Advances in Intelligent Systems and Computing

Volume 1202

AF806055

The series "Advances in Intelligent Systems and Computing" contains publications on theory, applications, and design methods of Intelligent Systems and Intelligent Computing. Virtually all disciplines such as engineering, natural sciences, computer and information science, ICT, economics, business, e-commerce, environment, healthcare, life science are covered. The list of topics spans all the areas of modern intelligent systems and computing such as: computational intelligence, soft computing including neural networks, fuzzy systems, evolutionary computing and the fusion of these paradigms, social intelligence, ambient intelligence, computational neuroscience, artificial life, virtual worlds and society, cognitive science and systems, Perception and Vision, DNA and immune based systems, self-organizing and adaptive systems, e-Learning and teaching, human-centered and human-centric computing, recommender systems, intelligent control, robotics and mechatronics including human-machine teaming, knowledge-based paradigms, learning paradigms, machine ethics, intelligent data analysis, knowledge management, intelligent agents, intelligent decision making and support, intelligent network security, trust management, interactive entertainment, Web intelligence and multimedia.

The publications within "Advances in Intelligent Systems and Computing" are primarily proceedings of important conferences, symposia and congresses. They cover significant recent developments in the field, both of a foundational and applicable character. An important characteristic feature of the series is the short publication time and world-wide distribution. This permits a rapid and broad dissemination of research results.

**** Indexing: The books of this series are submitted to ISI Proceedings, EI-Compendex, DBLP, SCOPUS, Google Scholar and Springerlink ****

More information about this series at http://www.springer.com/series/11156

Giuseppe Di Bucchianico ·
Cliff Sungsoo Shin · Scott Shim ·
Shuichi Fukuda · Gianni Montagna ·
Cristina Carvalho
Editors

Advances in Industrial Design

Proceedings of the AHFE 2020 Virtual Conferences on Design for Inclusion, Affective and Pleasurable Design, Interdisciplinary Practice in Industrial Design, Kansei Engineering, and Human Factors for Apparel and Textile Engineering, July 16–20, 2020, USA

Springer

Editors
Giuseppe Di Bucchianico
Dipartimento di Architettura
Università degli Studi "G.d'Annunzio"
Pescara, Pescara, Italy

Cliff Sungsoo Shin
Department of Industrial Design
University of Illinois at Urbana Champaign
Champaign, IL, USA

Scott Shim
University of Notre Dame
Indiana, IN, USA

Shuichi Fukuda
System Design and Management
Keio University
Tokyo, Japan

Gianni Montagna
Faculty of Architecture
University of Lisbon
Lisbon, Portugal

Cristina Carvalho
Faculty of Architecture
University of Lisbon
Lisbon, Portugal

ISSN 2194-5357 ISSN 2194-5365 (electronic)
Advances in Intelligent Systems and Computing
ISBN 978-3-030-51193-7 ISBN 978-3-030-51194-4 (eBook)
https://doi.org/10.1007/978-3-030-51194-4

This Springer imprint is published by the registered company Springer Nature Switzerland AG
The registered company address is: Gewerbestrasse 11, 6330 Cham, Switzerland

Advances in Human Factors and Ergonomics 2020

AHFE 2020 Series Editors

Tareq Z. Ahram, Florida, USA
Waldemar Karwowski, Florida, USA

11th International Conference on Applied Human Factors and Ergonomics and the Affiliated Conferences

Proceedings of the AHFE 2020 Virtual Conferences on Design for Inclusion, Affective and Pleasurable Design, Interdisciplinary Practice in Industrial Design, Kansei Engineering, and Human Factors for Apparel and Textile Engineering, July 16–20, 2020, USA

Advances in Neuroergonomics and Cognitive Engineering	Hasan Ayaz and Umer Asgher
Advances in Industrial Design	Giuseppe Di Bucchianico, Cliff Sungsoo Shin, Scott Shim, Shuichi Fukuda, Gianni Montagna and Cristina Carvalho
Advances in Ergonomics in Design	Francisco Rebelo and Marcelo Soares
Advances in Safety Management and Human Performance	Pedro M. Arezes and Ronald L. Boring
Advances in Human Factors and Ergonomics in Healthcare and Medical Devices	Jay Kalra and Nancy J. Lightner
Advances in Simulation and Digital Human Modeling	Daniel N Cassenti, Sofia Scataglini, Sudhakar L. Rajulu and Julia L. Wright
Advances in Human Factors and Systems Interaction	Isabel L. Nunes
Advances in the Human Side of Service Engineering	Jim Spohrer and Christine Leitner
Advances in Human Factors, Business Management and Leadership	Jussi Ilari Kantola, Salman Nazir and Vesa Salminen
Advances in Human Factors in Robots, Drones and Unmanned Systems	Matteo Zallio
Advances in Human Factors in Cybersecurity	Isabella Corradini, Enrico Nardelli and Tareq Ahram

(continued)

(continued)

Advances in Human Factors in Training, Education, and Learning Sciences	Salman Nazir, Tareq Ahram and Waldemar Karwowski
Advances in Human Aspects of Transportation	Neville Stanton
Advances in Artificial Intelligence, Software and Systems Engineering	Tareq Ahram
Advances in Human Factors in Architecture, Sustainable Urban Planning and Infrastructure	Jerzy Charytonowicz
Advances in Physical, Social & Occupational Ergonomics	Waldemar Karwowski, Ravindra S. Goonetilleke, Shuping Xiong, Richard H.M. Goossens and Atsuo Murata
Advances in Manufacturing, Production Management and Process Control	Beata Mrugalska, Stefan Trzcielinski, Waldemar Karwowski, Massimo Di Nicolantonio and Emilio Rossi
Advances in Usability, User Experience, Wearable and Assistive Technology	Tareq Ahram and Christianne Falcão
Advances in Creativity, Innovation, Entrepreneurship and Communication of Design	Evangelos Markopoulos, Ravindra S. Goonetilleke, Amic G. Ho and Yan Luximon

Preface

Industrial design has been one of the fastest growing and demanding areas which made significant contributions to our lives and societies. Traditionally, industrial designers focus on the aesthetics of products' form and style as well as ergonomics. This tradition has expanded its boundary to other subject matters such as service design, human–computer interaction and user experience design, sustainable design and virtual & augment reality. This is a strong indication that interdisciplinary collaboration across disciplines became an essential and critical practice in design.

This book covers five unique but overlapping design topics, namely: design for inclusion, interdisciplinary practice in industrial design, affective and pleasurable design, Kansei engineering and human factors for apparel and textile engineering.

This book has two underlying messages, the emerging importance of the social issue of inclusion and human diversity in contemporary society, and the increasing awareness that there is no such thing as a 'standard human being.' The first question has to do with inequality and social disparity as a necessary milestone toward economic revival, among other things, while a vital role in the strategies adopted by the European Union's Horizon 2020 Framework Program is played by the ones that focus on strengthening equality, participation and accessibility for all to goods, services and what Dahrendorf called 'life chances.' What this means is that the issue of social inclusion of diversity and for equality is firmly on political agendas all over the world, not least because of increasing awareness that new visions, new strategies, new tools and new approaches are needed, if we are to tackle the challenges arising from recent phenomena of economic and cultural globalization, demographic change, economic migration from poorer countries and an aging population in wealthier countries, phenomena that are destined to upset the entire planet's micro- and macroeconomic and social structures in years to come. The second issue tackled in this book is more technical in nature, since the paradigm change from 'designing for standards' and 'inclusive products and service design' to the enlightened awareness that there are no such concepts to fit the standard human being, this has immediate, direct repercussions on the specialized dimension of designing. The realization is at last taking hold not only that those individuals are physically, psychologically and culturally 'diverse,' but they also have widely diversified skills, abilities, aspirations and desires that make each one of us unique and not at all replicable. Since the diversity of individuals is the rule, not the exception, it makes sense to consider it as a resource, not as a limiting factor or a restriction on design, while equality between individuals, communities and peoples should be treated as fundamental strategic inputs to the sustainable development of contemporary society, where everybody should have the same opportunities to experience places, products and services. Numerous design approaches have been adopted to facilitate social and cultural inclusion in recent decades: design for disability, universal design, inclusive design and design for all. All of these philosophies, approaches and methodologies aim to build value on all aspects of human diversity, from psychophysical to cultural issues, and to offer equal opportunity to everyone in order to experience places, products, services and systems. With this in mind, this book sets out to forge a climate conducive to discussion and comparison between these approaches, without any prejudice in favor or against any one of them, but attempting to identify the elements they hold in common and to build each one's heritage of originality, because we are convinced that the true resource of design for inclusion may well be found in this very diversity of opinions.

The Interdisciplinary Practice in Industrial Design (IPID) aims at exploring and broadening the practice in industrial Interdisciplinary design process. The conference includes discussions on 1) a theoretical investigation as well as professional practice to foster interdisciplinary collaboration across disciplines, 2) design projects through interdisciplinary collaboration, 3) design process with external public and private sector partners with a solid record of interdisciplinary development experience and 4) design methods and techniques to investigate productive and effective interdisciplinary collaboration in design.

This book also focuses on an apparel and textile engineering along with positive emotional approach in product, service and system design and emphasizes aesthetics and enjoyment in user experience. This book provides dissemination and exchange of scientific information on the theoretical and practical areas of affective and pleasurable design for research experts and industry practitioners from multidisciplinary backgrounds, including industrial designers, emotion designers, ethnographers, human–computer interaction researchers, human factors engineers, interaction designers, mobile product designers and vehicle system designers.

A total of seventeen sections are presented in this book:

Section 10 Interdisciplinary Approach to Design Manufacturing
Section 11 Human Factors Research Methods for Hand and Tool Design
Section 12 Kansei Engineering
Section 13 Human Factors for Textile and Society
Section 14 Apparel Design and Textile Engineering
Section 15 Coordinating for Emotional Satisfaction
Section 16 Affective Value and Kawaii Engineering
Section 17 Sensory Engineering and Emotion

Sections 1–5 cover topics in design for inclusion, Sections 6–11 cover topics in interdisciplinary practice in industrial design, Section 12 covers topics in Kansei engineering, Sections 13 and 14 cover topics in textile design and engineering, and Sections 15–17 cover topics in affective and pleasurable design.

Each section contains research papers that have been reviewed by members of the International Editorial Board. Our sincere thanks and appreciation go to the Board Members as listed below:

S. Heffernan, New Zealand
A. Marr, UK
C. Morais, Portugal
F. Moreira da Silva, Portugal
D. Ramos, Portugal
M. Sbordone, Italy

Affective and Pleasurable Design

A. Aoussat, France
S. Bahn, Korea
L. Chen, Taiwan
K. Cho, Korea
S. Choi, USA
D. Coelho, Portugal
O. Demirbilek, Australia
M. Feil, USA
A. Freivalds, USA
Q. Gao, China
R. Goonetilleke, Hong Kong
B. Henson, UK
A. Ho, Hong Kong
W. Hwang, Korea
Y. Ji, Korea
E. Jung, Korea
J. Kim, Korea
K. Kim, Korea
K. Kotani, Japan
S. Minel, France
K. Morimoto, Japan
M. Ohkura, Japan
T. Park, Korea
P. Rau, China
S. Schutte, Sweden
D. Shin, USA
A. Warell, Sweden
M. Yun, Korea

Kansei Engineering

A. Lokman, Malaysia
Y. Matsubara, Japan
T. Tsuchiya, Japan

Special thanks to Professors Mitsuo Nagamachi and Shigekazu Ishihara for their valuable contribution to the research track focusing on Kansei engineering.

We hope that this book, which is the international state of the art in the industrial design, will be a valuable source of theoretical and applied knowledge enabling human-centered design of variety of products, services and systems for global markets.

July 2020

Giuseppe Di Bucchianico
Cliff Sungsoo Shin
Scott Shim
Shuichi Fukuda
Gianni Montagna
Cristina Carvalho

Contents

Designing for Inclusion in the Information Society

Design for Inclusion in the Living Environment

Interdisciplinary Design Education

Interdisciplinary Approach to Industrial Design

Coordinating for Emotional Satisfaction

Affective Value and Kawaii Engineering

Products and Services for an Inclusive Society

Human Centered Design Methodologies Applied to Complex Research Projects: First Results of the PLEINAIR Project

Giuseppe Mincolelli(✉), Gian Andrea Giacobone, Silvia Imbesi, and Michele Marchi

TekneHub – Dipartimento di Architettura, Università degli Studi di Ferrara, via Saragat 13, 44124 Ferrara, Italy
giuseppe.mincolelli@unife.it

Abstract. One of the data that national and international organizations – including the World Health Organization (WHO) – are observing is regarding people's sedentary lifestyle, of all age groups and without distinction of ethnicity, social or territorial characteristics. To try to give concrete and possible solutions, the PLEINAIR (Free and Inclusive Parks in Networks for Recreational and Physical Intergenerational Activity) was developed; a research project funded under the action 1.2.2 POR FESR 2014–2020. The project started in July 2019 and will last 24 months. Outdoor parks and recreational areas are the scope of the project, which aims to improve and design inclusive contexts to promote the adoption of active lifestyles for all and for all age groups. The operational tools through which we will try to provide real solutions are the OSOs (Outdoor Smart Objects): urban furniture, ludic tools or customizable inclusive strategies to encourage physical activity, conviviality and socialization among people.

Keywords: Human Centered Design · Co-design · Health · Wellness · Smart object outdoor

1 Introduction

"*…physical inactivity is worldwide the fourth most important mortality risk factor and causes 6% of all deaths. About 3.2 million people die each year because they are not active enough. Insufficiently active people have a mortality risk of 20% to 30% higher than those engaged in at least half an hour of moderate-intensity physical activity in most days of the week*" [1].

Above an extract of the data transmitted by the World Health Organization. The data are worrying and constantly growing [2]. According to the WHO, the decrease in physical activity is due to several factors: some caused by the individual behavior of the person and others caused by the – real and/or perceived – context where we live. The latter is perceived as unsafe and with widespread violence (especially in the evening/night hours), with a massive presence of vehicular traffic that brings bad air quality; lastly, it is perceived as not very accessible and inclusive with regard to primary infrastructures such as, for example, sidewalks, sports and recreational facilities.

G. Di Bucchianico et al. (Eds.): AHFE 2020, AISC 1202, pp. 3–9, 2020.
https://doi.org/10.1007/978-3-030-51194-4_1

Unfortunately, even at national level, in Italy, the data do not change significantly. According to the latest ISTAT data [3] in the two-year period 2017–2018, in Italy an estimated 2 million and 130 thousand children and adolescents are overweight, equal to 25.2% of the population aged 3–17 years (28.5% in 2010–2011). Although registering a drop on the number of overweight children compared to the previous two-year period, the data concerning the lack of physical activity in children and adolescents remains worrisome.

Among the different options and solutions that the WHO proposes, 3 priorities out of 5 [4] identify the promotion of physical activity as a primary element. Specifically, the Priority area 2 is focused on children, the Priority area 3 on adults and the Priority area 4 on the elderly. Therefore, the mission that the WHO [4] proposes is to promote physical activity with the consequent reduction of sedentary behaviors, to create and develop inclusive environments for people of different ages, cultures, sex and disabilities, that support physical activity in public and private spaces.

In this scenario, a scientific-industrial research team has developed the PLEINAIR project. The research is underway and will end in July 2021. The project is addressed to multigenerational users with the goal of setting up outdoor recreational and ludic environments that allow the synergistic use by people with skills and characteristics even very different from each other. The objective is to encourage physical activity with customized and profiled motivational strategies based on the specific characteristics and peculiarities of each person who will want to freely enjoy the park and its equipment. This objective will be possible thanks to the Outdoor Smart Objects (OSOs) project: sensors/actuators to be integrated with existing equipment or to be realized from scratch. The OSOs will also be able to monitor (thanks to IA and IoTM technologies) a series of useful parameters for assessing people's motor well-being and to trigger new and innovative ways of socio-recreational interaction between people and ludic equipment with the aim of improving people's quality of life (physical and emotional).

2 Methodology

The research project is complex since the objectives are broad and cross-cutting. Indeed, the research team has set both the real planning of the OSOs (to reach a Technological Maturity Level equal to 6 [5]) and the research of innovative virtuous dynamics that can arise in urban areas among users with very different characteristics (age, sex, culture, ability/disability, etc.) from each other. The goal is also to understand the deep and latent needs that lead different people to enjoy an outdoor park; understanding what are the possible transversal activities between the different users in order to be able to develop profiled motivational strategies based on their emotional, functional and social needs.

To achieve the goal, a highly trained and specialized scientific-industrial research team has been created.

The research group includes: the partners DataRiver, CIRI SDV, Future Technology Lab, TekneHub and A.I.A.S. The latter two have a specific expertise regarding Human Centered Design approaches to conceive and design inclusive and accessible services, environments and interfaces. Moreover, TekneHub also manages the specific aspects of OSOs design and interfaces, in terms of appearance, interactivity, inclusiveness and ergonomics. A.I.A.S, thanks to their experience gained over the years, has great skills in terms of both Assistive Technologies and vulnerable users, with specific reference to people with disabilities and the elderly, thanks to the management and coordination – through agreements – of day care centers or residential buildings. DataRiver, CIRI SDV and Future Technology Lab have an expertise in the fields of IoT/IoMT, sensors, as well as data management, Machine Learning/Artificial Intelligence, neuro-biomechanics of movement and analysis and development of motor behavior. The industrial partners, that are important for their real experience of industrial and experimental operations are: Ergotek s.r.l. (specialized in the production of wooden seats and furniture), Sarba s.p.a (producer of playground equipment and urban furniture), mHealth Technologies S.r.l (healthcare furniture sector). The methodology that is applied to the research is aimed at satisfying people's needs; they are the main actors of the project. Therefore, design research tools are adopted together with the Human Centered Design (HCD) approach. One of the biggest problems when working within a complex multidisciplinary team as in the PLEINAIR project, may be the one concerning a problem of language and communication between different expertise. As verified and confirmed by positive results within the Habitat project [6] (composed by some scientific-industrial partners that are also part of PLEINAIR), following HCD approaches [7] and employing application tools such as Quality Function Deployment [8] (QFD) or co-design [9, 10], can simplify the communication between partners and make the various planning phases more fluid and clear.

The PLEINAIR project is underway and some research workshops have already been carried out. The first saw the participation of the entire research partnership (academic and industrial) and had as objective, the focalization of project objectives and the beginning of the production of the common and shared material for the subsequent research phases (Fig. 1).

Fig. 1. View of the first workshop

The day had a playful-experimental but very structured nature. The event was divided into two distinct parts; within the first part, through tools such as brainstorming and cart-sorting [11], some graphic cards were distributed which represented national and international best practices regarding auxiliary objects and equipment that can be found in outdoor parks. For convenience of discussion, the audience was divided into groups of maximum 5–6 units. Each group was composed by at least one reference person belonging to one of the scientific partners in order to have widespread cross-cutting and sectorial competences. The goal was to carry out a competitive benchmarking between the different solutions, being able to gather the strengths and weaknesses in a PLEINAIR perspective. In the second part, we discussed which activities could be multi-generational, examining their weak and virtuous points and the possible changes to be made/integrated. The other two workshops were structured differently from the first one; in addition to the presence of some scientific partners, those workshops had the active participation of some Public Administrations of the Emilia Romagna Region (Bologna, Castenaso, Modena) through the components of the Municipal Offices responsible for the management of urban garden areas. The days were useful to understand stakeholders' needs and expectations regarding the topics covered by PLEINAIR. An active debate with the PAs was essential in order to understand the real existing problems concerning playgrounds and to know the virtuous (or not) ongoing activities. These events were also divided into two distinct parts.

The first part was focused on the state of the art of parks; in fact, considerations were expressed on the outdoor park environment (accessibility, maintenance, management,…), the equipment and auxiliary objects (seats, games, carousels,…) and any existing technologies such as, for example, IoT or monitoring of biomedical data on motor exercises. The second part was structured on the basis of the conclusions resulting from the first workshop. Taking into account the best practices, we observed the spill-over effects in terms of management, maintenance, inclusiveness, identity and affiliation with regard to the local communities. All with the help of graphic synoptic tables, gradually completed thanks to the contribution of all the participants. Each participant had the opportunity to monitor the progress of the work; this resulted in two immediate advantages (Fig. 2).

Fig. 2. View of the second workshop

The first one, was to not repeat the same considerations made by other people in such a way as not to waste energy and time; the second one, allowed a beginning of acquaintance between the different PAs and the project partners. Being able to understand each other, discuss and put yourself in other's shoes (in terms of needs, expectations but also obstacles and inconvenience) increases the awareness and responsibility towards a common project.

3 Results

The project has just started, therefore the results are partial. However, from the early stages of the research, it has been possible to notice that the themes introduced within PLEINAIR are very current in the governance of the involved PAs. During the three meetings, we observed – by applying user-centered approaches from the early phases of the research (intended both as users and as staff in charge of the management and maintenance of the park) – how it was possible to produce useful and shared material between all partners (academic and industrial) and project stakeholders. Therefore, we had the confirmation that working with HCD design tools leads to good results.

The first workshop led to the development of around 70 best practices, then divided into 9 macro-areas:

- Infographics with ground signage
- customized gymkhana playgrounds
- swings
- cognitive exercises with board
- physical activity with gym equipment
- yoga
- seats
- target shooting
- memory games

A unique code was assigned to each single activity (part of the different macro areas); then its usability was assessed on the basis of the typological characteristics of three users (child, adult and elderly). Then, a database of possible activities and viable solutions for the analyzed three categories of users – for now considered multi-generational – had been created. These activities and solutions were assessed with regard to the level of inclusiveness and the following temporary and permanent disabilities/abilities of possible users of an outdoor park, were taken into account:

- wheelchair user
- mom with stroller
- elderly person with walking stick
- visually impaired or blind
- person with mild cognitive disability
- deaf person

The activities that obtained the highest score are those that met both the requirement regarding the possible use by multigenerational users and a good inclusiveness for people with disabilities. Being able to extrapolate a series of activities considered multi-generational and inclusive for people with high skills in terms of ergonomics, design, sensors, technology, social relations, etc., has represented an important result for PLEINAIR. These activities will allow a more efficient work in the subsequent research phases.

The second and third meeting led to the drafting of about 80 needs expressed by the involved PAs with regard to an equipped outdoor park. Furthermore, as for the territories and parks managed by the PAs, we discussed on the real and concrete effects of the best practices developed during the first workshop. These evaluations will be pivotal starting points for the subsequent research phases.

4 Conclusion

The work carried out so far appears to be a good starting point for dealing with the subsequent research phases. The meetings have shown, as explained in the literature [12], that setting up an urban space with inclusive and accessible equipment for different users, in addition to reactivate social bonds within the neighborhoods, brings real benefit (emotional, motor and social) to the involved users.

As for the elderly, for example, the possibility of being involved in various activities with young people, allows them to increase their self-esteem, the perceived well-being and to reduce the stress. On the other hand, as for children, a positive attitude towards the elderly and a greater understanding of the ageing process, can be highlighted.

In conclusion, the objectives of the PLEINAIR project of improving people's lifestyle can certainly be achieved thanks to the creation of new and innovative socio-recreational dynamics that allow to combine different needs of different users.

References

1. WHO report: physical activity drafted by the Italian Ministry of Health - Department of Public Health and Innovation - General Directorate of European and International Relations by Katia Demofonti - Office III. Report n. 384, February 2014
2. Italian Ministry of Health. General Directorate of Health Prevention Office 8. Guidelines on physical activity for different age groups and with reference to physiological and physio-pathological conditions and specific population subgroups (2017)
3. Press release - lifestyles of children and teenagers- reference period: years 2017–2018, publication date: 29 October 2019
4. Physical activity strategy for the WHO European Region 2016–2025 © World Health Organization (2015)
5. Héder, M.: From NASA to EU: the evolution of the TRL scale in public sector innovation. Innov. J. **22**, 1–23 (2017)

6. Borelli, E., Paolini, G., Antoniazzi, F., Barbiroli, M., Benassi, F., Chesani, F., Chiari, L., Fantini, M., Fuschini, F., Galassi, A., Giacobone, G.A., Imbesi, S., Licciardello, M., Loreti, D., Marchi, M., Masotti, D., Mello, P., Mellone, S., Mincolelli, G., Raffaelli, C., Roffia, L., Cinotti, T.S., Tacconi, C., Tamburini, P., Zoli, M., Costanzo, A.: HABITAT: an IoT solution for independent elderly. In: Pau, G., Salerno, V.M., You, I., Sharma, V. (eds.) Sensors 2019, vol. 19, no. 5, p. 1258 (2019). In Special Issue "Internet of Things for Smart Homes" 1–23
7. Mincolelli, G., Imbesi, S., Marchi, M.: Design for the active ageing and autonomy: the role of industrial design in the development of the "Habitat" IOT project in advances in design for inclusion. In: Di Bucchianico, G., Kercher, P. (eds.) Proceedings of the AHFE 2017 International Conference on Design for Inclusion, 17–21 July 2017. The Westin Bonaventure Hotel, Los Angeles, California, USA, pp. 88–97 (2017)
8. Franceschini, F.: Quality Function Deployment, Franceschini Fiorenzo. Il Sole 24 Ore, Libro, HOEPLI.it; Il Sole 24 Ore; HOELPI, Milano, Italy (2003)
9. Mincolelli, G., Imbesi, S., Marchi, M., Giacobone, G.A.: New Domestic Healthcare. Co-designing assistive technologies for autonomous ageing at home. Des. J. **22**(sup1), 503–516 (2019)
10. Rizzo, F.: Strategie di co-design. Teorie, metodi e strumenti per progettare con gli utenti. Franco Angeli, Milano (2009)
11. Hudson, W.: Playing your cards right: getting the most from card sorting for navigation design. Interact. HCI High. Educ. **12**(5), 56–58 (2005)
12. Cartmel, J., Radford, K., Bell, K., Golenko, X., Fitzgerald, A.: Hearing children's voices in intergenerational learning and practice. In: Kernan, M., Cortellesi, G. (ed.) Intergenerational Learning in Practice, Chapter 5, pp. 196–213. Taylor and Francis, London (2019)

Envisioning the Future and Going Back: A Human-Centered Strategy to Develop the Styling of a Sports Car

Gian Andrea Giacobone(✉) and Giuseppe Mincolelli

Department of Architecture, University of Ferrara,
Via Ghiara 36, 44212 Ferrara, Italy
{gianandrea.giacobone,giuseppe.mincolelli}@unife.it

Abstract. Developing a vehicle means uncertainty because the competitiveness of global market implies investing high sums in advance without knowing if the final solution will be successful after its launch. Therefore, automotive styling must be well executed to reduce management risks and adopting a Human-Centered approach at the early stages of the process is one of the best strategies to avoid negative outcomes. In particular, the research used the Vision in Product Design, a strategic model to envision the concept in the future to anticipate future trends or upcoming wishes. However, since ViP generates a prediction because its context factors belong to the future, the research would risk of not completely matching costumer expectations because of future uncertainty. Therefore, the research pushed the intent of the vision backward, in a mediated context, where the character of the vision was merged with the real expectations of people gathered in the current context.

Keywords: Human-centered design · Vision in Product · Automotive styling · Conceptual design · Research methods

1 Introduction

Automotive styling is one of the key roles of the automotive development program and it takes place in the early stages of the design process [1]. This specialized discipline is specifically implicated in giving a form to the car, encompassing packages and negotiating engineering constraints such as size, ergonomics, performance and so on [2].

The intellectual decisions about technical and morphological aspects of the vehicle are carefully selected at that point. It delineates the right strategy for achieving the expected accomplishments of the company, in terms of revenues and return of investments [3]. This is because developing a vehicle means uncertainty [4] due to the competitiveness of the global market, which pushes carmakers to invest high sums in advance towards a shorter time-to-market process [5], without knowing if the final solution will be successful after its launch [6]. For this reason, automotive styling must be well executed in order to reduce management risks, but also to ensure that the design strategy and its final solution coincide properly with the customer expectations.

G. Di Bucchianico et al. (Eds.): AHFE 2020, AISC 1202, pp. 10–17, 2020.
https://doi.org/10.1007/978-3-030-51194-4_2

Considering this, adopting a Human-Centered approach at the early stages of the process is one of the best and successful strategies to avoid negative outcomes. Since the Human-Centered approach accommodates human needs, capabilities, and ways of behaving [7], the styling strategy is able to achieve the expected desirability of final product by determining its qualitative level of the total product experience through its appearance, functions, and operations [8].

2 The Project

This paper reports the activities of a specific research integrated in a multidisciplinary project called "Automotive Academy: a learning by doing project for innovation engineering automotive" which is supported by Emilia Romagna Region, Italy. The project is related to the field of automotive and its network involves many industries, research centers and four universities located on the territory. The research aims to develop the concept of a hybrid sports car, which aspires to become as a R&D platform and demonstrator for future technical or technological innovations – in terms of sustainable and high performance solutions – that will be investigated by the network in the coming years. In particular, this contribution refers to the discipline of automotive styling. Its goal is to envision and conceptualize both architecture and styling of the expected sports car by defining a successful design strategy to develop its final concept.

3 Methodology

In the automotive industry, the full-vehicle development program is defined by five main phases: the definition phase, the concept phase, the pre-development, the series development and early stages of the production phase [9]. The activities of this contribution were related to automotive styling and they touched the first two steps. Particular emphasis was placed on the concept phase, especially on the transition from the first visions or ideas (delineated by the product strategy) to the detailed concept solutions of the vehicle's styling and architecture.

This research conducted the concept phase by using a human-centered methodology, in particular envisioning a future concept solution by using the Vision in Product Design (ViP). The model was selected because it was perfectly suitable for structuring idea-generating throughout the styling process, while supporting design team in taking significant decisions for the product development.

ViP is a human-centered methodology that enables innovation by envisioning new ideas for the future [10]. The model produces strategic design decisions for a project starting from the analysis of people in their related context. Though, instead of simply solve present-day problems – which people every day require to be satisfied by functional solutions – ViP tries to discover new design opportunities (for life of people) that can be ideated in the future ahead. The process studies the current evolution of socio-cultural models [11] to understand the possible tomorrow by establishing the *raison d'être* of the final design, namely the intrinsic and anthropologic meaning of the artefact, which justify its existence in the new world [12]. The framework is articulated in a step-by-step process.

ViP is composed by two main areas that divide the model into two different design activities: the left side represents the process of *deconstruction* and it refers to the analysis of the current state of art; the right side is related to the process of *designing*, in which the vision is generated. Each area is divided into three main levels of investigation: product, interaction and context level. Instead, the entire vision process is divided into nine stages where the first one is addressed to the left side of the model (the existing context), whilst the other eight phases are completely dedicated to the right side. The outcome is represented by the advance concept.

4 Design Process and Results

According to the scope of the overall project, the final vehicle was related to the sports segment, especially to the track-day cars. Within the ViP, every activity was based on an existing car model, the KTM X-Bow, which was chosen as a primary case study to design the advanced concept according to the same sports car segment.

The vehicle of KTM was selected because its properties perfectly coincided to the requirements of the project but also because the phase of user analysis was performed through an event promoted directly by the same carmaker. Considering this, the analysis was conducted at a specific race event completely dedicated to the X-Bow model [13]. The activities regarded the observation of people while they were interacting with the X-Bow in its context of use. Moreover, the outputs of the user analysis were fundamental to investigate the left side of ViP, the "deconstructing" phase. They permitted to better understand the world of today that motivated the raison of being of the X-Bow in terms of product qualities and vehicle characteristics.

4.1 Generating the Vision

In the right side of the ViP model, the process envisioned a specific domain in which the advanced concept would be designed. To encourage the vision in diverging itself from the current context, neither constraints nor restrictions were included during this design process.

The selected time span of the vision was ten years ahead in order to anticipate new emerging automotive trends. The research gathered several context factors associated with the domain of mobility, which were used to generate the context of tomorrow. The most important factors were correlated each other and successively clustered together based on their affinities. All the factors were turned into a unified and coherent structure that explained the relationship among those separate elements. This process generated four different parallel clusters, which indicated four potential opportunities for designing the project. The scenarios were represented in a four-square matrix, where each square indicated a statement, namely a potential way in which the advanced concept would be designed. The two main axes represented the opposing forces among those scenarios and hence, produced four different statements:

- Top-left: "Resistance in changing", which indicates a context in which people resist against the technological innovation because they want to maintain their independence while driving and they prefer to have an individual relation whit their vehicles;
- Top-right: "Be a good citizen", which denotes a context where people what to preserve the status quo by passively accepting the emerging technology;
- Bottom-left: "Lead the changing process", which expresses a context in which people actively change the traditional transportation model of the cars in more shared and convivial mobility;
- Bottom-right: "Be part of the system", which represents a context where people passively follow social changes and technological innovation (for instance electrification, on-demand services and autonomous driving), losing at the same time their enthusiasm and independence in favor of an efficient mobility.

According to the goal of the project, the research selected the top-left square: "Resistance in changing". This ideology was translated in the statement for the vision, which was defined as: "I want people to experience mobility that expresses themselves as an act of resistance". The statement was the first point to discover the main human-product interactions that the concept vehicle would expect to produce.

Through the use of the analogies, the characters and the properties of the car were discovered. Storytelling facilitated the process of envisioning the advanced concept because it helped to identify those interactions. The process produced the correct analogy that fully matched the selected statement. It was: "It feels like making graffiti on the wall". This because the sensation of producing graffiti on the wall is recognized as a rebel act against a particular ideology or system. Indeed, the selected area "Resistance in changing" represents the resistance against the upcoming technological development presented by the opposing force "Be part of the system", in which technology prevails on automotive. For this reason, the analogy translates the car itself in a rebel expression or dissent, which is capable of escaping from the representation of a functional and anonymous object by providing to its owner a specific status and by maintaining its subjective experience of driving.

The analogy delineated the two main product characters, which were successively used to define the technical and styling properties of the advanced concept. The two primary product qualities were *secretive* and *adversative* because the two elements expressed perfectly the intrinsic meaning of the analogy. In fact, making graffiti on the wall is a dissenting act, which means that the behavior is intrinsically *adversative* because it acts as an opposing force or resistance against the prevailing system. The same behavior produces secretiveness because the action covers the identity of its producer even the drawing is visible to all.

4.2 The Advanced Concept

According to the two selected vehicle characters, the scenario envisioned a hidden and silent car, which can be driven only in the night or darkness as a symbol of rebelliousness against the daylight routine of an efficient and technological mobility. The selected powertrain is fully electric because it produces fast acceleration without sound.

The drivetrain is equipped with four independent electric motors, which are able to express an adverse way of driving through the four-wheel drive layout. The ingress/egress is located in the rear of the vehicle to cover the identity of the passengers. The occupant package is also hidden by an obscure coachwork because its configuration eliminates the traditional outlines of the doors from the vehicle. Therefore, it obscures both the occupant package and the vehicle's driving orientation, providing at the same time a unique secretive styling.

The cockpit covers the driver's identity as well by eliminating windshield and every lateral or rear window from the car. Consequently, the interface drives the vehicle only through digital screens, improving driver's perception by using the augmented reality generated by front cameras. The car features a three-seater configuration to balance the power of rebelliousness among the occupants. However, the driver's is positioned in the middle, whilst the other ones are slightly positioned behind it, so that the *adversative* character can imprint to the primary driver more activism by giving it the leadership of the rebel act of driving. This experience is also accentuated by maintaining the traditional steering hub: an anatomic steering wheel is designed to transfer every vehicle's behavior directly to the driver's hands such as the standard airplane joy-stick.

At the end of the process, the vision produced deliberately a visionary concept. Indeed, using ViP enabled the research to think in possibilities instead of constraints in order to come up with innovative and original idea. Since people tend to reason in terms of current problems and thus, their insights might be rooted in the situation they are in at that moment, ViP detaches the process from the constraints of present, forcing the research strategy to design the expected car in a novel and future reality.

However, since ViP generates a prediction because its context factors belong to the future, the research would risk of not completely matching costumer expectations because of future uncertainty. In addition, since the concept is designed for the future, the result could be also not accepted by the current audience because it could risk of striving with its socio-cultural environment due to the MAYA (Most Advanced Yet Acceptable), the principle coined by Raymond Loewy [14] to explain that every project needs to find a strong balance between novelty and typically.

4.3 The Mediated Context

Considering the above considerations, the concept was reframed and pushed its intent backward in a mediated context, in which the vision was capable of combining itself with the analysis of the KTM already delineated during the interaction with the X-Bow [13]. The advanced concept was not the final result but, on the contrary, became a source of inspiration for turning the vision into an integrated solution between the two selected characters of the vision and the technical properties of the X-Bow. In particular, the functional and morphological features of the advanced concept were utilized as guidelines to characterize the identity and the styling of the final concept. Whilst, the KTM's case study were used to define the technical specifications of the expected car.

Considering this, ViP conferred specific styling to the final car. The *adversative* character produced the sharpened style lines of the coachwork. Instead, the *secretive* aspect delineated the car's aerodynamics by hiding the two lateral air intakes in the front and another one under the chassis. This configuration allowed to cover the traditional outline of the air channel without affecting on the vehicle's downforce.

Regarding the powertrain, the final vehicle integrated the rear mid-engine layout of X-Bow with the *adversative* character of the vision by equipping it with a four-wheel drivetrain and a parallel hybrid system. In this case, this specific configuration maintains the mechanical gearbox and the sound of the traditional driving experience provided by the engine and, at the same time, it is able to provide different power settings based on the context of use. Moreover, the electric power source improves the efficiency of both fuel consumption and carbon emission, because it reduces engine size and optimizes its power source to fulfil the desired power demand [15] (Fig. 1).

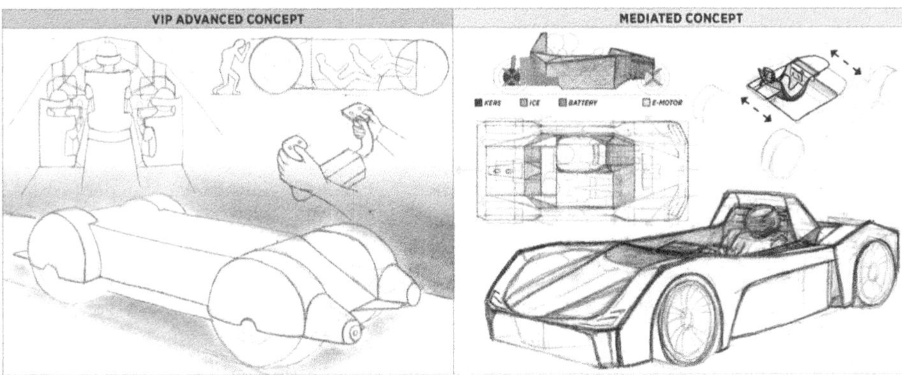

Fig. 1. On the left the advanced concept, on the right the mediated context.

The same character turned the three-seat layout of the advanced concept into an active driving seat. The latter, provided to the car a peculiar driving configuration because it could float from the traditional lateral seating position to the center driving system. In fact, the seating configuration is an asymmetrical two-seater, which decreased the track of the car but also it guaranteed much more involvement in driving by allowing the driver to experience the car in different modalities according to a specific context of use. The character of *adversative* configured also the driving interface toward a racing-oriented experience, conserving the pure driving sensation already provided by the X-Bow. All the features were integrated in the final solution of the expected concept.

5 Conclusion

In conclusion, ViP was a useful approach to plan a design strategy for developing an innovative vehicle for the future. The method was able to foresee new coming automotive trends, which became significant opportunities for the strategy to anticipate new potential social demands. The outcomes of ViP, indeed, were a significant source of inspiration for the concept phase because they were able to structure the styling strategy by designing a novel and original concept vehicle in response of the world of tomorrow. In particular, ViP drove the styling strategy of the overall integrated process, primary focusing on people's interactions and behaviors (experienced through the car) instead of technical specifications. The approach was able to bring out both the characters and the identity of the car, especially defining its styling and functions. Instead, the mediated context was necessary to support ViP in translating the vision into a feasible and tangible concept. The characters of the vision were processed with the technical specifications of the existing car model that were delineated during the analysis of the current context. At the end, the final result accomplished the research goal by elaborating the expected vehicle, producing at the same time, several technical and styling guidelines for the multidisciplinary network.

References

1. Feldinger, U.E., Kleemann, S., Vietor, T.: Automotive styling: supporting engineering-styling convergence through surface-centric knowledge based engineering. In: Maier, A., Škec, S., Harrison, K., Kokkolaras, M., Oehmen, J., Fadel, G., Salustri, F., Van der Loos, M. (eds.) DS 87-4 Proceedings of the 21st International Conference on Engineering Design (ICED 17), Vancouver, Canada, 21–25 August 2017. Design Methods and Tools, vol 4, pp. 139–148. The Design Society, Glasgow (2017)
2. Van Grondelle, E., Brand de Groot, S.: Discovering the meaning of form by exploded sketching. In: Bohemia, E., Buck, L., Eriksen, K., Kovacevic, A., Ovesen, N., Tollestrup, C. (eds.) Proceedings of the 18th International Conference on Engineering and Product Design Education (E&PDE 2016), Aalborg, Denmark, 8–9 September 2016. Design Education: Collaboration and Cross-Disciplinarity, pp. 374–379. The Design Society, Glasgow (2016)
3. Bhise, V.D.: Automotive Product Development: A Systems Engineering Implementation. CRC Press (2017). https://doi.org/10.1201/9781315119502
4. Brem, A.: The Boundaries of Innovation and Entrepreneurship. Gabler, Wiesbaden (2008). https://doi.org/10.1007/978-3-8349-9679-4
5. Brunner, H., Hirz, M., Wurzer, A.: Considering influences of sustainable production in early automotive engineering phases. Ann. Fac. Eng. Hunedoara Int. J. Eng. **15**, 85–90 (2017)
6. Gerhard, D., Brem, A., Voigt, K.I.: Product development in the automotive industry: crucial success drivers for technological innovations. Int. J. Technol. Mark. **3**, 203–222 (2008). https://doi.org/10.1504/IJTMKT.2008.019922
7. Norman, D.A.: The Design of Everyday Things, Revised and Expanded Edition. The MIT Press, Cambridge (2014)
8. Hekkert, P., Schifferstein, H.N.J.: Introducing product experience. In: Schifferstein, H.N.J., Hekkert, P.B.T.-P.E. (eds.) Product Experience, pp. 1–8. Elsevier, San Diego (2008). https://doi.org/10.1016/B978-008045089-6.50003-4
9. Brunner, H., Rossbacher, P., Hirz, M.: Sustainable product development: provision of information in early automotive engineering phases. In: MOTSP 2017 - International Conference Management of Technology (2017)
10. Van Boeijen, A., Daalhuizen, J., Zijlstra, J., Van Der Schoor, R.: Delft Design Guide. Bis Publisher, Amsterdam (2014). https://doi.org/10.1017/CBO9781107415324.004
11. Verganti, R.: Design, meanings and radical innovation: a meta-model and a research agenda. J. Prod. Innov. Manag. **25**, 436–456 (2008). https://doi.org/10.1111/j.1540-5885.2008.00313.x

12. Hekkert, P., Van Dijk, M.: VIP Vision in Design: A Guidebook for Innovators. Bis Publisher, Amsterdam (2011)
13. Giacobone, G.A., Mincolelli, G.: Human-centered design and quality function deployment: understanding needs on a multidisciplinary automotive research. In: Di Bucchianico, G. (ed.) Advances in Intelligent Systems and Computing, pp. 57–68. Springer (2020). https://doi.org/10.1007/978-3-030-20444-0_6
14. Loewy, R.: Never Leave Well Enough Alone. The John Hopkins University Press, Baltimore (1950)
15. Jain, S., Kumar, L.: 31 - fundamentals of power electronics controlled electric propulsion. In: Rashid, M.H. (ed.) Power Electronics Handbook, 4th edn, pp. 1023–1065. Butterworth-Heinemann, Oxford (2018)

The DfA Quality Label: Design for All Practical Evaluation Experiences for Inclusive Design Solutions

Giuseppe Di Bucchianico(✉)

University of Chieti-Pescara, Viale Pindaro 42, 65129 Pescara, Italy
pepetto@unich.it

Abstract. In order to give everyone equal opportunities to benefit from environments, products and services, thus favoring social inclusion through the enhancement of human diversity, in recent decades various design approaches have been developed, with common objectives, but different principles and methodologies. Among these, Design for All proposes a holistic approach to the project, focusing on the individual as a whole, trying to satisfy everyone's needs, from requirements to desires and aspirations. To do this it is also necessary to involve the entire supply chain of the decision. In an effort to investigate and experiment in practice this approach, for some years the DfA Italia association has issued a DfA Quality Label for product, service, environment or system, developed according to the Design for All approach. It is assigned through a particularly complex process, and declined with respect to two different levels. The highest level is assigned to design solutions meeting all the DfA principles and requirements; the lower one is assigned when not all the principles and requirements are fully satisfied (but can be, following some specific suggestions).

Keywords: Design for All · Quality Label · Evaluation process · Inclusive Design · Human diversity

1 Design for Inclusion Approaches for an Increasing Human Diversity

Contemporary society is in continuous and rapid transformation. New and growing socio-demographic phenomena (aging, dissolution of family and social ties, migratory phenomena) suggest for the next decades a society that, if left to itself, will be marked by a widespread social malaise, due to the increase in non-self-sufficiencies (with all the economic, social security, welfare and anthropological consequences that this entails) and by a dangerous weakening of interpersonal relationships (from individual alienation to mistrust towards those who are different).

These phenomena are attempted to respond to many with the idea that places of life must increasingly take on the role of real social and cultural melting pots, that is places where it is possible to actively build a positive amalgam of different elements (ethnic, religious, cultural, experiential, etc.), and in which all people, regardless of age, sex,

G. Di Bucchianico et al. (Eds.): AHFE 2020, AISC 1202, pp. 18–23, 2020.
https://doi.org/10.1007/978-3-030-51194-4_3

culture, physical and psycho-sensorial skills, must have the same opportunities to develop a life that responds to their own expectations and to decide on your business, home or lifestyle according to your desires and aspirations, consolidating relationships of trust and esteem with others.

The growing human diversity of contemporary society, therefore, can no longer be ignored, and indeed, if properly managed, respected and valued, it can represent a resource for the development of the social and cultural contexts to come, both at macro scale of communities, both at the micro scale of interpersonal relationships.

To do this it is necessary that all actors (social, economic, political and cultural) do their part, consciously contributing to improving the use of places and contexts of life and work. In fact, it was realized that the change of pace must first of all be of a cultural type: rigid and extraneous prescriptive rules do not manage to affect common thinking and therefore behaviors as specific and strategic actions of social awareness and conscious participation.

In this, the world of design has done its part, developing various design approaches in the last decades, such as Universal Design, Inclusive Design and Design for All: these, although having common objectives and reference principles, differ above all from a methodological point of view.

In particular, Design for All (DfA) was the last in chronological order to appear on the international scene. For this reason it had to compare with the two previous ones, acquiring their basic objectives, but immediately seeking their own holistic dimension in the transversal nature of the various disciplinary contributions to the project and a specific attention to the qualitative and subjective aspects, in contrast to the exasperated specialization proposed by scientific disciplines, attentive to quantitative and statistical aspects [1]. The topicality of the DfA lies above all in involving, with a participatory and conscious approach, the entire supply chain of the decision: it aims more than others to encourage social inclusion in a creative way, through paths that actively involve the greatest number of interested people directly or indirectly to the topics dealt with from time to time. An "inclusive" and holistic approach to the project, therefore that places the individual at the center at large, with an emphasis on the political and philosophical, rather than methodological, dimension of the design issues. In this way it becomes a formidable tool for competitive relaunch, precisely because it aims to intercept new types of users of goods and services hitherto "excluded", and above all it aims to convince "everyone" that every project action carried out in the direction of inclusion manages to improve the lives of even those who are apparently not directly involved.

To achieve these objectives, every good DfA project has its own strategy of participation and awareness of the issues addressed, thus transforming each time into an opportunity to disseminate the DfA principles.

Over the years, an academic, professional and institutional community has consolidated around the Design for All approach, gathered around national associations (which since 1993 have been operating under the aegis of the European association EIDD-Design for All Europe) who actively contribute to promoting the principles of the DfA through specific initiatives. Among these is the Italian association Design for All Italia with the DfA Quality Label.

2 The DfA Quality Label

In an attempt to actually investigate and experiment with the DfA approach, the Design for All Italy association has been issuing a DfA Quality label for some years.

It was born in 2009 from an idea of three members of the association[1]. Intentionally avoiding the language and approaches of the many design awards, which only recognize the value of the final product, the DfA Quality Mark aims instead to recognize the value of the entire design process, which is identified in the theory and in the Design for All practice. The holistic and systemic dimension of Design for All, in fact, in addition to requiring the synergistic work of a team of specialists from different disciplines (designers, engineers, ergonomists, sociologists, economists) also requires continuous discussion with users and decision-makers at all stages of the process, an activity which therefore assumes a decisive role in the success of the project.

It is therefore not a prize, but a certification of DfA qualities, in the idea of stimulating, through this activity, a specific awareness in the public opinion towards the themes of Design for All.

The DfA Quality Label is aimed at companies and entities, public or private, who have developed a specific product, service, environment or system, according to the Design for All approach.

2.1 The Advantages of the DfA Quality Label

The advantages of the Quality Mark are manifold (for companies, public clients and professionals) as evidenced by the same initial document presenting the initiative.

In particular, the advantages for companies are different: a DfA project extends the market (because it attracts new customers and loyalty the occasional ones), it increases its competitiveness (because it differs from the competition by acquiring visibility and reinforcing its image thanks to the innovative solutions and concrete social commitment), it increases profits (because it responds with a single investment to the different needs to which multiple projects should respond).

The advantages for public procurement concerns the reduction of social tensions through the conscious and declared inclusion of the excluded population groups, the acquisition of greater visibility on one's territory (thanks also to the attention that DfA initiatives arouse above all in the media local), the opportunity to differentiate from the neighboring local administrations, the return in terms of reputation of the financed project successes.

Finally, the advantages for professionals are represented by the satisfaction and stimulus to work on innovative contents and creative solutions, by the opportunity to specialize on a transversal theme (and therefore not related to a specific sector), by the possibility of gaining visibility through a ethical-social design commitment.

[1] The original idea was from Daniela Gilardelli, Avril Accolla and Gino Bandini Buti.

2.2 The Process of Assigning the DfA Quality Label

The DfA Quality Mark spreads with respect to two different levels (Fig. 1). The highest level is that of DfA-Quality, assigned in response to compliance with all DfA principles and requirements relating to the type of assignment of the assigned product. The DfA-Start is instead assigned when not all the principles and requirements are fully satisfied, but recognizes the DfA intent of the project and its realization, which is expressed in the observance of the fundamental principles of the discipline and in the satisfaction of some significant requirements.

Fig. 1. The different levels of the DfA Quality Label

The Label is assigned through a particularly articulated process, which required the adoption of a large and articulated Regulation, which governs all the following steps:

1 The applicant (company or institution, private or public) submits to the Design for All Italy association (or rather to a special Quality Label Committee within the association) a specific request, accompanied by suitable documentation, relating to a product, service, environment or system, which has already been implemented, and which is believed to have been done according to the Design for All approach, i.e. verifying the DfA requirements with respect to the target user (made up of all those who wish to benefit from and have a reasonable probability of using the product, environment or system autonomously).

2 The Quality Label Committee carries out a first assessment of eligibility and verification of suitability: through the analysis of the material attached to the application, in order to define compliance with what is required in the regulation and to check if there are no obvious contradictions between the description of the product/system and the DfA features boasted.

3 The Quality Label Committee appoints two figures among DfA Italia members: a "manager of the procedure", who manages the development of the subsequent phases; a "person in charge of the orientation document", who, through a transversal reading of the reference sector of the product to be evaluated, draws up a document in which the DfA generic principles and requirements (of physicality, perception and sensoriality, understanding, process) are applied to the specific product sector of the product to be evaluated). This document will act as a guide for the subsequent evaluation phases.

4 The "manager of the procedure" organizes the materials, coordinating the work of a Jury, appointed by the Quality Label Committee and external to the association,

made up of three well-known personalities and experts from DfA. The "manager of the procedure" will participate in the jury meetings, but without the right to vote. In particular, the Jury, having directly viewed the product/environment/system to be assessed, and having read the "orientation document", collectively verifies the consistency between the product features and the DfA requirements set out in the "orientation document", so effectively decreeing autonomously (with explicit motivation) the attribution or otherwise of the DfA Quality Label.

5 A specific event is periodically organized by the Design for All Italia association for the assignment of DfA Quality Labels: these are events that, also through the collaboration of the main communication channels, contribute to the dissemination of Design for All even among non-insiders (Fig. 2).

Fig. 2. One of the recent events for the DfA Quality Labels assignments, Milan 2016

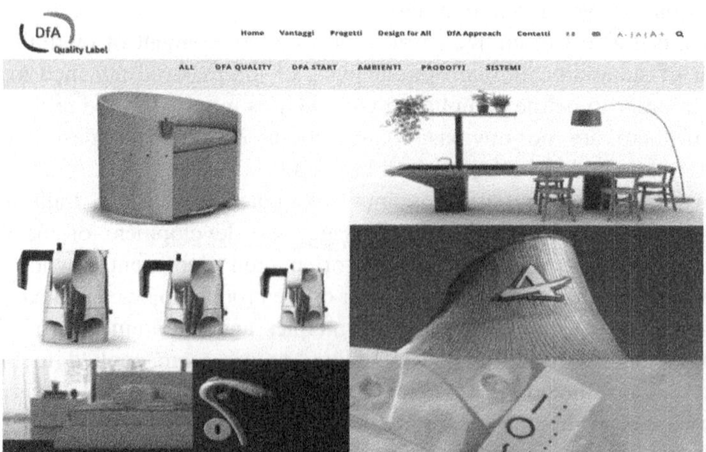

Fig. 3. A page of the DfA Quality label website, with some of the awarded products.

3 Conclusions

In 10 years the Design for All Italia association has awarded 12 quality Labels. In particular, 6 DfA-Quality and 6 DfA-Start (Fig. 3). The assignment events, held almost every two years in contexts of absolute prestige, attracted a varied audience, made up of both journalists, experts and curious people or people simply interested in the themes of design for inclusion.

In addition, during the years of activation of the DfA Quality Label, construction of a growing series of good examples of DfA products, environments and services was started in the most varied production sectors: from domestic kitchens to environments, from labels for the care of textile garments to motorized seats for indoor use. By definition, in fact, Design for All shuns any attempt to schematize and standardize the design process. In this regard, a few years ago Luigi Bandini Buti wondered if the disciplines of the project as they were divided into professions and specialized chairs put a brake on the development of projects: In his opinion, however, to respond to the multiple needs expressed by human diversity and from the complexity of reality, design would require a transversal and holistic approach. In the same way, he considered it unlikely to build rigid and schematic evaluation systems, which claim to evaluate the inclusive dimension of Design for All projects in a strictly scientific and quantitative manner, which instead require a critical and holistic reflection on the qualitative aspects of the solutions adopted.

The DfA Quality Label is trying to move precisely in this direction: it is a complex challenge to face, but which reflects the complexity of society and the individuals who are part of it.

Reference

1. Bandini Buti, L.: Prefazione. In: Accolla, A. (ed.) Design for All. Franco Angeli, Milano (2009)

Social Innovation in Fashion Design: Can Design Provide Opportunities of Inclusion to Refugees in Greece?

Giovanni Maria Conti(✉) and Maria Angeliki Panagiotidou(✉)

Politecnico di Milano, Milan, Italy
giovanni.conti@polimi.it

Abstract. Nearly 60.000 war refugees remain stranded in Greece [1], without any plan of amelioration for their living conditions. Design thinking strategy combined with social innovation could generate new scenarios of inclusion in the Greek society through the exchange of culture, knowledge and practical skills. In order to examine this hypothesis, members of Politecnico di Milano collaborated with Ankaa, a social NGO that offers vocational training to refugees in Athens. A one-month pilot period of workshops was organized to share practices of higher education in design of clothes and to awaken the creativity of participants by handmade processes. This period provided feedback for a long term educational offer designed specifically for this group of people. The aftermath planning included also a marketing strategy and entrepreneurial new scenarios to strengthen the organization, generate profit and finally to create a dialogue between the Greek society and this marginalized group.

Keywords: Social innovation · Clothes · Design · Inclusion · Education · Co-design

1 How Can Social Innovation Be Associated to Clothes Design

Social innovation is linked to bottom-up [2] solutions that focus on community participation, harnessing the skills and talents of the participants. This model [3] can be attached to educational opportunities of marginalized groups within the society, such as the refugees, in order to create social change. Education is seen as a tool to create new opportunities. Because this process relies on the initiative of the group and their involvement in strategic parts that may demand creativity, synthetic mindset, logistics, marketing and social skills, the educational offer challenges constantly the participants to think holistically and to take various roles. What is also important is the dynamics within the group. The students are seen as possible innovators and are given the opportunity of cross-disciplinary collaborations with local entrepreneurs thus, broadening their network and becoming members of the society.

This model is quite innovative for the fashion system, but the last decade can be found a lot of small enterprises and organizations that combine educational and professional offers for marginalized groups by making them vital part of their operating

G. Di Bucchianico et al. (Eds.): AHFE 2020, AISC 1202, pp. 24–31, 2020.
https://doi.org/10.1007/978-3-030-51194-4_4

system. Aligned with the principles of social sustainability, ethic production and transparency designers and companies collaborate with small communities of artisans and marginalized groups to strengthen their skills and then to offer them employment. Each model focuses on different skills and processes of learning depending on the skills of the group and the final collection of products that they create and sell. In this way, the students enhance their knowledge, get professional experience, improve their lives and finally earn their living while the facilitators gain new partners and can gradually scale up their business. Such an example is the I was a sari project where the designer Federica Vacca did co-design workshops with Indian artisans to create contemporary products from old saris, thus creating a new business model and economic boost to the local market and bringing them in contact with the European market [4] (Fig. 1).

Fig. 1. Two of the students of Ankaa and participants of the pilot project.

2 Primary Research in the Ankaa Organisation

Ankaa is a nonprofit educational center whose goal is to offer to refugees empowerment and opportunities of education, employment and social inclusion [5]. Their structure is focused on vocational training in various skills, while the last months have been created a professional production in collaboration with exterior partners. In order to design the one moth pilot project, a primary research was first done to understand their specific needs and prospects. Five visits were arranged where was assessed the space and tools available and most importantly some interviews were held. The director, coordinators and existing tailor teacher were asked separately some basic questions about their goal, processes, aspirations and notes from their experience of working with the refugees.

From their descriptions were highlighted also some characteristics of the refugee students. The fact that they are in a vulnerable situation facing daily problems in the camp is a fundamental aspect of the organization of the classes. The instability in students' life preoccupies a flexible schedule and the creation of a peaceful environment. Among them, there are beginners as well as experienced tailors with high skills. A mix class can give the opportunity to the advanced ones to become tutors and to the new ones to progress faster. By discussing with the existing teacher and assessing the space of the tailoring was examined the structure of the lessons and the final products. The advanced students were already creating clothes and accessories that were well made, but luck contemporary aesthetic and use of colors and materials.

The time spent in Ankaa prior and during the workshops showed the potential of the tailoring department and the possible ameliorations. Similarly, to the case studies seen on clothes design and social innovation, there was a need of alignment with the current western market and employment of creativity within the courses. Moreover, it was clear that the students were very introvert and used in traditional methods of learning by prioritizing doing over thinking. A last crucial factor was the absence of a common language to communicate, thus creating the need of visual tools.

3 The Pilot Project

The general structure of Ankaa followed the bottom-up logic and innovative thinking, but the class of tailoring, lacked professional and contemporary elements of production as well as the use of personal intuition and decision making of the students. To find adequate ways to solve this issue was held a one-month pilot workshop with ten students of tailoring using handicrafts and design thinking strategy.

3.1 The Goal

The final goal of this project is to offer to the participants the opportunity to become active members of the Greek society. Training in clothes construction is firstly a tool to gain practical skills, experience, find employment and broaden ones network. The holistic approach of clothes design encloses also a research part, knowledge of history, colors, fabrics, trends and handicraft techniques that are linked to cultural awareness and non-verbal messages. By familiarizing with these concepts, the students can become accustomed to the western culture, but also use the graphic signs on clothes as tools to express their personality as well. Thus, the creation of clothes can lead to cultural awareness and provide them the opportunity to get a voice in the society.

The practical goal of the pilot project was to meet the participants, assess their skills and the potential they have as a group and redesign a long-term educational plan. The primary goal of the training offer would be to enhance their skills and employability opportunities. The second aim would be to empower them and awaken their intuition by introducing the design thinking approach and urging them to take an active role in the classes. The final goal was to use the cultural background of the participants as a source of inspirations and find creative ways to incorporate it in the design process to promote their cultural heritage. The promotion of students and the products made can create awareness to the public and lessen the polarization that exists between them and the Greek society (Fig. 2).

Fig. 2. During the pilot project we collaborated with Fahima (on the right) that is the tailoring teacher of Ankaa and the rest of the students to co-create original patterns as in a team.

3.2 The Structure

The structure of the pilot project was flexible and broad in order to suit the participants needs, challenge their minds and urge them to take an active role and express themselves through the process of creation. The nurture of critical thinking presupposes a vast vocabulary of tools and experiences that can generate new answers and ideas on complex problems. This idea came initially from prior experience in architectural studies that follow the Bauhaus model and secondary research on co-design [6]. Introducing a variety of techniques that may seem unrelated to clothes, such as paper collage, can awaken the imagination of the students and make rethink the way that they use their hands, abilities and personality.

In all of the processes was accentuated the use of geometry as method to redefine the human body, question the standards and to finally create new patterns of clothes by doing synthesis of shapes and not by following traditional guides. The geometrical thinking together with eastern culture and western contemporary aesthetic were the main axis that influenced the choice of tools, processes and final aesthetic of the products created.

3.3 The Tools

The framework of this session was chosen to be wide and to include an introduction of different tools and means used in design disciplines, with the aim to create many stimuli and open the students' mindset. Secondary sources [7] like visual presentations of contemporary examples were firstly used in order to introduce the participants to the needs of the market and the western culture. Before passing to construction techniques, the participants were taught sketching and collage techniques that would provide a visual tool of communication and enable them to express their own thoughts. Then various handicraft traditional tools were implemented to urge the students to use their creativity and see various processes that can be associated to clothes construction such as patchwork, embroidery, hand sewing, block print and geometrical pattern making. The techniques that were chosen were all handmade to emphasize on the handicraft value and tradition and find links between the eastern culture and the contemporary European market. At the end were used also sewing machines in order to finish the products and use practices that are aligned with the recent manufacturing processes.

3.4 Occupational Therapy and Handicrafts

The choice of handicraft as tools occurred also because of its psychological effects on the makers and it's link to occupational therapy. Repetitive practices that presuppose the coordination of hands, mind and heart have a healing aspect on the makers [8]. This process necessitates a deep concentration, thus soothing of the mind from negative thoughts. Observing the participants during the workshop there was a peaceful atmosphere and each one seemed detached from the reality, expressing visually their thoughts on the cloth. This process in clothes was inspired by research from ancient groups of weavers, to the quilting bees [9] of the 19th century and the contemporary movement of craftivism [10]. In all of these examples groups of people gather regularly not to work, but to co-create something while socialising and expressing their personality on the cloth. In the recent example of craftivism, this action is linked also to common political and social beliefs of the group and the creation of messages through their crafts. All these can provide the opportunity not simply to learn a new skill, but to use this tool as a healing process and as the medium to communicate messages to the society and show the human point of view of the war.

3.5 The Practical Outcome

During the first lessons, each one was creating individually an artwork, whether it was a sketch or an embroidered fabric by using specific tools given and personal experiences as an inspiration. This lead to some very powerful images that contain social and political messages that if spread can aware people about this social phenomenon of displacement and losing one's home.

The final goal of the trials of various techniques was to observe how the students can use them synthetically to create a garment. All of the processes were collaborative, from making sketches and mockups on the mannequin to the decision-making and division of the work of the final products. The result was a collection of caftans and shirts with handmade block prints and embroidery details representing the common aesthetic of the group.

3.6 Conclusions from the Pilot Project

One of the biggest challenges was to understand the differences in terms of mentality, experiences, education, culture and social life that played a major role in the participation of the students. Coming from a design field we took experimentation and expression for granted, while for them were unknown terms. The participants were all used in being productive and doing manual jobs, thus needing a very specific context and examples in order to unlock their intuition and to start feeling comfortable expressing personal opinions. Even the shyest people became active participants in the group when they were given visual tools and slowly understood the process of co-creating in a group. The goal of the participants is to find employment, thus being efficient workers, but by experiencing this new approached they understood the variety of processes that can be linked to clothes apart from sewing, the value of artisan products, the role of design and the satisfactory feeling of expressing through this process.

4 Long Term Educational System

The long-term training offer that started this January aims at empowering the participants and at their alignment with the recent professional practices. Visual tools such as samples, presentations, sketches and guides of techniques were created in order to facilitate the communication. The main tools taught are pattern making and sewing in order to facilitate the employability, while added handmade processes are incorporated on each design.

In parallel with the courses is held a professional production within Ankaa with the aim to create a collection, generate profit and empower the organization. Thus, the courser of tailoring serves as the preparation of the students for their recruitment in the production space, while providing also the ground of experiments and research of aesthetic of the final products. Because of that the students are given the opportunity to experience all the phases from the creation to the finishing and branding while collaborating with more local designers, organizations and suppliers.

5 Future Scenarios and Marketing Strategy

The final goal of this venture is to create links of communication between the refugees and the Greek and Western society. Growing the network of Ankaa is fundamental for the creation of more opportunities. Weather this means involving more local actors in the courses and production such as local artisans or by creating a dynamic relationship with the audience. Starting from the virtual reflection of the organization, it should emphasize on the narration of personal stories of the participants and then present the courses and products made in order to create empathy [11] and engage the audience. Using network-focused design [12] to create durable connections requires the fostering of a meaningful relationship through media that interfere with the daily routine of people. By sharing honest information about the people involved, their motives, values and daily activities within Ankaa the audience would be more interested to look the products that are made. Most importantly, they may be even interested to join this effort and share their skills and experiences, thus broadening the network of active members with shared values (Fig. 3).

Fig. 3. Part of the pilot project was to experiment on handicraft techniques such as blockprint, but also to create processes that would enable the active participation and expression of all the students. Handicrafts and artistic methods provided a more positive atmosphere where people felt more confident and socialized, thus giving us more feedback.

References

1. UNHCR operational portal for refugee situations. https://data2.unhcr.org
2. Betts, A., Bloom, L., Weaver, N.: Refugee Innovation: Humanitarian Innovation that Starts with Communities. Refugee Studies Centre (2015)
3. Saul, J., Social Innovation, Inc.: 5 Strategies for Driving Business Growth Through Social Change. Wiley, San Francisco (2010)
4. Company I was a Sari. http://www.iwasasari.com
5. The Ankaa Project. https://www.ankaaproject.org
6. Meroni, A., Selloni, D., Rossi, M.: Massive Codesign. Franco Angeli, Italy (2018)
7. Thittichai, K.: Experimental Textiles. Anova Books, Milan (2009)
8. Monson, N.: Craft to Heal: Soothing Your Soul with Sewing, Painting, and Other Pastimes Hats Office Books (2005)
9. Gordon, B.: Textiles: the Whole Story: Uses, Meanings, Significance. Thames & Hudson, London (2013)
10. Corbett, S.: How to be a Craftivist. Unbound Publishing, London (2017)
11. Barnes, C., Blake, H., Pinder, D.: Creating and Delivering Your Value Proposition: Managing Customer Experience for Profit. Kogan Page Publishers, London (2009)
12. Rubino, S.C., Hazenberg, W., Huisman, M.: Meta Products: Meaningful Design for our Connected World. Bis Publishers, Amsterdam (2011)

Fashion and Inclusive Design: Assistive Technologies Applied to Clothing

Grazyella C. O. Aguiar[1(✉)], Leonardo M. Rincon[1], Miguel A. F. Carvalho[2], Christian Mailer[1], Fabieli D. Breier[1], and Fernando B. Ferreira[2]

[1] Department of Textile Engineering, Federal University of Santa Catarina, Blumenau, Brazil
grazyella.oliveira@ufsc.br
[2] Department of Textile Engineering, University of Minho, Guimaraes, Portugal

Abstract. The present work reports the results of a research project carried out by the Textile Engineering Department at the Federal University of Santa Catarina (Blumenau, Brazil). The project's main objective was to develop inclusive apparel textiles, meeting the different bodies and biotypes of people with or without disabilities. Contributing to the development of assistive technologies applied to clothing. Thus, in 2018, an inclusive fashion collection was developed for people with physical disabilities, co-produced with the members of ABLUDEF - Blumenau Association of Physically Disabled. The methodology used was based in a bibliographic research, deepening the approached themes in the project, as design and inclusive fashion, assistive technologies, and accessibility, as well as field research, conducted with ABLUDEF members. Among the final results, the garments made for the project were highlighted and modeled at the 6th South Brazil Awards for Inclusive Fashion, in Florianopolis (Brazil).

Keywords: Inclusive design · Fashion for all · Assistive technologies

1 Introduction

Today's fashion is becoming increasingly democratic in response to different styles, tastes and economic powers. Although fashion companies are concerned with serving different segments and consumer profiles, there is still a lack of studies and product development for people with disabilities.

Attempting to meet these market demands, the research project "Fashion is for everyone: a study of the principles of Universal Design applied to clothing" was created and developed by the Textile Engineering Course at Federal University of Santa Catarina (UFSC), Blumenau, Brazil, in partnership with the Department of Textile Engineering of University of Minho (UM), Guimaraes, Portugal.

The project, carried out in 2018, aimed to develop a collection of inclusive clothing for people with physical disabilities, in co-production with members of the Blumenau Association of Physically Disabled (ABLUDEF), based on a reflection of the different dimensions of sustainability. Among the different pillars of sustainability, the social

G. Di Bucchianico et al. (Eds.): AHFE 2020, AISC 1202, pp. 32–37, 2020.
https://doi.org/10.1007/978-3-030-51194-4_5

and the cultural were highlighted. Social sustainability was contemplated here through the development of products for people with physical disabilities and the involvement of other co-participating institutions. Cultural sustainability excelled in considering art as one of the aesthetic expressions inherent in human creation.

The methodology used in the project, applied bibliographic and field research, striving to strengthen pertinent concepts, such as inclusion, accessibility, inclusive fashion, universal design, sustainability, as well as Brazilian art featuring a Brazilian artist chosen as an inspirational theme of the collection.

2 Applied Methodology and Obtained Results

As mentioned before, the work presented here is the result of the research project developed in 2018, in which two volunteer students participated on a development team. One from the Textile Engineering course, involved in the field research, design of the collection, study of pattern design and study of the assembly process; and the other, a Control and Automation Engineering, involved in the successive tests to verify the feasibility and reliability of the technology, assembling all electronic devices, performing the programming and designing the application layout.

The objective of the project was to work in an integrated way with the ABLUDEF members, trying to meet all the demands observed for the creation of the collection. Thus, the partnership with the professionals and associates of the institution involved in the project was fundamental to create a relational, co-production and co-creation work system. Based on this principle, a set of actions were developed, allowing the relationship between the activities and the knowledge built throughout the year. Initially, it was considered the development of a collection targeting the needs of wheelchair users, incorporating in the garments technology to detect falls, although during the implementation of the project, it was realized that these features of clothing could also help others with different types of special needs, allowing to increase the target market, proposing products that could benefit more people.

According to the data from the 2010 Population Census, released by the Brazilian Institute of Geography and Statistics (IBGE), it is estimated that approximately 1/4 of the Brazilian population has some type of disability: "More than 45.6 million Brazilians reported having a disability. [...] The number represents 23.9% of the country's population" [1].

The methodology used in the project favored bibliographic and field research, with the use of participant observation sociological research, that consists of the actual participation of knowledge in the life of the community or group. The observer/researcher assumes, at least to some extent, the role of a group member. Therefore, it is a technique by which one comes to know the life of a group from within itself. Interviews were conducted with the director, teacher and some members of the ABLUDEF Association. To be able to detect the real needs of the subjects involved, a questionnaire with mixed questions (close and open), was applied.

By means of contact with ABLUDEF members, it was identified that the people who suffered the most from falls were those with physical disabilities with malfunction or paralysis of the lower and/or upper limbs, making it difficult to move. According to reports, falls were often frequent both in the street and at home. Most of the respondents used some kind of equipment that helped them to get around, such as walking canes, crutches and walkers, and most of them lived alone. When asked what could be improved in clothing, some answers guided the development of the collection, such as: pieces that would help them during the task of dressing, turning it easier, making them more autonomous; pieces without zippers, since some had difficulties in their hands; pieces slightly larger and more comfortable; and pieces that would not limit their locomotion.

For the development of the collection, it was used the planning methodology described by Treptow [2] as theoretical support, as well as the premises of Universal Design. As a result of the research, a set of garments were created, which endeavored to somehow meet the real demands of the users.

Employing the theories of Universal Design, products should be usable for everyone regardless of their disability. To do so, the necklines of the blouses were more open and had the same height in the front and in the back, making it easier to wear. For example, a person with visual impairment would not experience the situation of failing dressing it, avoiding the possibility of ending using it backwards. All seams of the garments had a clean finish, reducing discomfort with skin contact.

Trousers had an elastic waistband, allowing greater comfort to the wearer as it will grow as needed and stitched sideways for easy wearing. Most members sit for a long time, thus, a rigid, buttoned waistband ends up irritating the belly and sides, region that accumulates fat over time. Trousers were designed wider to disguise the use of a diaper, if the wearer needs it. The pattern design of trousers was looser and straighter, aiming to avoid arrested movements and block blood circulation of lower limbs. In terms of materials used, all pants were made with functional materials, donated by Invel Medicinal Clothing, featuring a unique patented technology from Bioceramica® MIG3®, infrared irradiator. This innovation incorporated into the tissue stimulates the production of nitric oxide (NO), soluble gas that has vasodilating and bronchodilating properties, which provides better blood circulation, bringing benefits to the user. This way, the whole leg is stimulated while wearing the pants. This population sits for long periods, resulting in low mobility of the lower limbs, and most have problems with blood circulation, causing problems such as edema (swelling) and even wounds.

A long vest was also made and sublimated, printed with the word "Palhaça" (female clown), by Gonçalo Borges, teacher, speaker and inclusive visual artist. Gonçalo paints his works with his mouth and his feet. He is currently a member of the Mouth and Foot Painters Association (APBP).

The key piece is a jacket with a Wi-Fi microcontroller inside to detect falls of users with poor mobility in the upper and lower limbs. The sensors, connected to a portable cell phone battery and a Wi-Fi network, were incorporated after the recognition of its relevance from the contact with ABLUDEF members, referring the constant number of falls. "Palhaças" (two female clowns) were printed on the back of the jacket. The artwork represents the affection between mother and daughter.

The fall detection system consists of a gyroscope/accelerometer, microcontroller with Wi-Fi connection and a buzzer. Through the gyroscope/accelerometer data and calculations performed by the microcontroller, it is possible to determine if the user has suffered a fall, as it accurately detects a fall in motion. If there is a fall, the electronic device beeps and sends an alert message via Wi-Fi to the desired contact (friend, family member, caregiver, among others). The buzzer alerts people close to the user if an unforeseen event occurs. In case of accident, the monitoring needs to be done remotely, the message alert causes a previously registered contact's mobile phone to receive a message via the application.

All components are assembled and attached together to occupy the smallest possible space and were inserted in a 3D printed box, providing greater flexibility and robustness for the garment. The electronic devices are powered by a common Power Bank battery (portable charger), aimed to charge mobile phones, offering great power to the device and autonomy to the user. The jacket, device, and application interface are presented in Fig. 1.

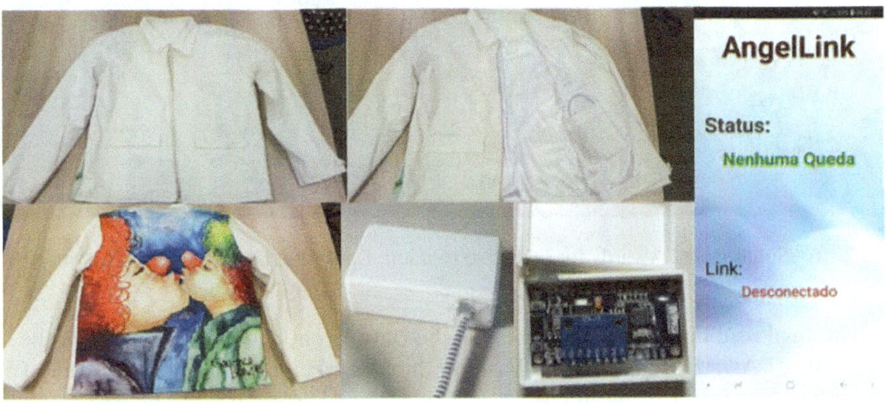

Fig. 1. Jacket equipped with falls detection system, representing the front view; the interior view (showing the pocket where the box with the device is inserted); the back view (with the print "Palhaças" by Gonçalo Borges); the 3D printed box; the electronic devices inside the box; and the application interface.

During the research, several tests were performed to collect information related to the movement of individuals during their daily life (walking, running and climbing stairs) and simulations of falls, in order to isolate and differentiate patterns present in each of these actions. The graphics obtained were compared by analyzing the difference between the patterns. In the end, it was concluded that the calculation of the parameters Sum Vector Magnitude (SVM), Euler angle, Gravity-weighted SVM (GSVM), Gravity-weighted Differential SVM of Acceleration (GDSVM) were essential for the correct determination of a real fall.

The project was submitted, classified and presented at the 6th South Brazil Awards for Inclusive Fashion. The exhibition took place on November 28, 2018, in the city of Florianópolis, Santa Catarina (Brazil). The three looks produced in the collection can be observed in Fig. 2, during its presentation in the event by ABLUDEF members.

Fig. 2. 6th South Brazil Awards for Inclusive Fashion. Looks modeled by ABLUDEF members.

The final results of the project include the runway show of the collection at the 6th South Brazil Awards for Inclusive Fashion; honors received from the Blumenau City Council; achievement of the finalist category of the 3rd Creative Brazil Award, in the Design category; reports in the mainstream media; nomination for the Academic Merit Commendation; support from an alderman; and work published on social media and broadcasted on Blumenau Legislative Television.

3 Conclusion

The main objectives of this project were achieved, highlighting the possibility of working in an integrated manner, providing interaction between university and community. The project offered the scholars involved greater knowledge about the different dimensions of sustainability, Brazilian and regional art, as well as provoking the perception of social responsibility in the diversity of our world.

Finally, the project has contributed to a greater conscious of the market demand and possibility of a building a more egalitarian society. This project intended to develop products that target people, especially those with disabilities, considering their references and social representations, with the objective to provide them greater autonomy and consequently better quality of life. Corroborating with this idea, is the quote from the teacher, artist and speaker Gonçalo Borges [5]: "It is not the law that will protect the disabled. It is awareness."

Acknowledgments. We would like to acknowledge the Department of Textile Engineering at Federal University of Santa Catarina (Brazil) and 2C2T-Science Center for Textile Technology from University of Minho (Portugal). This work is financed by FEDER funds through the Competitive Factors Operational Program (COMPETE) POCI-01-0145-FEDER-007136 and by national funds through FCT-Portuguese Foundation for Science and Technology, under the project UID/CTM/000264.

References

1. https://www.ibge.gov.br/en/statistics/social/population/22836-2020-census-censo4.html. Accessed 14th April 2015
2. Treptow, D.: Inventing fashion: Collection planning, 4 edn. D. Treptow, Brusque (2009)
3. Borges, G.: History. https://goncaloborges.com.br/historia/. Accessed 14 April 2015

Analysis of the Sensorial Characteristics of Materials for Inclusive Design

Stefania Camplone(✉) and Emidio Antonio Villani

University of Chieti-Pescara, Viale Pindaro, 42, 65127 Pescara, Italy
stefania.camplone@unich.it,
emidioantoniovillani@gmail.com

Abstract. In recent years the study of materials has extended from technical-engineering characteristics to sensory and perceptual ones. The material of an interface, in fact, also assumes a communicative value as well as contributing to improving or promoting the interaction of individuals with products. This also considering the different levels of reaction to sensory stimuli that individuals can have, and therefore the different perceptions, also in relation to their socio-cultural experience. Here we present the results of an educational laboratory which had as its theme the analysis of materials for industrial design and their sensorial characteristics, with particular attention to inclusive design. In particular, the analysis process initially envisages a reading of the characteristics of the materials through the students' descriptive skills related to the visual, tactile and acoustic dimensions, with reference to their use in the conformation of a product. Subsequently, the objects analyzed are also broken down into their main parts, while an original graphic summary rendering allows to evaluate the relative weight of each sensory aspect in the perception of the products, also in relation to human diversity.

Keywords: Materials · Expressive-sensorial features · Design for inclusion

1 Introduction

Every day we come into contact with different products that transmit their expressive qualities to us, and that we recognize through different perceptual levels. The products, in turn, can change their expressive ability through the materials that compose them. Thinking about the material is an educational laboratory based on the taxonomic research experience of sensory materials, which intends to build their characterization through the sensorial experience of the user.

The materials, in fact, are capable of activating our visual, tactile, olfactory and auditory channels, generating a new sensorial and communicative dimension which is particularly effective for identifying new qualities, aimed at a more inclusive project. The importance of the role played by the materials in the relationship with the "human factor" of the users, however, does not seem supported by the development of as many analysis and evaluation tools at the service of those who then have the responsibility of selecting the most suitable material for a given project that is the designer.

G. Di Bucchianico et al. (Eds.): AHFE 2020, AISC 1202, pp. 38–44, 2020.
https://doi.org/10.1007/978-3-030-51194-4_6

The interest of research stems from these assumptions and, more generally, from a simple consideration: the materials can, through their sensorial dimension, stimulate creative thinking, to respond to particular project specifications. Furthermore, a further consideration should be made.

The evolution of technology has made it possible to modify the characteristics of a material, designing it to your liking and also loading the products with sensory, expressive and symbolic qualities. The interpretation of the product by its possible users, and the consequent attribution of sensorial, communicative and symbolic values, is strictly connected to the way the senses are stimulated: it is certainly the sense that first "understands" the objects. It is evident that a product, through the material, tends to relate to all the senses, but it is equally understandable how the communicative aspect is instead determined by simple comparative physical-technological observations that describe its "character". The symbolic aspect is different, so that the product or material can be charged with "affective values" generated by the user's experience: he constructs his own interpretation based on the context, lifestyles, values, which they are different for each individual [1].

In general, the user expresses his evaluation of the product with which he interacts with three categories of judgment: aesthetic impression, semantic interpretation and symbolic associations. The assessment is conveyed through the senses which, based on the stimulations received, produce sensations that the brain processes in perceptions, which in turn are connected to the individual's experiential field. However, sensory experiences inevitably generate associations and meanings for products and materials, which can be traced back to one's personal history based on the cognitive process of associations and memories, for this reason, a material can generate attraction or aversion towards an artifact and therefore arouse an emotion. Since emotion is one of the major purchasing factors of a product, it is understandable how the choice of finishing materials is decisive for contributing to the creation of a more inclusive product and at the same time how fundamental it is for its success on the market.

2 Objectives: Choose Materials with Awareness

The paper presents the results of research conducted within the course materials for the design carried out in the Department of Architecture of Pescara. The main aim of the research was to develop students' sensitivity towards materials towards a more conscious choice. In this sense, the research conducted on samples of materials has on the one hand allowed to expand the students' vocabulary to describe the expressive-sensorial characteristics of products and materials, on the other to intercept sensations and emotions in the hypothesis of a more inclusive project.

The knowledge and awareness of the sensitive characteristics of the materials, in fact, aims at a design that is also more aware and more attentive to the needs of each individual, based on his or her sensory experience and cognitive abilities.

3 Method: The Educational Workshop

The educational workshop "think about the metter" used a singular and original methodology, divided into two distinct phases. On the one hand, the material is explored through its expressive and sensorial character capable of having a significant impact on the preferences and tastes of users of environments and products. On the other, through a backward process of decomposition of the product, up to the identification of its constituent materials, with the aim of understanding both the functional characteristics and the relationships that they establish with the users through the involvement of their own sensory qualities.

The investigation process develops in two phases. A first phase, intended as the "exploration" of samples of materials, the purpose of which is to verify how much it is possible to attribute to the materials a property related to sight, touch, hearing, highlighting even the most intangible characters (such as expressive, associative and emotional) and with the aim of outlining a profile. The second phase, of "experimentation", finds as the field of application that referred to the objects of use. The choice of the product by the students made it possible to verify, through their emotional experience, whether the material-sensory profiles identified previously corresponded to those generated during their relationship with the chosen objects.

The investigations, carried out on products of use and their materials, highlight sensory and symbolic-communicative aspects of considerable interest, with which it is possible to elaborate quantitative indicators of expressiveness, which can be collected in a table of sensory properties and emotional materials.

3.1 First Phase: Explore the Materials

In the exploratory phase, attention is paid to the analysis of materials, deciphering their characteristics through the senses. The sensory characteristics, in fact, derive from the interaction between the material and the user, who perceives its expressive qualities through polarity contrasts. In the first phase, therefore, groups of students developed a pair of tables, the first "sensory" and the second "expressive".

The first problem was that of being able to add the surfaces of a selection of materials in reference to the senses of sight, touch and hearing. To this end, each "sensory table" is structured to accommodate nine samples of materials, three for each sensorial aspect (photometric, tactile, acoustic). For each of these aspects, it is necessary to choose a pair of polarity contrasts on a list of possible adjectives or pairs of opposites (Table 1), with respect to which subsequently select three different materials that were characterized by these opposite sensory characteristics, according to a scale of increasing values.

Furthermore, since the selection of the materials to be analyzed must necessarily be dissociated from any shape or object in order not to be conditioned by the evaluation, the samples have been uniformed in shape and size (9 $\times$ 7 cm). The olfactory aspect and taste are not part of this investigation, as factors that do not contribute to providing useful values for this research.

Table 1. Polarity contrast

Polarity contrast		
Photometric aspect	Tactile aspect	Acoustic aspect
Translucent-Transparent	Smooth-Rough	Loud-Quiet
Hard-Soft	Cold-Hot	Severe-Acute
Reflective-Absorbent	Light-Heavy	Uniform-Variable
Natural-Artificial	Hard-Soft	Harmonic-Tune
Heterogeneous-Homogeneous	Tenacious-Yielding	Reflective-Absorbent
Color-Neutral	Velvety-Rough	Melodic and dissonant
Polished-Satin	Malleable-Drive	Weak-Ringing

A further attribute of each "sensory table" (Fig. 1) was also to associate a specific symbolic quality and meaning that each of the selected materials expresses. In this, words similar to those used to describe people's characters have been used: they are therefore expressive properties (rich, poor, attractive, chaotic, traditional, futuristic, sober, sophisticated, charismatic, playful, lively, ancient delicate, daily, calm, cutting, pleasant, irritating, extravagant, simple, slimy, hi-tech…), and therefore external, associated with materials rather than physical properties, which instead are intrinsic to the material.

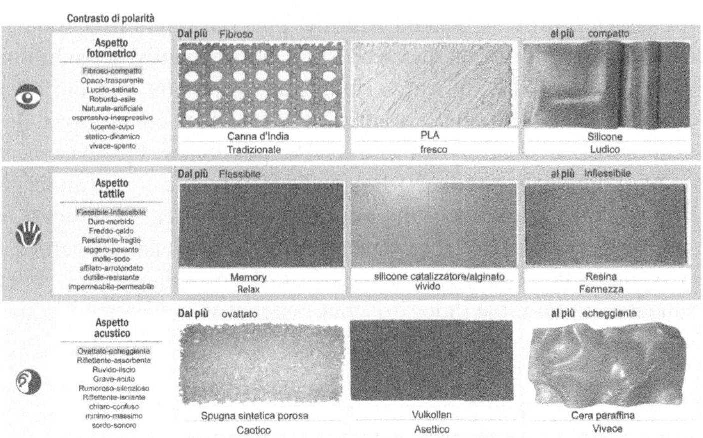

Fig. 1. Sensory table. Attribution of sensory characteristics to materials.

After comparing the chosen materials through polarity contrasts, the research process proceeds in analyzing the material from a communicative and symbolic point of view.

The expressive table (Fig. 2) instead analyzes the individual materials trying to associate some characters to each. Indeed, a material can evoke thoughts, feelings, associations. They "mean" different things in different situations related to the context of use, the way they are used, with respect to influences and influences of the experience lived by the user. Here the ability to associate each individual through their own field of experience comes into play.

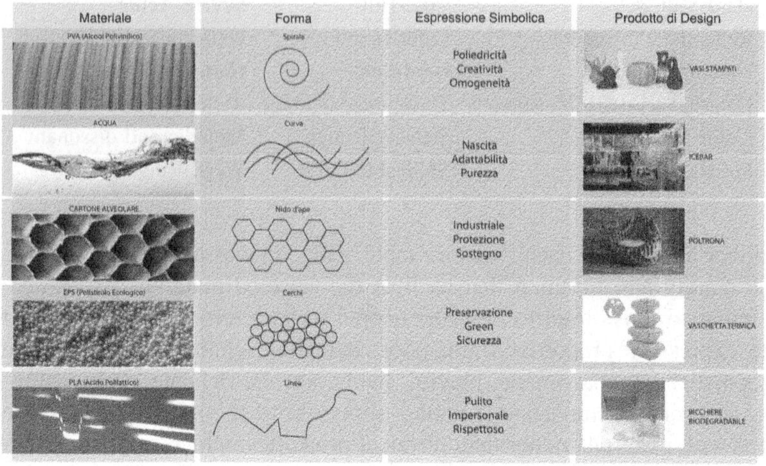

Fig. 2. Expressive table. Attribution of symbolic characters and association to materials.

The first character to be associated with the material refers to a simple geometric shape, which refers directly to its characteristics. Subsequently, a "symbolic" expression is attributed to the material: a description or expression that the brain suggests through terminological associations resulting from mental connections. It is evident that these elaborations are entirely subjective. The possibility, however, of involving a large sample of individuals (students) has made it possible to attribute greater "scientific" reliability to the selection of the information collected, to group it, measure it, interpret it and analyze it in the table of sensory and emotional properties.

Finally, on the same "expressive table", we ask to associate the material with a product, to understand how the "idea" of that material is concretized in the mind and experience of the observer.

3.2 Second Phase: From the Object to the Materials

In the second phase of activity of the didactic laboratory, the analysis process instead foresees a backward observation path, starting from a finished product to reach the material with which it was made. The exercise is particularly useful for understanding the various stages of construction, the production processes of the materials used and the surfaces.

After breaking down the product into its individual parts, the "technological table" (Fig. 3) describes each component through the material and the production process.

Finally, a "table of expressive and sensorial characteristics" analyzes the individual parts also from the point of view of their sensorial and emotional properties (Fig. 4), attributing to them a numerical value (from one to ten). In particular, the sensorial characteristics are again divided into photometric, tactile and acoustic characteristics, each referring to certain groups of adjectives. The expressive characteristics, on the other hand, are linked to symbolic associations, and therefore strictly connected to the individual's experience, his experiential field and the relative context that surrounds him. For each component, therefore, an "interpretation" of the material is attempted, quantifying, through a real numerical scale, the value of each adjective taken into consideration (Fig. 4).

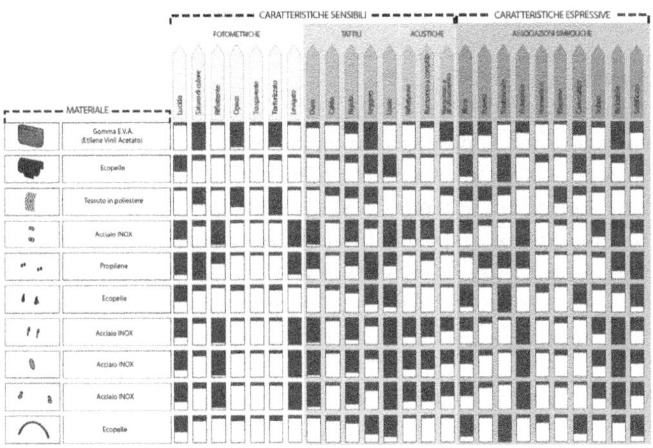

Fig. 3. Technological table.

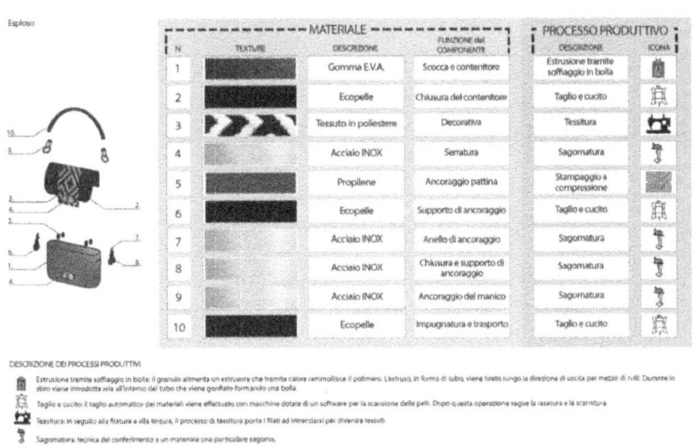

Fig. 4. Table of expressive and sensorial characteristics.

4 Results

The main result of the study was a considerable collection of information relating to different materials. These have been systematized in order to build a first database relating to sensory experiences and the production processes associated with them.

The experience conducted highlights how each individual perceives the materials differently in relation to their own experience. The attempt was to be able to conduct a comparison between the numerous associations, also to be analyzed statistically, to be useful to designers, also and above all in relation to human diversity, with all the experiences it expresses.

5 Conclusions

The educational workshop is proposed as an experience of knowledge of materials and their particular expressive and sensorial qualities.

In particular, the experience has proved particularly useful for students and their sensitivity to the importance that a careful and conscious choice of materials, even from an expressive and sensorial point of view, plays for the product development process. This especially in a scenario attentive to human diversity, both physiological and cultural.

Acknowledgments. This contribution reports the experiences conducted in the context of a Course in Design Materials, carried out at the Department of Architecture of the University "G. d'Annunzio" in Chieti-Pescara, held by Prof. S. Camplone, with the operational support of Emidio Antonio Villani. This contribution was written by Stefania Camplone (paragraphs 2, 3 and 4) and by Emidio Antonio Villani (paragraphs 1 and 5).

Reference

1. Ashby, M., Johnson, K.: Materials and Design. Butterworth-Heinemann, Oxford (2002)

User Friendly Eco Bag for Everyday Life

Stefania Camplone(✉), Giuseppe Di Bucchianico, and Grazia Ventrella

University of Chieti-Pescara, Viale Pindaro 42, 65127 Pescara, Italy
{Stefania.Camplone,Pepetto}@unich.it,
Grazia_Ventrella@hotmail.com

Abstract. The bag is a multi-tasking accessory for women, which combines their vain dimension with the utilitarian one related to different female personalities, the narcissistic character and the need to carry some "tools", which satisfies and adapts to their changing needs during the day. A bag is an accessory that "helps" but that, at the same time, "confuses" because of some problems, including the difficulty of finding objects inside, where everything is often stored in a disorderly way. Starting from these assumptions, here is the result of a research developed at the Department of Architecture of the University of Chieti-Pescara as part of a master's degree thesis. In particular, specific analyzes were conducted on the types of bags most used by a wide variability of female users in different contexts of use. The research ends with the concept design of a storage bag that responds to the main female needs, with particular attention to the realization through the use of environmentally sustainable 3D printable materials.

Keywords: Design for inclusion · Eco bag · 3D printed

1 Introduction

The bag is an indispensable and multi-tasking accessory, a combination of aesthetic taste and functional value, used by over 3 billion women[1] to carry with them, every day, a wide range of objects at any time and place.

At the same time, however, bags are not always simple to use. An "unorganized" bag leads to a reduction in usable space, a waste of time, etc. A German study, in fact, has calculated that on average each woman spends 76 days of her life looking for one of the many things she keeps in her handbag. A further investigation, conducted by a private company[2] out of about 1,500 women in order to find out the habits of a woman with her bag, confirmed the previous data in excess (81 days, which for almost a fifth of respondents double), also discovering that 34.5% of the women interviewed spend at least five minutes a day in an attempt to find their car keys, while only 3.7% of women declare to know where to find everything.

---

[1] In questo momento nel mondo ci sono 3,64 miliardi di donne [1].

[2] It is the Pilot Pen Australia Company.

G. Di Bucchianico et al. (Eds.): AHFE 2020, AISC 1202, pp. 45–51, 2020.
https://doi.org/10.1007/978-3-030-51194-4_7

Add to this, that the constant use of an accessory that has variable weight and bearing can cause postural problems for those who carry it.

Although this is an apparently trivial matter, facilitating women in the most comfortable and effective use of a frequently used accessory such as their bags certainly contributes to improving their daily lives, also in reference to frequent situations that may already be in themselves uncomfortable (in low light conditions, with busy hands, etc.).

Starting from these questions, a project research was carried out at the Department of Architecture of the University of Chieti-Pescara (Italy) in order to identify a possible design solution to reduce the time taken by a woman in "managing" a bag, that is, in finding easily the objects carried inside.

2 Method

The research was divided into successive phases. In particular, starting from a preliminary investigation and from a research on the state of the art, an in-depth fact-finding survey was carried out on the use of bags by women, in order to be able to define a significant framework of product requirements.

2.1 The Preliminary Investigation

In order to bring out the problems related to the use of bags in a more structured way, a preliminary investigation was carried out on a sample of 10 women, randomly chosen with different ages.

By asking them the question "What are the problems related to the use of your bag?" (Dwelling in this way on the most usual and daily gestures), some common problems have emerged: difficulty in finding the most common objects (keys, handkerchiefs, lighter, etc.), accumulation of receipts and other small objects, earphones that get tangled, embarrassment in making visible some objects (sanitary napkins, etc.), coins that fall from the wallet are lost in the bag, number of objects or compartments, dew on the surface of bottles with cold water that bathes the other objects, non-compliant dimensions to occasionally contain bulky objects (tablets, books, etc.), excessive weight, uncomfortable handles and shoulder straps (they leave marks, damage the hair, etc.), incorrect posture, accumulation of dirt, etc.

Basically the main problems therefore refer to the "internal organization" and to the "cleaning" of the bag: at the first, it has also been observed that each woman takes care of herself individually by using pouches recognizable by color, size and different material to the touch; to provide the second, the most common solution is to periodically schedule an internal cleaning of the bag to get rid of unnecessary items (receipts, pieces of paper or other).

2.2 The State of the Art and Typological Classifications

After identifying some main problems attributable to the lack of organization and the portability of the generalities of the bags, a first survey was carried out on the state of the art related to the different types of bags. This made it possible to identify four main categories:

– the hand bag, which do not include a shoulder strap;
– the shoulder bag: comfortable and capacious, it is characterized by a rigid double handle or a single longer one or by the wrap-around shoulder strap;
– the cross body bag: with an adjustable strap to distribute the weight, choosing the most comfortable height, and freeing the arms and shoulders;
– the bag at the waist: known as the "pouch", it is small in size, has an adjustable strap that allows it to be fixed at the waist, freeing the arms and hands.

Further typological classifications proposed by the market refer both to the moment of use (day or evening), and to the shape and size: small, medium, large, top box, rectangular, half moon (or hobo), horse saddle (or saddle), up to the "frame" with a vintage flavor, the "baguette", the "bucket", to finish the "quilted" and "weekend bag", which is nothing more than a small suitcase (Fig. 1).

Fig. 1. Types of hand, shoulder and waist bags.

The research on the state of the art also analyzed the main systems for organizing the contents of a bag, such as the arrangement of large and small objects, interchangeable systems of the components of a bag and the application of 3D printing on accessories fashion.

2.3 The Fact-Finding Survey on the Use of Bags

Subsequently, a fact-finding survey was designed and carried out on a sample of users through direct observation. The investigation focused mainly on the problems that emerged from the preliminary investigation, first of all the difficulty in finding objects in a bag. In particular, the objective was to acquire objective data on the opinions expressed by users on the practicality of use of their bags, in relation to different places and contexts, and on the specific contents of the bags used (also keeping to hypothesize a relationship between the content and the posture assumed in the transport of the bag).

The reference sample of the fact-finding survey consists of 26 adult women (ages ranging from 25–50 years) with different social statuses (singles, girlfriends or mothers) and different lifestyles, identified through a chat group, which allowed also to plan individual meetings: the survey was designed by associating a multiple choice questionnaire with direct observation.

The data were collected in March 2019 in two successive steps.

Initially women were asked to empty and photograph the contents of their bag (Fig. 2), used during the day, without eliminating any object that was inside it (for example, the presence of receipts, cards and handkerchiefs is was of great importance although it might have seemed irrelevant or "an element of personal disorder" to the women in question). This first step of the investigation immediately allowed to build a first list of the most frequent objects that women carry on a daily basis.

Fig. 2. Contents of the bag of some women interviewed.

Subsequently, each woman in the sample was asked to fill in a personal data sheet, a summary table of the contents of her bag, as well as indicating any problems and needs related both to the content of her bag and to the bag itself.

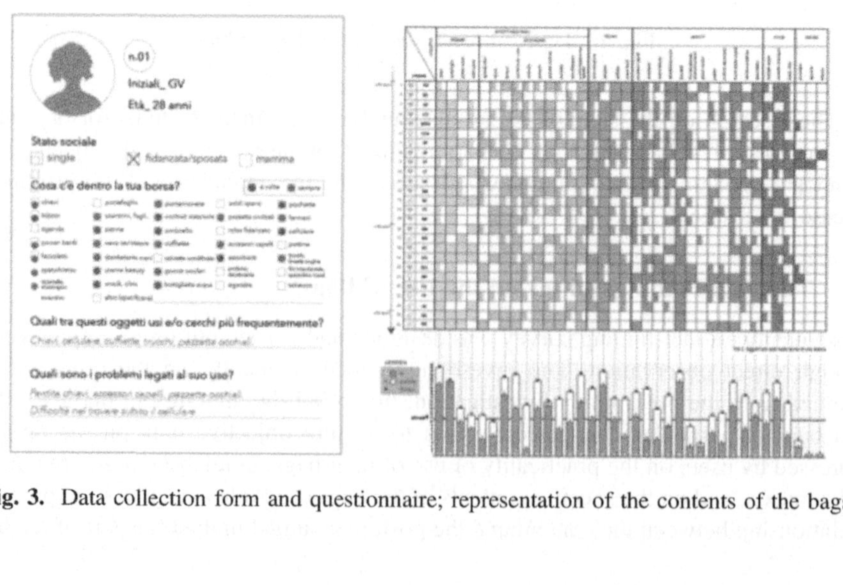

Fig. 3. Data collection form and questionnaire; representation of the contents of the bags.

The collected data were analyzed and graphed, grouping the objects contained in the bags according to specific categories (primary and secondary personal effects, techno, beauty, food and smoke) (Fig. 3). This made it possible to identify a sufficiently objective list of the objects most present in the bags and those most frequently searched for in the bags.

Finally, the investigation ended with a direct and passive observation (photos and videos) of a sufficiently large sample of 250 women while carrying their bag in specific contexts and times. The latter survey was useful for understanding the relationship between bags and places or which types of bags are mainly used in specific contexts (Fig. 4).

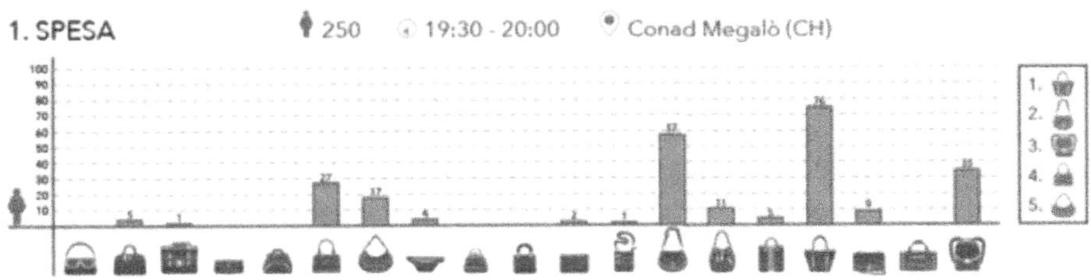

Fig. 4. Relationship between types of bags and contexts of use: shopping at the supermarket.

By interrelating the data collected, it was possible to identify the bag models adaptable to multiple contexts and therefore which are the most used bag models in the various contexts: "Saddle", "Tote" and "Backpack", each portable in a different way: on one shoulder, cross body, on both shoulders.

2.4 The Product Requirements

The fact-finding survey on the use of bags, carried out using some of the tools most used by traditional ergonomics to evaluate the usability of products, has allowed us to acquire useful information also with reference to the various possible ways of "interpreting" the accessory bag from a variable sample of women in different contexts of use. Human diversity, which the main design approaches for inclusion (Inclusive Design and Design for All) aim to enhance as a possible product innovation strategy, is also expressed through the choice of a fashion accessory in relation to both one's personal needs (of practicality and functionality), both to one's tastes, desires and aspirations.

The survey also clarified which are the most common ways of transporting bags, what are the most popular objects within the bags (in relation to contexts and times of use) and especially those that are "searched for" with more frequency by women in their bags.

This allowed to define a sufficiently large framework of actual product requirements, in relation to size, maneuverability and ease of use, perception and pleasantness (visual to tactile).

3 Results

The research subsequently developed a new bag system that favors the identification and organization of the objects to be brought along, grouping them by specific categories relating to frequency of use and size of the objects. It is a highly customized system, to meet the most diverse needs of possible users.

Subsequently, the research conducted a specific study on the possible product scenarios. Among the various innovative technologies investigated, 3D printing appears to be one of the most interesting options, as it allows you to create a customized product relatively quickly, with attention to detail and at relatively low prices. In particular, the most suitable technology has been identified in 3D SLS (Selective Laser Sintering) printing. It is an additive production process (classified as a powder bed casting technique), in which the heat of a laser melts a polymer powder with extreme three-dimensional precision, until the realization of a product completely defined in detail. It is a technology that is still in the testing phase but already widely used. For materials, attention was paid to environmentally sustainable, recyclable materials, which are also compatible with food products and which can be subjected to a natural dyeing process, to reduce their environmental impact.

4 Conclusions

Design for inclusion allows to pursue high levels of innovation even in consolidated sectors such as that of fashion accessories. Attention to human diversity, considered as an opportunity for the project, in fact favors the development of unprecedented design solutions, starting from fruitive needs, even apparently trivial.

In this, also the methods and techniques of investigation of ergonomics for the industrial product, if used with attention to the different strategies of use and to the needs of different users, prove to be extraordinary observation tools, in this case finalized to facilitate the female gender (but not only) in the use of one of the most useful and personal fashion accessories.

5 Credits

The research was developed as part of a thesis in industrial design at the Department of Architecture of Pescara (University of Chieti-Pescara, Italy). Thesis title: *Eco bag for everyday life. Organized system of objects in 3D printing with eco-sustainable materials*: supervisor prof. S. Camplone, correlator prof. G. Di Bucchianico, graduating G. Ventrella.

This paper, which summarizes the objectives, the phases of implementation and the main results obtained, was written by G. Di Bucchianico (paragraphs 1), G. Ventrella (paragraph 2, except 2.4), S. Camplone (paragraph 2.4, 3 and 4).

Reference

1. United Nations. Department of Economic and Social Affairs, Population Division (2019)

Design for Inclusion: Future Strategies and Challenges

Exploring Boundaries and Synergies Between Inclusive Design and Service Design

Daniele Busciantella-Ricci[1(✉)], Libertad Rizo-Corona[2], and Carlos Aceves-Gonzalez[2]

[1] Design Research Lab, Department of Humanities, University of Trento, Trento, Italy
d.busciantellaricci@unitn.it
[2] Ergonomics Research Center, Universidad de Guadalajara, Guadalajara, Mexico
libertad.rizo.din@gmail.com, c.aceves@academicos.udg.mx

Abstract. Service design is assuming a strategic role in contemporary society. From a design for inclusion perspective, this paper tries to outline a theoretical reflection based on the assumption that describes the possibility to identify four categories for understanding how service design and inclusive design can establish a relationship, i.e. (i) inclusive service design; (ii) design for inclusive services; (iii) service design for inclusion; (iv) inclusive design for service design. After a systematic literature review, the authors provide theoretical case studies from deliberately selected academic papers for each category. The result is the identification of a set of relationships between design for inclusion and service design to be studied in the inclusive design research field. Discussion and conclusions underline how these relationships can be helpful to practitioners in design and design researchers that would like to orient their activities understanding multiple ways of relating service design and design for inclusion.

Keywords: Design for inclusion · Service design · Inclusive service design

1 Introduction

Service design is assuming a strategic role in public and private organizations and, as it is being applied in several fields such as education, healthcare, policy making, and for spreading social innovation. Therefore, significant attention is required in the principles and approaches that drive the design process. As Sangiorgi [1] points out, the evolution of service design from service systems to value constellation and service ecosystem presents new challenges in practising service design. This requires "new design strategies and principles" [1]. In this sense, and with an emphasis of the attention to the human behaviour, capacity and needs, it is also possible to observe a structured shift in terms of meaning and approaches from service design to 'design for services' (cf. [2]; [3]). However, in the last decade, principles of service design among the practitioners are still anchored to some common points. Indeed, in 2009 Mager [4] describing the service design principles highlighted that it is a work with a holistic approach; it is an

G. Di Bucchianico et al. (Eds.): AHFE 2020, AISC 1202, pp. 55–61, 2020.
https://doi.org/10.1007/978-3-030-51194-4_8

interdisciplinary and co-creative work; it is based on the visual thinking and it has a radical approach. Successively, Stickdorn and Schneider [5] identified five principles of what they coined as service design thinking; it is user-centred, co-creative, sequencing, evidencing, holistic. Recently, Stickdorn [6] proposed six service design principles i.e. (i) human-centred, (ii) collaborative, (iii) iterative, (iv) sequential, (v) real, and (vi) holistic (cf: [6]).

From a ‘design for inclusion’ perspective, the first two principles might be the most significant. However, ‘human-centred’ is not necessarily inclusive and in service design is not always described as seen in inclusive design. ‘Human-centred’ replaced ‘user-centred’ in the late version with the aim of express the inclusion of all members, customers as well as staff in the organization [6]. Even so, the notion of the user remains mainly as a customer. The authors of this paper argue that it is still not clear how the two principles should guarantee a real inclusive approach in the whole design process. Despite “contemporary service design is described as a collaborative and inclusive practice” [7], recent studies [8] introduced the “service inclusion” paradigm as an egalitarian system that “provides customers with fair access to a service”. This emphasizes that “service exclusion can be significantly reduced by designing service solutions that respond to human diversity and by making resources available to consumers that facilitate access to services” [8]. Also, [8] underline that “service inclusion is multifaceted and comprises four key pillars”; these pillars are (i) enabling opportunities; (ii) offering choice; (iii) relieving suffering; (iv) fostering happiness. Thus, how service design recognizes the ‘diversity’ as a value? Furthermore, are the practitioners ready to be truly inclusive? These questions have driven the authors in searching how design for inclusion approaches (such as inclusive design) can establish a fruitful relationship with the service design principles and practice.

1.1 The Inclusive Service Design Approaches

Recent multidisciplinary studies in the field of human-factors allowed the identification of the inclusive service design (ISD) approach [9]. It comprises and taking advantage of the principles and methods of the domains of ergonomics, inclusive design and service design.

Following this direction, the authors of this paper try to outline a theoretical reflection for orienting researchers, practitioners and all those are engaged in interdisciplinary activities that require an inclusive approach in designing for and through services. The authors introduce an assumption that describes the possibility to identify four categories for understanding how service design and inclusive design can establish a reasonable relationship:

(i) inclusive service design; it means principles and methods of inclusive design, ergonomics and service design are used for designing services;
(ii) design for inclusive services; it is the design for an inclusive design result that is an inclusive service rather than focusing on the methods and the theoretical framework of the design process;

(iii) service design for inclusion (or service design for all); it means using service design for democratizing design or at least democratizing service design as a strategic tool for inclusion;
(iv) inclusive design for service design; it is the inclusive process of designing services through the inclusive design field of knowledge; in this case, the inclusive design attitude, praxis, approach and methods are used to design services.

Therefore, the aim of this paper is outlining the status quo of the categories for a first understanding of the efficacy of the framework.

2 Methodological Approach

Intending to understand this framework, the authors started a systematic literature review investigating specific sets of keywords in different search engines. The analysis was performed on the Web adopting "Google Scholar", "Scopus" and "Open Knowledge Maps" as search engines. "Inclusive service design", "inclusive design" and "service design", "design thinking" and "inclusive service", "service design" and "social inclusion" were the set of keywords the authors searched in the web engines respecting the use of the quotation marks for each single search session. No additional filters (e.g. date ranges) were selected except for Google Scholar where the default filter "citations" was deselected.

First, the results in searching for one set of keywords in Google Scholar were qualitatively filtered by the researcher by selecting relevant papers between the first 50 and 100 results. Successively, the selected articles have been in more depth analyzed through the following parameters:

- Relevance; how much the source is conceptually relevant with one of the hypothesized categories;
- Reliability; how much the source is reliable; the reliability value is determined by a subjective evaluation that considers the source of the item; e.g. a paper from a journal could be more reliable than a conference paper; also, the authors and their backgrounds have been considered as indicators for the reliability of the source;
- Impact; how much the contents of the source are or potentially are impactful on this research.

These parameters were used in a matrix for evaluating the single paper previously selected with a subjective score. Thus, for each selected paper, the three parameters were assessed with a value between 1 and 3. After this process, the same set of keywords were searched in Scopus. The researcher qualitatively filtered the results by selecting relevant papers. This search engine was used to confirm or add relevant papers among those found as relevant in google Scholar. Relevant papers that were found only in Scopus and not in Google Scholar were evaluated with the same matrix and parameters and same evaluation system. Finally, the same set of keywords were searched in Open Knowledge Maps following the same procedure as exposed before for the previous search engines. The duplicates were considered only one time.

The final selection is obtained by those papers had the highest score. Also, the matrix used for evaluating the relevant papers reports a section for additional cases. The authors used it for reporting cases and papers that did not appear in the search engines but are considered relevant from the authors perspective and backgrounds.

As a first investigation step on the framework, the authors limited the search process with keywords that do not explicitly report terms such as “design for all” and “universal design”. This is one of the next steps identified after the research presented in this paper. As an additional limit of this work, search engines such as Google Scholar, create indexes according to the capacity of the source to be identified by the Google algorithm. This means that relevant papers that are not indexed by the algorithm may not have been detected. It is also acknowledged that important contributions might have been disregarded due to the filter of the researchers’ perception.

3 Results

The result is the identification of a set of relationships between design for inclusion and service design to be studied in the inclusive design research field. Table 1 reports, for each set of keywords the total number of the found papers; the amount of the relevant papers; and the number of the papers were assessed with a score between 6 and 9. Table 2 reports the most relevant paper(s) for each hypothesized category.

Table 1. Number of papers for each set of keywords.

Set of keywords	N. of papers on Scholar/Scopus/Open Knowledge map	N. of relevant papers	N. of papers with a score between 6 and 9
“inclusive service design”	87/9/2	19	14
“inclusive design” and “service design”	928/89/3	51	18
“design thinking” and “inclusive service”	53/0/0	15	12
“service design” and “social inclusion”	3050/21/6	16	14

Table 2. The most relevant papers (score between 6 and 9) for each category.

Categories	Most relevant papers
(i) inclusive service design	Aceves-Gonzalez [9, 10], Bue Lintho and Begnum [11]
(ii) design for inclusive services	/
(iii) service design for inclusion	Blomkvist and Holmlid [7], Bridge [12]
(iv) inclusive design for service design	Darzentas and Darzentas [13], Parker et al. [14], Nickpour et al. [15], Liu [16]

4 Discussion

As a confirmation of the hypothesis, the first category is fully represented by the work of Aceves-Gonzalez [9, 10] and Bue [11, 17]. In the first case, the relationship between service design and inclusive design is the holistic combination among their principles and methods within the domain of the ergonomics. Providing knowledge and tools for the evaluation, design, and improvement of inclusive services is the primary goal of the "inclusive service design" (ISD) approach. A similar and explicit taxonomy was found in Parker et al. [14] and in Bue [11, 17]. However, Parker et al. [14] do not explicitly define "inclusive service design", and the approach is more oriented to use an inclusive design approach for designing services that consequently are more inclusive, which locates that paper within the fourth category presented in this paper.

On the other hand, Bue's work [11, 17] is oriented to understand how integrating universal design methodologies in service design. They also proposed a definition of a service that "is universally designed when its customer journeys are usable to all people, to the greatest extent possible, without the need for adaptation or specialized design apart from choosing preferred touchpoints" [11]. Both Aceves-Gonzalez [9, 10] and Bue [11, 17] addressed the need for SD to provide a more inclusive approach for including potentially excluded users and taking into consideration the diversity as a value in the design process.

The results of the second category did not produce any items which had the previously mentioned characteristics and reached a score higher than 5. However, a significant number of items were found, which almost complete the aspects of this category but failing to develop a process from a design perspective, even though having, as a result, an inclusive service.

One of the most relevant paper for the third category is the Blomkvist and Holmlid's [7] work; they emphasize that "contemporary service design is described as a collaborative and inclusive practice". Thus, from their perspective, emerges an approach in considering service design as an inclusive practice because it is based on collaborative practices. Also, the set of keywords 'service design' and 'social inclusion' produced many relevant results that fit into the third category. Among these, the work of Bridge [12] resulted in one of the most relevant about the category description. This kind of results acknowledged services and service design as a tool to alleviate social issues which at its core are inclusion issues, although the main focus of the work was not on service design. About the fourth category Darzentas [13] in a paper related to the context of self-services, point out that "an inclusive Design for All approach to self-service can be a driver for service innovation". In this case, the philosophy of design for all is adopted as a driver for including the needs of vulnerable customers in the service design process. This is a representative case for the fourth category, where a design for all attitude is adopted to contemplate more inclusive services.

About valid results outside the categories framework, many of the identified items were services related and directed to a vulnerable user profile although not from a service design perspective [18]. Also, another emergent issue was the different meanings of inclusion by area of knowledge and the need for the prevalence of a more comprehensive and robust concept of inclusion in design, highly relevant in service design. Which could have stronger consideration of economic issues of vulnerability for the users which has already been acknowledged in the work of Fisk et al. [8] and could be complemented with views from the sociology field where besides the economic also cultural, civic and interpersonal aspects are included [19]. This has also been recognized by Busciantella-Ricci et al. [20] as potential areas of analysis in inclusive service design similarly proposing them as "domains".

5 Conclusions

In conclusion, the identified set of relationships can be helpful to practitioners in design and design research that would like to orient their activities by giving a common ground for understanding multiple ways of relating service design and design for inclusion. However, to propose the categories as facilitators for the analysis and comprehension of the relations between service design and inclusive design the limits of these categories as well as their names might be clarified. Indeed, while the first, the third and fourth categories present papers to support the hypothesized definitions, the second category is not sufficiently supported. It means that the definition of this category need a strong reframe; or it means that the scenario of this category is still not enough investigated, or reported by the same practitioners. The reframing of the categories meaning is undergoing work. Also, in the course of the search, several projects which are not within the four categories have been found. These projects embody inclusive service design as outcomes and processes although not from the perspective of design. Finally, future work should include a more significant discussion on the literature that was identified in the course of the search process.

References

1. Sangiorgi, D., Patrício, L., Fisk, R.: Designing for interdependence, participation and emergence in complex service systems. In: Sangiorgi, D., Prendiville, A. (eds.) Designing for Service: Key Issues and New Directions, pp. 49–64. Bloomsbury Publishing, London (2017)
2. Meroni, A., Sangiorgi, D.: Design for Services. Routledge, London (2011)
3. Sangiorgi, D., Prendiville, A. (eds.): Designing for Service: Key Issues and New Directions. Bloomsbury Publishing, London (2017)
4. Mager, B.: Service design as an emerging field. In: Miettinen, S., Koivisto, M. (eds.) Designing Services with Innovative Methods, pp. 28–43. University of Art and Design, Helsinki (2009)
5. Stickdorn, M., Schneider, J.: This is Service Design Thinking: Basics - Tools - Cases. BIS Publishers, Amsterdam (2011)
6. Stickdorn, M., Hormess, M.E., Lawrence, A., Schneider, J.: This is Service Design Doing. O'Reilly Media, Sebastapol (2018)
7. Blomkvist, J., Holmlid, S.: Service designers on including stakeholders in service prototyping. In: Proceedings of INCLUDE 2011 Conference, pp. 615–624. Helen Hamlyn Centre for Design, Royal College of Art, London (2011)

8. Fisk, R.P., Dean, A.M., Alkire, L., Joubert, A., Previte, J., Robertson, N., Rosenbaum, M.S.: Design for service inclusion: creating inclusive service systems by 2050. J. Serv. Manag. **29**, 834–858 (2018)
9. Aceves-Gonzalez, C.: The application and development of inclusive service design in the context of a bus service. Doctoral dissertation (© Carlos Aceves Gonzalez) (2014)
10. Aceves-González, C., Cook, S., May, A.: Improving bus travel through inclusive service design. In: Soares, M.M., Rebelo, F. (eds.) Ergonomics in Design: Methods and Techniques, pp. 431–444. CRC Press, Boca Raton (2016)
11. Bue Lintho, O., Begnum, M.: Towards inclusive service design in the digital society: current practices and future recommendations. In: DS 91: Proceedings of NordDesign 2018, Linköping, Sweden, 14–17 August (2018)
12. Bridge, C.: Citizen centric service in the Australian department of human services: the department's experience in engaging the community in co-design of government service delivery and developments in e-government services. Austr. J. Publ. Admin. **71**, 167–177 (2012)
13. Darzentas, J.S., Darzentas, J.: Accessible self-service: a driver for innovation in service design. In: ServDes, 2014 Service Future, Proceedings of the fourth Service Design and Service Innovation Conference, Lancaster University, United Kingdom, 9–11 April 2014, pp. 143–153. Linköping University Electronic Press (2014)
14. Parker, C.J., May, A., Mitchell, V., Burrows, A.: Capturing volunteered information for inclusive service design: potential benefits and challenges. Des. J. **16**(2), 197–218 (2013)
15. Nickpour, F., Jordan, P.W., Dong, H.: Inclusive bus travel: a psychosocial approach. In: Langdon, P., Clarkson, J., Robinson, P., Lazar, J., Heylighen, A. (eds.) Designing Inclusive Systems, pp. 13–22. Springer, London (2012)
16. Liu, Y.: Exploring the application of Inclusive Design to the improvement of healthcare services. Doctoral dissertation, University of Cambridge (2018)
17. Bue, O.L.: Inclusive service design: current practices and future recommendations-a contribution to ensuring service design methodology is flexible, sensitive and adaptable to wide user groups. Master's thesis, NTNU (2017)
18. Lazar, J., Wentz, B., Akeley, C., Almuhim, M., Barmoy, S., Beavan, P., Beck, C., Blair, A., Bortz, A., Bradley, B., Carter, M., Crouch, D., Dehmer, G., Gorman, M., Gregory, C., Lanier, E., McIntee, A., Nelson Jr., R., Ritgert, D., Rogers Jr., R., Rosenwald, S., Sullivan, S., Wells, J., Willis, C., Wingo-Jones, K., Yatto, T.: Equal access to information? Evaluating the accessibility of public library web sites in the State of Maryland. In: Langdon, P., Clarkson, J., Robinson, P., Lazar, J., Heylighen, A. (eds.) Designing Inclusive Systems, pp. 185–194. Springer, London (2012)
19. Stewart, M., Reutter, L., Makwarimba, E., Veenstra, G., Love, R., Raphael, D.: Left out: perspectives on social exclusion and inclusion across income groups. Health Sociol. Rev. **17** (1), 78–94 (2008)
20. Busciantella-Ricci, D., Rinaldi, A., Tosi, F.: Supporting inclusive approaches in service design with netnography. In: Di Bucchianico, G., (ed.) Advances in Design for Inclusion, International Conference on Applied Human Factors and Ergonomics, pp. 290–301. Springer, Cham (2018)

Evaluating Possibilities in Chinese Market and Media for Strategic Design for Sustainable Development and Inclusion

Avril Accolla(✉), Ziquan Wang, and Huichan Liu

Design and Innovation, Tongji University, Shanghai, People's Republic of China
avril@avrildesign.com

Abstract. China is constantly changing, swiftly. The vastity of the country and the numerous challenges that it's facing simultaneously, allow for disparities and contradictions. In the last decades, China had impressive economic growth (the 2018 GDP of 9030.95 billion CNY, increased 2.6 times compared with 2009) and pulled out of poverty millions (from 98.99 million in 2012 to 16.6 million in 2018). In the last years, a growing population could tackle not only needs but quite some aspirations. Now that China has a thriving market and it's establishing its soft power, how is human diversity perceived in the society at large and the market? How are the market and its stakeholders following the growing suggestions of the government policies on sub mentioned topics of the last decade? Would it be possible to detect a role for strategic design for sustainable development in the commercial sectors? Can singular happening facts be escalated through the support of a design-driven approach to sustainability and inclusion? The authors attempt to highlight meaningful events in the mainstream media and the market and, combining them with government's and private institutions' actions, to understand and evaluate a possible framework for further strategic design action.

Keywords: Strategic design for sustainable development · Design for All · Human Diversity · New China · Inclusion · Market & media

1 The Speed and Quality of Change

In the last decades, China has shown a great speed of not only growth in urban population and household income, but also of meaningful change.

Planning, industrialization and uprising opportunities have triggered the outburst of several cities, for example, Chongqing witnessed a population growth of over 14 times in the last 10 years. In 2019 the Ranking of the New First-tier Cities[1] [36] (Table 1), shows an average increase in the disposable income exceeding 2,5 times between 2008 and 2018.

Shenzhen, some decades ago an unknown small town [6], has become "China's Silicon Valley of Hardware", with a nominal GDP of 2.42 trillion CNY in 2018, the

[1] Beijing, Shanghai, Guangzhou, Shenzhen are the 'Big 4' First-tier cities.

G. Di Bucchianico et al. (Eds.): AHFE 2020, AISC 1202, pp. 62–68, 2020.
https://doi.org/10.1007/978-3-030-51194-4_9

third-largest after Beijing and Shanghai, a market capitalization of 2.5 trillion US$ (30-11-2018) and a Stock Exchange (SZSE) which is the 8th largest in the world [7].

Table 1. Top five New First-Tier Cities. The data come from the Statistical Yearbook of each city: Chongqing 2019 [31], Hangzhou 2019 [32], Chengdu 2019 [33], Wuhan 2018 [34], Xian 2019 [35].

2008–2018	Urban resident population (ten thousand)	Gross domestic product (100 million CNY)	Per capita annual disposable income of urban household (CNY)
Chongqing	1419.09–20301.59	5817.55–20363.19	13321–34889
Hangzhou	340.76–515.04	4788.9748–13509.1508	24104–61172
Chengdu	807.86–1194.05	3944.9148–15342.7716	16942.62–42147.86
Wuhan	–	4115.51–13410.34	16712.44–43405.00
Xian	554.73–725.58	2318.14–8349.86	15207–38729

China's impressive economic growth (since it began to open up and reform its economy in 1978, GDP growth has averaged almost 10% a year [2]), with a GDP in 2018 of 903.095 billion CNY that has increased 2.6 times compared with 2009, also pulled out of poverty millions of people, from 98.99 million in 2012 to 16.6 million in 2018.

Within this great turmoil, the vastity of the country and the numerous challenges that it's facing simultaneously, allow for disparities and contradictions.

Medical services, even though in constant evolution and now following the comprehensive New Health Reform started in 2019 [37], see an unbalanced development among different regions and Provinces. Issues like the decline of capital-output rate, the decline of pension replacement rates and income gaps are relevant and current [5]. As one of the results, people's ANAs, and the actual capability to answer to them, varies considerably.

1.1 Old Countryside

China is surely not the only country ageing at a fast pace, it is interesting though to look into its specific way of ageing. The percentage of the population of over 60 and over-60 years old is smaller, and further decreasing, in the mega-cities: for example, it is around 10% in Beijing, Chengdu, Chongqing, Guangzhou, Shanghai, Shenzhen, Tianjin and Wuhan. Due to the work force's migration from rural to urban areas, the population of 60 and over-60 years old is larger in rural areas, and is going to increase even more: by 2030 will be 21.8% in rural areas and 14.8% in urban ones [11].

Abilities, Necessities and Aspirations of the ageing population may vary considerably between rural and urban areas considering the disparity in service development and access, in education and lifestyle in general.

2 Disposable Income, Aspirations and Soft Power

2.1 People's Purchase Power

Chinese growing economy allowed the majority of the population to answer to its basic needs: in 2017 the value set[2] in urban areas was 541 CNY/month and in rural areas 358 CNY/month [17]. In that same year, the average disposable income was 36.396,2 CNY/year (about 3.033 CNY/month) in urban areas, and 13.432,4 CNY/year (about 1.119 CNY/month) in rural areas [14]. With poverty[3] incidence of 3,1% in 2017 and 1,7% in 2018 and a disposable income's growth of 54% from 2013 to 2018 [14], an increasing part of the Country's population can start tackling not only needs but also some aspirations.

Nevertheless, there are relevant differences in disposable income of Chinese residents. Within the 5 levels of residents' income[4], in 2018 the lowest income level was one-tenth of the highest one, and in urban areas the average disposable income was 2,6 times the rural areas' one [14]. Moreover, due to the millions of rural migrants, 'cosmopolitan villages' still exist within the same city, even in largely internationalized ones [15, 16].

This creates deep differences in the A.N.A.s of the residents coming from different areas of the same city.

2.2 Country's Soft Power

China, since engaging with the WTO in 2001, reviews laws on intellectual property (IP) rights and enhances incentives on IP creation, bringing IP to the national strategy status in 2008, pushing for acceleration in 2015 and 2016[5] [8]. European companies in China have positively noted the enhancement on IP[6]; in the first half of 2019, the foreign invention patent applications were 78,000 with a year-on-year increase of 8.6% and the trademark applications were 127,000 with an increase of 15.4% [9].

Innovative industries are developing fast: from 2000 to 2017, Chinese IP royalties growth is the third in the world with an average annual rate of 20.1%, going from 1,28 to 28,66 billion US$ [3].

China, with the second-largest GDP, the largest manufacturing economy and goods' export in the world, is massively investing in R&D with the second-largest budget spending after the USA. In 2018 Chinese patent applications are around 1,54 million, 46.4% of the global total, ranking first in the world [1]. From 2000 to 2016, China's R&D budget growth is 27.4% of the world's, the USA's is 29.5% [3]. In more

---

[2] According to the Minimum Standard of Living Security set by China in 2017.

[3] According to the National Rural Poverty Standards.

[4] According to the Chinese Bureau of Statistics' division.

[5] Respectively with the: "National Intellectual Property Strategy Outline", "Several Opinions on Accelerating the Construction of a Powerful Country of Intellectual Property in the New Situation", the "Thirteenth Five-Year Plan" for National Intellectual Property Protection and Use.

[6] The European Business Chamber of Commerce's 2019 Business Confidence Survey on 585 EU companies, shows about 60% positive feedback on IP rights protection.

detail, the R&D budget represents the 2,2% of the GDP, South Korea's is the 4,3%, with an expenditure per capita of 388 US$, USA's is 1.590 US$ and Israel 1.867 US$ [37]: a relevant potentiality of further growth.

In terms of international impact, for example, in 2018, Chinese online games export was 9,59 billion US $ with a year-on-year increase of 15.8%, the domestic one is at 5,3%. The Beijing company ByteDance, developer of the video-sharing social networking Douyin Byte Bea and its version for overseas Tik Tok, bought the American short video platform Musical.ly, and in October 2018 became the highest monthly download application in the USA [10].

3 Human Diversity Perception, Market Cases and Trends

China, home to 56 ethnic groups and hundreds of dialects, is one of the most homogenous countries in the world. The average index of fractionalization (the probability that two randomly selected individuals in a community are from different ethnic groups) for China is 0.06 [24–26]. The Han is the majority of the population (the 91.51% with 1.2259 billion people) and it is quite similar in both culture and language [13]. Ethnic/cultural diversity is mainly considered a resource in local tourism.

Disables' rights are recognized and protected by the Constitution and specific laws tackling the inclusion of disables in the work-life and education since decades. Nevertheless, the overall participation of daily life is still scarce, but growing. In the "Digital Technology for the Disabled Empowerment Award" submitted by the UNESCO in Paris, Tencent is the first award-winning organization in Asia and the first award-winning company in the world [42], highlighting its support to the disable community with product-services *for All*: the phone QQ, which feature several accessibility features, and the mini-program for the "Guangdong Provincial Affairs" supporting residents in applying for more than 500 services, which feature several accessibility features.

The market is responsive to diverse ANAs, segmented and/or new. In the example of men's cosmetics, the market reached 3.2 billion CNY in 2016 [18], with a growth rate from 2016 to 2019 of 13.5%, higher than the world average of 5.8%. The top three brands of men's beauty are Chinese brands, established less than six years prior: "尊蓝 (respect blue)", "风雨 (wind and rain)", and "左颜右色 (left and right)" [18]. During the Double 11 campaign 2019, the first day of pre-sale increased by 56%.

Design as a strategic tool for successful sustainable development is found in a growing number of cases, alongside the cases where the lack of design in both its strategic role and in its implementation one is noticeable.

In the case of the mobile phones for the silver market [20], the focus is still mainly on some quite basic characteristics: oversized buttons, oversized font, high volume in operations, extra durable battery [21]. Nevertheless, the sales on Tmall of these 2G networks phones, have overtaken the prominent models (such as Huawei P30pro, Redmi NOTE7, Oppo and Vivo) to be second only to Apple [20]. However, due to the basic offer and the lack of a design approach, their popularity is declining: in 2018, the sales of basic elderly mobile phones fell by 31%. It is quite interesting to note that simultaneously the number of smartphones purchased by the elderly increased by

12.5% compared to 2017 [23]: it is a purchaser awareness and ability which creates an interesting opportunity for a Design for All approach.

三顿半 (Three Meals and a Half) founded in 2015 in Changsha, sells online high-quality instant coffee, coffee bags and coffee-related products [27]. In 2019 Double 11 Campaign the turnover has been 10 times the one of 2018. 三顿半 surpassed Nestlé in sales at a serving price that is 8 times higher. It is a Brand based on a strategic design approach and a unique selling proposition which combines: i. an innovative technology which allows high-quality taste and convenience in use everywhere; ii. a package designed to be the symbol of the Brand and to trigger emotional-based behaviour in the target consumers; iii. a thorough sustainable approach with 100% recyclable materials and a "return plan", in partnership with offline retailers (up to now 29 institutions and spaces in 17 cities), for the clients who are willing to recycle responsibly and be rewarded. The brand is so successful that at the end of 2019 has completed the A + application, has attracted funds for tens of millions of CNY, and is planning to open its firsts boutique offline [27].

This case, among others, shows the awareness and maturity of some areas of the Chinese market and society for sustainability design-driven business.

4 Conclusions

China has currently a relevant and meaningful role as a member of the international community and such ongoing cooperation allows for relevant and impactful exchanges, both ways, in diverse areas: not only economic, but also social, cultural and scientific. Accomplished the economic strength, China is now tackling the enhancement of its Soft Power and its Diversities, socio-economic, cultural and ethnic, will play an interesting role.

The Country's socio-economic development, as here summarized, and the presented market and media cases, show a possible readiness to welcome the offer of businesses oriented to an ethical approach and to answering Human Diversity's A.N.A.s. Both the independent and the supported initiatives to raise awareness and knowledge about Human Diversity show a society possibly ready not only to welcome an evolution towards it, but also to push for it.

A strategic design approach to sustainability and inclusion, like for example in the cases of *Tencent* and *Three Meals and a Half*, could escalate in a more general trend. Strategic design for Human Diversity, equality and social inclusion through innovation is a right answer to the simultaneous challenges, disparities and contradictions the Country currently faces. At the pace China evolves, we could soon be witnessing the rise of a new mainstream trend, with relevant impacts in industrial, commercial and institutional fields: the authors see a possible framework for further strategic design actions focusing on Sustainable Development.

References

1. Peng, X.: Xinhua International Review: China's Strengthening Intellectual Property Protection "Escorts" for Innovation and Development, 19 November 2019. http://www.xinhuanet.com/world/2019-11/19/c_1125248749.htm
2. The World Bank In China. https://www.worldbank.org/en/country/china/overview#1
3. How to See China's Scientific and Technological Progress? The World Has "Numbers", 02 November 2018. http://www.xinhuanet.com/politics/2018-11/02/c_1123650122.htm
4. The economy of China. https://en.wikipedia.org/wiki/Economy_of_China
5. Zhu, Y.: Tsinghua report: China's development losses due to imbalanced development are still in a high range, 26 April 2019. https://www.yicai.com/news/100175215.html
6. From Chasing the Times to Leading the Times. From the Miracle of Shenzhen's Development, 20 May 2018. http://www.xinhuanet.com/2018-05/20/c_1122860284.htm
7. Shenzhen. https://en.wikipedia.org/wiki/Shenzhen#Economy
8. China's Huge Progress in Protecting Intellectual Property Cannot Be Ignored, 30 March 2018. http://www.gov.cn/xinwen/2018-03/30/content_5278421.htm
9. China's IPR Protection Effectiveness Has been Widely Recognized, 24 July 2019. http://www.xinhuanet.com/politics/2019-07/24/c_1210212664.htm
10. The Dawn of The Explosive Growth of The Digital Cultural Industry Has Arrived, 19 August 2019. http://www.xinhuanet.com/tech/2019-08/19/c_1124891148.htm
11. World Health Organization. China country assessment report on ageing and health (2015)
12. Thomson, S.B., Wei, W.X., Swallow, P.: Equality and harmony: diversity management in China. Chin. Manag. Stud. (2019)
13. Churchill, S.A., Valenzuela, M.R., Sablah, W.: Ethnic diversity and firm performance: evidence from China's materials and industrial sectors. Empir. Econ. **53**(4), 1711–1731 (2017)
14. National Bureau of Statistics of China: China Statistical Yearbook-2019. China Statistics Press, Beijing (2019)
15. "Cure the City" or "Cure People"? Village In the City: "Ulcer" In a City that Cannot Be Cured? 16 January 2018. http://www.xinhuanet.com/politics/2018-01/16/c_1122264973.htm
16. Ye, X.: Beisi Village where small white-collar workers gather. Beijing Evening News, 10 November 2016
17. Urban and Rural Minimum Living Standards, 1999–2017. https://www.unicef.cn/figure-214-dibao-standard-urban-and-rural-areas-19992017
18. "He economy" is coming, 2018 Chinese men's beauty and skincare consumption trend report, 18 September 2018. https://zhuanlan.zhihu.com/p/44819978
19. Chinese Men's Beauty Consumption May Enter a Period of Golden Explosion, 06 November 2019. http://www.xinhuanet.com/fashion/2019-11/06/c_1125194919.htm
20. Tu, Y.: The Elderly Cell Phone is Fierce! Online Sales Surpass Huawei and Xiaomi, Designed for More than 200 Million Elderly in China, 06 June 2019. https://xueqiu.com/2378813576/127873423
21. The Elderly Cell Phone. https://baike.baidu.com/item/%E8%80%81%E4%BA%BA%E6%9C%BA/7169263?fr=zhidao_relate
22. Ren, Z.: We bought 5 elderly machines. Hengping decided to go home and send an elder one for the New Year! 17 January 2019. https://www.cnmo.com/reviews/653735.html
23. Suning Releases Chongyang Big Data: Elderly Cell Phone Sales Decline, "tide" Elderly Love iPhone, 15 October 2018. https://tech.huanqiu.com/article/9CaKrnKdFX6

24. Alesina, A., Zhuravskaya, E.: Segregation and the quality of government in a cross-section of countries. Am. Econ. Rev. **101**(5), 1872–1911 (2011)
25. Fearon, J.D.: Ethnic and cultural fractionalization by country. J. Econ. Growth **8**, 195–222 (2003)
26. Alesina, A., et al.: Fractionalization. J. Econ. Growth **8**(2), 155–194 (2003)
27. Zhang, Q.: Boutique coffee brand "three and a half" received tens of millions of financing, will open a flagship store in Shanghai next year, 19 November 2019. https://www.jiemian.com/article/3697059.html
28. Xu, Z.: 36Kr debut | "Three meal and a half" completed tens of millions of Series A financing, becoming the first domestic brand to top the Tmall Double 11 coffee list, 18 November 2019. https://36kr.com/p/5266829
29. Instant coffee collects countless fans! What marketing tricks did you use for three and a half? 10 December 2019. https://www.digitaling.com/articles/239322.html
30. Three and a half launch "Return Plan" to recycle users' empty coffee cans, 26 October 2019. http://dy.163.com/v2/article/detail/ESDI9FM60519FEEC.html
31. Chongqing Statistical Yearbook 2019. http://tjj.cq.gov.cn//tjnj/2019/zk/indexch.htm
32. Hangzhou Statistical Yearbook 2019, 23 October 2019. http://tjj.hangzhou.gov.cn/art/2019/10/23/art_1653175_39222973.html
33. Chengdu Statistical Yearbook 2019. http://www.cdstats.chengdu.gov.cn/uploadfiles/02010922/%E3%80%8A%E6%88%90%E9%83%BD%E7%BB%9F%E8%AE%A1%E5%B9%B4%E9%89%B4-2019%E3%80%8B%EF%BC%88%E5%85%A8%EF%BC%89.pdf
34. Wuhan Statistical Yearbook 2018. http://tjj.hubei.gov.cn/tjsj/sjkscx/tjnj/gsztj/whs/201911/P020191104653356795480.pdf
35. Xian Statistical Yearbook 2019. http://tjj.xa.gov.cn/tjnj/2019/zk/indexch.htm
36. New Tier 1 cities released in 2019: Chengdu ranks first, Kunming enters the list, 26 May 2019. https://www.yicai.com/news/100201553.html
37. UNESCO Institute for Statistics. List of R&D funds by country. http://uis.unesco.org/apps/visualisations/research-and-development-spending/
38. Reform of the Medical and Health System. http://www.gov.cn/zhengce/content/2019-06/04/content_5397350.htm?trs=1
39. Luo, Z.: China's population aged 60 and overreaches 241 million, accounting for 17.3% of the total population, 26 February 2018. http://www.xinhuanet.com/2018-02/26/c_1122456862.htm
40. Wong, C.K.S., Chung, K.H.M.: Work values of Chinese food service managers. Int. J. Contemp. Hosp. Manag. **15**, 66–75 (2003)
41. Began to return, for the road never travelled, 16 October 2019. https://mp.weixin.qq.com/s/ST9SDZWodOvtYff4DJpBxg
42. Tencent wins United Nations award! Important contributions to information accessibility!, 04 December 2018. http://www.siaa.org.cn/mmabu

Employing the Inclusive Design Process to Design for All

Jenna Mikus[1](✉), Victoria Høisæther[2], Carmen Martens[3], Ubaldo Spina[4], and Janice Rieger[1]

[1] QUT Design Lab, Queensland University of Technology, Brisbane 4000, Australia
{j.mikus,j.rieger}@qut.edu.au
[2] Design and Architecture Norway (DOGA), 0271 Oslo, Norway
vh@doga.no
[3] University of Hasselt, 3500 Hasselt, Belgium
carmen.martens@uhasselt.be
[4] Technologies Design and Materials European Research Centre (CETMA), 72100 Brindisi, Italy
ubaldo.spina@cetma.it

Abstract. The 2019 EIDD Design for All Europe Summer School in Viana do Castelo, Portugal brought together 20 international doctoral students and design professionals to explore and apply Design for All knowledge. The program culminated in a capstone design challenge, during which participants were divided into teams and asked to apply Inclusive Design (ID) principles to address Viana's urban planning issues. This paper presents the results of one of the four teams—outlining the design process, considerations, objectives, and outcomes. During this challenge, the team followed a prescribed ID process (based on the EIDD Design for All Europe-supported Inclusive Design framework [1] created by Design and Architecture Norway (DOGA) in collaboration with the Helen Hamlyn Centre for Design (HHCD) at the Royal College of Art (RCA)), testing its applicability. By engaging directly with lead user group members and relevant stakeholders, the team identified creative, pragmatic design solutions to meet design goals and innovate across people, planet, and profit. Thus, by applying ID as a people-centered strategy, participants created a conceptual urban design likely to result in sustainable innovation and resonate across demographics.

Keywords: Inclusive design · Universal design · Design for all · Human-centered strategy · Citizen involvement · Architecture · City planning

1 The Inclusive Design Process

From 9-13 September 2019, a capstone design challenge was conducted during the EIDD Design for All Europe Summer School held in Viana do Castelo, Portugal. The objective of the challenge was to work in interdisciplinary groups of three to four doctoral students and design professionals to address current Viana do Castelo (a.k.a. Viana) city planning issues with pragmatic Inclusive Design solutions, based on lead user insights. This paper presents the process and findings of one of the design teams.

G. Di Bucchianico et al. (Eds.): AHFE 2020, AISC 1202, pp. 69–76, 2020.
https://doi.org/10.1007/978-3-030-51194-4_10

1.1 An Inclusive Design Approach

"Inclusive Design is an approach to design and a business strategy. It aims to design mainstream products, services and environments that are accessible and attractive to the largest possible number of people." [1]

This definition of Inclusive Design comes from Design and Architecture Norway (DOGA), which published a book on Inclusive Design (ID). The EIDD Design Challenge required each group to employ a specified inclusive design approach as a prescriptive yet agile people-centered strategy for innovation. Part of employing this approach involved following an Inclusive Design process, which is supported by EIDD Design for All Europe and was created by DOGA in collaboration with the Helen Hamlyn Centre for Design (HHCD) at the Royal College of Art (RCA).

The process was to be used as a foundational truth for this Challenge—encouraging the design teams to position the design around the users, designing with and for them, while stepping through the four stages of the ID Process outlined in Fig. 1 below: Explore, Focus, Develop, and Deliver [1].

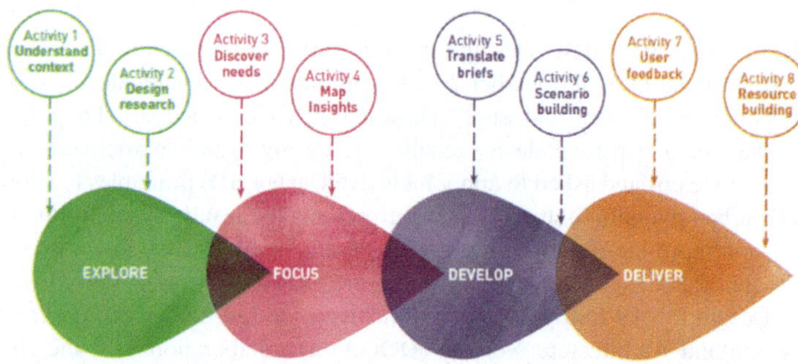

Fig. 1. DOGA's Inclusive Design Process [1], which depicts four 'teardrop'-shaped objects. Each 'teardrop' represents a stage in the Inclusive Design Process, contains two sub-activities, and feeds into the next from left to right like stepped chevrons. The first stage is Explore, the second Focus, the third Develop, and the fourth Deliver.

Due to the abbreviated 3-day time limitation, the team approached the Challenge using the DOGA 4-stage process [1] in a way that is appropriate for short-term innovation projects. Because early decisions inform end results, especially in projects with shorter engagement timelines, the team paid particular attention to the early Explore and Focus stages of the process. This provided contextual clarity and user insight up-front—understandings critical when generating new ideas for innovative projects—before developing and delivering our conceptual design.

1.2 Detailed Process Description

The team stepped through the 4 stages of the Inclusive Design Process (Fig. 1 above) by following the articulated activity and sub-activity protocol [1], noting specific actions taken, and highlighting the resulting output, as described below:

- **Stage 1: Explore**

  Activity & Sub-activity Protocol - Conduct Contextual Research (e.g., documenting literature, market, competitor research) and Design Research (e.g., conversing on-site to understand issues, select the lead users, and develop an initial plan or framework).

  Actions Taken - During the first day and a half, the team conducted tours and interviews throughout Viana do Castelo to choose the projects lead users and understand local needs and concerns. The term lead user refers to people who challenge the environment beyond an average, mainstream user. The team's task was to identify one or multiple lead users in this stage and learn about them. After meeting with the city architects and talking with people on the street, the team procured a map from the visitor's center and noted the featured slogan, "Quem gosta vem, quem ama fica" ("Those who like it, visit. Those who love it, stay."). The idea immediately resonated with the group, as most of the people they had met and spoken with were either pilgrims (i.e., tourists traveling through town on a religious pilgrimage) or tourism workers (i.e., locals working in the tourism industry). The team agreed that the primary lead users would be pilgrims and the secondary lead users would be tourism workers who would also likely benefit from design done for transient tourists. After speaking with their Portuguese-speaking colleagues, the team chose a team slogan of "Viana Fica Bem" ("Viana looks good" and makes all well)—encouraging an inclusive design experience for the short-term Viana-visiting pilgrim as well as the long-term inhabiting locals.

  Resulting Output - Lead user identification; Contextual understanding.

- **Stage 2: Focus**

  Activity & Sub-activity Protocol - Discover Needs (e.g., connecting and conversing meaningfully with users, adjusting data capture as needed) and Map Insights (e.g., capturing and analysing to highlight areas of opportunity).

  Actions Taken - Day 2 focused on finalizing and getting to know the needs of the identified lead user groups. After speaking with the pilgrims and tourism workers, the team collected the captured list of needs, collated them, and analyzed them to identify areas of opportunity. Special focus was put on understanding what the users truly needed and wanted, asking specific exploratory questions whenever possible to ensure full understanding of areas of concern. When translating the insights, the team noted needs across three primary themes: mobility (i.e., encouraging accessible and universal design), connection (i.e., linking nodal areas of activity around the city, people to each other, and people to place), and culture (i.e., embedding Viana-specific references in the urban landscape). These mapped insights became the team's primary design goals.

  Resulting Output - Need identification; Insight analysis.

- **Stage 3: Develop**

  Activity & Sub-activity Protocol - Translate Briefs (e.g., creating and selecting design needs crafted from themes identified in Stage 2) and Build Scenarios (e.g., creating user personas and profiles to work through design scenarios).
  Actions Taken - On Day 3, the team organized around the three identified themes of mobility, connection, and culture and brainstormed ways of designing to meet those as goals. While developing scenarios, the team recognized they were considering three types of infrastructural changes to attain people-centered solutions: physical, technological, and programmatic enhancements. The team also discovered that their conversations centered around two primary areas of the town—the train station to the north, a key area of entry to the partially "walled" city, and the boardwalk to the south, a vibrant area of community activity in modern buildings by local renown architects (e.g., Alvaro Siza Vieira, Eduardo Souto Moura). Both the train station and the waterfront were well-developed but disconnected, resulting in visitors not knowing about one when inhabiting the other. Consequently, the team's conceptual design focused on developing these geographic areas individually and in context for pilgrims (tourists) and tourism workers (locals) using physical, technological, and programmatic changes to enhance mobility, connection, and culture.
  Resulting Output - Thematic Design; Scenario Creation

- **Stage 4: Deliver**

  Activity & Sub-activity Protocol - Perform User Feedback (e.g., testing by presenting design suggestions to key stakeholders) and Resource Building (e.g., documenting images, insights, and ideas for future use).
  Actions Taken – Day 4 focused on Delivery. The team presented their findings and conceptual design suggestions to key stakeholders—a group of Viana city architects. Team members summarized the process followed, inclusive design concepts implemented, observations and lessons learned throughout, and global Inclusive Design examples they felt had potential for application in Viana. They discussed their chosen lead user groups (pilgrims and tourism workers); the rationale behind their chosen themes (mobility, connection, culture); and the emphasized geographic areas (train station, boardwalk, connecting street) and provided physical, technological, and programmatic changes that could be incorporated throughout Viana to meet these goals for the user groups. Stakeholder critiques were favorable and interested in the conceptual design suggestions. Many even mentioned how these ideas might be successful for the chosen lead users as well as other demographic groups. The final observation thus ended with a question: 'Could people-centered ID done for one or two lead user groups result in effective design for other user groups?'
  Resulting Output - Stakeholder Feedback; Resource Documentation

2 Design Considerations and Objectives

While following the ID process, the groups were asked to carefully consider lead user selection as well as sustainable innovation designs to ensure these topics would become design objectives and therefore a key success factor of the designs. Teams were asked to identify viable lead users that would represent typical Viana inhabitants and therefore have the ability to inform and benefit from the intended design. They were also asked to design for people, planet, and profit, if possible, to prompt what DOGA refers to as sustainable innovation [1].

2.1 Designing for Lead Users

As the ID process and other design thinking and Design for All best practice protocols state, it's important to designate targeted lead user(s) up front to understand how they can contribute to the design [1–4]. Figure 2 depicts some considerations by demographic, as noted by DOGA.

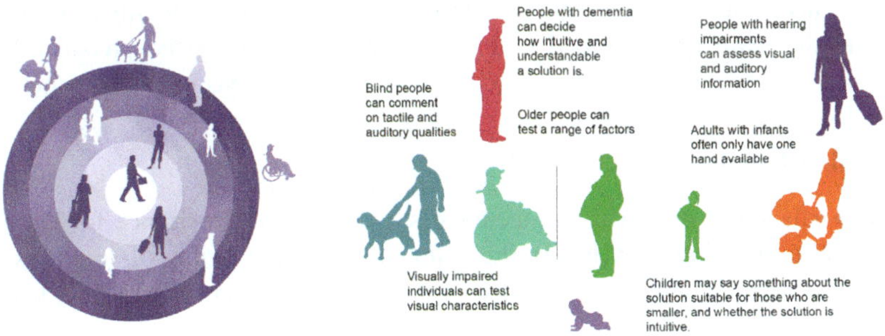

Fig. 2. DOGA's Lead User Definition Visuals [1], depicting a bulls-eye image of possible user groups (left) and examples of user groups with descriptive attributes (right). The bulls-eye diagram on the left (based on work by Professor Jeremy Myerson, RCA Helen Hamlyn Centre for Design) represents all possible user groups. 'Average users' are found in the center, and a variety of users are found around the perimeter. By aiming to design for the people on the outer edges, designers can include a broader market. The user groups depicted on the right are based on a DOGA visual that highlights possible design capabilities according to 'outer edge' user groups.

As mentioned above, after speaking with people in town, the team found two priority user groups in Viana: *El Camino* pilgrims (tourists traveling through town on a religious pilgrimage) and tourism workers (locals working in the tourism industry). These two groups were therefore designated as this team's lead users for this design challenge and the groups for whom they designed.

2.2 Designing for Sustainable Innovation

In addition to designing a city environment with and for users, teams were asked to achieve sustainable innovation in their designs. That is, team designs were required to balance triple bottom line aspects of social inclusion, environmental sustainability, and business opportunity (i.e., people, planet, profit).

These became prioritized considerations and success factors, in addition to addressing design considerations answering questions: Who, What, When, Where, Why, and How. By directly engaging with lead users and relevant cross-disciplinary stakeholders, the team answered these questions and identified solutions balancing social, environmental, and economic aspects with creativity and pragmatism. This prompted the creation of a socially sustainable and innovative design to benefit the people of Viana.

3 Final Conceptual Design

The team's final conceptual design stemmed from developing an understanding of the lead users' needs and concerns and defining goals around those identified needs. From that point, the team brainstormed potential solutions spanning the physical, technological, and programmatic and identified key geographic areas of improvement.

3.1 Design Results

To ensure balance of mobility, connectedness, and cultural experience for the pilgrim (tourist) and tourism worker (local), the team considered a variety of solutions. Some conceptual design examples by enhancement type can be found in Table 1 below.

Table 1. Example design ideas by infrastructural enhancement type.

Enhancement type	Conceptual design examples
Physical	Sidewalks and streets (e.g., rumble strips, smooth road) Sustainable energy sources (e.g., solar panels, wind turbines)
Technological	Elevated, accessible bike paths Tourist information kiosks Wearable device rentals for connected tourism Integrated train ticket systems
Programmatic	"Heart of Viana"-related cultural programs and branding Inclusive design policies and protocols Consistent indoor/outdoor signage and wayfinding

3.2 Suggested Next Steps

Because time was limited in this challenge, the team suggests conducting additional interviews and workshops to validate captured findings before finalizing and implementing a design. This should involve a variety of people within the pilgrim and tourism worker lead user groups to accommodate varying needs and preferences among members of the groups and to confirm validity. Consistent user engagement throughout the iterative design process and validation testing upon completion should also be incorporated to ensure satisfaction and build a sense of ownership.

In addition to engaging the lead user groups in more detail, the team also encourages conducting relevant cross-disciplinary stakeholder discussions before, during, and after implementation.

4 Conclusion

This design challenge highlighted how following an inclusive design process can help frame design research by gaining valuable insights from users and by translating those insights into innovative ideas and concepts. It ensures that design teams ask the right questions to a diverse group of participants and develop design solutions based on the feedback, thus resulting in true people-centered design.

By conducting research with two lead user groups—pilgrims and tourism workers—the team moved beyond functional problem-solving and experienced deeper engagement, trusted collaboration, and meaningful design. The design critique validated this process when the city architects hypothesized how the pitched design ideas might be successful for other demographic groups in addition to the chosen lead users.

This indeed proved the expanded market potential of Inclusive Design, as depicted in Fig. 3 and as described by DOGA: "Inclusive Design can help to increase market potential by widening the appeal of a design beyond the primary target market" [1].

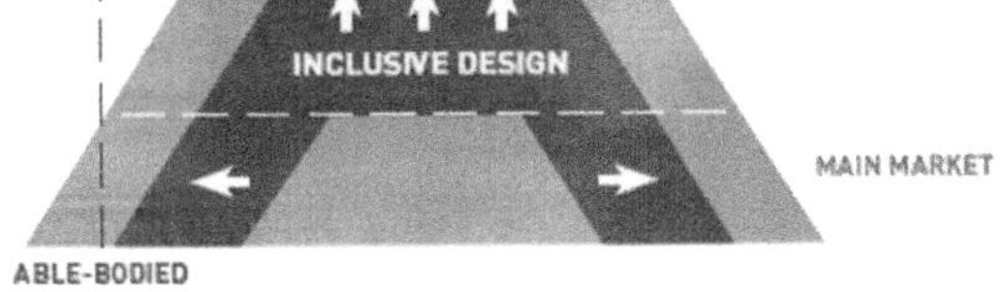

Fig. 3. DOGA's "Expand the Market Potential" visual [1] (based on Knut Nordby's 2003 Useability Pyramid [4]), depicting the main market (able-bodied) demographic on the bottom extending up to the personal assistance market (less able) at the top with expanded market potential building up and out from the center focus on Inclusive Design.

The findings from our research and design conclude that people-centered ID created for one lead user group can result in effective design for other user groups and therefore be a potential means to Design for All.

References

1. Design and Architecture Norway (DOGA) ed: Innovating with People-Inclusive Design and Arch. Design and Architecture Norway (DOGA) (2018)
2. Brown, T.: Design Thinking (2008)
3. Lidwell, W.: Universal Principles of Design, Revised and Updated. Rockport Publishers (2010)
4. Nordby, K.: Design for All: Shaping the end-users_Tel-eEurope: ETSI's involvement in laying the foundation for an all-inclusive eEurope. In: Presented at the Accessibility for All Conference, Nice, France, 27 March (2003)

Using AI to Enable Design for Diversity: A Perspective

Fang Li[1], Hua Dong[2], and Long Liu[1(✉)]

[1] Tongji University,
1239 Siping Road, Shanghai 200092, People's Republic of China
{FrankLiFang, LiuLong}@Tongji.edu.cn
[2] Loughborough University,
Epinal Way, Loughborough, Leicestershire LE11 3TU, UK
H.Dong@lboro.ac.uk

Abstract. Inclusive design focuses on diversity. The contextualized user-sensitive design framework of the interaction system needs to analyze and deal with complex diversity factors, which challenges the traditional design process, tools, and methods. Therefore, new technological progress is needed to provide more innovation potential. The authors point out that the design process of smart products is evolving in response to uncertainty. In the future, diversity-oriented design will tend to allocate design resources and values in an algorithmic way rather than the compromised unity solution. This paper analyzes the limitations and potential of the application of AI technology represented by deep learning in diversity-oriented design practice and design research, puts forward the goal and direction of further research, and discusses the critical links of AI-enabled diversity design in interdisciplinary research environment.

Keywords: Inclusive design · Diversity-oriented design · Artificial intelligence

1 Introduction

In order to make inclusive design sustainable in the competitive commercial environment, the authors' research focuses on the interdisciplinary study of inclusive design, economics, and technologies. As a design methodology initially driven by values, inclusive design advocates that design should pay more attention to the diversities of people. However, the existing design research generally focuses on how to realize and achieve inclusion in design. The lack of research on the causes and internal driving forces of inclusive design limits its expansion to broader and more universal economic and social fields.

Inspired by Matteo Bianchin and Ann Heylighen's concept of design fairness and justice [1, 2], we studied inclusiveness as an additional attribute of products from the perspective of supply/demand distribution in economics and give it economic definition and measurability. We also preliminarily summarized the economic causes of design exclusion, corresponding optimization strategies and sustainable constraints [3].

In view of the diversity of users and contexts in the human-computer interaction (HCI) system, we constructed a contextualized user-sensitive design framework

G. Di Bucchianico et al. (Eds.): AHFE 2020, AISC 1202, pp. 77–84, 2020.
https://doi.org/10.1007/978-3-030-51194-4_11

(Fig. 1) based on the analysis of the interaction between the behavior model and various elements. We studied the diversity and impact factors in the HCI system from the two primary dimensions of users and contexts. Our research regards the user's demands and capabilities as the functions of the dynamic interaction between the user's characteristics and the environment instead of fixed parameters, and proposes a new user research and analysis tool for user-sensitive design. Based on the contextualized user-sensitive design framework, the methods and system elements of user-sensitive design and contextualized design are studied, as well as the possibility of inclusive design by combining the two [4]. However, if designers want to customize their design to wrap each user's flexible demands and capabilities in different contexts, instead of summarizing diversity as a series of moderate equilibrium points, traditional design processes, tools, and methods are difficult to effectively analyze and deal with complex diversity factors and realize the inclusion of diversity required by contextualized user-sensitive design. Therefore, new technological advances are needed to provide more innovation potential.

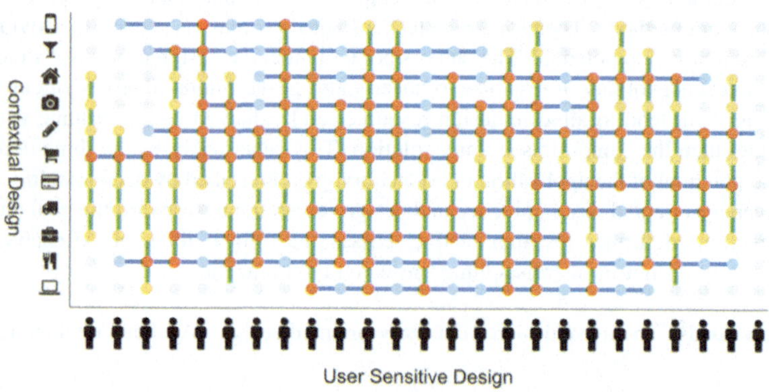

Fig. 1. The basic principle of contextualized user-sensitive design framework. It is to meet the different needs of users with different ability levels in the same scenario, and to ensure that the same user has the same experience in different scenarios [4].

2 The Shift of Design Process

Diversity-oriented design echoes the trend of technology, industry, and economy shifting from physical to digital. Traditionally, the physical design paradigm represented by architectural design and industrial design is to seek the best solution under constraints from the preset purpose, thus providing high-quality original design for mass production of high-cost physical products.

The double diamond model (Fig. 2) proposed by the British Design Council summarizes the basic paradigm of the design process described by the traditional design research theory [5]. The two significant diamonds respectively represent two processes. The former process is:

Discover: Start from the initial clear design challenges/objectives and seek to understand various possible causes/constraints through divergent research.

Define: Through analysis, clearly define the requirements and problems, thus determining the corresponding design principles and methods.

Based on the problems identified in the previous process, the latter process then:

Develop: Try various methods and means to meet the demands and solve the problems.

Delivery: Determine the final output and solutions.

The two diamonds which define the requirements and the solution have both relevance and relative independence. Multiple loop iterations may be required between different nodes or even the entire process. For designers, in specific design practices, many design processes are often directly informed of requirements through design briefing. These demands and problems determined by customers or clients may not have been carefully studied in the Discover and Define stages, or may have been obtained from the professional perspective of product managers or other roles and departments such as marketing. Designers need to decide whether to approve and accept these existing requirement definitions and/or communicate with the corresponding proposer of the requirements. Fortunately, for design researchers, the choice of research topics often has a certain degree of autonomy, and they can start from the research purpose, inject more energy into the Discover and Define stage, to make the research development and output more targeted.

However, the virtual design represented by interaction design, experience design, and service design is different. The new paradigm is to establish and match the most suitable solution for diversified needs and individual users under the preset target framework. The reproduction of digital products needs little attention due to little marginal cost, and its supply bottleneck lies in the richness and inclusiveness of design. It aims at

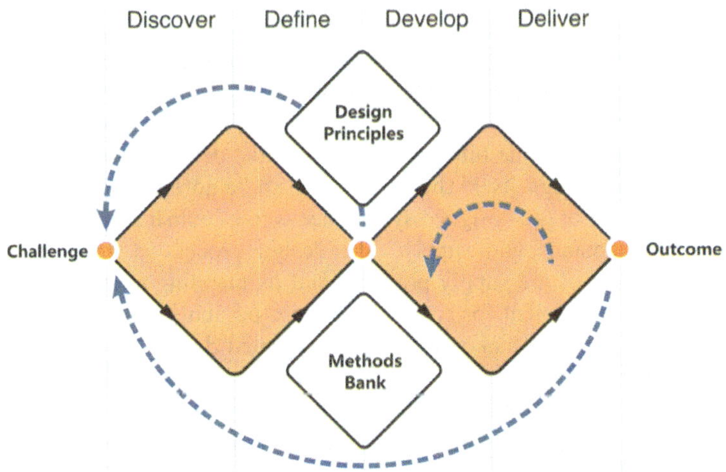

Fig. 2. The British Design Council's double diamond model [5].

establishing a mechanism to generate an uncertain design scheme according to uncertain needs, capabilities, and situations under the guidance of definite objectives (Fig. 3). This process makes it difficult for designers to cope with the challenges of diversity and uncertainty by relying on traditional design methods and tools that seek deterministic solutions. Traditional engineering methods that focus on accuracy and certainty have limitations in helping to solve design problems that are full of uncertainty and complexity. Qualitative research methods in sociology often have to spend much energy on the collection and analysis of unstructured data such as field investigation, which is challenging for large-scale applications in commercial design projects that focus on quick response and efficiency. Artificial Intelligence (AI) may be used to help define different problems, and even provide a variety of solutions/outcomes (Fig. 3).

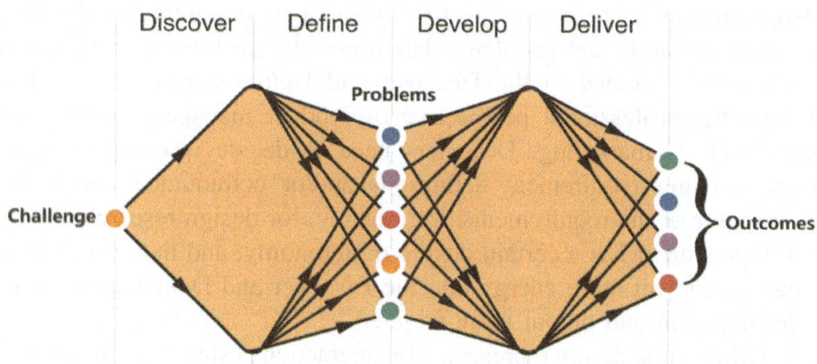

Fig. 3. Evolved design process of diversity-oriented design summarizes differentiated design problems and produces diversified solutions under the same design purpose.

3 Potential of AI

As low-cost mass customization becomes possible, the design methods to deal with diversity will rely more on inclusivity distribution instead of unity integration. Alternatively, to use the game theory terminology analogy, it is a more mixed strategy instead of a pure strategy. The future diversity-oriented design requires that the design output and even the design itself should become an "algorithm". For specific design purposes, the designer or the designed system needs to establish a mapping relationship between diverse demands and supply. The design process is the exploration and dynamic optimization of the supply potential and its mapping relationship. Designers are supposed to be competent for similar work in a particular field, but the enormous and complex demands may make the diversity-oriented design tasks extremely heavy and make designers struggle to cope with them. Moreover, designers' knowledge structure may not be sufficient to explore the characteristics of different users and different situations in various fields, which requires learning.

In recent years, machine learning technologies such as deep learning have proved their unique advantages in image recognition, NLP (Natural Language Processing), game, and other pattern recognition fields. Although some research has shown that the efficiency of deep learning systems in learning abstract and sophisticated rules is far less than that of human beings [6], sufficient data sets can make up and release the potential of deep learning to a great extent. With the migration and expansion of design research and practice to emerging fields, the types, forms, and accessibility of user and context data have increased significantly. More and more abundant data makes it possible for designers and design researchers to apply data-dependent AI technologies. Like image and speech recognition, if AI can learn the features of specific users/scenarios and their corresponding design patterns, and set corresponding design parameters accordingly, it can also provide reference designs for designers or provide alternative solutions for users. This kind of technology will greatly enhance the supply capacity of diversity-oriented design, thus improving the design inclusiveness of diversity requirements.

Existing deep learning and related technologies are still at an early stage in understanding; the results of training based on a large number of training examples are far from the abstract concepts summarized by human beings [7]. Deep learning classifiers are vulnerable to targeted input attacks. Some interferences that do not affect human perception will also significantly reduce the performance of machine-learning trained strategies [8]. Deep learning allows computers to construct complex concepts through simple concepts [9], but deep learning system's generalization and combination ability is still insufficient, so it often requires a huge training example set [10]. The professional experience of designers and design researchers can help optimize the design of deep learning systems for distributed representation of feature abstraction.

On the other hand, from AlphaGo's stunning performance in the Go chess field and DeepMind's open technical framework, it can be seen that the "black box" neural network can achieve performance beyond human level on specific tasks although it cannot directly help us to understand and explain the mysteries of the game more deeply. The "black box" may lead to some divergence of values. Some researchers worry that this state of "knowing what and not knowing why" is a significant shortcoming of machine learning. Lack of interpretability may be very harmful and unacceptable for some applications [11]. However, the Oriental thinking represented by Go chess is used to construct applied knowledge and experience with vague pragmatism before mastering the principle accurately. This kind of thinking is very consistent with the practical characteristics of design activities; it helps to promote new knowledge that is not clear and reliable to play a pioneering role in the design field, which is full of uncertainties. For example, deep learning solves problems from correlation rather than causality, which may not be a problem or shortcoming. We can argue that this makes it possible to explore and obtain feasible and better solutions through deep learning under the control of constraints.

The Deep Convolutional Generative Adversarial Networks (DCGANs) has shown its advantages in unsupervised learning [12]. DeepMind's self-supervised discovery model [13] and machine learning from long-tailed and open-ended distributed data have also made some progress [14]. However, the current deep learning system cannot draw open-ended inferences equivalent to the human-level based on the knowledge of the real-world [7]. Although design activities have a clear purpose, the openness and innovation of design results have always been one of the core values of design. The integration of machine learning technology in the design process needs to give full play to the respective advantages of designers and artificial intelligence, to avoid falling into a design cage of excessively focusing on known requirements [15].

4 Further Research

The primary purpose of further research is to improve the application level of inclusive design for diverse users and complex contexts, for example, to provide support for comprehensive human factors design research of a specific product in diversified usages. On the one hand, the research goal lies in the boundary of inclusive solutions expanded by the creativity of artificial design; on the other hand, machine learning can significantly improve the design efficiency of using existing knowledge to respond to differentiated demands.

According to the above research objectives, an ideal research environment should enable the establishment of interdisciplinary teams and platforms so that design research can be closely combined with relevant cutting-edge technology research, and an active division of task force can be carried out. Inclusive design research mainly focuses on exploring the technical, humanistic, and social reasons behind the existing problems and challenges, exploring the potential diverse needs in different contexts, and seeking strategies, methods, and tools to solve the problems, overcome the challenges and meet the needs. It means that designers need to create more design patterns that can be driven by algorithms. However, scientific and technological research provides more in-depth understandings, more effective technical means, and more possibilities for how to solve the problem of inclusiveness and realize these design patterns. Corresponding design practices can serve the technology transfer and commercialization of different scientific research outcomes, expand a broader range of application fields for artificial intelligence technology besides improving performance, and capture potential business opportunities.

5 Discussion

An inclusive design research from Cambridge University shows that there is an excellent potential for adaptation between AI-based Advanced Driver Assistance Systems and the diversity of drivers [16]. Some researchers try to introduce AI support to co-design [17], while others choose to apply inclusive design approaches to the development of AI products [18]. In the interdisciplinary research environment, we suggest that it may be necessary to focus on the application of machine learning in the design process. Firstly, through the establishment of a machine learning system with an interface for sensing user needs/capabilities and context features, the evaluation criteria and requirements of the target test set are studied so that machine learning can identify the design requirements and design problems corresponding to user and context features under a given design purpose. Secondly, together with researchers of artificial intelligence, the training models and feature representations of machine learning are optimized for different scenarios and tasks. Inclusive design tools, processes, and methods combined with machine learning technology are studied so that machine learning can identify corresponding design patterns or design options after inputting changing design requirements and problems. After reaching the milestone of phased research, new design tools, processes, and methods can be put into applied research in different sub-areas to summarize and establish corresponding domain knowledge and design pattern libraries. The outcomes of further research will provide reference for understanding inclusive design patterns applicable to different specific fields and the principles and mechanisms behind them, and improve the method of algorithmically providing appropriate design responses to a specific need.

References

1. Bianchin, M., Heylighen, A.: Fair by design. Addressing the paradox of inclusive design approaches. Des. J. **20**, S3162–S3170 (2017)
2. Bianchin, M., Heylighen, A.: Just design. Des. Stud. **54**, 1–22 (2018)
3. Li, F., Dong, H.: The economic explanation of inclusive design in different stages of product life time. In: Proceedings of the Design Society: International Conference on Engineering Design, vol. 1, pp. 2377–2386 (2019)
4. Li, F., Dong, H.: Achieving inclusion with contextualized user-sensitive design. In: Antona, M., Stephanidis, C. (eds.) 2019. LNCS, vol. 11572, pp. 113–132. Springer, Orlando (2019)
5. Design Council (UK). https://www.designcouncil.org.uk/news-opinion/what-framework-innovation-design-councils-evolved-double-diamond
6. Lake, B.M., Salakhutdinov, R., Tenenbaum, J.B.: Human-level concept learning through probabilistic program induction. Science **350**, 1332–1338 (2015)
7. Marcus, G.: Deep learning: a critical appraisal. arXiv e-prints (2018)
8. Huang, S., Papernot, N., Goodfellow, I., Duan, Y., Abbeel, P.: Adversarial attacks on neural network policies. arXiv e-prints (2017)
9. Goodfellow, I., Bengio, Y., Courville, A.: Deep Learning. The MIT Press, Cambridge (2016)
10. Lake, B., Baroni, M.: Still not systematic after all these years: on the compositional skills of sequence-to-sequence recurrent networks. In: 6th International Conference on Learning Representations, Vancouver, BC, Canada (2018)
11. Samek, W., Wiegand, T., Müller, K.-R.: Explainable artificial intelligence: understanding, visualizing and interpreting deep learning models. arXiv e-prints (2017)
12. Radford, A., Metz, L., Chintala, S.: Unsupervised representation learning with deep convolutional generative adversarial networks. arXiv e-prints (2015)
13. Gheshlaghi Azar, M., Piot, B., Avila Pires, B., Grill, J.-B., Altché, F., Munos, R.: World discovery models. arXiv e-prints (2019)
14. Liu, Z., Miao, Z., Zhan, X., Wang, J., Gong, B., Yu, S.X.: Large-scale long-tailed recognition in an open world. In: IEEE Conference on Computer Vision and Pattern Recognition, pp. 2537–2546. Computer Vision Foundation, Long Beach (2019)

15. Dong, H., Nickpour, F., McGinley, C.: Designing ergonomic data tools for designers. In: ICED 09 - the 17th International Conference on Engineering Design, pp. 53–63 (2009)
16. Caber, N., Langdon, P., Clarkson, P.J.: Designing adaptation in cars: an exploratory survey on drivers' usage of ADAS and car adaptations. In: Stanton, N. (ed.) AHFE International Conference on Human Factors in Transportation, vol. 964, pp. 95–106. Springer, Heidelberg (2020)
17. Walsh, G., Wronsky, E.: AI + co-design: developing a novel computer-supported approach to inclusive design. In: 22nd ACM Conference on Computer-Supported Cooperative Work and Social Computing, CSCW 2019, pp. 408–412. Association for Computing Machinery (2019)
18. Spencer, D.J., Poggi, J., Gheerawo, R.: Designing out stereotypes in artificial intelligence: involving users in the personality design of a digital assistant. In: 4th EAI International Conference on Smart Objects and Technologies for Social Good, GOODTECHS 2018, pp. 130–135. Association for Computing Machinery (2018)

Design for Inclusion and Diversity: Developing Social Competencies in Engineering Education

Claudia Lizette Garay-Rondero(✉), David Ernesto Salinas-Navarro, and Ericka Zulema Rodríguez Calvo

Escuela de Ingeniería y Ciencias, Tecnologico de Monterrey, Eugenio Garza Sada 2501, 64849 Monterrey, NL, Mexico
{clgaray, dsalinas, ezrodrig}@tec.mx

Abstract. This research proposes a novel method to design didactic products that favor the social inclusion of people with disabilities by creating learning experiences for engineering students that develop disciplinary and personal competencies. This manuscript briefly describes the notions of inclusion, diversity, and urgency to foster community equality, involving governments, companies, and society. We propose a methodology based on Experiential Learning, Challenged-based Learning, and Competency-Based Education to impart learning experiences that ignite awareness of social inclusion competencies and lead to improved learning outcomes. The methodology describes the design application for inclusivity, showing the interactions of students in different civil partnerships and institutions that support the needs of people with visual impairment, Down Syndrome, and Alzheimer's. The Results section discusses the participation of students in the learning experiences at host institutions that facilitated their involvement to understand social needs and then contribute to the design and production of supportive products.

Keywords: Design for inclusion · Social inclusion · Engineering education · Experiential learning · Educational innovation · Higher education

1 Introduction

The purpose of this work is to raise awareness among current undergraduate students of the social needs of vulnerable groups through innovative educational methods. This group (class) was conceived because the students have future responsibilities as decision-makers in organizations that incorporate inclusion, diversity, and accessibility at work in the production of products and services. We propose a methodology based on Experiential Learning (EL), Challenged-based Learning (CBL), and Competency-Based Education (CBE) to impart learning experiences that ignite awareness of social inclusion competencies and lead to improved learning outcomes.

Section 2 presents a seven-phase methodology. Then, Sect. 3 discusses the implementation of these phases by presenting a study case within the experiential learning space named "Lean Thinking Learning Space." The Results in Sect. 4 offer a brief description of the students' comments and reflections on their findings and involvement in the product design, production process, and implementation. Furthermore, the section

G. Di Bucchianico et al. (Eds.): AHFE 2020, AISC 1202, pp. 85–92, 2020.
https://doi.org/10.1007/978-3-030-51194-4_12

identifies opportunities to work with other various vulnerable groups and institutions for the benefit of increased social contributions. Finally, the conclusions about the students' engagement, achievements, and future work possibilities are presented to propose the next steps of this research work and highlight its value.

The main contribution of this educational innovation resides in the conceptualization and exemplification of an EL methodology that promotes inclusion and diversity in engineering education and develops awareness about the relevance of designing inclusive products in accommodating work environments.

2 Preparing Professionals to Design for Inclusion Through Experiential Learning Spaces

Inclusive design is based on four elements [1]: accessibility, adaptability, aesthetics, and affordability, without neglecting conditions that limit it, such as materials, time available, budget, and other operating factors of the space and product. In the daily life of people, one common problem is that the built environments, products, and spaces do not match the needs and expectations of their users, which impacts the quality of their lives [2, 3].

The need to guarantee an egalitarian society is largely in the hands of professionals involved in creating inclusive products and places that impact the social changes needed to create an inclusive society [4, 5]. For inclusive design, it is important, as some authors have mentioned [6–9], that future professionals develop these skills, considering not only the disciplinary aspects but the design principles as well. More important is that students develop these skills through teamwork, emphasizing the cultural and social aspects of real-life environments. This can be accomplished through collaborative learning [10].

Research has shown that learning based on developing knowledge through traditional instruction is less effective than using active learning techniques such as EL, CBL, and CBE. The primary reason is due to the skillsets that the students must develop for their future practice. These active learning methods tend to facilitate the development of the relevant attitudinal aspects and skills that the students need, increasing their social awareness, thus enabling them to generate solutions with a more inclusive approach [2, 11, 12]. The EL Kolb's model focuses on an experiential space in which activities are organized for the student [13], encouraging the students to face real-life issues and challenging them to find creative but appropriate solutions. The fact that their proposals might not be adequate, however, is part of the learning process [12, 14].

Although few studies have been conducted focusing on the impact of EL and the competencies related to inclusive design [7], the work that has been carried out involving architecture and design students stands out; however, there is less literature about the impact on engineering students. Nevertheless, the Lean-Thinking-Learning Space (LTLS) is focused on EL methodologies and can be used for the training of engineers. Here, students develop not only disciplinary competencies but also useful abilities to make a social impact, solving real-world problems [12].

The main objective of the LTLS is to foster social awareness and be a center for the development of inclusive competencies through learning experiences. The LTLS uses EL, CBL, and CBE procedures and methods, applied in seven phases, described as follows (see Fig. 1 and Fig. 2):

1. Students get in contact and interact with vulnerable people to identify their needs and requirements for inclusive-products design.
   1.1. The non-profit organization (NPO) and the professor define clear objectives, as well as the work schedule and assessment plan.
   1.2. The NPO actively participates in the development and evaluation of the students' abilities, skills, and competencies [15].
2. The students design prototypes and products that are inclusive and positively impact the functions and lives of disabled people.
3. The students test and validate their products with the users to obtain feedback.
4. With that feedback, the students re-engineer their products, applying engineering and manufacturing concepts, methods, and tools.
5. The stakeholders evaluate the prototype within the criteria established in step 3.
6. The students deliver and donate the manufactured products.
   6.1. The NPO and the professor schedule an event and invite all the people involved in the process, to raise awareness and empower the students.
   6.2. The formal closure of the project and its results are announced to the stakeholders.
7. The students reflect upon their work, and the evaluation committee assesses the students' competencies and learning outcomes.
   7.1. The students deliver a design portfolio of an inclusive project that includes documented evidence of their development of competencies, theoretical knowledge, and practical application in engineering, design, and sustainability. They also add an ethical discussion on the social impact they had.
   7.2. Finally, a discussion among the students is moderated by the professor to focus on improvements and express final thoughts.

An important feature of EL in the development of skills for inclusive design is the reflection and empathy that students develop for the special needs of people so that they seek to create an egalitarian society [2, 4, 11]. Professionals who generate inclusive designs for vulnerable groups have these characteristics [14]. For active learning to be successful, it is important not to set aside the development of these soft skills. The students must motivate themselves to learn these techniques, but keep in mind that doing this is more demanding for teachers and instructors than traditional teaching.

Following the above procedure provides an EL that directly impacts the training of professionals who work for more inclusive societies. Through this learning modality, the students prepare to address design problems from a more global perspective. They come to generate a more sensible and inclusive design proposal, which, in turn, leads to a more egalitarian society. This learning experience for the students results in generating synergy between the university and its community, laying the foundations for real social-impact projects.

3 Case Study

The methodology described in Sect. 4, has been applied on campus since 2018 during five consecutive semesters in an Industrial Engineering course, named *IN3038 Lab operations design and optimization.* This methodology has also been used in multi-disciplinary classes that include students from Industrial Design, where the objective is to develop "the ability to design a system, component or process to meet the desired needs within realistic limitations, such as economic, environmental, social, ethical, health and safety, manufacturing and sustainability" [15].

The main objective behind training in these competencies is for the student to focus on the development of a given product, not only from the design or manufacturing process point of view but also from a broader view of inclusion, diversity, and accessibility. The achievement of this objective is possible, thanks to the cooperating NPOs. The students are exposed to interdisciplinary collaboration, problem-solving, and critical thinking, abilities that will be of great importance in their near future.

The NPOs that actively participated in the design process, giving users' feedback; furthermore they received the donations, and the product follow-ups, are listed below. The NPOs are: educational institutions for the visually impaired; institutions focused on special-needs education for people with Autism, Down syndrome, and Alzheimer's disease; children's homes for mixed-aged girls and boys, and nursing/retirement homes.

Some of the most relevant designs developed through this methodology include wooden teaching toys, an educational kit for reading/writing in Braille, a slate for mathematical operations in Braille, and didactic objects to promote motor and cognitive skills for people with Autism, Down syndrome, and Alzheimer's disease.

Figure 1 illustrates some final products, and Fig. 2 shows the work process, delivery, and real-life product use from participants of the NPOs.

Fig. 1. Products designed for inclusion within the Lean Thinking Learning Space. Left: kit for reading/writing in Braille. Middle: didactic objects to promote motor and cognitive skills. Right: slate for mathematical operations in Braille.

Fig. 2. Left: testing the design for an inclusive product. Middle: assembly workstation for wooden didactic toys. Right: final delivery to the NPO.

This case study has presented the methodology employed in this research. This paper describes the testimony and evidence of the impact that the development of skills related to inclusive design had. The students observed, first-hand, people with various physical disabilities (the visually impaired, people with walking problems, older adults, those who lacked limbs, etc.) [4, 10]. After this experience, the students showed a high degree of empathy that inspired them to consider people with special needs in their product designs for vulnerable people.

Furthermore, it was observed that the students developed collaborative abilities and problem-solving and teamwork skills, which are fundamental for a successful, inclusively designed project; such products are inherently complex and require interdisciplinary knowledge. Therefore, it is essential to interact with people from different disciplines to generate a comprehensive design proposal. The development of knowledge also occurs through social interactions, so it is important to collaborate to learn and, of course, learn to collaborate [10, 12].

4 Results

Over 100 students have participated in this implemented methodology. They are mainly from the Engineering and Industrial Design program. There have been six academic staff members (Assistant and Associate Professors), multidisciplinary staff, and students involved in production and design, resulting in over 300 inclusive products manufactured and donated (See Sect. 3). Eight non-profit organizations have collaborated in the project, which has directly and indirectly impacted over 400 people.

Most of the participants have liked the approach of this program because it has facilitated the teaching-learning process, especially for the students who learn to combine practical skills with theoretical knowledge to tackle a real-life problem. Repeatedly, the assessments of the program have shown that the methodology has been popular among students. In the evaluations conducted among students after taking the course, some of the comments included: "The class was a challenge, but it was easy to understand, and there were plenty of opportunities", "The activities make you develop your abilities and work within a team", "The class was very cool and dynamic", "Such activity makes the course really interesting and a good way to apply practically the acquired knowledge obtained throughout my studies; it also motivated us to work and collaborate in teams", and "The atmosphere in the lab seemed like a real company".

Comments about the instructors included "The Professor has a vast theoretical and practical knowledge and was very accessible", "The Professor knows the topic and was very supportive of the students", "I like that the Professor knew how to explain the topics of the course and how to apply them in a practical setting", "Knows very well the main topics of the course and her teaching skills are great; I love the passion she used to teach", and "Very energetic class, the enthusiasm is contagious".

Beyond the popularity of the course, the students wrote reflections on the evaluations about the impact they had on other people. We observed that the students shifted their paradigm about disabilities because they felt involved in a close-knit community of classmates, academic staff, NPOs, and people with disabilities, all working for the same purposes, namely, inclusion, diversity, and accessibility. Some of the commentaries that stood out are:

"The most rewarding feeling for me is that this type of work directly benefited the elderly and people affected by Down syndrome and Alzheimer's disease that we encountered in the NPO; furthermore they will use our product, the one *we* designed; that is why this project helped us to be more sensitive in our surroundings", "The main goal of the project was to fulfill a production rate, but for me, it was more than that: it wasn't just about producing any product and passing the course, rather, it involved a deep understanding and analysis to determine who needed the product we designed. For them, this was a critical point, the social aspect of it".

Finally, some of the most satisfactory comments from the NPOs were: "The product that the students designed was excellent, teaching how to write and read in Braille", and "This product can break down barriers for people with disabilities [16]".

5 Conclusion

This research work proposes a novel teaching methodology focused on the development of competencies in inclusion and diversity. Furthermore, this method was validated through a case study based on experiential learning in a Lean Thinking Learning Space. Since 2018, the LTLS has developed and implemented steps for products and process designs characterized by inclusivity, diversity, and accessibility, and that directly benefit various groups of local people who suffer a range of disabilities.

The experiential learning in the LTLS by our students has resulted in their development of desired competencies (see Sect. 3), insights, shifts in paradigm, commitment, and inclusion (see Sect. 4), and this has also occurred among academic staff and personnel within the NPOs. It is noteworthy that in a traditional academic environment, these transversal skills are extremely difficult to develop; however, with the implementation of EL, CBL, and CBE methodologies, the students were able to improve them substantially.

Feedback from the students describes a great experience and their sincere commitment toward the community. The impact of this research work has positive implications for the training of students for inclusive design; it allows them to achieve greater satisfaction and improved outcomes in multidisciplinary learning, a crucial step in preparing future professionals for the labor market, which now requires a deep understanding of inclusivity, diversity, and accessibility in the designs of products and services.

In our research, we observed the impact of the final products in aiding individuals with special needs, thus having an immediate impact on their lives. Furthermore, we identified opportunities to work with other vulnerable groups and institutions to increase social contributions that can be made to the community.

For future work, we intend to keep monitoring students' feedback, conclusions, participation, and achievements to delineate the value and next steps of this research work. As an example, we believe that it is critical to strengthen the criteria within the methodology that assessed the social impact of the various students' projects. This can be achieved by applying experimental criteria to validate specified objectives such as product functionality, scalability, product lifetime, product follow-up in terms of usefulness (such as reading/writing Braille), and improvements.

It is also proposed for future work to implement a continual improvement process to validate the final product and scale it upward with the help of larger organizations and user groups. Doing this will allow the collection of vital information that can help other vulnerable groups. We suggest continuing to monitor the competencies of the alumni who experienced the LTLS to know how they are currently applying this methodology to their professional careers in terms of inclusive designs incorporating inclusivity, accessibility, and diversity. Data on the impact they have on everyday decisions such as the design and selection of products or processes, social entrepreneurship, recruitment of people with disabilities, among others, would be very useful, regardless of their roles in their current employment. Soon, the main goal of this methodology to have a larger impact on more students and academic staff, and even other institutions in other countries and regions, will be extended, starting first with other departments of this institution.

Some of the limitations found during the development of this research and implementation had to do with limited materials and manufacturing constraints and also the types of disabilities of the clients served by the NPOs. Ethical considerations and the safety of the stakeholders had to be considered, as well as the final product users, the academic staff, and the students. Deep knowledge of stakeholders' safety is crucial for the future success of the program. So far, the importance of Health and Safety regulations has been a key point when defining the scope of the project. As with any implementation of a new methodology, some academic-administrative frictions arose, specifically, regarding the time and implementation of the project, time constraints, and the interest of the NPOs in developing a new product.

Acknowledgment. The authors would like to acknowledge the technical support of Writing Lab, TecLabs, Tecnologico de Monterrey, Mexico, in the production of this work.

References

1. Sterling, M.C.: Universal design problem solving. J. Family Consum. Sci. **96**(2), 40–44 (2004)
2. Altay, B.: User-centered design through learner-centered instruction. Teach. High. Educ. **19**(2), 138–155 (2014)

3. Watchorna, V., Larkina, H., Angb, S., Hitch, D.: Strategies and effectiveness of teaching universal design in a cross-faculty setting. Teach. High. Educ. **18**(5), 477–490 (2013)
4. Mulligan, K., Calder, A., Mulligan, H.: Inclusive design in architectural practice: experiential learning of disability in architectural education. Disabil. Health J. **11**, 237–242 (2018)
5. Nicholl, B., Hosking, I.M., Elton, E.M., Lee, Y., Bell, J., Clarkson, P.J.: Inclusive design in the key stage 3 classroom: an investigation of teachers' understanding and implementation of user-centered design principles in design and technology. Int. J. Technol. Des. Educ. **23**, 921–938 (2013)
6. Hutter, L., Lawrence, M.H.: Promoting inclusive and accessible design in usability testing: a teaching case with users who are deaf. Commun. Des. Q. **6**(2), 21–30 (2018)
7. Kenning, G.: Reciprocal design: inclusive design approaches for people with late-stage dementia. Des. Health **2**(1), 142–162 (2018)
8. Chang, B., Tremblay, K., Dunbar, B.: An experiential approach to teaching universal design. Education **121**(1), 153–158 (2000)
9. Bang, M., Vossoughi, S.: Participatory design research and educational justice: studying learning and relations within social change making. Cogn. Instr. **34**(3), 173–193 (2016)
10. Hennessy, S., Murphy, P.: The potential for collaborative problem solving in design and technology. Int. J. Technol. Des. Educ. **9**, 1–36 (1999)
11. Blascoa, R., Blancoa, T., Marcoa, A., Berbegalb, A., Casasa, R.: Needs identification methodology for inclusive design. Behav. Inf. Technol. **35**(4), 304–318 (2016)
12. Garay-Rondero, C.L., Rodríguez, E.Z., Salinas-Navarro, D.E.: Experiential learning at lean-thinking-learning space. Int. J. Interact. Des. Manuf. **13**, 1129–1144 (2019)
13. Kolb, D.: Experiential Learning: Experience as the Source of Learning and Development. Prentice-Hall, New Jersey (1984)
14. Wharton, R., Parry, L.: The good, the bad, and the ugly: using experiential learning in the classroom. J. Scholarsh. Teach. Learn. **3**(2), 56–68 (2012)
15. Garay-Rondero, C.L., Rodríguez, E.Z., Salinas-Navarro, D.E.: Developing and assessing engineering competencies at experiential learning spaces. In: IEEE Frontiers in Education Conference (FIE), Covington, KY, USA, pp. 1–5 (2019). https://doi.org/10.1109/fie43999.2019.9028355
16. Lean Thinking Learning Space. https://www.youtube.com/watch?v=3fDT5Bab0FI&t

Design for Disability and Social Inclusion

Design for Inclusion in the Nautical Field. DfA Guideline for Small Motorboats

Jessica Lagatta(✉)

Department of Architecture, G. d'Annunzio University of Chieti-Pescara, Viale Pindaro 42, 65127 Pescara, Italy
jessica.lagatta@unich.it

Abstract. During the last decades a strong attention to the issues of human diversity, social inclusion and equality has developed in different fields and in the nautical one. This has allowed more and more users to approach the world of pleasure boating, including people with disability. Accordingly, several small motor boats are designed or adapted for different kinds of users. This state-of-art represents a breeding ground to employ Design for Inclusion approaches in this field. The research's aim here described is to define DfA-based guidelines for small motor boats, in particular. To obtain this result, the study has been conducted throughout different phases. During the early stages, different case studies have been selected, filed and analysed to comprehend what their strength and lacks about accessibility and ergonomic matter are. During the last stages, starting from this critical perspective, we have drawn up design guidelines based on the principles of "DfA Italia" for small motor boats.

Keywords: Design for inclusion · Design for all · Motorboats

1 Introduction

In the last years, ergonomic-based design theories are developing in order to respond to needs and desires of different people. They assume that if we consider human diversity from an inclusive point of view, it can represent a great resource rather than a problem. These new DfI "Design for Inclusion" approaches (e.g. Universal Design_UD, Design for All_DfA, ID_Inclusive Design) have already spread to several fields, including yacht design.

Today different sailing and motor yacht activities, previously considered as a prerogative of a social and economic elite, are currently practiced by people with different disability, age, culture, social status, etc. In the field of sailing sports we have different kinds of small boats that are designed according to the principles of some "Design for Inclusion" theories, like UD (e.g. Skud 18), DfA (Neo 495, Càpita), etc. Despite small motor boats are now widely used by all kind of people, there aren't case studies designed according to the disciplines of DfI, as is the case of the small sailing boat sector. However, at the same time, today there are many motor boats that are used by disabled, aged, young people, etc. This type of diffusion could bring the potential of a systematic application of the approaches of DfI also to the field of small motor boats.

G. Di Bucchianico et al. (Eds.): AHFE 2020, AISC 1202, pp. 95–102, 2020.
https://doi.org/10.1007/978-3-030-51194-4_13

2 Research Aim

Starting from the area of concern here described, we tried to comprehend what is the better approach of DfI to be applied in the nautical field. We hypothesise that DfA can represent the most suitable solution to meet the different and complex user needs and desires in this sector [1]. As a matter of fact, DfA approach provides for a constant relationship whit users and a holistic and well-structured design process. This allows a direct and continuous comparison with "human diversity" and, at the same time, the investigation of the issues most closely related to the motor boat ergonomic and performance aspects. Therefore, DfA approach could represent a competitive and more sustainable factor in economic, ethic and social terms [2]. Starting from this hypothesis, the research's main goal is to draw up design guidelines for motor boats, based on the DfA principles in particular.

3 Methodology and Research Phases

To obtain this aim, the research has been conducted through four main phases. In the first one, a critical-comparative analysis between the principal DfI approaches allowed us to understand that DfA is the most suitable one to respond to our problem areas. We particularly referred to "DfA Italia" principle. Since 2010, in order to certify the DfA quality of products, environments and systems, DfA Italia has established two labels: *DfA-Silver (Start)* and *DfA-Gold (Quality)*. The first one is awarded when the product, environments and systems observe the fundamental principles of this discipline and meets some significant requirements of it. The second one is awarded when they meet all the DfA principles and requirements. The principles to follow are summarized below: enhancing human diversity; promoting social inclusion and equality; being easy and pleasant to use for all potential users; not to marginalize in any physical or psychological manner; to be beautiful; to be socially, environmentally and economically sustainable; having the ultimate goal of improving the quality of life. The requirements that must also be verified are shared in the following four main categories: physicality, comprehension, perception and sensoriality, process. In Fig. 1 the specifications for each category are shown in summary [3].

PHYSICALITY	COMPREHENSION	PERCEPTION AND SENSORIALITY	PROCESS
• enabling prehensility • required minimum efforts, accuracy and skill • offering alternatives for handling and use • offering customized systems • using technical devices to meet the target user's needs and ability • correspondence to anthropometric differences • compliance with safety features • responding to diversified well-being needs	• self-explanatory use • use of semantic codes for the use understanding • formal/aesthetic quality • natural orientation's implementation • mutilsensory, transcultural and translingual communication • documentation for use	• color, tactile, sonorous contrasts • gustatory and olfactory stimulus • kinesthetic perception	• accuracy in the design approach and in the decision process • extension of the DfA principles to the whole value chain • public / promotional messages that promote the system, the product, the environment and the DfA philosophy

Fig. 1. DfA Italia requirements and their specifications.

Although these principles and requirements have been drawn up to certify the DfA quality of existing environments, products and systems, they can also represent a valid support to design and construct new ones. For this reason, we have been guided by them in drawing up DfA guidelines for small motor boats.

In the second phase, the state-of-the-art related to small motor boats has been studied to have a first overview of what has been done so far within the motor yacht field for disabled people. A selection of the most significant case studies has been carried out. In particular, 38 small motor boats (with a length less than ten meters) have been chosen based on users. They have been classified according to the following criteria: type 1_motor boats not designed for widened users[1] but that, thanks to the application of special aids or adaptations, are also used by people with motor disabilities; type 2_motor boats specifically designed and built for people with special motor or intellectual disabilities; type 3_motor boats not specifically designed for people with disabilities but, thanks to their spatial and distributional characteristics, currently used by user with limitations without any particular aid.

In the third phase, the 38 selected case studies have been catalogued to conduct an analysis of their main features. The data cataloguing has been carried out through FileMaker Pro 12®. This software allows the creation of *records* (data-sheet) easy to read and manage and statistical analyzes relating to case studies features. In particular, the following items have been examined and cataloged for each boat: dimensional characteristics (L.O.A., beam, displacement, motor, etc.); identification data (country and year of production, shipyard, type of user, crew members number, etc.); typological features (hull: monohull or catamaran; superstructure: open, half cabin or cabin cruiser; hull material: fiberglass, wood, aluminum, etc.); usage characteristics (pleasure: for daily trip, sport fishing, scuba diving, etc.; work: fishing, for rent, etc.; transportation: goods, people, bicycle, etc.); boarding and disembarking (mode and means by abled bodied and disabled users to access on board: gangway, crane, lift, etc.); accessibility features about: wheelhouse, toilet, cabin, etc.

Once case study boats data were collected, we proceeded to the quantitative and qualitative data-processing for every *records* item. This kind of analysis allowed us to understand, on the whole, what are the recurring features of the small motor boats considered. Taking into account the higher items percentage values for every data, we have the following recap: - *dimensional characteristics:* most of boats have L.O.A. between 4, 5 and 7 m and only one outboard engine; - *identification data*: most of the case studies are produced since 2005 in Europe, less than a quarter in the USA and a small percentage in Canada and New Zealand; the crew members number varies between 4 and 10; users type analysis shows that: 50% is type 3; 31% is type 1; only 18% is type 2. - *typological features*: monohull, open in fiberglass or aluminum; - *usage characteristics*: pleasure boat for daily trip and for sport fishing in particular; - *boarding and disembarking*: the access on board, in most cases, occurs by a gangway added to the stern or by a bow door opening. - *accessibility features*: frequently the cockpit and the wheelhouse is accessible by every user; it's not the same for the cabin and the toilet.

[1] Widened user is defined as the whole of: "autonomous users" «all people that, with their peculiarity, are able to perform a function independently from the help of other people» and "users with limitation" «that part of the All (target users) that has the most critical specificities to use the system autonomously» [2].

This study allowed us to understand what are the main features of the examined boats and the various design solutions currently adopted to meet the needs of every types of users. It also allowed to observe what are the accessibility and ergonomics problems and lacks. Starting from this critical perspective, during the fourth and last phase, the DfA guidelines, described in the following paragraph, were drawn up.

4 Design for All Guidelines for Small Motorboats

In accordance with the main features of the case studies described above, we consider only the boats that are: open, monohull, with a length between 4.5 m and 7 m, used for daily trip. Since DfA approach focuses its attention on users, in this paper we drew up design guidelines of the main boat elements with which the user interacts directly: access, seats and wheelhouse. In order to consider a boat "for all" it is necessary that most elements that compose it largely respond to the principles and requirements mentioned above. Obviously, it is impossible to describe in detail every requirement in this paper. We are here reporting only a summary of a wider account developed during the research.

Access Requirements for Boarding and Disembarking: access on board implies a certain degree of instability, especially if the boat is small. Users have to move from the mainland to an "artifact" that floats on a liquid and that is therefore in constant, albeit minimal, movement.

Physicality: systems of access on board have to be easily used by everybody. Several case studies have removable gangways, but they are heavy and cumbersome. Not every user has the physical capacity to move them before and after the access on board. In other cases a little crane helps users with physical disability to embark. This kind of technical solution cannot be considered "for all" as it is designed exclusively for a "particular user". It would be desirable that access on board was done in the same way for everyone, without the use of "special aids" such as crane or anything else. An adequate design solution could be, for example, a gangway that can be raised or lowered with very little effort (perhaps through hydraulic or electric movement), or an electronic retractable platform (e.g. in Easy Boat 21 case study [4]). Access on board frequently occurs through very close gangways. This requires a certain precision to go by foot or wheelchair, to avoid falling into the water. The width of the gangway should be designed: to allow everyone to pass without a high degree of precision; to respect the anthropometric differences of all users; to take on board: fishing equipment, strollers, bikes and all that is necessary to fully enjoy the boat trip. In order to make access safe for everyone, the platform that connects the quay to the boat should be long enough to guarantee a not too steep slope. In addition, vertical and horizontal support elements must be provided to prevent falls in water and reassure users during its use. Sometimes, in fact, the first problem to be solved in order to navigate is that of overcoming the fear of access on board. Making the access element structurally safe can help to board peacefully not only people with psycho-physical difficulties, but also e.g.: mums with strollers, children, elderly people, users with bicycles, suitcases or other goods and so on.

Perception and Sensoriality: The use of a different color between the gangway, the ground and the cockpit could help the user understand more immediately where the access phase on board begins and ends. Using a color in sharp contrast to the passage area along the edges of the walkway could help to understand the maximum limit within which people can go to stay safe. Tactile contrasts can be very useful to understand: where the surface is non-slip; where to cling with your hands to avoid falling; etc. Sound contrasts can be used too at the beginning and end of the opening and closing of the walkway to indicate when you can get on or off (for safety reasons and in case of blind users).

Comprehension: in the case studies, gangway aesthetic-formal quality is often not very appreciable. When gangway is used like an additional element of the boat, it does not reflect its style. Sometimes when it is an integral part of the boat (e.g. it can be opened from the bow or from the stern), it does not present a particular formal aesthetic care. It's important to design this "component" not only with the same architectural style of the boat but also with the same care in formal aspect. Moreover, all users must be able to understand how to close or open the gangway without particular difficulties, despite the design solutions adopted (electronic, mechanical, etc.). In general, we can say that, if properly designed, the morphology of the element, the chromatic and tactile contrasts, the formal aesthetic quality can allow a high level of self-explanation of use-operation and so a high level of comprehension.

Requirements for Seats: There are several factors that we have to consider when we design a seat for small motor boat. First of all, what the needs of the "users with limitation" are. In DfA approach these critical issues «are solved by taking into account the autonomous users, with the awareness that other users can employ ad hoc aid for their specific needs, which they already normally use» [2]. For example, a person with lower limbs disability should be able to access to the cockpit in the most autonomous way, through their wheelchair. Therefore, the cockpit spatial distribution should be designed to allow not only the access to wheelchairs but also the presence of comfortable seats for everyone.

Physicality: Given the difference between autonomous users and users with limitation, it is necessary to create a customized seating system inside the boat. This could consist of flexible or retractable elements that could allow a different design response to the needs of such a diversified user. For example, if there are users in wheelchairs an adequate space is necessary to contain them, as well as a system of "fixing" the wheels. Especially in small boats, a wheelchair could take up a lot of space by not allowing the presence of further seats. To remedy this drawback, different solutions could be used such as: reclosable or flexible seating systems inside the cockpit. Case study Silennis S.020 [5] could represent an example of good practice. Flexibility can be an adequate solution to respond at the different anthropometric dimensions of all users. Furthermore, seats size and position inside the boat should be designed to allow a comfortable side passage for: users both autonomous and in wheelchairs, for any movement of loads (such as e.g. fishing equipment, bags, backpacks, bicycles, etc.).

Perception and Sensoriality: chromatic contrasts could be very useful to indicate the points where the wheelchair can be fixed or those where the "transformable" seats can be found. An example of good practice on this issue is also the boat Silennis S.020. In fact, the moving parts which allow the flexibility of the system have been made with a black color, in stark contrast to the white support base. This lets you immediately understand which is the mobile part of the boat.

Comprehension: the possible design solutions described above should be easily understood by all users. Tactile and chromatic contrasts can help you to realize what the moving parts are, but in general the whole system must be designed with a good affordance. In closing, most of case studies seats do not have a good aesthetic-formal quality. It's important to design this element in such a way that it is not only functional but also beautiful. Furthermore, it should have the same style as the boat.

Requirements for the Wheelhouse: in almost all the case studies analyzed, the wheelhouse system consists of a "steering wheel" placed on a "column". Depending on the size of the boat, this can be positioned in a central or lateral area of the cockpit. The wheelhouse is an important part of the design of a boat "for all" because it can allow all users to be able to navigate in complete autonomy.

Physicality: prensility is fundamental when it comes to steering. The wheel must be designed in order to allow all users to grab it adequately, without any maneuverability problems. Not everyone, in fact, is able to grasp the wheel with the right force and with adequate prehensile ability (e.g. people who have disabilities in the upper limbs, etc.). This could involve risks especially in terms of safety during navigation and docking maneuvers. It is important that all the elements are non-slip and have an adequate size and shape to be correctly and comfortably grasped by anyone. When this is not possible, customization systems could be used. For example, joysticks, Sip/Puf systems, pedal steering, etc.; they are already used in small sailing yacht and could be transferred to the field of motor boat. In order to respond to the anthropometric differences of users, the wheelhouse and the seat should be flexible. Systems that allow you to raise and lower the column that supports the steering wheel or to adjust its inclination could be adopted. This could be more comfortable for all users including those who, being in a wheelchair, cannot vary the height of their seat and therefore of their arms. This flexibility would also allow to facilitate a broad view of the place where you are steering and a good comfort. In addition, the areas near wheelhouse should be sized to facilitate a comfortable passage to reach the command position for everyone. When you have a very small boat, space problems can be solved for example by designing rolling steering systems similar to the one used in the case study Pioneer Multi [6].

Perception, Sensoriality and Comprehension: wheelhouse should be designed to be immediately readable by everyone. The perceptual component is very important to make this happen. For example, creating color contrasts could help the user to clearly distinguish the elements that are on the dashboard. You could use very bright colors or particular illuminations for the buttons that are used most frequently. This differentiation could be accentuated through the use of materials that have a differentiated texture perceivable by touch. Sound contrasts could also prove very useful. The sounds, in fact, can accompany the warning lights in case of malfunction of some element of the boat. Or, when moored, they could perform a warning function when you get too close to the dock or some obstacle. Starting from this principle, new technologies have been introduced on the market that even allow navigation for blind people. They can navigate in complete autonomy thanks to the use of warning sounds from an acoustic instrument positioned in the bow of the boat. For example, the Homerus association, has been using this system for several years, allowing blind people not only to navigate independently on a sailing boat, but also to participate in regattas. The same technology or something conceptually similar could be transferred to motor boats to help all users, especially the less experienced ones, navigate with greater safety.

5 Results

Through this research, different results have been achieved. An important result is to create a wide case studies catalogue that made it possible to have a first state-of-the-art report of motor boats designed, not only for people with disabilities, but also for widened user. Moreover, the boats that meet the characteristics described in the selection criteria are increasing over time. For this reason the cataloguing has been structured in such a way that the *records* can be modified or added in the future. This will allow us to have a continuous update not only on the number of boats, but also on their technological, dimensional, architectural ergonomics characteristics, etc. which will be useful to analyze how they will develop over time. This represents, in itself, a newness in the nautical sector.

In line with the main thesis aim, the most important result is to develop some design guidelines for motor boats, based on the DfA Italia approach. The DfA guidelines here described refer to the main elements of the boat (access systems, seats, wheelhouse) with which users interact directly. We can consider these guidelines like an example for the design development of the remaining parts of the boat and a base from which to construct motor boats more and more able to respond holistically to the wishes of widened user and to allow social inclusion and equality through motor yacht practice.

Acknowledgments. This paper refers to a research developed by Ph.D. Jessica Lagatta at the Department of Architecture G.d'Annunzio University of Chieti-Pescara (Italy). The various paragraphs of the present paper can be considered the outcome of a common discussion with Prof. A. Vallicelli.

References

1. Lagatta, J., di Nicolantonio, M., Vallicelli, A.: Design for inclusion. Differences and similarities between DfA and UD in the field of sailing yacht design. Proc. Manuf. **3**, 2714–2721 (2015). Ahram, T., Karwowski, W. (eds.) Proceedings of the 6th AHFE Conference. Elsevier Procedia, Las Vegas
2. Accolla, A.: Design for All: il progetto per l'individuo reale. Franco Angeli, Milano (2009)
3. Design for All Italia; labels of quality. http://quality.dfaitalia.it/design-for-all-italia/wp-content/uploads/2015/12/Regolamento-marchi-italiano-2015-nuovo.pdf
4. Easyboat. http://www.velaetica.it/easyboat.html
5. Silennis S.020. http://www.silennis.com/
6. Pioneer Multy. https://pionerboat.com/pioner-multi/

Diary Methods Used in Research on Visually Impaired People

Dawoon Jeong and Sung H. Han(✉)

Industrial and Management Engineering, POSTECH, Cheongam-ro. 77, Pohang 37673, South Korea
{jdw0303, shan}@postech.ac.kr

Abstract. The diary method is a good way to figure out what practical needs visually impaired people (VIP) have based on their experience. The objective of this study is to investigate how to conduct the diary method in the research of visually impaired people through literature review and interviews. This study summarizes the research purpose of the existing studies, detailed methods such as input method and duration, and issues raised during the study. As a result, we derives considerations for researchers to perform efficient and effective diary methods. The research field for VIP needs a new type of the diary method applied with new technologies such as mobile, IoT, and blockchain.

Keywords: Diary method · Diary studies · Visually impaired people (VIP) · Visual impairment · Qualitative research

1 Introduction

There are about 285 million visually impaired people (VIP) [1]. Despite the fact VIP do their daily life like everyone else, they have a variety of difficulties in their daily lives with limited visual capabilities. Researchers need to figure out what problems VIP have in everyday life to give them practical help.

One good way to reveal the problems is a diary method. The diary method is a method for subjects to record their behavior, actions, thoughts, and experience. The method can be useful when researchers want to have a contextual understanding of the experience of a person. Of course, there are several representative methods of collecting qualitative data such as observations, surveys, and interviews. However, VIP are generally reluctant to participate in observations, fearing privacy violations [2]. For the researchers, observation is also labor intensive. Surveys are difficult to gather detailed experiences than the diary method. For an interview, it is difficult for participants to recall their past experiences within a limited time, and also too short to think of problems from the experiences that they are adopted to.

Although the diary method is a good way to collect VIP's experiences, conducting it without considering the characteristics of VIP is inefficient and ineffective to derive meaningful results. This study aims to investigate how to conduct the diary method in the research of VIP through literature review and interviews with researchers who have conducted the diary method for VIP.

G. Di Bucchianico et al. (Eds.): AHFE 2020, AISC 1202, pp. 103–109, 2020.
https://doi.org/10.1007/978-3-030-51194-4_14

2 Method

2.1 Literature Review

We reviewed the literature on how the diary method has been used by the researchers studying the behaviors of VIP. The book "What is diary method" [3] describes the diary method with literature published until 2014 that were gathered from Google Scholar. The research team searched literature featured the term 'diary method' or similar terms such as 'e-diary' or 'diary-interview method' in the title and/or abstract. Papers published since 2015 were collected from Google Scholar. The search strategy of this study is shown in Table 1. We removed 'double-blind', which is an experiment method for reducing or eliminating experimental biases.

Table 1. Search strategy

Strategy	Description
Research question	Identify how the diary method has been used in research about visually impaired people
Date	2015. 01. 01–2019. 12. 08
Database	Google Scholar
Language	English, Korean
Keywords (Synonyms)	Visually impaired (visual impairment, blind), diary method (diary study, diary studies, e-diary, diary-interview method)
Excluded keyword	Double-bind

2.2 Interviews

In-depth interviews were conducted with three researchers who had conducted a diary method for VIP. Two of them collected text diaries from the mobility-handicapped including VIP to understand their experiences when they walk [4]. The other gathered audio or text dairies from VIP to investigate VIP's requirements of speech synthesis when using smartphones [5]. This study conducted face-to-face interviews to investigate experiences of the diary method from the researcher's point of view. Interviewees were asked about the detailed methodology that they conducted, the issues raised during the study, and the points that the participants had difficulties with.

3 Results

3.1 Literature Review

For the literature before 2015, two papers were collected from the book [3]. For literature since 2015, 1650 documents were collected from Google Scholar. Among them, 11 papers were collected by checking title, abstract and contents. If a same researcher had written several documents with the same scope, only the most recent literature was analyzed. Therefore, we explored 13 literature for the research area, research objective, and detailed method (Table 2).

Table 2. Summarization of literature review

(Author(s), Year)	Objective	Tasks	Input method	Prior meeting	Post interview	Participants	Duration
(Papadopoulos and Scanlon 2002) [6]	Conduct qualitative research with audio diaries for VIP	Daily life	Audio			4	1-day
(Worth, 2009) [7]	Collect experiences of transitioning form adolescence to adult with audio diaries for VIP	Record experiences	Audio			22	
(Brewer et al. 2016) [8]	Investigate how old VIP create audio contents and the need to edit audio contents		Audio	O	O	5	2-week
(Metatla et al. 2016) [9]	Gain feedback and insights to help the next prototyping session		Audio			2–3	
(Shinohara and Wobbrock, 2016) [10]	Collect experiences that feel self-conscious or self-confident when facing assistive technology	Daily life			O	8	4-week
(Williams et al. 2016) [11]	Identify architectures that help or hinder blind pedestrian navigation	Walk		O		15	
(Jeong et al. 2017) [4]	Understand pedestrian experiences of the mobility handicapped	Walk in daily life	Text	O	O	8	1-week
(Królak et al. 2017) [12]	Identify accessibility issues in MOOCs for blind learners	Take 4-weeks coursera lectures	Text	O		7	4-week
(Zhao et al. 2017) [13]	To understand the effect of the enhanced photo-sharing feature and identify new challenges for current technologies to fully support their needs	Use the photo-sharing feature		O	O	6	1-week
(Balata et al. 2018) [14]	Observe long-term use of their revised navigation prototype	Use the navigation when traveling				3	2-week
(Zhao et al. 2018) [15]	Evaluate effectiveness and social acceptability of accessible bot in different real-life situations	Use accessible bot		O	O	6	1-week
(Lee et al. 2019) [5]	Investigate VIP's requirements of speech synthesis when using smartphones	Use smartphones with a speech synthesis speaker for general users	Text, audio	O	O	10	1-week
(Middleton and Byles 2019) [16]	Examine the everyday mobilities of young VIP in London	Walk or take public transportation	Video		O	8	

The research areas of the literature are divided into five categories as shown in Table 2. In each literature, 2–22 VIP participated in the experiments. The diary creation period for each study varied from 1-day to 4-weeks. There were audio recording, text writing, or video recording as an input method. Audio recording is the most commonly used input method. However, there is some literature not describing the input method (Table 3).

Table 3. Research area of the literature

Research area	Detail
Research methodology	Audio diary for VIP [6, 7], Participatory design [9]
Assistive technology	Overall assistive technology [5], Text-to-speech (TTS) technology [10], Navigation [11, 14]
User interface	Audio playing [5], Audio editor [8], MOOCs [12]
Waling/Mobility	VIP's experience during walking or using transportation [4, 11, 14, 16]
Social activity	Photo-sharing [13, 15]

Researchers contacted VIP before conducting the diary method to conduct interviews, pre-questionnaire or have training sessions. Post interviews were conducted to gain a deep understanding and further insights into the diaries.

There is no literature describing detailed analysis methods. Some literature classified diary contents and interview transcript conducting content analysis [6], inductive/deductive analysis [10], Burnard's method [13], or other methods [4, 5]. Middleton and Byles [16] collected the diaries of massive hours of video recordings. They thought it was impossible to check the entire footage to analyze. Instead, they made short films with only meaningful footage with participants and film makers to analyze.

3.2 Interviews

All interviewers held prior meetings with participants for preliminary surveys, diary/task description and training. They ensured that participants had no problem understanding what and how to carry out tasks and record diary. They responded that precautions were effectively communicated through the prior meeting. However, in prior meetings, the participants expressed difficulties in performing tasks on unused devices or in recording diaries every day.

They mentioned using audio, photo, or video as an input method for diaries would be better than text only. One interviewee said that receiving a diary as text only makes it difficult to understand the detailed context. Another mentioned that participants who were not familiar with a computer would have difficulties in typing. In case of audio recording, participants were easy to record but had difficulty delivering recorded files to researchers. They proposed the diary formats to increase participation and collect experiences in various situations. Examples include multiple choices that are easy to respond and forms that utilize mobile devices.

They conducted post interviews in the form of focused group interviews. They asked detailed context about received diaries due to lack of detailed information and rough writing style. They also checked how many people experienced similar events that were written/recorded in others' diary.

They conducted the diary method for one week. They all commented that a week is too short to produce meaningful results. One week is a short period of time for VIP to get used to new tasks and to discover the pain-points from the experiences that they are already adapted to. One researcher mentioned that if participants had to record diaries for a longer period, a post interviews would not be necessary.

4 Discussion

The research areas using the diary method were found to be smaller than the research areas of VIP. Assistive technologies for VIP and products/services used by VIP are more diverse than those investigated in this study. If the diary method is conducted in other research areas for which it has not yet been applied, the technologies/products/services can be further or newly developed from the VIP's perspective.

However, both VIP and researchers believed that the diary method is labor-intensive. Researchers should find an easier method for VIP to record diaries. The audio recording recommended by interviewees should be examined for its effectiveness. Nevertheless, it is not yet known which input method the VIP actually performs better.

In addition, diaries in the form of audio, photos, or videos are difficult to analyze, so it is necessary to come up with an efficient method to analyze them. TTS technology can be used to transcribe audio diaries. Technologies like deep learning can be used for transcribing photos or videos. Natural language processing can be used to classify the diaries.

For less distortion of memory and larger data collection, a new type of diary could be considered. Many VIP use their smartphones and IoT devices nowadays [17]. It is natural to expect that VIP can record diaries with mobile/IoT devices. Nevertheless, their blindness can expose them to a variety of risks. Before applying the new type of diary for VIP, researchers should need to determine its risks and possible solutions about it. For example, issues related to data security and reliability that may be of concern to the participants could be solved by technologies such as blockchain.

A total of 30% of reviewed literature and all interviewees conducted diary methods for 1-week. However, 1-week is a short period to collect significant experiences through the diary method without post interviews. It is necessary to define appropriate diary-recording period for gathering meaningful experiences.

5 Conclusion

This study examined the characteristics of the diary method conducted for VIP and explored the considerations when conducting the method for VIP. It would have been better to interview more researchers who conducted a diary method, but there were very few of them in Korea. In addition, we wanted to hear from the VIP who participated in the diary method, but could not. This is because it was difficult to obtain participants' information from other researchers due to research ethics. We hope this study will stimulate the use of a diary method in various VIP studies. For the future study, we will explore problems that VIP suffer from conducting various types of the diary method. Furthermore, we would design a diary method that can be effectively used by both researchers and VIP.

Acknowledgments. This research was supported by the MSIT (Ministry of Science and ICT), Korea, under the ITRC (Information Technology Research Center) support program (IITP-2018-0-01441) supervised by the IITP (Institute for Information & communications Technology Promotion) and by the National Research Foundation of Korea (NRF) grant funded by the Korea government (MSIP) (No. NRF-2019R1A2C1086366).

References

1. Pascolini, D., Mariotti, S.P.: Global estimates of visual impairment: 2010. Br. J. Ophthalmol. **96**(5), 614–618 (2011)
2. Okada, R., Yairi, I.: An indoor human behavior gathering system toward future support for visually impaired people. In: 15th International ACM SIGACCESS Conference on Computers and Accessibility, pp. 1–2. Association for Computing Machinery, New York (2013)
3. Bartlett, R., Milligan, C.: What is Diary Method? Bloomsbury Academic (2015)
4. Jeong, D.Y., Kwahk, J., Han, S.H., Park, J., Lee, M., Park, K., Kim, J.H., Jang, H., Koh, Y., Jeong, D.: Investigating pedestrian experiences of the mobility-challenged. In: Proceedings of Ergonomics Society of Korea, pp. 82–86. Ergonomics Society of Korea, Seoul (2017)
5. Lee, S., Park, Y., Kim, J.: Feautres of speech synthesizer on smartphone for the people with visual impairments. In: Proceedings of HCI Korea 2019, pp. 1243–1245. HCI Korea, Seoul (2019)
6. Papadopoulos, I., Scanlon, K.: The use of audio diaries in a study with visually impaired people. J. Vis. Impair. Blind. **96**(6), 456–459 (2002)
7. Worth, N.: Making use of audio diaries in research with young people: examining narrative participation and audience. Sociol. Res. Online **14**(4), 77–87 (2009)
8. Brewer, R.N., Cartwright, M., Karp, A., Pardo, B., Piper, A.M.: An approach to audio-only editing for visually impaired seniors. In: Proceedings of the 18th International ACM SIGACCESS Conference on Computers and Accessbility, pp. 307–308. Association for Computing Machinery, New York (2016)
9. Metatla, O., Martin, F., Parkinson, A., Bryan-Kinns, N., Stockman, T., Tanaka, A.: Audio-haptic interfaces for digital audio workstations. J. Multimodal User Interfaces **10**(3), 247–258 (2016)

10. Shinohara, K., Wobbrock, J.O.: Self-conscious or self-confident? A diary study conceptualizing the social accessibility of assistive technology. ACM Trans. Accessible Comput. **8** (2), 1–31 (2016)
11. Williams, M.A., Dubin, B., Amaefule, C., Nguyen, L., Abdolrahmani, A., Galbrath, C., Hurst, A., Kane, S.K.: Better supporting blind pedestrians and blind navigation technologies through accessible architecture. In: Designing Around People, pp. 237–246 (2016)
12. Królak, A., Chen, W., Sanderson, N.C., Kessel, S.: The accessbility of MOOCs for blind learners. In: Proceedings of the 19th International ACM SIGACCESS Conference on Computers and Accessbility, pp. 401–402. Association for Computing Machinery, New York (2017)
13. Zhao, Y., Wu, S., Reynolds, L., Azenkot, S.: The effect of computer-generated descriptions on photo-sharing experiences of people with visual impairments. In: Proceedings of the ACM on Human-Computer Interaction, vol. 1, pp. 1–22 (2017)
14. Balata, J., Mikovec, Z., Slavik, P.: Landmark-enhanced route itineraries for navigation of blind pedestrians in urban envrionment. J. Multimodal User Interfaces **12**(3), 181–198 (2018)
15. Zhao, Y., Wu, S., Reynolds, L., Azenkot, S.: A face recognition application for people with visual impairments: understanding use beyond the lab. In: Proceedings of the 2018 CHI Conference on Human Factors in Computing Systems, pp. 1–14. Association for Computing Machinery, New York (2018)
16. Middleton, J., Byles, H.: Interdependent temporalities and the everyday motilities of visually impaired young people. Geoforum **102**, 76–85 (2019)
17. Griffin-Shirley, N., Banda, D., Ajuwon, P., Cheon, J., Lee, J., Park, H., Lyngdoh, S.: A survey on the use of mobile applications for people who are visually impaired. J. Vis. Impair. Blind. **111**, 307–323 (2017)

Designing Accessible Maps on Mobile Devices for Blind and Visually Impaired Users

Tania Calle-Jimenez[1(✉)], Sergio Luján-Mora[2], Hugo Arias-Flores[3], Carlos Ramos-Galarza[5,6], and Isabel L. Nunes[4]

[1] Escuela Politécnica Nacional, Quito, Ecuador
tania.calle@epn.edu.ec
[2] University of Alicante, Alicante, Spain
sergio.lujan@ua.es
[3] Universidad Tecnológica Indoamérica, Ambato, Ecuador
hugoarias@uti.edu.ec
[4] Nova University of Lisbon, Monte de Caparica, Lisbon, Portugal
imn@fct.unl.pt
[5] Centro de Investigación en Mecatrónica y Sistemas Interactivos, Universidad Tecnológica Indoamérica, Quito, Ecuador
ps_carlosramos@hotmail.com
[6] Facultad de Psicología, Pontificia Universidad Católica del Ecuador, Quito, Ecuador

Abstract. Currently, there is a high percentage of people with visual disabilities worldwide. Sometimes, these people do not have access to the information of online geographic maps. A traditional map published on the Web does not provide the necessary alternative information for people with visual disabilities. This research presents a mobile application with geographic maps that provides the information in audio format to blind and visually impaired users. The application uses the Scalable Vector Graphics (SVG Tiny) format that allows additional information to be stored within the map code. SVG Tiny provides to screen readers the content inside of map code to transform text to audio. The tests were carried out through accessibility assessment tools, which showed a level of accessibility greater than 90%. In addition, tests were carried out with blind people, who were satisfied with the level of accessibility of the mobile application.

Keywords: Web accessibility · WCAG 2.1 · Accessible online maps · Mobile devices · Blind users · Visually impaired users

1 Introduction

Today, web applications include dynamic multimedia content. Anyone who has a computer or a smartphone with multimedia capabilities can access or produce web content. For example, a user can access social media, create and edit images, create access routes, among others. A specific case is the use of online maps in mobile applications, these maps allow users to be located in space. These tools are very useful when traveling or locating a particular place, orienting in a neighborhood. These examples are common uses of this type of content [1]. Most of this content is developed

G. Di Bucchianico et al. (Eds.): AHFE 2020, AISC 1202, pp. 110–116, 2020.
https://doi.org/10.1007/978-3-030-51194-4_15

for people without visual disabilities, who are also consumers and producers of dynamic multimedia content. People with visual disabilities also have the need to use online maps, that satisfy their needs and abilities [2, 3]. According to the World Health Organization, there are 285 million people with visual disabilities worldwide, of which 39 million are blind and 246 million have other vision disabilities [4].

Currently, people with visual disabilities can perform activities such as writing digital documents, navigate in Internet or sending and receiving emails. Screen reading software, speaking and braille devices, audio interpretation and conversion software allow people with visual disabilities to use computers, cell phones and other electronic devices independently [5]. Moreover, the web accessibility means that users with disabilities will be able to make better use of applications [6]. To solve the problems of accessibility, the Web Content Accessibility Guidelines (WCAG 2.1) can be used. WCAG 2.1 is the standard that is most often legally adopted via the ISO/IEC 40500 standard, in those countries that have legislation on web accessibility. There are three levels of accessibility compliance in WCAG 2.1: A, AA, and AAA [7]. Taking into account the guidelines of the web accessibility, this research aims to design accessible maps displayed on mobile devices for people with visual disabilities following the WCAG 2.1 and using de Scalable Vector Graphics (SVG Tiny) format [8].

The paper presents in Sect. 2 the related works, in Sect. 3 the method used for the design of the mobile application, in Sect. 4 the accessibility tests carried out, in Sect. 5 a discussion of the results and finally, conclusions and future work of this topic.

2 Related Work

The SVG format has been used for web content for some time to represent scalable images. This format is relevant for web applications because the images do not pixelate as the scale changes [8]. The authors in [9] show the use of SVG for the representation of mobile maps. This research presents an application to represent Geographic Information System (GIS) on mobile devices based on J2ME using the SVG format.

In [10], an application of SVG images based on data is proposed, this system creates an SVG image. This image was distributed to people with visual disabilities. In a later study [11], the authors explore different types of interactive maps "both physical and digital" in order to see the efficiency of the maps using SVG.

The authors in [12] analyze how screen readers recognize SVG format and the information that can be included within the images in that format. Additionally, the authors mention that maps with HTML content are better adapted to screen readers and included the WCAG 2.0 to developed maps. Subsequently, in [13], the authors present an accessible map of a subway system for people visual disabilities. This system is based on three components: the database component is responsible for storing web maps, the server component is responsible for converting a web map into an accessible web map, and the user component is responsible to deploy the accessible map on the Web.

In addition, in [14] a web application to provide a map of the University of Alicante in Spain is presented. This map is based on OpenStreetMap to obtain the buildings to compose the SVG image. This system is based on operations with PostGIS to locate the user and provide information on nearby places of the user location. Additionally, it makes an analysis based on the WCAG 2.0 [15] to check the level of accessibility of the application. Although the use of SVG is reviewed and developed in some researches and focused on accessibility, but an in-depth study of SVG Tiny [8] formats for mobile devices is not yet shown. Based on these related studies, this paper aims to design accessible maps with the SVG Tiny format for the Android platform based on WCAG 2.1.

3 Accessible Mobile Application Model

The Fig. 1 shows the model of the accessible mobile application, this presents a browser that allows communication between the website and the Text to Speech (TTS) engine of the mobile device. The mobile application connects with Google Maps to obtain the information of the coordinates. This location is sent to the server as a parameter of the HTTP request. This application gets the web page from the server and the page is analyzed and sent to the TTS engine. The website contains a database with the information of the maps in SVG Tiny format to offer the user an auditory response of the places on the map in which the users are browsing.

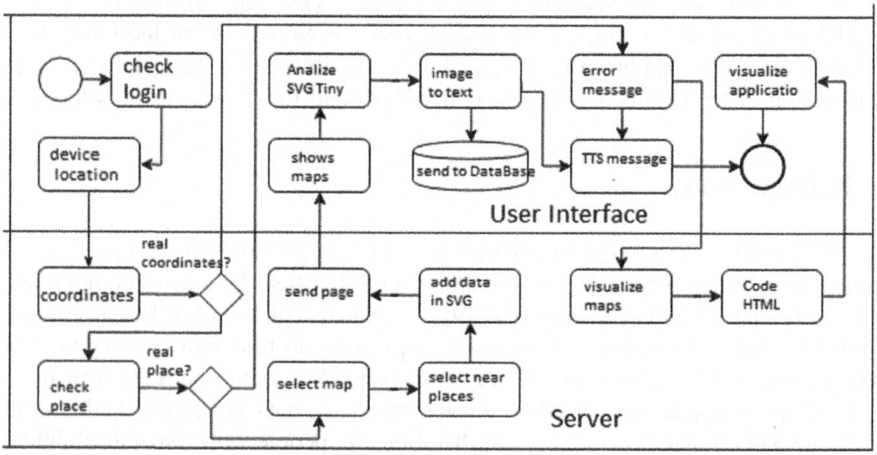

Fig. 1. Accessible mobile application model proposed

Subsequently, one of the main aims is to visualize accessible online maps that present information on mobile devices for people with visual disabilities. To achieve this, images based on SVG Tiny format will be used and WCAG 2.1. One GIS was used to design the maps to visualize and linked to the PostGIS spatial database, for the management of geographic data. The link allowed to store the map drawn in SVG Tiny in geographic data that can be accessed by the accessible mobile application.

Later, the design process draws the profile of the Google Maps, in order to use the coordinates in the PostGIS spatial database, which is related to the geographical points, and is compared with the GPS position captured by the mobile device. In addition, the spatial database allows verifying the map and the location of the user. The application allows the user to visualize the map and current position.

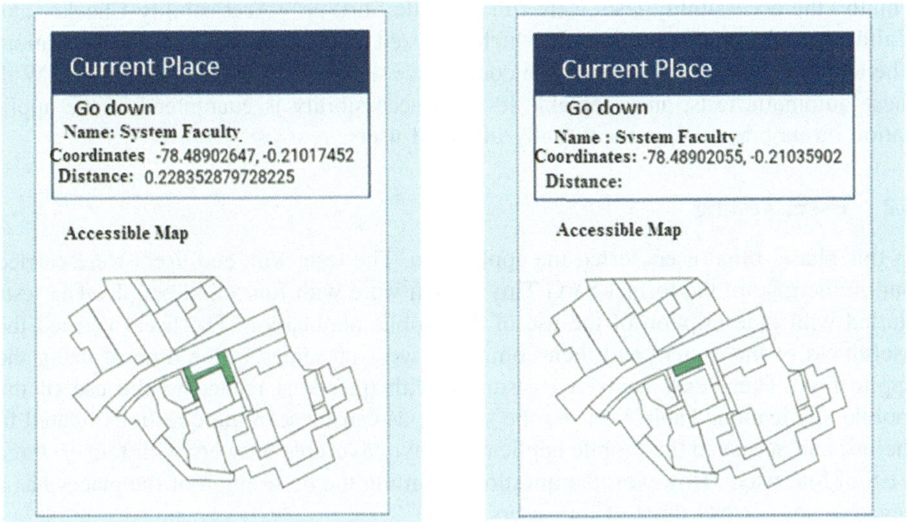

Fig. 2. Accessible mobile application

4 Testing

The assessment of this mobile application was through two types of web accessibility tests, with online tools and with end users. A synthesis of the tests performed is described below.

4.1 Web Accessibility Assessment with Tools

In this phase, three web tools were used WAVE, Examinator and PowerMapper. These tools evaluate that the applications comply with the WCAG 2.1. In addition, they are useful for software quality analysis. For this process, the tool entered to the mobile application and validated each of the web pages. This is intended to guarantee the accessibility of the mobile application.

The accessibility assessment of the mobile application with the WAVE tool showed zero errors and zero alerts, which means that the application complies with the WCAG 2.1. Examinator awards a score between one and ten as a qualification of the accessibility of the pages with respect to WCAG 2.1 and provides a detailed report of the tests performed. In this case, the accessibility assessment result had a rating of ten.

Finally, the accessibility tests were run with the SortSite - Accessibility Checker and Validator tool from PowerMapper, which analyzed all the web pages of the application. The tool presents a detailed report on compliance with the WCAG 2.1 guidelines. With these automatic tests, an acceptable level of accessibility is guaranteed of the application on mobile devices for visually impaired users.

4.2 Users Testing

In this phase, blind users tested the application. The tests with end users were carried out on the map in the format SVG Tiny, which were with four blind people. The tests started with a description of the use of the mobile application. The users verified the usefulness of the system and their comments were presented at the time of using the application. Then, they answered a survey with questions related to the use of the mobile application. Table 1 shows the survey, as can be seen the questions related to the use and access to the mobile application have favorable answers with four or three YES of four users. However, the question regarding the description of the places has a negative answer NO three of four users.

Table 1. Survey of blind users.

Questions	Answer	
	Yes	No
Do you consider that the mobile application displays an accessible map?	4	0
Do you consider that the mobile application clearly describes the places traveled?	1	3
Do you consider that mobile application is easy to use?	3	1
Do you consider that the information in audio format presented by the mobile application is clear?	4	0
Do you think the mobile application buttons worked correctly?	4	0
Would you recommend the mobile application to other blind users?	4	0

Table 2 presents the feedback comments of the blind users and the taken actions to respond to the comments, the users contributed to the improvement of the mobile application for a next version.

Table 2. Actions carried out in the mobile application regarding the suggestions of the testers.

Tester	Suggestion	Action
1	The mobile application is useful for walking alone and knowing where we are. One suggestion is to add a more representative name of the place. The mobile application could include a sign reader that would help us locate part of the name of the place	The authors add description to the places
2	The mobile application is useful to know where came from and where to go. The application should say the proper name of the places	The authors add description to the places
3	The application is interesting because does not consume so much resource and the map is clear and easy to use. One suggestion is the application can be manipulated from the headphones	Future works
4	The application meets the objective of showing nearby places. One recommendation is that the application have detailed description of what is in each building. In addition, the implementation of the Z axis within the application would extend the information of the places that have more than one floor	Future works

5 Discussion

The automated tests presented positive results, which guarantees that the page meets the AA criteria of the WCAG 2.1. Regarding the tests with end users, they mentioned that the mobile application is useful for accessing map information. In addition, the application helps users to know the place where is located on the map and for which place the user goes. Regarding the related works, this research shows one accessible mobile application using the SVG Tiny format, in compared with [13], the authors proposed web application displayed in computers. This application provides greater mobility for the visually impaired user, who can access map information in order to know the location in real time. In the same way that the accessibility tests were addressed in [14], successful results were reached with respect to users and accessibility assessment software. The mobile application to follow the accessibility guidelines of the WCAG 2.1 contributes to the development of new innovations to support people with visual disabilities as the authors do in [15]. In addition, the screen reader can access the information within the map. Although the mobile application informs users of the closest places, it is necessary the participation of an assistant to help the blind user to move through the place since a specific route is not given in real time. This limitation can be solved if a route system is implemented.

6 Conclusion

This research presents a mobile application, such as a first approach to an accessible online maps for people with visual disabilities. The research approach was to design an SVG Tiny map of a place and comply with the WCAG 2.1. The information included in the SVG Tiny allows the screem reader to communicate to the user the places in an auditory format. In addition, the application allows to determine the distances between the places and the mobile device. The tests were carried out through accessibility assessment tools, which showed a level of accessibility greater than 90%. In addition, tests were carried out with blind people, who were satisfied with the level of accessibility of the mobile application. For future work, this application can be complemented with the integration with other applications that provide a route service.

References

1. Kalbag, L., Pickering, H.: Accessibility for everyone (2017)
2. Park, K., Goh, T., So, H.J.: Toward accessible mobile application design: developing mobile application accessibility guidelines for people with visual impairment. HCI **4**, 31–38 (2014)
3. Henry, S.L., Abou-Zahra, S., Brewer, J.: The role of accessibility in a universal web.: In: Proceedings of the 11th Web for all Conference, pp. 1–4 (2014)
4. World Health Organization.: Visual Impairment and Blindness (2014). http://www.who.int/blindness/publications/globaldata/en/
5. Sandoval, C., Gómez, G.: Accesibilidad en web para personas con discapacidad visual. Síntesis Tecnol. **2**(1), 1–10 (2017)
6. World Wide Web Consortium.: Accessibility W3C Recommendation (2005). https://www.w3.org/standards/webdesign/accessibility
7. Word Wide Web Consortium.: Web Content Accessibility Guidelines (WCAG) 2.1 (2018). https://www.w3.org/TR/WCAG21/
8. Word Wide Web Consortium.: Scalable Vector Graphics (SVG) (2018). https://www.w3.org/TR/SVG2/
9. Binzhuo, W., Bin, X.: Mobile phone GIS based on mobile SVG. In: Proceedings IEEE International Geoscience and Remote Sensing Symposium, IGARSS 2005, vol. 2, p. 4 (2005)
10. Engel, C., Müller, E.F., Weber, G.: SVGPlott: an accessible tool to generate highly adaptable, accessible audio-tactile charts for and from blind and visually impaired people. In: Proceedings of the 12th ACM International Conference on PErvasive Technologies Related to Assistive Environments, pp. 186–195 (2019)
11. Ducasse, J., Brock, A.M., Jouffrais, C.: Accessible interactive maps for visually impaired users. In: Mobility of Visually Impaired People, pp. 537–584 (2018)
12. Calle-Jimenez, T., Luján-Mora, S.: Accessible map visualization prototype. In: Proceedings of the 13th Web for All Conference, pp. 1–2 (2016)
13. Calle-Jimenez, T., Eguez-Sarzosa, A., Luján-Mora, S.: Design of an architecture for accessible web maps for visually impaired users. In: International Conference on Applied Human Factors and Ergonomics, pp. 221–232 (2018)
14. Juan-Armero, S., Luján-Mora, S.: Using SVG to develop web maps for people with visual disabilities. Enfoque UTE **10**(2), 90–106 (2019)
15. World Wide Web Consortium.: Web content accessibility guidelines WCAG 2.0. W3C Recommendation (2008). https://www.w3.org/TR/WCAG20/

Balance of Considerations Between Technological and Social Factors: A Case Study on Smartrack™ for the Visually Impaired

Yi Lin Wong[1], Kin Wai Michael Siu[2(✉)], and Chi Hang Lo[2]

[1] Hong Kong Baptist University, Kowloon Tong, Kowloon, Hong Kong
elaineylwong@hkbu.edu.hk

[2] The Hong Kong Polytechnic University, Hunghom, Kowloon, Hong Kong
{m.siu, sdpaullo}@polyu.edu.hk

Abstract. Rebuilding inclusive facilities in old districts in metropolitan cities with high population density encounters a number of problems. It is necessary to find a solution to renew the city with minimal interruption of people's everyday life. Through a case study of Hong Kong, the paper addresses the conditions of the public facilities provided for the visually impaired persons (VIPs) in Hong Kong. VIPs' needs and concerns while using the facilities were investigated. Based on the findings, a new tactile guide path using smart technology, named Smartrack™, is designed. It is argued that supplementary actions have to be taken to help people with special needs to understand the issues of public privacy when smart technology is being brought to them. Inclusive researchers, designers, as well as educators, should collaboratively attain a balance among different considerations between technological and social factors.

Keywords: Smartrack · Visually impaired · Public design · Guiding path · Privacy

1 Introduction

People with special needs and disabilities encounter several problems in their daily life affecting their living quality. Although governments and non-government organizations (NGOs) are implementing different measures to help this group of people to adapt to the society, it is still difficult for most of the disabled persons to live like others without any accessibility issues in the city. Especially in metropolitan cities with high population density such as Hong Kong, the facilities are quite limited because of the land problem. Older districts, where the elderly dwell, should get the priority to be redesigned, and the facilities there should also be upgraded and improved for the needed. However, they are also the most difficult area to be rebuilt because the residents refused to move out, and the elderly's everyday life would be largely disturbed because of the construction work. Either the government would move all residents away from the districts and relocate them in newly developed towns away from the city center, or the residents would have to live near to the construction site. Many arguments and conflicts

G. Di Bucchianico et al. (Eds.): AHFE 2020, AISC 1202, pp. 117–123, 2020.
https://doi.org/10.1007/978-3-030-51194-4_16

would occur when the districts are about to rebuild and renew. Given this dilemma, it is necessary to develop facilities and renew the city with minimal interruption of people's everyday life.

In Hong Kong, one of the facilities that are in need to be rebuilt in the public areas is those designed for the visually impaired persons (VIPs) [1]. The needs of the VIPs and the inclusion of their daily activities to the society are often neglected [2]. In the previous studies, one of the most important observations was that VIPs would not use the tactile guide path on pedestrian streets [3]. They depend on their own to find their ways and other public facilities. It is necessary to redesign the facility so that the resources are not wasted and remained unused.

Through a case study of Hong Kong, as the needs of VIPs are seldom addressed [4], the study first explores the needs of Hong Kong's VIPs when navigating in a city and approaching pubic services. Based on the findings, a new smart tactile guide path for the VIPs is designed using smart technologies. The tactile guide path, named Smartrack™, is used as the major technological device to help the VIPs to find their ways in daily life. As smart technology is used, personal privacy should also be considered. The paper thus discusses the balance and tension between privacy in public spaces and smart public services experienced by the needed people. It is hoped that the study will be able to help VIPs to optimize their daily life and also educate VIPs and the related associations about public privacy when using smart technology.

2 Method

2.1 Case Study

Hong Kong has established policies and regulations addressing the issues of city inclusiveness in these few decades. Barrier-free design, which is also known as inclusive design and universal design, has already accepted by society. In this study, three old districts in Hong Kong were investigated, i.e., Sham Shui Po, Kwun Tong, and Kowloon City. They are the fourth, first and fifth densest districts in Hong Kong. During the study, the population densities of these three districts were 43,381, 57,530 and 41,802 people/km^2 [5]. Densely populated districts were chosen because the facilities were relatively old, and this was able to reveal more information about the VIPs' difficulties and needs in living in Hong Kong. Figure 1 shows the location of these three old districts.

The research and design team of the Public Design Lab at the Hong Kong Polytechnic University has been researching the needs of VIPs in different aspects of their lives. The study was conducted based on the trust between the research and design team and the VIPs. As VIPs were a marginalized and neglected group in society, they might not be open to talk to or make contact with other people. The trust developed was essential to gain reliable information from the VIPs.

Fig. 1. Location of the three old districts in this study.

2.2 Procedures

A non-government organization (NGO) in Hong Kong was approached to recruit VIPs to participate in the study. Sixteen VIPs (8 female and 8 male) living in the three old districts were recommended by the NGO. They had different levels of visual impairment, and most of them needed to use a cane to help them to navigate in the city. Their residential information was collected, and based on this piece of information, fifteen parks in these three old districts were chosen for investigation. These parks were close to or in the residential areas that the 16 VIPs lived in. The team's researchers and designers visited the parks, and the inclusive facilities were investigated. Photos were taken to record the provision of the facilities during observations.

After visiting the parks, VIPs were invited to go to the park that is near to their home. They started walking to the park from their home with researchers and designers. Their different processes of wayfinding of the VIPs were recorded. Semi-structured interviews were conducted to understand their needs. Questions like 'How familiar are you with the park we visited?', 'What kind of difficulties have you encountered in finding the way to parks and navigating in the city?', 'Is technology able to assist your daily life?', and 'How you understand public privacy?' were asked. The interviews were recorded, and notes were taken during the interviews.

3 Findings and Discussions

3.1 Facilities in the Parks

Limited facilities designed for the VIPs were found in the parks included in this study. The most common facility was the tactile guide path. Some parks also had a tactile map located at the entrance of the park. According to the Universal Accessibility guidelines, in order to ensure the VIPs to find the tactile map, the Architectural Services Department suggested that detectable clues such as tactile paths should be provided to lead users to the tactile map [6]. In the field observation, it was believed that the users should be able to find the tactile map following the tactile path. However, the color of some tactile guide path has a low contrast to that of the ground, and the VIPs might not unable to notice the path from a distance easily. Figure 2 shows an example of a tactile path turning at a right angle with low contrast to the ground.

Fig. 2. An example of a tactile path turning at a right angle with low contrast to the ground.

Tactile map was not a common facility that some smaller parks or sitting-out areas do not have the map, and it was not mandatory to install tactile maps in public spaces. Therefore, different designs of tactile maps could be found in the visited parks. Figures 3 shows two different designs. One at the left-hand side is to combine the textual and tactile information, and the other one is to illustrate the two information separately. The design of the tactile map was not standardized. It can be argued that the variation of the design did not facilitate VIPs to use the maps in different parks.

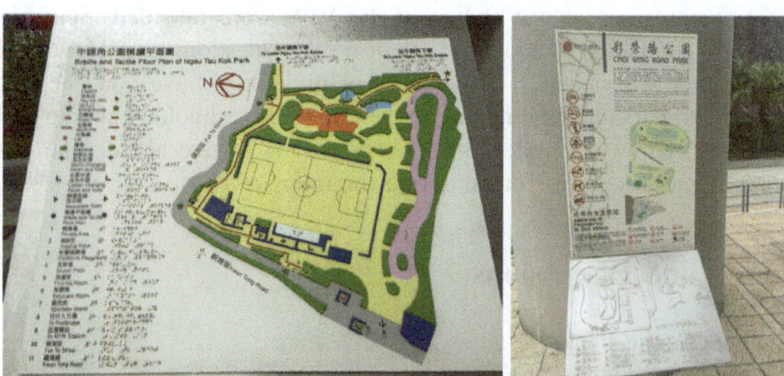

Fig. 3. Two different tactile map designs found in the parks.

3.2 Difficulties Encountered When Using the Tactile Path

The VIPs who participated in the study were able to find the park near their home. However, they reported that this was because they were familiar with the areas near to their home, and they knew very well where the park was and the road condition on the way to the park. Most of the VIPs did not use the tactile guide paths on the streets or in the park. In the interviews, they expressed that the current tactile guide paths were not user-friendly. It often brought people to turn at an acute angle like a robot. Besides, the VIPs also expressed that if they followed the tactile guide paths to move around, they would have to walk for an extra-long distance before reaching the facility or the destination, whereas the direct distance between the starting point and the destination is much shorter. Even they did not care about the long-distance walk and follow the tactile guide paths, it was not guaranteed that they could reach the facility they need. Many tactile guide paths led the VIPs in the wrong direction.

Therefore, the VIPs who participated in the study responded that most of the VIPs in Hong Kong did not use the tactile guide paths, and they tended to believe in their only memories and experiences when navigating in the city. They would only go to the places where they were familiar with, or they would need volunteers from the NGO to bring them to familiarize the route and the place before going there alone. They tended not to explore new places, as there were many unknown road conditions and dangers.

3.3 New Tactile Path for the VIPs: Smartrack™

The findings from the observations and interviews show that the design of the tactile path does not cater for the VIPs' needs. The VIPs participated in the study had a very limited social space in life without others' help. Considering Hong Kong's busy traffic on pedestrian roads, it is not practical if the construction work for building facilities for the VIPs would create major disruption. The design of the new tactile guide path for the VIPs should be succinct yet functional, i.e., meeting the needs of the VIPs.

In view of the issue, the research and design team has invented and designed a revolutionary tactile path, named Smartrack™, for guiding VIPs to walk on pavements safely (Fig. 4). The new design applies radio-frequency identification (RFID) technology and various tangible tactile forms. VIPs are expected to receive information related to the path through the vibration pulse from a specially designed cane and also the tactile sensation underfoot and at the tip of the newly designed cane.

With a thin RFID tag attached inside individual tactile path tiles, the sensor at the tip of the newly designed cane can recognize the tactile path within a range. Different kinds of vibration pulses will be generated at the cane handle to inform users to walk, turn or stop. Besides, at the top of the newly designed cane, there is a button to switch on or turn of the location mode. The location mode helps VIPs locate public facilities by using the Smartrack™ App in smartphones. Smartrack™ forms a network in the city, and direct VIPs to public facilities freely and inclusively.

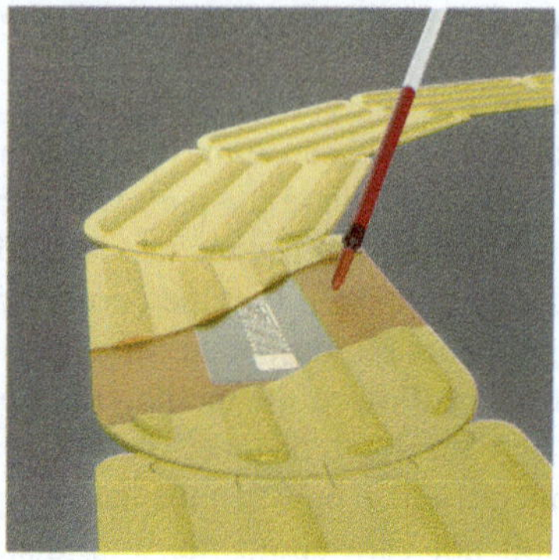

Fig. 4. Design of Smartrack™.

3.4 Issues of Public Privacy

The new design helps the VIPs to navigate in the city with confidence. However, while they enjoy the convenience provided by the design and the services, they, the users, also have to know about the privacy issues. However, most of the VIPs are not aware of the issues and unable to protect their privacy.

Many social activities happen in public spaces that are designed for public gathering and communication. These activities are not private, and also the space where the activities are held, are open for all [7, 8]. However, when each individual is considered in a public event, the issue of privacy emerges, especially in the technological era. For example, the advancement of location service with Wi-Fi and GPS in smartphones has disclosed an individual's activities at different locations to the mobile carriers, despite better mobile services provided to the smartphone users. Location privacy, according to Cremonini, Braghin, and Ardagna, is 'the right of individuals to decide how, when, and for which purposes their location information could be released to other parties' [9]. Lack of location privacy protection may lead to different kinds of attacks such as physical harassment and unsolicited advertising. Other technologies also collect personal information in public spaces while the city is advancing its services and facilities to be a smart city. While many voices from the general public express that the use of personal information should be used cautiously, people with special needs or disabilities are one of the community groups who's public privacy is less concerned. It may be due to the limited public information received by the group and the poorly established channels for the group to express opinions. While using the newly designed Smartrack™, the VIPs' privacy may be disclosed without being informed.

Considerations on the privacy of people with disabilities when providing public service should be given when smart technologies are involved. Education related to the knowledge of public privacy is essential to protect this group of people. A balance of consideration between technological and social factors should be achieved so that people can have different preferences and choices in daily life. This is the kind of equity and inclusiveness that every smart product should be claimed for the benefits of people with special needs.

4 Conclusions

The study adopted a case study approach to investigate the public facilities provided for the VIPs in Hong Kong. Sixteen VIPs were invited to take part in the study. Through field observations and interviews, VIPs' needs and concerns while using the facilities were investigated. Based on these findings, a new tactile guide path is designed, and it is named Smartrack™. Smartrack™ is able to allow the VIPs to navigate in the city confidently and conveniently, and it helps them to find their ways to the public facilities easily.

However, most of the VIPs are unaware of public privacy issues. It is argued that although through *smart* products and services are able to optimize the life quality of the people with special needs, their *privacy* may be disclosed and the users are unable to be notified. Supplementary actions have to be taken to educate them to understand the issues. Inclusive researchers, designers, as well as government officials and educators, should collaboratively attain a balance among different considerations between technological and social factors.

Acknowledgments. The authors wish to acknowledge the support provided by the Hong Kong Blind Union for the data collection and preparation of the paper. The authors thank The Hong Kong Polytechnic University for its manpower support for the research. The Eric C. Yim Endowed Professorship provided financial support for the data analysis. Mr. Maurice Kwok and Mr Yeung have participated in the development of the Smartrack™. (Note: The Hong Kong Polytechnic University solely owns the intellect property of Smartrack™).

References

1. Siu, K.W.M.: Accessible park environments and facilities for the visually impaired. Facilities **31**(13/14), 590–609 (2012)
2. Marmeleira, J., Laranjo, L., Marques, O., Pereira, C.: Physical activity patterns in adults who are blind as accessed by accelerometry. Adapt. Phys. Act. Q. **31**, 283–296 (2014)
3. Siu, K.W.M., Xiao, J.X., Wong, Y.L.: Inclusive design of open spaces for visually impaired persons: a comparative study of Beijing and Hong Kong. In: Di Bucchianico, G. (ed.) Advances in Design for Inclusion: Proceedings of the AHFE 2018 International Conference on Design for Inclusion, pp. 171–179. Springer, Orlando (2019)
4. Faruk, M., Ormerod, M., Newton, R., MacLennan, H., Abbas, M.Y.: Tactile paving a necessary intervention, but does it suit everyone? In: Bust, D. (ed.) Proceedings of the International Conference on Contemporary Ergonomics, pp. 303–308. Taylor & Francis, Nottingham (2008)
5. Census and Statistics Department: Hong Kong 2016 Population Census. Census and Statistics Department, Hong Kong (2016)
6. Universal Accessibility for External Areas. Open Spaces & Green Spaces. https://www.archsd.gov.hk/archsd/html/ua2/3_6_4.html
7. Aubock M, Cejka A.: Open space: the city. Verein Planbox, Wien (1996)
8. Siu, K.W.M.: Open for all: open spaces and new urban lifestyles. Int. J. Hum. **7**(12), 25–28 (2010)
9. Cremonini, M., Braghin, C., Ardagna, C.A.: Privacy on the internet. In: Vacca, J. (ed.) Computer and Information Security Handbook, 2 edn., pp. 739–753. Morgan Kaufmann Publishers/Elsevier, Waltham (2013)

Design and Development of an All-Terrain Wheelchair for Rural Areas in Mexico: The First Stage of the Project

Alberto Rossa-Sierra[1(✉)], Ana Paula Diaz-Pinal[1], Fabiola Cortes-Chavez[1], Felix Maldonado de la Fuente[2], and Gilberto Carrillo-Alvarado[3]

[1] Facultad de Ingeniería, Universidad Panamericana, Alvaro del Portillo 49, 45010 Zapopan, Jalisco, Mexico
{lurosa,0188968,fcortes}@up.edu.mx

[2] Facultad de Arquitectura y Urbanismo, Universidad de Chile, Portugal 84, Santiago de Chile, Chile
femaldon@uchile.cl

[3] Centro de Innovación en Diseño Industral y Manufactura, Universidad Don Bosco, Calle Plan del Pino, Cantón Venecia, Ciudadela Don Bosco, Soyapango, San Salvador, El Salvador
gilberto.carrillo@udb.edu.sv

Abstract. Estimates that around 10% of the world's population, some 650 million people have some disability. Studies indicate that about 10% of these people need a wheelchair to move. Therefore, estimates that 1% of the total population requires a wheelchair, about 65 million people worldwide. Calculations by the UN show that 80% of that amount lives in developing countries. In 2010 it was estimated that 20 million of those who needed a wheelchair to get around did not have one. In developing countries, moreover, chairs are required to allow people with disabilities to move through irregular surfaces since they either live in the countryside, or in cities where the infrastructure is not adequate, and there are no sidewalks, ramps or wheelchair facilities. From this problem arises the need for the design and development of an all-terrain wheelchair of low cost and simple manufacturing, which allows its reproduction in almost any developing country.

Keywords: Wheelchairs · Product design · Inclusive design

1 Introduction

As reported in a report by the World Health Organization [1] there is an estimate of approximately 70 million people in the world who require a wheelchair. Of these, only 15.5% can acquire it. The same report reflects, for example, that in India, 47.8% of people with disabilities have to crawl on their own, while 38.6% use a cane or crutch, and 9.7% do have no chance to move out of bed. 78% of people with motor disabilities treated in that country have never had access to a wheelchair.

G. Di Bucchianico et al. (Eds.): AHFE 2020, AISC 1202, pp. 124–130, 2020.
https://doi.org/10.1007/978-3-030-51194-4_17

At the Academy of Design and Innovation of Faculty of Engineering of Universidad Panamericana, México, we have been working in recent years with projects of application in vulnerable groups, such as the disabled motor. Derived from the projects previously developed [2, 3], we wanted to deepen the development of increasingly specific solutions to our environment. So we detected a severe lack of the development of proposals accessible to people with disabilities for wheelchairs who they can be used on dirt roads and cobblestones, such as rural or peripheral areas to large cities in our country, and that their final cost is as accessible to the bulk of the population in Mexico.

The General Law for the inclusion of persons with disabilities in Mexico of 2011, amended in 2018, defines physical disability as the sequel or malformation that derives from a condition in the neuromuscular system at the central or peripheral level, resulting in alterations in the control of movement and posture, and that by interacting with the barriers imposed by the social environment, it can prevent its full and active inclusion in society, on equal terms with others [4].

Official data in Mexico (2014) indicate that approximately 6% of the population lives with a disability (7.2 million inhabitants), with motor disability being the most frequent (57.5%) (Fig. 1). The same official data show that, of the aid used by people with motor disabilities, the use of wheelchairs is the most common (37%).

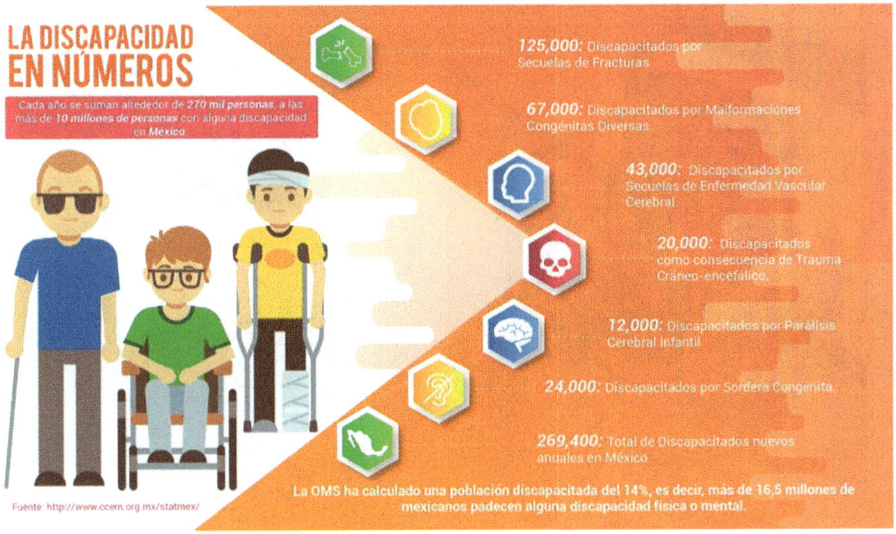

Fig. 1. Disability data in Mexico. http://edomex.gob.mx/dia_internacional_personas_discapacidad

People who have mobility-related difficulties form the population group whose most affected activity is displacement, because the construction of their immediate environment, such as housing, school, workplace, and transport - which according to the CIF (Classification International Functioning, Disability and Health) are part of the environmental factors - in most cases, they are not adapted so that people with this type of difficulty can move freely. One way to modify is the use of enablers, among which are: cane, wheelchair, walkers or crutches, and even the help of someone else [5].

Of every ten people with disabilities to walk or climb stairs, eight use some technical help, and two do not. This population group is the most benefited from the use of mechanical aids; Men outnumber women with this type of support [6]. In addition to the above, it is reported that 49% of people with disabilities live in rural areas, which makes this mobility even more complicated, not to mention that many of the inhabitants in urban areas are in settlements without streets or walkable sidewalks for their motor condition.

Although there are no precise data on the number of people who cannot afford a wheelchair in Mexico, following the WHO data, this amount can be significant. As an example, we will say that the cost of a standard wheelchair in Mexico is higher than the average salary of a worker for a month ($150 approx.). Additionally, it is the case that some people with disabilities save “for years”, and when it is possible to buy a wheelchair, they acquire one that is not the most appropriate to them (by size, the resistance of the chair, etc.). Still, it is the one that should last a lifetime, when the ideal would be to change it according to the needs of the user and their physical changes. Other times, the purchase or donation of a chair is accessed without; however, these do not resist the harsh use that environmental conditions in rural or marginalized areas impose, so they end up deteriorating it in a short period of product life (Fig. 2).

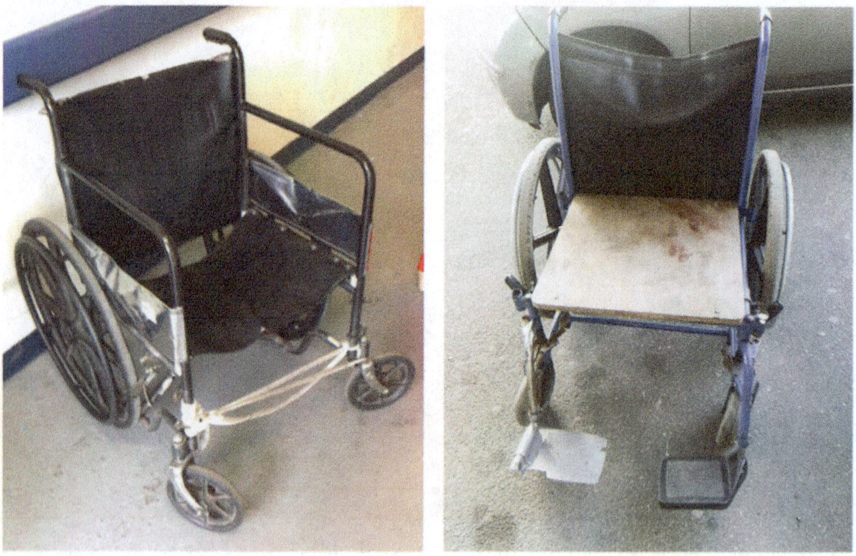

Fig. 2. Example of damaged wheelchairs for the use in hard environmental conditions.

2 Analysis of Existing Products

For the preparation of this project, we have searched for alternatives to the design of all-terrain wheelchairs or rural areas. However, much of what we founded escapes the economic possibilities of the bulk of the population; the most economical being national production models that starts at $660, reaching quotas for mechanized chairs with caterpillars that cost around $11,500.

After a search for similar products, some all-terrain wheelchairs were detected that fit our type of project. The first of these, is the project called "Safari seat", developed by Janna Deeble for Kenya and other East African countries. This proposal consists of an "all-terrain" chair made from bicycle parts [7]. The project is novel; however, it conditions to find bicycle parts and welding equipment for its manufacture in areas where it is not necessarily feasible to find these inputs. Its reported cost is GBP 200 ($260), and donations are through a financing platform for creative projects called "Kickstarter" (Fig. 3).

Fig. 3. "Safari seat" wheelchair project, developed for Kenya in 2015.

The second concept found that is of interest is called "GRIT Freedom chair". This chair mentions that MIT mechanical engineers have designed it, and among its main features is its robust construction; it has three wheels, the front has moved forward to improve stability [8]. Perhaps the most important feature is its impulse mechanism that is achieved based on two levers located on the sides of the chair (Fig. 4). The cost of this chair is $2,995, which already makes it is cost-prohibitive for the bulk of the population.

Fig. 4. "GRIT Freedom chair" wheelchair project, created by MIT mechanical engineers in 2012.

The third and final concept selected for being of interest for our purposes is the "Hippocampe" concept. Which is a chair developed in France by the VIPAMAT company and offers a tubular chair with curved shapes that have the characteristic of being able to exchange the wheels to be used in all-terrain (narrow pneumatic wheels) or sand (wide solid wheels) (Fig. 5).

The cost of this wheelchair is $3,750. The cost of this chair is entirely prohibitive for countries like Mexico.

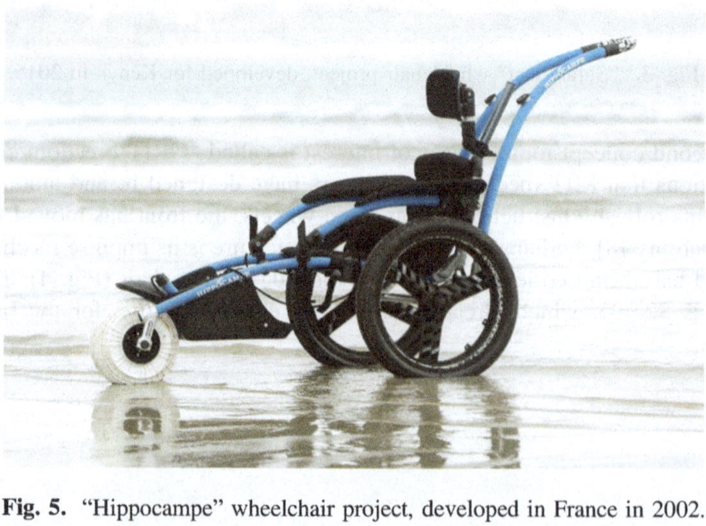

Fig. 5. "Hippocampe" wheelchair project, developed in France in 2002.

3 Proposal Design Development

The first action of the project will be to replicate the manufacture of an off-road chair to carry out field tests with it, including maneuverability tests, stress in use tests, and ease of assembly and transportation. For this, we chose the "GRIT Freedom chair" model, considering its structure and its impulse mechanism strong. It should be noted that even though the selected chair is promoted as an "open source" project, we can't find the production plans or assembly instructions online.

So we gave ourselves the task, using photographs and some general measures, to elaborate on the production plans that we will use for its manufacture. We take advantage of the realization of this model on a computer to perform simulation tests by finite element (FEA), to optimize the use of materials in our design proposal (Fig. 6).

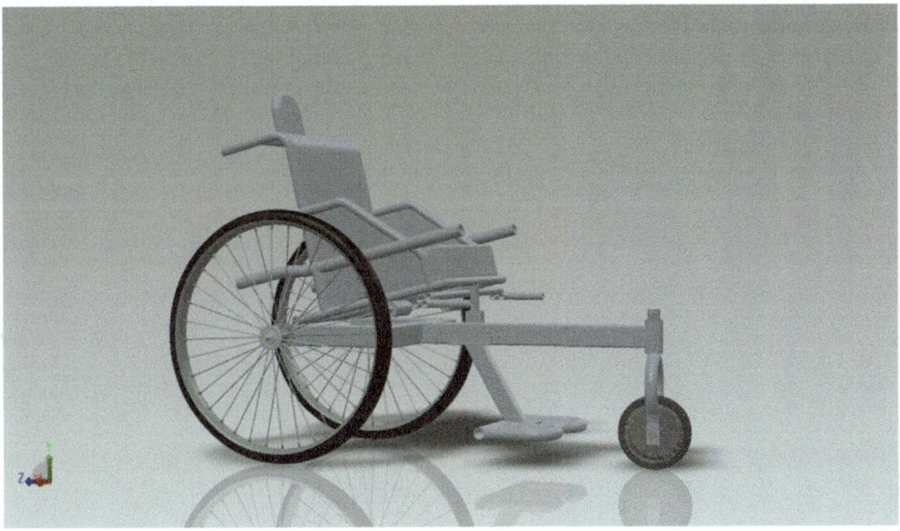

Fig. 6. Render of GRIT Freedom chair used for reproduction and tests.

4 Results

The manufacture and testing of the selected wheelchair demonstrated the capacity of its use, the ease of maneuverability, and limitations of the wheelchair. We have taken into account the opinion of the users and are currently working on the design phase of our design proposal, aiming to make our wheelchair lighter, improve its maneuverability, and, as far as possible, improve the final product price of the project.

Acknowledgments. An especial acknowledge to Universidad Panamericana, Campus Guadalajara, for the financial support received through the Research Found Program 2019–20. This research is part of the work of the Innovation Net for digital production in SMEs (Red IFADI) from the Iberoamerican Program for Science and Technology for Development (CYTED).

References

1. World Health Organization: Guidelines on the provision of manual wheelchairs in less resourced settings (2008)
2. Rossa-Sierra, A., Cortes-Chavez, F., Vidana-Zavala, D., Munno, M.G.T.: Design of a wheelchair for low-income countries, the second stage of a project. In: International Conference on Applied Human Factors and Ergonomics, pp. 301–309. Springer, Cham (2019)
3. Rossa-Sierra, A., Landa-Ávila, I.C., Jiménez-López, O.: Design of a low-cost wheelchair for open-source platform: first phase. In: International Conference on Applied Human Factors and Ergonomics, pp. 269–277. Springer, Cham (2018)
4. de la Federación, D.O.: Ley general para la Inclusión de las Personas con Discapacidad. Recuperado el, 21 Enero 2020 (2011)
5. Organisation mondiale de la santé, World Health Organization, & World Health Organization Staff. International classification of functioning, disability and health: ICF. World Health Organization (2001)
6. INEGI, I.: Las personas con discapacidad en México: Una visión al 2010 (2013)
7. https://www.kickstarter.com/projects/safariseat/safariseat-open-source-wheelchair-for-developing-c
8. https://www.gogrit.us/
9. https://www.vipamat.com/hippocampe-beach-and-all-terrain.php

Designing for Inclusion in the Information Society

Building Communities Through Digital Data Sharing

Soroosh Nazem(✉), Valeria Bruni, Elena Fabris, Alan Marcus, Barbara Melis, and Graziella Roccella

PLANET Holding Ltd., PLANET Group,
2 Portman Street, W1H 6DU London, UK
s.nazem@planetidea.it

Abstract. Social inclusion through data management is a new challenge in the urban transformation processes, which is highly relevant for the real estate sector. Digital technologies can accelerate the formation of accessible urban districts, where a district manager collects and analyzes data to provide feedback to inhabitants. Neighborhoods and urban areas, with their inhabitants, work as a platform connecting tangible and intangible elements.

Keywords: Social inclusion · Data collecting · Data management · Sharing information · Sharing services for citizens · Social innovation · Urban platform

1 The Role of Data Sharing in Creating Accessible Communities

The accessibility of all citizens to collective life is the key to transforming the population of a territory into a community. Communities are subjects capable of creating local value, and are the cornerstone for facing the challenges identified for current cities [5]. Social inclusion projects and their ability to create collective value is universally considered a best practice to creating strong communities [1].

Digital technologies can be leveraged to accelerate the formation of accessible and smart cities due to their ability to reinvent older services and allow for the creation of innovative new services [4]. In particular, digital media accelerates the sharing economy, inclusive economy, and circular economy processes that give substance to a community network.

Accessibility through the digital medium, which creates the community as described above, is critical for Affordable Housing interventions. In these districts, where economic resources of individual citizens are limited, being able to count on an expanded support network becomes a way to achieve a higher quality of life. In 2025 affordable housing projects worldwide are expected to involve 440 million families, or 1.6 billion people [3]. Real estate developers can play an important role, in collaboration with city governments and nonprofit organizations, "[…] to keep abreast of emerging solutions in construction techniques and materials, (…) to ensure an adequate flow of skilled labor, and consider new solutions in financing and innovative tenure models" [2]. Further, the real estate sector can ensure development projects are quality

G. Di Bucchianico et al. (Eds.): AHFE 2020, AISC 1202, pp. 133–138, 2020.
https://doi.org/10.1007/978-3-030-51194-4_18

built and within budget, but also through the application of innovative digital technologies (e.g. apps, IOT devices) they can improve the quality of life for residents.

As a real estate player, Planet Smart City[1] considers the community itself as a backbone of a platform capable of transferring and receiving information between its members through tools which specifically respond to their local needs (see Sect. 2). This approach requires real estate developers to devise a responsible strategy for governing the collection, use, and distribution of data to return useful information (see Sect. 3). This strategy must lead, to maximizing the benefits of both the virtual networks and networks of physical relationships to achieve social inclusion (see Sect. 4).

This approach will be described in detail below using an example of an intervention completed by Planet near Milan, Italy. It embraces the "Design for inclusion" philosophy by managing the data analysis of the experiences and needs of every individual, and can be applied for both brownfield and greenfield residential development projects.

2 Community and Data

As platforms, neighborhoods and urban areas, with their inhabitants, work in a dynamic nonlinear way [7]. The primitive and constitutive elements of the whole, taken in their own right, possess properties and functionalities that distinguish them in a unique and linear way. But when a large number of these primitive elements aggregate and form a system as a result of the relationships established between them, they begin to manifest themselves in the overall aggregate of properties and behaviors. The non-linear type, of which there was no trace in the atomic elements, therefore denotes the so-called emergent behaviors.

Similarly, real estate developers can consider their top-down interventions as structural, with the purpose of activating the most efficient systems. From this perspective, communities are the fuel of the system, interacting with (and in) the environment, with people activating new relations between the parts.

Thus, while acting top-down, Planet's design moves bottom-up, supporting people in interactions. Planet accelerates the growth of systems using social management and social design principles, inspired by the idea of Autonomy Respecting Assistance [6]. This approach enhances the ability of people and communities to self-determination. The resulting model is therefore a double platform, pairing the specific local onsite platforms (supported by social, tech and spatial infrastructures) with a digital platform (as a managerial and organizational global tool), the Planet App (see Fig. 1). In such a

[1] Planet Smart City designs and builds cities and neighbourhoods tackling the global affordable housing deficit, placing people at the heart of every project. In a multi-disciplinary team, the company integrates planning and architectural solutions, technological systems, ecosystemic resources and social innovation solutions to deliver high-quality low-cost homes, creating lasting value. Planet has the vision to offer everyone the opportunity to live in a better home and therefore works in partnership with developers worldwide to revitalise existing communities through smart technologies and social innovation programs. www.planetsmartcity.com.

model, community management teams assist the communities in acquiring capacities to self-determine their futures (see Sect. 4).

The key is to allow communities in reaching the right information to actually enable interactions with the system. In a data strategy perspective, this means considering emergent behaviors as fundamental data flows that can be intercepted, wrapped, aggregated and finally returned to communities in the form of useful and fully understandable information (see Fig. 1).

Planet believes, from an organizational perspective, that substituting the old transactional model for welfare with a model that is shared, collective, and relational will have a deeper impact and produce better results [8]. Through digitalization and new technologies, we can achieve this in ways that could potentially support large numbers of people. The main risk element of this innovation consists in building a strong enough strategy to ensure an "equilibrium of information". In other words, everyone in the community would have the same access to fully understandable information as a sort of social dividend.

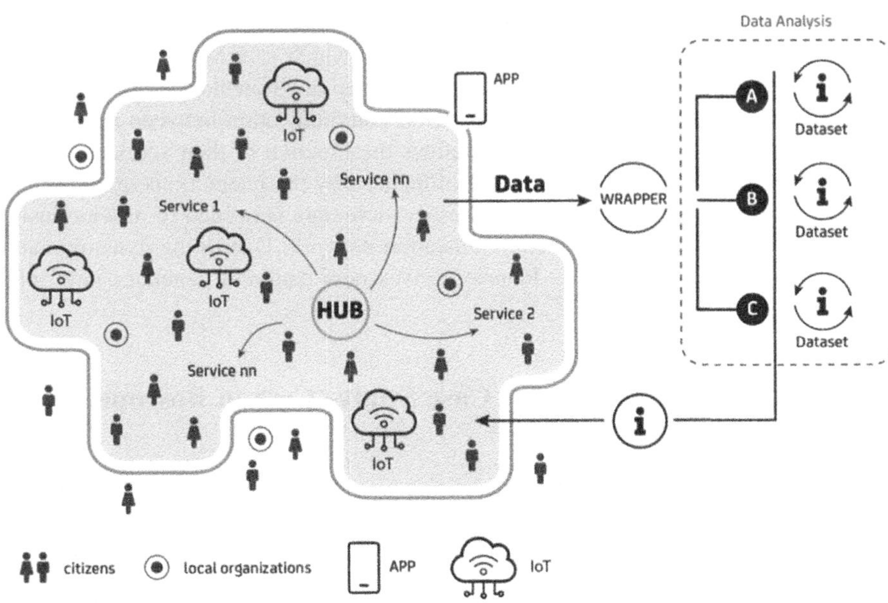

Fig. 1. Planet "accelerating impact at neighborhood scale: urban systems", (2019)

3 Planet's Data Strategy

In the digital age it has become essential for organizations to formally define their data strategy and address the increasingly pressing issue of data accessibility (Dwaync Gefferie, 2018), (Sas) [9], (SAS) [11], (Marr, 2019) [10]. Planet is therefore developing a detailed data strategy in order to provide the most efficient and optimized services to its communities; it consists of two components: Internal and External.

The Internal Data Strategy outlines how Planet can organize and manage data internally to improve its services. To achieve this, Planet will focus on how data flows through the organization and where each functional team is positioned in this flow. It further describes how each functional team handles data to ensure it is organized logically within the entire flow. For any given project Planet must be aware of its internal data flows throughout the entire process, from the beginning to the end.

The External Data Strategy is directly related to the communities where Planet works and offers its services. It must clearly outline the services that Planet provides communities with and all of the related interactions between them.

It is important to emphasize that communities serve Planet as both data sources and information consumers simultaneously. Planet is fueled by the data produced by its communities. By analyzing this data using methods defined by the Internal Data Strategy, Planet can generate useful information and eventually knowledge that can led to improving the quality of life for residents in its communities. The type of data collected can be evaluated and refined in an iterative process to maximize information and knowledge generation.

Planet provides a variety of services for residential communities. As a service provider, Planet must focus on two parameters: 1. which services are most useful to communities, and 2. how are these services connected and how do they influence each other. The majority of information is based on communication between services and not by the services acting alone. Planet defines the function of their services, but it is not clear how they are related as they are influenced by the interactions of community residents. To better understand these processes more data is needed to provide insight into the workings of the services communication network. Due to the dynamic nature of the data-information cycle, as Planet receives new data and generates new information, this network can evolve.

4 The "Quartiere Giardino" Case Study (Cesano Boscone, Italy)

Planet believes that the use of digital data must reinforce network development among inhabitants and within social systems. Planet applies this method at its "Quartiere Giardino" project (located near Milan - Italy) which from March 2018 Planet is transforming the existing area into a smart urban ecosystem altering and improving the lives of its 5,000 residents.

The interactions that inhabitants and city users have with their surroundings or with other inhabitants are being documented. This data is being made available to the residents themselves to support their decision-making. In fact, sharing data regarding 'what' and 'how' available services or products are used by residents, is useful for both the city administrator and the residents themselves. Further, when residents have access to other residents' personal data, interests, and routines they can more effectively forge social relationships and expand their networks. This consequently leads to the accelerated creation of spontaneous opportunities and support relationships. The main advantage for residents to live in an inclusive community is to benefit from a system of dense and integrated relational networks. The more numerous the connections are between networks and systems the greater the inclusive capacity of a community becomes, accelerating the ease and speed with which residents are able to find support and opportunities to solve their problems.

Planet is working towards these results through the work of a "community manager" who is directly engaged with a community. This figure's main role is to enable data sharing tools and to demonstrate the potential for networks to spontaneously generate value within a community's social systems.

For these reasons the work of the community manager at Quartiere Giardino has focused on the following:

- engaging inhabitants to activate new or renewed peer relationships. Social platforms (e.g. facebook, whatsapp) are reducing both the quality of real-world relationships and interpersonal skills. It is believed that their use improves their social networks without compromising the quality of their human relations. However, this is an illusion because virtual relationships do not create trust between people as effectively as real-world relationships do;
- amplify the value of collaboration and sharing, by addressing concrete needs such as: sharing and borrowing objects instead of owning them, exchange used books or objects, and share skills and abilities;
- facilitate access to the Planet App as the main data-generating and sharing platform for the community, and document the value of bridging the digital divide (e.g. elderly) and the opportunities generated by sharing data.

Lastly, Planet recommends that, to create an inclusive community, the following data are shared: personal data, service use and interaction data, data detailing interactions between residents and their environment, and interests and skills data. By harnessing this data, Planet aims to reduce the time needed to organize networks without compromising trust generation.

5 Conclusion

In Planet's experience data management, which provides residents with access to an enhanced community life, permits: co-design with inhabitants; making the best use of available technologies to improve people's quality of life; maintaining a database to evaluate the real impact of our work on community growth; distinguishing Planet from other real estate developers and to solidifying its position in the sector as innovators. It also allows Planet to produce a product that has more value than just homes; and, Planet to be more attractive on the real estate market.

Further, Planet recognizes the importance that data management plays in the current digital age. Many players are approaching the topic with opposing views; for example some are focusing on the protection of privacy (European directive 95/46/CE), while others are addressing the theory of auto-determination of populations through Data Cooperatives (Pentland, A. et al. 2019) [12]. Planet will consider all positions during the definition of its own data strategy.

The topics addressed represent core elements to Planet, and are being applied in their greenfield projects. For these types of developments where Planet starts and manages new communities from scratch, it's possible to build the online community of future residents even before the neighborhood is launched. For Planet's Smart City Laguna project, the strengths, weaknesses, opportunities, and threats of this specific situation are being evaluated.

Acknowledgments. The authors clarify that the positions expressed in the paper are the result of the constant comparison with all the colleagues in Planet, to whom a heartfelt thanks go. Furthermore, many thanks to Matthew Deluca that help us to make the paper content more clear.

References

1. Mulgan, G., Tucker, S., Ali, R., Sanders, B.: Social innovation. The Young Foundation (2007)
2. World Economic Forum: Making Affordable Housing a Reality in Cities. WEF Report (2019)
3. McKinsey Global Institute: A blueprint for addressing the global affordable housing challenge. McKinsey & Company (2014)
4. McKinsey Global Institute: Smart cities: digital solutions for a more livable future. McKinsey & Company (2018)
5. United Nations: New Urban Agenda. UN Report (2017)
6. Ellerman, D.: Autonomy-respecting assistance: toward an alternative theory of development assistance. In: Hayman, R., King, K., MacGrath, S. (eds.) A new approach to African Development? Internal and External Visions, Center of African Studies, p. 368. Edinbirg University, Edinburg (2007)
7. Sennett, R.: Building and Dwelling: Ethics for the City. Allen Lane (2018)
8. Cottam, H.: Radical help. How we can remake the relationships between us and revolutionise the welfare state, Virago (2018)
9. Dwayne: Become data driver or perish: why your company needs a data strategy and not just more data people (2018). https://www.forbes.com/sites/bernardmarr
10. Marr, B.: Why every company needs a data strategy for 2019 (2019). https://towardsdatascience.com
11. SAS: The 5 Essential Components of a Data Strategy. SAS Report
12. Pentland, et al: ITIF technology explainer: what are digital platforms? (2019). https://itif.org/publications/2018/10/12/itif-technology-explainer-what-are-digital-platforms

How to Democratize Internet of Things Devices: A Participatory Design Study to Improve Digital Literacy

Matteo Zallio[1(✉)], John McGrory[1,2], and Damon Berry[2]

[1] Autonomous Systems Laboratory, Stanford University, Lomita Mall, Stanford, CA, USA
matteo.zallio@Stanford.edu

[2] School of Electrical and Electronic Engineering, Technological University Dublin, Dublin, Ireland
{john.mcgrory,damon.berry}@TuDublin.ie

Abstract. The global introduction of affordable Internet of Things (IoT) devices offers an opportunity to empower a large variety of users with different needs. However, many off-the-shelf digital products are still not widely adopted by people who are hesitant technology users or by older adults, notwithstanding that the design and user-interaction of these devices is recognized to be user-friendly. Considering the habits and preferences of those who have used mainly analog devices for most of their lives, such users may encounter challenges such as digital illiteracy and technology anxiety when faced with newly-released digital IoT-based products. In view of the potential of IoT-based devices, how can we reduce the obstacles of a cohort with low digital literacy and technology anxiety and enable them to be equal participants in the digitalized world? This article shows the method and results achieved in a community-stakeholder workshop, developed through the participatory design methodology, aiming at brainstorming problems and scenarios through a focus group and a structured survey. The research activity focused on understanding factors to increase the usability of off-the-shelf IoT devices for hesitant users and identifying strategies for improving digital literacy and reducing technology anxiety.

A notable result was a series of feedback items pointing to the importance of facilitating educational experiences through learning resources to support individuals with different abilities, age, gender expression, to better adopt off-the-shelf IoT-based solutions. This fundamental first step of user experience research generated considerations for designing open-source learning tools to foster inclusive and more accessible use of IoT-based technologies for improving older people's daily activities.

Keywords: Participatory design · User experience research · Usability · Internet of Things · Smart ageing friendly environments · Inclusive design · Learning tools

G. Di Bucchianico et al. (Eds.): AHFE 2020, AISC 1202, pp. 139–150, 2020.
https://doi.org/10.1007/978-3-030-51194-4_19

1 Introduction

The ageing profiles of populations across the globe, from Europe to USA, to Asia, have consequences for different aspects of daily life [1].

The United Nations, with the Sustainable Development Goals (UN-SDG) [2] describe how health & well-being, sustainable cities & communities, innovation & infrastructure, represent a baseline for guaranteeing a better future for these populations.

These goals affect the use of different devices, services and solutions as well as the settings where humans spend most of their life: in the community, at work, and at home.

According to the World Health Organization [3], sustainable environments provide effective support for well-being & socialization. However, every user has a distinct sense of place [4] as well as different ways of interacting with the surrounding environments, objects, technology and to socialize and connect with others.

Technologies, in particular IoT-based devices, collaborative robotic applications and cloud-based systems have been released into the consumer market over the past decade. Such devices have a potential to foster independent living, to increase psychophysical health and advance older adults' context in physical and social activities [5, 6].

Inexperienced or tentative users may encounter challenges such as digital illiteracy and technology anxiety with newly released digital IoT-based products notwithstanding that the configuration (e.g. Plug & Play), design and user-interaction of IoT devices is considered to be user-friendly [7–9].

These challenges can be related to concerns about tolerance of error, system reliability, social exclusion burden, affordability, security and privacy of data generated about the users and their dwelling, and higher energy consumption [10–12].

The question arises: given the potential of IoT-based devices, how can the obstacles of poor digital literacy and technology anxiety be reduced to enable marginalized users to fully join the digitalized world?

The proposed research aims to investigate the factors that could lead to improvements in the use of off-the-shelf IoT devices for certain individuals through the participatory design methodology. Initial discussions with international experts in disciplines such as product inclusion, technology development and caregiving services were activated as a starting point for creating a community-stakeholder workshop.

The main goals of this study were to (1) involve experts working with older individuals in conversations and ethnographic research to identify problems encountered in the daily use and perception of IoT-based devices, (2) stimulate brainstorming of scenarios and assess features that could potentially support future solutions to identified problems, and (3) identify and test a feasible strategy that could lead to reduction and removal of digital illiteracy barriers and improve confidence in using IoT-based technologies.

2 Materials and Methodology

The impetus for investigating digital literacy and awareness of older adults in the use of IoT-based technologies arose from several years of research with stakeholders, as well as from various researchers and published academic reports [13, 14].

The methodology chosen to perform the initial exploratory research step centers on the participatory design research methodology. Participatory design is a democratic process for understanding and designing systems involving human work, based on the argument that people should be involved in the designs they will be using, and that all stakeholders have equal opportunities to provide inputs into the process [15, 16].

According to Spinuzzi, participatory design is research [17]. Participatory research is characterized by defining problems, ideas addressing problems and solutions in cooperation with and for people. Designers and researchers see themselves as facilitators who attempt to empower users in understanding their problems and framing ideas to solve needs [18].

The main activity of this participatory research was framed within a workshop, composed by a focus group and structured survey, deployed by conducting the participants (with already extensive working experience with senior citizens) in the mindset of older users, replicating some of their obstacles as a low digital literacy cohort.

Examples of this procedure refer to a partial communication of information regarding certain devices selected for the workshop, a lack of rigidity in progression of outcomes and using cross-disciplinary groups of participants.

The focus group procedure, which offers a deep collection and examination of detailed human experiences [19], was undertaken with the goal of analyzing views and perspectives regarding: (a) generalized causes that illuminate problems associated with newly released off-the-shelf IoT-based devices, (b) understand what specific problems people may encounter when using a voice assistant or a smart plug in conjunction with certain Activities of Daily Living (ADLs) and Instrumental Activities of Daily Living (IADLs), (c) brainstorm ideas on how those devices might be used to solve different key challenges and which aspects might be difficult to learn from different users.

The structured survey instrument [20] to investigate human phenomena, such as emotions and opinions, was used to obtain final remarks and feedback on the previously brainstormed ideas and to understand advice about future strategies for developing learning tools for empowering older individuals to benefit from IoT-based technologies. Those two complementary tools were jointly selected to enable a deeper understanding of stakeholders' views, that would otherwise be hard to achieve with single quantitative methods of analysis. The intention was to triangulate data from the focus group and structured survey in order to assure consistency of information; and to assure anonymity and confidentiality of participants (structured survey) by collecting insights that would be otherwise difficult to collect in a focus group.

2.1 Recruitment of Participants

The community-stakeholder workshop was included in the activities of the International Forum on Active Assisted Living, held in Aarhus, Denmark on September 2019.

A total amount of 45 participants, coming from several European countries, with different linguistic background, aged above 25 years old chose to participate. The group included healthcare sector specialists, design and engineering professionals and academic experts. Participants voluntary participated to a two-hour workshop session.

The activity was articulated within a two-layer participatory design research activity: a demonstration session, following a brainstorming of problem causes (focus group), and feedback evaluation (structured survey).

An introductory twenty-minute-long demonstration session on constraints and difficulties in learning how to use certain IoT-based devices (in order to enable participants to walk in older people's shoes) was provided by session facilitators, with two examples for voice assistants, smart lighting systems and wearable devices. Participants included a spectrum of stakeholders involved in work and life experiences with older individuals, who encountered difficulties in using IoT-based technologies. Subjects didn't receive any compensation for their participation and were able to drop the session at any time.

2.2 Data Collection

After the introductory session, participants were divided in to twelve working groups of three to four individuals each and were asked to define causes of problems regarding the use of IoT-based devices. The focus group activity was framed around two selected off-the-shelf IoT technologies: a voice assistant and a smart plug device.

In order to facilitate the participatory activity, each group received an infographic card (as shown in Fig. 1) where a series of ADLs and IADLs icons were shown to facilitate problems and use cases brainstorming. Each group was equipped with a whiteboard, post-its and pens to crystalize and note thoughts and statements through text, icons and symbols.

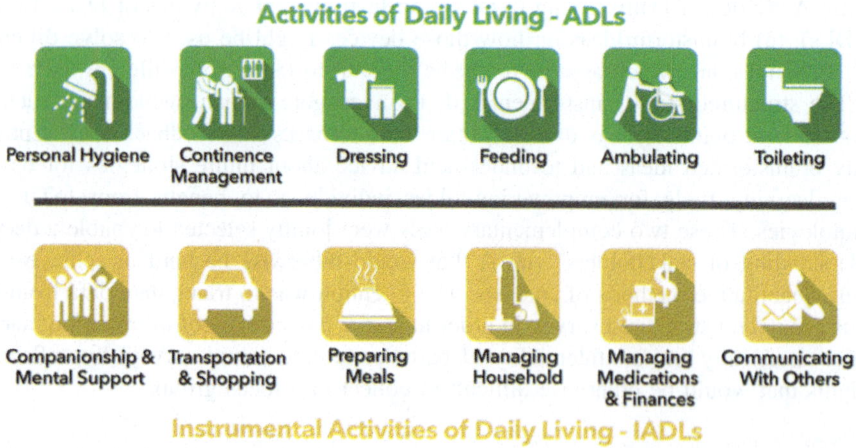

Fig. 1. Infographic representation of main ADLs and IADLs to facilitate the understanding of usage scenarios of IoT-based technologies: voice assistant device and a smart plug.

Mentorship from the research team was delivered to facilitate stakeholders in expressing as many feedback as possible and prototyping scenarios of use of the selected devices. Conceptualizing ideas was driven by the paradigm of quantity rather than quality in order to emphasize the understanding of certain features, use and operation of a voice assistant and a smart plug for people who have limited literacy skills regarding those devices.

In the second stage, participants were asked to provide feedback through a structured survey in order to clarify what type of learning resources would be useful to better interact with IoT-based devices, what sort of extended audience could benefit from those learning resources and how should those tools be developed and shared across the community.

3 Participatory Design Research

Out of twelve groups, respectively six provided feedback regarding the voice assistant device and six regarded the smart plug. The large quantity of information was condensed here into two infographics (Fig. 2, 3) where ADLs and IADLs were used as support criteria to facilitate the brainstorming activity.

3.1 Results and Feedback from the Focus Group: Voice Assistant

The researchers noticed throughout the session that stakeholders were pretty familiar with voice assistant devices and their knowledge of use and interaction with those IoT-based technologies was fair. However, the majority of participants expressed concerns regarding ease-of-use for older people. In virtually all six groups, participants created simple use cases for both ADLs and IADLs and experimented with verbal interactions in order to simulate a real-life context.

Some of the most prominent and common feedback for the use of voice-based assistance in relation to ADLs were.

1. "Receive step-by-step instructions and support with morning routine/orientation for people with early stage dementia, by considering a question/answer procedure to be facilitated by the device".
2. "Improve the conversation between end-users and family members to keep an active lifestyle; for example: 'you need to stand up', 'you need to walk', 'it's time for medication' or 'it's time for a tea with friends'".
3. "Serve as a home companion for considering different clothing tips according to weather forecast, trends in the market or seasonal preferred choices of the community".

Interestingly, in the field of IADLs, more curiosity, awareness and creative feedback was distilled from the focus group, enabling different interactions with other devices and other family members and caregivers out of the home environment.

1. "Support meal preparation, suggestions for different diets, vegetarian/vegan options and improve creativity in cooking, and as a result satisfaction in meal preparation or ordering from a local shop".
2. "Improve communications of certain topics among peers but also with chatbots that would lower the barrier and stigma of restricted or taboo topics".
3. "Suggest guidance using simple chains of instructions for learning how to improve DIY skills and small repairs at home".
4. "Talk and learn to users possible schedule time for personal activities, housekeeping, and connection of other personal devices to scheduled activities".

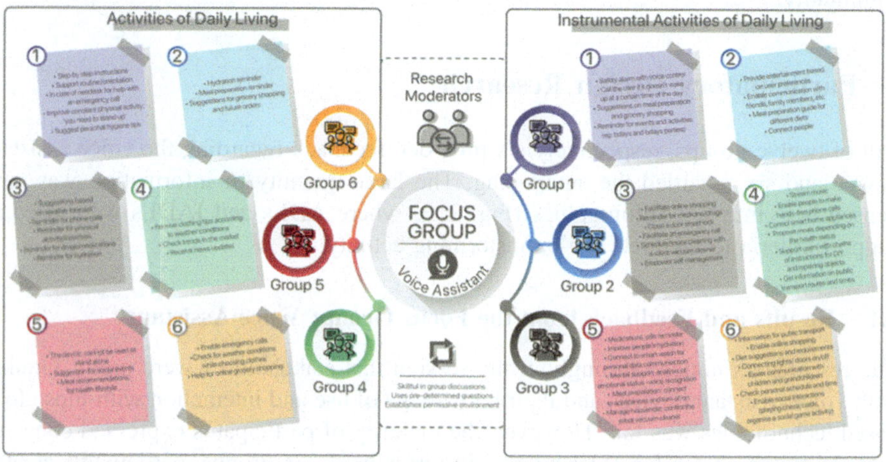

Fig. 2. Infographic with representation of feedback elicited from the focus group on voice assistant device in the context of use with ADLs (on the left) and IADLs (on the right).

Some of the most prominent and representative quotes collected during this research activity related mainly to three usability aspects: a) device reliability: use of sentences by the user that are not currently understandable by voice assistance devices, due to limitations of the Digital Audio Processing software; b) tolerance of error and easiness of deleting a process or an input that was erroneously initiated; c) lack of trust about the storage, processing and privacy of the audio registrations, (during various activities), in particular personal and intimate conversations in environments such as bathroom and bedroom.

3.2 Results and Feedback from the Focus Group: Smart Plug

The second half of participants focused mainly on understating scenarios associated with use of a smart plug that upgrades traditional home appliances in to connected, remotely controlled devices.

The majority of participants were not fully familiar with the potential offered by smart plugs, so workshop facilitators had to emphasize several functions during the initial explanatory session. For example, a lamp connected to a smart plug is controllable from smartphones or tablets and the switch on the lamp can be left unused.

As a result, many stakeholders developed several creative scenarios of use regarding ADLs and IADLs and expressed major concerns related to economic affordability, possible higher energy consumption and system reliability. Major feedback regarding the ADLs scenario focused mostly on feeding people and personal hygiene.

1. "Regularly control the use of personal electronic devices in the bathroom in order to detect usage patterns for personal hygiene".
2. "Detect energy consumption of kitchen appliances (refrigerator, kettle, oven, cooker) for detecting anomalies and misuse".
3. "Check through remote control of switches the activities performed by users, and if needed, set up notifications or alarms".
4. "Create ambient to follow daily routine".

Focusing on IADLs activities such as companionship and social support, managing household and communication with others were the most discussed by stakeholders.

1. "Find out easily how to control different devices, the energy consumption and schedule timing and zone activity in a reliable, user-friendly way".
2. "Understand usage patterns of appliances in the living room, kitchen, bedroom, bathroom and set up requests and alerts (via mail, texts, calls) primary to home users, then to family members and caregivers".
3. "Connect smart plugs with web-based services to foster productivity at home by connecting sensors and actuators that control devices automatically (according to use patterns)".
4. "Connect and turn off-on devices (TV, phone, tablet, computer, robot companions, cameras) to enable conversation and communication with others".

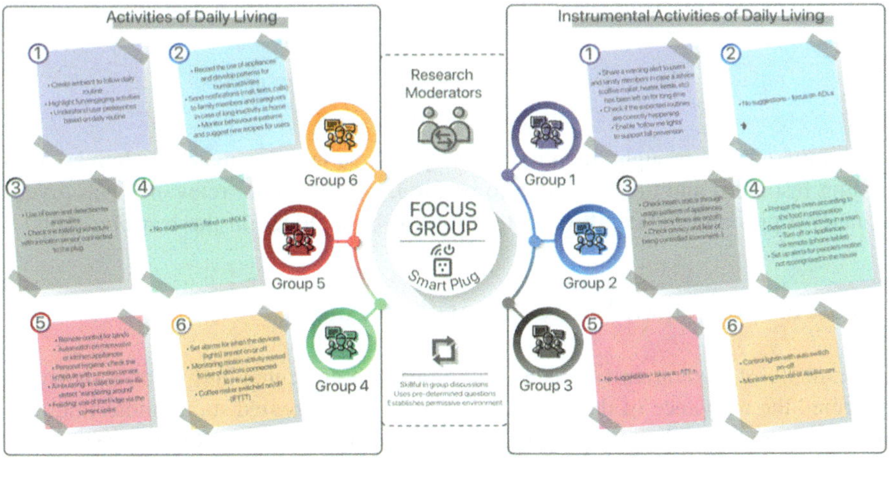

Fig. 3. Infographic with representation of feedback elicited from the focus group on smart plug device in the context of use with ADLs (on the left) and IADLs (on the right).

The scenario regarding the use of smart plug devices was slightly more complex, mostly because of the inner lack of knowledge of potential functions and capabilities of the selected device.

The cost of the device, compared to a traditional alternative with no Internet capability, was a particular topic of discussion among most of the groups. A further challenge posed by the group regarded the ease-of-use of function integration and the human-computer interaction process.

3.3 Structured Survey: Learning Tools

During the last stage of participatory design research, a structured on-line survey was submitted to participants. Three questions were submitted to participants with a Likert evaluation scale (1 to 10) to rank which solution should take priority and was most important. Wittink and Bayer [21] noted that a ten-point Likert scale can improve measurement reliability, reduce multicollinearity problems and minimize skewness in the distribution of the data. Clarifications were elicited on awareness-related concerns and around types of learning resource for better understanding the use of IoT-based devices.

The concluding survey was answered by a total of 20 of the workshop participants.

The survey was framed across three main questions (for more details refer to Zallio et al. [22]) respectively on: finding modalities to enable users to better understand the use and functionalities of off-the-shelf IoT-based devices, on identifying the major features that learning resources should include and also identifying potential audiences for those learning resources (Fig. 4).

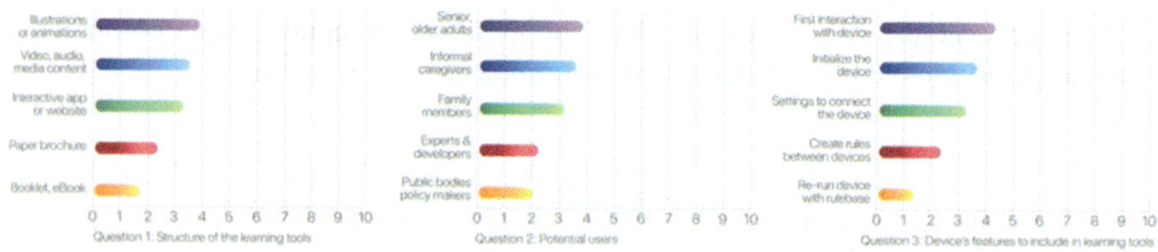

Fig. 4. Data visualization of the survey responses.

Major feedback brought attention to the quality of learning material, delivery of the learning resources and the languages to be used, while a minor relevance was given to the inner topics and the depth of instructions to be included in the learning tools.

4 Discussion: Open-Source Learning Tools

The results of the community-stakeholder workshop presented in the previous sections confirmed the relevance of the research question and presented some interesting topics for investigation in relation to easing digital illiteracy and technology anxiety.

The use of the chosen methodology (participatory design research) allowed the research team to mobilize a significant number of stakeholders with experience in dealing with older adults and elicit meaningful scenarios with constraints and potentialities for the use of certain IoT-based devices.

The focus group supported participants to 'dig deeply' in to different device features, to develop use cases and brainstorm constraints, issues and potentials in relation to daily activities.

Discussions allowed participants to frame contributions around system reliability, ability to learn an easy methodology to control the device, familiarization with interface and commands, the justification of higher costs and higher energy consumption compared to a traditional devices and difficulties in understanding privacy and data management of IoT-based technologies.

The participatory activity led stakeholders towards the common need of developing tools for improving digital literacy, such as interactive user manuals, educational resources and informational experiences.

The final results underlined the importance of developing a new form of learning experiences, that would be open-source (supported and perhaps created by the community of users), independently developed, with emphasis on different educational levels (from beginners, to experts), and easily accessible by different media (smartphones, tablets, tv, computers, voice assistant appliances, wearable devices).

In order to highlight the major features from the survey, a normalization of ratings by adjusting values measured on different scales to align to a normal distribution was performed.

According to each survey question a detailed visualization of normalized data was created (accessible at Zallio et al. [22]) and major responses underlined the importance of three main aspects:

1. The structure and media to use for developing and communicating the learning resources: graphical illustrations or animations (infographics and icons), video, audio and media content, and interactive mediums (application or website).
2. Enlarge the audience of possible users: from seniors, to informal caregivers, and family members who want to learn efficiently how to interact with a new device or simply receive user-friendly usage tips.
3. The main features that are encountered during a first interaction with the device (turning on-off-standby), initializing and setting up the device (log in-out options, settings, privacy), integration and connection of the device with others (Bluetooth-Wi-Fi pairing, data transferring).

All these criteria are based on the assumption that a large majority of people (90%) aged 55 and more have a Wi-Fi network at home and can easily access to the Internet, computers and smartphones, according to the Linkage Connect Technology Survey [23]. This ability to easily access information allowed workshop participants to be more confident in considering learning tools based on interactive mediums like apps with embedded audio and video recordings and infographic, animated content.

Data interpretation was grounded both in focus group and survey activities and the findings highlighted the relevant need to develop educational tools to foster technology skills and awareness, improve the ability to choose devices according to specific personal challenges, and to increase self-awareness capabilities to interact with a variety of off-the-shelf IoT-based devices.

5 Conclusions and Future Work

This user research generated a series of solution-oriented feedback that in the opinion of the authors could be useful to implement design and test of learning experience tools.

The main scope of this work was to create a strong methodological base of ethnographic research to offer grounded information for ideating, prototyping and testing learning solutions across different countries, languages, cultures and individuals.

The feedback collected through the participatory design research underlined the importance of developing resources to facilitate educational experiences to support individuals with different abilities, age, gender expression in conducting daily activities. These educational experiences have the potential to foster the adoption of off-the-shelf IoT-based devices for improving the performance of daily activities and to develop a better human-machine interaction process.

The next steps of project development will focus on deploying prototypes of training packages to foster learning experiences, based on interactive mediums, with multilingual platforms, through the use of graphical illustrations with a specific goal of enlarging the audience of potential users.

The main features that will be included in the learning resources will initially focus, as suggested from consultation, on three steps of human-machine interaction: turning the device on/off, general and privacy settings and technology integration.

As outcomes it is foreseen that those resources could lower cultural and language barriers through the inclusive use of off-the-shelf IoT-based technology, boost social inclusion with the open-source future creation of those learning resources, and in the longer term foster cross-sectoral cooperation among companies, educational facilities and institutions.

Acknowledgments. This research, as an impulse into a larger international project (Hands-on SHAFE funded under the ERASMUS+programme), was made possible thanks to the Active Assisted Living Forum, in Aarhus, Denmark on September 23–25, 2019. A team of researchers from Stanford University (USA) and Technological University Dublin (Ireland) developed the participatory design research and successfully run the collaborative workshop supported by the AAL Programme.

References

1. European Commission: Population ageing in Europe. https://bit.ly/2NcjHC7. Accessed 18 Jan 2020
2. United Nations: Sustainable Development Goals, August 2016. https://bit.ly/2d4dcA4. Accessed 20 Jan 2020
3. World Health Organisation: Measuring the age-friendliness of cities (2015). https://bit.ly/2ARrhjY. Accessed 20 Jan 2020
4. Jorgensen, B., Stedman, R.: Sense of place as an attachment: Lakeshore owners' attitudes toward their properties. J. Environ. Psychol. **21**, 233–248 (2001). https://doi.org/10.1006/jevp2001.0226
5. Task Force on Research and Development for Technology to Support Aging Adults. Emerging Technologies to Support an Aging Population; Committee on Technology of the National Science & Technology Council: Washington, DC, USA (2019). https://www.whitehouse.gov/wp-content/uploads/2019/03/Emerging-Tech-to-Support-Aging-2019.pdf. Accessed 20 Jan 2020
6. Peek, S.T., Luijkx, K.G., Rijnaard, M.D., Nieboer, M.E., Van der Voort, C.S., Aarts, S., Van Hoof, J., Vrijhoef, H.J., Wouters, E.J.: Older adults' reasons for using technology while aging in place. Gerontology **62**(2), 226–237 (2016)
7. Hoque, R., Sorwar, G.: Understanding factors influencing the adoption of mHealth by the elderly: an extension of the UTAUT model. Int. J. Med. Inf. **2017**(101), 75–84 (2017)
8. Vaportzis, E., Clausen, M.G., Gow, A.J.: Older adults perceptions of technology and barriers to interacting with tablet computers: a focus group study. Front. Psychol. **8**, 1687 (2017). https://doi.org/10.3389/fpsyg.2017.01687
9. Tsai, H.S., Shillair, R., Cotten, S.R.: Social support and "playing around": an examination of how older adults acquire digital literacy with tablet computers. J. Appl. Gerontol.: Off. J. Southern Gerontol. Soc. **36**(1), 29–55 (2017). https://doi.org/10.1177/0733464815609440
10. Martínez-Alcalá, C.I., Rosales-Lagarde, A., Alonso-Lavernia, M.Á., Ramírez-Salvador, J.Á., Jiménez-Rodríguez, B., Cepeda-Rebollar, R.M., López-Noguerola, J.S., Bautista-Díaz, M. L., Agis-Juárez, R.A.: Digital inclusion in older adults: a comparison between face-to-face and blended digital literacy workshops. Front. ICT **5**, 21 (2018). https://doi.org/10.3389/fict.2018.00021
11. Zallio, M., Fisk, M.J.: Smart Homes. In: Gu, D., Dupre, M. (eds.) Encyclopedia of Gerontology and Population Aging. Springer, Cham (2019)
12. Lorenz, A., Oppermann, R.: Mobile health monitoring for the elderly: designing for diversity. Pervasive Mob. Comput. **5**, 478–495 (2009)
13. Wang, S., Bolling, K., Mao, W., Reichstadt, J., Jeste, D., Kim H.C., Nebeker, C.: Technology to support aging in place: older adults' perspectives. In: Healthcare, vol. 7, no. 60 (2019). https://doi.org/10.3390/healthcare7020060
14. Schreurs, K., Quan-Haase, A., Martin, K.: Problematizing the digital literacy paradox in the context of older adults' ICT use: aging, media discourse, and self-determination. Can. J. Commun. **42**(2) (2017)
15. Muller, M.J., Kuhn, S.: Participatory design. Commun. ACM **36**(6), 24–28 (1993)
16. Muller, M.J.: PICTIVE—an exploration in participatory design. In: Robertson, S.P., Olson, G.M., Olson, J.S. (eds.) Proceedings of the SIGCHI Conference on Human Factors in Computing Systems (CHI 1991), pp. 225–231. ACM, New York (1991)
17. Spinuzzi, C.: The methodology of participatory design. Tech. Commun. **52**, 163–174 (2005)
18. Clement, A.: Computing at work: empowering action by "low-level" users. Commun. ACM **37**(1), 52–63 (1994)

19. Morgan, D.L., Spanish, M.T.: Focus groups: a new tool for qualitative research. Qual. Sociol. **7**, 253 (1984). https://doi.org/10.1007/BF00987314
20. Phillips, A.W.: Proper applications for surveys as a study methodology. Western J. Emerg. Med. **18**(1), 8–11 (2017). https://doi.org/10.5811/westjem.2016.11.32000
21. Wittink, D.R., Bayer, L.R.: The measurement imperative. Mark. Res. **6**, 14–23 (2003)
22. Zallio, M., McGrory, J., Berry, D.: How to democratize internet of things devices. A participatory design research (2020). https://arxiv.org/abs/2002.06308
23. Linkage Connect Technology Survey. https://linkageconnect.com/wp-content/uploads/2019/05/2019-Link-age-Connect-Technology-Study-Report.pdf. Accessed 30 Jan 2020

Talking About Inclusiveness: How Inclusive Are Transportation App Services?

Adriana Rodríguez-Villaseñor[1](✉), Patricia Reyna-López[1], Alber Duque-Álvarez[1], Carlos Aceves-González[2], and Libertad Rizo-Corona[2]

[1] Maestría en Ergonomía, Universidad de Guadalajara, Guadalajara, Mexico
{adriana.rodriguezv, patricia.reyna, alber.duque}@alumnos.udg.mx
[2] Centro de Investigaciones en Ergonomía, Universidad de Guadalajara, Guadalajara, Mexico
c.aceves@academicos.udg.mx, libertad.rizo.din@gmail.com

Abstract. This research aimed to know the perception of users with motor disabilities who need to use assistive devices about how inclusive it is one of the leading ride-hailing platforms. Transportation app services have become aware of the needs of people with disabilities and the difficulties they face to integrate into active life to satisfy their basic and recreational needs. Thus, companies behind these services have developed alternatives in their service to reach the goal of inclusiveness. However, the remaining question is if they achieve such a goal. Two surveys were applied to assess the service. Results suggest that whereas the app advertises the inclusiveness of its service, more than half of users are not receiving the required attention. Eighty per cent of the participants considered necessary the provision of a vehicle physically adequate to their needs with only 36% who responded to have traveled in suitable cars.

Keywords: Inclusive service · Transportation app · Interaction · Disabled people

1 Introduction

According to the Conventions on the Rights of Persons with Disabilities by the United Nations personal mobility of people with disabilities must be ensured in terms of shape, ease of access at an affordable cost [1]. Consequently, there is an increasingly widespread acceptance that efforts should be made to tackle the barriers to disabled people's participation and engagement in society [2]. Transportation systems help in this process since they allow people to connect with work, study, social life, and services [3]. The recent years have witnessed the rise of ride-hailing platforms (RHPs), these are services that match drivers with consumers who demand a ride [4]; services like UBER are used by people to move around spaces with autonomy. This service has enhanced awareness of the needs of vulnerable groups, identifying their requirements and in consequence, adapting their services by adjusting features according to users' needs and conveniences and create

G. Di Bucchianico et al. (Eds.): AHFE 2020, AISC 1202, pp. 151–156, 2020.
https://doi.org/10.1007/978-3-030-51194-4_20

specialized alternatives in its service to fulfill market gaps considering customers despite their conditions [5]. They pretend to increase accessibility in their services by providing options for as many people as possible. In pursuit of this, they have developed the WAV and Assist alternatives [6]. In Mexico, UBER offered just inclusive service Assist, which was created for people living with some type of disability, pregnant and elderly. They are matched with top-rated drivers who have been trained to provide extra assistance based on a program designed by the *National Integration Council for People with Disabilities* (*CONADIS* for its acronym in Spanish) and provided by *TELETÓN México* (a non-profit organization that seeks to serve people with disabilities, cancer and autism) [5]. In this training, drivers learn how to fold wheelchairs, manage crutches and walkers, also they learn inclusive language and practical advice for assistance of these people in case the passenger may need it; this service is provided in cars without appropriate modifications, however, are big enough to transport a wheelchair in the trunk [6, 7].

With all the aid characteristics added to the regular service in these new options, it is assumed that companies are reaching the goal of inclusiveness. However, are they actually achieving it? To address this question, it must be considered the expectations and users' perception of satisfaction regarding service. Perhaps, some situations are perceived as simple for people without motor disabilities but can turn complicated and restrict mobility to persons who have a disability.

A user relies on two standards to determine if the service has high quality. One is which his/her expectations on the service and an adequate or minimum acceptable service quality [8]. Expectations are assumptions that serve as a starting point to establish judgment by the user on service performance, accepting technical quality associated with tangible and intangible expectations [9]. The perception is the outcome after receiving a service. Users have expectations which cause the desired quality, but after receiving the service, the real quality is generated. The difference between real and expected quality results in quality perceived or perceived service [10], which is a matter of this study.

This research aimed to know the inclusiveness in the service provided by one of the leading platforms of a mobile transportation app by asking people with motor disabilities about their expectations, and their perception of critical situations they have faced throughout the journey's stages (shown in Fig. 1) using the service.

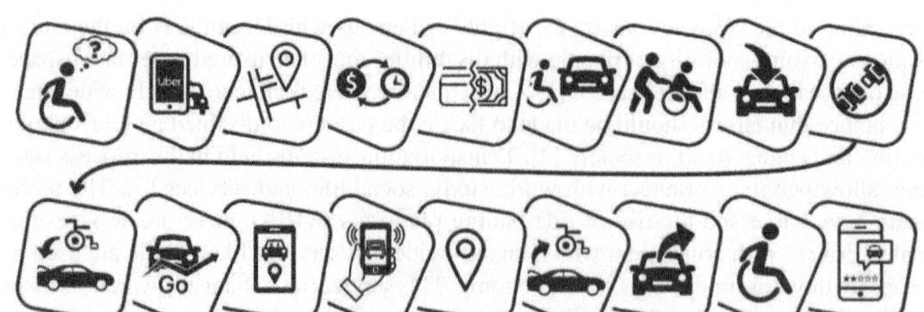

Fig. 1. Uber journey stages

2 Method

2.1 Participants

A total of 11 people with motor disability users of an assistive device like a wheelchair, a walking stick, crutches, or a walker. All the participants were UBER users, with at least a trip in one of the provided services. They gave consent for being included as part of the study. The surveys were responded by six women and five men aged between 17 and 42 years.

2.2 Instruments

Two surveys of the ServQual model were employed to evaluate the quality of UBER service. A Likert scale of 7 points was used, where one was the lowest and seven was the highest level of satisfaction and pleasantness according to their necessities [11]. Allowing to acknowledge and evaluate expectations and perceptions of users by giving an overview of the service and its inclusiveness.

The ServQual model is an instrument that evaluates customers' perspectives and perceptions about the quality of service of an organization [11], each section of this model consisted of 22 items arranged into five dimensions: 1) *reliability:* refers to the ability to execute the promised service reliably and carefully. That is, the company keeps its promises, about deliveries, service provision troubleshooting, and pricing, sensitivity, security, empathy, and tangible elements; 2) *sensitivity:* it is the willingness to help users to provide fast and adequate service. It refers to the attention and

Table 1. Expectations and perception questions

Expectations	
Tangible elements	Should Uber app be easy and intuitive?
Reliability	Should Uber drivers be skilled?
	Should Uber cars have an adequate design for the user?
Sensitivity	Should Uber drivers offer help as soon as they arrive at the starting point?
	Should Uber drivers always be willing to help?
Empathy	Are Uber cars best suited for your needs?
	Should Uber drivers give personalized attention?
User's perceptions	
Tangible elements	Is Uber app easy and intuitive?
Reliability	Was the driver skilled while helping you?
	Did Uber's car provide you with physical support on your trip?
Sensitivity	At the time your car arrived, did the driver offers you help?
	Was the Uber driver willing to help you?
Empathy	Is Uber's car right for you?
	Did the Uber driver give you personalized attention?

promptness when dealing with requests, answering questions and complaints from customers, and solving problems; 3) *security:* it is the knowledge and attention of employees and their ability to inspire credibility and trust; 4) *empathy:* refers to the level of individualized attention that companies offer their clients. It must be transmitted through a personalized service adapted to the client's taste and 5) *tangible elements:* it is the physical appearance, physical facilities, such as infrastructure, equipment, materials, personnel [12]. Table 1 comprises the items of the survey, where the highest discrepancies between expectations and users' perception appeared.

2.3 Procedure

First, the participants were contacted and invited by phone or email. Once they had provided verbal consent and agreed to participate in the study, an email with the link of the survey was sent to each participant. The survey was responded on the WEB by using Google Sheets. For data analysis, the scores 5, 6 and 7 of the Likert scale were unified as positive responses of the service attributes in the different aspects that were evaluated. The analysis was undertaken using SPSS V19.

3 Results

Results in Fig. 2 show the differences between participants' expectations and perceptions of the service. There is a difference between what users with a mobility impairment expected and what they have experienced. It highlights data about drivers' skills and help, where according to their experiences, only 27% of the participants mentioned being satisfied with the drivers' skills. Also, just 30% of the participants have received help from the drivers when picking them up.

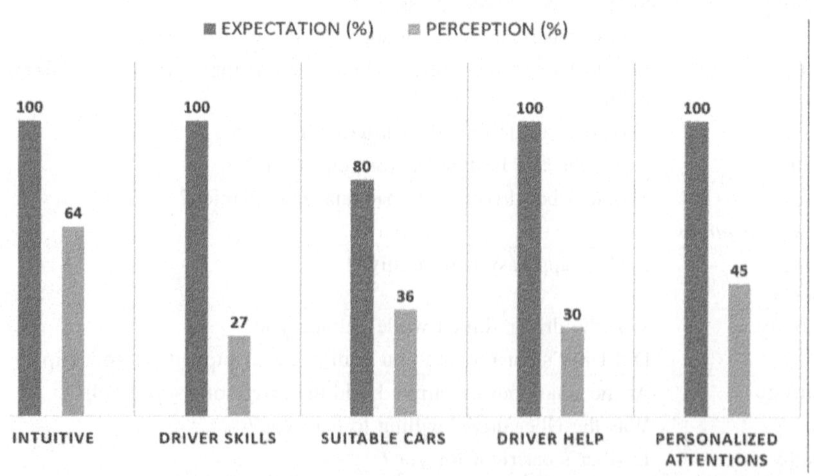

Fig. 2. Discrepancies between expectations and perceptions of participants.

4 Discussion and Conclusions

The users of mobile transportation app services seek a safe, fast and accessible way to move, whether they have a disability or not. It is crucial to guarantee truly inclusive transportation systems, especially to improve the accessibility of people with mobility impairments [3]. Some discrepancies were found between the advertised service of UBER and expectations and what the users with motor disabilities have experienced. Results suggest that the main differences are in the interaction with the app, the driver and the vehicle characteristics since they interfere in addressing their needs.

After the analysis, the higher gap appears in the ability of the driver and the help provided by them, which could be because not all the drivers have the training to assist people with disabilities; 64% of the users surveyed had not used UBER Assist since they did not know the service. It is relevant to consider that users also encounter difficulties when choosing the means of transport since the choice depends not only on tangible attributes but also on the experiences and perceptions of the previously received service [3]. The lack of training in matters of inclusion in UBER drivers could reduce the overall qualities of the service; since the assistive devices could be mistreated or even some user could be injured.

It is relevant to mention that under researchers' perspective arrange an extraordinary trip to test the service involves a high risk to them and compromises the investigation, because of the effect of selective memory. Therefore, they were directed to answer in retrospective by remembering their trips, and that could make them tend to value better the service of what it was, which is a limitation of this study.

The fact that only a third part of the respondents knew about UBER Assist suggests that the publicity of this service characteristic is limited. However, it is relevant to mention that providing an excellent inclusive transport service it is a matter of social interest and the responsibility is not only of who offers it, but it is also necessary that the government, the transport companies, and society get involved and work together for social welfare.

In conclusion, this study provides an overview of the difficulties faced by people with motor disabilities when using UBER service in Mexico. Although there are some advances and attempts to be inclusive, there is still a lot to do to meet the needs and expectations of users with the provided service. Companies that offer ride-hailing services should redouble their efforts to be more inclusive. Moreover, if civil organizations get along with the companies, they could train more UBER drivers, and these efforts could increase and facilitate the transportation of people regardless of their age or capabilities.

Although the results obtained by the ServQual used in this research exhibit a notable difference in what the user expects and what receives, it is suggested to do additional research, considering that this type of survey may have a reliability bias, due to the data that was left out, like the motives behind the given answers. It is also proposed to extend the research to a more significant number of participants with motor disabilities and incorporate additional participants that are within other vulnerable groups like are people living with other disabilities, pregnant, and elderly.

Acknowledgements. The authors extend their appreciation to many people who made this research possible, particularly the participants of the study.

References

1. Lawson, A.: United nations convention on the rights of persons with disabilities (CRPD). In: International and European Labour Law, no. December, pp. 455–461 (2018)
2. Unless, R., Act, P., Rose, W., If, T., Rose, W.: This is a repository copy of travel apps for disabled people: evaluating their contribution towards inclusive transport. White Rose Research Online URL for this paper: Version: Accepted Version Conference or Workshop Item: Matthews, B and Wardman (2015)
3. Márquez, L., Poveda, J.C., Vega, L.A.: Factors affecting personal autonomy and perceived accessibility of people with mobility impairments in an urban transportation choice context. J. Transp. Heal. **14**(June), 100583 (2019)
4. Lam, C.T., Liu, M.: More than Taxis with An App: How Ride-hailing Platforms Promote Market Efficiency (2018)
5. U. Assist: UberASSIST Un viaje hacia la inclusión (2020)
6. U. Gobernación: ¡Uber, Conadis y Teletón lanzan UberASSIST para asistir a las personas con discapacidad y adultos mayores! (2018)
7. U. Conadis: Uber, en alianza con CONADIS y Teletón, lanza uberASSIST en ocho nuevas ciudades de México (2018)
8. Peralta Montecinos, J.: Rol de las expectativas en el juicio de satisfacción y calidad percibida del servicio. Límite: Revista de Filosofía y Psicología (14), 195–214 (2006)
9. Marzo, J.C., Martínez-Tur, V., Ramos, J., Peiró, J.M.: La satisfacción del usuario desde el modelo de la confirmación de expectativas: Respuesta a algunos interrogantes. Psicothema **14**(4), 765–770 (2002)
10. Galviz, G.I.: Calidad en la gestión de Servicios (Fondo Editorial Biblioteca Universidad Rafael Urdaneta, ed.). Vereda del Lago, Maracaibo, Venezuela (2011)
11. Parasuraman, A., Zeithaml, V.A., Berry, L.L.: Servqual: a multiple-item scale for measuring consumer perc. J. Retail. **64**, 12 (1988)
12. Nishizawa, M.: Desarrollo del Modelo ServQual para la medición de la calidad del servicio en la empresa de publicidad Ayuda Experto. Perspectivas **34**, 181–209 (2014)

Emotional Design and Human Factors Design as Tool for Understanding Efficiency in Information Design Process at Medical Documents

Mariel Garcia-Hernadez[1(✉)], Fabiola Cortes-Chavez[2], Marco Marin-Alvarez[1], Alberto Rossa-Sierra[2], and Elvia Luz Gonzalez[3]

[1] Unidad Azcapotzalco, Universidad Autónoma Metropolitana, Av San Pablo Xalpa 180, 02200 Col. Reynosa Tamaulipas, CDMX, Mexico
a12173804149@alumno.azc.uam.mx,
marma@correo.azc.uam.mx

[2] Facultad de Ingeniería, Universidad Panamericana, Álvaro del Portillo 49, 45010 Zapopan, Jalisco, Mexico
{fcortes,lurosa}@up.edu.mx

[3] Centro de Investigaciones en Ergonomía, Universidad de Guadalajara, Calzada Independencia Norte No. 5075, Huentitán El Bajo, 44250 Guadalajara, Jalisco, Mexico
elvia.gmunoz@academicos.udg.mx

Abstract. This work will focus on improving the design that currently has medical documents, specifically results of blood chemistry used in Mexico, which are not designed to be understood by the different users who really interact with this product. It is important to propose a taxonomy of the different users that interact with the documents that correspond to the results of clinical analysis, in order to develop products that are focused on the cognitive requirements of each of them.

Keywords: Information design · Ergonomics · User centered design · Medical documents · Emotional design

1 Introduction

The study of ergonomics is essential, we must know the person who will be behind the product, it is necessary to know it and study its needs as its limitations to be able to understand what type of object needs, with all the characteristics and specifications that it requires. Each user is different, and it is necessary to group them according to their personalities, limitations, needs, etc. [1]. Currently, the literature shows that it is essential to develop products that are focused on the needs of the end-users, however there are still few products and services that have been developed focused on the needs of a specific user, thus causing an increase in the risk of errors and understanding of them.

G. Di Bucchianico et al. (Eds.): AHFE 2020, AISC 1202, pp. 157–163, 2020.
https://doi.org/10.1007/978-3-030-51194-4_21

This work will focus on improving the design that currently have medical documents specifically results of blood chemistry used in Mexico, which are not designed to be understood by the different users who really interact with this product. These contain information that only the doctor or health officer could understand, however, other users such as the patient and/or the patient's family member are not contemplated, which generates misunderstanding of the information that is being presented to them. It is important to propose a taxonomy of the different users that interact with the documents that correspond to the results of clinical analysis, in order to develop products that are focused on the cognitive requirements of each of them.

Ergonomics in the design plays a fundamental factor at this time, since the design of the information is not addressed in a way appropriate to the cognitive needs of all users, errors occur and loss of time for users, when they generate documents that do not invite reading; instructions that are not understood and generate frustration, ambiguous information and excessive fatigue for those who interact with these devices [2].

This is why this work seeks to demonstrate the importance of addressing the cognitive and communication requirements from the design strategy of the products previously mentioned to the end user. As well as the current information generated by chemical laboratories is inefficient, since it is directed only to users of health specialists (doctors, nurses, laboratories) and does not consider that there are other users of this product, such as the patient, and/or the patient's relative. Not being designed for all users, the product does not meet the needs of the latter, resulting in an inefficient design. Finally, it is shown that there are different types of users for this product.

With the results of this study, a design proposal was generated that can support information designers so that the algorithm found in the results can generate an extra scale, focused for the group of users who do not work within the area of the health, with this the designer will get a product that increases compression efficiency when considering all users.

2 ISO Standard

The ISO standards IEC 62366-1, 2015, ISO 13485 2016, in the area of medical devices and usability have clear rules about the importance of safety and specify the obligation of designers or developers to detect user needs before the design process [3].

3 Who Is the User

Speaking of who is the user? In a study by Syed Ghulam and Ian Robinson of Brunel University, a literature review of 556 articles was conducted to identify who were the users to which each document referred. With the information, we conclude that all the documents speak of different users and they were given the task of proposing a grouping of them, generating two large classes of users: main users that include the following groups: health professionals, patients, careers, people with special needs, trainers and students, and secondary users including groups: researchers, others, students and trainers. Each of these groups is divided into different types divided into subtypes, where family members were in a subtype. This means that family members will be considered low users [4, 5].

The study concludes: A formal definition of who the users are, are missing, Information about all of different users it's necessary, all users are different from each other, all users are important, and should be ranked according to their function [4, 5].

The object leads the user experience, Designers need to consider all the experience as behavioral, expressive and physiological feelings. Designer must visualize before and during the design process [4, 5].

The FDA agree that some errors coincide with the poor design of medical devices, designers need to give importance to Human factors to understand the real needs of users, during the design process [6].

4 Scenario Based Design Method

The scenarios are an integral part of the international standard for the usability engineering of medical devices (IEC 62366: 2007) [7].

The method analyzes how an object or interface will be used in the future and provides to all the types of user, the needs that must be met at each time and stage of the product. "The scenarios are based on observational studies and future projections with the analysis of data" [8].

5 Information Design Requirements and Cognitive Ergonomics

W we talk about design for usability, we refer to any decision that is aimed at developing design artifacts that take the user and their needs as a central axis [9]. The author states that the design for usability includes two user items; physical characteristics and cognitive characteristics. The author refers to physical characteristics to all those considerations that the designer must consider for the development of any design artifact that are related to the user's physical morphology; weight, height, age, etc. In terms of cognitive characteristics, they are those that refer to the skills, knowledge, experiences and beliefs of the user regarding the design artifact and the context in which it interacts with it [10]. A design oriented to usability has the purpose of covering both characteristics in correlation with the context of use and usability requirements of the design product. However, due to the object of research study we will focus specifically on the cognitive characteristics of the user.

The graphic interfaces, must cover the cognitive needs of their users [10]. These needs, are the following; reduce cognitive effort, minimize the error rate during reading, require a lower learning effort on the part of the user, simplify representational transformations of information, match the memory of intermediate states to solve problems, improve performance, productivity and efficiency to perform a task related to the graphical interface increase the reading and browsing comfort through the information that is presented in the graphic interface with respect to the user [9].

These needs can be included in three important requirements to be covered by the interfaces:

1. Be effective. When we talk about effectiveness in the interface we mean that the information it contains can be easily understood by the user.
2. Be usable. Referring to this term, we would like to point out that the interface is easy to use and that the elements that make it up do not generate visual noise or visual bumps during the reading of the interface.
3. Be nice. The interface must be aesthetically pleasing to the user, since by being so we can infer that it satisfies visually.

Emoticons are graphic representations of an easy expression or physical body; these forms are embedded in human communication [11]. These representations compensate for the lack of non-verbal information, [12]. Also, the use of emoticons increases the understanding of nonverbal of the meaning. Emoticons express more emotional information than written text [13]. The use of emoticons is associated with strengthening the message, so this could be useful for those users that doesn't have knowledge about medical or technicalities about items that can be seen in the results of clinical analysis terms. As a contribution of this research work, we designed a Likert scale using emoticons. On the Likert scale emoticons green, and orange yellow were used. The green color was used to represent the normal reference values of the elements presented in the results of clinical analyzes. Yellow was chosen to represent slightly higher or lower values than those that could be within normal range. On the other hand, orange was chosen to represent higher or much lower values than those represented by the yellow color. The choice of this color coding was taken from the Human Engineering Committee of the Association for the Advancement of Medical Instrumentation (AAMI), [14]. There is a color coding that is related to the different warning parameters in medical devices. This color coding is:

6 Hypothesis

Seeing from strategy design information to the end user, especially those who do not belong to the area of health, decrease their level of uncertainty receiving the results of clinical tests, if they are presented supported emoticons and color scales.

7 User

It was taken as the end user profile sheet laboratory results. The study considered that the participant was between 25 and 44 years old, with a minimum high school education, with a medium-medium high socioeconomic level, also that will reside in the city of Puebla and Guadalajara.

8 Experiment

The usability test aimed at the following points:

1. Compare which of the two editorial proposals was the most effective (in as for the understanding of the information) for the user.
2. Compare which of the two editorial proposals was the most pleasant for the user (reader).
3. Compare which of the two proposals was the one that generated less stress for the user.

For the development of the usability evaluation, the total user sample was divided into two groups, using the A/B Test tool.

The editorial proposals that were developed for this usability test were the following: (Fig. 1a, b).

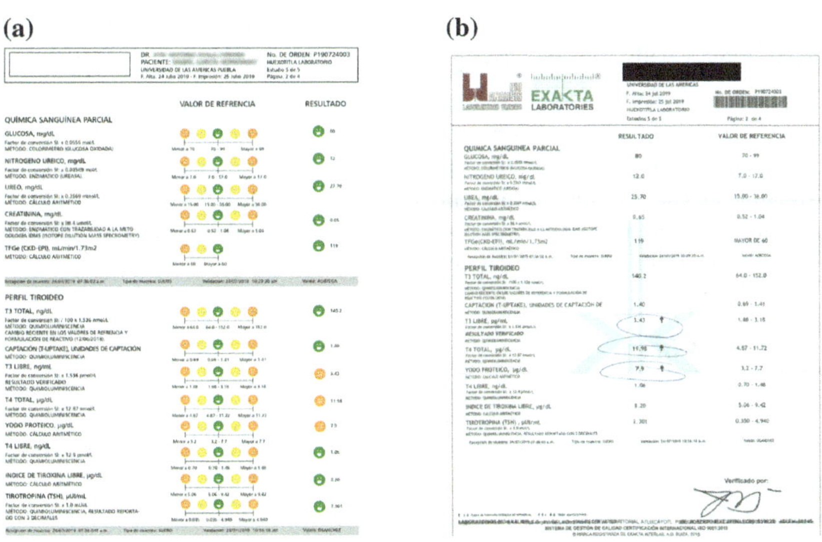

Fig. 1. **a**. Proposal "A". **b**. Proposal "B"

An electronic form was used through the survermonkey.com platform, one for the "A" proposal and the "B" proposal, the "A" being the re-designed design piece and the "B" proposal the original proposal.

9 Results

The results of the A/B test were as follows:

1. Variable "Effectiveness":
   Observations: We find that the proposal "A" is clearer than reading with 37.5% in the premise "totally agree". Likewise, the same editorial proposal presents a greater understanding between the numerical result and the level that level represents. As also, this proposal is easier to understand than "A".
2. Variable "Satisfaction":
   Observations: According to the items corresponding to this variable, the proposal "A" is the one that presents the highest qualification regarding the user's satisfaction with it. We can observe that this user is visually more pleasant to the "A" proposal than to the "B" proposal, just as he feels more satisfied with the presentation and arrangement of the information.
3. Variable "Stress Level":
   Observations: After analyzing the corresponding graphs this variable, we can determine that the proposal "A" generates much less stress to the user that the proposal "B". Being the proposal "A" the highest qualified with variables ranging between "very quiet" and "quiet", unlike the proposal "B", where the "uneasy" reagent ranges with 50% preference.

After presenting the previous results, we can establish that the proposal that presented a better level of efficiency, effectiveness and satisfaction was the "A". Therefore, we can say that the use of emoticons and color nomenclature within a scale helps the understanding of the information presented to the user. Also, it reduces the stress you may feel at the uncertainty of the results themselves and make more palatable the data contained in the result set of clinical analysis.

10 Future Works

In future projects it is recommended to work on the impact of emoticons on the patient's health status, this under the assumption that doctors need the patient to be optimistic about their health status. In this way, generating a state of less stress during the patient's recovery process.

References

1. Cañas, J., Wearns, Y.: Cognitive Ergonomics. Panamericana Medical Publishing House, Madrid (2001)
2. Czajkowski, K., Fitzgerald, S., Foster, I., Kesselman, C.: Grid information services for distributed resource sharing. In: 10th IEEE International Symposium on High Performance Distributed Computing, pp. 181–184. IEEE Press, New York (2001)
3. Derks, D., Bos, A.E.R., Von Grumbkow, J.: Emoticons and online message interpretation. Soc. Sci. Comput. Rev. **26**(3), 379–388 (2008)
4. Desmet, P., Hekkert, P.: Framework of product experience. Int. J. Des. **1**(1), 57–66 (2007)
5. Eilouti, B.: Scenario-based design: new applications in metamorphic architecture. Front. Archit. Res. **7**(4), 530–543 (2018)
6. Frascara, J.: ¿Qué es el diseño de información? Buenos Aires (2011)
7. ISO: A practical guide ISO 13485: 2016 Medical devices (2017)
8. Jordan, P.W.: Introduction to Usability. Taylor & Francis Group, Londres
9. Romero-Medina, A.: Cognitive Ergonomics and Usability. Ergonomics. Optional subject 5th year Degree in Psychology. University of Mursia, Murcia (2006)
10. Shah, S.G.S., Robinson, I.: Medical device technologies: who is the user? Int. J. Healthc. Technol. Manag. **9**(2), 181 (2008)
11. Thompson, D., Mackenzie, I.G., Leuthold, H., Filik, R.: Emotional responses to irony and emoticons in written language: evidence from EDA and facial EMG. Psychophysiology **53** (7), 1054–1062 (2016)
12. Ward, J.R., et al.: An analysis of medical device-related errors: prevalence and possible solutions. An analysis of medical device-related errors: prevalence and possible solutions 1902 (2009)
13. Walther, J.B., Addario, K.P.D.: The impact of emoticons on message interpretation in computer-mediated communication. Soc. Sci. Comput. Rev. **19**(3), 324–347 (2001)
14. Wiklund, M.: Medical Device and Equipment Design: Usability engineering and ergonomics. Taylor & Francis Group, Florida (1995)

Design for Inclusion in the Living Environment

Moving Beyond Visitor and Usability Studies: Co-designing Inclusion in Museums and Galleries

Janice Rieger(✉)

Queensland University of Technology (QUT), Brisbane 4100, Australia
j.rieger@qut.edu.au

Abstract. Museums and galleries have made efforts to be more inclusive over the last ten years, primarily through the emphasis on visitor studies, however they continue to have issues with making their environments and content accessible. This research addresses these issues and presents an alternative approach to creating inclusion in museum and galleries through co-design. By using co-design methods to actively engage people with differing abilities, this study creates new trajectories for inclusion that address the full spectrum of need and choice, for all users of the museum and gallery. Moving beyond visitor studies, the research presents new methods and strategies for museums and galleries when designing for inclusion. This paper presents key findings from case study research undertaken through the *Vis-ability* exhibition in Australia, to propose alternative ways of creating inclusion in museums and galleries, and how co-design can deepen our understanding of design for all.

Keywords: Abilities · Co-design · Design for all · Exhibition design · Inclusive design · Museums · Universal design

1 Introduction

Museums and galleries as public institutions and producers of cultural knowledge emphasise social cohesion and inclusion, and yet they still have issues with making their content and environments accessible to all. In this paper, I present key findings from the research undertaken in the Vis-ability exhibition at the QUT Art Museum, to propose alternative ways of creating inclusion in museums and galleries [1]. Through an extensive and intuitive consultation and research process with participants of differing abilities, I worked with a team of over 30 advisors to rethink exhibition design and the design of museum and galleries more generally. A co-design and multisensorial approach was undertaken to create 12 inclusive outputs that employed alternative mediums and technologies, such as tactile and audio experiences, and plain language summaries.

The key findings from this research offer a useful prompt to designers, curators, managers, educators and scholars to engage users with differing abilities more fully in this work in the future. The findings affirm not only the need for choice but argue for co-design methodologies to actively engage people with differing abilities to co-design

G. Di Bucchianico et al. (Eds.): AHFE 2020, AISC 1202, pp. 167–172, 2020.
https://doi.org/10.1007/978-3-030-51194-4_22

the spectrum of choices. This research shifts knowledge production around inclusion in the museum and gallery away from visitor and usability studies and towards the need to train museum and gallery staff in co-design processes. Additionally, this research shows inclusion in museums is more sustainable when it is co-designed with the communities the museum seeks to include, on an ongoing basis. Co-design has the capacity to create new trajectories for inclusion, that address the full spectrum of need and choice, for all users of the museum and gallery.

2 Co-designing Inclusion

As museums and galleries are increasingly adopting a visitor-centered approach, new methods for studying, designing and implementing this approach are needed [2–4]. Moving beyond visitor studies that often create personas or construct journey maps with imagined users, this research presents an alternative approach to creating inclusive museums through co-creation. For example, the Smithsonian recently developed a series of personas for a workshop exercise with the National Museum of Natural History (USA) to develop journey maps of imagined visitors [2]. Current literature indicates designers base their designs on what they already know (about themselves and their peers) or on stereotypical notions of others [5]. By continuing to imagine our visitors and users, we are designing for 'absent or even imaginary people' rather than with participating users [5, 6]. Co-design methods allow for an inclusive and participatory processes to enable designers and museum professionals to empathise with diverse 'real' users.

Studies that document co-designing with people with disabilities often focus on the person's disability and their inability to contribute to the co-designing process, but Rieger, Herssens and Strickfaden [7] in their co-design process, instead concentrate on the abilities of the participants and recognize their individual expertise. This requires a questioning of underlying assumptions and biases to freely explore, experiment and co-create.

Co-designing inclusion is also not only about the process, but about understanding design as a meta-category comprised of three elements. Fry expands [8, 9]:

1. the design object—the material or immaterial outcome of designing
2. the design process—the system, organization, conduct and activity of designing
3. the design agency—the designer, design instruction in any medium or mode of expression and the designed object as it acts on its world

Therefore, in co-designing for inclusion, the methods used must consider the object, process and the agency of *all* participants. Inclusion in museums is a complicated process, especially when considering the spectrum of choices and needs of all visitors and their differing embodiments [10].

3 Case Study

The data collected via interviews, workshops, visitor feedback and observations in this case study reinforced the need for choice and for a spectrum of choices that are multisensorial and scaffolded in a way that is accessible to all. This requires an understanding of multimodality and the multisensorial. Where multimodality is about representation and communication, whereas multisensoriality is about how we engage and interact with these differing modes [6, 7].

Vis-ability: Artworks from the QUT Art Collection, was exhibited at the QUT Art Museum, Brisbane, Australia from May 11–August 4, 2019. *Vis-ability* was conceived as a project to broaden understanding of the lived experiences of people with dis-abilities. *Vis-ability* presented key recent acquisitions to propose alternative ways of engaging with the QUT Art Museum Collection and to consider how multisensorial technologies can deepen our understanding and experience of art [11]. Several artworks were translated through alternative mediums such as tactile and audio experiences, to offer a fresh multisensorial experience of art.

The exhibition, its programming, exhibition catalogue and events were all created out of a co-designed process over a period of 9 months. As a part of this process, an intensive 3-day workshop was delivered that brought together over 30 people, inclusive of international experts from Belgium, Canada and France to co-design inclusive outputs for this exhibition. The 12 co-designed outputs were:

- 3 Audio Description Pods
- Augmented Reality Simulation Goggles
- Simulation video of a museum visitor who is blind
- Tactile Model based upon Catherine Parker's painting, *Present portal,* 2017.
- Soundscape based upon Catherine Parker's painting, *Present portal,* 2017.
- Touch/Descriptive Tours
- Sensitivity Staff Training
- Co-designed Public Programs
- Curriculum and Workshops for High School Groups
- Inclusive Exhibition Catalogue (with audio links and a plain language summary)

Translating visual art into haptic and aural experiences requires moving from one perceptual mode to another [12]. The co-creation process in *Vis-ability* involved many layers of interpretation and translation due to the many participants involved in the process [12].

In co-designing the three audio descriptions for *Vis-ability*, we realized that there is no one audio description that is correct. There are many different approaches and one artwork could result in any number of descriptive translations [13]. These audio descriptions were designed to be heard before the art was seen. This is why we chose to design pods to conceal the works of art. Audiences were encouraged to listen to the description of the artwork before looking through the viewing windows in the pods to see the work. This research reinforced an understanding that audio descriptions can add to the impact of an artwork, not just for audiences who are blind or have low vision, but for all audiences [13].

The online exhibition catalogue was co-designed with people who are blind to include accessible images, text, and audio files. It also included access to readable exhibition didactics and a text-only version of the catalogue for accessible software programs [1]. Through a consultation process with people with differing abilities we decided to also include a plain language summary of all the essays in the exhibition catalogue to be inclusive of all reading levels.

In an effort to also make the *Vis-ability* exhibition programs inclusive, regular workshops were held called *Making visible*, in which visitors engaged in a design process to create a small tactile artwork, based on an original work in the exhibition. Additionally, we conducted inclusive staff training and also co-designed new Touch/Descriptive tours to be inclusive of people who are blind or have low vision. As a result of these inclusive programs and tours a new co-facilitator/tour guide who is blind is now working at the QUT Art Museum. Over the course of this exhibition, visitors who identify as blind or vision impaired became regular visitors. Another unexpected outcome of these new tours and inclusive programs co-designed for *Vis-ability,* is that they are now a part of every exhibition at the QUT Art Museum. By co-designing inclusion across the museum, it has expanded and sustained these new audiences. The following are feedback from workshop participants and visitors, many who are blind and had not visited an art museum before:

> *The opening of the Vis-Ability exhibition was a culmination of all the learnings of the co-design workshops into a sensory delight – a tactile, auditory and visual experience. The exhibition enabled art to be more accessible, not just for those who had visual impairments (although this was the main benefit), but also enabled all who participated the opportunity to engage in art in a more meaningful and embodied way*—Workshop participant feedback

> *Thank you for all the information and again your workshop was really good and I know a lot of us were taken away with this workshop and on looking at other ways to communicate art to people who can't see the art.* – Visitor feedback

> *Love the soundscape, and Vis-ability artworks provide an opportunity for more people to feel/touch the pieces* – Visitor feedback

During the *Vis-ability* exhibition, 17 workshops were held, 557 visitors participated in the programs and 944 artworks were made. This exhibition also had one of the largest attendance records for a small university museum, with 5166 visitors.

This case study provides further evidence that if tactile and audio elements in a museum or gallery are not created *with* people with differing abilities, they are often not useable or meaningful for the very people they are designed for and become tokenistic [14, 15]. If the people who visit museums and galleries are not involved in materially shaping their experiences, designers risk producing a 'scenic' notion of inclusion and of the visitors themselves [16]. Through co-design methods, museums and galleries can move away from the 'paradox of participation' that often creates tokenistic inclusionary practices, towards co-creating inclusion [16, 17].

4 Conclusion

To create inclusion *with* visitors rather than *for* them, this requires moving beyond visitor studies that often create personas or construct journey maps with imagined users, towards a co-designing of inclusion. Through co-designing *Vis-ability* the expectation was that we could come to some kind of a consensus on how we could create inclusion in museums and galleries. Needless to say, this was not an outcome of the research. What did transpire was that most participants wanted choice and to be offered a spectrum of choices depending on their needs and desires with how to engage with the work, exhibitions, programs or even the museum architecture.

Through this research a shift occurred, which moved away from the concept of inclusion as universal, and towards inclusion as choice. For example, with the audio descriptions, some visitors wanted to sit down and put on headphones to listen. Other visitors wanted the choice to listen to the audio through their own devices whereas some visitors wanted the audio descriptions to filter into the exhibition spaces. In terms of the content of the audio descriptions, most visitors wanted the choice of different narratives—either with short explanations of the work or with longer curatorial narratives.

Designing for choice in museums and galleries can be a difficult task, however there are many approaches, tools and techniques which allow for the co-designing of a wide range of choices. The essential difference is that these choices *need* to be co-designed with the very visitors engaging with them. By moving beyond imagined visitors, and co-designing with real visitors, concepts around inclusion move away from universal design to design for all. This research contributes new knowledge to an identified gap in the literature on how to effectively facilitate co-designing for inclusion in museums and galleries. By putting people at the centre, this research focuses on co-design practices to create cultures of inclusion alongside people with differing abilities.

Acknowledgments. I gratefully acknowledge the contributions of the participants in this project, and the QUT Art Museum and their staff.

References

1. Vis-ability: Artworks from the QUT Art Collection, QUT Art Museum (2019). http://www.artmuseum.qut.edu.au/whats-on/2019/vis-ability.php
2. Villaespesa, E., Álvarez, A.: Visitor journey mapping at the Museo Nacional Thyssen-Bornemisza: bringing cross-departmental collaboration to build a holistic and integrated visitor experience. Museum Manag. Curatorship **35**, 1–8 (2019)
3. Samis, P., Michaelson, M.: Creating the Visitor-Centered Museum. Routledge, London (2016)
4. Roppola, T.: Designing for the Museum Visitor Experience. Routledge, New York (2013)
5. Strickfaden, M., Heylighen, A.: Who are they? In: Proceedings of Include 2009 (2009)
6. Rieger, J.: Doing dis/ordered mapping/s: embodying disability in the museum environment. Doctoral thesis, University of Alberta, Canada (2016)

7. Rieger, J., Herssens, J., Strickfaden, M.: Spatialising differently through ability and techné, CoDesign, pp. 1–17 (2018). https://doi.org/10.1080/15710882.2018.1531134
8. Fry, T.: Lexicon. EcoDesign Foundation (1998, unpublished)
9. Willis, A.M.: Ontological designing. Des. Philos. Pap. **4**(2), 69–92 (2006)
10. Rieger, J., Strickfaden, M.: Dis/ordered assemblages of disability in museums. In: Hadley, B.J., McDonald, D. (eds.) Routledge Handbook of Disability Arts, Culture, and Media, pp. 48–61. Routledge (Taylor & Francis Group), Abingdon (2018)
11. Dionysius, K., Rieger, J.: Introduction, Vis-ability: Artworks form the QUT Art Collection QUT Art Museum, Australia, pp. 6–18 (2019). ISBN 978-0-85856-010-9
12. Herssens, J., Rieger, J., Strickfaden, M.: Co-creating haptic and soundscape experiences, Vis-ability: Artworks form the QUT Art Collection, QUT Art Museum, Australia, pp. 19–33 (2019). ISBN 978-0-85856-010-9
13. Hadley, B., Dionysius, K., Boulton, S.: Audio descriptions, Vis-ability: Artworks form the QUT Art Collection QUT Art Museum, Australia, pp. 45–53 (2019). ISBN 978-0-85856-010-9
14. Candlin, F.: The dubious inheritance of touch: art history and museum access. J. Vis. Cult. **5**(2), 137–154 (2006)
15. Strickfaden, M., Vildieu, A.: On the quest for better communication through tactile images. J. Aesthetic Educ. (JAE) **48**(2), 105–122 (2014)
16. Jones, P.: Situating universal design architecture: designing with whom? Disabil. Rehabil. **36**, 1369–1374 (2014)
17. Rieger, J., Strickfaden, M.: Taken for granted: material relations between disability and codes/guidelines. Societies **6**(1), 6 (2016). https://doi.org/10.3390/soc6010006

Challenges of Universal Design in Coastal Resort Architecture and Planning

Agata Bonenberg(✉)

Faculty of Architecture, Poznan University of Technology,
ul. Berdychowo, Poznań, Poland
agata.bonenberg@put.poznan.pl

Abstract. Holiday time is one of the most important and pleasurable occasions when daily routines can be changed for healthier and beneficial for well-being: physical as well as psychological. The opportunity to both enjoy and use this time well is important for all, and also crucial for persons with disabilities and the elderly. In this article the author collects and presents specialized solutions in universal design for coastal resort architecture and planning, provides an assessment of a case study, and presents her own project for a the Międzyzdroje Promenada Zachodnia competition in Poland.

Keywords: Preventative care · Housing architecture · Disabilities · Universal design

1 Introduction

Holiday time is one of the most important and pleasurable occasions when daily routines can be changed for healthier and beneficial for well-being: physical as well as psychological. The opportunity to both enjoy and use this time well is important for all, but seems particularly crucial for persons with disabilities and the elderly. To fully take advantage of holiday possibilities, it is important for holiday destinations to provide well designed services and excellent accessibility; this article concentrates on challenges of accessibility of beach resorts (Figs. 1 and 2).

G. Di Bucchianico et al. (Eds.): AHFE 2020, AISC 1202, pp. 173–181, 2020.
https://doi.org/10.1007/978-3-030-51194-4_23

Fig. 1. Persons with disabilities benefit from experiences offered by holiday resorts providing architectural space and services, which have been adapted to their needs. Photo A. Bonenberg

Fig. 2. JOB beach chair: a wheelchair for the disabled that goes into water. JoB, an acronym for "Jamme'o bagno" (a captivating invitation in Neapolitan dialect to take a bath in the sea), is a specific product for the holidays, for the bathing of people with disabilities. Photo A. Bonenberg, Hersonissos

2 Aims and Methods

The aims of this publication are to:

- collect and present specialized solutions in universal design for coastal resort architecture and planning,
- assess coastal resort architecture and planning of a chosen case study (Creta Maris Beach Resort, Hersonissos, Greece),
- present the author's own project, which follows concepts of universal design for coastal resort architecture: Competition for Międzyzdroje Promenada Zachodnia, Poland.

Applied methods:

- analysis of a case study (Creta Maris Beach Resort, Hersonissos, Greece): a range of solutions supporting mobility within coastal resort architecture and planning researched through product presentation, description and author's photographs,
- bibliographical analysis,
- research by design: competition projects for Międzyzdroje Promenada Zachodnia, Poland, coastal resort with multiple pro-accessibility solutions.

3 Architecture, Aesthetics and Universal Design Solutions in Hotel Resort Architecture

A holiday resort can be defined as a self-contained recreational and commercial establishment providing most of a holidaymaker's needs, developed around a hotel or condominium and equipped with a wide range of amenities.[1] Amenities typically include entertainment, wellness, sports activities related to cultural or natural contexts; beaches, mountains, forests, historical towns or villages.

Architecture of resorts is designed to celebrate holidays: aesthetics usually underlines natural contexts, promotes wellbeing and celebrates the locality. This is why resort architecture around the world typically follows one of these architectural trends:

- contemporary vernacular architecture (contextualized architecture with local influences)
- traditional local architecture (merging with spatial context or recreating it)
- eco-architecture (demonstrating vicinity to nature)
- neo-modernist architecture with local influences (creates an image of global aesthetics)

Universal design, introducing a wide range of accessibility solutions, often has a "medical", "hospital-like" appearance. Double handrails, ramps, transportation platforms, interior elements may bring to mind healthcare institutions and infrastructures. For holiday resorts this kind of association is a negative one. As the aesthetics of architectural space play an important role both in the popularity and marketing strategies of resorts, it is important that universal design elements and solutions for persons with disabilities are applied in a way that they don't have a strong impact on aesthetical expression and quality of space. It is the author's belief, that they can become an inspiration for new interesting solutions in the architectural interiors and exteriors of holiday resorts.

4 Universal Design in Resort Architecture - Case Study

The case study presented in this publication pertains to Creta Maris Beach Resort in Hersonissos, Greece. It has been chosen for a wide range of different accommodation types available and a problem associated with the site's complex topography, which presented an important design challenge. The Resort opened in 1975 and initially centred around the main hotel building. A vast bungalow area to the north gave the experience of a small village set in a hilly landscape, with twisting paths, small piazzas and indigenous greenery contrasting with the white walls of small-scale buildings. The design goal was to provide an experience of Aegean countryside and architecture for a relatively large numbers of visitors with accommodated in 680 rooms, suites, and bungalows made available to them. The resort includes a conference centre, six

[1] Definition by the author, based on Collin Dictionary, "2015 State of the Industry". Developments, Wikipedia.

restaurants, seven bars, six outdoor and one indoor swimming pools, a waterpark, SPA, tennis courts, a mini soccer court, mini golf, and two theatres (Figs. 3 and 4).

Fig. 3. Creta Maris, Hersonissos, Greece, 2019. Photo: resort commercial materials.

Fig. 4. Electric wheelchair user friendly nonslip footpaths without steps or thresholds to enable electric car operation.

The necessity of adjusting the area to the needs of families with small kids, elderly persons and persons with disabilities occur in a planning resort's:

- master plan
- architecture and interiors of multiple buildings
- services provided and staff training.

These three elements are decisive when it comes to comfort, usability and accessibility of the resort for guests with mobility restrictions. Some possible shortcomings in architectural solutions aimed directly at persons with disabilities can be compensated with services provided, like easy and available transportation within resort boundaries or waiters' service on request in the restaurants.

5 Elements Implemented to Support Persons with Mobility and the Visual Perception of Architectural and Urban Spaces

The tables below present elements implemented to support persons with mobility problems shown together with the visual impact and perception of spaces resulting from these solutions. This juxtaposition shows that most of the technical and spatial solutions applied do not have a negative impact on visual quality of architectural space and have been designed in line with an overall aesthetics and atmosphere of place.

An important characteristic of the resort area is the hilly terrain, which constrains multi – levelled (terraced) location of spaces: external piazzas of various functions, pools, bars, hotel, bungalows and restaurants. The topography makes accessibility the major challenge and as a result a variety of external pathways, which transform into ramps and steps have been put in place (Tables 1 and 2).

Table 1. Resort plan – public areas

	Solutions for persons with disabilities; assessment (±)	Visual impact/perception of urban space with assessment (±)
1. Foopaths	+ Electric wheelchair friendly nonslip footpaths without steps or thresholds in place to enable electric car operation	+ Use of natural stone gives a vernacular appearance to pathways. Unregularly shaped stones create village-like aesthetics (pitched paths)
	− Absence of double handrail can in some cases decrease comfort of use	+ Inserting double handrail along all inclined paths in the resort would have a very strong visual impact
	− Intensively used surfaces get slippery along some fragments	
	+ Provision of sitting spaces on benches, niches or stone walls facilitates short breaks during walks	+ Benches, niches with sitting spaces give an impression of a friendly space encouraging to stay outside, teeming with people as real communities are
2. Piazzas and courtyards	+ Electric wheelchair user friendly nonslip surfaces without steps or thresholds in place to enable electric car operation	+ Use of natural stone gives a vernacular appearance to piazzas. Unregularly shaped stones create village-like aesthetics
	+ Provision of sitting spaces on benches, niches or stone walls facilitates short breaks during walks	+ Impression of friendly, inhabited piazzas designed in friendly, vernacular style
	+ Presence of greenery provides shadow, lowering temperatures around benches and other sitting areas. Improves air quality and humidification, especially during the hot summer season	+ Presence of well-kept greenery both along pathways, piazzas, courtyards and bungalow areas create very attractive landscapes
	− The topography can make it difficult to move around	+ Topography makes the resort's landscape varied and interesting with terraced hotel buildings and a random looking composition of bungalows
3. Beach	+ Beach path: portable, nonslip beach access matt with white stripes for people with visual impairments is one of the important elements of inclusive technology for coastal resorts for people with disabilities, walkers and elderly visitors or parents with strollers	− The use of the matts takes away the impression of a "wild" beach
		+ Location of matts can help define zones of the beach and indicate paths towards the beach entrances, improving orientation from the see
4. Waterpark	+ The inclusive waterpark unit with a shallow basin and manually operated water jets is easy to use by children and persons with mobility problems	+ Shallow basin has a positive impact on aesthetics of waterpark area
	− Water temperature in water jets cannot be adjusted	
	+ Waterpark area is accessible with ramp and stairs	+ Use of natural stone gives vernacular appearance
	+ Open external showers without thresholds are available	+ High quality of external shower design
	+ A fully accessible bathroom is available in the vicinity	
5. Bars	− Lack of lowered sections of counters at the main reception, and a number of beach bars, snack bars located in many locations in the Resort	+ The result of eliminating the lowered counter is a more compact form of the beach bar hut. A well designed hut with a lowered counter would have little impact on the architectural form

Table 2. Architecture and interior design

Hotel areas	Solutions for persons with disabilities with assessment (±)	Visual impact/perception of urban space with assessment (±)
1. Main entrance and lobby	+ The main entrance and hall are situated on one level, with an unobscured, vast space. The main entrance is protected from rain by a wide porch and recessed area	+ Vernacular formal solutions applied
2. Hotel communication	+ Accessibility ensured by stairs, lifts and ramps within hotel area	+ Universal design solutions do not have an impact on aesthetics and formal composition of spaces
3. Hotel bedrooms	+ The bedroom size makes it possible to arrange them as fully accessible	+ Universal design solutions do not have an impact on aesthetics and formal composition of spaces
	+ Height of the beds for an easy transfer to the bed directly from their powerchair	
4. Hotel entertainment areas	Accessibility ensured by stairs, lifts, ramps	+ Universal design solutions do not have an impact on aesthetics and formal composition of spaces
5. Restaurants	− 3 main restaurants with buffet service are problematic for guests with disabilities	
	+ 2 restaurants with table service serve persons with disabilities well	
	− 3 snack bars with counter service are problematic for guests with disabilities, as low-access counters are not in place	+ The result of eliminating the lowered counter is a more compact form of the bar. A well designed hut with a lowered counter would have little impact on the architectural form
6. Bungalow area	− Many bungalow entrances have thresholds or are accessible by stairs	
7. Retail area	+ Accessibility ensured by stairs and ramp	+ The solution does not have a negative impact on aesthetics and formal composition of space
8. SPA	+ Accessibility ensured by stairs and ramp	+ The solution does not have a negative impact on aesthetics and formal composition of space

6 Ramps and Passages

External ramps within the resort's boundaries are necessary for an electric vehicle service, for wheelchair users' accessibility and are beneficial for people pushing prams, pushchairs and baby carts. Gradients used within the resort vary, depending on natural topography. In heavily used areas they do not exceed 1:20. In all cases, clear flat unobstructed turning areas are present at the top and at bottom landings. In most cases in the vicinity of buildings the approach to a ramp is highlighted by the use of colour contrast, tone and texture change. The surfaces of ramps are non-slip, made mostly of natural stone or ceramic material to give a local, vernacular appearance. Lighting for paths and ramps comes from the sides to avoid shadow. Lamps have been positioned along the length of the pitched pathways and ramps and feature non-glare illumination. In the image below four selected ramp types are shown:

A., B. Ramped access provided at the amphitheatre, gently sloping with a gradient not exceeding 1:20, with firm surface suitable for wheelchair users.
C. Ramped access to the inclusive waterpark area
D. Ramp and continuous series of steps between two landings
F. Temporary beach ramp made of timber with a non-slippery surface (Figs. 5 and 6).

Fig. 5. Variety of external ramps at Creta Maris Beach Resort, Hersonissos, Greece, 2019 Photo A. Bonenberg.

Fig. 6. Variety of internal ramps at Creta Maris Beach Resort, Hersonissos, Greece, 2019 Photo A. Bonenberg.

Internal ramps and steps with ceramic or stone surfaces link the hotel' levels. Importantly, one can observe, that the use of ramps is very common for fully-abled people because of their comfort. The images above depict:

A. Ramp accompanying steps in the main hotel building providing access to guest rooms
B., C. Ramp accompanying steps in the main hotel building
D. Ramped access from the main lobby to the bar; steps available in the vicinity
E. Entertainment room equipped both with a ramp as well as steps.

Absence of continuous handrails on each side of the ramp is a common solution in the resort, which is understandable outside buildings, in long external paths. Even though people using manual wheelchairs or especially electric wheelchairs do not normally use handrails while using ramps, they sometimes may find handrails useful to steady themselves on a long ramp.

Ramp handrails prove very useful for persons with a visual impairment. A very good practice is the provision of seats along most of the routes and paths; especially those long and sloping. It provides an opportunity to take a rest during walks.

Still, circulation is kept clear of elements like lamps, bins, seats for the electric resort car service.

The downside of the resort's facilities for wheelchair users is lack of lower counter sections at the main reception, and at a number of beach bars and snack bars located in many locations at the resort. Lower section counters should be around 760 mm high, with 700 mm of vertical clear space underneath.

The resort's interior design is based on colour, tonal and textural contrast between floor surfaces, doorways. Shiny surfaces were avoided to minimize confusing reflections. It is in line with the vernacular style of the hotel, bungalow area and restaurants' architecture, where dark natural stone or natural ceramic materials are used for flooring and paving. Off-whites are used for plastered walls in the majority of interiors and hotel rooms. Floor surfaces are non-slip, door saddles and thresholds have been eliminated in the main hotel area.

Lighting has been installed at good levels and it minimizes shadows. The Visual Information system around the resort is clear, contrasted (white letters on a brown background) available in alternative formats (Braille in escalators).

Summing up, the case study has both strong and weaker points when it comes to comfort and accessibility. The popularity of the place among families with young children, persons with disabilities and elderly guests prove, that the spaces are adequately equipped for these groups of users.

7 Research Application

The author used the experiences within the scope of universal design in the "Międzyzdroje - Promenada Zachodnia" project made at Poznań University of Technology Faculty of Architecture's Architecture and Spatial Planning Department. The project included a hotel complex with shopping pavilions and was made as part of a competition for Brester Sp. Z o.o. Warszawa. The project's accessibility for disabled persons and the elderly is one of its primary attributes. Apart from standard facilities a special footways have been designed, facilitating wheelchair access to the sea to permit persons with disabilities to bathe. A number of new rehabilitation and therapy services will be available. The beach will also be accessible to those with impaired mobility through the use of special ramps and an inclined access way with ample space for necessary infrastructure such as bathing wheelchair hire and changing rooms with appropriate facilities. A (wheelchair friendly) path along a dune enhances the overall composition. It will span the entire coastal section of the investment (Fig. 7).

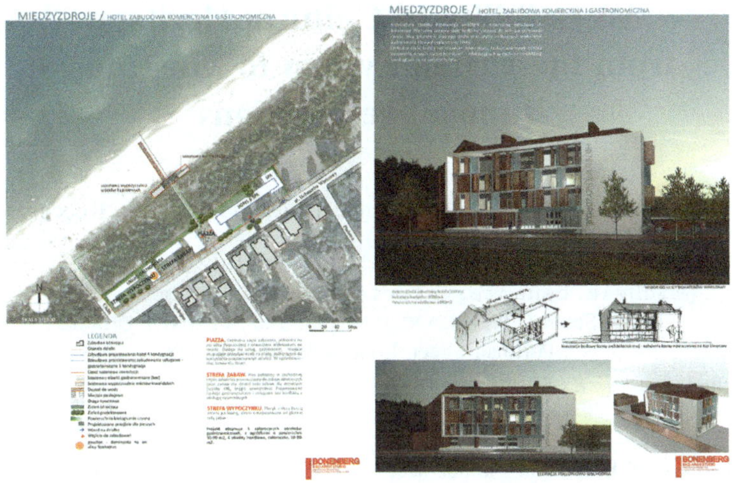

Fig. 7. "Międzyzdroje - Promenada Zachodnia" project made at Poznań University of Technology Faculty of Architecture's Architecture and Spatial Planning Department

References

1. Bonenberg, A., Zabłocki, M.: Residential architecture for health and longevity. Universal kitchen design and Architektura mieszkaniowa sprzyjająca zdrowiu i długowieczności. Projekt kuchni uniwersalnej [in] Space & FORM | Przestrzeń i FORMa 31_2017 edited by. https://doi.org/10.21005/pif.2017.31.b-02
2. Branowski, B., Zabłocki, M.: Kreacja i kontaminacja zasad projektowania i zasad konstrukcji w projektowaniu dla osób niepełnosprawnych, [w]: Ergonomia produktu. Ergonomiczne zasady projektowania produktów, edited by Jabłoński J., Wydawnictwo Politechniki Poznańskiej, Poznań (2006)
3. Mori, N.: Rough set approach to product design solution for the purposed "Kansei". The Science of Design Bulletin of the Japanese Society of Kansei Engineering (2002)
4. Pilch, A.: Poziom motywacji osób starszych a efekty rehabilitacji (Motivational predictors of successful rehabilitation in elderly patients). Fizjoterapia **19**(4), 17–25 (2011). ISSN 1230-8323. https://doi.org/10.2478/v10109-011-0029-z. Accessed 18 Nov 2017
5. Altman, N.: Healing springs: the ultimate guide to taking the waters: from hidden springs to the world's greatest spas. Inner Traditions/Bear & Company (2000)
6. Crebbin-Bailey, J., Harcup, J., Harrington, J.: The Spa Book: The Official Guide to Spa Therapy. Cengage Learning EMEA (2005)

Accessibility to Cultural Heritage from the Urban System to Museums. Innovative Solutions

Michela Benente[1(✉)], Valeria Minucciani[1], and Graziella Roccella[2]

[1] Politecnico di Torino, Viale Mattioli 39, 10125 Turin, Italy
{michela.benente, valeria.minucciani}@polito.it
[2] Planet Smart City, (PLANET Holding Ltd., PLANET Group), 2 Portman Street, London W1H 6DU, UK
g.roccella@planetsmartcity.com

Abstract. When talking about accessibility, it is essential to underline the different concepts and perceptions of disability: as the ideal man does not exist, we can say that we are all disabled. This paper focuses on the paradigmatic case of Cultural Heritage, whose accessibility is very complex because it includes all possible access difficulties. Yet there are still few cases in which it can be said that the Heritage is really accessible to all. In order to proceed to a true inclusive design, the first step is to know the status quo and recognize the elements that prevent a complete and wide accessibility. The results of the knowledge have to be mapped and communicated both to planners/designers and to citizens. In these phases, at least three very important functions are performed by the web: the availability of information, the interoperability, the dissemination of a "culture of inclusion".

Keywords: Cultural heritage accessibility · Inclusive welcoming · Open database

1 Introduction

Usually the term accessibility is associated with "the possibility of being visited by people with disabilities" [1].

It is therefore essential to stress the different concepts and perceptions of disability: since the ideal man does not exist, we can say that we are all disabled.

In particular, this paper focuses on the paradigmatic case of Cultural Heritage, whose accessibility is very complex because it includes all the possible difficulties of access.

The term accessibility refers to the physical component and specifically to people with disabilities and in particular mobility difficulties. It is necessary, instead, to shift the attention to a more inclusive approach, based on welcoming all possible users.

Therefore, a welcoming context, without architectural and cultural barriers that compromise the satisfaction of human needs, is the basis of the project. The intention to focus on a concept of accessibility not only compliant with the law but also functional, efficient, comfortable and above all qualitative, arises from this reflection.

G. Di Bucchianico et al. (Eds.): AHFE 2020, AISC 1202, pp. 182–188, 2020.
https://doi.org/10.1007/978-3-030-51194-4_24

In this sense, the principles of Universal Design, also recalled in the Convention on the Rights of Persons with Disabilities, adopted on 13 December 2006 by the ONU General Assembly, signed and ratified by the Italian State in 2009 [2] constitute a fundamental reference.

The Italian context, in general, presents few cases in which it can be said that Heritage is really accessible for all. The research field offers opportunities for reflection and comparison between different disciplines that should act in synergy to achieve effective project actions [3].

2 The Difficulty as an Opportunity

The first step of inclusive design is in knowing the status quo and recognizing the elements that preclude complete and wide accessibility. However, knowing and critically interpreting the surrounding situation is also a complex task: in fact, Cultural Heritage is always part of an urban (or suburban) and architectural system. Access to Cultural Heritage is a challenge for the planner/designer, whom might experiment several difficulties, however these constraints should be seen as an opportunity to improve projects. The sensibility of those who, at different levels (political, design, management) face with the project in the Cultural Heritage field affect the overall approach and, therefore, the effectiveness in terms of real reception of the users of the Heritage. The absence of sensitivity very often derives from the fact that "The biggest deficit that we have in our society and in the world right now is an empathy deficit. We are in great need of people being able to stand in somebody else's shoes and see the world through their eyes" [4]. The social role of those working in Cultural Heritage requires, more than in other fields, the ability to put themselves in the shoes of other people and to look at the world through their eyes, providing solutions as inclusive as possible. We need to switch our attention from a design practice that simply complies with regulations to a design attitude able to meet the needs of all the people. Accessibility to Cultural Heritage depends also on the welcoming that starts from the urban scale to the architectural one, looking at the elements as complex interrelated systems.

If the attention to accessibility that has featured many Italian cities [5] has partly triggered processes of awareness raising, it has also generated a specific attention to people with motorial, visual and hearing impairments effectively excluding other possible audiences. Furthermore, even in the most virtuous cases, the Plans didn't consider the urban system as a complex system, but as sets of points or portions of cities that are not connected to each other. This is more evident when looking at museums and the actions they have undertaken for accessibility, that didn't improve their relationship with the urban systems they belong to. The affiliation to different institutions and administrative organizations causes an absence of coordination both in the project actions and in the sharing of objectives related to heritage and cities.

3 From Law to Sensibility

Also for museum institutions, attention to specificity leads to the implementation of actions not aiming at inclusion, but strictly compliant with accessibility, exclusively as a response to specific standards and requests. Consequently, information about accessibility reflect this approach, simply presenting an on/off situation compared to the standard imposed by law. Looking at the multiplicity and heterogeneity of users, to their expectations and possible combinations of their needs, we need to act differently. The collection of information must be properly designed to meet the welcoming needs, to provide for changes and different combinations of needs that users can express.

Therefore, in order to be effective both in designing and using, information cannot be limited to the numerical data of compliance with standards (yes/no) but must describe the situations in detail. This enables a dynamic update of the database in case of new standards and above all, it allows wider queries responding also to those users whose needs are not protected by law or are not standardized.

Paying specific attention to restrict groups of users will ensure that they are not excluded, but also will lead to their isolation from other users. Dedicated entrances, activities and paths are a common practice derived by the respect of the rules rather than by a real sensibility. The inclusion, thanks to which the fruition takes place in a shared way, is achieved when, enlarging one's gaze, one conceives "environments, systems, products and services that can be used in an autonomous way by people with different needs and abilities, involving human diversity in the design process. This holistic and innovative approach constitutes a creative and ethical challenge to meet the needs of all users regardless of their physical, sensory and cognitive abilities" [6]. Furthermore, it is essential to consider the levels of behaviour as an important variable for the project. Among them: culture, society, motion, rest, info and knowledge [7].

Last but not least the project should consider the value of Cultural Heritage. The implemented actions are limited to allowing the physical access to artifacts or elements, rather than focusing on processes of value and identity sharing. This leads to the practice, still very common, of considering accessibility exclusively as the removal or overcoming of barriers [8]. The respect for the rules could conceal a "false" intention of inclusion, and often generates invasive interventions not pertinent to the historical context or, otherwise, nothing is achieved under the alibi of heritage conservation.

As mentioned above, working on the welcoming implies the need to know both the physical contexts in which one works and the potential users and their aspirations. The results of knowledge must therefore be mapped and communicated both to planners/designers and citizens. It should also be considered that there are at least three very important tools that the web can offer: the availability of information, the interoperability and the dissemination of a "culture of inclusion".

The current panorama shows the absence of a global vision that leads to the fragmentation of data collected by people and the creation of dedicated tools that are not put into one system. The complexity of the topic increases when quantitative data are added to qualitative data. If the first ones are objective, the second ones put us in front of the issues related to the subjective component, to the individual perception that are difficult to be quantified univocally. If the quality indicators have to measure satisfaction, gratification, well-being and the growth gained from experience: then they can motivate and/or direct project strategies, but on the other hand the complexity increases furthermore.

4 Shared Platforms and Accessibility

The planning of a cultural visit is normally preceded by the definition of itineraries, hours and things to do.

The consultation of websites is the usual way followed by direct contact, email and telephone, to verify first of all how the site can be reached and then the services, activities and timetables by transmitting the information with respect to needs and aspirations of individuals or groups. Although with different profiles, tourists or citizens are increasingly using digital applications to obtain useful information to meet their daily or travel needs.

Digital tools are increasingly important and also constitute, for sites that collect the impressions and the degree of satisfaction of users, a feedback of the services offered, both for those who provide the service and for those who want to use it and build their opinion and choices on the feedback of previous users. This is even more important for people with specific needs, not necessarily disabilities, who need such information. The latter very often use and evaluate the goods and services offered, sharing their experience with the community and offering a valuable information service that dedicated websites and apps don't provide. The information offered presents an attentive view of the specific and cross-cultural needs between cities and museums/cultural goods.

It is therefore obvious how important is to change the approach in creating shared platforms and implementing data collections, where the method requires the respect of a principle of univocally such that homogeneous and comparable data can be obtained. If properly designed, the database allows the creation of continuity between architecture and city in order to improve the accessibility and welcoming to cultural heritage.

4.1 Crowd Mapping for Cultural Heritage

The most important features of crowd-mapping, as a specific crowdsourcing initiative, are the precision of details, the collection of data in "almost real-time" and of course: accessibility [9]. These kind of instruments, although characterized by different levels of precision due to the non-professionalism of the people who populate them, are precious to address accessibility related to cultural heritage. Usually, crowd-mapping based platforms are thought to be used for coordinating helps in case of disaster, as it happened with Ushaidi in 2010, after Haiti earthquake. Nowadays there are other different uses for this kind of platforms [10] that aim at aggregating large amounts of user-generated inputs on a specific issue and thus creating a close to real-time intelligence. Regarding accessibility, a clear strategy is needed when beginning collecting data. The maps can be created either by filling in data on a specific platform or data can be entered from text messages and social media feeds combined with open geographic data. This different data sources, direct and indirect, can create confusion. The intent of data collection therefore should be very clear and based on the people needs. Digital Platforms are usually accessible via websites or mobile apps.

The accessibility and affordability of the information that people can get is crucial to provide them with the transportation means and the services they need to get to a cultural site, to provide support for their families and eventually to let them feel part of a community. When cultural events are in place, such as a music concert or a congress or also an exhibition, it often happens that spontaneous 'temporary' communities arise by making use of social media and apps. On the other way, local 'permanent' communities can also arise spontaneously but it is more usual that they are created by a top-down approach by one specific heritage site. Whatever is the approach, temporary or permanent, and wherever they are based, locally or virtually, these communities have the possibility to act in order to change the spaces and make the Cultural Heritage more accessible.

Some strategic actions that can be put into practice by them are for example the integration of innovative solutions, such as accessible digital platforms or the request for design methods based on Design for All principles [11]. In the end, all of these actions are based on the same mission: to improve the quality of living of people. Some tangible solutions that can be integrated in the field of cultural heritage are e.g. designing access roads to a cultural site, a library, a museum in a way that all the people can reach those site in an acceptable amount of time; and the inter-relation with something that can be done during the trip, the paths themselves could also be made "fun" by making use of 'Fun Theory' projects. These principles allow people to increase their education and skills, and also reduce waste of time for the things they need most. The result is that all people can achieve behaving in a better way and are also happier whatever their abilities are.

5 From Accessibility to Welcoming

To solve the problem of how to access cultural heritage it is necessary to think differently and shift the focus from accessibility to welcoming. Ultimately engaging people to collaborate to open platforms is a little step, we recognize, but it is one of the best we have, to begin creating places that are open, smart, sustainable, socially inclusive. This is a way to empower people to make a difference in their community.

If, as mentioned above, the principles of Universal Design should become the reference for the design of spaces and structures intended for use, the data collected must be able to allow outputs that guide designers in making choices in a global vision that respects users' needs and expectations. The basic principle is to think, plan and implement accessible, safe and comfortable environments and solutions. Therefore, accessibility or rather welcoming should become part of the common living culture. A concept integrated into all aspects of everyday life that improves everyone's life, simplifies interactions and promotes performance and possibilities to communicate, move and interact with others and the environment. The sharing and exchange of knowledge outlines a cultural model based on the principle of equal dignity of people and equal opportunities for all [12]. Only with an intellectual approach based on these principles we can think of developing sustainable territories, resources and services, also from a social point of view. It is therefore essential to act both by raising awareness of designers and institutions and by intervening in the field of training and research. Creating awareness and preparation on these sheets in the younger generations of designers is unfortunately not yet considered a specific field of training and research to be integrated with consolidated knowledge [13] (Fig. 1).

Fig. 1. "A mile in my shoes" Empathy Museum (photo: http://www.empathymuseum.com).

References

1. Accessibilità. https://www.garzantilinguistica.it/ricerca/?q=accessibilità
2. Law n. 18/2009. https://www.gazzettaufficiale.it/gunewsletter/dettaglio.jsp?service=1&datagu=2009-03-14&task=dettaglio&numgu=61&redaz=009G0027&tmstp=1237200828355
3. Linee guida per l'accessibilità dei paesaggi vitivinicoli del Piemonte. https://www.paesaggivitivinicoli.it/wp-content/uploads/2019/06/Linee-Guida-2018-1.pdf
4. Empaty museum. http://www.empathymuseum.com
5. Plans for the Elimination of Architectural Barriers (P.E.B.A.), Laws n. 41/1986 e n. 104/1992. http://netwees.altervista.org/il-peba/il-peba-riferimenti-normativi
6. The 7 principles. http://universaldesign.ie/What-is-Universal-Design/The-7-Principles/
7. DFA training and consultancy. http://www.dfaitalia.it/servizi/

8. Linee Guida per il superamento delle barriere architettoniche nei luoghi di interesse culturale. https://www.beniculturali.it/mibac/multimedia/MiBAC/documents/1311244354128_plugin-LINEE_GUIDA_PER_IL_SUPERAMENTO_DELLE_BARRIERE_ARCHITETTONICHE.pdf
9. da Silva Lima, N., Caldas Leite, J.P., Cardoso de Paiva, A., Marcia Oliveira Maia, I., Corrêa Silva, A., Braz Junior, G., de Souza Baptista, C.: Mobile Application for Crowdmapping Accessibility Places and Generation of Accessible Routes, Orlando (2018)
10. What Are Digital Platforms? http://www2.itif.org/2018-tech-explainer-digital-platforms.pdf?_ga=2.226863903.368629717.1579699608-1926710858.1579699608
11. The EIDD Stockholm Declaration 2004. http://dfaeurope.eu/what-is-dfa/dfa-documents/the-eidd-stockholm-declaration-2004/10/
12. The sustainable development goals. https://sustainabledevelopment.un.org/content/documents/21252030%20Agenda%20for%20Sustainable%20Development%20web.pdf
13. Benente, M., Azzolino, M.C., Lacirignola, A.: Accessibilità e fruibilità nei luoghi di interesse culturale, Roma (2016)

Inclusive Museums: From Physical Accessibility to Cultural Appropriation

Michela Benente and Valeria Minucciani(✉)

Politecnico di Torino, Viale Mattioli 39, 10125 Turin, Italy
{michela.benente, valeria.minucciani}@polito.it

Abstract. Cultural Heritage belongs to all people, and its preservation is dedicated to the cultural growth and human promotion, then it must to be available to all citizens. Today, sensitivity has increased on this issue, and over time different levels of accessibility to Cultural Heritage have been recognized: social, financial, physical, perceptual and cognitive, cultural accessibility – but the final stage is related to "appropriation", when people get the Cultural Heritage not only from the intellectual point of view but also from the emotional one, really benefiting from it. In museums, as confined places where the communication of heritage is the main goal, it is clear that physical accessibility is only a prerequisite for being able to really appropriate the contents. Several guidelines have been developed, but the sensitivity of curators and designers (from the museological and museographic point of view) is crucial. This paper aims to discuss these issues, with particular reference to a wider concept of accessibility.

Keywords: Museum experience · Cultural inclusion · Personal meaning · Mutual understanding · Cultural exchange

1 Introduction

The debate on accessibility in general has gradually gained attention in recent decades, although undoubtedly its consciousness rose after the Second World War, as shown by the Declaration of Disabled People's Rights adopted in 1948 by United Nations General Assembly [1]. Over time, several researches developed the concept and involved directly industrial and architectural design: in 1985 the American architect Ronald Mace coined the term *Universal Design* (further developed in 1997 by a working group of architects, designers and researchers) also called, in Europe, Design for all. This approach, focusing on the peculiarities and needs of human beings, that may change in the course of the life, aims to find inclusive solutions more than dedicated solutions to disabled people.

More recently, with the International Classification of Functioning, Disability and Health 11 (ICF), developed in 2001 by the World Health Organization, the concept of disability has been extended from the medical model to the bio-psycho-social one, shifting the attention from the *person* to the *environment*, which barriers and handicap are created by. Nevertheless, a sort of misunderstanding persists, according to which

G. Di Bucchianico et al. (Eds.): AHFE 2020, AISC 1202, pp. 189–195, 2020.
https://doi.org/10.1007/978-3-030-51194-4_25

accessibility solutions are devoted to people with disabilities, while they actually increase accessibility (i.e. quality, safety, comfort) for all.

In general, the term accessibility means the ability to access everywhere, as well as the ability of everyone to use/enjoy a service or a product or an experience regardless of any kind of personal condition: it implies that user can be comfortably autonomous and safe [2].

2 Accessibility and Cultural Heritage

The concept is enriched with much more complex meanings when the service or product to which it refers is the enjoyment of Cultural Heritage: it belongs to all people, and its preservation is dedicated to the cultural growth and human promotion of all. It is therefore not acceptable for someone to be excluded from it, for various reasons.

As the Convention on the Value of Cultural Heritage for Society (Faro Convention, 2005) emphasizes the role of heritage with regard to human rights and democracy, promoting a wider understanding of communities and society and representing meanings and values, it is clear that ensuring accessibility to cultural heritage cannot be limited to provide its "comfortably autonomous and safe use". Today, access to Heritage is an issue on which sensitivity has increased, and several scholars focused on standards and guidelines to improve it [3], but it should be remembered that, unlike urban spaces, accessibility to Cultural Heritage is not limited to the possibility of physically accessing it, since various conditions must be met in order to really benefit from it. This is clearly stated by the ICOMOS Ename Charter (ICOMOS 2008) that highlights not only the physical but also the intellectual access to Cultural Heritage sites. Then a wider accessibility concept has been in several occasions proposed [4], putting a greater emphasis on the accessibility to *knowledge* and *experience*.

And yet, albeit with interesting exceptions, accessibility to Cultural Heritage has remained unbalanced towards the physical aspects. However, it is not enough to consider also difficulties in sensory perceptions, as well as cognitive or psychological impairments - although it is a remarkable advancement. It is important to change the perspective, reflecting in terms of *normal people* because it is precisely *normal* for everyone to have limitations and difficulties - just as it is normal for everyone to feel comfort or discomfort in certain situations or places.

3 Accessibility in Museums

The reference to museums, as places expressly designed to make Cultural Heritage accessible in broad sense, is particularly significant because they are confined places where the overall arrangement is expressly designed to communicate contents and to make the encounter with Cultural Heritage easy and fruitful.

The term *encounter* is crucial: in fact, in every encounter a mutual understanding is needed, as well as an exchange. If it is true that the museum visitor should adopt an attentive and concentrated attitude, it is equally true that the museum, with all the resources at its disposal, should make its contents comprehensible, interesting and attractive. It is therefore clear that a first stage of this process is reaching physical, sensory and perceptive accessibility (without which there can be no encounter), but the museum's mission cannot end there.

Conversely, it has constituted till now the only concern, or almost the only one, about museum accessibility.

With reference to Italian case, several measures have been taken, as well as recommendations in support of accessibility to Cultural Heritage. In summary, Italian legislation on Accessibility is considered very complete among western countries: it originates in 1971 and in 1989 some rules *result oriented* have been introduced, based on a more modern performance approach. Referring to museums. the Ministerial Decree of May, 10th 2001 "Atto di indirizzo sui criteri tecnico-scientifici e sugli standard di funzionamento e sviluppo dei musei", regulates a number of requirements and in particular, in the Chap. 7 dedicated to "Relations of the Museum with the Public and its Services", clearly specifies the activities that must be ensured to all visitors, not limited to the mere, physical accessibility of the building and exhibits. In particular, this Decree underlines the removal of any barrier or obstacle of any nature that could prevent or limit the access and the use at all levels.

The Ministerial Directive November 2007 sets out some general principles and criteria to promote accessibility and enjoyment of the Cultural Heritage, underlining the following ones: knowledge of the needs of different public with reference to the specific heritage we are concerned with, and conditions of its accessibility; training of staff; welcoming to different public; availability of exhaustive information; adoption of systems that can compensate for accessibility shortcomings not surmountable by direct actions.

Over the last few years, overcoming architectural, cognitive and sensory barriers has actually been one of the most important interventions of Ministry (Ministero dei Beni e delle Attività culturali e del Turismo): the 2008 Ministerial Decree "Linee guida per il superamento delle Barriere Architettoniche nei luoghi di interesse culturale" ("Guidelines for overcoming Architectural Barriers in Places of Cultural Interest") identifies several design criteria to facilitate the accessibility to Cultural Heritage: Orientation, Overcoming of distances, Overcoming of differences in level, Use of space and equipment. The document devotes specific attention to the exhibition spaces, that must perform a real mediation between heritage and visitor, and should facilitate the reading and understanding. The document offers many suggestions, mainly focusing on the design of physical elements and spaces, but also considering the effectiveness of information and communication in general and the accessibility to contents and referring to the important issues of monitoring and maintenance of interventions.

Finally, in 2018, the "Guidelines for Plans for Elimination of Architectural Barriers (P.E.B.A) in Museums, Archaeological Sites and Parks" were published. These plans, which will have to be adopted, include a very detailed set of requirements aimed at making the museum comfortable, starting from how it can be reached. Several recommendations concern the availability of information, and great attention is paid to the furnishings (heights, colors, etc.) as well as to services aimed at different kinds of needs. The plan also envisages special training for the staff, who must be prepared to meet the different difficulties of visitors.

A substantial part of the plan is dedicated to the museum experience, where perceptive and cognitive needs are highlighted, and museum fatigue is also considered. With regard to texts and captions (which must, of course, be accessible also to people with sight or hearing difficulties) their clarity, legibility, comprehensibility and hierarchisation have to be carefully planned, according to a consistent communication strategy. Recommendations on safety, maintenance and monitoring complete the document.

It must be acknowledged that these guidelines are very advanced and comprehensive, yet a strong imbalance towards physical accessibility requirements (here also intended as sensory and perceptual ones) is still evident and the recommendations regarding quality parameters are as much shareable as generic. This, far from appearing as a criticism, is rather the demonstration that the specific sensitivity for the accessibility to Cultural Heritage cannot be enclosed in any normative cage, however advanced it may be.

As a proof of Italy's sensitivity to these issues, ICOM Italy established on 2007 a thematic commission for museum accessibility and within the 25th ICOM General Conference (Kyoto, 2019) organized "Talking about accessibility, inclusion, usability: a discussion for a new International Committee" in order to open a discussion on the opportunity to establish an International Committee for Accessibility.

4 Museums as Inclusive Places

Addressing the accessibility to Cultural Heritage in an innovative way requires a change of perspective. The principle that heritage is at the service of the people, and not vice versa, has to be embraced. In museums, this means the centrality of the visitor and not of the object.

Focusing on humans, with their cultural understandings and abilities, should provide for a flexible approach by museum professionals, avoiding ready-made and generalized solutions: the museum is the *mise-en-scène of the encounter between the heritage and its heir.* However, for this to happen, a number of different and multi-faceted conditions must be met, which show how complex the possibility of really accessing to cultural Heritage is. Today, at international level, there is a growing concern on the "social, cultural and intellectual accessibility" for all [5], and, over time, different levels of accessibility to Cultural Heritage have been recognized:

1) *Social accessibility.* It refers to social role or condition of people. For instance, people working hardly could renounce to visit CH due to lack of time, but also detained people, or sick persons, or elderly, cannot;
2) *Financial accessibility.* Visiting cultural sites implies economical means, as well as travelling to reach them: only a limited class of people can afford it:
3) *Physical accessibility.* It relates to the practical access, offering to visitors a sensorial *experience.* Of course this aspect concerns particularly the physically disabled people, but not only;
4) *Perceptual* and *Cognitive accessibility.* "The phase of understanding culture and heritage … has to do with the perceptiveness of the receiver…." [6]. The main

obstacle is related to people with serious difficulties in cognitive process (see dyslexic persons), but other cases could occur.

5) *Cultural accessibility*. "The levels of mental accessibility and perceptiveness are closely linked to the receiver's educational background, way of living and habitual mode of intellectual operation" [6]. Furthermore, in case of different cultural contexts, visitors should have the necessary information and basis to understand them. This is the crucial phase, that's the accessibility to contents.
6) *Appropriational accessibility*. If previous phases are satisfied, the last one can occur: appropriation. This is the real and complete access to CH, that become a personal asset and wealth. People will never forget it again, but on the contrary they'll contribute to its protection, conservation and divulgation. This phases implies both an intellectual appropriation and an emotional one. "This final moment represents the apex of accessibility, the ultimate goal to be aimed for by policy makers in the field" [6].

As we can see, Physical Accessibility is only one of several phases, nor the first one, although it is still considered as the main concern. Actually, it is a prerequisite for being able to really appropriate the contents.

Figure 1 also shows that the experience inside the museum is constantly conditioned by the particular atmosphere and emotional qualities of the exhibition space (since the museum is a physical place). This component can strongly affect reactions and personal involvement of visitors, and shouldn't be underestimated. Space is always emotional (in the sense that it induces reactions of an emotional nature), then museums cannot be neutral [7]. Without a complete access, that's without the appropriation, Cultural Heritage cannot really be inherited, preserved and passed down.

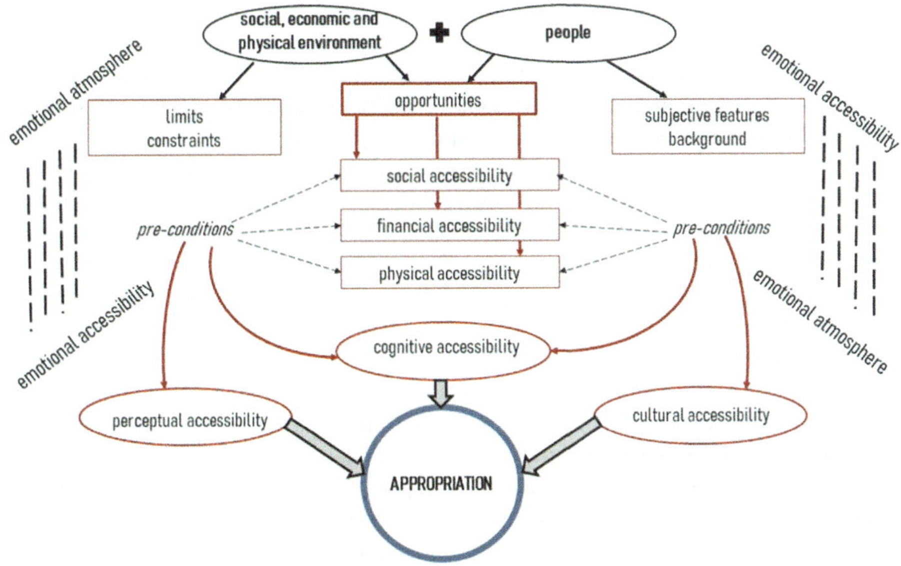

Fig. 1. The accessibility to Cultural Heritage process.

From this point of view, it becomes manifest that, in order to really comply with all the accessibility phases, a regulatory system, however advanced and complete it may be, cannot be effective: it can be used to comply with certain phases of accessibility, but the last ones that have been listed above depend strongly on the sensitivity of the designer (from the museological and museographical point of view). For example, the cultural roots of people can lead to different interpretative processes: not only the exhibited objects, but even the display solutions could mean different things to different people [8].

So when does the appropriation process take place? Of course, it is a highly personal and not standardisable experience. However, even if it has not yet been codified in its complexity, some points for reflection are presented here, in relation to meaningfulness in visiting museums.

Appropriation occurs when an aspect of the exhibition (an object, a display) comes into direct and deep contact with the most intimate and personal spheres of the visitor. By way of non-exhaustive example:

1) when it creates a link with experiences from his past
2) when it creates a link with a person who is important to him
3) when it offers the answer to a question which he has always considered important
4) when it connects to one of his dreams or hopes
5) when, on the other hand, it is linked to one of his fears
6) when it generates surprise and wonder in him.

Surely many other cases could occur. But these show how the visit is first of all a *transformative experience* for the visitor, who leaves the museum holding something (because he has appropriated it): it could be a notion, an emotion or a competence that will remain in him and which he will be able to spend on a next occasion, to continue his process of growth and personal development (which is very different from one visitor to another).

It is therefore clear how the textual communication apparatus of museums, but also their non-verbal communication, seem to have seldom taken these aspects into account.

A new museology is trying to change this perspective.

A museum can exclude much more a person without a cultural background than a physically disabled person; it can be much more incomprehensible to children than to the blinds. It can be inaccessible to all, finally, if it does not try to get in touch with everyone, generating the virtuous process of appropriation. Then, as mentioned above, at the core of the museum, paradoxical as it may seem, *there are people and not objects*. At the same time, curator vigilance is required: his mission becomes even more complex because, not intrusively, he has nevertheless to avoid that the exhibition tells scientifically wrong or historically distorted stories. It is not true that "the object speaks for itself", on the contrary *objects are silent*, and indeed when they seem to be talking they can lie [9]: the role of the curator therefore remains even more fundamental and irreplaceable. Furthermore, he should be supported by a sensitive, passionate and motivated staff: this is a challenge in many museums but at the same time it is a resource that too many times has been under-used avoiding experimentation and involvement of employees.

It should be further mentioned that the museum visit is a transformative experience not only for the visitor but also for the museum: the visitors' feedback - which today we have many ways to receive, to check and to evaluate in real time – not only acts as mouthpiece and is a formidable tool for museum experience improvement, but also it perfectly represents the mutual understanding and the cultural exchange mentioned above.

A new research season must be launched, where the compliance with quantitative or binary (on/off) parameters has to be overcome and replaced by the definition and evaluation of qualitative requirements, to improve more and more the educational value of museums and their impact on the personal growth of citizens.

In this new season, accessibility to Cultural Heritage is a wide topic, much more focused on cultural, social and emotional aspects than on physical and perceptive ones: because after having accompanied the visitor in front of the artwork it is evident that *physical access is not enough*; after he has learned and understood the content of the artwork, it is clear that *cognitive and cultural access is not enough.* He must leave the museum richer than when he entered, because the artwork will have become his own.

References

1. Evcil, A.N.: Raising awareness about accessibility. Procedia Soc. Behav. Sci. **47**, 490–494 (2012)
2. Tzanavara, A.: Museum and people with disabilities: accessibility - education - social inclusion. Ph.D. thesis, University of Aegean, Corinth (2013). http://thesis.ekt.gr/thesisBookReader/
3. Donika, G.: Rethinking accessibility to cultural heritage: sensing archaeology. In: Modern Conservation. National Committee at ICOMOS, Srbija, Belgrade (2016)
4. Sørmoen, O.: Accessibility to Cultural Heritage: Nordic Perspectives. Nordic Council of Ministers. TemaNord, Copenhagen (2009)
5. Pouhere Taonga Act.: Providing for Physical Access to Heritage Places. New Zealand Historic Places Trust, Pouhere Taonga (2011)
6. Bogiantzidis, N., Deffner, A., et al.: Accessibility to culture and heritage: designing for all. In: Conference AESOP 2015. Czech Technical University, Faculty of Architecture (2015)
7. Zumthor, P.: Atmospheres: Architectural Environments, Surrounding Objects. Birkhäuser, Basilea (2006)
8. Roberts, D.A.: Collections Management for Museums. Museum Documentation Association, Cambridge (1988)
9. Crew, S.R., Sims, J.E.: Locating authenticity: fragments of a dialogue. In: Karo, I., Lavine, S. D. (eds.) Exhibiting Cultures. The Poetics and Politics of Museum Display. Smithsonian Institution Press, Washington (1991)

Bathroom Handrail Requirements for Use by the Elderly

Yoshiaki Goto[1], Satoshi Kose[2(✉)], Shinji Tanaka[3], and Hajime Yada[4]

[1] Okayama University of Science,
1-1, Ridaicho, Kita-ku, Okayama 700-0005, Japan
goto@archi.ous.ac.jp
[2] Building Research Institute, 1, Tachihara, Tsukuba, Ibaraki 305-0802, Japan
skose@gakushikai.jp
[3] Comprehensive Housing R&D Institute, Sekisui House, Ltd.,
6-6-4, Kabutodai, Kizugawa, Kyoto 619-0224, Japan
stanaka@gz.sekisuihouse.co.jp
[4] Almetax Manufacturing Co., Ltd.,
1-1-30, Oyodonaka, Kita-ku, Osaka 531-0076, Japan
yadarin@almetax.co.jp

Abstract. Getting in and out of a bathtub during bathing can be difficult as people grow older, but Japanese people prefer to go into the tub. Experiments were conducted on the effectiveness of handrail positioning to assist their bathing activities. The height of a tub edge (measured from the outside) was changed along with different positioning of a handrail, both vertical and horizontal. General finding was that subjects preferred the floor height between 350 mm and 450 mm for the tub edge. The preferred handrail height they grabbed was between 760 mm and 1060 mm from the floor, with acceptable handrail position just above the edge or 100 mm inside the tub. Observation of subjects' behavior and past experiences suggest that seating area such as bath board be provided for frail seniors to sit down to get in and out of the tub rather than forcing them to step across.

Keywords: Bathroom · Bathtub · Handrail · Position · Seniors

1 Introduction

In Japan, bathroom used to be smaller, with a small and deep bathtub. In that context, getting in and out of a bathtub during bathing can be difficult as people grow older. Traditional bathroom design in Japan is that there is a wet area outside the tub where people wash bodies – which can be used as roll-in area if only shower is to be used – but Japanese people prefer to get into the tub. In this context, proper handrail installation to assist them getting in and out of the tub is vital for the comfort and safety of the users. Acceptable handrail position for toilets was already examined and reported elsewhere [4], but bathroom handrails require different functions than toilet handrails, and they needed further investigation.

G. Di Bucchianico et al. (Eds.): AHFE 2020, AISC 1202, pp. 196–203, 2020.
https://doi.org/10.1007/978-3-030-51194-4_26

Experiments were conducted on the effectiveness of handrail positioning to assist bathing activities of senior subjects. The height of the tub edge from the outside floor was changed along with different positioning of a handrail, both vertical and horizontal, and preferred height of the tub edge of was investigated as well as preferred positioning of the handrails.

2 Experiment with Dry Bathtub

2.1 Experimental Setup A

Depth of the inside of the tub was fixed to be 550 mm, and the outside was set at seven different dimensions between 100 mm and 550 mm (100, 250, 300, 350, 400, 450, 550, lower from the tub edge, respectively). The subjects stepped in and out of the tub as simulated bathing. Their behavior was video recorded from four directions, i.e., front, back, side and from above, to analyze how they moved and used the handrail. They were also asked what they felt about the ease of stepping in and out, with 5-point scale rating (from −2 to +2).

Twenty-six senior subjects participated in the experiment without warm water in the tub. Characteristics of the subjects are given in Table 1.

Table 1. Characteristics of experimental subjects with dry bathtub.

	Number of subjects	Age range (average)	Stature range (average) in mm
Male	10	65–84 (74.8)	149.7–171.5 (158.8)
Female	16	57–79 (70.7)	141.5–156.5 (148.7)
Total	26	57–84 (72.3)	141.5–171.5 (152.6)

For the experiment, fiber-reinforced plastic bathtub (with dimensions 715 W 1100 L × 550 H) was cut open for ease of observation form the side (see Fig. 1). The height of the outside floor was changed with adjustment of wooden deck. Nine vertical handrails with 34 mm diameter and length 1500 mm were mounted on the side wall. Nine horizontal handrails were also used as a separate experiment B (it was impossible to conduct both experiments simultaneously).

Fig. 1. Experimental subject getting out of cutout bathtub.

2.2 Experimental Results A

Preferred height of tub edge was around 350 mm and 450 mm, depending on the direction as shown in Fig. 2. Observation of the subject's behavior revealed that majority used both hands if the outside height is 100 mm, i.e., height difference is 450 mm between inside and outside, but with moderate height differences, 18 or more subjects used only one hand. Some subjects changed which handrail to grasp depending on the height difference, but some only changed the grabbing height (Fig. 3a and 3b).

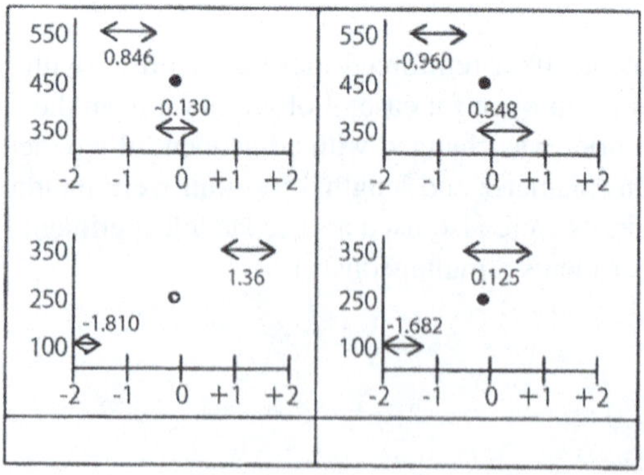

Fig. 2. Subjective evaluation on the ease of getting in and out of the tub. Horizontal axis shows preference of 5-point scale rating (from very poor to very comfortable). Vertical axis is the height of the tub edge from the outside floor.

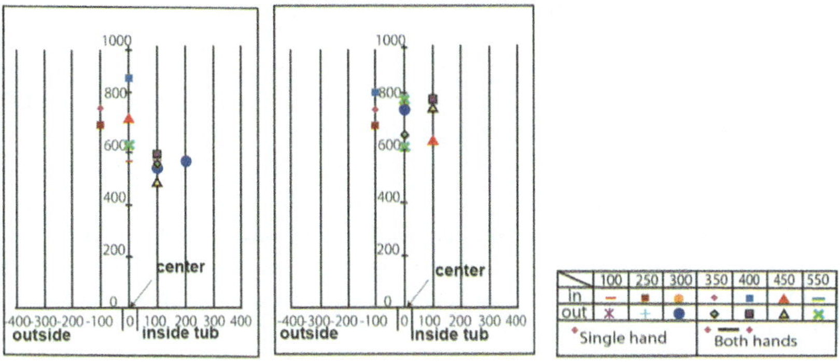

Fig. 3a. Change of grasp point by subjects using single hand, getting in and out. Horizontal axis shows the distance measured from the center of the tub (left side shows outside of tub). Vertical axis is the height measured from the top of the tub edge.

2.3 Discussion A

Most preferred outside tub edge height is 450 mm. However, acceptable lower height was different depending on the direction: 350 mm is accepted when getting in, but 250 mm was also accepted when getting out. It might be because 250 mm will allow subjects to secure their posture more quickly outside. Firm handrail grasp may also be contributing. Since the subjects' posture became less stable when getting in, it will be better to limit the outside height of the tub edge between 350 mm and 450 mm.

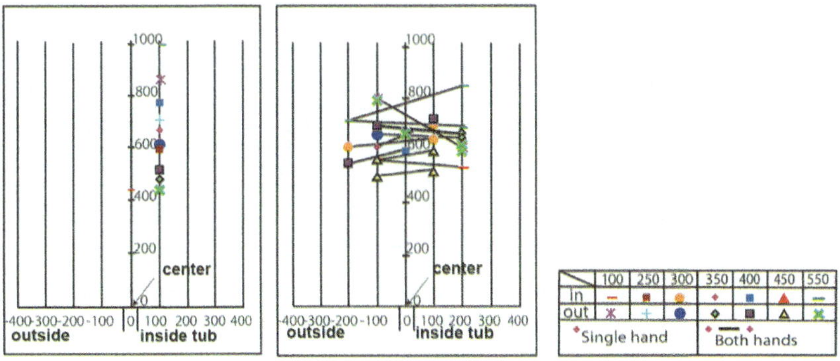

Fig. 3b. Left figure shows change of grasp point height by subjects using single hand, getting in and out (without changing the rail to get in and out). Right figure shows subjects using both hands during getting in and out.

2.4 Experiments B with Vertical and Horizontal Rails

With the previous results in mind, a further set of experiments was conducted with the condition of outside floor 400 mm lower from the tub edge, for both vertical and horizontal handrails.

2.5 Experimental Results B

The grasped points observed during the experiments B are shown in Figs. 4 and 5. Figures 4 are results for vertical handrails, getting in and out respectively, and Figs. 5 are for horizontal handrails, getting in and out.

When subjects used vertical handrail with only one hand, their choice tended to be the vertical handrail just above the edge (center one) or next ones (inside or outside). Grabbed height was around 1000 mm and 1200 mm from the outside floor.

With horizontal handrails, most subjects chose 800 mm or 900 mm in height from the outside floor as they got in. One subject who used both hands chose the highest one, i.e., 1300 mm.

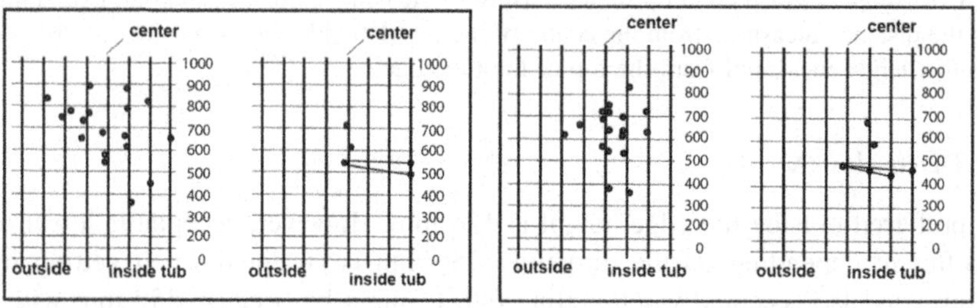

Fig. 4. Vertical handrail grasp point during getting in (Left Box) and getting out (Right box). Left side in each box is for one hand use, and the right side for both hands.

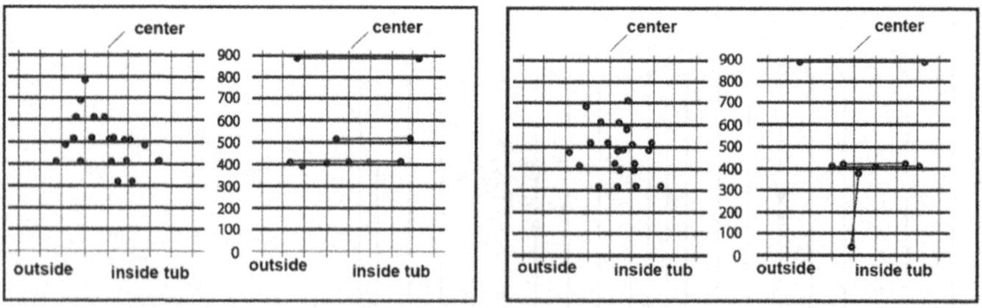

Fig. 5. Horizontal handrail grasp point during getting in (Left Box) and getting out (Right box). Left side in each box is for one hand use, and the right side for both hands.

2.6 Discussion B

Most grabbed handrails during getting in were just above the tub edge (center one) or the next ones. Most grabbed height for the center one was 600 mm above the edge, while other heights ranged between 500 mm and 700 mm. During getting out, preferred handrail was the same as getting in, preferred height being between 500 mm and 700 mm. Judgement by the subjects revealed that vertical handrails 200 mm apart from the center are unusable on both directions.

There were variations in the choices among subjects. Some of the reasons may be that they do not use handrail during their daily bathing activities at home.

3 Experiment with Wet Bathtub

3.1 Experimental Setup

In order to narrow down the range of preferred position of handrail installation, experiments were conducted with bathtub filled with warm water (Fig. 6). Seventeen subjects participated in the experiment as shown in Table 2. Bathtub of the size 600 W × 1050 L × 530 D was used, with outside tub edge height 360 mm. Five vertical handrails were installed on the wall (handrail 3 is above the tub edge center, handrail 1 is 200 mm inside, handrail 2 is 100 mm inside, handrail 4 is 100 mm outside, and handrail 5 is 200 mm outside). Video recording was conducted to check the grabbed position of the handrail.

Table 2. Characteristics of experimental subjects with wet bathtub.

	Number of subjects	Age range (average)	Stature range (average) in mm
Male	9	65–86 (71.8)	150.5–171.7 (160.5)
Female	8	64–74 (69.0)	144.0–157.0 (150.6)
Total	17	64–86 (70.5)	144.0–171.7 (155.9)

Fig. 6. Experimental subject using the handrail with warm water in the tub, facing the wall.

3.2 Experimental Results

Since previous experiments revealed overall allowable positions, subjective evaluation was mainly examined. Figures 7 and 8 show the result of preference between handrail 3 or handrail 2 versus others. Handrail 2 (100 mm inside the tub) was significantly better than 1 and 4. Choice of handrail 3 was also significant compared to 1 and 5.

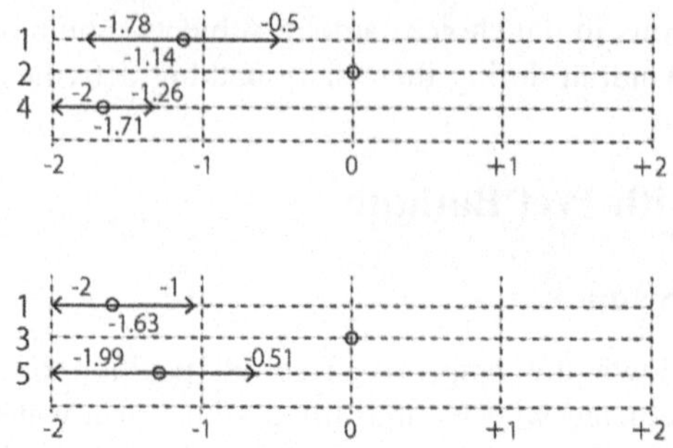

Fig. 7. Subjective evaluation on the ease of getting in the tub. Horizontal axis shows preference of 5-point rating, optimal handrail rating set to 0.

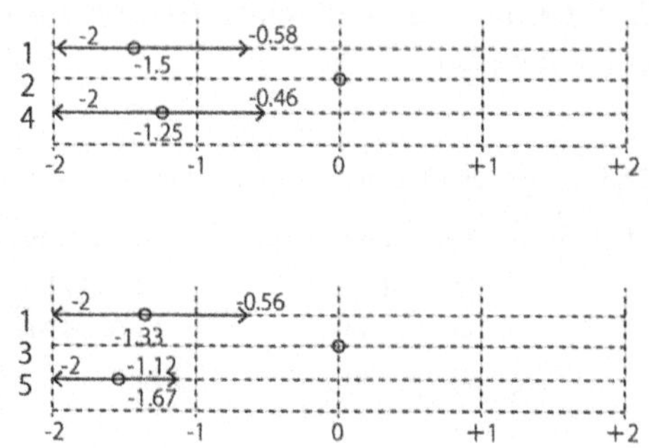

Fig. 8. Subjective evaluation on the ease of getting out of the tub. Horizontal axis shows preference of 5-point rating, optimal handrail rating set to 0.

3.3 Discussion

It is generally understood that handrail be placed to support one's final posture after completion of the action, rather than its start. When stepping over the bathtub, it means that the handrail be placed further on the other side than the near side. During the experiment, however, many subjects grabbed the handrail nearer to one's self, as shown in Fig. 7 and Fig. 8. It might be because senior subjects stood facing the wall and moved parallel to the wall, whose behavior is different from youngsters who tended to walk without facing the wall. The seniors might feel that posture more secure than otherwise. As to the grabbed height, most subjects chose same or similar height, between 500 mm and 700 mm above the tub edge during getting in and getting out.

4 Conclusion

Taking into all the results into consideration, proposed handrail position is just above the tub edge or 100 mm inside the tub, with the height range covering between 760 mm and 1060 mm from the floor. However, sitting down and standing up activities will also occur in the wet area outside the tub within a Japanese bathroom, and it means that additional handrails will be required.

The observation of subjects' behavior and past experiences suggest that seating area at the edge or some arrangements such as a bath board be advisable for frail seniors to sit down during getting in and out of the tub rather than forcing the users to walk into the tub.

Acknowledgements. The experiments were conducted as part of joint research between the Building Research Institute, the Ministry of Construction, and the Comprehensive Housing R&D Institute, Sekisui House, Ltd. Summaries of these researches were reported at the Architectural Institute of Japan Annual meetings in 1992 [1, 2] and 1993 [3].

References

1. Yada, H., Goto, Y., Tanaka, S., Ukitya, N., Kose, S.: Bathroom handrail requirements for use by the elderly. Part 1. Basic experiment on allowable height of bathtub upstand. In: Summaries of Technical Papers of Annual Meeting of AIJ 1992 -Architectural Design-, pp. 941–942 (1992). (in Japanese)
2. Goto, Y., Yada, H., Tanaka, S., Kose, S., Ukiya, N.: Bathroom handrail requirements for use by the elderly. Part 2. Basic experiment on preferred position and layout of handrails. In: Summaries of Technical Papers of Annual Meeting of AIJ 1992 -Architectural Design-, pp. 943–944 (1992). (in Japanese)
3. Goto, Y., Yada, H., Kose, S., Ukiya, N., Kikuchi, Y.: Bathroom handrail requirements for use by the elderly. Part 3. Preferred position and layout of handrails for bathtub. In: Summaries of Technical Papers of Annual Meeting of AIJ 1993 -Architectural Design-, pp. 931–932 (1993). (in Japanese)
4. Kose, S., Sugimoto, Y., Goto, Y.: Acceptable handrail position for use by the elderly. In: Designing for Everyone - Proceedings of the Eleventh Congress of the International Ergonomics Association, pp. 1116–1118 (1991)

Beyond Walking: Improving Urban Mobility Equity in the Age of Information

Dadi An, Jiangchao Wang(✉), Puxian Wang, Yuheng Yang(✉), Yunchen Pu, Haorui Ke, and Yuqin Chen

School of Art Design and Media, East China University of Science and Technology, 130 Meilong Road, Xuhui District, Shanghai, China
61915633@qq.com

Abstract. With the ubiquitous cellular network and smart devices, information and communication technologies (ICTs) have significantly influenced our everyday mobility. However, there is still lack of accessibility information for limited mobility groups, including old, disabled public transit users and disabled pedestrians. This paper aims to present some practical inspirations to overcome the informational barriers for disabled people to be involved in urban activities. A review and an analytical explanation of current solutions is carried out to explore the possibility of ICTs application in the context of Asian high-density cities. The ubiquitous ICTs offer a huge potential to help and liberate people (disabled or older) who face challenges with an access to a certain minimum level of mobility.

Keywords: ICTs · Public transport · Mobility equity · Inclusive design

1 Lack of Support for Limited Mobility

In China, there are large number of disabled and aging people. People with limited mobility, such as physical disable and those with other constrains on using public transport are badly in need of effective, ease to use travel information [1]. This is related to whether the establishment of barrier free facilities is reasonable, whether it will bring inconvenience to this kind of people in daily life, which is also an urgent problem to be solved in the construction of urban humanity.

Based on the research methodology of anthropology and environmental behavior, an observation on over 20 wheelchair users have been carried out in Shanghai. It is obvious that the range of daily activity of disabled people is severely limited.

Although there are remarkable developments of infrastructure construction in Shanghai, for instance, 18 metro lines with 233 stations within central city, the lack of accessibility for disable is still a universal phenomenon. Lifts and other universal entrances are only one quarter of all metro entrances. During the interview, it is widely believed that disabled people are eager for more opportunities to enjoy the use of public transportation to get a broader range of mobility. However, disabled people's opportunity for urban experience are inaccessible for lack of information support about the accessibility and pre-planning their route, making their metro usage less inclined.

G. Di Bucchianico et al. (Eds.): AHFE 2020, AISC 1202, pp. 204–209, 2020.
https://doi.org/10.1007/978-3-030-51194-4_27

2 Mobility Takes an Informational Turn

With the popularization and wide use of smart phones and mobile Internet, it is inevitable to consider the impacts that pervasive mobile information has on the composition of everyday urban life [2], especially daily mobility. Particularly, there are wide arguments that mobility impairments have difficulties to obtain information that provide them accessible public transport network. According to the age of information, locative media, and mobility, mobile communication technologies promote urban mapping become mobile mapping, a collaborative mapping platform applied on a smartphone.

2.1 Mapping

Mapping technologies have been applied in planning and research in urbanism since the 1960s [3]. The method of storing information in electronic map generally uses vector image storage. The map scale can be enlarged, reduced or rotated without affecting the display effect. In the early stage, the map scale can not be enlarged or reduced by using bitmap storage. Modern electronic map software generally uses geographic information system to store and transmit map data, as well as other information systems.

2.2 Mobile Mapping

Locative-based service can enable ordinary people to tie bit of media and information to specific locations in the physical world, marking up the built environment with personal notations [4]. Electronic urban mapping can empower users to not only read the map, but also write on the map. Specialized e-map platform operators, such as Google map, are free, labeled, API open, dynamic and real-time. Value added service websites on the electronic map operation platform. They integrate the map function into their own website to enhance the function; cooperate with some platform operators to feedback rich POIs information.

3 Urban Wheelchair Map

3.1 Technologies

With the rapid development and integration of Internet, new generation mobile communication and 3S technology, electronic map and Internet map emphasize the social service of spatial information for the public, which has become the mainstream trend of geographic information technology and electronic applications. On the other hand, urban infrastructure is constantly optimized. With the development of the city, the number of facilities for the disabled is increasing. For these people, accurate facility data is needed to provide map services.

Thus, we describe a concept of solutions based on possible technologies, including LBS (Location-based service), AR (Augmented reality), and crowdsourcing, a smartphone application prototype is developed to support wheelchair users with customized information service and access (Fig. 1). Besides, the advantages and disadvantages of barrier free facilities in the city are summarized through the social function and message function of the app. In view of these shortcomings, design or improve barrier free facilities, improve the utilization rate and reasonable degree of barrier free facilities.

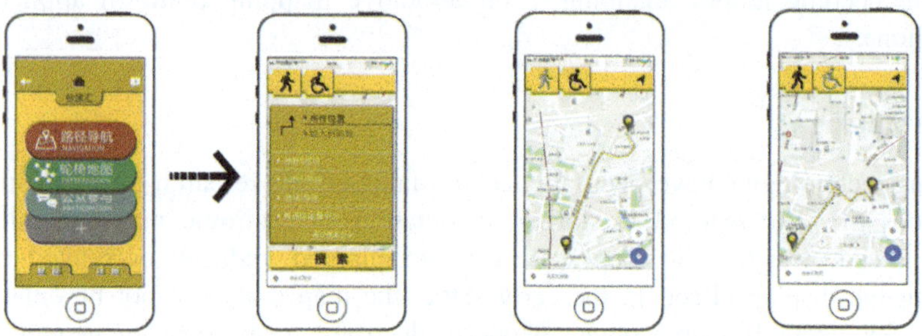

Fig. 1. Typical interfaces of the prototype

3.2 Prototype of Functions

The purpose of urban wheelchair map is to serve the disabled and the disabled, including line planning, social networking, additional functions, online communities and other services. Additional functions such as online cloud doctor, intimate taxi, hi chatting, social networking, rehabilitation supplies purchase and other supporting services are included as well.

Path Navigation. Path navigation pays attention to the travel problems of people who are inconvenient to move (especially those who use wheelchairs, strollers, and luggage), in the form of mobile application or applet, the elevator location, ramp and other information are embedded in the mobile phone map, the accessibility of the destination is analyzed, and the barrier free path map is provided to provide people with effective and accurate map information service and travel route planning, with two path navigation modes of wheelchair and walking.

Public Participation. Users can take photos and upload the improper public facilities around them, such as the broken blind path, the broken elevator, the damaged or blocked ramp, etc. The positioning information of the photos can be uploaded to the urban management platform, providing the basis for the accurate urban management and maintenance of relevant departments (Fig. 2). The basic data, location data and other information of urban wheelchair map users can also provide meaningful data support for planning management departments, urban research, accessibility research and other academic institutions on the premise of privacy protection.

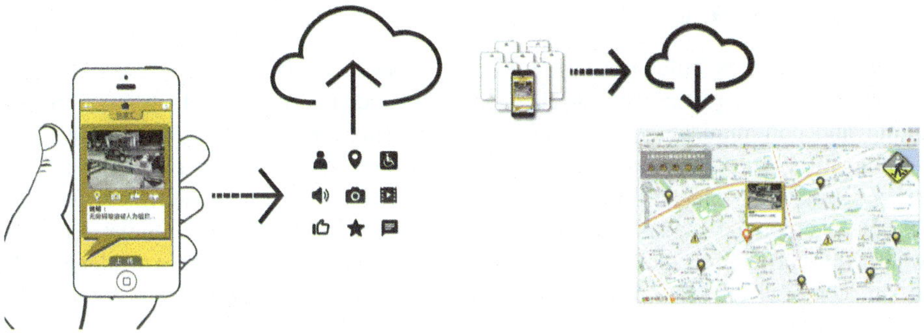

Fig. 2. Public participation based on LBS

Online Doctor. Because of physical reasons, you can contact the nearest hospital through the online cloud doctor, a sub project in the applet, and make an appointment with the hospital. Health care services will be also available.

Intimate Taxi. Due to physical reasons, the taxi has a safe range, and the taxi outside the safe range must be accompanied by relatives to avoid accidents when traveling alone. Join the emergency call to let their families know the situation and location information in the first time.

"Hi Chatting". Online social media provides the disabled a wider platform of communication on their everyday life. In addition to basic dating features, shopping, publishing, exchanging, such as rehabilitation products, can also found in this platform.

4 Discussion

4.1 Mobility Equity

The concept of Urban Wheelchair Map is expected to impact users (people with mobility impairments), built environments, and mobility equity. Physical mobility plays a key role in social inclusion [5]. With the support of access information, people with impairments can define their trips in advance and travel in an optimal way. In the social perspective, all users are empowered to participate the benefit of public transport. Particularly, by uploading photos and descriptions on the platform of LBS and crowdsourcing, citizens are encouraged to release the real-time problems and to engage creating better walking environments. Besides, civil participation with the combination of multiple source, such as crowd, open and social media, the paper concludes that with the increasingly importance of ICTs in urban daily life, accessibility information, inclusive design and online civil participation are emergence strategies in promoting urban walkability and have a huge potential to improve social equity.

4.2 Crowd Sourcing

Mobile mapping is influenced by open source and technologies of crowd sourcing. Collaborative and participant-based mapping has become essential features in constituting and communicating urban mapping [6]. This emergence of volunteered geographic information form a new integrated resource by combining data uploaded by users. Urban Wheelchair Map provides the identification of locations of universal entrances of public transport and their access to the public, as well as the locations where the barrier-free facilities are under urgent repairs. Volunteers can identify the urban accessibility problems on a base layer map to provide customized data source for limited mobility.

New data sources have emerged that are related to ICTs, and given the governments' awareness, urban data have become increasingly available to the public and academia and have broken through the obstacles for quantitative urban studies [7]. The platform of participation makes map users turn to be map makers, data consumers become data producers, providing mobility patterns or pedestrian flows for accessible maps. In the case of protecting the privacy of the disabled and the people with mobility disabilities, data statistics and regular release are conducted for the use of small programs to provide reliable data support for relevant workers and attract more people to engage in relevant industries.

5 Limitations and Implications

The applet is designed in a hierarchical structure. Select Baidu API to obtain map attribute configuration and basic map operation (mark creation, clearing, scribing, drawing encapsulation, homing, amplification, reduction, compass, click, select point, screenshot, calculate longitude and latitude distance, coordinate conversion, etc.). There are a lot of codes about map scattered into business logic, redundant and repetitive content. If there is a problem, it is not convenient for the unified modification of the problem. Everyone needs to learn some API related to map use when using, and some people need to maintain it after encapsulation. Map switching requires a lot of modification to support multiple map applications.

The above problems can be solved by using the form of small program for urban wheelchair map. China has a huge number of disabled, aging population and the growth rate of disabled people are increasing year by year. These are the problems that China has been facing at this stage. The demand for service procedures such as Urban wheelchair map is large and increasing year by year.

Furthermore, it is not only a solution to promote inclusive mobility but also a tool to develop municipal administration and the construction of infrastructure. ICTs play a major role to combine technologies with urban designers, policy makers, and citizens. Meanwhile, it is obvious that mobility accessibility, design for all, and civil participation become key strategies in promoting equal rights and equal opportunity for disabled people within the built environment of the information society [8]. With the development of modern urban construction, people have higher and higher requirements for the design and fine management of urban public space, especially for the vulnerable groups in urban environmental design, which can fully reflect the civilization, inclusiveness, age friendliness and internationalization of a city or a country.

Acknowledgments. We would like to acknowledge the support by the Shanghai Pujiang Program (18PJC022), Shanghai Summit Discipline in Design, and the Fundamental Research Funds for the Central Universities. We would also like to thank the anonymous referees who provided useful insights on the earlier version of this paper.

References

1. May, A., Parker, C.J., Taylor, N., Ross, T.: Evaluating a concept design of a crowd-sourced 'mashup' providing ease-of-access information for people with limited mobility. Transp. Res. Part C **49**, 103–113 (2014)
2. Kelley, M.J.: Urban experience takes an informational turn: mobile internet usage and the unevenness of geosocial activity. GeoJournal **79**(1), 15–29 (2014)
3. e Silva, A.D.S., Sheller, M.: Mobility and Locative Media: Mobile Communication in Hybrid Spaces, pp. 167–187. Routledge, Abingdon (2014)
4. Al-Kodmany, K.: Sentient city: ubiquitous computing, architecture, and the future of urban space. J. Urban Technol. **19**(3), 137–140 (2012)
5. Lucas, K.: Transport and social exclusion: where are we now? Transp. Policy **20**, 105–113 (2012)
6. Gordon, E., e Silva, A.D.S.: Net Locality: Why Location Matters in a Networked World (2011)
7. An, D., Tong, X., Liu, K., Chan, E.H.W.: Understanding the impact of built environment on metro ridership using open source in Shanghai. J. Cities **93**, 177–187 (2019)
8. Toboso, M.: Rethinking disability in Amartya Sen's approach: ICT and equality of opportunity. Ethics Inf. Technol. **13**(2), 107–118 (2011)

Acceptable Handrail Diameter for Use by the Elderly

Satoshi Kose[1(✉)], Yoshihide Sugimoto[2], and Yoshiaki Goto[3]

[1] Building Research Insitute, 1, Tachihara, Tsukuba, Ibaraki 305-0802, Japan
skose@gakushikai.jp
[2] Naka Corporation, 39, Shinmachi, Yashio 340-0807, Japan
shimura-s@naka-kogyo.co.jp
[3] Okayama University of Science, 1-1, Ridaicho, Kita-ku, Okayama 700-0005, Japan
goto@archi.ous.ac.jp

Abstract. Psychological experiments were conducted to determine allowable range of handrail diameter for use by the elderly on stairs and in the toilets. The results were cross-examined with ergonomic standpoint. The present findings are that: 1) Simple grab of handrails yields the optimal diameter of 40 mm; 2) Simultaneous application of body weight against the handrail indicates 36 mm or slimmer as the optimal diameter. The latter result is more important as seniors need stability and reliability when using handrails. These findings are in accordance with the guidelines in Sweden, New Zealand, and England & Wales, where diameters between 30 and 50 mm are unanimously accepted. Present research results point toward a revised recommendation in Japan on handrail diameters to secure safer traverse on stairs.

Keywords: Handrail · Grabrail · Diameter · Senior · Stair · Toilet

1 Introduction

Japan is growing older at an amazing speed. A highly aged society with more than a quarter of the population being 65 years of age or over is already reached. This necessitates preparation in every aspect of the society. Building design can never be exempted from this change. Although the previous design guidelines have rarely taken into consideration the needs of the aged persons, it has become inevitable to do so from now onwards, because buildings will last many years once they are constructed. The needs of the aged persons come mainly from their deteriorating ability to cope with the functional strength. Lower limbs are particularly vulnerable to such deterioration. The authors aimed at the establishment of guidelines of building design toward the aged society [9]. There has rarely been any theoretical basis that supports or denies the Japanese recommendations which have simply been imported from abroad. The authors tried to solve questions in a quantitative way to convince architects and designers. Among many, handrails are the most important provisions to prevent accidental falls and resulting injury (especially on stairs). Practical guidelines should be presented for their installation. The present paper describes the result of experiments on

G. Di Bucchianico et al. (Eds.): AHFE 2020, AISC 1202, pp. 210–217, 2020.
https://doi.org/10.1007/978-3-030-51194-4_28

handrail diameter for stairs and in other situations, previously presented as progress reports in Japanese only [1, 2].

To secure safety and to eliminate drawbacks of using stairs, handrail provisions become vital to assist unhindered traverse. However, appropriate information on the selection of handrail design has never been transmitted to the designers. Some design examples on stair handrails suggest, for example, that ones with 45 mm diameter or greater are preferable, with the allowable maximum being 65 mm in diameter. The shape of the section is left to the choice of the designers as well. In the market, larger sections are preferred, as it reduces the need to have more supporting hardware, or of craftsmen's work. It is quite different from the regulations or guidelines in other countries where the range of diameter being between 30 mm and 50 mm [5, 7, 8]. They also suggest that the section be of a shape for easy grasping. The recommendation is based on experiments. In order to solve the questions and establish recommendations, experiments were conducted.

2 Experiment A

2.1 Experimental Procedures

As a starting, subjective evaluation of handrail diameter was conducted with various dimensions, from 25 mm to 60 mm (11 different ones from 25, 30, 34, 35, 38, 40, 42, 43, 45, 50, up to 60). The subjects were randomly exposed to experimental specimens, and they evaluated their thickness in 5 rank score as they grabbed the rail. The subjects were covered of their eyes with an eye mask so that visual information will not be employed during evaluation (see Fig. 1).

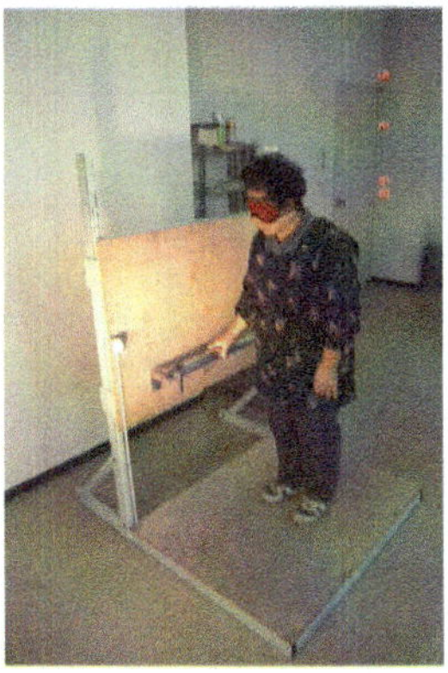

Fig. 1. A female subject with an eye mask, grabbing a horizontal handrail.

Next, in order to simulate actual handrail use situations, subjects were requested to lean backwards, holding the rail to keep their stability. They then evaluated the subjective preference (reliability) of the handrail diameter. The specimens were tested from thick to slim, i.e., from 60 mm and onwards (only nine rails were tested: 60, 50, 45, 43, 40, 38, 34, 30, and 25).

Thirdly, the distance between the handrail and the wall was tested for the ease of use, from 25 to 50 mm with the increment of 5 mm. The evaluation was done regarding the physical contact (with grasped hand in vertical position) and subjective evaluation (whether they feel the gap is reasonable enough).

2.2 Experimental Subjects

Thirteen senior subjects (seven males and six females) and 14 youngsters (seven males and seven females) participated in the experiment. The handrail was placed near to the height of each subject's waist, the easiest position to grasp it, with the arm and hand naturally hung from the shoulder.

2.3 Results

As to psychological evaluation of handrail diameter, each subject gave similar response as shown in Fig. 2. From the figures, evaluation was standardized and the areas between medium preference to optimum were assumed as allowable range. The results were added among all subjects (Fig. 3). The average diameter then becomes 40.6 mm, and this is not related with age or sex. This is however related with size of the hand, the regression coefficient being 0.35 (Fig. 4). The regression coefficient is 0.56 for aged subjects only.

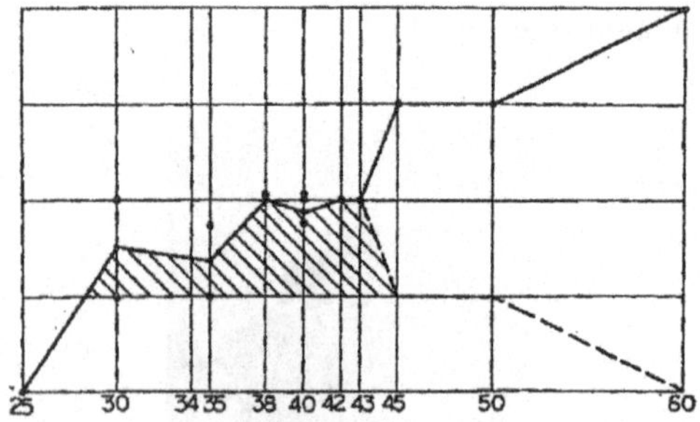

Fig. 2. Typical response by a female subject. Horizontal axis shows diameter in mm, and vertical axis means 5-point rating, bottom line shows too slim, middle optimal and top line too thick. Shaded area was assumed to be acceptable.

The second experiments yielded a different result. As shown in Fig. 5, subjects did not reject slimmer handrails as unacceptable. Rather, thicker ones do not ensure reliable grip. The summed-up figure (Fig. 6) drawn through similar procedures as above suggests that average diameter be slimmer than 35.7 mm. The slimmer handrails received, however, a little negative response as to their psychological reliability. The subjects seem to have felt some fear of the handrails collapsing.

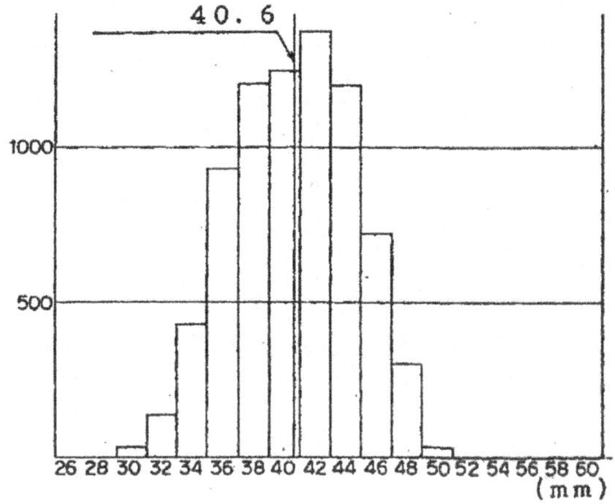

Fig. 3. Accumulated response from all subjects as acceptable diameter (i.e., not too slim, not too thick). Average was 40.6 mm.

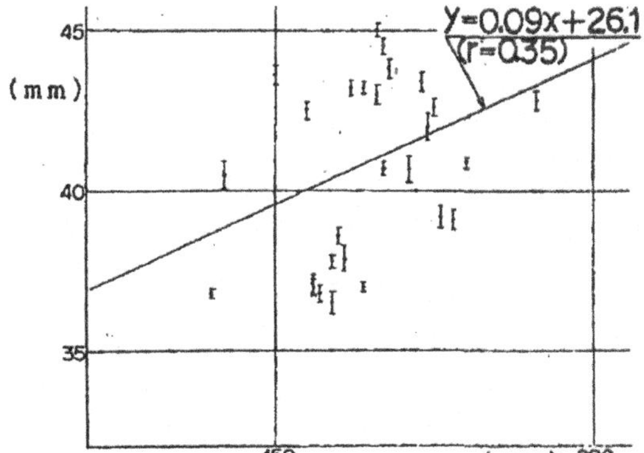

Fig. 4. Relationship between hand size (horizontal axis in mm) and optimal diameter (vertical axis in mm).

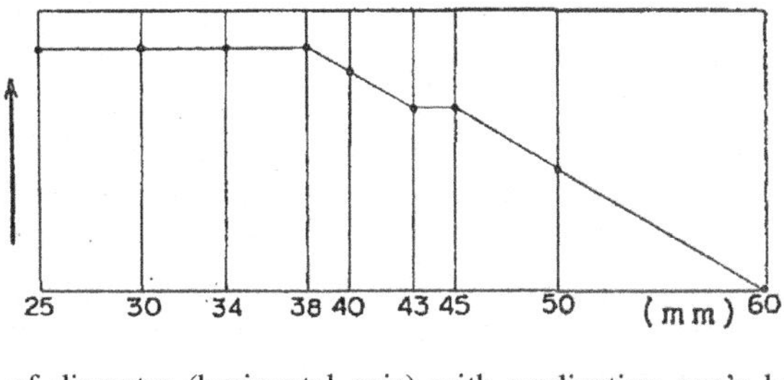

Fig. 5. Evaluation of diameter (horizontal axis) with application one's body weight leaning backwards. This is an example of a female subject. Top side means acceptable, bottom unacceptable.

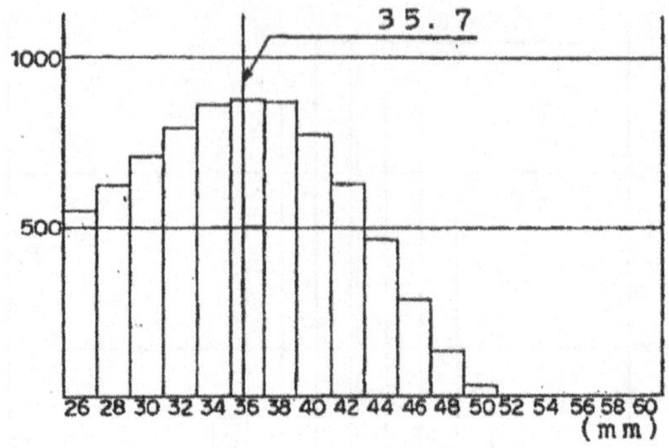

Fig. 6. Accumulate response as acceptable diameter from all subjects with leaning backwards. Averaged diameter is 35.7 mm.

The third experiment on acceptable distance between wall and a handrail yielded a result of 37.4 mm (sigma = 3.4) for physical contact, and 32.0 mm (sigma = 3.9) for psychological evaluation.

3 Experiment B

3.1 Experimental Procedures

Based on the results of experiment A, further experiments were conducted under more practical settings, i.e., horizontal handrail for walking, three handrail arrangements for toilets, and stair handrail. Subjects evaluated the thickness, and distance from the wall (physical and psychological). Sixteen handrail specimens were used (10, 15, 20, 23, 25, 28, 30, 32, 34, 36, 38, 40, 42, 45, 50, and 60 mm in diameter). Height and distance from the toilet stool were set at the subject's preference [6]. For horizontal handrail, subjects applied one's full weight with firm grip.

3.2 Results

Firm grip of horizontal handrail resulted in 31.9 mm diameter as the average compared to 40.6 mm without full weight application in experiment A (see Fig. 7). The difference of result may be due to how the instruction on weight application was interpreted by the subjects.

Results on toilet handrail for three arrangements, i.e., vertical, horizontal and inclined 45° are given in Fig. 8. There was more variance on the diameter for horizontal handrail than others. It may be due to its difficulty in the direction of force application (push up) while others are more in the form of pulling up.

Results on stair handrail is given in Fig. 9. The average was 37 mm, which is consistent with other handrail arrangement.

The evaluation of the gap between the wall and the handrail gave similar results among handrails for horizontal walking, toilet and stair. Figure 10 gives the result of

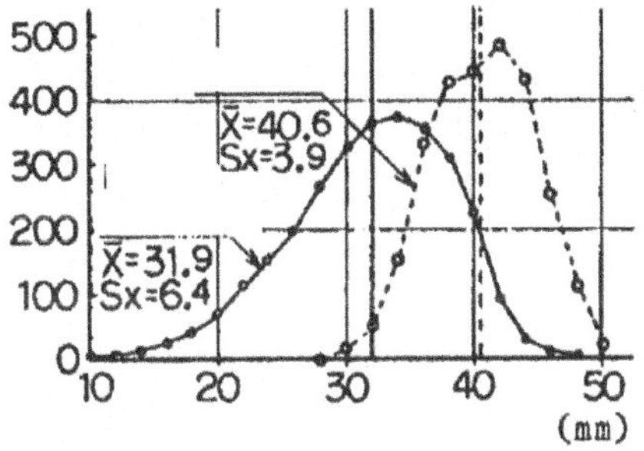

Fig. 7. Difference in accumulated response between slight grabbing condition (dotted line) vs firm grip (solid line) for horizontal rail. Slimmer diameter was chosen with firm grip. Horizontal axis is diameter in mm, vertical axis frequencies.

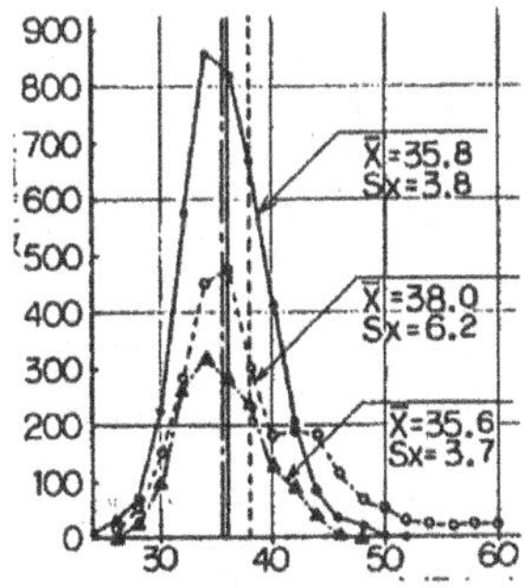

Fig. 8. (Left) Difference in accumulated response on optimal diameter for toilet handrails of three arrangements: vertical, horizontal and inclined 45°. Average is 35.8 mm for vertical, 38 mm for horizontal, and 35.6 mm for inclined.

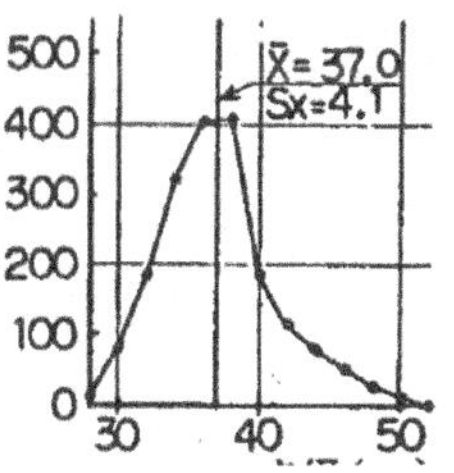

Fig. 9. (Right) Accumulated response on optimal diameter for stair handrails.

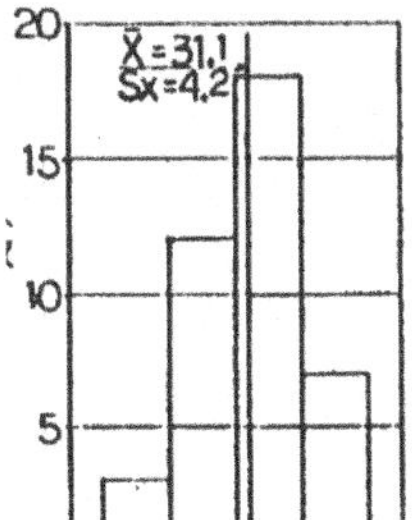

Fig. 10. (Left) Frequency distribution on acceptable handrail distance from the wall (mm).

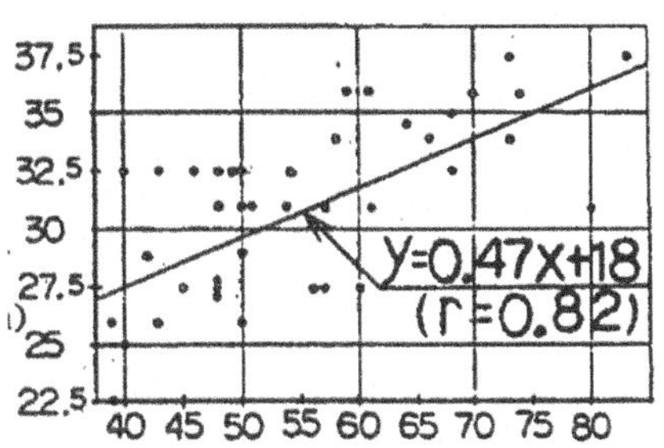

Fig. 11. (Right) Relationship between handrail clearance and body weight. Horizontal axis shows body weight of the subjects (in kg) and vertical axis is distance from the wall (mm).

physical evaluation. There is good relationship between the gap and the weight of the subjects (Fig. 11). Seniors gave similar psychological evaluation among different conditions, but younger adults preferred somewhat wider gap, particularly for vertical and inclined handrails.

3.3 Discussion

If the horizontal handrail is to be used just as a guide, for example in corridor, 40 mm diameter would be acceptable. If it were to prevent the user from falling, slimmer handrails such as 32 mm would give firmer grip (see Fig. 7). For toilet handrails, 36 mm would be acceptable. Stair handrails can be 37 mm. Gap between handrail and the wall be better 40 mm for comfort, but can be compromised to be 36 mm.

4 Conclusion

Experimental results suggested that the diameter be between 32 mm and 40 mm, depending on the user needs. A wider gap between the wall and handrail is sometimes difficult to ensure, particularly for narrow circulation spaces in dwellings, but it is advisable to have 36 mm for easier use.

Acknowledgements. The experiments were conducted as part of joint research between the Building Research Institute of the Ministry of Construction, the Naka Corporation Technical Laboratory, and the Comprehensive Housing R&D Institute of the Sekisui House, Ltd. Summaries of these researches were reported at the Architectural Institute of Japan Annual meetings in 1989 [1] and 1990 [2]. The former part was also orally presented at 22nd International Congress of Applied Psychology [3] and at the annual meeting of Japan Ergonomics Society Kanto Chapter [4].

References

1. Kose, S., Goto, Y., Sugimoto, Y.: Handrail requirements for use by the elderly. Part 1. Basic experiment on handrail diameter. In: Summaries of Technical Papers of Annual Meeting of AIJ 1989 -Architectural Design-, pp. 799–800 (1989). (in Japanese)
2. Kose, S., Sugimoto, Y., Goto, Y., Konishi, T.: Handrail requirements for use by the elderly. Part 2. Allowable diameter and necessary clearance from the wall. In: Summaries of Technical Papers of Annual Meeting of AIJ 1990 -Architectural Design-, pp. 697–698 (1990). (in Japanese)
3. Kose, S., Sugimoto, Y., Goto, Y.: Acceptable handrail diameter for use by the elderly. In: 22nd International Congress of Applied Psychology '90 Kyoto Abstracts, p. 385 (1990)
4. Kose, S.: Evaluation of handrail diameter for toilets and stairs. In: Presentation Abstracts of Japan Ergonomics Society Kanto Chapter 1989, pp. 105–106 (1989). (in Japanese)
5. Hall, N.B., Bennet, E.W.: Empirical assessment of handrail diameter. J. Appl. Psychol. **40**(6), 381–382 (1956)
6. Kose, S., Sugimoto, Y., Goto, Y.: Acceptable handrail position for use by the elderly. In: Designing for Everyone - Proceedings of the Eleventh Congress of the International Ergonomics Association, pp. 1116–1118 (1991)

7. Statens Planverks Forfratteningssamling: SBN (Svensk Bygg Norm) 1980 Utgava 2, p. 337 (1980)
8. Standard Association of New Zealand: NZS4121-1985. Code of Practice for design for access and use of buildings and facilities by disabled persons, p. 16 (1985)
9. Kose, S.: Design Guidelines of Dwellings for the Ageing Society: Japanese Approach Toward Universal Design. CIB World Building Congress, April 2001, Wellington, New Zealand (2001)

A Holistic Approach to Inclusive Mobility Under Resilience and Sustainability Goals: A Pilot Project in Brazil

Magaly Romao[1(✉)] and Anabela Simoes[2]

[1] Prefeitura Municipal de Rio Verde, Rio Verde, Brazil
magaly.romao.jau@gmail.com
[2] LUSOFONA University, Lisbon, Portugal
anabela.simoes@ulusofona.pt

Abstract. Today's societies are dependent on increasingly complex and interdependent safety-critical systems with often poorly understood vulnerabilities. If their provision of service is cut off or drastically reduced, the dependent activities will either fail or perform at a lower level. Transport systems are a good example of such propensity to natural and man-made disruptions. Due to their complexity, uncertainty, and safety criticality, Mobility offers a high number of opportunities for research and innovation, particularly in the present technological era. Thus, objectives like disaster risk reduction, sustainable development strategies, and sustainable transport, are nowadays included as well. The holistic approach is based on the principle of promoting inclusive mobility without creating any negative impact in other areas of each targeted community, thus requires an integrated perspective of the work towards inclusion under safety, sustainability and resilience conditions.

Keywords: Transport · Mobility · Inclusion · Resilience · Sustainability

1 Introduction

Today's societies are dependent on increasingly complex and interdependent safety-critical systems, such as, electric power, transportation, water supply, information and communication, among others. Furthermore, a continuous state of crisis appears to be the new norm for city life in the 21st century. Unexpected incidences, both natural and manmade, are occurring in a more prevalent basis and on a larger scale. While it is impossible to predict which type of disruption will come next, we must anticipate, monitor, respond and learn from past events in order to protect people and their settlements.

Transport systems must be safe, secure, sustainable and inclusive accommodating everyone's mobility, particularly those with special needs and during crises. However, they are a good example of such propensity to natural and man-made disruptions [1]. Due to their complexity, uncertainty, and safety criticality, Mobility offers a high number of opportunities for research and innovation, particularly in the present technological era.

G. Di Bucchianico et al. (Eds.): AHFE 2020, AISC 1202, pp. 218–224, 2020.
https://doi.org/10.1007/978-3-030-51194-4_29

The UN has launched recently the 17 Development Goals, being the 11th Goal the trigger of this proposal, as it targets cities and every human settlement to be inclusive, safe, resilient and sustainable [2]. In accordance, Simon et al. [3], and Yamagata and Maruyama [4] highlighted the increasing importance of urban issues towards resilient and sustainable communities. Thus, objectives like disaster risk reduction, sustainable development strategies, and sustainable transport, are nowadays included as well. The present epidemic outbreak is an example of the needs for community resilience as a way to be prepared to resist, absorb, adapt to and recover from acute shocks and chronic stresses keeping critical services functioning. Furthermore, every system should be monitored, and operators should learn from on-going processes through the community and cross-regional collaboration, to increase adaptive abilities and strengthen preparedness by anticipating and responding to future challenges [5].

Transport systems must be safe, sustainable and inclusive in accommodating everyone, especially those with special needs, and particularly during crises. Any emergency evacuation will be made easier if operators do not encounter barriers in logistics, communication and transportation aspects in their rescue efforts.

Furthermore, important demographic changes in nowadays society, together with the fast and intensive technological development, as well as concerns about new threats and stresses, are leading to new realities. Thus, the concept of Resilience has become an important tool to enhance a community's emergency preparedness and response as its impact on sustainability and inclusive mobility is enormous.

2 A Pilot Project: Rio Verde

An ongoing project focusing on inclusive mobility relies on Resilience and Sustainability as overarching concepts to develop an integrated project that will allow to take into consideration the following main target and principle: to promote inclusive mobility within the scope of flexible solutions that adjust to the continuous dynamics of communities. The principle of avoiding that any isolated improvement in any subsystem within the community will compromise any target in any other subsystem is being taken into consideration.

This project is being carried out in Rio Verde, (State of GOIAS) being promoted and supported by the City Government targeting two urban areas: Transport and Environment. While this remains a relatively limited scope, it is expected to produce useful guidance and contribution towards that end [4]. Today's societies are dependent on increasingly complex and interdependent safety-critical systems with often poorly understood vulnerabilities. If their provision of service is cut off or drastically reduced, the dependent activities will either fail or perform at a lower level [6].

The recent concerns about the environment are leading decision-makers to promote public transport, shared mobility schemes and greener modes. However, few infrastructures around the world are designed to be safe and inclusive enough to accommodate every user groups. So far, decision-makers, operators and designers have not considered accessibility as a priority in transport planning. Superimposed on this lack of awareness, the increased volume of migration and tourism movements, as well as impending technological developments, all require resilience and inclusive mobility be built into the planning, operations, deployment of transport systems and infrastructure design right from the start.

3 The Holistic Approach

The holistic approach is based on the principle of promoting inclusive mobility without creating any negative impact in other areas of each targeted community. Either involving the design and development of new urban areas, or a new operation line of public transport, or based on retrofitting existing urban areas or elements of an existing transport system, the holistic approach requires an integrated perspective of the work towards inclusion under safety, sustainability and resilience conditions. A safe and inclusive mobility is required for all within a sustainable and resilient community. With this objective in mind, the project is centered on the development of the following main tools to be used in further applied projects under the proposed holistic approach: (1) the conceptual framework presiding at the holistic approach and its development, (2) the project methodology with its different steps to be followed towards the project goals; and (3) the maturity model, which is necessary to be followed along to any further applied project by means of assessing the defined maturity stages in order to ensure that each reached target will be stable.

Projects on Urban Resilience, Sustainable Cities and Inclusive Transport are more and more required. As a result of the existing interdependencies between and within subsystems, such services are extremely sensitive to incidents, technical failures, disruptions, natural disasters or other threats. Transport systems must be safe and inclusive in accommodating everyone, especially those with special needs, and particularly during crises. Any emergency operation will be made easier if operators do not encounter barriers in logistics, communication and transportation aspects in their rescue efforts.

The recent concerns about the environment are leading decision-makers to promote public transport, shared mobility schemes and greener transport modes. However, few infrastructures around the world are designed to be safe and inclusive enough to accommodate every user groups. So far, decision-makers, operators and designers have not considered accessibility as a priority in transport planning. Superimposed on this lack of awareness, the increased volume of migration and tourism movements, as well as impending technological developments, all require resilience, sustainability and inclusive mobility be built into the planning, operations, deployment of transport systems and infrastructure design right from the start.

4 Relationships to the Existing Body of Knowledge

As referred above, Simon et al. [3] highlight the increasing importance of urban issues to make cities and human settlements inclusive, safe, resilient and sustainable. Thus, objectives like disaster risk reduction, sustainable development strategies, and sustainable transport, should be included as well. Actually, a continuous state of crisis appears to be the new norm for city life in the 21st century. Unexpected incidences, both natural and manmade, are occurring in a more prevalent basis and on a larger scale. While it is impossible to predict which type of disruption will come next, anticipation, monitoring, responding and learning from past events are required.

Transport systems are expected to be safe, reliable, inclusive, sustainable and resilient. However, they become vulnerable once they operate as open systems sharing the space with other subsystems (e.g. private cars with public buses) and users (e.g. pedestrians and cyclists) leading to the system complexity. As their operation is dynamic, the level of uncertainty is high, thus increasing the system vulnerability. Vulnerability has been defined as the degree to which a system, or part of a system, may react adversely during the occurrence of a hazardous event [9]. System vulnerability in transport refers to its susceptibility to extreme strains, which is related to the potential for the propagation of shocks within a network [9]. The term is also understood as: (1) the differences in the degree of damage incurred from natural hazards suffered by an individual, a whole community, a city, or an entire region [10]; and (2) the system's ability to cope with the social and economic liability of the resulting event. From the human perspective, people become "vulnerable" if their limited access to resources is the most critical factor in obtaining a secure livelihood or recovering effectively from a disaster. Resilience and vulnerability are two different concepts although linked and complementary in understanding and explaining the systems and people's response to change, shocks and disruptions. Both concepts appear together in the frame of disaster risk research and management [11].

Two important and recent projects addressing the communities' resilience are: (a) the Smart Mature Resilient Project (SMR) [8] which aimed at matching the needs of European cities for resilience with tangible policy proposals to enhance their maturity and capacity to absorb, resist and bounce back from shocks; and (b) the Resilience in Action project [12], which was started in 2013 and involve 100 cities under a City Resilience Framework. The framework expresses the relationship and complementarity of four subsystems: Health and Wellbeing, addressing people; Infrastructure and Environment, addressing the place; Economy and Society, addressing the organization; and Leadership and Strategy, addressing knowledge-related required by management and coordination. Some of the ideas mentioned in this framework have inspired the creation of the proposed Conceptual Framework Towards a Resilient Community.

The recent wide recognition of the importance of urban Sustainability represents an important prerequisite for progress towards that objective. However, the worldwide difference in terms of political and financial resources among countries and cities within the same country compromise the sustainability goal. According to Simon et al. (2016) [3], sustainability is itself a complex and contested notion, containing diverse elements, some relatively easy to measure and others more qualitative. Moreover, like development, sustainability has the triple characteristics of being simultaneously a normative aspiration, a state of being and the means of attaining that state. The same authors define the following three key dimensions of sustainable cities: accessibility, greenness and fairness.

Accessibility is generally defined as: (1) the ability of people to reach goods or services as measured by their availability in terms of physical space, affordability and appropriateness; and (2) the provision of services and facilities, job opportunities, education and housing, as well as the means of reaching them. In urban terms, density is considered one factor affecting accessibility, together with connectivity, diversity and intensity. Finally, Waters, J. (in Simon et al., 2016) [3] define Accessibility as the ability of individuals to participate in necessary or desired activities for the wellbeing of humanity. However, the concept is here viewed as an individual ability instead of a quality of the city or ever urban system. Furthermore, accessibility is rather applied to physical systems than to information and communication systems, per instance. Thus, the concept of Inclusion, here adopted, provides the notion of every system for all potential users.

The urgency of such approach is expressed in the inclusion of a specifically urban Goal (N. 11) to make cities and human settlements inclusive, safe, resilient and sustainable in the frame of the 17 Sustainable Development Goals (SDGs) adopted by the UN at the 2015 General Assembly. These SDGs Goals apply to all countries, regardless of per capita income or position on the Human Development Index. As time is running, the present exploratory project to pave the way for new applied research projects targeting a safe and inclusive mobility for all within resilient and sustainable communities, seems to be relevant and timely welcome.

5 The Conceptual Framework

The areas to be addressed focus on complex and safety-critical subsystems, within which the vulnerabilities and identified frailties or the existing barriers to an inclusive mobility represent the targets to be overcome. The sources of variability and uncertainty represent the main factors to predict a system functioning and its level of physical exposure, as well as the continuity of critical services in the case of an undesirable and unexpected event. Uncertainty and variability, associated with complexity, can make risk management in the transport sector equally complicated [13]. As Leveson [14] pointed out, accidents within complex environments tend to produce unpredicted chain reactions, the domino effect of which could rapidly reach intolerable proportions. Depending on the transport infrastructure (rigid, like railways systems, or flexible, like the road transport system), it is possible to maintain the service following any disruption, by either replacing or complementing a temporary disrupted transport system by changing the vehicle(s) or route(s).

Addressing the above referred problem statements, a conceptual framework has been developed following a holistic approach highlighting a dedicated transport segment. This multidisciplinary tool will guide: (a) the analysis to identify and assess the system vulnerabilities and their impacts on community activities, together with the system interdependencies and critical processes that operate across multiple boundaries; (b) the transformative project and the related actions towards the project goals; (c) the development of a management platform for different transport modes; (d) the resources and their deployment capabilities towards an inclusive, safe, sustainable and resilient community.

In addition, two axes will complete the actions towards a barrier-free, safe, sustainable and resilient community: (a) an educational program towards the development of a Safety Culture highlighting the shared values, actions, and behaviors demonstrating a commitment to safety over competing goals and demands [7]; (b) The development of a maturity model to accompany each step of the project paving the way towards a consolidated resilience [8].

6 Next Steps

The next steps for action are: (a) Focus Groups interviews with citizens of different ages and mobility-related functional abilities; (b) Focus Groups interviews with experts on mobility impairment; (c) the analysis towards the identification of the urban vulnerabilities and mobility-related barriers; (d) the development of the simulation scenarios for testing different mobility solutions, especially addressing the human side of their impacts in the society; (e) the development of the transport-related management platform under a human-system integration approach; (f) the education action towards the development of a safety culture allowing for a permanent attitude regarding safety; (g) the development of the maturity model and its different steps and related assessment; (h) the assessment of the proposed solutions in simulated testing scenarios; (i) the implementation of the integrated solutions towards an inclusive, safe, sustainable and resilient community.

References

1. Mattson, L.-G., Jenelius, E.: Vulnerability and resilience of transport systems: a discussion of recent research. Transp. Res. Part A **81**, 16–34 (2015)
2. United Nations Development Goals. https://www.un.org/sustainabledevelopment/cities
3. Simon, D., Palmer, H., Parnell, S., Waters, J.: Rethinking Sustainable Cities: Accessible, Green and Fair. Policy Press, University of Bristol, Bristol (2016)
4. Yamagata, Y., Maruyama, H. (eds.): Urban Resilience: A Transformative Approach. Springer, Cham (2016)
5. Vugrin, E., Warren, D., Ehlen, M., Camphouse, C.: A framework for assessing resilience of infrastructure and economic systems. In: Gopalakrishnan, K., Peeta, S. (eds.) Sustainable and Resilient Critical Infrastructure Systems: Simulation, Modelling and Intelligent Engineering, pp 77–116. Springer, Heidelberg (2010)
6. Bellini, E., Nesi, P., Coconea, L., Ferreira, P., Simões, A., Candelieri, A.: Towards resilience operationalization in urban transport systems: the RESOLUTE approach. In: Walls, L., Revle, M., Bedford, T. (eds.) Risk, Reliability and Safety: Innovating Theory and Practice, pp 2110–2118. Taylor & Francis Group, London (2017). ISBN 978-1-138-02997-2
7. Morrow, S., Coplen, M.: Safety culture: a significant influence on safety in transportation. U.S. DOT John A. Volpe National Transportation Systems Center, Cambridge, MA 02142-1093 (2017). http://www.fra.dot.gov
8. Smart Mature Resilient Project. http://smr-project.eu/home/. Accessed Aug 2018
9. Proag, V.: The concept of vulnerability and resilience. In: Proceedings of the 4th International Conference on Building Resilience, Salford Quays, UK (2014)
10. Fekete, A., Hufschmidt, G., Kruse, S.: Benefits and challenges of resilience and vulnerability for disaster risk management. Int. J. Disaster Risk Sci. **5**, 3–20 (2014). https://doi.org/10.1007/s13753-014-0008-3
11. Reggiani, A.: Network resilience for transport security: some methodological considerations. Transp. Policy **28**, 63–68 (2013)

12. Kete, N.: City Resilience Framework. The Rockefeller Foundation, Arup (2015). https://assets.rockefellerfoundation.org/app/uploads/20160105134829/100RC-Resilience-Framework2015.pdf. Accessed 2015
13. Marais, K., Dulac, N., Leveson, N.: Beyond normal accidents and high reliability organizations: the need for an alternative approach to safety in complex systems. In: Engineering Systems Division Symposium Proceedings. MIT, Cambridge (2007)
14. Leveson, N.: A new accident model for engineering safer systems. Saf. Sci. **42**, 237–270 (2004)

Interdisciplinary Design Education

"Grit" Assessment of First Year Design Students

Benjamin Bush(✉) and Chris Arnold

School of Industrial and Graphic Design, College of Architecture, Design and Construction, Auburn University, Auburn, AL, USA
bushbem@auburn.edu

Abstract. Author Angela Duckworth's book "Grit" [1] raises interesting questions about the idea of how perseverance and passion play into the development of young design students. Talented students are well appreciated but sometimes talent isn't enough to address complex design problems, sometimes one needs hustle to overcome obstacles. This study uses Duckworth's "Grit Test" and "Activity Test" to quantify the level of "grit" of a student population entering the first year of an intensive industrial design program. The resulting data was used to identify and track "gritty" students through the summer and fall semesters. While grit is an incredibly important attribute for students to have in the industrial design profession, subsequent assessment reveals the grittiest students did not necessarily improve academically over the two semesters. This paper identifies the methods of research and speculates how grit applies to students beginning their studies within an established industrial design curriculum.

Keywords: Grit · Talent · Perseverance · Passion

1 Methods

A test identical to that offered by Angela Duckworth in her book "Grit" [1], was administered to 66 first time design students. This test consists of 10 questions. Odd numbered questions inquire about the individuals' passions while even numbered questions probe one's internal perseverance. Responses were gathered using a five-point Likert scale range from "Not like me at all" to "Very much like me". Based on responses, total scores were averaged to provide an overall "Grit Score".

This test was given at the beginning of a twelve-week grouping of summer classes that serves as a 'gateway' to the professional curriculum within a four-year American public university. Duckworth's book posits that grit can also be identified through repeated performance and improvement in a challenging activity. To obtain this information from the pool of available design students, the following questions were asked in addition to Duckworth's survey: "What events did you compete in during high school, how many years did you engage in this activity, and what awards did you receive?" The intent of this question was to measure grit beyond the student's subjective opinion. If a student was able to stick with an activity over multiple years it can be concluded that he or she has notable levels of grit.

G. Di Bucchianico et al. (Eds.): AHFE 2020, AISC 1202, pp. 227–233, 2020.
https://doi.org/10.1007/978-3-030-51194-4_30

The data from this survey was used to track this cohort of students over the course of seven months, from the beginning of the Summer semester to the end of the Fall semester. At the conclusion of the Fall semester, the students' grit scores were compared against their academic score to see if grittiness, based on this measure, played an important part in their improvement as young designers (Fig. 1).

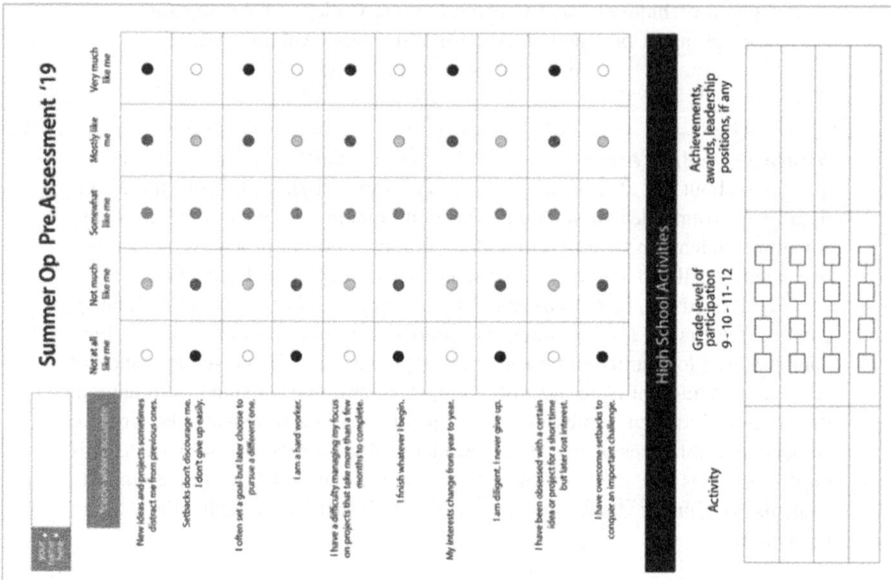

Fig. 1. Angela Duckworth's grit test with activities question

2 Why Administer the Grit Test Now?

2.1 The Summer Studio as a Gateway Course

The "Summer Studio", which occurs after the first two semesters of the students' freshman year, serves as a gateway to the industrial design professional program. The students are evaluated on their learning and skill performance on tasks related to technical drawing, perspective drawing, and presentation renderings skills, all completed by traditional hand-tool based techniques. Each module of the three-class grouping lasts 10 class meetings, with each class meeting lasting approximately seven hours. Instructions and demonstrations are given to the cohort. Each set of newly taught skills are measured through assignments which build on one another. Novice students are expected to display adequate capability on a new skill in only a matter of hours. Based on previous studies [2] students who have design experience from previous schooling have a noted advantage. At the end of the summer semester, the highest-ranking students (typically forty to forty-five students), based on academic performances, are offered entry into the professional program. Based on previous years, the retention rate for students entering this professional program is roughly 95%. This percentage supports the effectiveness of this method to identify students who will successfully progress through the program until graduation.

2.2 Effort Counts Twice

Grit is understood as an important attribute to investigate for two reasons. Firstly, the majority of students who enter this professional program have not had the opportunity to acquire design skills prior to pursuing this degree. If they are to be offered the opportunity to continue in the program, they will have to combine their inherent talent with perseverance. Talent is a tricky thing. Angela Duckworth describes talent like this [1]:

> Talent is how quickly your skills improve when you invest effort. The outcome of talent and effort produce skill. Once you have skill that's not it, practitioners still must apply effort to that skills to gain achievement. In her eyes, effort counts twice!

$$\text{Talent x}\,\textit{Effort} = \text{ Skill } \rightarrow \text{ Skills x}\,\textit{Effort} = \text{ Achievement.} \tag{1}$$

Since most students lack design skills upon entering our program, willingness, effort, and "grit" are believed to be more reliable attributes to identify and develop.

2.3 Grit is Transferable

Secondly, "grit" is a transferable attribute. Students have varied educational foundations, life experiences, and socioeconomic influences. A privileged young adult with five years of violin experience could exhibit the same amount of "grit" as a lesser advantaged student with five years of football experience. While the talent in a fine motor skill like instrumentation may not translate well to gross motor skill in activity like football, the drive to persevere and engage in deliberate practice is much more common.

3 Hypothesis

Students with elevated levels of grit will consistently improve their academic performance, relative to their peers, as they progress through the sophomore fall semester of the industrial design curriculum.

4 Grit Test Outcomes

Out of the 66 students surveyed, only forty-five (63%) subsequently gained entry into the program. While it is believed that grit can play an important role in the success of these students, their starting points aren't all equal. Even if a student has phenomenal amounts of "grit", if he or she doesn't have the time to develop that talent, he or she will not make it into the program due to enrollment constraints.

4.1 Cumulative Grit Score

Students who appear in the chart below were determined to have the highest "grit" scores relative to their peers. Students that have been given an open star next to their name, as noted in the following figure, scored exceptionally high on both "personal perseverance" and "internal motivation" questions. This star system will be used in subsequent figures (Figs. 2 and 3).

1	Student ZZ ☆	10	Student D ☆	19	Student 3	28	Student DD
2	Student L	11	Student W ☆	20	Student 10	29	Student T
3	Student R ☆	12	Student H ☆	21	Student J	30	Student NN
4	Student G ☆	13	Student 2 ☆	22	Student FF	31	Student V
5	Student UU ☆	14	Student C	23	Student Y	32	Student O
6	Student U ☆	15	Student MM	24	Student QQ	33	Student VV
7	Student N ☆	16	Student OO	25	Student Z ☆	34	Student KK ☆
8	Student S ☆	17	Student 11 ☆	26	Student BB ☆	35	Student Q
9	Student B ☆	18	Student II ☆	27	Student HH	36	Student P

Fig. 2. Beginning of summer semester - cumulative grit score

		10	Student D ☆			28	Student DD
2	Student L	11	Student W ☆			29	Student T
		12	Student H ☆	21	Student J		
4	Student G ☆					31	Student V
5	Student UU ☆	14	Student C	23	Student Y		
6	Student U ☆			24	Student QQ	33	Student VV
7	Student N ☆	16	Student OO	25	Student Z ☆		
		17	Student 11 ☆	26	Student BB ☆	35	Student Q
						36	Student P

Fig. 3. Accepted to the program - cumulative grit score

Both internal perseverance and personal passion are important factors in meeting Duckworth's definition of grittiness. Twenty-one out of thirty-six students (58.3%) who ranked highest in the *Internal Perseverance Score* (not shown) were accepted into the program while nineteen out of thirty-six students (52.7%) scored among the highest in the *Personal Passions Score* (not shown) were accepted into the program. It is believed that the small variation between the two suggests that grit gained from life experiences is better suited for overcoming the challenges of the summer semester 'gateway' curriculum.

4.2 Activity Score

Students in the program demonstrate participation in a wide range of extracurricular activities prior to their admission to the university, including: water polo, drum line, art classes, and national honor societies. Researchers chose to assign additional value to such extracurricular activities as there is consensus that they add much to the student's educational experience. For students who participated in an activity for four years, an additional three (3) points were awarded. If a student participated in an event for only one (1) year, even if he or she won an award as a result of that activity, no additional points were awarded. The goal of this survey is to identify grit, not short-term achievement. Students receiving a total of six (6) points have been given a black background in the following figure, those totaling five (5) points in dark grey, totals of four (4) points are in grey, totals of three (3) points in light grey, totals of two (2) points are in cross hatch, total of (1) point are in back slashes.

Twenty-three out of thirty-six (63.9%) students listed in the *Activity Score* data set were accepted into the program. Of these twenty-three students, seventeen (73.9%) also appear in the cumulative grit score.

Figure 4 suggest that development of grit through previous experiences may be a better indicator of future success as it pertains to industrial design studies. The industrial design program focuses on a "learning through doing" approach. Future studios will issue design briefs that increase in scope and complexity. Successful students will need to reflect back upon past successes and failures to adequately address difficult design briefs. It is believed that students who have a history of overcoming challenges are more likely to assign themselves high "grit" scores not because they felt they were gritty in the moment - instead, they had the real-life experience to justify the score they would report.

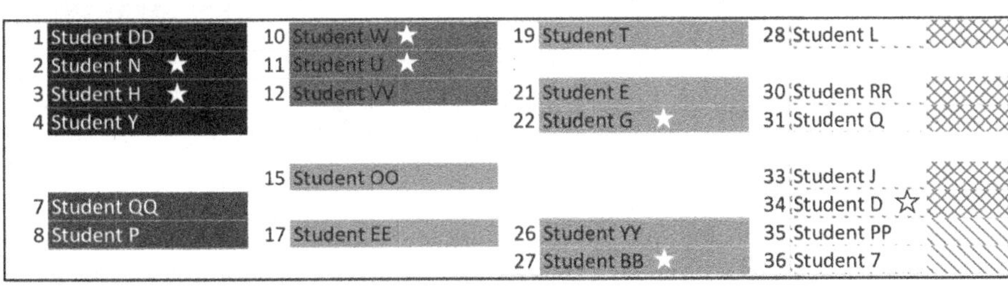

Fig. 4. Accepted to the program - high activity score

5 Fall Semester Outcomes

With the initial test complete, and gritty students identified, the hypothesis would be tested in a more typical design studio environment. Would gritty students continue to improve relative to their peers as they progress through their sophomore year? Given the higher success rate of the *Activity Grit Score*, opposed to the *Cumulative Grit Score*, this data set was used as the primary metric to track gritty students for the remainder of the school year.

5.1 Semester Results at Midterm

At spring midterm, a large number of gritty students received higher studio-based scores. A strong clustering at the top is encouraging for the use of "grit" as a predictor, but it also means that there is no other place to go but down.

It's important to note that the fall semester curriculum is not simply a continuation from the summer 'gateway' experience. Formerly, students were asked to apply one new skill to that directly related to a previous learned skill. The assignments of the Fall Semester "foundations" studio ask the students to apply a wider range of skills that are not as easily coupled with one another.

This is not to say that grit doesn't apply in this space. There are still numerous opportunities where placing extra effort into an assignment will yield a greater result.

The goal of the study is to notice improvement as the student continues to add their gritty effort to their design works. At this point, Fig. 5 serves as a baseline for future measurement.

1 Student Q	10 Student OO	19 Student 14	28 Student 7
2 Student VV	11 Student D	20 Student 5	29 Student AA
3 Student N	12 Student RR	21 Student K	30 Student 9
4 Student 12	13 Student Y	22 Student T	31 Student BB
5 Student W	14 Student EE	23 Student H	32 Student L
6 Student J	15 Student SS	24 Student CC	33 Student M
7 Student YY	16 Student P	25 Student A	34 Student WW
8 Student X	17 Student G	26 Student C	35 Student DD
9 Student U	18 Student XX	27 Student Z	36 Student PP

Fig. 5. Midterm ranking with activity score

5.2 Semester Results at Final

The results at the end of the fall semester were a surprising. There is a clustering of identified gritty students within the top-ten (80%). However, the grittiest students made no considerable change. Referencing the six most gritty students (those identified by black, dark grey, and grey backgrounds) from the Midterm ranking, three obtained higher academic rankings while the other three students obtained a lower ranking at the conclusion of the semester. Of the nineteen (19) gritty students identified in the final rank, nine (9) increased in rank, nine (9) decreased in rank, and one (1) remained stationary (Fig. 6).

1 Student VV	10 Student Y	19 Student F	28 Student N
2 Student OO	11 Student K	20 Student 9	29 Student PP
3 Student 12	12 Student X	21 Student G	30 Student Z
4 Student W	13 Student YY	22 Student 6	31 Student DD
5 Student T	14 Student C	23 Student A	32 Student 8
6 Student J	15 Student XX	24 Student H	33 Student U
7 Student Q	16 Student RR	25 Student LL	34 Student 5
8 Student 14	17 Student P	26 Student EE	35 Student TT
9 Student D	18 Student SS	27 Student BB	36 Student CC

Fig. 6. Final ranking analyzed with activity score

6 Conclusions

When it comes down to it, the researchers base hypothesis appears to be false. The grittiest students in the cohort, as evaluated by Duckworth's metrics, did not improve as the semester progressed. It is believed that there are three probable causes for this deviation from the expected result; the curriculum itself, student life structure, and a limited timeline for "grit" to emerge as an influencing factor.

Perhaps the programs' fall semester of the Industrial Design curriculum is not solely conquered through relentless grit. Students are adding foundational design skills like composition, contrast, and balance to the concepts gained during the summer 'gateway' sequence. The strength of the fall curriculum is that it bounces back and forth between two-dimensional design and three-dimensional design. While this approach helps students gain a more wholistic understanding of design, it can also be challenging to apply the learning outcomes from one project to another.

It is possible that there are more factors at play than just a student within a classroom. These young adults could still be adjusting to managing their own schedule and social calendar. An abundance of high school activities and accomplishments could indicate that these students' parents provided large amounts of structure and encouragement to their children. Part of participating in a professional program carries the weight of transitioning youthful tendencies to responsible habits.

The last cause is the very reason for this study. The goal is to see student improvement over time, knowing that long-term success comes from making multiple steps in the right direction. As students' progress through the curriculum they will be presented with more comprehensive, complex design challenges. Perhaps differences in the subsequent Spring semester will be a better place for so called "gritty" students to flourish? The projects are longer and fewer new skills and concepts are introduced on a weekly basis.

References

1. Duckworth, A.: Grit: The Power of Passion and Perseverance. Scribner, New York (2016)
2. Bush, B.: The "Grit" Test Applied to Summer Semester Industrial Design Students. https://youtu.be/rWqv-SM0OQw

Identity, Wayfinding and Construction for an Underserved Population in Ecuador

David Smith(✉)

Auburn University, Auburn, AL, USA
drs0049@auburn.edu

Abstract. Socially responsible design is a topic of significant interest to designers in all design disciplines. Socially responsible design manifests itself in multiple ways as designers explore how to integrate the concept into their practice. This paper is a case study about a particular manifestation of socially responsible design. It documents a collaborative effort between educators in graphic design and building science attempting to impact a poor neighborhood in Quito, Ecuador. Important elements of the project were that it accommodates a service learning class for students, that it utilizes local building methods, and that it meets a genuine need of the community. The project serves as an example of interdisciplinary collaboration finding a unique way to integrate socially responsible design into practice and as an educational tool.

Keywords: Socially responsible design · Social design · Graphic design · Building science · Ethical responsibility · Interdisciplinary · Underserved

1 Introduction

Design in the service of a better world comes with different names, different concerns and different tactics. This category of design is sometimes called design for good, green design, socially conscious design, socially responsible design, or as Milton Glaser prefers, *designism* [1].

While the names are varied, the subject matter is even more diverse. An early crusader in the push for socially conscious design was industrial designer Victor Papanek. An outspoken leader on the importance of considering the human condition, sustainability and the environment, Papanek said of design in the 21st century, "It will be understood that no design stands on its own: all design has social, ecological and environmental consequences that need to be evaluated and discussed in a common form" [2]. Graphic Design's preeminent professional organization, AIGA, offers designers a way to participant in socially responsible design through their 'Design for Good' program. Designers are encouraged to get involved in various social concerns in their community ranging from domestic violence to water access [3]. Others frame the idea of socially responsible design in terms of activism. In his book, *Design Activism: Beautiful Strangeness for a Sustainable World*, Alastair Fuad-Luke identifies the variety of forms design activism takes as types of capital. While the global notion of 'capitalism' has come to dominate economic and political thinking, Fuad-Luke believes activism can exert an effect on multiple types of capitals. His five capitals framework

G. Di Bucchianico et al. (Eds.): AHFE 2020, AISC 1202, pp. 234–239, 2020.
https://doi.org/10.1007/978-3-030-51194-4_31

includes natural, human, social, manufactured and financial [4]. This viewpoint gives designers a virtual endless list of ways to inject socially responsible design into their practice. An example of a possibly more extreme version of design activism can be seen in the work of *Adbusters*, a magazine without ads acting as a subversive challenge to how design is used in society. Adbusters coined a new social movement named 'culture jamming' which in their view was taking the storytelling, culture-generating power back from the commercial and corporate forces [5].

It is easy to understand why designers would want to apply their skills to lessen world ills like poverty and overconsumption, but no one is there to tell them in practical terms, how to make socially conscious design part of their day-to-day work life [1]. Design education seems like the best place to expose budding designers to social design and how they might include it in practice. History shows us that students have a hunger for work that can make a positive impact on the world. According to Ida Kamilla Lie in her article, *'Make Us More Useful to Society!': The Scandinavian Design Students' Organization (SDO) and Socially Responsible Design, 1967–1973*, Nordic design students were dissatisfied with the state of design practice and education at that time. They wanted their education to be more in line with addressing what they felt were urgent challenges of society [6]. Fast forwarding to today, the School of Visual Arts in NYC has addressed social design in a big way offering a MFA degree in Design for Social Innovation. Program founder, Cheryl Heller emphasizes that design is ultimately not valuable if it does not consider how it interacts with people in the real world [7]. Another vehicle higher education often uses for exposing students to social needs is service learning. Service learning is the real-world application of classroom knowledge into community settings, and this can be applied by many different academic disciplines, not just design [8].

We can see that socially conscious design is important to the design disciplines, and that design education can play a role in integrating it into practice. The following is an examination of an interdisciplinary effort between educators in graphic design and building science from Auburn University who take a service learning class to Quito, Ecuador. There, they undertake a community project which serves as an example of service learning that includes socially responsible design.

2 Community Need

Ecuador's population is approaching 17 million with over 20% below the poverty level. High poverty and income inequality mostly affect the indigenous, mixed race, and rural populations. The government has increased its social spending to ease these problems, but critics question the efficiency and implementation of its national development plan [9]. According to the World Bank Poverty Report, Ecuador is a poor country, measured by the number of people who cannot afford to purchase a basic basket of goods. Although the oil boom of the 1970s led to unprecedented growth, poverty remains pervasive [10]. 64.2% of the country's population live in urban areas, with Quito being the capital and home to 1.848 million citizens. As the capital, Quito bears the brunt of protests over economic issues. In the summer of 2018 citizens protested over the rapid increase of 'extreme' poverty from 7.9% to 9% [11].

As recently as October of 2019 protesters took to the streets in Quito over governmental austerity measures because of national debt.

In the high elevations of the Andes Mountains the city of Quito lies in a valley. The heart of the city is where most of the governmental offices, services, and wealthier citizens live. While the poor can be found all over Quito, as the population spreads up in elevation away from the main lower city areas, the neighborhoods become poorer.

One of those neighborhoods is Aida-Leon, which is home to a small church named Iglesia Bautista Esperanza Eterna. This church provides services of growing importance to the children of Aida-Leon, but with extremely limited resources the prospect of helping more children seemed hopeless.

3 Primary Objectives

Iglesia Bautista Esperanza Eterna provides facilities for two vital community services. The first is an after-school program for elementary and middle-school children. School is only in session for a half-day and most parents need to be out working or looking for work when their children arrive home. The afterschool program provides a safe place for children to go when they are let out of school. It includes nutritional snacks/meals, recreation and additional learning opportunities. The second service provided is daycare for pre-school children. The church only provides the space for the daycare while the government provides a very small budget for limited staff and supplies. Any expansion or improvements to the daycare must come from the limited resources of the Aida-Leon community. By American standards, this community is considered underserved as there is no welfare or alternatives for the poor families of Aida-Leon. There exists a great need to expand both of these programs as they continue to meet more and more community needs, but the existing small building housing these services with mostly volunteer workers was at full capacity.

SIFAT (Servants in Faith and Appropriate Technology) is an American ministry that brings assistance to projects helping communities create solutions that can eventually be maintained by the community without requiring additional relief or funding. SIFAT with church, daycare and community leaders came up with a plan to expand both programs. This would require: 1) Moving the daycare to an old unused building on the property after making it usable; 2) Build a new building adjacent to existing building for expansion of the after-school program; and 3) Renovating the existing building to better support the two programs. Volunteer teams have been visiting the site over the last 2 years donating supplies and their time as labor. The project is making good progress but still unfinished.

It was through SIFAT that Auburn University Professor of Building Science, Dr. Scott Kramer, began bringing students to the project as volunteer teams twice a year for a service learning course. The primary objectives of the course are to:1) Provide students with a real-world connection to people from a different culture; 2) Meet a deep community need; and 3) Utilize local building methods. Among the ways students made a cultural connection with people was the worksite itself. A Quito based site foreman and building engineer worked side by side with students on the new building. Ecuadorian meals were eaten together prepared by local volunteers. The students also had opportunities to interact with the children and families benefiting from the services. We've already established that this work will have a positive impact on the community by allowing the daycare and after-school programs to reach more children and their families. The last objective of using local building methods was facilitated by the local site foreman and building engineer. No heavy equipment is used and the concrete and bricks are locally made.

4 Design Overlay

In the Fall of 2018, Dr. Kramer invited Assistant Professor of Graphic Design, David Smith to join their service learning class in Quito in hopes they might enrich the work with graphic design. There was no predetermined need for design, but the goal was to see how graphic design might enhance the ongoing construction work. The objective of Prof. Smith's first trip was to assess where design might have the greatest and most practical impact on the project. Interviews with the church leaders, daycare center director, and volunteers were conducted, as well as visits to similar daycare and afterschool programs in other poor Quito neighborhoods. Many ideas were considered, such as interior or exterior murals, posters with health or educational messages, and building identification signs. It was determined that building identification and wayfinding signs would have the most useful impact.

There were several factors that led to the decision of choosing signage. First, the ongoing construction on the property and location changes of the services led to some confusion about where to find the right service, particularly for those new to the services. Secondly, the property itself is an odd shaped lot with buildings hugging two of the property lines and a big basketball court in the middle. Fencing surrounds the whole property with additional fencing around the individual buildings. The fencing is important to protect young children being cared for on the premises. However, this layout of fencing and building configuration was confusing and difficult to negotiate. Lastly, there is a lack of clear building identification, especially for the church. Two distinct community services are being offered, the after-school program through the church and the daycare program through the community, sanctioned by the government, and in cooperation with the church. Additionally, the church has worship services and other activities in their main meeting area. The church had no effective signage to identity the church or its after-school program. The daycare had an existing sign but it was badly faded, unreadable in spots and out of date. Also, the location of the daycare sign was directly in front of the church further convoluting what service is actually in the building.

The strategy for clarifying the location of the services was to make a distinction between what was church oriented and what was daycare oriented. This required a unique visual identity for both entities. The daycare was to use standardized graphics provided by the Ecuadorian government which specified logos, typography, colors and sizes. The church had no such system, logotype or typography that consistently expressed the church name. Photographs and measurements were taken so plans could be made for the signage. Before leaving Ecuador Assistant Professor David Smith of Graphic Design and Assistant Professor Alan Bugg of Building Science wrote a grant proposal to fund the signage project. A week after returning to Auburn University they learned they had received a small grant from the McWhorter Fund for Excellence to fund the project.

Back in Auburn specifics of the design including production and installation methods had to be worked out. A logotype was developed for the church as an umbrella visual identity. The logo became part of a simple wayfinding signage system to help direct people from the front of the church for worship activities or to the after-school program in the back. The daycare signage was a much easier solution since it was now in its own building. A sign identifying the daycare would be erected near the fence entrance in front of their building, and a second identity sign was planned to go near the road, since the daycare was at the back of the property. All design work was submitted to and approved by the church and daycare leaders in Quito before going into production.

In the United States the cost of permanent signage is very expensive, but using local building methods found in Quito was the solution. In Quito it was observed that most small businesses would wrap vinyl banner material around a frame of steel tubing to serve as their permanent signage. We produced the banner material in Auburn ensuring the details were correct. The banners were relatively light and flexible making them easy to transport with our personal luggage rather than shipping. The tubing would be fabricated by local tradesmen in Quito.

Spring of 2019 brought a new service learning class of students to Quito who would not only continue to work on the new building but also signage for the project. Working with a local Aida-Leon metal tradesman, production included wrapping vinyl banners around metal tubing and then setting the sign posts in concrete or affixing them to existing concrete structures.

5 Conclusion

How does this signage project serve as an example of socially responsible design in practice and as an educational tool? A review of the experiential and physical outcomes can help us understand.

Student Outcomes. Students experienced cultural exchanges with the people of Aida-Leon. They worked with them and ate with them. In some cases, they even visited homes of families in Aida-Leon. Students learned local building methods without the benefit of heavy equipment giving them further cultural perspective. Students took part in work as a 'service' believed to be meeting a real community need. The experience went beyond the student's chosen discipline of building science and stepped into the discipline of communication design giving them an interdisciplinary experience.

Community Outcomes. Leaders and volunteers participated in making decisions about how design would affect their offerings to the community. Real physical changes took place in the form of building construction and signage, expanding the ability to help more children and families. Local tradesman and workers were paid to help with the project.

Socially Responsible Design. This design project was conducted with the purpose of helping a community to better offer social services through visual communication. Amplifying distinction of services will help the community better understand where to find daycare and after-school help. This level of clarity in visual communication and production is not typically available to an underserved grassroots effort. The work was implemented based on the input and desires of the community. This project was conducted in an interdisciplinary educational context demonstrating socially responsible design in practice.

References

1. Scalin, N., Taute, M.: The Design Activist's Handbook: How Design. Simon and Schuster, Blue Ash (2012). pp. 14, 19
2. Papanek, V.: The Green Imperative, p. 48. Thames and Hudson, New York (1995)
3. AIGA | the professional association for design. Design for Good (2020). https://www.aiga.org/design-for-good/. Accessed 31 Jan 2020
4. Fuad-Luke, A.: Design activism, p. 6. Earthscan, London (2009)
5. Lasn, K.: Design anarchy. Adbusters Media Foundation, Vancouver (2006). Chap. 3
6. Lie, I.: 'Make us more useful to society!': the Scandinavian design students' organization (SDO) and socially responsible design, 1967–1973. Des. Cult. **8**(3), 327–361 (2016). https://doi.org/10.1080/17547075.2016.1223363
7. Mattioli, G.: Social design: a discipline in its own right - metropolis (2020). https://www.metropolismag.com/design/design-education/social-design-a-discipline-in-its-own-right/. Accessed 31 Jan 2020
8. Wolpert-Gawron, H.: What the heck is service learning? (2016). https://www.edutopia.org/blog/what-heck-service-learning-heather-wolpert-gawron. Accessed 31 Jan 2020
9. CIA, gov.: South America: Ecuador—The World Factbook - Central Intelligence Agency (2020). https://www.cia.gov/library/publications/the-world-factbook/geos/ec.html. Accessed 1 Feb 2020
10. Ecuador poverty report: World Bank, Washington, D.C., p. ix (1996)
11. Telesurenglish.net: Ecuador: Poverty Increases by 3% in Just Six Months (2018). https://www.telesurenglish.net/news/Ecuador-Poverty-Increases-by-3-in-Just-Six-Months-20180717-0013.html. Accessed 1 Feb 2020

ID Case Study: The Value of Intellectual Property in Innovative Product Development

Joyce Thomas(✉)

Auburn University, Auburn, USA
jkt0022@auburn.edu

Abstract. A new category in the vacuum cleaner industry was created in 1979 when Black & Decker introduced its first Dustbuster hand vac. B&D sold more than a million Dustbusters in that first year and inspired competitors to rapidly flood the market with knockoffs. Enterprising manufacturers in Asia reverse-engineered the Dustbuster and tried to sell their reproductions in the U.S. market. B&D embraced its intellectual property and successfully litigated imitators because of patents covering its unique appearance and configuration.

Eureka saw this new market category as an opportunity to embrace emerging Asia manufacturing, first creating its own hand vac that did not infringe on B&D's patents and then following up with a new product by adding a handle and floor nozzle to introduce the very first light weight battery-operated upright vacuum cleaner.

This case study discusses the value of intellectual property in innovative product development.

Keywords: Intellectual property · Product development · Design patent · Utility patent · Innovation

1 Introduction

In 1979, Black & Decker (B&D) introduced its first Dustbuster, a battery-operated handheld vacuum cleaner, and created a new category in the vacuum cleaner industry with its visual type form and technology. B&D sold more than a million Dustbusters in that first year and inspired competitors to rapidly flood the market with similar products and knockoffs. Dustbuster became nearly a generic term for this type of product.

In the same timeframe, the U.S. and China reestablished diplomatic relations that had long been disrupted and signed a bilateral trade agreement that gave a start to a rapid growth of trade between the two nations [1]. The recognition that China needed Western technology and investment created opportunities for foreign businesses that wanted to set up manufacturing there. Enterprising manufacturers there reverse-engineered the Dustbuster and tried to sell their reproductions in the U.S. market. B&D embraced its intellectual property and successfully litigated imitators because of design and utility patents covering its unique appearance and configuration.

A manufacturer in Hong Kong approached Eureka (Electrolux) with a Dustbuster imitation. Anxious to explore manufacturing in Asia, Eureka charged their design team to create a "Me Too" product for Eureka, which neatly sidestepped the patents Black &

G. Di Bucchianico et al. (Eds.): AHFE 2020, AISC 1202, pp. 240–247, 2020.
https://doi.org/10.1007/978-3-030-51194-4_32

Decker was fiercely protecting. The Mini Mite had limited commercial success; however, in a follow-up product development exploration when their design team added a handle and a floor nozzle to create a new product, Eureka scooped the market with the introduction of the very first light weight battery-operated upright vacuum cleaner. By coupling the motor unit to the nozzle with a bridge structure and placing the filter into the removable cup, a uniquely patentable product was created. Originally sold as the QuickUp, this new design was more than a simple product extension of the Dustbuster genre.

When Black & Decker entered this new product category, their original design accidentally copied those unique features that were now covered by Eureka's utility and design patents. Challenged by Eureka with patent infringement, Black & Decker settled out of court and paid royalties on their product through its life cycle. This market subsequently became one of the fastest growing segments of the floor care industry.

This case study discusses the value of intellectual property in innovative product development.

2 U.S. Patent Law – Protecting Intellectual Property from 1790–Present

Federal patent laws protecting an inventor's intellectual property were established in 1790 in the United States and underwent a number of changes over the next 160 plus years [2]. Modern patent law in the U.S. dates to the Patent Act of 1952 which requires an inventor not only to describe her/his invention, but also requires that the new and useful aspects of the invention be non-obvious. "The subject matter sought to be patented must be sufficiently different from what has been used or described before that it may be said to be non-obvious to a person having ordinary skill in the area of technology related to the invention." [3]. This aspect was added to keep inventors from taking ownership of a pool of knowledge that was known in a particular field.

According to the US Patent and Trademark Office a patent grants the inventor property rights for 20 years from the date the application was filed. This gives the inventor "the right to *exclude others* from making, using, offer for sale, or selling" the invention in the U.S. or "importing" it into the U.S. [3].

Pending United States Patent Applications were not published for access by the general public until March 15, 2001, as provided for by the 1999 American Inventors Protection Act amendment [2]. Since 2001, applications are now published 18 months from the earliest filing date [3].

Until quite recently, U.S. patent law favored the "*first to invent*" which included: 1) conception of the product, device, service and 2) reduction to practice – developing/prototyping the idea and discovering that the theory works. While the patent filing date set an "official" invention date, examiners also reviewed other signed, dated and witnessed materials to determine the actual conception date. The date of reduction to practice was then also critical in the resolution of priority between inventors. Patents were required be filed within 1-year of disclosure of the invention to the public or the invention passed into the public domain.

Provisional filing of patents was introduced in 1994 [2]. This allows an inventor one year from the date the provisional patent is applied for to convert it to a patent application. Provided the provisional application sufficiently describes the invention as filed for in the subsequent patent application, it will set the effective filing date for the invention.

In March of 2013 the U.S. patent law shifted to a "*first inventor to file*" system similar to many European countries, maintaining the requirements for reduction to practice and filing within 1 year of public disclosure.

Why does this matter to this case study? The patent for the QuickUp was filed in the time frame prior to the publication of applications so at that time in history there was no disclosure of pending applications in the Patent and Trademark Office.

3 Case Study

3.1 Innovation in the 1980's – Times Were Different Then!

Times were very different in the 1980's from our current experience. There was NO internet. Not only did this create a challenge for introducing new products, it also influenced the ability to research existing products and patents to make sure that innovations provide new and useful aspects of that are non-obvious to a person of ordinary skill. Paper copies of new (and old) patents were purchased and circulated through the engineering and design departments. Patent applications for new inventions were not published until the patent actually issued – thus providing the opportunity for accidental "copying" when similar innovations were being developed by different companies at similar times.

This was also a less litigious era when it came to enforcement of IP rights. Between 1980 and 1996 the number of patent infringement cases that were initiated in the U.S. ranged from about 800 per year to approximately 2000 per year. That number began climbing and peaked at about 3000 cases in 2004, but then spiked up to 5000 between 2010–2012. According to Quinn [4] "for the time period from 1980 to 2011 the number of trials remained remarkably constant, averaging 100.03 trials per year. In FY 2012 the number of trials spiked to 139".

In the 80's, consumer awareness of new and existing products was literally by word-of-mouth and through traditional media advertising. Product feedback from consumers to manufacturers mostly funneled through the independent organizations that sold their products. Vac shops selling Eureka's products were a good source; they heard the customer and sorted out the legitimate complaints and suggestions. The problem was that it was mostly verbal communication from the dealer to the sales branch or service department, and then again, usually verbal back to headquarters. "We acted on a lot of info, for sure, but how far up the chain information got depended on who was talking to who and how good a talker they were. Of course, we got letters as well – some phone calls – and there was plenty of anecdotal feedback from local experience. But no doubt the internet – from online customer reviews to guys on YouTube – have really changed the feedback loop." – Scott Bolbock [5].

Consumer and product trends were disseminated through industry publications (actual print magazines) – most of which have disappeared from publication, but some continue in online versions e.g. Home Appliance World [6]. Copies of these publications circulated throughout companies, passed person-to-person or through the mailroom.

Eureka also relied on focus groups to help them understand the wants, needs and desires of consumers. In the early 1980's, the product characteristic most desired in the vacuum cleaner industry was power to deep clean carpets. Uprights and canister vacs began advertising amps and horsepower (HP) ratings while the ASTM.org was developing methodology to measure performance characteristics that manufacturers hoped would give consumers confidence in products prior to purchase.

Carpeting had been a main stay in American homes – ramping up from about 6 million square yards sold in 1951 to almost 400 million yards in 1968. In the 70's shags and synthetics boomed, but by the start of the 80's consumers began to be savvier about homeownership and the value of turning a profit on their homes as they relocated for new jobs. By the late 80's moving into the 90's hardwood floors became more popular [7]. As the population became more transient and more women moved into the workforce, rising from 18 million in 1950 to 66 million in 2000 [8], the preferences of consumers began to shift towards lightweight products that completed the job quickly, making it *look* cleaner. This set the stage for the opportunities of battery-operated products.

3.2 Hand Vac Innovation – from Dustbuster to Mini Mite

"Borne out of the ashes of a marketplace failure, Dustbuster created a new mass-market category as it took the Black & Decker brand from the workshop into the home. Dustbuster has continued to lead the category for thirty years" [9].

In 1974, B&D built upon their history of innovation and work with NASA on battery-operated products for their moon mission and introduced one of the first consumer products to be powered by rechargeable batteries. Their Mod 4 system of power tool products included a hand vac, flashlight and other products that were all powered by the same battery filled handle that was moved from device to device. Although clearly innovative and ultimately very useful, consumers were not ready to adopt rechargeable batteries as a reliable power source since their primary previous experience was with disposable batteries in toys.

Seeking to understand why this product was not resonating with users, B&D discovered in their own research that "92% of women users surveyed in the Mod 4 aftermath were highly satisfied with the Spot-vac" [9]. This moved the focus of product usage to the home and thus the Dustbuster was born as a product concept in early 1977. Product development, prototyping, consumer research all ensued and ultimately the Dustbuster patent was filed December of 1978 and issued in December of 1980. Production began in January 1979 (Fig. 1). B&D found themselves in the position of needing to fight in the U.S. courts for injunctions to protect their product from imitations. The onslaught of other rechargeable knockoffs didn't begin until manufacturing really started taking hold on the Chinese mainland. Then, all of a sudden, you could get dozens of them.

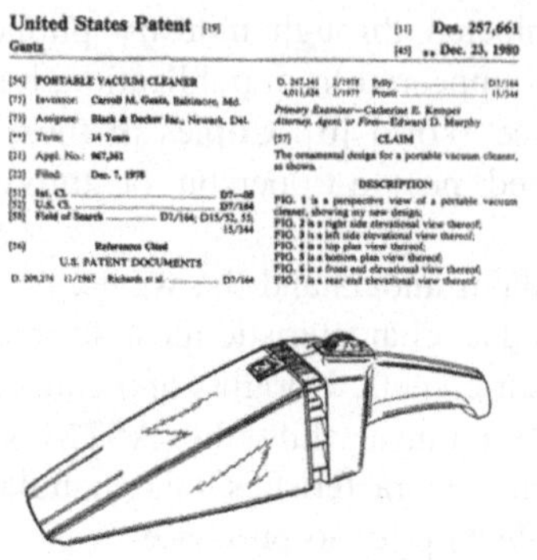

Fig. 1. B&D U.S. Patent Des. 257,661

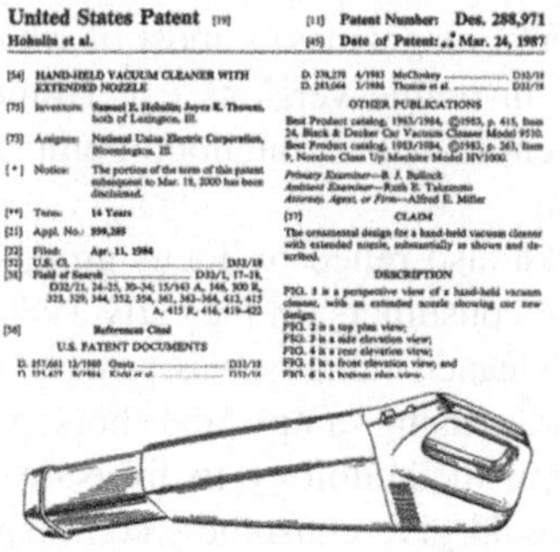

Fig. 2. MiniMite U.S. Patent Des. 288,971

The Dustbuster hit the market at the right moment – it met the needs of consumers in the changing marketplace. B&D was literally the only player in the rechargeable category. Up to that point, Eureka, and likely all the major players, considered them to be toys or novelties.

Eureka held back on jumping into this market, concerned about the customer satisfaction of the products and the potential damage to their brand. So, they surveyed consumers who owned Dustbusters. Finding that the satisfaction was relatively high, that was no longer an issue. Anxious to explore manufacturing in Asia, when a manufacturer in Hong Kong approached Eureka with a Dustbuster imitation that had an injunction prohibiting its sale in the U.S., Eureka charged their designer to create a hand vac for Eureka that "designed around" the Black & Decker patents. Eureka launched the Mini Mite hand vac in 1984, and while this was during the period that China was in the process of opening to the West, it was produced in Hong Kong for the first couple years until their manufacturer built a factory in Shenzhen. A U.S. patent was filed in April 1984 and issued March 1987. The Mini Mite allowed Eureka to enter this market that was continuing to expand and was a catalyst for their future new product innovation (Fig. 2).

3.3 Innovating a New Product Category – Lightweight Battery-Operated Upright Vacuums

In the designer's original exploration for the Mini Mite, ideation sketches included a wand (suction tube) and nozzle attachment that reached the floor from the hand vac when the user was in a standing position. At that moment, the hand vac product development was a priority and this idea was passed over.

However, in a follow-up product exploration after the Mini Mite went into production, the design team literally added a handle and a floor nozzle from another vacuum cleaner to a Mini Mite and tried it out. Designers, product coordinators and model makers all took it home to give it a try and discovered that the relatively low power of the Mini Mite motor wasn't enough, but all agreed that the prototype was a

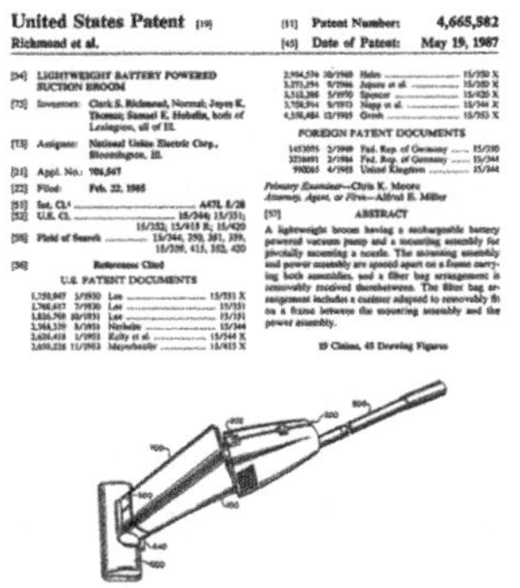

Fig. 3. United States Patent 4,665,582

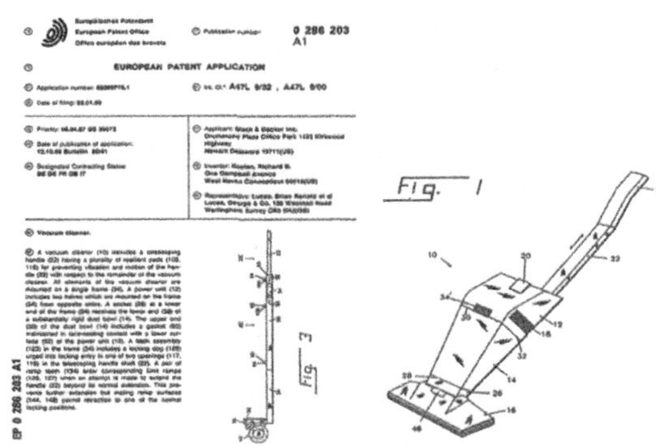

Fig. 4. B&D European Patent Application

start on a great idea. Through iterative product development, the designer coupled the motor unit to the nozzle with a bridge and placed the filter into the removable cup that was supported on the bridge, creating a uniquely patentable product. A patent application was filed in Feb 1985 and issued May 1987 (Fig. 3).

Focus group research showed that consumers agreed and so Eureka scooped the market with the introduction of the very first light weight battery-operated upright vacuum cleaner. Originally sold as the QuickUp, this new design was more than a simple product extension of the battery-operated hand vac genre.

When Black & Decker entered this product category, their original design accidentally copied those unique features that were now covered by Eureka's utility and design patents that were not published until May 1987. A B&D European patent application for their product filed in January 1988 has a priority date from U.S. Patent Application filed April 1987 (Fig. 4). Challenged by Eureka with patent infringement, Black & Decker settled out of court and paid royalties on their product through its life cycle – roughly about three years. They eventually launched a new platform that designed around the QuickUp patents.

Scott Bolbock, product planning at Eureka told me "I don't know if we appreciated at the time the challenges of introducing a new category to the market. B&D had blazed the trail for rechargeable hand vacs, but QuickUp was all alone. Corded 120v stick vacs were dead as a category. Consumers were naturally dubious of battery's power to do the job, the length of time they'd run, and the overall integrity of the rechargeable concept" [5].

Following the introduction of this product and the rough sales starts, Bolbock addressed the need to include marketing launch costs in the Investment Request for new products in his conversations with Swedish management (Electrolux). "We only planned for the development and production tools and depended on the operating budget to support any new launches."

In the mid 1980's Eureka still did much of their business through independents, catalogues, and catalog showrooms. It was relatively easy to get the initial distribution of the products, but without a real campaign to explain why people needed something they didn't understand, the movement was slow. Eureka's marketing department created a clever promo tape with a butler and maid dancing with the QuickUp. It was shot on 16 mm film and initially distributed on Fairchild 8 mm players for point of purchase display in some stores. In retrospect these promos were probably good for the small number of people that saw them, but not very effective on a national scale. Eureka struggled in gaining sales until this industry category began to take off with other competitors entering the market where they could compete on price.

In 1999, the original tooling was approaching the end of its useful life. Eureka took the opportunity to upgrade the product with new features and styling. This resulted in a new design patent that extended protection for this design for another 14 years. The footprint of the new product was limited by the existing carton size to not change shipping container quantities from China. Some additional technical changes were made to add a handgrip on the body so that it could be used as a hand vac and changes in the area of the filtration that resulted in a new utility patent for the filter. In production, the revised version didn't have enough filter area to be effective, so the product was reverted back to the original filter design several years later.

Battery-operated products subsequently became one of the fastest growing segments of the floor care industry. It is estimated that 150 million Dustbusters have been sold in the past 40 years through multiple product iterations/designs with close to $6 billion in sales for B&D [9]. The QuickUp/BrushUp product family line incorporates both battery-operated models and corded versions in two design iterations and is estimated to have sold nearly 9 million units totaling nearly $350 million in retail sales in the U.S. and Europe.

Bolbock says "In general, I'd say that up to 2000, we were the major player in that segment. The downside of that was that people weren't really looking for them and bought by accident. Electrolux (Eureka's parent company) did the Ergo Rapido and over the next decade everyone got in and it became a category" [5]. After 36 years, the BrushUp is still on the market in the U.S. in its 2nd generation form.

3.4 Author's Note

Joyce Thomas is an innovator and educator with a passion for design. Starting her career as a professional industrial designer for Eureka/Electrolux, she worked across interdisciplinary boundaries and integrated marketing, engineering, and consumer needs into creative products for 30 years before becoming an educator. This case study illustrates products for which she was the primary designer and is listed as inventor on the patents.

Scott Bolbock was a product coordinator and manager for Eureka and worked with Thomas and the engineering, design and manufacturing teams on these projects. He retired from Electrolux as Senior Director, Marketing, and is currently a director of Continuing Education at Heartland College in Normal, Illinois.

References

1. U.S. Trade with China: Selected Resources. Library of Congress. https://guides.loc.gov/us-trade-with-china
2. A brief History of the Patent Laws of the United States (2014). https://ladas.com/a-brief-history-of-the-patent-law-of-the-united-states-2/
3. United States Patent and Trademark Office: General Information Concerning Patents. https://www.uspto.gov/patents-getting-started/general-information-concerning-patents#heading-2
4. Quinn, G.: IPWatchDog.com "The Rise of Patent Litigation in America: 1980 – 2012" (2013). https://www.ipwatchdog.com/2013/04/09/the-rise-of-patent-litigation-in-america-1980-2012/id=38910/
5. Bolbock, S.: In personal correspondence with the author, 2 December 2019 & 14 January 2020
6. Home Appliance World. https://www.homeappliancesworld.com
7. Kueber, P.: Wall-to-wall carpeting history from the 1950s to today — an exclusive interview with Emily Morrow, Shaw Floors (2012). https://retrorenovation.com/2012/01/16/wall-to-wall-carpeting-history-from-the-1950s-to-today-an-exclusive-interview-with-emily-morrow-shaw-floors/
8. Toossi, M.: A century of change: The US labor force, 1950–2050. Monthly Lab. Rev. 125 (2002): 15. https://www.bls.gov/opub/mlr/2002/05/art2full.pdf
9. Gantz, C., Umbach, S.: The Little Vac that Roared: How Dustbuster "Cleaned Up" in the Market Place, IDSA Catalyst Case Studies (2009). http://www.idsa.org/images/pdfs/case_studies_2009_blackdecker.pdf

PET: Polyethylene Terephthalate – The Ubiquitous 500 ml Water Bottle

Clark Lundell(✉) and Joyce Thomas

Auburn University, Auburn, USA
lundece@auburn.edu

Abstract. Only 10% of worldwide PET production is recycled. PET is associated with Single Use Plastics, used a few min and discarded. A PET bottle in a landfill takes 500 years to decompose. That same bottle in the ocean will break down into micro plastics that last 500 years. Micro-plastics are consumed by organisms as part of a food chain, which includes us!

Since 2009, Coca-Cola distributed more than 35 billion PET bottles worldwide and introduced a PlantBottle packaging that converts sugars in plants into ingredients for fully recyclable PET plastic bottles.

China stopped buying 2/3's of the world's recyclable plastic in 2018.

Ideonella sakaiensis is a bacterium capable of producing enzymes that consume PET plastic as its' sole carbon and energy source. Producers of PET bottles are encouraging recycling, pursuing plant based PET and bacterial consumption of PET to promote decomposition in landfills and oceans as an alternative to recycling.

Keywords: PET · Recycling · Environment · Single Use Plastics

1 Introduction

The PET bottle was patented in 1973 by Nathaniel Wyeth, an American mechanical engineer and brother of the famous painter Andrew Wyeth [1].

PET is a plastic resin polyester formed by combining two monomers: modified ethylene glycol and purified terephthalic acid. It is made into plastic sheet used to create plastic PET pellets that are thermoformed by injection and blow molding into a bottle shaped negative mold producing the ubiquitous 500 ml Water Bottle.

The idea of selling bottled water was clever. The first commercially distributed water in America was bottled and sold by Jackson's Spa in Boston in 1767. Drinkers of bottled spa waters believed the water at mineral springs had therapeutic properties and that bathing in or drinking the water could treat common ailments. Who would have thought that you could charge money for something freely available, but also turn it into a multi-billion-dollar business? [2] Crystal Geyser bottled water is the best inexpensive bottled water, given issues of production cost, plastic recyclability, amount of material used, taste and lack of contaminates in the water [3]. There are some places where bottled water is the only option for clean drinking water, but elsewhere, it's just trendy and hugely wasteful.

G. Di Bucchianico et al. (Eds.): AHFE 2020, AISC 1202, pp. 248–254, 2020.
https://doi.org/10.1007/978-3-030-51194-4_33

2 PET Bottles

PET bottles are identified by the number 1 inside a triangular symbol embossed often on the bottom of the bottle. PET is also used in: clothing using DACRON or recycled PET fiber, 3D printer filaments, carpet and shampoo microbeads. A 500 ml bottle contains 16.9 fluid oz. and a standard soft drink can is 12.0 fluid oz (355 ml). You must not to expose PET bottles to 100+ degree F environments. They can leach endocrine (hormone) disruptors like antimony, which can cause developmental disorders in children.

Often the PET bottle plastic is different from the plastic screw cap used to seal it which has different recycling requirements. Caps are generally High-Density Polyethylene (HDPE - #2) or Foamed Polypropylene (#5) with PP on bottom. It is a harder plastic and has a higher melting point than PET. The bottle cap is often removed by rapidly crushing the PET bottle, which results in the cap being blown off separating it from the PET bottle to be recycled through its own unique process.

In the USA 50 lb of PET plastic passes through a typical household each year and only 30% of PET produced each year is recycled. In Europe 50% is recycled. PET has the burn characteristics of soft coal.

Bottling makes up 30% of the world PET production, 20 million metric tons. A metric ton equals 2,200 lbs. 60 million metric tons of PET are produced each year worldwide and readily recycled back into bottles, clothing and carpet. Unfortunately, less than 10% of worldwide PET production is recycled. PET is often associated with Single Use Plastics, used once and discarded, i.e. bottles, straws, food packaging, and bags. These products are used a few min in their life cycle and discarded to exist hundreds of years on and in the planet. Because PET is food safe, it is used throughout the food delivery system. 300 million metric tons of plastic are produced worldwide each year. 50% of the total plastic production is for single use products.

A PET bottle covered in a landfill can take 500+ years to decompose. That same bottle, born on a storm water and river conveyor to the ocean, will ultimately break down into a micro plastic that becomes part of five worldwide ocean gyres and can last in that state for 500+ years. Micro plastics are consumed by organisms that become part of the universal food chain, which includes us!!! Over one million marine animals die each year as an outcome of contact with or ingestion of plastic.

2.1 Recycling Plastics

Since 1950, the birth date of modern plastics production, 8 billion metric tons of plastic have been produced. This level of production is the third largest by pound of material manufactured by humankind in history only after steel and concrete. It is anticipated that by 2050 the total amount of plastic produced will be 12 billion metric tons.

Of the 300 million metric tons of worldwide plastic produced 10% is recycled, 10% burned, and 80% is discarded. Many of the products (textiles, carpet, bottles, etc.) produced from recycled material ultimately are discarded. Those materials become part of a landfill or of the 8 million metric tons of plastic that winds up in the oceans each year. This is enough to make four rows of plastic grocery bags filled with plastic, which can line the worldwide ocean seashore for its entire length (700,000 miles).

Once a designated “recycling bin” for the world’s post-consumer plastic scrap, China said no more when it instituted a ban on post-consumer plastic imports in 2018. In its first year, Operation National Sword led to near-total elimination of plastic scrap imports in 24 categories, including plastics such as PET, PE, and PVC. This approach will allow China to ramp up manufacturing of biodegradable and alternate products to replace single-use plastics. The ban comes on the heels of China announcing that its mega-dump, the largest in the country is full 25 years ahead of schedule. The Jiangcungou landfill in Shaanxi Province is the size of 100 football fields and was built to receive 2,500 tons of waste per day. Instead, the landfill received 10,000 tons of waste per day.

While China is one of the largest waste producers in the world, the USA produces more waste per person than China. Canada is the largest producer of waste in the world with estimates that it produces 36.1 metric tons (about 80,000 lb) of waste per day. In total, Canada produces 1.33 billion metric tons of waste per year. In 2015, the top partner countries from which Canada imported plastic or rubber include USA, China, Japan, Germany and Mexico [4].

2.2 Regulation of Single Use Plastics

There are three states in the USA that have banned single-use plastic bags. California was the first, followed by Hawaii then New York, whose ban will go into effect in March of 2020. Numerous cities and counties across the country have also banned single-use plastic.

China, one of the world’s largest consumer of plastics, announced its plan to ban single-use plastics across the country [5]. The proposal, aims to reduce plastic waste that has become ubiquitous in Chinese cities, promote recycling and green packaging and curb their overflowing garbage dumps. Single-use and non-degradable plastic bags will be banned in major Chinese cities by the end of 2020 and across the country by 2022. In addition, production of single-use straws and disposable plastic tableware along with household products containing microbeads (shampoo, toothpaste) will be banned by the end of 2020 and sales of such products will be banned two years after that. By 2025 the proposal, introduced by both the National Development and Reform Commission, and the Ministry of Ecology and Environment would ban all single-use plastics across the country. The phased approach will allow time for China to ramp up manufacturing of biodegradable and alternate products to replace single-use plastics.

Beverage companies in India are worried about a possible ban on single-use plastic and seeking clarity on whether this will have any impact on the polyethylene terephthalate (PET) bottles that they use. Prime Minister Narendra Modi, is leading efforts to scrap such plastics by 2022. In 2019 Modi urged Indians and government agencies to “take the first big step” toward eliminating single-use plastic [6]. Bisleri International, the country’s largest packaged water maker, said existing laws only allow the use of PET and glass for packaging water, and claims “Almost all PET is recycled in India. Authorities need to step up incentives for recycling” [7].

India’s primary packaging is completely recyclable and does not consist of single-use (non-recyclable) plastic, according to Coca-Cola India vice president for public affairs and communications. Lightweight packaging is being redesigned to reduce the quantity of plastic used in every bottle. Beverage makers and industry representatives are talking with the government on PET and single-use plastic. Consumption is already declining so switching to glass or cans entirely is not an option because both involve higher costs, leading to reduced profitability. Airlines, hotels and other companies have begun reducing use of single-use plastic. Air India has said it will stop single-use plastic on most of its flights and replace 200 ml water bottles with 1,500 ml ones on all flights. Delhi International Airport Ltd plans to be a single-use plastic free airport by the end of 2020. This includes water, juice and soft drink bottles. Fifty-five Indian airports are single-use plastic free now according to the Airports Authority of India (AAI) [6].

PET is not a single-use plastic (as defined in India) and is extensively recyclable. Size does not hamper recyclability. Statements like this bring into question the interpretation of "single use". Does it refer to a material's characteristics or products made from any type material? The government is working on a national blueprint to eliminate use of plastic bags, small bottles and straws, which is over and above the ban on single-use plastic such as bags. The expected ban may specify penalties for violations.

PepsiCo is aiming to make its packaging 100% recyclable, compostable or biodegradable by 2025. They are working to reduce the use of plastic across packaging, and make beverage PET bottles lighter in order to use less plastic. An estimate by the Central Pollution Control Board (CPCB) suggests that India generates about 25,940 tons of plastic every year. Coca-Cola and PepsiCo India officials said they recycle all their PET bottles and are leveraging technology to minimize the use of plastic [6].

2.3 Environment and Aspiration

A whole host of environmental problems come with our need to carry around an aspirational peer group bottle. Only half of bottled water is sourced from a natural spring, and the rest is tap water. When it's actual spring water, the transportation cost involved in getting it from the spring to the shelves can be as much as making and bottling the water itself. Consider the California droughts in the last few years. It takes three liters of water to manufacture a single liter of bottled water. California residents were ordered to restrict water use in the face of drought. Nestle continued to bottle and ship California's natural spring water at a rate of 36 million gallons a year, consuming 1 billion gallons of water in the process [8].

Is the age of the plastic water bottle ending? According to data from the Guardian [9], one million plastic bottles are sold around the globe every single min. Annual consumption is set to top half a trillion by 2021. Alternative solutions are needed.

2.4 A New Recyclable Future?

There is confusion about what is single-use plastic and clarity is required. The term 'single-use' is defined differently in the western world and includes any plastic material, recyclable or not, used once and discarded. Pressure is mounting globally on the beverage industry to reduce its use of plastic. For soft-drink makers Coca-Cola and PepsiCo, the production of PET bottle sizes from 250 ml to 600 ml constitute 50% of annual sales.

Coca-Cola will sell Dasani water in aluminum cans in the U.S. instead of plastic bottles. Rival PepsiCo said it will do the same for its Aquafina water. Coca-Cola had also said it will experiment with hybrid bottles, those made with a blend of recycled plastic and plant-based materials.

Since 2009, Coca-Cola has distributed more than 35 billion bottles in nearly 40 countries using its current version of PlantBottle packaging, made from up to 30% plant-based materials [10]. Coke estimates that the use of PlantBottle packaging has helped save annual emissions of more than 315,000 metric tons of carbon dioxide. PlantBottle packaging uses patented technology that converts natural sugars found in plants (sugarcane and waste from the sugarcane manufacturing process) into the ingredients for making fully recyclable PET plastic bottles. Both materials meet Coke's criteria as plant-based ingredients for PlantBottle material. These principles include demonstrating improved environmental and social performance. PlantBottle packaging is more expensive than conventional PET. Coke feels the market value and appeal is worth absorbing the added cost and does not pass this on to the consumer. The packaging looks, functions and recycles like traditional PET but has a lighter footprint on the planet and its resources.

The reason we have a problem with plastic pollution is convenience. Trying to change the delivery mechanism is going to be very difficult. There are other initiative's that are taking steps to create solutions.

One solution to improve the recyclability of the existing products is to find ways to cause PET to biodegrade more quickly. Ideonella sakaiensis is a bacterium capable of breaking down and consuming PET [11]. This bacteria uses two novel enzymes, PETase and MHETase, to decompose the plastic into two environmentally benign substances which serve as its' main food source. A colony of Ideonella sakaiensis could completely degrade a low-grade plastic water bottle in six weeks. Higher-grade PET products would require heating and cooling to weaken it before bacteria could start eating. The bacteria could also be used to reduce waste during plastics manufacturing.

Ari Jónsson, a product design student in Iceland, has combined powdered agar, a substance made from red algae, with water to fashion a new biodegradable bottle. When water is added to agar powder, the result is a jelly-like material that can be shaped into a bottle by a mold being cooled into a solid in a refrigerator. As long as it is filled with water, the bottle keeps its shape; as soon as it is empty it will begin to decompose. The substance can be reheated into a jelly and made into a brand-new bottle [12]. Jónsson's invention is unlikely to go into mass production soon as Agar tears too easily. Inventions like this, can provide solutions to problems posed by traditional plastic containers.

A new bottled water company, Cove is using a bottle made from a biopolymer called PHA. If the bottle ends up in a compost bin, landfill or ocean–it will fully biodegrade, made by bacteria through a fermentation process, is not new, but it has only recently become economical to produce. Startups, like the Bay Area-based Full Cycle Bioplastics, use food waste to feed the bacteria, solving two problems at once. Cove's bottles are created from biomass, but the company is also working to make the same product from captured greenhouse gases. This product could be carbon neutral and even carbon negative [13]. The company also plans to source water near the markets where it sells. Cove, which is a public benefit corporation, has guidelines that say it will not source from areas that are currently in a drought.

VeganBottle is a 100% biodegradable bottle made from sugar cane manufactured by Lyspackaging, a French packaging company. This solution combines sugar cane bagasse (the fiber/pulp that remains after extracting the sugar cane juice) with other plant-based products to produce material that is suitable for holding water, and other liquids including oil, beer or even perfume. Lyspackaging currently produces traditional plastic bottles, is planning to move half of its production to the biodegradable alternative, aiming to reach 80% of production by 2020 [14].

Paper Water Bottle® (PWB) is an eco-friendly solution. The water containment is made of a compostable pulp shell made from sustainable combinations of plant-based fibers. PWB has been awarded 16 global patents. PWB began in 2005, when creator, Jim Warner, a global designer and brand creator, took a walk with his son in the city. His son asked, "What do you do for a living?" Warner replied, "I design and make bottles and packaging for people." His son looked down at litter on the sidewalk and asked, "You make trash?" [15].

Recognizing the reality of this innocent question, Warner was determined to address the growing concern over plastic and its disposition in landfills.

3 Conclusion

Patented in 1973, the polyethylene terephthalate (PET) bottle makes up 50 lb of waste (water bottles and other food packaging) that passes through a typical household in the U.S. (125 million) each year. While 30% of PET produced each year in the U.S. is recycled, less than 10% of worldwide PET production is recycled. A 500 ml bottle contains 16.9 fluid oz. while a standard soft drink can is 12 oz. (355 ml).

PET is associated with Single Use Plastics, used a few min in its' lifecycle and discarded. The number 1 inside a triangular symbol identifies PET bottles. It takes 77 million years to make fossil fuels from which PET is produced and 30 min to be actively used and discarded as a water bottle. A PET bottle in a landfill takes 500+ years to decompose. That same bottle in the ocean will break down into micro plastics and last in that state for 500+ years. Micro plastics are consumed by organisms as part of a food chain, which includes us! Additionally, if PET is exposed to 100+ degree environments, consumption of contained liquids can cause developmental disorders.

Recycling facilities are facing mounting challenges because China stopped buying 2/3 s of world's recyclable plastic in 2018. Acknowledging the impact of PET on the environment, Coca-Cola has introduced PlantBottle packaging that converts sugars in plants into ingredients for making fully recyclable PET plastic bottles. Coca-Cola, Pepsi Co and Dr. Pepper are launching an "Every Bottle Back" initiative to keep bottles in recycling bins instead of oceans and landfills.

Producers of products contained in PET bottles are encouraging recycling which should reduce demand on landfills. They are also pursuing plant based PET and bacterial consumption of PET to promote decomposition in landfills as an alternative to recycling. Other alternatives are still needed, such as better methods of creating biodegradable bottles, and better consumer understanding of how single use plastics affect their environment.

References

1. Fowler, G., Wyeth, NC.: Inventor, Dies at 78: Developed the Plastic Soda Bottle, New York Times (1990). https://www.nytimes.com/1990/07/07/obituaries/n-c-wyeth-inventor-dies-at-78-developed-the-plastic-soda-bottle.html
2. Kelly, D.: Convenient Products That Are Destroying the World, GRUNGE (2017). https://www.grunge.com/47174/convenient-products-destroying-world/?utm_campaign=clip
3. Monnier, MN.: The Best Bottled Water, Reviews.com (2019). https://www.reviews.com/bottled-water/
4. Byrnes, H., Feolich, T.: Canada produces the most waste in the world. The US ranks third, USA Today (2019). https://www.usatoday.com/story/money/2019/07/12/canada-united-states-worlds-biggest-producers-of-waste/39534923/
5. Nace, T.: China To Ban All Single – Use Plastics, Forbes (2020). https://www.forbes.com/sites/trevornace/2020/01/20/china-to-ban-all-single-use-plastics/#48928fac7293
6. Mcdermid, B.: India's PM Modi exhorts nation to end usage of single-use plastic by 2022, Reuters (2020). https://www.reuters.com/article/us-india-pollution-plastic/indias-pm-modi-exhorts-nation-to-end-usage-of-single-use-plastic-by-2022-idUSKBN1WH29C
7. Bhushan, R.: Beverages fear pain as government steps up to push out single use plastic, The Economic Times (2019). https://economictimes.indiatimes.com/industry/indl-goods/svs/paper-/-wood-/-glass/-plastic/-marbles/beverages-fear-pain-as-government-steps-up-to-push-out-single-use-plastic/articleshow/71002245.cms?from=mdr
8. Wilson, J.: Nestlé is still taking national forest water for its Arrowhead label, with feds' help, Palm Springs Desert Sun (2019). https://www.desertsun.com/story/news/environment/2019/06/13/nestle-still-taking-public-forest-water-its-arrowhead-label-feds-help/1362211001/
9. Laville, S., Taylor, M.: A million bottles a minute: world's plastic binge 'as dangerous as climate change? The Guardian (2017). https://www.theguardian.com/environment/2017/jun/28/a-million-a-minute-worlds-plastic-bottle-binge-as-dangerous-as-climate-change
10. Meyers, E.: Coca-Cola debuts 100% plant-based bottle as companies go eco-friendly, Christian Science Monitor (2015). https://www.csmonitor.com/Business/The-Bite/2015/0604/Coca-Cola-debuts-100-plant-based-bottle-as-companies-go-eco-friendly
11. Yoshida, S., et al.: A bacterium that degrades and assimilates poly (ethylene terephthalate). Science **351**(6278), 1196–1199 (2016)
12. Morby, A.: Ari Jónsson uses algae to create biodegradable water bottles, Dezeen (2016). https://www.dezeen.com/2016/03/20/ari-jonsson-algae-biodegradable-water-bottles-iceland/
13. Peters, A.: This new water company's biodegradable bottle melts away if it ends up in the ocean, Fast Company (2019). https://www.fastcompany.com/90304442/this-new-water-companys-compostable-bottle-fully-biodegrades-in-the-ocean
14. Cane, G., von Herranz, A.G.: Introducing the First Biodegradable Bottle Made From Sugar, PESET | Digital for Good (2017). https://en.reset.org/blog/introducing-first-biodegradable-bottle-made-sugar-cane-11212017
15. Core77 Design Awards 2016, Runner Up Packaging Award, Core77 (2016). https://designawards.core77.com/Packaging/46847/Paper-Water-Bottle

Leveraging Design Competitions to Enhance Learning Goals and Student Motivation Within a Studio Project

Wendell Wilson(✉) and Young Mi Choi

Georgia Institute of Technology, Atlanta, USA
wendell.wilson@design.gatech.edu

Abstract. Industrial Designers traditionally use models to visualize and test designs. As fabrication processes shift from analog shop equipment to digital technologies, ID education must adjust curriculums. As a result, graduates often have gaps in their training. Accordingly, the School of Industrial Design at Georgia Tech. created a new course to bridge this gap between digital and traditional modelmaking.

This course has consistently featured the project that is the focus of this paper. Recently, a competition was leveraged so that project deliverables would satisfy the entry requirements while the prizes could add value to successful project completion.

The results of this most recent project were noticeably improved. Clearly leveraging the competition motivated students and enhanced the learning goals inherent in the project. Through a comparison of the results of this project over multiple semesters, it will be shown that competitions can serve as a useful motivational element in undergraduate design education.

Keywords: Industrial design · Design education · Modelmaking · Design competition · 3D printing

1 Introduction

Industrial Designers have traditionally used 3D models to visualize, communicate and test their designs. Whereas in the past, Industrial Design models were often utilized for promotional photography and marketing purposes, models are primarily used today for testing and validation. The tools and processes used to fabricate these models has shifted from traditional analog shop equipment and tools to digital technologies such as 3D printing, laser cutting and other rapid prototyping processes. As these new processes emerge, Industrial Design education has had to make decisions regarding specific skills that can be included within curriculums given the realities of course scheduling. Many ID graduates end up with noticeable gaps in their model making training and skills as a result of the tradeoffs that have been made. These gaps can manifest in poor decision making regarding materials and processes used to model designs or/and compromises in the quality of models themselves. Recognizing these potential shortcomings, the School of Industrial Design (SID) at the Georgia Institute of Technology created an elective course intended to supplement the skills being covered

G. Di Bucchianico et al. (Eds.): AHFE 2020, AISC 1202, pp. 255–260, 2020.
https://doi.org/10.1007/978-3-030-51194-4_34

in undergraduate studio courses. This new course, "Advanced Model Making", was intended to bridge the gap that has become apparent between teaching digital and traditional model making skills in current undergraduate ID education.

This course has been offered for 3 consecutive semesters thus far. The project that is the focus of this paper has been featured as a part of all 3 iterations of this course. The basic project timeframe and nature of the final deliverables has not changed substantially. In the most recent iteration, a national competition sponsored by *Wacom* and *SolidProfessor* was leveraged so that the deliverables produced by students over the course of the project would meet the entry requirements for the competition. The substantial prizes promoted by the sponsoring companies added perceived value to successful project completion. The only modifications necessary to piggyback the assignment onto the competition included a narrowing of the project focus from a "head or body mounted product" to a "Halloween themed mask" and the additional stipulation that students provide proof-of-entry (to the competition) as a part of their final project documentation.

In addition to the success and notoriety associated with one of the student's work being recognized as the winner of the competition, it was noted that the overall quality of most student work associated with this project was noticeably improved over that of previous course iterations.

Through a general comparison of the results of this project between that of two previous iterations that did not involve a competition and those of the most recent course where the project deliverables were entered in a competition, it will be shown that leveraging available competitions can be a useful motivational element in undergraduate education.

2 Background

The goal of this paper is to present observations of the differences in the results of a selected modeling project undertaken by 3 successive classes. In the most recent iteration of this class, the project was "piggybacked" onto a national design competition having similar goals as the class assignment. It is believed that by leveraging the competition, with the associated opportunities for public recognition and substantial awards, the quality of final project deliverables produced by the students can be significantly enhanced.

The class was structured as a design studio, which uses project based assignments to introduce learning objectives. Some past observations have indicated that competition can spur improved creativity in project work [1]. Project work presented as part of an internally organized competition [2], judged by a panel of peers and invited judges, can enhance learning and outcomes. The approach may help to improve design, engineering skills as well as presentation, communication and other soft skills.

Other cases have considered the submission of class project work to fully external design competitions [3]. External competition approach may have both positive and negative impacts on learning. Some positives include improving students' awareness of the intellectual and commercial potential of their ideas. It can help to improve the real-world relevancy of projects that are tackled in educational settings. Some negatives include possible delayed learning of basic skills and theory due to excessive focus on the competition's success criteria. The judging standards of different competitions may be mixed, and programs may risk losing control of necessary evaluation standards. Additionally, student's intellectual property may not be well protected when submitted externally.

Provided that the potential drawbacks are accounted for and the learning objectives are appropriately defined, the experience gained through competition appears to generally be an asset. It can provide additional confidence. Students can gain a higher sense of accomplishment not only through winning/placing in a competition, but also via feedback from other points of view they may not always get from a class-only project.

3 Method

This study was conducted within an undergraduate elective course in the School of Industrial Design at the Georgia Institute of Technology. The class, typically taken by 2nd or 3rd year SID students is focused on introducing and familiarizing students with advanced modelmaking techniques that are either not typically covered within studio classes (or to a greater degree than what can be addressed within the course of a studio project). Techniques typically covered in this class include body scanning & laser cutting, vacuum forming, 3D printing, kinematic development and urethane casting. The project that this the subject of this paper involved the students designing a 3D printed mask to fit their face as represented by a 3D scan performed in our body scanning lab.

Realizing that it is a simple matter to 3D print a single model of a CAD design, students were required to create their designs as a 3-piece assembly rather than a single part, forcing students to consider appropriate assembly features in their designs as well as accounting for the tolerances associated with 3D printing. To fine tune their designs, a two-step process was prescribed whereby (1) students would print trial parts to ascertain the success of their assembly and dimensional assumptions, making adjustments to their CAD models and (2) a second set of final prints representing the refined design. Towards this goal of using an iterative process, the School of Industrial Design provided a budget for the 3D printing of parts that could be used by the students in making the multiple prints. Additionally, a number of 3D printers are located throughout the SID that can be also be used by students without expense.

A competition sponsored by *SolidProfessor* and *Wacom* was announced shortly before the outset of this project. The competition involved the design of "Halloween themed" masks that could be 3D printed. Since the scope of the completion was similar to that of the planned assignment, the assigned project was "piggybacked" onto the entry requirements of the competition.

Each student was tasked with designing a mask in sketch format. Subsequently designs were modeled in CAD. Assembly features to permit assembly of the masks in 3 pieces were developed and incorporated into their CAD models. Some in-class discussion took place in which proposed assemblies were presented and potential problems were considered. Discussion of optimal orientation of parts during the 3D printing process also took place.

Students then developed plans for how they would troubleshoot their proposed assembly methods without printing the entire mask assembly – the goal being to minimize the time & cost of printing while still being able to effectively test their assumptions. Trial parts were printed that were subsequently presented to the class by the students. Discussion of problems and successes helped students make decisions regarding changes to their designs. Students revised their CAD models accordingly and printed their final mask components. Assembled masks were presented to the class, displayed on head form fixtures previously fabricated using laser cut components based on the 3D scans of each student's head. Students then submitted their CAD data to the design competition. One student's work was awarded the 1st place designation – receiving both publicity and a prize of substantial value.

Observing that the student work completed for the competition seemed to be, as a group, of higher quality than that of previous iterations of the project, it was speculated that layering the competitive element onto the project requirements had served to more highly motivate students. This paper focuses on the analysis of results from 3 consecutive iterations of this 3D printing project to ascertain whether the element of competition might be an effective motivational element.

4 Results and Discussion

Preconceptions about what we might discover were deliberately avoided at the outset of the study but it was presumed, at least by this author that the results from the project associated with a competition would be superior to those resulting from previous iterations of the assignment which had no competitive element.

Figures 1 and 2 depict the models fabricated as a part of this project from 3 different semesters that the course has been offered.

Although it is somewhat subjective, a comparison of the models resulting from this project can be made according to general considerations. Following are the general topics of comparison that were employed:

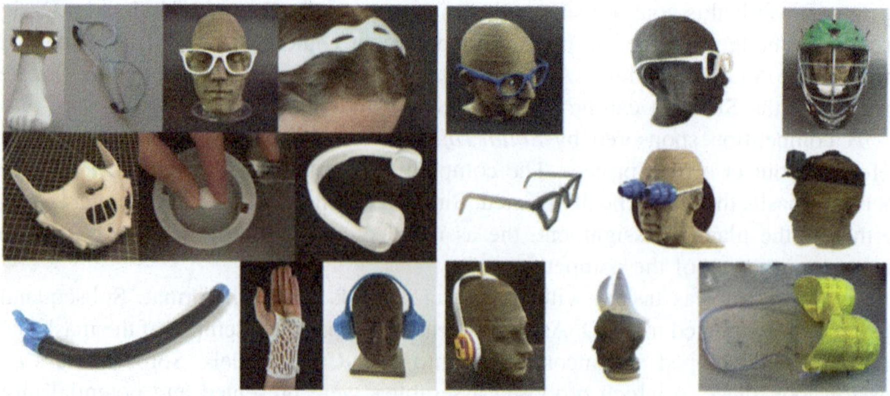

Fig. 1. Final 3D printed models course #1

Full Face Coverage: Whether or not the models provide coverage of the user's face; more coverage equating to larger, more elaborate designs?

Did not Meet Project Scope: Did the models met the assigned project scope?

Cohesive Aesthetic: Is the consistent application of a visual theme apparent in the final model?

Fig. 2. Final 3D printed models course #3

Appropriate Assembly: A 3-piece assembly was assigned. How appropriate for the overall design were the assembly techniques that were utilized? Was the final model divided into 3 pieces without further consideration? Scored on a scale from 1–3 with 1 representing little consideration of assembly techniques and 3 representing use of sophisticated assembly techniques. In general, assemblies comprised of parts merely divided into multiple pieces and glued together without additional design provisions received a lower score. Assemblies which used sophisticated assembly features integrated into individual components received higher scores.

Level of Effort: Did the final design require minimal effort to complete or was it more elaborate in design? Scored on scale from 1–3 with 1 representing minimum effort and 3 representing an elaborate design. In general, smaller assemblies comprised of parts having simple geometry received a lower score. Larger more complex assemblies received higher scores.

As illustrated by the comparison shown in Table 1, the results of Class #3 (featuring the design competition), the resulting models show greater consistency.

Table 1. Comparison of project deliverables over 3 semesters

Class #:	# of models:	Full face coverage:	Did not meet scope:	Sophisticated design:	Comprehensive asthetic:	Appropriate assembly:
1	10	1 *(10%)*	1 ***(10%)***	3 *(30%)*	2 *(20%)*	4 *(40%)*
2	9	0 *(0%)*	1 *(11.1%)*	2 *(22.2%)*	1 *(11.1%)*	5 *(55.5%)*
3 *(Competition)*	12	10 ***(83%)***	2 *(16.6%)*	5 ***(41.6%)***	8 ***(66.6%)***	8 ***(66.6%)***

5 Conclusions and Recommendations

The improvements in final project deliverables clearly demonstrate that leveraging the competition as a part of the assignment served to better motivate students and potentially enhance the learning goals inherent in the project.

While the sample size of this study was too small to draw firm conclusions, several opportunities for improvement of future projects of this nature can be identified:

- Narrowing the scope of final deliverables to a more specific category provides a more objective means of comparison and clearly simplifies project evaluation (i.e. mask-to-mask vs. headphones-to-respirator).
- Effort should be made to identify additional competitions that might be even better suited for enhancing course assignments – particularly for studio courses that typically focus more on overall design rather than only physical modeling.
- Promotion of competitive successes within the academic environment adds motivation for future student participation so effort should be made to publicize the results of this project to prospective students.

References

1. Lin, R.L.: Teaching achievement study of creative design competitions. In: International Conference on Genetic and Evolutionary Computing, pp. 72–79. Springer, Cham, November 2016
2. Chua, Y.L., Koh, Y.Y.: Internal competition in engineering education—a case study of project design competition in UNITEN. Adv. Sci. Lett. **23**(2), 708–711 (2017)
3. Wang, S.F.: The impact of design competition on design education. In: IOP Conference Series: Materials Science and Engineering, vol. 573, no. 1, p. 012060. IOP Publishing, July 2019

Strategic Pathways for Interdisciplinary Design Education

William Bullock(✉)

University of Illinois, Urbana-Champaign, USA
wbullock@illinois.edu

Abstract. Pathways for understanding critical components of interdisciplinary design education are necessary in the expanding universe of design. As the design of interactive products, environments, systems, and services expands, the shift has been profound for an academy historically immersed in product design education. New disciplines think and work differently. Facilities are often outdated and are not well-designed or equipped for interactive learning. Some collaborative partners are overly bureaucratic, isolated, and often have low budgets. Understanding the critical components of interdisciplinary design education is key to the development of new and more effective models. The author draws from over four decades of experience in education conducting interdisciplinary product design studies for industry.

Keywords: Interdisciplinary · Design education · Industrial design

1 Introduction

There is a long tradition of the design of physical artifacts we use, from architecture and civil engineering structures to industrial design for everyday products (Moggridge 2007). Industrial design has as much work as it ever did but the physical design's form and styling are not currently seen as the most central feature of a design by business (marketing) and engineering and even by the customer. UI and UX design are bringing about the merger of the physical, digital, and the mechanical. Because of the digitization of products, the tendency is for the physical design of a product to receive less emphasis. The design of digital technology products now focuses less on designing a beautiful or utilitarian physical object and more on designing our interactions with it. In spite of the reports of its demise, in the words of one of my colleagues, "industrial design is alive and kicking." Industrial designers are in demand, but the profession does not have the respect today that it had in years past when form (aesthetics) and styling were a central focus. Regardless of the importance of the digital interface, the physical design of the product still has to be designed.

The major stakeholders, engineering, business and design, are also facing change. Traditionally siloed and working independently as part of a step-by-step design development process, no one was responsible for overseeing all three domains. This has begun to change, however. Academic institutions have embraced design thinking in curriculums across the academy. This has created an unprecedented interest in design thinking methods along with interest in for interdisciplinary collaborations for mutual

G. Di Bucchianico et al. (Eds.): AHFE 2020, AISC 1202, pp. 261–267, 2020.
https://doi.org/10.1007/978-3-030-51194-4_35

benefit. It has also opened the door for industrial design to learn from and to share our knowledge in human-centered design principles and aesthetics.

Industrial design is ideally positioned to lead in integrating this interdisciplinary effort: Industrial designers work in all three domains and are responsible for creating a user-centric interactive product experience that delights customers. With the focus on UI/UX design, our sister profession graphic design, historically limited to 2D design application at the periphery of product development, is increasingly moving to a more prominent role in UI/UX design.

2 Industrial Design's Design Thinking Challenge

In response to this new paradigm shift to design thinking, industrial design education must adapt. New knowledge is being generated at an accelerating rate. Coding and new software tools must be understood. The challenge is how do we integrate this new expertise from stakeholder domains while maintaining a strong design core curriculum for our students (Nae 2017). And while tools are important, we also need to find new philosophical meaning in the ways we respond to the living environment where making and living come together.

Industrial design has traditionally served as a catalyst by working across disciplines as part of an interdisciplinary product design and development process. A product's beauty or form quality may not be the most important aspect of a design but looks are often the deciding factor on whether the product is purchased.

With limited budgets and faculty resources and the need to keep pace with developments in industry, most industrial design programs embrace interdisciplinary education. This usually takes the form of a collaborative studio where industry clients agree to act as clients on a studio design assignment assigned by the faculty. This is accompanied by a financial gift or gifts such as materials or equipment donations to the school. Faculty assign and manage the project and industry professionals are invited in at regular intervals to critique student work. For a long-standing example of this consider the "automotive schools"—institutions supported by automobile manufacturers who hire their graduates almost exclusively. Most university industrial design programs are involved in such symbiotic relationships with industry partly to obtain supplemental support, and to provide students with opportunities to address the human, technical, and manufacturing factors that have to be balanced in product design.

I spent the better part of my career as an industrial design educator at major research universities championing models of interdisciplinary design education. The goal of these programs is to introduce core competencies in interconnected fields. This involves inviting students and colleagues from different disciplines to collaborate on real life product research and development projects. While collaborations routinely take the form of undergraduate studio projects through industry collaborations, my research focuses on proprietary product design research and development studies for industry.

In my current lab at the University of Illinois at Urbana-Champaign, teaching and research are comingled. Interdisciplinary teams of advanced students are invited to work on collaborative projects under the direction of faculty researchers to solve real-world design problems. The rationale for the lab is that companies can no longer compete based on technology alone. To be successful, user needs must be met by creating innovative products that delight customers. For this reason, my lab links user-centered/customer focused and technology-driven disciplines to work toward common goals and solutions. This experience helps prepare students for leadership positions in collaborative product development. The disciplines integral to this effort are industrial design, graphic design, architecture and marketing, given their user-centered focus, and engineering and the sciences to support the integration of technology and manufacturing innovation. Our design process methodology combines customer, technology, business and marketing research with structured analysis, interdisciplinary brainstorming, rapid prototyping and evaluation. Research deliverables include development of smart products and technology transfer, strategic product and systems concepts, sustainable product development, ethnographic studies, trends forecasting, ergonomic verification testing and envisioning new products and systems. Research guides companies in articulating the future or addressing immediate needs through a fresh independent perspective.

Storytelling is a particularly effective technique used on interdisciplinary projects. Memorable stories (think Velcro; Penicillin and others) delight with surprising or novel outcomes. I use storytelling in my lab in two ways: First, for students to learn important case studies about the product design and development and as way to present research to clients; and, second, as a way to approach the human-centric basis of design thinking. Typically student teams develop a user persona for the user being designed for.

While story telling techniques are readily available online and in a myriad of books, I was recently invited to an informative workshop on storytelling by Dr. Shawn Jordan, Associate Professor of Engineering at Arizona State University (Storytelling in Design and Engineering). He suggested the story is a journey over time: How it was before? (Departure); What changed? (Difficulty/Discovery-over time or abruptly); How is it now? (Return); What does it mean? (Return). He also recommended that in preparing to tell your story that you visualize it by illustrating it with quick sketches or icons plus writing down the main points of your story.

3 Opportunities and Challenges

Collaborating on sponsored projects through funded research studies makes sense. Teaching and research can be comingled while sharing teaching responsibility with knowledgeable collaborators. It is also fun. Witnessing the excitement and commitment of students on these projects is a win win for everyone. Faculty learn about the collaborative research process and the success of various theoretical approaches, students receive course credit and valuable experience in research-driven innovation from faculty mentors, teaming collaboration and leadership while working on real projects. Industry clients receive unvarnished deliverables in the form of information including concepts based on an independent perspective unencumbered by the day-to-day realities of the corporate world.

Interdisciplinary collaborations can be challenging. Collaborations with colleagues takes more time, much more time than teaching alone. Presentations, advising, lesson preparation, and grading are all done jointly, in addition to client meetings. Scheduling weekly early evening class during the heart of the week provides the best opportunity for students from different majors to participate. So, while saving time is not one of the benefits, these collaborations provide expertise from other disciplines and overall positive learning experience for students.

The industry-funded research model is time consuming and difficult. A streamlined research approval process is needed. The process used at state-supported universities is typically modeled on significant research funding from the federal government. The proposal submission and approval process and the back in forth between company lawyers and university legal system is much too cumbersome to provide a quick turnaround necessary to meet industry needs. Industry funding for industrial design product research and development studies is typically not large, usually in the range of $50,000 or less for a semester-long research project and more typically half that. Overhead charged to projects, typically between 50 and 60%, reduces the available project budget by half or more. This leaves too little to purchase release time from teaching in order do the research. Hence, research industry collaborations usually occur in addition to full-time teaching. On larger federal grants funds are usually sufficient to provide faculty release time from teaching by funding replacement instructors.

While I enjoy working with colleagues at any stage of their career, early-career collaborators typically bring expertise in emerging areas along with enthusiasm, dedication, and interest. I find myself both mentor and mentee in these collaborations. I learn much or more from them as they do from me. I have been fortunate to work with good partners, and this experience more than makes up for the extra time required to develop and manage interdisciplinary projects. Such projects can help early career tenure track faculty develop their own research agenda.

The interdisciplinary teaching strategy I have found that works best is one of mentor-apprentice. The advantage of mentoring as an approach is that it empowers students to take responsibility for their own education and thus to develop responsibility for developing life-long learning skills. I have progressed from telling students what they should do to advising them on options to consider and why. While I still prime the pump by providing overall project planning, I involve students in finalizing schedules along with deliverables and deadlines. I take the time to explain why a particular project is a good experience and what they will learn. Usually students are highly motivated as they "apply" to take this class as an elective. The group is also vetted by submitting their resumes to build a qualified and motivated group representing a balanced mix of human- and technical-centered participants responsive to the needs of the project brief.

I find interdisciplinary projects generally work best for advanced undergraduates and graduate students who have the necessary skill sets and design project experience. There are exceptions of course, but inexperienced undergraduates can quickly get lost on teams unless detailed frequent guidance and direction are provided. As one might expect motivation is key, and I favor interest and motivation over talent. I have found that some students like while others abhor working on teams–often because they had a bad team experience in another class.

Interdisciplinary projects and research collaborations for creative product development projects can also provide badly needed funding. While funding for research in the arts pales in comparison to that of the sciences, opportunities for industrial design are improving. Industrial design is now being invited to collaborate on sponsored research projects at the proposal stage rather than afterward. I have received calls from well-meaning colleagues over the years inviting me to work on a project after it was funded as in, "ID had been included in the proposal, ostensibly to improve the chances of funding collaborative research" and now it was time to deliver. A good sign for our profession is that RFPs are beginning to couple theoretical discovery along with beneficial application, and this favors applied research that is in the traditional domain of most industrial designers.

Current budget models may not support interdisciplinary work. As higher education curricula have embraced design thinking, disciplines such as business are teaching courses such as product design and development, innovation, and design entrepreneurship. Engineering is teaching design thinking and perspective drawing. The humanities have design labs. Engineering and business have sizable enrollments, deep pockets, and faculty resources. Some are hiring industrial design faculty and recent graduates to teach design classes. Industrial design faculty, ostensibly due to their expertise, are also in demand to teach courses or workshops in design thinking in these areas. Many of these design courses can and should be taught by the design thinkers and doers in industrial design.

State of the art space designed and equipped with the tools to support interdisciplinary teaching and research is essential. Timely feedback on interdisciplinary projects is important. My colleagues and I require weekly progress reports in class. Teams meet with faculty researchers as needed. Students generally hold each other accountable but not always. Students understandably do not like complaining about their classmates, so determining the causes of team underperformance can be challenging. Once a team's underperformance becomes obvious, however, addressing the issue with the entire team in a private meeting usually works. It may seem obvious, but lack of candid communication and feedback between members is often the root cause of team underperformance; i.e., failing to show up for meetings or not delivering on promised work without explanation.

At the end of the semester I ask the students to complete an Individual and Peer Evaluation, in which they evaluate their performance and the performance of other team members, identifying their respective contributions to the project, initiative, and productive functioning of the team.

About three decades ago my graduates began to be hired to design interactions for increasingly smart products in the content on user screens, i.e., the entire experience. One of our graduates was in charge of interface design for one of the top major appliance manufacturers. Around the same time I attended a presentation at an IDSA conference by Bill Moggridge, Co-founder of IDEO. Bill talked about his experience in designing the first laptop computer and in spite of all the neat user features he had designed into the product including a clever latch and other features, when a working prototype arrived he forgot all about the physical design and felt he was being "sucked down into the screen". His experience with the product had very little to do with the physical design, and he needed to learn how to design the software piece. And thus the field of interaction design was born.

4 Strengths of Industrial Design

Industrial design gets it right. One of the strengths of good industrial design education programs is that we prepare graduates with general design skills and design process knowledge and experience that allows them to apply their knowledge more broadly. For example, every year at our annual Career Fair, companies tell me that they prefer hiring industrial design graduates in these emerging fields of interaction, interface and experience design. Reasons cited include the understanding of the user-centered process of design, ability to do research in search of unmet user needs, and visualization of 3D form using CAD and other software skills. One interesting comment was that "screens are no longer flat"–they deliver three-dimensional information in addition to graphics so understanding perspective and form-giving can be a decided asset in designing 3-D virtual worlds.

Industrial design graduates can apply their knowledge and skill in so many different ways. While preparation for entry level positions in the profession is one of our goals, industrial design undergraduate education has always been about providing students with general knowledge in the sciences, humanities, social sciences, and art and design while empowering them to learn how to learn and participate as an informed citizen of a democratic society. For example, the 2000 IDSA member directory listed over 80 specialty employment areas for members including business and industrial products, business development, consumer products, design explorations, design resources, design strategy, environments, furniture, medical and scientific products, packaging and graphics (including graphical user interface, software: web and multimedia design), product development process (including design research, ecological design, ethnographic research, universal design) along with transportation.

My own experience as a practicing designer along with that of many of my colleagues in industry bears out this diversity of opportunity in the field. In my work in corporations, design consultancies and as an independent consultant, I have been involved in the design of products (electronics, fitness and appliances), furniture, exhibits, point of purchase displays, equipment, packaging, graphics, illustration, aircraft human factors, human factors research, expert witness, in addition to being an educator.

So history tells us that industrial designers have been working in what is now touted as "new" areas of opportunity such as interaction design, experience design (UX), interface design (UI), design research and strategy from the beginning and in fact have helped found and shape these fields. What has changed is focus and opportunity. The recognition by industry that the user needs must be met through better design has given rise to the awareness of design thinking as an approach that puts the needs of user first in the creative design process. This reflects what I have been teaching my students for more than four decades. In fact, one of the challenges in education is to teach an 18-year-old the concept of designing for others as opposed to themselves. We approach design problems by trying to understand the user and their needs. This problem-solving process is more method and philosophy than exact science. I have personally witnessed some awful designs recommended by human factors experts based on empirical studies. Equally, industrial designers can produce bad results in spite of their best efforts. Where scientific/technical reasoning and the art of design intersect is perhaps the zone where the optimal outcomes can be found. The goal is to understand the user and to uncover unmet or unrecognized user needs. If we are successful, the product sells. If not it is soon forgotten.

The process we follow is an iterative one with a goal of creating a new and/or improved solutions by design. We are design thinkers and importantly design doers. We are by tradition the "form meisters" or giver of form and we expect our students to be skilled in form giving through an understanding of basic design principles and aesthetics. We want our students to design the user experience, whether that results in a physical product or a virtual or interface-based one.

5 Conclusion

So the good news is that industrial design remains relevant and important. Change remains a challenge and to stay relevant, we will need to continue to strengthen our interdisciplinary education and research initiatives, particularly with engineering and business (marketing). Industrial design is ideally suited to bring these domains together through research and education. But to support this work, we must press for budget models and facilities that make interdisciplinary collaboration possible and appropriate resources to fund it.

Industrial design has traditionally served as a catalyst by bringing disciplines together as part of an interdisciplinary product design and development process. One of the industrial designer's key strengths is the ability to visualize the product's form and usability. A product's beauty or form quality may not be the most important aspect of a design; however, a product's look and aesthetics are often the deciding factor on whether or not the product is successful. We can and should continue in these leadership roles even as the design universe expands. We should not abandon our role in determining the new physical beauty, aesthetics, and style to design thinkers.

References

Nae, H.-J.: An interdisciplinary design education framework. Des. J. **20**(suppl. 1), S835–S847 (2017). https://doi.org/10.1080/14606925.2017.1353030

Moggridge, B.: Designing Interactions. MIT Press, Cambridge (2007)

Cagan, J., Vogel, C.: Creating Breakthrough Products: Innovation from Product Planning to Program Approval. Prentice Hall PTR, Englewood Cliffs (2002)

A Student Driven Model for Creative Pedagogy

Angie Fahlman(✉) and Lois Frankel(✉)

School of Industrial Design, Carleton University, Ottawa, ON, Canada
angiefahlman@cmail.carleton.ca,
lois.frankel@carleton.ca

Abstract. This literature review discusses the possibility of students being the driving force of an interdisciplinary research educational model for creative pedagogy. Current models lack the robust interdisciplinary and individually oriented approach for disciplines in the field of creative studies such as Graphic Design, where the students are the driving force for their educational methods and assessments. This paper posits that, if the education system is designed for the average learner, then the system is designed for none. The purpose of this literature review is to initiate discussions about customized learning plans for the graphic design discipline and to provide ideas about how to progress towards a more user-driven interdisciplinary approach for accomplishing this goal. This review seeks to shed light on the possibilities of individual learning plans across creative disciplines. Finally, the paper synthesizes these diverse ideas into an early stage interdisciplinary model for developing a student-driven learning model for creative pedagogy.

Keywords: Interdisciplinary · Pedagogy · Stakeholder · Beneficiary · Taxonomies

1 Introduction

Present design education is outdated and needs to move in a more progressive direction [1]. Discussions among the design and education community should be initiated in order to address the flaws of designing for average users. The "flaw of averages," the conceptual pitfall that traps so many designs, process and product in underperformance [2].

Imagine for a minute if learners, industry representatives, educational professionals and educators could contribute to a more cohesive learning model that would be adaptable according to each learner's preference. In essence it could be an individual-based learning model configured to help each learner to excel, learn and digest information. Presently, many members of the learning population benefit from individual education plans (IEP) [3] that accommodate according to how they learn. Individual education plans are used by school for individuals who require accommodations due to learning disabilities. The IEP allows these learners to adjust their learning environments to foster better outcomes. This concept is supported by Augsten and Gekeler [4] who suggest that, in this new era, each student has to be evaluated on a more individual basis. This paper intends to shed light on how interdisciplinary perspectives might help shape an individual-based learning model for creative pedagogy.

G. Di Bucchianico et al. (Eds.): AHFE 2020, AISC 1202, pp. 268–276, 2020.
https://doi.org/10.1007/978-3-030-51194-4_36

This literature review attempts to create a better understanding of the educational landscape at a post-secondary level by providing a brief description of past learning theories and discussing how they may contribute to the design of a new model. The paper introduces interdisciplinary stakeholders, their roles and their possible involvement in the process. It emphasizes each stakeholder's unique perspective and how they may enrich the end model. Potential obstacles and negative aspects are discussed. Finally, future considerations for this model are touched upon.

2 Background

2.1 Pedagogy Theorist Overview

This paper provides a brief overview and critique of four key learning theories which will be synthesized to help shape a new student driven learning model. The four learning theories include: Kolb's Experiential Learning Process and Learning Style Inventory (LSI) [5], Gardner's Multiple Intelligence Theory [6], Bloom's Taxonomy [7, 8], and Universal Lesson Plan (UDL) [9]. Thereafter the paper will introduce the stakeholders and provide descriptions of their roles and their interdisciplinary contributions.

2.2 Kolb's Experiential Learning Process

David Kolb was an educational theorist with a background in psychology who produced an experiential learning process model and a twelve-item self-assessment tool called the Learning Style Inventory (LSI). Kolb divides the learning process into a four-stage cycle. In addition to the learning process model, Kolb's twelve-item assessment tool assists learners in identifying the way they learn from their experiences. The LSI tool produces a score for each of the four stages in the experiential learning process, that results in an overall categorization of their learning style [5] (Fig. 1).

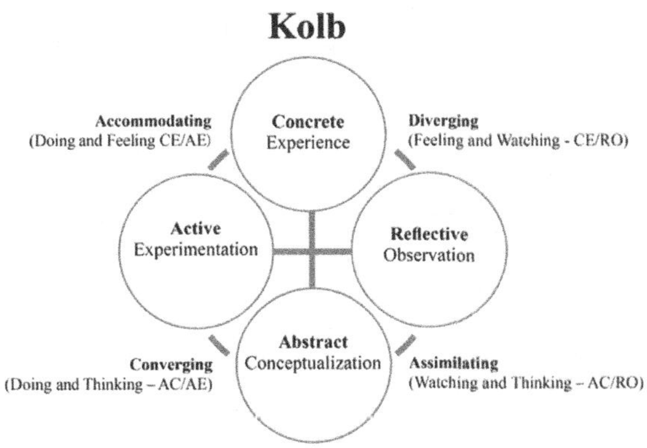

Fig. 1. Kolb's experimental process diagram [1]

Kolb's psychology background contributes to his insights into the cognitive processes of learning. His theories discuss the way a learner experiences learning. Although the average path for a learner is acknowledged by this paper it is not embraced as a superior method by some [10]. Another note for concern is that it does not accommodate learners who evolve throughout the education process or learners who develop compensation methods due to a variety of factors and concludes that model evaluations should be taken at multiple times [10]. There was no mention of how often the student should take the test or at which stages of the learning process.

2.3 Gardner's Theory of Multiple Intelligence

Howard Gardner is a practitioner of education with a background in developmental psychology. Gardner's Theory of Multiple Intelligence proposes that learners will each have their own set of intellectual strengths and weaknesses. The theory recognizes that each learner has a different and unique collection of abilities and talents. Gardner's theory introduces the notion that each person possesses capacities in eight of the sets of intelligence. According to this theory, most individuals have stronger and weaker intelligence categories, which may be due to environmental influences and exposures (Fig. 2 shows an example) [6, 11].

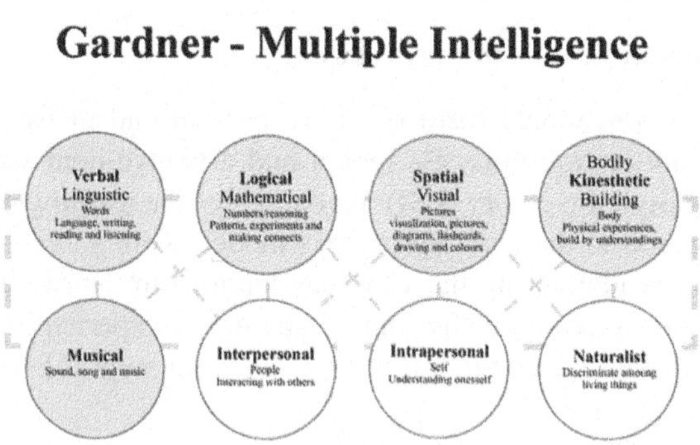

Fig. 2. Gardner's multiple intelligence diagram [2]

Based on this multiple-intelligence theory, a learner at the post-secondary level will already have a set of established intellectual strengths and weaknesses, which may provide a rationale for supporting the concept that an individual-based learning model could possibly benefit the learner. It may be beneficial to develop a tool for identifying a learner's multiple intelligence domains and his or her weaknesses and strengths. There may be a relationship between the categories in Gardner's theory and those in Kolb's twelve-item assessment tool which if combined may contribute to a more robust individual-based learning model. It would also be beneficial if this test was administered within every semester as a learner's intelligence could evolve, as Gardner suggests [6, 11].

2.4 Bloom's Taxonomy

Benjamin Bloom, who was an educational psychologist, and a group of colleagues, introduced the concept/language of classifying cognitive functioning by dividing learning into three different domains. Each domain is broken further into varying levels of ability, where the lower level should be addressed before the higher levels can be achieved. Bloom's cognitive domain model is based on the use of verbs. This theory breaks down how people learn and provides educators with a sequence for learning. The objective of Bloom's Taxonomy is to align educational goals, curricula (differentiate content), and assessments in hopes of producing attainable learning outcomes [7, 8].

Bloom's theoretical perspective may contribute to Kolb's and Gardner's theories by providing descriptors and, if applied properly, could set up achievable goals and provide measures for success (Fig. 3).

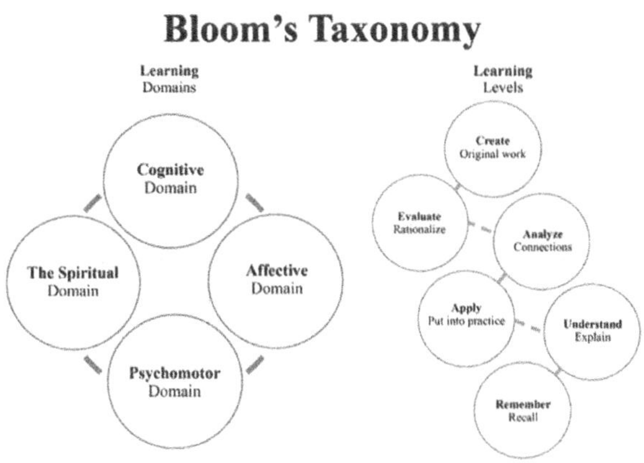

Fig. 3. Bloom's taxonomy diagram [7, 8]

In the education industry there are critics of the mentioned learning theories. There is a community of thought that categorization methods place learners in a section that may produce limitations and self-esteem issues, plus possibly hinder expectations [10]. In contrast, design educator, Ken Friedman supports research through a pedagogical program of self-learning [12]. Both views are valid and should be explored further.

2.5 UDL (Universal Lesson Plan)

Some post-secondary educational institutions have adopted the universal lesson plan, called Bridge, Pre-Assessment, Post-Assessment, or Participatory Learning, (BOPPPS) for generating a more cohesive learning experience. According to the literature, the lesson plan structure should contain all six sections for every lesson. Lessons plans are generally created based on the curriculum topic and administered according to one of the three lesson plan types, Psychomotor (Skills), Affective (Attitudes, feelings & values), and Cognitive (Knowledge) (Fig. 4 shows an example) [9].

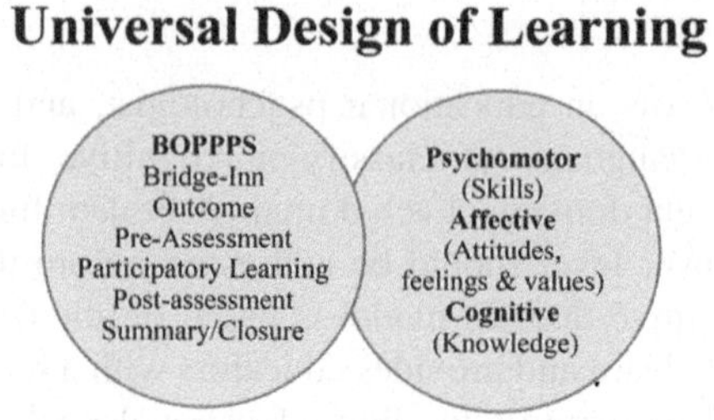

Fig. 4. UDL – universal design of learning diagram [4]

The universal lesson plan is a basic practice that provides guidelines for preparing class curriculum for a general student population, not individual learners. It appears that this model could help promote comprehension by adjusting the lesson plan type per individual learner. Overall, it reinforces curriculum lesson development and the assessment process.

2.6 Outlining Stakeholders Plan

The ultimate stakeholder – the student – directly benefits from an approach that permits them to customize their own learning plan. Mcdonagh [13] states that the end-user's input helps shape its own design. Mcdonagh [13] proposes that the design should develop in a way that transforms the user into an active participant in the design process. Moreno and Villalba [14] supports this view by stating that the role of the educator is changing into one of facilitation, which is more of a guide who allows the learner to contribute to their own experience and directly benefit from his or her involvement.

Moreno and Villalba [14] advocate for interdisciplinary collaboration in the design of curricula and promote this approach as beneficial to all the stakeholders. The inclusion of stakeholder could produce more sound outcomes, industry from various disciplines, institutions, facilitators and finally, the end-user [14]. By fostering the inclusion of interdisciplinary perspectives in the development of a new model, the education system could become a pioneer for this holistic approach [14].

The primary stakeholders are based on the existing system structure in the post-secondary institutions, which include faculty that oversee course development and execution. Their positions provide unique perspectives, allowing for insight into the possible obstacles, student struggles and achievements.

Secondary stakeholders would be industry representatives who would be consulted to determine any gaps or missing skills. This group would provide an interesting perspective on how they interpret the students learns outside in the workforce. Dubberly [15] identifies the importance of involving the community members who contribute to the discipline in the education process.

Tertiary stakeholders would consist of support staff from the administration, advising and counselling. These stakeholders have direct interactions with the students and therefore have inside knowledge of student struggles and navigating difficulties. They would give insight into process and implementation perspectives.

3 Discussion

This paper will synthesize the four learning theories, plus introduce the inclusion of interdisciplinary perspectives in terms of stakeholders in order to help shape a student driven learning model for creative pedagogy. Dubberly [15] urges the teaching community to begin discussions on re-inventing design education in order to stay relevant. This new approach, the fluid model, may or may not be a significant factor in student success.

3.1 The Fluid Model

The new proposed fluid model incorporates portions of the four learning theories, then introduces the inclusion of stakeholders to create a more cohesive and adaptable educational model (Fig. 5).

The model begins by illustrating who the stakeholders are. The post-secondary educational institution, it's faculty, along with the creative industry, creates rich and current curriculum initiatives. The student also participates in these discussions through course reviews and feedback. The support faculty devises a seamless administrative process, which takes into account the specialized student's plan. The student is the primary focus and beneficiary that selects methods in which they would like to receive instructions and be assessed, based on Gardener's theory. It should be a common practice that the student is not only the learner but also contributes based on their direct experience with the product [16].

Both Kolb and Bloom's theories are absorbed within the UDL theory and utilized through instructional delivery. It is proposed that the student be given the opportunity to adjust their methods each semester to ensure they have control of their learning experience.

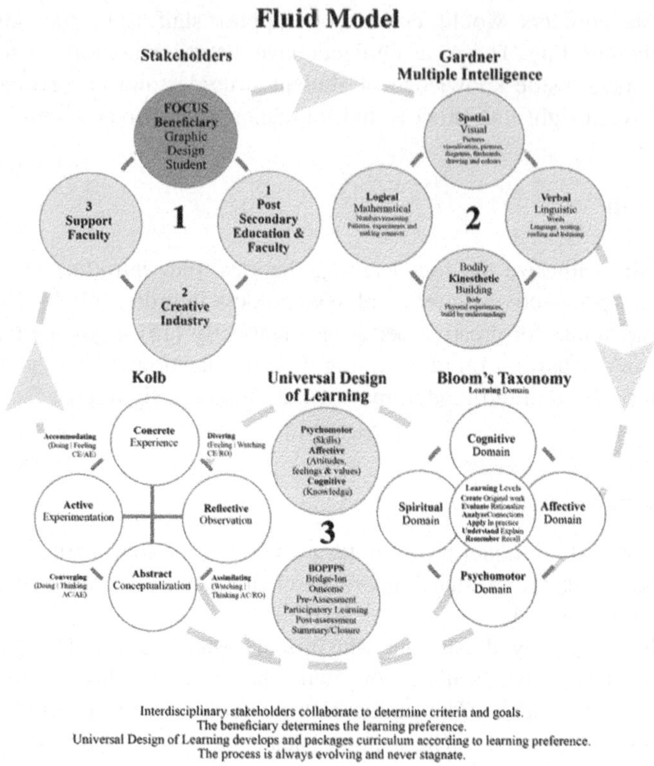

Fig. 5. Model

In the education industry there are critics of the mentioned learning theories. There is a community of thought that categorization methods place learners in a section that may produce limitations and self-esteem issues, plus possibly hinder expectations [10]. In contrast, design educator, Ken Friedman supports research through a pedagogical program of self-learning [12]. Both views are valid and should be explored further.

4 Future Research and Gaps

Both this interdisciplinary approach for a new learning model for creative pedagogy could open up avenues for improving learning and teaching dynamics. The continuous flow of interdisciplinary perspectives from all stakeholders may foster better relationships, promote knowledge progression and improve the educational environment. Without continuous prompting from education institutions, industry and students the existing process will remain in place without challenge [1].

In addition, there could be resistance from within the educational community, industry and learners, due to the unknown effects that the new model might have. Another point to consider is the level of complexity that may arise due to the network of contextual relationships, which may make the situation more intricate and therefore more laborious [14].

5 Conclusion

This paper began by providing a brief overview of four learning theories. Then stakeholders were introduced to provide insight into their roles and interdisciplinary perspectives. The learning theories where synthesized and stakeholders were integrated into the process in order to produce a new learning model that was student driven. Future considerations were mentioned along with the possibility of adapting this model to future frameworks and other learning disciplines outside of the creative learning environment.

In order to have a better and more in-depth view of this topic a greater literature review will have to be conducted. It should be expressed that there is plethora of materials on this topic and yet none have managed to have a large enough impact or produced a drastic enough change. The education community needs to ponder why is this so? What are the barriers that are stopping the progress to a more individual-based learning platform? Is it due to the unknown, funding, lack or insight or resources? These and many more questions need to be addressed in order to remove the obstacles and forge to a new and higher level of education.

References

1. Design Education Manifesto, ICOGRADA. A proposal for the future of design education (2011). http://www.icograda.org/uploads/resources/IcogradaEducationManifesto.pdf. Accessed Oct 2000
2. de Neufville, R., Scholtes, S.: Flexibility in Engineering Design
3. OCDSB. p. 187 (2011). https://ocdsb.ca/cms/One.aspx?portalId=55478&pageId=238021. Accessed Dec 2019
4. Augsten, A., Gekeler, M.: From a master of crafts to a facilitator of innovation. How the increasing importance of creative collaboration requires new ways of teaching design. Des. J. **20**, S1058–S1071 (2017). https://doi.org/10.1080/14606925.2017.1353049
5. Kolb, A.Y., Kolb, D.A.: The learning way: meta-cognitive aspects of experiential learning. In: Simulation and Gaming, vol. 40, pp. 297–327 (2009). https://doi.org/10.1177/1046878108325713
6. Gardner, H.: Intelligence Reframed: Multiple Intelligences for the 21st Century, Basic Books. ProQuest Ebook Central (2000). https://ebookcentral-proquestcom.proxy.library.carleton.ca/lib/oculcarleton-ebooks/detail.action?docID=903998
7. Chapman, A.: Bloom's Taxonomy – learning domains (2006). http://www.businessbalss.com/bloomstaxonomyoflearningdomains.htm. Accessed 21 Aug 2009

8. Anderson, L.W., Krathwohl, D.R., Bloom, B.S.: A taxonomy for learning, teaching, and assessing: A revision of Bloom's taxonomy of educational objectives. Allyn & Bacon (2001). findaddressphonenumbers.com/Address-of-Allyn-and-Bacon-0810
9. CAST Universal Design for Learning Guidelines version 2.2. (2018). https://iswnetwork.ca. Accessed Feb 2018
10. Webb, M.: A definitive critique of experiential learning theory (2003). http://web.archive.org/web/20061109200729/http://www.cc.ysu.edu/~mnwebb/critique/TheCritique_final2_wtp.pdf. Accessed July
11. Gardner, H.: Multiple Intelligences after twenty years. In: Paper presented at the American Educational Research Association, Chicago, Illinois, 21 April 2003 (2003)
12. Friedman, K.: Creating design knowledge: from research into practice. In: IDATER 2000 Conference, pp. 13–15 (2019)
13. Mcdonagh, D.: Design research: methods and perspectives: brenda laurel (editor). Des. J. **7**, 55–56 (2004). https://doi.org/10.2752/146069204790718510
14. Moreno, L.A., Villalba, E.R.: Transdisciplinary design: tamed complexity through new collaboration. Strat. Des. Res. J. **11**, 42–50 (2018). https://doi.org/10.4013/sdrj.2018.111.07
15. Dubberly Design Office 10 Models of Teaching + Learning, Pages 9–10 (2009)
16. Michaelson, M.: Creating the Visitor-Centered Museum. Routledge, New York (2017)

Steps to Safer and More Sustainable Industrial Design Studios

George K. Chow(✉)

University of Houston, Houston, USA
georgekchow@gmail.com

Abstract. In many industrial design studios different materials, methods and processes are used which are not sustainable or eco-friendly and are often toxic or hazardous for students to be exposed to. Foam board, high-density urethane foams, blue and pink insulation foam, Bondo, spray adhesive, epoxy and thermoset plastics are some examples of materials that are commonly used in the modeling of concepts and prototypes but are neither healthy or sustainable. An effort has been made by many educators to encourage students to create designs which are eco-friendly and sustainable but the materials, methods and processes in which students use to create their designs still remains largely non-environmentally friendly and toxic. This paper explores steps, materials and methods that were utilized in a sophomore industrial design studio case study to provide a healthier, more environmentally friendly setting for students to generate their designs.

Keywords: Sustainability · Industrial design · Education · Materials · Safety

1 Toxic and Unsustainable Materials and Modeling Methods

Industrial design students are trained to use a wide variety of materials and processes when ideating, developing prototypes and constructing final models. Unfortunately, the majority of the materials and processes that are used in many studios are highly toxic and/or not environmentally friendly [1]. For example, in the ideation and prototyping phase of a project, students will generate 3-D physical sketch models often using blue or pink insulation foam to represent their designs. They will cut the foam with band saws, use rasps to roughly shape it and use sand paper to further smooth out the surfaces. The refinement process often generates thousands of tiny particles that can be easily inhaled especially if the students are not wearing any protective respiratory masks. Sometimes students will use heated electrical wire cutters to melt wire-thin paths through the foam. The melting process emits toxic fumes which can also be inhaled. The foam is made of polystyrene, a petrol-based plastic which is not easily recyclable in the U.S., and the manufacturing process depletes non-renewable natural resources. Using high-density polyurethane foam modeling materials such as "yellow" foam and RenShape are even more toxic than the blue or pink insulation foams as they have found the particulates to be highly carcinogenic [2]. Many schools do not use "yellow" foam or RenShape as often due to their higher cost and toxicity, however, they are still utilized occasionally because of the superior modeling fidelity that can be

G. Di Bucchianico et al. (Eds.): AHFE 2020, AISC 1202, pp. 277–283, 2020.
https://doi.org/10.1007/978-3-030-51194-4_37

achieved with the materials. Once the foam models have been shaped, students will usually apply joint compound and Bondo to the exterior and sand it to achieve a smooth finish before painting. Both the joint compound and Bondo are also toxic especially if the strong fumes or sanded particles are inhaled. Primer is applied before painting the models and both the primer and paint are poisonous if the fumes and sanded particles are inhaled. When it comes time for the final presentation, students are often required to make a poster or presentation board which is usually mounted on foam board for rigidity. The most common foam boards are made of two sheets of white coated paper with a polystyrene foam layer in the middle. Most recycling facilities are not set up to recycle polystyrene foam and to make matters worse, the foam layer is adhered to the coated papers making the separation of materials for recycling even more difficult [3]. In order to adhere the paper posters onto foam board, a generous amount of spray adhesive is commonly used, resulting in noxious odors and harmful fumes filling the air [4].

The students are not at fault with the use of these materials and processes because industrial design educators are teaching them to use these industry-standard materials and processes. We need to do much better as educators and as an industry and offer designers materials and processes that are eco-friendly and not harmful to our health. This paper explores simple steps that were taken by industrial design educators to use more sustainable and less-toxic or non-toxic materials, methods and processes in a sophomore level industrial design studio course taught in the Fall semester of 2019.

2 Assign Eco-Friendly Projects

On the first day of class, course syllabi are customarily distributed to students to help them understand the overall objectives, content and schedule of a course. The course syllabus for this case study studio was six pages long. Printed out, it would have been three-double sided pages per syllabus multiplied by 30 (28 students and two professors) which is a total of 90 pieces of paper. To save on paper and ink, the instructors opted to provide the syllabus to the students digitally instead. This was not only for the material savings but also to set an example to the students to try to be as sustainable as possible.

The first project that the instructors gave to the students was to design and build a cardboard chair that could hold up a minimum of 250 lbs. utilizing only corrugated cardboard with no glue, tape or fasteners. The project in itself was an exercise in sustainable design as the finished product concepts were entirely recyclable and biodegradable. The students were encouraged to utilize digital sketching instead of traditional sketching on paper to save paper during the ideation phase of the project. When it came time to build quarter scale and half scale 3-D sketch models of their best concepts, the prototypes were also made from cardboard. Students had the option to accent their final cardboard chair model with an eco-friendly, biodegradable paint that they could purchase or make themselves such as milk, chalk or flour/corn starch based paint. These natural paints use natural raw ingredients including water, plant oils, resins, plant dyes, essential oils and natural minerals such as clay, chalk, talcum, milk casein, natural latex, beeswax, earth and mineral dyes. Students could also purchase and utilize some of the most environmentally friendly, non-toxic natural paints from a wide variety of suppliers including BioShield, Auro, Green Planet Paints, Livos, Weather-Bos, Silacote, Anna Sova, The Real Milk Paint Company, and Master's Blend [5].

To encourage the students to use as little cardboard as possible, they were challenged to design the lightest chair that could still hold up 250 lbs. The winning design from the competition weighed only 2.2 lbs, less than 1% of the total weight that it could support.

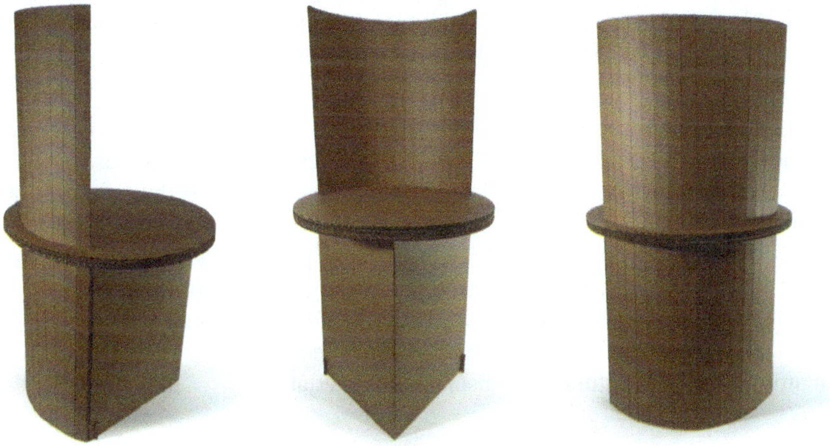

Fig. 1. "Helio", the lightest cardboard chair in the class (2.2 lbs.). Design and photos by Sarah Tran. From left to right: ¾ back perspective view, front view, back view.

3 Use More Sustainable Materials and Processes

A second project that was given to the students was to design and build an eco-friendly case for an existing 7 × 10" e-tablet (such as an iPad) that would protect it from a 2-story fall (approximately 22 ft.). To minimize the amount of materials used, students were challenged with a competition to see whose case would be both the lightest and strongest. The students were encouraged to think about ease of assembly and disassembly so that their products could be easily disassembled into different materials to better facilitate closed-loop recycling [6]. The students were also encouraged to research different eco-friendly materials and processes to utilize for their concepts. After researching existing materials and methods, students conducted their own interesting experiments. One of the most successful experiments was a paper pulp molding process where recycled paper cup holders were pulverized, mixed together with only water, salt and cornstarch and molded over a form to create the shells of the case [7]. This process yielded concepts that were bio degradable, recyclable, non-toxic and strong.

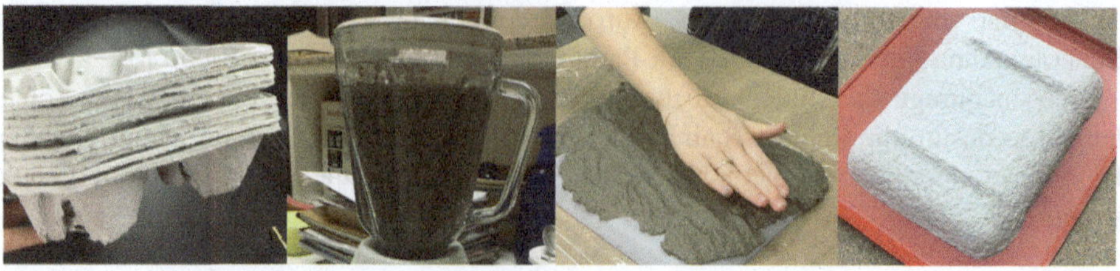

Fig. 2. Paper pulp molding process utilizing recycled cup holders. From left to right: recycled paper cup holders, pulverizing holders with blender, applying paper pulp over mold and drying the case shell. Process and photos by Anne-Elisabeth Baker.

Students were encouraged to utilize cardboard, paper, cloth and other natural, biodegradable or recyclable materials to create sketch models and test prototypes. Some of the students lined the interiors of their cases with polyethylene foam and/or cork. Polyethylene foam is 100% recyclable and cork is a natural, renewable, recyclable and biodegradable material [8]. One student placed triangular and trapezoidal pieces of cork on the outside as a protective covering for the case. The negative space areas in between the cork pieces allowed users to fold the cover origami-style into a display stand. For clasps, hinges, or liners, some of the students used wool felt which is also a strong, durable, natural and sustainable material.

Fig. 3. Left: "Sustain", a paper pulp molded e-tablet case by Anne-Elisabeth Baker. Right: E-tablet display configuration with cork lining.

Some students made their cases using a fiberglass layup process, however, instead of using traditional fiberglass and epoxy, they opted to use linen and EcoPoxy, a combination of more sustainable materials. Linen is a natural cloth material made from flax plant fibers and EcoPoxy is an epoxy with the highest bio-renewable content currently on the market. Some of EcoPoxy's products are made with 70% bio-based content and the company is working towards making their products 100% bio-based [9].

Fig. 4. Linen and EcoPoxy layup process. From left to right: Linen cloth, cloth draped over molds and covered with EcoPoxy, lightly sanded case shells to achieve distressed look. Process and photos by Jacob Roberts.

Fig. 5. From left to right: "Fiber", a linen cloth and EcoPoxy e-tablet case by Jacob Roberts, display configuration and clasp straps used to securely hold "A" frame display configuration.

Once the cases were completed, the students placed 7 × 10" pieces of ¼" thick glass in each case to represent the size, weight, and thickness of a typical e-tablet, and dropped them from a 2nd floor (21.5 ft.) of a college building. 93% of the students' cases passed the 2nd story drop test. Those that passed the 2nd floor drop test dropped their cases again from the 3rd floor (43 ft.). 52% of the class passed the 3rd floor drop and those that passed dropped it again from the 4th floor (64.5 ft.). 44% of the class passed the 4th floor drop test. There were no more floors to drop the cases from so the tests ended at that point. The lightest case to endure a 4th floor drop test weighed 13.2 oz (0.825 lbs.) without the glass in it. The ability of the cases to protect the glass "e-tablets" from such high drops showed that it is possible to use sustainable materials and methods to achieve and even exceed design goals.

4 Sustainable Presentations and Process Books

At the conclusion of each project there is typically a final presentation where students are required to create poster boards showcasing their design concepts. Oftentimes posters are mounted onto non-recyclable foam boards with copious amounts of toxic spray adhesive. Depending on where the students spray the adhesive, it can become a sticky mess and the smell from the fumes is noxious if not sprayed in a proper spray booth or location with proper ventilation. In order to make the studio class safer and more sustainable, students were instructed to forgo mounting their posters onto foam board. Instead, the paper posters were pinned directly into the walls and recycled after the presentations.

Process books are another common final deliverable which can cost a lot of money, use a considerable amount of ink and many pieces of paper if printed out. An average length process book can be 20–40 pages, but sometimes more depending on the project. If all 28 students printed out their process books double-sided, 280–60 pieces of paper would have been used. To save paper, ink, binding materials, and cost, students were instructed to submit their process books electronically as PDFs instead. Other benefits of electronic submissions include long-term record storage, high quality image and color reproduction and the ability for instructors to quickly view, give feedback, and grade electronically.

5 Conclusion

An effort to develop more environmentally friendly and sustainable products has gained much traction and acceptance in the industrial field in general, however, the materials and model-making processes utilized in the ideation, prototyping and presentation phases still remains largely unsustainable and unhealthy for practitioners and students. A transformational shift needs to occur in the industry and in education towards utilizing safer, sustainable materials and methods for generating concepts. In the education realm, instructors can start by assigning more sustainability minded projects- not only with the final deliverables but also addressing sustainability issues in the projects such as ease of assembly/disassembly for closed-loop recycling, biodegradable or recyclable material selection and giving constraints to reduce overall material use. Instructors can also encourage and challenge students to investigate and utilize more sustainable materials and methods like the e-tablet case paper pulp molding example. Collectively, a tremendous amount of paper, ink, and money can be saved simply by switching to digital sketching, syllabi and process books. Use of non-recyclable materials like foam board and toxic spray adhesives can be eliminated from studio environments altogether. Finally, further research needs to be conducted to discover and develop safer and more sustainable materials and methods for designers to utilize. If everyone implements some of these steps, we can begin to make our studios, design processes and profession safer and more sustainable in the near future.

References

1. Hallgrimsson, B.: Prototyping and Modelmaking for Product Design. Laurence King Publishing, London (2020)
2. American Cancer Society, https://www.cancer.org
3. Recycling Partnership, https://recyclingpartnership.org
4. United States Consumer Product Safety Commission, https://www.cpsc.gov/
5. Earth Easy, https://learn.eartheasy.com/guides/non-toxic-paints/

6. McDonough, W., Braungart, M.: Cradle to Cradle: Remaking the Way We Make Things. Vintage, London (2019)
7. Thompson, R.: Manufacturing Processes for Design Professionals. Thames & Hudson, London (2015)
8. Lefteri, C.: Materials for Inspirational Design. RotoVision, Mies, Switzerland (2006)
9. EcoPoxy, https://www.ecopoxy.com

Project Pilot Run 2.ZeRo

Adam Feld(✉)

421 E. Lewis Street, Lafayette, LA 70503, USA

Abstract. Educators have a responsibility to prepare students for their profession through a building of methods and skills and understanding, increasing in complexity, over the course of a four-year degree. However, educators cannot be expected to know everything and are sometimes limited in course content based on factors such as class size, studio level, curriculum, and the total amount of credit hours. It is the educator's responsibility to their students to develop advanced student research experiences (ASRE's) such as professional student work experience. Sponsored or funded projects are great ways to provide students an understanding of work for a client; these experiences are where educational pedagogy and professional practice converge. Although working for a client is a positive experience for students, it is the educator's responsibility to maintain rigor and studio standards. This paper will follow a third year industrial design sponsored project and highlight the areas of pedagogy and profession.

Keywords: Educational pedagogy · Professional practice · Sponsored project · Sewn goods · RedBull · E-Sports · Pattern making

1 Introduction

It can be said that eSports Gaming began in the early 70's with pong and asteroid tournaments. Since then, it has grown to different platforms and a multitude of games. One byproduct of eSports is live streaming games for others to watch. Live streaming has become so popular that in 2015, 89 million viewers watched over 3.7 billion hours of eSports gaming and "…as an overall category, gaming has more YouTube followers than the news, movies and education combined [1]". Most of the live streaming occurs from a fixed location, where gamers have equipment to live stream themselves playing videogames as well as their face and voice. While gamers are live streaming, they also have a chat open so viewers can comment and receive responses from the gamer in real-time. When gamers are on the road at tradeshows, tournaments, or conventions their fans still want a live stream but at these times, the gamer needs to have the ability to transport their equipment, usually in some type of bag. This paper followed a RedBull Sponsored project in the Spring'19 Semester. In this project, students developed concepts for the means of safely and efficiently carrying live streaming equipment for an eSports gamer. One such gamer is ZeRo, the best Super Smash Bro's player on the Wii U. He has difficulties when live-streaming on the go. For ZeRo, live streaming on the go requires his assistant to hold all equipment; a video capture device and microphone, power pack(s), Live U (a device the combines data streams), and four

G. Di Bucchianico et al. (Eds.): AHFE 2020, AISC 1202, pp. 284–290, 2020.
https://doi.org/10.1007/978-3-030-51194-4_38

Wi-Fi routers all of which are on different cellular networks. All of this equipment is placed in a backpack. However, the backpack, though designed for a gamer, is not adequately equipped to handle ZeRo's or his assistant's needs. This designed, gaming backpack does not accommodate the scale of the different pieces of equipment as well as does not provide the means of "seeing" the live stream as it goes into the cloud. In this project, fourteen students worked with ZeRo as a primary research source, for their inspiration, conceptualizations and design work to develop concepts for a new, wearable, live-streaming container. This was done through sketching, basic mock-up and pattern making through paper to finally develop a concept prototype that was to be tested by ZeRo, himself. In addition, the students received feedback from a local sewn goods design professional who consulted on student concepts as well as how to develop specific visualizations for the project and provide appropriate feedback to ensure each design was manufacturable.

2 Development

Fig. 1. ZeRo with student reviewing initial concept sketches

This phase of the project held: the analysis/generation of form and brand language, initial client interaction, ideation sketching, refined sketching, presentations sketching, physical mockup, and initial pattern. The main goal of the project was to develop a prototype as a one-off from manufacturing that would be suitable for the client. It is important to note that each milestone had a combination of instructor of note and either ZeRo or the industry expert or all three. Development of the sewn good product began similar to standard industrial design projects. Students were provided with a project brief as well as a bio of ZeRo and his needs. In order to optimize face-to-face interactions with ZeRo, students developed initial materials for review at his first visit to the studio. There were three initial assignments; 1) to develop image reference boards (mood boards) to describe what RedBull means to each student, 2) to analyze physical products, produced by or in partner with RedBull, and 3) to develop 25 ideation sketches based on the information provided in the bio as well as assignments 1 and 2. Ideation sketches are loose, freehand sketches in two-point perspective with basic value and shading and typically 3–5 per page. At the initial meeting, ZeRo discussed his issues with his current equipment and demonstrated how the backpack will be packed for any live streaming event. After the demonstration, ZeRo sat with each student and reviewed each their work. Students then reflected on their feedback from ZeRo and used that information to progress to refined sketches. Many of the ideation sketches were validated by the discussion and demonstration but all students continued to push their ideation sketches further based on their one-on-one conversations. For all students, working for a client and not for themselves was an eye opener. Many students thought their ideation sketches would be successfully accepted by ZeRo and for some, this was not the case. This was a wonderful exercise, for the students, to remove themselves from the project and design for another person.

After the ideation phase came the translation into digital and beginning physical modeling phase. In this phase, students worked in vector software to develop a "rough draft" orthographic drawing of their proposed design. This activity teaches how to "see" your design and translate it from a 2-point perspective drawing to flat views. Orthographic view drawings force students to make critical decisions on the scale of the design and feature placement. In addition, the orthographic drawings assisted students in the initial pattern development. Patterns are flat drawings of all different parts and shapes that, when assembled, create a 3-dimensional product. Few of the students, with the exception of one or two, have ever developed a pattern. Because of this, the digital orthographic drawing provided a base or starting point. Students reverse engineered their orthographic drawings for their pattern and used that to develop a ½ scale mockup. Mockups are a physical model of the proposed design. They typically are used for overall scale and proportion. For theses mockups, students used their pattern and cut out their pieces in craft paper. Craft paper does not promote organic form but allowed for very crisp and clean "box-like" forms for the initial mockup. It was not expensive and students could explore trouble spots in their design with less worry about cost. Because the pattern was developed as vector artwork, students were able to use the laser cutter to cut out their parts. Laser cutting their pattern gave students the confidence to continue to try to perfect their pattern prior to moving onto the next phase of the project. They could laser cut, assemble, review, and laser cut again, usually within the class period. This review process was critical for student growth. Critiques of sketches typically have an emotional connection since the student developed the sketches by hand. Digital work has an emotional disconnect which allowed students to more critically review their work and push their designs further. With the ½ scale mockups complete, students then presented to the industry professional, Patti Dunn, owner and operator at Tchoup Industries. She reviewed the work and provided feedback as to how each design would be manufactured. After the review she lectured on how her designs are fabricated and tips and tricks for successful designs.

3 Fabrication

Fig. 2. Students sewing concepts

After creating the ½ scale mockups and receiving feedback from the instructor and industry professional, students reworked their pattern and began working in fabric. This next version of the design would be full scale and in duck canvas. Duck canvas is monetarily appropriate as well as having attributes similar to the final material, ballistic nylon. It is here that it should be noted that at the beginning of this phase of the project, the majority of the class did not know how to sew. Lack of sewing knowledge was not a deterrent as those who did know how to sew were able to pass on the knowledge. This was an excellent opportunity for classmates to become teachers and pass on knowledge. All students in the class would become proficient at sewing by the end of the project. Students used the laser cutter to develop the ½ scale models and were eager to do the same with the full-scale mockups. Using the laser cutter allowed the students to take risks and update their pattern quickly. Many of the students developed multiple full-scale mockups to perfect their patterns.

Once the full-scale mockups were complete, students then needed to check back in with ZeRo and show him their progress. ZeRo was unable to physically be co-located with the students during this critique however; he was able to use his home live streaming equipment to video chat with the class. Each student developed a PDF that contained the project thus far; it had the sketches, ½ scale mockup, and full-scale mockup with details highlighted. Students then sat, one-on-one, in a video chat with ZeRo and he reviewed their designs and offered suggestions for improvement. After the one-on-one chats the class sat down and discussed their individual feedback as well as how they, as a group, developed their projects and any difficulties they encountered and how they overcame them.

4 Final Designs

Fig. 3. Final design and poster

With the full-scale mockup complete, the updated pattern prepared, and the feedback from ZeRo, students began on their final model. Prior to building their model, students needed to estimate how much material they would need so the instructor could order the industry standard material for the final design. Specifying materials for purchase was a new concept for students, what if they ordered too little, too much? What if they had to change their design? Students were concerned about the cost of the material and were hesitant to make an order. Once the hesitation was over, the materials were purchased and shipped to the studio. When students received their materials, they quickly realized that with new materials come new properties than that on which they practiced. Students quickly overcame that challenge and also realized that the new material worked very well on the laser cutter. Since students had worked with the laser cutter for their previous models, the final model's pattern was developed and produced quickly. While the students were completing their final designs, they were also producing their final visuals for the project. The final visual was a poster that documented the entire project from beginning to final design.

Once everything was complete, i.e. the due date, students brought their designs to a local business and set up their final show. Since students were designing for ZeRo, he came to the final presentation and select a student's design that would be considered the "winner" of the project. That design would then be handed off to the industry professional, Patti Dunn, for any adjustments and a specification sheet for the design to be produced by a professional model maker. At the final presentation, students had the opportunity to present to ZeRo as well as Patti Dunn. After all students had presented, ZeRo and Patti privately reviewed each design and had a discussion on which was a good fit for the overall project direction as well as which would be the most manufacturable.

5 Project Conclusions

It can be argued that a sewngoods project can be a standard type of project in Industrial Design. However, not many university programs offer projects in that area. The instructor of note's job is to teach skills, methods, and principles of design for the creation of products but this was a product the instructor was not as familiar with having run only one sewngoods project previously. Project Pilot Run 2.0 was the second sewngoods project the instructor had run. In the first project, the instructor was a novice but learned from the industry professional as well as through student work. In the second project, the industry professional affirmed the information disseminated in class. As Edgar Dale discusses in his Cone of Experience (also known as the cone of learning), people remember 10% of what they hear, 30% of what they see, and 90% of what they do [2]. Since this was the second time the instructor held a sewngoods project, the ability to disseminate information was faster and students progressed further in the project. It is not always necessary for faculty to be an expert in everything. Sponsored projects allow for a collaboration that is not always possible in non-sponsored projects and allows the instructor the opportunity to be both be learner and facilitator of information and resources. For students, this project provided the chance to work for a client and learn how to remove themselves from their work. The heavy use of the laser cutter was unexpected and afforded students the confidence to perfect their designs.

References

1. Casselman, B.: FiveThirtyEight.com. Resistance Is Futile: ESports Is Massive… and Growing. ESPN. ESPN Internet Ventures, 22 May 2015. https://www.espn.com/espn/story/_/id/13059210/esports-massive-industry-growing
2. Rob Segers Blogspot. http://robsegers.blogspot.com/2013/03/edgar-dale-cone-of-experience.html

Industrial Design + Horticulture: A Collaborative Approach to Greenhouse Design

Carlton R. Lay(✉)

Auburn University, Auburn, USA
cr10030@auburn.edu

Abstract. Creating collaborative environments between design and the sciences has the potential to solve problems in new and creative ways through the design process, a process unfamiliar or unknown to many scientific fields. This type of collaboration has been an ongoing pursuit, with a recent instance being a cross disciplinary workshop between Schools of Industrial Design and Horticulture. This unique workshop consisted of a studio-style multidisciplinary workshop in which professors guided students to generate innovative concepts for greenhouse cooling. The focus of this particular collaboration was the creation of energy-efficient, long-term, and cost-effective solutions for greenhouse designs that allow year-round food production in both food-poor locations and temperate climates that are home to a significant portion of greenhouse farming. Opportunity for creative and novel solutions to cooling are vitally important and timely due the negative impact of climate change on food production in many at-risk and low-income regions.

Keywords: Food production · Design · Systems design · Design collaboration · Infrastructure · Greenhouse · Temperature control · Climate · Microclimates

1 Introduction

As regional climates evolve over time, food production methods must evolve accordingly in order to provide access to nutritious food for a rapidly growing global population in a consistent and efficient way. A shortage of food in multiple locales throughout the world is exacerbated by floods, droughts and rising CO_2 levels which reduce production levels and the nutritional value of foods [1]. Despite technological improvements that increase particular crop yields, extreme weather events have caused significant yield reductions [2].

In addition, not all climates that are home to large population areas are suited for sustainable food production as will be shown below. One method to alleviate changes in the planet's productivity of food resources and increase yield in poor growing areas is to provide protection for crops while increasing use efficiencies of natural resources, particularly water and nutrients. Controlled-environment agriculture (CEA) is a type of protected agriculture that allows for a high level of control over the growing environment and production inputs. The most popular form of CEA, which has been used

G. Di Bucchianico et al. (Eds.): AHFE 2020, AISC 1202, pp. 291–298, 2020.
https://doi.org/10.1007/978-3-030-51194-4_39

for generations, is greenhouse technology. Controlled-environment greenhouses are defined as greenhouses that "use fans, evaporative cooling systems, heating, or other electro-mechanical measures to control plant growing conditions." [3].

Greenhouses serve many useful purposes for enhancing the production of a variety of crops including control of temperature, humidity, light, and insect populations. Greenhouses provide a more consistent temperature across day and night by creating an enclosed environment with heat retention characteristics. The amount of light that is allowed to pass into the enclosed greenhouse and be converted to infrared radiation within the structure can be controlled through window glazing, window materials, window angles, shading techniques, etc. It is the enclosed nature of greenhouses that allows for the introduction of controlled humidity, light, warmth and other climate characteristics, as well as electricity and support structures for introducing active systems that enhance plant growth.

An ideal example of utilizing greenhouses for extremely productive "precision farming" is in the Netherlands. "The Netherlands is a small, densely populated country, with more than 1,300 inhabitants per square mile. It's bereft of almost every resource long thought to be necessary for large-scale agriculture. Yet it's the globe's number two exporter of food as measured by value, second only to the United States, which has 270 times its landmass." [4].

The Netherlands has accomplished this productivity and efficiency through significant improvement in greenhouse design for both plant and animal production. Although the climate in which the Netherlands is located provides significantly less sunlight throughout the year than what is optimal, or even possible, for growing, they have made their country highly productive through better design of climate specific greenhouses [5].

2 The Problem to Solve

The aptly-named "greenhouse effect" can be both friend and foe in temperate regions, passively warming greenhouses in cooler seasons but trapping unwanted heat in warmer seasons. Some of the detrimental effects of undesirably high temperatures on plants include reduced stem strength, reduced flower size, reduced leaf size, delayed flowering, flower bud abortion, and reduced growth rate [6].

The overheating problem of greenhouses is especially pronounced in the regions where they are most needed. The World Bank reports that in tropical regions containing many of the poorest countries, climate change impacts on agricultural productivity are expected to be particularly harmful. The vulnerability of these countries is also especially likely to be acute in light of technological, resource, and institutional constraints [7]. Common cooling techniques that are less expensive than traditional air conditioning systems most often include evaporative cooling or passive ventilation. Evaporative cooling becomes less effective as humidity increases, making it highly inefficient for tropical regions. Passive ventilation can be both inexpensive and effective to lower greenhouse temperatures closer to the exterior temperatures but has decreased effectiveness as size of the greenhouse increases [8]. In addition, passive venting requires venting within taller roof structures in order for the efficiencies of the natural "chimney effect" to be effective enough to achieve the appropriate air exchange [9].

Cooling a greenhouse to appropriate growing temperatures in the summer is more difficult and cost-prohibitive in warm regions than warming a greenhouse in the winter to appropriate growing temperatures. There is no cooling source as easily accessible as the sun, used for heating. With finite resources, it will take innovation, development of new technologies and innovative approaches to utilizing a variety of existing technologies to meet the world's food demand.

3 An Opportunity for a Novel Approach

The information above is but a sampling of the in-depth knowledge and broad scope that professionals, and professors, within a specific scientific field have, and that collaborations between design and other sciences, can benefit from. Creating this collaborative environment between design and the sciences has the potential to solve problems in new and creative ways through a process that is unfamiliar or unknown to many scientific fields.

Historically, it has been uncommon for certain disciplines and industries to seek input or support from the design disciplines. The lack of connection between certain disciplines/industries and creative disciplines have likely hindered the development of solutions to many problems and challenges that are perceived as "givens". Therefore, the collaboration itself is unique and innovative. There is great opportunity in forming cross disciplinary relationships between the design discipline and "un-designed" industries with elegant and efficient systems as the goal. Agriculture and food production is one of many disciplines that has lacked integrated design input. Agriculture and its systems of growing have, for the most part, either developed organically over centuries or have experienced recent, dramatic and rapid change due to scientific advancement in genetic engineering and chemical development. Our effort seeks to be innovative in its multidisciplinarity as it takes advantage of both the deep agricultural knowledge within the field of horticulture and the unique combination of creativity and critical thinking in the phased process of industrial design.

An initial, more informal collaboration was formed between an industrial design studio and a professor of horticulture. This collaboration utilized the professor of horticulture's thorough knowledge of aquaponics growing systems and the professor of design's guidance in design methodologies to teach and aid students in the design of small scale aquaponics systems (the raising of both plants and fish in a symbiotic relationship in which fish provide nutrients for the plants and plants clean the water for the fish). The design brief asked the students to design the systems for the "millennial" market, the fastest growing and highest spending segment of the population on small scale food gardening in the United States [10] (Fig. 1).

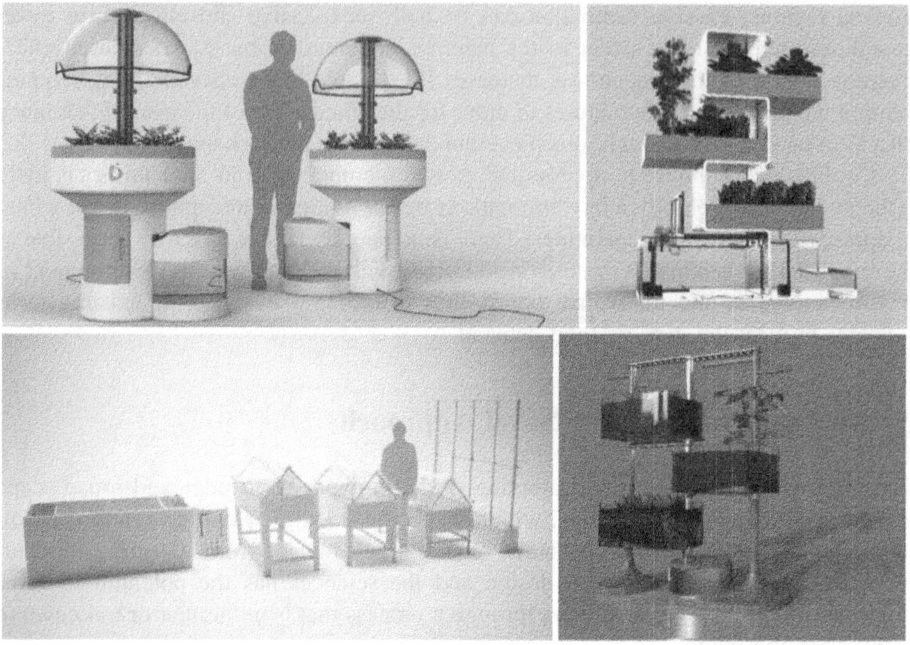

Fig. 1. Aquaponics growing systems for millennials by Hoyt, Chen, Lakeman & Masaschi

The success of this informal collaboration led to the desire for a more structured and formal collaboration to solve other challenges, including the overheating of greenhouses during summer months. The plan for a formal collaboration led to the securing of a College of Architecture Design and Construction Seed Grant which provided funds for a cross disciplinary workshop between Horticulture and Industrial Design. This unique collaboration consisted of a studio-style multidisciplinary workshop in which professors guided students to generate innovative concepts for greenhouse cooling, explored through scale prototyping. This workshop took place during the Fall semester of 2019. A collaboration between students of horticulture and students of industrial design allowed for both disciplines to learn from each other and benefit from each other's particular strengths and knowledge. The collaborative processes of research, idea generation, understanding of plant needs as well as user needs, physical prototyping, applied scientific testing of prototypes, data collection and analysis allowed both disciplines to share their unique strengths, enhancing the learning of all students and professors.

The primary goal of our ongoing collaboration is currently to create energy efficient, cost effective solutions for greenhouse designs that allow year round food production in some of the most food-poor locations on earth. Greenhouses in warm and humid climates depend on high cost active cooling systems to remove excess heat in summer months, preventing smaller farming businesses and poorer regions from utilizing the productivity of controlled environment greenhouse growing throughout the year. As solutions to the cooling problem have yet to be developed for warm and humid climates through methods that don't require significant labor and upfront cost or long term maintenance costs, the opportunity for creative and novel solutions to cooling are opportune and timely due to the increased awareness of the negative impact of climate change on the food production in these regions most affected as mentioned above.

A variety of methods for idea generation were introduced to the participants in the first exploratory phase of the multidisciplinary workshop. One common practice within the design field that the students found interesting and utilized heavily during the workshop is the study of other categories of products, scientific fields, nature, etc.… that are not necessarily related to the problem or situation at hand but may have similar traits, needs or outcomes. This "borrowing" from other disciplines or situations is often called analogous inspiration [11]. For example, many types of manmade and natural systems require cooling and particular methods have been developed or discovered for each of the following: computers utilize heat sinks, motors utilize strategic air movement and lubricants, protective equipment utilizes chemical reactions, heat shields utilize ceramics, insects utilize thermoelectricity [12], to name a few. For this particular challenge, the guiding of students toward creative thinking through analogous inspiration led to innovative methods of cooling greenhouses, as well as other environments

Of the five teams in the workshop, concepts from two of the teams, were most intriguing and proven most effective through their prototyping and will be explored further, beyond the workshop at a larger scale and with higher craft and precision, giving more accurate results of the effectiveness of each concept. One concept to be further explored, with a more sophisticated and robust build, utilized enhanced natural convection through a process of dehumidification and re-humidification of air, initially inspired by Middle Eastern architecture (Fig. 2). The second concept to be explored, in a variety of forms, involved using expanded CO_2 gas to cool individual microclimates within the greenhouse, rather than the entire greenhouse (Fig. 3).

Fig. 2. Construction of enhanced natural convection greenhouse

Fig. 3. Construction of CO_2 cooled microclimates within greenhouse

4 Next Steps

Continued development, refinement, and testing of the concepts will likely be done in multiple phases based on both the type of concept to be tested as well as internal and external funding support. We anticipate that any concepts that will be generated may fall into either of two categories:

1. Cooling methods that can be tested through retrofitting or placing into existing greenhouses a prototype of the concept.
2. Cooling methods that will require a redesign of the greenhouse itself.

Through the Horticulture department, access is available to a variety of greenhouses in which to test the concepts that fall into the first category. As the cost of designing and constructing a new type of greenhouse will likely be at much greater expense, funding for concepts falling into category two will be sought through external sources. Dividing the concepts into these categories will allow management of costs and priorities throughout the design and development as funding becomes available.

5 Significance and Impact

There are over 40,000 farms that utilize greenhouses for growing crops in the United States, totaling over one billion square feet of growing area. This totals close to a $15 billion annual market value [13]. For many of these commercial greenhouses, either production is slowed during the summer months due to the high greenhouse temperatures or great reoccurring expense is absorbed to maintain productivity.

Efficiencies, or lack thereof, in food production is a global issue. Climate change, geographical location, population densities, supply chain difficulties in transporting food long distances, land with depleted nutrient levels, all contribute toward uneven distribution of quality food around the globe. Rising global temperatures are altering climatic zones around the planet, with consequences for food and water security, local economies, and public health [14]. Those who will be hit the hardest by these changes will be located closer to the expanding tropics and semi-arid zones north and south of the equator [15].

If new methods for cooling greenhouses for warmer climates, like those in the climate zones mentioned above, during the hottest summer months allow maintained plant productivity at lower cost, annual crop yield can be significantly increased. If cooling systems can be developed that do not compromise insect and pest inhibitors as many passive ventilations systems do, the crops that are grown in these greenhouses can be grown without pesticides, therefore generating healthier foods, healthier consumers, and healthier work environments.

Greenhouses not only provide a controlled environment for growing plants in existing soils, which in some regions may be inadequate for food production due to lack of nutrients, pollution, soil type, and other limiting factors, but can also provide a stable and controllable environment for non-traditional forms of food production that do not require nutrients from existing soils. Aquaponics - defined above, and hydroponics in which nutrients, specific for each plant type can be directly added to the water supply for the plants, both eliminate the need for soil based growth and are effective systems when combined with greenhouses that are appropriately designed for the corresponding climates in which they are located.

As the significant development of greenhouse systems accomplished in the Netherlands for their specific climate conditions has allowed a naturally minimally productive country to be the second larger exporter of food by value, the design and development of greenhouses better suited and more strategically designed for warmer climates could have far reaching impact. The ability to control the environments within a greenhouse precisely and appropriately according to the climate in which it is located, specifically, addressing the widespread problem of excessive heat in warmer seasons, will allow food production increases, reduced water usage, reduced pesticide and herbicide usage, and in locations not naturally conducive to sustainable food production. The greenhouse systems and their continued development initiated though the collaboration between the schools of Industrial Design and Horticulture, have considerable potential to allow increased productivity in a significant portion of the world.

References

1. USEPA: Climate Impacts on Food Supply (2017). https://19january2017snapshot.epa.gov/climate-impacts/climate-impacts-agriculture-and-food-supply_html
2. Hatfield, J., Takle, G., Grotjahn, R., Holden, P., Izaurralde, R.C., Mader, T., Marshall, E., Liverman, D.: Agriculture. Climate Change Impacts in the United States: The Third National Climate Assessment, J. M. Melillo, Terese (T.C.) Richmond, and G. W. Yohe, Eds., U.S. Global Change Research Program, 150–174. (2014)
3. Hickman, G.W.: International Greenhouse Vegetable Production Statistics. Cuesta Roble Greenhouse Consultants (2016)
4. Viviano, F.: National Geographic (2017) https://www.nationalgeographic.com/magazine/2017/09/holland-agriculture-sustainable-farming/
5. Locatelli, L.: Photographer: National Geographic (2017). https://www.nationalgeographic.com/magazine/2017/09/holland-agriculture-sustainable-farming/
6. Kesler, J.R., Jr.: Dealing with the Heat in Southern Greenhouses. Auburn University

7. Kurukulasuriya, P., Rosenthal, S.: Climate Change and Agriculture: A Review of Impacts and Adaptations. Environment department papers; no. 91. Climate change series. World Bank, Washington, DC. ©World Bank. https://openknowledge.worldbank.org/handle/10986/16616 License: CC BY 3.0 IGO (2013)
8. Valensi, S.: Ventilation in Greenhouses. https://azrom.com/technology-highlights/ventilation-in-greenhouses/
9. DPI. https://www.dpi.nsw.gov.au/agriculture/horticulture/greenhouse/structures-and-technology/ventilation
10. AgWeek (2019). https://www.agweek.com/lifestyle/home-and-garden/4678306-are-millennials-future-gardening-or-are-they-throwing-trowel
11. IDEO.org. http://www.designkit.org/methods/6
12. Dume, B.: Insects Stay Cool with Thermoelectricity. Biophysics - Research Update (2003). https://physicsworld.com/a/insects-stay-cool-with-thermoelectricity/
13. Kopf, A.: Helping Farmers Increase Profits (2017). https://medium.com/agrilyst/lets-talk-about-market-size-316842f1ab27
14. Jones, N.: Redrawing the Map: How the World's Climate Zones Are Shifting. Yale Environment 360. Yale School of Forestry & Environmental Studies (2018)
15. Freedman, A., Allen, M.: Climate change is redrawing maps. Axios (2018). https://www.axios.com/climate-change-extreme-weather-shifts-f6c866ce-6f6c-465e-a5e3-005e15564ce5.html

Designer Thinking, A Unique Combination of Cognitive Modes

Paul Skaggs(✉)

Brigham Young University, Provo, UT 84602, USA
Paul_skaggs@byu.edu

Abstract. The theory of cognitive modes states that individuals have unique and dominant ways of receiving, processing, and using information and that these styles influence behaviors and abilities. These abilities are natural talents. Every profession uses different. The work someone is most likely to enjoy and be successful in is work that uses his or her specific abilities. Research identifies three dominant cognitive modes for industrial designers: creative thinking, visual thinking, and adaptive thinking. This paper describes a series of experiences developed to help assess creative, visual, and adaptive thinking in students that are considering the industrial design program at Brigham Young University.

Keywords: Design thinking · Divergent thinking · Visual thinking · Adaptive thinking · Creative thinking · Aptitudes · Tolerance for ambiguity

1 Introduction

The industrial design program at Brigham Young University is a limited-enrollment program necessitated by fixed resources and accreditation-required studio teaching methodologies (NASAD 2016a, b). The student interest in the program is greater than what the resources allow. Forty students are accepted into the industrial design pre-program each year and fifteen are accepted into the professional program. This selection process necessitated a way to determine who would move on to the professional program. A first-year foundation course curriculum was developed to assess students for three dominant thinking styles; creative, visual and adaptive thinking. A specific course was designed to help determine if a student had the core competencies for industrial design.

2 Purpose

The purpose of this paper is to describe the methodology for development of a course curriculum to assess a student's proclivity for industrial design and admittance into the program at Brigham Young University.

G. Di Bucchianico et al. (Eds.): AHFE 2020, AISC 1202, pp. 299–305, 2020.
https://doi.org/10.1007/978-3-030-51194-4_40

3 Method

Aptitudes are innate talents—special abilities for doing, or learning to do, certain tasks naturally. These measured traits are highly stable over long-term periods. Every occupation uses certain aptitudes. The work someone is most likely to enjoy and be successful in is work that uses his or her specific aptitudes (Hoffman 2002). For example, if someone is an industrial designer but does not possess aptitudes relevant to industrial design, his or her work might seem unrewarding, perplexing, difficult, and unpleasant.

To determine industrial-designer aptitudes, a survey of fifty-seven design managers, or persons responsible for hiring industrial designers, was conducted. The managers were asked what aptitudes an industrial designer needs in order to be successful. This survey returned 285 responses. A review of the responses grouped duplicate items and removed items that did not fit the definition of an aptitude, such as attitudes, special knowledge, or a teachable skill. The final list of aptitudes included eighty-three separate statements.

The aptitudes were categorized in order to organize and clarify the results and to determine dominant cognitive modes. (Skaggs 2002). In addition to the survey, a literature review was also undertaken. As a result, the industrial-designer aptitudes were organized under these three dominant cognitive modes: creative thinking, visual thinking, and adaptive thinking.

4 Cognitive Modes

Creative thinking is the ability to generate new and useful ideas (Sternberg and Lubart 1995; Runco 1991). Creative thinking is separated into four subcategories: fluency, flexibility, originality, and elaboration. Fluency is the ability to generate lots of ideas. A flexible thinker can look at a problem from different points of view. Original ideas are new to a domain. Elaboration is the ability to communicate the idea and its details. This is an important ability for industrial designers who typically are hired for their ability to generate and evaluate new ideas.

Visual thinking is the ability to perceive the visual world accurately—the ability to not only receive visual information but also to process, manipulate, and apply the information. Silverman (1997) suggests that less than 30% of the population primarily uses visual/spatial thinking over other forms of thinking, such as auditory or kinesthetic.

Visual acuity is the ability to mentally hold and compare images, orient or rotate images, and manipulate images, such as folding up a flat pattern or assembling parts (Newcombe and Frick 2010). Visual language is the way visual thinkers communicate ideas. They tend to think with pictures; their "alphabet" is visual (Silverman 1997). Sketching is a quick and intuitive method for visual thinkers to communicate ideas (Ware 2008). This is an important skill for industrial designers who are hired for their ability to create, develop and effectively communicate idea.

Adaptive thinking suggests a certain lack of rigidity in thought processes. More specifically it refers to a person's tolerance for ambiguity and adaptability and his or her ability to make intuitive judgments (Gelb 1998). Design is a messy and complex process that makes some people uneasy; but designers are often energized by it (MacDonald 1970). This is important to industrial designers who work at the fuzzy end of product development, where many factors are undefined and can change rapidly, and where there is a resistance to closure. A designer must have a willingness to embrace and work with ambiguity, paradox, and uncertainty.

Taking into consideration these three cognitive modes, a course was developed that would more effectively determine a student's proclivity for industrial design, specifically the ability to think creatively, visually, and adaptively.

5 The Course

The course "Structured Creativity" was created. The course is a seven-week (block) class in the first-year foundation series of four pre-professional industrial design classes. The class was designed for faculty and students to assess aptitudes in aspects of creative, visual and adaptive thinking.

6 Assignments

The scope of this paper will not allow for the description and discussion of all the assignments in the course but will define a few assignments in detail. Learning outcomes were developed using elements of recognized and validated tests for creative, visual, and adaptive thinking: the Torrance Test of Creative Thinking (TTCT), Visual Thinking test (Cooper and Shepard 1984) and the Tolerance for Ambiguity test (Budner 1962).

Most of the assignments are not focused on a single aspect of a single category but evaluate several items across categories. For example, an assignment may test a creative attribute of alternative uses and at the same time assess the visual thinking attribute of representation and an adaptive thinking attribute of categorization or functional fixedness. Almost all the adaptive thinking aptitudes are assessed by assignments and students' responses to those tasks in the other two categories.

7 Creative Thinking

The Torrance Test of Creative Thinking (TTCT) is the most used standardized assessment tool for creativity. The test is scored on the four sub-categories of creativity: (1) fluency—the number of responses, (2) flexibility—the variety of the responses, (3) originality—the uniqueness of the responses, and (4) elaboration—the amount of clarity and detail of the ideas (Torrance 1992).

7.1 Assignments in Creative Thinking

Fluency Assignment. The assignment is called the Snowball Fight. Students are given twenty minutes to generate as many snowball fight product concepts as possible. They are instructed to create one idea per 8.5 × 11 page. At the end of the twenty minutes the pages are collected and counted for fluency, the number of ideas or pages. The concept pages are then reviewed to determine the different categories of ideas; this number represents flexibility. Categories include ideas in; snowball making, throwing/shooting, storage, protection, cold abatement, vehicles, forts, teamwork, and so on.

Fluency Assessment. The snowball fight assignment is one of five different assignments to assess fluency. The assessment shows that students' results are consistent for fluency over all five assignments.

Flexibility Assignment. The assignment is called the Lip Balm Container. The test requires participants to generate several possible uses for common objects, for this assignment a lip balm container was chosen. The assignment is described as follows: "Take thirty minutes to explore as many different uses for a lip balm container as possible. Communicate your ideas as simply and clearly as possible on an 8.5″ × 11″ page."

Flexibility Assessment. Flexibility is demonstrated by looking at the lip balm container alternative uses on different levels, first as a container for other items, second level looking at the container not as a container but as a form, and third looking at the container as parts of a whole.

Originality and Usefulness Assessment. If we define creativity as the generation of new and useful ideas, both aspects of the definition are constraints that must be considered. The narrow target of creative ideas will be missed if one is overly practical or on the other hand overly imaginative.

Elaboration Assessment. The prerequisite to the structured creativity course is a design sketching course training the students how to effectively communicate ideas through sketching. The assessment of elaboration is on how clearly the student can communicate their ideas visually.

8 Visual Thinking

The Cooper and Shepard (1984) investigations into the development of visual [spatial] skills led to theories that divide the ability to visualize into subsets. The aspects most relevant to industrial design are listed in descending order according to difficulty: (1) The ability to hold and compare images in your imagination; (2) The ability to orient 3D objects mentally; (3) The ability to mentally deconstruct and manipulate components of a 3D object.

8.1 Assignments in Visual Thinking

Hold and Compare Assignment. This assignment asks the students to view several objects and then sketch those objects from memory. The first series is composed of is several cylinders. Students are asked to recall the height to diameter proportions and compare the proportions of each object in relation to each other. The objects include; a 1-gallon paint can, a Pringles potato chip cannister, a toilet paper roll, a coffee mug, and a tuna fish can.

Hold and Compare Assessment. The assessment is for accuracy in each objects proportion and in their relationship to each other.

Orientation Assessment. This assessment is called Cube Snakes and is a test created by Cooper and Shepard (1984). The students are shown a slide with two images. The student is required to mentally orient one image to see if it matches the other object on the page. Eight different pages are shown, and the students have seven seconds to record their answer.

Manipulation Assignment. The assignment is a flat pattern assignment, one aspect of manipulation. The students are given an image of three geometric shapes. They are asked to create a flat pattern of the three shapes but not to test the pattern. The flat patterns are cut out and assembled in class.

Manipulation Assessment. The assessment is simple; is the form complete and the correct proportion?

9 Adaptive Thinking

Adaptive thinking is the ability to perceive uncertainties, contradictory issues which may be difficult to understand, as well as information with vague, contrary or multiple meanings in a neutral and open way (Mclain 2015). Bochner (1965) categorized attributes of individuals who are intolerant of ambiguity. Bochner's attributes for intolerance were rewritten to tolerance for ambiguity and more specific to industrial design. The list is as follows; (1) comfortable with uncertainty; (2) a low fear response to the change; (3) tolerance for fluctuating stimuli; (4) and delaying selection from multiple opportunities. These aptitudes are not assessed with specific assignments but with evaluating the students in all their assignments, approaches, and attitudes toward assignments that require adaptive thinking.

9.1 Assessments in Adaptive Thinking

Comfortable with Uncertainty Assessment. To move forward with uncertainty, decisions need to be made. These important decisions can be made using intuitive judgement. According to Agor (1986 p. 39–41) intuition is not a guess; good intuitive decisions are based in part on "input from facts and experiences, combined and integrated with a well-honed sensitivity and openness to other clues."

A low Fear Response to Change. Designers seem to have a natural desire to change things—to make things easier, better, or sometimes just different. The designer is not resistant to change because they know that change brings unfamiliar problems and designers like to solve problems.

Tolerance for Fluctuating Stimuli Assessment. In the work world of an industrial designer schedules are shifting, budgets expand and contract, and product requirements change. Designers must work with changing parameters.

Delaying Selection from Multiple Solutions Assessment. Designers like to explore lots of ideas and so have a willingness to defer judgment and seek alternative ideas, solutions, or conclusions. This is called resistance to closure and requires adaptive thinking.

10 Submission and Overall Evaluation

The final assessment for admittance into the professional program requires a portfolio of work from the first semester set of four block (seven-week) courses. The courses are Visual Literacy, Visualization Methods, Structured Creativity, and Form & Surface Development. All the faculty that teach the first-year foundation courses rank the portfolios. They all have components that help to determine, at a deeper level, a student's creative-, visual-, and adaptive-thinking abilities.

11 Summary

In trying to assess a student's aptitude for industrial design, we have found that looking for these traits of creative, visual, and adaptive thinkers in the professional program applicants has proven valuable. The program has had more success with the new assessment than with the more generic portfolio review, as indicated by the higher level of work and engagement in the program classes. These results are also shown in the higher quality of the professional-level application portfolios. This quality of work was validated by a visit of NASAD accreditors as part of our program accreditation review. The visitors' report says, "The level of proficiency in the first year of the professional program was observed to be especially high." We attribute this level of quality in the professional program to finding students with strong aptitudes in creative, visual, and adaptive thinking as a part of the application process.

References

Agor, W.: How top executives use their intuition to make important decisions. Bus. Horiz. **29**(1), 49–53 (1986)

Bochner, S.: Defining intolerance of ambiguity. Psychol. Rec. **15**, 393–400 (1965). http://doi.org/10.1007/BF03393605

Budner, S.: Intolerance of ambiguity as a personality variable. J. Pers. **30**, 29–50 (1962)

Cooper, L., Shepard, R.: Turning something over in the mind. Sci. Am. **251**(6), 106–115 (1984)

Gelb, M.: How to Think Like Leonardo da Vinci. Dell Publishing, New York (1998)

Guilford, J.P., Hoepfner, R.: The Analysis of Intelligence. McGraw-Hill, New York (1971)

Hoffman, E.: Psychological Testing at Work: How to Use, Interpret, and Get the Most Out of the Newest Tests in Personality, Learning Style, Aptitudes, Interests, and More!. McGraw-Hill, New York (2002)

Mac Donald Jr., A.P.: Revised scale for ambiguity tolerance: reliability and validity. Psychol. Rep. **26**(3), 791–798 (1970)

McLain, D.L., Kefallonitis, E., Armani, K.: Ambiguity tolerance in organizations: definitional clarification and perspectives on future research. Front. Psychol. **6**, 344 (2015). http://doi.org/10.3389/fpsyg

NASAD.: National Association of Schools of Art and Design, Brigham Young University Visitors Report. Reston, VA (2016a)

NASAD. 2016. National Association of Schools of Art and Design Handbook, Standards and Guidelines, 106. Reston, VA (2016b)

Newcombe, N., Frick, A.: Early education for spatial intelligence: why, what, and how. Mind Brain Educ. **4**(3), 102–111 (2010)

Runco, M.: Divergent Thinking (Creativity Research), vol. xi, p. 199. Ablex Publishing, Norwood (1991)

Silverman, L.: Finding and Serving Gifted Children with Learning Disabilities. Address to the Colorado Parent Institute, Denver (1997)

Skaggs, P.: Aptitudes for industrial design. Master's thesis, Rochester Institute of Technology (2002)

Sternberg, R., Lubart, T.: Defying the Crowd: Cultivating Creativity in a Culture of Conformity. Free Press, New York (1995)

Suzuki, H., Hiraki, K.: Constraints and their relaxation in the processes of insight. Cogn. Stud. Bull. Jpn. Cogn. Sci. Soc. **5**(2) (1998). http://wsd.irc.aoyama.ac.jp/hiblog/suzuki/files/2009/07/etlreport97.pdf

Torrance, P.: A national climate for creativity and invention. Gifted Child Today **15**(1), 10–14 (1992)

Ware, C.: Visual Thinking: For Design. Morgan Kaufmann, Burlington (2008)

Interdisciplinary Approach to Industrial Design

Redefining User Needs

James Rudolph(✉)

Department of Design, University of Notre Dame, South Bend, IN, USA
jrudolp2@nd.edu

Abstract. There is consensus in the field of medical device development that successful product development requires a deep understanding of the user and their needs. Despite the importance placed on identifying user needs, from practitioners and academics alike, there is little agreement on what constitutes a user need. This is problematic for medical device development teams, as they are responsible for defining user interface requirements and ensuring they meet the needs of intended users. The author proposes a framework for defining user needs for medical device development to improve consistency and to optimize the development of user interface requirements.

Keywords: Human factors · User needs · Medical device development · Product development · User interface requirements · Industrial design · Design strategy · User research · Design research · Ergonomics

1 Introduction

There is consensus in the medical device industry that successful product development requires a deep understanding of users and their needs [3, 7, 8, 12, 13]. A review of prevalent human factors and medical device development literature supports this perspective. Money, et al., in their research, state: "The importance and value of focusing on user needs has been recognized as having a number of health related benefits including; improved patient safety, improved compliance and health outcomes, and higher levels of patient and user satisfaction [10]." There are many similar examples published in articles and books related to human factors, design research, and device development.

Making explicit the connection between user needs and product solutions is not only desirable from a product development and business perspective, but also mandated by international medical device regulations. From the U.S. Food and Drug Administration's (FDA) design input section of the Quality System Regulation (21 CFR 820.30(c)): "Each manufacturer shall establish and maintain procedures to ensure that the design requirements relating to a device are appropriate and address the intended use of the device, including the needs of the user and patient [4]." Despite the importance placed on identifying user needs, from practitioners, regulators, and academics alike, there is little agreement on what constitutes a user need. If identifying user needs is an important part of designing useful, safe, and effective medical devices, there should be a common understanding of how we define user needs.

G. Di Bucchianico et al. (Eds.): AHFE 2020, AISC 1202, pp. 309–315, 2020.
https://doi.org/10.1007/978-3-030-51194-4_41

2 Background

In medical device development, the identification of user needs serves multiple purposes. First, it ensures the development team is working on a product or service solution that will provide value—the product will enable a user to accomplish a goal, solve a problem, or complete a task. Second, identifying and documenting successful user needs provides actionable insights that can inform appropriate product solutions, and ultimately be translated into user interface requirements [14]. This second objective ensures the development team is providing traceable and verifiable user interface requirements, and documenting their efforts in a way that meets national and international medical device regulations, including ISO 13485 and the FDA's design control requirements [2, 4].

To bring a successful medical device to market, of course, there may be many other needs to consider throughout the development process, including clinical, business, marketing, and other stakeholder needs. The process for identifying these stakeholder needs, however, may require different definitions, research methods, and validation processes.

3 Methodology

The author conducted a literature review of prevalent human factors and design research publications, including the FDA's guidance documentation, to identify prevailing and disparate definitions of user needs. In addition, an online survey was conducted to uncover definitions of user needs in current use amongst practitioners in the medical device development industry. Survey participants included 15 individuals with backgrounds in engineering, industrial design, and human factors. Definitions for user needs in the available literature and the survey results were analyzed for similarities and differences between content, descriptions, and examples used to communicate user needs.

Analysis. A visual display of results from the literature review and survey was created, following a process described by Miles et al. [9], which allowed for efficient review, analysis, and comparison of user needs definitions. Definitions were arranged into categories according to content similarities, differences, and examples provided. The analysis provided input for creating a practical way of thinking about user needs in medical device development. The author proposes a theoretical framework for understanding and defining user needs in medical device development. The framework has practical implications for development teams, including documentation efficiencies, consistency between development firms and manufacturers, and most importantly, the ability to provide a direct link (traceability) between user interface requirements and user needs.

4 Results

The author analyzed 14 different definitions of user needs across various media platforms, including published journal articles and books, websites, government guidance documents, and international standards. The results were then coded according to key words, definitions, and primary examples used (if available).

A cursory review of the published literature highlights the discrepancies amongst user needs definitions. See examples in Table 1. The samples selected for inclusion in the table represent the range of disparate definitions. Some of the more prominent publications acknowledged the importance of understanding user needs, but did not include a definition for a user need or a synonymous phrase (e.g., user needs statement, user problem, user point-of-view) [1, 17]. For example, in the Association for the Advancement of Medical Device's guidance document, HE75:2009, the section on user needs outlines two key recommendations: 1) Prioritize user input, and 2) Do not rely exclusively on "thought leaders." There is no description of what constitutes a user need. Adding to the confusion amongst various definitions, user needs are often described using different terminology, including user needs statements, user requirements, marketing requirements, user needs requirements, and user preferences, amongst other terms and phrases.

Table 1. Published or online definitions of *user needs*.

	User needs definitions
1	User needs are requirements that add value to a product, service, or environment for a user. Capturing user needs is a process of engaging users to understand their problems, process, goals, and preferences [15]
2	Properly articulating "a need" creates a three-part concise statement of desired outcomes, including: 1) A direction (increase/decrease/ensure/avoid); 2) A target (comfort, trauma, ease of use); and 3) A context (day in the life step/activity) [12]
3	A need statement is a concise description of a goal (what the device needs to do) but does not indicate how to achieve the goal [12]
4	Traditional needs statements have 3 components: 1) a user, 2) a need, and 3) a goal. These are then combined following the pattern [A user] needs [need] in order to accomplish [goal] [11]

A similar trend was identified in the results of the online survey. An overview of selected definitions can be found in Table 2. The samples outlined below were selected to showcase the range of how current practitioners understand user needs.

Table 2. Survey participant definitions of *user needs.*

	Survey results – user need definitions
1	Basic necessities required by the user to carry out an action
2	A requirement of the user (not a business need) that does not define a solution
3	Articulated or unarticulated insight that helps a user accomplish their goal
4	User needs allow the user to successfully manipulate and understand the product technology, as intended by the manufacturer

The definitions discovered through this research highlight the disparity in current use. The core differences between definitions related to two primary content elements: 1) the relationship of a user need to a product or solution; and 2) the conflation of goals and tasks as synonymous.

User Needs and Product Solutions. One category of user needs definitions sought to separate the solution from the user need. In these examples, the authors tended to define the need as a goal or "desired outcome" [11, 12, 16], and the goals were described as independent of the user need. In one example, from the FDA's HE75, there is a clear expressed need, but no indication of a product or solution: "The user needs information regarding important changes in the patient's condition… [1]."

On the other hand, there were a group of definitions that tied the user need directly to a product or solution. The definitions that fell into this category defined the need as a task or specific step in the process of interacting with a device. The user needs, in these examples, were dependent on and defined by the solution provided. For example: "Waveform scale markings should allow the user to accurately estimate the associated parameter values with sufficient resolution to meet user needs [1]." This example expresses the need in a way that can easily translate to a well-defined user interface requirement, one that can be verified through testing or human factors standards.

Conflating Tasks and Goals. The conflation of tasks and goals may be linked to how one interprets the relationship between a user need and a product or solution. If a solution is required to define a user need, for example, then one can presumably define needs more specifically and relate the need directly to a task required to complete a job. Conversely, if the solution is to remain independent of the need, one must consider higher-level user goals (e.g., what is the user trying to accomplish?).

There are several other variables provided in the definitions that may require additional research, including alternative terms and phrases for user needs, the examples used to describe user needs, and the specific terminology of the definitions. To provide a useful structural framework for product development teams, however, I will focus on the two core discrepancies identified in the research to date.

5 Discussion

Analysis of available user needs definitions in published literature, online resources, and from survey respondents indicate a disparate range of how we understand user needs. User needs are often used to describe a multitude of considerations, including user goals, desires, challenges, preferences, and tasks required to successfully interact with a product. The initial analysis uncovered two primary trends, outlined above. The two prominent definitions provide an opportunity for creating a practical framework for product development teams.

User Needs Framework. Two primary categories of user needs emerged through this research: 1) Motivational or Goal-oriented Needs and 2) Experiential or Task-based needs. Both categories are important for product developments teams as both contribute to a more thorough understanding of prospective users. It is not a matter of which definition is more accurate, but a matter of when it is most prudent to identify each type of user need in the development process.

Motivational Needs. A common theme amongst user needs definitions included the description of a user need as a goal. The goals, as several definitions indicated, do not prescribe any specific solution, but rather communicate an individual's "desired outcome" [11, 12, 15]. As one author explained: "A need statement is a concise description of a goal but does not indicate how to achieve the goal [12]." Motivational needs are defined by high-level goals that a user aspires to accomplish, including career, social, socio-economic, aesthetic, and individual self-fulfillment goals. At first glance, these types of needs may seem ambiguous, but motivational needs can also be specific. For example, a physician who specializes in interventional cardiology may aspire to reduce his or her restenosis rates or procedure times.

Understanding motivational needs is important during the Strategy phase of product development, when the development team is determining the right product to bring to market. As much of the published research suggests, it is important for the development team to understand why users aspire to accomplish the identified goals [7, 10, 12, 16]. Once the underlying, often latent, motivational needs are understood, they provide direction and inspiration for ideation, problem solving, and the creation of alternative solutions. Solutions that best meet the intended user groups' motivational needs should be considered, in conjunction with appropriate stakeholder needs, for refinement and testing. Motivational needs inspire and inform product solutions, but they do not demand a particular solution. As such, they do not provide a clear link between user needs and user interface requirements.

Experiential Needs. As several of the analyzed definitions pointed out, some user needs are intrinsically connected to a user's experience interacting with a given product or solution. *Experiential needs* can be defined as tasks the user must complete to successfully interact with a device, including physical and cognitive related tasks. For example, using the case of the interventional cardiologist, an experiential need would be to gain access to the femoral artery before inserting and positioning a guidewire or stent.

In medical device development, experiential needs are best identified during the development process, after a general solution has been determined. A task analysis is conducted to document the proposed step-by-step process required to successfully interact with the device, and each step in the interaction process becomes a user need. These critical steps, or experiential user needs, then inform specific design solutions. They also enable the team to optimize the solution for use through iterative testing and refinement. The proposed design solution must enable a user to safely and effectively complete each task. The user interface requirements (or design specifications) can then be traced to the user need, and importantly, can be verified through inspection, demonstration, testing, or analysis [2, 13].

Both motivational needs and experiential needs, as defined in this framework, are critical to developing successful medical devices. Motivational needs ensure development teams work on medical devices that are valuable and focused on meaningful problems, while experiential needs ensure the solutions are optimized for safety and usability. An overview of the proposed structural framework, including examples, is provided in Table 3.

Table 3. Structural framework for user needs.

	Strategy <<<	>>> Development
Objective	To define the optimal product and/or service to bring to market	To implement and optimize the selected product for production and launch
User needs	Motivational (Goal-oriented)	Experiential (Task-based)
Definition	Higher level goals the user hopes to accomplish, including personal, emotional, social, cognitive, and service related goals	Lower level tasks the user must complete to successfully interact with a device or system, including physical and cognitive related tasks
User need examples	A physician hopes to conduct a successful coronary intervention surgery	The physician must gain access to the femoral artery to position a guidewire
	An individual hopes to lose weight to maintain his or her health and well-being	The individual must research and purchase foods that contribute to a health, low-fat diet

6 Conclusions

This research highlights the structural discrepancies amongst existing definitions for user needs. Additional research is required to understand perspectives from a wider audience, including professional domains outside medical device development. Future research should also consider alternative naming conventions for user needs, including user needs statements, latent needs, unmet needs, user insights, and alternative terminology not considered by the author of this study. Finally, the proposed framework will need to be implemented in a professional development environment if we are to truly understand its practicality, usefulness, and legitimacy.

Ultimately, understanding user needs is critical to developing medical devices that enhance our natural human capabilities, provide safe and effective interactions, and increase a user's overall satisfaction with medical devices. This paper proposes a theoretical framework that may have important implications for medical device product development teams, including the effective creation and documentation of clearly defined user needs and the translation of user needs into effective user interface requirements.

References

1. Association for the Advancement of Medical Instrumentation: Human Factors Engineering - Design of Medical Devices. ANSI/AAMI HE75:2009, Arlington, VA (2009)
2. Association for the Advancement of Medical Instrumentation: Medical Devices – Part 1: Application of Usability Engineering to Medical Devices. ANSI/AAMI/IEC 62366-1:2015, Arlington, VA (2015)
3. Food and Drug Administration (FDA): Applying Human Factors and Usability Engineering to Medical Devices: Guidance for Industry and Food and Drug Administration Staff. CDRH, Maryland (2011)
4. Food and Drug Administration (FDA): Quality System Regulation – Design Controls, 21CFR820.30 (2019)
5. ISO: IEC 623661-1:2015: Medical Devices – Part 1: Application of Usability Engineering to Medical Devices (2015)
6. ISO: A practical guide ISO 13485: 2016 Medical devices (2017)
7. Martin, J.L., Murphy, E., Crowe, J.A., Norris, B.J.: Capturing user requirements in medical device development: the role of ergonomics. Physiol. Meas. **27**, 49–62 (2006)
8. Martin, J.L., Barnett, J.: Integrating the results of user research into medical device development: insights from a case study. BMC Med. Inform. Decis. Mak. **12**, 74 (2012)
9. Miles, M.B., Huberman, A.M., Saldana, J.: Qualitative Data Analysis: A Methods Sourcebook, 2nd edn. Sage, Washington, D.C. (1994)
10. Money, A.G., Barnett, J., Kuljis, J., Craven, M.P., Martin, J.L., Young, T.: The role of the user within the medical device design and development process: medical device manufacturer's perspectives. BMC Med. Inform. Decis. Mak. **11**, 15 (2011)
11. Nielsen Norman Group. https://www.nngroup.com/articles/user-need-statements/
12. Privitera, M.B.: Contextual Inquiry for Medical Device Design. Elsevier, Waltham (2015)
13. Privitera, M.B., Evans, M., Southee, D.: Human factors in the design of medical devices – approaches to meeting international standards in the European Union and USA. Appl. Ergon. **59**, 251–263 (2017)
14. Privitera, M.B., Murray, D.L.: Applied ergonomics: determining user needs in medical device design. In: 31st Annual International Conference of the IEEE EMBS, pp. 5606–5608. IEEE Press, New York (2009)
15. Simplicable. https://simplicable.com/new/user-needs
16. Ulwick, A.: Turn customer input into innovation. Harvard Bus. Rev. **102**, 91–97 (2002)
17. Wickens, C.D., Lee, J., Liu, Y., Becker, S.G.: An Introduction to Human Factors Engineering. Pearson Prentice Hall, Upper Saddle River (2004)

The Role of Design in Consumer Behaviour: How Design Can Influence Consumer Decision Making at a Point of Purchase

Ruzbeh Irani(✉) and Lois Frankel(✉)

School of Industrial Design, Carleton University,
1125 Colonel by Dr, Ottawa, ON K1S 5B6, Canada
ruzbehirani@cmail.carleton.com,
lois.frankel@cunet.carleton.ca

Abstract. Numerous disciplines are attempting to discover how consumers behave at a point of purchase (POP). One factor that is significant to the development of the content behind what influences consumers' behaviour is design. Existing design and brand research will be explored; this literature review reveals that these traditional research methods do not produce the intended target behaviour. New forms of research that provide an interdisciplinary approach, such as the Design with Intent Method (DWI Method), Neuromarketing, emotional design or sensory design applications, may allow for more authentic results to help influence consumer behavior. If designers can design with the intent to change consumer behavior using more genuine methods of research, then behavior can be influenced and predicted. By exploring numerous examples of how design has provoked emotion and influenced consumer behavior, this literature review suggests that design influences consumer decision making at the point of purchase.

Keywords: Decision making · Design with Intent Method · Consumer behavior · Point of purchase · Sensory aspects · Neuromarketing · Emotional design

1 Introduction

Why do people buy what they buy? Numerous disciplines are attempting to answer this question. Design can play an important role in developing the content behind what influences people's behavior as consumers [1]. This literature review explores the contributions of various disciplines from an interdisciplinary lens.

According to the literature, design and brand research that is used, such as 'Buy Me Research' [2] and advertising through displaying information [3] has not been producing desirable outcomes. Buy Me Research is a preliminary method used for designing a point of purchase (POP). It investigates how the consumer behaves through interviews and questionnaires in stores at the POP or in the consumer's home [2]. Although this type of research arouses tendencies from the consumer as they know that they are being targeted with advertising, it also produces resistance to messages at the POP [2]. Rogers [6] notes that the reason traditional methods of research are misleading

G. Di Bucchianico et al. (Eds.): AHFE 2020, AISC 1202, pp. 316–322, 2020.
https://doi.org/10.1007/978-3-030-51194-4_42

and inaccurate is because there is a separation between the 'rational mindset' during interviews and the 'subconscious browsing' state of the consumer at the POP. Ninety-Five percent of people's decisions are made subconsciously [6], hence the inaccuracies in such traditional approaches to consumer behavior research.

There are only a few scientific studies suggesting that by simply labeling fast food with the calories on all menus, consumer behavior will lean more towards a healthy alternative [3]. Unfortunately, this is not the case with the research done by Lisa J Harnack and Biran Elbel. Their research suggests that what consumers say and actually do are different [3]. Both use methods that inevitably produce superficial results due to their inability to replicate the user's authentic subconscious behavior at the POP. New forms of research that provide an interdisciplinary approach, such as the Design with Intent Method (DWI Method) and Neuromarketing allow for more authentic results to help influence consumer behavior [4, 5]. The DWI Method brings forth a goal of the design process which is to modify or redesign the system so that it influences consumer decision making towards a target behavior [4]. Neuromarketing provides tools for information to be extracted for design research purposes [5]. Design can influence people's emotions; without emotion, they can't make decisions [6]. Emotions can operate at a subconscious level during times of purchasing and at a POP, which is why design plays a significant role in a consumer's decisions [6]. If designers can design with the intent to change consumer behavior using more genuine methods of research, then behavior may be influenced and predicted. By exploring numerous examples of how design has provoked emotion and influenced consumer behavior, this literature review suggests that design influences consumer decision making at the point of purchase.

2 Background

The results of traditional consumer behaviour research, such as buy me 'Buy Me' research, have been found to be inaccurate and misleading [2]. Buy Me research was conducted by Research International as an initial study where consumers were observed and interviewed in a supermarket setting. Six types of products were compared, detergents, spreads, coffee, shampoo, jams and yogurts. [2] These categories were used to consider the influence design had on different brands and other POP decision influencers such as promotions, price, merchandising display and brand memories [2]. One of the participants said during an interview that packaging doesn't influence her decision at a POP [2]. "I don't care what it's in, brown paper bag as long as the products are OK. I must say I find that very attractive; I must say I do like these bottles. I'm very drawn to the packaging of Bonne Maman because I think they're very nice. I think that pack is just, you know, that would put me off . . . The packaging is better, isn't it, and I would think the inside would be better" [2]. On one hand, the consumer says they aren't influenced by the packaging but on the other, they have numerous subjective comments about packaging and branding. This contradiction in consumer behavior and their perception of their behavior seems to be consistent in traditional research methods. Unfortunately, participants of the study only provide superficial data that does not aid the consumer's decision-making process [2].

The reason for this is the separation between the rational mindset of the consumer during interviews vs the submissive browsing state of the consumer at the POP [2].

Why do people buy what they buy? It is easy to assume that it is the product of advertising, but advertising just information isn't enough [3, 7]. Research done by Lisa J Harnack and Brian Elbel in calorie food labeling suggests that simply advertising the number of calories on a menu does not change the consumer's behavior, rather it informs them [3, 7]. Their research reports that consumers acknowledge the labeling and its relevance but on its own, the information is not enticing enough to influence the consumer's behavior to a lower calorie alternative. One of the studies was conducted in numerous fast food joints, where subjects were asked to provide their receipt to the researcher after making a purchase (with and without calorie labeling on menus) [3]. Some of the subjects said that calorie labeling did influence their behavior although, their receipts suggested no change in calorie consumption between menus with labels and ones without. One purpose of this literature review is to note this inconsistency in consumer behavior and the perception the consumer has of their own behavior to identify a possible opportunity for design to play a role. The inconsistency suggests that consumers want their behavior to change if the outcome is in their favor.

One approach to help with the dilemma of influencing consumer behavior is the Design with Intent Method (DWI). DWI focuses on the importance of the user's behavior and its relation to an operation, or where the user's behavior is strategically altered [4]. "The goal of the design process is to modify or redesign the system to achieve this: to influence the user's behavior to a particular 'target behavior'. The method is a 'suggestion tool', inspiring design solutions as suggested techniques, with examples applicable to particular target behaviors" [4]. The DWI method has a structure involving eight lenses to approach a problem from: architecture, error proofing, interaction, ludic perceptual, cognitive, Machiavellian and security [4].

3 How Decision Making Really Works

Nowadays there is no significant difference in quality or price; people's choices are based on more than traditional variables [8]. Mediatization has influenced consumer decisions to be based more on a product's ability to amplify personal attributes like identity play, gaining social currency and self-branding [8]. These attributes can make the decision-making process difficult for consumers [8]. Hanna Willman, notes in her article, 'The Future of Consumer Decision Making', there is a direct correlation between the complexity of the decision task and the choices consumers make [8]. The subjective importance of choice significantly affects how much effort consumers put into making that choice [8]. Willman utilizes the following heuristics to simplify the consumer's decision-making process [8]: satisficing, lexicographic, eliminating by aspects, frequency of good and bad features and finally, equal weight. Satisficing is the first satisfactory choice made by the consumer and tends to be what the consumer ends up purchasing. Lexicographic; the most important feature to the consumer will be chosen while they neglect any products without that feature. Eliminating by aspects is used when the consumer works through the alternatives one by one but with a certain attribute in mind. Once a product meets the attribute in mind, the second attribute is then considered by the consumer while continuing to search through the alternatives. Frequency of good and bad features involve the consumer adding the negative and positive attributes respectively, the negative attributes are then subtracted from the positive for each product. The one with the highest score will be the product of choice. Finally, the equal weight heuristic; all products and their alternatives' are assigned a value and are added, the highest score is the one the consumer selects.

These heuristics fit into story maps of how the consumer contemplates the decision to purchase certain products over alternatives. Similar to how the film industry uses story mapping to illustrate the motivations, aspirations and fears of the character, designers also use story mapping to illustrate the process that the consumer may go through with a particular product. It may be used as a tool to increase the ease of navigation of a process by breaking down its complexities. This interdisciplinary tool may be used for opportunistic change in understanding these heuristics. People tend to use these heuristics to simplify the task of decision making for non-important choices.

If designers can map the process (the heuristic) a consumer uses to simplify their decision, that may open up opportunity for new methods to be used to influence consumer behavior by enhancing the heuristic used. As an example, if calorie labeling isn't an influencing factor on the consumer's decision, by having designers develop an understanding of the heuristic used during the fast-food purchase, they may be able to further reduce the complexity of the consumer's decision. If consumer decisions have a certain complexity then designers could utilize the POP's surroundings to develop a strategy to change consumer behavior. When designers undergo a product design process, the consideration for the user context could play a significant role in how successful the product is. Context is a significant driving force when it comes to decision making [8]. If the process to make a decision is too difficult it can procreate negative emotions [8]. If the complications of the consumer's decision can be solved with the aid of the POP's contextual environment, then the recognition of the consumers' heuristic of decision making during the POP is critical to the consumer's emotional state.

4 Neuromarketing and Emotional Design

Neuromarketing is a marketing communication field that involves neuroscience and marketing [5]. Every year nine out of ten new products fail; approximately one hundred million dollars are wasted in market research. The reason for this is because traditional methods of marketing fail to pay attention to consumer's unconscious emotional experiences [5]. Consumers often have a rationale for their purchase decision although their emotions play a significant role [1]. Sensory stimuli can also influence consumer's emotions and as a result, change the nature of the consumer's behavior [1]. If the appropriate sensory stimuli are applied at the right time, it can calm, relax, de-stress, energize, improve mood and influence decision making [1]. This can eradicate the perceived difficulty of a decision. As Fig. 1 illustrates, the contextual environment, if designed correctly, may evoke emotion in the consumer which could then reduce the difficulty of the heuristic the consumer is using to make their purchase decision. Consumers rely more on their emotions when a decision is difficult, when information is limited and when they believe that their emotions are relevant to the purchase [8]. Point of Purchase Advertising International found that sensory elements can be capitalized on as seventy-five percent of buying decisions are made at a POP [9]. Not only do sensory aspects consider the emotions of the consumer, but they are also a practical application for profits. In fact, combinations of senses can have a "super-addictive effect", while improving the consumer's ability to remember the brand [9].

Color, touch, smell and sound all affect consumers' behavior [1]. Research evidence suggests that color influences people's moods and manifests itself in our decisions [1]. Colors like blue can lower the blood pressure of a consumer, while yellow is generally the first color consumers see [1]. When the color of lighting is considered in an environment, it could affect how long a consumer stays there. Neuromarketeers at Google performed a study to identify what shades of blue would generate more clicks by the user [5]. By adopting a specific shade of blue in Google's search engine, Google was able to increase its annual revenue by two hundred million dollars [5]. Some of the literature focusing on food quality and consumer preferences found that colour intensity was most impactful when it came to the consumer's overall liking of the food [10].

Touch is a sense that is predominately used by consumers to influence their decision when purchasing clothes or furniture. When holding a piece of clothing, grasping the fibers to feel how soft it is, weighing it from one hand to the other or pushing down on a couch to feel it's resistance, consumers are fueling their emotions to make the most rational choice they can [1]. Research conducted by Joann Peck, Associate Professor of Marketing at the University of Wisconsin-Madison School of Business, suggests moderating the effects of an impulse purchase on impulse-purchasing behaviour with design considerations for NFT (Need For Touch) customers [11]. NFT customers' preferences and motivations in response to a product are based on extracting information through touch. There are two components of NFT, the autotelic component of NFT will be a focus and relates to touch as hedonic-oriented response seeking through fun, arousal, sensory stimulation, and enjoyment [11]. Peck examines individual differences in NFT of customers and how it affects their impulsive-buying behaviour at a grocery store [11]. There were two types of subjects, first, customers who had a high autotelic NFT and the other were those who had a low autotelic NFT [11]. The study discovered that those with higher NFT were buying significantly more impulsively.

Smell is one of the most powerful influencers of all the senses because it allows the brain to be occupied with other thoughts the consumer may have [1] at the POP. Smell is most closely linked to emotions because it fast tracks signals to the limbic system and links emotions with memories. Lavender essence increases the consumers' time and money spent [1]. 'Feminine scents' doubled the sales of women's clothes and the same for men with 'masculine scents' [1]. Olfactory stimuli are the second most 'liked' when it comes to influencing consumer behaviour [9], this influence can be seen through a study done by The New York Times that found that eighty-one percent of consumers prefer to choose food and beverages based on their smell rather than visual appearance [9]. This is an important insight that companies keep in mind when using smell to reinforce their brand identity. The smell consumers' encounter when walking into Subway may seem like a byproduct of freshly baked bread but the reality is that the innocent and subtle scent is intentionally used to develop a 'scent brand' [12]. Scent brands are becoming common among retailers [12].

Finally, the role that sound plays influencing people's behaviour is significant to the rationale of their purchase. Sound can evoke the taste, texture and freshness of a product. The 'crinkle' of a bag of chips or corn flakes may suggest the quality of the snack to the user [9]. Although, sound isn't one of the leading senses when it comes to influencing consumer behaviour, it can be used or combined with other sensory stimuli to create an influential factor at the POP.

5 Discussion

Research conducted could consider the consumer's mental state to produce more authentic results. Traditional research methods may simply end up advertising what the consumer should be concerned about [3, 7], rather than designing to suggest that the product under consideration could solve the purchaser's problem. As POP environments change and as consumers evolve with their shopping experiences, research to understand consumer behaviour needs to advance as well. Design research should not provoke disingenuous behavior but rather provoke authentic consumer behavior using tools to analyze and synthesize information.

There are critiques suggesting that the influence of emotional design and neuromarketing may mislead consumers to overspend or purchase products that may not be of any use to them. This may be a cause of concern especially in times of economic crisis and wealth inequality gaps. Statistics Canada has claimed that in 2018 the annual household savings was only 1.4%, the lowest rate since 2005 [13]. If consumers' decision making is constantly influenced by factors such as neuromarketing and emotional design, then when does the consumer actually purchase a product based on their own decision, with no external influence? The rebuttal to this concern is, that all consumers are influenced by some sort of external factor in their lives; therefore, products that are designed intentionally to provoke a purchase share no ethical difference from the consumer's other unplanned external factors. The reality is that consumers are going to continue to purchase products in the future; the question is whether or not designers should influence consumers' behaviour to exploit profit or to benefit the consumer.

6 Conclusion

The role of design in consumer behaviour has influenced consumer decision making at the point of purchase by tackling this problem from an interdisciplinary lens. The Design with Intent Method (DWI Method), Neuromarketing, emotional design and sensory aspect application can all be used to tap into the subconscious/unconscious mind to help influence a purchase decision at a POP. By understanding how decision making really works using Willman's heuristics to simplify the consumer decision making process, designers could create a story map for the consumer and apply these interdisciplinary approaches to the consumer's mapped process. Consumers have a multitude of external factors affecting their decisions, especially at a POP because of their constraints when shopping in a retail store (time, state of mind, etc.). Since there are many aspects affecting the consumer at a POP, it is necessary to understand consumer decision making as an interdisciplinary design problem that cannot be solved using traditional monodisciplinary methods.

References

1. Soars, B.: Driving sales through shoppers' sense of sound, sight, smell and touch. Int. J. Retail Distrib. Manag. **37**, 286–298 (2009)
2. Connolly, A.: How does design affect decisions at the point of sale? J. Brand Manag. **4**, 100–107 (1996)
3. Elbel, B., Kersh, R., Brescoll, V.L., Dixon, L.B.: Calorie labeling and food choices: a first look at the effects on low-income people in New York City. Health Aff. **28**, w1110–w1121 (2009)
4. Lockton, D., Harrison, D., Stanton, N.A.: The design with intent method: a design tool for influencing user behavior. Appl. Ergon. **41**, 382–392 (2010)
5. Wu, T.: Neuromarketing: The new science of consumer decisions. https://www.ted.com/talks/terry_wu_neuromarketing_the_new_science_of_consumer_decision
6. Rogers, D.: The Science of Shopping and Future of Retail (2013). https://www.youtube.com/watch?v=qVJAr5eGOKA
7. Harnack, L.J., French, S.A.: Effect of point-of-purchase calorie labeling on restaurant and cafeteria food choices: a review of the literature. Int. J. Behav. Nutr. Phys. Act. **5**, 3–8 (2008)
8. Willman-Iivarinen, H.: The future of consumer decision making. Eur. J. Futur. Res. **5**, 14 (2017)
9. Week, D.: Sensory design in packaging (2015)
10. Raz, C., Piper, D., Haller, R., Nicod, H., Dusart, N., Giboreau, A.: From sensory marketing to sensory design: how to drive formulation using consumers' input? Food Qual. Prefer. **19**, 719–726 (2008)
11. Peck, J., Childers, T.L.: If I touch it I have to have it: individual and environmental influences on impulse purchasing. J. Bus. Res. **59**, 765–769 (2006)
12. Lockton, D., Harrison, D., Stanton, N.A.: Design with Intent: 101 Patterns for Influencing Behaviour Through Design v.1.0. Equifine, Windsor (2010)
13. Liew, C.: Canadians Are Awful at Saving Money: Here's How You Can Improve (2019)

Designing an Experience Ecosystem on Campus

Scott Shim(✉)

University of Notre Dame, 237 West Lake Hall, Notre Dame, IN 46556, USA
scottshim@hotmail.com

Abstract. This paper summarizes an inclusive pedagogical experience from the perspectives of design educators and students. It is a case study of a collaborative capstone course at an established design program at a major university and a narrative of a sponsored project to redesign a service/experience-driven ecosystem on campus. Non-linear journey of interdisciplinary teams showcases how contextual experts from diverse background were able to embrace unfamiliar challenges. The participants in this study include senior level undergraduate students from various majors.

The author will discuss inclusive pedagogical examples from a 14-week capstone studio course that was defined by three key design phases: Design Intent and Opportunity, Conceptualization, and Implementation, each employing its own relevant tools, methodologies, and design approaches. This collaborative design development course inspired participants to engage in design learning from a holistic, user-centered, and interdisciplinary approach.

Keywords: Service design · Experience design · Interdisciplinary · Design education · Collaborative design

1 Introduction

Today, the opportunity spectrum to implement creativity is broadening from a traditional manifestation of objects to human needs that can be explored through alternative resources. Designers are now presented with more complex problems that have to be approached holistically and with human-centered intentions. These projects frequently require empathetic sensibility with a sequential ecosystem in mind.

> *"No product is an island. A product is more than the product. It is a cohesive, integrated set of experiences. Think through all of the stages of a product or service – from initial intentions through final reflections, from first usage to help, service, and maintenance. Make them all work together seamlessly."*
>
> — *Don Norman, inventor of the term "User Experience"* [1]

Such an immersive shift is a result of blended expansion in the emerging areas of industrial design, design research, UX design, interaction design, social innovation, entrepreneurship, and more. Consequently, Human Centered Design (HCD) has rooted itself in creative strategies and is now being implemented in the opportunity assessment process. Shim [2] states, "*Utilizing a methodology arising from HCD's natural heritage, User Experience (UX)/Service Design brings together divergent stakeholders in*

G. Di Bucchianico et al. (Eds.): AHFE 2020, AISC 1202, pp. 323–329, 2020.
https://doi.org/10.1007/978-3-030-51194-4_43

a dynamic, iterative process that accelerates innovative thinking and making." User Experience (UX)/Service Design is now perceived as a common thread in offering students unique opportunities to engage in innovative creativity in academia. It has delivered emerging practice for collective inclusion and furthermore provides a fundamental Segway into traditional core competencies (such as ideation, visualization and prototyping) in design. Stickdorn and Schneider [3] confirms the value of the balance and writes *"the combination of a general understanding at the top end of the Ts, and an additional deep expertise at the bottom end of the Ts lead to a valuable collaborations that trigger useful service concepts and ensure their implementation."*

2 Context

The Collaborative Innovation program at the University of Notre Dame is a minor program in its 5th year and was conceived as a dynamic catalyst for process-based, cross-disciplinary collaboration between various academic departments in Liberal Arts and wider interests across the university. The initial constituency of undergraduate students was drawn from within Liberal Arts, but it is has now expanded to include students from business, science and engineering, all of whom are involved in bringing successful solutions to complex problems. Building a strong core competency in design thinking and collaborative innovation, it is intended to fulfill growing demand for this skill among these various communities, attracting broad and diverse student enrollment and re-establishing the primacy of a humanistic, collaborative approach to the complex and integrated challenges that face our rapidly changing world. The Collaborative Innovation program has placed humanization mindset as an overarching approach in an effort to increase contextual relevance. As Mootee [4] states, *"great design is only as good as it's ability to connect and forge relationships with the people who ultimately use it."*

The Capstone course in the Minor curriculum, titled 'Collaborative Design Development' (CDD), offers an engaging experience to develop innovative user experience solutions through an association of expertise from business, political science, entrepreneurship, engineering, science, anthropology, graphic and industrial design students and faculty. Collaborating teams of graduate and undergraduate students engaging in two procedural cycles, beginning with a framing of need and concluding with comprehensive proof of concept, and a contextually resolved product or system. All collaborative team members engaged throughout the research and developmental process. Each participant also shared rotating leadership responsibilities, providing direction within their specific areas of expertise and in the context of a sequential course outline.

3 Project Overview

The project topic in Collaborative Design Development class was to future cast the Huddle Mart, which is a convenience store on campus close to lecture halls and student residential housing. It serves approximately four thousand students daily and is accessible from early morning to late night for snacks, medicine, and other common necessities (Fig. 1).

Fig. 1. Current Huddle Mart

The Huddle Mart is operated by the Campus Dining Services which also operates 18 different dining/food related units on campus. The Director of the Campus Dining Services approached the class as a sponsor in assessing the future opportunities of campus dining, specifically the Huddle Mart. Three teams were made up from 15 students with diverse majors and background that were enrolled in this class, and had 14 weeks to work on this fully funded project. The midterm and final presentations were attended by the senior administrators who were directly involved with the campus experience.

4 Framing the Direction

The director of Campus Dining Services (CDS) kicked-off the class with a presentation on the operation of Huddle Mart. Students were able to gain valuable insight based on the quantitative information and expert opinions of how the industry is influenced by available technology used by external competitors. It was clear the Campus Dining Services (CDS) were driving the conversation around re-designing the physical space of the Huddle Mart and had pre-determined to incorporate the proven technologies now available on the market. One particular technology shared during class was the no checkout ecosystem in an Amazon Go store. According to Bishop, 'the system is logging the items as the shopper goes along, which eliminates the need to go through a traditional check-out line. When customers exit the store through a "transition area," the system senses that they're leaving, adds up the items and charges their Amazon account.' As an inspirational check-in, the class decided to take a class trip to Chicago and experience the customer journey first-hand (Fig. 2).

Fig. 2. Class trip to Amazon Go store in Chicago

Students were excited to witness the state-of-the-art facility and the technological advancements at Amazon Go, but felt the technology driven improvements would require a massive overhaul in infrastructure and maintenance requirements. As student groups began to collect information through generative literature research on convenience stores and consumer behaviors, we began to identify a specific niche personality that isn't usually displayed in a typical commercial space; college campus life style. The groups started looking into the details of this sector using the AEIOU (Activities, Environments, Interactions, Objects and Users) research method. There was a strong consensus among students that the unique experience should be the driver in this innovation effort, and that additional elements would support the logistics of execution. This paper will now describe one of three proposal outcomes in detail, and walk through the holistic approach in establishing a User Experience (UX)/Service design concept.

Student Persona

After reviewing and synthesizing the findings from the Intercept interviews, Behavior mapping and Photo Study, three categories were identified as core characters that reflect college student lifestyle; Control, Time Value and Social Snacking.

Control

The ability to control options and food intake, in choosing the amount of money spent, the healthiness of food, the customization of food, and time when items are purchased.

Students need to feel that they are getting value for the food items they purchase and want their purchase behaviors to be a result of their needs and desires, not the constraints imposed by Campus Dining or the Huddle itself.

Time-Value

Purchasing experiences at the Huddle do not have to be as fast as physically possible (ex: Amazon Go), but students need to feel that the purchasing process is efficient and does not waste time or involve needless steps. Speed is important, but above all, students must see that time was not unnecessarily spent in purchasing items, whether it be between classes or just before settling down for a long study session.

Social Snacking
The social norms present on campus dictate a student's purchasing and eating behaviors.

Pressures to eat healthy melt away when others in a group also engage in unhealthy behaviors. Snacking, especially late-night, can also become a social activity and an easy way to spend time with friends. Snacking also becomes important for students, either by themselves or in a group, or when settling down for study sessions. The needs and desires of students change as the varying social situations change throughout the day or week.

5 Ideation and Validation

Further validation indicated 'Time Value' as the most reflective necessity for students dealing with food consumption on campus. The team focused on creating an integrated solution and ended up with two distinctive ideas; one focusing on a delivery system and the other focusing on integration of space (Fig. 3).

Fig. 3. Two final concepts in Time Value category

Both ideas had pros and cons, but the design team focused on journey map to elaborate 'valuable concepts into reliable and systematic solutions that are actionable for future successful implementation' Kumar [5]. During this process, the design team identified a natural scene that is quite common on college campus; thousands of students walking from lecture halls to lecture halls. This was a eureka moment in that the team saw these moving individuals as part of the delivery ecosystem. When the design team identified a clever payment transfer portal that combines a university charge platform and commercial digital payment method, they came up with a disruptive experience that utilizes students' existing habits on campus called 'RELAY'- a delivery service whereby students can utilize other students' daily routes for pick-up and delivery in exchange for flex points; an extension of campus meal plans that are accepted only on select on-campus retail food service establishments. The design team have identified that utilizing flex points as a means of supplying delivery compensation, allows for a redistribution of cash to those that have less flex points and require more (thus incentivizing more spending and consumption).

6 Final Proposal

Our institution sits on about 1,200 acres in a suburban campus and the walking time from one end to the other is approximately 40 min. It's not a big campus but with 12,000 students taking classes and walking through campus for extracurricular activities, human traffic is available for a good period of the day. The design team identified this daily phenomenon as a core principle in proposing a cost-effective delivery system in the college campus environment. After multiple testing explorations on campus, the design team was confident that if an ecosystem offered benefits the deliverers, students would gladly participate in this self-sustaining service (Fig. 4).

Fig. 4. Prototyping and testing the experience

The RELAY model works to consider both the deliverer and customer at every step of the way. In doing so, we are able to build accountability within the framework of the system to ensure quality and efficiency (Fig. 5).

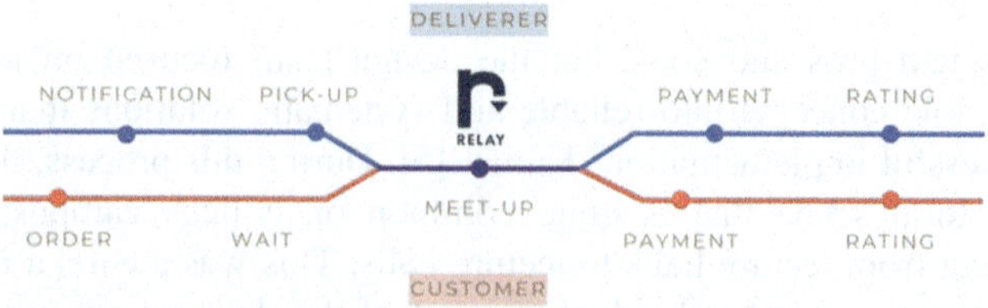

Fig. 5. Deliver and Customer journey map

As a way to clarify the procedural components of the ecosystem, the design team prototyped a delivery App for communication purposes, and a meeting station Kiosk. The App function includes order placement, connecting customers with deliverers, satellite tracker, notifier and payment methods. Kiosk has a sensor recognition that informs the customer when the delivery has arrived. As Verschoor [6] wrote *"Business is aimed at creating value, and as the context changes, business becomes part of a dynamic system in which value creation is achieved through engagement and vision"*, this project demonstrates how the User Experience (UX)/Service design education can contribute in emerging practice of value creation and further assist converging toward collaboration. The Collaboration with the Campus Dining Services (CDS) brought creative empowerment to light and students gained exposure to organic processes that can often result in disruptive innovations. The ideas from this class are currently being reviewed by the Student Affairs group for potential implementation in the near future.

7 Conclusions

Students from across the campus have recognized the distinctive attraction of User Experience/Service Design as it serves as a creative toolbox for interdisciplinary collaboration. Many institutions have revisited a pedagogy to humanize the content that prepares them to begin immediate and productive careers across a wide variety of industries, where collaborative design thinking is valued at a premium. Stickdorn and Schneider [4] write *"understanding the value and the nature of relations between people and other people, between people and things, between people and organizations, and between organizations of different kind are now understood to be central to designing services."* There is reason to believe contextual creativity can be inclusively embedded in the core of higher education rather than left as an exclusive study. We're hoping these students promote human-centered approach and cross-collegiate collaboration, resulting in the spread of divergent thinking across the entire academic community.

References

1. Definition of User Experience Design. https://www.interaction-design.org/literature/topics/ux-design
2. Shim, S.: Design education as an inclusive pedagogy. In: AHFE 2018 International Conference Proceedings (2018)
3. Strickdorn, M., Schneider, J.: This is Service Design Thinking. Wiley, Hoboken (2011)
4. Mootee, I.: Design Thinking for Strategic Innovation. Wiley, Hoboken (2013)
5. Kumar, V.: 101 Design Methods. Wiley, Hoboken (2013)
6. Verschoor, C.: Change Ahead. BIS Publishers, Amsterdam (2015)

Ageless Design: A Design Method of Product Longevity for Cross-Generation

Yong-Gyun Ghim[1](✉) and Cliff Shin[2]

[1] School of Design, University of Cincinnati, Cincinnati, OH, USA
ghimyn@uc.edu
[2] School of Art + Design, University of Illinois, Champaign, IL, USA
thecliff@illinois.edu

Abstract. Product longevity is a method for sustainable production and consumption in the current consumerist culture driven by planned obsolescence. While product aesthetics affect the product lifetime, existing approaches from industrial design perspectives are limited. Timeless design has been mentioned as one of the design strategies for product longevity. However, timeless design has limitations because it is largely focused on a product's appearance and lacks consideration of varying needs for different age groups. This paper introduces ageless design as a more comprehensive method with an inclusive perspective and usability added to timeless design. Ageless design aims to address the needs of multiple generations as well as to make a product accessible and easy to use regardless of age so that a product can not only be used across generations but also endure long years of use. Subsequently, this paper examines the differing VOCs for multiple generations and builds a framework for ageless design. A representative product of ageless design is analyzed based on the framework, then the elements and the guidelines for ageless design are verified.

Keywords: Sustainable design · Product longevity · Timeless design · Ageless design · Industrial design

1 Introduction

With an ever-increasing concern on the environment and more awareness of the general public on sustainable consumption, manufacturing corporations and designers are asked to endeavor in increasing a product's lifecycle. Researchers have mentioned timeless design as one of the strategies for product longevity because a long-lasting design has a high probability to become a part of users' lifestyles both aesthetically and practically. However, timeless design has certain limitations. It lacks consideration of wider audiences. For example, a new voice of consumers (VOC), particularly from millennials and generation Z, has become increasingly important in product development that includes product planning, manufacturing process, point of purchase, and post-purchase. Unlike older generations who embraced materialistic value as a measure of success, younger consumers see materialism as a belief that possessions lead to happiness. It is now necessary for designers to understand the various VOCs to better understand product longevity from their perspectives and re-define product longevity

G. Di Bucchianico et al. (Eds.): AHFE 2020, AISC 1202, pp. 330–337, 2020.
https://doi.org/10.1007/978-3-030-51194-4_44

through the lens of industrial designers. This paper investigates timeless design from the product lifecycle perspective and proposes "ageless design" as a way of complementing timeless design for a longer product lifetime.

2 Product Life Cycle Design

Design for product longevity, which seeks to increase the lifespan of a product, is a way of sustainable development. Increased longevity helps reduce waste as well as carbon emission [1]. Strategies of increasing a product's lifetime, both physical and emotional, have been studied and proposed by many researchers. However, design strategies for product longevity have been largely developed around engineering design aspects [2], while studies focused on industrial design perspectives are very limited. For example, Table 1 shows several design strategies from existing literature [2–6] that are grouped into specific disciplines based on general practices in the industry.

Table 1. Design strategies for product longevity.

Engineering Design (technical aspects)	Robust design Modular design: Design for upgradability Design for adaptability Design for maintenance	Design for disassembly Design for repair Design for reuse Design for refurbishment Design for remanufacture
Both Disciplines	Material selection	Product attachment
Industrial Design (aesthetic, emotional)	Timeless design Superior appearance	Sensory design

2.1 Product Aesthetics and Sustainability

The aesthetic quality is one of the most important factors in products [7] and it has a fundamental role in product sustainability [6]. In our current consumerist culture, consumer products are frequently updated and become quickly obsolete, and product appearance has been a purposeful tool in that regard. At the same time, there have been various efforts from designers to make longer-lasting products through careful consideration of the appearance and materiality. In short, a product's aesthetics either support a long-term use or discourage it.

Planned obsolescence, a business strategy to increase profits through stimulating repetitive consumption, deliberately shortens the usable life of a product [8, 9]. Products are constantly changed and product aesthetics are reduced to mere styling that follows short-lived trends. Progression of technology, frequent updates, and changes of design combined with attractive and distinctive style make products obsolete and lead to premature disposal [7, 10, 11]. Some aesthetic elements serving this purpose are high contrast, sharp shape, shiny or vivid color, highly polished surface [10, 12].

Zafarmand et al. proposed seven categories of aesthetic attributes that contribute to product sustainability: 1) aesthetic durability - continuation of aesthetic appeal during product lifespan; 2) aesthetic upgradeability and modularity - easy update of a style or functions through modular design; 3) simplicity and minimalism - minimizing the number of parts and materials; 4) logicality and functionality - functional integration of a product's form in a clear and logical way; 5) natural forms and materials - drawing forms and materials from nature such as biomimicry; 6) local aesthetic and cultural identity - locally designed and produced products fitting to local users' tastes; 7) individuality and diversity - aesthetics based on individual or a group's taste [7].

In particular, aesthetic durability achieved through neutral design or timeless design will encourage users to retain products for longer periods. Design for simplicity and "less is more" are considered as guiding principles in that direction: Dieter Rams pointed out unobtrusiveness, honesty, unfashionableness, and little design as principles for good design which is also long-lasting and environmental-friendly [13]. Logicality and functionality also support the simplicity attribute through clear communication of function without unnecessary aesthetic elements [7, 14]. Finally, individuality and diversity can be understood as a user-centered design approach that listens to users and meets their needs.

2.2 Timeless Design for Product Longevity

Timeless design has been mentioned by multiple researchers as one of the design strategies contributing to product longevity [4, 7, 11, 14, 15]. Timelessness is a quality of not being affected by time, so not becoming obsolescent after a long period of time. In the design industry, it refers to a design that has been beloved by users for decades or even a century, without going through major changes or modifications. For example, Mont Blanc had celebrated 100-year-old design for its Meisterstück Pen.

Lobos proposes four strategies for timeless design: appearance, product efficiency, materials selection, and user experience (Fig. 1). Simple and classic appearance can endure a long period of time without going out of style, so the product can be used longer. Lobos also points out a clear and honest communication of function as an element of timelessness in products. Product efficiency is eliminating any superfluous features or unnecessary components. For materials selection, designers should consider their durability, recyclability, and environmental impact during extraction and manufacturing, while minimizing additional processes like painting. A product that provides joy and pleasure in addition to good function and usability connects to users at a stronger level and makes it be used longer [14].

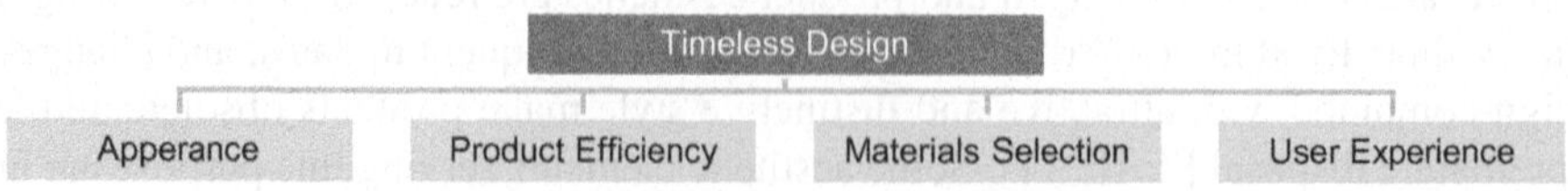

Fig. 1. Four strategies for timeless design.

However, there are certain limitations and ambiguity of timeless design for product longevity. Firstly, users have different views and opinions toward timeless design, particularly between the younger and the older generation. Timeless design is normally associated with classic design, anti-fashion, simplicity, and minimalism [7, 9, 14], but it lacks a clear definition of aesthetic attributes. Specific needs and expectations for a product also vary according to different age groups. A product for longevity needs to be inclusive of multiple age groups at any point in its life cycle, addressing the varying needs of multiple generations whether it is functionality, usability or aesthetics. Secondly, timeless design does not explicitly address usability aspects of a product, which becomes critical as the user becomes old. As a long-lasting product can be used over a few decades, the same user can go through multiple life stages, for example from an adolescent to a mid-age and to an elderly. That means a product purchased today should remain equally usable and aesthetically pleasing to the same user after a long period of time regardless of the user's age.

3 Ageless Design

We need another dimension to complement timeless design for product longevity. This study seeks to strengthen the design methods for product longevity by considering a wider spectrum of age groups both at the time of purchase and over the course of product usage. It is necessary to first understand the characteristics of each generation.

3.1 Cross-Generational Consideration

A generation is defined as "a set of historical events and related phenomena that creates a distinct generational gap" [16]. As a new VOC has been getting more attention and critical to the market, it is necessary to understand better. Authors see Millennials and Generation Z as new VOC. Authors think that new VOC is critical because they interact with the old generation, which can be approached as cross-generation consideration.

Although each generation has a distinctive character due to different historical events, it does not mean that each generation is not connected. Gen X and Millennials share many values, such as a need for empowerment, life balance, and environmental concern [17]. Millennials and Gen Z share similar experiences of economic crisis, which causes those two generations practical and digital/device native. The parents of Millennials are typically baby boomers and the parents of Gen Z are Gen Xers. Their constant interactions between generations can be a great influence on each other. Therefore, inclusiveness is a critical element. It has two different meanings. One is that design would be able to expand and reach throughout different generations. The other is users from generation to generation would want to embrace the design into their lifestyles due to various reasons (i.e. usability, durability, or aesthetics).

3.2 Inclusive Approach for Product Longevity

Ageless design is an alternative design method that complements timeless design for product longevity. It seeks to appeal to both young and old generations, to be accessible and easy to use regardless of age, to support durability, and to minimize wear and tear. Therefore, a product can not only be used across generations but also endure long years of use (Fig. 2). Ageless means "not growing old or showing the effects of age", while timeless is "not restricted to a particular time" [18]. Compared to the timeless which tends to focus on the aesthetic quality, the meaning of ageless conveys a more human quality that it is more suitable to deal with concepts related to both a product's and human aging. When the design of a product is ageless, it means that a person's age and subsequent physical capabilities or emotional needs are not affected by the design.

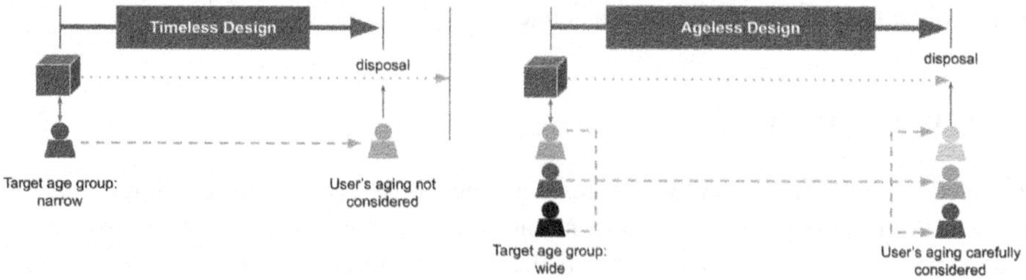

Fig. 2. Comparison of timeless design and ageless design

Authors propose a framework of ageless design based on the aforementioned four strategies of timeless design [14] and the three major areas of industrial design practice, which are form, CMF (colors, materials, finishes), and usability. Product efficiency from Lobos' four strategies is included under the form because placement and number of outer components is one of formal qualities of a product while features and inner components are generally out of the scope of industrial design practice. Authors also use the term usability instead of user experience in order to prevent any confusion resulting from ambiguity with the term. Table 2 shows the framework with elements comprising each focus area and subsequent guidelines that support product longevity. For example, the overall formal language should communicate function; applying diverse colors reflecting individuality will appeal to wider generations; easy-to-use and adaptable touchpoints enable the use of a product across generations.

3.3 Case Study

The Kitchenaid Stand Mixer is a classic example of ageless design. Since its launch in 1919, the stand mixer has been around for over 100 years without a major design change. Authors analyzed the stand mixer based on the ageless design framework (Fig. 3).

Form. When a mother prepares a meal using the stand mixer and her kids helping her, the stand mixer is being spontaneously exposed to the younger generation. It is

Table 2. Framework for ageless design.

Area of Focus	Elements	Guidelines
Form	Overall formal language	Communication of function Enduring aesthetics
	Outer components	Minimal components
CMF	Color	Individuality & diversity
	Materials	Durability, Graceful aging
	Finishes	Pleasurable surface texture
Usability	Functionality	Consistency / efficiency of operation Ease of maintenance
	Touchpoint	Ease of control, Adaptability

Fig. 3. Kitchenaid Stand Mixer

typically a positive experience and is how the stand mixer gets introduced to her kids. In other words, the kitchen environment is given to her kids. From those children's perspectives, aesthetics are defined as inherent, stable properties of an object that produce a pleasurable experience in perceivers [19, 20]. Aesthetic beauty leads to desirable outcomes such as an immediate desire to own [21] and an increased desire to display or care for a product [22].

CMF. Kitchenaid has been responding to the new VOC from Millennials. Kitchenaid has utilized CMF in order to increase the number of choices that the user can choose from. According to Kitchenaid, they have introduced options of color, engraving, and bowls whereas Kitchenaid had originally launched the stand mixer in white [23]. Historically, the stand mixers have been high gloss finished. However, Kitchenaid has introduced matte finishes in order to respond to the new VOC.

Usability. The stand mixers have many attachments to help cooking preparation easier. According to Kitchenaid, there are 51 attachments for the stand mixers. These attachments can increase usability not only for the younger generation for making ravioli but also for the older generation for simply slicing ingredients, which leads to the use of those attachments for everyday meal preparation.

4 Conclusion and Discussion

Product longevity is a complicated matter. In addition to various challenges to each field, it has to deal with sophisticated behaviors of consumers as well as conflicting dynamics among manufacturers, markets, and policymakers. There is no perfect formula that would untangle all the issues. At the same time, it is undeniable that design has a great potential to affect consumers, the business, and the environment, particularly when the scale of designed products is considered. More thoughtful consideration about product appearance and usability to increase a product life span will bring a positive impact to the environment. To that end, ageless design is proposed as a more structured and improved method for industrial designers as well as product managers. Ageless design is inclusive and user-centered: not only does it appeal to multiple generations, but also endures a long period of time through enabling a product to be equally usable regardless of the age of the user.

This research is an attempt to lay a ground of ageless design. Based on the literature review, a proposition of ageless design with a guiding framework was discussed, then it was examined through a case study. However, more robust research is necessary to verify the validity of the theory. For example, an exercise of designing a product based on ageless design guidelines outlined in this research and testing it with people at different age groups will provide better understanding of the theory.

References

1. Cooper, T.: The value of longevity: product quality and sustainable consumption. In: Global Research Forum on Sustainable Consumption and Production Workshop, Rio de Janeiro, Brazil, pp. 1–9 (2012)
2. Haines-Gadd, M., Chapman, J., Lloyd, P., Mason, J., Aliakseyeu, D.: Emotional durability design nine - a tool for product longevity. Sustainability **10**(6), 1948 (2018)
3. Den Hollander, M., Bakker, C., Jan Hultink, E.: Product design in a circular economy: development of a typology of key concepts and terms. J. Ind. Ecol. **21**(3), 517–525 (2017)
4. Rose, C., Ishii, K., Stevels, A.: Influencing design to improve product end-of-life stage. Res. Eng. Des. **13**, 83–93 (2002)
5. Van Nes, N., Cramer, J.: Influencing product lifetime through product design. Bus. Strategy Environ. **14**, 286–299 (2005)
6. Vezzoli, C.: Design for Environmental Sustainability. Springer, London (2018)
7. Zafarmand, S., Sugiyama, K., Watanabe, M.: Aesthetic and sustainability: the aesthetic attributes promoting product sustainability. J. Sust. Prod. Des. **3**, 173–186 (2003)
8. Bartels, B., Ermel, U., Pecht, M., Sandborn, P.: Strategies to the Prediction, Mitigation and Management of Product Obsolescence. Wiley, Hoboken (2012)
9. Guiltinan, J.: Creative destruction and destructive creations: environmental ethics and planned obsolescence. J. Bus. Ethics **89**, 19–28 (2009)
10. Cooper, R.: Ethics and altruism: what constitutes socially responsible design? Des. Manag. Rev. **16**(3), 10–18 (2005)
11. Page, T.: Product attachment and replacement: implications for sustainable design. Int. J. Sustain. Des. **2**(3), 265–282 (2014)
12. Walker, S.: Experiment in sustainable product design. J. Sustain. Prod. Des. **7**, 41–50 (1998)

13. Lovell, S.: Dieter Rams: As Little Design as Possible. Phaidon, London (2011)
14. Lobos, A.: Timelessness in sustainable product design. In: 9th International Conference on Design & Emotion, pp. 169–176 (2014)
15. Niinimäki, K., Koskinen, I.: I love this dress, it makes me feel beautiful! Empathic knowledge in sustainable design. Des. J. **14**(2), 165–186 (2011)
16. Parry, E., Urwin, P.: Generational differences in work values: a review of theory and evidence. Int. J. Manag. Rev. **13**, 79–96 (2011)
17. Matlay, H., Szamosi, L.T.: Just what are tomorrow's SME employees looking for? Education + Training (2006)
18. Merriam-Webster. http://merriam-webster.com
19. Arnheim, R.: Art and Visual Perception, vol. 197, pp. 96–161. University of California Press, Berkeley (1954)
20. Gombrich, E.: Art history and psychology in vienna fifty years ago. Art J. **44**, 162–164 (1984)
21. Norman, D.: Emotional Design: Why We Love (or Hate) Everyday Things. Basic Books, New York (2005)
22. Bloch, P.H.: Seeking the ideal form: product design and consumer response. J. Mark. **59**, 16–29 (1995)
23. Lidwell, W., Manacsa, G.: Deconstructing product design: exploring the form, function, usability, sustainability, and commercial success of 100 amazing products. Rockport Pub (2011)

Analysis of Design Elements of Automatic Household Food Waste Composting Machine Based on Perceptual Engineering

Chuyi Zhao and Huabin Wang(✉)

School of Design, South China University of Technology,
Guangzhou 510006, China
hbwang@suct.edu.cn

Abstract. With the improvement of people's living standards, the pursuit of material life has also been continuously improved. The improvement of life has also brought new kitchen waste and realized local resource utilization. Kitchen waste composting is of great significance for sustainable development. The product's styling design of the machine is an important design element for achieving householdization. The analysis of its design elements such as color, material, surface texture, and styling technology through the analysis of perceptual engineering provides a reference for product design and production.

Keywords: Automatic household food composting machine · Design elements · Perceptual engineering

1 Introduction

With the implementation of national policies, the civic spirit, cultural level, and quality of life have continued to improve, people's living standards have improved, and the requirements for the spiritual experience of products have become higher and higher, and their awareness of environmental protection and green development has become stronger. Consumers have shifted from a single function to diversification. The introduction of policies such as waste classification and the fight against poverty, and the on-site recycling of food waste are one of the necessary means for sustainable development.

Food waste disposal products have emerged at the historic moment, and household automatic food waste composting machines, in addition to solving the problem of food waste disposal, achieve self-production and sales, and sustainable recycling, is one of the first choices for domestic and rural users. The color, material, surface texture, and styling process of composting machines are important design elements for the domesticization of products, and they have an important force in promoting the more domestic and widespread products.

G. Di Bucchianico et al. (Eds.): AHFE 2020, AISC 1202, pp. 338–345, 2020.
https://doi.org/10.1007/978-3-030-51194-4_45

2 Perceptual Semantic Analysis of Household Automatic Food Waste Composting Machine

Perceptual engineering is a science that prioritizes human needs, transforming the user's emotions and intentions into designed product development technologies, and quantifies the sensibility through engineering. The method of perceptual engineering is applied to household automatic kitchen waste composting machines. Analyze its semantic characteristics in the design.

2.1 Household Automatic food Waste Composting Machine Figures

At home and abroad, the treatment methods of kitchen waste mainly include crushing and direct discharge, composting, making feed, anaerobic gas generation, incineration, landfill, etc. Compared with the methods mentioned above, the use of high-temperature aerobic composting treatment Technology, which converts kitchen waste into organic fertilizer that can be absorbed and used by plants. Residents put fertilizer in the plant land of their own gardens, which is very suitable for household waste treatment. Household food composting machine, there are higher requirements in automation, insulation and heating capacity, aeration and deodorization capacity.

In the current food waste treatment methods, most of them are limited to a relatively limited scope, such as technology, materials and modeling structures, and ignore the interaction between human and machine environment, which brings people the inconvenience of handling food waste. But also caused a waste of resources (Fig. 1).

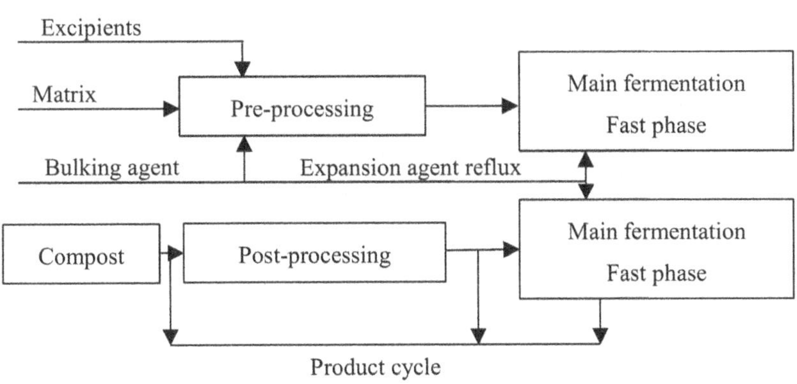

Fig. 1. Composting process

2.2 Preliminary Analysis of Perceptual Intention

Design elements of household automatic food waste composting machine. Perceptual intent semantics. Through internet, newspapers, magazines, advertisements and other media terms, refer to more than 200 adjectives in the "Cognitive Study of Modeling

Intent Attribute Adjective Clusters". Finally, a questionnaire survey was conducted (63 design related personnel and 43 irrelevant design personnel). A total of 109 questionnaires and 3 invalid questionnaires were retrieved. The user selected, through his own experience, the intention and meaning of the home automatic kitchen waste composting machine.

Emotional intentional semantics with more than one third of the votes were selected, and 15 groups of intentional semantics were finally selected as shown in Table 1.

Table 1. Perceptual intention semantics

Mark number	Intentional meaning	Ticket number	Mark number	Intentional meaning	Ticket number
1	Fine	24	9	Sanitary	33
2	Strong	26	10	Cozy	35
3	Rounded	23	11	Green	26
4	Streamlined	27	12	Modern	35
5	Elegant	21	13	Fresh	25
6	Professional	32	14	Atmospheric	21
7	Quiet	25	15	Domestic	27
8	Durable	29			

2.3 Representative Semantic Phrases

The sentimental intention semantics of the design elements of the household automatic food waste composting machine are too complicated. In order to select representative semantic phrases, a second questionnaire survey was conducted on the basis of selecting 15 groups of semantic meanings (70 people were involved in the design, and the design was not (Relevant staff of 36 people), a total of 107 questionnaires were retrieved, and one invalid questionnaire was used. 15 groups of intentional meanings were listed as a matrix of 15 * 15, allowing users to choose the most suitable description of emotional intentions.

According to the multiple response analysis matrix, analyze its response rate and penetration rate, and combine the chi-square goodness-of-fit test to analyze whether the selection ratio of each item has a significant difference. As shown in Table 2, the top two votes for each group are selected. The intentional semantic combination specifically compares the differences through the response rate or penetration rate. Compared with other combinations, the response rate and penetration rate are significantly higher.

Table 2. Multiple response analysis

Intentional meaning	Intentional meaning 1	Ticket number	Intentional meaning 2	Ticket number
Durable	Professional	29	Modern	24
Quiet	Strong	25	Modern	21
Professional	Durable	29	Strong	23
Legant	Cozy	21	Strong	18
Streamlined	Durable	18	Fresh	17
Rounded	Atmospheric	28	Strong	26
Strong	Rounded	26	Quiet	25
Fine	Green	28		
Sanitary	Cozy	25	Domestic	25
Cozy	Sanitary	25	Elegant	21
Green	Fine	28	Quiet	19
Modern	Durable	24	Quiet	21
Fresh	Rounded	23	Atmospheric	20
Atmospheric	Rounded	28	Domestic	27
Domestic	Atmospheric	27	Sanitary	25

The representative semantic phrases with a relatively high proportion are selected from the following: professional durable, green delicate, comfortable and sanitary, and atmospheric smooth.

3 Analysis of Design Elements of Household Automatic Food Waste Composting Machine

In recent years, kitchen waste treatment has only become popular. The products on the market do not have a more extensive and mature design. The design elements of food waste treatment products must be combined with home appliances for design thinking.

As shown in Fig. 2, using the hierarchical category classification method, from 0 perceptual to n perceptual, a perceptual tree diagram of a household automatic food waste composting machine is established, and its feelings and product technology are analyzed from n perceptual. Analyze and make a framework guide.

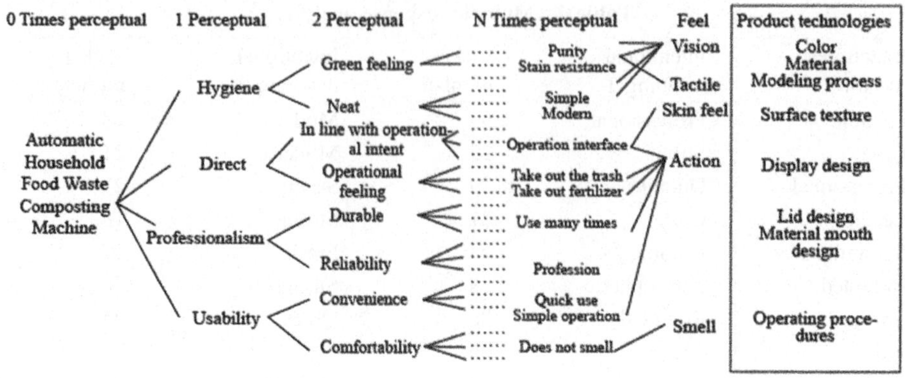

Fig. 2. Hierarchical taxonomy

3.1 Unity and Diversity of Colors

The color design of home appliances has gradually been enriched. On the one hand, the people's living standards and the level of aesthetic ability have improved. On the other hand, the influence of popular colors has become greater and greater, and the use of popular colors in the design of home appliances has become more widespread. The bold use of color also provides more creative space for design color.

As shown in Table 3, it is an approximate example of the color of home appliances in the existing market. Color is used to express consumer emotions and capture consumers' spiritual pursuit. The use of home appliances can no longer meet the needs of monochrome, and more colors are favored.

Table 3. Example of appliance color

Color	Color intent
White series	Clean and tidy, it is also a widely used color in home appliances
Silver series	Simple and atmospheric
Gray series	Deep and calm, sense of technology, modernity
Red series	Fashion, passion, gentleness, fullness
Blue series	Neat, pure and fresh
Mixed color series	Rich color mix became part of furniture decoration

The matching of colors is to seek the unity of colors on the basis of diversification. In the color design of composting machines, it is necessary to seek to combine colors with structural modeling to achieve the unity and diversity of colors in modeling. From the representative meaning of perceptual engineering It seems that consumers prefer simple, clean, fresh, and atmospheric color intent.

3.2 Materials and Surface Texture

The choice of materials and surface textures is limited by various conditions such as cost. The environment in which kitchen waste composters are used is mostly outdoor environments, which have high requirements for strength and corrosion resistance. Their own characteristics and use process require materials and surfaces. The texture is easy to clean and keeps it clean and hygienic.

As shown in Table 4, the characteristics of the materials are briefly explained. The choice of materials is also closely related to the modeling process, and is not limited to the use of the same material in the design of home appliances. So that the product is not only cold supplies, but also a special feeling.

Table 4. Examples of home appliance materials

Material		Features
Metal	Galvanized	Compared with electrogalvanized, hot galvanized has better corrosion resistance, radiation protection, oxidation resistance and heat dissipation
	Zincizing	The price is more expensive, the appearance is high, the strength is strong, the corrosion resistance, not easy to rust, and the production process is more complicated
Plastic	ABS	Durable, non-rusty, brittle, relatively expensive, and easy to surface treat
	PP	Good toughness, low cost, non-wear resistance, and difficult surface treatment such as screen printing and painting

The choice of surface texture can have more choices and more performance in the case of limited materials. It can highlight the charm of texture while saving costs. It is often combined with color to emit its unique feelings and intentions for the product. Its charm is also reflected in its unity of denseness and sparseness. Do not blindly pursue overly full textures. Relaxation is the aesthetic pursuit of modern consumers. It is gradually applied to the design of home appliances and is compatible with home decoration (Table 5).

Table 5. Examples of surface texture of home appliances

Surface texture	Features
Spray paint	Good surface finish, easy to clean, not easy to rust
Spray	More advanced technology, energy saving and efficient, safe and reliable, smooth texture, bright color
Brushed metal	Brushing can make non-mirror metallic luster on metal surface
Lacquered glass	Spraying, roller coating, screen printing or spray coating on the glass surface makes the glass extremely decorative and expressive
Laminator	Very strong decorative and formability

3.3 Modeling Process

The overall structure of a household food waste composting machine consists of a display screen, an inner bucket, a stirring rod, a motor, thermal insulation cotton, an aeration device, a bucket lid, and an outer bucket.

The cavity position of the entire structure is roughly divided into front-rear layout type and left-right layout type. The two layout forms are suitable for each internal structure, and the cavity position determines the further design of the molding process. The left and right layout types are respectively the inner bucket on the left display and the device on the right, the inner bucket on the right display and the device on the left.

About 88% of people are accustomed to using their right hand. During the process of using a food composting machine, dump the food waste into the inner bucket, which may cause some residual garbage to remain on the operation surface. If the display is placed on the right side, the residue on the countertop will have a certain degree of influence on the display screen, which will affect the use experience. The inner barrel is located on the right side of the display and the device is located on the left. The cavity

Table 6. Commercial food composting machines

Food waste composting machines on the market	Cavity layout	Height	Modeling process
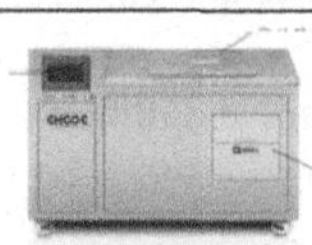 Xipu environmental protection	Left and right layout type	800mm	Metal welding
 Yongerjia Environmental Protection	Front and rear layout type	600mm	Monolithic injection
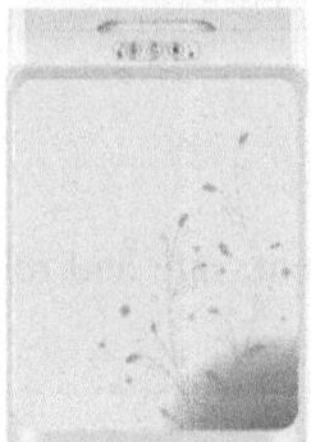 Yongerjia Environmental Protection	Front and rear layout type	580mm	Monolithic injection

layout is better. The commercially available products are short and the display is mostly vertical. Surface, it is not convenient to read data, and the layout on the table is better (Table 6).

The high cost performance of household products is the first factor for consumers to consider. Based on the impact of processing technology and material costs, the overall design is divided into two parts. The upper part is a cover upper body design. The shape is more and more complex, involving flip covers and displays. The screen and the streamlined part of the model, so the injection molding process is better; the lower part of the model is simpler, the overall account is relatively large, the cost of using the injection molding process is higher, the sheet metal stamping process cost is lower, can achieve a smaller design, can be sprayed The material paint meets the product design requirements. This type of styling process selection can achieve faster product production flow, reduce production costs, and improve product cost performance.

4 Conclusion

This article uses the method of perceptual engineering to analyze and summarize the design elements of the color, material, surface texture and modeling technology of the household food waste composting machine. Based on the actual situation of the domestic food waste composting machine, it gives an overall overview and advice on the design elements. Those who pay more attention to the professionalism of this type of home appliances do not have a clear intentional preference for the design elements shown above. In the design and production, they should pay more attention to the perceptual intentions of consumers, professional and durable, green and delicate, comfortable and hygienic. These representative meanings of the type and the atmospheric smooth type provide some ideas for design and production.

References

1. Guo, Y., Liu, M., Zeng, Q., Zhang, B., Huang, X., Yang, H.: Improved design and test of fermentation tanks for rural household kitchen waste treatment. Mod. Agric. Equip. (03), 29–33 (2017)
2. Wang, Y.: Research and application of CMF trend for home smart washing machines. Harbin University of Science and Technology (2015)
3. Wang, Y., Zhang, L.: Analysis of home intelligent washing machine modeling based on perceptual imagery. Ind. Des. (10), 112–113 (2016)
4. Liu, M., Guo, Y., Wei, J., Mai, Y., Zeng, Q.: Design of automatic kitchen waste composting machine for households. Mod. Agric. Equip. (02), 34–38+42 (2018)
5. Liu, Y.: Research on the design of mechanical products based on perceptual engineering and rough set. Lanzhou University of Technology (2014)
6. Yan, L.: Research on visual color matching based on Kansei engineering. In: Proceedings of the 2nd International Conference on Contemporary Education, Social Sciences and Ecological Studies (CESSES 2019) (2019)
7. Yu, P., Chen, J., Li, Y.: Design of a small household solar compost box. Food Mach. **34**(09), 121–125+200 (2018)

Have the Display Illumination Design Misled Customers on Color Perception? A Study on Differences of Color Caused by LED Lighting on Leather Products

Cong Xu[1], Chih-Fu Wu[2(✉)], Dan-Dan Xu[1], and Ying-Shan Lai[2]

[1] The Graduate Institute of Design Science, Tatung University, No. 40, Sec. 3, Zhongshan N. Rd., Taipei City 104, Taiwan
[2] The Department of Industrial Design, Tatung University, No. 40, Sec. 3, Zhongshan N. Rd., Taipei City 104, Taiwan
wcf@gm.ttu.edu.tw

Abstract. This research aims at identifying the differences of color perception by applying 6 types of LED lighting on 4 types of leather finishing. The 6 lighting types include three levels of correlated color temperature (CCT) multiplied by two illuminance levels. The 4 types of leather finishing include red and brown, with grains and without grains. The dependent variables were blackness, saturation and hue of 3 characteristics encoded for each color in the Natural Color System. A total of 15 volunteers participated the experimental study. The results reveal that CCT, illumination and color have significant effects on the perception of color differences. At the same time, a three-factor interactive effect of grains, CCT and color, also shows significant effects. The results of this study can provide references for the display lighting design of leather products and the color design of leather products under different lighting conditions.

Keywords: Display illumination design · LED lighting · Differences of color · Leather products · Correlated color temperature

1 Introduction

Currently, Light emitting diode (LED) has been widely applied to display lighting in commercial retail space due to its higher power efficiency, longer usable lifetime, lower maintenance cost and easier adjustment and control. However, like other light sources, LED light source can also cause people to see the wrong color of illuminated object, and the color difference caused by it could be even bigger [1]. Thus, that illumination of LED light could cause cognitive deviation on the color of displayed goods should be fully considered when designing display lighting in retail stores.

Now, LED lights are usually adjusted in three aspects: color rendering, illuminance and correlated color temperature (CCT). In terms of color rendering of light source, IESNA has suggested that CRI applied to retail stores should be at least 80–90 [2]. Besides, since the CRI of LED lights is usually set at factory, it is rarely studied as a control variable in experiments [3]. Also, studies have found that the illuminance and

G. Di Bucchianico et al. (Eds.): AHFE 2020, AISC 1202, pp. 346–353, 2020.
https://doi.org/10.1007/978-3-030-51194-4_46

CCT of LED light source will both affect human visual perception [4]. Human pleasure, arousal and dominance will be stronger as illuminance increases. When people have a strong feeling of pleasure and arousal, his dominance of space will also get stronger, therefore causing customers to stay for a longer period in consuming environment or increasing consumption probability [5]. Higher CCT presents colder light source color, usually applied to office space to increase people's attention; Lower CCT produces warmer light source color, usually applied to public places such as commercial space to create a more relaxed atmosphere [6]. The study of Pousset et al. also supports the argument that different CCT and illuminance will affect the perception of object's color [7].

Leather has been a popular material since ancient times. Different cutting methods and surface treatment enable its wide applications to manufacturing fields as clothing, shoes & hats, luggage, daily necessities, furniture and industrial products. In order to promote sales, many shops now often use LED light source to closely illuminate leather products in the method of accent lighting or vertical lighting, so as to have a good display. But leather is a material with its own characteristics. The microstructure of its surface forms specific texture types, further causing the reflection of light to form specific luster and directional flow [8]. So it is worth having a further discussion on the conditions of cognitive deviation of color caused by LED-illuminated leather products. In conclusion, the present study was designed to examine if leather is illuminated with different LED light source, namely different combinations of CCT and illuminance, how human's cognitive deviation of leather color caused thereby will be affected in three aspects of Natural Color System (NCS) Color Circle: blackness, saturation and hue. The study will also provide essential reference for designers when they deal with display illumination design of leather products.

2 Materials and Methods

2.1 Participants

A total of 16 volunteers were recruited for the experiment. To ensure its validity, all subjects were tested for color blindness, color weakness, poor concentration & focus, and Snellen eye test. 15 of them gained corrected binocular vision result of 0.8 or more in Snellen eye test with no color blindness, color weakness and concentration & focus problem. Demographic details of the 15 subjects are as follows: 8 males, 7 females, age range 22–49 years, average age at 32.02, SD = 5.52.

2.2 Experimental Design

Design of Light Source Combination. Physical stores often set the correlated color temperature of display illumination at 2700 K–4000 K [9]. And Mark Karlen said in his work that 2700 K–4000 K was suitable for illumination in residential buildings, hotels, shopping malls and display while 4100 K–6000 K could be used in special lighting occasions when identifying color is of great significance [10]. Therefore, three CCT levels were selected in this experiment: 2700 K warm, 4000 K neutral and

6000 K cold. And based on the previous discussion, the color rendering index of LED light source set at Ra = 85 was selected.

According to CNS (Chinese National Standards) illuminance measurements and JIS (Japanese Industrial Standards) illuminance measurements, 750 lx is applicable to the indoor illumination in big shopping malls, wholesale stores, boutiques, stationery shops and exclusive stores while 1500 lx is suggested for counters, display and windows, and therefore the illuminance levels required for the experiment 750 lx and 750 lx were decided.

Therefore, 6 light sources of the color rendering index Ra = 85 were set as CCT × illuminance: 2700 K × 750 lx, 4000 K × 750 lx, 6000 K × 750 lx, 2700 K × 1500 lx, 4000 k × 1500 lx, 6000 K × 1500 lx.

Design of Light Box and Objects. To meet the experimental requirements, a wooden light box with the same size as standard color controller light box TILO-T65 has been customized for the experiment. A set of LED light beads is arranged at the top of the box, which can emit six types of light sources by the combination of 3 levels CCT and 2 types illuminance, and vertically illuminate the objects placed under it.

Meanwhile, Chan-di Liu (2013) has pointed out that there are 11 leather colors commonly used on the market at present. However, the most popular color for women is black and red, while the most common one on the market is black and brown; in terms of leather texture, Lichee pattern is the most common one used on the surface of fashion bags. Chen et al., through sampling statistics, found that people prefers white above all the other leather colors, followed by black, red and brown; small Lichee grains are preferred above all the other leather grains, followed by no grains, middle Lichee grains, large Lichee grains and convex grains [11]. Based on the above discoveries and interviews with experts, plus considering the purpose of this study which is aimed to examine how LED lighting affects the color difference of leather, after contract with suppliers, we have got four 2 cm × 4 cm cowhide leather samples in standard colors in the NCS color system: two piece of S3060-R red samples and two piece of S5030-Y60R brown samples, with same color pieces mist-sprayed with small Lichee grains and no grains.

The selection of comparison colors is centered with the two standard color stimuli: S3060-R red and S5030-Y60R brown. 12 comparison colors were selected for each stimuli in the three aspects: blackness, hue and saturation.

Variables. Independent variables of this study include different levels of correlated color temperature (2700 K, 4000 K, 6000 K) and illuminance of LED lighting (750 lx, 1500 lx), leather grains (no grains, grains) and colors (red, brown), with the dependent variables are saturation, blackness and hue of the leather under the influence of the four independent variables.

2.3 Procedure

15 participants were tested in an independent light-pollution-free laboratory. Before the test was administered, the researcher explained the experimental objectives and procedures to the subjects. Subjects compared the four objects, namely red and brown leather with and without grains, under 6 different illumination combination

(CCT × illuminance) in the left LED light box, and then selected the color that was most similar to the standard stimuli color swatch in the right TILO-T65 light box. These selected colors were then recorded by researchers regarding their blackness, saturation and hue for statistic analysis.

3 Results and Discussion

A total of 24 sets of comparative color swatch selection tests were performed by 15 subjects under three levels of correlated color temperatures, two levels of illuminance, two types of surface textures, and two colors. Under the experimental conditions of each group, the values of the blackness, saturation and hue of the comparative swatches selected by each subject were recorded and sorted using SPSS V26 for quantitative analysis. MAVONA was adopted in the study to explore relevant color difference.

The result indicates that CCT and color have significant effects on blackness, saturation, and hue, illumination has a significant effect on blackness and saturation. And there is no significant effect on blackness, saturation, and hue as change of grains. However, CCT and grains have a two-factor interactive effect on blackness and saturation; CCT and color also have a two-factor interactive effect on blackness; Besides, CCT, grains, and color have a three-factor interactive effects on blackness and saturation. Post-hoc test was then performed, in order to reveal the main effects of the three independent variables: correlated color temperature, illuminance and color.

The effect of three levels of CCT on the blackness of leather materials was 6000 K > 4000 K > 2700 K, indicating that when the CCT was adjusted from a low correlated color temperature of 2700 K to a medium one of 4000 K and then increased to 6000 K, the blackness of the leather material would increase accordingly, further indicating that when CCT got higher, the naked eyes will recognize a deeper color compared with lower CCT. Moreover, the effect of illuminance on blackness was 750 lx > 1500 lx, which means that compared with 750 lx, 1500 lx will significantly reduce the blackness and appear to be brighter to naked eyes. In terms of color, LED-illuminated red leather showed a greater difference in blackness than that of LED-illuminated brown leather.

Judging from saturation of the dependent variables, the effect of CCT on saturation was 2700 K > 4000 K > 6000 K, indicating that compared with the medium and high color temperature, 2700 K would significantly increase leather saturation, 6000 K would significantly reduce saturation and 4000 K was in-between. The effect of illuminance on saturation was that 1500 lx displayed higher color saturation than 750 lx. Meanwhile, compared with red, brown can also show higher saturation under LED light source.

The result of post-hoc tests also shows that changes in CCT and color both will also cause significant difference of color differences of hue. In terms of the color difference of hue, according to NCS color circle, if the deviation value of hue is smaller than the value of the stimuli hue, it means the color is deviating towards the right side of NCS circle, if the deviation value of hue is bigger than the value of the stimuli hue, then the color is deviating towards the left, where the colors are distributed on the circle yellow, brown, red and blue, from left to right [12]. Based on the means and SD of hue deviation of red and brown leather under the six different light sources presented in Table 1, it is not difficult to find that at 1500 lx, red will deviate toward blue, and such deviation of 2700 K will be significantly greater than that at 6000 K, with $F(2) = 5.75$, $p = .005$. Meanwhile, LED-illuminated brown leather will deviate toward yellow, but the change of various combinations of different CCT and illuminance did not lead to significant differences among color differences. In addition, comparing red leather with brown leather, it is found that under the six light sources, the color difference of hue of red was significantly greater than that of brown.

Table 1. Means and SD of the colors' hue under the 6 light sources

Colors	Standard light	2700 K		4000 K		6000 K	
		750 lx	1500 lx	750 lx	1500 lx	750 lx	1500 lx
Red	0	−6 (5.623)	-7^{*} (4.661)	−4.67 (5.713)	−6 (4.983)	−4.33 (6.261)	-2.67^{*} (5.833)
Brown	−60	−8.67 (3.457)	−9.33 (2.537)	−8.33 (3.790)	−9 (3.051)	−7.33 (5.208)	−9.33 (2.537)

$^{*}p < .01$

The study also performed post-hoc tests on the significance of independent variables at various levels under interactive effects. In terms of saturation, the interactive effect between color temperature and grains was that only at a medium CCT of 4000 K, the color difference of leather without grains turned out to be significantly higher than that of leather with grains. As for leather without grains, there was no significant difference in the color difference of saturation caused by low and medium color temperatures, while the change in saturation caused by low and medium color temperatures were both significantly greater than high color temperature. However, as for leather with grains, there was no significant difference in saturation difference caused by medium and high color temperature, but the saturation difference caused by low color temperature was significantly greater than that of medium and high color temperatures. From the perspective of blackness, in the case of leather without grains, the high correlated color temperature showed a noticeably darker color than the medium and low color temperatures, but there was no significant difference between the middle and low color temperatures. In the case of leather with grains, there was no significant difference in color difference caused by blackness at high and medium color temperature, which both were significantly greater than that at low color temperature. In addition, the interactive effect of color temperature and color on blackness showed that the blackness of red was remarkably greater than that of brown at the levels of 4000 K and 6000 K. And the effect of CCT on blackness of red was that high color temperature > medium color temperature > low color temperature. However, in the case of brown, the effect of color temperature on blackness was slightly different from that of red, namely, high color temperature was still remarkably greater than low color temperature, but there was no significant difference between medium and high color temperature, as well as no difference between medium and low color temperature.

The further analysis revealed the comparison results of three-factor interactive effects among color temperature, grains, and color at various levels, indicating that at each level of combination of CCT and grains, the effect of color on the color difference of saturation all turned out to be red < brown, which was consistent with the main effect of color on saturation, and the comparison results in saturation between red and brown at each level under the effect of two-factor of color temperature and color. In addition, at 4000 K and the color being red, leather without grains had a significantly greater effect on saturation than leather with grains. From the perspective of color temperature, with color being red, the low color temperature had a significantly greater effect on saturation difference than high color temperature. In the case of leather without grains, there was no remarkable difference between low and medium color temperature in saturation difference. However, in the case of leather with grains, the effect changed to that medium and high color temperature showed no remarkable difference in color difference of saturation. Color being brown, the effect of color temperature on saturation was 2700 K > 4000 K > 6000 K regardless of grains.

From the perspective of blackness, under three-factor interactive effect of color temperature, grains, and color, at all levels, the color effect all turned out to be red was significantly greater than brown, indicating that red show greater difference of blackness than brown under the effect of color temperature and grains at all levels. Only with red leather at 4000 K did grains had a significant effect on blackness, indicating that leather without grains had less effect than that with grains. Under the four levels of combination among red, brown, grains and no grains, there showed a tendency that the high color temperature has a remarkably greater effect on blackness than the low color temperature, where change in grains level had a major effect on blackness at medium color temperature. For instance, as for red leather without grains, the medium color temperature had less effect on blackness than the high color temperature, while for red leather with grains there was no significant difference on blackness at the medium color temperature and high color temperature, which were both greater than that at the low color temperature. Meanwhile, as for brown leather without grains, there was no significant difference on blackness among medium, low and high color temperature, which went the same for brown leather with grains.

The following trends can be estimated from the above experimental results that CCT and color have a significant effect on the difference of blackness, saturation, and hue of the leather material under LED illumination; illuminance has a significant effect on blackness and saturation; grains has a significant effect on color difference of leather only by its interactive effect between color temperature and among color temperature and color. From the above mentioned of interactive effect between CCT and color, it can be found that at 4000 K and 6000 K, the effect of red on blackness is significantly greater than that of brown, and the change of CCT has no effect on the independent variable-color. And the change of red and brown, under two-factor interactive effect, causes a slight change in the effect of color difference of blackness at three color temperatures. Therefore, this study claims that the importance of the four independent variables affecting the color difference of leather is color > correlated color temperature > illuminance > grains, which also confirms the founding of Dou et al. that visual effect caused by color temperature is greater than that caused by illuminance [1].

4 Conclusion

A total of 15 subjects were invited to perform 24 set of color difference comparison experiments under the conditions of three levels of color temperatures, two levels of illuminance, two types of surface textures, and two colors. The experimental results showed that the magnitude of the visual effect on the color difference of leather was: color, color temperature, illuminance, and grains. In addition, under the six light sources of different color temperature and illuminance combination, although the color difference of leather did not reach an unit of the NCS color, the hue of red leather had a tendency to turn blue and the hue of brown one turn yellow, which was recognizable by naked eyes. In addition, red had a deviation tendency that was significantly greater than brown. Therefore, in terms of display lighting, it is particularly necessary to consider carefully whether to place red leather products and brown ones together under the same light source for display, so as to avoid unexpected display effect. Overall, low color temperature or high illuminance both enhance saturation, while high color temperature or low illumination will relatively reduce it. Besides, high color temperature or low illumination will obviously deepen leather blackness, while low color temperature or high illumination will cause the opposite. The combination of color temperature and illuminance has no interactive effect on color difference. Therefore, the study suggests in order to avoid misleading consumers' visual perception of color saturation, LED light source of 4000 K × 1500 lx should be considered in leather products display. It can ensure that the color of leather products is not over-bright or under-saturated, and the brightness of displayed products can be appropriately enhanced.

References

1. Dou, X., Wu, C.-F., Lin, K.-C., Liou, J.-J.: What color does the consumer see? perceived color differences in plastic products in an LED-Lit environment. Sustainability **11**(21), 5985 (2019)
2. Park, N., Farr, C.A.: The effects of lighting on consumers' emotions and behavioral intentions in a retail environment: a cross-cultural comparison. J. Inter. Des. **33**(1), 17–32 (2007)
3. Sun, C.-C., Lee, T.-X., Ma, S.-H., Lee, Y.-L., Huang, S.-M.: Precise optical modeling for LED lighting verified by cross correlation in the midfield region. Opt. Lett. **31**(14), 2193–2195 (2006)
4. Hawes, B.K., Brunyé, T.T., Mahoney, C.R., Sullivan, J.M., Aall, C.D.: Effects of four workplace lighting technologies on perception, cognition and affective state. Int. J. Ind. Ergon. **42**(1), 122–128 (2012)
5. Iwata, T., Saito, Y.: The effects on productivity of lighting conditions in rest space. In: Proceedings of the 10th International Conference on Indoor Air Quality and Climate, Beijing, pp. 454–459 (2005)
6. Paschotta, R.: Encyclopedia of Laser Physics and Technology, 1st edn. Wiley Online Library, Berlin (2008)

7. Pousset, N., Obein, G., Razet, A.: Visual experiment on LED lighting quality with color quality scale colored samples. In: CIE 2010: Lighting Quality and Energy Efficiency, Vienne, pp. 722–729 (2010)
8. Anderson, B.L.: Visual perception of materials and surfaces. Curr. Biol. **21**(24), R978–R983 (2011)
9. Sayigh, A.: Sustainability, Energy and Architecture: Case Studies in Realizing Green Buildings. Academic Press, Cambridge (2013)
10. Karlen, M., Spangler, C., Benya, J.R.: Lighting design basics. Wiley, New Jersey (2004)
11. Chen, T.L., Chou, Y.K., Hsu, A.: Consumer lifestyles and preferences for sofa leather. Adv. Mater. Res. **650**, 687–691 (2013)
12. Hard, A., Sivik, L., Tonnquist, G.: NCS, natural color system—from concept to research and applications. Part I. Color Res. Appl. **21**(3), 180–205 (1996)

The Development Trend of Personalized Bank Card Design from a New Perspective

Xinxiong Liu and Jingyang Du(✉)

Huazhong University of Science and Technology, Wuhan 430074, China
duer22@126.com

Abstract. In recent years, bank card business has become a very important profit growth point of major commercial banks, but the homogenization of bank card is becoming more and more serious, and the competition for bank card market is becoming more and more fierce. In order to attract more customers, banks have launched personalized bank cards. Personalized bank cards not only have different functions, but also pay more attention to improving the internal attributes of products compared with ordinary bank cards. This paper analyzes the design characteristics of the existing personalized bank card products of the major banks, and forecasts the development trend of the personalized bank card design combining with various factors in the new perspective.

Keywords: Personalized bank card design · Product design

1 Personalized Bank Card from a New Perspective

In recent years, personalized bank cards have been changing. In the early years, single corporate joint bank cards and event joint bank cards have gradually enriched, not only with the theme of color personalized credit cards, but also with special materials based personalized bank cards. From these phenomena, when developing new bankcard products, bankcard has begun to pay attention to the cultural meaning, design aesthetics and other factors of bankcard, and also pay more attention to user experience in design. Today's personalized bank card is not only more and more beautiful, but also more and more meaningful.

Individuality refers to the individual and special nature of one thing different from other things. [1] Personalized bank card, which is different from ordinary bank card, is specially designed with unique theme and additional special technology. In recent years, from the perspective of the bank card products of card issuers, the personalized bank card functions of various banks have gradually converged. From the obvious tendency of product homogeneity, banks began to work intensively, entering the stage of characteristic and personalized development, gradually focusing on stimulating the consumption enthusiasm of cardholders, cultivating the brand loyalty of cardholders, and starting to break through the tradition for a specific customer group simple payment concept, pursuit of multi-functional additional services and product personality differences.

The personalized bank card from a new perspective is a bank card made for a specific group of people with a unique design theme and special technology. Some

G. Di Bucchianico et al. (Eds.): AHFE 2020, AISC 1202, pp. 354–359, 2020.
https://doi.org/10.1007/978-3-030-51194-4_47

personalized bank cards also have the characteristics of limited time and limited quantity. Generally speaking, personalized bank card is different from conventional products. The characteristics of the product include: the particularity of customer group, the particularity of card theme, the uncertainty of issuing time, and the randomness of issuing time.

2 Preparation Analysis of the Design Features of Personalized Bank Card from a New Perspective

Personalized bank card has the characteristics of the times. When analyzing the design features, we must select the contemporary and representative bank card for analysis.

2.1 Pay More Attention to the Design Theme of Personalized Bank Card from a New Perspective

The theme has always been the basic idea contained in literary works. Complex works, the subject of which has polysemy, often embodies the unity of multiple subjects [2]. The theme of personalized bank card refers to the basic idea expressed by personalized bank card, which takes the card face as the final form of expression and reflects the user's use needs, social needs and emotional needs through various design means.

Themes that Fit the Bank's Image. Most banks have roughly the same classification of bank cards, but due to the different nature of banks, when banks issue new personalized bank cards, they will have a bias in the choice of topics, which is not only to stabilize the bank's established image in the user's mind, but also to promote the bank's image. China Merchants Bank is a joint-stock commercial bank held by enterprise legal person, while industrial and Commercial Bank of China is a state-owned bank under the central management. They have the same classification of bank cards, but the theme of personalized bank cards is different. Most of the personalized bank card themes of China Merchants Bank choose animation, games, football, etc., while most of the industrial and Commercial Bank of China choose medical care, armaments, celebrity quotes, etc. From the choice of theme style, the former is more lively and younger, while the latter is more stable and mature. In the new perspective, the design theme of personalized bank card should be close to the image characteristics of the bank itself.

Personalized Themes for Target Customer Groups. The specific target customer group of personalized bank card is one of its personalized characteristics. From the perspective of product design, the human centered design concept is the main goal of product system design at present. In the design of personalized bank card, we need to pay special attention to the needs and consumption characteristics of the target users. Zunshi credit card is a credit card product issued by Agricultural Bank of China for the public. This product is designed with Chinese style and carries forward the traditional virtue of respecting teachers and valuing the way. Customers can choose one of them to handle, and can provide a series of honorable courtesy such as children's consumption management and happy holiday. Industrial and Commercial Bank of China has also designed teachers' credit cards for the majority of teaching staff. In addition, there are also overseas students' home-based credit cards and medical credit cards designed for the staff groups of all kinds of medical and health institutions across the country. Under the new perspective, the design of personalized bank card pays more attention to the choice and grasp of the target group.

2.2 Pay More Attention to the Visual Design of Personalized Bank Card from a New Perspective

Layout of Necessary Design Elements. Although bank card is a tool of daily life consumption, its security level is high, and there are international general standards in form. At present, in addition to some special forms of cards on the market, the size of the bank cards circulating on the market at home and abroad are the same, and the same size design has many advantages. First, it is convenient for the cardholder to carry. Second, it can be designed for the fixed size of the bank card matching ATM and POS machine. In the design of traditional bank card, there are requirements for the necessary design information on the front and back. The front information: Bank logo, chip, bank card organization, card number, and the back is other information. In the design of personalized bank card, in order to better display the card face, simplify the front card face information, put the card number on the back, greatly avoiding the impact of the card number on the display of the front design scheme. Under the new perspective, the necessary design elements of the front and back sides of the personalized bank card are adjusted according to the design of the card surface. Industrial and Commercial Bank of China cosmic constellation card won the "best safe payment card" award with its unique design concept and innovative technology, becoming the only product in China's banking industry to win this award in 2018. The expression of the twelve constellations uses the simple design concept, only keeping the most concise star orbit chart composed of points and lines, highlighting the personality and trend of the card, and creating the back offset card number information in the domestic industry for the first time, so as to make the artistic expression of the whole card more complete.

Diversified Design Style. In strategic marketing management, Mark E Parry mentioned that product attributes can be divided into internal attributes, external attributes and performance attributes. [3] The design of personalized bank card is mainly reflected in the card face. The card face design is graphic design, most of which are static. Diversified design style is one of the methods to enhance the personalized. When the card face rotates, the koi's tail will start to swim, as if it is playing a dynamic video. It not only attracts a large number of users to open cards visually, but also achieves the propaganda function.

2.3 Other Design Features of Personalized Bank Card from a New Perspective

The application and performance of new materials and processes in personalized bank card design. In recent years, in order to better express the design of personalized bank card, the bank applies new technology to business card printing, which not only enhances the design effect of card surface, but also improves the value of bank card. Most of the personalized bank cards are made of PVC. In order to improve the product value and enhance the design effect, some of the personalized bank cards are made of special materials. Industrial and Commercial Bank of China's Jurassic themed bank card is made of titanium alloy. Compared with the ordinary PVC card, the titanium alloy card has higher cost and huge cost difference. At the same time, the unique design enhances the personality of the bank card. The raised Jurassic logo in the middle and the steel like texture around all add the label of "personality" and "characteristic" to the bank card. That is to say, through the design, on the one hand, it reflects the rarity and collection value of the personalized bank card, on the other hand, it reflects the special identity of the cardholder. In addition, there are many personalized bank cards using new technology and new materials on the market, such as mirror card, water drop card, fragrance card, etc.

The card face of the personalized bank card and the coordination of the design peripheral. In addition to focusing on the product itself, product packaging and display are also part of product series design. The whole product series can better reflect the specialty of the product itself, and also can bring better experience to users. At present, most banks have the function of applying for bank card online, and some personalized bank card application web pages are specially designed. In addition, other products matching with main products are particularly important. The Doraemon fan card issued by China CITIC Bank is not only the basic master card, but also the opposite sex card. Pokémon fan card has launched a limited edition Pokémon backpack or Pokémon travel wash and storage set of four. The design of these peripheral products is one of the means to maximize the effect of auxiliary design.

Personalized bank card should pay attention to many aspects of cooperation to achieve better marketing results. The success of any product is not single, so is the design of personalized bank card. With the improvement of user's design literacy and aesthetic, the requirements for personalized bank cards are gradually higher. In addition to visually attracting users, personalized bank card should also pay attention to other aspects of design. Excellent product copy-writing can better express the design concept of the product, and appropriate marketing plan can further improve the best performance of the design plan. China Merchants Bank launched the "angry birds" parent-child bank card for parent-child families, with great care in the card surface design, and designed the corresponding card surface for different roles in the family. Four of the small bird cards are made of color card base materials, with pearlescent backing and 3D eye popping technology, so that the card surface is reversed and shiny, and the big eyes of the birds are particularly lifelike, and specially designed the official card Web application page. China CITIC Bank issued seven Yanka series, which are of different colors. Each color has different product meaning and unique product advertising language. The official website has made a unique interface design for Yanka series, and also added interactive design to make users more integrated into the product series when applying. In addition to the slogans of each color, the bottom of each color also shows the voice of some users, enhances the emotional resonance of users for the product, and improves the degree of personalization of the product.

3 The Development of Personalized Bank Card Design from a New Perspective

The personalized bank card under the new perspective can meet the market demand no longer through unilateral promotion. In addition to the user demand and market segmentation, it pays more attention to the design and the application in multiple fields. Bank is a financial institution established according to law to manage money and credit business. The corporate image must always be positive. As one of the daily tools, the bank card is closely related to people's life, and it carries the mission of corporate image. At the same time of personalization, the personalized bank card should always keep such original intention, which also needs to pay more attention to the personalized owner in the design The choice of questions.

With the information and intelligence of life, the function of the bank applying for the card independently appears. The personalized bank card can not only consider the card surface in actual use, but also pay attention to the card surface display of the application bank card in the web interface and mobile phone interface, as well as the consideration of interactive design, including the visual design of the surrounding materials of the bank card. Comprehensiveness and integrity are also one of the means to improve the product personalization.

In recent years, the design discipline has gradually developed from a single discipline research to a common integration of cross disciplines, the design style has developed from a single discipline to a diversified one, the two-dimensional visual performance has gradually expanded to three-dimensional, and the personalized bank card has paid more attention to the multi-faceted integration of design concepts in the design.

4 Conclusion

Through the analysis of the development of personalized bank card and the collection of various factors of personalized bank card, it can be concluded that the design trend of personalized bank card in the new perspective needs to be reflected in the following aspects: in the development of new products, pay attention to avoid product homogenization, and study the target customer group; in the theme selection, on the basis of positive guarantee In the design, we should pay attention to the integration of various design concepts, as well as the use of design styles and new techniques; in the production, the use of special materials and new technologies helps to express the design scheme. The trend of personalized bank card design from a new perspective promotes "personalization" through comprehensive design in many aspects, gradually forming characteristic products and brands, solving the problem of homogeneity, and achieving the goal of creating excellent and easy-to-use personalized bank card products.

References

1. Chinese code. Personality. [EB/OL]. https://www.zdic.net/
2. Chinese Dictionary Compilation Office: New Edition of Modern Chinese Dictionary. Sichuan Dictionary Press, Sichuan (2018)
3. Perry, M.E.: Strategic Marketing Management. Trans. Li Yisong. China Finance and Economics Press, Beijing (2003)

A Conceptual Model for Designing Sustainable City Development

Amira Ashraf(✉) and Thomas Garvey

School of Industrial Design, Carleton University, Ottawa, Canada
amiraashraf@cmail.carleton.ca

Abstract. This literature review highlights key factors of sustainable cities by looking at the World Design Capital (WDC) programme of the World Design Organization (WDO) which is an international organization representing the profession of industrial design. Exploring the WDC, helps recognize cities for their effective use of design. This literature review explores successful design strategies, which could be utilized as preliminary steps towards sustainability and suggests a conceptual model for sustainable city development.

Keywords: Interdisciplinary design · Sustainable cities · UN SDGs · Development policy

1 Introduction

Cities are looking forward for new ways to develop their standards for sustainable development. However, the current industrial model of efficiency in most cities does not allow for enough sustainable practices. The current rates of production continue to accelerate the use of resources and energy. Today, approximately 55% of the world's population are situated in urban areas. This proportion is expected to increase to 68% by 2050. Per the United Nations data, projections show that urbanization coupled with the overall growth of the world's population could potentially add another 2.5 billion people to urban areas by 2050, with close to 90% of this increase developed in Asia and Africa [1]. With sensible policies and changes, it is possible to slow the rate of decline in resources without depleting them drastically and improve the quality of life for the increasing population.

Design strategies are being explored today to assist with sustainable development with many organizations such as the United Nations (UN) and the World Design Organization (WDO). The World Design Capital (WDC) is one of the WDO programmes, which recognizes cities for their effective use of design to drive economic, social, cultural, and environmental development [2]. While these WDC cities are awarded for successfully utilizing design, their success could be used as ideal models for future cities. As a result, it may help to look at which steps have already been taken into consideration in these cities. Through sustainable design practices, this literature review explores some of the preliminary factors that cities have considered in their processes towards establishing successful development.

G. Di Bucchianico et al. (Eds.): AHFE 2020, AISC 1202, pp. 360–369, 2020.
https://doi.org/10.1007/978-3-030-51194-4_48

This paper will look at sustainability at a conceptual level for sustainable city development. This concept acknowledges that the city demands economic, social, cultural, political, and environmental objectives for the perpetuity of a city. Therefore, the concept of a sustainable city will be looked through the lens of these five aforementioned objectives and perspectives of sustainability. This paper analyzes literature in the area of sustainable cities, and the WDC programme, and synthesizes sources to highlight patterns and themes in achievement. It looks at stakeholders involved in sustainable development and the collaboration between partners such as governments, private and public institutions, and communities and their part in the global movement. Delving into examples such as design influence, infrastructure, and policies, this paper highlights some of the key factors needed for sustainable cities.

From previous WDC cities, a lot can be learned from the recognizable cities making huge advancements in innovation and sustainability. From WDC cities to these recognized sustainable cities, and emerging smart cities, there is a window to look for design strategy opportunities which could be a potential catalyst for developing cities that are considering the next step for change.

This paper looks at the United Nations Sustainable Development Goal (UN SDG) 11, and the WDC programme, delving deeper into sustainability and design components while considering the interdisciplinary aspects. Additionally, the perpetuity of sustainable development is discussed for continuous growth. Furthermore, a conceptual model highlighting key considerations for sustainable development is presented. This model consists of five factors that highlight a few aspects drawn out from the analysis of information presented in the paper. These aspects are prevalent in the information in the main analysis and are chosen based on reoccurrence and/or importance. With this research, the purpose of this model is to add value to the organization as it gives a preliminary insight on the factors involved for sustainable development and to understand the issues better.

2 Background

2.1 World Design Organization (WDO)

The WDO is an international non-governmental organization and worldwide society that represents the profession of industrial design. The organization is the highest body for the voice of industrial design. Formerly known as the International Council of Societies of Industrial Design (ICSID), the organization was founded in 1957 from a group of international organizations. In 2017, they were renamed as the World Design Organization, advocating for 'Design for a Better World', and encouraging and sharing knowledge of industrial design-driven innovation to strengthen the economic, social, cultural and environmental quality of life. Today, WDO includes over 150 member organizations [3]. The WDO suggests that their "programmes are designed to enable collaborations among the global industrial design community, to identify challenges and opportunities and to address some of the world's most significant economic, social and environmental challenges, from a design perspective" [3]. The WDO has consultative status with the UN and has embraced the UN Sustainable Development Goals

(SDGs) as a framework for action by engaging the design community in collaborative efforts and executing international programming. The SDGs framework has the capacity to simultaneously offer positive social and environmental influence and furthermore, financial return and brand value [2]. The UN SDGs are a global set of 17 goals that were designated by UN member countries to solve some of humanity's substantial challenges by year 2030. Out of the 17 goals, the WDO has identified seven SDGs which are particularly relevant to the industrial design community, with one of them being Goal 11 – Sustainable Cities and Communities [2].

2.2 United Nations Sustainable Development Goals (SDGs) – Goal 11

In a global partnership, the 17 SDGs are an urgent call for action by all countries – developed and developing. "They recognize that ending poverty and other deprivations must go hand-in-hand with strategies that improve health and education, reduce inequality, and spur economic growth – all while tackling climate change and working to preserve our oceans and forests" [4]. The UN SDGs are increasingly being adopted by the world, "not just by governments, but also by civil society, academia, and the private sector, as a way to break down silos and effectively collaborate around a common and universal development agenda" [2].

The goals have a number of targets and indicators. The targets are qualitative descriptors and the indicators are quantitative measurements. Sustainable development Goal 11, aims to make all cities and human settlements inclusive, safe, resilient and sustainable. With 10 targets and 15 indicators, by 2030, it aims to ensure access for all to adequate, safe, and affordable housing and basic services; accessible and sustainable transport systems; inclusive and sustainable urbanization and capacity for participation; sustainable human settlement planning; and to strengthen efforts to protect the world's cultural and natural heritage. Moreover, UN SDG 11 intends to substantially decrease the direct economic losses to global gross domestic product caused by disasters, reduce the adverse per capita environmental impact of cities including air quality and other waste management, and to provide universal access to safe and inclusive green and public spaces.

Furthermore, UN SDG 11 aims to support positive economic, social and environmental links between urban and rural areas by strengthening national and regional development planning, and substantially increasing the number of cities and human settlements by adopting and implementing integrated policies. It aims towards inclusion, resource efficiency, mitigation, adaptation to climate change, and resilience to disasters. It also aims to develop, implement and support the least developed countries, through financial and technical assistance, in building sustainable and resilient buildings utilizing local materials [4]. UN SDG 11 is an important goal for developing and developed cities, therefore it should be an essential consideration towards further city development.

2.3 World Design Capital (WDC)

Through a year-long programme of events, the WDC city, which is designated every two years, showcases best practices and accomplishments in sustainable design-led urban policy and innovation that enhance the lives of citizens [5]. Establishing themselves as prominent design hubs, which is one of the real strengths of the WDC designation, these cities continue to communicate the importance of design as a tool to address some of the biggest urban challenges long after their initial designation year. These cities join an international network where they can learn and share innovative design-led urban revitalization programmes and strategies. Through this, the cities can build a global reputation as a leading centre of design, creativity, and innovation [5].

A recent graduate thesis written by Asfand looks at how the WDC programme can better incorporate the UN SDG 11 in future cities. Asfand [6] suggests that strategies can be explored and utilized to assist with sustainable development, and the WDC programme should assist cities in need of development. Asfand suggests that UN SDG 11 be integrated with the WDC programme considering a framework of 10 factors. The 10 factors are: knowledge-based resources; funding support requirement; case studies; WDC programme re-evaluation; WDC cities selection approach; design linkage to UN SDG 11 approach; design linkage to sustainable development; designers' arbitrative approach; role of citizens; and government responsibility [6]. Some of these factors can be considered essential towards sustainable city development such as design linkage to UN SDG 11 and government responsibility. He suggests that the WDC programme should aim towards linking developing cities that need design for various forms of sustainable development. Through sustainable design initiatives, there are opportunities for a future step towards implementing sustainable development and the WDC programme within developing nations.

3 From Sustainability to Sustainable Cities

Sustainability, let alone sustainable development, is a contested concept with a wide selection of meanings. Egger states that "[at] a conceptual level, sustainability relates to the maintenance or improvement of the integrated natural systems that collectively comprise life on our planet" [7]. Considering that the dominance of people and their actions has the most influence on the global environment, human population can be said to be central to these natural systems. "The Earth's capacity to support people is determined by natural constraints and by human choices concerning economics, environment, culture (including values and politics) and demography" [7]. Egger goes on to say that "Human carrying capacity of the Earth is therefore dynamic and uncertain. Sustainability must be concerned with the level of impact on the earth by the human population" [7]. Several researchers have pointed out that sustainability could be examined in different dimensions, such as economic, social, cultural, political, and environmental [8]. In order to achieve long-term sustainability, all different aspects of sustainability must be addressed.

Egger [2] states that, "the model of the sustainable city manifests itself in countless different forms depending upon the area's history, culture, economic base, climate, geography and politics to name just a few of the determining variables. Sustainability must predominantly address enduring human purpose as the core of any city development strategy". Although the measurement of the performance of a city is the human condition of its inhabitants, it is essential to examine other factors to determine its sustainability in regards to being able to maintain over time, a high quality of life for its inhabitants without compromising the human condition in other local and global regions. If its structure and operation acknowledge the indivisibility of the planet, only then can a city be recognized to be sustainable [7]. He continues to state that the concept of the sustainable city "must recognize that the city needs to meet social, environmental, political and cultural objectives as well as economic and physical ones" [7]. He continues to say that "the key attributes of such a city… include equitable access to basic services, beauty in its art and architecture, creativity to optimize human potential, resource efficiency and minimal ecological impact, ease of contact, mobility, integrated and compact communities and diversity" [7].

4 Helsinki as the 2012 World Design Capital

The WDC programme is a celebration of the cities enforcing design strategy development tools in transforming their cities and securing post-event legacy, such as improvements in infrastructure and social-economic conditions. Previous WDCs have included Turin (2008), Seoul (2010), Helsinki (2012), Cape Town (2014), and Taipei (2016), followed by Mexico City (2018). These cities utilized design to work towards greater planned urban development. Asfand states that "[the] cultivation of design as a force allowed the cities to remain flexible on the development procedures as it gave them the opportunity to constantly question strategies and ideas rather than being constrained through the territories of architecture, landscape and engineering concepts" [6]. Therefore, the WDC programme does not limit the cities to adhere to all specific guidelines, so cities are able to explore different avenues that may elicit innovation and breakthroughs. The criteria for selecting the WDC city, which can be found on the WDO's website, looks at the city's existing design assets, the city's plans for using design to renew itself, and the ways in which citizens benefit from the use of design in city planning. Furthermore, the criteria examines the capacity to organize and finance a year-long programme of design events, the mobilization and participation of the wider design community, large sectors of the population and visitors, and finally, the fostering of a wider dialogue on urban revitalization strategies. Qualifying cities bid for the title of the WDC, and a WDC selection committee selects which city to award over several months of evaluation.

The city of Helsinki is one of the primary examples that was devoted to using design as its creative power for transformation. The strong emphasis on design culminated in 2012, when Helsinki was designated as the WDC by the WDO. Apart from its famous designers and products, the bid by Helsinki not only emphasized the reach of the design legacy of Helsinki, but also presented something new which would appeal to the WDO and help strengthen and reinforce the WDC brand [8].

The idea of 'Open Helsinki- Embedding Design in Life' WDC 2012 went beyond the concept of design from goods to include services and systems. One of the objectives was providing user-oriented municipal services and to strengthen resident participation and their responsibility. Other aims of the project were to make the city "attractive, functional, and enjoyable; to create interesting, beautiful, and functional urban residential areas and environment; and to provide educational services that people need" [8]. Liu goes on to state that the "WDC in Helsinki is not just about image but about intervening in everyday life, enhancing quality of life, boosting economic indicators, and raising awareness about the use of design in all aspects of urban development" [8].

5 Interdisciplinarity

When working on issues with design and sustainability, it is important to address and collaborate with various views. Understanding different perspectives is the first step to knowing how to deal with that design possibility. The WDC programme is still young with each city and group of stakeholders taking on different approaches. For design and sustainable cities, it is necessary for governance to be considered in the long-term. Having a legacy plan for events can provide justification for public expenditure, acquire local support, and encourage other cities to bid for events [8].

Similar to establishing the WDC, the process towards sustainable cities is as much of an interdisciplinary framework, as it facilitates the participation and involvement of numerous stakeholders. It involves designers and end-users in the co-creation of products and services that contribute to the enhancements of the quality of life in cities. The concept of an open city, highlights a new form of urban governance providing user-oriented municipal services as well as reinforcing resident participation and their responsibility in the transformation of the city.

Open Helsinki relates to popular concepts such as co-design, design education and open data [3]. In its initiative for co-design, Helsinki involved the community in designing places, products and/or services. Pekka Timonen, executive director of WDC Helsinki 2012, said in an interview "a design-driven city is a user-centric city" [8]. In regards to design education, the WDC Helsinki 2012 improved possibilities and tools to provide design education to youth through school and recreational activities. Open Helsinki's initiative for open data, allowed for positive cooperation in the region between municipalities and across sectors such as state government bodies [8]. Making public information accessible for free promotes stakeholders and communities to collaborate and stimulate innovation. Being transparent increases citizen involvement since they have access to the city's decisions and data resources. The open access to information can lead to new services and businesses as it may advance research and development [8].

The WDC Helsinki 2012 represented a huge cooperation process between cities, the national government, corporate partners, the international design community, universities and so forth. The National Government with its support and funding laid the solid ground for the process involving dozens of NGOs joining the initiative. The WDC Helsinki 2012 brought together the international design community and organized nearly 80 international seminars, conferences, and encounters in the Helsinki metropolitan area [3]. The WDC Helsinki 2012 provided "four directions of impacts: (1) improving the organizing incapacity of Helsinki; (2) making Helsinki as a city more attractive internationally; (3) encouraging city departments to adopt design methods and to make city smarter; and (4) improving the competitiveness of international and local companies based in Helsinki" [8].

6 Legacy Plan

Liu states that the legacy of the WDC title on cities has become a prime focus. However, the perpetuity beyond the WDC designated year varies considerably from city to city due to the levels of investment, especially from the public sector. The issue of legacy and sustainability impacts is critical for cities to thrive and to continue being a model for design cities. There are several definitions of legacy. Liu highlights one definition of legacy as "all planned and unplanned, positive and negative, tangible and intangible structures created for and by a design event that remain longer than the event itself" [8]. Without a legacy plan, the investments are less likely to pay back and due to budget restraints, the legacy plan needs to be very focused. The legacy plan called Design Driven City (DDC) was implemented by the main task force of WDC Helsinki [8].

The primary initiative of DDC was to develop the city with design methods and find approaches to bring the public sector close to the design field. This was executed through the innovation of diversified design activities performed around the city which provided local communities an opportunity to participate in activities, and understand the WDC programme. These activities include idea forums, design exhibitions and open identity Helsinki workshops. Moreover, the DDC project hired city designers and city staff members who worked alongside one another in the process of urban development. Liu states that "the former mainly aims at promoting Helsinki internationally, while the latter seeks to firmly place design at the core of public service development and strengthen the user-oriented urban service design method" [8]. He continues to say that "the former helped the latter to deepen their understanding of citizens' needs and encourages residents to play a more active role in city decision-making and development. The task of the city designers is to determine the areas where the design can be applied and determine the method of applying the design in these areas" [8].

The WDC Helsinki 2012 and DDC project both share the same bottom-line: using design to help better shape cities. Liu states that "DDC has two concrete aims: (1) strengthening the use of design in city development initiatives and supporting these with know-how and expertise (DDC has its own design experts, called City Designers); and (2) educating the public sector on the significance of design, and participating in public discourse surrounding it" [8]. Distinctive from other WDC cities such as WDC Seoul 2010, which focused on the 'hard legacy' (i.e., building up infrastructures), WDC Helsinki 2012 concentrated on the 'soft legacy', specifically picking up some representative cases, which could be achieved firmly to ensure the effectiveness of the project [8]. "Such design strategies resulted in Helsinki emerging as an internationally appealing design city that used design as an apparatus to instill awareness of social, economic and environmental development issues to the local public" [6].

7 Conceptual Model for Sustainable City Development

The following is a conceptual model consisting of five factors that highlight a few aspects drawn out from the analysis of the information presented earlier. These aspects were prevalent in the information presented earlier, and were chosen based on reoccurrence or importance. These aspects were then categorized under the five types of sustainability that are needed to be met for the concept of a sustainable city as stated by Egger. Together, an overall system can be designed to be partially sustainable. The concept for a sustainable city development acknowledges that the city demands economic, social, cultural, political, and environmental objectives for perpetuity of a city. The factors outlined below highlight a few aspects to consider.

7.1 Economic Sustainability

Economic success can sometimes be correlated to quality of life in a city. Having equitable access to basic services, resources, transport systems, and affordable housing are a few economic indicators. At the same time, to substantially decrease direct economic losses, disaster prevention is valuable. In order to support positive economic, social, and environment links between urban and rural areas, national and regional development planning needs to be strengthened.

7.2 Social Sustainability

With creativity optimizing human potential, there must be continuous communication within diverse integrated and compact communities. Knowledge-based resources and open access to information helps allow for new services and business with potential, to advance research and development. Social sustainability should involve inclusive and sustainable urbanization and capacity for participation. The role of citizens is essential to integrate sustainable settlement planning. It is fundamental to enhance the use of design in city development initiatives and supporting these with know-how and expertise. The public sector must educate on the significance of design, encouraging participation in public discourse surrounding it.

7.3 Cultural Sustainability

Cities thrive with the beauty in their art and architecture. Efforts should be encouraged to protect the world's cultural and natural heritage. The WDC and DDC projects have made the city of Helsinki more attractive internationally. At the same time, the projects encourage city departments to adopt design methods and to make cities smart. Design places a vital role in city development, much like the DDC project, that have their own design experts called City Designers.

7.4 Political Sustainability

Politics play an influential role towards sustainable city development. For successful development, it is as much an interdisciplinary collaboration as it is an immense cooperation between cities, the national government, corporate partners, the international design community, universities and so forth. These stakeholders should consider devoting the use of design as their creative power for transformation. Government responsibility is significant as it can substantially increase the number of cities that adopt and implement integrated policies. At the same time, levels of investment and funding support majorly affect the scope of any project. Politics may also improve the competitiveness of international and local companies and improve the organizing capacity. Furthermore, legacy plans ensure that investments are more likely to pay back.

7.5 Environmental Sustainability

There must be minimal ecological impact while linking design to sustainable development. This also includes universal access to safe and inclusive green and public spaces. It should aim to reduce the adverse per capita environmental impact of cities including air quality and other waste management. It should plan towards inclusion, resource efficiency, mitigation, and adaptation to climate change, and resilience to disasters. Additionally, it should develop, implement, and support least developed countries through financial and technical assistance, in building sustainable and resilient infrastructure utilizing local materials.

8 Conclusion

While society tends to see the problems of today more acutely, we need to be open to future possibilities, and it will take time to make changes to how we design. Although a completely sustainable city is a concept right now, sufficient sustainability is possible. This paper has developed a model of sustainable city development by undertaking a preliminary analysis of the factors involved. To achieve long-term sustainability, all different aspects of sustainability must be addressed utilizing design strategies at the forefront for prosperous development. As indicated in the conceptual model above, these key factors can facilitate sustainable city development. This hopes to present recommendations for developing and developed cities mostly in an ever-changing environment.

The UN SDGs, and the WDC programme, have contributed to our understanding about sustainable city development; however, there are limitations to this paper. Although the factors proposed in the analysis model are valuable, the generalizability to other cities can be questioned. Unfortunately, this analysis was deficient in examples of cities recognized as sustainable, as it looked specifically at only one WDC city in detail. Currently, there have only been six WDC cities and due to the programme being young, there is not enough sufficient information about past WDC cities. Similarly, a few WDC cities have carried out legacy programs and systematically produced measurable data of the outcome of the legacy. This prevents concrete comparative analysis and judgment of the success of WDC Helsinki 2012 to be conducted at this stage.

Therefore, this literature review indicates some components for future study, including but not limited to, quantitative data, investigating trends, and common matters that affect sustainable city development. Furthermore, other practices of recognized cities which are making huge advancements in innovation and sustainability should also be examined and considered.

References

1. 68% of the world population projected to live in urban areas by 2050, says UN | UN DESA Department of Economic and Social Affairs (n.d.). https://www.un.org/development/desa/en/news/population/2018-revision-of-world-urbanization-prospects.html
2. WDO. United Nations Sustainable Development Goals (n.d.). https://wdo.org/about/vision-mission/un-sdgs/
3. WDO. (n.d). https://wdo.org/
4. Sustainable Development Goals. (n.d.). https://sustainabledevelopment.un.org/?menu=1300
5. WDO | Programmes | World Design Capital. (n.d.). https://wdo.org/programmes/wdc/
6. Asfand, M.: How can World Design Capital (WDC) programme better incorporate United Nation's Sustainable Development Goal 11 (SDG 11) in future strategies? (2019). https://curve.carleton.ca/167299e9-53e6-48d7-a28d-8af2f87719ec
7. Egger, S.: Determining a sustainable city model. Environ. Model. Softw. **21**(9), 1235–1246 (2006). https://doi.org/10.1016/j.envsoft.2005.04.012
8. Liu, Y.: Legacy planning and event sustainability: Helsinki as the 2012 world design capital. Sustainability **10**(7), 2453 (2018). https://doi.org/10.3390/su10072453

The Challenges of Interdisciplinary Participation and Anti-oppressive Principles

Dawson Clark(✉) and Lois Frankel(✉)

Carleton University School of Industrial Design, 1125 Colonel By Drive, Ottawa, ON, Canada
{dawson.clark,lois.frankel}@carleton.ca

Abstract. Championed as an emancipatory, bottom-up approach, participatory design (PD) continues to be a popular method for collaborative and interdisciplinary designers. PD approaches not only emphasize, but require for their success, the engagement of key stakeholders throughout the design process, including end-users, interdisciplinary team members and subject matter experts. This literature review examines contemporary perspectives on PD, highlighting participation and challenges to PD approaches, before introducing the reader to the principles of Anti-Oppressive Practice (AOP). Reframing challenges to PD approaches as issues of participation allows for the consideration of AOP principles to meet these challenges.

Keywords: Participatory design · Interdisciplinary design

1 Introduction

Participatory design (PD) discourse has only recently gotten around to addressing participation as a matter of it's concern [1], and through this belated address it has been made evident that there is a need for PD to take responsibility for the concept of 'participation' in the design and execution of its methods. This paper builds on literature within the field of PD calling for participation to be a critically considered aspect of collaborative design projects and suggests another discipline that may assist with this task.

To discuss the concept of participation with the realm of PD, the discipline is first defined by its aspects and scope within the design discourse. Secondly, contemporary perspectives on participation are examined along with challenges encountered by PD practitioners in both research and practice. These challenges and perspectives combine in the discussion to demonstrate how challenges facing PD approaches are in fact issues of participation that the designer who wishes to utilize a participatory approach must address.

The principles and goals of anti-oppressive practice (AOP) provide a new perspective for PD to consider issues of participation. AOP emphasizes the understanding of power structures and interpersonal relationships within the field of social work [2] and draws a number of parallels with the emancipatory goals espoused by PD proselytizers. A brief overview of AOP allows the reader to reach a cursory level of understanding for their potential in PD.

G. Di Bucchianico et al. (Eds.): AHFE 2020, AISC 1202, pp. 370–377, 2020.
https://doi.org/10.1007/978-3-030-51194-4_49

The concepts of PD and AOP connect through what they can both offer one another in the face of their critics. With both disciplines acknowledging a need for greater reflexivity in their respective practice, there exists a potential for both disciplines to grow stronger through collaboration. The complexity and nuance of both disciplines calls for active collaboration between their respective experts across contexts and problem spaces.

2 Participatory Design

For the purposes of this paper, it is important to develop a contextual understanding of what entails a participatory design approach. Halskov and Hansen identified five fundamental aspects of participatory design (Table 1) that serve as useful criteria for judging whether an approach can be considered participatory. The various terms used to describe collaborative design processes such as co-design, co creation, and user lead design [3] can thus be grouped under the term participatory design so long as they encompass the fundamental aspects as laid out by Halskov and Hansen.

Table 1. Fundamental aspects of participatory design approaches [1]

Aspect	Description
Politics	People who are affected by a decision should have an opportunity to influence it
People	People play critical roles in design by being experts in their own lives
Context	The use situation is the fundamental starting point for the design process
Methods	Methods are means for users to gain influence in design processes
Product	The goal of participation is to design alternatives, improving the quality of life

2.1 Participation in the Context of Participatory Design

Though there is no formally agreed upon definition for what constitutes participation in the application of PD methods, authors do tend to agree that the concept of participation is worth critical examination [1, 4, 5]. The purpose of this section is to examine the relevant literature and build a shared understanding towards the contemporary perspectives on participation in PD.

Borrowing from Actor-Network Theory (ANT) Andersen et al. place participation as a 'matter-of concern' within PD practice, pointing out that participation is dependent on the context of the project (partial) and that the outcomes of participation are synthesized out of necessity of their communication (overtaken) [4]. The example given by the authors is that of a case worker being interviewed as part of a design project. The children that the case worker has worked with are thus partially included through that interview, though their participation is overtaken by the case worker who synthesizes their experience. These key aspects of participation require that participation be considered on a project-to-project basis and designed for carefully.

To create an infrastructure around the discourse of PD practice and the issues of participation, Saad-Sulonen, Eriksson, Halskov et al. propose a temporal perspective on the matter [5]. They outline five distinct lenses through which PD practitioners may view the matter of participation (Table 2). Combined, these perspectives on PD show an increasing demand for PD practitioners to critically consider how they will design for, and view, participation.

Table 2. Temporal lenses for participation [5]

Lens type	Description
Phasic	Participation is considered during different phases of a project, and adapted iteratively
Emergent	Participation is considered as a continuous and unfolding phenomenon
Retrospective	Understanding and interpretation of how participation changes after the end of a project
Prospective	Exploration of future participation in terms of sustaining after project end or enabling it to emerge post project
Long-term	Stretched view of participation, looking to the past and the future to influence the present

2.2 Challenges to Participatory Design

The cross-contextual nature of PD applications adds to the complexity of understanding challenges facing participatory designers. PD approaches are used across a variety of contexts, including healthcare [6], design for people living with a disability [7], design for the 90% [8, 9], and business [10]. Each of these contexts naturally comes with its own barriers to participation and by extension, challenges to implementing PD practices. Luckily, the discourse on PD includes authors who have been brave enough to not only discuss, but to seek out barriers to successful implementations of PD.

Hussain, Sanders and Steiner noted four broad challenges to participation through their experience co-designing in rural Cambodia. These challenges include human aspects, sociocultural aspects, organizational aspects and temporal & financial aspects [8]. Their findings mirror those of Drain, Shekar and Grigg who found they had to develop several competencies within their participants in the developing world in order for effective codesign to take place [9].

Anti Pirrenen, synthesized a total of 20 barrier-enabler pairs impacting codesign spread throughout the service design process [11]. The concept of barriers and enablers was also utilized by Kleinsmann and Valkenburg who classified codesign challenges on an organizational level. Together the two papers highlight the importance of interdisciplinary communication, building mutual understanding, flexible methodologies, shared vocabulary and continuity friendly practice as key aspects of successful codesign projects in the realm of service design and business.

Traditional PD tend to require a level of cognitive and physical ability that may not always be realistic when working with people living with a disability. Hendriks, Slegers and Duysburgh note that PD practitioners who wish to work in these areas must heavily modify their methods to ensure they can be used effectively [7]. There is also a need for PD practitioners in this field to share effective, considered methodologies that take into account participation, ethical challenges and the adaption of existing techniques [7].

What the literature shows, is that regardless of context, PD approaches are by no means simple to implement. The challenges facing those who wish to engender stakeholder participation are numerous and inconducive to 'turnkey' applications of PD methods.

3 Anti-oppressive Principles

Anti-Oppressive Practice (AOP) is an integral element of social work practice, first originating in the 1970s [12]. Touted as a process based on the complexity of changing social interactions and relationships, AOP is concerned with addressing and mitigating power imbalances in the implementation of social work practice [2].

Table 3 outlines key principles for Anti-Oppressive principles as stated by Burke and Harrison [2].

Table 3. Key principles for anti-oppressive practice

Anti-oppressive practice...
Is flexible without losing focus
Includes the views of oppressed individuals and groups
Challenges and changes existing ideas of practice
Can analyze the oppressive nature of organizational culture and its impact on practice
Includes continuous reflection and evaluation of practice
Has multidimensional change strategies which incorporate the concepts of networking, user involvement, partnership and participation
Has a critical analysis of the issues of power, both personal and structural

These principles are echoed by Potts and Brown, two anti-oppressive researchers who add to the discussion through their research-based lens and highlight that from their perspective anti-oppressive research is social justice and resistance in both process and outcome [13]. Anti-oppressive principles call for the recognition of knowledge as being socially constructed and a result of complex power structures and social relationships [13]. Critics of AOP call for a greater reflexivity in its practice and a further embrace of the complexity of human experience [12], though, as with any body of knowledge that has existed for some time, criticism is generally aimed at poor applications that have strayed from the guiding principles, rather than with the principles themselves.

To summarize, AOP is concerned with highlighting, acknowledging and working within the complexity generated by power imbalances, social networks and organizations. The parallels between the emancipatory goals of PD and the principles of AOP make a case for greater collaboration between the two bodies of knowledge and are further explored in the discussion.

4 Discussion

4.1 Challenges as Issues of Participation

The assumption that any method or tool denotes participation is dangerous, and demonstrates a lack of consideration for the 'people' and 'politics' aspects of PD. This paper argues that challenges faced by PD practitioners are in fact issues of participation. Accepting the ANT perspective on participation developed by Andersen et al. requires PD practitioners to consider participation on a project – to – project basis, and thus, tailor their methods to ensure they are capable of engendering project-specific participation. Table 4 outlines some of the challenges to PD as laid out in the literature and illustrates how they can be reframed as issues of participation.

Table 4. Challenges to PD reframed as issues of participation

Aspect of PD [1]	Challenges to PD	Challenge reframed as issue of participation
Politics	Conflicting Goals and Expectations [11]	Who oversees outlining the goals for this project? How are they agreed upon?
People	Physical Impairment of Participants [7, 8]	How can we tailor our approach to better serve the abilities of our participants?
Context	Accessing Remote or Restricted Areas [6, 8, 9]	How important is geographic location for participation in this project?
Methods	Lack of Design Knowledge [8, 9]	How much (if any) participant training will be required to ensure effective participation?
Product	Design freedom within the task [10]	What is the best way to leverage participant experience in the design process, within the constraints of this project?

As demonstrated by the table, choosing a PD approach implies a responsibility towards the stakeholder's researchers choose to engage, and by keeping participation as a central matter of concern, practitioners can better fulfill that responsibility. This, combined with selecting the appropriate temporal perspective from Table 2, allows for the PD practitioner to more responsibly address project specific participation. This focus on participants calls for a thorough understanding of the impacts of choices made when designing PD approaches, and it is here where the perspectives and principles of anti-oppressive social work becomes particularly relevant.

4.2 Anti-oppressive Principles to Address Issues of Participation

The principles of AOP as outlined by Burke and Harrison already begin to emerge when we reframe challenges to PD as challenges of user participation. Mere exposure to these principles is by no means sufficient to label a particular PD approach as 'Anti-Oppressive', but it does provide another lens through which practitioners can reflect on their own methodologies. Table 5 further illustrates this link between issues of participation and AOP principles.

Table 5. Issues of participation and relevant AOP principles

Issues of participation	Relevant AOP principle
Who oversees outlining the goals for this project? How are they agreed upon?	Has a critical analysis of the issues of power, both personal and structural
How can we tailor our approach to better serve the abilities of our participants?	Challenging and changing existing ideas of practice
How important is geographic location for participation in this project?	Flexibility without losing focus
How much (if any) participant training will be required to ensure effective participation?	Includes continuous reflection and evaluation of practice
What is the best way to leverage participant experience in the design process, within the constraints of this project?	Has multidimensional change strategies which incorporate the concepts of networking, user involvement, partnership and participation

Table 5 demonstrates how AOP expertise is directly relevant to addressing issues of participation that arise in the development of PD methodologies. It is unlikely that PD practitioners set out to oppress or further disenfranchise their participants. Rather, negative consequences of PD may arise when the designer does not treat participation as a matter of concern. Incorporating individuals with AOP knowledge in the design of PD methodologies may help to ensure that participation is treated as a matter of concern in a responsible and ethical manner throughout the project.

4.3 Bridging Through Reflexivity

Critiques and reflections on both AOP and PD have called for a greater reflexivity in the practice of both disciplines [12, 14]. In this case, reflexivity is the increased consideration of relationships, complexity and the power dynamics of knowledge structures in practice [14]. Where PD may be lacking in its understanding of power dynamics and social networks, critics of AOP have highlighted concerns with adherence to a lens that blocks out complexity in favor of a framework of oppression [12]. It may be possible then, that collaboration between these disciplines across different contexts and application may strengthen one where the other lacks and vice versa. PD approaches are well documented in their efforts to deal with complexity [4, 6, 8] and AOP is inherently concerned with understanding power dynamics and social networks [2, 13]. There is an opportunity for further research and collaboration in this direction, and the results of this collaboration may prove to be symbiotic.

5 Conclusion

This paper adds another voice to the participatory design discourse regarding matters of participation and suggests an interdisciplinary collaboration amongst the disciplines of PD and AOP. Designers must take participation into account when developing collaborative design methodologies and be constantly asking how they can act responsibly on behalf of participants. Introducing AOP as a discipline that concerns itself with the principles that PD must increasingly embrace serves as an introduction between two large schools of knowledge. Moving forward, PD and AOP practitioners may build further shared understanding and evolve both disciplines through a collaborative approach towards reflexivity between and within these disciplines.

References

1. Halskov, K., Hansen, N.B.: The diversity of participatory design research practice at PDC 2002-2012. Int. J. Hum Comput Stud. **74**, 81–92 (2015). https://doi.org/10.1016/j.ijhcs.2014.09.003
2. Burke, B., Harrison, P.: Anti-oppressive practice. In: Barrett, S., Komaromy, C., Robb, M., Anita, R. (eds.) Communication, Relationships and Care, pp. 131–138. Routledge, London (2004)
3. Sánchez de la Guía, L., Puyuelo Cazorla, M., De-Miguel-Molina, B.: Terms and meanings of "participation" in product design: From "user involvement" to "co-design". Des. J. **20**, S4539–S4551 (2017). https://doi.org/10.1080/14606925.2017.1352951
4. Andersen, L.B., Danholt, P., Halskov, K., Hansen, N.B., Lauritsen, P.: Participation as a matter of concern in participatory design. CoDesign **11**, 250–261 (2015). https://doi.org/10.1080/15710882.2015.1081246
5. Saad-Sulonen, J., Eriksson, E., Halskov, K., Karasti, H., Vines, J.: Unfolding participation over time: temporal lenses in participatory design. CoDesign **14**, 4–16 (2018). https://doi.org/10.1080/15710882.2018.1426773
6. Donetto, S., Pierri, P., Tsianakas, V., Robert, G.: Experiencebased co-design and healthcare improvement: Realizing participatory design in the public sector. Des. J. **18**, 227–248 (2015). https://doi.org/10.2752/175630615X14212498964312
7. Hendriks, N., Slegers, K., Duysburgh, P.: Codesign with people living with cognitive or sensory impairments: a case for method stories and uniqueness. CoDesign **11**, 70–82 (2015). https://doi.org/10.1080/15710882.2015.1020316
8. Hussain, S., Sanders, E.B.N., Steinert, M.: Participatory design with marginalized people in developing countries: challenges and opportunities experienced in a field study in Cambodia. Int. J. Des. **6**, 91–109 (2012)
9. Drain, A., Shekar, A., Grigg, N.: Participatory design with people with disability in rural cambodia: the creativity challenge. Des. J. **21**, 685–706 (2018). https://doi.org/10.1080/14606925.2018.1488923
10. Kleinsmann, M., Valkenburg, R.: Barriers and enablers for creating shared understanding in co-design projects. Des. Stud. **29**, 369–386 (2008). https://doi.org/10.1016/j.destud.2008.03.003
11. Pirinen, A.: The barriers and enablers of co-design for services boundary-crossing collaboration and organisational change. Int. J. Des. **10**, 27–42 (2016)

12. de Montigny, G.: Beyond anti-oppressive practice: investigating reflexive social relations. J. Progress. Hum. Serv. **22**, 8–30 (2011). https://doi.org/10.1080/10428232.2011.564982
13. Potts, K., Brown, L.: Becoming an anti-oppressive researcher. In: Brown, L., Strega, S. (eds.) Research as Resistance: Critical, Indigenous, and Anti-Oppressive Approaches, pp. 255–283. Canadian Scholars Press, Toronto (2005)
14. Pihkala, S., Karasti, H.: Reflexive engagement - enacting reflexivity in design and for "participation in plural". In: ACM International Conference Proceeding Series, pp. 21–30 (2016). https://doi.org/10.1145/2940299.2940302

A Dimensional Themes' Matrix for Identifying Design Problem Structure

Dabin Lee, Sangjin Joo, Soyoon Park, Hwang Kim, and KwanMyung Kim(✉)

Graduate School of Creative Design Engineering, Ulsan National Institute of Science and Technology, 50 UNIST-gil, Eonyang-eup, Ulju-gun, Ulsan 104-904, Republic of Korea
{kkang58322, jimmy928, hwangkim, kmyung}@unist.ac.kr, yoonyoon1014@gmail.com

Abstract. Designers use various design methods and tools to understand user needs, identify using process and context, and set a design direction in the design process. Generally, Human-Centered Design (HCD) process requires the process of classifying and interpreting a vast amount of raw user data into meaningful information. Designers frequently use affinity diagrams to transform the data into meaningful information. However, it is difficult to get rational insights only within the affinity diagram, since it is challenging to synthetically analyze the derived keywords. This paper presents a new method of analyzing user data called 'dimensional themes' matrix'. The matrix enables designers to understand the design problem structure by positioning user data on the intersections of two keyword dimensions. It provides rationales to insights by enhancing the connectivity of the keyword themes, which is difficult to capture with the original affinity diagram.

Keywords: Human-Centered Design · UX research · Dimensional themes' matrix · Affinity diagram · Design problem structure

1 Introduction

In Human-Centered Design approach, designers use various user research methods to identify user data, such as user needs and contexts [1]. In this process, they frequently use interviews and observations. Interviews help moderator identify problems in an explicit knowledge level through communication with users. In observations, researchers identify implicit needs in a behavioral knowledge level by observing the actual behavior of users. Designers can understand user needs and contexts by systematically analyzing data related to the users. It helps define design problems and set solution direction. To extract meaningful information from a vast amount of data, feasible and effective tools and methods are needed. As such, designers use affinity diagrams to analyze qualitative data inductively. The affinity diagram is a tool that clusters data based on similarities and codes keywords to each group [2]. Using the affinity diagram, designers determine the implicit information and hierarchical links of data [3]. Afterward, designers are required to understand and interpret user context,

G. Di Bucchianico et al. (Eds.): AHFE 2020, AISC 1202, pp. 378–384, 2020.
https://doi.org/10.1007/978-3-030-51194-4_50

user needs, and problems, so-called insights, from the affinity diagram to benefit the user by increasing product usability [4]. In many cases, however, it is a limitation to draw insights only within the affinity diagram. It flattens multidimensional data into unidimensional groups, which makes it hard to define the connectivity among keywords [5]. For a conclusive interpretation of data, a new method for analyzing multidimensionality of data is required. In this paper, we present a tool called 'dimension themes' matrix' to overcome the limitation of the affinity diagram and efficiently interpret the data from user research.

2 Methodology

We proceed with the smart helmet project for the safety of construction site-workers. In construction sites, mostly happen fatal falls, fall-related injuries, and risks of fallen objects. Besides, most of the construction sites are large, which makes it difficult for safety supervisors to manage individual site-workers. The main function of the proposed smart helmet is to collect brainwaves from users' foreheads and send to a central server through wireless ad hoc networks. The aim is to enable safety supervisors recognize the physical and psychological health conditions of site-workers in a real-time situation. To understand user needs and circumstances of the construction site, we conducted user research through interviews with three stakeholder groups, and arrange observations. The data collected from user research were initially analyzed through an affinity diagram and subsequently, the dimension themes' matrix was designed to systematically interpret the processed data. The detail of the methods and tools followed in this study is presented in the following sections.

2.1 User Study

2.1.1 Interviews

The interviews were held with three stakeholder groups: an inspector, two safety supervisors, and four site-workers. As shown in Table 1, the basic information related to the groups is presented.

Table 1. Basic information of interviewees

	Inspector	Safety supervisor	Site-worker
Number of interviewees	1	2	4
Main job	Construction site management	Filed supervision	Construction work
Term of office	Over 20 years	Over 10 years	Over 15 years

The main contents of the interviews were related to the management of the construction site and the usability of safety helmets. All the voice conversations were recorded instantaneously and later transcribed. Figure 1 shows the location and situation of the interviews.

(a) (b) (c)

Fig. 1. Focus group interview with three stakeholders (a) safety supervisors, (b) inspector, and (c) site-workers.

2.1.2 Observation

After conducting the interviews, we visited the construction sites for contextual inquiry. The observation divided into two phases from inquiry to lunchtime, and afternoon work to wrapping up. Two main works were performed simultaneously in the construction site. Design team divided into two groups with two people and observed both work sites, and changed observing site after lunchtime. Using shadowing and fly on the wall, which are specific methods of observation, the entire working process had observed. In the interim, certain actions that site-workers perform, such as extremely bending their waist, were photographed and some of the uncertain points were recorded through memo and drawings. Figure 2 shows the activities of the workers at the site.

(a) (b) (c)

Fig. 2. Photographed observation data from (a) inquiry, (b) working to (c) wrapping up.

2.2 Data Analysis and Interpretation

2.2.1 Affinity Diagram

Upon the completion of data collection, we printed all raw data, which include transcribed data and field study photos to conduct affinity diagram. It requires the process of classifying and grouping all raw data based on similarities and coding keywords for each group [6]. In Fig. 3, the initial keyword data mapping using an affinity diagram is shown.

Fig. 3. Initial keyword-data mapping using affinity diagram.

Using the affinity diagram, it was difficult to sufficiently and reliably classifying the data. There were two main reasons for this problem, the duality of data and the lack of connectivity of keywords from the affinity diagram. Specifically, certain data could belong to more than one keyword, which debases the validity of data positioning. It supports that the data contains a multidimensional context, and it hinders the logical classification of the data within the affinity diagram. We concluded that this affinity diagram is difficult to identify the structural design problems that the data involved. To overcome such a problem, a method was required that can logically map the keywords and data. Figure 4 shows the limitation of the affinity diagram for data positioning.

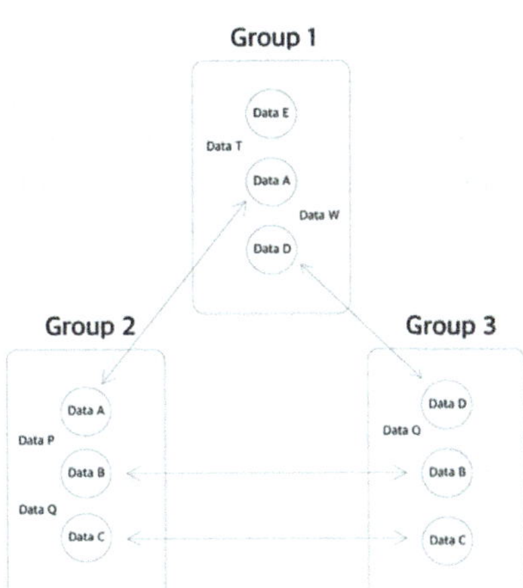

Fig. 4. Limitation of affinity diagram identified by the design team.

2.3 Dimensional Themes' Matrix

At first, our goal was to find connectivity among each derived keyword. These keywords were divided into two themes: situations occurring during fieldwork and user needs related to the usability of helmets (See Table 2).

Table 2. Keyword themes: situation-based and needs-based.

Situation	Needs
Inquiry	Appearance
Using heavy equipment	Site management
	Communication
Small-scale work	Helmet managing
Big-scale work	Emergency
Lunch	Sweat
Snack	Condition
Wrapping up	Tool
Education	
Rest	

Then, we put needs-based keywords on row and situation-based keywords on the column to make a two-dimensional matrix. Situation-based keywords, especially, were listed in chronological order. In the matrix, we could make cells that the two keywords, one from the row and the other from the column, produce by intersecting each other. For each cell, we positioned every data that correspond to both keywords.

3 Results

The mapping process systematically demonstrates the problems that the data potentially possess. With the development of the matrix, we could analyze the problems and needs that the user faces more reliably. It was able to match the problems that were mentioned in the interviews and actual situations where the problem arises. Also, by comparing the amount of data contained in each cell, we could identify the importance of specific needs that occur from a particular situation. Certain cells that contain a large amount of data has highly marked in terms of significance, since the situation and need that the cell indicates judged to be mentioned frequently from various users. Subsequently, we determined critical factors that should be considered intensively during the project. Some of the questions and insights were figured out from each cell. Figure 5 shows dimensional themes' matrix we made.

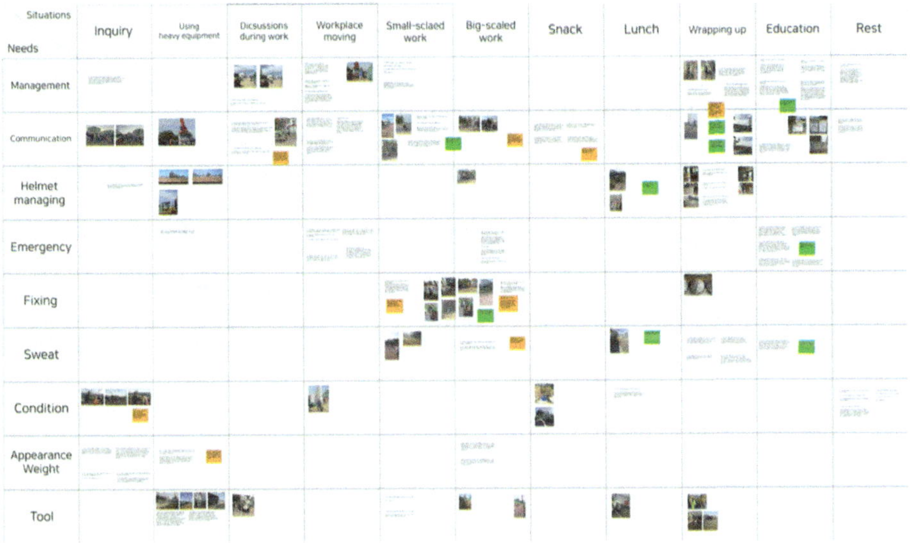

Fig. 5. Dimensional theme's matrix, situation-based theme keywords on row and needs-based theme keywords on column.

4 Conclusion and Discussion

In this paper, we presented an effective method for analyzing user research data with a case design project. The method provided us a means to systematic analysis of affinity diagram results. To identify structural design problems from user research data, it is required to place multi-dimensional user research data on a map and demonstrate the connections among them. In order to achieve this, we proceeded with a stereoscopic processing of the data that were flattened after conducting the affinity diagram. The dimensional themes' matrix was useful to position and integrate the multi-layered data on the map. This helped us comprehend the user needs and contexts simultaneously. Accordingly, it provided determination of complex problem structures that the data possess, which had been difficult to figure out with the affinity diagram. The dimensional themes' matrix that we used maps the keywords and the user research data in two-dimensional matrix. As the number of keyword themes increases more than two, we may need more complex matrices. For example, with three themes, there can be three matrices built with three 1×1 combinations of the themes. An efficient way to draw a high level of dimensional matrix needs to be further discussed. We expect that this method can be used for supplementing the limitation of affinity diagram method and enhancing its practicality.

Acknowledgements. The work was supported by the UNIST funded product development project (1.190110.01).

References

1. Zoltowski, C.B., Oakes, W.C., Cardella, M.E.: Students' ways of experiencing human-centered design. J. Eng. Educ. **101**, 28–59 (2012)
2. Plain, C.: Build an affinity for KJ method. Qual. Prog. **40**, 88 (2007)
3. Holtzblatt, K., Wendell, J.B., Wood, S.: Rapid Contextual Design: A How-to Guide to Key Techniques for User-Centered Design. Elsevier, Amsterdam (2004)
4. Babbar, S., Behara, R., White, E.: Mapping product usability. Int. J. Oper. Prod. Manag. (2002)
5. Alloway Jr., J.A.: Be prepared with an affinity diagram. Qual. Prog. **30**, 75 (1997)
6. Algozzine, B., Haselden, P.G.: Tips for teaching: use of affinity diagrams as instructional tools in inclusive classrooms. Prev. Sch. Fail. Altern. Educ. Child. Youth **47**, 187–189 (2003)

Reconstructing Social Identity Through Physical Attractiveness

Young Ae Kim[1(✉)] and Qiuwen Li[2]

[1] School of Design, South Dakota State University, 117 Grove Hall, Brookings, SD 57007, USA
YoungAe.Kim@sdstate.edu
[2] Department of Art and Art History, Santa Clara University, 500 El Camino Real, Santa Clara, CA 95053, USA
Qli2@scu.edu

Abstract. Attractiveness is an attribute of an individual's interest and desire for someone or something that outlines affiliative behaviors in human interaction. Physical attractiveness is the aesthetic perception of the physical traits of a human-based on individuals' subjective preferences and/or cultural and social differences. It has a significant contribution to how people are judged in sexual attractiveness, physical attractiveness, physique, social opportunities, relationships, and more.

In a previous study, we discovered that physical appearance was a significant factor of happiness for young adults, as they desire flawless images of themselves in social media [1]. The psychological driving forces of improving physical attractiveness are mainly two-fold – 1) it comes from one's eagerness to satisfy the needs of beauty and shape themselves as "the ideal one in mind", and 2) it comes from the pressure of the external, such as public opinion, employment and so on.

The aim of this present research is to address the relationship between social identity and physical attractiveness trends based on Eastern and Western cultural differences. Social media was used as a measure of social identity because it is a primary interaction tool for current young adults. The research displays significant differences between Eastern and Western cultures in terms of preferences for cosmetic interventions. Comparisons from two Instagram hashtags, #beautiful and #happiness, indicated significant bias of social media images that represent the preferences of cosmetic interventions by culture.

Keywords: Attractiveness · Happy · Happiness · Social identity · Self-worth · Self-esteem · Vanity validation · Identity crisis · Cosmetic medicine · Cosmetic surgery · Dermatology · Body image · Problematic social media use (PSMU) · Fear of missing out (FOMO) · Social media

1 Introduction

Social media provides a massive amount of daily interactions with photographs. More than 200 million users are using Google Photos every month and there are more than 1.6 billion animations and 24 billion selfies [2]. Frequently, people are taking photos

G. Di Bucchianico et al. (Eds.): AHFE 2020, AISC 1202, pp. 385–393, 2020.
https://doi.org/10.1007/978-3-030-51194-4_51

and sharing them with not only friends and family, but also on social media, creating countless contents to search for ourselves. However, there is a disconnection between real life and photos in social media. This gap increases significantly with the use of filters that are provided through social media platforms such as Snapchat and Instagram, as well as photograph editing apps which allow individuals to manipulate their photos with a couple of simple finger gestures.

Social media has become a window into how we perceive ourselves and has been influencing beauty trends. It creates the opportunity for any users to display the best version of themselves to the public and provides several societal benefits – dissemination of knowledge, discussion of new ideas, and community building. The desire to present oneself as flawless has led to the era of photo manipulation tools. There are a flood of applications allowing users to enhance the appearances of their body shapes and facial features. In fact, everyone wants to look perfect with a fast and magical fix instead of going to the gym or on a diet with long-term commitment because people want to look the best in photographs of oneself.

The purpose of this research is to contribute to the existing body of research on the uses of social media in relation to attitudes toward cosmetic surgery. The study aims are as follows; 1) to examine what types of photographs of oneself are posted most frequently in Eastern and Western culture on Instagram and how they are framed, 2) to analyze if each culture has different attitudes toward types of cosmetic surgery, and 3) to observe the relationship between the social media photograph of oneself and the attitudes toward types of cosmetic surgery.

Snapchat Dysmorphia. The cumulative availability of photo filters in apps have intense psychological effects [3]. The selfie age produced a concept called, "Snapchat Dysmorphia." It refers to the replication of filtered body images through cosmetic procedures without the filters and/or photo manipulation apps.

Snapchat, Instagram, and Snow filters enhance one's features flawlessly. It gives a desirable look from one's perspective that fulfills the beauty standards of society. It is an ongoing discussion whether the impacts are positive or negative on self-esteem and self-worth with use of these apps. The biggest problem is people are failing to remember what they actually look like. There was a time that we all knew celebrities and models in ads were edited [4]. Now, people are not only comparing themselves to these spotless celebrities and models images but also to the intentional or unintentional fake self in social media.

The pervasiveness filtered images on social media can decrease an individual's level of self-worth and self-esteem. This makes an individual feel inadequate for not looking a certain way in the real world, which may activate Body Dysmorphic Disorder (BDD) [5]. Life experience or personality traits are other sources that influence BDD risk; however, in this research we are primarily addressing social media photographs of oneself as a risk factor. In 2017, more than 55% of plastic surgeons had patients who wanted to improve their appearance toward the filtered selfies of themselves, which has grown 13% since 2014 [4]. Millennials and Gen Z are trying to achieve the perfect version of themselves in real life by seeking out cosmetic interventions as these images in social media become the norm. This causes insecurities of an individual's physical appearance from weight to aging.

There are many ways to measure self-worth and self-esteem in life. However, it is psychologically harmful to allow self-worth to be contingent upon appearance. This mental illness, categorized in the obsessive-compulsive spectrum, is extremely common and rapidly increasing as millennials and Gen Z are influenced by social medias through daily use.

Real Life vs Photos of Ourselves. The reality is no one is looking at an individual as closely in real life as with selfies. Camera angles determine the way an individual looks in selfies. It is not hard to find the instructions on how to take the best selfies from YouTube and/or Google. Social influencers are honest about how selfies look different depending on the angle of the camera. In fact, photos do not represent the correct depictions of how people look and easily dwell on little imperfections that do not appear in real life. For instance, a camera angle below eye level will show parts of one's face that normally are not seen when looking at eye level. In other words, people do not see others from the same angles as in photos.

Social media filters and photo editing apps are leading people to work on particular areas of individuals' faces, such as eyes, jawline, the middle part of lip, and more (See Fig. 1). In fact, people are looking for targeted cosmetic procedures just like these editing tools [6]. Looking at filtered photographs of oneself encourages impractical beauty standards and expectations of what can be attained. Expectations are getting higher because an individual can self-study their look through these apps and filters.

Fig. 1. BeautyPlus App – Photo filters for selfies.

There is no doubt that social media is setting trends for millennials and Gen Z in terms of the desired self-image of an individual. However, trends are continually changing. What we do and/or desire today will be different 10 years from now. The problem is that we do not know what the outcomes of cosmetic surgeries will be. It will take time to understand how particular surgeries will affect an individual because it is different than instant editing filters and/or apps with no side effects or permanent damage. Thus, the physical alteration should not be considered lightly by wanting to get more likes and receive invisible validations on social media from anonymous populations. Instead, it must be a long-term decision process. This study is interested in how different types of framed photos of oneself are associated with the attitudes toward types of cosmetic surgeries by different cultures.

Self-worth and Self-esteem. Social media can help people access information about a wide range of people and encourage boundless networking opportunities. However, social media has a significantly negative link to several things, such as anxiety, narcissism, lack of social skills, and higher levels of loneliness due to frequent use. This is because stories on social media are optimistic, confident, and celebratory. It seems everyone around an individual has a great lifestyle, is living a dream life, and has wonderful relationships. As a matter of fact, it only encourages an individual to compare oneself to others, who are displaying better versions of themselves on social media. Comparing oneself with these positive aspects has negative influences on self-worth, self-esteem, and relationships because people are constantly seeking validation through social media likes, not in real life [7].

Sharing the best version of oneself in social media has become a way of getting likes, shares, and validations from others, which creates the illusion of social engagement and popularity with disguising one's genuine personality. Higher activity of social media involvement discourages one's self-worth and self-esteem. In other words, viewing social media profiles with confident content has a negative effect on well-being because of an upward comparison on lifestyle and lively social network. In addition, receiving vicious social support while being exposed to continuously increasing social comparison can damage one's self-worth and self-esteem further.

Fear-Of-Missing-Out (FOMO). According to the 2017 Match Singles in America study, social media has influenced people's dating lives. 57% of singles admit social media has initiated their FOMO [8]. It refers to a pervasive apprehension on one's social anxiety caused by a desire to stay connected with what others are doing [9]. It is inevitable that people feel obligated to check for updates on social media once they interact with it for an extended period of time. It can develop to a compulsive concern that one might neglect various opportunities for social interaction, satisfying events, and a unique experience.

Throughout the 1970s and 1980s, scholars recognized a tendency to think about what others are thinking and doing and have understood an apprehension around missing out and comparing their experiences to that of others' [10]. It is a fundamental human motivation and need. People fulfill this need by pursuing social groups that provide the feeling of belonging. Social media provides a digital portal to access affiliation groups and enables the exercise of keeping in touch. It may cause an emotional reaction of being out of touch with real life due to the lack of activity of connecting with affiliation groups on social media. It is the same feeling where losing a friend with a high level of attachment can cause a negative emotion of grieving as if a part of oneself was damaged [11].

FOMO is a strong predictor of how people frequently use social media [12]. People who have a greater FOMO experience tend to use more diverse types of social media more frequently. People are routinely exposed to social media feeds of friends' daily lives and often develop nervous feelings, asking themselves, "Why am I not doing things that they are doing?" Paradoxically, the more an individual checking up on others' social media, the more one finds events that they are missing out on. It is a vicious cycle that strengthens itself through the use of social media as a means of diminishing the anxiety, which is also another source of FOMO.

In this study, we anticipate that the frequency of social media use is an active influence of one's current beauty standards and cumulative desire of cosmetic interventions. Our focus is on unveiling the factors that predict types of photographs of oneself from different cultures.

Social and Cultural as a Predictor of Cosmetic Surgery (Physical Attractiveness). The dissatisfaction of body image influences an individual's thoughts on one's body, which involves divergent feelings between one's real self and the filtered version [13]. The research found that teasing, media influence, and body dissatisfaction of physical appearance was the strongest predictor of interest in cosmetic surgery [14]. In contrast, people who were satisfied with their body image were not likely to desire changes to their body [15]. It proves that body dissatisfaction is a mediator between the perception of one's own body and considerations to change it. People with negative perceptions of their body will likely become displeased with their appearance, resulting in exploring methods of improving their physical appearance with cosmetic surgery [16]. In addition, teasing and bullying experiences motivate people to consider cosmetic surgery to fix their body dissatisfaction. Frequency of interacting with people who have cosmetic surgery experience increases the likelihood of people's desire of cosmetic procedure [17]. In other words, the more interactions one has with cosmetics, either positive or negative, the more likely one will have cosmetic intervention.

People have perceptions of beauty standards based on the society they live in because it is innate within the social and cultural context [18]. The media is a primary method in displaying the messages of beauty ideals, which affect perceptions of individuals. Much of the historic research on attitudes towards cosmetic procedure were focused on being influential in determining behavior, not indicating behavior itself. Therefore, this study aims to examine the effect types of photographs in Instagram have on the desire for different types of cosmetic intervention in Western and Eastern culture. The hypothesis is that viewing images (either facial, upper body, or entire body) of someone on Instagram will increase a desire for certain types of cosmetic intervention.

2 Method

2.1 Procedure

We examined a quantitative observational cross-sectional study. We convenience sampling. The study population consisted of female users in Instagram from two different locations – Japan and the United States. The inclusion criteria for the participants were as follows: females from both countries between 18 and 35 years old. A sample size of 400 was calculated using a 95% CI and 0.05 degrees of freedom. The data was collected through Instagram during the day. The data was extracted for statistical analysis in SPSS.

The Instagram users were fully informed about the goals and purposes of the study and have participated voluntarily. They were able to refuse to share their images for this study at any given time. Informed consent was required at the beginning process of gathering images from the individual Instagram users.

We gathered the 400 most recent Instagram posts of images of individuals between January 20, 2020 and compared with the 2018 International Society of Aesthetic Plastic Surgery Report. The 2019 ISAPS 'International Survey on Aesthetic/Cosmetic Surgery' statistics were examined by country, taking into account the underlying international base population according to the official United Nations population estimates. The number of surgeons was used to determine the number of procedures performed per surgeon.

The Instagram posts were coded for facial, upper body, and entire body of the photograph of oneself which were tagged with the hashtag #Happy and/or #Happiness. For the reliability analysis, we randomly selected 10% of the material. The analysis generally indicated reliable measurement. Krippendorff's α values were acceptable all (all values, $\alpha > .76$).

2.2 Variables

Instagram Image Types. The selfie types were coded as present when the Instagram post included an explicit reference of either face, upper body, or entire body. Images with hashtags of #happy and #happiness are considered only as a presentation of one's desire and satisfaction.

Emotions. The following emotional expressions were coded when they were visually or verbally present: Happiness/Happy (e.g., smiling faces, proud/confident faces and gesture).

Demographics. A total of 400 most recent Instagram image posts from both groups, 200 United States females and 200 Japan females, were downloaded. The age of the individual was coded. We coded whether the post included face, upper body, or full body of oneself.

Data Analysis. We calculated descriptive statistics with percentage values and 95% confidence intervals based on bootstrapping techniques. The percentage values are significantly different when the confidence interval does not overlap. The confidence interval lower and upper limits are reported after the percentage value.

3 Results

This research findings indicate that the use of social media may be correlated with amplified acceptance of particular types of cosmetic surgery. These findings can assist three areas as follows; 1) the future social psychologists discussion regarding awareness of self-worth through social media, 2) future patient-physician dialogue regarding cosmetic surgery perceptions, which vary by social media or photo editing application use, and 3) the attitude towards the types of cosmetic procedure. Overall social media is a significant driver of cosmetic interventions. It is important to recognize the latent societal benefits of social media, such as the dissemination of knowledge, discussion of new ideas, and community building, however, it is also important to recognize the human nature to engage in comparison, which encourages negative influences on self-worth by comparing oneself to others (e.g. superior).

The highest demand countries for surgical cosmetic procedures are the United States and Brazil followed by Japan. There was an explicit reference to the preferences of physical appearance where people want to improve, which are significantly different by culture. Within the search of hashtags, such as #happy and #happiness, physical appearance was a strong predictor of one's happiness.

In the United States, the distribution among types of Instagram images (e.g. Face/Head, Upper Body, and Entire Body) are almost equally desired to post. The types of cosmetic procedures are also desired evenly on their body. Americans considered the breast and overall body image more important than face area alone. Their beauty interest is in the overall frame rather than a particular area of a body even within face/head area (see Table 1).

Table 1. Instagram image post and surgical cosmetic procedure comparison.

Instagram images			Cosmetic procedure (Surgical only)				
	USA	Japan		USA	USA surgeons	Japan	Japan surgeons
Face/Head	48	183	Face/Head Eyelid Facelift Fat Grafting Rhinoplasty	366,781 (total) 119,784 73,805 49,413 48,222		207,104 (total) 141,596 27,210 7,326 28,644	
Upper Body	79	11	Breast	602,003		6,168	
Entire Body	73	10	Entire Body	523,599		11,021	
Total	200	200	Total	1,492,383	7,009	224,293	2,663

In contrast, Japanese strongly desired to post face and head photography of themselves and nearly did not care to display their upper body nor entire body in the Instagram post. The same trend was observed in the surgical cosmetic procedure preferences where Japanese have drastically high aspiration to have facial/head area of cosmetic procedures than other areas. In particular, they preferred eyelid surgery the most, which demonstrates their specific face area of interest instead of the whole face and/or its structure. In fact, Japanese preferred to have the facial cosmetic procedure 18 times more likely desire than having a cosmetic procedure on their other body areas. The breast was the least interested area to improve their body image. They were 19 times more likely to post facial specific images than images including their upper or entire body.

The physical attractiveness was the dominant image types in Instagram under the hashtags of #happy and #happiness as proved from our previous study. The appearance of physically attractive body images on Instagram is a strong predictor to young adult themselves being happy on social media.

4 Discussion

Procedures can be predicted by making observations of the trends of social media and taking the particular demographic and geographic factors into consideration. It is important to understand social media has taken a vast role of communication and trends of the current generation (e.g. millennials and Gen Z). It is a direct route to reach them without any filters or disruptions. In fact, social media has become the window to express and share their desires. Anybody can access this shared information anywhere, which means it became a social norm and standard.

The findings display the distinctive differences in preferred body images between Western and Eastern culture despite sharing the same social media platform. The perception of attractiveness keeps changing over time; for example, the large rounded eyes with double eyelids with voluminous lower eyelids are preferred in Eastern culture instead of single-folded eyelids with almond-shaped eyes in Asia. Physical attractiveness conveys positive advantages cross culturally. However, we should remember that everyone has a moderate level of physical attractiveness and one's attraction nurtures and strengthens naturally. There is no doubt that being physically attractive is highly desirable, which can be obtained naturally [19]. However, it weakens in self-worth and devalues other important aspects (e.g. relationship, achievement, experience, etc.) of one's quality life.

This research suggests that beauty standards and preferences are highly correlated with cultural influence, which is demonstrated both in social media and attitude towards types of cosmetic procedures. We hope that millennials and Gen Z recognize that desired artificial interventions are influenced by social media and find the value of other qualities of life beyond the flawless image of themselves.

5 Limitations

The results of the study provide support in understanding the psychology and influence of social media on attractiveness and cosmetic procedures. This study is specifically meaningful to understand body image concerns that are portrayed through social media and the preferences in types of cosmetic procedures from two different cultures. It is possible that an individual will likely have negative satisfactions from cosmetic treatments if one's motivations were purely from social media influence [15]. It is important to deliver an awareness of social media image influences on one's behaviors and body image in order to increase psychosocial performance. It is necessary to implement media literacy in the educational system for young adults to reduce the negative perception of their body image and to diminish cosmetic procedure opportunities to improve their dissatisfactory body image.

References

1. Kim, Y.A.: Emotional evidence: influences on happiness from the frequent positive visual exposure. In: International Conference on Applied Human Factors and Ergonomics, pp. 227–237. Springer, Cham (2018)
2. Sabharwal, A.: Google Photos: one year, 200 million users, and a whole lot of selfies. Google Blog, 27 (2016)
3. McLean, S.A., Paxton, S.J., Wertheim, E.H., Masters, J.: Photoshopping the selfie: self photo editing and photo investment are associated with body dissatisfaction in adolescent girls. Int. J. Eat. Disord. **48**(8), 1132–1140 (2015)
4. Pleasures, A.: Beauty Sick How The Cultural Obsession with Appearance Hurts Girls and Women
5. Phillips, K.A., Pinto, A., Jain, S.: Self-esteem in body dysmorphic disorder. Body Image **1**(4), 385–390 (2004)
6. Economides, J.M., Fan, K.L., Pittman, T.A.: An analysis of plastic surgeons' social media use and perceptions. Aesthetic Surg. J. **39**(7), 794–802 (2019)
7. Silva, C.: Social media's impact on self-esteem. The Huffington Post, 22 February 2017
8. The Single in America (2017). https://www.match.com/dnws/cpx/en-us/singlesinamerica/2017/#technicallydating
9. Przybylski, A.K., Murayama, K., DeHaan, C.R., Gladwell, V.: Motivational, emotional, and behavioral correlates of fear of missing out. Comput. Hum. Behav. **29**(4), 1841–1848 (2013)
10. Simon, J.: Love: addiction or road to self-realization, a second look. Am. J. Psychoanal. **42**(3), 253–263 (1982)
11. Belk, R.W.: Possessions and the extended self. J. Consum. Res. **15**(2), 139–168 (1988)
12. Franchina, V., Vanden Abeele, M., Van Rooij, A.J., Lo Coco, G., De Marez, L.: Fear of missing out as a predictor of problematic social media use and phubbing behavior among Flemish adolescents. Int. J. Environ. Res. Public Health **15**(10), 2319 (2018)
13. Grogan, S.: Body Image: Understanding Body Dissatisfaction in Men, Women and Children. Routledge, London (2016)
14. Markey, C.N., Markey, P.M.: Correlates of young women's interest in obtaining cosmetic surgery. Sex Roles **61**(3–4), 158–166 (2009)
15. Lee, H.H., Damhorst, M.L., Paff Ogle, J.: Body satisfaction and attitude theory: linkages with normative compliance and behaviors undertaken to change the body. Fam. Consum. Sci. Res. J. **37**(4), 466–488 (2009)
16. Markey, C.N., Markey, P.M.: Relations between body image and dieting behaviors: an examination of gender differences. Sex Roles **53**(7–8), 519–530 (2005)
17. Brown, A., Furnham, A., Glanville, L., Swami, V.: Factors that affect the likelihood of undergoing cosmetic surgery. Aesthetic Surg. J. **27**(5), 501–508 (2007)
18. Thompson, J.K., Heinberg, L.J., Altabe, M., Tantleff-Dunn, S.: Exacting Beauty: Theory, Assessment, and Treatment of Body Image Disturbance. American Psychological Association, Washington, DC (1999)
19. Li, N.P., Valentine, K.A., Patel, L.: Mate preferences in the US and Singapore: A cross-cultural test of the mate preference priority model. Personality Individ. Differ. **50**(2), 291–294 (2011)

Happiness Participatory Media: Cultural Differences in Happiness on Instagram

Qiuwen Li[1(✉)] and Young Ae Kim[2]

[1] Department of Art and Art History, Santa Clara University, 500 El Camino Real, Santa Clara, CA 95053, USA
Qli2@scu.edu

[2] School of Design, South Dakota State University, 117 Grove Hall, Brookings, SD 57007, USA
YoungAe.Kim@sdstate.edu

Abstract. The Happiness Participatory Media analyzes Instagram data for visual cues with emotional expression (happiness), hashtags (words), gender, age, and locations, in order to get a clearer picture of how people portray their happiness from different cultures. This research looked at the actual patterns of the dataset to see how it is different and/or similar to the generalized idea of happiness today.

In our previous research, we were curious about how to approach theoretically social media images in general and to understand one's emotional expression through social media, in particular, Instagram. We incorporated an ensemble of computational methods from machine learning, image processing, and information design to extract useful psychological indicators from photographic data. Hence, we discovered that the daily communication method has been changed drastically, which heavily influenced individuals' perceptions of what they considered to be happy. In fact, physical appearance was by far a more important factor for young adults' happiness than relationships, achievement, or travel/experience (Li & Kim 2019).

Happiness Participatory Media centralized disparate archives of happiness expressions through Instagram from 3 countries in the international locations Seoul, Korea (four seasons), Naha Japan (one season), and New York (four seasons) and Los Angeles (one season) in the United States using a mix of theoretical, artistic, and quantitative methods. This rich media visualization will be assembled with thousands of photos to reveal interesting patterns and allow participants to navigate the whole set of 4,000 photos (1000 images of each location) with hashtags of #beautiful, #relationship, #achievement, and #travel. In addition, we will address social media as not only a function of emotional expression through images of oneself, but also opportunities to display a perfect version of oneself with a desired lifestyle. The project brought to light the development of participatory media and the diverse social & cultural contexts.

Keywords: Happiness · Instagram · Cultures · Participatory media · Visualization · Lifestyle · Emotion · Social media · Social network · Social comparison · Self-concept · Emotional responding · Body image · Vanity validation

G. Di Bucchianico et al. (Eds.): AHFE 2020, AISC 1202, pp. 394–402, 2020.
https://doi.org/10.1007/978-3-030-51194-4_52

1 Introduction

Social media is part of people's daily lives and is changing how we communicate and connect with others around us. According to Instagram's marketing report in 2020, more than 1 billion Instagram accounts worldwide are active every single month and more than 500 million Instagram accounts worldwide are active every single day [1]. For many Instagram users, they check their own and others' feeds and stories multiple times a day. They see others posting smiling selfies, travel photos, and personal/professional accomplishments. The eMarketer predicts that users will spend an average of 28 min per day on Instagram in 2020 [2].

The use of social media platforms, such as Instagram, gives users easy access to more information and affects how they approach their lives. In previous research, we discovered the preference for communication types through smart devices and what are the direct influences on sensing the feelings of others [3]. In addition, we found that the daily communication method has been changed drastically, which influenced heavily on individuals' perceptions of what they considered to be happy. In fact, physical appearance was by far an important factor of young adults' happiness than relationships, achievement, or travel/experience [3]. We have been focused on emotional visual signals with described words and how they influence individuals' perceptions, value of life, presentation of themselves, and communication method changes.

The Happiness Participatory Media analyzes Instagram data onto visual cues for emotional expression (happiness), hashtags (words), gender, age, and locations, in order to understand visual methods of portray one's happiness from different cultures and climate environment. We analyze the patterns of the dataset to create theoretical essays to deliberate visual methods (types of photographs of individuals) how it is different or similar to the generalized idea of happiness today within. The photo filters from social media platforms and photo editing apps were not concerns in this study in terms of how well flawlessly display themselves.

Positive Reinforcement. All actions we think and say and do have consequences that influence ourselves and others. In behavioral psychology, positive reinforcement occurs when a stimulus or reward is presented as a consequence of a behavior and increases the likelihood of the behavior recurring [4]. One example of positive reinforcement of everyday life, a mother gives her son ice cream (reinforcing stimulus) for doing homework (good behavior). Virtual compliments (e.g., views, likes, and comments) on social media are different types of positive reinforcement. Research demonstrates that people feel less lonely and happier when they receive messages from close friends on social media [5]. Each view, "like", and comment, increases people's feeling of self-worth and becomes dependent on positive reinforcement.

Social Comparison Theory. First proposed by social psychologist Leon Festinger in 1954. The theory explains how individuals evaluate their own opinion and abilities by comparing them with others in order to reduce uncertainty and learn how to define themselves [6].

Comparing and contrasting one's life with others often happens to adolescents and young adults and is part of their identity formation [7]. The social comparison between oneself and others on social media can cause people to feel bad about themselves and magnifies the negative impact. Social media envy happens when they compare themselves to celebrities or others who have achieved unusually high levels of social success; causing negative psychological and behavioral outcomes.

Happiness in Social Network. The Framingham Heart Study, a multigenerational study, was initiated in 1948 with 4739 participants on happiness and found that the happiness of others with whom a person is connected will affect their happiness. Furthermore, an individual's happiness is associated with happiness of people up to three degrees removed from them in the social distance. Social distance refers to the closest social distance between alter and ego. The study shows that a person is 15% more likely to be happy is their friends or family members (social distance 1) are happy (instead of unhappy). At social distances of 3° of separation, a person is still being happier, but, the percentage increase is only 6% [8] (see Fig. 1). In other words, a friend's friend, even if someone never met, can have a huge impact on a person's happiness. Figure 2 shows how the relative importance of distance affects happiness. According to the statistics, friends, spouses, siblings, and neighbors can significantly influence an individual's happiness, but only if they live close to the person [8].

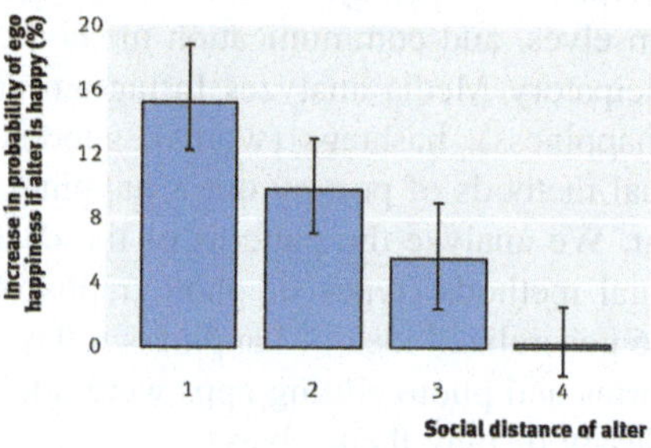

Fig. 1. Social distance and happiness in the Framingham social network.

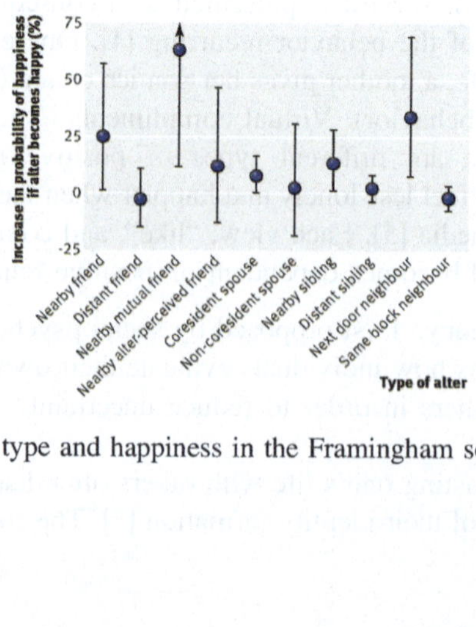

Fig. 2. Alter type and happiness in the Framingham social network.

Happiness and Different Types of Happiness. People view things differently in life, including happiness. According to Alan Goldman, an American philosopher and Professor of Philosophy, happiness is an emotion and contains all features of multi-component states about emotion. Different emotional states include judgments, feelings (pleasant sensations), physical symptoms (smiling, etc.), and behavioral dispositions (energy, openness) [9]. The study also classified happiness into four types: sensation; attitude; and feeling. The first type is happiness as a sensation, which has both a mental and/or physical cause (e.g., sexual pleasure or the pleasure of eating). The second type is happiness as a kind of intentional attitude, which takes pleasure in numerous objects and activities and can be felt in various ways. The third is also called "pure feeling pleasure", which is the feeling of achievement or unexpected change for the better [10].

Cultural Differences in Self-concept. Akey element in understanding cultural influences in self-concepts is distinguishing the difference between interdependence and independence [11]. Western culture emphasizes personal achievement more than group consensus, because Westerners tend to regard themselves as independent individuals. In Western culture, people tend to focus on emotional expression. Moreover, high arousal emotions (e.g., joy and anger) are valued and promoted more than low arousal emotions (e.g., sadness and reflection) [12].

Eastern culture, in contrast to the independent cultural background, tends to think about themselves in terms of interconnectedness with group members (" interdependent "). Asians tend to focus on the external factors or surroundings of the individual, with a special emphasis on the concept of humans as social beings. In the context of interdependent Asian cultures, what the group focuses on is relatively more important than what the individual focus. Furthermore, individual emotional control is highly valued in Asian culture, especially those related to negative emotions and social separation emotions [13].

Gender Shape Emotion Response. In addition to cultural differences, gender also influences how people feel about emotions, including the intensity and regulatory strategies. Data collection indicated that both American and Japanese men reported a higher frequency than did women for disengaged positive emotions, but, the pattern was reversed for engaged positive emotions [14] (see Fig. 3).

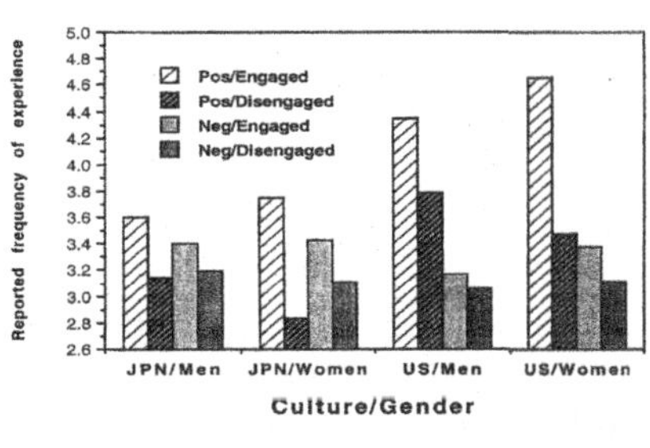

Fig. 3. Reported frequency of experiencing four types of emotions for Japanese and American men and women.

Photo Taking vs. Sharing on Social Media. Information sharing with others is a foundational desire for people. Photo-taking is one method of sharing an individual's personal experience with others, which encourages the engagement of positive (enjoyable) experiences. Taking photos can inspire positive (enjoyable) and negative (unpleasant) experiences [15], however, it decreases one's positive experience if an individual is taking pictures with the intention to share on social media, due to the rise of anxiety or self-presentational concern [16].

2 Method

2.1 The Project Team

The project presentation has three major contents; 1) observation of the demographics of people taking pictures of themselves with the best representation of their interests and perception; 2) the visualizations of revealing patterns; 3) an interactive application in both 2D and 3D platforms that allows interactions and exploration of 4000 images associated with four hashtags (#beautiful, #relationship, #achievement, and #travel) as well as discovering new patterns.

The Team is multi-disciplinary with a wide range of backgrounds, including human factor psychologists, visual designers, programmers, data scientists from China, Korea, and the U.S. The project was coordinated by Young Ae Kim and Qiuwen Li who were also responsible for creative direction and visualizations.

2.2 Data Collection

The first step was to create a dataset of images of four hashtags from each location, which required many steps. Each Instagram post has many hashtags in addition to the four research hashtags and often they overlap because Instagram allows thirty hashtags per post. A closer examination reveals that there are two types of images, images taken by someone else and images taken by oneself (selfies). For this project, it was not a concern whether it is taken by oneself or not. We focused on areas of four cities with different climate environments, including Seoul, South Korea (four seasons), Naha, Japan (one season), New York City (four seasons) and Los Angeles, California (one season), because we wanted to collect images and data with the same conditions in order to improve reliability and validity of the data. The project team inspected all photos collected to make sure images were tagged accurately in order to keep the data visually comparable.

2.3 Computer Analysis

The data set of 4000 images will be analyzed by the different types of images, including location, climate differences, face size and orientation, how one's body framed within the picture (e.g. full body, upper body, or face specific) emotional expression, presence of smile, eyes are closed or open and more. We will be comparing the measured images of characteristics between cities, genders, and ages. These measurements will be implemented in both VR and AR applications (see Fig. 5 and Fig. 6) to allow users t filter the image database by any combination of hashtags or an individual hashtag as well as any types of images of characteristics (see Fig. 4).

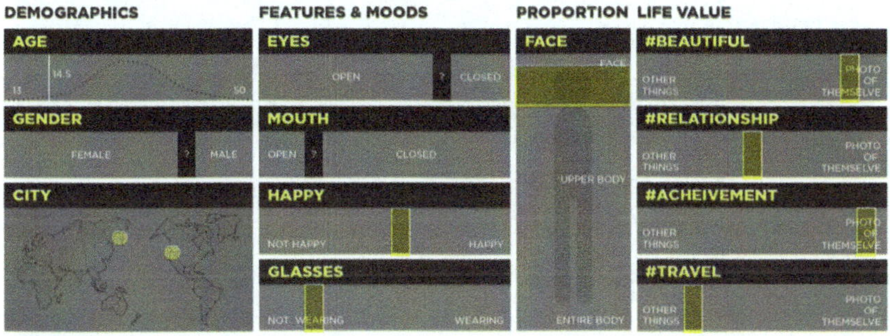

Fig. 4. VR User Controller Interface.

3 Outputs

3.1 The 3D Visualization Environment with Virtual Reality (VR)

VR technology provides an engaging learning experience for users. It will allow users to access a big data set in an interactive manner. Happiness Participatory Media VR allows for self-guided exploration of the dataset. As a user enters the VR space, they will be exposed to space with hashtags on Instagram that represent individuals' happiness, such as #beautiful, #relationship, #achievement and #travel. A user can get closer to each hashtag and choose a specific hashtag to access further information that is involved with the chosen hashtag. Users can select any hashtags floating around them, and as they choose a particular hashtag, a group of images associated with a chosen hashtag from 4,000 images and mean scores of datasets will be displayed. Figure 5 displays the VR environment and Fig. 6 demonstrates the interface after a user chose a specific keyword. The VR experience will provide a multi-dimensional environment with the most memorable data visualization experience to explore the difference and similarity of how people portray their happiness from different cultures.

3.2 The 2D Visualization Platform with Augmented Reality (AR) Experience

AR experiences expose users to learning and exploring environments through what they find interesting. Creativity and curiosity will be fostered for both student and general public users. Happiness Participatory Media AR is the perfect way to get students' imagination flying. Participants will have interaction with 30 exhibited images of the most representable image from the 4,000 dataset through Artivive app, which provides moving images of each image that sparks users' creativity. Figure 7 is a demonstration of how it works.

Fig. 5. The Virtual Reality Environment User Interface

Fig. 6. User Interface of each Instagram hashtag.

Fig. 7. Artivive App AR User Interface.

4 Limitations and Future Direction

In previous research, we discovered the preference of communication types through smart devices and what are the direct influences on sensing the feeling of others (Li & Kim 2018). In addition, we found that the daily communication method has been changed drastically; which influenced heavily on individuals' perceptions of what they considered to be happy. In fact, physical appearance was by far a more important factor for young adults' happiness than relationship, achievement, or travel/experience (Li & Kim 2019). We have been focused on emotional visual signals and words and how they influence individuals' perceptions, value shift, and communication method changes.

The Happiness Participatory Media project will be a window to reach a broad public (non-specialist audience), allowing them to filter the image database by any combination of hashtags, individual hashtags, or any type of images of characteristics. It will also allow participants to understand innate complications associated with certain hashtags. By examining a large sample in diverse geographical locations at the same time period, we can avoid focusing on an individual's agendas. Instead, we can observe patterns at large and provide a third-person point of view to see how social media describes meanings and values of life. We need to keep in mind that this study is limited to users of smartphones in active social media platforms; Instagram, in particular. The statistics we used are within four chosen cities among ages between 13–50 years old.

All photo data of this research has constraints and affordance due to the standardized appearance, format and resolution of posted images (612 × 612 pixels) on Instagram, which are edited by its filters and photo editing apps. The control experience has advantages for researchers, because it allows researchers to examine a variety of scopes and ranges of patterns and fast locational mapping. The image format of Instagram also helps to create a standard foundation for comparing the multitude of photos.

References

1. Facebook for Business. https://www.facebook.com/business/marketing/instagram
2. Instagram Stats That Matter to Marketers in 2020. https://blog.hootsuite.com/instagram-statistics/
3. Li, Q., Kim, Y.A.: Happiness on Instagram–content analysis and engagement based on attention theory. In: International Conference on Applied Human Factors and Ergonomics, pp. 13–20. Springer, Cham, July 2019
4. Flora, S.R.: Power of Reinforcement (SUNY Series, Alternatives in Psychology). State University of New York Press, Albany (2004)
5. Burke, M., Kraut, R., Marlow, C.: Social capital on Facebook: differentiating uses and users. In: Proceedings of the SIGCHI Conference on Human Factors in Computing Systems, pp. 571–580, May 2011
6. Festinger, L.: A theory of social comparison processes. Hum. Relat. **7**(2), 117–140 (1954)
7. Social Media Comparison & Teen Mental Health. https://www.newportacademy.com/resources/empowering-teens/theory-of-social-comparison/
8. Fowler, J.H., Christakis, N.A.: Dynamic spread of happiness in a large social network: longitudinal analysis over 20 years in the Framingham heart study. BMJ **337**, a2338 (2008)
9. Goldman, A.H.: Happiness is an emotion. J. Ethics **21**(1), 1–16 (2017)
10. Goldman, A.H.: Life's Values: Pleasure, Happiness, Well-Being, and Meaning. Oxford University Press, Oxford (2019)
11. Mesquita, B., Karasawa, M.: Different emotional lives. Cogn. Emot. **16**(1), 127–141 (2002)
12. Lim, N.: Cultural differences in emotion: differences in emotional arousal level between the East and the West. Integr. Med. Res. **5**(2), 105–109 (2016)
13. Li, Q., Li, Z.: Influence of culture diversities in emotional control and emotional reaction. J. Jiyuan Vocat. Tech. Coll. China (2019)
14. Kitayama, S., Markus, H.R., Kurokawa, M.: Culture, emotion, and well-being: good feelings in Japan and the United States. Cogn. Emot. **14**(1), 93–124 (2000)

15. Diehl, K., Zauberman, G., Barasch, A.: How taking photos increases enjoyment of experiences. J. Pers. Soc. Psychol. **111**(2), 119 (2016)
16. Barasch, A., Zauberman, G., Diehl, K.: How the intention to share can undermine enjoyment: photo-taking goals and evaluation of experiences. J. Consum. Res. **44**(6), 1220–1237 (2018)

Collaboration to Improve Experience in Hospital Environments

Kayla Daigle(✉) and Lois Frankel

School of Industrial Design, Carleton University, Ottawa, Canada
{KaylaDaigle, LoisFrankel}@carleton.ca

Abstract. Hospital environments are continuously responding to external pressures and societal expectations for healthcare. This results in increased demands for practitioners and hospital staff, which greatly impact the quality of care for patients. The need for interdisciplinary collaboration is emerging within the public sector in response to these changes. This literature review aims to identify the ways that patients and practitioners could or have been engaged with co-design methods, to improve the user experience within stressful environments, specifically in hospitals. In order to frame the current pain points within a hospital, this literature review also identifies current issues within the environment, and previously explored solutions relative to patient and practitioner experience. Furthermore, the different disciplines that can likely be involved in improving the experience within this context through interdisciplinary collaboration will be addressed, specifically, ties to architecture, sensory design, psychology and cognitive science, and social sciences.

Keywords: Interdisciplinary · Co-design · Collaboration · Healthcare · User experience · Environmental design · Sustainable

1 Introduction

The current design of hospital environments has been linked to fatigue, anxiety, and other psychological stressors for both patients and their practitioners which impact the overall comfort, experience, and usability of these spaces [1]. This suggests that there could be an opportunity to engage patients and practitioners in an interdisciplinary design research approach to improving the design of hospital environments. This literature explores the ways that patients and practitioners could or have been engaged with co-design methods, to improve user experiences within stressful environments, specifically in hospitals. More public sector institutions are expressing their interest in the incorporation of design theory and practice, such as user experience processes and measures in their research and development [2, 3]. This provides an opportunity for designers, key stakeholders and end-users to work collaboratively to identify and improve upon current barriers within hospital environments. There are six main aims of providing quality healthcare; (1) safety, (2) timeliness, (3) effectiveness, (4) efficiency, (5) equitability, (6) patient-centeredness [4]. These aims focus on the current experience and interactions of patients, practitioners, and hospital staff, otherwise described as the "core" population within the emergency department [5].

G. Di Bucchianico et al. (Eds.): AHFE 2020, AISC 1202, pp. 403–409, 2020.
https://doi.org/10.1007/978-3-030-51194-4_53

Due to the nature of emergency departments, the physical and psychosocial environment is a dynamic, high-traffic and chaotic space, which leads to error and potential for accidents [5, 6]. The complexity of these spaces requires collaborative processes that can address and adapt to changes within the environment. Often, collaborators within healthcare environments include architects, government agencies, hospital administration, and environmental psychologists [7, 8]. It is crucial for design researchers to involve end-users and key stakeholders throughout the entire process, taking into account their current pain points, challenges, and wishes for an improved hospital experience. These collaborative methods have the potential to give a voice to people with lived experiences and to provide real-world solutions that could improve their interaction and experience within this context.

In order to frame the current pain points within a hospital, this literature review identifies current issues within the environment, and previously explored solutions relative to patient and practitioner experience. Drawing upon previously explored co-design methods may provide researchers with a baseline of activities that have been successful in similar contexts, along with a framework to follow when starting a new project. By identifying the structure and context of healthcare environments, design researchers may establish collaborative activities and co-design methodology that best fits the task at hand.

2 Background

The dynamic and continuous nature of hospital environments has been proven to affect patients and practitioners in many ways. Burnout and stress are key contributors to the experiences of patients and practitioners within this context [1, 9]. Studies have shown that over 40% of nurses reported high levels of burnout and fatigue due to their workplace environment [1, 10], while other studies have reported that almost 46% of physicians in the United States feel the effects of burnout due to work activities [9]. Meriam-Webster dictionary defines burnout as the "exhaustion of physical or emotional strength or motivation usually as the result of prolonged stress or frustration" [11]. This burnout affect has been directly linked to lower patient satisfaction levels, reduced health outcomes, and increased public sector costs [1, 9].

According to a national survey conducted in 2011, 63 percent of physicians indicated rising levels of stress in their workplace [9]. Healthcare researchers have defined stress as "a cluster of physiological and psychological responses to situations which threaten well-being [12, 13]. With increases to external pressures (policies, regulations, finances) and changes to social influences, healthcare systems have faced challenges to their quality of care for patients and practitioners [14]. As patients and hospital staff continue to be affected by changes within the public sector, so too does their experience with the environment change to adapt to increasing pressure. As Bodenheimer (2014) stated, "physician dissatisfaction [is seen] as an early warning sign of a healthcare system creating barriers to high-quality practice" [9]. It is only when all parts within the healthcare system work in cohesion, that patient and practitioner experience and interaction can be seen as good.

2.1 Hospital Environments

Hospital environments are often the source of conflicts in patient and practitioner experiences. The physical, inanimate environment has been proven to impact health-care providers, patients, and visitors [8, 15]. It may be valuable for research to consider why and how the environment shapes experience and interaction within the space.

The physical space within a hospital is designed mainly for utility and basic ergonomic considerations for provider workflow and patient care. Studies have shown that worn-out physical structures and architecture can influence user perceptions, and through incremental change, can improve patient health and visitor experience [3, 16]. Further research has demonstrated the influence that natural materials and exposure have on patient care quality safety, and recovery from stress, suggesting that urbanized designs hinder these effects [12, 17].

Factors such as noise and wayfinding also contribute to the perceived comfort and ease patients and their providers experience. Noise interference (pollution) has been shown to interfere with provider workflow and patient comfort, privacy, and recovery, resulting in higher levels of dissatisfaction among these populations [1, 18]. Aside from noise, wayfinding is a key factor in the burnout and stress effects experienced by patients, visitors, and hospital staff. These effects have been linked to impacts in health and recovery, as well as delays in patients obtaining medical treatment [19]. The environment could be constructed in a way as to reduce these negative effects.

2.2 Patient and Practitioner Experience

Hospital environments are dynamic spaces which foster interaction and communication between patients, visitors, practitioners and other staff. User experiences within this context should be assessed. Stress and burnout effects have been associated with reduced patient and practitioner satisfaction, safety, and quality of care [1]. These effects are often the result of emotional exhaustion, depression/anxiety, lack of personal accomplishment, cognitive weariness, overwork, change in attitude, and lack of effectiveness [1, 7, 18].

The perceived comfort and experience for patients and practitioners are the result of the physical environment and its influence on a users' interaction. The literature suggests that design and architectural choices within these spaces could mitigate some of the risks of burnout and stress. Physical structures, spatial arrangement, and materials all seem to influence the quality of care provided to patients [18]. Design intervention has the potential to provide for improved user experience and interaction in hospital settings.

2.3 Defining Good Experience

When identifying what contributes to patient and practitioner experience, it is important to consider what a good psychological experience is. Research demonstrates that there are many factors which affect these experiences, however, there seem to be few studies which define what good patient and/or practitioner experience represents. Common themes emerge in what patients expect and what practitioners require. Patients desire communication and relationships with quality-care providers while being at ease within their space, while providers often want an environment that provides a calm space that facilitates ease of workflow [1, 16]. A positive environment supports the psychological and social well-being of its users, while providing structure and adaptability to changes within the hospital context [16, 20]. The environment should offer a sense of security, safety, and peace for all people involved.

Factors such as age, gender, physical and mental health status, cognitive status, culture, education, and residential history all influence the perceived comfort and satisfaction of people within a given environment [12]. Because of this, interdisciplinary collaboration may be used to identify solutions to existing environmental challenges that exist within hospitals.

3 Collaboration in Healthcare

Collaboration such as co-design processes in healthcare environments may be key in identifying real conflicts within this context. Co-design can be defined as "the creativity of designers and people not trained in design working together in the design development process", whereas a collaborative process is "an approach that engages with stakeholders and end-users to meet their needs" [14]. Collaborative methods allow hospital staff, patients, and practitioners to reflect on their lived experiences in identifying areas to improve and implement change [2]. As Bodenheimer and Sinsky (2014) suggest, "Health care is a relationship between those who provide care and those who seek care, a relationship that can only thrive if it is symbiotic, benefiting both parties" [9]. This symbiotic relationship may serve as a means of communication between different stakeholders and end-users, with the goal of developing better processes and quality of care [3].

Many collaborative approaches have proven to be successful with the design of healthcare environments and are recognized by a variety of terms; participatory design, service design, experience design, evidence-based design [3, 7, 14]. Each of these methods carries its own merit and role in interdisciplinary collaboration in hospitals.

3.1 Interdisciplinary Collaboration

Literature provides evidence that interdisciplinary practices within hospital settings are emerging. A common term used within healthcare research is interprofessional collaboration, which focuses "on the mostly face-to-face interaction between mainly physicians and nurses that occurs in the conduct of their work" [6, 8]. This definition seems to suggest that design researchers can align with other key specialists as a way to develop sustainable solutions within healthcare environments. These key collaborators often include architects, construction managers, interior designers, and other healthcare personnel [8]. The use of a methodology that incorporates many subject matter experts and specialists, both external and internal to the context, advocates for a more informed evaluation of the built environment [4, 20]. Those involved in such a collaborative process should ask questions along the way to direct their focus and gain insight into key areas. These questions, as suggested by a previous study, are concerned with the task being accomplished, how to recognize improvement, and what change can create this improvement [21].

In the literature, it seems that within these collaborative approaches, communication is key; the use of understandable jargon among professionals is essential so that open communication can occur [20]. Through collaboration and co-design activities, patients and practitioners are given a voice through which they can address their concerns and direct change within this setting.

Successful collaborative processes have allowed healthcare professionals to create lasting change within their workplace environment, while improving patient care as a result. Jones (2013) identifies *simplexity* as a harness to "generate a series of responses to successive levels of problem inquiry" through group ideation and methodology [3]. This process of strategic planning in conjunction with collaboration is ideal for ideation, exploration, collaborative innovation, and project planning [3].

3.2 Challenges in Collaboration

Though there have been many collaboration successes within hospitals, there are also a number of challenges to consider. A major consideration is the diffusion of power within a group. Due to the nature of practitioner, physician, and nurse workflow, it is often difficult for members of this group to move between an 'expert' role into the role of a group member of equal power, which is important in collaborative processes [2]. It is important for facilitators to attend to group dynamics, while providing communication strategies to encourage a shared understanding [8].

4 Discussion

The findings of this literature review provide insights into collaborative processes and frameworks that could be applied within healthcare environments. By looking at previous case studies and methodologies, researchers may benefit from a framework that best identifies current barriers within this context, as shown in Fig. 1, while giving power to end-users and stakeholders to implement change. Open-communication and dynamic power structures have been shown to encourage full participation and creation of ideas. This collaboration leads to improved systems within healthcare environments.

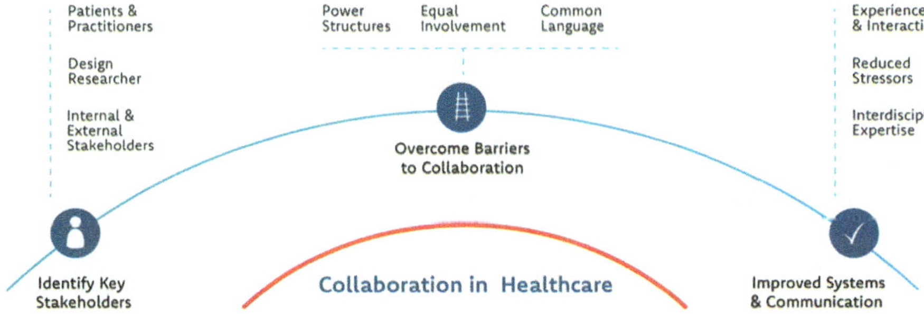

Fig. 1. Design framework for collaboration in healthcare environments

5 Conclusion

This review focused on identifying the ways that interdisciplinary collaboration may be used in the context of healthcare to reduce the effects of burnout and stress, experienced by patients and practitioners because of the physical environment. It notes that hospital environments today are linked to feelings of fatigue, anxiety and other psychological stressors. These factors affect a persons' overall comfort, experience, and usability within these settings. This suggests that interdisciplinary collaboration and co-design activities may foster open-communication and interaction among key stakeholders and end users, as a way to implement change. Therefore, this literature review presents the value of collectively engaging stakeholders in design research activities, approaches and methods, as a way to better understand the current context and develop solutions for these real-world problems.

References

1. Williams, E., Halbesleben, J., Manwell, L., McMurray, J., Rabatin, J., Rashid, A., Linzer, M.: The Effect of Workplace Health Care Worker Stress and Burnout on Patient Outcomes. Presented at the 9 November 2011. https://doi.org/10.1201/b11219-13
2. Donetto, S., Pierri, P., Tsianakas, V., Robert, G.: Experiencebased co-design and healthcare improvement: realizing participatory design in the public sector. Des. J. **18**, 227–248 (2015). https://doi.org/10.2752/175630615X14212498964312
3. Jones, P.H.: Design for Care : Innovating Health Care Experience. Rosenfeld Media, Brooklyn (2013)
4. Chartier, L.B., Cheng, A.H.Y., Stang, A.S., Vaillancourt, S.: Quality improvement primer part 1: preparing for a quality improvement project in the emergency department. Can. J. Emerg. Med. **20**, 104–111 (2018). https://doi.org/10.1017/cem.2017.361
5. Wears, R.L., Perry, S.J.: Human factors and ergonomics in the emergency department. Ann. Emerg. Med. **40**, 206–212 (2002). https://doi.org/10.1067/mem.2002.124900
6. Dean, M., Gill, R., Barbour, J.B.: Health communication "let's sit forward": investigating interprofessional communication, collaboration, professional roles, and physical space at EmergiCare (2016). https://doi.org/10.1080/10410236.2015.1089457
7. Sternberg, E.M.: Healing Spaces: The Science of Place and Well-Being. Belknap Press of Harvard University Press, Cambridge (2009)
8. Keys, Y., Silverman, S.R., Evans, J.: Identification of tools and techniques to enhance interdisciplinary collaboration during design and construction projects. Heal. Environ. Res. Des. J. **10**, 28–38 (2017). https://doi.org/10.1177/1937586716684135
9. Bodenheimer, T., Sinsky, C.: From triple to quadruple aim: care of the patient. Ann. Fam. Med. **12**, 573–576 (2014). https://doi.org/10.1370/afm.1713.Center
10. Aiken, L.H., Clarke, S.P., Sloane, D.M., Sochalski, J.A., Busse, R., Clarke, H., Giovannetti, P., Hunt, J., Rafferty, A.M., Shamian, J.: Nurses' reports on hospital care in five countries. Health Aff. **20**, 43–53 (2001). https://doi.org/10.1377/hlthaff.20.3.43
11. Burnout | Definition of Burnout by Merriam-Webster. https://www.merriam-webster.com/dictionary/burnout. Accessed 01 Dec 2019
12. Thake, L.C., Bambling, M., Edirippulige, S., Marx, E.: A Psychoevolutionary approach to identifying preferred nature scenes with potential to provide restoration from stress. Heal. Environ. Res. Des. **10**, 111–124 (2017). https://doi.org/10.1177/1937586717705085

13. Baum, A., Grunberg, N.E., Singer, J.E.: The use of psychological and neuroendocrinological measurements in the study of stress. Heal. Psychol. **1**, 217–236 (1982). https://doi.org/10.1037/0278-6133.1.3.217
14. Reay, S., Collier, G., Kennedy-Good, J., Old, A., Douglas, R., Bill, A.: Designing the future of healthcare together: prototyping a hospital co-design space. CoDesign **13**, 227–244 (2017). https://doi.org/10.1080/15710882.2016.1160127
15. Alvarado, C.: The Physical Environment in Health Care. Presented at the 9 November 2011. https://doi.org/10.1201/b11219-20
16. Lindgren, B.-M., Molin, J., Lundström, M., Strömbäck, M., Salander Renberg, E., Ringnér, A.: Does a new spatial design in psychiatric inpatient care influence patients' and staff's perception of their care/working environment? A study protocol of a pilot study using a single-system experimental design. Pilot Feasibility Stud. **4**, 1–7 (2018). https://doi.org/10.1186/s40814-018-0383-4
17. Velarde, M.D., Fry, G., Tveit, M.: Health effects of viewing landscapes - landscape types in environmental psychology. Urban For. Urban Green. **6**, 199–212 (2007). https://doi.org/10.1016/j.ufug.2007.07.001
18. Harris, D.: A material world: a comparative study of flooring material influence on patient safety, satisfaction, and quality of care. J. Inter. Des. **42**, 85–104 (2017). https://doi.org/10.1111/joid.12100
19. Mustikawati, T., Yatmo, Y.A., Atmodiwirjo, P.: Understanding wayfinding experience of hospital visitor through tours and maps analysis. Environ. Proc. J. **2**, 149 (2017). https://doi.org/10.21834/e-bpj.v2i6.992
20. Lacanna, G., Wagenaar, C., Avermaete, T., Swami, V.: Evaluating the psychosocial impact of indoor public spaces in complex healthcare settings. Heal. Environ. Res. Des. J. **12**, 11–30 (2018). https://doi.org/10.1177/1937586718812439
21. Chartier, L.B., Stang, A.S., Vaillancourt, S., Cheng, A.H.Y.: Quality improvement primer part 2: executing a quality improvement project in the emergency department. CJEM. **20**, 532–538 (2018). https://doi.org/10.1017/cem.2017.393

Research on Emotional Interaction Design in Landscape Installation

Jiang Yuhao(✉)

College of Design, Hunan University,
Pailou Road, Changsha 410082, Hunan, China
jyh0313@outlook.com

Abstract. Study the connotation, value and design methods of emotional interaction design in landscape installations, let people generate emotion-level interactions in their interactions with landscape installations. Through case analysis, and summarizing the methods of integrating emotional interaction design into landscape installation, emotional interaction in landscape installation is divided into three levels: instinct layer, behavior layer, and the reflection layer. It is considered that incorporating emotional interaction design into landscape installations is an effective way to improve the attractiveness of landscapes and meet the emotional needs of users and proposes an emotional interaction design method for landscape installations.

Keywords: Landscape installation · Emotional design · Interaction design

1 Analyze the Emotional Interaction of Landscape Installations

1.1 The Emergence of Emotional Interaction Design in Landscape Installations

In 1984, Bill Moggridg first proposed the concept of interaction design. From the user's perspective, interaction design is a technology that makes products effective and delights users [1]. Initially, interaction design was promoted in the computer field with the concept of "human-computer interaction". In the development of the following forty years, interaction design has evolved from meeting people's simple use needs to provide users with a better experience, and then to meet users' emotional needs in the experience. Research fields of interaction design have also evolved from products and software interfaces to other disciplines. The emergence of emotional interaction design is not only the development of this discipline but also a change in design thinking.

According to Kevin Lynch: "In urban scenes, people are more than simple observers. Together with other participants, people also become part of the scene [2]. It can be seen that people should not just be viewers. Instead, it should be a participant who interacts with the surrounding landscape to achieve a high-level spiritual and emotional experience. Landscape installations are an important medium for people to interact with the landscape. Therefore, the establishment of interactive relationships between people and landscape installations is an important basis for good emotional

G. Di Bucchianico et al. (Eds.): AHFE 2020, AISC 1202, pp. 410–417, 2020.
https://doi.org/10.1007/978-3-030-51194-4_54

interaction. At present, in the field of landscape design, emotional interactive design is mainly focused on the practical research of different types of landscape such as waterfronts; The application of system control and software development in landscape design; And the landscape design methods in emotional interactive design.

1.2 The Connotation of Emotional Interaction Design in Landscape Installation

In psychology, emotion is the attitude and experience of whether or not objective things meet their own needs. It is a complex and stable physiological evaluation. Emotion is one of the unique psychological phenomena of human beings. People can feel the emotion from the surrounding things. Emotional interaction is a design method of emotional dialogue between designer, product and user. It is based on the user's emotional needs and generates emotional resonance through the user's interaction with the product.

The landscape installation is usually an important part of the landscape space and is the carrier for expressing the connotation of the landscape site. In landscape installations, emotion is high-level information transmission, and emotional interaction is a non-material interaction state. It conveys emotional information in the process of people's experience and interaction, awakening people's deep emotions and memories. An excellent landscape installation should be touching, able to convey emotions and recall memories, resonate with people spiritually, and introduce people to deep thinking.

1.3 Features of Emotional Interaction in Landscape Installations

Themed. Emotional interaction in landscape installations has a distinct theme. The theme is the main idea of the landscape installation, such as saving resources, cherishing time, national equality, and so on. In the design step, determining the theme is a key step in experience design. Taking the theme as the core, through the organization of the experience process and the surrounding environment, people can experience the theme of the landscape installation in the interaction process.

Participateable. Emotional interactions in landscape installations are participatory. An important prerequisite for experience design is human participation. At the same time, only when people have an interactive relationship with the landscape installation, can they deeply understand the theme of the landscape installation to produce an emotional experience.

Different. Since emotion is a person's subjective psychological feeling, each person's emotional interaction in landscape installations is different. In emotional psychology, the generation of emotion is restricted by three conditions: stimulus, physiological and cognitive factors, of which cognitive factors are the key factors that determine emotions. M.B. Arnold believes that the evaluation and measurement of external environmental impact is the direct cause of emotions. Different assessments of the same stimulus produce different emotional responses. Everyone's experience and cultural background are different, and the emotional feelings and reactions to the stimulus of landscape installation are also different.

2 Content of Emotional Interaction in Landscape Installation

Norman divides the user's emotional experience into three levels: instinct, behavior, and reflection layer [3]. Based on Norman's theory, analyze the emotional interaction design of landscape installations from three levels of instinct, behavior, and reflection, as shown in Fig. 1.

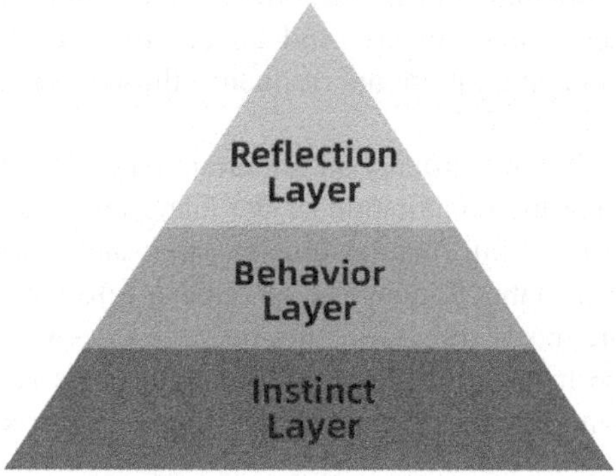

Fig. 1. Three levels of landscape installation emotional interaction design

2.1 Instinct Layer

Humans are very sensitive to the strong signals of the external environment, so the reaction of the instinct layer is formed. The emotional interaction of the instinct layer is related to the first response, which is called sensory stimulation. By stimulating people's sensory systems, people can experience direct emotional experiences, including vision, touch, hearing, smell, taste and so on.

Vision is the most important channel for people to receive external stimuli. The visual interaction of landscape installations is reflected in the properties of the landscape installation itself, that is, the shape, color, light and shadow, and the spatial scale characteristics of the environment in which the landscape installation is located. For example, the round shape without edges makes people feel friendly and the curves are more dynamic than straight lines; The colors with higher saturation give people a lively feeling; Soft and weak light and shadow render a calm and romantic atmosphere, and bright and flashing light and shadow render a dynamic and lively atmosphere. Space, where the landscape installation is located, is an important place for spontaneous and social activities to take place in the city. The appropriate spatial scale makes people feel comfortable and provides necessary conditions for people's interaction [4]. Rough and hard materials such as steel and concrete give an industrial feel, and smooth and soft fabrics give a delicate feeling; The auditory sense is the sound during the interaction with the landscape installation. Wind, rain, bird calls and soft music in nature can make people feel relax; Smell and taste are usually linked through synesthesia. Fresh taste can make people happy and sweet taste can evoke sweet memories.

The design goal of the instinctive layer of the landscape installation is to attract participants to generate positive psychological feelings through sensory stimulation, create a pleasant interactive space, and pave the way for a good emotional experience. As shown in Fig. 2, the "domino effect" landscape installation consists of 120 dominoes. Each group of dominoes has different colors and lighting effects. The sound characteristics of each group of installations are different: vocals, percussion, marimba, balafon, flute and so on. When players topple the dominos in a long cascade, the resulting multi-colored domino orchestra plays a unique musical work in a colorful and luminous bow of pastel hues. The "domino effect" landscape installation attracts people to interact with the installation through visual and auditory stimuli at the instinct level, thereby bringing participants a sense of fun and relaxation.

Fig. 2. "Domino effect" landscape installation

2.2 Behavior Layer

The behavior layer is related to usage. In landscape installations, the emotional interaction at the behavioral level is reflected in the interaction model. The simple and easy-to-understand interaction mode allows participants to easily participate in the interaction of landscape installations and experience the fun of communicating with the landscape installations [5]. Emotional interaction modes of the landscape installation in the behavior layer can be divided into two types: manipulation type and capture type.

Manipulative interaction means that humans actively change the state of the landscape installation to obtain an emotional experience. As shown in Fig. 3, The design of the landscape installation "Around About" is inspired by the Karesansui. A piece of gravel was laid on the ground. By artificially pushing the device that rotates around the center point, the shape of the gravel can be continuously changed to generate different shapes, thereby enabling participants to realize the artistic conception of Karesansui. With the rapid development of science and technology, more and more advanced electronic technologies are used to realize the interaction of landscape installations. Therefore, the capture interaction type that captures human actions, sounds and other behaviors has gradually become the mainstream. As shown in Fig. 4, Dune is the landscape installation that interacts with human behavior. The hybrid of nature and technology is composed of hundreds of fibers that brighten according to the sounds and motions of passing visitors. The design of the behavior layer should give landscape installation interaction more entertainment and fun.

Fig. 3. "Around About" landscape installation

Fig. 4. "Dune" landscape installation

2.3 Reflection Layer

The experience that can reach the reflection layer can give users an unforgettable impression and allow people to have emotional communication with landscape installations. Therefore, in the emotional design of landscape installations, the goal is to allow users to have a good experience, and also have an impact on the participants after the interaction [6]. Incorporating realistic meaning, traditional culture, and the spirit of the times in the design of landscape installations can leave a more perceptual impression on users. Through the spontaneous association, resonance and reflection of participants, the highest level of interactive experience can be generated [7]. As shown in Fig. 5, the "love" interactive landscape installation is formed by a colored board to form a love shape, and a tall arrow forms an arrow to penetrate the heart. The separate channel for men and women symbolizes the journey of love. When the participants reach the middle position from the two channels together, the pressure sensor is triggered to light up the overall device. The device attracts people to interact with it through bright colors and light effects. The simple and inadvertent way of interaction and the changes in the lights surprise people, which causes the participants to resonate emotionally, let them feel the sweetness of love, and create a reflection layer emotional interactive experience.

Fig. 5. "Love" landscape installation

3 Value and Method of Emotional Interaction Design in Landscape Installations

3.1 The Value of Emotional Interaction in Landscape Installations

Increase the Attractiveness of the Landscape. At present, most landscape installations can only be used for viewing, and people cannot communicate effectively with them, making the experience tedious. By enhancing the interactivity of landscape installations and making the landscape space lively, people can interact with it effectively, thereby obtaining the resonance and emotional experience of the reflection layer, which greatly improves the attractiveness of the landscape.

Meet the Emotional Needs of Users. Real design can move people's hearts, it can convey emotions, recall memories and surprise people. Let people feel the emotional interaction during the experience [8]. Through the emotional interaction design of the instinct layer, behavior layer, and reflection layer, the landscape installation enables people to communicate with the landscape environment, so that participants have a good experience and emotional satisfaction.

3.2 Design Method of Emotional Interaction in Landscape Installation

Because landscape installations are in fixed locations and changing spaces, landscape installation design methods are unique. The landscape installation needs to think from the perspective of emotional interaction and how to bring high-quality experience and design quality to the participants.

Dynamic Sensory Stimulation Enhances the Emotional Experience. The external environment stimulates our sensory system, which produces a direct emotional experience. Landscape installations exist in a changing environment, so dynamic sensory stimuli can strengthen participants' emotions. At the beginning of the design, the designer should determine the multi-channel sensory stimulation method suitable for the venue according to the characteristics of the venue. At the same time, it is also necessary to consider the impact of changes in space and time on the landscape installation. For example, we cannot only consider the sensory stimuli brought by the landscape installation during the day, but also the effect at night. When the season changes, the plants in the environment where the landscape installation is located will change in form and color. The landscape installation should make changes for the changes in the space in which it is located, forming dynamic sensory stimuli, and strengthening the participants' emotional experience.

Shape Emotions with Simple and Fun Interactive Mode. Emotions are vague and illusory, but appropriate technology can play a role in emotional expression and emotional stimulation [9]. With the integration of computer vision, speech recognition systems, and computer technologies into interaction design, more and more novel interaction methods have emerged. In the landscape installation of emotional interaction design, a simple interaction method can shorten the distance between a person and a landscape installation. At the same time, interesting and unexpected ways of interaction can create a strong sense of excitement. When the landscape installation can communicate and interact with people, the user's sense of experience and happiness will increase.

Use Metaphors to Build Emotional Bonds. Metaphors in landscape installations translate the theme through the language of design. Designers can extract the emotional memories that people have formed during long-term life experiences, and express them through the materials, shapes, and interaction methods of landscape installations, which resonate with the participants.

4 Conclusion

With the advent of the experience economy era, user experience and emotional design have become the focus of designers' attention. Landscape installations are the main place for people to communicate in landscape space. Traditional ornamental landscape installations have not been able to meet the increasing needs of people's emotional experience. More and more designers have begun to pay attention to people's emotional needs and hope that people will have a good user experience. Taking emotional interaction as the starting point, new perspectives and methods are proposed for the design of landscape installations to enhance the emotional communication between landscape installations and people.

References

1. Copper, A.: Interaction Design Essence. Publishing House of Electronics Industry, Beijing (2008)
2. Kevin, L.: The Image of the City. Huaxia Publishing House, Beijing (2001)
3. Norman, A.D.: Emotional Design. Publishing House of Electronics Industry, Beijing (2005)
4. Jan, G.: Life Between Building. China Architecture & Building Press, Beijing (2002)

5. Yu, J.H., Fu, T., Wu, X.: Emotional interaction design method of furniture. Package Eng. **12**, 49 (2019). (in Chinese)
6. Lan, Y.Q., Liu, P.: Research on emotionalization of interactive products based on user experience. Package Eng. **12**, 23–28 (2019). (in Chinese)
7. Edie, T.A.: Design for Emotion. Posts & Telecom Press, Beijing (2014)
8. Fu, X.Z., Hu, J.C., Zheng, Y.L.: Mobile Desing. Publishing House of Electronics Industry, Beijing (2013). (in Chinese)
9. Sun, X.X., Jin, W.K.: Design of emotion interaction in mobile applications. Package Eng. **14**, 51–54 (2014). (in Chinese)

Exploring the Potential Uses of Ocean Plastic and Public Engagement Activities for Raising Awareness

Xingyu Tao(✉)

Edinburgh Napier University, 10 Colinton Rd,
Edinburgh EH10 5DT, Scotland, UK
x.tao@napier.ac.uk

Abstract. This research explored the qualities and properties of ocean plastic as an accessible material in design and examined public perception by conducting public engagement activities. From addressing the needs of ocean plastic challenge by reviewing the literature, the author finds complex meaning in ecological and economic agendas of ocean plastic. The laboratory experimentation with recycling ocean plastic includes mainly Polyethylene, Polypropylene and Polystyrene, which were collected on the beaches in Scotland. The context of ocean plastic is reframed through working with design students to raise the awareness of ocean plastic and encourage them to take actions to recycle ocean plastic in the future.

Keywords: Ocean plastic · Public engagement · Recycling

1 Introduction

Plastic is prevalent in the marine environment and is increasingly attracting attention from academic, governance and media since it affects ecology and socio-economy negatively [14, 15, 34]. Scholars have done geographical and quantitative research, but the research on the exploration of either ocean plastic as an accessible material or public perception towards ocean plastic has not been thoroughly examined. This study aims to address this gap by conducting interdisciplinary research related to design, material science and public engagement. Firstly, the paper will review the existing literature and theories on ocean plastic and public recognitions on this issue. It will analyse the destructive impacts that ocean plastic has caused ecologically and economically. Since public awareness is a critical aspect in this study, the author seeks to investigate the studies of environmental (ocean plastic) learning activities that other researchers did in the past, which will pave the way for the public engagement activities in this study. Then this practice-based research will explore possible ways to recycle ocean plastic aesthetically as a new valuable material and introduce it to pilot study participants – design students whose attitudes to ocean plastic will be analysed and examined.

G. Di Bucchianico et al. (Eds.): AHFE 2020, AISC 1202, pp. 418–425, 2020.
https://doi.org/10.1007/978-3-030-51194-4_55

2 Ocean Plastic

Plastics are ubiquitous, and the mass production of plastics has caused environmental issues – plastic waste. Hartley et al. state, "Pairwise comparisons indicated that extensive use of plastic in everyday products and packaging and the behaviour of the public were perceived as the most important causes of marine litter" [17]. There are different ways for continental plastic to enter the ocean – flowing into watercourses by storm-water runoff or going straight into coastal waters [9]. According to Dauvergne, laundry, tyre erosion, and city life cause plastic pollution, which is spiralling upwards and accounting for approximately one-quarter of plastic entering into the oceans [11]. Governance is challenging due to "the properties of plastic itself, including its longevity, toxicity, malleability, and the propensity to disintegrate into microplastics" [11]. An estimated 80% of marine litter originates from land-based sources, and plastics are the main component [1, 37]. There are at least 8 million tonnes of plastics entering into the ocean every year quantity-wise [20]. Because most plastics are not biodegradable, quantities of the plastics in the ocean are increasing [1]. As a result, according to Ocean Conservancy's estimation, there were over 150 million tonnes of plastic waste in the ocean in 2015 [35, 36]. The amount of plastic leaking into the ocean would double in five years from 2010, from around 8 million to 16 million tonnes if trends continued [20, 33]. Moreover, a calculation given by the Ellen MacArthur Foundation suggested that there could be more plastic (by weight) than fish by 2050 [12].

Marine litter is found in all the ocean compartments [28], and plastics are considered as one of the most severe problems due to their properties [27], posing a vital threat to the ecosystem. Besides, plastics destroys coral [26], spread invasive species [23] as well as cause accidents with ships [6]. It is reported that around 700 marine organisms have encountered marine litter, and plastic was responsible for entanglement and ingestion for most cases (92%) [14]. The encounter also includes habitat destruction, transport and bioaccumulation of contaminants [41]. Hastings and Potts state, "… litter can impact and deteriorate a range of natural functions that provide on-going social and economic benefits." [18] A calculation provided in a 2014 UN report estimated that plastic waste damage to the marine ecosystem equates to approximately US$13 billion, even going as far as to state that this is likely underestimated [49]. Marine environments as well as human and ecosystem health all require protection from plastic pollution [42], and one of the reasons is that microplastics absorb, release chemicals [29], become bioavailable after ingestion [44], and consequently bioaccumulate in food chain [29].

Marine litters also have an impact on declining in tourism as they are washed ashore and appearing on beaches [22, 31]. A study done by Ballance, Ryan and Turpie indicated, "85% of both tourists and residents would not visit beaches… If they had more than two items of debris per metre, and if the beaches had more than ten large items of debris per metre, 97% of all visitors would not visit them." [2] The geography of tourism, especially "where tourism forms the mainstay of the local economy", could have been significantly affected [18].

3 Public Perceptions

Human behaviour is the sole source of marine litter, which plays an essential role in both contributing to and mitigating the issues, and changing perceptions and behaviour is vital to engage society effectively and in developing and implementing widely acceptable solutions to reduce litter inputs into marine and coastal system [16, 17, 24, 38, 39]. Therefore, actions to the problem lie on land [16, 17], which means, solving ocean plastic problem requires effort from every aspect of society and everywhere in the world. Environment education (EE) has been considered as a key role in shaping the understanding of individuals and communities about the complex relationship with the natural world, and in raising knowledge, attitudes, and behaviours to improve and protect the environment [40, 46]. Hidalgo-Ruz and Thiel engaged 8–16 years old students (total 983) in activities about microplastics in the marine environment [19]. The participants found it very interesting and fun and would take part in other activities about environmental issues in the future [19]. Research with school students has positive results – they realise environment issues [7, 21], tend to take actions in an ecologically responsible manner [13], and can influence others' perceptions and behaviours [10]. Therefore, the reviews indicate that it can help foster children's knowledge, attitudes, and behaviours related to environmental problems to participate in EE activities [16, 17].

It seems to work well to engage public on local issues when combining formulations and implementation of solutions [3], as Biasutti also believes that integrating theoretical learning and practical activities may be more effective to increase the environmental knowledge and change attitudes and behaviours [5]. Riemer, Lynes and Hickman's research suggested, "It is important for youth to be engaged in a manner that fosters active citizenship so that they develop a sense of environmental-social responsibility that is sustained into adulthood" [43]. Education for Sustainable Development (ESD) also requires learner-centred and interactive methodologies for teaching methods [8, 32, 45] to raise environmental perceptions [25]. Besides, Ballantyne and Packer's research shows that short-term environmental learning experiences may foster pro-environmental attitudes and behaviours as well as proposes a strategy for facilitating post-visit sustainable actions, which can be recognised as long-term impacts [4].

4 Materials and Methods

The project is interdisciplinary research by combining material science and product design skills. In order to gain a better understanding of the ocean plastic situation, the author went to the beaches in the Isle of Harris (Scotland) for a field trip (see Fig. 1). Samples were collected from the beaches and numbered (n = 24), and then examined and processed in the polymer laboratory at Edinburgh Napier University (see Fig. 2). Design students also participated in workshops to interact with ocean plastic. Design students were chosen as participants because they are potential designers and are more likely to use ocean plastic in future designs or to design more sustainably.

Fig. 1. Plastic waste on the beach – the Isle of Harris, Scotland

Fig. 2. List of samples

The samples were examined for the types of plastic as some of them did not contain recycling codes. The result shows that most of the pieces collected are Polyethylene (PE, 46%) and Polypropylene (PP, 46%) with only two pieces of Polystyrene (PS). They are colourful, however, with fishy odour and contaminants such as grinding, organisms and sand embedded. Ocean plastic recycling is challenging as it requires to clean the contaminants from the plastic, and it will not obtain a thoroughly cleaned result. After the pre-treatment process, the samples were sawn or granulated to smaller pieces for further operations.

Compression and extrusion were employed as recycling methods. The granules were compressed to blocks of different sizes, colour combinations, and material combinations. The remaining contaminants after cleaning and the sea smell are still noticeable on the blocks, which are considered as the uniqueness of the post- ocean plastic products. The project also adopted laser-cut, which is a common method used by designers. The blocks were cut to different shapes, such as highland cow, leaf, initials, and 3D puzzle, and some of them just remain the block shape as a coaster or jewellery. The colour combination experiment proposed a potential use of ocean plastic for lighting. The remnants were mixed and compressed in the end in order to achieve a no-waste project.

3D printing is another approach often used by designers and increasingly widely employed in other industries. Polylactic Acid (PLA) is the primary material used as 3D printing filament. PLA polymer is biodegradable, and it degrades rapidly at high humidity and temperatures (55–70 °C) [30]. However, it still takes time to degrade and costs raw material and energy to manufacture. This research experimented with extruding ocean plastic and using it on 3D printer. Firstly, the research extruded the filaments by using 100% ocean plastic (HDPE, ø = 1.75 mm). They work well on the 3D printing pen; however, they do not work on the PRUSA I3 MK3 3D printer. The material melts and comes out of the nozzle but is not capable of being printed more than "blobs". 1:1 proportion of ocean plastic (HDPE) and PLA were mixed and extruded to be printed afterwards. The model started to slide on the printer bed after printing halfway through. The filament still works well on the pen (see Fig. 3.).

Fig. 3. Potential uses of ocean plastic – lighting, tourism, children toy, jewellery, 3D printing etc.

Two pilot studies (n = 80 and n = 75) took place with second- and third-year design students at Edinburgh Napier University. The participants were given a lecture about ocean plastic and this project, and then they interacted with the samples made from the previous experimentations by the researcher in this project. They took part in a survey on www.menti.com and commented on the changes in their opinions about ocean plastic before and after the participation, and if they would consider using ocean plastic as a material in their future designs.

5 Discussion

This study explored different ways to recycle ocean plastic from a designer point of view. Ocean plastic is not as perfect or shiny as raw plastic after being in the marine environment for a long time, but it still can be applied to many fields. As mentioned earlier in the paper, the contaminants remained on the post- ocean plastic products can be converted to uniqueness, which is also recognised as exclusivity or rareness. These features are preferred to create the impression of ownership in the tourism industry (i.e. tourist experience activities), therefore, to create a place-person relationship to represent of a specific place [47, 48]. Apart from the potential use in tourism, the simpleness of recycling ocean plastic shown in this study indicates that it is accessible for the public to recycle plastic by using everyday tools, such as blender (for shredding), oven and sandwich maker (for melting and compressing). Lighting design is another proposed use of ocean plastic. The research also shows the potential use of ocean plastic for 3D printing. It is still under the exploration stage, and the researcher is currently experimenting with making it mature enough to use, to reduce the amount of raw material and energy used to produce PLA. With regards to raising public awareness, the project got positive feedback from the participants. Most of them reported that their participation changed their attitudes towards ocean plastic, and they never thought about using ocean plastic in their design before but would consider using it in the future. One of the participants started to use ocean plastic in a design project after the activity, and he thinks the activity opened up a new world for him to design more sustainably. A small number of students expressed their concerns about whether these potential applications of ocean plastic would end up being in the landfill and finally in the ocean again.

6 Conclusion

The purpose of this research was to explore the potential uses of ocean plastic and to examine public perceptions of and engagement with ocean plastic and post- ocean plastic products. The findings suggest that some methods of recycling ocean plastic are accessible and achievable to the public as the author herself does not have a material science background, which suggests that public can use alternative tools to create their post- ocean plastic items instead of working in a laboratory. Besides, the results of the engagement activities with potential designers reveal that the ocean plastic learning experiences, to some extent, can improve participants' knowledge and shape their attitudes and behaviours. The public is willing to be involved and take actions to protect the ocean. Other applications of ocean plastic and public engagement activities with the public with different backgrounds will be explored in the near future.

References

1. Andrady, A.: Microplastics in the marine environment. Mar. Pollut. Bull. **62**(8), 1596–1605 (2011)
2. Ballance, A., et al.: How much is a clean beach worth? The impact of litter on beach users in the Cape Peninsula, South Africa. S. Afr. J. Sci. **96**(5), 210–230 (2000)
3. Ballantyne, R., et al.: school environmental education programme impacts upon Student and family learning: a case study analysis. Environ. Educ. Res. **7**(1), 23–37 (2001)
4. Ballantyne, R., Packer, J.: Using tourism free-choice learning experiences to promote environmentally sustainable behaviour: the role of post-visit 'action resources'. Environ. Educ. Res. **17**(2), 201–215 (2011)

5. Biasutti, M.: An intensive programme on education for sustainable development: the participants' experience. Environ. Educ. Res. **21**(5), 734–752 (2015)
6. Cho, D.: Challenges to marine debris management in Korea. Coast. Manage. **33**(4), 389–409 (2005)
7. Cohen, S., Horm-Wingerd, D.: Children and the environment. Environ. Behav. **25**(1), 103–120 (1993)
8. Corney, G., Reid, A.: Student teachers' learning about subject matter and pedagogy in education for sustainable development. Environ. Educ. Res. **13**(1), 33–54 (2007)
9. Cozar, A., et al.: Plastic debris in the open ocean. Proc. Natl. Acad. Sci. **111**(28), 10239–10244 (2014)
10. Damerell, P., et al.: Child-orientated environmental education influences adult knowledge and household behaviour. Environ. Res. Lett. **8**(1), 015016 (2013)
11. Dauvergne, P.: Why is the global governance of plastic failing the oceans? Glob. Environ. Change **51**, 22–31 (2018)
12. Ellen MacArthur Foundation: New Plastics Economy, Catalysing Action. Cowes (2017)
13. Evans, G., et al.: Young children's environmental attitudes and behaviors. Environ. Behav. **39**(5), 635–658 (2007)
14. Gall, S., Thompson, R.: The impact of debris on marine life. Mar. Pollut. Bull. **92**(1–2), 170–179 (2015)
15. Gregory, M.: Environmental implications of plastic debris in marine settings—entanglement, ingestion, smothering, hangers-on, hitch-hiking and alien invasions. Philos. Trans. R. Soc. B Biol. Sci. **364**(1526), 2013–2025 (2009)
16. Hartley, B., et al.: Turning the tide on trash: empowering european educators and school students to tackle marine litter. Mar. Policy **96**, 227–234 (2018)
17. Hartley, B., et al.: Exploring public views on marine litter in europe: perceived causes, consequences and pathways to change. Mar. Pollut. Bull. **133**, 945–955 (2018)
18. Hastings, E., Potts, T.: Marine litter: progress in developing an integrated policy approach in Scotland. Mar. Policy. **42**, 49–55 (2013)
19. Hidalgo-Ruz, V., Thiel, M.: Distribution and abundance of small plastic debris on beaches in the SE Pacific (Chile): a study supported by a citizen science project. Mar. Environ. Res. **87–88**, 12–18 (2013)
20. Jambeck, J., et al.: Plastic waste inputs from land into the ocean. Science **347**(6223), 768–771 (2015)
21. Kahn, P., Lourenço, O.: Water, air, fire, and earth: a developmental study in Portugal of environmental moral reasoning. Environ. Behav. **34**(4), 405–430 (2002)
22. Keswani, A., et al.: Microbial hitchhikers on marine plastic debris: human exposure risks at bathing waters and beach environments. Mar. Environ. Res. **118**, 10–19 (2016)
23. Kiessling, T., et al.: Marine litter as habitat and dispersal vector. In: Bergmann, M., Gutow, L., Klages, M. (eds.) Marine Anthropogenic Litter, pp. 141–181. Springer, Cham (2015)
24. Klöckner, C.: A comprehensive model of the psychology of environmental behaviour—a meta-analysis. Glob. Environ. Change **23**(5), 1028–1038 (2013)
25. Kollmuss, A., Agyeman, J.: Mind the gap: why do people act environmentally and what are the barriers to pro-environmental behavior? Environ. Educ. Res. **8**(3), 239–260 (2002)
26. Kühn, S., et al.: Deleterious effects of litter on marine life. In: Bergmann, M., Gutow, L., Klages, M. (eds.) Marine Anthropogenic Litter, pp. 75–116. Springer, Cham (2015)
27. Law, K., Thompson, R.: Microplastics in the seas. Science **345**(6193), 144–145 (2014)
28. Law, K.: Plastics in the marine environment. Ann. Rev. Mar. Sci. **9**(1), 205–229 (2017)
29. Li, W., et al.: Plastic waste in the marine environment: a review of sources, occurrence and effects. Sci. Total Environ. **566–567**, 333–349 (2016)

30. Lunt, J.: Large-scale production, properties and commercial applications of polylactic acid polymers. Polym. Degrad. Stab. **59**(1–3), 145–152 (1998)
31. McIlgorm, A., et al.: Understanding the economic benefits and costs of controlling marine debris in the APEC region (MRC 02/2007), Coffs Harbour (2008)
32. McNaughton, M.: Implementing education for sustainable development in schools: learning from teachers' reflections. Environ. Educ. Res. **18**(6), 765–782 (2012)
33. Mortillaro, N.: Humans have produced 8.3 billion tonnes of plastic, researchers say. http://www.cbc.ca/amp/1.4210279
34. Mouat, J., et al.: Economic impacts of marine litter (2010)
35. Ocean Conservancy: International coastal cleanup: 2015 report (2015)
36. Ocean Conservancy: Stemming the tide: land-based strategies for a plastic-free ocean (2015)
37. OSPAR Commission Press Release: Prevention is better that the cure when tackling marine litter in the North East Atlantic, London (2014)
38. Pahl, S., et al.: Channelling passion for the ocean towards plastic pollution. Nat. Hum. Behav. **1**(10), 697–699 (2017)
39. Pahl, S., Wyles, K.: The human dimension: how social and behavioural research methods can help address microplastics in the environment. Anal. Methods **9**(9), 1404–1411 (2017)
40. Palmer, J.: Environmental Education in the 21st Century. Routledge, London (1998)
41. Pettipas, S., et al.: A Canadian policy framework to mitigate plastic marine pollution. Mar. Policy **68**, 117–122 (2016)
42. Raubenheimer, K., McIlgorm, A.: Can the Basel and Stockholm conventions provide a global framework to reduce the impact of marine plastic litter? Mar. Policy **96**, 285–290 (2018)
43. Riemer, M., et al.: A model for developing and accessing youth-based environmental engagement programmes. Environ. Educ. Res. **20**(4), 552–574 (2013)
44. Schneider, F., et al.: Collected marine litter — a growing waste challenge. Mar. Pollut. Bull. **128**, 162–174 (2018)
45. Scoullos, M., Malotidi, V.: Handbook on Methods Used in Environmental Education and Education for Sustainable Development. Athens (2004)
46. Stapp, W.: The concept of environmental education. Environ. Educ. **1**(1), 30–31 (1969)
47. Swanson, K.: Souvenir, Tourists, and Tourism. Wiley, New York (2012)
48. Swanson, K., Timothy, D.: Souvenirs: icons of meaning, commercialization and commoditization. Tour. Manag. **33**(3), 489–499 (2012)
49. United Nations Environment Programme (UNEP): UNEP Year Book Emerging Issues Update: Plastic Debris in the Ocean (2014)

Interdisciplinary Approach to Digital Design

Application of Typical Technical Features in Industry 4.0 for Product Innovation Design

Huabin Wang(✉) and Zhanwen Gan(✉)

South China University of Technology, Panyu District, Guangzhou, Guangdong Province, China
{377381343,290675111}@qq.com

Abstract. Objective: The research on the impact of advanced electronic information technology on industrial design innovation and application that has appeared since Industry 4.0, combined with the inspiration of "Made in China 2025", provides a new innovation path for product design in the era of Industry 4.0. Methods: A conceptual model was established based on industry 4.0, the typical technical features of industry 4.0 were expounded and summarized in blocks, and the mechanism of technology's effect on product innovation design was studied by case analysis. Results: Verifying the rationality and universality of the conceptual model, and proved the influence of typical technology characteristics in the 4.0 era. Conclusion: The analysis of the functional mechanism of technology characteristics in the 4.0 era and the induction of innovative design ideas have certain guiding significance for future enterprises to apply technology to design.

Keywords: Industry 4.0 · Advanced electronic information technology · Case analysis · Product innovation design

1 Overview of the History of Industry 4.0

As early as 2012, General Electric The company once proposed the "Industrial Internet", which aims to use highly developed Internet to strengthen electronic information technology and promote the rapid transformation of industrial models. This is similar to the German "Industry 4.0". At the same time, "Industry 4.0" It has also begun to appear in Chinese media on a large scale and has become a very hot topic, but at this time, China, which is in the stage of parallel development of Industry 2.0 and 3.0, does not have the strong industrial foundation of Germany, nor the United States that can lead the world. The development of innovative electronic information technology, according to its own national conditions, the concept of "Made in China 2025" was proposed, focusing on re-planning the development direction of China's manufacturing industry, increasing the development of advanced electronic information technology, and solving product quality problems. And strengthen industrial basic capabilities and move

This paper is funded by the Humanities and social sciences Research project of the Ministry of Education, No. 20YJA760079.

G. Di Bucchianico et al. (Eds.): AHFE 2020, AISC 1202, pp. 429–436, 2020.
https://doi.org/10.1007/978-3-030-51194-4_56

towards "smart factories" and "connected factories". 2.0 and 3.0, does not have the strong industrial foundation of Germany, nor the United States that can lead the world. The development of innovative electronic information technology, according to its own national conditions, the concept of "Made in China 2025" was proposed, focusing on re-planning the development direction of China's manufacturing industry, increasing the development of advanced electronic information technology, and solving product quality problems. And strengthen industrial basic capabilities and move towards "smart factories" and "connected factories".

2 Analysis of Typical Technical Characteristics of Industry 4.0

At present, the development speed of Industry 4.0 in countries around the world is amazing. Germany has set off a wave of transformation, and other countries have followed suit. They have also increased investment in the development of advanced electronic information technology, so "technology" has become the industry 4.0 Symbol, such as industry Internet of things, Cloud computing, industrial big data, industrial robots, 3d printing, etc. are all new high-techs that have emerged quickly after the introduction of Industry 4.0, but it is worth noting that the technology referred to here is no longer a single technology, but It has risen to the integration of technology and technology, and then formed the three typical technical characteristics of Industry 4.0, which we call highly automated, highly informatized, and highly networked (Table 1).

Table 1. Analysis of typical technical characteristics of Industry 4.0.

General characteristics	Covered technology types	Features
Highly automated	Intelligent robot, artificial intelligence	Towards a "smart factory" to increase production efficiency on a large scale
Highly informative	Big data, virtual reality technology, augmented reality technology	Data information acquisition and processing has become one of the core of enterprise development
Highly networked	CPS system, cloud technology, 5 g communi- cation network	Everything connected

2.1 Highly Automated

Automation once led the innovation in the era of Industry 2.0. Today's Industry 4.0 still has great room for development. According to the development trend of technology, automation will be presented in the actual product design, production and processing with intelligent features. It is a big enabler for achieving high automation.

Intelligent Robot. The intelligent robot is based on a powerful central processing unit, which is equivalent to the "brain" of the robot. As far as the intelligent robot itself is concerned, it is already a carrier of multiple technologies, involving the fusion of multiple sensor information, navigation and positioning, path planning, robot vision, intelligent control, human-machine interface technology, etc. With the enhancement of computer technology, these intelligent robots are also continuously improving, and gradually have sensory, interactive, and even autonomy. The application of a large number of intelligent robots will greatly enhance the softness of automated production lines and allow simultaneous production of multiple product lines, which provides the possibility for the emergence of "unmanned factories" and also lays the foundation for the promotion of "smart factories".

Artificial Intelligence. Artificial Intelligence, abbreviated as AI. It is a new technological science that researches and develops theories, methods, technologies and application systems for simulating, extending and expanding human intelligence. Even though the current application of artificial intelligence technology in the manufacturing industry is still weak, its prospects are unlimited.

For a complete manufacturing process, intelligent robots are generally responsible for parts processing, assembly, and transportation. When it comes to product quality inspection, the application of artificial intelligence technology can greatly reduce the investment in human costs. Artificial intelligence technology can Through its own machine learning principles, it learns from the process of continuously inspecting products and improves the accuracy of quality inspection, and the production efficiency of the company is also improved.

2.2 Highly Informative

Looking at the development of informatization from the perspective of Industry 3.0, on the one hand, it has indeed accelerated the development of more new technologies, but to some extent, it has weakened the development of manufacturing. Computer technology at that time seemed incompatible with traditional manufacturing. As a result, U. S. internal manufacturing has become hollow, and Industry 4.0 is aiming to better apply information technology to industrial manufacturing.

Big data, which is the basis of intelligent manufacturing, has applications in mass customization in manufacturing, including data collection, data management, order management, intelligent manufacturing, and custom platforms. The core is custom platforms. Custom data reaches a certain order of magnitude. Big data applications can be realized.

In addition to the benefits of big data technology, the development of highly informatization is an important dimension is the introduction of virtual reality technology and augmented reality technology.

In other words, it can integrate the real world and virtual world information to assist our daily life and work. No one in the "factory" is actually not all involved in manufacturing. Many links in the production line still need to be operated by people. In order to reduce the burden on people, augmented reality technology can assist workers to work, thereby greatly improving production effectiveness.

2.3 Highly Networked

The key technology in Industry 4.0 is to use the Cyber-Physical System (CPS) to make the supply, manufacturing, and sales information in production digital and intelligent, and finally achieve a fast, effective, and personalized product supply.

With the advent of cloud technology, cps has also been more perfected. Cloud technology refers to unifying a series of resources such as hardware, software, and network within a wide area network or local area network to achieve data calculation, Store, Processing and shared Kind of Hosting Technology. In the early days of development, most of the CPS will be stored in local space, with computer-assisted equipment as a carrier, but there are problems such as small capacity, difficult backup, and inconvenience of movement. Cloud technology helps CPS solve these problems After it connects the physical device to the Internet, it also uploads in the "cloud". As long as the server can operate normally, the uploaded data can be permanently saved, and can be edited and used anytime, anywhere. cps + cloud technology has also become High-quality, advanced solutions produced by enterprises.

In addition, in recent years, the development of the 5th-Generation mobile communication network, that is, 5G technology has begun to enter all aspects of production and life, which means higher production efficiency for Industry 4.0, making CPS technology More highly integrated upgrades, big data, artificial intelligence, and cloud computing will be well blessed into this system at 5G network speeds, providing a new business model for the manufacturing industry.

3 Enlightenment of Industry 4.0 on Product Innovation Design

Today, technology has been Moving towards a highly automated, informatized, and networked structure in the 4.0 era, it provides unlimited possibilities for the development of the current manufacturing industry, and the manufacturing industry is closely related to the current product design. To a certain extent, it also affects the product design process, giving the product design model more innovation possibilities.

3.1 Highly Automated-Improve the Efficiency of Design Creative Output

Take Alibaba as an example. As an Internet company, they are also full of ambitions for these technologies. To this end, they invested a lot of money to set up the Alibaba Intelligent Design Lab to study the design of automation. At the Alibaba User Experience Design Forum Its artificial intelligence "designer" Luban was announced that it will be renamed "Luban" and launched as a platform product for global users. Its product features include one-click generation of posters, intelligent typography, design expansion, intelligent creation, etc. It reduces the tedious work of many e-commerce users in packaging design and layout of products in their stores, and can also serve the Taobao e-commerce platform with the highest market share in the world (Table 2).

In contrast, artificial intelligence has now become the core of this process, and its objects have risen to Internet products. Ali's "Luban" robot can design 8,000 posters in

Table 2. Comparison of the conventional human drawing method and artificial intelligence drawing method based on e-commerce posters.

	Design approach	Features	Average output speed	Average cost
Graphic Designer	Routine manpower mapping	Design is highly subjective, but requires multiple revisions	1800 s/sheet	50–150 yuan/sheet
"Luban" robot	Artificial intelligence plotting	Through a lot of learning typography design, a large number of creative design effects can be generated	0.000125 s/sheet	Non-VIP: ¥9.9/piece VIP: ¥0.0099/piece

one second, which is more productive than human designers. Full, so can we judge whether artificial intelligence can promote product innovation and design? Compared with human subjective design work, this is obviously inconceivable. But from another perspective, Luban robot as a design The platform, that is, a service system, is not only for Ali itself, but also for users of major products. It is extremely innovative. The innovation here does not refer to how much innovation the designed poster will have, but it can reduce a lot of tedious work for designers, so that people who can't design can also use this platform to create and have certain productivity, and can directly serve other products under their own, becoming an innovative service design.

3.2 Highly Informatization-Using Reliable Data to Aid Design and Production

In the traditional industrial design phase, it is generally necessary to analyze the project requirements-user data research-provide multiple solutions-digital implementation of the solution-transfer the determined solution to the manufacturing team for large-scale production, etc., which takes a lot of effort and time to determine the best feasible product for the client company. In the Internet era, the high level of informationization will relieve these problems to a certain extent for enterprises.

Many companies can use big data to Accurately locate the real needs of users, or can collect a large amount of user information, and this information is extremely valuable for the company's work in the design process (Table 3).

Taking the development of shoe companies as an example, Aokang International, as the leading domestic men's shoe giant, first introduced a new generation of foot shape measuring instrument in 2017, so that the user's foot shape 3D data can be obtained in 15 s, and automatically linked with The shoe shape database established by the company is used for matching, and the generated data is transmitted to the factory for customized production, providing users with the appropriate shoe shape.

Table 3. Aokang international and Shuangchi footwear technology application analysis

	Adopting technology	Application phase	Features	Output Efficiency
Aokang International	Big data analysis	User data collection	Powerful database comparison to design shoe shapes for users	Data comparison in 15 s
Shuangchi shoes	Virtual reality technology assists in data collection	User data collection	Multi-dimensional recording of user's foot shape and exercise status, more reliable design	Real-time (0 s) logging data

In addition, Shuangchi Shoes, which focuses on intelligent manufacturing, uses its self-developed Shuangchi vr measurement device to collect static information on consumer foot shape data, and also obtains dynamic information about the pace of the user's walking., Inside and outside walking, inside and outside walking, etc. are effectively stored and transmitted to the foot database. Compared to traditional foot measuring instruments, the biggest feature of vr measuring equipment is to bring users into the virtual reality scene, so that users have physical body. The immersive sports experience allows them to show their true foot shape and exercise state in a relaxed state, while improving the fun of data collection, while ensuring that the data is more effective.

3.3 Highly Networked-Connecting Full Links of Design and Manufacture

The foundation of highly networked is the Internet of Things, and the core of the Internet of Things is CPS with cloud technology. This definition is only for industrial manufacturing, and for product innovation and design, it is far beyond the Internet of Things technology itself.

The application value of the Internet of Things lies in the establishment of a networked scenario that opens up the data links between various people and devices, thereby helping a customer group to meet their needs or pain. This gives product design workers some Design ideas for IoT products (Table 4).

For example, in many work and life scenarios, not every participating member can reach the scene. For those users who want to know or control a scene but cannot visit the scene in person, they can use remote access to the scene to help them collect with sensors. This data will network the equipment to be controlled, thereby helping them to remotely understand or control the situation on site.

Device service is another innovative design method. After the device is networked, users can access and use the corresponding device anytime, anywhere, and at the same time, the administrator can also maintain it.

In addition to controlling devices remotely, the Internet of Things builds a variety of networked devices into a digital network, which can help users in the scene to obtain a better experience. Users only need to access this live network in the field, and they can Swim among them.

Table 4 Networking applied to design link analysis

Design phase	Action form	Features
Analyze project needs	IoT collaboration	The direction of product design will join the concept of "Internet of Things"
User data research	Cloud storage data	Designers can use user data for in-depth design anytime, anywhere
Provide multiple solutions	IoT collaboration	Users can participate in the design scene
Digital implementation of the solution	IoT collaboration	Collaboration between designers is more authentic
Manufacturing	IoT collaboration	The management of the production workshop is more standardized, and the efficiency is rapidly improved

4 Conclusion

Obviously, under the era of Industry 4.0, global industrial manufacturing has developed rapidly under the support of many advanced technologies, and China has, without exception, started to move to the leading level of global manufacturing. However, the road to "Made in China 2025" is still far away. This first requires us to advance the application of technology to the product innovation design process, reduce the pressure on design work, increase the output ratio of design work, and highly integrate and effectively use information, analyze effective information, and summarize innovation. Design ideas, and add technology to the physical product or service system, and only when we properly integrate the technical characteristics of Industry 4.0 into product innovation design, the development and transformation of enterprises can be effectively carried out, "Made in China 2025" It will be closer to us.

References

1. Jing, S.: Industry 4.0 based on "Made in Germany" and its inspiration to China's innovation. J. Chongqing Three Gorges Univ. **35**(01), 107–114 (2019)
2. Li, Z.: Analysis of new technology in industry 4.0. Sci. Technol. Innov. (03), 44–45 (2019)
3. Wang, H.: Research on product design concept under the background of "Internet of Things". Design (09), 179–180 (2013)
4. Jian, X.: Comparative study of "Made in China 2025" and Germany" Industry 4.0". China Collective Econ. (10), 86–87 (2019)
5. Liu, Y., Liu, Q.: The role of industrial design in the transformation and upgrading of manufacturing industry from the perspective of Industry 4.0. Packag. Eng. **39**(08), 113–116 (2018)
6. Lu, N.: On the innovation of productive service industry under the deep integration of Industry 4.0 and "Internet+". Economist (05), 27–28 (2017)

7. Li, D.: Internet+ promotes the promotion of China's Industry 4.0 strategy. Smart Factory (11), 55–58,64. (2016)
8. Chen, S.: Footwear Industry Design 4.0 Innovator in the Big Data Era (2017). https://www.sohu.com/a/211628057_283830
9. Yu, D.: A new starting point for shoe companies' change—foot type big data. Chin. Foreign Footwear Ind. (02), 82–87 (2018)
10. Qin,W., Chen,S., Peng, M.: Recent advances in industrial internet: Insights and challenges. Digit. Commun. Netw. (2019)
11. Gulati, N., Kaur, P.D.: Towards socially enabled internet of industrial things: architecture, semantic model and relationship managemen. Ad Hoc Netw. (2019)
12. Anonymous: Industrial Internet groups to collaborate on cloud computing. IoT Res. Control Eng. (2019)

Effects of Industry 4.0 on Human Factors/Ergonomics Design in 21st Century

Manutchanok Jongprasithporn[1(✉)], Nantakrit Yodpijit[2], Chanakamon Phaisanthanaphark[2], Yotsuda Buranasing[2], and Teppakorn Sittiwanchai[2]

[1] Department of Industrial Engineering, Faculty of Engineering, King Mongkut's Institute of Technology Ladkrabang, Bangkok, Thailand
manutchanok.jo@kmitl.ac.th, mjongpra@gmail.com
[2] Center for Innovation in Human Factors Engineering and Ergonomics (CIHFE^2), Department of Industrial Engineering, Faculty of Engineering, King Mongkut's University of Technology North Bangkok, Bangkok, Thailand

Abstract. The fourth industrial revolution as known as Industry 4.0 becomes a big part of the digitized economy in the 21st century. Industry 4.0 considerably drives the entire supply chain through digital infrastructures and significantly changes work systems. There are many new jobs and work opportunities available for workers with different skill levels due to society and technological changes. Understanding user needs and new technologies is a key factor for the design and development of products and systems where human exists. It is to ensure that the operation and maintenance of such systems are compatible with human abilities and limitations. The purposes of this paper are to explore theoretical and empirical concepts of Industry 4.0 influencing the design of work systems and present factors affecting human factors/ergonomics design in terms of physical, cognitive and organizational perspectives.

Keywords: Human Factors/Ergonomics · Industry 4.0 · Physical, cognitive and organizational perspectives

1 Introduction

Nowadays, the business competition is quite high. Companies must find appropriate ways to develop business in order to increase productivity, work efficiency, and cost reduction. Technology is one of major roles to help companies with high success. Therefore, industries have been changed with significant utilization of computers and automated systems since 2011 into the fourth industrial revolution (Industry 4.0) [1]. Industry 4.0 applies different technologies to change human living and create the business opportunities. Internets and technologies are integrated in the digital and physical world for monitoring and controlling in manufacturing systems for the improvements of productivity and work efficiency [1] and [2]. The main concept of Industry 4.0 is smart factory [1] and [3]. Before Industrial 4.0 implementation, workers are required to understand work systems and their responsibility. As such, understanding the effects of industry 4.0 is very important to help companies achieve business success.

G. Di Bucchianico et al. (Eds.): AHFE 2020, AISC 1202, pp. 437–443, 2020.
https://doi.org/10.1007/978-3-030-51194-4_57

Human Factors/Ergonomics (HF/E) concerns with the relationship between the user, equipment, and environment or fitting a job to a person. It is to ensure that the operation and maintenance of such systems are compatible with human abilities and limitations. HF/E is defined by The International Ergonomics Association (IEA), it concerned with the collaboration between humans and other elements of system for development quality life and overall system performance [4]. The field of HF/E is characterized into three domains of specialization: physical, cognitive, and organizational [4–6] and [7]. Understanding user/operator needs and new technologies is a key factor for the design and development of products and systems where human exists. Proper HF/E design leads to the improvement of work systems and the prevention of repetitive strain injuries and other musculoskeletal disorders in Industry 4.0 era.

Due to significant changes in activities of daily living, new technologies have been designed and developed to meet user needs. The main purposes of this paper are (1) to explore theoretical and empirical concepts of Industry 4.0 influencing the de-sign of work systems, and (2) to present factors affecting HF/E design in terms of physical, cognitive and organizational perspectives.

2 Methodology

This research concentrates on Industry 4.0 and HF/E characteristics. The project has been conducted with the following research methods.

2.1 Background of Industry 4.0

In sum, the first industrial revolution refers to steam power in term of decreasing worker and animal workload. The second industrial revolution involves electrical power for mass production. The third industrial revolution concerns with computer and automation. The fourth industrial revolution as known as Cyber-Physical Systems (CPS) and defines by the use of internet and technology to integrate the digital and physical world. It applies the technologies in all parts of factory as called smart factory to support entire supply chain and meet customer demand. Industry 4.0 is the technology revolution. It consists of Internet of Thing (IoT), Cloud, Big Data, Augmented Reality (AR), Additive Manufacturing (AM) and Autonomous Robots as shown in Table 1 [1–3] and [8].

Table 1. The definition of the technologies enabling Industry 4.0.

Technology	Definition	Reference
IoT	The use of internet for connecting human-human, human-machine, and machine to machine	[1–3, 10, 11] and [12]
Cloud	The storage of data that can assess in any time and place. Several small business, use Cloud for reducing cost as IT infrastructure	[1–3, 11] and [12]

(continued)

Table 1. (*continued*)

Technology	Definition	Reference
Big Data	Huge information that keep the various data for analysis. It helps for decision making and business opportunities	[1–3, 11] and [12]
AR	Technology that simulate the virtual objects in real environment	[1–3, 11] and [12]
Additive manufacturing	Making 3D objects from the drawing file for the simulation of the prototype	[1–3, 11] and [12]
Autonomous robots	The use of autonomous robot for decreasing human workload. It can work instead of human in hazardous industrial or monotonous work	[1–3, 11] and [12]

2.2 Overview on HF/E

HF/E is defined as the relationship between human, machine, environment, and its interaction in order to develop human well-being and system performance [4]. As noted by [5] and [7], HF/E concerns human beings, human capabilities and limitations for appropriate work systems design. The configurations of HF/E include (1) Physical Ergonomics deals with physical activity and human body (2) Cognitive Ergonomics concerns with metal processing for supporting human and machine functions, and (3) Organizational Ergonomics focuses on organizational structures, teamwork, and society relationship and communication [4–6], and [7].

Work system is the aggregation of various components that interact with each other to achieve the same goal [5]. Human–machine systems have been implemented in work systems for many decades. Human-machine systems have three domains as follows: (1) manual system considers of hand tools and other equipment using human physical energy for operating, (2) mechanical system uses human for controlling devices, and (3) automated system is unmanned working environment to get the job done [5, 6] and [7].

2.3 Data Analysis

From previous studies, this paper follows main pillars of Industry 4.0, as noted by [10], and three characteristics of HF/E: physical, cognitive and organizational, as defined by [4–6] and [7].

The analysis of the qualitative data of this research has three phases. The first phase focuses on reviewing and highlighting concepts of Industry 4.0. The second phase completes interviews on potential parties from various businesses. The third phase identifies the factors affecting HE/E design and refines the findings by the experts.

3 Results

3.1 Concepts of Industry 4.0

Industry 4.0 refers to Cyber-Physical System (CPS). It is the connection between physical world such as tools, machines, people, and cyber world via internet system, and the use of technologies for communication and real-time information sharing. Industry 4.0 has also focused on monitoring and controlling functions on machines and human for work agility, work efficiency, productivity, human error reduction and better decision making [2] and [10]. In Industry 4.0, the manufacturing system can produce products and provide services based on real-time customer demands using technologies to meet customer needs [2, 3, 8, 11] and [12]. The six pillars of Industry 4.0 are given in Table 2 [1–3, 11] and [12].

Table 2. The definition of the six pillars in Industry 4.0

Pillar of Industry 4.0	Definition	Reference
Virtualization	Technology that simulates physical world into digital form	[2, 10] and [13]
Inter collaboration	The connecting of components in system via internet	[2, 3, 10, 11] and [13]
Decentralization	The connecting of components in system though internet for appropriate decision making	[1–3, 10, 11] and [13]
Real-time communication	The real-time communication system that makes better decision making	[2, 3, 10, 11] and [13]
Service capability	The use of internet for internal and external organization management	[2] and [10]
Modularities	The flexible system that can change the condition by putting a module in or replace it	[2] and [10]

3.2 Interview of the Potential Parties

The data analysis is carried out through interviews with managers in higher management level. In-depth interviews are used to provide a broader view of the influence of Industry 4.0 on HF/E in companies. The managers explain the impact of Industry 4.0 on the design of HF/E in three following areas:

Physical HF/E

The implementation of Industry 4.0 concerns operator fatigue minimization, time reduction, productivity increase, cost saving, and work efficiency improvement. Benefits from applying physical HF/E include the decrease of sick leave of workers from hard working and the minimization of accidents from worker recklessness.

Cognitive HF/E
Most workers believe that if workers do not have enough knowledge and skills of new technologies, the company will lay off the workers. Workers do not have motivation for dealing with new technologies. Therefore, companies have to educate workers and put the right man on the right job.

Organizational HF/E
Work pattern is changed by integration new technologies. The companies have to change and determine the boundaries and roles of departments for supporting business operations.

3.3 Results from Qualitative Data Analysis

Physical HF/E
Autonomous robot is implemented in manufacturing systems for mass production. It supports human limitation such as working in dangerous places and lifting heavy parts, [2, 8] and [13]. The application of VR for plant layout simulation to improve workplace safety and productivity [2, 3, 8] and [13].

Wearable technologies such as smart phone, smart watch and smart glasses provide operator's real-time information on health condition and location. These are very useful for investigating operator's behavior, posture and activity while working [2, 8] and [13].

Cognitive HF/E
VR deals with simulation to make job safe and easy to perform. It is useful for decision making and real-time problem solving [2, 8, 10, 12] and [13]. VR reduces language limitation, work difficulty and stress, and supports migrant and old workers by displaying tasks in picture and large letters of working information [2, 3, 8] and [13]. Wearable technologies are used to investigate workers' stress at work [2] and [13].

Job allocation can be performed based upon skills of workers that are analyzed using Big Data features for matching worker skills to job tasks. Workers must have special skills and proper knowledge of new technologies, so they can work effectively with minimum stress [2, 3] and [13].

Organizational HF/E
After implementing new technologies, the organizational structure must be changed in the ways of work operation and management to fit with the technologies. It is very critical to identify the duties and organizational climates clearly. A new design of work system will contribute participation of all departments and adjust overall organizational structure including responsibilities and communications [2, 3, 8] and [13].

From the qualitative data analysis above, Industry 4.0 factors affecting HF/E design in the all three perspectives are shown in Table 3.

Table 3. The effects of Industry 4.0 on HF/E design.

Pillars of Industry 4.0	Physical	Cognitive	Organization	Application	Reference
Virtualization	✓	✓	✓	AR, and AM	[1–3, 9, 10] and [13]
Intercollaboration	✓	✓	✓	IoT, AR and Autonomous robot	[2, 3, 9–11] and [13]
Decentralization		✓	✓	Big data and IoT	[2, 3, 9–11] and [13]
Real-time communication	–	✓	✓	IoT, Big data and Cloud	[1–3, 9–11] and [13]
Service capability	–	✓	✓	IoT	[1–3, 9–12] and [13]
Modularities	✓	✓	✓	Cloud and IoT	[2, 3, 9–11] and [13]

Table 4 illustrates how industry 4.0 affects HF/E design on physical, cognitive, and organizational perspectives.

Table 4. The effects of industrial revolution on HF/E

Industrial revolution	Physical	Cognitive	Organization	Reference
Industry 1.0	• Human workload	–	–	[1–3, 8] and [13]
Industry 2.0	• Human workload • Fatigue	• Stress	• Profitability	[1–3, 8] and [13]
Industry 3.0	• Human workload • Fatigue • Comfortability	• Stress	• Profitability • Productivity • Work efficiency	[1–3, 8] and [13]
Industry 4.0	• Human workload • Fatigue • Comfortability • Safety	• Stress • Decision-making	• Structure of organization • Productivity • Work efficiency	[1–3, 8, 10, 11] and [13]

4 Conclusion and Discussion

Findings from this study present the effects of Industry 4.0 on HF/E design based on the six pillars (IoT, Cloud, Big Data, AR, AM, and Autonomous Robots) of Industry 4.0 and three perspectives (physical, cognitive and organizational) of HF/E. Industry 4.0 factors affect physical HF/E are Virtualization, Intercollaboration and Modularities with using IoT, AR, AM, Cloud and Autonomous Robot. The factors influencing cognitive and organizational HF/E are Virtualization, Intercollaboration, Decentralization, Real-time Communication, Service Capability and Modularities with using IoT, AR, AM, Big Data, Cloud and Autonomous Robot. The findings from this study supports a premise that Industry 4.0 is a central role in HF/E design influencing physical, cognitive and organizational approaches significantly and changes of HF/E design must be made substantially in 21st Century.

References

1. Frank, A.G., Dalenogare, L.S., Ayala, N.F.: Industry 4.0 technologies implementation patterns in manufacturing companies. Int. J. Prod. Econ. **210**, 15–26 (2019)
2. Alcácer, V., Cruz-Machado, V.: Scanning the Industry 4.0: a literature review on technologies for manufacturing systems. Eng. Sci. Technol. Int. J. **22**, 899–919 (2019)
3. Büchi, G., Cugno, M., Castagnoli, R.: Smart factory performance and Industry 4.0. Technol. Forecast. Soc. Change. **150**, 119790 (2020)
4. International Ergonomics Association. https://www.iea.cc/whats/
5. Sanders, M.S., McCormick, E.J.: Human Factors in Engineering and design. Christopher Rogers and Tom Holton, Singapore (1993)
6. Tayyari, F., Smith, J.L.: Occupational Ergonomics Principles and applications. T.J. Press Ltd., Padstow, Cornwall (1997)
7. Chengalur, S.N., Bernard, T.E., Rodgers, S.H.: Kodak's Ergonomic Design for People at Work. Wiley, New York (2004)
8. Pinzone, M., Albè, F., Orlandelli, D., Barletta, I., Berlin, C., Johansson, B., Taisch, M.: A framework for operative and social sustainability functionalities in human-centric cyber-physical production systems. Comput. Ind. Eng. **139**, 105132 (2019)
9. Norros, L.: Developing internet of thing/ergonomics as a design discipline. Appl. Ergon. **45**, 61–71 (2014)
10. Manavalan, E., Jayakrishna, K.: A review of internet of things (IoT) embedded sustainable supply chain for Industry 4.0 requirements. Comput. Ind. Eng. **127**, 925–953 (2019)
11. Culot, G., Nassimbeni, G., Orzes, G., Sartor, M.: Behind the definition of Industry 4.0: analysis and open questions. Int. J. Prod. Econ. 107617 (2020)
12. Hofmann, E., Rüsch, M.: Industry 4.0 and the current status as well as future prospects on logistics. Comput. Ind. **89**, 23–34 (2017)
13. Kadir, B.A., Broberg, O., Da Conceição, C.S.: Current research and future perspectives on human factors and ergonomics in Industry 4.0. Comput. Ind. Eng. **137**, 106004 (2019)

Human-Machine Interaction Design Process for Complex Product Systems as Spacesuits

Gang Liu, Li Huang, Mengyin Jiang(✉), and Yi Su

Tsinghua University, 5th Floor of Sancaitang Building, Chengfu Road, Haidian District, Beijing, China
jiangmy@mail.tsinghua.edu.cn

Abstract. The research is an exploratory case study of the lunar landing suit. The purpose is to explore innovative prototype solutions to improve the Human-Machine Interaction (HMI) efficiency of the lunar landing suit based on human factor ergonomics simulations, simulation experiments, and modern industrial design theories.

The HMI design in Complex Product System (CoPS) such as spacesuit is mainly based on functional requirements. The main principle is to apply human-centered HMI symbiosis as the design guiding ideology. The goal is to improve people's efficiency in system control, including reducing operation response time and improving operation accuracy. The result is the establishment of a new framework of HMI design for the spacesuit, which can effectively support the design practice. Based on this framework, a new conceptual rendering of the spacesuit is designed.

Keywords: Complex product system · Human machine interaction · Design framework · Spacesuit · Visualization

1 Introduction

China's lunar exploration project, since 2004, is named "Chang'e project". The project is divided into three phases: "unmanned lunar exploration", "manned lunar landing" and "establishment of lunar base" [1]. The first phase – the unmanned lunar exploration – can be divided into three stages: circling, landing and returning [2]. It is expected that China's unmanned lunar exploration will achieve the final step of "circling, landing and returning" in 2020, with the return of the unmanned lunar lander back to Earth [1, 3]. Next, China is expecting to begin the second phase of lunar exploration – manned lunar landing. Therefore, the development of the lunar landing suit has reached a critical stage.

China has not yet developed a mature space suit for landing on the moon, but has developed an extravehicular space suit for flying in the sky. The successful development of extravehicular space suit has accumulated experience for the future development of space suit for landing on the moon, including experience in human-machine interaction (HMI) design. The extravehicular spacesuit is for astronauts, and the design of "human centered design" is the principle of human-machine interaction design. The main goal of the human-machine interaction design of the lunar landing suit is to

G. Di Bucchianico et al. (Eds.): AHFE 2020, AISC 1202, pp. 444–450, 2020.
https://doi.org/10.1007/978-3-030-51194-4_58

increase the safety of the operation process, enhance the comfort, reduce the operation response time, improve the operation accuracy, and increase the efficiency of information transmission.

At present, only the United States has successfully landed on the moon for many times with the Apollo program [4]. In 1966, the Soviet Union once carried out the plan of landing on the moon by developing the Кречет-94 spacesuit specifically used for lunar walking [5]. After 40 years, the United States released a new lunar landing suit as an upgrade of the Apollo spacesuit in October 2019 and plans to return to the moon in 2024 [6] (Fig. 1).

Fig. 1. Soviet and American moon landing suits (Left-Кречет-94, Middle-Apollo, Right-New spacesuit of US)

The technicality of the whole HMI design is quite different between the United States and the Soviet. The Apollo program in the United States uses soft spacesuits, while the Soviet spacesuits adopt a semi-rigid design. The United States spacesuit consist of different parts worn individually. The Soviet space suit can be opened and worn from the back. At the same time, a ring-shaped safety device is added around the spacesuit to ensure the safety of the personnel during walking. From the perspective of human-machine interaction between the wearer and the spacesuit, the interaction process of spacesuits from the United States is more refined, but the wearing process is more complex due to the nature of the soft suit. After putting on the American spacesuit, one can carry out relatively complex operations. Soviet spacesuits, on the other hand, are easy to wear and take off, but not easy to make fine movements. The two kinds of spacesuits use different design technologies and have different strengths. Chinese designers can benefit and learn from both designs.

2 Design Content for HMI

Manned space system is a typical human-machine interdependent system that is complex and composed of both human and equipment. The core task of this system is to ensure the safety of the human and to complete task activities. The HMI of spacesuit is implemented in the following three parts, including the key nodes of HMI design (Fig. 2):

- Pressure clothing: including airtight clothing, trunk, arms, legs, gloves, helmets, boots and movable joints.
- Life support system: heat control, oxygen supply, food and water supply, metabolite removal, emergency life protection, astronaut real-time physiological state recording system, etc.
- Information control and communication guarantee: including intelligent power supply, communication with the outside world, data display and human-computer interaction, intelligent control, etc.

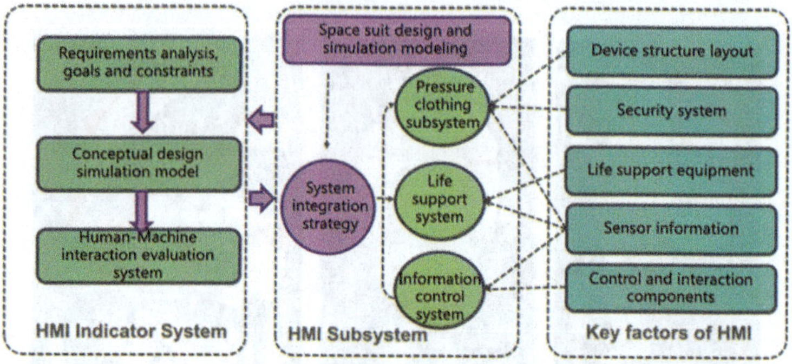

Fig. 2. Research framework for Human-Machine Interaction (HMI) design of lunar spacesuit

The basic HMI functions of the extravehicular service include: 1. Electrical control; 2. Air pressure and liquid temperature control. These two parts are used to regulate the safety protection system and life support system of the extravehicular service. In the design process of the lunar landing suit, the existing HMI operations are classified, a spacesuit with adaptive flexibility is constructed, and the conceptual design scheme of new materials based on the mechanical counter pressure (MCP) is studied. The HMI design of spacesuit can explore new forms of visual interaction, and multi-dimensional human-machine interaction can be carried out through wearable devices, augmented reality glasses, vibrations, visual prompts, and more. Through the application of new HMI technology, it is hoped that more HMI functions will be added to the lunar landing suit and the overall weight of the system module will be reduced (Table 1).

Table 1. Main contents of human-machine interaction design

Num.	Category	Contents
1	Basic human dimensions	All parts of the body, especially the upper limb reach area simulation analysis, upper limb operation limit area, reach area, flexible area calculation
2	Hardware interface interaction button	Handle rotation, stretching, toggle switch, key switch, switch protection cover, button identification, positioning, operation, reset, etc., accuracy of action steps and operation efficiency
3	Software interface interaction elements	Software user interface interaction elements shape, size, material, click impact, voice interaction, entity interaction, physiological signal interaction, touch sensitivity

Space within the spacesuit is relatively limited, astronauts need to complete a large number of tasks in a narrow space, which puts forward high requirements for the rational layout of HMI functions. Good HMI design requires the combined effort and communication between hardware design engineers, software programmers and task planning engineers. The main parts of HMI research include: the hardware HMI design of the operation interface of the extravehicular spacesuit, the HMI of the operation information interface, and the process planning design of the operation task. In addition, HMI design should also include communication and social activities that care for the psychological wellbeing of the astronauts.

3 Design Process for HMI

The success of space activities is the result of the seamless cooperation between human and machine. Human behavior and task demand are the main focus of the designers in the design process of aerospace products. They should further combine behavior and demand with the technical requirements, and put forward a reasonable HMI architecture scheme that can be presented visually for astronauts, researchers, developers and other stakeholders to review and optimize.

The study of human factors is to take people as the influencing factors in the system, and to study human perception, decision-making and operation processes.

As a typical complex product system (CoPS) [7], the EIIV (Empathy-Insight-Imagination-Visualization) frame demonstrated in Fig. 3 can be used as the process framework for HMI design in spacesuits [8]. The main steps of EIIV framework include: 1. Sort out the design requirements and all kinds of information; 2. Focus on the needs of insight tasks. 3. Look for possible solutions for innovation points. 4. Refine the determined solution to form the determined solution.

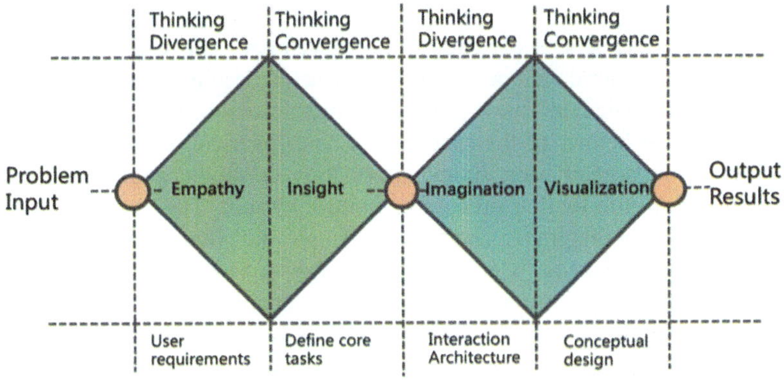

Fig. 3. The EIIV design framework for the complex product system of Lunar spacesuit

Manned space flight requires a complex life support system, which is highly HMI based. It may be necessary to use technologies such as flexible joints,bioelectric control, posture recognition, augmented reality glasses, body vibration, visual cues to achieve multi-dimensional interaction between human and the machine. The research on the HMI design of extravehicular spacesuit not only needs to solve the functional problems, but also pay more attention to user experience to reduce the cognitive stress on astronauts, and to increase their operation efficiency, accuracy and comfort. Main contents of human-machine interaction design are shown in Table 2.

Table 2. Main contents of human-machine interaction design

Num.	Category	Contents
1	Ergonomics	Dynamic comfort joint design/augmented reality flexible interface design/ergonomic analysis model/exoskeleton technology and mechanical back pressure system/bionic structure joint
2	Flexible sensor	Voice controlled interaction/flexible environmental indicator sensor/flexible tactile and pressure sensor/flexible information display device
3	Interactive component	Gas-liquid console layout and design/control component layout/physical operation button layout/sealing mechanism and safety redundant backup
4	Smart helmet	Communication module based on interactive control/micro-projection display technology based on EEG, muscle, and eye movement signal recognition, virtual reality display, transparent display/front window OLED display technology and text display technology/aptamer layout

4 Conceptual Design of HMI

China's lunar landing suit is currently being developed, and China's extravehicular spacesuit "Feitian" was developed with reference to designs from US and Russian (Soviet) spacesuits. "Feitian" have a unique bionic structure with bending joints and good mobility. The helmet of the spacesuit is hemispherical and has a wide field of vision. "Feitian" spacesuit uses digital information communication technology with integrated personal information processing control system. The gas-liquid console of the spacesuit is highly integrative and easy to operate and control. OLED technology is applied to the display screen, which is more energy-saving and efficient [8].

New forms of interaction and new electronic technologies are being explored for the new generation of moon-landing-spacesuits to improve its functionality and reduce the overall weight. The design process of the spacesuit can benefit from the visualization of conceptual designs. For example, new interactive possibilities such as the integrated arm display HMI module can be explored (Fig. 4).

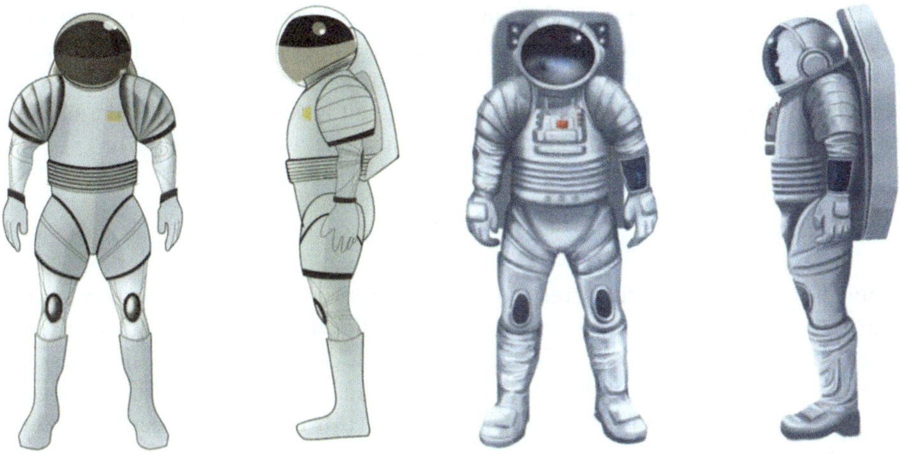

Fig. 4. Conceptual renderings of human-computer interaction design for lunar landing suit

The lunar landing spacesuit is a subsystem to the lunar module. Thus, the design of the spacesuit needs to consider its integration with the lunar module, and make full use of assistive technologies such as touch screen and voice recognition. Digital display screen can be used to complete tasks such as monitoring activities, video communication, immersive exercise, entertainment and more to increase the richness of information within the limited space, and to care for the psychological wellbeing of the astronauts.

Fig. 5. Conceptual rendering of HMI design between lunar lander and lunar spacesuit

The new concept design of the lunar module includes features such as reducing physical buttons, and increasing the use of touch screen and voice activated interactions. Physical buttons are only used as redundancy backup. Two airlocks are added to store the spacesuit. This conceptual lunar landing suit is designed to enter the suit from the back, which can be worn and taken off by just one astronaut (Fig. 5).

5 Conclusion

Manned lunar landing requires very complex engineering in the manned space system. The "astronaut" is an important system factor and thus the HMI design of the spacesuit is very important. Based on the concept of "human centered design", this study proposes that the adoption of the "EIIV" process framework for the HMI design of spacesuits for a variety of possible purposes. The HMI design of spacesuit should be based on the functional requirements, and the main goal is to improve safety, efficiency and comfort.

Acknowledgments. This research is funded by the National Key Laboratory of Human Factors Engineering open project fund, Project number: 20181450148.

References

1. Hejun, L., Xiangnan, N., Xin, Z.: Progress of Chang'e 5 probe project. J. Aerosp. China **17** (01), 72 (2016)
2. Weiren, W.: A good beginning of Chang'e project (Part 1). J. Aerosp. China **11**(01), 8–11 (2010)
3. Weiren, W.: A good beginning of Chang'e project (Part 2). J. Aerosp. China **11**(02), 5–9 (2010)
4. Apollo Lunar Suit. https://airandspace.si.edu
5. Skoog, A.I., Abramov, I.P.: The Soviet/Russian spacesuit history. J. Acta Astronaut. **60**(12) 1–10 (2006)
6. NASA releases the next generation of lunar landing suits. 美航天局发布下一代登月宇航服. http://www.xinhuanet.com
7. Hobday, M.: Product complexity, innovation and industrial organization. J. Res Policy **26**(6), 689–710 (1998)
8. Gang. L.: Human-Machine Interaction Design Process for Complex Product Systems. 复杂产品系统人机交互设计框架研究. D. Tsinghua University. Beijing (2019)
9. He, S., Liu, X., Li, T.: Design of airtight bearings in feitian EVA space suit. J. Space Med. Med. Eng. **26**, 387–390 (2013)

Integration of Multi-domain-Simulation into the Design of Cyber-Physical Systems

Tanja Steinbach(✉), Yübo Wang, Vladimir Kutscher, and Reiner Anderl

Otto-Berndt-Straße 2, 64287 Darmstadt, Germany
steinbach@dik.tu-darmstadt.de

Abstract. The development of smart systems is a multidisciplinary process that is gaining in importance. The share of software integrated in a smart product will continue to increase in the further course of Industrie 4.0 and the increased use of cyber-physical systems. At the same time, a prolonged time-to-market will not be accepted, so that the development of this kind of products must be further accelerated. This can be achieved by increasing the parallelization of the product development phases using supportive methods and tools.

Within the scope of the presented work, a concept for the integration of the corresponding embedded software into design of a smart product is presented and validated. This should enhance the interaction of the mechanical and software engineering of cyber-physical systems and support the parallelization of the development. Thus, on the one hand, the behavior of a system based on the original software can be considered in early phases of development. On the other hand, the development of embedded software can be assisted long before the first physical prototype.

The increased integration and parallelization of the development of smarter systems creates synergies that can positively influence not only the time-to-market but also the traditional goals of cost and quality.

Keywords: Industrie 4.0 · Cyber-physical systems · Embedded software

1 Introduction

The development towards more complex and individualized products, smaller batch volumes and intensified global competition sets high demands on industrial production. The consequences are increasingly complex, automated and networked manufacturing processes [1, 11]. Industrie 4.0 offers the potential to master these challenges by creating dynamic, real-time optimized, decentralized and cross-company value creation networks [2, 12].

The key technologies here are cyber-physical systems (CPS) and their interconnection via Internet of Things as well as their digital twin [3]. CPS cause an extensive elimination of the separation between virtual and real world [4]. CPS enables value creation systems to communicate with each other automatically and to determine the optimal production process independently. The division of the factory into small autonomous units therefore enables decentralized and self-controlling production,

G. Di Bucchianico et al. (Eds.): AHFE 2020, AISC 1202, pp. 451–456, 2020.
https://doi.org/10.1007/978-3-030-51194-4_59

which is referred to as a revolution from the classic production process to an *intelligent factory* or *smart factory* [5].

This global competition, as well as the increasing demand for individualized products, are the developments that make it necessary for companies to switch to Industrie 4.0. It stands for holistic integrated individualized virtual product development and flexibility of the production processes.

2 State of the Art and Challenges in Industrie 4.0

2.1 Challenges in Industrie 4.0

Industrie 4.0 is an essential topic, especially for manufacturing companies. Nevertheless, many companies shy away from the costs involved and the uncertainty of the benefits. Small and medium-sized enterprises in particular have not yet made much progress in the industrial revolution [6]. Since the reasons for these inhibitions lie largely in unsuspectingness [7], a tool for visualizing and illustrating the benefits of change can bring significant progress. This is to be built up in the course of the project, within the framework of which this work is also being developed. The mentioned tool is a virtual factory based on combination with multi-domain co-simulation systems, which represent not only the optical but also the functional properties of the corresponding real factory.

2.2 Multi-domain Co-simulation

Due to the specialization of the different simulation software, extremely accurate and realistic results are delivered. On the other hand, the individual focal points also have the disadvantage that the simulation programs are structured differently and therefore work with different simulation languages.

For this reason, it is necessary to create a separate model for each simulation program, which can be read and used by it. [8]. This makes it clear that a centrally controlled simulation, which uses the individual focal points of the different software, but can nevertheless work with one model or system, would be of great benefit.

An overall system that would be too complex for simulation as a whole is broken down into subsystems. For each of these subsystems, a model is then developed in the corresponding special program, which is then simulated. The results of the individual sub-simulations are coordinated and exchanged within the framework of multi-domain co-simulation via the co-simulation environment.

Depending on the availability of computing and storage capacities, as well as licenses, the simulations can run in parallel or sequentially. With parallel co-simulation, the total computing time is usually only slightly longer than the simulation time of the slowest model. Systems of co-simulations can be real-time capable or non-real-time capable. This means that a real hardware component can be integrated into a real-time capable system, which can be used, for example, as a test device. The main problem with such real-time co-simulations is that latency times through communication, calculations or data preparation lead to distortions in the representation of the real overall system [9].

3 Objectives and Conceptual Structure

3.1 Introduction to Robotics

Robot models can be divided into five different classes. These are the service robot, the mobile robot, the micro- and nanorobot and the humanoid robot, as well as the aforementioned industrial robot. Regarding to DIN 8373, a robot is an automatically controlled, freely programmable manipulator, programmable in three or more axes, which can be fixed or movable [10]. An articulated arm robot, which is used in industry, mainly for manufacturing processes is the focus of this paper.

3.2 Objectives of Multi-domain Co-simulation of Robots

The main objective is the development of a multi-domain co-simulation of a production robot for the connection of a trajectory control. After the implementation of the start and target values, as well as environmental conditions of the virtual space by the domain of the virtual environment, the optimal path of a previously modelled robot arm should be calculated (see Fig. 1). The movement of the robot arm along this path should be simulated and the corresponding results should be returned to the virtual reality environment. These results can be used to evaluate the properties of the robot arm and the task being performed.

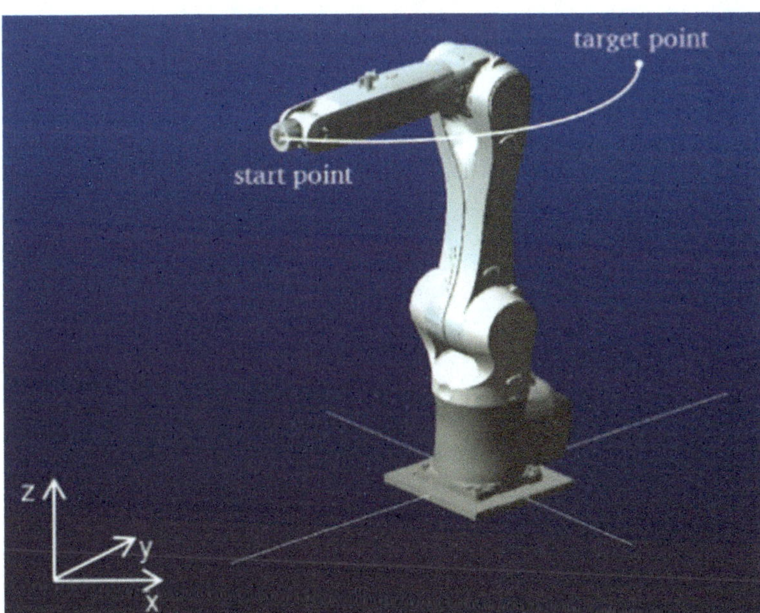

Fig. 1. Trajectory curve of the KR 10 R 1100 Agilus Sixx by KUKA

The aim of these objectives is to support companies in forcing the challenges mentioned above by using a platform and a virtual reality environment to advance the development of a CPS by using the functionality of a robot production cell in combination with a multi-domain co-simulation. Specifically, an existing or new factory is to be built in a virtual environment, including all individual production elements such as machines or conveyor belts. These will not only be reproduced realistically, but their functionality will also be simulated realistically. This allows a factory to be viewed and inspected before it is set up or rebuilt.

An additional aim of these objectives is to force predictive maintenance, which is based on the realistic multi-body simulation of machines and systems and enables faults or failures to be detected before they occur. With the help of this principle, production costs can be greatly reduced.

For the outlook, by networking the virtual factory with the real factory, including sensors, it will be possible to check the factory at any time and from any location and, in the next step, also to control it.

3.3 Structure of the Multi-domain Co-simulation

Regarding the objectives, a conceptual structure of a multi-domain co-simulation is presented in Fig. 3. The used programs are: Unity, which is one of the leading software for virtualization and animation; Matlab, which is one of the leading software for multi-paradigm numerical computing; MSC Adams, which is one of the leading software for multi-body-simulation.

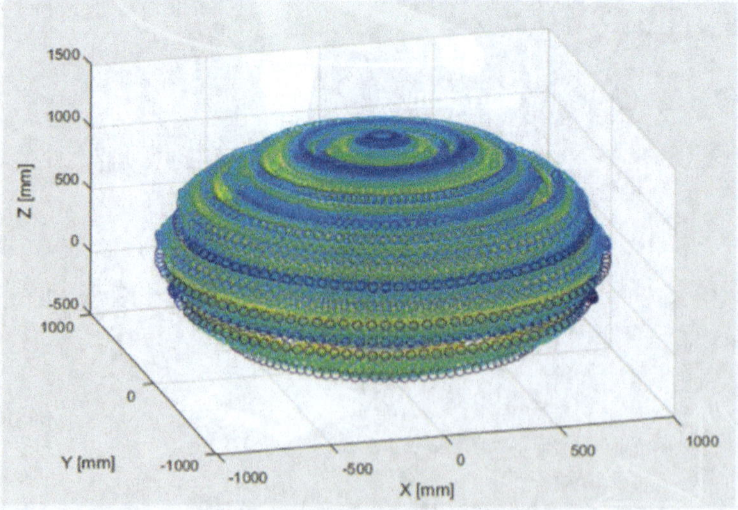

Fig. 2. Robot condition grid

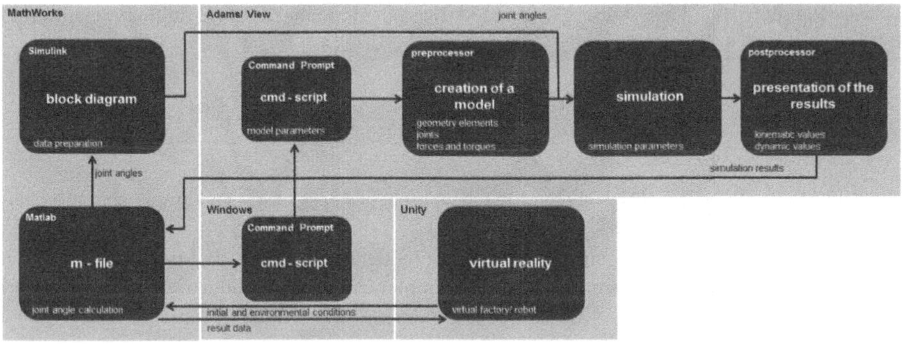

Fig. 3. Process of multi-domain co-simulation

The process starts with the implementation of the initial and final conditions and the boundary conditions. These are selected in Virtual Reality, which is represented by a robot KR 10 R 1100 Agilus Sixx by KUKA or a whole virtual factory could be selected, and passed to Matlab via the program Unity.

A script in Matlab, the so-called m-file, prepares these input values for further processing. While the input values are available as points in the robot condition grid (see Fig. 2), MSC Adams requires input data in the form of angular sizes of the six joints of the robot arm.

In Matlab, the optimal path from the starting point selected in Unity to the target point is calculated and converted via a robot condition grid for each point of the trajectory curve into the individual joint angles, which can be processed by MSC Adams. When calculating the optimal path, possible obstacles must be taken into account in addition to the goal achievement.

After the robot model is available in the preprocessor of MSC Adams, the joint angles are implemented by Matlab. MSC Adams simulates the given movement of the robot with all kinematic and kinetic properties and returns the results of the simulation to Matlab, where they are further processed and returned to Unity for visualization. Additionally, in case of model change request Windows Command can be used for the implementation of the robot model.

4 Summary and Outlook

In this paper a process of multi-domain co-simulation with an integrated optimal trajectory control for a production robot is developed. The trajectory control is based on a robot condition grid. Therefore, the fundamental basement for a VR environment to advance the development of a CPS by using the functionality of a robot production cell in combination with a multi-domain co-simulation is established.

Consequently, the extension from one single robot to a holistic virtualization of a production environment. Therefore, new factories and production resources are able to be commissioned in a virtual environment. Herein, it is necessary to extend the research effort in simulation of a robot Programmable Logic Controller.

Acknowledgements. This work was funded by the Hessian LOEWE initiative within the Software-Factory 4.0 project.

References

1. Anderl, R.: Industrie 4.0 – technological approaches use cases and implementation. at - Automatisierungstechnik **63**, 753–765 (2015)
2. Anderl, R., Picard, A., Wang, Y., Fleischer, J., Dosch, S., Klee, B., Bauer, J.: Guideline Industrie 4.0 – guiding principles for the implementation of Industrie 4.0 in small and medium sized businesses, VDMA Forum Industrie 4.0, Frankfurt (2015). ISBN: 978-3-8163-0687-0
3. Bauernhansl, T.: Die Vierte Industrielle Revolution – Der Weg in ein wert-schaffendes Produktionsparadigma. In: Vogel-Heuser, B., Bauernhansl, T., ten Hompel, M. (Hrsg.) Handbuch Industrie 4.0: Bd. 4: Allgemeine Grundlagen, 2. Aufl. pp. 1–31. Springer Vieweg, Berlin (2017)
4. Hinrichsen, S., Jasperneite, J.: Industrie 4.0 – Begriff. Stand der Umset-zung und kritische Würdigung Betriebspraxis & Arbeitsforschung **2**, 45–47 (2013)
5. Umsetzungsempfehlungen für das Zukunftsprojekt Industrie 4.0: Abschlussbericht des Arbeitskreises Industrie 4.0, Hrsg. Henning Kagermann, Wolfgang Wahlster, Johannes Helbig (2012) https://www.iosb.fraunhofer.de/servlet/is/21752/Umsetzungsempfehlungen_Industrie_4.0_final_2012-10-02.pdf?command=downloadContent&filename=Umsetzungsempfehlungen_Industrie_4.0_final_2012-10-02.pdf. Accessed 09 Jan 2019
6. Demary, V., et al.: Digitalisierung und Mittelstand. Eine Metastudie. Unter Mitarbeit von Barbara Engels, Klaus-Heiner Röhl und Christian Rusche. Köln: Institut der deutschen Wirtschaft Medien GmbH (IW-Analysen, Nr. 109) (2016). http://hdl.handle.net/10419/157156
7. Leyh, C., Bley, K.: Digitalisierung: Chance oder Risiko für den deutschen Mittelstand? – Eine Studie ausgewählter Unternehmen. HMD **53**(1), 29–41 (2016). https://doi.org/10.1365/s40702-015-0197-2
8. Voss, V.: Wiederverwendbare Simulationsmodelle für die domänen- und disziplinübergreifende Produktentwicklung. Zugl.: Stuttgart University, Diss. (2012). Heimsheim: Jost-Jetter (ISW/IPA Forschung und Praxis, 192) (2012)
9. Zehetner, J., et al.: Echtzeit-Co-Simulation für die Regelung eines Motorprüfstands. In: ATZ - Automobiltechnische Zeitschrift 116. Jahrgang. Online verfügbar unter www.springerprofessional.de/ATZ. Accessed 15 Oct 2019
10. Abele, E.: Werkzeugmaschinen und Industrieroboter. Vorlesungsskript. Technische Universität Darmstadt, Darmstadt. Institut für Produktionsmanagement, Technologie und Werkzeugmaschinen (2014)
11. Kagermann, H., Wahlster, W., Helbig, J.: Umsetzungsempfehlungen für das Zukunftsprojekt Industrie 4.0: Abschlussbericht des Arbeitskreises Industrie 4.0, Hrsg. acatech - Deutsche Akademie der Technikwissenschaften e.V. (2013)
12. Plattform Industrie 4.0: Umsetzungsstrategie Industrie 4.0: Ergebnisbericht der Plattform Industrie 4.0, Hrsg. Bundesverband Informationswirtschaft, Telekommunikation und neue Medien e. V., Verband Deutscher Maschinen- und Anlagenbau e.V., Zentralverband Elektrotechnik- und Elektronikindustrie (2015)

Integrated Keyword Mapping Process: Bridging User Research to Style Concept Development

Sangjin Joo, Dabin Lee, Soyoon Park, Hwang Kim, and KwanMyung Kim(✉)

Graduate School of Creative Design Engineering,
Ulsan National Institute of Science and Technology, 104-904, 50 UNIST-gil,
Eonyang-eup, Ulju-gun, Ulsan, Republic of Korea
{jimmy928,kkang58322,hwangkim,kmyung}@unist.ac.kr,
yoonyoon1014@gmail.com

Abstract. In this paper, we introduce 'Integrated Keyword Mapping' that will be helpful to connect user research data to concept visualization. Designers commonly experience a *broken chain situation,* a logic lacking situation when they jump into idea sketch phase after user research. This appears to happen due to the non-existence of appropriate methods that can be used for translating user research data to visual language. In our design process of a brainwave detecting smart hard hat, we conducted user research through user observations and focus group interviews. When we translated the user research result to visual, we met a similar situation. To successfully link the research data to the design, we invented the method, with which we drew out concept keywords not only from the research data but also from the designer's interpretation. We found that this provides a logical connection between the user research and the design outcomes.

Keywords: Design method · Integrated keyword mapping · Broken-chain situation

1 Introduction

Designers commonly expect that they can draw out logical design outcomes while they are employing Human-Centered Design methods. However, in some cases, the design outcome may lose logical connection with the initial user research results. We call this *'Broken-chain situation'*. Lindemann shows that the mental contents of the designer often play a role in design errors [1]. Thus, in order to navigate away from such errors, the design methods are often used; to reduces the size of the mental step required in the design process [2]. Our design team carried out a design project of a brainwave detecting smart hard hat where we discovered ourselves causing a Broken-chain situation. This design project, currently ongoing in UNIST, began in June 2019. At the beginning of the design process, we conducted thick user research with shadowing – an observational method and focus group interviews, leading the research data to various design insights. We started the design development based on this user research data. In

G. Di Bucchianico et al. (Eds.): AHFE 2020, AISC 1202, pp. 457–463, 2020.
https://doi.org/10.1007/978-3-030-51194-4_60

this paper, we address a problematic situation in the design development and a substitute design method that has a potential of preventing such situations.

2 Hard-Hat Style Design

In our design project, we firstly conducted user research to understand the user and identify design problems and solution directions. To generate style concepts, we then generated style concept keywords based on our interpretation of the user research data. However, we found that the initial style concepts of the helmet represented by drawings had a weak connection with the user research outcomes. We assumed that this happened because designers' interpretation was made subjectively and institutively, which is very common in actual design practice. In order to logically connect the research outcomes to style concepts, we conducted another session of keyword mapping. This attempt included not only designer interpretation-based keywords but also keywords derived from the user research data.

2.1 Initial Method

Initially we conducted a keyword generation session based on our knowledge of the users (Fig. 1). This process was administrated by only the members who participated in the user research process – observation and interviews. Because we all had a good understanding of our research data, the keyword generation was not a difficult task, rather quick and easy. In this process, we came up with some of the product brands that presented relevant atmosphere to our desired outcome. Based on the key brands, we branched keywords out while we considered the style direction of the smart hard hat.

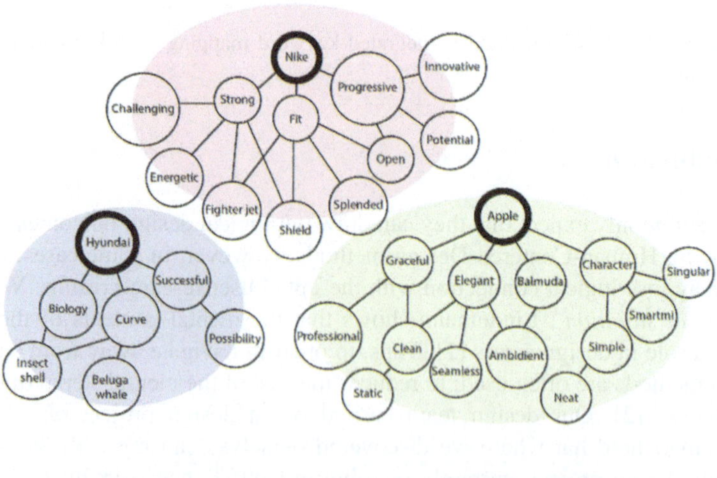

Fig. 1. Keyword generation for the hard-hat design

We generated style concepts through rough sketches and grouped them into 10 styles (Fig. 2). It was difficult to narrow down the style concepts into one direction. We discussed the sketch results and came to a mutual understanding that the sketch did not represent the overall context of the user. This finding led us to revisit the process and check how the design went wrong. We discovered that the keywords generated were vague and could be translated into many different emotions. This appeared to result in diverse visual translations. This is because emotion by its nature is implicit, and inexperienced researchers may have a problem to uncover the verbal descriptors to represent the emotional state [3]. After the discussion, we concluded that the keywords did not have a reasonable link to the user research data, and therefore decided to conduct a keyword generation process again.

Fig. 2. Initial style concept sketches

2.2 Broken-Chain Situation

In the design process it is necessary for the designers to keep in mind of the user spectrum, needs, and wants. Contextual Design approach, which defines what user needs to address, what the system should do, and how it should be structured, based on the criterion of data gathered from the users, is one way to keep user's experience on priority [4]. To gain an objective perspective on the target user, it is essential to conduct a user research. The designers need to stay with the research result during the whole design process to create a user-friendly outcome. However, in some situations, designers may lose a link of the research data to the design outcome. This situation happed in our design team.

We named this 'Broken-chain situation' (Fig. 3). This is a situation when the designers proceed to a design result, they lose the link of the initial user research data to the design outcomes. In Fig. 3, as designers proceed with the design process, the design outcome can be developed to 'result B,' which has low relevance to the research data compared to 'result A.' Its low relevance means a broken connection between the user research data and the design outcomes. This situation would make designers produce design outcomes that are irrelevant to the user research data. This situation has cost us time, which was invaluable to most of the design process.

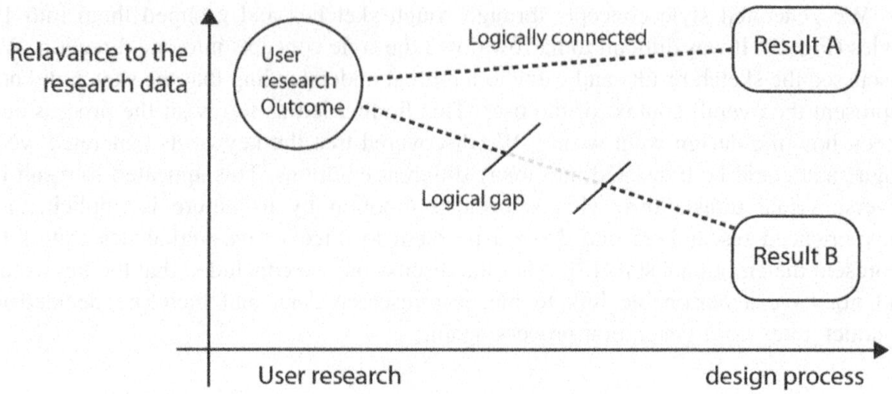

Fig. 3. Broken-chain situation

2.3 Integrated Keyword Mapping

Designers need to consider the users while designing because involving users in design leads to satisfying designs [5]. To do that, designers utilize various user-focused methods. In our design process, we utilized focus group interviews with a wide range of stakeholders and observations – shadowing and fly on the wall.

We generated keywords from the original focus group data. To successfully regenerate the keywords from the user data, we reviewed the interview scripts and photos from user observations. After reviewing through the research data, we generated keywords based on the actual data, while deciding reasons of the generation.

Table 1. Keywords extracted from the user research data

Keyword	Data	Reasoning
Insurance	*I would have ended in an accident if it weren't for the hat*	hard hat works as an insurance
Diligence	*We usually come to the site around 6 in the morning*	It is a cultural virtue to be diligent
Equal	*It's not the military. You don't go on ordering workers*	It's an equal relationship rather than subordinates and superiors
Veteran	*I started working since 1990 until 2016*	know-how of the work usually inherits to the next generation of workers
Strong	User photo of lifting heavy objects	Tough work that requires strengths

Table 1 shows the process and reasoning of the keyword generation. In this process, we individually reviewed the data and came up with potential keywords along with the associated user research data, and reasons that support the validity of the keywords. The process produced more than 100 keywords. After that, we screened out keywords that had weak reasons. Finally, we constructed a keyword map by linking and grouping the remaining keywords according to their meanings.

The co-evolution of problem solution is one theory that suggests that iteration between problem space and solution space is important in creative design [6]. Because we discovered the keywords from the user data, they are considered to belong to the problem space. After mapping keywords into different groups, we noticed that the map was incomplete and lacking because we felt that there were no keywords indicating designers' holistic interpretation based on their own experience with the users. Usually, designers tend to capture solution directions immediately when they are exposed to design problems as Darker (1965) mentioned about 'primary generator' [7]. Therefore, keywords proposed by designers based on their own interpretation can be considered to belong to the solution space.

We supplemented the keyword map with the keywords derived from our holistic interpretation about the users and the context (see words in red in Fig. 4). Finally, because this keyword mapping was made to support designers to create style concepts, we added visually reminiscent keywords; keywords that can remind us of visual moods of the style direction such as 'whales,' 'Admiral Yi,' 'knights,' etc. We called this method 'Integrated Keyword Mapping,' a method where we collected keywords from three different bases; user research data, designers' interpretation and visual reminiscence, and constructed them as a single map.

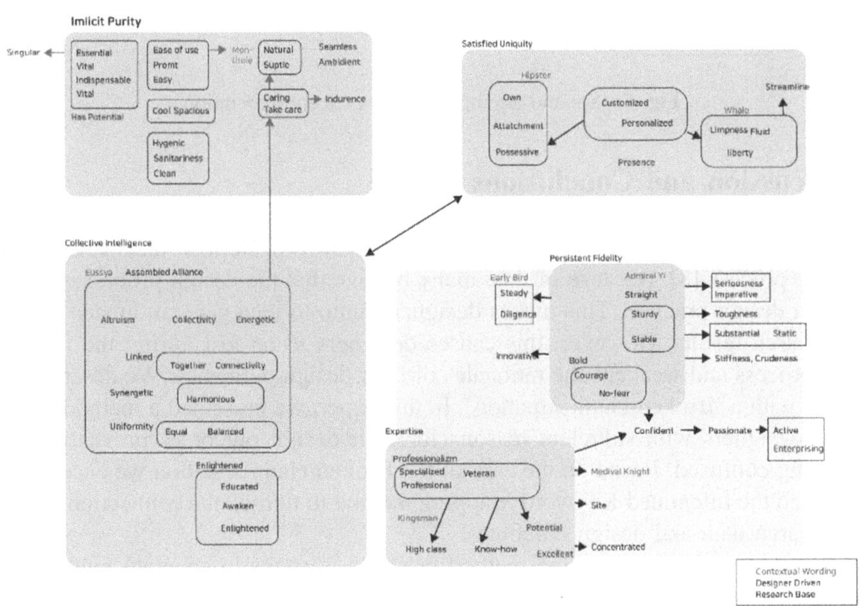

Fig. 4. Integrated Keyword Map

3 Results

We generated additional style concepts (Fig. 5) based on the Integrated Keyword Map. The sketches were grouped into 7 different concepts. By using the Integrated Keyword Map, we generated design outcomes without producing a process gap by going through well-defined steps starting from the user research data. This also helped us explain how and why our design outcomes were generated in such a way without harming our creativity.

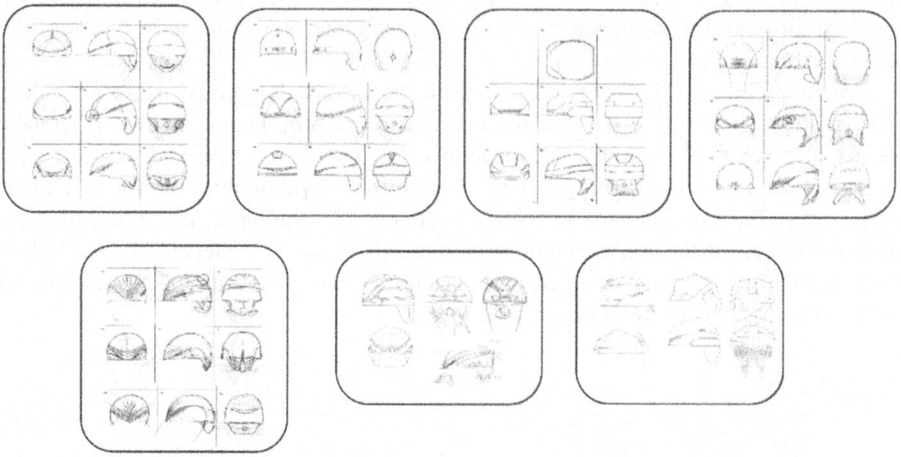

Fig. 5. Second attempt of style concept generation

4 Discussion and Conclusions

It is said that generating design ideas is involved in a mysterious mechanism in our cognitive process [8]. Because of this, many believe that the design process cannot be logical in design practice. This makes designers jump to idea generation step ignoring user research results. However, this causes designers to be lost during the idea generation process and weakens the rationales of their design outcomes. We described this situation with a 'Broken-chain situation.' In this paper, we presented a method that can provide designers with sufficient reasons for their design outcomes preventing them from being confused. Based on the case of the Broken-chain situation we encountered, we devised the Integrated Keyword Mapping method to tighten the connection between user research data and design outcomes.

Integrated Keyword Mapping method provides a means to generate multi-layered keywords from user research data, designers' holistic interpretation and visual reminiscence. The method helped us link the style concepts to the user research data. With this process we were able to convert the research data (text and pictures of users) to exterior sketches smoothly. We believe that this method can contribute to filling the process gap occurring frequently in current user-centered design methods.

Acknowledgements. The work was supported by the UNIST funded product development project (1.190110.01).

References

1. Lindemann, U.: Human Behaviour in Design. Individuals, Teams, Tools. Springer, Heidelberg (2003)
2. French, M.J., Gravdahl, J., French, M.: Conceptual Design for Engineers. Springer, Heidelberg (1985)
3. Lokman, A.M., Kamaruddin, K.A.: Kansei affinity cluster for affective product design. In: International Conference on User Science and Engineering (i-USEr), pp. 38–43. IEEE (2010)
4. Beyer, H., Holtzblatt, K.: Contextual design. Interactions **6**, 32–42 (1999)
5. Abras, C., Maloney-Krichmar, D., Preece, J.: User-centered design. In: Bainbridge, W. (ed.) Encyclopedia of Human-Computer Interaction. Sage Publications, Thousand Oaks vol. 37, pp. 445–456 (2004)
6. Dorst, K., Cross, N.: Creativity in the design process: co-evolution of problem–solution. Des. Stud. **22**, 425–437 (2001)
7. Darke, J.: The primary generator and the design process. Des. Stud. **1**, 36–44 (1979)
8. Lawson, B.: How Designers Think. Routledge, London (2006)

Emotional Interaction Design for Driverless Food Truck Under Campus Scenarios

Ming Yan[1,2], Jianmin Wang[1(✉)], and Wenjuan Wang[1]

[1] Shenzhen Research Institute of Sun Yat-Sen University, Shenzhen, China
wangjianmin@tonji.edu.cn

[2] Car Interaction Design Lab, College of Arts and Media, Tongji University, Shanghai, China

Abstract. By understanding the background and significance of the development of driverless food truck, the trend of vehicle modeling and its emotional interaction are analyzed. The use scene and function of the truck were positioned, and the modeling and design elements of the truck in the future were summarized. Use industry Knowledge of design and interaction design to build prototypes of smart driverless food truck. The online and offline methods are used to plan the driving environment and the vehicle driving routes. Design the exterior appearance of the vehicle docking station, and analyze the usage scenarios between the driverless food truck and the consumer to rationally plan the vehicle location, the information required by the user, and the operation mode. This paper combines industrial design with emotional design, service design, and perceptual engineering to design an emotional interactive system for driverless food truck under autonomous driving technology. So as to deeply consider and explore the interdisciplinary practice of industrial design.

Keywords: Autopilot · Driverless food truck · Modelling · Industrial design · Emotional interaction design

1 Introduction

Nowadays, in the urban scenic areas and campus scenarios, there is a large pedestrian flow, and the existing vending machines have a common program of merchandise sales and labor-intensive. This has led to increased demand for mobile machines for vending in public spaces. With the development of driverless technology, innovative combinations of vending machines and the technology become achievable.

This paper analyzes the autopilot technology and promotes them to be combined with sales and tourism economic industries. The design not only includes relevant driving assistance information from vehicles, warehouse goods and vehicle docking stations, but also effectively communicates online with users. Focusing on buyers, occupants and other pedestrians, the vehicle's vending system is more intelligent.

Shenzhen Collaborative Innovation and Technology Program - International Science and Technology Cooperation Project; National Social Science Fund(19FYSB040); Tongji University's Excellence Experimental Teaching Project.

G. Di Bucchianico et al. (Eds.): AHFE 2020, AISC 1202, pp. 464–471, 2020.
https://doi.org/10.1007/978-3-030-51194-4_61

Improve product operability, safety and Usability. The purpose of this paper is to push the design of the emotional interactive system of mobile vending vehicle under the autonomous driving technology to the "conventional operation" mode, and provide mobile vending vehicle for sale in different regions such as tourist attractions, industrial parks, campuses, amusement parks and even the entire city.

2 Design Survey and User Analysis

The design is user-centric, focusing on the actual needs of users and improving the product experience. By determining the pain points and needs of users when using existing products, study their thinking habits, required functions and usage scenarios. Summarize the functional architecture of the service system to provide a basis for the vehicle's styling design [1].

2.1 User Research

The survey is divided into three different stages: quantitative research, qualitative research and results analysis. Quantitative analysis refers to distributing questionnaires to target groups via the Internet, analyzing larger samples and understanding user expectations [2]. Qualitative analysis refers to recording user usage process through user interviews and observations, analyzing a small number of specific samples and discovering the opinions and suggestions of different types of users [3]. Finally, through the research in the two stages, the user experience map and service blueprint during the entire process are explored, and the functional architecture of the system is designed.

2.2 Investigation and Analysis

According to user research, the sales scenarios that trucks should have can be divided into four processes: goods preparation, goods distribution, sales at the predetermined location, and return to the storehouse after the sale. Among them, the process of user participation is concentrated in the stage of reaching the predetermined place for sale. This stage can be divided into four stages: finding a sales vehicle before shopping, understanding the sales process and information of the sales vehicle, purchasing goods, and experiencing after-sales service (As shown in Fig. 1).

3 Driverless Food Truck Design

Summarize the driverless food truck with automatic driving function in different countries in the existing market, conduct a practical analysis on them, and improve them according to user needs. Propose the direction of product improvement, including the improvement of the interactive interface, the upward movement of the shipping site and the demand for the development of online programs and docking sites.

Fig. 1. Product usage scenarios

3.1 Product Positioning

This truck is positioned as a minibus in the first-tier city scenic spot, 6–8 seats, pure electric, short-distance travel, with the functions of selling goods and picking up passengers for sightseeing. Supported by self-driving technology, multiple people can make appointments and make online purchases at the same time, realizing the user's shopping needs and emotional care for co-touring. The system sets corresponding docking stations, charging sites, and warehouses in the fixed tour area to meet its operational requirements. Create a safe, comfortable and convenient user experience for users.

3.2 Modeling Design

Carry out detailed positioning of the product design plan from the vehicle structure, vehicle size, functional positioning. Structure positioning [4]: The main function of this product is to make short trips in scenic spots and campuses, so it is positioned as a mini car. Size positioning: Since the average small bus can carry between 8 and 20 people, but this sales vehicle is suitable for places with large passenger flow, so its volume should be reduced to facilitate its shuttle in scenic spots to develop services. The size is about 7000 mm $\times$ 2500 mm $\times$ 3050 mm. Driving function positioning: The cruising range is about 100 km, which can meet the daily driving. The maximum speed is located at 60 km/h, which can be used in areas with large traffic. Carrying function positioning: the function of satisfying the storage and purchase of passengers' purchase items required by users for daily travel.

By planning the layout of the vehicle, it can be divided into the following four parts: (1) passenger seating space and seats (2) product storage and collection space (3) door structure (4) the exterior shape of the vehicle.

The overall layout of the interior of the vehicle is shown in the following figure. On the left is a hand-drawn sketch of the internal layout. From which, a plan that is more in line with function and structure is selected and marked as light blue. According to the plane structure of the sales vehicle 7000 mm $\times$ 2500 mm, it is divided into two areas on average, which are the user pickup area (pink) and the ride area (blue), in which a sale and pickup machine is placed in the pickup area. The area uses half-packed ring seats that can accommodate up to 8 people at the same time. The open area between the pickup area and the riding area is set up for special users who are in wheelchairs or wish to stand on the vehicle and enjoy the view (Figs. 2, 3, 4, 5).

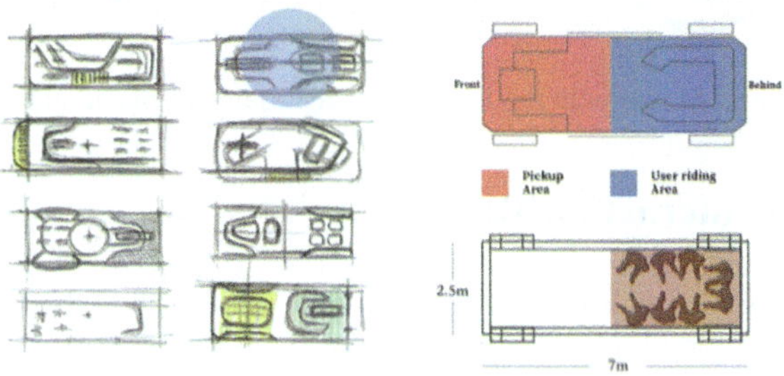

Fig. 2. Driverless food truck interior layout

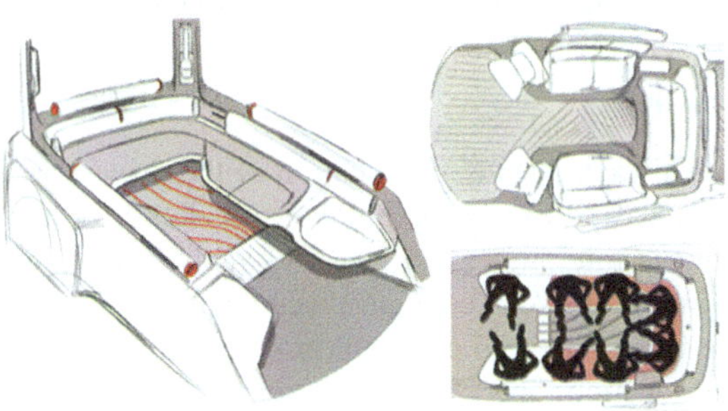

Fig. 3. Passenger seating space and seat

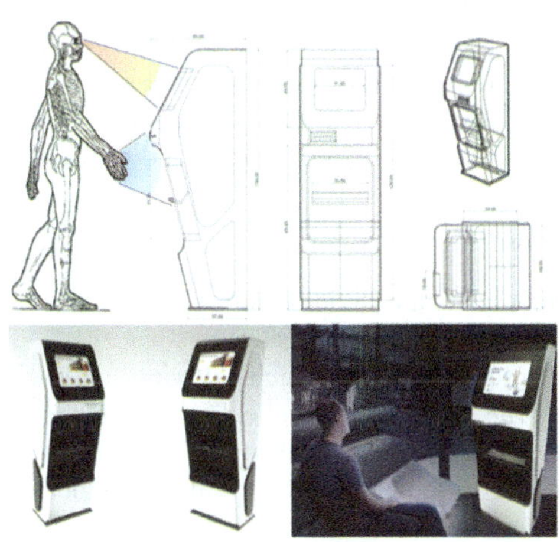

Fig. 4. Use scene & three views of sales and pickup space

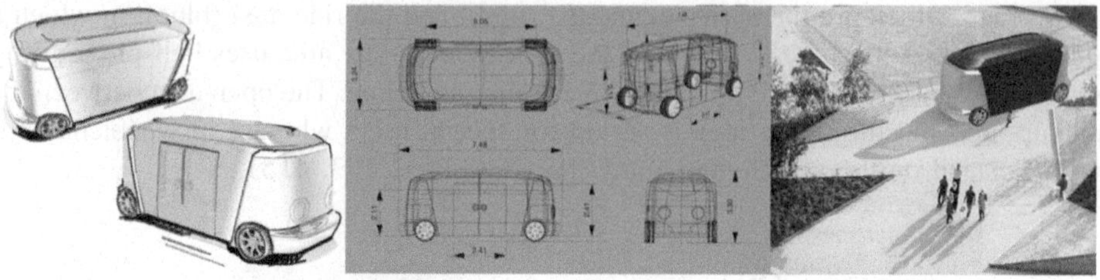

Fig. 5. Vehicle modeling & Three views & Scene graph

4 Emotional Interaction Design

The future driverless car is not only a controllable vehicle, but also a "friend" who can establish emotional communication with people [5]. The services of this system include the following six parts, namely service equipment providers, users, retailers, trucks, online programs and databases. Truck sales and online programs are the media of the entire service system, linking users, equipment providers and retailers, and sharing information through the database in the background to realize their respective interests.

4.1 Online Interaction

The mobile APP can always display the location of the driverless food truck, which is convenient for users to find shopping points, and obtain online information about product quality, temperature, inventory, etc., to realize the function of booking in advance before buying, using scan code and face payment, After the success, the APP can display the four stages of cargo preparation stage, cargo delivery stage, sales stage at the predetermined location, and carrying passengers according to the running status of the sales vehicle. In addition, the online data platform is also used to ensure the safety of driving. It uses GPS technology and data statistics background to allow the machine to actively find someone or guide the user to the nearest site to pick up the goods, improving the service experience.

After purchasing the product, if there are problems such as product quality, or failures are encountered during the purchase process, such as the machine is not actively shipped or the inventory expires. Users can enter the after-sales service through the APP, making the service more direct and efficient.

4.2 Offline Interaction

Offline interactive components are divided into multiple parts such as road markings, vehicle docking stations, and off-board hardware.

During the whole process of sales vehicles driving, all roads will be marked with corresponding signs on the road surface, indicating that this is the designated route for unmanned sales vehicles, reminding tourists to pay attention to avoid. Within the scope of the docking station, the pavement marking will also indicate the real-time status of the station through text and color.

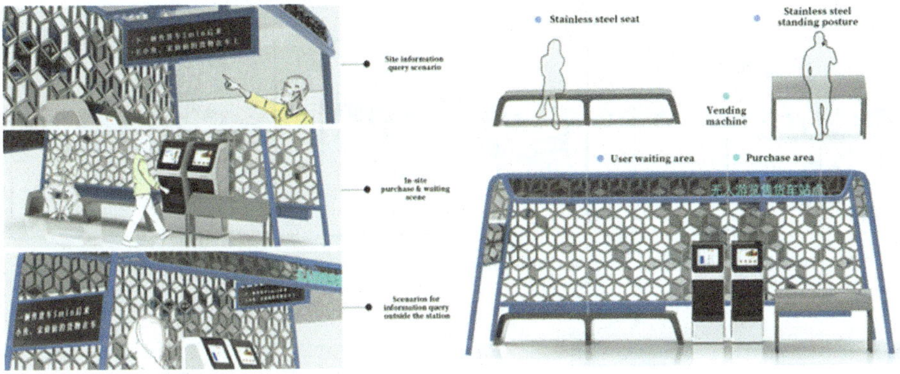

Fig. 6. Dock site area division and usage scenarios

According to the user's functional requirements for the site, it is divided into four parts: user rest area, information prompt area, shelter area and shopping area (Fig. 6).

In order to form better communication between vehicles and people, this driverless food truck will use LED matrix and holographic technology to realize the grouping, pattern and digital projection of lights to remind pedestrians and system administrators on the road of the vehicle's operating state [6], according to different states change (Figs. 7 and 8).

Fig. 7. Emotional interactive display outside the vehicle

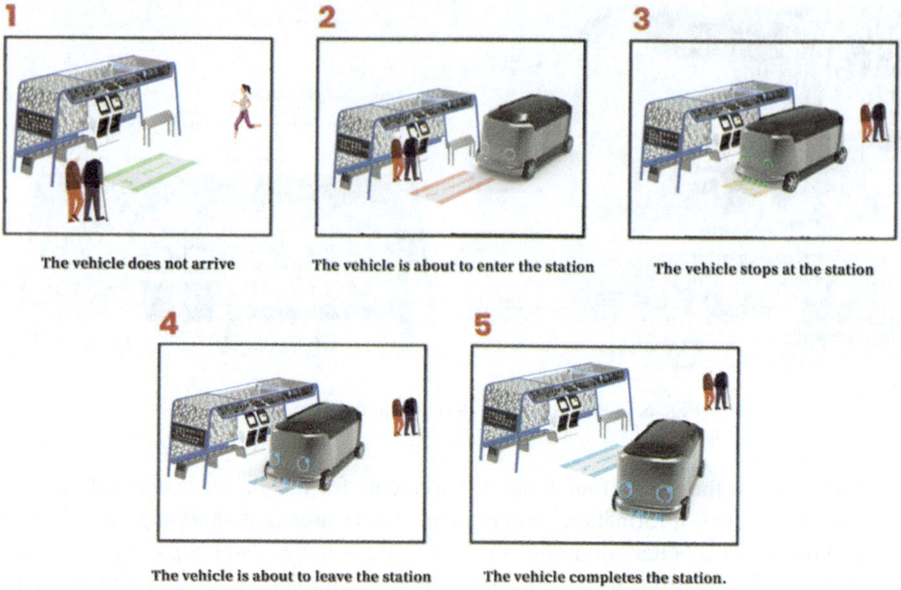

Fig. 8. The system's emotional service design

5 Conclusions and Future Studies

This design uses online and offline interactive interaction methods, detailed design of vehicle layout and styling, road markings, and truck docking stations, and specific positioning of the overall service system in combination with lights and other interactive components of the body. The system's emotional service design is divided into five typical system scenarios: the vehicle does not arrive, the vehicle is about to enter the station, the vehicle stops at the station, the vehicle is about to leave the station, and the vehicle completes the station. Through the different forms of car lights, the electronic display information of the bumper part, the color of the site ground and the site display information, the overall service provided by the system is carefully displayed.

In future research work, we will continue to deepen or adopt different theoretical support and research methods, so as to more objectively and deeply develop the research on the emotional interactive system of unmanned tour sales vehicles under automatic driving technology.

References

1. Barry, R.G.: Norwegian contributions to Arctic environmental sciences from the 1880s to the third International Polar Year. J. Adv. Polar Sci. **27**(01), 1–7 (2016)
2. Zhang, X., Tse, W.W., Victoria, S.: Improved properties of the big five inventory and the rosenberg self-esteem scale in the expanded format relative to the likert format. J. Front. Psychol. **10**, 1286 (2019)

3. Dwey, A.: How to Illustrate and Design Concept Cars M. Vekce Publishing Lid. (2017)
4. Delarge, C.A.: Story telling as a critical success factor in design processes and outcomes design. Manag. Rev. **2**, 76–79 (2004)
5. Kim, H.K., Han, S.H., Park, J., Park, J.: Identifying affect elements based on a conceptual model of affect: a case study on a smartphone. J. Int. J. Ind. Ergon. **53**, 193–204 (2016)
6. Chang, Y.M., Chen, C.W.: Kansei assessment of the constituent elements and the overall interrelations in car steering wheel design. J. Int. Ind. Ergon. **56**, 9705 (2016)

Interdisciplinary Approach to Design Innovation

An Interdisciplinary Approach to Accessible Museum Exhibitions

Carla Ayukawa(✉) and Lois Frankel

School of Industrial Design, Carleton University, Ottawa, ON, Canada
carla.ayukawa@carleton.ca

Abstract. As museums continue to satisfy their legal requirements to deliver accessible services for people with sensory, mobility and cognitive impairments, there is a growing parallel concern to provide visitors with exhibitions that expand their social identity and user experience. This literature review explores the possibility of a new approach that could improve accessible services and visitor experience through emerging technologies in the areas of personalization, rapid prototyping, and ability-based design. This proposed approach considers an interdisciplinary co-design method that combines technological and disability expertise with a visitor-centered approach to accessible museum exhibitions. This literature review explains how it could create an interface which respects the visitor's abilities, previous knowledge, and preferences and at the same time, provide a unique and engaging social experiences.

Keywords: Exhibition design · Personalization · Ability-based · Co-design · Constructivist approach · Participatory design · Visitor-centered · Museum identities

1 Introduction

This literature review explores the possibility of a new approach to developing museum exhibitions. It presents perspectives from the literature that show that such an approach may be timely and crucial; it addresses how the limitations of the current system in its ability to offer accessible designs gives rise to diminishing visitor experience. The literature addresses philosophies governing this approach that are applicable to areas that aim to create improved and meaningful user-experiences.

According to the literature, the delivery of accessible services and environments has become a focus for institutions and governments. Museums have responded by developing accessibility standards and guidelines (Standards) that define the visitor and their accommodations have incorporated accessible design into their exhibitions. However, the traditional exhibition development approach leads to designs that sometimes sacrifice a visitor's museum experience [1, 2].

This paper examines the cause of the diminishing visitor experience and through a literature review, finds support for an approach to accessible museum development that merges ability-based design processes and technologies, visitor-centered approaches, and personalization.

G. Di Bucchianico et al. (Eds.): AHFE 2020, AISC 1202, pp. 475–483, 2020.
https://doi.org/10.1007/978-3-030-51194-4_62

2 Background

2.1 Museum Exhibition Development – Four Hundred Years of Tradition

Museums have a reputation as the holders of enlightenment and culture for the benefit of society [3]. Museums, over the last four hundred years, have evolved from personal collections to public collections with a mandate to curate, organize, interpret and present exhibitions which communicate a story or idea [4, 5]. They are careful when making changes and as a result the museum exhibition tradition creates a unique ecosystem for museum exhibition development [1].

Traditional exhibition development process is referred to as information-centered or a didactic form of engagement where the museum offers two things to the visitor: their collection and their knowledge. It begins with a core idea and involves curatorial research. The curator, who evaluates, selects and researches the museum's object collection, would develop an exhibition brief that includes a core idea, theme and storyline. An interpretive planner organizes the curatorial information, develops the exhibition treatment and writes the content. Together they create the exhibition brief and interpretive plan. The exhibition designers use the interpretive plan to translate the exhibition content into a 3D environment [4, 6, 7] (see Fig. 1).

Over the last few decades, this information-centered approach evolved in part due to the visitors' desire for less didactic and more physical, social, and sensory experiences. Through research into visitor motivations, the visitor experience was formulated, which later led to defining a visitor-centered approach for museum exhibitions [8, 9].

Fig. 1. This figure outlines a traditional hierarchical system with an Information-Centered Approach to exhibition development. Standards tend to be included as part of the exhibition design phase.

2.2 Inclusive Access to Exhibitions

Today, most museums cater to a broad visitor profile and are legally required to be accessible to people with sensory, mobility and cognitive impairments. Most organizations use Standards and accessibility accommodations as a solution to creating accessible exhibitions is based on the medical model of disability [9].

One of the methods that Standards use is "multi-modal formats" of the subject content. A museum's traditional didactic content is translated into various sensory modalities, which could include haptic, visual and auditory information (e.g. Braille, video with closed captions and audio). It could also include hierarchies of text to accommodate different levels of cognitive load and language skills [10, 11].

The literature discusses mindsets that can arise toward the Standards. They are sometimes considered to be a "box that requires ticking" and to limit an exhibition designer's creative freedom [2]. This results in accommodations as add-ons to a exhibition's budget or an after-thought in an interpretive plan which leads to the problem of competing accommodations [12]. Museums can move away from creating competing accommodations and their effects, by changing their mindset about disability from the medical model to a model that focuses on ability [13]. Museums can create exhibitions that automatically conform to the visitors' abilities [14].

3 Approaches to Accessible Museum Exhibitions

The traditional method of creating accessible museum exhibitions is not sustainable [15]. Institutions must realize that in order to create meaningful advances towards engaging, valued, and accessible exhibitions, they must make substantial investments towards changing their attitudes. An information-centered approach could become a visitor-centered approach. The disability model could evolve to an ability-based model. The following literature review explores relevant contexts that may contribute to a new approach to accessible museum exhibitions by addressing the development of a visitor-centered and ability-based exhibition while also improving the user-experience.

3.1 The Visitor-Centered Approach

The creation of the visitor-centered approach was driven by the realization that the exhibition content may not be the primary driver of people's motivation to visit the museum. Traditional visitor categorization using quantitative variables: age, level of education, gender, social arrangement (family, school group, tourists) or frequency of visit was sometimes the driver of the messaging and content of an interpretation plan [1, 8].

A study of museum visitors revealed motivations went beyond just their affinity for the subject being presented. Museum visitors indicated that they are often only vaguely aware of what is on display when they make their decision to visit. The research by Falk (2009) led to his creation of five museum visitor identities: 1) Explorer; 2) Facilitator; 3) Experience Hunter; 4) Professional/Hobbyist; and, 5) Recharger and recognized that the visitors craved social sharing and personal "recharging" [8].

Museums that embrace a visitor-centered approach recognize the constructivist method of learning and use it to their advantage [16, 17]. This change also created a mindset open to the idea that visitors may prefer a multisensory-multimodal engagement with the exhibition content [18, 19].

3.2 Visitor-Centered Personalization Approach

Personalization in museums emerged from mobile interactive experiences where a person's preferences are digitally recorded, retrieved, and frames the delivery of information for a personal experience. It is sometimes created from a model that uses the user's behaviour to suggest a custom experience based on their inferred pace, theme or style to empower the visitor to raise their engagement in their museum experience [18, 20]. It expands on the constructivist approach by using personalization technology to maintain a dynamic model of the visitor's preferences based on their previous and real-time behaviours [21].

An application of a visitor-centered constructivist approach using personalization is the High Museum of Art's use of a dating app framework to determine their visitors' art preferences. Visitors "swiped" through the museum's collection to find artworks they like and were directed to their locations in the museum via a custom map. By understanding the visitors' art preferences the museum's personalization software was able to direct them to the art – in essence to curate a personalized exhibition experience [22].

By having a connection to the visitors' preferences, personalization can transform an exhibition from a place of information to an enrichment of personal identity.

3.3 Using Co-design to Create Visitor-Centered Personalized Museum Exhibitions

Using a variety of emerging technologies such as radio-frequency identification (RFID) tags and rapid-prototyping, a research group, *the meSch Project*, created a visitor-centered museum exhibition by using co-design practices to interrupt the traditional exhibition development process [23]. These specific practices, that are part of the co-design decision-making within a participatory design method brought together curators, interpretive planners, exhibition designers, and actors from many other disciplines such as computer science, engineering, social science, interaction and product design along with cultural heritage professionals and technologists. Through co-design they each played a role in the knowledge development and created new concepts to address the problem. Technology played a major part in building their storylines by using a tangible user interfaces (TUI).

One of the values of the process was social value. This was due to the teambuilding and the relationships that were forged. The results of this exercise is of great importance when considering the change in group dynamics that would have been necessary to bridge the traditional hierarchy of exhibition development into a collaborative scheme [23]. The collaboration resulted in the design of an environment that allowed for the application of personalization to the development of the exhibition content. This personalization technology was manifested in TUI's that delivered the exhibition content to the visitor in their preferred format.

3.4 Using Co-design with Disability Community

In the case of the *Welcome Collection* and the Helen Hamlyn Centre for Design, Royal College of Art, experts in the field explored a solution to create accessible museum exhibitions by applying co-design and user-centered design methodologies to the exhibition design process. The researchers employed user-journey mapping with a variety of participants who presented different disabilities in order to reveal positive and negative experiences with an exhibit. Later, visitors collaborated with designers to develop and test prototypes, which were incorporated into new, less prescriptive accessibility guidelines. The researchers identified that although the personal preferences, personalities, and difference in disability types made agreeing on an approach to implementing a universal design difficult, the benefits of using a co-design method for including visitors with disabilities increased awareness among the museum staff [2].

3.5 Digital Strategies for Personalized Learning

The literature also explores an appropriation of ability-based design, rapid-prototyping and visitor/user centered holistic approach in creating a learning environment; it considers all of the complexities of the disabled experience to support an individual toward self-expression, rather than the medical model based on assistance and intervention [24]. This approach was examined in Wilson et al.'s project that observed children with autism to personalize their own content and learning experience. It applies participatory design and a constructivist mindset to a user-centered approach in order to motivate a child with autism and improve their user-experience [25].

This project overcame similar challenges associated with traditional information-centered exhibitions and the disability model. Its results reinforce the possibilities of success for a new approach to accessible museum exhibition development.

3.6 Summary

The literature review of the different contexts and approaches reveals a common thread that if carefully appropriated, could result in a new approach to exhibition development.

4 Discussion

The advances of the Internet allow museums to look beyond their information-centered approach to one that is visitor-centered. Museums can change their exhibition development ecosystem by leveraging advances in technology and reframing the model of disability design to an ability-based design. This discussion seeks to synthesize the insights from the literature review to form a new ecosystem that improves visitor experience and provides accessible exhibitions.

4.1 Proposed Accessible Exhibition Development Ecosystem

The proposed accessible exhibition development ecosystem could appropriate ability-based design concepts, personalization, and, rapid-prototyping technologies and use a visitor-centered approach. This would allow the visitor greater control over their experience through self-determination and preference to the type and depth of content [8, 14, 21].

The adoption of an ability-based model determines the method of engagement. In this case, accommodation is not an add-on to the exhibition—the ability-based design could conform to the visitor preferences [2].

The technological nature of ability-based design allows for personalization that supports the visitor-centered approach. Visitors may become empowered with the confidence that their personal construct of knowledge through prior experiences is acknowledged and valued [17]. For some, their curiosity could be boosted through a custom "self-curated" museum tour based on their pre-formatted interests [22, 25]. Through personalization, the exhibition experience will be able to suggest new areas of interest that could offer the visitor with more choices and deepen their social sharing and identity. The personalization would also recognize the visitor's individual identity and deliver the content in the preferred format (audio, visual, didactic and/or tactile) and cognitive level (beginner to expert) (see Fig. 2).

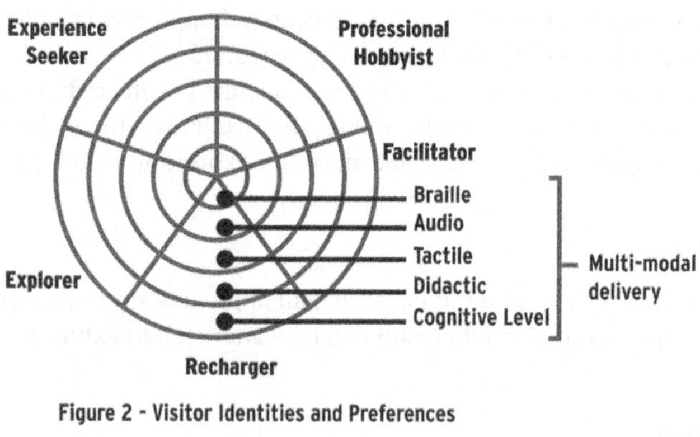

Fig. 2. A museum visitor can have 5 different identities (adapted from Falk, 2009). Each identity can have different preferences or abilities for the multi-modal delivery of their content.

Creating this new visitor-centered ecosystem would require collaboration that would differ from the traditional hierarchical exhibition development team and would be more participatory and include the accessibility community and technology disciplines. The use of rapid-prototyping techniques could test ability-based designs with people of various abilities in order to determine, with curators and interpretive planners, the specific content to be included in the interpretive plan. The technology could be proposed, built, tested and modified as ideas take shape. The process could require a curator to use a larger depth of knowledge of the museum's collection. For the

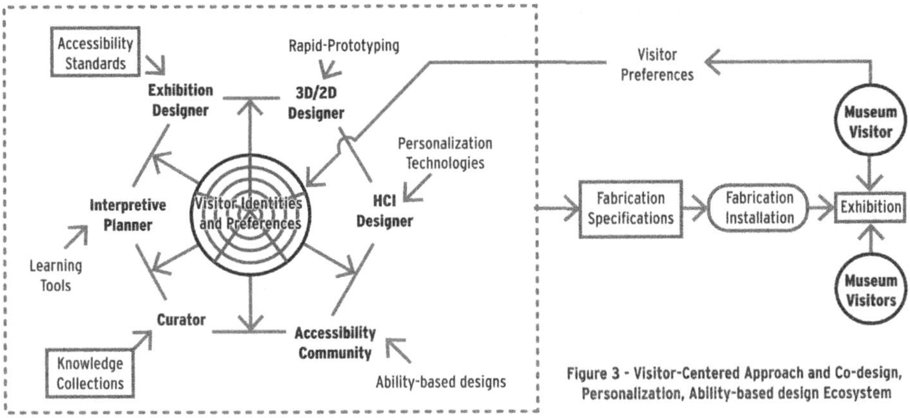

Fig. 3. This figure outlines a proposed approach for a Visitor-Centered Approach that includes a co-design method to integrate new knowledge to generate concepts and ideas for accessible designs.

exhibition designer, knowing that it could be possible to cater to each visitor may open the doors to greater flexibility and creativity (see Fig. 3).

5 Conclusion

The points drawn from the literature review provided background to create a framework for further research into creating accessible exhibitions with improved visitor engagement and experience. By understanding the importance of the visitor preferences and motivations when developing exhibitions and employing the emerging technologies that make personalization possible, the ability to create sustainable, accessible, and pleasurable museum exhibitions without sacrificing visitor experience may be possible.

Future research could explore this framework through clinical exercises that include community members with different abilities as described in the following sections.

5.1 Co-design and Interdisciplinary Participatory Design

Further research into the co-design method could include participants with visual impairments along with museum professionals, technologists, and theatrical and literary artists. It could include TUIs, emerging personalization and rapid-prototyping technology. Participants would contribute their unique knowledge when designing the interfaces as the exhibition theme is developed. They could evaluate potential objects, texts or audio-descriptions and explore other expressive mediums. At the same time, the curators could contribute how to best express the exhibition theme and ensure contextual accuracy.

5.2 Visitor-Centered Approach and Ability-Based Design

Testing the visitor-centered approach, the museum visitor(s) could share their level of interest, type of interaction and their method of interface with the museum's exhibition based on their abilities and preferences.

References

1. Samis, P., Michaelson, M.: Creating the Visitor-Centered Museum. Routledge, New York (2017)
2. Hesseldahl, K., McGinley, C., Monk, G.: Using design thinking to develop new methods of inclusive exhibition making. In: Craddock, G., Doran, C., McNutt, L., Dónal, R. (eds.) Studies in Health Technology and Informatics, pp. 151–160. IOS Press (2018). https://doi.org/10.3233/978-1-61499-923-2-151
3. Dudley, S.H. (ed.): Museum Materialities: Objects, Engagements Interpretations. Routledge, New York (2010)
4. Hughes, P.: Exhibition design. Laurence King Publishing, London (2015)
5. Lorenc, J., Skolnick, L., Berger, C.: What is Exhibition Design. RotoVision, Hove (2007)
6. Mayrand, Y., Nicks, J., Spencer, H.: The Manual of Museum Exhibitions. Rowman & Littlefield Publishers Inc., Connor Hill (2001)
7. Summers, J.: Creating exhibits that engage: a manual for museums and historical organizations. Rowman and Littlefield, Lanham (2018)
8. Falk, J.H.: Identity and the Museum Visitor Experience. Left Coast Press, Walnut Creek (2009)
9. Osborne, F., Luby, E.M., Fogarty, C.: An Accessibility Adventure: Teaching and Learning Exhibition Design Through the Lens of Disability Studies (2014)
10. Ingenium: Ingenium Accessibility, Ottawa (2018)
11. Smithsonian: Smithsonian Guidelines for Accessible Exhibition Design., Washington, D.C. (2012)
12. Cachia, A.: Talking Blind: Disability, Access, and the Discursive Turn. Disabil. Stud. Q. 33, (2013). https://doi.org/10.18061/dsq.v33i3.3758
13. Gomez, J.L., Langdon, P.M., Bichard, J.A., Clarkson, P.J.: Designing accessible workplaces for visually impaired people. In: P. M. Langdon et al. (ed.) Inclusive Designing. pp. 269–279. University of Cambridge, Cambridge (2014). https://doi.org/10.1007/978-3-319-05095-9
14. Wobbrock, J.O., Kane, S.K., Gajos, K.Z., Harada, S., Froehlich, J.: Ability-based design: concept, principles and examples. ACM Trans. Access. Comput. **3**, 1–27 (2011). https://doi.org/10.1145/1952383.1952384
15. Rieger, J., Strickfaden, M.: Taken for granted: material relations between disability and codes/guidelines. Societies **6**, 6 (2016). https://doi.org/10.3390/soc6010006
16. Harlow, D.B., Skinner, R.K.: Supporting visitor-centered learning through practice-based facilitation. J. Museum Educ. **44**, 298–309 (2019). https://doi.org/10.1080/10598650.2019.1590682
17. Jones, N.: Charting a new course: the craft of curation and the visitor-centered Museum. Collections. **10**, 218–229 (2014). https://doi.org/10.1177/155019061401000206
18. Rothberg, M., Reich, C.: Making Museum Exhibits Accessible for All: Approaches to Multi-Modal Exhibit Personalization, pp. 1–18 (2014)

19. Claisse, C., Ciolfi, L., Petrelli, D.: Containers of Stories: using co-design and digital augmentation to empower the museum community and create novel experiences of heritage at a house museum. Des. J. **20**, S2906–S2918 (2017). https://doi.org/10.1080/14606925.2017.1352801
20. Hashemi, S.H., Kamps, J.: Exploiting behavioral user models for point of interest recommendation in smart museums. New Rev. Hypermedia Multimed. **24**, 228–261 (2018). https://doi.org/10.1080/13614568.2018.1525436
21. Not, E., Petrelli, D.: Blending customisation, context-awareness and adaptivity for personalised tangible interaction in cultural heritage. Int. J. Hum Comput Stud. **114**, 3–19 (2018). https://doi.org/10.1016/j.ijhcs.2018.01.001
22. Forbes, J., Rucket, I.: A Museum 'Dating' App Helps Visitors Fall in Love with Art. https://www.aam-us.org/2019/07/11/a-museum-dating-app-helps-visitors-fall-in-love-with-art/. Accessed 20 Nov 2019
23. Avram, G., Ciolfi, L., Maye, L.: Creating tangible interactions with cultural heritage: lessons learned from a large scale, long term co-design project. CoDesign **00**, 1–16 (2019). https://doi.org/10.1080/15710882.2019.1596288
24. Fraunberger, C., Good, J., Parés, N.: Autism and technology beyond assistance & intervention. In: Conference Human Factors Computer System – Proceedings, pp. 3373–3378 (2016). https://doi.org/10.1145/2851581.2856494
25. Wilson, C., Brereton, M., Ploderer, B., Sitbon, L., Saggers, B.: Digital strategies for supporting strengths - And interests-based learning with children with autism. In: ASSETS 2017 - Proceedings 19th International ACM SIGACCESS Conference Computer Accessibility, pp. 52–61 (2017). https://doi.org/10.1145/3132525.3132553

An Interdisciplinary Framework for Designing Adaptive Snowsports

Lindsay McCauley(✉) and Lois Frankel

School of Industrial Design, Carleton University, Ottawa, ON, Canada
{Lindsay.McCauley,Lois.Frankel}@carleton.ca

Abstract. This literature review lays the groundwork for the past, present, and future of adaptive snowsports. The paper acknowledges the design challenges of inclusivity and individualism for adaptive snowsports equipment. In addition to surveying literature on adaptive snowsports design, various design approaches and case studies are explored to expand the discussion about different perspectives and techniques for achieving design solutions that meet the needs of many people. By engaging in a literature review, the paper presents an initial interdisciplinary recommendation that could influence designers and researchers in contributing to adaptive snowsports solutions that address these challenges to ultimately meet the needs of many.

Keywords: Adaptive snowsports · Interdisciplinary approach · Inclusive design approach · Individualized design approach

1 Introduction

Participating in adaptive snowsports, such as sit skiing, track skiing, or visually impaired skiing, can be a freeing experience for persons with disabilities [1–4]. Having the opportunity for people of all abilities to participate in winter sports can foster social inclusion [2]. According to a 2010 study by Nasuti and Temple, "skiing can positively influence self-esteem, physical self-worth, standing balance, and gross motor function among individuals with a disability" [3]. Fortunately, adaptive programs, such as Special Olympics International [5] and Canadian Adaptive Snowsports (CADS) [6] provide downhill skiing opportunities to children and adults with cognitive and physical disabilities. Over the years, a range of adaptive equipment has been designed to support diverse needs, such as for persons who are visually or hearing impaired, amputees, paraplegics, quadriplegics, as well as those who have spina bifida, cerebral palsy, cognitive disabilities, or loss of coordination, balance, and mobility [7]. Each skier has unique needs. Adaptive equipment is often used differently from person-to-person and constantly requires modifications [6]. Park et al. (2014) note that, for designers, adaptive snowsports can present design challenges, as no disabilities are the same. A unique paradox of inclusivity and individualism arises in the literature, where designers must consider the collective needs of many and the exceptional needs of individuals [8]. In the instance of designing a sit ski, a designer may want to aim to develop a device that is inclusive, accessible, and usable for as many people as possible. In contrast, a designer applying an individualized approach may want to create a

G. Di Bucchianico et al. (Eds.): AHFE 2020, AISC 1202, pp. 484–490, 2020.
https://doi.org/10.1007/978-3-030-51194-4_63

unique, custom-made sit ski to provide specialized fit. The paradox occurs in the design of adaptive snowsports equipment when combining the two approaches to ultimately meet the needs of as many people as possible. This literature review aims to identify how various perspectives and case studies may help designers navigate the paradox. It hypothesizes that an interdisciplinary approach may help create guidelines for designers when working in the area of adaptive snowsports equipment by integrating different methods and techniques.

2 Design and Disability/Ability

According to Statistics Canada (2017), one in seven Canadians have reported living with a disability [9]. Many people will experience some form of disability within their lifetime, be it permanent, temporary, episodic, or situational [10]. Designers have a responsibility to consider the diversity in user needs and abilities during the design process. The following are a couple approaches for design and disability/ability.

Ability-Based Design. Many approaches for designing for persons with disability centralize disability. In other words, traditional 'design for disability' approaches seek solutions to compensate for the limitations of disabled users. By definition, the word 'disability' indicates the lack of ability and focuses on what a person cannot do. Wobbrick et al. (2011) introduce a refocused perspective called 'ability-based design' [11]. Rather than asking "what disability does a person have?", ability-based design asks "what can a person do?" [11]. Approaching design from an ability-based perspective seeks to leverage the full range of human potential. For example, in 2019, a CADS volunteer recognized that a skier in the program was having difficulty pulling an adjustment cord attached to their sit ski. Due to limited dexterity from progressive multiple sclerosis, the skier was not able to firmly grip the cord. The CADS volunteer applied an ability-based approach and explored cord solutions that could work with the skier's abilities. Inspired by life-preserver ripcords, a small ball was attached to the end of the cord for quick and effective use [6].

Design *For* and Design *With* Users. Design *for* users and design *with* users are two approaches that emphasize the needs of end users. Design *for* users occurs when users are considered during the design process in an effort to make user-friendly products, services, and experiences [12]. Studying users through interviews, observations, and focus groups can lead to insights about needs, desires, problems, and dreams [13]. Seeking user input can help designers gain empathy, which is important for understanding alternative perspectives. Kouprie and Visser (2009) propose a framework for supporting empathic approaches for design. In overview, the framework recommends that designers attempt to gain a greater appreciation of the user's world, memories, and experiences. By developing empathy, the designer may better understand how to design for the user [14].

With that being said, empathy is not equivalent to lived experience. Designer, Liz Sanders, is known for saying "we are all experts of our own experience" [15]. In other words, despite being trained professionals, designers must recognize that other people may be more knowledgeable in certain situations. Sanders is an expert in co-creation or participatory design, which is a method of designing with users. Co-creation can be defined as "any act of collective creativity, i.e., creativity that is shared by two or more people" [15]. Design *with* users suggests that end-users become co-designers, offering expertise throughout the design process [16]. This might be achieved through workshops or generative sessions that include end-users in creation activities using materials provided by designers and researchers. Designing with users and other stakeholders often forms interdisciplinary teams, which may contribute to richer, more diverse outcomes.

3 Adaptive Snowsports

Cost of Adaptive Snowsports Equipment. The literature presents remarkable information about the cost of adaptive snowsports equipment noting that participating at any level, elite or beginner, can be expensive [6, 17, 18]. A mono-ski averages $7000 Canadian dollars [19], while a pair of racing skis can reach upwards of $1400 Canadian dollars [20]. The price difference highlights that adaptive skiing is approximately five times more expensive than traditional skiing, without including other costs such as additional gear and lift tickets. High price points may lead to barriers of access for many skiers [21]. In 2014, Scope released a study showing that individuals living in the United Kingdom with disabilities have a higher cost of living than people without disabilities, called the "financial penalty of disability" [22]. People with disabilities generally have to pay for more everyday expenses, products, specialist items, and services. Since many skiers with disabilities are already burdened with a financial penalty, adaptive snowsports equipment should be made accessible by minimizing economic barriers.

Benaso et al. (2018) examine potential causes for the high price points for adaptive ski equipment. In 2018, Dr. Jaimie Borisoff was interviewed to discuss whether manufacturers consider consumers' financial means during the design process. Dr. Borisoff oversees the creation, design, and production of adaptive equipment at British Columbia Institute of Technology (BCIT). In the interview, he indicates that his team always considers cost when designing new equipment "because there is no point in making something if no one can afford it" [23]. Rather, the cost stems from other factors, such as small production runs, customization, and complex materials, parts, and engineering [24]. Sit skis are individually manufactured, usually with materials that have high strength to weight ratios [25]. Fortunately, many adaptive snowsports programs offer scholarships or grants for their students to reduce the price of ski lessons and gear [5, 6]. Skiers in adaptive programs are often able to borrow, trade, rent, or buy used equipment [6]. Skiers who decide to purchase their own equipment may require additional customization or desire higher performing gear [23].

Design of Adaptive Snowsports Equipment. Barreda points out that adaptive snowsports equipment is typically designed to be adjustable. Popular sit skis, such as the Mountain Man and Dynamique offer a range of set up options to enhance the fit based on the skier's dimensions [19]. In order to serve a breath of abilities and performance levels, adjustability is an important design consideration. Understanding how different skiers interact with adaptive equipment can help optimize safety, human well-being and overall system performance [26]. Similar to a ski boot, adaptive snowsports equipment should be snug and comfortable. To increase fit and comfort, sit ski manufacturers like Tessier offer custom-made seats that are molded to the contours of the skier [25]. In other cases, foam, gel, or cushion inserts may be used [27]. Due to equipment availability and costs, fitting each skier with custom-made equipment or molded sit ski buckets is often unattainable for adaptive programs. Instead, CADS uses materials that are readily available, like pool noodles, foam, and duct tape to tailor equipment to each skier [6].

4 Case Studies: Inclusive Approach and Individualized Approach

Inclusive Approach to Adaptive Equipment. In 2016, Microsoft Corporation published its own inclusive design principles. Microsoft defines inclusive design as "a methodology…that enables and draws on the full range of human diversity. Most importantly, this means including and learning from people with a range of perspectives" [28]. Microsoft practices three main principles of inclusive design: (1) Recognize exclusion, (2) Solve for one, extend to many, (3) Learn from diversity. These principles are reflected in the design process of the Microsoft Xbox Adaptive Controller [28]. Traditional Xbox controllers required users to have two hands and finger dexterity. As a result, users who are unable to interact with the controller are excluded from playing Xbox. Microsoft recognized these physical barriers of use and designed the Xbox Adaptive Controller. Users with widespread abilities and needs were recruited to participate in Microsoft's design process. Learning from diversity and other perspectives ultimately led Microsoft to the final design of the adaptive controller that features two oversized buttons that are easy to hit with any appendage. Arguably, the most important features are the nineteen ports located at the top of the controller, allowing users to plug in any specialized controllers, such as joysticks, adaptive surfaces, or sip and puff sensors. The device serves as an inclusive tool that can easily be modified to support individual needs.

Individualized Approach to Adaptive Equipment. Another case in the literature is that of a charity-based firm in the United Kingdom, Remap, which provides a practical example of an individualized approach to design adaptive equipment [29]. Volunteers at Remap collaborate with persons with physical disabilities for free to provide one-off solutions and equipment to help individuals achieve independence and a better quality of life. Remap describes a successful project where they helped a woman named Rosie adapt her every-day wheelchair for the "Snowdon Push". The "Snowdon Push" is a challenge that takes place at Mount Snowdon, the highest mountain in England and Wales. Rosie's team members were required to push, pull, climb, and wheel Rosie to the summit and then return back down. Rosie's wheelchair needed to be modified to sustain steep, mixed surfaced terrain, as well as to be easy for her teammates to bring her up and down the mountain. Through in-person discussions and collaboration, Remap volunteers and Rosie decided to attach a front wheel for steering and braking the wheelchair, along with pulling straps, and a rear pushing and lifting frame. During training sessions, adjustments were made to suit all members of the team. After the challenge, Rosie's modified wheelchair was disassembled to return to its original state as an every-day wheelchair. Due to its volunteer nature and personalized, iterative process, Remap is a fantastic example of an individualized, decentralized, quality-of-life approach.

5 Discussion

In integrating the insights from the literature it seems that adaptive snowsports equipment could be approached with both inclusivity and individualism in mind, which may acknowledge the universal problem of designing for everyone's unique differences. Essentially, no two individuals are alike. This paradox encourages the design of barrier-free systems, but also the integration of customized supports to ensure that people of all abilities gain equitable opportunities to participate in winter snowsports. This concept is ability-based, incorporating principles from Wobbrock, et al.

As discussed, empathy is also a valuable asset for designers, especially when designing for skiers with disabilities [14]. Learning to be empathetic may help increase a designer's ability to understand different perspectives and needs of users. Nevertheless, empathy is not equivalent to lived experiences. When a designer relies on empathetic approaches, they risk making assumptions about the user [30]. To reduce or avoid making assumptions, stakeholders should be integrated within the design process. In the context of adaptive snowsports, design *with* users recommends that designers, users, manufacturers, and other stakeholders collaborate in order to achieve more appropriate and usable equipment [15]. Design is becoming more of a participatory activity; this involves including people from the beginning and throughout the whole process. Respecting and including people with lived experiences in the design process, may result in interdisciplinarity. Working in interdisciplinary teams increases the diversity of knowledge and experiences shared, which may lead to enriched results that are more inclusive.

In terms of individualization, adaptive snowsports equipment should be designed to incorporate a tolerance for future modification and customization. Anticipate individualization, rather than impede it. It is not uncommon for products to be designed so that only manufacturers can service the product [31]. As a result, product owners may be barred from repairing or making modifications on their own. These limitations create physical and financial barriers to access. An individualized approach to adaptive snowsports equipment should allow for ease of maintenance and customization. For example, instead of using rivets as a permanent fastening system, use bolts so that users have the ability to make their own adjustments.

6 Conclusion

This paper reviewed literature about inclusive, individualized, accessible, and ability-based design. It focused on adaptive snowsports, such as sit skiing and track skiing, which are popular activities with therapeutic and rehabilitative potential, for mind and body. Skiers with disabilities using adaptive equipment can vary significantly in abilities, from limited mobility, to cognitive impairments, to a lack of eyesight or hearing. As a result, the literature points to the importance of addressing diverse user needs in the design of adaptive equipment. Utilizing an inclusive approach, which seeks to embrace human diversity, may help designers create adaptive snowsports equipment and opportunities that are usable for as many people as possible. Even so, some users have exceptional needs that cannot be addressed through inclusive design and require one-of-a-kind design solutions. In such cases, an individualized approach was shown to support unique, custom-made designs.

Recommending that designers should apply both approaches is contradictory, as inclusivity suggests design for all, while individualism suggests design for one. In an exploration of how this paradox could be navigated, the paper provides recommendations, derived from synthesizing the insights from the literature that designers should strive to provide solutions that address both approaches simultaneously to meet the needs of many. Collaborating with adaptive skiers, manufacturers, and various other stakeholders can offer interdisciplinary perspectives, skills, and experiences. An interdisciplinary team is more representative of the diverse needs, experiences, and abilities of adaptive skiers compared to a single designer. Designing with other people, such as non-designers, can lead to enriched outcomes. By learning from diversity, adaptive equipment should be designed to extend to many. On the other hand, adaptive equipment should be designed to allow people other than designers or manufacturers to individualize or adjust their equipment. For example, rather than using manufacturer-specific parts, designers should use universal elements that are readily available to the general public. Instead of using rivets that are one-time-use, use bolts, which are readily available and easy to change. Ultimately, by designing for ability, rather than disability, human potential and diversity can be leveraged to ultimately meet the needs of many people.

References

1. Laskowski, E.R.: Snow skiing for the physically disabled. Mayo Clin. Proc. **66**, 160–172 (1991)
2. Lape, E.C., Katz, J.N., Losina, E., Kerman, H.M., Gedman, M.A., Blauwet, C.A.: Participant-reported benefits of involvement in an adaptive sports program: a qualitative study. Pm&r. **10**, 507–515 (2017)
3. Nasuti, G., Temple, V.A.: The risks and benefits of snow sports for people with disabilities: a review of the literature. Int. J. Rehabil. Res. **33**, 193–198 (2010)
4. Miller, M.A., Groah, S.L., Cifu, D.X.: Poster 173 Psychologic benefit of a community-based adaptive ski program: a survey. Arch. Phys. Med. Rehabil. **76**, 1073 (1995)

5. SpecialOlympics: Alpine Skiing. https://www.specialolympics.org/our-work/sports/alpine-skiing
6. Canadian Adaptive Snowsports: CADS (2019). https://www.cads.ski/
7. SkiCanada: Accessible Skiing Information: Canadian Adaptive Snowsports (CADS): Go Skiing Go Snowboarding. https://www.skicanada.org/ready/accessible-skiing-information/
8. Park, J., Morris, K., Stannard, C., Hamilton, W.: Design for many, design for me: universal design for apparel products. Des. J. **17**, 267–290 (2014)
9. Statistics Canada: Infographic: Canadian Survey on Disability (2017). https://www150.statcan.gc.ca/n1/pub/11-627-m/11-627-m2017008-eng.htm
10. Bradshaw, E.: Disabilities. UWIRE Text. p. 1. Gale Academic OneFile (2019). https://link-gale-com.proxy.library.carleton.ca/apps/doc/A582478960/AONE?u=ocul_carleton&sid=AONE&xid=1086ba06. Accessed 1 Feb 2020
11. Wobbrock, J.O., Kane, S.K., Gajos, K.Z., Harada, S., Froehlich, J.: Ability-based design. ACM Trans. Access. Comput. **3**, 1–27 (2011)
12. Nielsen, L.: Personas - User Focused Design. Human–Computer Interaction Series (2019)
13. Laurel, B.: Design Research: Methods and Perspectives. MIT Press, Cambridge (2003)
14. Kouprie, M., Visser, F.S.: A framework for empathy in design: stepping into and out of the users life. J. Eng. Des. **20**, 437–448 (2009)
15. Sanders, L., Stappers, P.J.: Editorial. CoDesign **10**, 1–4 (2014)
16. Carlson, C., Peake, W., Joiner, J.: Letting context speak. Commun. Des. Q. Rev. **2**, 34–39 (2014)
17. Canadian Paralympic Committee: Paralympic Sport Development Fund helps sit skiers reach new heights at Rocky Mountain Adaptive, https://paralympic.ca/news/paralympic-sport-development-fund-helps-sit-skiers-reach-new-heights-rocky-mountain-adaptive
18. Lazzaro, J.J.: Adaptive technologies for learning & work environments. American Library Association, Chicago (2001)
19. Barreda, K.: Adaptive Skiing. https://adaptiveskiing.net/
20. Race Place: Race Place Ski Shop - Your source for Ski Racing Equipment & Supplies. https://the-raceplace.com/
21. Russell, R., Kutin, J., Stewart, M.S., Rankin, G.: Money minded Impact Report 2017 (2017)
22. Scope.: Priced out: ending the financial penalty of disability by 2020. https://blog.scope.org.uk/2014/04/15/priced-out-ending-the-financial-penalty-of-disability-by-2020/
23. Benaso, A., Johnston, L., McGauley, L., Malo, J.: Breaking barriers how access to adaptive skiing changes lives (2018). https://bcitnews.com/2018/12/16/breaking-barriers-how-access-to-adaptive-skiing-changes-lives/
24. Houle, M.H.: The Cost Of Disability: Or, Why We Can't Have Nice Things (2017). https://wheresyourdog.com/2016/02/06/the-cost-of-disability-or-why-we-cant-have-nice-things/
25. Tessier: Outdoor Adaptive Sports Equipment (2019). http://www.dualski.com/en/
26. Woodson, W.E.: Human Factors Design Handbook: Information and Guidelines for the Design of Systems, Facilities, Equipment, and Products for Human Use. McGraw-Hill Book Company, New York (1981)
27. Cooper, R.A.: Rehabilitation Engineering Applied to Mobility and Manipulation. Institute of Physics Pub. (1995)
28. Microsoft Design (2019). https://www.microsoft.com/design/inclusive/
29. Remap. The Snowdon Push (2018). https://www.remap.org.uk/2130/the-snowdon-push/
30. Goodman, E., Kuniavsky, M., Moed, A.: Observing the user experience: a practitioners guide to user research. Morgan Kaufmann, Amsterdam (2012)
31. Cooper, T.: Longer lasting products: alternatives to the throwaway society. Gower, Farnham (2016)

Case Study of Cultural Design Innovation Enabling Brand Development

Huabin Wang(✉) and Lei Han(✉)

South China University of Technology,
Panyu District, Guangzhou, Guangdong Province, China
377381343@qq.com, 1279456545@qq.com

Abstract. Under the policy-driven background, cultural design innovation has broad prospects. Based on the user experience level, this paper divides cultural design innovation into visual perception level cultural design innovation, behavioral operation level cultural design innovation, and spiritual connotation level cultural design innovation, and selects The Palace Museum, LKK Design Company and MUJI as three categories of cases for analysis. The path of cultural design innovation is obtained: discovering the typical appearance characteristics of popular cultural IP, establishing the relationship between cultural practices and behavior levels in real life, and injecting a culture consistent with long-term development into the brand idea. This paper analyzes the path of cultural design innovation through case study, with a view to bringing some guidance to the development of related enterprises.

Keywords: Cultural design innovation · Enterprise development · Case study

1 Design Innovation Is an Important Driving Force for Enterprise Development

Design innovation is of great significance for enterprises to improve their competitiveness and create more profits. Design can help companies create new product concepts, enhance user experience, and appropriately configure related technologies in products to create differentiated products [1].

The market is constantly changing, enterprises only in advance accurately predict market trends, continuous innovation in the market competition in an invincible position. Because of the important role of design for enterprises, design-driven innovation has become an important link in the enterprise's innovation behavior. The Italian scholar Verganti found through research on successful enterprise products that the semantic and design expressions delivered by the products of these companies are extraordinary novelty, even exceeding the product's own functions and technological innovations. He called such a strategy "design-driven innovation" [2].

This paper is funded by the Humanities and social sciences Research project of the Ministry of Education, No. 20YJA760079.

G. Di Bucchianico et al. (Eds.): AHFE 2020, AISC 1202, pp. 491–497, 2020.
https://doi.org/10.1007/978-3-030-51194-4_64

In 2010, the Ministry of Industry and Information Technology, the Ministry of Education and other departments jointly issued "a number of guidance on the development of industrial design", clearly pointed out that the industrial design industry is an important part of the productive service industry, its level of development is one of the important signs of industrial competitiveness. This report indicates that vigorously promoting the development of the construction of industrial design industry is of great significance for enriching the product variety and enhancing the added value of the products, and can promote the establishment of independent brands in the market, effectively enhance China's industrial competitiveness [3].

2 Cultural Design Innovation Has Broad Market Prospects

With the consumer market product variety more and more rich, increasing consumption demand, consumers in daily life no longer meet the general consumption demand, but in all aspects of life need a higher level of spiritual satisfaction, the pursuit of a can fit the status of consumer status, value identity, personality image of symbolic consumption, symbolic consumption. Data show that China's annual potential size of cultural consumption is 4.7 trillion yuan, the actual consumption is only about 1 trillion yuan [4].

The potential consumption gap of nearly 4 trillion yuan shows that the cultural consumption market is still a "blue sea" with great potential, and cultural design innovation as the biggest driving force of cultural market will play an increasingly important role. The reports of the 17th and 18th National Congresses of the Communist Party of China have clearly stated that "the cultural industry should be vigorously developed" and that "the cultural industry should become the pillar industry of the national economy" [5]. To vigorously develop the cultural industry, the main grasp is to vigorously develop cultural and creative industries, to build a high visibility, reputation and competitiveness of cultural and creative products brand.

3 Hierarchical Types of Cultural Design Innovation

Cultural design innovation is divided into three categories according to the different levels of cultural symbols acting on the experience level.

3.1 Visual Perception Level Cultural Design Innovation

Visual perception level cultural design innovation refers to the representative object of culture as the cultural symbol, and the design by the characteristics of its shape, color, texture and so on. The outstanding advantage of objects as cultural symbols is that we can catch the eyes of consumers with the visual characteristics of culture, so as to attract consumption. At present, most of the tourism creative products in the market have adopted such a cultural design innovation method, The Palace Museum is a typical case.

Cultural creation of the imperial palace has a unique advantage in the context of the "Forbidden City", the most profound cultural IP in China. Chinese people are curious about the mysterious Forbidden City since ancient times. The Forbidden City Culture and Innovation excavates various characteristics of the Forbidden City to make it a cultural symbol, and then designs the cultural symbol to be reasonably implanted in the product and enters the modern life.

The note holders of the Cultural creation of the imperial palace use the shapes of the guards and cats in the Forbidden City as cultural symbols. This design makes the note holders both functional and cultural. The guards often appear in a standing posture, and cats in the Forbidden City often appear in court paintings in a lazy posture, so their images are very suitable for desktop decoration, and this is related to the use of note pads and the crowd coincidentally. The note holder designed by the Palace Museum has been greatly loved by consumers and has achieved good commercial value. These cultural and creative products have brought unprecedented vitality to The palace museum, with more than 8,700 cultural and creative products and a turnover of more than 1 billion yua [6]. According to relevant reports, the sales of Cultural creation of the imperial palace increased year by year: from 600 million yuan in 2013 to 1 billion yuan in 2016, and reached 1.5 billion yuan in sales by 2017 [7] (Fig. 1).

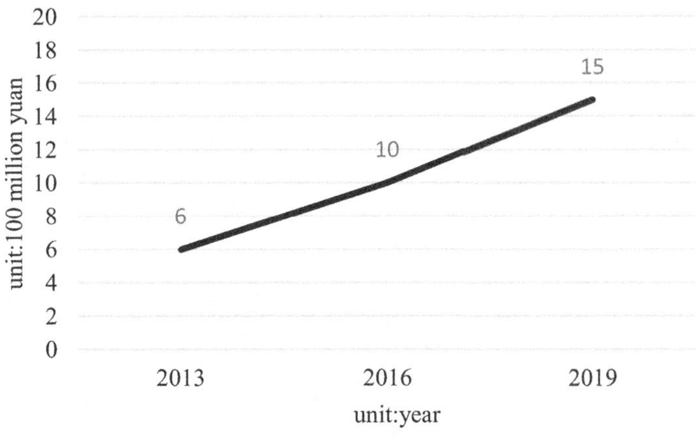

Fig. 1. Sales of cultural creation of the imperial palace

3.2 Behavioral Operation Level Cultural Design Innovation

Behavioral operation level cultural design innovation is to use behavior rituals related to cultural conventions as cultural symbols, and use their operation methods to design. The outstanding advantage of using behaviors corresponding to cultural conventions as design symbols is that it can give the product a vivid experience in operation, and it is easier to evoke empathy from users. Unlike the visual impact brought by the cultural symbols of utensils, the sense of cultural belonging evoked by behavioral actions has a greater psychological impact on people.

"God Luck" toothpick box is based on the ancient blessing draw as a cultural symbol. The toothpick box is shaken as if it is shaking the sign holder. The action of the user taking out the toothpick is like drawing a bamboo stick in the sign holder, which is lively and interesting. In order to match the product's blessing meaning, the designer also uses the shape image of the Temple of Heaven as a component of the toothpick box. The red color of the inner box strengthens the traditional sense of belonging to the blessing and prayer. The entire toothpick box expresses the cultural symbols of blessing vividly. When the user uses the simple gestures, he evokes the induction of traditional blessing customs hidden in his body, so a little action has emotions, and the cultural symbols make the toothpick box It has meaning that can't be separated from people. This toothpick box won the Red Dot Award for outstanding appeal during the design phase.

It is LKK Design Company's ingenious use of cultural symbols and design languages. The company's products are more and more loved by users, and it has developed into a well-known design company in the industry [8].

3.3 Spiritual Connotation Level Cultural Design Innovation

Cultural design innovation at the level of spiritual connotation is to inject cultural connotation into the brand concept, to make the brand product show a continual design language, and to achieve the purpose of conveying cultural connotation by bringing users the reflection and perception after use. The prominent advantage of using cultural connotation as the brand concept is that it gives the brand products a source of vitality and family genes.

MUJI emerged during the second oil crisis during the Japanese economic downturn [9]. It takes Zen culture as the brand's cultural symbol and regards the "out of nothing" pursued by Zen culture as a product claim. It believes that too much decoration will distract users from focusing on the product itself. The application of design is extremely restrained. This style quickly became popular in Japan. Over time, MUJI's brand image that combines simple beauty and good functions has been deeply imprinted in users' hearts. As MUJI went out of Japan, it once again attracted consumers with its simple and unique design style, and captured the hearts of the people of the world through its Zen culture, and this design language that symbolizes Japanese Zen culture also spread throughout the world, and since then MUJI has become a symbol of Japanese culture.

MUJI's lazy sofa has been a popular household product since its birth in 2002. From the outside, it is just a sponge without shape, but it can be changed in shape to provide users with comfortable support. Lying or sitting on one's own will, no matter which posture is comfortable. This sofa, which has no shape but can change its shape to meet the needs of users, is the embodiment of the brand concept of "nothing is found out of nothing", that is, there is no shape but there are infinite shapes. There are not many restrictions to better use the product as you wish.

MUJI maintains such an excellent spiritual level of cultural design innovation, which makes it enjoy a high brand reputation and good sales performance in the world. From 2014 to 2018, MUJI's turnover was 220 billion yen, 260 billion yen, 310 billion yen, 330 billion yen, and 370 billion yen [10]. From the data point of view, the sales of MUJI have increased year by year, which proves that the brand has good vitality (Fig. 2).

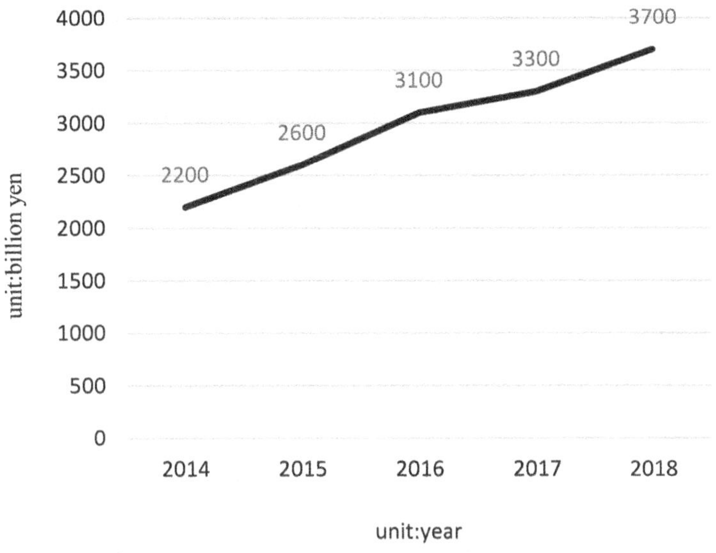

Fig. 2. Turnover of MUJI

4 The Path of Cultural Design Innovation

According to the research summary of three types of cultural design innovation cases, we can see that different types of culture are suitable for refining different cultural symbols, and the corresponding design innovation paths are also different.

Visual perception level cultural design innovation can achieve better market effects by selecting popular cultural IP as the background. Popular cultural IP's modeling, color, texture and other characteristics often have a high recognition in the market, a large fan base, and a high degree of market acceptance. As long as the interesting image is refined, a usage scene consistent with the image can be used as a text Create products into the market. In order for the brand to achieve sustainable development, it is best to choose a cultural IP with a large body and more typical utensils. In this case, there are many cultural symbols that can be designed. Such cultural design innovation and its path are suitable for the development of tourism cultural and creative products.

The key to cultural design innovation at the behavioral operation level is to find customary rituals with obvious characteristics of behavioral actions, extract the behaviors that are deeply embedded in the cultural group in the customary rituals, and associate such behaviors with real-life product use scenarios, look for similar action elements, and then guide by surface features to make the culture at the behavioral operation level perceptible to users. The surface features of the product and the style of the operation result are unified and clear, making the user experience stronger and more susceptible to cultural touch.

Spiritual connotation level cultural design innovation acts on the highest level of the user experience level. It has long-term stability and is the type of cultural design innovation that has the most profound impact on products. Cultural design innovation at the level of spiritual connotation requires brands to choose the right culture based on market trends and their own development goals, and to use this culture as the brand concept to gradually become the core driving force for brand products. Cultural design innovation at the level of spiritual connotation usually transforms cultural connotation into a suitable design language through product semantics such as metaphor.

5 Summary

Cultural design innovation can be divided into three categories according to the types of cultural symbols: visual perception level, behavior operation level, and spiritual connotation level. Based on this classification method, this article selects three typical cases for analysis, and tries to find out the corresponding design methods that cultural design innovation should use in design practice and its strengths and weaknesses in the market through case studies. For cultural design innovation at the level of intuitive experience, a suitable design method should be to choose a cultural background with high popularity and easy promotion, and choose appropriate utensils as cultural symbols. For cultural design innovation at the behavior operation level, the appropriate design method should be the abstract ritual-level custom rituals as cultural symbols, and then find products with similar operation methods in modern objects, fuse the two, and at the same time, make the products easy to learn. It is more sexual and needs to give users prompts at the level of intuitive perception. For cultural design innovation at the level of spiritual connotation, the appropriate design method is to use the realm of cultural pursuit as the brand's design language.

References

1. Chen, X.: Research on design-driven innovation mechanism and design pattern evolution. Zhejiang University (2011)
2. Verganti, R: Design as brokering of languages: the role of designers in the innovation strategies of Italian firms. Des. Manag. J. 34–42(2003)
3. "Several Guiding Opinions on Promoting the Development of Industrial Design". Liaoning Building Materials, pp. 20–21(2010)
4. Hao, N.: Theoretical research and practice of cultural and creative product design. Industrial Design, pp. 73–76(2016)
5. Du, F.: Programmatic documents for continued development of socialism with chinese characteristics——on the new thoughts, new judgments, new requirements, and new deployments in the report of the eighteenth national congress of the CPC. J. Harbin Inst. Technol. (Soc. Sci. Edn.), 4–19 (2013)

6. Jin, Q., Zhang, X.: Research on cultural and creative product development of museums. J. Xi'an Univ. Archit. Technol. (Soc. Sci. Edn.), 42–46 (2016)
7. Chai, X.: Research on the spreading strategy of cultural creative products in the new media era. Shaanxi Normal University (2019)
8. Ge, X.: LKK Design company new power of industrial design. Yingcai, pp. 78–79(2008)
9. Zhang, M.: Thinking of design details caused by the minimalism of "MUJI". Anhui Jianzhu University (2015)
10. Lu, C.: Development status, problems and countermeasures of MUJI under the background of economic globalization: based on SWOT analysis framework. Market Modernization, 9–10 (2019)

Innovation's DNA in Postmodern Society Age

Ekaterina Barcellos(✉), Galdenoro Botura Jr., and Lívia Inglesis

Sao Paulo State University, Av. Engº Luiz Edmundo Carrijo Coube, Vargem Limpa 14-01, Bauru 17033360, Brazil
kettymoda@gmail.com, galdenoro@gmail.com, liviaibarcellos@gmail.com

Abstract. The concept of innovation is still used as a synonym for creativity. Despite the overvaluation of creative ideas to achieve goals, logical systems have become even more essential for the digital society. A brief demonstration of the innovators' DNA throughout history indicated that people with different skills had multifaceted elements, in a synergy that allowed them to glimpse different perspectives to extract unusual solutions. From Da Vinci to Bezos and Musk, innovators have an intelligence factor, added to the creative imagination assets but mixed with technological knowledge. The tangency between intelligence and creativity is defining to obtain innovation, confirmed from the past to the current context. Therefore, the innovative process requires specific skills, which combine logical factors with creativity. Especially in the postmodern context, the rational and technical base added to Design methodologies and strategic thinking provides greater and better resources to generate innovation.

Keywords: Innovation · Design methodology · Creativity · Intelligence · Engineering and systems

1 Introduction

From the analysis of periods and elements of history, mainly in the last 500 years, it is possible to conclude that there are convergences between the intellectual profile and the thought models of inventors and innovators, and variables that determine these similarities.

At first, the character of the innovators stands out as eminently curious and questioning, a natural characteristic of the human being, as cited by [1], analyzing Da Vinci, Einstein, Jobs, and others. Natural curiosity acts as a fuel that links philosophy to empiricism [3], leads to scientific reflection, evidence, and practical attempts, which cause changes in understanding, actions, and results in innovation [4]. It is intuitive as mentioned by [2], "*the tendency to think of new and better ways of doing things and experiencing them in practice.*" New perspectives of understanding favor the proposition of ideas and problem-solutions [5], necessary in the diverse-adverse contemporary world. The concept of "*Innovation is not a new phenomenon. Arguably, it is as*

Sao Paulo State University, UNESP - Universidade Estadual Paulista Júlio de Mesquita Filho, Bauru – Sao Paulo, Brazil.

G. Di Bucchianico et al. (Eds.): AHFE 2020, AISC 1202, pp. 498–505, 2020.
https://doi.org/10.1007/978-3-030-51194-4_65

old as mankind itself' [2]. Its origin stems from the human being's search for survival motivated by curiosity and natural issues and by the desire to obtain more satisfactory benefits and conditions [2–4]. It took centuries to establish the concept of innovation as we know it, considered a highly strategic and effective asset, contradicting past considerations [4, 6]. To improve this understanding, a comparative graph of the use of the terms "invention" x "innovation" was created by authors to measure their difference and relevance over the course of historical evolution. The references were verified by citations found in books using Google Books online tool Ngram Viewer.

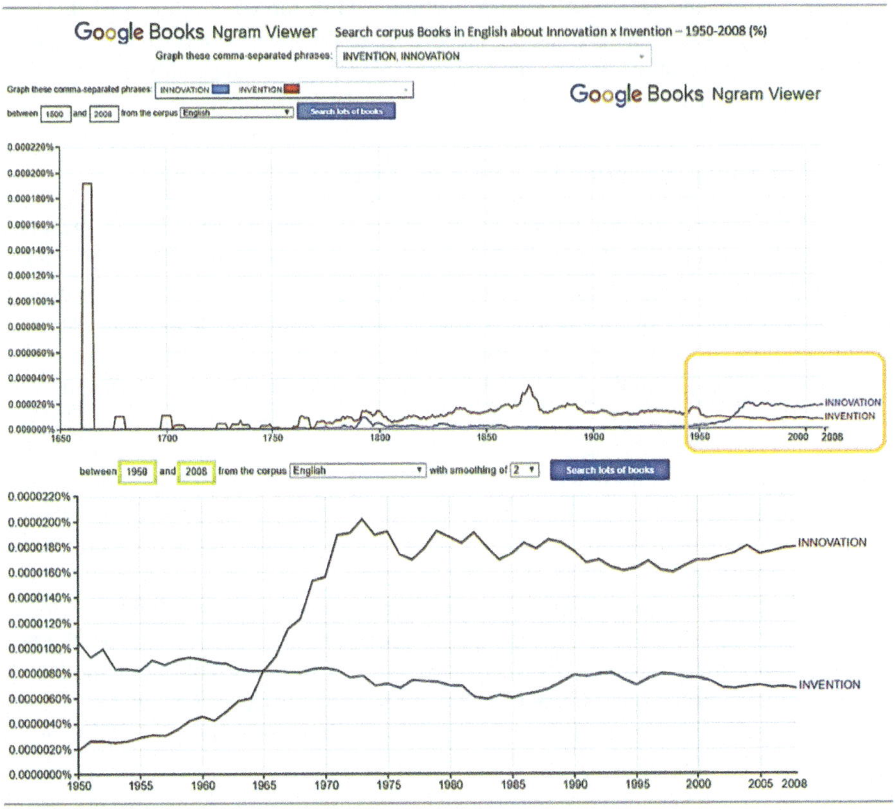

Fig. 1. Comparative scale between terms "invention" x "innovation" from 1650–2008. Source: search in corpus Books (English) by Google Ngram Viewer tool, developed by authors (2020)

The index reading is available from 1500 onwards. Data between 1500 and 1650 was despised due to index zero, explained by the Middle Age period (476 to 1453) when both terms were totally ignored or prohibited. As Fig. 1 demonstrates, both concepts suffered different assignments and different scales of value [4]. The context of each era effectively interfered and changed its judgment and importance. According to [4] descriptions, with the advent of the Scientific Revolution, only the term "invention" started to make sense and began to be mentioned in relevant publications.

The act of inventing was, therefore, overvalued, and a surprising interest was reached between 1660–1670, while the term innovation was despised and offensive, considered vulgar and opportunistic at the time. Due to the changes based on the context of each phase, [4] categorized innovation by behavioral socio-economic, and cultural relevance bias. He divided it into three distinct phases by epistemological meaning: 1. "Prohibition" - from the 1st millennium to the 19th century; 2. "Instrument" - from the end of the 19th century until the middle of the 20th century. In this phase, after the Second World War, it came to be understood as instrument due to the influence of the army, war technologies, and the rising of digital tech in Silicon Valley; 3. "Value" - from the mid-20th century to today. After the development of the R&D status, a strategic value change occurs. According to this new understanding, endorsed by [2] and [4], innovation got the character of development and public policy. In the last two decades, it was consolidated and reached its current relevance.

1.1 Is Innovation generated by a kind of DNA?

The Innovator's DNA, a book written by [6], brings a comprehension that suggests five fundamental competencies for an individual to become an innovative professional: associating, questioning, observing, interconnecting networks, and experimenting. However, there is no way to guarantee that someone innovates using this pattern of actions, as there are other elements that interfere and determine the achievement of innovative results. Certainly, in the past, after the Middle Age, an invention was a difficult challenge, given many adverse conditions and motivation absence [3]. When investigating variables throughout human history, up to the present moment, it is possible to list factors that determine the attribution of a person as an innovator. As explained by [4], the context is one of them, as it determines society's configuration in its demands to accept innovations' proposal and also defines how it will be adopted by individuals [2, 4].

According to [7], several theorists associate creativity and intelligence with the emotional attitude of inspiration, empathy, purpose, and motivation. Some of the most relevant among them, such as [5, 8, 9] and [10], who studied thought models for problem-solving, point out that a possible innovative DNA carries many potential variables in addition to creativity and intelligence, such as capacities and skills, or also flexibility and fluency of ideas [7]. Other additives stand out, as association and connection between thoughts [8] persistence and resilience [5]. The latter two are seen as mandatory criteria by the renowned scientists [5], in order to tirelessly persist through many tests and trials and to be able to deal with a succession of errors until innovate. To better investigate a possible DNA, Fig. 2 shows a careful selection of 25 innovative inventors indicated as the most brilliant ever, referenced in lists and indexes: starting in ancient Greece, with pre-Socratic philosophers, and ending with the main digital innovators of the 21st century. The data were compiled from MIT Invention Index research (2011). Other bases consulted were "Thinking Creative Index 2017"; "The Best Americans" (2005); "The 100 Greatest Britons" (2002); "The Best Italians Ever"; "Unsure Besten" (Our Best); "The Best French of All"; "The best Soviet", "The Best Brazilian ever". The most prominent in the Top 10 list, in each country, were included, as long as they were really relevant innovators in the rest of the world.

Most of the elected in this list have varied knowledge and technical mastery in their areas of knowledge and many other skills. They also showed interest in more than one scientific, cultural, or artistic field. A kind of natural form of enhanced associative creative intelligence, multi-directional thinking, brain flexibility, and knowledge.

Fig. 2. List of the 25 most innovative personalities throughout history. Source: [15], 2020.

Also noted after the 20th century is the aptitude for engineering mixed with design (as Da Vinci worked). Creative projects and entertainment bring the dominance on digital systems in the 21st century, perceived in most recent innovators by the ability to deal with disruptive technical performance products, services, and experiences.

1.2 Intelligence and Creativity

The indistinct sum of the skills demonstrated by these notables denotes an important indicator for composing an innovative DNA. It is necessary to investigate the intelligence profile of these personalities who stood out, as models and inspiration to understand their creative thinking models and imagination. By delimiting their characteristics, it is possible to perceive similarities that indicate a "standard", regarding their competencies and abilities. The most innovative elected representatives indicated a majority tendency towards increased knowledge in logical sciences, health, and earth sciences, engineering associated with physics and chemistry.

Research published by MIT [11] corroborates and attest to the point that individuals who have a greater interest in science, mathematics, computing, and engineering form a profile of greater potential to become innovators. Unfortunately, the same statistics showed lower academic interest from female representatives, in inventive careers such as science, technology, mathematics, or engineering. The same who claim female empowerment but avoid choosing this path. Totally similar results were noted during surveys to indicate strategic guidelines to motivate the culture of innovation for developing countries. As an example, Brazilian reports measured in collaboration with the MIT Lab and the National Confederation of Industries of Brazil, in 2018 and 2019, showed discouraging data on female disinterest in these areas of knowledge.

Most theorists on this subject, such as [2, 3, 6–10] and [12] reaffirm the notion that innovative ideas are generated by the conjunction of elements or qualities that 'facilitate' the creative mental process. Meanwhile, Einstein said that "*intelligence is not knowledge, but imagination.*" Therefore, since he is one of the most intelligent individuals in all history, highlighted in Fig. 2 list as one of the greatest innovators, it can be concluded that intelligence connects with creativity and imagination, and the result of that sum are innovative ideas and solutions. It is about imagining and accomplishing by ability, skill, competence, and adequacy to innovate [8, 9]. Among other aspects, it involves perception and cognitive dimension [2, 3] but it also depends on a type of intelligent association and connection of ideas through creative [7], analogical and logical thinking, a structured space for problems and critical thinking, provided by lateral plus vertical thinking [12], divergent plus convergent thinking [9].

In addition to the motivation and emotion needs, the cultural context and aspects of personality [8], becoming an innovator is linked to flexible and fluid thinking [7]. Also, the ability to change the perspectives of knowledge and action, expanding it incessantly in the right direction, focusing a problem-solving situation, achieving goals and objective.

Their noticed interest in more than one scientific, cultural, or artistic field works as a kind of organic form of enhanced associative creative intelligence, multi-directional thinking, brain flexibility to deal with problems [13]. This transversal creative thinking is as compared to the way designers use to think. So, this bias gave rise to the Design Thinking method [14]; therefore, designers are usually less involved with requirements and techniques.

By perceiving the current diversified and multidirectional scenario, we see a clear direction for the revaluation of logical training. Computer science and all refined digital knowledge alters performance and results in innovation. A greater reinforcement is notable in systems and logical areas, such as IT engineering, engineering design, associated with the application of models and Design methodologies directed to creative thinking [13, 15] and other tools like Design Thinking [14].

Figure 3 shows the last analysis of knowledge profiles of the 10 largest leaders in innovation in the world indicated by Forbes. Of these ten, eight are leads in front of the most innovative companies in the world, from a list of twenty most valuable companies globally in 2019.

Forbes

Rank	Name	Knowledge and graduation area	Company	Company Innovation Premium
1	Jeff Bezos	computer science and electrical engineering.	Amazon	71.49
1	Elon Musk	physics. energy physics.	Tesla Motors	72.16
3	Mark Zuckerberg	psychology computer science	Facebook	61.84
4	Marc Benioff	Business Administration. computer programmer	Salesforce.com	80.7
5	Reed Hastings	computer science	Netflix	72.16
6	Satya Nadella	electronics and commun engineering computer science	Microsoft	31.59
7	Shantanu Narayen	computer science	Adobe	64.43
8	Tim Cook	industrial engineering Business Administration.	Apple	11.58
9	Arne Sorenson	lawyer	Marriott International	53.66
10	Larry Page	computer engineering computer science	Alphabet	33.87
	Sergey Brin	computer science mathematic		

Fig. 3. Ten most Innovative Leaders by Forbes List 2019. Source: adapted by [15], from www.forbes.com/lists/innovative leaders/(2020)

From the presented list, surrounded by innovators classified as the best in 2019, it can be seen that 90% of them come from technological, logical and pragmatic areas of science, especially computer science and engineering fields.

When verifying the most valuable companies in the world, the same pattern is repeated with the additive that there are founding partners, CEOs and teams in disruptive companies that present the clear sum between logic and creativity. They are leaders qualified or graduated in technologies or engineering, such as electrical, mechanical, or engineering design, system computer science and industrial design, product design, system design and similar. So, this composition enhanced the results obtained in innovation. This aspect strongly confirms what this article is investigating.

2 Conclusion

It can be concluded that creativity and imagination should not be separated from associative logical intelligence. Lateral thinking must be followed by vertical thinking [12], and divergent thoughts must also be concluded by convergent ones [9]. Intelligence and creativity are the right compositions to generate an innovation DNA pattern, without competing for the domain. This means that the brain is designed to solve problems and will intuitively move to search for the solution choosing the side of completeness, by associating new ideas and insights. This also means that the creative association seeks the deposit of a missing scope, of more technical, practical, and logical knowledge, as it represents a balance to enhance the imagination and creative capacity, and thus, achieve the best and most innovative results possible. Creative multidirectional thinking grows with the partnership of the logical and pragmatic domain of engineering and systems. To integrate them in partnership is to achieve the ideal form of a holistic balance. The objective is to conduct this process supported by models of systemic design processes integrated with technical knowledge to create disruptive solutions adapted to the socio-economic and cultural context of current societies.

Acknowledgments. The researcher wants to thank the support of FAPESP - Foundation of Support and Research of the State of Sao Paulo, which provided resources that generated this article by process Nr 2016/11169-4.

References

1. Isaacson, W.: Leonardo da Vinci. Simon & Schuster, New York (2018)
2. Faberberg, J.: Innovation: a Guide to the Literature. In: Fagerberg, J., Mowery, D., Nelson, R. (eds.) The Oxford Handbook of Innovation, pp. 1–26. Oxford University Press, New Yok (2004)
3. Gino, F.: The Business Case for Curiosity. Sept.–Oct. 2018 Issue, Harward Business Review, Boston, MA, https://hbr.org/2018/09/curiosity (2018)
4. Godin, B.: Innovation Contested: The Idea of Innovation Over the Centuries. Routledge - Taylor & Francis, New York (2015)
5. Newell, A., Shaw, J., Simon, H.: The processes of creative thinking. In: Gruber, H., Terrel, G., Wertheimer, M. (eds.) Contemporary Approaches to Creative Thinking. Atherton, New York (1962)
6. Dyer, J., Gregersen, H., Christensen, C.: Innovator's DNA: Mastering the Five Skills of Disruptive Innovators. Harvard Business Review, Boston, MA (2019)
7. Alencar, E.S.: O Processo da Criatividade. Produção de ideias e Técnicas Criativas. Editora Makron Books, São Paulo, Brazil (2000)
8. Kneller, G.F.: Art and Science of Creativity. International Thomson Publishing, Boston (1965)
9. Guilford, J.P.: The Nature of Human Intelligence. McGraw-Hill, New York (1967)
10. De Bono, E.: The Mechanism of Mind. Simon and Schuster, New York (1969)
11. MIT Lemelson. https://lemelson.mit.edu/news/2011
12. De Bono, E.: Lateral Thinking: Creativity Step by Step. Harper Collins, New York (2015)

13. Barcellos, E.E.I., Botura Jr., G.: Design process: interaction methodology between design and engineering as an innovation factor. In: Gestão e Tecnologia de Projetos, vol. 14, no. 2, pp. 65–74, São Carlos, São Paulo, Brazil (2019). http://dx.doi.org/10.11606/gtp.v14i2.137709
14. Barcellos, E.E.I., Botura Jr., G.: Design thinking: user-centered multidisciplinary methodology based on people and innovation. In: Advances in Human Factors, Business Management, and Leadership, AHFE 2017, vol. 594, pp. 173–182. Springer International Publishing, New York (2018)
15. Barcellos, E.E.I., Botura Jr., S., J.C.P.: Design Methodology for Innovation Generation. (Thesis) São Paulo State University. Faculdade de Artes, Arquitetura e Comunicação - Universidade Estadual Paulista Júlio de Mesquita F.º- UNESP, Bauru, São Paulo, Brazil (2020)

Exploration and Practice of Industry-Education Integration in Design Schools

Wei Ding, Xinyao Huang, and Dadi An(✉)

School of Art Design and Media, East China University of Science and Technology, 130 Meilong Rd., Xuhui District, Shanghai, China
61915633@qq.com

Abstract. With the country attaching great importance to the development and employment prospects of academy, the Chinese State Council has issued a series of policies on the integration of industrial education to promote cooperative academy, broaden the areas and deepen the level of cooperation between academy and enterprises, as a measure to better match the demand of supply majors and industries. Similar to the classical Triple Helix Model (THM), Industry-education Integration (IEI) also emphasizes the leading role of the integration of educational resources, key enterprises and other social organizations in the process of innovation activities. According to the professional features of design education and relevant industry advantages, THM can be diversified updated. This paper considers the practical design education experience in home furnishing, intelligent manufacturing, cultural innovation, platform operation, rural revitalization and other industries, and looks into the future development prospects by summarizing three main modes of IEI.

Keywords: Industry-Education Integration (IEI) · Design education · Triple Helix Model (THM) · Innovation

1 Background

Recently, Chinese State Council vigorously promotes the integration of industry, education and research, which is an important way for academy to train and bring up innovative talents [1]. IEI pays more attention to the “integration” development, the purpose is to realize the transformation of the transformation of education itself and industry, so as to better realize economic development and construct the new education system.

THM comes from the background of New American Economy, and the application of Chinese design academy should be combined with Chinese specific national conditions to extend the original concept (Fig. 1). It links universities, industry and government among knowledge fusion of university, industry and government. This interconnected linkage system ultimately promotes the economic innovation and development [2].

In the era of innovation-driven development and design education, IEI has been widely used in academy to integrate enterprises, universities and society closely. IEI promotes the economic development of society and realizes the collaborative sharing of

G. Di Bucchianico et al. (Eds.): AHFE 2020, AISC 1202, pp. 506–513, 2020.
https://doi.org/10.1007/978-3-030-51194-4_66

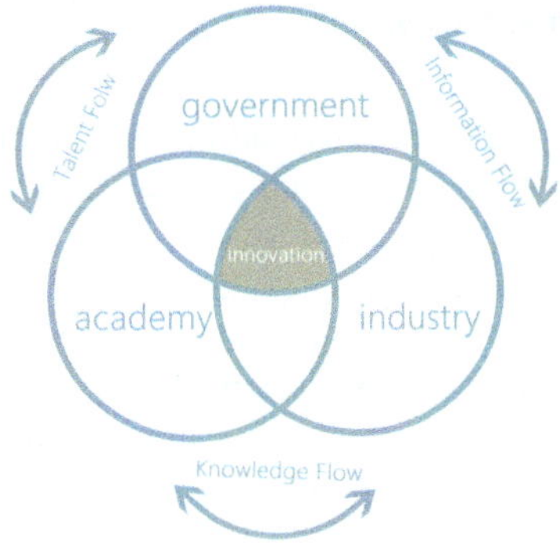

Fig. 1. Triple Helix Model

knowledge. Meanwhile, it improves the quality of talent training and expands the cooperative network so as to promote the cultivation of interdisciplinary and innovative technical talents by providing students more practice opportunities.

2 IEI in Design Education

With the rapid development of Internet industry, there is a mismatch between design education and the talent needs of enterprises. It is difficult to employ talents with corresponding capacity, while the academy curriculum system is not closely related to social needs. Training programs pay more attention to basic theoretical knowledge without practical ability. As a practical and innovative discipline, design is in urgent need of expanding students' horizon and practical ability. Firstly, experienced designers from industry are rarely to get a lecturership, cutting-edge design concepts cannot be quickly spread to academy. Second, the cultivation of talents is not in proportion to the social demand. Third, training programs do not match the new social and economic development needs.

Therefore, design education should consider how to cultivate talents with the development needs of the society. It is necessary use the idea of IEI to connect the talents of academy with the needs of industry, so as to cultivate application-oriented design talents.

3 Three Modes of IEI

IEI is an important tool of cultivating applied talents in design education to better promote the fusion of talent supply and demand [3]. Based on years of experiences on exploration and practice in design education, we summarized three IEI modes in current Chinese context.

3.1 University-Government Dominated Mode

Rural Revitalization. With the creativity and inspiration of academy, design for rural revitalization is held to serve the revival and activation of village culture [4]. Through the intelligence and technical support of academy, it is possible to revitalize rural production, life and ecology, and to enhance the value of rural culture. Meanwhile, the village and residents can build local identity and rural revitalization by stimulating the village "endogenous development" by design. There are significant effects on activating village "self-hematopoietic" ability and more systematic, comprehensive, and profound impact on the future development of Chinese rural (Fig. 2).

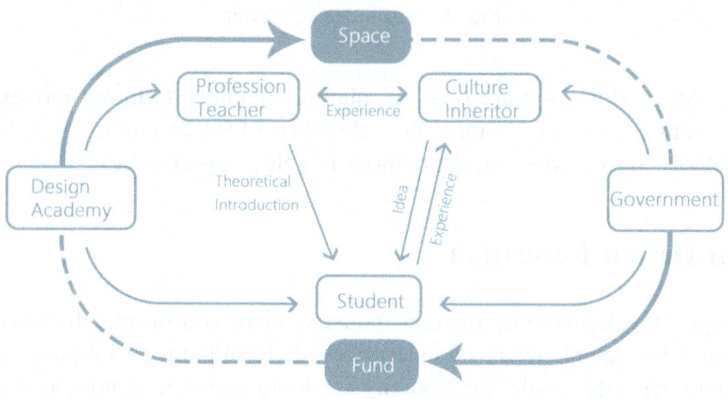

Fig. 2. University-government dominated mode

In view of the current development situation and regional brand construction of Songxi, Fujian, a design team from ECUST began to combine design discipline and innovation force to provide effective support for the poverty alleviation plan of Songxi county. Series of activities such as "Thousand Years Songxi Hundred Years Sugarcane Cultural Tourism Festival", as well as local specialty brand planning and design (Fig. 3) has been released to provide Songxi county with a distinctive regional characteristics of products, packages and culture.

Urban Renewal. Cooperation between design academy and government can also play and important role in urban renewal by design. It is well know that there exist various sites to be renewed in city centers, including aged living community and industrial heritage. Urban governments can provide development platform for design students' mental play by integrating the existing technology and excellent creative, as well as various funds to increase investment in research and raise the possibility of transformation of scientific and technological achievement to promote people's livelihood, environments, living service, sustainable development, etc.

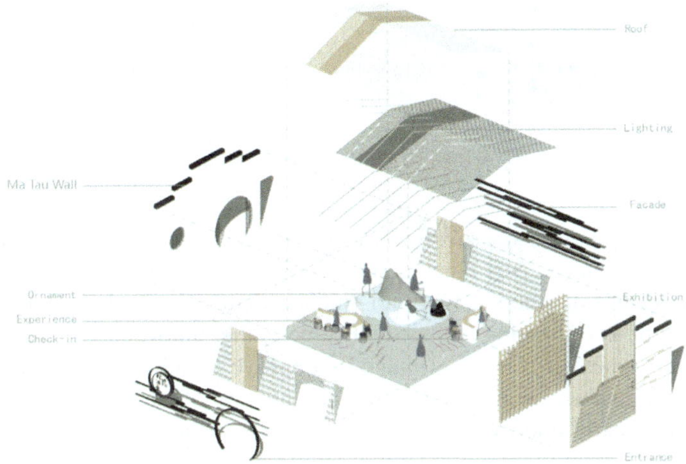

Fig. 3. Design for Songxi Tea House

3.2 University-Enterprise Dominated Mode

Design Competition. Enterprises provide design demand and subject to arouse students' imagination and creativity. In return, academy can provide enterprises with rapid design scheme, and greatly enrich enterprises' design strategy, which is conducive to the new product rapid hatch. This mode produce clear design demand, Business oriented training makes students involved in theory and practice of knowledge to the enterprise in manufacturing industry through the latest design and trend [5]. By organizing design competitions, enterprises can provide the fastest design scheme and discover excellent and distinctive ideas, as well as design talents. Enterprises can maximize and expand the frontier vision to plan the development trend in advance and better serve the economic development [6] (Figs. 4 and 5).

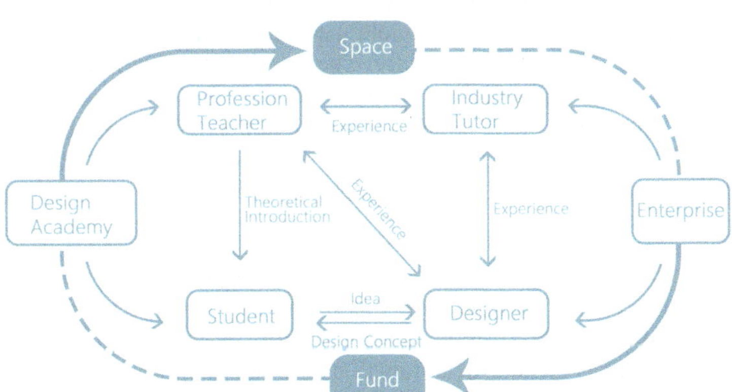

Fig. 4. University-enterprise dominated IEI mode

Fig. 5. Winning-works of OPPLE - ECUST design competition

Workshop. Workshop also provides practice tutor, guidance and wisdom. Students and enterprise technical personnel communicate in the workshop, turn individual learning into groups, teachers taught into guidance, enterprise technical personnel gain design wisdom and thinking so as to promote the innovation of enterprise [7]. Workshop can establish corporate image to improve brand quality with higher economic and social value. On the other hand, introduction of design competition into the classroom can greatly improve students' enthusiasm for design.

3.3 University-Science Park Dominated Mode

Creative Incubation. An innovation incubation platform promotes to explore the value of creativity to transform industrial capital into real social value [8]. The platform includes content discovery, community sharing, cultural and creative investment, brand IP incubation, e-commerce marketing, property rights protection and so on, completing a full series of creative work from generation to production to sales, so as to realize the co-creation, sharing and using of projects, talents, technologies and brands. With the combination of online and offline LCS (Lab + Club + Shop) dual education mode, course, business practices, incubation, linking to the whole process innovation entrepreneurship education, eventually to build platform for the industry belt around ECUST, explore a new mode of regional economy and industry, and strive to become Shanghai IEI culture innovation belt, study and research of wen gen belt, explore Chinese a complete industry chain including design and marketing, forming a mode of promotion can be copied (Fig. 6).

It is an effect way to realize the integration of university knowledge with various industries and technologies, and to promote the modernization of higher education so as to integrate theory into practice rapidly. The creative incubation platform not only gives the design community a fair opportunity, but also completes the crossover integration of design, technology, brand and production in the real sense. Driven by the large platform, the characteristic leading trend of the design industry has been greatly developed, which also promotes the development of the integration of industry and city.

Creative Community. Creative community is a composition of nearby residents, creative class and related sectors. In comparison with creative industry center, the creative community is more exoteric and diversiform. With a fusion of daily life and the creative space, it has certain social function to promote the development of the city. Creative community in China plays an important role in design industry transformation. the formation of a community can activate the cultural confidence of the native

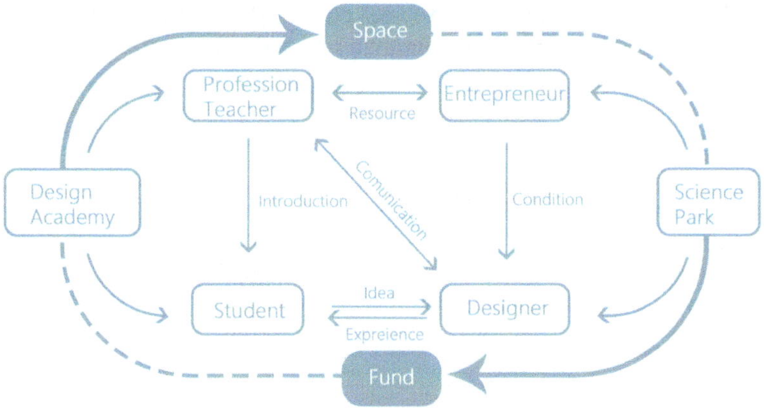

Fig. 6. University-science park dominated mode

community, promote the city's cultural evolution, receiving financial support for the local culture, can not only establish local characteristic culture brands, but to promote cultural construction and development of additional industry.

In respect of practice, ECUST adopts the concept of community economic thinking and the support of high-density, diversified communities in Shanghai with scientific, educational and cultural resources of universities, it is feasible and effective to build an "Innovation and Entrepreneurship Platform of ECUST Industrial Design Belt" to serve Xuhui district of Shanghai.

4 Triple Helix Model and Design Education

IEI modes enrich the connotation of design education, illustrates the trend for future development. Firstly, the IEI close the relationship between education and society, make the education built on real social background and help the social demand closely connected with the talent we need. Second, through the "integration of science and technology", get through the barriers between the professional to achieve the all-round development of talent. Third, through "innovation and entrepreneurship", students' thinking and practical ability can be cultivated, and design education can be guided through the construction of a new design thinking framework to enable design talents to have innovative and entrepreneurial thinking. Fourth, through the "knowledge system" to construct a new system and research direction, from "knowledge" to "ability" is the most important thing in the training of design talent. Fifth, through "discipline development", build key areas of new disciplines. In today's omnimedia field, information, people and things are closely linked. Based on the new changes in art design and industry, discipline development will also form a new discipline system.

Originally, innovation plays a role like an armchair strategy and become a barrier between different disciplines and majors in universities, between colleges and social enterprises, where enterprises could only develop innovation in the areas that they are good at. Today's model (Fig. 7) of collaborative development is a giant axe that breaks the invisible wall. The academy and society remove design barriers by tearing down walls, thus benefiting from each other through interaction in the process of innovation.

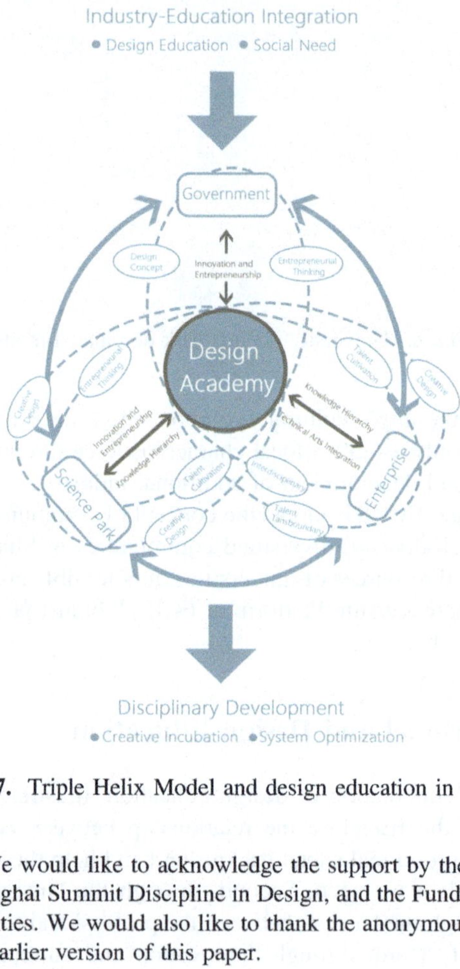

Fig. 7. Triple Helix Model and design education in China

Acknowledgments. We would like to acknowledge the support by the Shanghai Pujiang Program (18PJC022), Shanghai Summit Discipline in Design, and the Fundamental Research Funds for the Central Universities. We would also like to thank the anonymous referees who provided useful insights on the earlier version of this paper.

References

1. Li, Y.: Research on the system environment and optimization of the integrated development of education in China (2018). (in Chinese)
2. Leydesdorff, L.A., Etzkowitz, H.D.: Introduction: Universities in the Global Knowledge Economy (1997)
3. Yao, R.: Research on evaluation system of the applied undergraduate college integration of production and education based on stakeholder theory (2018)
4. Zhao, W., Zhang, Y.: The role and function of government in industry-university-research cooperative education (2014). (in Chinese)

5. Zou, S.: Study on the construction path of the ecological circle of production and teaching integration – a case of A College (2016)
6. Santomauro, M.: Research Activities at CEFRIEL, a University-Industry Cooperation Example (1991)
7. Pfeiffer, A., Uckelmann, D.: Open Digital Lab for You – Laboratory-based learning scenarios in education, research and qualification (2019)
8. Cabral-Cardoso, C.: A bridge over troubled water: the difficult crossing from academia to the 'real world' of industry (2000)

Research and Practice of Whole Process Product Design Innovation Theory System Based on Service Design Thinking

Wei Ding, Defang Chen(✉), and Junnan Ye

East China University of Science and Technology,
130 Meilong Road, Shanghai, China
2351962834@qq.com

Abstract. With the coming of the information age, industrial design has entered the 4.0 era. The circulation of resources among designers, enterprises, suppliers and universities has been accelerating. At the same time, design has gradually developed from a single discipline to a multidisciplinary. The purpose of service design thinking is to mobilize the resources of designers, enterprises, multidisciplinary and all parties to co-create, so that the creativity gradually changes from closed innovation to open innovation. Therefore, under the background of the current era, "design" has gradually changed from simple product design to a full process service design, including demand analysis, product design, product landing, commercial operation and so on. Consequently, a scientific and systematic design service system can help designers and design enterprises to carry out efficient and low-risk design innovation, so as to enhance their competitiveness.

Based on the practical cases of MOMA company, the whole process closed-loop service design innovation theoretical system of "four chain integration and seven step innovation" was put forwarded. Four chains include product chain, brand chain, interaction chain and service chain.

Keywords: Product design · Service design · Four chain integration and seven step innovation

1 Theoretical Research on Service Design and Design Service System at Home and Abroad

According to Gao Ying, the specific difference between service design and traditional design in concept lies in that it is not only the design of things, but also the design of the way of using things, even the design of the desire to use things. The design dictionary published by the International Design Research Association defines it as: "service design sets the function and shape of services from the perspective of customers type. Its goal is to ensure that the service interface is useful, available and desired by customers, while the service provider feels that it is effective, efficient and identifiable."

Industrial Design Researcher.

G. Di Bucchianico et al. (Eds.): AHFE 2020, AISC 1202, pp. 514–518, 2020.
https://doi.org/10.1007/978-3-030-51194-4_67

Researchers at home and abroad define most of the objects served by service design as individual users, and this paper will explore the use of service design thinking to mobilize the resources of designers, enterprises, multi-disciplinary and all parties to create together, so as to gradually change the creativity from closed innovation to open innovation. Using service design thinking, taking product design process as service object, establishing a scientific and systematic design service system can help designers and design enterprises to carry out high-efficiency and low-risk design innovation, so as to enhance their competitiveness.

2 Current Design Service Background

2.1 Advantages of China's Product Design Services in the Current Era

Since the 21st century, China's design industry has developed rapidly. With the continuous improvement of national design awareness, more and more domestic design resources, and the continuous output of innovative talents from the design education industry to the society, the scale of China's design industry is becoming larger and large. The design services are beginning to be subdivided, which ask new requirements for the scientific methods of design.

2.2 Disadvantages of China's Product Design Service in the Current Era

With the development of product design and service industry, all kinds of defects and deficiencies are gradually emerging. Because of China's large population base and grate design demand, the process and methods of product design services need to be summarized scientifically and verified through practice. Otherwise, the design speed can not keep up with the customer's requirements, and the design quality can not meet the customer's requirements.

3 The Whole Process Closed-Loop Service Design Innovation Theoretical System of "Four Chain Integration and Seven Step Innovation"

Based on the practical cases of MOMA industrial design company, this paper will put forward the whole process closed-loop service design innovation theoretical system of "Four Chain Integration and Seven Step Innovation".

In this system, "four chains" refer to product chain, brand chain, interactive chain and service chain. Seven steps refer to insight, definition, discovery, design, transformation, integration and communication (Fig. 1).

The first chain in the four chain integration is the product chain. Product design are the core of the system, and the other three are all around product design. Nowadays, the product design is not only about one hardware product, but also a combination of software products and hardware products. The combination of industrial products and Internet products makes the product chain more important and complete.

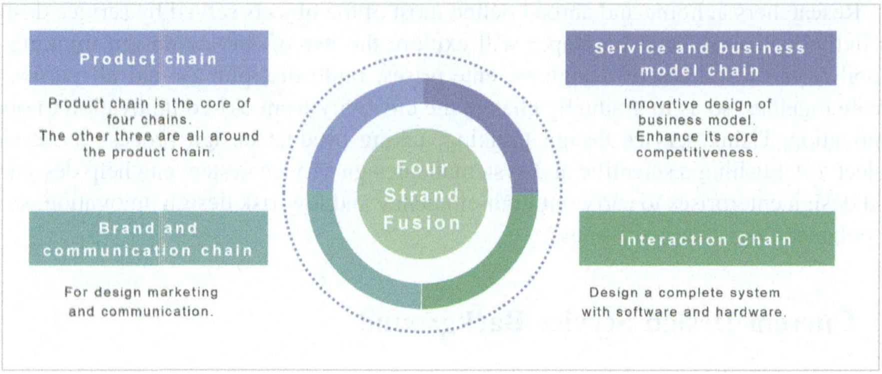

Fig. 1. Diagram of four chain fusion relationship

The second chain is service design and business model chain. In the past, we lacked business model chain in the field of system business model innovation, which is a whole process service design, aiming to provide customers with more trustworthy whole process services and business model innovation design, so as to enhance their core competitiveness. Of course, the business model is also based on the product design, around the product characteristics and market demand, business model innovation.

The third chain is interaction and user experience chain. Now most of the products have begun to shift from simple hardware products to intelligent products, that means product design is from the separation of hardware and software to the combination of hardware and software. In such a background, the central problem behind all these phenomena is the Internet of things. Through sensor technology and big data, the design becomes a complete system combining software and hardware, including products and services. The role of designers in the Internet of things is to connect products and the Internet with design. Through the combination of hardware products and software interaction, designers can provide users or customers with convenient and operable design experience, so as to make the design more valuable.

The fourth chain is brand and communication chain. Product designers need to have the knowledge reserve and ability of marketing and communication, and have the consideration of marketing and market at the beginning of design. Therefore, through the integration of four chains, we can maintain the value of product design, so that product design can give full play to its original goal of design (Fig. 2).

The first step in the seven step innovation is insight. Designers need to observe today's society, business and people. They need to find problems and the pain points of design. To explore new problems, through scientific methods of research, so as to tap the real demand, and then demand and market research. This is the whole process of insight.

The second step is to define. The important task of today's design is to define the products we really need today. This requires the insights in the first step to define the design according to the needs. Before defining the design, we should first understand the needs in all aspects, including target population, demand positioning, market research, etc. After understanding the demand, analyze the market situation of the demand. After the above foundation is determined, you can start to conceive the design prototype.

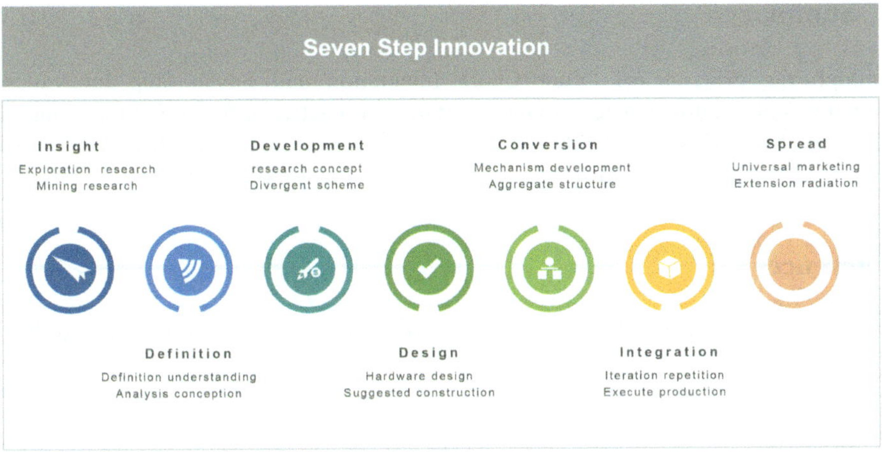

Fig. 2. Seven step innovation diagram

The third step is to develop. According to the design definition, designers can develop the software and hardware prototype and generate low fidelity product design and interaction design. Using the low fidelity model, designers can carry out the feasibility test. The test needs ordinary users and relevant professionals, so that it can carry out the test in an all-round way and improve the reliability of the test.

The fourth step is design. The design should visually express the problems found and the needs generated by the problems, and refine the development prototype. There are two basic qualities for designers. The first one is image thinking, that requires designers have the ability to transform the prototype into a high fidelity prototype; the second is role introduction, that is, the attitude of thinking problems from the perspective of consumers.

The fifth step is transformation. Designers should transform the design language into public programming language, with scientific design language and context. Designers should be more professional and design theory should be more standardized. The successful design cases will be transformed into design theoretical results, and then the next design practice under the guidance of theory will form a complete theory practice theory closed-loop.

The sixth step is to realize the integration of supply chain. The whole supply chain is from supplier to designer, and then to marketing needs. Designers need to connect these elements and understand the whole design process, so as to provide customers with more worry free design services.

The seventh step is spread. Designers and design teams should spread their designs through their own marketing and packaging capabilities, and gain strong communication focus. Related personnel can make use of the power of communication to fully embody the value of design.

4 Summary

Through the principle of "Four Chain Integration and Seven Step Innovation", designers and design agencies can carry out design practice according to this principle, transform design theory in the design process, guide the next practice with the theory of successful design cases, and form a good design closed-loop in the current era.

References

1. Gao, Y., Xu, X.: Service design: a new concept of contemporary design. J. Literat. Art Res. **6** (2014)
2. Erlhoff, M., Marshall, T.: Perspectives on Design Terminology. Design Dictionary, Boston (2008)

Product Design Education for Circular Economy

José Vicente[1,2(✉)]

[1] Faculdade de Artes e Letras, DA, Universidade da Beira Interior, R. Marquês de Ávila e Bolama, 6201-001 Covilhã, Portugal
jmanvicente@gmail.com

[2] Faculdade de Arquitetura, CIAUD, Universidade de Lisboa, R. Sá Nogueira, 1349-055 Lisbon, Portugal

Abstract. Design has continually developed new approaches to find the most appropriate solutions to the growing environmental and social problems. At the same time higher education courses have tried to adapt their curricula accordingly. The most recently proposed model is circular economy. It reinforces the idea of a paradigm shift to a system of closed loops where there is no waste.

This article develops a state of the art on the integration of sustainability in product design in higher education and its evolution to embrace circular economy. This analysis includes identifying past experiences, which contents are addressed, what methodologies are used, what type of approach (focused or dispersed) and what are the needs for teaching staff. This paper tries to identify gaps in order to purpose better solution for circular economy integration.

Keywords: Design education · Product design · Sustainable design · Circular economy · Circular design

1 Introduction

Humanity is at an historic moment, full of particularities and challenges. Our success as a species has brought us significant benefits, like an exponential demographic growth, but that has been based on a development paradigm fully supported on economic growth. This brings enormous pressure on natural resources and ecosystems, creating environmental problems of various scales and complexities, of which climate change is the most significant. Associated with these problems, but also with many other inefficiencies of contemporary societies, social inequalities continue to increase. On the other hand, we live in a period in which Humanity is more capable than ever to face challenges, in which technology plays an increasing role and where design is assumed, in a multifaceted and growing way, as a discipline with the potential to positively contribute to global development. Design is considered one of the main responsible for the circularity of products [1] and, together with technology, innovation and the responsibility of production and consumption, is seen as one of its main facilitators [2].

But for designers to be able to practice sustainability in an effective way it is necessary that they receive solid training. It is very important for designers learn about sustainability, social responsibility and the circular economy (CE), namely its key

G. Di Bucchianico et al. (Eds.): AHFE 2020, AISC 1202, pp. 519–525, 2020.
https://doi.org/10.1007/978-3-030-51194-4_68

concepts [3] in order to take advantage of this opportunity and make a difference. It is then necessary to know the teaching of product design and know how it has been adapting and creating knowledge to develop a capacity in the area of sustainability. It will be important to identify the relevant and innovative skills and content, but also which pedagogical practices are the most appropriate for this creation of skills.

2 Methodology

The work exposed in this article is based on a methodology of critical review of literature, based on the analysis of articles on product design education for sustainability, in its various stages of development.

3 Circular Economy

The multitude of summits, strategies, declarations and international protocols have contributed positively to promote a solution to the environmental problems but have always fell short of solving it entirely. The most recent strategy presented by the European Commission [4] is CE, which aims to help citizens and companies to make a transition to a stronger and more sustainable economy, proposing to close the production and consumption cycles by increasing the recycling and reuse, extracting the minimum amount of raw materials, promoting energy savings and reducing greenhouse gases [5]. In the European guidelines one can note that design is considered in a direct and relevant way as a tool to support the development of this plan, differently to what happened in the previous plans.

The changes in mentality and the pressures that the business and industrial companies have been suffering due to economic crises, pollution and resource depletion [6] make the concept of CE, which is relatively new, gain attention and traction, as it is seen as a way to continue to increase prosperity while reducing dependence on materials and energy, managing to change the current paradigm of production and consumption of linear economy [7].

CE advocates to close and slow the cycles, that is, tries to replicate nature, where the water and nutrient cycles are abundant. In our daily lives, we have to move away from the linear optics of producing-using-throwing away [8] and thus interpreting this new CE strategy as the new paradigm of sustainable development, while still distinguishing the two concepts [9].

But we must not forget that the third pillar of sustainability has to be addressed. In the production and consumption system, the social function can be worked through corporate social responsibility [10]. Knowing that the activity of organizations generates positive and negative impacts at the economic, social and environmental level and that in a context of globalization, companies are increasingly under the jurisdiction of laws of various States, under the scrutiny of a more demanding public opinion, social responsibility is an instrument through which companies can contribute to a fairer society and a cleaner environment [11].

It is also important to bear in mind that the essential basis for the implementation of sustainability is education. This happens in general training and, more importantly, in education in the area of product design. CE has been very advocated, but it is not yet being widely practiced [12], for this it is necessary to facilitate the education of this theme.

4 Evolution of Product Design Education Towards CE

Higher education institutions have an important role in the promotion of a paradigm shift towards sustainable development. For that to happen they need to go beyond the simple integrations of key ideas in the course contents [13]. And need to overcome the constrain of having curricula that are already packed with other important content [14].

4.1 Ecodesign and Design for Sustainability Education

Looking at the international panorama of product/industrial design education, most of the courses have mandatory or optional curricular units (CU) to integrate sustainability, but there are also several institutions that present 1st cycle degrees totally guided by this theme. The majority of the CU addresses contents in the scope of ecodesign, focusing on life cycle analysis, lower impact materials or recycled/reused material [14].

The development of ecodesign CU, even when based on the same matrix of contents like the PROMISE Ecodesign Manual [15], have to take into account several variables of the local context, like profession background, experience, learning style, local economic developments and cultural context [16].

The ecodesign CU are structured with both theoretical and practical dimensions, with contents ranging from general information regarding sustainable development to specific operative design decisions. The operationalization of this content, similar to the professional context, is made through the use of guides, strategies and tools like "Ecodesign in Higher Education Guide" [17] or other qualitative and quantitative tools. Some programs use student-led seminar to promote the widespread dissemination of the contents throughout the entire class. To enhance the motivation of students in the apprehension of this type of content and development of skills, some universities establish contact with real companies to frame the projects/exercises being created, which proves to be a win-win situation [18].

Most approaches focus on the integration of concepts only related to the environment (ecodesign), but it has been known for a long time that extending the scope of sustainability to integrate social aspects, although a challenge, results in a richer learning process, with greater credibility and acceptance [19].

So that the high number of elements related to sustainability does not cause confusion and dissatisfaction in students during their learning process, the evaluation process has to be simplified, focusing on the students' ability to identify options for sustainable product concepts and to be able to evaluate these options [19]. Also, since the complexity of sustainability may present added difficulties in conceptualizing solutions, an approach based in systems-thinking, interdisciplinary and anticipatory presents improvement in students' skills [20]. To ease the learning curve of sustainability, literature states that it should be introduced early in the courses, it should have a bigger importance in the course, that it should be linked with the design studio CU and this contents should be made an expectation rather than an option [21, 22].

The integration of sustainability in product design courses requires a method that links theory with design projects, especially with real-life partnerships with companies, so that students can develop their own case studies in a transdisciplinary way [24].

Several authors indicate that due to the increasing complexity and multidisciplinary of design for sustainability, that integrates criteria from social sciences, renewable energy, corporate social responsibility, social innovation, etc., these contents should not be constrained in a dedicated CU, but should be divided in more than one [20] or dispersed throughout the entire course [24]. Others argue that only through a focused design for sustainability approach throughout the entire program a truly paradigm change can come [25].

The effective transition from ecodesign to design for sustainability pedagogies required comprehensive resources, but also faculty with expertise [23, 24], and this expertise should be connected with professional experience of faculty and fully disseminated among students, since the lack of knowledge in designers is one of the major barriers identified to practicing sustainable design [26]. It should also be noted that the support and autonomy given to faculty may be very important to integrate sustainability in their teaching [21], and this can be a problematic issue, since universities are usually underfunded, and staff is burdened with bureaucratic activities.

The integration of design for sustainability in design practice is still not mainstream, but, nevertheless, students need to be prepared with a range of skills to support and promote a fully sustainable practice, for when that may occur [27].

4.2 Design for Circular Economy Education

Since the affirmation of CE that design for sustainability has started to embrace this approach. It presents a more robust relation between the environmental problem-solution and the economy and, from the design perspective, addresses the problems in a more pragmatic way, base in a clearer and more synthetic model [28].

During the last few years practical guidelines have been developed for design practice [29]. But this new approach presents another challenge related with the possibility of creating new business models, meaning designers should widen their multidisciplinary approach. Also, the perspective of reverse cycles, creates more complexity and depth to the life cycle approach that design for sustainability incorporated from ecodesign.

Also, the new designer archetypes, which are not exclusive product creators, but designers of holistic solutions, have to integrate a wide set of skills to meet the diverse requirements of the CE. Business stakeholders are encouraged to seek the development of such competencies in their product development teams to drive successful implementation of sustainability strategies [6].

Design for CE contents should contribute to understand this new model and its implications for a new economic paradigm. They should promote a multidisciplinary approach, bridging different stakeholders, namely through the development of case studies with companies [30]. Besides some of the contents already referred it is essential that students are fully aware of all design for CE principles and strategies. Regarding the methods, some of the more recent and successful results in the area of design education for CE points to curricular units based on a practical methodology, with a strong design studio nature [31]. For the implementation of this approach several tools were already developed by Ellen MacArthur Foundation and other have been adapted by other sources.

5 Conclusion

Despite the discussion on whether sustainability should be addressed in a specific CU or across the entire design studios CU, or even throughout the entire course has not yet had a conclusive outcome due to the very different contexts where the courses are taught, it's possible to reason that the best solution would be the development on an entire course with a product design for CE matrix. This would be a solution to the problem of having a sustainability CU packed with contents, and without removing to much hours for the regular contents. However, due to the lack of awareness or training from part of the teaching staff it is still difficult to integrate these skills smoothly.

To enable the implementation of design for CE contents in regular product design degrees a pedagogical manual should be developed, structured with the theoretical and practical contents, packed a set of case studies and with a robust set of tools specific for learning design for CE. This manual can be adapted from Design for Sustainability but need to respond to CE specificities.

References

1. European Environment Agency: Circular by design-Products in the circular economy. EEA Report, No. 6/2017. European Environment Agency (2017)
2. Bocken, N., Pauw, I., Bakker, C., Grinten, B.: Product design and business model strategies for a circular economy. J. Ind. Prod. Eng. **33**(5), 308–320 (2016)
3. Hollander, M., Bakker, C., Hultink, E.J.: Product design in a circular economy: development of a typology of key concepts and terms. J. Ind. Ecol. **21**(3), 517–525 (2017)
4. EC. Closing the loop–An EU action plan for the circular economy. Com, 614. Brussels: European Commission (2015)
5. Camocho, D., Vicente, J., Ferreira, A.M.: Meeting the circular economy agenda: supporting tools for a new strategic design practice. Convergências - Revista de Investigação e Ensino das Artes, vol. XII, no. 24 (2019)
6. De Los Rios, I.C., Charnley, F.J.: Skills and capabilities for a sustainable and circular economy: the changing role of design. J. Cleaner Prod. **160**, 109–122 (2016)
7. Ellen MacArthur Foundation: Towards the Circular Economy. Ellen MacArthur Foundation, vol. 1, pp. 1–96 (2015)
8. Stahel, W.R.: Circular economy. Nat. Int. J. Sci. **531**, 4 (2016)
9. Geissdoerfer, M., Savaget, P., Bocken, N.M.P., Hultink, E.J.: The circular economy a new sustainability paradigm? J. Cleaner Prod. **143**, 757–768 (2017)

10. Vicente, J., Frazão, R., Rocha, C., Silva, F.M.: The integration of social criteria in sustainable design for furniture. In: Proceedings of European Roundtable on Sustainable Consumption and Production 2010, Delft, The Netherlands (2010)
11. UNEP: Guidelines for social life cycle assessment of products. UNEP (2009)
12. Andrews, D.: The circular economy, design thinking and education for sustainability. Local Econ. **30**(3), 305–315 (2015)
13. Tilbury, D.: Higher education for sustainability: a global overview of commitment and progress. High. Educ. World **4**, 18–28 (2011)
14. Ramirez, M.: Sustainability in industrial design education: a worldwide survey. In: Connected 2007 International Conference on Design Education, Sydney (2007)
15. Brezet, J.C., Horst, T.v.d., Riele, H.t.: PROMISE Handleiding voor milieugerichte productontwikkeling. Den Haag: NOTA/SDU (1994)
16. Diehl, J.C., Brezet, J.C.: International ecodesign education: personalised design knowledge transfer. In: ICED 05 International Conference on Engineering Design, Melbourne (2005)
17. Doorsselaer, K., Verhulst, E.: Ecodesign in Higher Education-Guide. OVAM (2013)
18. Badni, K., Coles, R.: A review of the recent eco-design education initiative for industrial design and technology undergraduates at Loughborough University. J. Des. Technol. Educ. **8**(2), 15 (2009)
19. Boks, C., Diehl, J.C.: Integration of sustainability in regular courses: experiences in industrial design engineering. J. Cleaner Prod. **14**, 932–939 (2006)
20. Banerjee, S., Upadhyay, P., Punekar, R.M.: Teaching design for sustainability for socioeconomic ecosystems—three case studies. In: Chakrabarti, A., (ed.) Research into Design for a Connected World. Smart Innovation, Systems and Technologies, vol, 135. Springer, Singapore (2019)
21. Chamberlin, L., Greenham, T., Davis, S.: Fractures in the faculty: the state of sustainable design teaching in the UK. In: EPDE2018 20th International Conference on Engineering and Product Design Education, London (2018)
22. Dias, A.C., Almendra, R., Silva, F.M.d.: Teaching of sustainability in industrial/product design courses in Portugal. In: Simpósio Brasileiro de Design Sustentável + 4th International Symposium on Sustainable Design, Porto Alegre (2013)
23. Dewulf, K., Wever, R., Boks, C., Baller, C., D'hulster, F.: Sustainability in design engineering education; experiences in Northern Europe. In: EcoDesign 2009: 6th International Symposium on Environmentally Conscious Design and Inverse Manufacturing, Japan (2009)
24. Bhamra, T., Lofthouse, V., Norman, E.: EcoDesign education strategies: a recent initiative for industrial design and technology undergraduates at Loughborough University. J. Des. Technol. Educ. **7**(2), 133–141 (2009)
25. Fiorentino, C.: Education on design for sustainability: focus and integration vs. fragmentation. In: Designa 2012 International Conference on Design Research, Covilhã (2012)
26. Blincoe, K., Fuad-Luke, A., Spangenber, J.H., Thomson, M., Holgren, D., Jaschke, K., Ainsworth, T., Tylka, K.: DEEDS: a teaching and learning resource to help mainstream sustainability into everyday design teaching and professional practice. Int. J. Innov. Sustain. Dev. **4**, 1 (2009)
27. Lofthouse, V.: Preparing the way for mainstream sustainable product design–methods and approaches to inspire change. Form Academic, vol. 5, no. 1 (2012)
28. MacArthur, E.: Towards the Circular Economy, Economic and Business Rationale for an Accelerated Transition. Ellen MacArthur Foundation, Cowes (2013)
29. Bovea, M.D., Pérez-Belis, V.: Identifying design guidelines to meet the circular economy principles: a case study on electric and electronic equipment. J. Environ. Manage. **228**, 483–494 (2018)

30. Celades, I., Ros, T., Rocha, C., Camocho, D., Schmidt, K., Pamminger, R., Schmidt, S., Kalleitner-Huber, M., Veral, S., Ruiz, L., Mulet, E., Chulvi, V.: KATCH_e: introducing circular economy into higher-education design curricula-overview of training needs, state of the art, trends and policies. In: 18th European Roundtable on Sustainable Consumption and Production, Greece (2017)
31. Ramirez, M.: Industrial design education and the circular economy: experiences from UNSW Sydney. In: Paper presented at the Unmaking Waste 2018, Adelaide, Australia (2018)

Interdisciplinary Approach to Design Manufacturing

Analysis of the Components of the Scene System and Its Characteristics

Yan Wang(✉) and Jiadong Han

School of Art Design and Media, East China University of Science and Technology (ECUST), M. BOX 286, No. 130 Meilong Road, Xuhui District, Shanghai 200237, China
54756432@qq.com

Abstract. This paper analyses the elements and characteristics of scene design, so that designers can establish the theoretical framework and work-flow of scene design. Systematic analysis was used to analyze the factors, relationships and mechanisms. This paper puts forward that scene consists of objective scene, emotional psychology and scene interaction. We need to be more systematic in defining scene system design and use it to guide our design, adhere to the design concept of people-oriented, let people, things and fields be unified in sensibility and logic, release the value of scene system design.

Keywords: Scene design · Explicit interaction · Invisible interaction · Neutrality · Reciprocity and conflict

1 Introduction

"This flower dies with you when you have not seen it. When you see this flower, its color suddenly turns bright and you will know that it is not outside your heart," said Wang Yangming, a great scholar of the Chinese Ming Dynasty, in Chuan Xi Lu. This statement explains the various elements of the scenes system and their relationship. Scene system, as its name suggests, refers to the environment that stimulates input into a person and causes human behavior and emotional interaction. Scene system consists of three core elements: objective environment, emotional psychology and scene interaction.

2 Elements of Scene System

There is a step-by-step process in the cognition of the scene system. At first, we equated the scene system with the objective environment in which people act, and focused more on how to create the objective environment, focused less on the people act in the environment, on the real time interaction between people and the environment. But in modern China, technology is constantly evolving, social forms are constantly being deconstructed and reconstructed, and more importantly, both the objective environment and human behavior can be recorded in real time. The scene system, which is composed of objective environment and emotional psychology and scene

G. Di Bucchianico et al. (Eds.): AHFE 2020, AISC 1202, pp. 529–536, 2020.
https://doi.org/10.1007/978-3-030-51194-4_69

interaction, becomes a dynamical system that can be fully perceived every minute. The scenario design around "scene" is a response to this rapid change. Design is no longer just an explicit visual tool. "Using design and making design a core capability of an organization and key part in value innovation require a changing role of design". As a systematic methodology, scenario design is expanding and reshaping as the concept of scene evolves [1].

2.1 Objective Environment

The objective environment refers to the natural environment, space, objects, people (groups), social culture and so on that constitutes the occurrence of things. As the name implies, the objective environment exists objectively. It doesn't change because of people's sadness and happiness. There are a few key points to note when understanding an objective environment:

The First Point: The basic influence of natural environment on human beings. We often take the natural environment as the starting point of thinking, and at the same time we ignore the existence of the objective environment. We often talk about the importance of the natural environment and put it aside when we design. For example, when we are working on the design, we don't think about the change of seasons, and it's hard to feel the change of day and night. In fact, the light/temperature that changes with the seasons/days defines our design in a more fundamental way. The emphasis on the natural environment is not an idea, but an ability to fully understand the natural environment, and to build a bridge from the natural environment to the scheme design.

Point 2: The limitations of objects on human behavior, in our design process, our interpretation is more about the object's care for our emotions and support for our actions, but in fact, the objects support our actions, while they strictly limit our actions. It constantly reduces the possibilities, which is very important, for example, you're not welcome to walk barefoot on the concrete floor, and the sofa enclosure in the meeting area suggests that people who are not involved should not be approached.

Point 3: Social Culture is an objective existence, as objective as the natural environment, sometimes it's a rigid existence, its existence is no less hard than the rock. There are a variety of laws, institutions, and customs that constantly regulate people's behavior. In China, for example, it is self-evident that leaders/ elders/ teachers take the lead, there are clear physical boundaries between men and women. In the scenario design, if there is no understanding of these social culture and be able to walk with ease between, is bound to hit a brick wall.

2.2 Emotional Psychology

Emotional psychology refers to people in an objective environment to receive stimulation and the emotional response is formed by sensory perception and perceptual judgment. There are two processes involved, one is to feel the external things and the present state of one's body through one's sense, in which case we often have a subconscious stress response. The second is to make perceptual judgments about the information we receive, and it doesn't just reflect the signals we receive. By contrast with our own memories, information is integrated into the existing knowledge framework, and formed judgments. What's more, our brains have the ability to process information automatically without us noticing it, which greatly improves our productivity and quality of life. An open door, for example, appears as a parallelogram in our retina, but we perceive it to be a rectangle with four right angles. This information is either directly or transmuted into our bodies, and we form our own emotional response to it. We are constantly constructing our own emotional psychology, which constitutes our rich and colorful social life.

2.3 Scene Interaction

Scene interaction refers to the dynamic change process of environment and emotion with the development of behavior. This can be expressed by an example, the couple broke up, saw the sofa that was bought together and it remind them the emotional injury, so the girl throws out the sofa and feel better. The emotion changes with the situation, the situation changes with the behavior, the behavior changes with the emotion, which constitutes a complete interactive process. The view of process thinks behavior as a long-term and dynamic course of action rather than a fixed state of action.

3 The Characteristics of the Scene System

3.1 Relevance

A situation is essentially an expression of a relationship. Objective environment, emotional psychology and scene interaction are interrelated with each other. From another point of view, we can further divide the objective environment and emotional psychology into three categories: people, objects and fields. The relation among people, things and fields constitutes the scene system. The situation here transcends the realm of physical concepts. It becomes the field to analyze the psychological scene system of the characters' the social relationship. For example, when we eat at KFC, we find that there are two dining scenes at KFC, one is made up of a table and several chairs, chairs are around the table and are very close to each other. It's for friends and family to sit down and eat together, such a dining scene represents a close personal relationship. The other is made up of a long table by the window and several individual high chairs. There is a certain distance between the high chairs, which is suitable for eating alone. Such a scene system represents a different social relationship. Social relations include family, economic, political and cultural relations. The family scene system is related to the family relationship; the bank scene system is related to the economic relationship. Whether from the internal relevance or external relevance, only when the relevance exists, space will become the scene system. About social relations, metaphors and/or analogies encourage consideration of how to apply existing ideas to new situations [2].

3.2 Field

In the scene system, there are all kinds of visible objects, and also a large number of invisible objective exists. With the rapid development of science and technology, microwave signals, Wi-Fi signals, Bluetooth signals, short-and long-wave signals, and so on, are all around us. These electromagnetic waves are transmitting all kinds of information all the time, though we can't see or touch them, their absence will seriously affect our lives. Good Wi-Fi is now almost a requirement for both space and interior design. In addition to electromagnetic waves, there are sound waves, airflow, temperature, light, and so on, and these dynamic factors are also very important to the scene system. For example, when we design a grand theater, under this scene system, we have to consider the effect of sound waves, how to let the audience sitting in all directions to maximize the enjoyment of sound, the designers need to consider the transmission characteristics of sound waves. These invisible objective existence, visible things and the social relations based on history form the field characteristics of the scene system. This requires us not only to see the scene system with our eyes, but also to think about the scene system with our brains. The scene system is not presented at a certain moment, it "not only discovering but constructing the reality of a design situation" [3].

4 The Evolving Characteristics of the Scene System's Element

The situation system, which is composed of objective environment, emotional psychology and scene interaction, is driven by both self-change and other-change. With the occurrence of self-change and other-change, the scene system is constantly changing, the objective environment, emotional psychology and scene interaction form the new objective environment, new emotional psychology and new scene interaction (Fig. 1).

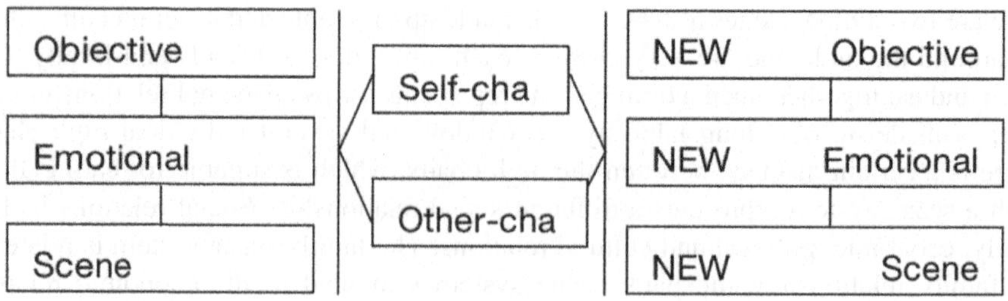

Fig. 1. The evolution of the scene system

4.1 Self-change

When we talk about the objective environment, we tend to put our eyes on the 3D space and consider it, ignoring the influence of "time" on the objective environment. Einstein said in theory of relativity: "Time, space and matter cannot be separated. No matter how which dimension space, it cannot be separated from time. Without time there is no space." The objective environment is diachronic, because of the existence of time, the scene system is changing all the time, and the old scene system is changing into the new one. Many things we thought would last forever are finally lost in the dust of history.

4.2 Other-change

Other-change refers to changes that have taken place under the influence of external factors, which we are most focused on and good at. Responding to external changes is the basic ability to survive as a living organism. In the process of other-change, what is more troublesome is the choice of whether to follow the old rules, by following the old rules, we can greatly reduce costs/increase efficiency, but it's good for only slow-moving societies. In a rapidly changing society, there will be repeated setbacks. For example, before, the company's average survival time is 5 years, we can spend half a year to decorate the office space well. Today, many companies survive less than a year, making it impossible to design office space the way it used to be. In addition, highly emotional human beings are fickle, this makes emotions often one of the triggers for situation change.

5 The Relationship Between Scene System's Elements

As a dynamic system, there must be various relationships among the elements, how to analyze these relations and carry on the design based on them is very important to the scene system design. scene system design classifies the relationship among scene system's elements into three states of neutrality, reciprocity and conflict, and the three states can be changed under certain conditions, they are not invariable. Under the influence of these dynamic factors, the dynamic characteristics of the scene system are constructed, this requires us to look at the scene system in a dynamic way. The design scene system is not a display of a certain time; the design scene system is one kind of embodiment of process.

5.1 Neutrality

Neutrality originally meant being impartial and not inclined to favor one side or the other. Here, its meaning can be extended to describe the weak correlation between two elements, each maintaining its original state. There is almost no interference with each other. This kind of relation widely exists in the objective environment, if we look closely at our home, we see this relationship between a bowl and a sofa, a wardrobe and a kitchen knife, although they all belong to the same objective environment – our home

– they are neutral and weakly correlated with each other. When we design, do not try to carry on the unified planning layout to the objects that show neutral relations, the result of doing so is to make things worse.

5.2 Reciprocity

Reciprocity refers to a mutually reinforcing relationship. When we design, we should pay particular attention to the design of the two elements that exist in this relationship, when we design for one side in a reciprocal state, the other side also benefits. We can save resources, improve efficiency, and we can play the design of the maximum value, form a virtuous circle. The famous Japanese architect Tadao Andou is very good at grasping the influence of light on architecture. In fact, there is such a reciprocal relationship between light and shade and architecture, good architects design buildings according to light, shadow, light and shade, and the top design masters, they design the light and shadow through the architectural. Tadao Andou's masterpiece, the Church of light, solved the lighting problem by dividing the outside light into cross shaped walls to form a cross of light inside the church, it is also a good foil to the holy, clear, pure and shocking atmosphere of the church. It is the mutual achievement of architecture and light that gives us a taste of the master's brilliance.

5.3 Conflict

Conflict refers to a relationship of mutual hostility, for example, air switches and household appliances cannot work simultaneously, when the air switch begins to work, all the household appliances stop working. In addition, in the bedroom, the husband and wife have different light demands at night, one wants to read a book, the other wants to sleep, this is the obvious conflict. In fact, clear positive and negative relationships are everywhere in scene system design, we can only choose the best among the worst. In the practice of design, we can hone our feeling ability of scene system and learning ability.

6 The Interactive Process of the Scene System

The interactive process between the scene system's elements, divided by whether or not they attract significant attention, it can be divided into explicit interaction and implicit interaction.

6.1 Explicit Interaction

Explicit interaction is an interaction in which the user needs to be distracted. With the development of Mobile Internet and multimedia technology, people's time becomes more and more fragmented. Whether you're working or studying, you can always be distracted by a message on your phone, this in turn reflects the high speed/efficiency with which information is being transmitted. In this environment, architectural space and interior design began to merge with multimedia technology and the Internet, we can see that today's large shopping malls are full of large and small LED screen, the store owners urgently need consumers to be able to receive the information in the first time and turn it into purchase behavior. In such a business situation, users always bear the risk of being distracted, the essence of multimedia lies in its dynamic visual, such interaction is the dominant interaction of the situation, focus ads take full advantage of this point, successfully attracted the attention of office buildings and other elevator crowds. Scene systems, especially in a business scene system, merchants are full aware of human nature on the basis. They are competing fiercely for attention.

6.2 Implicit Interaction

Implicit interaction is to complete the interaction when the user is not aware of the situation. Naoto Fukasawa advocates "unconscious design," also known as "intuitive design". He believed that unconscious design could remove clutter and bring things back to their roots. It is a reflection and extension of the primitive relationship between human and objects. It goes by instinct, not by middle process. "Intuitive people tend to understand the world in terms of the big picture and the patterns, their learning and problem solving process is automatic, not intentional. They synthesize the data and identify connections to provide "non-conscious understanding of the rules and principles that governs a situation" [4]. Implicit interaction is the fusion of scene system and behavior, and it doesn't need people to think too much. It is able to see what you need from the beginning and use the most reasonable design language to reach the value that makes life flow like water, without any trace.

In the scene system of smart homes, when you wake up in the morning, the curtains open for you, the bread is baked in the toaster, the shirts are ironed for you, such an interaction does not need to interrupt the user's attention. It can predict user's need, and provide the most suitable solution for the user silently, it makes life easier, the problem-solving process occurs naturally and efficiently, this is the value of implicit interaction. Almost by definition, exploratory research concerns situations where you do not yet know what patterns of relationships to expect and what the time frame might be [5].

7 Conclusion

In these rapidly evolving times, the trend of scene system design is becoming more and more obvious, new technologies make it possible for everything to be connected, the frequency, depth, and results of these connections and understandings have been greatly enhanced. Scene system design is no longer a transformation of single physical space. We need to be more systematic in defining scene system design and use it to guide our design, adhere to the design concept of people-oriented, let people, things and fields be unified in sensibility and logic, release the value of scene system design.

References

1. Koomans, M., Hilders, C.: Design-driven leadership for value innovation in healthcare. J. Des. Manage. J. **11**, 43–57 (2016)
2. Read, G.J.M., Salmon, P.M., Goode, N., Lenné, M.G.: A sociotechnical design toolkit for bridging the gap between systems-based analyses and system design. J. Hum. Factors Ergon. Manuf. Serv. Ind. **28**, 327–341 (2018)
3. Schön, D.A.: Designing as reflective conversation with the materials of a design situation. J. Knowl. Based Syst. **5**, 3–14 (1992)
4. Suh, J., Cho, J.Y.: Analyzing individual differences in creative performance: a case study on the combinational ideation method in the interior design process. J. Inter. Des. **43**, 9–23 (2018)
5. Spector, P.E.: Do not cross me: optimizing the use of cross-sectional designs. J. Bus. Psychol. **34**, 125–137 (2019)

Sound Design for Electric Vehicles: Lessons and Requirements Learned from a Chinese Automobile Manufacturer

Yang Zhao[1], Hua Dong[2], and Long Liu[1(✉)]

[1] College of Design and Innovation, Tongji University, Shanghai 200092, China
zhaoyangisvip@163.com, liulong@tongji.edu.cn
[2] School of Design and Creative Arts, Loughborough University, Loughborough LE113TU, UK
h.dong@lboro.ac.uk

Abstract. This paper studied the sound design for Electric Vehicles (EVs). As an emerging product with great potential of application, EVs alongside their sound issues have attracted attention from different stakeholders. In this paper, the lessons and requirements of EVs' sound-related practices from a Chinese automobile company were identified by using a What-How-Why lens as an interview framework to collect descriptive data from ten EVs' sound design practitioners. The purpose of the research is to highlight the EVs' sound design practitioners' insights and to extend these to address EVs' sound-related issues in design. The findings are based on a qualitative content analysis which could help EVs' sound design practice.

Keywords: Sound design · Electric vehicles · Experiences · Requirements

1 Introduction

In recent years, Electric Vehicles (EVs) are increasingly appearing on the road due to its environment-friendly nature. Global plug-in vehicle deliveries reached 21 million units for 2018, 64% higher than for 2017 [1]. As China has identified EVs as one of the country's seven strategic emerging industries and has implemented a series of measures to promote their development [2], the growth of EVs sales in China has become the highest in the world, with sales reaching 12 million in 2018, which accounts for 56% of all plug-in sales worldwide [1]. However, as an emerging product, EVs have many new characteristics in comparison with traditional combustion engine vehicles [3]. Changes in sound are one of the new characteristics, which make EVs significantly silent. This has both positive and negative impact, because 'quiet is good; silence may not be' [4].

In fact, the changes of EVs have not only influenced consumers' safety, sense of security as well as the product experience [5], but also aroused governments and automobile companies' concerns and actions to tackle the problems emerging with the EVs' sound. Experiments conducted by Refaat Hanna [6] and Nicolas Misdariis et al. [7] show that the noises of wind and tire/road friction from EVs are not loud enough to signify the presence of an EV to other road users. At low speed, in particular, the

G. Di Bucchianico et al. (Eds.): AHFE 2020, AISC 1202, pp. 537–543, 2020.
https://doi.org/10.1007/978-3-030-51194-4_70

incident rate is twice as that of combustion engine cars [8]. Sound issues also affect people's perception, emotion, purchasing decision, response, and experience of privacy [9–12], in that people have different experiences due to the differences in pitch and tone of surrounding sounds [13].

Some actions have been taken by governments and automobile companies [14] to confront the EV's sound-related issues. The European Parliament required that all new silent EVs must be equipped with acoustic alerting systems by July 2019 [15]. The Chinese government also published a Technical Requirements for slow driving EVs [16], in order to mitigate risks posed by EVs. Other actions for better vehicles' quietude and acoustic interiors have been applied by automobile manufacturers like Toyota, Mercedes-Benz, Volkswagen, BMW, etc. for advertising their cars' different design [12]. Besides the silence of vehicles, the intelligentised Human-Vehicle-Voice-Interaction system is also being promoted extensively [17].

In the stakeholders' line of actions, the EVs' sound-related safety issues and marketing tactics are dominated. What is missing, however, is the hidden experience that automobile industry should confront with the EVs' sound issues. Therefore, this paper will contribute to a special perspective to EVs' sound and research: (1) To apply the Golden Circle model to develop a WHW lens to conduct a set of semi-structured interviews with ten EVs' sound design practitioners from a special Chinese automobile company. (2) To analyse the interview data, and (3) to propose a descriptive picture of the actions and experiences for EVs' sound design.

The objectives of this study lie in: (1) To identify the advantages and disadvantages related to EVs' sound from the perspective of automobile sound design practitioners, with the consideration of their current design practice. (2) To identify how to enhance the auditory experience through sound design as well as how to evaluate it, and (3) to clarify why good sound design is needed for EVs. The results of this study could provide reference to the automobile sound researchers and sound designers.

2 Method

Semi-structured interviews were undertaken (45-min duration each) with ten EVs' sound design practitioners from a large automobile company in China. The company is in the top 50 tier position in the 2019 Fortune Global world's 500 largest companies [18] and is the top 10 among the global automobile companies. In particular, the company is one of the leading electric vehicle manufacturers in China, which was ranked among the top three companies in terms of EVs sales across the country in 2018 [19]. In addition to this, the company is setting the sound of EVs as an important tactic to improve the overall product experience.

Interviewees were filtered by 3 criteria: (1) Representing a wide spectrum of expertise from different departments relating to sound/noise and user experience. (2) Knowing details of sound design. (3) Having at least three years' working experience in the company. Interviews were taken place by adopting a Why-How-What (WHW) structure, which acted as a lens to explore the innovation situations of EVs' sound. The WHW lens was developed based on the Golden Circle model [20] in businesses to reflect the process of innovation, which starts with Why (i.e. why you do what you do?), followed by How (i.e. how you do it?) and finishing with What (i.e. what are your products/services?) [21]. The interview protocol comprised of three themes: (1) Why does the automobile company pay attention to EVs' sound? (2) How does the automobile company design and evaluate the sound of EVs? (3) What EVs' sound do users like in the eyes of automobile sound design practitioners?

2.1 Data Analysis

The interviews were transcribed in full and an inductive qualitative content analysis was undertaken with the help of the NVivo 12 software (QSR International, Pty Ltd.). Content analysis aims to categorize data (e.g. words, phrases, sentences, or paragraphs) with identical meanings as aggregated attribute, consequence and value, and entails a reading of a body of texts, images, and symbolic matter derived from the interviewees' opinions [22]. An iterative and systematic coding approach was adopted, combined and (re)-grouped into meaningful categories (See the coding example in Fig. 1). Figure 2 reflects the mapping relationships of data collection, data analysis and the results.

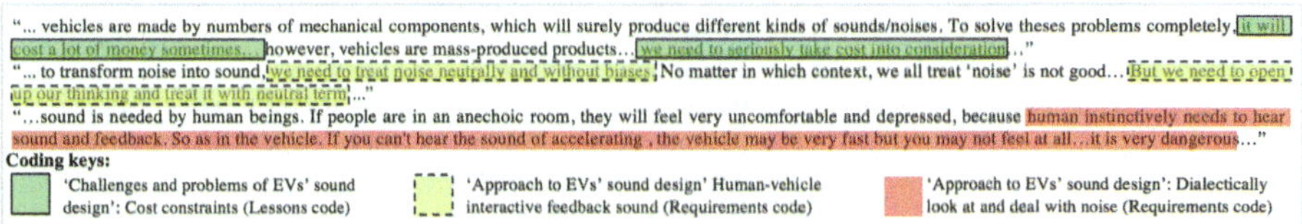

Fig. 1. An example of how the content was coded

Results of data analysis		Strategy of content data analysis		Strategy of data collection	Example
Lessons(L) and Requirements(R)		Attribute(A)-Consequence(C)-Value(V)		WHW lens: Why – How - What	
L1: Sound as an EVs' saling point to promotion	→	V	→	Why	EVs' sound pressure level, sound quality.
L2: Pros and cons of EVs in sound	→	A	→	Why	The cabin is suitable for the tuning of sound.
L3: Current works related to EVs' sound design	→	C	→	How	Improving/optimizing the quality sound.
L4: Departments involved in EVs' sound design	→	C	→	How	Electronic, Human Machine Interaction.
L5: Current EVs' sound design planning	→	V	→	What	Noise control, sound immersion.
L6: Challenges in EVs' sound design	→	A	→	What	Costs, empirical based sound design.
L7: EVs' sound evaluation method	→	C	→	How	Subjective scoring, leadership oriented.
R1: Design requirements for EVs' sound	→	C	→	How	Design philosophy, follow human nature.
R2: Evaluation requirements of EVs' sound	→	C	→	How	Pleasantness, users autonomy.
R3: New dimensions of EVs' sound evaluation	→	V	→	Why	User based and Quantity based evaluation.

Fig. 2. Mapping relationships of data collection, data analysis and the results

3 Results

The primary objective of this study was to empirically explore the experiences of EVs' sound design in a particular Chinese automobile company. Overall, two key findings, 'lessons' and 'requirements', were identified together with several sub-factors. Within the 'lessons' finding, 7 sub-factors (L1-L7, as shown in Table 1) were identified (with a total of 24 insights). For the 'requirements', 3 sub-factors (R1-R3), together with 20 insights, were identified, shown as descriptions of examples, comments and interpretations.

Table 1. Lessons and requirements emerged from the data

Findings	Sub-factors	Examples, comments, and interpretations
Lessons	**L1:** Sound as EVs' selling point	**L1-1:** Advertising EVs' sound pressure level
		L1-2: Advertising EVs' sound quality, e.g. "the sound of closing doors"
		L1-3: Advertising EVs' audio system, e.g. "equipped with Bose speakers"
	L2: Pros and cons of EVs in sound	**L2-1:** The cabin of the vehicle is very suitable for the tuning of sound
		L2-2: More opportunities for sound design brought in though the intelligentisation of automobile
		L2-3: Too much of tranquillity
		L2-4: The high-frequency band noise of the EVs' engine was intensified, which cause discomfort to consumers
		L2-5: The wind noise and tyre/road noise of EVs tend to more prominent for the absence of masking effects of the combustion engine
		L2-6: Low voice recognition rate, inaccurate semantic understanding and limited feedback capability of human-vehicle auditory interactive system negatively impacts user's experience
	L3: Current works related to EVs' sound design	**L3-1:** Developing and optimising vehicles' noise performance
		L3-2: Improving and optimizing the quality of some individual sounds, e.g. horn sound, the sound of opening and closing doors, lifting sound of windows
		L3-3: Eliminating abnormal noise
	L4: Departments involved	**L4-1:** Noise, Vibration and Harshness (NVH); Electronic; Human-machine interaction; Market research and analysis; Engine
	L5: Planning of EVs' sound design	**L5-1:** Noise control: "continue to reduce the noise intensity in the cabin"
		L5-2: Further improvement of EVs' sound quality
		L5-3: Creating a sense of sound immersion for users
	L6: Challenges in EVs' sound design	**L6-1:** The restrictions of sources on designing and delivering EVs' sound
		L6-2: The extensive homogenization of vehicles' sound due to the suppliers' convergence of automobile components
		L6-3: The automobile companies' rights to design the sound of EVs are relatively limited due to the greater decision-making rights the suppliers hold
		L6-4: Empirically-based sound design rarely focusing on customers' needs
		L6-5: Doctrinal thinking in dealing with sound-related practical issues lacks flexibility, e.g. "the tactic to noise is to eliminate rather than optimize it into user wanted sound"
	L7: EVs' sound evaluation methods	**L7-1:** Subjective scoring
		L7-2: Comparing with another brand vehicle
		L7-3: Decision-making process dominated by leaders' opinions
Requirements	**R1:** Design requirements for EVs' sound	**R1-1:** Philosophy based design, e.g. "Confucianism"
		R1-2: Following the evolutionary law of human hearing
		R1-3: Application of bionics in EVs' sound design, e.g. imitating the horse sounds
		R1-4: A systemic design framework for EVs' sound design is needed
		R1-5: Human-vehicle interaction-based sound design
		R1-6: Relevance of sound feedback semantics and the user's queries
		R1-7: Delivery priority of sound in a certain situation
		R1-8: Intelligently customise sound content
		R1-9: Acoustic principles-based sound design
		R1-10: Multi-sensory based sound design
		R1-11: Defining automobile brand identity by sound
		R1-12: Contextualised design of sound, i.e. "EVs' sound cannot exist alone, they must exist in some specific contexts"
	R2: Evaluation requirements of EVs' sound	**R2-1:** Consistency of information conveyed by sound and users' perception, experience
		R2-2: Pleasantness
		R2-3: Users' autonomy
		R2-4: Robustness. "whether there is a remedial action when an accident occurs"
	R3: New directions of evaluation	**R3-1:** Quantity-based evaluation. "…should be evaluated with a combination of subjective and objective methods"

4 Discussion

Some refreshed insights could be formulated based on the bottom-up empirical investigation, some of which are highly relevant to previous research findings. Some insights are newly emerging with contribution to the existing EVs' sound design and research.

4.1 Lessons and Implications

Some sub-factors identified were similar to those advocated by previous research (L1-L3, L6, L7, R2, R3). Concerning to the findings of sound as a selling point, Karin Bijsterveld et al. [12] claim that automobile manufacturers have been using sound as a promotional highlight since the 1920s. Nevertheless, these insights: advertising EVs' sound pressure level and audio system, etc. provide specific practical operation methods in the 21st-century context. Although the pros and cons of EVs in sound found in this study have been mentioned extensively in previous academic studies [8, 12, 23], it is still valuable for it was certified by the front-line practitioners who engaged in EVs' sound for many years. Concerning the tranquility nature of EVs, it was recognized as both pros and cons for the reason that previous researches claimed tranquility poses potential safety hazards to other road users [2, 24]. In light of this, this study classified it as a disadvantage feature of EVs' sound.

In terms of the excellent acoustic tuning feature of cabin ambient, it was a piece of practical based experience and was seldom appeared previously. Furthermore, previous researches generally regard brand, customer and legal requirements as the first factors to be considered in the delivery of right sounds for vehicles. These were also highly stressed by the design practitioners involved in this study, e.g. the automobile company have a number of requirements and interests in using sound to define brand image and identity. Moreover, the evaluation methods of EVs' sound were mentioned both in previous researches and this study. Previous researches evaluated EVs' sound both in qualitative and quantitative methods, e.g. by using subjective Jury tests [25] and sharpness vs. time curve analyzing method [26] to evaluate the sound. However, some specific commonly used practical evaluation methods were revealed, e.g. the method of comparison to other brand vehicles, leaders' opinions tended to dominate the decision-making process.

Some insights have not been retrieved and seen in existing research publications (L4-L6). In the automobile company, the EVs' sound design involved extensive departments, as many as five. This kind of cross-department collaboration is a double-edged sword, which on one hand enriches the refreshed ideas of sound design. On the other hand, challenges the unification and standardization of sound design. In addition to noise controlling and sound quality improvement, the automobile company has intended to create users' sound immersion, nevertheless, the process is still an ongoing plan and has not been implemented. In terms of EVs' sound design challenges, the automobile company has limited sources to support professionals' dedication (e.g. to investigate what kind of sounds users prefer), on the contrary, too much attention is being paid to solve existing noise problems. Besides, the delivery of sound for a better ambient acoustic quality requires higher cost (e.g. more expensive acoustic package materials). Similarly, the converging suppliers of automobile components lead to less sound design schemas and challenge the differentiated sound design possibilities. On top of that, the doctrinal and inherent thinking patterns limited the emergence of fresh ideas to EVs' sound design (L6).

4.2 Requirements and Implications

This study categorized the 'requirements' into three factors (R1-R3), some of which were mentioned previously, whilst some are newly uncovered.

The 12 design requirements for EVs' sound, four (R1-4, R1-5, R1-9, R1-11) were previously described by Mark et al., Nees et al., and Watson et al. [25, 27, 28]. In contrast, nine are newly uncovered, which are novelty and exclusive. These insights can be generally divided into two layers: the philosophy layer and the practicality layer. For the design philosophy layer, it is concluded that EVs' sound design ought to follow the insights of (1) the concepts of 'Confucianism' (R1-1) (2) the evolutionary law of human hearing (R1-2), and (3) a dialectical perspective to the disposal of noise. Practically, for an EVs' sound design project, six new requirements are identified: R1-3, R1-6, R1-7, R1-8, R1-10, R1-12. As to the EVs' sound evaluation requirements, this study demonstrates some specific and practical requirements compared with previous research [28]: R2-1, R2-3, R2-4, e.g. whether all of the nature of alarm (duration, intensity) based sounds are in the same dimension in the process of evaluation.

5 Conclusions

Some of the 44 emerged insights are newly uncovered in this empirical industrial experiences-based study, which will bring in more nuanced research and design themes and directions for EVs' sound researchers and practitioners. In order to facilitate the success of EVs to meet the goals for environmental protection, to solve the safety issues, and to enhance EVs' user experience, academic researches should consider their research topics alongside the industrial lessons and requirements for effective EVs promotion. Furthermore, based on the fertile research outcomes, the automobile industries are encouraged to put forward more practice-based insights for academia in order that both sides could maintain a mutually enforceable relationship for better EVs' application.

References

1. Global EV Sales for the 1st Half of 2019. http://www.ev-volumes.com/country/total-world-plug-in-vehicle-volumes
2. Liu, Z., et al.: Domestic and foreign present situation and the tendency of electric vehicles. Electr. Power Constr. **36**(7), 25–32 (2015)
3. Nilsson, M.: Electric Vehicles: An Interview Study Investigating the Phenomenon of Range Anxiety. http://www.elvire.eu
4. Norman, D.A.: The Design of Future Things. Basic books, New York (2007)

5. Van der Auweraer, H., et al.: Electric vehicle exterior sound and sound source design for increased safety. In: Inter-Noise and Noise-Control Congress and Conference Proceedings, Institute of Noise Control Engineering (2011)
6. Hanna, R.: Incidence of pedestrian and bicyclist crashes by hybrid electric passenger vehicles. In: Electric Vehicles (2009)
7. Misdariis, N., et al.: Do electric cars have to make noise? An emblematic opportunity for designing sounds and soundscapes. In: Acoustics (2012)
8. Fagerlönn, J., et al.: Acoustic vehicle alerting systems: will they affect the acceptance of electric vehicles? In: Proceedings of the Audio Mostly 2018 on Sound in Immersion and Emotion (2018)
9. Västfjäll, D., Kleiner, M.: Emotion in product sound design. In: Proceedings of Journées De sign Sonore (2002)
10. Lageat, T., Czellar, S., Laurent, G.: Engineering hedonic attributes to generate perceptions of luxury: consumer perception of an everyday sound. Mark. Lett. **14**(2), 97–109 (2003)
11. Özcan, E., van Egmond, R.: Product Sound Design: An Inter-Disciplinary Approach (2009)
12. Bijsterveld, K., et al.: Sound and Safe: A History of Listening Behind the Wheel. Oxford University Press, Oxford (2014)
13. Everest, F.A., Pohlmann, K.C.: Master Handbook of Acoustics. Wiley, Hoboken (2015)
14. Genuit, K., Fiebig, A.: Sound design of electric vehicles: challenges and risks. In: Inter-Noise and Noise-Control Congress and Conference Proceedings, Institute of Noise Control Engineering (2014)
15. Regulation (EU) No 540/2014: The European Parliament and of The Council on the sound. Level of motor vehicles and of replacement silencing systems, and amending Directive 2007/46/EC and repealing Directive 70/157/EEC Text with EEA relevance (2014)
16. Standardization: Technical Requirements for Acoustic Vehicle Alerting System of Electric Vehicles Running at Low Speed (2016)
17. Luxury experience. https://rw.roewe.com.cn/marvel/innerhtml/experience
18. Fortune Global 500 (2019). https://fortune.com/global500/2019
19. Statista, Electric vehicle production forecast (2019). https://www.statista.com
20. Sinek, S.: Start With Why: How Great Leaders Inspire Everyone to Take Action. Penguin, London (2009)
21. Sinek, S.: Find Your Why: A Practical Guide for Discovering Purpose for you and Your Team. Gramedia Pustaka Utama, Jakarta (2019)
22. Krippendorff, K.: Content Analysis: An Introduction to its Methodology. Sage publications, California (2018)
23. Bodden, M.: Principles of active sound design for electric vehicles. In: Inter-Noise and Noise-Control Congress and Conference Proceedings, Institute of Noise Control Engineering (2016)
24. Parizet, E., et al: Warning sounds for electric vehicles. In: Fisita World Automotive Congress (2014)
25. Allman Ward, M., et al.: Designing and delivering the right sound for quiet vehicles. In: Inter-Noise and Noise-Control Congress and Conference Proceedings, Institute of Noise Control Engineering (2014)
26. Lei, L., et al.: Study on the sound quality evaluation of the automobile door closure. Chin. J. Autom. Eng. **2**(02), 129–133 (2012)
27. Nees, M.A., Walker, B.N.: Auditory displays for in-vehicle technologies. Rev. Hum. Factors Ergon. **7**(1), 58–99 (2011)
28. Watson, M.O., Sanderson, P.M.: Designing for attention with sound: challenges and extensions to ecological interface design. Hum. Factors **49**(2), 331–346 (2007)

Design of Doll's Arm Based on Shape Memory Alloy Actuator

Danying Liu[1], Xifan Ding[1](✉), Mingjiang Jin[2], and Shanang He[1]

[1] School of Design, Shanghai Jiao Tong University, Shanghai 200240, China
ccding@sjtu.edu.cn
[2] School of Materials Science and Engineering, Shanghai Jiao Tong University, Shanghai 200240, China

Abstract. In this paper, the application background and current status of shape memory alloy were briefly reviewed. And a 0.08 mm shape memory alloy (SMA) microwires were applied in driving device of a new doll arm design. By using the driving function of SMA, the feasibility in the design process of smart dolls was demonstrated and the system was tested to achieve statistics the parameters of electric actuator. Furthermore, taking into account the artistic and technological value of dolls, it is of high value of the movable and controllable doll arm device based on shape memory alloy micro actuator, and this work also provides ideas for the transformation design of traditional dolls to intelligent dolls.

Keywords: Shape memory alloys · Micro actuator · Intelligent · Doll joint

1 Introduction

In the mid-1980s, the concept of smart material was put out to describe the material system which was integrated with perception, drive and information processing [1]. Although these concepts have not been unified in the literature, the research on them has never been interrupted. In the 13th Five-Year Plan Outline of China, "Research and Development and Application of Key New Materials" is listed in the tenth place of "100 Major Projects to Start" and smart materials such as shape memory alloys and self-repairing materials are included.

Although the research on shape memory alloys is developed continuously, only a few patents have been applied commercially due to material limitations and lack of knowledge about its design and related tools. The market entry projects are largely confined to applications such as biomedicine, aerospace, robotics and so on [2]. A few mass-produced shape memory alloy applications in the market include underwear, eye frame, mobile phone antenna, SMA pneumatic valve for car seat and auto-focusing module on smart phone [3].

The rapid development of frontier science and technology is constantly impacting on traditional art design. For designers, the birth of new materials and related technologies can largely promote the emergence of new design schemes. In recent years, there are more and more research data about smart actuator of shape memory alloys. This paper is to verify the feasibility of using this technology to design intelligent dolls.

G. Di Bucchianico et al. (Eds.): AHFE 2020, AISC 1202, pp. 544–550, 2020.
https://doi.org/10.1007/978-3-030-51194-4_71

2 Summary of Shape Memory Alloys

Shape memory alloys (SMA) is a kind of intelligent material which can remember the original shape. When stimulated by specific stimuli such as thermomechanical or magnetic changes, it can "remember" or retain the previous shape, which is called shape memory effect [3]. Shape memory effect is a major characteristic of shape memory alloy, which is originated from thermoelastic martensite transformation. Another important characteristic is pseudoelasticity (also known as superelasticity), which is characterized by large recoverable strain (generally up to 8%), that is, it can apply stress to the material to provide large recoverable deformation at a relatively constant stress level, which is a special hysteretic energy dissipation energy [4, 5]. Because of its excellent deformation ability, shape memory alloy has been widely used in medical, aerospace, modern vehicles, robots and other fields [6–9].

In recent years, with the rapid development of modern science and technology, the performance requirements of shape memory alloy are higher and higher in various fields, and SMA is increasingly used in intelligent micro actuators [10]. The purpose of miniaturization is to develop a actuator with high mechanical properties in a limited space. Using SMA material in the microactuators has the advantages of excellent output power per unit mass, lightweight, silent drive, simple structure, etc. Kheirikhah et al. [11] Proposed that SMA actuator can be used as a substitute for small robots due to its excellent power quality ratio and simple structure. Modabberifar and others [12] introduced a kind of fixture similar to gecko. It uses shape memory alloy wire to drive, which is cheaper, smaller, lighter and simpler than the existing fixture. It can be installed on the robot control arm, can also be used for taking and placing operations or as a manual clamp. Of course, there are also some challenges in current SMA driving elements, such as high energy consumption, limited fatigue life and low response speed [13].

Since the 1980s, SMA has been widely used in the application of human robot. Compared with the industrial manipulator, in the human robot, the requirements for the joint are focused on whether the configuration of the freedom of the robot arm and the mechanical structure are in line with the freedom and joint structure of the human arm. The driving of the joint should meet the requirements of large output power, small size and low power consumption. At the same time, the arm should have high positioning accuracy, high safety and low inertia Advantage. These requirements are just in line with the characteristics of SMA.

3 Design and Evaluation of Doll Arm Based on SMA Smart Actuator

3.1 Design Analysis

In this design, the emotional needs of the players are the focus of the doll design. Firstly, we classify the functions of the doll based on the user's needs, with the help of Maslow's needs theory and Kano model, and draws the conclusion that the movable joints and modeling of the doll are a basic type of needs for consumers, and the intelligent design of the doll is that the designers are achieving the goal of consumers' needs for the doll The potential requirements mined after the movable design belong to the expected requirements. The requirements are converted into functions. According to the functional design structure, material calculation and selection are carried out according to the design requirements. Finally, the production verification and evaluation process is carried out to complete the production of the doll joint.

The realization of human figure interaction and aesthetic function has a positive effect on the establishment of emotional connection between players and human figures. One of the ways to realize human figure interaction is to establish the connection of its internal structure. The traditional ball-joint doll is mainly controlled by two solid high elastic rubber bundles. And two arms share one (Fig. 1). The spherical part and bayonet on the joint enable each joint to be fixed after completing the action. Taking the arm joint as an example, the main part of hand lifting is the spherical joint connected with the lower arm. The main advantage of traditional doll using high elastic rubber band to connect various parts and complete the action is cheap, and using its structure to simply realize the mobility of the doll, but precisely because of the simplicity of rubber band connection, the mobility of the doll can' t be precisely controlled, and the action also has a fixed angle, which is not flexible. When we think about how to make the human figures move, we can easily associate with the robot man. The application of SMA actuator in the robot gives us great inspiration. The original exquisite spherical joint of the doll guarantees its greater mobility, while the hollow structure provides the space for internal driving and placement. Therefore, it is feasible to use the principle of robot actuator for reference to realize the design of the movable doll.

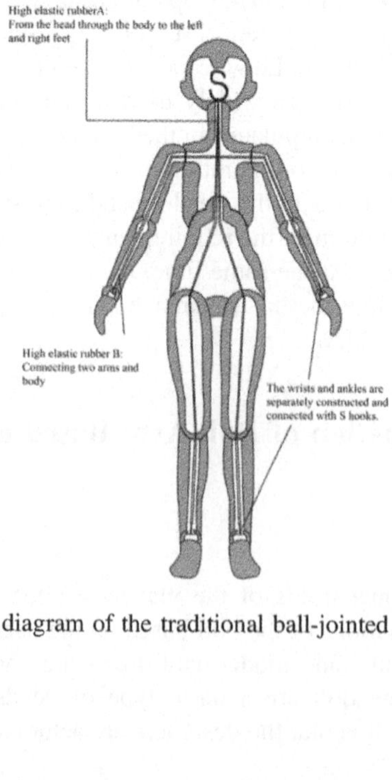

Fig. 1. The schematic diagram of the traditional ball-jointed doll's internal structure

3.2 Action Description and Driving System Design

Based on the arm structure of the traditional doll, taking the lifting-lowering action as an example, two shape memory alloy wires are used to connect the upper and lower arms instead of the high elastic rubber bundles, and the swing arm motion is realized by electric drive. Usually there are two kinds of reset mechanism available, one is spring reset, the other is double wire pull reset. Considering the narrow space requirement inside the arm, this scheme chooses double wire pull as reset mechanism. The design scheme and schematic diagram are shown in Fig. 2. In the twin-memory alloy wire drawing system, SMA1 is superelastic wire, showing superelastic properties, while SMA2 is shape memory alloy, playing a driving role by its shape memory effect. The driving wire is driven by electric heating. When the current is turned on, the driving wire deforms and makes the arm straighten. When the current is turned off, the superelastic wire rebounds and the arm restores to the bending position.

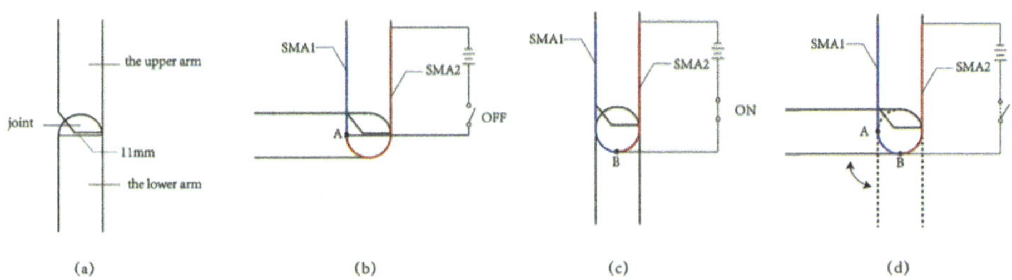

Fig. 2. Schematic diagram of joint action driven by SMA: (a) shows the joint structure, (b) shows that when the current is off, the joint is in a 90° bending state, the driving wire is driven by electric heating, when the current is on, the driving wire deforms, and when the upper arm is fixed, the lower arm takes the elbow center as the origin, completing a 90° rotation action, so as to straighten the joint and reach the state of (c), when the current is off, SMA1 rebounds and the joint returns to the bending position, as shown in (d).

In the process of completing the design drawing, the designer's requirements for materials will be gradually clear, which is also the selection process of material physical properties. According to the wire selection criteria of M. Mertmann and G. Vergani [14], in this design, the arm joint diameter of a doll with a common size of 1/4 (about 45 cm in height) is 11 mm, and the lift angle is 90 degrees. The lifting-lowering action of the arm can be switched by using shape memory alloy wire. The deformation of this action is $\frac{1}{4}$ circumference, so it can be concluded that the strain of the driving element wire should reach $\Delta l = \frac{1}{4}\pi D = \frac{1}{4}\pi \times 11 = 8.635\,\text{mm}$. According to the recoverable deformation of the SMA wire drive element, the length of the wire is: $l = \frac{\pi D}{4} \times \frac{100}{8} = 8.635 \times \frac{100}{8} = 107.9375\,\text{mm}$.

According to the calculation, the force needed to realize the swing arm movement is about F = 150 g = 0.15 kg = 1.47 N. Then the driving force of shape memory alloy wire is estimated at 300 MPa, and the cross-sectional area of wire is selected as $S = \frac{1.47\text{N}}{3 \times 10^8 \text{N/m}^2} = 4.9 \times 10^{-9}\text{m}^2 = 4.9 \times 10^{-3}\text{mm}^2$

The diameter of shape memory alloy wire used in bulk system is $d = 2 \times \sqrt{\frac{S}{\pi}} =$ 0.08 mm and its weight is 3.44 g (Fig. 3).

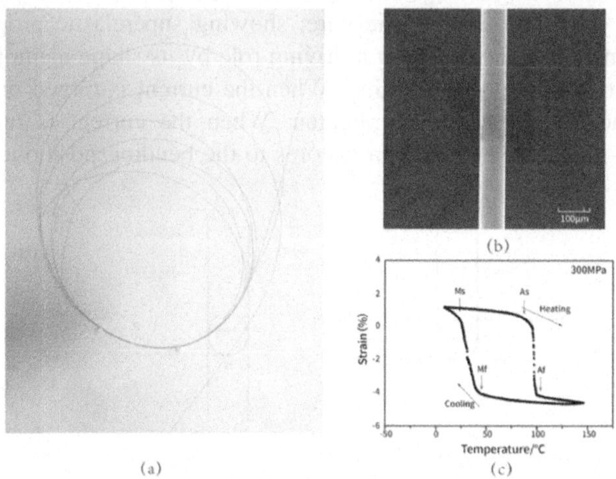

Fig. 3. 80 μm shape memory alloy wire used in this design: (a) is the picture under the macro state of the alloy wire, (b) is the picture under the micro state of the alloy wire, (c) shows the transformation temperature of 80 μm shape memory alloy wire under constant load of 300 mpa.

According to the calculation results, material requirements are put forward and corresponding materials are obtained. This step is the process of selecting, improving or developing materials. In this experiment, the diameter and bending angle of the joint determine the length and diameter of the required alloy wire. The phase transformation temperature and superelasticity of shape memory alloy wire are affected by the diameter, so it is an important step in this design to calculate the required wire diameter.

In order to verify the feasibility of the scheme, a simple joint imitating device is made. Double wires are arranged according to the design drawing. After power on, the device realizes the cyclic action of bending, straightening and bending, as shown in Fig. 4.

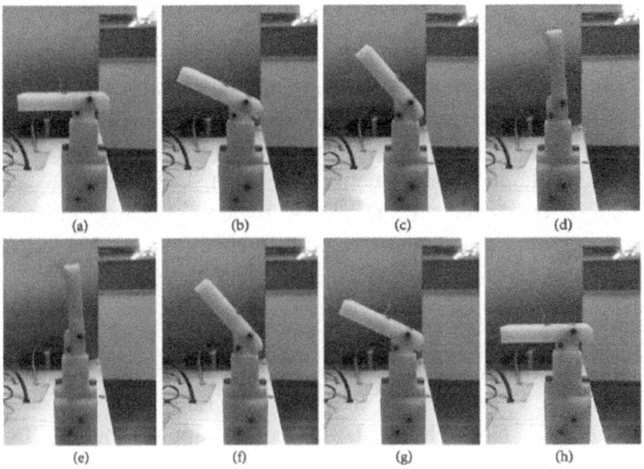

Fig. 4. SMA electric actuator simulates the bending-straightening-bending motion of arm

Shape memory alloy is driven by different heating methods. In this system, electric heating is selected to drive, and the driving current is related to the macro scale and working environment of the wire. In this paper, the current required for driving is tested by gradually loading the current, the loading rate is 10 mA/s, and the evolution of joint shape is observed. The experimental results show that when the load current of the wire rises to 100 mA, as shown in Fig. 4(b), the joint starts to rotate; when the current reaches 150 mA, the joint turns to (c); when the load current reaches 200 mA, the joint completes 90° rotation, as shown in (d). (e)–(h) is the cooling process, and the joint is in (f) state when the power is off for 0.1 s; when the power is off for 0.3 s, the joint returns to its original state when it reaches (g), 0.5 s, as shown in (h). In this experiment, the driving current of 80 μm shape memory alloy wire is 200 mA and the operating power of full load is 0.16 W, which can meet the design requirements.

4 Conclusions

In this paper, the working principle and application status of shape memory alloy are briefly reviewed, especially its development prospect in the field of new intelligent drive. A joint driven by shape memory alloy wire was designed. The structure of the system was constructed by using double wire pull system. The diameter of the wire was 80 μm, the length was 108 mm, and the weight was 3.44 g. The driving current of 80 μm shape memory alloy wire is 200 mA and the operating power of full load is 0.16 W.

As a summary, shape memory alloy micro smart actuator was introduced into the design of the articulation of the doll arm and obtains a series of parameters. The system has the advantages of reliable and simple structure, low energy consumption and intelligent mobility. At the same time, the original artistic sense of the doll is well retained. This design provides a new application direction for SMA and provides an idea for the design of smart doll.

References

1. Shoufeng, Z., Xuejian, Z.: On the development of intellectualization and functionalization of materials in the "Intelligent Age". Northern Arch. **3**(04), 71–73 (2018). (in Chinese)
2. Kauffman, G.B., Mayo, I., et al.: The story of nitinol: the serendipitous discovery of the memory metal and its applications. Chem. Educ. **2**(2), 1–21 (1997)
3. Mohd, J.J., Leary, M., Subic, A., et al.: A review of shape memory alloy research, applications and opportunities. Mater. Des. **1980–2015**(56), 1078–1113 (2014)
4. Sheliang, W., Xiang, Z., Junqiang, Z., et al.: Finite element analysis on the phrase transformation pseudo-elasticity of shape memory alloy. Eng. Mech. **24**(10), 180–185 (2007). (in Chinese)
5. Oliveira, J.P., Miranda, R.M., Braz Fernandes, F.M.: Welding and joining of NiTi shape memory alloys: a review. Prog. Mater Sci. **88**, 412–466 (2017)
6. Auricchio, F., Boatti, E., Conti, M.: SMA biomedical applications. In: Lecce, L., Concilio, A., (eds.) Shape Memory Alloy Engineering. Butterworth-Heinemann, Boston (2015)
7. Singh, K., Chopra, I.: Design of an improved shape memory alloy actuator for rotor blade tracking. Smart Struct. Mater. **4701**, 244–266 (2002)
8. Strittmatter, J., Gümpel, P.: Long-time stability of Ni-Ti-shape memory alloys for automotive safety systems. Mater. Eng. Perform. **20**, 506–510 (2011)
9. Kuribayashi, K.: Millimeter size joint actuator using shape memory alloy. In: IEEE Micro Electro Mechanical Systerms, Proceedings, An Investigation of Micro Structures, Sensors, Actuators, Machines and Robots, pp. 139–144 (1989)
10. Hu, M., Hongliang, Y., Xiaobin, S.: Research progress of nanocrystalline NiTi shape memory alloys. Mater. Rep. **33**(13), 2237–2242 (2019). (in Chinese)
11. Kheirikhah, M.M., Rabiee, S., Edalat, M.E.: A review of shape memory alloy actuators in robotics. In: Robot Soccer World Cup, pp. 206–217. Springer, Heidelberg (2011)
12. Modabberifar, M., Spenko, M.: A shape memory alloy-actuated gecko-inspired robotic gripper. Sens. Actuators, A **276**, 76–82 (2018)
13. Nespoli, A., Besseghini, S., Pittaccio, S., et al.: The high potential of shape memory alloys in developing miniature mechanical devices: a review on shape memory alloy mini-actuators. J. Sens. Actuators, A **158**(1), 149–160 (2010)
14. Mertmann, M., Vergani, G.: Design and application of shape memory actuators. Eur. Phys. J. Special Top. **158**(1), 221–230 (2008)

Integration of Consumer Electronics Products Designed in Mexico

Carlos Garnier-Ortiz(✉), Alberto Rossa-Sierra, and Fabiola Cortes-Chavez

Universidad Panamericana, Facultad de Ingeniería, Álvaro del Portillo 49, 45010 Zapopan, Jalisco, Mexico
{cgarnier, lurosa, fcortes}@up.edu.mx

Abstract. In Mexico, Jalisco has developed along four decades an important infrastructure to initially manufacture electronic components, and over time, being able to transition to a design sector in a joint effort between government, electronics industry and universities. As a result, product design should be included to deliver complete consumer electronics solutions while taking in consideration diverse aspects such as functionality, ease of use, brand positioning, aesthetics and emotional aspects. However, unlike transnational corporations this has not occurred to the electronic sector in Mexico. This paper will intent to explain the missing link between consumer electronics companies that are not beneficiating from the insertion of industrial design to generate complete solutions.

Keywords: Product design · Technology based products · Consumer electronics · Design management

1 Introduction

For decades, Mexico's productive sectors have undergone the transformation of agribusinesses into the industrialized production of goods. Given Mexico strategic geographic position, this makes the country a very competitive destination for producing and exporting goods and services.

It is important to mention the efforts Mexico has made in order to cease being an unattractive country for foreign investment, and to become a strategically differentiated country not based in cost strategies, but on innovation and technology development. In this way, the electronics sector in the state of Jalisco has become an important priority for local and federal government in the country.

The arrival of transnational companies since the early 80s, and the subsequent creation of companies of Mexican origin in the areas of electronics and information technology, gave birth to relevant projects that began with manufacturing processes of electronic components, but which within a few years demanded the experience to create digital applications and the development of intelligent systems in the areas of

G. Di Bucchianico et al. (Eds.): AHFE 2020, AISC 1202, pp. 551–557, 2020.
https://doi.org/10.1007/978-3-030-51194-4_72

software and firmware generating significant exports besides the manufacturing services, highlighting an important differentiating strategy not based in the offering of low costs, but supported by innovation and design as a strategic tool for the economic development of Jalisco.

On the other hand, it is since the 1950s that Mexico has had activities and presence of product design as a professional activity. In the following years, the foundation of design centers and the first industrial design schools leveraged the country's commitment to training professionals in design and innovation. During the 70s to 90s the opening of specialized careers in design areas at universities among the country flourished.

Product design in Jalisco has been rich in exploring and developing in areas such as point of purchase, packaging and particularly in the furniture and fashion sectors (jewelry, footwear and clothing design), generating experts in these areas that have met the needs of national and international markets to make evident the existence of a design sector in the state. However, in terms of complexity, knowledge and specific skills, designing furniture or footwear represents a different degree of complexity and experience than the required to design and develop electronic products in an era in which mass production, complex market strategies and new generations of consumers are demanding products that must be useful, beautiful and with a high emotional sense to generate the right experiences between consumer products and their users.

This particular issue results relevant as according to the Mexican Institute for Competitiveness, it is estimated that in Mexico about 7% of GPD came from the creative industries in the last 10 years and that it has grown by 0.18% in this period. It is expected that if they grow according to their potential, creative industries could have an increase in the coming years between 3 and 4% per year [1].

2 Integrating Product Design to Consumer Electronics Designed in Jalisco

2.1 The Electronic Sector in Jalisco

Talking about the electronic sector involves mentioning the development and manufacture of appliances or components that process some kind of information, and although different regions in Mexico share a vocation for the development of technologically based products, as in north of the country with companies such as Whirlpool, Sony or LG, or in the central region where companies such as GE, Mabe or Applica are located or have been established for mentioning just a few examples, it is undoubtedly in Jalisco where the development of components for the generation of electronic products has its greatest exponential and development at the service of domestic and international companies, while covering any of the six major subsectors of this industry: computing and office, electronic components, communications, audio and video, measuring instruments, control, navigation and electronic medical equipment, and magnetic and optical media [2].

As the years passed, Jalisco's productive capacity in electronics aspects and the talent developed in universities led the local government to name this region as the

Mexican Silicon Valley, alluding to its importance and growth while compared to its counterpart in California in the United Stated of America.

Part of the current challenges for Jalisco´s electronics industry includes further development of the software industry, and assuming that successful clusters such as California's Silicon Valley or that of Finland (where Nokia predominates), are underpinned by a growing domestic market, which demands its products and services [3], including among others, product development via innovation and design services.

2.2 Investment in Jalisco's Electronic Sector

According to data from the National Institute of Statistics and Geography in the state of Jalisco operate 104 manufacturing plants of electronic products, some of them are large transnational companies such as Flextronics, Foxconn, HP, Jabil, Motorola, Continental and Sanmina [4], and if we talk about the profitability of this ecosystem, according to information from the Ministry of Economy, from 2002 to 2012 Jalisco received 2,057 million dollars of foreign investment in that industry, just above states such as Chihuahua with 1,819 million and Baja California with 1,734 million [4], both also with a long vocation in this activity, in which the main difference while compared to Jalisco lies in the experience in the design of multidisciplinary solutions (electronic engineering, mechanical engineering and industrial design) for these markets.

2.3 Relevance of the Participation of the Experts in Innovation and Design in the Electronic Areas

Innovation and design experts are required to have the skills and understanding to know how to execute products appropriate to the technological needs of the moment and to be able to take advantage of the enormous infrastructure developed in the state to position the design and innovation of technology-based products as a key area within Jalisco's creative industries.

However, taking as an example the case of European small and medium sized enterprises, the role of the designer and the management's appreciation of design are not developed in them. In any case, European companies have stopped trying to compete with low prices to focus on the development of high-value products and services using design and design management, but is still difficult to use design and this circumstance is assumed as a lack of design management [5]. While there is a tendency to favor design as a way to promote the innovation expert as an ally for innovation, design management is continually neglected by relegating the design as a mere aesthetical production and enduring the idea, that occasionally collaborating with designers is enough [6], so the following two aspects are clear here:

1. Lack of knowledge of design management concepts related to product design.
2. Designers show difficulty to implement a value model in their daily professional practice in which their professional skills are understood and appreciated.

This means that the skills of the product designer are not recognized and valued within companies even tough design is a strategic tool that companies can use to achieve competitive development, however most of them do not use design as a strategy in their favor [7]. Being in this way, investment in product design is beneficial for the development of companies depending on how and when it is integrated, as involving this profession in its development and research processes has a significant and positive influence in their economical results and the perception of these companies, which can be perceived in the performance shown by their designers based on the following skills according to Gemser: generation of appearance concepts, ease of use, manufacturing feasibility, efficient use of materials and functionality [8].

2.4 Asia as an Example

Countries like Japan were able after the Second World War, to revive their economies stablishing a product development strategy (electronics + industrial design), or as in the cases of China and Korea, both countries with a limited tradition of designing products but with a rich manufacturing heritage as in the particular case of China, but that when compared to Mexico showed a clear delay in terms of industrial design applied to technological developments. But in these examples, China and Korea managed to educate young generations in innovation and design, generating engagement with domestic and foreign companies established in these countries and that while being supported by government incentives and stimuli, got to achieve an economy based in the support and presence of international brands of Chinese and Korean products such as in the cases of Huawei or LG, which not only showed a correct performance and functionality in their products, but also boasted a contemporary design that has been internationally accepted and recognized by their customers.

Small businesses can differentiate their products through design. An innovative design offers a competitive advantage [9], and this is important, because it implies, among other aspects, the need for companies to adjust their expectations and to understand the role that design has in generating innovation of products and services applied to consumer electronics, because it is not only about creating functional and aesthetical gadgets, but solutions that are aligned with the users' needs, technologically feasible, able to be manufactured and that can result in an adequate cost as Asia is proving us that can be done.

3 Method and Materials

3.1 The Experiment

With the intention of understanding and validating the skills of innovation and design experts and their appreciation within research and development teams, where the work of product designers is assumed within the electronics industry in various occasions by engineers, an experiment will be conducted within two types of development teams: one with only engineers and other with designers + engineers (Fig. 1).

The issue of the product development is determined to be an electronic voting machine, in part because it is a product that not everyone has interacted with, but which requires the understanding of users and the aspects of secrecy of voting. According to the Electoral and Citizen Participation Institute of the State of Jalisco, an electronic

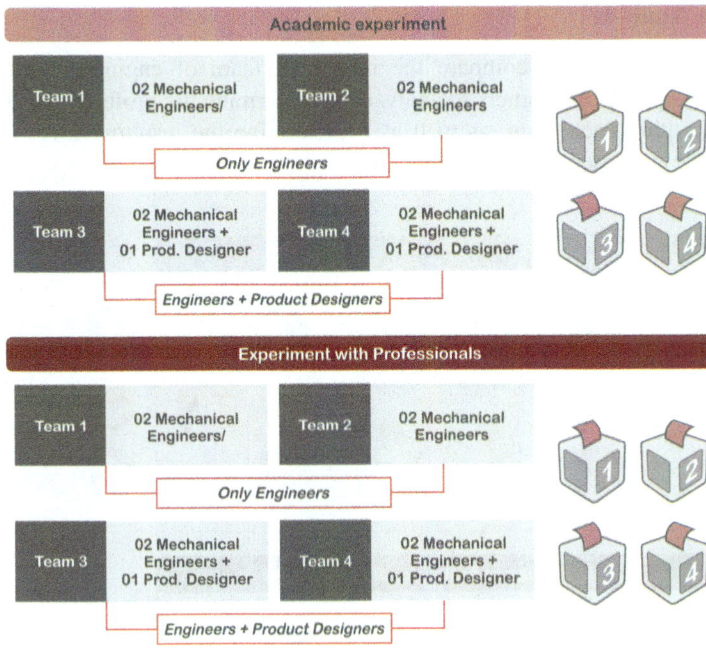

Fig. 1. Conformation of innovation cells for experiment with students and professionals.

voting machine is a device through which votes emitted by citizens on an election day, are received, recorded and counted in an electronic manner [10].

A design brief will be developed and delivered to the development teams specifying the general characteristics of the device. A package of 3D models, which will include battery, PCB, thermal printer and touch screen will be delivered as well.

3.2 Participants and Procedure

Students and professionals will be the subject of the study. In the case of students, the experiment will occur within two universities in the metropolitan area of the city of Guadalajara. Two student teams will be conformed only by engineers (groups 1 and 2), while a second set of groups are intended to be composed by engineers and product designers (groups 3 and 4). The exercise will be guided as a class topic within the students' academic courses.

The project with professionals will occur in similar conditions but inside of product development facilities of locally based companies, in the city of Guadalajara. The procedure will be to invite two innovation cells formed by a team of engineers (mechanical and/or electronic) to conform groups 1 and 2, and a second set of teams including product designers as well as engineers (groups 3 and 4). These can be seen on Fig. 1.

Each team will deliver their conceptualization process and a 1:1 scale mock up made out of carton, foam board or a similar material.

3.3 Data Analysis

The experiment seeks to compare the results of team of engineers with those of engineers and product designers to analyze and determine the professional skills of the developers within each team, as well as to determine the resulting level in terms of quality of the proposed solutions (Fig. 2).

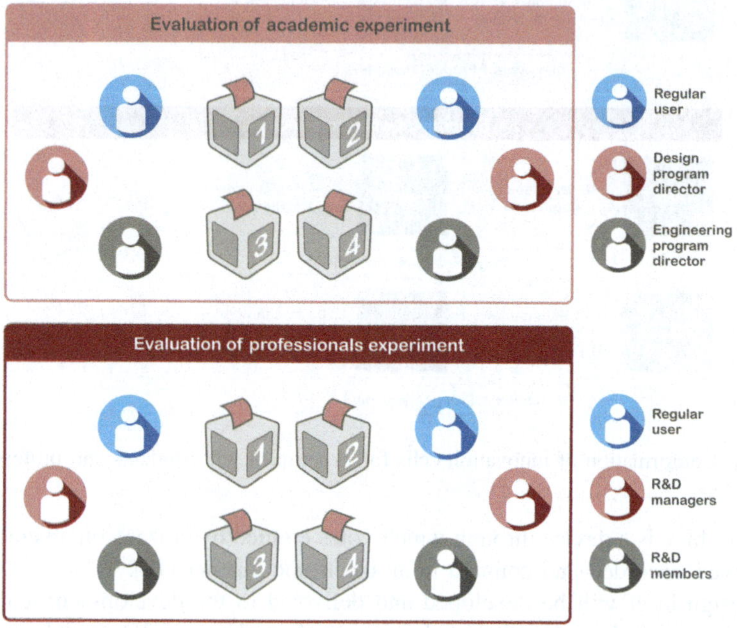

Fig. 2. Evaluating teams for experiment with students and professionals.

Results will be evaluated by individuals who can make use of an electronic voting machine (regular users), academic program directors, research and development managers and R&D team members using the expert appraisal process, by dividing the results into two parts, one for the academic results and a second one for the professionals.

4 Future Work

After the experiment, the analysis of the results will help to develop a tool to document and generate a model of participation for designers to be included in the development of technology-based products in a consistent and systematic way.

References

1. Quesada, J.: México, Meca del diseño industrial. Forbes México (2016). https://www.forbes.com.mx/forbes-life/mexico-meca-del-diseno-mundial/
2. ProMéxico. Diagnóstico sectorial de la industria eletrónica (2014). http://www.promexico.gob.mx/documentos/diagnosticos-sectoriales/electronico.pdf
3. Jaén Jiménez, B.: Los nuevos retos de la industria electronica en Jalisco. Gaceta UDG (2018). http://www.gaceta.udg.mx/Hemeroteca/paginas/314/314-10.pdf
4. Morales, R.: Jalisco es líder en la inversión electronica. El Economista (2012). https://www.eleconomista.com.mx/empresas/Jalisco-es-lider-en-la-inversion-electronica-20140112-0075.html
5. Best, K.: Design Management and Business in Europe: A Closer Look (2016)
6. Borja de Mozota, B.: The Four Powers of Design: A Value Model in Design Management (2016)
7. Kotter, P.: Design: A Powerful but Neglected Strategic Tool (2015)
8. Gemser, G.: How Integrating Industrial Design in the Product Development Process Impacts on Company Performance (2014)
9. Hodges, N.J.: Innovation by Design (2018)
10. IEPC Jalisco. ¿Qué es la urna electronica? (2012). http://www.iepcjalisco.org.mx/urna-electronica/que-es-la-urna-electronica

Extracting Contour Shape of Passenger Car Form in Front Quarter View Based on Form Similarity Judgement by Young Chinese Consumers

Chunrong Liu(✉), Yuchun Shi, and Minghua Zhang

Shanghai Jiao Tong University, Shanghai 200240, China
cheeronliu@sjtu.edu.cn

Abstract. To contribute to consumer-oriented car design practice, how a reduced contour shape is extracted to reveal the underlying vital parts of triple-compartment passenger car form in front quarter view is explored based on young Chinese consumers' form similarity judgement. By defining 25 character lines for 87 representative form samples and segmenting points on character lines, coordinate value data of these points are recorded and slope values of all 207 segments are calculated. Furthermore, factor analysis is completed and 27 common factors are extracted. The findings show that (1) the passenger car form in front quarter view can be extracted and presented as a reduced but recognizable contour shape with 28.816% loss of information on the form in front quarter view; and (2) the contour shape is mainly consisted of character lines defining design elements in side view such as side windows, beltline/shoulder line, A-pillar and C-pillar, roof and skirt area, and those defining design elements in front view such as headlight, and the upper and lower grille.

Keywords: Consumer/user research · Passenger car form in front quarter view · Contour shape extraction · Form similarity judgement

1 Introduction

From brand diversity and production/sale volume points of view, passenger car market in the Chinese mainland is likely the most intensely competitive market all around the world, where young Chinese consumers as the main force for car consumption [1] propose a higher request of a passenger car form and its symbolic value. Understanding their perception of passenger car form, especially the form in front quarter view, the most important angle of view [1], so as to meet their mental appeals, is a key point for automaker to be the winner in fiercer competition.

In this study, in order to explore main design elements in passenger car form in front quarter view configured by a set of character lines which are usually used to describe passenger car form [1–7], young Chinese consumers are invited to estimate on the similarity between 130 form samples from mainstream brands and product lines. And representative passenger car forms in front quarter are then selected by cluster analysis. By factor analysis with acquired data on all character lines for all representative samples, the contour shape of passenger car form in front quarter view is extracted.

G. Di Bucchianico et al. (Eds.): AHFE 2020, AISC 1202, pp. 558–564, 2020.
https://doi.org/10.1007/978-3-030-51194-4_73

2 Methods

2.1 Form Similarity Judgement and Representative Samples

Experimental Preparation. A total of 130 pictures of the triple-compartment passenger car form in front quarter view with engine displacements from 1.5 L to 2.4 L are collected in the Chinese mainland's car market, involving in products from 31 mainstream automobile brands including Beijing Benz, Beijing-Hyundai, BYD, BMW-Brilliance, Changan, Changan Ford, Changan-Mazda, Chery, Dongfeng, Dongfeng-Citroen, Dongfeng-Honda, Dongfeng-Nissan, Dongfeng-Peugeot, Dongfeng-Renault, Dongfeng-Yueda-Kia, FAW, FAW Audi, FAW-Mazda, FAW-Toyota, FAW-VW, GAC, GAC-Honda, GAC-Toyota, Great Wall, Qoros, SAIC-GM Chevrolet, SAIC-GM Buick, SAIC-GM-Wuling, SAIC MG, SAIC Roewe and SAIC Volkswagen. All pictures are then processed and randomly marked with serial numbers for input into a self-developed grouping tool.

Form Similarity Judgement Experiment. By the grouping tool, Subjects, 18 to 30, are invited to compare and estimate on the similarity between all samples, resulting in 30 pieces of valid data on similarity matrix which are then computed into an averaged similarity matrix.

Cluster Analysis and Sample Screening. The averaged similarity matrix data is analyzed by hierarchical cluster analysis. With resulted dendrogram, it is reasonable to classify all samples into 7 categories according to the relevant cluster analysis principle [8]. The K-means cluster analysis is executed. Among the initial 130 form samples, 87 representative samples are selected for the following data acquisition due to their relatively shorter distance values to the center of their corresponding categories.

2.2 Definition of Character Lines and Segments

Character Lines. Although there are some losses in form information, the character lines are usually used to describe the passenger car form in design studies. Since "the least cost includes the most information" [3], the character lines can provide a relatively complete description and expression of form. In this study, the character lines of passenger car form in front quarter view are defined and composed of 25 descriptive lines that outline main parts of form in quarter view. Considering differences in partial features such as headlight, fog light, grille, hood, shoulder and skirt area, the versatility for character lines is discussed to ensure that all representative form samples may be described and configured into a set of character lines. The final definition of character lines is shown in Fig. 1 and the labelling for these character lines is listed in Table 1.

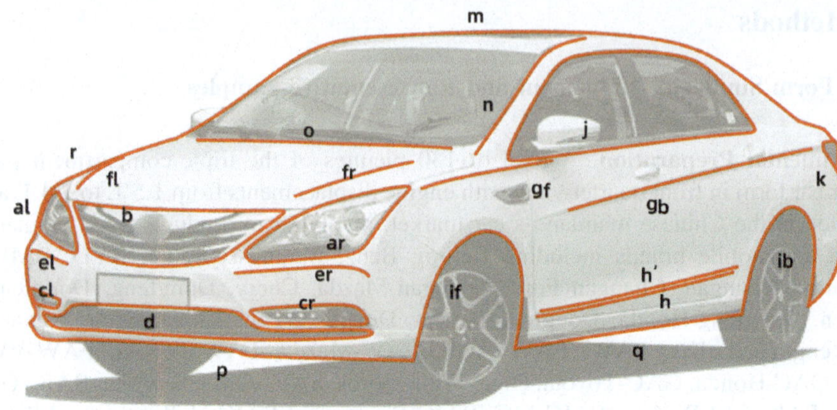

Fig. 1. Definition of character lines.

Table 1. Labelling of character lines for different form features.

Parts	Character lines' labelling letters
Headlights	al, ar
The upper grille	b
Fog lights	cl, cr
The lower grille	d
The transitional edge on front bump	el, er
The transitional edge on hood	fl, fr
Beltline and shoulder line	gf, gb
Skirt area	h, h'
Wheel arches	if, ib
Side window	j
Taillight	k
Rear end	l
Outline of right A-pillar to rear deck	m
Left A-pillar	n
The upper edge of hood	o
Bottom line of front bump	p
Bottom of side body	q
Outline of hood	r

Edit Points and Segments on Character Lines. A set of edit points is used to guide the flow of a character line while the character line, which is partitioned by edit points on it with appropriate space, draws the outline of specific form feature. The more complicated a form feature, the more edit points its character line calls, e.g., to define the upper grille which is labeled with 'b', 22 edit points are employed and every edit point is marked with the letter and its serial number, i.e., 'b1', 'b2', …, 'b21', 'b22', respectively. Apparently two adjacent edit points draw a segment named, e.g., ‘b12’, ‘b23’ and so on, and 22 edit points divide fully closed character line ‘b’ into 22 segments as shown in Fig. 2. In this way, a total of 207 segments are defined finally on all 25 character lines as illustrated in Fig. 2 and Fig. 3.

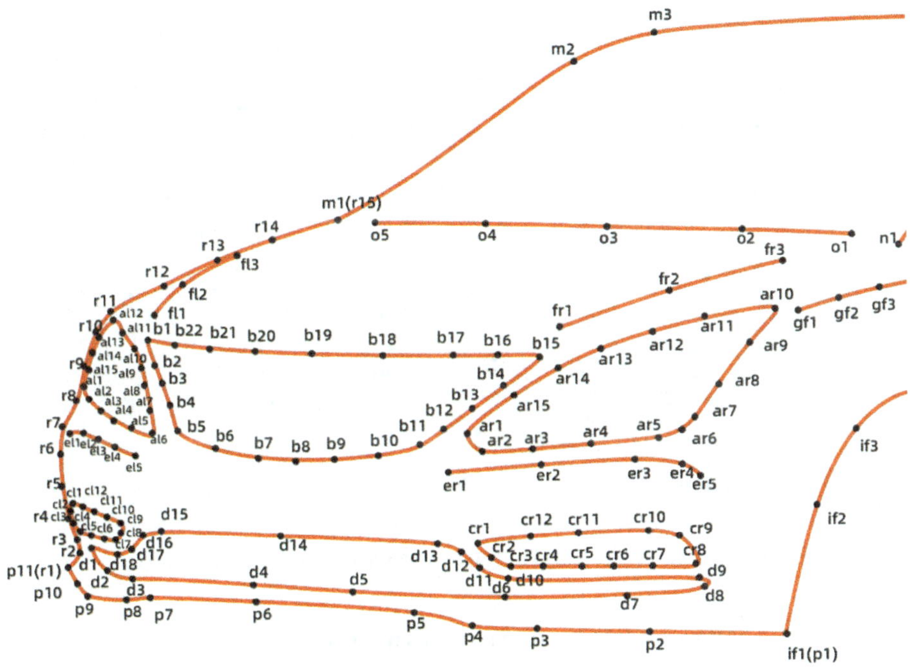

Fig. 2. Edit points and segments on character lines (the front part).

2.3 Coordinate Values of Edit Points

All pictures of representative form samples are imported into the same orthographic viewport for data acquisition of coordinate values of edit points. These pictures are scaled to the uniform length and positioned to align the center of front wheel with the same point in the viewport. The edit points are moved to adjust and fit flow of character lines of specific form features in corresponding form picture. The coordinate value data of all start points and end points of 207 segments are then recorded.

2.4 Slope Values of Segments

The slope of a segment reflects the degree of its own inclination and it can be calculated by the ratio of the difference between y-coordinate values divided by the difference between x-coordinate values of its end point and start point. All slope values of 207 segments in a form sample are obtained in this way and the slope values of all segments of 87 representative samples are calculated and recorded into a complete piece of slope value data.

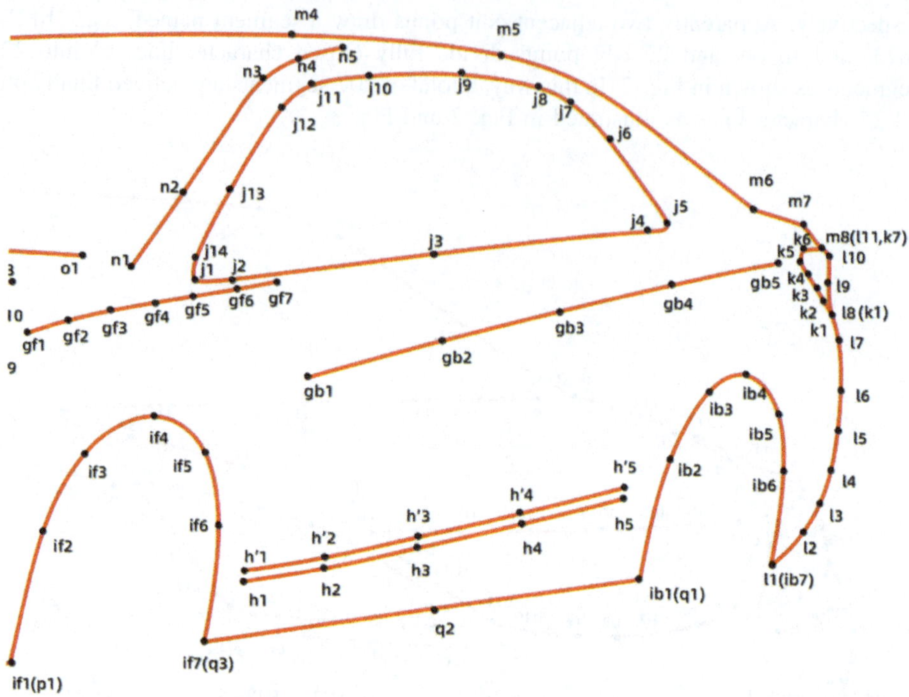

Fig. 3. Edit points and segments on character lines (the rear part).

2.5 Factor Analysis and Contour Shape Extraction

With the principal component method and 'varimax rotation' option, factor analysis procedures are performed five times with the slope value data as input and by setting eigenvalue is 'greater than 1', 'greater than 2', 'greater than 3', 'greater than 4' and 'greater than 5', respectively. By reproducing the segments in configuration of character lines according to factor analysis results, it is found that most of form information (94.779% and 83.182%) are retained when eigenvalues are set to be 'greater than 1' and 'greater than 2', while the retained form information (43.171%) is too less to express a recognizable contour shape of passenger car form in front quarter view when eigenvalue is set to be 'greater than 5'. By comparing the reproduced contours related to other two eigenvalue settings, the contour shape corresponding to eigenvalue 'greater than 3' is finally used as shown in Fig. 4, which has 28.816% loss of form information and is composed of 136 retained segments determining 27 common factors.

Fig. 4. The extracted contour shape.

3 Conclusion

In this study, the contour shape is extracted as a reduced expression of passenger car form in front quarter view by quantitative analyses based on young Chinese consumers' perception of and similarity judgement on the triple-compartment passenger car form in front quarter view.

The findings show that the reduced but recognizable contour shape can be extracted and described by 136 segments on 25 character lines with 28.816% loss of information on passenger car form in front view. The extracted contour shape is mainly consisted of character lines defining design elements in side view such as side windows, beltline/shoulder line, A-pillar and C-pillar, roof and skirt area, and those defining design elements in front view such as headlight, and the upper and lower grille.

From the extracted contour shape, it is observed that the design elements in side view are more completely remained than those in front view, implying the focus for designing passenger car form in front quarter view.

References

1. Liu, C.: Design Strategy Development for Product Innovation. Shanghai Jiao Tong University Press, Shanghai (2015). (in Chinese)
2. McCormack, J.P., Cagan, J.: Speaking the Buick language: capturing, understanding, and exploring brand identify with shape grammars. Des. Stud. **1**, 1–29 (2004)
3. Zhao, D., Zhao, J.: Automobile form feature and feature line. Packag. Eng. **28**, 115–117 (2007). (in Chinese)
4. Hu, W., Zhao, J., Zhao, D.: Study on styling image of vehicle based on form feature lines. China Mech. Eng. **20**, 496–500 (2009). (in Chinese)
5. Chen, Z.: A feature-line-based research of TOYOTA Brand Automobile Styling Evolution. Master's thesis, Guangdong University of Technology, Guangdong (2013). (in Chinese)
6. Liu, C., Zhang, M.: Extracting contour shape of passenger car form in rear view based on form similarity judgement by young Chinese consumers. In: Shin, C. (ed.) Advances in Interdisciplinary Practice in Industrial Design. AHFE 2019. Advances in Intelligent Systems and Computing, vol. 968, pp. 137–145. Springer, Cham (2020)

7. Liu, C., Xu, Q.: Extracting contour shape of passenger car form in front view based on form similarity judgement by young Chinese consumers. In: Marcus, A., Wang, W. (ed.) Design, User Experience, and Usability. Application Domains. HCII 2019. Lecture Notes in Computer Science, vol. 11585, pp. 74–84. Springer, Cham (2019)
8. Zhang, W.: Advanced Course of Statistical Analysis in SPSS. Higher Education Press, Beijing (2004). (in Chinese)

Human Factors Research Methods for Hand and Tool Design

Lend a Hand for 3D Scans: Scanning Methodology and Data Collection for Tool and Glove Design

Bethany Juhnke[1], Colleen Pokorny[1], Linsey Griffin[1(✉)], and Susan Sokolowski[2]

[1] Human Dimensioning Laboratory, College of Design, University of Minnesota, 1985 Buford Drive, St. Paul, MN 55108, USA
{juhnk006, pokor033, lgriffin}@umn.edu
[2] School of Art + Design, College of Design, University of Oregon, 70 NW Couch Street, Portland, OR 97209, USA
ssokolow@uoregon.edu

Abstract. Methods to conduct large-scale anthropometric studies to capture civilian measurements are inefficient and expensive. Industrial engineering principles were applied to improve the data capture process to build comprehensive datasets. The goal was to transform the raw materials (the participant) into a tangible product (anthropometric data) with minimal waste (time, equipment, and space). Traditional elements of an anthropometric study were evaluated based on *how* the study was conducted. Developed methods were applied to a study capturing scans of 398 participants over 7 days. Participants continually flowed through the study stations and completed it in 23.09 min on average. The study cost $34.18 per participant, compared to a traditional anthropometric study cost of $46.95 per participant. The results present the value of applying industrial engineering principles to anthropometric study design to improve the quality and accessibility of data used for human factors analyses and product design.

Keywords: Anthropometry · Human factors · Industrial engineering

1 Introduction

Current methods to conduct large-scale anthropometric studies to capture civilian measurements are inefficient and expensive. Trained personnel must prepare, recruit, measure, capture, and analyze data from willing participants to create the necessary databases of civilian anthropometric measurements [1]. Unfortunately, inefficient methods to conduct anthropometric studies continue to limit the scope of datasets, ultimately limiting the downstream impact to improve human factors analyses and product design through industrial engineering processes. More efficient methods are

Research was supported by the Minnesota Agricultural Experiment Station, and the National Institute of Food and Agriculture, U.S. Department of Agriculture, Hatch under MIN-53-088.

G. Di Bucchianico et al. (Eds.): AHFE 2020, AISC 1202, pp. 567–574, 2020.
https://doi.org/10.1007/978-3-030-51194-4_74

needed to reduce the cost and time required to create anthropometric databases. By applying industrial engineering principles [2] to the data capture process, we can improve the process to capture comprehensive anthropometric databases for the purposes of improving human factors analyses and ultimately product designs.

2 Literature Review

2.1 Anthropometric Studies

Many anthropometric datasets are limited in scope and insufficient for product design of specific products, such as gloves and tools for hands [1, 3–6]. Additional anthropometric studies fill knowledge gaps for specific regions of the body, including; the hands [6–9], the waist-hip-thigh region [10], and the breast shape and movement [11]. Recently, new scanning technologies have been used to capture the dynamic change of the human body to impact product designs [6, 10]. However, there are no known studies to develop more efficient methods to capture large-scale civilian anthropometric data.

2.2 Human Factors

Human variability (size, age, gender, race, etc.) impacts how individuals interact within a system and is the focus of a human factors analysis [12]. Workspaces and tools must be appropriately designed for the user to be most effective [12]. Inadequate designs are the most common sources of error, fatigue, and discomfort [12]. Anthropometric studies provide the quantitative data to improve workspace designs and tools based on the dimensions of the body [12]. The reverse may also be true; evaluating the human factors of an anthropometric study can improve study design and the ability to capture data. This study focuses on methods to improve the workspace and tools with which large datasets of anthropometric data are captured.

2.3 Industrial Engineering

Principles of industrial engineering focus on improving processes through evaluating the integration of people, materials, information, and equipment [13]. The evaluation focuses on transforming raw materials into a final product by reducing waste (time, material, labor, etc.) at specific procedural steps and holistically. Human factors analyses of workspaces and tools inform industrial engineering process evaluations.

To apply industrial engineering principles to an anthropometric study, we considered the final product to be one participant's set of anthropometric data. The dataset from one participant includes demographic and background information, manual measurements, and 3D hand scans. Our goal was to transform the raw materials (the participant) into a tangible product (anthropometric data) with minimal waste (time, equipment, and space). The production line needed to accommodate researchers and equipment to continuously capture the datasets efficiently and cost effectively.

The anthropometric dataset can be studied to draw conclusions about the body for the purposes of human factors and product design.

3 Methods

The traditional elements of an anthropometric study are recruiting, consent, demographics, landmarking, and capturing data [14]. Anthropometric studies are labor intensive and time consuming. Our objective was to develop an efficient method for a large-scale anthropometric study using industrial engineering principles. This new methodology was applied and tested as a large-scale anthropometric study conducted at the Minnesota State Fair in the Driven to Discover (D2D) Barn.

The D2D event draws a large civilian population from across the state. The D2D study goal was to gather anthropometric data from 300 participants over 7 days using the Artec Leo, a 3D wireless structured light scanner. The research team was allotted a 13' × 12' space. The suggested time per participant was 15 min, therefore, efficiency was critical to the success of the study.

The research process was broken into three segments: recruitment, demographics, and background; landmarking and manual measurements; and 3D scanning. Methods developed focused on efficiency and cost-effectiveness with the constraints of time, space, and participant numbers. This industrial engineering approach informed the *how* of each segment within an anthropometric study.

4 Results

Applying industrial engineering principles to the design of anthropometric studies resulted in a large-scale anthropometric database at minimal cost. Overall, 398 individuals were scanned in two hand positions for the D2D study. The study occurred over 7 days for 41 active study hours. The study space accommodated 4 participants at a time, participants completed the study in 23.09 min on average, and financial cost per participant was $34.18.

4.1 Time Analysis

The D2D study maximized the available space, prioritizing the scanning area. The spatial layout, scanning area and radius, and participant path are shown in Fig. 1. Five researchers were required per shift and were assigned to specific stations.

Survey, Landmarking and Manual Measurement. Verbal consent was obtained from participants to reduce the recruiting and enrollment time and eliminated the need for printed consent forms. Next, participants completed an iPad survey about demographics and background information. Landmarks were applied using washable markers, which were more efficient than previous methods [9]. Manual measurements were captured to standardize data with existing datasets [1, 3]. The average amount of time to complete the survey, landmarking, and manual measurement stations can be seen in Fig. 2.

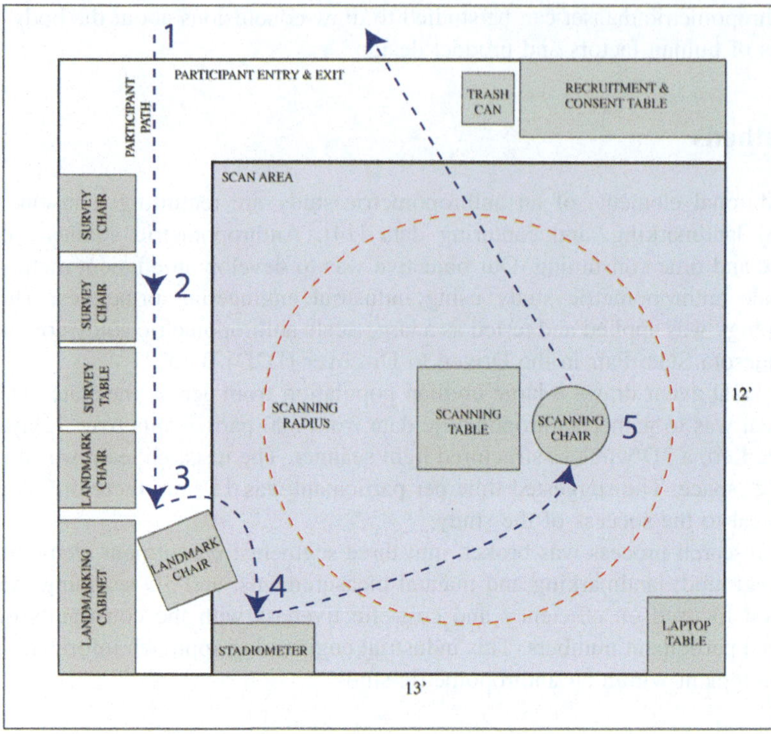

Fig. 1. D2D Study Layout and Participant Path. The participant was recruited and completed consent (1), completed a survey (2), their hand was landmarked (3), manual measurements were collected (4), and their hand was scanned (5).

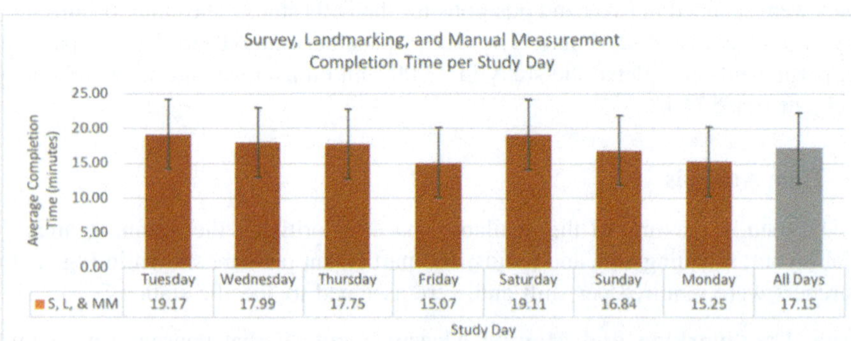

Fig. 2. Average survey, landmarking and manual measurements completion time for each day of the study with the minimum and maximum times denoted by the error bars.

Scanning. An efficient scanning method was developed to capture two hand positions per participant. The average time to complete the scan sequence per day and average overall scan times can be seen in Fig. 3. Scanning time was reduced towards the end of the study as the research team became more familiar with their roles.

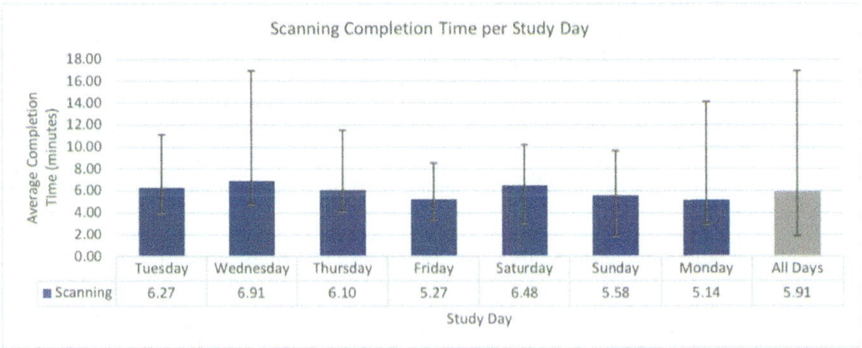

Fig. 3. Average scanning completion time for each day of the study with the minimum and maximum times denoted by the error bars.

Overall Study Time. The survey, landmarking and manual measurement segment of the study took 17.15 min on average to complete and the scanning segment took 5.91 min on average. The overall time for participants to complete the study took 23.09 min on average.

4.2 Financial Analysis

Overall, 398 individuals participated in the study over a 7-day period. The research and development, the D2D study costs, and a theorized laboratory study were calculated for a financial comparison between this study design and a traditional anthropometric study. Research and development costs can be seen in Table 1. D2D study costs can be seen in Table 2.

Table 1. The research and development costs to develop the D2D study protocol based on improving efficiency and reduce costs.

Study Research and Development Costs	
Item	Expense
Supplies	$503.65
Researcher Time (2 researchers @ 100 hours @ $20 per hour)	$4,000.00
TOTAL R&D COST	$4,503.65

Table 2. The costs to conduct the D2D study broken down to the cost per participant.

D2D Study Costs	
Item	Expense
D2D Fees	$7,000.00
Researcher Time (5 researchers @ 48 h @ $20 per hour)	$4,800.00
Equipment to Conduct Study	$1,804.68
TOTAL COSTS	$13,605.00
Cost Per Participant (398) over 7 days	$34.18

Table 3. The cost to conduct a traditional anthropometric study in a laboratory setting broken down per participant.

Traditional Anthropometric Study Costs Based on D2D Study Design	
Item	Expense
Recruitment Costs (@ 7.5 minutes per participant + $250 for supplies)	$1,806.28
Researcher Time (1 researcher @ 8 hrs/day @ $20/hour for 44.5 days)	$7,114.00
Participant Reimbursement ($20 per participant)	$7,960.00
Equipment to Conduct Study	$1,804.68
TOTAL COSTS	$18,685.39
Cost Per Participant (398) over 44.5 days	$46.95

Comparison to Traditional Anthropometric Study. One advantage of the D2D study was the lack of no-show participants. A no-show rate of 36.07% was calculated from a previous study [10]. Applying the no-show rate to the D2D study, 623 slots would need to be scheduled to capture 398 datasets. Scheduling participants every 30 min for 7 hours per day would require 44.5 days to complete the study. The traditional anthropometric study analysis is presented in Table 3.

The financial analysis of the D2D study does not account for prototyping supplies and equipment (computers, scanners, etc.) already available in an anthropometric laboratory. Also, the traditional anthropometric timeline of 44.5 days does not consider staff morale when considering the length of the study [15]. Overall, the D2D study design was more efficient at capturing anthropometric data and less expensive per participant when compared to traditional anthropometric research studies. An event such as a state fair also draws a wider demographic of people than are recruited to an academic research laboratory. The results show that by applying industrial engineering principles and accounting for human factors in study design, large-scale anthropometric studies can be executed at minimal costs to researchers.

5 Discussion and Limitations

The time goal for the D2D study was 15 minutes per participant, but ultimately, the study took on average 23.09 minutes per participant, despite attempts to ensure each of the three major stations took five minutes each. Even though we did not meet this goal, we easily met our recruitment goal of 300 participants, ending with a total of 398 participants. Several factors were identified that impacted efficiency throughout the process and could be improved upon for future studies. Landmark, understanding the limitations of technology in an unpredictable environment, and incorporating a more robust researcher training are recommended to improve study efficiency.

This study shows the value in applying industrial engineering principles into the design of an anthropometric study. By considering constraints that affect efficiency and cost, a method was developed that efficiently captured anthropometric data from a large civilian population at minimal cost to researchers. This method can be used to create a comprehensive anthropometric database that improves human factors analyses and provides accessibility of data to improve product design.

References

1. Robinette, K.M., Blackwell, S., Daanen, H., Boehmer, M., Fleming, S.: Civilian American and European Surface Anthropometry Resource (CAESAR), Final Report, vol. 1. Summary. Sytronics Inc. Dayton, Ohio (2002)
2. Salvendy, G. (ed.): Handbook of Industrial Engineering, 2nd edn. Wiley, New York (1992)
3. Paquette, S.: Anthropometric Survey (ANSUR) II Pilot Study: Methods and Summary Statistics. Anthrotch, US Army Natick Soldier Research, Development and Engineering Center (2009)
4. Size USA. http://www.sizeusa.com/
5. Kuebler, T., Luebke, A., Campbell, J., Guenzel, T.: Size North America–The New North American anthropometric survey. In: International Conference on Human-Computer Interaction, pp. 88–98 (2019)
6. Griffin, L., Kim, N., Carufel, R., Sokolowski, S., Lee, H., Seifert, E.: Dimensions of the dynamic hand: implications for glove design, fit, and sizing. In: International Conference on Applied Human Factors and Ergonomics, pp. 38–48. Springer, Cham (2018)
7. Griffin, L., Sokolowski, S., Seifert, E.: Process considerations in 3D hand anthropometric data collection. In: Proceedings of 3DBODY.TECH 2018-9th International Conference and Exhibition on 3D Body Scanning and Processing Technologies, Lugano, Switzerland, pp. 121–130 (2018)
8. Sokolowski, S.L., Griffin, L., Chandrasekhar, S.: Current technology landscape for collecting hand anthropometric data. In: Proceedings of 3DBODY.TECH 2018 - 9th International Conference and Exhibition on 3D Body Scanning and Processing Technologies, Lugano, Switzerland, pp. 142–153 (2018)
9. Griffin, L., Sokolowski, S., Lee, H., Seifert, E., Kim, N., Carufel, R.: Methods and tools for 3D measurement of hands and feet. In: Chung, W., Shin, C. (eds.) Advances in Interdisciplinary Practice in Industrial Design. AHFE 2018. Advances in Intelligent Systems and Computing, vol. 790, pp. 49–58. Springer, Cham (2018)

10. Griffin, L.A., Juhnke, B., Seifert, E., Pokorny, C., Doran, K.: Method to capture and analyze the waist-hip-thigh body region of seated-standing 3D scans. In: Proceedings of 3DBODY. TECH 2019 - 10th International Conference and Exhibition on 3D Body Scanning and Processing Technologies, Lugano, Switzerland, pp. 254–265 (2019)
11. Pei, J., Griffin, L., Fan, J., Ashdown, S.P.: The detection of the upper boundary of breasts using 4D scanning technology. In: Proceedings of 3DBODY.TECH 2019 - 10th International Conference and Exhibition on 3D Body Scanning and Processing Technologies, Lugano, Switzerland, pp. 266–271 (2019)
12. Wickens, C.D., Lee, J.D., Liu, Y., Gordon Becker, S.E.: An Introduction to Human Factors Engineering. Pearson, Upper Saddle River (2004)
13. Womack, J.P., Jones, D.T.: Beyond Toyota: how to root out waste and pursue perfection. Harvard Bus. Rev. **74**, 140–158 (1996)
14. National Health and Nutrition Examination Survey (NHANES): Anthropometry Procedures Manual. Center for Disease Control, January 2009. https://www.cdc.gov/nchs/data/nhanes/nhanes_09_10/BodyMeasures_09.pdf
15. Vincent, K.B., Kasperski, S.J., Caldeira, K.M., Garnier-Dykstra, L.M., Pinchevsky, G.M., O'Grady, K.E., Arria, A.M.: Maintaining superior follow-up rates in a longitudinal study: experiences from the college life study. Int. J. Multiple Res. Approach. **6**(1), 4406 (2012)

Qualitative Survey Methodology and Data Collection for Performance Glove Design and Fit

Susan L. Sokolowski[1(✉)], Linsey Griffin[2], Bethany Juhnke[2], Colleen Pokorny[2], and Chrissy Bettencourt[1]

[1] Sports Product Design, University of Oregon, 70 NW Couch Street, Portland, OR 97209, USA
{ssokolow, chrissyb}@uoregon.edu

[2] University of Minnesota, Design, Housing and Apparel, 1985 Buford Avenue, St. Paul, MN 55108, USA
{lgriffin, bjuhnke, pokor033}@umn.edu

Abstract. Performance gloves worn for work, sport and thermoregulation are known to have fit challenges because there is a lack of accurate and relevant civilian anthropometric data for product manufacturers. Fit challenges also exist because manufacturers neglect to communicate with users at the point-of-purchase and during the design process about their experiences and needs with glove fit. Poor fitting gloves can interfere with sensory information, accuracy, protection, mobility and blood flow. There is a considerable opportunity to improve glove fit. This qualitative study addressed the issue of communicating with users during the design process, by collecting fit data directly from users about their experiences with performance gloves used for work, sport and thermoregulation. Results from the study uncovered that both men and women have fit challenges across all performance glove market segments and simple user check-points during the design process could help to develop better performing products.

Keywords: Qualitative survey · Glove design · Glove fit · Work · Sport · Thermoregulation

1 Introduction

Several studies have reported the need of properly fitting users who rely on performance gloves [1–5]. Fit challenges exist for performance glove users because there is a lack of accurate and relevant anthropometric data available to product manufacturers [2]. Fit challenges also exist because manufacturers neglect to communicate with users at the point-of-purchase and during the design process regarding their fit needs. Poor fitting gloves can interfere with sensory information, accuracy, protection, mobility and blood flow. As an example, the US Bureau of Labor Statistics, reported that one million occupational workers were treated with acute hand injuries in 2014 [6]. These injuries occurred because users experienced products that fit and functioned poorly or chose to not wear protection at all because it hindered performance due to fit [6]. There is an

G. Di Bucchianico et al. (Eds.): AHFE 2020, AISC 1202, pp. 575–581, 2020.
https://doi.org/10.1007/978-3-030-51194-4_75

opportunity to improve glove fit during the product design process. This qualitative study collected fit data directly from users regarding performance gloves used for work, sport and thermoregulation (protection from hot, cold and wet environmental conditions).

2 Background

2.1 Glove Market Segments

In the U.S., the performance glove market can be broken into three market segments - work, sport and thermoregulatory (Table 1):

Table 1. Glove market segments and typical end uses.

Work gloves	Sport gloves	Thermoregulatory gloves
Construction	Baseball	Cold environments
Landscaping/gardening	Boxing	Hot environments
Medical	Cycling	Wet environments
Restaurant	Football	
Other	Golf	
	Hockey	
	Lacrosse	
	Skiing/snowboarding	
	Soccer goalkeeping	
	Weightlifting	
	Other	

2.2 Fit Issues Due to the Lack of Civilian Hand Anthropometric Data

For gloves that are available in the U.S. market, there is a lack of civilian hand anthropometric data to inform fit [5]. The majority of anthropometric available in the U.S. is from military personnel [5]. The data is also limiting, as there are typically only three measures collected - hand circumference, breadth, and length [5]. With the absence of data from civilians and limited measures, it is nearly impossible for glove manufacturers to accurately define how glove patterns should be developed and sized across a sizing system.

2.3 Fit Issues Due to How Manufacturers Communicate Glove Sizing at Point-of-Purchase

In addition to the absence of appropriate anthropometric data, the methods used by manufacturers to size users into gloves are varied, confusing and not standardized [5]. There are several issues with consumer-facing communication about sizing which makes it challenging for anyone to find gloves that fit. The issues include:

Poorly Explained Measures. Glove sizing charts do not clearly direct consumers on how to measure their hands, in order to size themselves into the right gloves. As an example, the palm circumference between manufacturers is not standardized [5]. Some show it as a circumferential measurement at the first and fifth finger interdigital web, others show it at the knuckles or somewhere in-between the knuckles and interdigital web. Some manufacturers do not specify where to measure the circumference at all [5].

Inconsistent Measures. Between manufacturers, glove sizing systems have different measurement scales and require consumers to use different types of tools (e.g., metal rulers, soft plastic tape measures, paper templates) to measure their hands. The variety of tools can be confusing for users that purchase hand wear between different manufacturers [5].

Not Enough Measures. Similarly to Sect. 2.2, consumers are usually directed to capture very few measures of their hand – likely because there are very few anthropometric measures to reference [5]. Some manufacturers only require one measurement (e.g., hand length) to select a glove size [5]. In order to fit all five fingers and the palm, more than two to three measures are needed to fit users effectively.

2.4 Fit Issues Due to the Design Process

The authors have found that a major contributing factor to glove fit - is the design process that manufacturers use. Many manufacturers neglect to interface with users to collect fit information on state-of-the-art products at market, wear test data of products during concept development or wear testing of a full-size range with a large subject sample. In all, there is an opportunity to improve the process.

3 Methodology

A Qualtrics survey was developed and administered via iPad to subjects that volunteered at the University of Minnesota's Driven to Discover (D2D) program, at the 2019 Minnesota State Fair. The event provided a unique data collection opportunity, as it drew-in a large civilian population from across the state of Minnesota and beyond, over seven days. The qualitative survey collected information pertaining to sex and fit experiences with gloves used for work, sport and thermoregulation (e.g., hot, cold, wet environments). Other qualitative and 3D hand scan data were collected in the study but will not be addressed in this paper. Subjects were asked about performance glove fit experience, through a visual tool that indicated 17 locations of fit (A-Q) on the hand (Fig. 1). The visual tool separated glove fit locations into lengths, depths and widths.

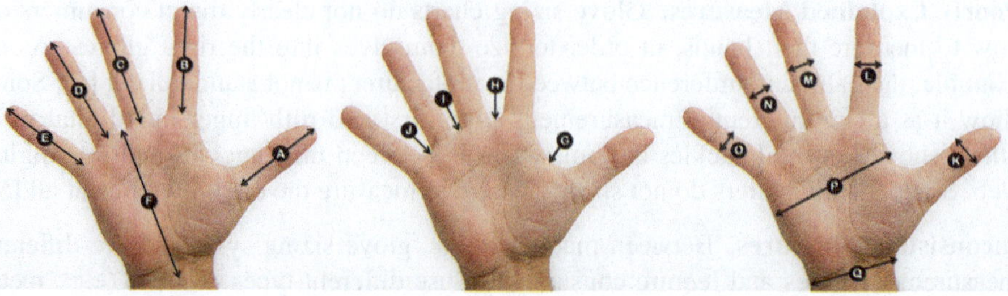

Fig. 1. Visual fit tool provided to subjects that participated in the qualitative survey. The 17 glove fit locations (A-Q) included lengths *(left)*, depths *(center)* and widths *(right)*.

4 Results

417 subjects (167 men and 250 women) participated in the study. Data were analyzed between sexes and glove market segment. Work glove fit data (125 subjects) represented construction, landscaping/gardening, medical, restaurant and other professions and were analyzed for fit (too big/long and too small/short). Sport glove data (100 subjects), for baseball, boxing, cycling, football, golf, hockey, lacrosse, skiing/snowboarding, soccer goalkeeping, weightlifting and other activities were also analyzed for fit (too big/long and too small/short). Data collected from users that required gloves for thermoregulation (hot, cold, wet protection) were also analyzed for fit from 302 subjects. The percentage of subjects who reported fit issues at the locations outlined in Fig. 1 were calculated (Table 2). The percentages were then converted into "heatmaps" (Fig. 2) to illustrate how subjects perceived glove fit at the 17 different locations on the hand. Darker areas indicated the locations that were frequently reported as either too big/long or too small/short, and lighter ones indicated locations which were less frequently reported as too big/long or too small/short.

Table 2. Percentages of subjects who reported gloves being too big/long or small/short by location on the hand. Locations are pictured in Fig. 1 *(above)*.

Location	Men big/long	Women big/long	Men small/short	Women small/short
A	7.84%	9.51%	4.09%	5.68%
B	9.89%	11.69%	4.97%	6.53%
C	9.70%	10.44%	6.39%	7.67%
D	9.70%	10.65%	5.15%	6.11%
E	10.63%	12.00%	5.33%	5.11%
F	2.61%	2.59%	6.04%	5.82%
G	3.92%	3.93%	7.46%	4.83%
H	3.17%	3.41%	9.59%	6.53%
I	2.99%	3.41%	9.24%	7.24%
J	2.80%	3.62%	9.41%	7.10%
K	2.99%	2.90%	2.49%	2.56%
L	2.43%	1.96%	1.78%	2.70%

(continued)

Table 2. (*continued*)

Location	Men big/long	Women big/long	Men small/short	Women small/short
M	2.80%	2.28%	1.60%	3.13%
N	2.61%	2.28%	1.95%	2.84%
O	2.43%	2.28%	1.95%	2.13%
P	3.73%	2.90%	6.22%	4.40%
Q	5.41%	4.34%	3.55%	2.41%
None	13.99%	9.72%	12.61%	16.48%
Other	0.37%	0.10%	0.18%	0.71%

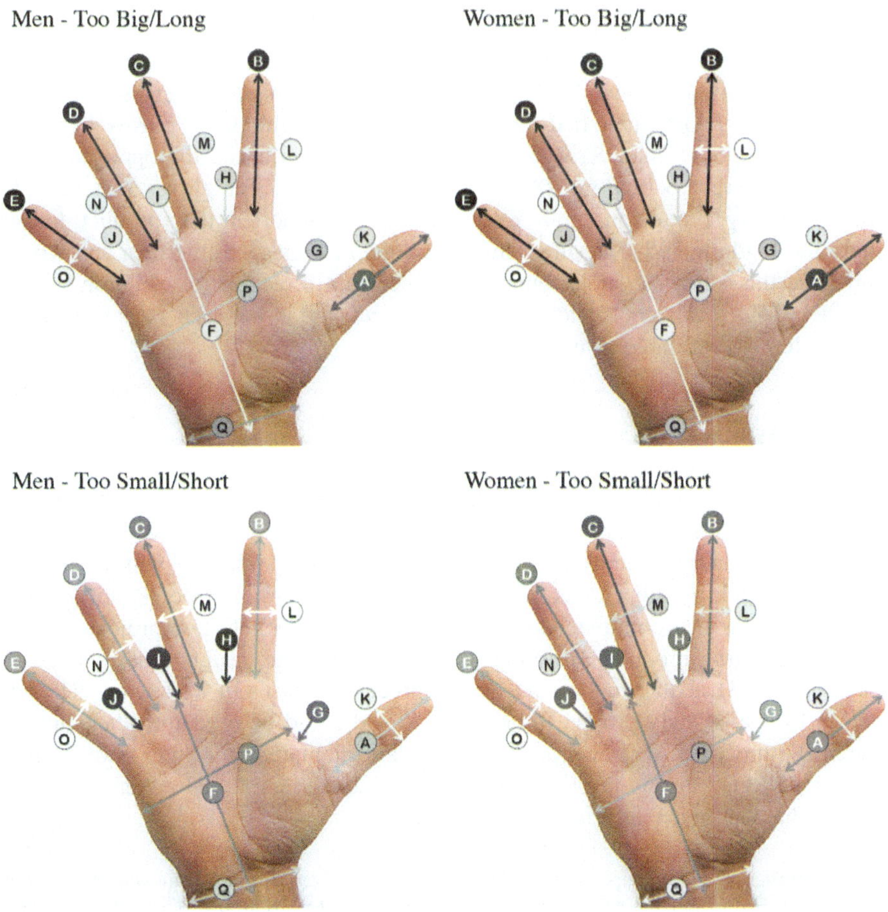

Fig. 2. Subjects (by sex) reported whether gloves were either too big/long *(top)* or too small/short *(bottom)* across different areas of the hand. Darker areas indicate more reports of issues and lighter areas indicate less reports of issues.

In all, both sexes in the study had issues with glove finger length (being too long) across all glove market segments. Men reported that they had more issues with glove finger crotch depth than women. Palm length/width and finger breadth had the least amount of reported issues (Table 3).

Table 3. Percentages of subjects who reported gloves being too big/long or small/short in at least one area on the hand, by market segment.

Use	Men too big/long	Women too big/long	Men too small/short	Women too small/short
Work	83.33%	92.58%	88.41%	78.91%
Sport	81.91%	88.79%	88.96%	82.47%
Thermoregulation	88.29%	89.58%	85.98%	85.22%

When looking at the fit data by market segment, the most compelling result was that women reported their work gloves to be 92.58% too big/long versus 78.91% too small/short. For sport, men had slightly more issues with their gloves being too small/short (88.96% versus 81.91%), where women experienced the opposite – their sport gloves were too big/long (88.79% versus 82.47%). For gloves that provided thermoregulatory protection, both men and women reported that product fit was slightly too big/long. What is surprising is that the fit issues that the men and women reported in this study are not swinging in one clear direction (too big/long or too small/short). What this could indicate is that subjects are in between sizes or there are not enough sizes available to users.

5 Conclusion

The findings support that along with collecting better hand anthropometric data (which can be a costly process) and communicating better glove sizing procedures to consumers, there is an opportunity during the design process to interface with users to get better fit feedback. Feedback can occur anywhere in the design process, but the most effective way to incorporate feedback is to include it 1) before a project starts to understand existing product, 2) during the project once the design concept has been confirmed and 3) at the end of the design process to test out a size set. The tool developed for this study proved to be effective and could be used for other glove fit research. Future work will also analyze the other qualitative attributes and 3D hand scan data collected during the D2D program, to uncover additional patterns and relationships that can better inform glove design and fit.

Acknowledgments. This paper is based upon work supported by the Minnesota Agricultural Experiment Station, and the National Institute of Food and Agriculture, U.S. Department of Agriculture, Hatch under MIN-53-088.

References

1. Campbell, R.B.: Patterns of Female Firefighter Injuries on the Fireground. National Fire Protection Association, Data and Analytics Division (2017)
2. Corriveau, S.: Applying ergonomics to industrial glove design. Occup. Health Saf. **83**(1), 8–10 (2014)
3. Health and Safety of Women in Construction (HASWIC) Workgroup and Occupational Safety and Health Administration: Women in the Construction Workplace - Providing Equitable Safety and Health Protection. http://web.archive.org/web/20190713042919/, https://www.osha.gov/doc/accsh/haswicformal.html
4. MacDonald, L.A.: Training Outline: Ergonomics in the Building Trades. Training Materials Developed for the Women in Building Trades, Jamaica Plain, Massachusetts (1990)
5. Sokolowski, S.L, Griffin, L.: Women's leather protective work wear gloves: a comparative pilot study between 3D hand scans, product specifications and sizing. In: ASTM 11th Symposium on Performance of Protective Clothing and Equipment: Innovative Solutions to Evolving Challenges, October 2019
6. U.S. Bureau of Labor Statistics: Injuries, Illnesses, and Fatalities. http://web.archive.org/web/20190928110357/, https://www.bls.gov/iif/

How Sport-Specific Hand Positioning Can Inform Glove and Mitt Design

Susan L. Sokolowski(✉), Carly Anderson, and Sarah Klecker

Sports Product Design, University of Oregon, 70 NW Couch Street, Portland, OR 97209, USA
{ssokolow, canders6, sklecker}@uoregon.edu

Abstract. There are thousands of hand wear products available to athletes in the marketplace, but very few have considered how sport-specific hand positioning informs design. Most are designed with a 2D front and back view. When designers consider sport-specific positioning of the hand, products can be developed in a 360° manner, to dynamically address how the hand moves and interfaces with the environment and other equipment in sport. For this paper, the Performance Hand Wear and Tools Product Innovation Framework was used by graduate students to design a glove and mitt, based on sport-specific hand positioning for female biathletes and winter night runners.

Keywords: Sport product design · Gloves · Mitts · Sport hand positioning

1 Introduction

Athletes wear gloves, wraps, braces and mitts to provide abrasion resistance, impact protection, visibility, grip and thermal comfort while training or competing in sport. There are thousands of hand wear products available for athletes on the global market, but very few have considered how sport-specific hand positioning informs design. Most are designed with a 2D palmar and backhand view, without consideration of the sides or dynamic motion of the hand during sport. When designers consider sport-specific positioning of the hand, products can be created in a 360° manner and each part of the product can be functionally designed more effectively. For this paper, graduate students in a sports equipment design studio used the Performance Hand Wear and Tools Product Innovation Framework (Table 1), to design gloves, based on sport-specific hand positioning needs of female biathletes and winter night runners [1].

G. Di Bucchianico et al. (Eds.): AHFE 2020, AISC 1202, pp. 582–588, 2020.
https://doi.org/10.1007/978-3-030-51194-4_76

Table 1. Hand Wear and Tools Product Innovation Framework [1].

Phase One	Phase Two	Phase Three
Strategies for understanding the hand wear & tool project background	Strategies for defining the user's 3D & 2D hand	Strategies for hand wear & tool product innovation
–Historical & state-of-the-art product anatomy –Technology landscape –Market landscape –User needs	–Hand anatomy –Hand physiology/biomechanics –User injuries –Hand landmarking –3D hand scanning –3D hand scan processing –Hand anthropometry –3D hand forms –2D hand croquis	–Ideation drawing –Materials identification –Product prototyping –Product patterning & form –Product tech drawings –Fitting & modeling on the 3D hand –Product validation

2 Background

2.1 General Sport Hand Wear Design Process and Challenges

The design process used for sports equipment is dependent on if the product is soft or hard goods based. If the product is soft goods based (made up of soft textile materials), an apparel design process will be used [2]. If the product is hard goods based (made of hard or molded parts), the product will be created with an industrial design process [2]. For most sport gloves, they are soft goods based with small molded parts, therefore they typically take-on an apparel design process – where the designer will research, ideate and create 2D technical flat drawings (front and back views) that are sent to the factory for product development. This method has short comings as the hands in sport need 360° functionality, and the product is only designed in 2D.

2.2 The Sport of Biathlon and Hand Wear Needs

Biathlon is a winter sport that combines cross-country skiing and rifle shooting [3]. Participants in a typical biathlon ski between 1 to 3 miles, and then stop to shoot targets [4]. Depending on specific competition rules, missed shots will result in extra distance or time being added to the athlete's finish [4]. The demands of the hand in biathlon require quick movements. In the cross-country portion of the event, the hands are used to push ski poles, to quickly propel forward. When shooting, the athlete must swiftly release their poles, grab their rifle and make five shooting attempts before skiing on to the next location. Since the sport is in a cold environment, it is essential that hand wear provides the athlete warmth, along with grip and dexterity when engaging with equipment.

Most biathletes wear cross country ski gloves (Fig. 1), or liners from winter gloves [5]. These gloves are bulky and do not provide the feel needed to move from ski pole to rifle efficiently.

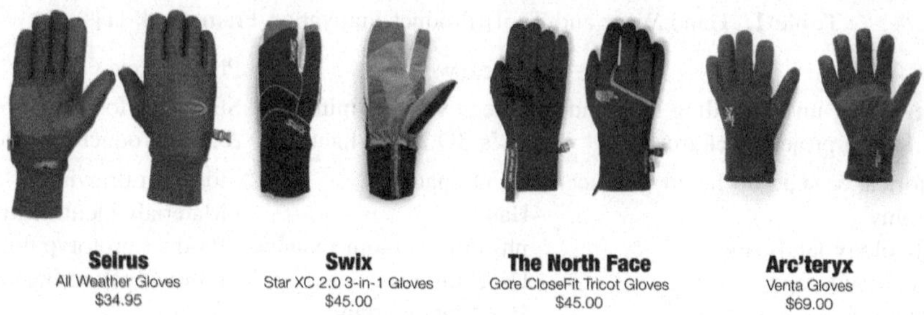

Fig. 1. Cross country glove market research.

2.3 Winter Night Running and Hand Wear Needs

Winter night running presents complicated thermoregulatory challenges for the athlete. As the runner steps into the cold, the body begins losing heat and vasoconstriction occurs [6]. With continued activity, the runner's core temperature warms and blood flow is redirected back to the muscles and skin [6] If the runner is wearing too much insulation, they may begin to overheat. If sweat is trapped against the body by layers of warm clothing, then as soon as core temperature begins to drop after running, the athlete will become chilled due to convective heat loss [6]. If evaporation of sweat is allowed to occur during the entire run, the body will stay comfortable [6]. Therefore, the most important job of hand wear designed for winter running should be to manage the conflicting demands of warming and cooling.

Along with thermoregulatory needs, decreased daylight during winter means that runners will likely run in the dark, and visibility can be a safety concern. The involvement of the hands and wrists in nighttime visibility is highlighted by research from 3 M [7]. It was found that humans are wired to recognize biomotion (patterns similar to the way a human or animal moves), and that the placement of reflective materials on major body joints/parts (e.g., wrist and hands) was more important than the brightness or color of material in making someone visible [7].

Finally, running throughout the winter is often a means to marathon racing in the spring. Marathon training requires ingesting gels and fluids to keep the body energized and hydrated during long runs. Thus, there is a need for hand wear to include a secure water bottle grip and a pinching dexterity for opening gel packets during runs.

There is a myriad of winter running hand wear options in the marketplace, but none of them truly consider the all of the needs outlined. There are either too thin and do not provide enough warmth or are mitt options that can keep the hands too warm as the runner heats-up. The few options that do provide variation in coverage and thermal protection generally do so by having a wind resistant cover that goes over the fingers that can fold back at the knuckles, leaving a simple glove silhouette. While this option is better, the fingers aren't the best place on the hand to allow venting. 65–70% of sweat flow in the hand is concentrated on the dorsal surface, making that the best place to promote cooling (as opposed to the fingers) [8].

3 Hand Wear Design Research Methodology

From understanding the sport, athlete needs and product backgrounds, students were tasked to develop methodologies to collect sport-specific dynamic hand data that could be used to design better products.

3.1 Women's Biathlon Design Research

For this project the student worked with an elite biathlete to 3D scan their rifle, to develop a research tool, to discover the functional zones of the right and left hands (Figs. 2 and 3). With the rifle scan a 3D model was created using *Rhinoceros 3D,* and it was then laser cut in cardboard and assembled. The model was then painted, so contact data could be transferred to the hands of six subjects. Through this approach, it was determined that asymmetric gloves were needed in order to optimize the thermoregulatory, dexterity and grip requirements of the hand for the sport of biathlon.

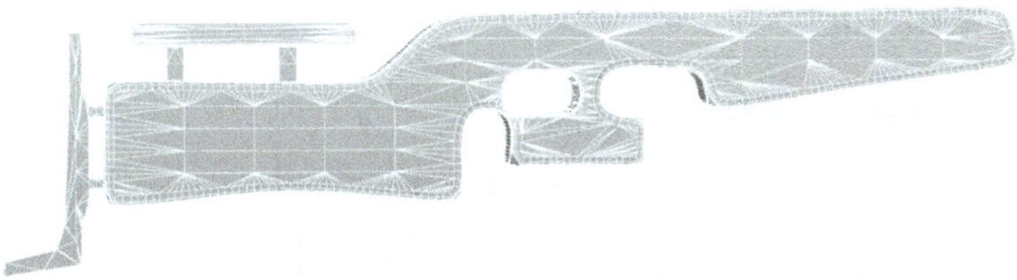

Fig. 2. 3D file of biathlon rifle.

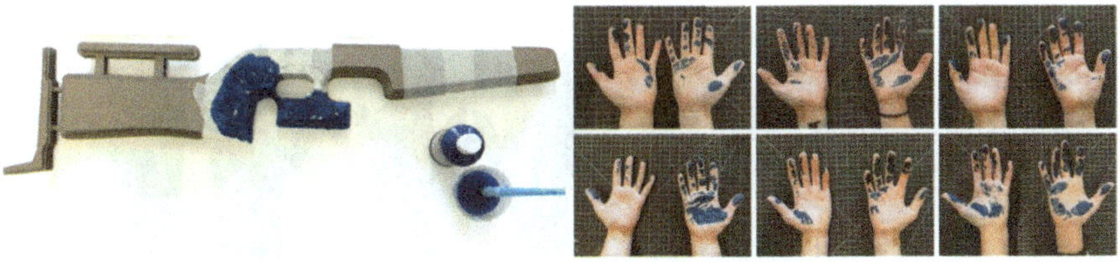

Fig. 3. Research process to understand points of contact.

3.2 Women's Winter Night Running Design Research

For the women's winter running project, a nighttime study was conducted with retro-reflective tape on a basic glove to determine the most functional arrangement of reflectivity in motion from various angles (Fig. 4). Reflective markers were placed along the metacarpal body/dorsal surface of the hand, distal and proximal phalanges, metacarpal heads, top of the wrist, and lateral edges of the hand at the thumb and 5th finger edges (Fig. 4). The gloves were then worn on a treadmill in a darkened room. Video was taken from the front, back, and side views with light shining directly from the camera to stimulate the retroreflective markings. Each video was broken down into frames and analyzed to see which parts of the hand were visible most frequently from each angle. A water bottle grip study was also conducted with paint to analyze points of contact (Fig. 4). To do this the student covered a traditional 22 oz water bottle used by runners with tape and white paint. Wearing a black latex glove, the student grabbed the bottle while running. The white paint left markings on the hand in areas key to the gripping the bottle.

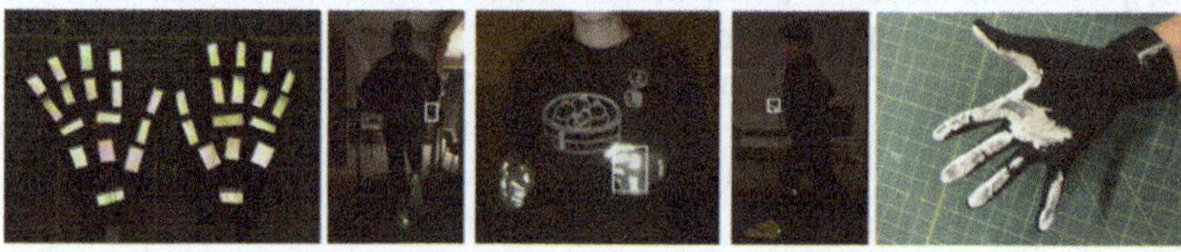

Fig. 4. Running reflective and points of contact study.

4 Design Results

4.1 Women's Biathlon Glove Design

For the women's biathlon project, asymmetric right and left gloves were developed with articulated fingers and insulation that considered optimal dexterity and warmth (Fig. 5). Suede material was mapped to the palm where the research showed points of contact to improve grip. To complete the project, the student developed technical drawings, took product photography and created a digital presentation of the process.

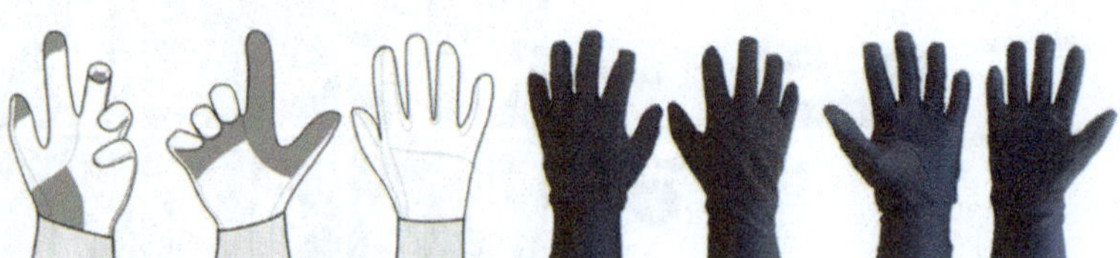

Fig. 5. Women's biathlon glove showing asymmetric design.

4.2 Women's Winter Night Running Mitt

A thermal "lobster" mitt was developed that allows for a balance of dexterity for gel-opening and bottle grip, while maintaining insulation. Based on thermoregulation research, it was determined that optimal ventilation placement was on the dorsal surface of the hand to enable the fingers to stay comfortable. Point of contact research showed that grip was needed on the fingers and thumb (palm side), as well as in-between the thumb and index finger web. Based on the reflectivity assessment, reflective detailing was placed along fifth finger side and dorsal surface of the mitt

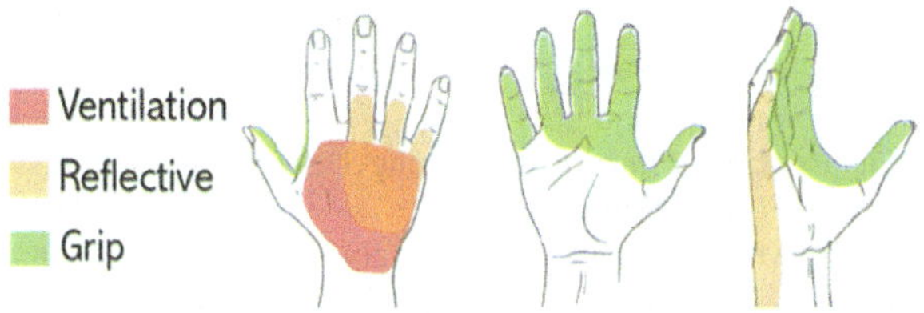

Fig. 6. Hand mapping of ventilation, grip, and reflective details.

Fig. 7. Women's winter running mitt design.

(from the third to fifth finger), to optimize visibility. A hand map in Fig. 6 shows where the respective elements were placed.

The final design features a lightweight stretch fleece on the palm, around the wrist cuff, and on the gussets between the fingers (Fig. 7). This allows for mobility and dexterity in grip as well as a secure fit. There is a quilted water and windproof shell on the top of the mitt to prevent moisture from chilling the body, with a zip vent to allow heat to escape if necessary. Sticky silicone material was placed on the areas defined by the grip test to best allow a secure water bottle grab. Finally, reflective materials were placed in the areas outlined by the research for optimal runner visibility and safety.

5 Conclusion

By understanding dynamic sport-specific hand positions, designers can carefully define shape and 360° functional product needs. Future work could explore using similar research methods for other sporting activities or occupations where performance hand wear and tools are required.

References

1. Sokolowski, S.L.: The development of a performance hand wear and tools product innovation framework. Fashion and Textiles, January 2020
2. Sokolowski, S.L.: Sports industry meets academia: The pedagogical development of an MS degree program in sports product design. Technol. Innov. **20**(3), 165–177 (2019)
3. Vaughan, K.: Taking photographs of winter sports action indoors. PSA J. **70**(8), 33–35 (2004)

4. Grabianowski, E.: How biathlon works - Biathlon basics (2019). https://adventure.howstuffworks.com/outdoor-activities/snow-sports/biathlon1.htm
5. Wray, B.: Personal Communication (2019)
6. Watkins, S.M., Dunne, L.E.: Functional Clothing Design: From Sportswear to Spacesuits. Fairchild Books, New York (2015)
7. Sayer, J.R., Mefford, M.L.: High visibility safety apparel and nighttime conspicuity of pedestrians in work zones. J. Saf. Res. **35**(5), 537–546 (2004)
8. Taylor, N.S., Van Den Heuvel, A., Caldwell, J., Taylor, E.A.: The Roles of Hands and Feet in Temperature Regulation in Hot and Cold Environments (2009)

A Visible Functional Grasp to Measure the Complete Hand

Colleen Pokorny, Bethany Juhnke, and Linsey Griffin(✉)

Human Dimensioning Laboratory, College of Design, University of Minnesota, 1985 Buford Avenue, St. Paul, MN 55108, USA
{pokor033, juhnk006, lgriffin}@umn.edu

Abstract. Glove and tools are designed to protect our hands, but manufacturers are limited by the available anthropometric hand data, which fails to reflect functional measurement changes of the hand while performing tasks. Advancements in 3D scanning technology have improved the ability to capture data, but minimal research has focused on capturing functional hand dimensions. The purpose of this study was to develop a comprehensive protocol to capture the dorsal and palmar side of the hand in functional positions across a large population using the Artec Leo. The development of this protocol considered the following elements; scanning technology, hand positions, hand support apparatuses, scanning platforms, and standardization across a population. Two functional hand positions, splayed and grasp, were selected based on clear visibility of the palmar side. The protocol and final scans contributed to a robust anthropometric database to improve the design, fit, and function of products for hands.

Keywords: Functional hand positions · Hand anthropometric database · 3D scanning

1 Introduction

The hand is one of the most complex parts of the human body and allows humans to perform a wide variety of tasks. Gloves and tools are designed to protect our hands during work and play. However, glove and tool manufacturers are limited by the available anthropometric hand data, often only using three measurements from a single hand position to inform their designs [1]. There are no established functional hand positions or methods to obtain data about other hand positions. Therefore, ill-fitting gloves and tools are a common problem for many trades of work, which increases the risk of hand injuries [1]. The hand is dynamic, and products designed using limited measurements do not accurately reflect real-world usage. It is critical to develop more robust anthropometric sizing systems of the hand in functional positions so that designers and engineers to create innovative designs for diverse populations.

Research was supported by the Minnesota Agricultural Experiment Station, and the National Institute of Food and Agriculture, U.S. Department of Agriculture, Hatch under MIN-53-088.

G. Di Bucchianico et al. (Eds.): AHFE 2020, AISC 1202, pp. 589–596, 2020.
https://doi.org/10.1007/978-3-030-51194-4_77

Advancements in 3D scanning technology have improved the ability to capture data of the hand [2], but minimal research has focused on capturing functional hand dimensions [1]. Previously developed methods for capturing functional hand positions are limited in scope, primarily capturing data from the dorsal side of the hand [3, 4]. Methods are needed to capture not only functional hand positions but to measure both the dorsal and palmar side of the hand. This data is key to understanding the complexities of the hand to build robust anthropometric sizing systems. The purpose of this study was to develop a comprehensive protocol to capture the dorsal and palmar side of the hand in two functional hand positions across a large population using the Artec Leo, a light-based 3D scanner.

2 Literature Review

The small size of the hand, coupled with shape change when performing tasks, makes anthropometric studies of functional hand positions challenging to conduct. Previous studies were instrumental in developing new methods to capture the complexities of the hand with 3D scans [5–7]. Functional hand positions and standardization methods across a population were key components to inform methodology development. As new scanning technologies improve the quality of data captured, they must be checked against existing methodologies to confirm viability. Therefore, previous 3D scanning protocols for hands were reviewed and tested.

2.1 Functional Hand Positions

The variability of hand shape and sizing impacts product design. The first studies to evaluate glove design against anthropometric measurements focused on the hand's dorsal side [3, 4]. As 3D scanning technology advanced, new methodologies to capture the hand were developed. Previous research developed a protocol to scan hands in three positions with the Occipital Structure Sensor [1]. While this study captured the palmar side of the hand, limitations in the scanning technology resulted in poor visibility of palmar details and landmarks. New methods need to be developed to capture the full complexity of functional hand positions. Our approach focused on capturing clear views of the dorsal and palmar sides of the hand in functional hand positions.

2.2 Standardizing Across a Population

In addition to capturing clear views of the palmar side of the hand in functional positions, considerations must be made to ensure the position is comparable across a population. A splayed hand position is standard in traditional anthropometric measurements of the hand [8, 9]. 3D scanning studies have previously standardized the splayed hand position through the use of scanning platforms, such as plexiglass frames, to support the hand [1, 7]. However, methods to capture other functional hand positions are not standardized across a population. Previous studies scanned participants' hands while gripping a glass cylinder, using the same sized glass for all participants [1, 7]. These studies were limited in that the final shape of the hand was dependent on the object being held. Support apparatus sizing must be standardized to account for the natural size of the participant's hand. Our work explored methods needed to standardize a functional grasp across a population, taking into account the unique size of each participant's hand.

3 Methods

The development of a new protocol to capture the dorsal and palmar side of the hand in functional hand positions considered the following elements; scanning technology, hand positions, hand support apparatuses, scanning platforms, and standardization across a population. These elements were tested and developed in conjunction as each one impacted the other, as seen in Fig. 1, to derive a new protocol.

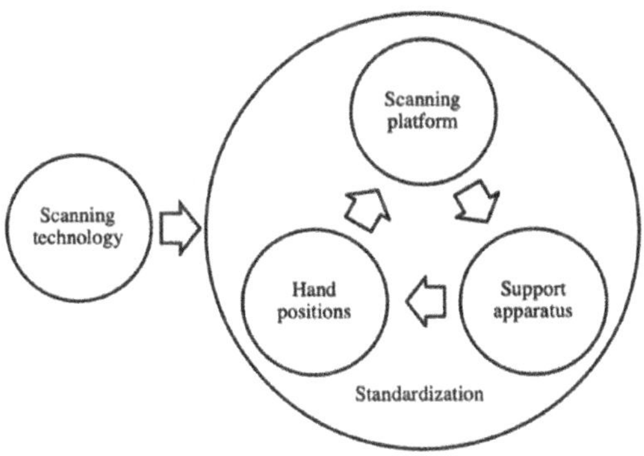

Fig. 1. Elements considered in the development of a new protocol.

3.1 Scanning Technology

First, the Artec Leo was selected as scanning technology for protocol development. The Artec Leo is a wireless handheld 3D scanner using structured light scanning technology. The scanner captures the shape, texture, and color of objects in high resolution [10]. The Artec Leo was selected due to its advanced capabilities and flexibility to capture objects of different sizes and varying detail.

Next, the Artec Leo was tested with an existing functional hand protocol in which the participant held a glass cylinder and was scanned with the Occipital Structure Sensor scanner [1]. However, the structured light of the Artec Leo refracted through the glass surface, resulting in scan artifacts and revealed a lack of compatibility with existing methods. Based on this initial testing, the protocol was redesigned with new hand positions and support apparatuses to improve compatibility with the new scanner.

3.2 Variables of Method Compatibility

Capturing both the palmar and dorsal sides of the hand is necessary to analyze dynamic hand change for product design. The palmar side of the hand is difficult to capture due to obstructed views in functional positions. Four elements were tested simultaneously to develop a new protocol with the Artec Leo; hand positions, support apparatuses, scanning platforms, and standardization across a population.

Hand positions and support apparatuses were evaluated for the visibility of the palmar side of the hand in scans and the ability to standardize support apparatuses across a population. Previous hand positions and support apparatuses [1] were tested as well as free-standing hand positions [9]. Lastly, new support apparatuses were developed and tested to complement additional functional hand positions.

The position of the participant's hand and arm angle is critical to obtaining scans of both sides of the hand. Testing began with existing scanning platforms based on previous studies [1, 7]. New scanning platforms were also developed and evaluated to assist with ease of scanning, ergonomics of the participant, and standardization of hand positions. The angle of the participant's arm on scanning platforms was also evaluated based on the ability to capture the space between the fingers and palmar side.

Lastly, testing considered the ability to standardize the hand positions and support apparatuses across a population. The variability in hand size across a population warrants the development of hand supports of various sizes to capture functional positions accurately. Selecting a support apparatus based on individual hand size achieves a natural hand shape compared to using a single sized support apparatus for an entire population as per previous studies [1]. The results of testing hand positions, support apparatuses, scanning platforms, and standardization is listed in Table 1.

4 Results

A protocol to capture 3D scans of functional hand positions using the Artec Leo was the result of this study. A 45° scanning platform with a plexiglass plane was selected to support the arm angle and hand position. Two functional hand positions, splayed and grasp, were selected based on clear visibility of the palmar side and landmarks in scans. Mesh cylindrical hand support apparatuses were developed and sized to standardize the functional grasp based on the participant's cylindrical grasp measurement. The final scans contributed to a robust anthropometric database to improve the design, fit, and function of products for hands. Details of the final positions, scanning platform and support apparatus can be found in Table 2 and images in Fig. 2.

4.1 Standardization

Standardizing the participant's position ensured the arm and wrist position were the same across all scans. First, the participant's elbow was placed on the scanning platform cross-beam with the forearm resting flat against the plexiglass plane. The height of the cross-beams was adjusted, so roughly 4″ of forearm extended past the end of the plexiglass plane.

Splayed Hand. To standardize the splayed hand position, the participant splayed their fingers apart as wide as comfortable to clearly view the space between the fingers, as previously tested [1, 7, 9]. Next, the participant's hand, fingers, and wrist were positioned flat against the clipboard, straightening out the fingers to comfortably create a flat surface. Once the wrist angle and hand position were aligned, the clipboard was removed, and the hand was scanned (Fig. 2a).

Table 1. Results of hand position testing

Functional hand position	Position and support	Result	Standardization
Splayed hand	Arm bent at 90° with the elbow on table. No forearm support	Issues with viewing angles and ergonomics of scanner. Microshifts in participant's hand	Arm angle
	45° scanning platform. Hand extended past plexiglass plane	Clear scans. Improved ergonomics for scanner and stability of participant's arm	Scanning platform angle, hand, arm, and finger position
Gripping a glass	Glass cylinder	Light refractions in scans caused poor visibility	Cost prohibitive, not easy to acquire
	Plexiglass cylinders	Light refractions in scans caused poor visibility	Inexpensive, available in multiple circumferences
Fist	45° scanning platform	Restricted views of palmar side. No delineation between fingers	Based on each participant's hand size
Pen	45° scanning platform. Pen	Restricted views of palmar side. Difficult to extract fingertip measurements	Same size pen across population
Turning a knob	45° scanning platform. Metal hoop held then removed for scan	Shifts in hand after hoop removed	Inexpensive, available in multiple circumferences
	45° scanning platform. Metal hoop in varying sizes	Clear view of palmar side. Not a natural position	Inexpensive, available in multiple circumferences
Grasp position	45° scanning platform. Plexiglass cylinder held then removed for scan	Shifts in hand after cylinder removed	Inexpensive, available in multiple circumferences
	45° scanning platform. Mesh cylinders were made from various material	Varying issues with visibility of palm, based on mesh thickness and opening size, issues with the amount of hand support. Difficult to shape. Improved by removing excess mesh for best visibility and hand support. Natural functional hand position	Inexpensive, possible based on ½ increments due to mesh opening sizes

Table 2. Details of final functional hand positions

Functional hand position	Position	Support apparatus
Both positions	Arm held at 45° for scanner ergonomics	45° scanning platform with plexiglass plane to support forearm modified from [5]
Splayed hand	Fully opened hand with 4″ of forearm extended past plexiglass plane	N/A
Grasp position	Grasping mesh cylinder with fingertips and thumb aligned along 2″ gap	Cylinder of 2 × 2 easy-to-form galvanized steel wire cloth with opening size of 0.459″. 2″ gap between fingertips of 1st and longest fingers. Excess mesh removed for visibility. Dipped in 1 coat Plasti Dip primer and 3 coats Plasti Dip white

Functional Grasp. The grasp apparatus design was sized for circumferences of 7″–10″ at ½″ intervals. Excess mesh was removed to improve visibility. Each cylindrical pattern was symmetrical across a midline for use by both hand dominances. Lastly, each apparatus was color-coded with tabs indicating the circumference and holding direction of the apparatus. The hand was positioned around the grasp apparatus, and the wrist was aligned using a clipboard, as per the splayed hand. The clipboard was removed, and the hand was scanned (Fig. 2c).

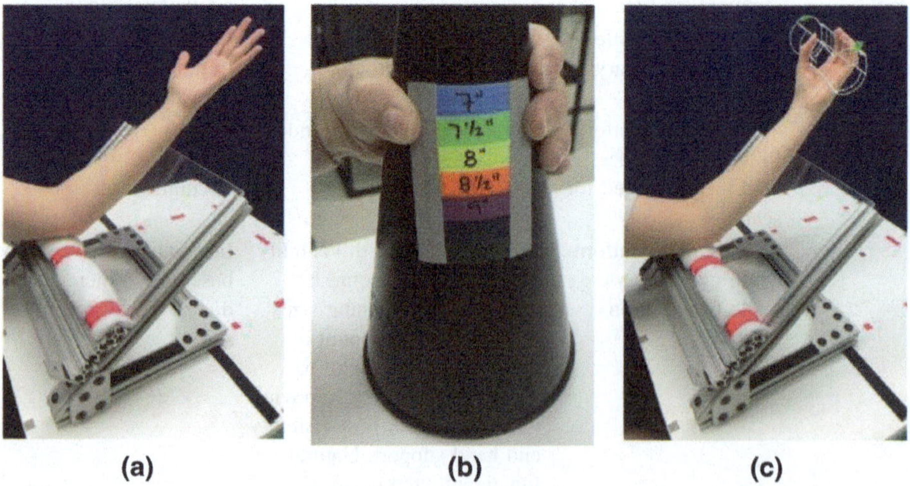

Fig. 2. (a) Splayed hand position. (b) Cylindrical grasp measurement tool. (c) Functional grasp position.

Cylindrical Grasp Measurement. A cylindrical grasp measurement tool [11] was developed to standardize the size of the functional grasp apparatus across a population. Circumferences were color-coded with tape corresponding to the grasp apparatuses' circumferences. Participants wrapped their dominant hand around the cone and adjusted their hand until their first and longest fingers created a 1½″–2″ gap between fingertips (Fig. 2b). This measurement allowed for the accurate selection of the grasp apparatus.

5 Discussion

The complexities of the hand require researchers to examine its form, shape, and measurements from multiple perspectives. Methods need to be continually improved to encourage more robust, inclusive datasets of the hand. This research contributes to the growing body of knowledge for 3D hand scanning and improves upon previous studies by offering a method that enables 3D hand scans of functional hand positions across a varied population and accounts for each individual's hand size. This procedure was developed and utilized for a large anthropometric survey of hands, where nearly 400 subjects participated. Future research includes developing methods to stabilize the hand to eliminate sway and incorporating improved landmarking methods to aid in analysis.

References

1. Griffin, L., Kim, N., Carufel, R., Sokolowski, S., Lee, H., Seifert, E.: Dimensions of the dynamic hand: implications for glove design, fit, and sizing. In: International Conference on Applied Human Factors and Ergonomics, pp. 38–48. Springer, Cham, July 2018
2. Li, Z., Chang, C.C., Dempsey, P.G., Ouyang, L., Duan, J.: Validation of a three-dimensional hand scanning and dimension extraction method with dimension data. Ergonomics **51**(11), 1672–1692 (2008)
3. Nasir, S.H., Troynikov, O., Watson, C.: Skin deformation behavior during hand movements and their impact on functional sports glove design. Procedia Eng. **112**, 92–97 (2015)
4. Williams, G., Hodder, S., Torrens, G., Hodgson, T.: The integration of anthropometry into computer aided design to manufacture and evaluate protective handwear. In: Human Factors in Design, Safety, and Management, Delft, pp. 275–288 (2005)
5. Griffin, L., Sokolowski, S., Seifert, E.: Process considerations in 3D hand anthropometric data collection. In: Proceedings of 3DBODY.TECH 2018-9th International Conference and Exhibition on 3D Body Scanning and Processing Technologies, Lugano, Switzerland, 16–17 October 2018, pp. 121–130 (2018)
6. Sokolowski, S.L., Griffin, L., Chandrasekhar, S.: Current technology landscape for collecting hand anthropometric data. In: Proceedings of 3DBODY.TECH 2018-9th International Conference and Exhibition on 3D Body Scanning and Processing Technologies, Lugano, Switzerland, 16–17 October 2018, pp. 142–153 (2018)
7. Griffin, L., Sokolowski, S., Lee, H., Seifert, E., Kim, N., Carufel, R.: Methods and tools for 3D measurement of hands and feet. In: Chung, W., Shin, C. (eds.) Advances in Interdisciplinary Practice in Industrial Design. AHFE 2018. Advances in Intelligent Systems and Computing, vol. 790, pp. 49–58. Springer, Cham (2018)

8. Robinette, K.M., Blackwell, S., Daanen, H., Boehmer, M., Fleming, S.: Civilian American and European Surface Anthropometry Resource (CAESAR), Final Report, vol. 1. Summary. Syntronics Inc Dayton OH (2002)
9. Seifert, E., Pokorny, C., Griffin, L., Holschuh, B.: Validation of artec eva for hand anthropometric data collection. In: Proceedings of the 2019 Conference of the International Textile and Apparel Association, Las Vegas, NV, USA, October 2019
10. Artec Leo Specifications. https://www.artec3d.com/
11. Vergara Monedero, M., Sancho Brú, J.L., Gracia Ibáñez, V., Pérez González, A.: An introductory study of common grasps used by adults during performance of activities of daily living. J. Hand Ther. **27**(3), 225–234 (2014)

Investigation of 3D Functional Grip Shape to Design Products for Dog Walking and Hiking

Susan L. Sokolowski(✉), Chrissy Bettencourt, and Jennifer Null

Sports Product Design, University of Oregon, 70 NW Couch Street, Portland, OR 97209, USA
{ssokolow, chrissyb, jcrawfo7}@uoregon.edu

Abstract. For product designers, the hand is a challenging part of the body to innovate products for. Anthropometrically, the availability of hand data relevant to civilians are sparse, as most data are from military personnel. In addition, data are typically of static hands which do not include task-related or functional knowledge, like grip shape. Inappropriate data could misinform product shape, design features, fit, performance, and safety. This paper explored how 3D functional hand grip shape could be captured from civilians and inform the design of products for dog walking and hiking.

Keywords: 3D functional grip · Equipment design · Design process

1 Introduction

The human hand is a complex system of bones, joints, muscles, ligaments and nerves that work in unison to perform tasks and interface with equipment. When using equipment, the hand must hold, grip and direct with ease and accuracy, so tasks are conducted safely. The hand is a challenging part of the body to innovate products for, as the availability of anthropometric data relevant to civilians' hands are sparse, as most data are collected from military personnel. In addition, data are typically of static hands which does not include task-related or functional knowledge. Inappropriate data could misinform product shape, design features, fit, performance, and safety. This paper explored how 3D functional hand grip shape could be captured from civilians and inform the design of products for dog walking and hiking.

2 Background

2.1 General Hand Anatomy and Biomechanics

As a part of the upper limb, the human hand is an intricate assemblage of bones, ligaments, joints, muscles, tendons, nerves, veins, skin, and nails, to allow for a wide array of tasks [1]. Comprised of 27 bones, 123 ligaments, 29 joints and 34 muscles, the structural foundation of the hand allows for grasping, pulling, abducting, adducting, pushing, and holding [1]. Nerves stem from three major branches: median, ulnar, and

G. Di Bucchianico et al. (Eds.): AHFE 2020, AISC 1202, pp. 597–604, 2020.
https://doi.org/10.1007/978-3-030-51194-4_78

radial, and an elaborate vascular network (arteries and veins) transport blood in and out of the hand [1, 2]. The skin and nails provide protection and is the surface from where anthropometric data can be collected [1].

2.2 Dog Walking

Dogs are walked for a variety of reasons, including elimination, mental and physical stimulation, socialization, and training [3]. Dog walking involves a person holding a leash in either hand or attached to their body by a clip, and the other end either being lassoed around the dog's neck or clipped to a harness or collar at the dog's neck. Dog owners and professional dog walkers may use any leash, harness, or collar of their choosing, and they may hold the leash in any way they wish. However, animal shelter volunteers must abide by a set of rules to ensure the safety of both the volunteer and the dog. For example, the Oregon Humane Society (OHS) in Portland, Oregon requires that all volunteers use the approved Be Bop USA slip lead leash [4]. Volunteers must use the "safety hold," (Fig. 1), where the leash loops around the walker's thumb and is gripped by four fingers in the palm of the hand. The hold ensures that the leash is secured in the hand in case the dog pulls on the leash, but it can be easily dropped in case of an emergency.

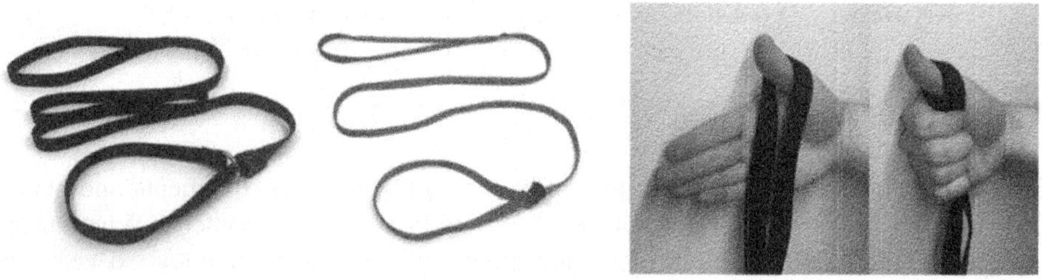

Fig. 1. Large *(left)* and small *(center) Be Bop USA* slip lead dog leashes used by OHS volunteers. The "safety hold" is required by all shelter volunteers walking dogs at the OHS *(right)*.

The OHS shelter may house up to 130 dogs, that are each walked a minimum of two times per day by staff and volunteers [5]. Nearly 2,000 volunteers participate in dog walking activities at the OHS [6]. It is estimated that of the 78 million dogs owned in the US, 23% were obtained from an animal shelter or humane society, and approximately 3.3 million dogs enter one of the 3,500 animal shelters in the US every year [7, 8].

Unlike dog owners, volunteer dog walkers are unfamiliar with the dogs they are walking. The dogs may be unpredictable, pulling hard on the leash if startled or excited. Behavior specialist, Tanya Roberts, describes how many volunteers complain about fatigue in their hands after walking dogs and difficulty gripping such a thin leash [4]. Softer, easier to grip leashes are often requested by volunteers [4].

2.3 Hiking

Hiking is defined as taking a long walk, typically along a trail or through the woods. Many individuals choose to hike with hiking poles [9]. Hiking (or trekking) poles are commonly used to prevent injury or provide stability and assistance on steep uphill and downhill walks. A hiking pole consists of a hand grip (often accompanied by a hand strap), a stick section made of wood, aluminum or other light weight material, and tips at ground level made of metal with optional rubber covers that are used depending on the ground conditions (e.g., rocks, ice, pavement).

Hiking poles come in two styles: adjustable or fixed height. To determine proper adjustable pole fit, the hiker should stand with their elbows at a 90-degree angle and adjusts the poles to the height where the tips are on the ground next to their feet. It is recommended to adjust the poles slightly shorter when going uphill and longer when going downhill [10]. Despite the number of hiking poles available on the market, there is little difference in shape and function of the hand grips (Fig. 2). The grips also do not consider the change of hand position when hiking on a flat surface versus uphill or downhill.

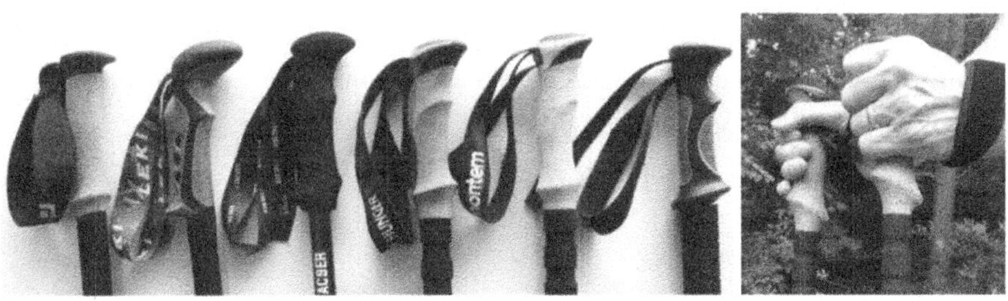

Fig. 2. Hiking pole grips vary slightly in design *(left)* [11] and are modeled only for flat surface hiking *(right)* [9].

Fig. 3. Steep downhill hiking hand positions [13–15].

The lack of support for changing hand positions is particularly noticeable when the angle of decline becomes too steep to continue using the grip as intended (Fig. 3). When hiking downhill the poles are used as a breaking-system, which means higher forces are being exerted onto the arms, wrists, and hands [10]. When the hand grip does not appropriately fit the hand, pain can be caused by unnecessary pressure, friction on the palm and irritation of the ulnar nerve [12].

3 Innovation Methodology

As discussed in the background, both dog walking and hiking equipment must fit a wide range of hand sizes and take on a variety of tasks (e.g., holding, pulling, and push-off). By using the *Performance Hand Wear and Tools Product Innovation Framework*, developed by Sokolowski, graduate students were tasked to develop a method of collecting 3D functional hand grip shape to design better fitting and performing equipment [16]. By working with an in-class convenience sample, procedures were developed where subjects individually modeled 3D functional grip forms with "air-dry clay." The clay forms were 3D scanned and compared digitally to sculpt "average" forms. The refined forms were 3D printed and integrated with other product parts that each student designed for their individual user needs. To complete the project, technical drawings, product photography and digital presentations were created.

4 Data Collection and Design Results

4.1 Dog Walking Results

For the dog walking project, an ergonomic leash handle was created that considered safety and security needed by volunteers at OHS. Subjects squeezed the "air-dry clay" in the "safety hold" position (Fig. 4). The forms were then 3D scanned with a *Structure* scanner and overlaid to create an average form. The average form was then digitally sculpted using *AutoDesk Meshmixer*, refined in *Rhinoceros*, and 3D printed to confirm shape and proportion (Fig. 5). While the finger shape varied significantly from person to person, the palm shape was easily combined to create a form that conformed to each subject's hand shape (Fig. 6).

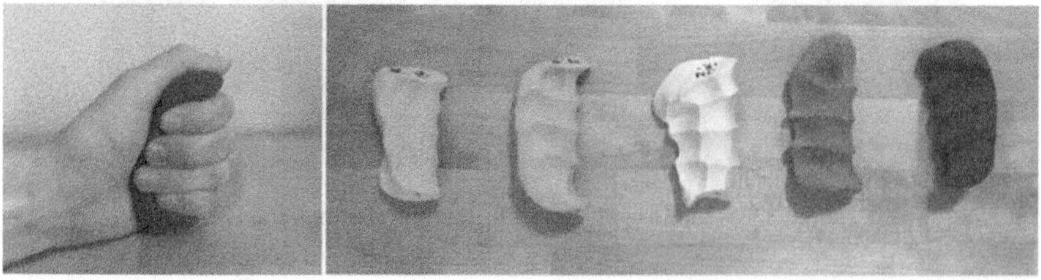

Fig. 4. Subjects created personal grip forms by squeezing "air-dry clay."

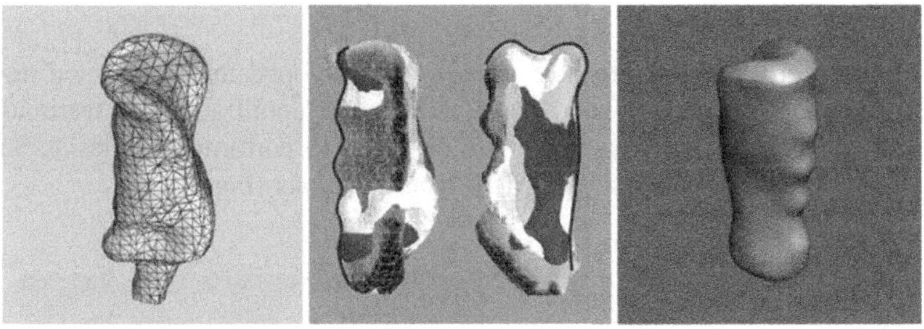

Fig. 5. 3D scans of clay forms *(left)* were overlaid *(center)* to create an average form *(right)*.

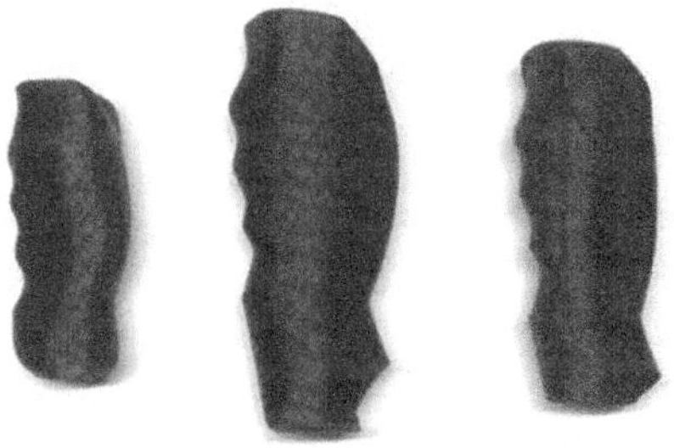

Fig. 6. 3D printed prototypes show refined shape and proportion.

The final leash design called the "Bobbie of Silverton," was inspired by the Japanese lean manufacturing term, "poka yoke," which means mistake-proofing [17]. The leash grip (Fig. 7), accommodates and encourages the "safety hold", while discouraging poor holding mechanics where the walker wraps the leash around their wrist. The leash fits comfortably in the right or left hand and can be easily transferred between hands.

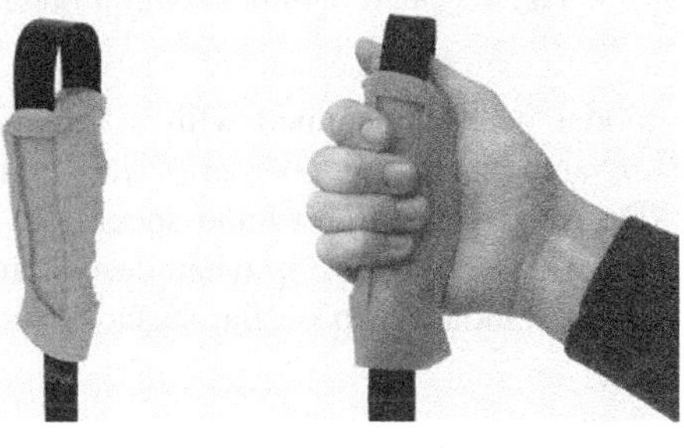

Fig. 7. "Bobbie of Silverton" dog leash, designed for shelter volunteers at the OHS.

4.2 Hiking Results

For the hiking project, trekking pole grips were developed that considered downhill activities or the steep decline hand position. Flat and uphill activities were not considered as the standard grips available are designed to perform these tasks. Subjects were asked to create clay imprints of both hand positions (Fig. 8).

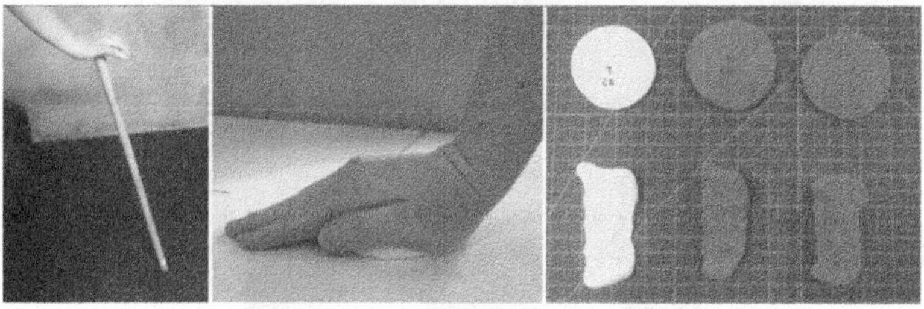

Fig. 8. Subject replicating downhill break position using a 1.25″ dowel (*left*) and creating palm imprint with clay (*center*). Palm imprint and standard hand grip clay forms (*right*).

The individual clay palm and standard grip imprints were then combined to create one unified new grip design (Fig. 9).

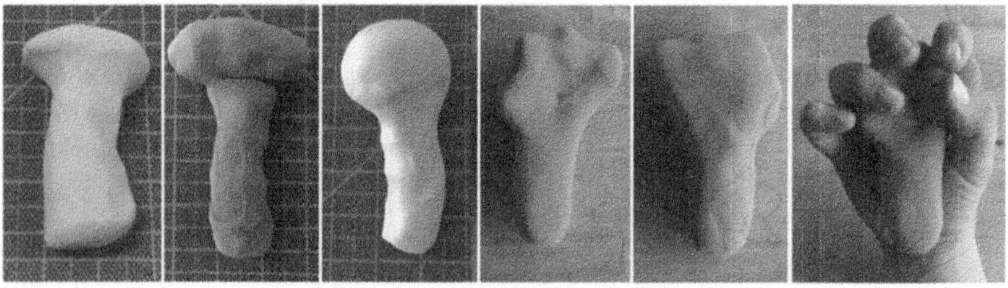

Fig. 9. Unification of clay grip imprints.

The final clay model was 3D scanned with a *Structure* scanner, sculpted using *AutoDesk Meshmixer,* and the hiking pole grip was created using *Rhinoceros*. The design, called the "Descent," is right/left hand specific, where the thumb, index, and middle fingers are engaged to hold the grip when descending a steep decline. The base of the grip is used like a standard grip on flat surfaces and uphill inclines (Fig. 10).

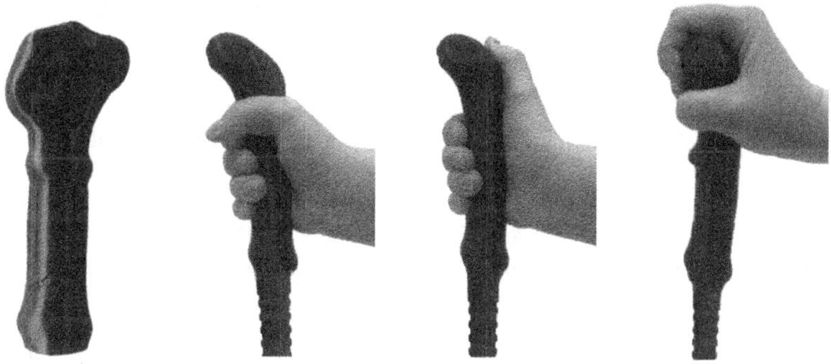

Fig. 10. The "Descent" hiking pole grip.

5 Conclusion

By using 3D clay forms and scanning, the graduate students were quickly able to define grip shape and functional details needed for performance equipment. Future work could explore other activities that require equipment used with the hands and/or include a larger sample size to statistically define forms and sizing across a population.

References

1. Sokolowski, S., Griffin, L., Carufel, R., Kim, N.: Drawing hands for glove design: does the data match-up? In: Chung, W., Shin, C.S. (eds.) Advances in Interdisciplinary Practice in Industrial Design, pp. 68–77. Springer, Cham (2019)
2. LaBat, K.L., Ryan, K.S.: Human Body: A Wearable Product Designer's Guide. CRC Press, Boca Raton (2019)
3. The Animal Foundation: The importance of walking your dog. https://animalfoundation.com/whats-going-on/blog/importance-walking-your-dog
4. Roberts, T.: Personal Communication (2019)
5. Oregon Humane Society: Walking the dogs. https://www.oregonhumane.org/walking-the-dogs
6. Oregon Humane Society: Volunteer at OHS. https://www.oregonhumane.org/get-involved/volunteer
7. ASPCA: Pet statistics. https://www.aspca.org/animal-homelessness/shelter-intake-and-surrender/pet-statistics
8. The Humane Society of the United States: Pets by the numbers. https://www.humanesociety.org/resources/pets-numbers
9. Tobalski, J.: Wandering rose travels - Five reasons you should consider hiking poles. https://www.wanderingrosetravels.com/trip-activities/hiking/five-reasons-consider-hiking-poles/
10. Bumgardner, W., Fogoros, R.N.: How to use trekking poles for stability. https://www.verywellfit.com/how-to-use-hiking-and-trekking-poles-for-stability-4145313
11. Alexa, R.: Tactical - What are the best lightweight trekking poles for hiking and backpacking. https://www.tactical.com/best-lightweight-trekking-poles-hiking-backpacking/
12. Nall, R., Reed-Guy, L., Morrison, W.: Ulnar nerve palsy (disfunction). https://www.healthline.com/health/ulnar-nerve-dysfunction

13. Wenatchee outdoors: trekking poles – The why and how. https://wenatcheeoutdoors.org/2019/05/20/trekking-poles-the-why-the-how/
14. Holmes, J.: Cool of the wild - How to use trekking poles. https://coolofthewild.com/how-to-use-trekking-poles/
15. Johnson, D.: Backpacking light - Gossamer gear lighttrek 3 trekking pole review. https://backpackinglight.com/gg_lightrek3/
16. Sokolowski, S.L.: The Development of a Performance Hand Wear and Tools Product Innovation Framework. Fashion and Textiles, January 2020
17. Lin, A.: Lean in real life - 10 examples of Poka Yoke in daily life. https://tulip.co/blog/lean-manufacturing/poka-yoke-examples-everyday-life/

Human Factors Design of Digital Information Systems for Inspection Body (IB) in Industry 4.0 Era

Manutchanok Jongprasithporn[1(✉)], Nantakrit Yodpijit[2], Danitar Tontoh[2], and Teppakorn Sittiwanchai[2]

[1] Department of Industrial Engineering, Faculty of Engineering, King Mongkut's Institute of Technology Ladkrabang, Bangkok, Thailand
manutchanok.jo@kmitl.ac.th, mjongpra@gmail.com

[2] Center for Innovation in Human Factors Engineering and Ergonomics (CIHFE^2), Department of Industrial Engineering, Faculty of Engineering, King Mongkut's University of Technology North Bangkok, Bangkok, Thailand

Abstract. Industry 4.0 highly considers new industrial products and services platforms where various emerging technologies are combining to implement for digital solutions. Digital transformation plays a significant role in the disruption of the traditional information systems and drives the business to digitalization. The Inspection Body (IB) conducts technical assessments, generally by providing factory/business inspection services on behalf of private customers or their parent organizations to provide them with industrial product certification and/or information on compliance with legislation, standards, specifications, and contractual commitments. Importantly, data accuracy, traceability, and confidentiality controls are critical for IB operations where the digital information systems can significantly enhance the planning, managing, and monitoring of the IB's Management Information System (MIS). This paper presents both theoretical and practical applications of human factors product design and development to make a useful and easy to use digital information systems for IB. Major methods for human factors design includes the focus of potential users, the empirical measurement of usage, and the iterative design where the simulated, prototype, and real models are created. Limitations and future work for human factors design for digital information systems in other industrial businesses are also discussed.

Keywords: Human factors design · Digital information systems · Inspection Body (IB) · Management Information System (MIS)

1 Introduction

Industry 4.0 is mostly considered as the fourth industrial revolution as the trend towards automation and data exchange in industrial products and services platforms [1]. In fact, Industry 4.0 comprises areas that are not categorized as an industry, such as smart cities, for instance. Various emerging technologies and principle concepts

G. Di Bucchianico et al. (Eds.): AHFE 2020, AISC 1202, pp. 605–612, 2020.
https://doi.org/10.1007/978-3-030-51194-4_79

boosting businesses to identify and implementing Industry 4.0 solutions are currently working together to implement for digital solutions, as follows [2]:

1. Interconnection is the ability of systems to connect and communicate with each other via networking infrastructure.
2. Information transparency is defined by the business information in both financial and non-financial information. The information transparency encourages the stakeholders' trust in business operations. Insufficient information provision can be misperceived and may result in the unjustified expectations of interested parties [3].
3. Technical assistance is an ability of systems to support human operators in information visualizing and conducting the tasks that unsafe, unpleasant, and exhaustive for humans.
4. Decentralized approach is the ability to perform, and the independent decision is possible to be made in the digital systems. Exclusively in the case of exceptions, interferences and conflicts are escalated to the upper human operator level.

The digital information systems support the operation of Industry 4.0 in both off-site and on-site in which all business documents can systematically retain, maintain, and retrieve documents in real-time. The novel system is designed with an emphasis on the requirement of stakeholders proposing at the longevity data storage and ease of document access [4, 5].

IB is a private organization for conformity evaluation of products/processes either based on defined or general requirements. This evaluation aims at describing the functionality and any concerned characteristics of the investigated target, such as safety, reliability, capability, etc. The IB is accredited by the Accreditation Body (AB) under ISO/IEC 17020 [6]. In connection with AB, the IB can be classified into three types, A, B, and C. The type determines the relationship between the IB and the investigated target.

Since the importantly, data accuracy, traceability, and confidentiality controls are critical for IB operations where the principle concepts of Industry 4.0 (interconnection, information transparency, technical assistance, and decentralized) and digital document information systems can significantly enhance the planning, managing, and monitoring of the IB. The purpose of this research project is to present the proposed method focusing on theoretical and practical approaches of human factors product design and the development of digital information systems of IB.

2 Methodology

The ISO 9241-210 [7] provides specifications and guidance for human-centered design principles and activities throughout the life cycle of computer based interactive systems. Technical human factors and ergonomics issues are addressed in this ISO to conceive the understanding of relevance and importance of the overall design process. This standard is concerned with approaches in which both hardware and software elements of interactive systems can improve the human-system interaction of such systems are referred to as products, systems, or services. The design principles under this standard are as follows.

1. The understanding of users (stakeholder), tasks, and environments
2. The involvement of users.
3. The user-centered evaluation is a tool to drive and refine the design.
4. The design team comprises multidisciplinary experiences and aspects.

This method adopts the design principles and activities of ISO 9241 for the design and development of the Industry 4.0 digital information systems of IB. The methodology consists of four steps, including 1. understanding and specifying the context of use, 2. specifying the user requirements, 3. producing design solutions, as presented in Fig. 1. Since the suitable design cannot typically be accomplished without iteration, repeating a sequence of steps until the desired outcome is obtained can also reduce the uncertainty during the design of the system progressively, the iteration of the proposed represents by the dashed line in Fig. 1.

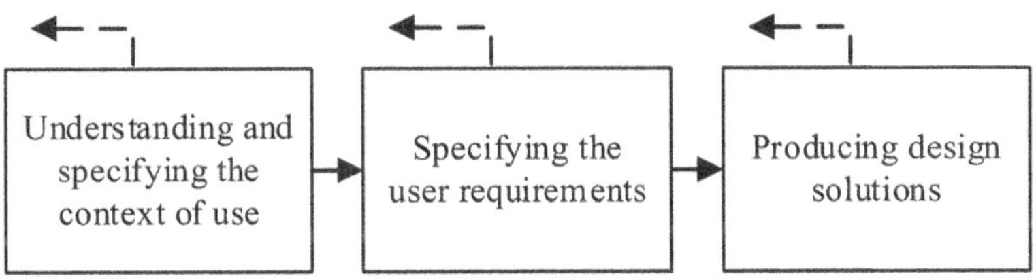

Fig. 1. Design framework of digital information systems

2.1 Understanding and Specifying the Context of Use

The context of use is defined by the users, tasks, organizational environment, technical environment, and physical environment from the system of IB. This process collects and reviews the current manual systems of IB. The outcomes reveal the needs, issues, and limitations of the users on the existing operational system. Moreover, feedback from customer and customer service records is also taken into consideration at this stage. The result of the context of use can represent the intention of the system and provide the scope with detail to support the requirements, design, and evaluation activities. The context of use comprises the following elements:

1. The description and demographic of stakeholder.
2. The goals of the organization and the operational task of users.
3. The identified relevant environments of the system.

2.2 Specifying the User Requirements

Consistently specifying user requirements can support the design and development of interactive systems. The user requirements specify the required interaction to achieve the intended outcomes of any system interaction and user-related quality requirements in terms of effectiveness, efficiency, and satisfaction. This process aims at identifying user needs and specifying the functional and other requirements for the digital information systems of IB. This activity creates a clear statement of user needs regarding the context of use and business objectives in previous, in order to comply with human-centered design principles. Since this system will become available in the main operation of IB and replace the current hand-operated system, the design of the operational process must include all stakeholders to optimize both the organizational and technical systems. This step uses the semi-structured interviews of all stakeholders and the review of relevant documents to investigate the user requirements. The guidelines for the interviews are as follows:

A: Role and responsibility on the digital information systems
B: Required Features
C: Definition of "Good systems"
D: The value-added functions of systems
E: Desired performance of systems
F: Acceptable downtime for the maintenance period
G: Average tasks per day

2.3 Producing Design Solutions

The human-centered design intends to deliver a pleasant experience to the users with consideration along the design process, where design solutions are the key to achieve this aspiration. This step establishes the drawing on the context of use and state of the art of the developed systems.

3 Results

3.1 Understanding and Specifying the Context of Use

Description and Demographic of Stakeholders: The case study is the accredited IB under ISO/IEC 17020. The stakeholder of digital information systems consists of experienced operators and executives from the different sections include 1. Customer Service, 2. Technical Expert, 3. Factory Inspection, 4. Financing and Accounting, and 5. Top Management. Figure 2 illustrates the organization chart of IB.

Goals of the Organization: The operation focuses on neutrality, compliance, confidentiality, and customer experience to provides the factory inspection services and deliver audit reports to the customer.

Relevant Environments of the System:
Technical environment:

Hardware: For users including desktop, Laptop with networking, and internet connectivity. Cloud services provider for IB's infrastructure.

Software: Any operating systems with web browsers.

Materials: Documents listed in quality management systems and other financing and accounting documents.

Location: In the office, remote work, and on-site customer's factory

Social and Cultural Aspects:

Organizational Structure: Functional structure

Work Practices: Ethics, good logical, active self-management, written communication, critical & Creative thinking, resourcefulness, diplomacy skills

Attitudes: Passionate continuous improvement, ability to adapt to many environments, strong desire for efficiency of organization

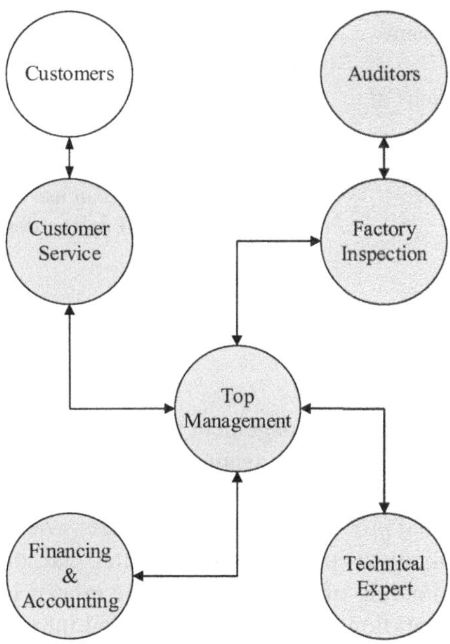

Fig. 2. Organization chart of IB

3.2 Specifying the User Requirements

Identifying User and Other Stakeholder Needs: The interview affects the user requirements. Therefore, the interview focuses on one group is user operation role; the conversation uses around 30 min as the open-ended question. The participants are the experienced operators and executives of IB with experience in 3–15 years in IB business. Table 1 presents the acquired user requirement.

Table 1. The user requirements.

Question	Answer
A: Role and responsibility on the digital information systems	New users of IB corporations, who coordinates and proceed business a customer. They are use paper and cellphones
B: Required Features	Job administration, communication within teams, and documents and records management
C: Definition of "Good systems"	Good traceability, real-time notification, and flexible files attachment and download
D: The value-added functions of systems	Provide the step-by-step user manual with accurate detail of each functions, images and related references
E: Desired performance of systems	Can access the system everywhere and every time
F: Acceptable downtime for maintenance period	One day. The team can use the alternate solution such as cooperate email during downtime
G: Average tasks per day	1–10 tasks

3.3 Producing Design Solutions

The proposed digital information management systems of IB emphasizes the essential features, including task administration, team communication, and document management. The concepts of these digital information systems are based on delivering benefits of data flow and operational processes. The process starts when the customer fills in the job request form and submit to customer service. After that, the systems will assist in implementing and coordinating of each step with the related individuals/departments. Figure 3 illustrates the workflow of the proposed digital information systems for IB.

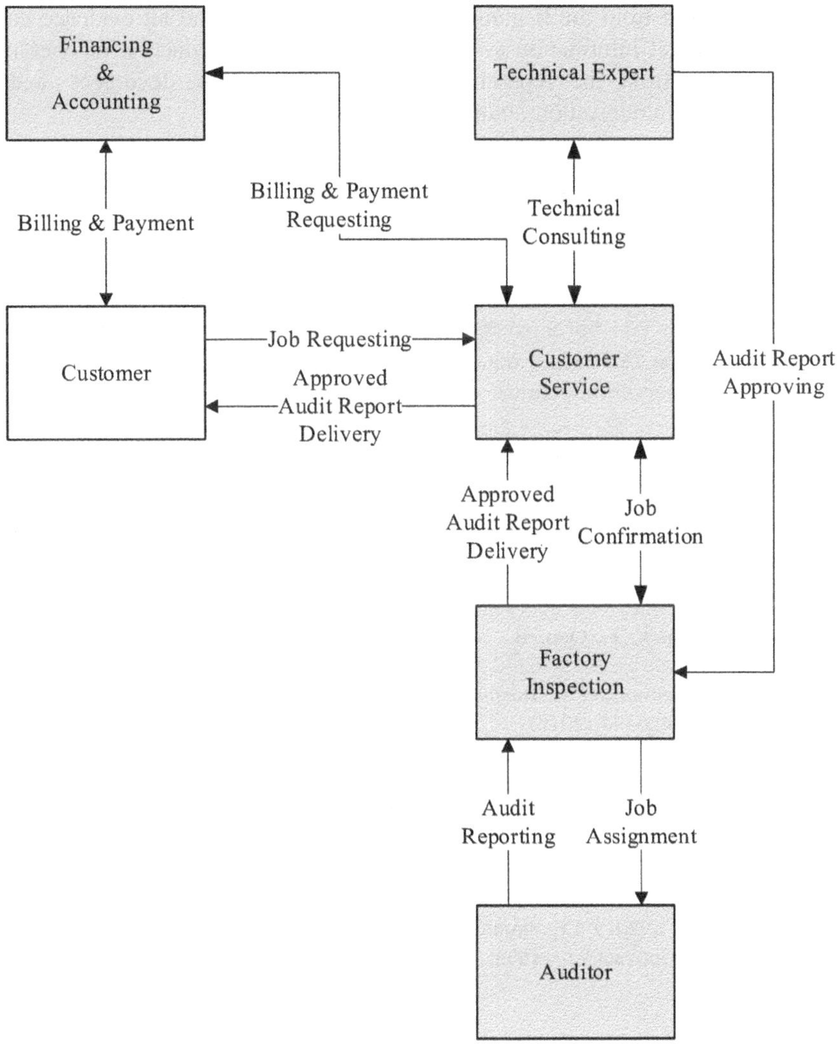

Fig. 3. Business operation flow of the systems

4 Conclusions and Discussions

This paper presents the implementation of ISO 9241-210 on IB's digital information systems. The essential results from the understanding and specifying the user requirements are synthesized and used for producing human factors product design solutions. The effective implementation of digital information management can support the operation and lead to cost-saving solution of the organization. This can make every job to be visualized from the beginning until finished work and all evidence can be followed. The digital information systems also promote the reduction of operational errors, especially from the repetitive process and overload document, and the improvement of communication ability in the organization.

5 Limitations and Future Research

The evaluation and modification after the first implementation of digital information systems for IB need to be taken in the near future. It is expecting that the digital information systems can be expanded to support the more complex operation with massive data volume. Advanced database, infrastructure, and networking technology will be suitable to serve these future demands.

References

1. Marr, B.: Why Everyone Must Get Ready for The 4th Industrial Revolution. https://www.forbes.com/sites/bernardmarr/2016/04/05/why-everyone-must-get-ready-for-4th-industrial-revolution
2. Hermann, M., Pentek, T., Otto, B.: Design Principles for Industrie 4.0 Scenarios, Hawaii, USA (2016)
3. Kundeliene, K., Leitoniene, S.: Business Information Transparency: Causes and Evaluation Possibilities, pp. 340–344 (2015)
4. Hesham, S.A., Issa, M.B., Maha, D.A.: Investigation of Document Management Systems in Small Size Construction Companies in Jordan, pp. 3–9 (2017)
5. Jones, S.: eGovernment document management system: a case analysis of risk and reward. Int. J. Inform. Manage. **32**, 396–400 (2012)
6. ISO.: ISO/IEC 17020:2012 Conformity assessment - Requirements for the operation of various types of bodies performing inspection. https://www.iso.org/standard/52994.html
7. ISO.: ISO 9241-210:2019 Ergonomics of human-system interaction - Part 210: Human-centred design for interactive systems, https://www.iso.org/standard/77520.html

Kansei Engineering

Pilot Study of Kansei Human Machine Interface for Prevention System of Unintended Acceleration

Toshio Tsuchiya(✉), Kanon Watanabe, Mizuki Hirano, Kotaro Kuma, and Shintaro Wakamatsu

Faculty of Economics, Shimonoseki City University, 2-1-1 Daigakucho, Shimonoseki, Yamaguchi 751 0834, Japan
tsuchiya@shimonoseki-cu.ac.jp, {s172063, s170149, s171068, s171205}@eco.shimonoseki-cu.ac.jp

Abstract. This pilot study focuses on Unintended Acceleration which is a driver's experience full and unexpected acceleration with a pedal misapplication. The causes of such accidents have been investigated in a great deal of engineering reports from various viewpoints of mechanical or operational contexts. The innovation of human machine interface is expected to have a potential to contribute for reducing such accidents. The primary purpose of the previous studies is to identify the misapplications which causes the accidents by unintended acceleration and to develop the accident prevention mechanism by hardware and software approaches. The existing researches contributed to develop the suppressing the accident caused by a misapplication. However, the emotional aspect associated with driver's desire to the active safety has not been noticed enough on the studies. This study started by motivation to consider the possibility of use the active safety system which introduced Kansei Interface. In order to solve the problem, the overview of unintended acceleration will be surveyed at the beginning of this study. Then, the existing prevention systems will be examined in the meaning of effectiveness and limitation. Kansei interface in order to prevent the accidents will be considered based on the survey. To develop the prevention system using Kansei interface as the final objective of our study, we will propose a rapid prototyping method to simulate the pedaling error.

Keywords: Kansei engineering · Unintended acceleration · Human machine interface

1 Introduction

Accidents of Unintended Acceleration (UA) by pedaling misapplications and as phenomenon caused by erroneously at the time to drive a car becomes a serious issue in Japan [1]. The Japanese Minister of Land, Infrastructure and Transport announced that the installation of automatic brakes will be made completely obligatory for domestic new cars from November, 2021 [2]. EU is, prior to this, going to make obligatory introduction of collision damage reducing devices in Europe from early 2020. The passive safety systems

G. Di Bucchianico et al. (Eds.): AHFE 2020, AISC 1202, pp. 615–620, 2020.
https://doi.org/10.1007/978-3-030-51194-4_80

such as the automatic driving systems, the automatic brakes, the suppressing accelerations and the gear lock mechanisms, managed for avoiding the accident are still not spread, because it must be implemented into the mechanical system of the vehicle. In fact, the accident of the pedal misapplication changes in around from 6,000 to 7,000 cases a year in Japan [3]. In addition, the accidents by UA cannot be completely prevented because the safe equipment is not perfect in general. This study intends to develop a system to prevent serious accidents by active safety mechanisms composed by smart sensors and devices. This system would be implemented without a mechanical device in the drive system of the automobile, then it cannot exactly prevent the accident by braking operation. It is however expected that the interface supports the driver by oneself to control appropriate operations.

Regarding with design of the systems, Kansei interface is introduced in order that the driver turns attention into the system intentionally. A driving support system generally tends to be ignored because it is unavoidable to annoy the driver and habituation. The main purpose of the system is to prevent the accident caused by unintended acceleration, at same time it attempts to turn the driver's attention into the response of the system by using Kansei interface. This study aims to introduce human machine self and social interfaces as Kansei interface [4]. The human machine self-interface is that the user promotes conversations with oneself through the interface. It is a characteristic to be caused by the internal experience of the user including a sense of unity by sympathizing with the confirmation of feelings to the value of the product and feelings such as accomplishment, fulfilment, simulated or intellectual experience. The human machine social interface promotes the conversation with the people that constitutes society through interface. The interface is representative of applications concerned with social medias. The prototype of the system is composed as a rapid prototyping technology by utilizing smart sensors and devices.

2 Advanced Safety Vehicle

An advanced safety vehicle (ASV) is the car which implemented a system supporting the safe driving using advanced technologies. ASV promotion plan by The Japanese Minister of Land, Infrastructure and Transport announced three fundamental principles and a direction of the approach of the driving support for the practical use, the development and the spread promotion. The principals are shown as for just supporting to the driver (not for automatic driving), ensuring acceptability of the driver and ensuring acceptability of the society. The interesting point is the acceptability of the driver that is to need to consider usability and reliable in the interface of the system. It can be said that ASV should be required an emotional level interface to be accepted by the driver not only utilities. This is a prime motivation to star this study.

Automobile manufacturers and equipment makers have provided the advanced safety devices such as emergency braking systems and suppressing acceleration devices [5]. Besides the devices, intelligent driving support systems such as Advanced Driver Assistance Systems (ADAS) are also indicated significantly effective to the accident prevention. In these systems, the emergency braking systems and the suppressing acceleration devices are associated with the preventing accidents by pedal misapplication. Installation of the systems for the new car in Japan is 45.5% and 35.9% respectively in 2015, and it is expected to increase currently. The accident caused by the pedal misapplication is not reduced in spite of the diffusing the prevention systems.

Practical Living Information Online Network System (PIO-NET) received 142 queries about ASV, 119(83.8%) of them were related to emergency braking systems. There were the examples that accidents caused the emergency brake didn't operate and that brakes operated suddenly without obstacle [5]. Also, malfunctions of ASV were reported as variable of cases. ASV essentially purposes to reduce damage by accidents, so it is not for preventing accidents by every situation. For example, an emergency braking system operates when it detected an obstacle and the collision could not be avoided. The first and the most important operation of the system is to inform the risk of the accident to the driver, because conditions of working the prevention operation is not consistent and deferent between equipment manufacturers. It can be said that the current ASV just works when the driver could not avoid the accident and is not to prevent completely. We can see therefore that the interface to inform the appropriate circumstance and the foreseeing about the risk of the accident.

3 Unintended Acceleration by Pedal Misapplication

Unintended acceleration, which is generally caused by pedal misapplication, typically begins when the driver first enters the car and is initializing a driving cycle while shifting from Park to a drive gear (Drive or Reverse) [6]. Pedal misapplications - where the right foot contacts the accelerator instead of the brake that was intended - have been linked to these accidents [1]. The pedal application error analysis by NHTSA resulted regarding with driver age as a U-shaped function shows significant formed by 16 to 20 age group and over 76 age group from both the North Carolina database and the news media reports [7]. NHTSA also pointed the overrepresentation of female drivers due to shorter stature as one of the reasons to increase the likelihood of a pedal misapplication. Schmidt examined the North Carolina Police Accident Report database to identify the contributing factors by pedal disapplication [1]. It is considered that UA by pedal misapplication is caused by variable errors in hitting the pedal and biases in the aiming direction. The factors that such accidents occur is variously conceivable, but in any case, the foot placement errors are suggested about a direct factor of the pedal misapplication.

To explore accident causes by pedal misapplications can contribute to develop prevention systems for the problem. Institute for Traffic Accident Research and Data Analysis (ITARDA) reported an analysis of driving behaviors of elderly drivers that carry a high risk of leading to accidents by pedal misapplications from literatures and original database [8]. This report summarized causes of the accidents due to pedal misapplication by elderly drivers have the following characteristics, location, maneuver, object of damage, getting flustered and sitting posture. The report concluded from the research to prevent pedal misapplication accidents as; to use the creep phenomenon when driving in parking lots and similar locations, to be cautious of pedestrians and vehicles emerging from unexpected places in parking lots and similar locations, to check on one's driving once again owing to the decline in their physical abilities and flexibility and concentrate on driving.

4 Kansei Interface to Prevent Accident by UA

From the investigations to consider and to presume accident patterns, we can be certain that a prevention system of pedal misapplications would be able to be composed into in-vehicle interface implemented with accident prediction mechanisms. This induces us into a consideration of Kansei interface to prevent the accidents by pedal misapplications. The points to resolve the problem would be represented as methods to inform appropriate signals to the driver and to predict the status of risk. The problem of prediction would be given by the previous studies to investigate the accident situations involving UA. This study considers the problem of informing to the driver.

The vehicle manufacturers provide several in-vehicle interfaces to notify driving information. The interfaces usually employ multimodal devices combined with display, sound or voice. The combination of the modals is used according to the informing situation. EyeSight provided SUBARU is an integrated ASV system to support safety drive [9]. The system includes the collision avoidance functions, the touring assist utilities and the warning and informing interface. The interface includes sound and visual alarm to notify case of unstable driving, lane departure and preceding car departure. The notification is however equipped without the accident prediction. It would also have difficulty to implement the accident prevent notification to the interface such as habituation. An accident occurs not often. So, the user doesn't be informed during driving the car eventually. EyeSight employs the situations as unstable driving, lane departure and preceding car departure which are very often in daily driving. On the other hand, UA by pedal misapplication is almost not happened. It is a most important difficulty to compose the interface into an accident prevention mechanism.

A useful solution is to introduce interactive interfaces to promote to attract driver's attention. Suzuki has developed Multi-information display for a model of K-car. On the interface, a character of the rabbit says hello When the engine is started (Fig, 1). The interface tells the driver about various vehicle information, including careless mistakes of the driver and fuel consumption meter, by animations, voice and sounds. It also plays melodies and messages associated on the day of celebrations, birthdays and seasonal events. This interactive interface gives meaningful suggestions in order to solve the issue of habituation.

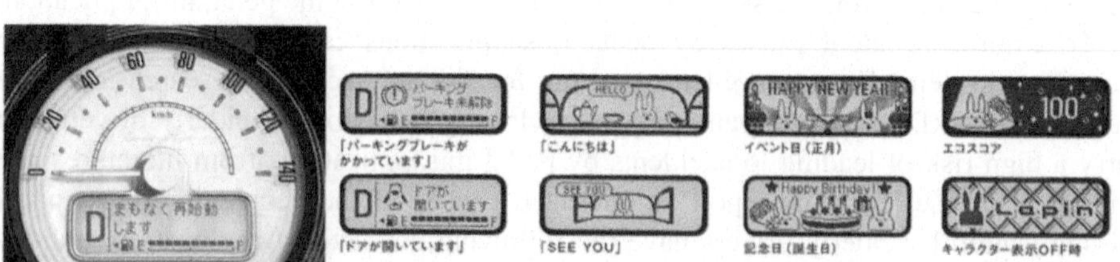

Fig. 1. Multi-information display produced by SUZUKI for a model of K-car. https://www.suzuki.co.jp/car/lapin/interior/

We are considering that Kansei interface is also a useful methodology. Kansei interface is understood as comprising emotional aspects directing to compose it like intuitive and desirable to use [4]. Kansei interface has several strategies to induce users' attention into the system. The main concept is defined as interface like to use unintentionally which is basically satisfied easy to see, easy to understand and easy to use. Inoue et al. introduced the interface elements to induce like to use as Human-machine Interface Caused by Itself, Human-machine Self Interface and Human-machine Social Interface. The characteristics of the interface are as follows. The interface is designed to promote interactions between the user and the product. The user is attracted by internal UX from the interface such as a sense of the accomplishment, a pseudo experience of the achievement, a confirmation of the intelligent experience, a sympathize with the product meaning and so on. The interface is explained that users accelerate interactions with people form a society through the interface. Then, it is recognized that Kansei engineering should be required to apply.

In order to develop Kansei interface for a prevention system by UA, we attempt to make an experiment by a prototyping. The prototype system is composed by a simplified driving simulator, a sensing system of driving information and an interface prototyping system. The driving simulator is composed by store-bought handle, brake and accelerator devices which can be connected to a personal computer by USB cables. The personal computer displays graphics of driving scene interlocking driving operations. In order to compose sensing systems, HOTMOCK, produced by Holon Create [10], is employed as a prototyping tool. It is integrated kit with sensor devices, controller board and device manager software and interface prototyping software. This experiment will simulate potential ideas of Kansei interface to evaluate them by the rapid prototyping method. The prototype is evaluated in the means of usability and Kansei (Fig. 2).

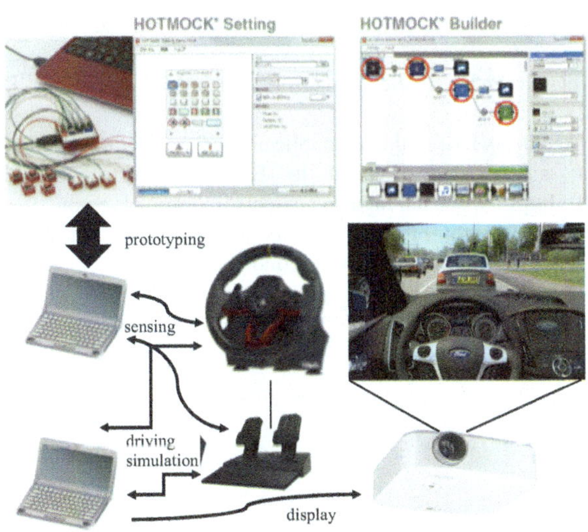

Fig. 2. Rapid prototyping using driving simulator and HOTMOCK. https://www.hotmock.com/

5 Conclusion

This study focuses on Unintended Acceleration with a pedal misapplication. Literatures which investigated the causes of the accidents by UA were examined. The various researches identified the misapplications which causes the accidents by UA and to indicated typification of the misapplications. In order to contribute for reducing the accidents, approaches by human machine interfaces were investigated. The problem of habituation is the most difficult to resolve for developing a prevention system. The existing products to contributed accident prevention were examined in the means of emotional effect to the driver. Kansei interface for preventing the accident by unintended acceleration was considered based on the survey. To develop the system using Kansei interface, we proposed a rapid proto-typing method to simulate the pedaling error. For future works, we are planning to test usability and effect of the driver's emotion by using the composed prototype.

References

1. Schmidt, R.A.: Unintended acceleration: a review of human factors contributions. Hum. Factors **31**(3), 345–364 (1989)
2. Japanese Ministry of Land, Infrastructure, Transport and Tourism. http://www.mlit.go.jp/report/press/jidosha07_hh_000327.html
3. Kazumitsu, S., et al.: Analysis of the recognition, action properties of the elderly person for the prevent of accidents caused by misapplication on Accelerator and Brake. Technical report, International Association of Traffic and Safety Sciences (2016)
4. Inoue, K.: Interface design textbook, 2nd edn. Maruzen Publishing, Tokyo (2019)
5. National Consumer Affairs Center of Japan. http://www.kokusen.go.jp/news/data/n-20180118_1.html
6. Schmidt, R.A., Young, D.E.: Cars gone wild: the major contributor to unintended acceleration in automobiles is pedal error. Front. Psychol. (2010). http://www.frontiersin.org/fpsyg.2010.00209
7. Locco, K.H., et al.: Pedal application errors. Technical report No. DOT HS 811 597, NHTSA (2012)
8. Hirakawa, A.: Accidents Due to the Misapplication of Accelerator and Brake pedals - For the Prevention of an Accident Typical to Elderly Drivers∼. ITARDA Information No. 124 (2018). https://www.itarda.or.jp/contents/92/info124_e.pdf
9. SUBARU, Subaru EyeSight. https://www.subaru.com/engineering/EyeSight.html
10. Holon Create, HOTMOCK. https://www.hotmock.com/

Meaning and Approach of New Product Designing Through Kansei Engineering

Yuuki Shigemoto(✉)

Department of International Liberal Arts,
International College of Arts and Sciences, Fukuoka Women's University,
1-1-1 Kasumigaoka, Higashi-ku, Fukuoka 813-8529, Japan
y.shigemoto@fwu.ac.jp

Abstract. This paper reopens Kansei Engineering (KE) that Nagamachi has founded and developed since 1970s from the perspective of design management. The frameworks of innovation of meaning and design thinking provide a new description of Nagamachi's practice that translates unawakened future meanings of consumers' life into product design. The product designing through KE therefore maximises its performance only when combined with solid design thinking, following the logic of science and utilising technology. In this sense, his hitherto works should actually be called Kansei Design. Furthermore, design thinking that consists of intuitive and analytical thinking also provides an opportunity for considering future application of artificial intelligence (AI) to KE approach. The two types of thinking are bridged by "abductive reasoning" of which development in programming is expected for increasing automated Kansei Design system.

Keywords: Kansei Engineering · Kansei Design · Design thinking · Innovation of meaning · Design management

1 Introduction

Kansei Engineering (KE, hereafter) is the powerful approach for designing innovative product and its subsequent service that often yield absolute emotional bond between product and consumers. However, the fact that there are still quite a few companies that apply the established methods and procedures of KE but nonetheless face market failures indicates the space for considering possible conditions that enable the KE approach to maximise its performance.

One of such conditions is spotted by the perspective of design management, which has been emerging as an academic discipline since the late 1970s. By nature, design is the action of sense making [1]. The philosophical rhetoric has been taken into the innovation research that has been recently paying attention to the *Innovation of Meaning* (IoM) (also known as design-driven innovation) in comparison to technological innovation [2]. Subsequently, the perspective of design research brings a new explanation to the distinction between successful and unsuccessful product development by KE; the absolute principle of KE is enquiring into inner sense of consumers who "feel" something desired for their lives but they cannot clearly envision what the

G. Di Bucchianico et al. (Eds.): AHFE 2020, AISC 1202, pp. 621–626, 2020.
https://doi.org/10.1007/978-3-030-51194-4_81

awkward senses are. In other words, the nature of KE is to catch the unawakened future meaning of consumers' life, and subsequently, cristalyse it by following the logic of science and utilising technology [3]. Consequently, as in the age of artificial intelligence (AI, hereafter), this paper addresses the emergent question; "How can we design the artificial intelligence that can design *Kansei*?".

The research question has been addressed for being challenged by empirical studies in the future. The main aim of this paper is to encourage an extensive discussion regarding a possible direction of future research and practice of KE. Despite the plethora of attempts by computer scientists for developing the system of inductive, deductive and abductive reasonings, a relationship between KE and AI has been paid little attention to. The two realms may be well bridged by the frameworks of IoM and design thinking, which come from the growing research area of design. The new insight is expected to shed light of the philosophy behind the approach of KE that is only able to maximise its performance with a solid design thinking (or design attitude).

2 Design Context Behind Kansei Engineering

Since 1970s, Mitsuo Nagamachi (Professor Emeritus at Hiroshima University, Japan) has developed a unique methodology regarding product development by combining psychology and engineering, and his research and practice have kept updating our life. In 1990s, the methodology and its academic expansion gained global attention as Kansei Engineering (KE), introduced by his paper "Kansei engineering: a new ergonomic consumer-oriented technology for product development" [4].

The Japanese term *Kansei*, which remains tricky to be translated into English (neither into any other languages), may often cause confusion amongst researchers including both native and non-native Japanese speakers. Considering the discussions around the etymological origin and possible definitions for *Kansei* [5–8], this paper defines *Kansei* as the concept referring to man's affective movement and cognitive response that are driven by surrounding environments, inspired by Shigemoto's [9] description; "the essence of the methodology of Kansei engineering is to capture consumers' affective movements that involve both positive and negative feelings, desires, attitudes, thoughts, and so forth towards product. In other words, Kansei engineering in nature targets all possible senses that consumers could feel in interaction with products (p. 32)". Thus, KE is premised to start from seeking to understand meanings of products that consumers are tacitly wishing to find in their lives. Without this perspective, the following processes of KE approach do not always work well.

It is also notable that Kansei Engineers usually collaborate with technical experts for R&D. Subsequently, the most important role of Kansei Engineers is translation of the meaning of a product which is sensed by consumers' *Kansei* into the specifications of design [3, 10, 11]. Their insights and future visions around consumers' life are crystalised by the technology that collaborative companies can offer. Thus, during the product development, Kansei engineers are required to understand discourses of more than engineers; there are specific techniques for observation and interpretation of consumer desire (the discourses of marketers and consumers), for collaboration with industrial experts (the discourses of engineers and managers) and for providing product and service to customers. In particular, it is crucially important to identify what truly is meaningful to consumers – this paper would like to call this stage as "meaning seeking". The similar perspective is shared as "need finding" amongst design thinkers in mechanical engineering (see [12]). The meaning seeking in consumers' life has been traditionally vocational to designers. But today, business experts are also regarded as benefitted by thinking like designers [13].

3 Artificial Intelligence in Kansei Engineering

The whole process of designing a new product by Kansei Engineering (KE) approach goes through the similar procedures of deep learning (namely, inductive and deductive logics) by the artificial intelligence (AI). Consequently, application of AI into KE product design has been continuously attempted (e.g. [14–18]). Whilst the computer-aided inference achieves the extraction and optimisation of consumers' *Kansei*, it is an exciting challenge to develop the AI that is capable of supervising the whole KE designing process. The practice of taking consumers' *Kansei*, which reflects socio-cultural systems and thus changes time by time, into product development is deemed to encounter various "wicked problems" that are the problems that have no definitive formulation, no single right answer or no single right path to reach an answer. And it is design thinking that has dealt with such wicked problems [19–22].

Design thinking is structured by the combination of intuitive thinking (i.e. sense seeking for the future) and analytical thinking (i.e. sense making by logic) (Martin 2009). This is what enables the KE methodology to be best utilised and sustains the design of *Kansei*. On the one hand, KE leverages intuitive and analytical thinking for interpreting and connecting consumers' *Kansei* with elements of product features (i.e. design elements) throughout new product design, respectively. On the other hand, the core of design thinking is "abductive reasoning" that mediates the intuitive and

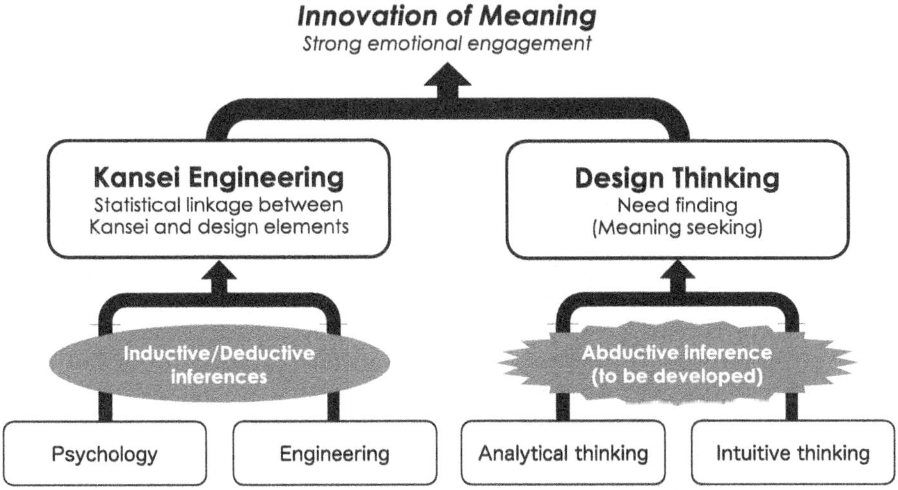

Fig. 1. Kansei Engineering and artificial intelligence from the design management perspective

analytical thinking. Thus, establishment of the abductive inference engine would potentially enable AI to achieve *Kansei* designing that will result in the innovation of meaning (IoM) (see Fig. 1). Given this, the coming challenge for combining KE and AI research will be to develop better abductive programmes (see [23, 24]). For this objective, more collaborations amongst engineers, computer scientists, scholars in design and management should be encouraged.

4 Kansei Design – Critique to Nagamachi's Kansei Engineering

Whilst the author would like to re-appreciate Nagamachi's achievement of establishing Kansei Engineering, this paper is humbly made critical to the fact that Nagamachi named his research and practice as "Kansei Engineering". What he has done should actually be called "Kansei Design" that is only achieved by the methodology of Kansei Engineering with decent design thinking. It is assumed that the true value of Nagamachi's product development through Kansei Engineering is in fact beyond the domain of engineering. What he has been doing is rather a holistic design of affective meaning that conveys new value to consumers. However, sufficient languages and frameworks to describe his work had neither been in the field of engineering nor in management. The new insight from design management has provided a space for a better explanation of KE approach, and for a possibility for developing the AI towards bringing about an automated-design system that crystalises consumers' *Kansei*. Today, a product solely cannot be completed its value. Subsequent service and socio-cultural structures of the markets should be considered as a part of generating a new meaning of the product. In this respect, more and more open innovations across academic and vocational disciplines are expected, and design management may take a key role to consider the future of product design; as Hands [25] says that "the remit of design management has expanded considerably, moving towards engaging critically with global challenges, demonstrating its value far beyond the delineated traditional boundaries of production and commerce" (p. 1).

5 Conclusion

This paper has reviewed Nagamachi's Kansei Engineering (KE) from the perspective of design management. In short, Kansei Engineering aims for quantitative understanding and embodiment of human senses (*Kansei*) in product and service that make new senses (*Meaning*). In fact, the design practice that is sustained by the established methodology of KE and design thinking should be called Kansei Design. This new insight highlighted by the frameworks of innovation of meaning (IoM) and design thinking also provided an opportunity for considering future application of artificial intelligence (AI) to KE approach. As well as inductive and deductive logics, abductive reasoning in programming is expected for developing the automated Kansei Design system that grasps what consumers are truly seeking for – in the age where consumers themselves are not aware of their inner desires. Further investigation of Kansei Design is expected to foresee an emergence of a new managerial strategies that may increase consumer-oriented society where product and service provide not only utilitarian benefit but also hedonically joyful experiences.

References

1. Krippendorff, K.: On the essential contexts of artifacts or on the proposition that "design is making sense (of things)". Des. Issues **5**, 9–39 (1989)
2. Verganti, R.: Design-Driven Innovation. Harvard Business Press, Boston (2009)
3. Nagamachi, M.: Kansei/Affective Engineering. CRC Press, Boca Raton (2010)
4. Nagamachi, M.: Kansei engineering: a new ergonomic consumer-oriented technology for product development. Int. J. Ind. Ergon. **15**, 3–11 (1995)
5. Alexandros, S., Sotirios, P., Dimitrios, K.: Affective engineering: towards a consumer culture theory approach to kansei engineering. In: Proceedings of the 2011 Conference on Designing Pleasurable Products and Interfaces, p. 23. ACM (2011)
6. Lee, S., Harada, A., Stappers, P.J.: Pleasure with products: design based on Kansei. Pleasure Prod.: Beyond Usabil. 219–229 (2002)
7. Levy, P.: Beyond Kansei engineering: the emancipation of Kansei design. Int. J. Des. **7** (2013)
8. Lokman, A.M.: Design & emotion: the Kansei engineering methodology. Malays. J. Comput. **1**, 1–11 (2010)
9. Shigemoto, Y.: Designing emotional product design: when design management combines engineering and marketing. In: Fukuda, S. (ed.) Advances in Affective and Pleasurable Design, Proceedings of the AHFE 2019, pp. 28–39. Springer (2019)
10. Nagamachi, M.: Kansei Inobeshon (Kansei Innovation). Kaibundo (2014). (in Japanese)
11. Nagamachi, M., Ishihara, S., Nishino, T., Matsubara, Y., Tsuchiya, T., Kanda, T., Inoue, K.: Shohinkaihatsu to Kansei (Product development and Kansei). Kaibundo (2005). (in Japanese)
12. Plattner, H., Meinel, C., Leifer, L.: Design Thinking: Understand–Improve–Apply. Springer, Heidelberg (2010)
13. Verganti, R.: Design thinkers think like managers. She Ji: J. Des. Econ. Innov. **3**, 100–102 (2017)
14. Chen, J.-S., Wang, K.-C., Liang, J.-C.: A hybrid Kansei design expert system using artificial intelligence. In: Pacific Rim International Conference on Artificial Intelligence, pp. 971–976. Springer (2008)
15. Li, Y., Zhu, L.: Extracting knowledge for product form design by using multiobjective optimisation and rough sets. J. Adv. Mech. Des. Syst. Manuf. **14**, JAMDSM0009 (2020)
16. Matsubara, Y., Nagamachi, M.: Hybrid Kansei engineering system and design support. Int. J. Ind. Ergon. **19**, 81–92 (1997)
17. Nagamachi, M.: History of Kansei engineering and application of artificial intelligence. In: International Conference on Applied Human Factors and Ergonomics, pp. 357–368. Springer (2017)
18. Yang, S.-M., Nagamachi, M., Lee, S.-Y.: Rule-based inference model for the Kansei engineering system. Int. J. Ind. Ergon. **24**, 459–471 (1999)
19. Buchanan, R.: Wicked problems in design thinking. Des. Issues **8**, 5–21 (1992)
20. Coyne, R.: Wicked problems revisited. Des. Stud. **26**, 5–17 (2005)
21. Farrell, R., Hooker, C.: Design, science and wicked problems. Des. Stud. **34**, 681–705 (2013)

22. Rittel, H.W., Webber, M.M.: Dilemmas in a general theory of planning. Policy Sci. **4**, 155–169 (1973)
23. Denecker, M., Kakas, A.: Abduction in logic programming. In: Computational logic: Logic Programming and Beyond, pp. 402–436. Springer (2002)
24. Kakas, A.C., Kowalski, R.A., Toni, F.: The role of abduction in logic programming. Handb. Log. Artif. Intell. Log. Program. **5**, 235–324 (1998)
25. Hands, D.: Design Management: The Essential Handbook. Kogan Page Publishers, London (2017)

Real, Augmented, Virtual, and Robotic Animals in Smart and Playable Cities

Anton Nijholt(✉)

Human Media Interaction, University of Twente,
PO Box 217, 7500 AE Enschede, The Netherlands
a.nijholt@utwente.nl

Abstract. Animals are part of city life. Many animal species live in the city. On the street, in buildings, in our homes, in vacant lots, parks, cemeteries, and in gardens. In smart cities the life of domestic and working animals will be augmented with digital technology. Urban wildlife will be confronted with digital technology as well. Robotic and virtual animals will also become part of street life. We survey the developments that lead to this future city life in which humans, augmented animals, robotic animals and virtual animals will co-exist.

Keywords: Animal-computer interaction · Smart cities · Playable cities · Companion animals · Working animals · Robotic animals · Virtual animals

1 Introduction

Our future urban environments are smart. There will be sensors and actuators that monitor our activities and suggest or tell us what to do or force us to behave in a particular way. Research on smart and playable cities mentions the use of sensors and actuators to improve a city's efficiency, that is, make the city smart by collecting information about the behavior of its inhabitants and use it to improve its infrastructure, traffic control, public transport, waste management, sustainability, and energy consumption. Usually, when considering a city's smartness, let alone a city's playfulness [1], it does not include how in the physical real-world situation children, disabled persons, and non-human animals take part in daily urban life. Neither are these categories of city dwellers involved in the decision and design process. In this short paper we will focus on the non-human animals and how they will co-exist with humans in the smart city. We give a broad interpretation to the concept of non-human animals.

First of all, animals can be augmented with wearable technology, sensors and actuators, even with implants. This allows them to interact with smart technology embedded in the urban environment. They can be monitored, controlled, and supported in their activities. Useful, especially for working animals and their handlers, and for pets and their owners. Moreover, they can become nodes in the Internet of Things, so that their behavior can also be monitored on a more global level and decisions can be made that relate to a certain species, a city district, a park, or to a certain community.

Secondly, we can expect that in the near future we have robopets (robotic pets) in our urban environments. People will keep robopets and take them for a stroll on the

G. Di Bucchianico et al. (Eds.): AHFE 2020, AISC 1202, pp. 627–633, 2020.
https://doi.org/10.1007/978-3-030-51194-4_82

streets or in a park. The robopets keep them company, allow some social interaction (both with the robopet and other robopet owners), and keeps them healthy by staying active. The robopet can tell its owner that it is time to go for a walk or that it wants to engage in a playful activity. We can also expect robotic working animals in the urban environment. Robotic delivery dogs, robotic police dogs, and when necessary, robotic search and rescue dogs will be on our streets. For real dogs, dog sitting and dog walking can be done by a smart robotic dog as well.

Thirdly, augmented reality allows us to see and have physical interactions with animated virtual animals that are projected on walls, streets, and buildings. It is also possible to have virtual animal activity overlaid on the camera view of the real world that we see on our smartphone, or, much more immersive, to have this animal activity made visible on a head-mounted augmented reality device, glasses, or contact lenses. In this augmented reality world we co-exist with natural living animals, augmented living animals, and animated virtual animals that are aware of their environment and interact with living animals and human city dwellers.

In this paper we provide an overview of the various natural, augmented, robotic, and virtual animals that will co-exist with their human inhabitants of the urban environments.

2 Living and Augmented Animals in Smart and Playable Cities

Animals living in our smart cities will be confronted with digital technology. This can lead to humorous situations where a speed-tracking camera gets confused by a pigeon (Fig. 1, left) or two security cameras are used in a non-intended way (Fig. 1, right). Smart technology can also help to decrease the number of birds killed by colliding with buildings, decide where to build shelters for stray dogs, or keep animals away from certain places using artificially produced scents. Smart technology can also help to detect urban wildlife that needs to be protected or hunted and killed.

Fig. 1. Non-intended use of a city's smart technology

But of course we must first of all think of how smart technology can help keep our pets and working animals healthy, how smart technology can help living with our pets more comfortable, and how to improve the performance of our working animals. Dogs and cats are our traditional pets, but, of course, fish in aquariums, hamsters, stick insects, rats, rabbits, birds, snakes, and chickens can be held as pets as well. Not all of them are companion animals in the sense that we can have them accompany us when we leave our house and take them out on the streets.

When we talk about working animals it is mainly about dogs: assistance dogs, sniffing dogs, and search and rescue dogs. In developing countries horses, goats, elephants, bears, and snakes are given tasks that cannot be done by humans, or can be done cheaper if animals are used. There are also many examples where animals are used for entertainment.

In the next subsection we shortly discuss how companion animals and their human companions can profit from digital technology, making their life more enjoyable, comfortable, social, and healthy. The subsequent subsection will address smart technology for performance-oriented working animals.

2.1 Augmented Companion Animals and Pets

Digital toys that draw the interest of pets and engage them in playful activities have been designed. Playgrounds for dogs or parks can be designed in such a way that digital technology can introduce dog-personalized challenges. Parks can be equipped with smart technology and social media coverage that makes it interesting for dogs and dog owners to be there, to engage with other dogs, or to have social interaction and community building between dog owners.

Animals, whether they are pet animals or working animals, can be equipped with smart devices, usually hidden, and without their consent, in a wearable: a collar, a 'vest', or even a head-mounted device. It makes animals part of the Internet of Things (or the Internet of Animals). In the case of pet animals this is mainly done to monitor animals, see what they are interested in, monitoring their health condition, and learn about their preferences and social interactions [2]. On-body smart technology can include microphones or touch-sensitive actuators to interact remotely with an animal. With the help of a vibrotactile vest an animal can remotely be hugged (Fig. 2, left) [3]. Vibration patterns can simulate a hug or a caress. Dogs can learn commands that are mediated by haptics (Fig. 2, middle) [4]. Facial recognition technology for dogs has been investigated as well [5]. It can help to find lost dogs.

In playable city research [1] the audience that is addressed consists largely of well-educated and creative young people. However, smart technology makes it possible to introduce playful urban environments for animals as well.

Fig. 2. Pet animals and working animals. Hugging a pet chicken (left. Commanding a dog with haptics (middle). Eagle trained to intercept drones (right)

2.2 Augmented Working Animals

We can distinguish different kinds of 'working' animals. We can have animals that are augmented with digital technology to increase their performance. On-body sensors and actuators can assist them in their tasks and the communication with human handlers. Smart collars and smart vests for animals have been introduced with the aim to have them following commands, to control them and to provide them with suggestions how to perform their task. Animal senses that are useful in performing tasks that can or will not be done by humans can be augmented with smart technology. Smart collars, on-body microphones, and haptic actuators can help to communicate with dogs in search and rescue operations. Dog-drone communication makes it possible to monitor a dog while it performs a task in an impassable environment. Dogs can be trained to follow a drone [6]. Eagles can be trained to intercept a drone (Fig. 2, right) [7]. With more advanced technology, perhaps learning from animal senses, we can expect that the tasks that nowadays are given to working animals can be done by robotic animal-like robots.

3 Robotic and Virtual Animals in Smart and Playable Cities

Animal-like robots have been on the market for many years. There are robotic pets with sensors and actuators that make them interactive in socially appropriate and appreciated ways. On the other hand, there are also animal-like mobile robots that perform tasks that used to be done by humans or living animals. In addition, we can distinguish virtual animals that can appear in an urban environment for entertainment, or for educational or training purposes. Research results on making cars autonomous (self-driving cars), can as well be used to make augmented and robotic animals artificially intelligent and autonomous.

3.1 Robopets and Robotic Working Animals

Robopets can improve the quality of life for seniors by showing affection and by stimulating social and physical activity [8]. Aibo is a robot dog with computer vision, facial recognition and touch sensors that detect when it is hugged. Robotic cats and seals have been introduced as well. In nursing homes, they are used to help seniors with dementia. By providing them with more intelligence and knowledge of the environment in which they have to perform, they can act more autonomously and assist their owners outside their homes, go outside to look for playful situations or social interactions with living or other robotic dogs. Real dogs will meet with robotic dogs (Fig. 3, left), they can have a robotic dog as a dog sitter, or get confused when meeting delivery robots on the street (Fig. 3, middle).

Robotic animals can perform useful functions. When talking about robotic working animals we should consider whether tasks of living working animals should be replaced by robotic animals. Obviously, robopets that improve the quality of life of seniors can be considered as working animals as well. Walking a robotic dog can have many advantages over walking an actual dog. Real birds can be chased away from an airport by a robotic bird (Fig. 3, right), rather than by a trained bird of prey. We can attempt to replace the police's sniffing dog with a robot dog that has an 'electronic nose'. Robot dogs can assist policemen in their tasks and act as 'mobile remote observation devices' or they can be used to spread or restrain a crowd.

Fig. 3. Dog meets a robotic dog (left). Dog meets a delivery robot (middle). Robotic bird scaring other birds away from an airport (right)

3.2 Virtual Animals in Smart and Playable Cities

We provide three examples of virtual animals that are designed so that they become immersed in our urban environments, rather than being displayed on our smartphones.

Some years ago, a London bus stop was equipped with a camera showing in real-time the street view, but overlaid with a fast-approaching tiger, scaring people looking at the billboard (Fig. 4, left). In order to have the tiger moving in this environment in a natural way the overlay had to be done such that the tiger avoids collisions with the fixed street furniture. A model had to be made of the layout of the street and its buildings. People started to 'interact' with the tiger once they discovered how to have their own movements captured by the bus stop's camera.

Urbanimals [9] is a project implemented on the streets in Bristol (UK) in 2015. Animated animals (Fig. 4, middle) aware of their environment, including approaching and interested humans, were projected on walls and streets and invited passers-by to play, compete or otherwise interact with them. Also, in this case, the environment had first to be modeled in 3D in order to give the various animals (kangaroo, beetle, dolphin, rabbit) environment-aware movements, such as climbing the stairs or avoiding collisions with objects.

Although not yet implemented in an urban environment, in [10] research is reported on walking a virtual dog. The study focuses on the interaction behaviors of dog handler, dog and a passerby (Fig. 4, right) in a shared physical space. The dog handler is wearing an augmented reality head-mounted device. The passerby was present in two conditions: with and without a head-mounted device.

Fig. 4. Augmented reality tiger on a billboard (left). Animated and interactive kangaroo projected on a wall (middle). Walking a dog in augmented reality (right)

4 Conclusions

Our future smart urban environments include human and non-human animals. The focus of this paper is on non-human animals. How can human animals expect to live with non-human urban animals, whether they are living, augmented, robotic, or virtual? Non-human artificial animals will be part of our daily life activities. We can have artificial pets and meet them on our streets. There can be artificial working animals as well, maybe taking care that we humans behave in a preferred manner. We, and living animals have to co-exist with artificial and augmented animals and artificial and augmented humans. In this short paper we introduced examples of augmented, robotic, and virtual animals that will appear in our future smart urban environments. The aim of this short survey is to create awareness of this urban animal life in the near future.

References

1. Nijholt, A. (ed.): Playable Cities: The City as a Digital Playground. Series: Gaming Media and Social Effects. Springer, Singapore (2016)
2. Bekoff, M.: Hidden tales of yellow snow: what a dog's nose knows - making sense of scents. Psychology Today, June 2009. https://www.psychologytoday.com/sg/blog/animal-emotions/200906/hidden-tales-yellow-snow-what-dogs-nose-knows-making-sense-scents
3. Lee, P., et al.: A mobile pet wearable computer and mixed reality system for human–poultry interaction through the internet. Pers. Ubiquit. Comput. **10**(5), 301–317 (2006)
4. Golan, Y., et al.: Dogs can understand haptic communication. In: Proceedings Sixth International Conference on Animal-Computer Interaction (ACI 2019). ACM, New York (2019) https://doi.org/10.1145/3371049.3371066
5. Kumar, S., Kumar Singh, S.: Monitoring of pet animal in smart cities using animal biometrics. Future Gener. Comput. Syst. **83**, 553–563 (2018)

6. Foster, M., et al.: Preliminary evaluation of dog-drone technological interfaces: challenges and opportunities. In: Proceedings Sixth International Conference on Animal-Computer Interaction (ACI 2019). ACM, New York (2019) https://doi.org/10.1145/3371049.3371065
7. Castle, S.: Dutch firm trains eagles to take down high-tech prey: drones. NY Times, 28 May 2016. https://www.nytimes.com/2016/05/29/world/europe/drones-eagles.html
8. Abbott, R., et al.: How do "robopets" impact the health and well-being of residents in care homes? A systematic review of qualitative and quantitative evidence. Older People Nurs. **14** (3), 1–25 (2019). https://doi.org/10.1111/opn.12239
9. Dobiesz, S., Grajper, A.: Animating the static. case study of the project "urbanimals". In: Proceedings of 34th International Conference on Education and Research in Computer Aided Architectural Design in Europe (eCAADe 34), vol. 1, pp. 691–697. University of Oulu, Oulu (2016)
10. Norouzi, N., et al.: Walking your virtual dog: analysis of awareness and proxemics with simulated support animals in augmented reality. In: Proceedings of IEEE International Symposium on Mixed and Augmented Reality (ISMAR), pp. 253–264. IEEE Press, New York (2019)

On the Role of "Muscle Memory" in Interaction Design

Miao Liu and Xingchun Yang(✉)

East China University of Science and Technology,
Shanghai 200237, People's Republic of China
183787975@qq.com, 1379917938@qq.com

Abstract. "Muscle memory" refers to the human body's muscles have memory effects. After the same action is repeated many times, the muscles will form conditioned reflexes. With the development of screen technology, after more and more physical buttons are replaced with screen operations, the structure of the machine no longer follows its functions to design, especially some button designs. This poses new challenges to the accuracy, timeliness, and reduction of mis-operation rates. This article studies different interface designs and proposes several interface design principles that are consistent with muscle memory. Based on this principle, an APP launcher based on "muscle memory" is produced to verify whether the criterion is reasonable. This article links "muscle memory" with interoperability, and aims to create, build, and facilitate convenient relationships between people and products and services. A feasible direction is proposed for exploring better human interaction experience.

Keywords: Muscle memory · Electromyograph · Interaction design

1 "Muscle Memory" Discussed in this Article

1.1 Definition of "Muscle Memory"

Muscle memory is a form of procedural memory that involves consolidating a specific motor task into memory through repetition. When a movement is repeated over time, a long-term muscle memory is created for that task, eventually allowing it to be performed with little to no conscious effort. This process decreases the need for attention and creates maximum efficiency within the motor and memory systems [1].

Muscle memory consolidation involves the continuous evolution of neural processes after practicing a task has stopped. The exact mechanism of motor memory consolidation within the brain is controversial. However, most theories assume that there is a general redistribution of information across the brain from encoding to consolidation. Hebb's rule states that "synaptic connectivity changes as a function of repetitive firing". In this case, that would mean that the high amount of stimulation coming from practicing a movement would cause the repetition of firing in certain motor networks, presumably leading to an increase in the efficiency of exciting these motor networks over time [2].

G. Di Bucchianico et al. (Eds.): AHFE 2020, AISC 1202, pp. 634–640, 2020.
https://doi.org/10.1007/978-3-030-51194-4_83

1.2 "Muscle Memory" Discussed in this Article

The academic controversy of "muscle memory" is whether muscles can store memory. The focus of this article is not to explore the answer to this question, but that the existence of the "muscle memory" phenomenon is true, which has been confirmed. One of the earliest and most notable studies regarding the retention of motor skills was by Hill, Rejall, and Thorndike, who showed savings in relearning typing skills after a 25-year period with no practice [3]. Findings related to the retention of learned motor skills have been continuously replicated in studies, suggesting that through subsequent practice, motor learning is stored in the brain as memory. This is why performing skills such as riding a bike or driving a car are effortlessly and 'subconsciously' executed, even if someone had not performed these skills in a long period of time [3].

2 Case Studies on the Use of "Muscle Memory" in Interaction Design

In actual life, there are actually many cases of "muscle memory". Recalling the time when you just practiced typing when you first came into contact with a computer, we need to find every letter, and many people will type each letter. However, when you are familiar with blind typing, you can type the characters you need quickly without looking at the keyboard. Comfortably, not just rely on the brain to remember keys, your finger is familiar with the position of the keyboard, for example, for a new keyboard need to re-adapt to a period of time in order to improve accuracy and speed. After mastering the blind typing, it is much more efficient than the previous stage of finding and hitting. Blind typing omits a search step. In accordance with the principles of interaction design, it is an optimization of the interaction process, and the step of finding this step has become the information that the user has learned through learning. This is an important difference between interactions that incorporate "muscle memory" and other interaction designs.

If some operations should be designed as muscle memory but not designed, the user will be very resistant to operation. For example, the design of the part in the lower right corner of the picture requires clicking the first-level menu and then the second-level menu, which lacks the consistency of operation and breaks Connectivity between the primary and secondary menus. This hinders the production of "muscle memory" and can cause discomfort to the user experience (Fig.1).

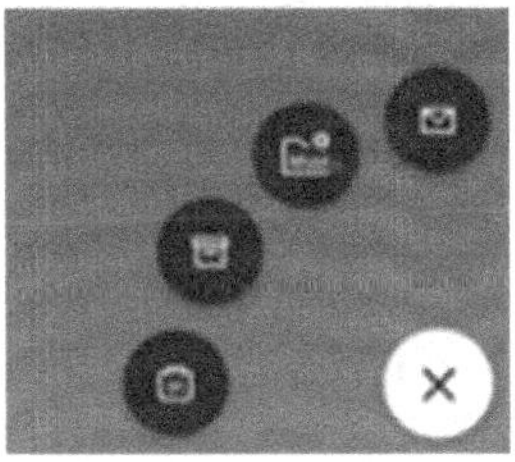

Fig. 1. Examples of lack of muscle memory design

3 Discussion on Interactive Model Based on "Muscle Memory"

3.1 Research on User Model Based on "Muscle Memory"

Cognitive-perceptual-learning-memory is a typical interaction step that conforms to "muscle memory" and is inseparable from memory. The following aspects should be considered:

(1) **User Cognitive Model**
Human cognitive processes are performed under the coordination of different types of memory. Information is first perceived by people through audiovisual channels, and then processed by the perception processor to selectively perform memory processing or experience extraction. These characteristics provide some design guidelines for interaction design. For example, during the design process, a large amount of information should be classified and organized according to mutual relationships. The design of the interactive interface should be concise, to avoid the distraction and interference of irrelevant task information to the user's attention, and to show personality. The design can also significantly enhance the user's impression of design details, which will help to extract the memory experience.

(2) **User perception model**
The user perception model is summarized based on human perception channels. After receiving the information through the human sensory channel, it enters the sensory processor for processing. In order to reduce the burden of identification and memory of the user of the graphical symbol user interface, the teaching design of the interface, from layout to menu design, should be considered in accordance with the user's perceptual characteristics. For example, most users are right-handed, so it is more convenient to arrange the control operation group on the right side of the display area; according to the primary and secondary principles of the interface layout, leave the upper-left area where the user's attention is most concentrated to the display area.

(3) **User fault tolerance model**
There are no users who can go wrong. User errors are often caused by two reasons, one is internal, intuitive user to operate, misuse, incoherent thinking, or excessive accumulation of information processing tasks simultaneously, such as fatigue mistakes; and second, the machine does not meet the system design process Users' cognitive laws, memory processing mechanisms, and experience habits lead users to make mistakes. Therefore, when the interface design should research the characteristics of the user's own error-prone, and avoid trying to reduce the chance of user error.

(4) **User learning model**
A good human-computer interaction interface should be self-evident. After the user contacts, it will be a short time to learn a little operation, and it does not need too high learning costs. The interaction designer's job is to study the learning process of the target user, understand the user's processing and memory

mechanism for information, and capture the user's subconscious reactions and actions that are difficult to perceive, so as to improve the method of interaction design to reduce learning difficulties. For example, compared to the existing design methods and metaphors for learning materials and user interface elements in favor of information processing and memory factor analysis and control.

3.2 Interaction Principle Based on "Muscle Memory"

"Muscle memory" reflects an automatic memory that does not require conscious participation. The characteristic of this kind of memory is that people don't realize deeply that they have this kind of memory, and they don't subconsciously extract this kind of memory, but it is manifested in the operation of specific tasks. In the process of human-machine interaction, designers should fully consider this mechanism and think about how to use this mechanism to accelerate the rate of information acquisition by humans, so as to propose effective design methods, and then users can easily and accurately with minimal awareness Burden to complete the current task. Based on this consideration, the author proposes the following interaction design principles based on "muscle memory":

(1) **Consistency.** Consistency refers to the design factor used in the design to be consistent with existing experience and knowledge structure, taking into account the most common concepts and user experience, designed to meet the user's needs and preferences of products, reduce the user's learning cost.
(2) **Multi-sensory experience.** Make full use of new technologies in the design process, using sound, video, gestures, expressions, eye movements, vibrations and even odors to provide a multi-sensory experience, which can provide users with multiple channels to quickly obtain information and feedback, which is more conducive to "muscle "Memory".
(3) **Fault tolerance, recovery and feedback.** Different users have different cognitive styles and abilities, and this level of influence should be fully considered in the design. Designers should allow and encourage users to practice using their subconscious actions and intuition, which requires the system to be fault tolerant and recoverable. Able to monitor and report errors, give prompts, and cancel recovery of operations that have been performed, giving reasonable feedback.
(4) **Simplify interaction.** For the user's frequent affairs, analyze the execution steps in the task, find and eliminate redundant actions, and then merge or reduce unnecessary operations according to the person's past experience and cognitive habits, and realize the automatic execution of these operations. This simplifies the interaction process, reduces user information processing time and cognitive load, and improves interaction efficiency and quality.

4 Experimental Verification-Quick Launcher

Based on the above-mentioned principles, the author conducted an experimental investigation on a quick launcher software to verify the reliability of the proposed principles. The interface of the quick launcher is shown in the figure below. After clicking any area of the screen, four folders appear. According to the direction of finger movement, the folder in the corresponding direction is activated to enter the next level menu. When activated, there is a vibration feedback lower level menu. There is a folder or corresponding software in it, and you can open the corresponding folder or software by moving your finger in the corresponding direction again to complete the operation of opening specific software (Fig. 2).

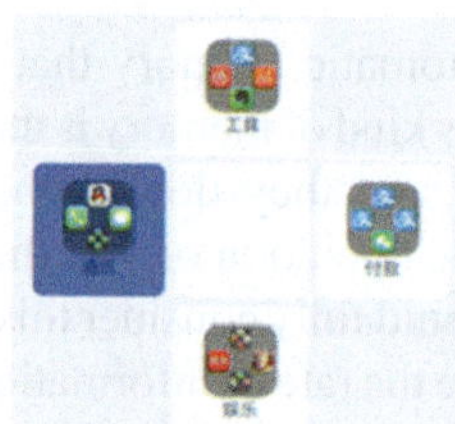

Fig. 2. Quick launcher interface

4.1 Experiment: Validate Interaction Principles Based on "Muscle Memory"

Research Purposes. Verify that the interaction principle based on "muscle memory" is effective.

Research Method. Subjects: 5, both graduate students, between the ages of 22–25 years old, right-handed, hand, eye health, East China University of Technology, and has not previously participated in similar experiments, also unused software used in the experiment.

Participants specified operations on a mobile phone software such as Quick Launcher according to the task list. The task list was designed to include two processes, namely (1) familiarization process: familiarize with the current operation and understand the basic software operation principle. (2) Learning process: Learn a single task in the list, repeat a single operation several times, the purpose is to let it quickly form muscle memory. (3) Application process: After the previous learning process, the participants have generated basic "muscle memory" and then applied the operation.

The interaction process is divided into five levels: (1) startup operation: press the screen to display the folder (2) first-level interaction: select the first-level menu after pressing (3) second-level interaction: enter the first-level menu and select two Level menu (4) Level three interaction: Enter level two menu and select level three menu (5) Level four interaction: enter level three menu and select driver menu. The task list is divided into 7 groups from top to bottom, and the tasks are divided into 7 levels. Among them, task 1 and task 2 verify the directional difference between the secondary interactions, and the directions are right and upper. Task 3 verifies the secondary interactions. There is continuity between the second-level interaction and the third-level interaction; task 4 verifies the second-level interaction, and there is no continuity between the second-level interaction and the third-level interaction; task 5 verifies the third-level interaction, the third-level interaction and the fourth-level interaction. There is no continuity between them; task 6 verifies the three-level interaction, and there is continuity between the three-level interaction and the fourth-level interaction. Task 7 verifies four levels of interaction. In order to reduce the experimental error caused by gradually becoming familiar with the software, the subjects were required to randomly select tasks for execution. Participants recorded the screen during the operation, and used the electromyograph to monitor the flexor hallucis flexor, brachial abductor and adductor hallucis on the palm of the operator in real time to collect EMG data. And record the completion in the task list.

Research Result

The analysis of the EMG data of the 35 groups during the learning process showed the following phenomenon (FIg. 3):

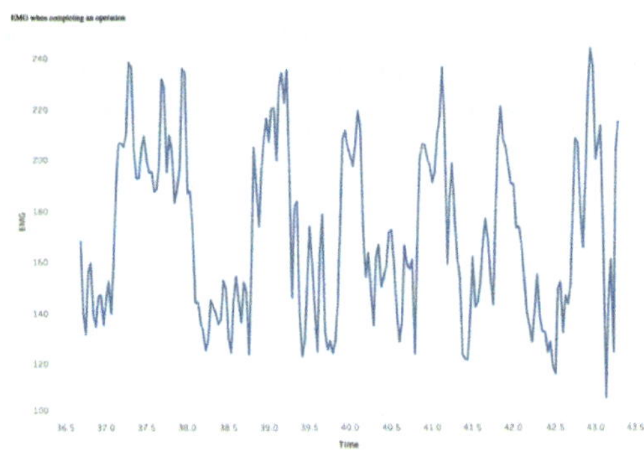

Fig. 3. EMG when completing an operation

When learning a new task, that is, completing the learning process of an operation, the curve period decreases with time, and in the video recorded on the screen, repeating the same operation becomes more and more smooth. The phenomenon indicates that there is a "muscle memory" phenomenon in the touch operation of the quick launcher (Table 1).

Table 1. Task operation accuracy table.

	Tester1	Tester2	Tester3	Tester4	Tester5
Task1	100.00%	97.06%	97.06%	94.12%	100.00%
Task2	93.75%	96.88%	90.63%	84.38%	93.75%
Task3	87.50%	81.25%	93.75%	81.25%	90.63%
Task4	93.75%	96.88%	100.00%	81.25%	93.75%
Task5	96.88%	100.00%	96.88%	100.00%	93.75%
Task6	75.00%	84.38%	84.38%	93.75%	96.88%
Task7	84.38%	87.50%	96.88%	90.63%	90.63%

(1) The comparison between Task1 and Task2 shows that the operation is directional, and the correct rate of the right sliding operation is higher than the up-sliding operation. This is consistent with the habitual movement of thumb sliding operation, and is consistent with the existing experience and knowledge structure.
(2) It can be seen from the table that the average accuracy rate of Task3 is lower than Task4, and the average accuracy rate of Task5 is lower than Task6. Combined with the video, it is necessary to think about when to stop when there is continuity between two levels of interactive actions It increases the information that users need to process, increases the cognitive burden for users, and blocks the continuity of operations.
(3) Task5 although it has also the third-level interaction, but its high accuracy rate, the decomposition of its operation, the first slide left, then right slide, after the habit, left slide right slide operations can be combined into one operation, simplifying the interaction, in line based on "muscle memory "Interaction principle.
(4) Vertical comparison. As the number of menu levels increases, the accuracy rate tends to decrease. Displaying too many menus brings more content for the user to learn, which increases the user's information processing time and cognitive load.

5 Conclusion

"Muscle memory" can evoke the individual's experience to automatically affect the current task, and the existence of the "muscle memory" phenomenon plays an important role in the process of human-product interaction. Understanding and understanding of "muscle memory" helps to create more Good human-machine relationship. Through the design of relevant experiments, this paper reveals the role of "muscle memory" in interaction design, and proposes interaction principles based on "muscle memory", hoping to provide more natural guidelines for interaction. In this paper, muscle memory in biology is introduced into interaction design. The principles and experimental design of the conclusion are bound to have many shortcomings, and it needs to be improved in future research.

Acknowledgments. We thank everyone who provided valuable suggestions and feedback during the writing of this paper. Thanks to the supported by Shanghai Summit Discipline in Design - DA17014.

References

1. Krakauer, J.W., Shadmehr, R.: Consolidation of motor memory. Trends Neurosci. **29**(1), 58–64 (2006)
2. Ma, L., et al.: Changes in regional activity are accompanied with changes in inter-regional connectivity during 4 weeks motor learning. Brain Res. **1318**, 64–76 (2010)
3. Lee, D.T., Schmidt, A.R.: Motor Control and Learning: A Behavioural Emphasis, 4th edn. Human Kinetics, Windsor (2005)

UX Design for Mobile Application of E-Commerce Site by Using Kansei Interface

Yuiko Mimura(✉), Toshio Tsuchiya, Kaho Moriyama, Kanna Murata, and Sana Takasuka

Faculty of Economics, Shimonoseki City University, 2-1-1 Daigakucho, Shimonoseki, Yamaguchi 751 8510, Japan
{tsuchiya, s170185, s171186, kas170096}@eco.shimonoseki-cu.ac.jp, tsuchiya@shimonoseki-cu.ac.jp

Abstract. This paper focuses on UX design procedures and aims to develop Kansei interface in order to improve usability of mobile applications. This study is intended to examine about contribution of Kansei interface to UX design. Kansei interface is understood as comprising emotional aspects directing to compose it like intuitive and compelling to use. The characteristics of Kansei interface are explained at the beginning of this article. Especially, human machine social interface is focused on as a principal element of Kansei Interface. Then, the designing mobile interface is illustrated as an example of UX design. UX design processes explore a customer journey map by user investigation. The application of Kansei interface is examined for the extracted activities at touch points. The human machine social interface implemented as Kansei interface is investigated by user questionnaires. The influence on the UX of Kansei interface is discussed from the experiment.

Keywords: Kansei engineering · UX design · Mobile application

1 Introduction

According to Ministry of Economy, Trade and Industry, the market size of the electronic commerce for consumers (BtoC) of 2018 in Japan is approximately 18 trillion JPY. The market size spreads at the growth rate more than 5% and is expected that it will rise in the future. The ratio of the E-commerce market size for the transaction market size is more than 6%, it is relatively low currently, and is considered to grow up more [1]. It may be said that the spread to small companies is indispensable. An application of the smartphone as a plat-form of E-Commerce is important to increase the users of the net shopping site. So, the large-scale EC malls targeted the expansion of sales by utilizing the big data of enclosure the application users. For this reason, the sites made much account of usability of the application particularly UX design.

On the other hand, the enclosure of the user is difficult in the small-scale sites, and they cannot provide enough UX like the EC mall. This study attempts to develop a UX design for mobile phone applications. A human-machine social interface, which is a kind of Kansei interface [2, 3], would be implemented into the application. The human-machine social interface includes the users' purchasing tendency data obtained from

G. Di Bucchianico et al. (Eds.): AHFE 2020, AISC 1202, pp. 641–647, 2020.
https://doi.org/10.1007/978-3-030-51194-4_84

web scraping on the net. The web scraping extracts information from web sites and is a process to convert the information into analyzable structured data. The need for web scraping is increasing, and it's getting convenient to get the information easily using it.

The UI composed by UX design will be tested by the task analyses using a prototyping method. In this study, the candidates of utility for prototypes are examined in the meaning of Kansei interface. Human machine social interface will be implemented into the prototype. Kansei interface is explained before getting to main part of this study.

2 Kansei Interface

The first scholar to give attention to Kansei interface was Kageyu Noro. He described The Kansei interface as the highest level upon physical interface and cognitive interface [2]. The physical interface is to have evaluation points of the perception level before the recognition, for example, easy to hear, easy to push or easy to see. The cognitive interface is caused by process of thinking such as memory and reasoning. These two levels of the interface are evaluated by functions, efficiency and minimizing of physical and cognitional load. On the other hand, the Kansei Interface is required an emotional characteristic different from those two levels. Norman outlined the essence of three level theory of human behavior as Reaction (visceral), Routine (behavioral) and Reflection levels [4]. Kansei interface is recognized as one of design methods of Reflection level. He defined reflection level as the highest level of interfaces at which the system continually monitors its own operations [5]. The interface is pointed both the highest level of analysis and the weakest in today's systems.

Interface design that can be intuitive to use anyone is required by the widespread use of multi-functional information communication devices typified smartphones. A new trend of the interface includes objects appealing to the five senses, borrowing from game elements and arousing emotions especially "lovely" or "enjoyable". It became the cause that the interfaces for representing the emotion of the user increased by User-generated contents and Consumer-generated medias of the spread of social medias. For the spread of smartphones, it can be thought that the active and affirmative emotion that the user wants to use of influences as well as evolutions of the functional features. It can be said that such emotional features of the highest level on Reflection corresponds to Kansei interface. The lower level interfaces such as Reaction and Routine are required to be evaluated by the aspects of functionality, efficiency and the physical or cognitive load. In contrast, Kansei interface should be estimated by emotional elements. Then, it is important to define the features of Kansei interface to represent its emotional characteristics.

2.1 Intuitive User Interface

Inoue et al. investigated the problem of representing the characteristics and the meaning of the intuitive interface from a questionnaire survey [3]. The research suggested the relationship between "user-friendly" and "intuitive" interface. The results of the questionnaire explored that an intuitive interface is user-friendly but user-friendly interface is not necessarily intuitive. This research was compiled progressively and arrived at 10 rules [6].

Rule1. Emphasis and Restraint of displaying (Simplify Display)

It becomes simple, highly visible UI by reducing the number of UI elements.

Rule2. Visibly Simple (Simplify Display)

A design simplifying several UI elements by cognitive organization is required. There are well known elements including organization such as proximity, similarity, closure and common fate.

Rule3. Operation by Body Motion (Simplify operation)

UX using the natural body motion in life such as the weight movement, the utterance, the tap, the push which are called as NUE(Natural User Experience) motioned unconsciously. NUI(Natural User Interface) is an concept applied NUE to the operation. It is not necessary an intentional operation, so NUI can realize intuitive operations.

Rule4. Automatic Operation (Simplify operation)

There is contradiction between simple physical operation and the multi-functionality of the device. Therefore, UI automating sequential operations should be designed by simple operations. The method of simplify the operation is such as to replace the function with physical buttons or icons as cognitive representations.

Rule5. Use of Visual Metaphor (Imaginable Operation)

When icons or contents of the information appliance are represented as metaphors copied from the real world, the devices become intuitive for the recognition.

Rule6. Use of Operational Metaphor (Imaginable Operation)

UI simulated the physical operation has high affinity with user retrieval. The user can operate the e-book without being conscious which is designed as similar to physical operation to turn up page of the real book with a finger.

Rule7. Mapping Display into Operation (Interface Without Conscious in Operation)

The user images the operation only watched display by the recognition of the sense level promptly.

Rule8. Use of Direct Manipulation (Interface Without Conscious in Operation)

The user can operate by "direct manipulation" to operate the objects and contents directly without a user being conscious of UI.

Rule9. Real and Prompt Feedback (Interface Without Conscious in Operation)

Real and prompt feedback makes the interface Intuitive, consequently the user can operate intuitively without conscious.

Rule10. Consistency of Display and Operation (Interface Without Conscious in Operation)

Display and operation of UI are required consistency to be able to do the same UX between software and between hardware. It is the method to promote organization that makes the interface intuitive using consistency of visual metaphors, operational metaphors, mapping and feedback expressions.

2.2 Interface Like to Use Unintentionally

Intuitive User Interface is an approach to improve the interface more user-friendly. Although, a concept of the operation to induce enjoyment or like to use includes the intuitive user interfaces. The interface like to use unintentionally, namely user cannot help using, must need the requirement of easy to see, easy to understand and easy to use as basic features of Reaction and Routine level. Inoue et al. introduced the elements that the user can't help using the interface [6].

Human-Machine Interface Caused by Itself
This interface is designed to promote interactions between the user and the product. The elements of the interface need to acceptable for the user. A physical specification, for example, comfortable touch feeling, bring the feeling which the user wants to use. A futuristic impression which means the media including advanced technology implicates the users like to use. A feedback from the interface, such as a trackability according to the operation, the sound, the animation and the 3D graphic, is also induce user's preferable feeling. It is to conclude that the interface like to use is brought by a design of coordination between the user's comfortableness and the appropriate feedback [7].

Human-Machine Self Interface
The concept of Human-machine self-interface can be explained that the user promotes interaction with the product by oneself [3]. The user is attracted by internal UX from the interface such as senses of the accomplishment, pseudo experiences of the achievement, confirmations of the intelligent experience, sympathizes with the product meaning and so on. This kind of the interfaces have characteristics such as contents of discovery to satisfy the user's intellectual desire, mechanisms to learn the user's demand and representations to reflect the user's mind by character designs. To design Human-machine self-interface is required to examine and identify the relationship between the user's mental state and the product.

Human-Machine Social Interface
The aim of Human-machine social interface is explained that the user accelerates interactions with people form a society through the interface [3]. This interface has two aspects. One is explained as that the user enjoys to use the interface for boasting of one's properties or experiences through social network services. The other is associated with idea of "Gamification". Gamification is for the use of video game elements in non-gaming systems to improve UX and user engagement [8, 9]. The problem of introducing Human-machine social interface is to identify the relationship between members of the society shaped as applications or network services. Then, the interface should reflect the identified relationship. Designing of Human-machine social interface needs to represent irrational users' psychological aspect such as behavior, passion, emotion, value and so on, so it is recognized that Kansei engineering should be required to apply.

3 Kansei Interface for UX design

Kansei interface would be able to affect the consciousness of the user to enjoy the interface itself. In Kansei interface, the design of the interface cannot be clearly discriminated with its contents and applications because the purpose of the interface might be reflected the utility of the product. Therefore, it is effective to apply UX design into Kansei interface.

This study attempts to design Kansei interface for an example of smartphone application by using UX design processes as a pilot study. A ticket sales E-commerce site is selected for the object of the application. At the first step, experiences brought from the application were extracted by a focus group interview. Then, the extracted experiences were expressed into a customer journey map. This experiment collected ten phases to carry users' activities to the application and touch points of the phases were defined. Then, forty-two potential elements as countermeasures of the utility for the application were extracted. Finally, we selected eight ideas among the elements as the candidates which might be implemented by Kansei interface (Table 1).

Table 1. Candidates of utility for ticket sales site to be designed as Kansei interface extracted from customer journey map.

No.	Candidate of utility	Phase
1	Recommendation utility such as indies introduction, the hot artist	Reservation
2	Schedule management, calendar utility such as the live schedule or ticketing	Until concert day
3	The timetable utility that can manage the appearance time of the artist in festivals	
4	SNS utility that can communicate between users of the application	
5	Enquire utility for the popularity of goods sales at the venue represented by good!	
6	Real-time information utility that the situation and waiting time understand at the product sales and the entrance	Concert day – until entrance
7	Automatic brightness adjustment utility to light a screen automatically when the user shows the electronic ticket	Concert day – until curtain time
8	Traceability utility that can manage the history, which means schedule, the start, the curtain time, highlight video, and so on of the event	After concert

Then, we investigated relationships between user experience and emotion brought from the extracted potential utilities. A questionnaire was performed to examine user preferences for the utilities. We gathered thirty-nine respondents whose attributes are in Table 2. The results of the user preferences for the candidates are shown in Fig. 1. The number of utilities is according to Table 1. The resulted favorite utilities are No. 2, 3 and 6, which are related to functional elements. The not favorite utilities are No. 1 and

4, which are regarding user's preferences. These results indicate that Kansei interface should be designed as relating the functional futures to user's emotional preferences such as stress-less usability, entertainment elements like enjoyable display or personification, gamification and so on. This experiment suggested the process of UX design would be applied Kansei interface.

Table 2. Attributes of respondent who answered questionnaire of user preference of application utility.

Frequency of ticket purchase (paper or electronic)		
Not used	Sometimes (around 1/year)	Often use
18	16	5
Frequency of electronic ticket purchase		
Not used	Sometimes (around 1/year)	Often use
21	14	3

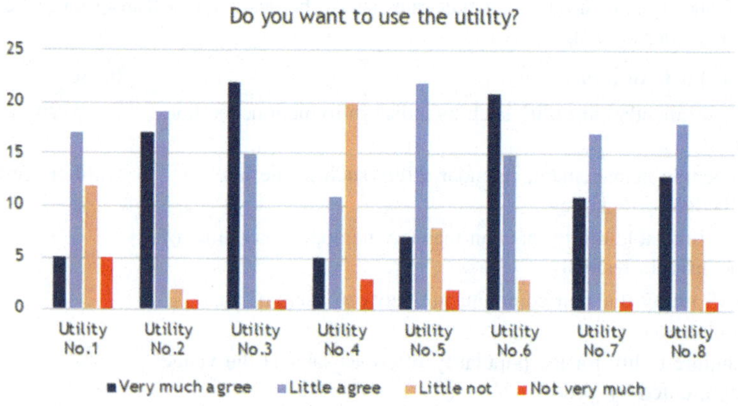

Fig. 1. The results of questionnaire to investigate favorite utilities for Kansei interface.

4 Conclusion

This paper illustrated the UX design procedures aiming to develop a Kansei interface. The concept of Kansei interface was defined as comprising emotional aspects directing to compose it like intuitive and desirable to use. The characteristics of Kansei interface was represented as the reference of UX design. Then, Human-machine self and social interfaces and Gamification were focused on. Then, the designing mobile interface was shown as an example of UX design. UX design of E-commerce site explored a customer journey map by user investigations. The application of Kansei interface was examined for the extracted activities at the touch points. The expected future work is to build Kansei interface to represent the relationship between application utilities and user's emotional aspects.

References

1. Ministry of Economy, Trade and Industry Japan. https://www.meti.go.jp/press/2019/05/20190516002/20190516002.html
2. Noro, K.: Illustrated ergonomics. Japanese Standards Association (1990)
3. Inoue, K.: Interface Design Textbook, 2nd edn. Maruzen Publishing, Chiyoda City (2019)
4. Norman, D.A.: Emotional Design: Why We Love (or Hate) Everyday Things. Basic Books, New York (2004)
5. Norman, D.A., et al.: Affect and machine design: lessons for the development of autonomous machines. IBM Syst. J. **42**(1), 38–44 (2003)
6. Inoue, K., et al.: Proposal of 10 rules to intuitive interface design methodology. In: The 62nd Annual Conference of JSSD 2015, June 2015
7. Inoue, K., et al.: Consideration of fascinating interface design. In: The 59th Annual Conference of JSSD 2012, June 2012
8. Deterding, S., et al.: Gamification: using game design elements in non-gaming contexts. In: Proceedings of CHI 2011, pp. 2425–2428, May 2011
9. Saitou, A.: What is Game-nics. Gentousha (2007)

Research on the Color of Physiotherapy Equipment - Example of the Lower Limb Equipment

Peng Jyun Liu[1,2](✉)

[1] Institute of Applied Arts, National Chiao Tung University, 1001 University Road, Hsinchu 300, Taiwan, ROC
a804338@gmail.com

[2] Department of Creative Product Design, Asia University, 500, Lioufeng Rd., Wufeng, Taichung 41354, Taiwan, ROC

Abstract. The global population is aging, creating a gradual increase in patients with chronic diseases. According to the theory of color therapy, color is able to increase positive feelings and enhance the effectiveness of treatment. This study, based on Kansei Engineering theory, explores the effect of color on use of Physical therapy equipment. The purpose of this study is to reinforce the patient's positive feelings and thus improve the efficiency of treatment. Firstly, relevant adjectives were found from references. Find five experts were sought out to select main adjectives and colours for feelings relating to lower limb physical therapy equipment images, get 6 sets of vocabulary and 12 colors. Secondly, a questionnaire survey was conducted through semantic differences method to further find the colours which relate closest to positive images of lower limb physical therapy equipment, valid sample of 334 people. Finally, according to the results, the lower limb physical therapy equipment recoloured, followed by another questionnaire survey, valid sample of 418 people. The results of this study show that yellow green (C30-M0-Y60-K0), red (C0-M55-Y30-K0), blue green (C60-M0-Y25-K0) color can enhance user rehabilitation have a positive feeling.

Keywords: Color · Physiotherapy equipment · Kansei Engineering · Color image

1 Introduction

Taiwan's aging population has resulted in more and more patients with stroke and chronic diseases. At the same time, the demand for rehabilitation has gradually increased, resulting in an increase in the number of rehabilitation patients. Taiwan's Census and Statistics Department (2016) points out that an average of 31 people develop cerebrovascular diseases every day. In addition, the proportion of patients aged 45 to 64 is the highest in the outpatient population of rehabilitation medicine, accounting for 33.4% [1]. In February 2014, the central health insurance department of the Ministry of health and welfare pointed out that the whole Taiwan spent a total of 500 billion yuan on health insurance in one year, of which the application for rehabilitation payment was

G. Di Bucchianico et al. (Eds.): AHFE 2020, AISC 1202, pp. 648–661, 2020.
https://doi.org/10.1007/978-3-030-51194-4_85

about 15 billion yuan, accounting for 3%–4% of the total expenditure. The medical management team of the health insurance department said that the growth rate of health insurance and rehabilitation medical expenses from 2008 to 2011 was 6%–7%, and in 2015 the number of Chinese rehabilitation outpatient clinics reached 1,733,450 and the number of hospitalized patients was 12,445, as shown in Error: Reference source not found. Wei [2] pointed out that the health insurance department has calculated that the health insurance expenditure for rehabilitation treatment has increased by 5% to 6% every year. In order to reduce the waste of health care resources, the health care department will take primary control over rehabilitation medicine. In the future, if the frequency of rehabilitation treatment for patients in medical units is too high, or if the duration and proportion of rehabilitation treatment are too long, health care will not be able to provide benefits [3].

Li and Lin [4] proposed that in order to improve the rehabilitation efficiency and make effective use of health care resources, if rehabilitation patients have good rehabilitation motivation, it will be conducive to achieving the maximum independent function of daily life, improve the quality of life and reduce the burden on families and society. During the rehabilitation treatment, the rehabilitation equipment can help the patients to get appropriate assistance for their health recovery during the rehabilitation. However, how to increase the patients' concentration, motivation and positive energy in using the rehabilitation equipment to improve the rehabilitation efficiency, reduce the frequency of rehabilitation treatment, and make the patients recover successfully as soon as possible are the problems that should be discussed in the rehabilitation road. The purpose of this study is to increase the positive energy (such as active, positive, optimistic, etc.) of patients through discussing the color plan of rehabilitation equipment, thus improving the rehabilitation efficiency and restoring health as soon as possible. At the same time, it can also reduce the heavy burden of health insurance funds, so that the whole population can enjoy good health insurance resources and medical environment.

Rehabilitation is a long and hard process. In terms of physical therapy, the basic course of treatment is 3–6 months, even lasting more than one year [5]. Three months after stroke discharge is a golden recovery period, and six months to one year is a golden rehabilitation period. However, if the recovery condition is poor, it may be necessary to enter the long-term care phase [6]. Active rehabilitation treatment during the golden period can restore a certain degree of walking ability and self-care. If rehabilitation treatment is stopped after 6 months, it will usually deteriorate again. However, as long as rehabilitation continues, even if stroke lasts for more than one year, the patient's condition may be improved again [7].

With the aging of the population and the promotion of policies, the scale of the medical device market is continuously expanding. Statistics from the European Union Medical Device Commission show that the United States, the European Union and Japan have accounted for more than 80% of the global medical device market. Among them, the United States is the world's largest producer and consumer of medical devices, accounting for more than 40% of the world's consumption [8]. It can be seen from this that the increasing market of medical equipment is directly proportional to the growth of elderly patients and patients with chronic diseases. Physical therapy is the most common treatment among rehabilitation patients. Therefore, this type of rehabilitation equipment will become one of the major medical equipment to be produced in the future.

Pleasant equipment is helpful for treatment, and at the same time, it can improve the therapeutic effect of patients and help patients to recover their daily life functions as soon as possible [9]. The ancient Egyptian murals have developed the theory of color therapy, and the therapeutic results have achieved obvious results [10]. Colors can bring people different feelings, such as beauty, ugliness, relaxation, stress, etc. [11]. For example, blue helps calm the mood of patients, while green has hope and the feeling of life, with soothing and auxiliary functions [12]. In the shape and function design of rehabilitation equipment, patients can feel safe. Color in appearance has certain influence on emotional expression of rehabilitation equipment [13]. At present, the common color matching of rehabilitation equipment is achromatic and pure color, mostly using gray, white, black, red, blue, green and other colors. Generally, cool colors such as green and blue are more common in medical devices, because green makes eyes more comfortable and can relieve patients' negative emotions such as tension and uneasiness, while blue gives people a quiet and calm feeling and can help patients control their emotions [14]. It can be seen that color is also one of the most important factors in the design of rehabilitation equipment.

Color therapy can be traced back to murals in Egyptian pyramids, acupuncture therapy in ancient China, and the theory and application of color therapy proposed by India's chakra energy center. At present, some doctors use color to help treat diseases. For example, Peter Mandel combines color with acupuncture, and uses special light pens to transmit color to various acupuncture points, proving that the treatment method has obvious effects [14]. Zhang Zhenlin and Shangguan Zhaoyi mentioned different diseases in their books and suggested usable colors, which can find the most effective way in the treatment process. For example, the effective colors for headache and hypertension are blue and green. The effective colors for allergy are indigo and soft orange. In 1982, an American study showed that blue light had a significant effect on the treatment of rheumatoid arthritis. Color therapy is also used to treat Alzheimer's disease, attention deficit and depression, allowing patients to have a variety of treatments that do not require medication [15]; Taichung Children's General Hospital (2015) specially organized a "Happy Sand Boy, Paint for Summer" activity, inviting volunteers and students to paint on the walls of the rehabilitation treatment area. Through color art activities, the positive and optimistic mood of patients can be restored, so that patients can put positive and positive feelings into the rehabilitation treatment [16].

The sensory organs of human receiving external information are 87% visual, 7% auditory, 6% olfactory, tactile and gustatory [17], it is mainly visual. Visual senses will be stimulated by information to produce perception and memory. When these memories and perceptual interactions are linked, cognition and related emotional significance will be produced [18]. The reaction to color is mostly derived from physiological feeling and psychological experience, while the memory, impression, symbol, association, experience, traditional habits, etc. when using color will affect our feelings to color [19]. People often associate colors with descriptive words for various colors with emotional associations [20], and endow feelings with colors and characteristics, thus forming different psychological feelings [21]. Guo and Ke [22] pointed out that color can not only express emotions, but also bring people different emotions and feelings in spirit.

Kansei Engineering is most commonly used to study people's feelings brought by products. It is a Consumer-oriented product development technology. It takes the emotional response and cognition of users as the research basis, and constructs products that meet consumers' sensory or psychological needs through statistical analysis and computer application, including product shape, color, material, size, function, etc. In a nutshell, the aim is to transform the consumer's sensibility or inner feelings into product design elements [23], as shown in Error: Reference source not found. The scope of application is very wide. Bo [24] took ALESSI products as an example to discuss and understand the semantics of product shape. In the research, the different designs of ALESSI are analyzed through Kansei Engineering, and the feeling of the product to users is discussed. Among the six emotional symbols used in ALESSI design, children's symbols are most popular with consumers. The study has sorted out six main influencing factors, which can serve as guidelines for future product design. Guo [25] conducted research on the design of space atmosphere for chemotherapy in Zhanghua Christian Hospital. The atmosphere feeling of the chemotherapy room environment in the hospital was discussed. The results showed that the reporting space of the hospital should be open, eye-catching and friendly. The waiting and discussion space is full and quiet. However, the treatment space must be warm and private. Dong [26] took cervical vertebra therapeutic instrument as an example to analyze the application of Kansei Engineering in home medical rehabilitation equipment. The experimental results show that these four groups of words, durable, safe, friendly and modern, are the main factors that affect the design of the cervical therapeutic apparatus. The conclusion can provide a reference for the design of rehabilitation related equipment (Fig. 1).

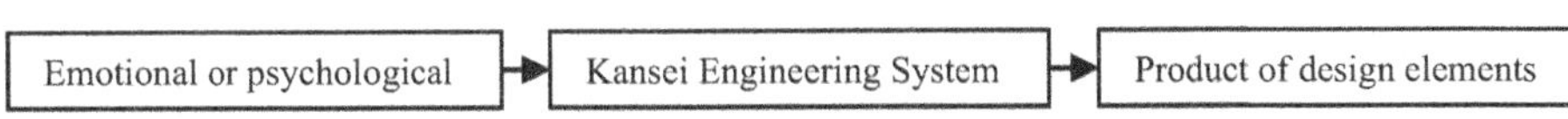

Fig. 1. Kansei Engineering System (KES) (Nagamachi 1995)

2 Research Method

The design process of this study is divided into five stages, including: (1) On-site visits to rehabilitation treatment centers; (2) Selection of rehabilitation-related descriptive words; (3) Color selection of rehabilitation imagery; (4) SD survey of rehabilitation related color image preference; (5) SD survey of rehabilitation equipment color plan.

Phase 1 is a rehabilitation treatment center visits, due to the regional and time constraints of the research team, purposive sampling was adopted to conduct field observation in three hospitals in central Taiwan. The selection of hospitals is based on the fact that the rehabilitation treatment center of the hospital has been established for more than 20 years, and the number and mode of medical treatment are also considered. Therefore, regional hospitals and regional hospitals are selected as the main targets. Before the on-site visit, communicate with the hospital in advance so as to conduct investigation and photographing during the outpatient service without affecting the patient and infringing the patient's privacy. At this stage, the investigation focuses on the types of existing rehabilitation equipment and the color matching of the existing rehabilitation equipment. In addition, it is necessary to know what kind of rehabilitation equipment patients will spend a long time watching during rehabilitation, so as to serve as the test sample carrier for the experiment.

In the second stage of the collection and selection of positive descriptive words related to rehabilitation, a total of 5 physiotherapists and 2 functional therapists with over 3 years of rehabilitation experience were invited by way of expert discussion. Rehabilitation-related descriptive words use 54 descriptive words (Table 1) listed in Dong [26] research related to product modeling. First, 54 pairs of descriptive words are made into word cards, and experts screen and integrate inappropriate and similar descriptive words. Finally, voting and discussion are conducted by consensus, and adjective pairs suitable for rehabilitation imagery are selected as SD survey evaluation scales.

Table 1. Adjective Pairs for Product Modeling (source: Dong [26])

Harmonious-conflicting	Safe-dangerous	Female-male	Delicate-coarse
Passionate-insipid	Nervous-relaxed	Steady-light	Exaggerated-reserved
Natural-artificial	Conventional-alternative	Bright-gloomy	Beautiful-ugly
Streamlined-straight	Rich-monotonous	Comfortable-uncomfortable	High-level-low-level
Innovative-conservative	Plain-affected	Smart-clumsy	Quiet-noisy
Bright-dark	Heavy-thin	Pure-dirty	Rational-emotional
Round-sharp	Implicit-overt	Warm-cold	Happy-sad
Happy-dull	Positive-negative	Popular-Individual	Solid-fragile
Enjoyable-uncomfortable	Calm-excited	Light-heavy	Friendly-cold
Simple-cumbersome	Excited-composed	Fixed-mobile	Environmentally friendly-polluting
Soft-hard	Lively-inflexible	Excited-quiet	Expensive-cheap
Standard-unordered	Healthy-Weak	Smooth-rough	Coordinated-abrupt
Anxious-peaceful	Modern-Traditional	Advanced-backward	Durable-Vulnerable
Clean-dirty	Efficient-inefficient		

The third stage is the color selection of rehabilitation imagery, which is mainly to find out the color matching colors used in rehabilitation equipment. The color selection in this stage is based on PCCS 129a color ticket produced by Japan Color Research in 2005 as the tested sample. A panel of experts was held to invite 3 experts with more than 4 years of design experience and 2 physiotherapists with more than 5 years of practice experience, totaling 5.

In the fourth stage of rehabilitation, SD survey on color image preference was conducted. The questionnaire design in this stage was divided into two parts, including basic data and color survey. The basic data report includes the background, gender, age, place of residence and how long the subject has received rehabilitation treatment. The color investigation part is the SD evaluation relationship between color and adjectives (Table 2). The sample number of the questionnaire is based on 1,745,895 outpatients and inpatients of the whole rehabilitation clinic and 9,872 practitioners of the whole rehabilitation therapist in 2015, with a total of 1,755,767. The confidence level of 95% and the sampling error of ±5 are input through the Creative Research Systems statistical website, and 384 copies need to be distributed in total. Subjects selected patients with rehabilitation for more than 3 months and rehabilitation therapists as subjects. The measuring scale adopts Likert 5 scale, and the questionnaire is distributed in the form of network questionnaire.

Table 2. Content example of rehabilitation imagery color SD questionnaire

		1	2	3	4	5	
	Healthy	☐	☐	☐	☐	☐	Weak
	Comfortable	☐	☐	☐	☐	☐	Uncomfortable
	Modern	☐	☐	☐	☐	☐	Traditional
C0-M55-Y30-K0	Efficient	☐	☐	☐	☐	☐	LowEffective

In the fifth stage, according to the colors obtained from the survey results in the fourth stage, the rehabilitation equipment will be color matched again without changing the overall design of the equipment. There are two types of color matching: the main color and the auxiliary color. The color matching principle is that the proportion of the main color is more than the auxiliary color. The color plan is located at the position where the vision is more frequently watched. In addition, achromatic color (black and white) will be included for color matching. Adobe Photoshop CS6 software is used to simulate the color matching tool in this study. The color planning vehicle is based on the on-the-spot investigation results of the rehabilitation center and uses the formal principle of beauty as a reference for color planning, and 20 sets of color matching samples are carried out on each carrier.

Considering that the number of samples is too large for the examinees to fill in and answer, a total of 5 designers with more than 3 years of design experience were invited to carry out preliminary color matching sample screening. In the form of expert discussion, the research team provided 20 sets of color matching paper samples of each carrier for each expert to review. In order to avoid the paper samples not being clear enough, the computer film projection method was used to provide expert reference, and the sample selection was decided by consensus through discussion. The experts obtained 8 samples and 6 pairs of descriptive words from the color survey of rehabilitation equipment through expert discussion (Table 3). The number of questionnaires distributed at this stage is 384 (confidence level 95% and sampling error of ±5). The subjects are all rehabilitation therapists and rehabilitation patients who have been undergoing rehabilitation therapy for more than 3 months. The measurement scale is Likert 5 scale, and the questionnaire is distributed in the form of network questionnaire.

Table 3. Results of expert screening samples

Number	A2	A3	A14
Color carrier	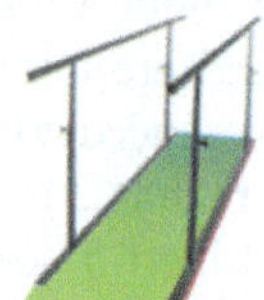		
Color matching principle	Symmetry Gradient	Symmetry Harmony	Symmetry
Number	B2	B12	B16
Color carrier	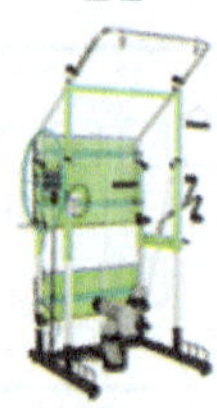	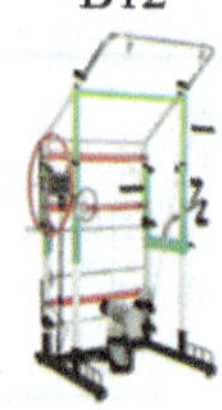	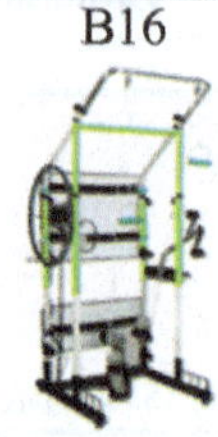
Color matching principle	Symmetry Harmony	Symmetry Gradient Compared	Symmetry Harmony

3 Research Results

According to the types of indoor rehabilitation equipment used in physical therapy in the rehabilitation medical center in Taichung, we know that the most common rehabilitation equipment are percutaneous electrical stimulator, intermediate frequency vector interference wave, neck traction, waist traction, parallel rods, followed by Xinwangai traction machine. According to the investigation results of the usage modes and time-consuming of each device, when performing vector interference wave and percutaneous electrical stimulator therapy, the patient will gaze at the device itself in a sitting posture, but does not need to spend too much attention on the device, while the parallel rod and the Xinwangai tractor are devices that the rehabilitation patient needs to gaze at for a long time. The results of the rehabilitation equipment color matching survey showed that the commonly used colors are pure color and achromatic color, and the used colors are black, white, blue, green, etc. Therefore, this study will use parallel rods and Xinwangai tractor as carrier to study the color plan (Table 4).

Table 4. Results of on-site visits to rehabilitation medical centers using rehabilitation equipment. (Photo data source: draw for this study)

Equipment name	Parallel rod		Xinwangai tractor	
			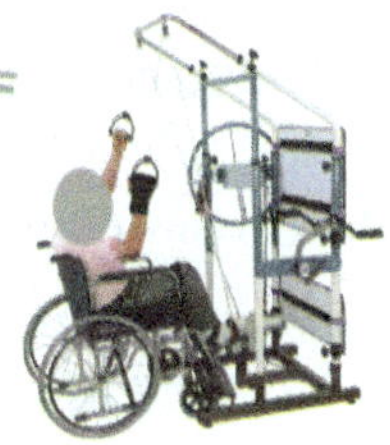	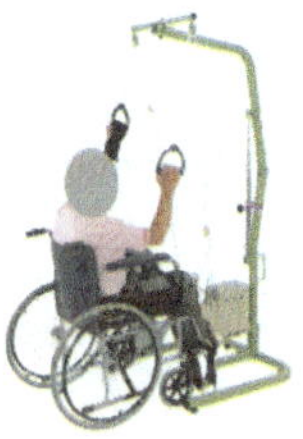
Category	A	B	A	B
Tone Used	Pure tone		Pure tone, light grey tone, achromatic color	
Color Used (CMYK)	Silver (C0-M0-Y0-K25), Green (C100-M0-Y100-K0), Blue (C100-M100-Y0-K0), Black (C0-M0-Y0-K100).		White (C0-M0-Y0-K0), Dark cyan (C100-M0-Y0-K45), Black (C0-M0-Y0-K100), Khaki (C0-M4-Y42-K6).	

For the selection of rehabilitation imagery descriptive words, experts will not be suitable for deletion, and the first suitable descriptive words will be selected for the second time. In the selection process, there is no limit on the number of experts to select and the relevant descriptive words suitable for rehabilitation will be selected by consensus. According to the results of the discussion, six groups of adjectives were obtained, namely, healthy-weak, comfortable-uncomfortable, modern-traditional, efficient-inefficient, safe-dangerous, rich-monotonous (Table 5).

Table 5. Selection results of experts on rehabilitation imagery descriptive vocabulary

Healthy-Weak	Modern-Traditional	Safe-Dangerous
Comfortable-uncomfortable	Efficient-inefficient	Rich-monotonous

According to the color screening results of rehabilitation imagery, the most appropriate color tone is light color tone, followed by bright color tone, and The third is soft tones (Table 6). Reference source not found. Therefore, 12 colors (Fig. 2) in the light color tone in PCCS are adopted as the color tone of this research.

In order to accurately adopt expert validity for verification, the rehabilitation color imagery SD questionnaire design invited a total of 3 experts with design background to examine the appropriateness of the topic content according to the research purpose, and to correct the color block size. In the pre-test questionnaire, there were 16 males and 20 females, and 36 questionnaire were collected. The reliability Cronbach's Alpha value of 0.95 is good. In the pre-test questionnaire, it was found that the subjects were too close to colors, resulting in visual fatigue. In order to avoid the visual fatigue and color confusion of the subjects, the correction of color ticket items in the questionnaire is arranged in a relative color manner, for example, the first color is lt2 and the second color is lt14.

Table 6. Expert tone sorting results

Color number	p (light)	lt (Shallow)	b (Bright)	ltg (light grey)	v (pure)	g (grey)	d (chaos)	dp (deep)	dk (dark)
Sorting	5	1	2	4	3	7	6	8	9

Fig. 2. 12 colors of lt (light tone)

The official questionnaire was distributed for a total of 11 days, 390 copies were distributed and 334 copies were recovered, with a recovery rate of 86.9%. The questionnaire recovery rate should be at least 50% to be sufficient for report analysis. The recovery rate exceeding 70% is already a very good amount of information (Qiu 2002). The confidence level of the questionnaire was 95%, the sampling error was 5.36, the Cronbach's Alpha value of reliability was 0.92 as good, KMO verification was 0.85 > 0.8 as good (Kaiser 1974) (Table 5, 6, 7 and 8), Bartlett's spherical verification significance was 0.00 as significant. This questionnaire has good reliability and validity, and the data of the questionnaire were suitable for analysis. The results of the questionnaire were analyzed by SPSS 20 software, and the average, standard deviation and mode were calculated. SPSS 20 was used to analyze the survey results at this stage. Color among adjectives was less than the median in the results at this stage. Therefore, unilateral evaluation was adopted. The evaluation was based on the criteria of average value <2.9, standard deviation <1, and mode <2 to find colors close to rehabilitation image adjectives. The survey results are shown in Table 7.

According to the overall survey results in Table 6, lt10 yellow-green (C30-M0-Y60-K0) is closest to healthy, comfortable and modern images. Lt2 red (C0-M55-Y30-K0) is closest to modern and abundant images. Lt14 turquoise (C60-M0-Y25-K0) is closest to the image of safety.

During the SD survey of rehabilitation equipment color plan, 3 experts with design background were invited to carry out expert validity verification. During the verification process, it was suggested that the original color matching of color matching carrier and the samples after color matching should be randomly sorted. Before the questionnaire, the subjects included 21 males and 13 females, and 34 were collected. The reliability Cronbach's Alpha value is 0.93.A total of 418 questionnaires were distributed and 418 were recovered, with a recovery rate of 100%. The questionnaire was distributed for a total of 6 days. 95% confidence level and sampling error is 4.59, Cronbach's Alpha value is 0.80 > 0.8 or above is good, KMO test is 0.95 > 0.8 or above is good, Bartlett's sphere test significance is 0.00 significant, representing that the questionnaire has reached good reliability and validity, and the average, standard deviation and mode of the questionnaire are analyzed.

Table 7. Analysis of investigation results of rehabilitation color imagery

	lt 2	lt 4	lt 6	lt 8	lt 10	lt 12	lt 14	lt 16	lt 18	lt 20	lt 22	lt 24
Healthy-Weak	2.65 1.24 2	2.70 1.10 2	2.80 1.20 4	2.49 1.17 2	1.96 0.97 1	2.38 1.17 2	2.30 1.19 1	2.54 1.17 2	2.76 1.23 2	3.26 1.19 4	3.23 1.15 4	2.95 1.14 3
Comfortable-uncomfortable	2.58 1.15 2	2.75 1.08 3	2.63 1.12 3	2.40 1.12 2	1.94 0.86 2	2.14 1.00 1	2.32 1.16 1	2.42 1.13 2	2.71 1.19 3	3.12 1.19 4	3.21 1.15 3	2.74 1.13 3
Modern-Traditional	2.16 0.95 2	2.70 1.16 3	2.75 1.18 3	2.49 1.13 2	2.07 0.91 1	2.63 1.20 3	2.49 1.17 3	2.51 1.21 2	2.69 1.20 3	2.82 1.26 3	2.93 1.25 3	2.70 1.18 3
Efficient-Inefficient	2.62 1.18 3	2.90 1.13 3	2.94 1.15 3	2.61 1.16 3	2.46 1.10 2	2.30 1.01 3	2.50 1.18 2	2.52 1.16 3	2.69 1.17 2	3.13 1.18 4	3.13 1.13 3	2.98 1.22 3
Safe-Dangerous	2.98 1.27 3	2.93 1.06 3	2.76 1.10 3	2.55 1.14 3	2.26 1.11 2	2.17 1.00 1	2.09 0.99 1	2.31 1.17 1	2.50 1.18 3	3.09 1.12 3	3.11 1.09 3	3.05 1.11 3
Rich-Monotonic	2.21 0.95 2	2.70 1.10 3	2.87 1.11 3	2.51 1.18 2	2.40 1.16 2	2.61 1.19 3	2.59 1.17 3	2.65 1.12 3	2.87 1.23 3	3.09 1.22 3	3.11 1.18 3	2.75 1.14 3

Note:

1.Selection is based on average < 2.9, standard deviation < 1, and mode < 2.

2.The numerical value of each pair of adjectives, the first column is the average; the second column is the standard deviation.

Table 8. Color plan of rehabilitation equipment-overall analysis results

	受测样本1	受测样本2	受测样本3	受测样本4	受测样本5	受测样本6	受测样本7	受测样本8
Healthy-Weak	4.35 0.95 5.00	1.66 0.75 1	1.75 0.78 2	2.04 0.96 2	3.87 0.97 3	1.67 0.83 1	1.88 0.85 2	1.69 0.74 2
Comfortable-uncomfortable	4.23 0.96 5	1.73 0.79 2	1.70 0.74 2	2.00 0.97 2	3.83 0.95 3	1.67 0.79 1	1.82 0.87 2	1.77 0.72 2
Modern-Traditional	4.34 0.91 5	1.76 0.86 1	1.62 0.71 1	2.04 0.96 2	4.16 0.91 4	1.66 0.78 1	1.71 0.76 2	1.69 0.75 2
Efficient-Inefficient	4.16 0.95 5	1.82 0.85 2	1.72 0.76 1	2.04 0.97 2	4.18 0.93 5	1.71 0.83 1	1.77 0.84 1	1.81 0.81 2
Safe-Dangerous	2.24 0.99 2	4.45 0.88 5	4.47 0.83 5	4.21 0.96 5	2.13 0.97 2	4.48 0.77 5	4.39 0.89 5	4.41 0.92 5
Rich-Monotonic	4.48 0.89 5	1.78 0.92 1	1.71 0.78 1	2.09 0.97 2	4.11 0.89 4	1.68 0.83 1	1.76 0.77 2	1.77 0.81 2

Note:

1.selection is based on average < 2.5, standard deviation < 1, and mode < 2. The dangerous-safe words are placed upside down, so the samples selected in this image are based on average < 4.5, standard deviation < 1, and mode > 4.

2.The numerical value of each pair of adjectives, the first column is the average; the second column is the standard deviation.

The results of the survey were analyzed by SPSS 20 to find color plan samples of rehabilitation equipment close to rehabilitation imagery. In this stage, the selected samples were selected unilaterally for each adjective selection. Since the positive descriptive words are all less than the median in the results of this stage, unilateral evaluation was adopted. According to the average value <2.5, standard deviation <1, and mode <2, in addition, the dangerous-safe words in the questionnaire are placed upside down in order to verify the credibility of the examinees. Therefore, the samples selected in this imagery are based on the average value <4.5, standard deviation <1, and mode >4. From Table 8, we know that sample 1 and sample 5 have no imagery that are close to healthy, comfortable, modern, efficient, safe and rich. The overall survey result Table 8 is the closest to the rehabilitation imagery, followed by sample 2 and sample 3.

4 Conclusions and Suggestions

Based on the survey results of systematic and scientific research on the positive color imagery of rehabilitation equipment, the research conclusions are as follows:

1. Color of Rehabilitation Imagery
   (1) According to the overall results, the most suitable color for rehabilitation imagery is lt10 yellow green (C30-M0-Y60-K0), followed by lt2 red (C0-M55-Y30-K0) and lt14 blue green (C60-M0-Y25-K0). Therefore, this result will be used for subsequent color planning.
   (2) In terms of the background of the subjects, the colors most suitable for rehabilitation imagery in the rehabilitation patient population are lt10 yellow green (C30-M0-Y60-K0) and lt2 red (C0-M55-Y30-K0), followed by lt12 green (C50-M0-Y50-K0) and lt14 cyan (C60-M0-Y25-K0). Among the rehabilitation therapists, lt2 red (C0-M55-Y30-K0), lt4 red orange (C0-M50-Y40-K0), followed by lt24 purplish red (C0-M50-Y10-K0) are the most suitable rehabilitation imagery. The lt2 red color (C0-M55-Y30-K0) was found to be the closest to the reconstructed imagery between the rehabilitation patients and the rehabilitation therapists.
   (3) In terms of gender, the most suitable color for rehabilitation imagery in male population is lt10 yellow green (C30-M0-Y60-K0), followed by lt14 blue green (C60-M0-Y25-K0).The most suitable colors for rehabilitation imageey in female population are lt10 yellow green (C30-M0-Y60-K0), lt2 red (C0-M55-Y30-K0), followed by lt14 blue green (C60-M0-Y25-K0).Lt10 yellow-green (C30-M0-Y60-K0) and lt14 blue-green (C60-M0-Y25-K0) are found to be the closest to the reconstructed imagery between male and female populations.
   (4) The results of this stage show that the overall standard deviation is large, which indicates that the opinions of the subjects are scattered to a large extent. There are two possible reasons for this estimation: 1. Color difference is caused by different displays of computers or 3C products; 2. The standard deviation will be indirectly affected by the different subjective consciousness and preferences of each person's feeling for color. Finally, the degree of concentration of the participants' opinions is checked by mode.

2. Color Plan for Rehabilitation Equipment
   (1) The overall color matching sample that most conforms to the rehabilitation image is sample 6, followed by sample 2, sample 3 and sample 7. The results show that the color matching principles are mainly symmetry, harmony and gradation.
   (2) According to the background of the subject, the most suitable rehabilitation imagery in the rehabilitation patient population is sample 6, followed by sample 8, sample 2, sample 3 and sample 7. Among the rehabilitation therapists, the most suitable imagery is sample 6, followed by sample 7, sample 2, sample 3 and sample 8. Sample 6 is found to be the most consistent with the rehabilitation imagery in the tested population.
   (3) According to gender, the most suitable rehabilitation imagery in male population are sample 2, sample 3 and sample 6, followed by sample 8 and sample 7. In the female population, the most suitable rehabilitation imagery is sample 6, followed by sample 2, sample 7, sample 3 and sample 8. Sample 6 is found to be the most consistent with the rehabilitation image in the gender group.
   (4) In terms of color matching carrier A and B, samples 2 and 3 are the most suitable rehabilitation images in color matching carrier A; In color matching carrier B, the most suitable rehabilitation imagery is sample 6, followed by sample 7 and sample 8.
   (5) The overall results of SD survey at this stage found that the subjects' perception of the original color matching of the carrier selected in this study was not close to the rehabilitation imagery. In addition, when analyzing various ethnic groups, it is found that the opinions of the subjects on sample 4 are scattered to a large extent, possibly because the tested samples are less discriminating. In addition, the dangerous-safe adjectives were put upside down in the questionnaire, and the standard deviation was found to be greater than 1 in the analysis of various ethnic groups. The possible reason was that some subjects did not answer in detail.

Although the results of this study are the same as those of green and blue used in the color matching of rehabilitation equipment sorted out in the literature, the color matching area of rehabilitation equipment currently on the market is mainly occupied by achromatic color, and the green (100, 0, 100, 0) and blue (100, 100, 0, 0) used are mainly in pure tones. Through expert screening and a large number of questionnaire survey results, this study shows that the color matching color suitable for rehabilitation equipment is light tones. The colors are lt10 yellow green (C30-M0-Y60-K0), lt2 red (C0-M55-Y30-K0) and lt14 blue green (C60-M0-Y25-K0), respectively, which are more suitable for color matching of rehabilitation equipment. From the research results, we can know that color does affect psychological feelings, and the research hypothesis is verified. The research results can be used to improve the color matching of current rehabilitation equipment, increase the positive feeling of patients or therapists for rehabilitation treatment, improve the treatment effect of patients, and further reduce the cost of health insurance payment. At the same time, the research results can provide reference for future medical or color-related research, hoping to improve the current medical color matching, so that patients can be treated in a better environment, so that patients can shorten the treatment time and return to daily life as soon as possible.

This study is limited in time and can only conduct preliminary research and discussion. There are still some deficiencies. There is still an opportunity for further research on this research direction in the future. Suggestions for future research are as follows:

1. Increase the number of on-site visits to the field: This study is currently aimed at regional hospitals and abbotsford regional hospital. In the future, teaching hospitals, medical centers, clinics, etc. can be observed in depth. Because rehabilitation patients in different medical hospitals do not have the same symptoms, for example, rehabilitation treatment of patients with different symptoms can be observed in teaching hospitals and medical centers.
2. Questionnaire distribution method: due to the color difference caused by different screen displays during network distribution, the questionnaire can be distributed through paper in the future. For fear that the pictures in the paper questionnaire are too small or unclear, a chart card produced by professional printing method can be attached or a computer projection method can be used to provide the reference for the examinee to answer.
3. Using scientific measuring instruments: Because everyone has different feelings for color, in order to increase the accuracy of the research, eye movement instrument experiments can be used to obtain more objective research results and verification.

Acknowledgments. Thanks to Taiwan's Ministry of Science and Technology for funding to attend the international conference. Thanks to Chiao Tung University and Asian University for conducting this research, Relocation assistance.

References

1. Su, Z., Chen, S., Wang, T., Wang, Y., Lian, Y.: Classification of diseases in rehabilitation medical services: recent experience of a medical center in China. Taiwan J. Rehabil. Med. **38** (4), 229–236 (2010)
2. Wei, Y.: Rehabilitation frequency is too high, health insurance will not pay. Freedom Times, Taipei, Taiwan (2014). http://news.ltn.com.tw/news/life/paper/756262. Accessed 05 Sept 2016
3. Dong, N.: Those with a high frequency of rehabilitation will not pay for future health insurance. TVBS News. Taipei, Taiwan (2014). http://news.tvbs.com.tw/local/522316. Accessed 20 Sept 2016
4. Li, P., Lin, Q.: Explore the motivation and obstacles of rehabilitation patients engaged in rehabilitation. J. Welfare Technol. Serv. Manage. **4**(2), 277–278 (2016)
5. Pan, J.: A new doctor in pain: Dr. Pain, a musculoskeletal pain expert, will take you to find the source of pain, relieve the pain, and get rid of the pain. Sancai Publishing, Taipei (2016). ISBN 9789863427353
6. Lin, Y.: Golden period within 3 months after stroke repair. Health Medical Network, Taipei, Taiwan (2014). http://health.ettoday.net/news/382072. Accessed 22 Mar 2017
7. Zhu, X.: 6 ideas, stroke rehabilitation is no longer difficult. Health Magazine Issue 187, Taipei, Taiwan (2014). http://www.commonhealth.com.tw/article/article.action?nid=68459&page=2. Accessed 22 Mar 2017

8. Zhang, L.: 2016 domestic and foreign medical device industry market size and company rankings. China Business Industry Research Institute, Taipei, Taiwan (2016). http://m.askci.com/news/dxf/20160808/15112451379.shtml. Accessed 20 Sept 2016
9. Li, C., Huang, L., Chen, M.: Investigation on the use of upper limb rehabilitation products in stroke patients. Sci. Technol. J. Hum. Soc. Sci. **19**(2), 103–114 (2010)
10. Guan, Z.: The magic of color: make good use of color to enhance the interest and health of life (Originally by Carin Hick). Think Tank Culture Press, Taibei City (1998). ISBN13 9789578396388
11. Lin, W.: Research on the influence of digital color configuration on attention and relaxation. Master's Thesis, Department of Information Management, National Changhua Normal University, Changhua (2015)
12. Zhang, X., Zhang, M.: Emotional design of medical equipment. Packag. Eng. **29**(6), 157–159 (2008)
13. Guo, J.: Design and simulation of lower limb rehabilitation equipment based on ergonomics. Taiyuan University of Technology Master's Thesis in Design Art, China (2010)
14. Yu, H.: Humanity research in medical product design. Art Des.: Theor. Edn. (1), 109–110 (2014)
15. Li, R.: Color therapy can help sleep, but also treat diseases. Epoch Times (2013). http://www.epochtimes.com.tw/n61567/%E8%89%B2%E5%BD%A9%E7%99%82%E6%B3%95%E8%83%BD%E5%8A%A9%E7%9C%A0-%E4%B9%9F%E8%83%BD%E6%B2%BB%E7%99%82%E7%96%BE%E7%97%85.html. Accessed 14 Mar 2017
16. Huang, X.: Children's general hospital adds color to rehabilitation. Health Medical Network, Taichung, Taiwan (2015). http://www.healthnews.com.tw/news/article/22416/?act. Accessed 20 Sept 2016
17. Wodsworth, R.H.: Basics of audio and visual system design. National Audio-visual Association, Washington, D.C. (1983)
18. Zheng, Y.: Research on color recognition in older people. Master's thesis, Department of Industrial Design, National Yunlin University of Science and Technology, Yunlin, Taiwan (2001)
19. Zheng, G., Lin, P.: Color scheme. Yifengtang Publishing, Taipei (1987). ISBN 9979789494427
20. Gao, G.: Investigation and research on aging color image and preference. Master's Thesis, Institute of Industrial Design, Tunghai University, Taichung, Taiwan (2002)
21. Shi, S.: Architectural environment color design. Shuxin Publishing House, Taipei (1991). ISBN13 9789575313425
22. Guo, Z., Ke, M.: Research on the user's visual perception and emotions with different combinations of wall tones and furniture colors-taking the waiting room of a dentist clinic as an example. J. Des. Environ. **15**, 53–70 (2014)
23. Mitsuo, N.: Kansei engineering: a new ergonomic consumer-oriented technology for product development. Int. J. Ind. Ergon. **15**(1), 3–11 (1995)
24. Bo, R.: Exploration and cognition of product shape semantics-taking ALESSI products as an example. Master's Thesis of Institute of Design and Creation, Ming Chuan University, Taipei (2005)
25. Guo, R.: Research on the atmosphere design of chemotherapy space-a case study of Changhua Christian Hospital. Degree Thesis of Institute of Innovation and Design, Taipei University of Science and Technology, Taipei (2011)
26. Dong, C.: Research on the application of perceptual engineering in home medical rehabilitation equipment. Harbin Engineering University Master Lu Wen, China (2014)

Young Chinese Consumers' Perception of Passenger Car Form in Side View Analyzed with Quantitative and Qualitative Methods

Chunrong Liu, Kui Gao, and Qichang He(✉)

Shanghai Jiao Tong University, Shanghai 200240, China
{cheeronliu,heqc}@sjtu.edu.cn, 13913599877@163.com

Abstract. It explores the perception of the triple-compartment passenger car form in side view of young Chinese consumers/users based on their estimate on the similarity of 130 form samples. Nine form categories are classified by cluster analysis and consumers' perceptual map is plotted by multidimensional scaling, respectively. The gradual form change related to the distribution of form samples in the perceptual map is further analyzed qualitatively. The result shows that young Chinese consumers perceive and judge in their cognitive process the passenger car form in side view and its variation by both overall and local form features. The former are features including the length of vehicle and overall vehicle's posture, while the latter are ones such as the ratio of the width of front side door to that of rear side door, the inclination of beltline, and the transitional relationship between C-pillar and rear deck.

Keywords: Passenger car form in side view · Consumer/user research · The perceptual mapping · Quantitative and qualitative analyses

1 Introduction

The past decade has seen the boom in auto consumption to the Chinese mainland that auto manufacturing and sales both revealed a general trend of dramatic increase and the Chinese mainland ranked first in car productions and sales for many years [1]. In 2018, 23,709,782 passenger cars were sold in the Chinese mainland [2]. The passenger car product has been a mobility need for daily lives of the Chinese consumers, among whom the young consumers begin to be in main force role as passenger cars' consumers tend to be younger obviously [3].

It was found that the appearance of passenger car is ranked the forth place among a dozen of key factors influencing Chinese consumers' purchase decisions [4] and that when Chinese consumers observe and perceive passenger car form, the side view is the second most important point of view that reflects the design intent and product family features while the front quarter view is the greatest concern [5, 6]. Meanwhile, it is proved that there exists significant difference between designers' and users' product form perception [7]. It is neccesary for designers to understand the characteristics of young consumers' perception of passenger car form in different views for the creative ideation process.

G. Di Bucchianico et al. (Eds.): AHFE 2020, AISC 1202, pp. 662–668, 2020.
https://doi.org/10.1007/978-3-030-51194-4_86

Based on exploring form similarity judgement by young Chinese consumers/users, the perceptual characteristics of passenger car form in front view [8], in rear view [9], in front quarter view [10] and in rear quarter view [11] are revealed respectively. In this study, the characteristics of young Chinese consumers' perception of passenger car form in side view are probed, in an attempt to help designers understand young Chinese consumers' morphological awareness of passenger car form in side view.

2 Quantitative and Qualitative Analyses Procedure

2.1 Stimuli

Firstly, 130 pictures are collected from websites, presenting main passenger car forms in side view with engine displacements from 1.5 L to 2.4 L, which are in main force role in the current Chinese mainland's family car market [12] and manufactured as production cars in Chinese market by home-grown and joint-ventured motor corporations which, respectively, hold such 25 brands as FAW, Dongfeng, SAIC, GAC, Changan, Chery, BYD, FAW-Toyota, FAW-VW, FAW-Mazda, FAW Audi, SAIC-VW, GAC-Honda, Dongfeng-Honda, Dongfeng-Nissan, Dongfeng-Peugeot, Dongfeng-Citroen, Dongfeng-Renault, Beijing-Hyundai, BBAC (Beijing Benz), Changan-Mazda, Changan Ford, BMW-Brilliance, SAIC-GM Buick and SAIC-GM Chevrolet.

Secondly, all pictures are selected in a way that ensures they were visually at as approximately similar viewing angle as possible, and are then processed in specific aspects to reduce the possibility that these aspects distract the subject from the passenger car form presented in the pictures [10].

Finally, all samples in pictures of passenger car form in side view are marked randomly with serial numbers for the follow-up form similarity judgement survey.

2.2 Form Similarity Judgement Experiment

60 young consumers/users, including 32 male and 28 female subjects with age of 20 to 45, are invited to estimate on the form similarity between 130 passenger car samples presented in the corresponding pictures with a self-developed application tool (as shown in Fig. 1), where a subject carries out grouping tasks in seven turns to evaluate the form similarities until all samples are put into eight subgroups, resulting in a piece of similarity matrix data.

As a result, 60 pieces of valid data are acquired, and an averaged similarity matrix is calculated by averaging all similarity matrices.

2.3 The Perceptional Mapping and Qualitative Analysis

By analyzing the dissimilarity matrix transformed from the averaged similarity matrix with multidimensional scaling method, subjects' perceptual map of passenger car form in side view is plotted, visually mapping the form samples in subjects' cognitive space.

In the perceptional map, every sample is located with accurate (x, y) coordinate value resulted from multidimensional scaling.

With help of related car styling and design knowledge, authors' team members observe and analyze in a noncontinuous way the latent regularities, facing the wall with sample pictures printed in A4 size and stuck approximately to it according to their (x, y) coordinate values, i.e., their corresponding locations as presented in the perceptual map. As an output of individual exploration and group discussion, main clues on form distribution and change are traced and recorded.

3 Result of Qualitative Analysis

3.1 Form Changes on the Horizontal Dimension

By comparing and analyzing the form samples of passenger car form suck on the wall and form change along the horizontal direction, the regularities of the morphological transitional features of passenger car form in side view from left to right along the x axis in the perceptual map are explored and summarized as follows.

Firstly, there is a certain form change clue observed that form samples located toward the left end of the x axis have longer overall length than those located toward the right end of the x axis.

Secondly, there is a certain form change clue observed that form samples located toward the left end of the x axis have shorter front overhang (as illustrated on the left in

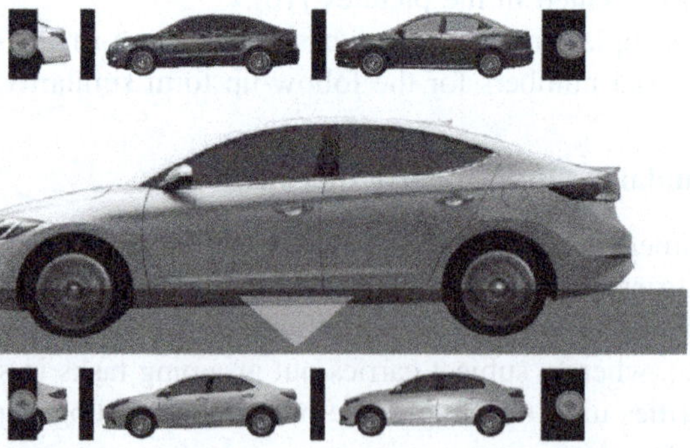

Fig. 1. The interface of the application tool used for form similarity judgement.

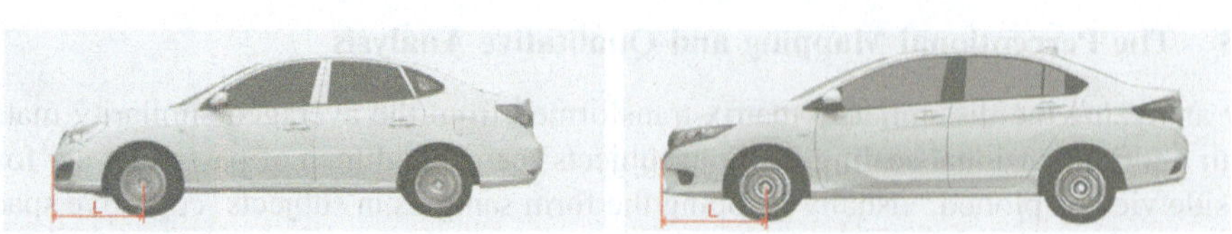

Fig. 2. The feature of front overhang length.

Fig. 2) than those located toward the right end of the x axis, which tend to have longer front overhang (as illustrated on the right in Fig. 2).

Thirdly, there is a certain form change clue observed that for form samples located toward the left end of the x axis, the front door has nearly identical width to rear door while for those toward the right end of the x axis, the front door is distinctly wider than rear door.

Fourthly, there is a certain form change clue observed that for form samples located toward the left end of the x axis, the C-pillar appears more probably thinner while being more probably thicker for those located toward the right end of the x axis.

Fifthly, there is a certain form change clue observed that for form samples located toward the left end of the x axis, the end area of side window is more probably to cross the parting line of rear door (as illustrated on the left in Fig. 3) while being more probably to be enclosed in rear door (as illustrated on the right in Fig. 3) for those located toward the right end of the x axis.

Sixthly, there is a certain form change clue observed that for form samples located toward the left end of the x axis, the side window is completely edged with chrome trims (as illustrated on the left in Fig. 4) while for those located toward the right end of the x axis, the side window is only edged with chrome trim along the top edge of doors or without any chrome trim along its boundary (as illustrated on the right in Fig. 4).

Seventhly, there is a certain form change clue observed that for form samples located toward the left end of the x axis, the distance of scuttle to front fender in side view is more probably greater (as illustrated on the left in Fig. 5) while for those

Fig. 3. The feature of relationship between the end area of side window and rear door.

Fig. 4. The feature of chrome trims.

Fig. 5. The feature of vertical distance of scuttle to front fender in side view.

located toward the right end of the x axis, that distance is more probably shorter (as illustrated on the left in Fig. 5).

3.2 Form Changes on the Vertical Dimension

By comparing and analyzing the form samples of passenger car form suck on the wall and form change along the vertical direction, the regularities of the morphological transitional features of passenger car form in side view from the top end to the bottom end along the y axis are explored and summarized as follows.

Firstly, there is a certain form change clue observed that for form samples located toward the bottom end of the y axis, the overall vehicle is more apparent in posture of dive (as illustrated on the left in Fig. 6) than form samples located toward the top end of the y axis (as illustrated on the right in Fig. 6).

Secondly, there is a certain form change clue observed that for form samples located toward the bottom end of the y axis, the beltline is generally in simple and integrative shapes (as illustrated on the left in Fig. 7) while for those located toward the top end of the y axis, the beltline is more probably in complicated and varied shapes (as illustrated on the right in Fig. 7).

Thirdly, there is a certain form change clue observed that for form samples located toward the bottom end of the y axis, the profile of C-pillar and rear deck looks more probably flat and stationary (as illustrated on the left in Fig. 8) while it is more probably smooth and as if moving (as illustrated on the right in Fig. 8) for those located toward the top end of the y axis.

Fourthly, there is a certain form change clue observed that for form samples located toward the bottom end of the y axis, the outline of the end area of side window looks more probably in a turn-back shape (as illustrated on the left in Fig. 9) while it looks more probably in an extension shape (as illustrated on the right in Fig. 9) for those located toward the top end of the y axis.

Fifthly, there is a certain form change clue observed that for form samples located toward the bottom end of the y axis, the side skirt area is generally less curvatured

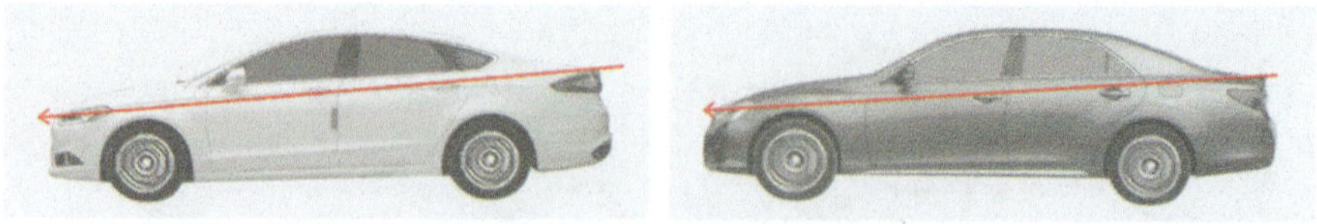

Fig. 6. The feature of overall vehicle posture.

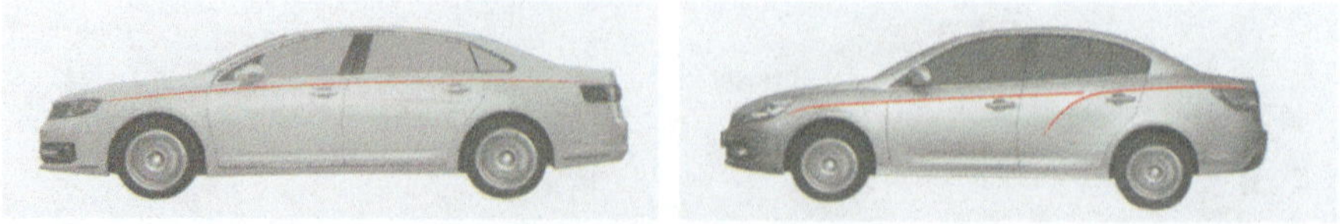

Fig. 7. The feature of beltline.

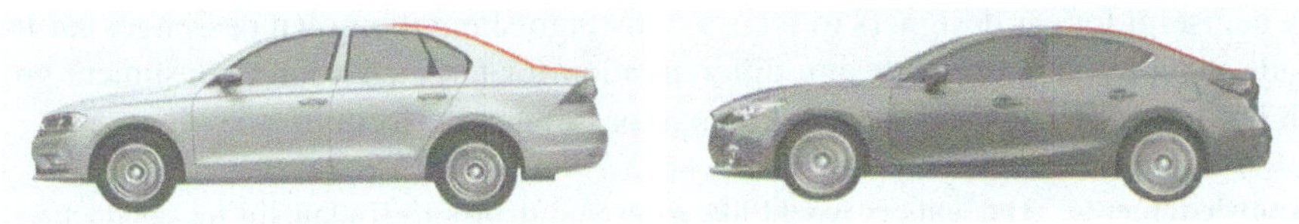

Fig. 8. The feature of profile of C-pillar and rear deck.

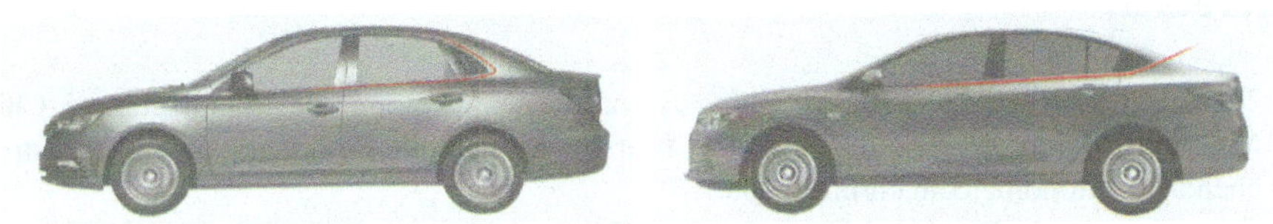

Fig. 9. The feature of the outline of the end area of side window.

Fig. 10. The feature of the side skirt area.

surface (as illustrated on the left in Fig. 10) while it is featured with larger bulging degree (as illustrated on the right in Fig. 10) for those located toward the top end of the y axis.

4 Conclusion

As young Chinese consumers begin to be in main force role in the fiercely competitive auto market in the Chinese mainland where auto consumption remains room to grow after two-year slightly slowing-down, it is unavoidable problem for automakers to understand the perceptual characteristics of passenger car form's dissimilarity of young Chinese consumers/users and form's variance implied in their morphological cognition.

By quantitative multidimensional scaling and qualitatively analyzing morphological shifts hidden in the perceptual map, it concludes that the characteristics of consumers' perception of passenger car form in side view can be mapped on the horizontal and the vertical dimensions in the perceptual map by dimension reduction, presenting the differential form features and gradual changes along the horizontal and vertical directions, respectively, and involving in both overall and local form features.

As the side view is the second most important point of view when Chinese consumers/users observe and perceive passenger car form, the characteristics of young Chinese consumers' perception of passenger car form in side view unveiled in this study may be useful for car designers to focus on distinguishing their own passenger car form in side view by intentional design differentiation for targeted young consumers group with both overall and local form features in passenger car form in side view.

Acknowledgments. The authors would like to extend thanks to Kuikui Jin for conducting the form similarity judgement survey in this study.

References

1. The Annual Report in 2017 by China Association of Automobile Manufacturers: China Maintains to Stay World No. 1 for Nine Years in Automobile Production and Sales. https://chejiahao.autohome.com.cn/info/2168516
2. Sales Volumes of Passenger Car in Various Countries/Districts in the World in 2018. http://www.caam.org.cn/chn/4/cate_40/con_5225632.html
3. The Influence of Slowing Growth of Chinese Macroeconomics in 2016 on Automobile Sales and Characteristics, Demand and Capacity of Consumer Populations. http://www.chyxx.com/industry/201612/476907.html
4. Yang, F.: Study on influence of passenger car's form on consumers' purchase decision. Master's thesis, Shanghai Jiao Tong University, Shanghai (2016). (in Chinese)
5. Liu, C.: Design strategy development for product innovation (Revised Edition). Shanghai Jiao Tong University Press, Shanghai (2019). (in Chinese)
6. Liu, G., Huang, D., Liu, C.: Preference of car form feature based on DEMATEL method. China Pack. Ind. **24**, 34–36 (2014). (in Chinese)
7. Hsu, S., Chuang, M., Chang, C.: A semantic differential study of designers' and users' product form perception. Int. J. Ind. Erg. **25**, 375–391 (2000)
8. Liu, C., Ding, X., Xie, Y., Jin, Y.: Young consumers' perception of the similarity between the passenger car forms in front view. Pack. Eng. **39**(24), 5–10 (2018). (in Chinese)
9. Liu, C., Jin, Y., Ding, X., Xie, Y.: Young chinese consumers' perception of passenger car form in rear view. In: Marcus, A., Wang, W. (eds.) Design, User Experience, and Usability: Users, Contexts and Case Studies. DUXU 2018. LNCS, vol. 10920. Springer, Cham (2018)
10. Liu, C., Gao, K.: Young Chinese users' perception of passenger car form in front quarter view analyzed with quantitative and qualitative methods. In: Ahram, T., Falcão, C. (eds.) Advances in Usability and User Experience. AHFE 2019. Advances in Intelligent Systems and Computing, vol. 972. Springer, Cham (2020)
11. Liu, C., Xie, Y., Jin, Y., Ding, X.: Understanding young chinese consumers' perception of passenger car form in rear quarter view by integrating quantitative and qualitative analyses. In: Fukuda, S. (ed.) Advances in Affective and Pleasurable Design. AHFE 2018. Advances in Intelligent Systems and Computing, vol. 774. Springer, Cham (2019)
12. 2016' Report on Local Brands in China's Car Market. http://www.motortrend.com.cn/umbraco/surface/Article/ArticleDetail/5279

Examining the Incidence of Intuitiveness and Cognitive-Affective States in the Kansei and Rating of Mobile Apps

Sandrine Fischer(✉)

San Francisco, CA, USA
sandrine.fischer@ieee.org

Abstract. Kansei analyses have predominantly been used for modeling users' affective responses on physical products. We propose a method for expanding Kansei attributes to measure user satisfaction for digital products. By leveraging constructs of intuitive use and cognitive-affective states of learning, along with online product reviews, we build a catalog of interaction Kansei. Finally, we demonstrate how this catalog can be used to determine which attributes are most relevant for measuring user satisfaction of mobile applications.

Keywords: Kansei analysis · Intuitive use · Mobile applications · Prior knowledge effects · Cognitive-affective states · User satisfaction · Learning

1 Purpose

The Kansei of a product designates the multidimensional emotional response, impression, or feeling its properties (e.g., design, material) and qualities (e.g., utility, safety) impart on people. Kansei are typically words like adjectives and adverbs that express how users feel about products. Such words can be derived from product manuals, advertisements, and specialized articles. They can also be obtained from end-users themselves, through customer interviews or opinions posted online [1]. A more recent approach is to analyze academic papers related to Kansei [2, 3], since such literature conveniently compiles a variety of definitions and descriptions.

One issue with mining Kansei publications, though, is that they almost exclusively only consider objects and devices that are physical, e.g., clothing, household appliances, tools, and electronics. Such focus restricts the spectrum of Kansei words to physical characteristics, such as aesthetics, appeal, or sturdiness. Should this approach be extended to products that are digital, though, it may benefit from specifying Kansei related to how people interact with software and user interfaces. This would mean incorporating words from literature specializing in user behavior and cognition. To explore this view, we implemented a method for extracting product Kansei based on constructs from the domain of user interaction.

S. Fischer—Independent Researcher.

G. Di Bucchianico et al. (Eds.): AHFE 2020, AISC 1202, pp. 669–675, 2020.
https://doi.org/10.1007/978-3-030-51194-4_87

2 Method

The first goal of the method is to build a catalog of Kansei attributes and words that reflects constructs of user interaction. Prime candidates for addition to the catalog are constructs with a demonstrated impact on the use of digital products (or environments), but are yet to be documented as Kansei. In Sect. 3, we discuss in more detail the selection of two such constructs.

The method, shown in Fig. 1, surfaces the Kansei words pertaining to a construct of interest by collecting and processing the content of related academic articles. Since terminology used by scholars sometimes differ from words used by end-users, a consolidation task is necessary to reconcile the disparity of language. This requires finding related words such as synonyms and antonyms, either manually, or with the aid of a thesaurus [4] and/or lexical database [2, 3]. Here, an expert eye is needed to ensure that synonyms and antonyms stay true to the construct while grouping them into meaningful higher-order Kansei attributes. In this way, one can assemble an ever-growing Kansei catalog that accommodates key constructs of user interaction.

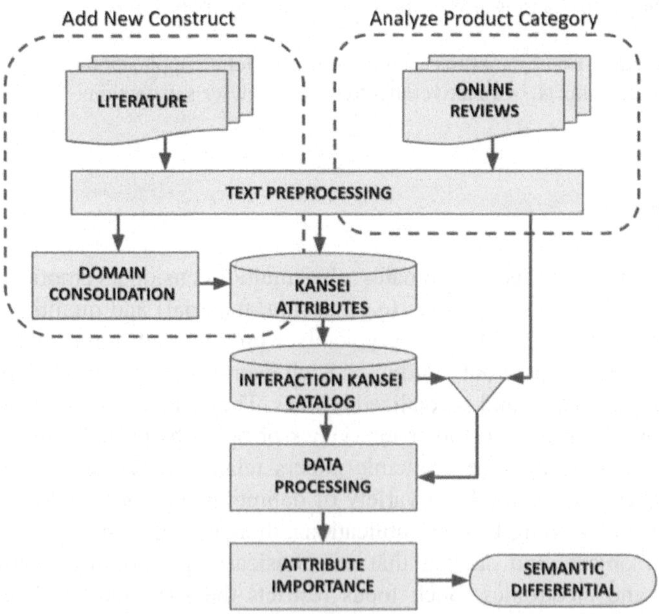

Fig. 1. Two-part method for building a catalog of Kansei words and selecting those that drive product satisfaction. In the two upper boxes, Kansei words and attributes are extracted from academic literature, organized in the Kansei catalog, and reconciled with words commonly found in end-user opinions. In the bottom half, the Kansei catalog is analyzed per product category in order to select which attributes drive user satisfaction.

In traditional Kansei Engineering analyses, the extraction of Kansei words and attributes is a basis for maximizing positive feelings and greater product satisfaction.

Thus, the second goal of our method is to identify, of all possible product responses listed in the catalog, those that correspond to positive reviews. The method provides a process to determine the Kansei attributes most highly correlated with positive ratings of a given product (bottom portion of Fig. 1) by analyzing online reviews of its category (upper-right of Fig. 1). This process interprets the rating of a review as a measure of user satisfaction, and uses a logistic regression classifier to select the attributes that drive greater satisfaction in that category. An application is presented in Sect. 4 using online reviews for a variety of mobile applications (apps).

3 Interaction Kansei Catalog

Thanks to a decoupling of function and form, digital products support greater multi-functionality than physical ones. More functionality should not mean more work, however, as users want digital products that are intuitive, i.e., they should take minimal effort, practice, and training to use. Intuitive use (or *intuitiveness*) is explained by a cognitive mechanism called *knowledge transfer*. In a nutshell, prior knowledge is subconsciously transferred to analog tasks, making their processing effortless. When knowledge is either missing or incompatible, cognitive mechanisms like interpretation, analysis, and learning take over. This makes processing not only effortful and prone to error, but also more conscious and explainable.

Literature on intuitive use details the behavioral [5], cognitive [6, 7], and subjective [8] repercussions of knowledge transfer on users. But views on the affective side of intuitiveness remain anecdotal. Studies on cognitive-affective states of learning [9, 10] provide the kind of structure required to frame the affective repercussions of trying new products. This research documented that a positive state of engagement occurs when learning tasks are mildly challenging. However, when learning requirements are too great, a shift occurs towards negative affective states, such as confusion, frustration, and disengagement.

Granted, digital products vary greatly in the prior knowledge required for their use, and discrepancies are likely to affect users' impressions, be it behavioral, cognitive, subjective, or affective. Therefore, traditional Kansei terms may benefit from incorporating the two constructs of intuitive use and cognitive-affective states of learning. Using research papers from 2008 to 2015 [11], we extracted the most common adjectives from these constructs and grouped their synonyms/antonyms into attributes. At the time of writing, our Kansei catalog, summarized in Table 1, contained 297 words grouped into 18 attributes.

Table 1. Sample of the interaction Kansei catalog. Kansei attributes for the constructs of intuitive use and cognitive-affective states of learning have been combined with the set of traditional Kansei terms. Note that each attribute is also a member of its corresponding positive Kansei, not shown because of redundancy.

Construct	Attributes	Kansei (positive)	Kansei (negative)
Intuitive use	Intuitive	Apparent, anticipated	Unaware, illogical
	Familiar	Similar, novel	Unconventional, unusual
	Knowledgeable	Proficient, trained	Non-expert, naïve
	Easy	Efficient, effortless	Difficult, slow
	Correct	Accurate, successful	Inexact, wrong
	Simple	Basic, elementary	Complex, multiple
Cognitive-affective states	Relaxing	Calming, eased	Frustrating, stressful
	Engaging	Amusing, fascinating	Banal, boring
Traditional Kansei	Appealing	Attractive, likeable	Bulky, worthless
	Comfortable	Gentle, smooth	Tough, loud
	Handy	Accessible	Inaccessible
	Aesthetic	Delicate, beautiful	Ugly, inelegant
	Compact	Small, short	Whole, complete
	Innovative	Futuristic, modern	Dull, old-fashioned
	Correct	Right, accurate, true	Wrong, inaccurate,
	Practical	Convenient, manageable	Inconvenient, useless
	Technological	High tech, advanced	Classical, basic
	Reliable	Dependable, safe	Weak, imprecise

4 Validity

A corpus of user reviews was gathered for 16 categories of iOS and Android mobile apps. These categories were selected to span three levels of prior knowledge on the user's part:

- **Basic (intended for broad audiences):** bucket list, travel journaling, point of sale, shopping, parking, meal delivery
- **Intermediate:** job boards, resume builders, app store news, travel planning, fasting, car sharing
- **Advanced (designed for specialized audiences):** personal finance trackers, business analytics, marketing applications, electric car control

Each category contained at least 5 apps, for which up to two years' worth of online reviews from users in the USA were collected. After filtering out reviews with fewer than seven words, sentences were segmented and part-of-speech tagging was applied to extract adjectives and adverbs that could serve as potential Kansei words. These candidates were then mapped to the catalog outlined in Table 1. Subsequent data processing consisted of dropping attributes that did not appear in at least three reviews, and removing all sentences that did not contain at least one attribute. Since the attribute "easy" appeared in a larger number of reviews, we removed all sentences where it was the only attribute. The remaining sentences were assigned the star rating of their parent review for the subsequent modeling steps.

Reviews for the Basic, Intermediate, and Advanced levels were modeled separately. For each level, a binary target variable was created by assigning ratings with 5 stars to a positive class and ratings with 3 or fewer stars to a negative class. Ratings with 4 stars were excluded to create a greater separation between positive and negative reviews. Because 5-star reviews outnumbered the others, the datasets were balanced by randomly down-sampling the positive reviews to match the number of negative reviews. The final number of rows for each dataset was 2 * 5685 for Basic, 2 * 8470 for Intermediate, and 2 * 3514 for Advanced.

The feature weights w_k, as reported by each of the three models, were used to determine which attributes more strongly predict a higher/lower rating. To avoid over-representing attributes that do not appear in many reviews, the frequency f_k of each attribute was measured and a weighted feature importance i_k was calculated as w_k * f_k. Figure 2 shows the top attributes for each category along with their weighted feature importance.

Basic		Intermediate		Advanced	
i_k	**Attribute**	i_k	**Attribute**	i_k	**Attribute**
0.154	easy	0.329	easy	0.199	intuitive
0.149	handy	0.157	practical	0.191	easy
0.091	appealing	0.115	simple	0.145	practical
0.062	simple	0.082	reliable	0.08	correct
0.060	practical	0.036	handy	0.077	simple
0.035	engaging	0.035	intuitive	0.055	reliable
0.025	compact	0.018	relaxing	0.041	handy
0.022	aesthetic	0.015	engaging	0.013	appealing
0.020	correct	0.011	familiar	0.011	knowledgeable
0.014	reliable	0.01	comfortable	0.011	engaging
0.014	comfortable	0.009	correct	0.005	comfortable
0.012	innovative	0.008	appealing	0.005	familiar
0.011	intuitive	0.007	compact	0.004	compact

Fig. 2. Weighted feature importance i_k for attributes in each of the three models. Only attributes with the largest magnitudes are shown.

Overall, it is clear from Fig. 2 that the two added constructs play a role in user satisfaction and should be considered in the Kansei of mobile apps. The importance of some attributes, such as *easy*, *simple*, *practical*, and *comfortable*, remained large across all three levels. Attributes such as *aesthetic* and *innovative* were mildly important for Basic apps and disappeared in the Kansei of Intermediate and Advanced apps. Among the attributes with varying importance, the biggest movers were *handy*, *engaging*, and *intuitive*. The attributes *handy* and *engaging* decreased from left to right, indicating that they are more important for Basic apps, and less so for Advanced ones. Conversely, the attribute *intuitive* increased from left to right, indicating it is most important to the satisfaction of Advanced apps.

By no means does this suggest that Basic apps *are* innovative, or that Advanced apps are the *most* intuitive. Rather, because they are not expected to be innovative, users may appreciate Basic apps with elements of innovativeness. Conversely, because Advanced apps involve specialized and less-standardized knowledge, users felt much more satisfaction upon finding that some were intuitive to use. This view echoes [12] and [13], who stated that products with low knowledge requirements (or whose domain is familiar) have the least to gain from being made more intuitive. Still, such products could gain from more creativity or originality, for example by adding elements of surprise or engagement. In fact, we found *engaged* to be more important for Basic apps than for Advanced ones. Again, Basic apps require common knowledge and little learning. This is consistent with [10]'s dynamics of cognitive-affective states, in that users who undergo less learning are more likely to experience a state of engagement.

5 Outlook

In this paper, a method was proposed for expanding traditional Kansei to constructs of user interaction, along with a principled way for determining attributes relevant to a given product category. The two added constructs primarily focused on how first-time usage affects user experience. Applying this method to a range of mobile apps revealed that users are more satisfied when apps feel "easy to use", whereas the aesthetic attribute is much less important. While being engaging, intuitive, or practical all sound like desirable product qualities, their actual relevance may fluctuate with the level of prior knowledge required by the users. We conclude that Kansei can achieve a finer understanding of user satisfaction and overall product ranking by considering constructs of users' prior knowledge.

Next steps include supplementing the Kansei catalog with constructs such as usability, persuasiveness, and playfulness. Application-wise, there is potential for our method to shortlist attributes or descriptors for questionnaires administered during Kansei analyses and user tests.

References

1. Li, Z., Tian, Z.G., Wang, J.W., Wang, M.W.: Extraction of affective responses from customer reviews: an opinion mining and machine learning approach. Int. J. Comput. Integr. Manuf. (2019). https://doi.org/10.1080/0951192X.2019.1571240
2. Wang, W., Li, Z., Tian, Z., Wang, J.W., Cheng, M.N.: Extracting and summarizing affective features and responses from online product descriptions and reviews: a Kansei text mining approach. Eng. Appl. Artif. Intell. **73**, 149–162 (2018)
3. Wang, W., Wang, J.W., Li, A.Z., Tian, Z., Tsui, E.: Multiple affective attribute classification of online customer product reviews a heuristic deep learning method for supporting Kansei engineering. Eng. Appl. Artif. Intell. **85**, 33–45 (2019)
4. Kobayashi, H., Ota, S.: The semantic network of KANSEI words. In: SMC 2000, pp. 690–694 (2000)
5. Blackler, A., Popovic, V., Mahar, D.: Investigating users' intuitive interaction with complex artefacts. Appl. Ergon. **41**(1), 72–92 (2010)
6. Fischer, S., Itoh, M., Inagaki, T.: Prior schemata transfer as an account for assessing the intuitive use of new technology. Appl. Ergon. **46**, 8–20 (2014)

7. Still, J., Still, M.L., Grgic, J.: Designing intuitive interactions: exploring performance and reflection measures. Interact. Comput. **27**(3), 271–286 (2015)
8. Diefenbach, S., Ullrich, D.: An experience perspective on intuitive interaction: central components and the special effect of domain transfer distance. Interact. Comput. **27**(3), 210–234 (2015)
9. Baker, R.S.J.D., D'Mello, S.K., Rodrigo, M.M.T., Graesser, A.C.: Better to be frustrated than bored: the incidence, persistence, and impact of learners' cognitive–affective states during interactions with three different computer-based learning environments. Int. J. Hum.-Comput. Stud. **68**, 223–241 (2010)
10. D'Mello, S.K., Graesser, A.C.: Dynamics of affective states during complex learning. Learn. Instr. **22**, 145–157 (2011)
11. Fischer, S.: List of intuitive use and learning research papers, 2008–2015. ResearchGate (2017). https://www.researchgate.net/project/list-of-intuitive-use-and-learning-research-papers
12. Fischer, S., Itoh, M., Inagaki, T.: Screening prototype features in terms of intuitive use: design considerations and proof of concept. Interact. Comput. **27**(3), 256–270 (2015)
13. Fischer, S.: Designing intuitive products in an agile world. Intuitive interaction: research and application (2018)

Research on Hospital Self-service Machine Interaction Design Based on Kansei Engineering

Zhengyu Shi[1], Huifang Shang[1(✉)], Chuanshun Wang[2], Zhengyu Wang[1], and Meiyu Zhou[1]

[1] School of Art Design and Media, East China University of Science and Technology, Shanghai 200237, China
945308090@qq.com, 753501966@qq.com, hfshang@ecust.edu.cn, zhoutc2003@163.com

[2] Shanghai Xian Dai Architectural Design (Group) Co., Ltd., Shanghai 200237, China
2079226881@qq.com

Abstract. In top tertiary hospitals situated in big cities in China, long registration wait times, payment and short treatment times are common problems due to the considerable number of patients. For this reason, self-service terminals are introduced by hospitals to shorten waiting time and improve patients' medical experience. In order to improve the quality of outpatient service reasonably and effectively, this paper used the qualitative and quantitative methods to study the interface of self-service terminal. The research combined Kansei Engineering theory with Service Design theory to make the perceptual evaluation of self-service terminal interface and improve the service content. Firstly, the perceptual vocabularies were collected from the literature of medical self-service system research, and the interface samples were collected from the self-service terminals used by public hospitals in Beijing, Shanghai and other big cities. Secondly, through the discussion of the expert group, the words with no obvious characterization were deleted, then the questionnaire method was used to screen out the high-frequency words and representative interfaces. Thirdly, the vocabularies and samples were integrated into a perceptual evaluation scale of the interface of self-service terminals. The research used factor analysis method to process the collected data from the questionnaire survey and summarized the significant perceptual vocabularies. Finally, combining with the field research of the hospitals, this paper discussed and analyzed the interface elements of the self-service system, and summarized an improvement strategy. It is to improve the readability of the interface, optimize the logic of interface interaction and strengthen the interface guidance.

Keywords: Kansei Engineering · Hospital self-service machine · Service design · Human-computer interaction

G. Di Bucchianico et al. (Eds.): AHFE 2020, AISC 1202, pp. 676–688, 2020.
https://doi.org/10.1007/978-3-030-51194-4_88

1 Introduction

1.1 Research Background

In hospital outpatient service, patients gradually increased the attention to service experience, and service quality has high influence on clinical process. Which is mainly demonstrated in whether the problem of hospital service process can be solved reasonably, whether the hospital's long waiting time and short time of treatment problem can be solved effectively, and whether the medical and health service development level can meet the people's demands etc. In view of the problem of hospital treatment service, large public hospitals build an "Internet + Intelligent medical" system and put corresponding self-service equipment to improve the service efficiency and meet patients' needs.

1.2 Research Status

According to the latest researches, many research teams in China and abroad have completed the framework of the theoretical system of self-medical-service and put it into practice. Self-medical service is an intelligent medical service system, which is concentrated on patient needs, through its foundation of hardware and software. In self-service and mobile terminals, the intelligent medical treatment is mainly the research and realization of human-computer interaction technology [1, 2]. The self-service equipment in hospital is mainly supported by HIS information system, and self-service machine and information screen as the carrier. Through Socket, Web Service and other interfaces, the self-service equipment interacts with LIS medical insurance, unionpay and other heterogeneous system data, and adopts DDN line to realize the docking with the online bank [3].

According to *Survey report on National Health Informatization (2019),* the self-service equipment in Chinese hospitals have provided 11 functions, such as registration, reservation service, queuing and stationing, and satisfaction evaluation. Among the tertiary hospitals, the highest proportion of service function opening is as follows: reservation service is 98.6%, self-service registration service is 82.5%, and queue-to-station service is 63.5%. In the secondary hospitals, the highest proportion of service function opening is as follows: booking service is 44.7%, satisfaction evaluation is 43.6% and queuing station to station service is 42.6% [4]. With the use of self-service equipment, the patient treatment process is convenient, and the work efficiency of hospital is also improved. On mobile devices, such as "Spring Rain Doctor", "WeDoctor", "Ping An Health Cloud" and other apps, users can get more medical information and treatment services.

Through the research and application of self-service equipment, foreign hospitals not only provide services, but also increase patients' medical experience. Lyu et al. optimized the design of community medical service machines to improve the efficiency, inclusiveness and accuracy of their services [5]. Megan et al. studied the public health information of hospitals and concluded that the dissemination of health education services and preventive screening through self-service devices could bring benefits to patients' health [6]. Markill et al. customized the self-service terminal of the health center for hospital patients based on their cognition of the medical environment [7].

1.3 Purpose and Significance

According to the literature on hospital self-service terminals and the survey of three hospitals in Shanghai, in large public hospitals, intelligent service devices are used less often by patients, mainly because of the availability of interactive interface, low acceptance degree of elderly users, maintenance and update of service equipment, etc. [8]. In this paper, through the combination of Kansei Engineering and Service design theory, from the three elements of man, machine and environment, research and analysis for hospital self-service equipment, targets are providing self-service interaction design strategies for patient preferences, optimizing the self-help service process and perfecting the self-help service system when people see a doctor.

2 Research Theory

2.1 Kansei Engineering

"Kansei Engineering" [9] was first proposed by Mr. Kenichi Yamamoto, who elaborated on Kansei Engineering in his *Automobile Culture Theory* published in 1986. Kansei Engineering is a technology that combines perceptual cognition with engineering theory. By analyzing people's perceptual cognition and preference for products, and combining qualitative reasoning with quantitative analysis, this technology calculates and processes the perceptual information of product demand. Research data can effectively combine perceptual cognitive information with product design elements to obtain product design trends [10].

2.2 Service Design

Service design is a design activities with the purpose to improve user experience and service quality, which effectively integrates and organizes the information exchange, infrastructure application and other relevant elements in the process of interaction between consumers and suppliers from a multi-dimensional disciplinary perspective [11]. The ultimate goal of service design is to design activities to meet human needs. In the design process, it adopts the participatory cooperation mode to let stakeholders in the service chain participate together and pays attention to the nodes of the service process to accurately grasp the operation rhythm of the service process. Service design adheres to the theoretical principle of serving people and makes the service system reasonable and tangible. In the process of promoting the sound and smooth service, it pays attention to the user's experience [12].

2.3 Application of Service Design and Kansei Engineering

Carreira et al. applied the method of extended Kansei Engineering to improve the design of travel services by combining user experience requirements (ER), user experience compositions (EC) and product elements in product service systems (PSS) [13]. Chen et al. designed and studied international express service based on Kansei Engineering theory, and provided new service contents for its service system

[14]. Wang et al. studied the catering service design by combining Theory of Inventive Problem Solving (TRIZ) and Quality Functional Deployment (QFD), and tested the usability of the design results [15]. Based on the combined application of Kansei Engineering and Service design, this paper studies the self-service terminals of hospitals by referring to the above practice methods.

3 Methodology

This paper mainly studies the interaction design of hospital self-service machine by using the theoretical method of Kansei Engineering. The test method mainly adopts questionnaire method, semantic difference method and statistical method. The optimization analysis method mainly adopts the factor analysis method to extract the most influential perceptual vocabularies. Using the theory of Service Design, this paper analyzes the problems and difficulties of patients in the human-computer interaction process and summarizes the corresponding service requirements. Through the deployment of service demand and perceptual vocabularies, the corresponding service content is researched and planned.

3.1 Collection and Selection of Perceptual Vocabularies

Firstly, this study summarized the interface of the hospital self-service machines that have been put into use in the market, and obtained the relevant attributes of the interface, including interface style, interface function, interface content, interface operation and interface experience. In this paper, perceptual vocabularies of interface description were extracted from the literature of hospital self-service terminal research, descriptive vocabularies of hospital service equipment were extracted from medical journals, and oral presentation vocabularies were obtained from user evaluation, and 38 perceptual vocabularies were preliminarily obtained, whose classification is shown in Table 1. On the basis of the collected perceptual vocabularies, an expert group was set up to discuss and classify the perceptual vocabularies. Secondly, the study conducted a questionnaire survey to the audience to obtain relevant feedback. The conclusions obtained from the two were summarized. Finally, six perceptual words were screened out, and six antisense words were selected to match with them to obtain a representative perceptual vocabularies group (As shown in Table 1).

3.2 Collection and Screening of Interface Samples

Through field research and network recommendation collection, the study obtained 25 interface samples of hospital self-service machines, mainly including the service equipment used by public hospitals in Beijing, Shanghai, Ningbo and other regions. Through the analysis and discussion of the expert group, which was mainly composed of interface experts and hospital patients, the interface samples were classified and summarized based on the interface style and form. Finally, through the process of screening and deleting similar samples, six different representative interface samples were obtained, including the service terminal interface used by affiliated hospital of

Table 1. Perceptual vocabularies selection of hospital self-service machine interface.

NO.	Interface Style	Interface Functions	Interface Content	Interface Operation	Interface Experience
1	Simple	Nonprofit	Ordered	Automated	Friendly
2	Traditional	Full-featured	Public	Convenient	Kind
3	Quasi-materialization	Intelligent	Clearly	Simple	Detailed
4	Visualization	Efficient	Accurate	Safe	Reliable
5	Flat	Effective	Understandable	Easy to learn	Smoothly
6	Characteristic	Purposeful	Intuitive	Easy to use	Humanized
7	High color contrast	Universal	Shared	Empirical	Rapid
8			professional	Normative	Guiding
Filter	Simple-Complex	Full-featured-Sparse	Precise-Rough	Understandable-Doubtful	Convenient-Troubled	Humanized-Mechanized

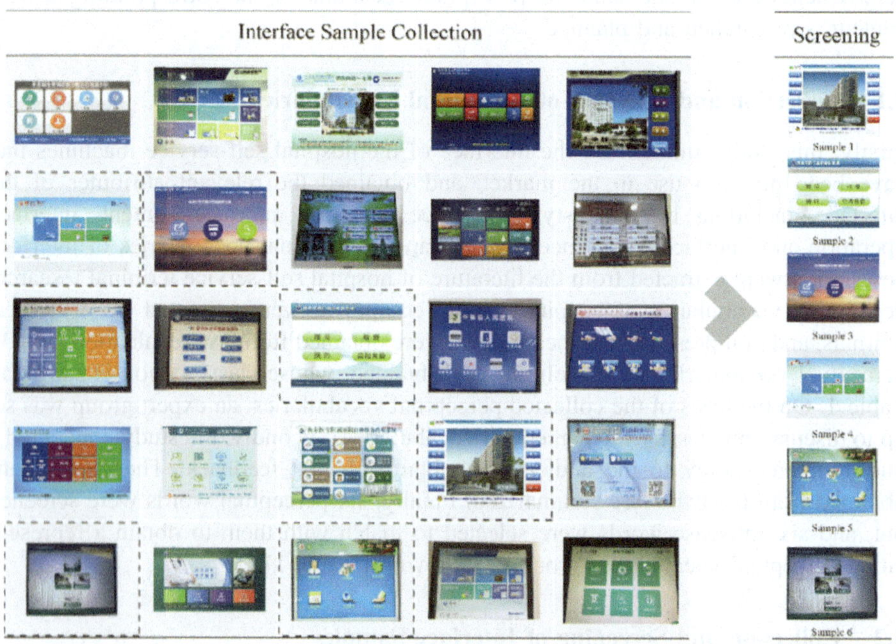

Fig. 1. Sample screening

Ningbo university medical college, Shanghai Ruijin hospital, Peking University Shougang hospital, Shanghai Huashan hospital, the Sixth people's hospital and Zhongshan hospital (As shown in Fig. 1).

3.3 Establishment of Multidimensional Perceptual Matrix

In this study, six groups of perceptual vocabularies and six interface samples about the interface of hospital self-service machine were obtained to form the perceptual scale. Based on Semantic Differential (SD) [16], each interface sample was accompanied by an intention space composed of six groups of perceptual terms, and a multidimensional perceptual matrix was established. In this study, the semantic difference method was used to establish a 5-order SD rating scale (As shown in Fig. 2).

Perceptual Words	Corresponding Score	Perceptual Words	Representative Sample
Complex	-2 -1 0 1 2	Simple	
Sparse	-2 -1 0 1 2	Full-featured	
Rough	-2 -1 0 1 2	Precise	
Doubtful	-2 -1 0 1 2	Understandable	
Troubled	-2 -1 0 1 2	Convenient	
Mechanized	-2 -1 0 1 2	Humanized	

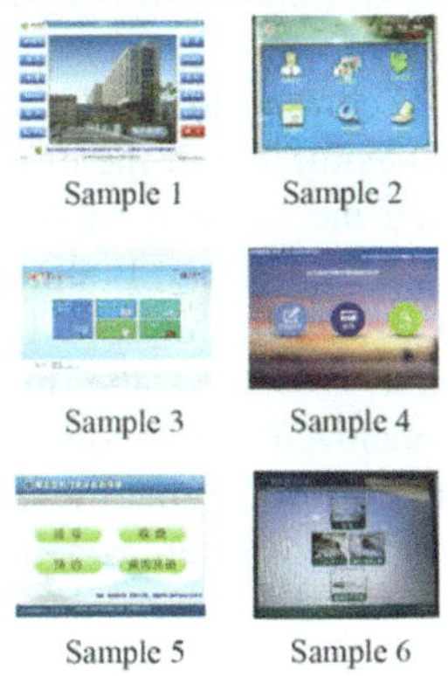

Fig. 2. SD rating scale

In the questionnaire survey, the experiment divided each group of perceptual image vocabularies evaluation of the interface into five levels, with 0 as the dividing line, and the two ends of the scale respectively indicated the degree of participants' perception of images. For example, for the evaluation score of "Simple" and "Complex", the left end is "Complex" with a score of −2, "Relatively Complex" with a score of −1, and the middle part is "No Special Feeling" with a score of 0, "Relatively Simple" with a score of 1, and "Simple" with a score of 2. The study used the SD rating scale to test the perceptual image of the interface samples of the subjects and obtained the corresponding evaluation data.

3.4 Questionnaire Data Statistics

This questionnaire test selected 94 participants who had self-service treatment experience in local hospitals in Shanghai, such as Huashan hospital affiliated to Fudan university, the Eighth people's hospital and the Sixth people's hospital, etc. The main characteristics of them were the education level of undergraduate and above, and their ability to understand intelligent machines was relatively high. The study conducted questionnaires on the subjects and obtained relevant data. After statistics, the mean

Table 2. Statistical table of perceptual vocabularies of interface samples

Vocabulary/sample	Sample 1	Sample 2	Sample 3	Sample 4	Sample 5	Sample 6
Simple-Complex	0.585	0.745	0.872	0.851	0.883	0.660
Full-featured-Sparse	0.798	0.777	0.830	0.755	0.649	0.734
Precise-Rough	0.617	0.691	0.915	0.777	0.681	0.734
Understandable-Doubtful	0.840	0.957	0.936	0.840	0.809	0.787
Convenient-Troubled	0.542	0.819	0.819	0.789	0.957	0.723
Humanized-Mechanized	0.500	0.649	0.830	0.766	0.649	0.691

value of perceptual images of the representative samples was obtained (As shown in Table 2).

According to the data in Table 2, samples 5 and 3 scored higher on the perceptual word "simple". In the evaluation of "Full-featured" perceptual vocabularies, sample 3 scored the highest; In the evaluation of "Precise" perceptual vocabularies, sample 3 scored the highest; In the evaluation of "Understandable" perceptual words, samples 2 and 3 are the highest; In the evaluation of "Convenient" perceptual vocabularies, sample 5 scored the highest; Sample 3 scored higher on the perceptual word "Humanized". After data evaluation and comprehensive consideration, it was preliminary concluded that the interface style of sample 3 was the most popular among users among the 6 samples.

4 Analysis

4.1 Questionnaire Data Analysis

The main analysis method is factor analysis [17], and many factors affecting the interface of hospital self-service machine are reduced to a few independent comprehensive factors through data statistical calculation.

In this study, the mean value of perceptual images representing interface samples was input into SPSS, and the data were processed and analyzed. Table 3 shows the statistical table of variance explanation (Statistical Table of Total Variance Explanation). Eigenvalue greater than 1 is the most common method to determine the number of factors, that is, factors with eigenvalues greater than 1 are retained and factors with eigenvalues less than 1 are omitted [18]. In the initial eigenvalue column, it is shown that the eigenvalue of factor 1 is 3.169, far greater than 1, and has a significant effect. The eigenvalue of factor 2 is 2.000, which is also greater than 1. However, the eigenvalues of other factors are all less than 1, and the influence degree is small, so there is no need to consider. In addition, the cumulative contribution rate of 86.161% can be obtained by selecting factor 1 and factor 2, that is, these two common factors can explain the total variance of about 86.161%, indicating that factor 1 and factor 2 have the most significant influence among the six factors and are significant factors.

Table 3. Statistical table of variance

Factor	Initial eigenvalue			Intercept the load of the sum of squares			Cyclic sum of squares loading		
	Total	Variation %	Cumulative %	Total	Variation %	Cumulative %	Total	Variation %	Cumulative %
1	3.169	52.822	52.822	3.169	52.822	52.822	3.030	50.503	50.503
2	2.000	33.339	86.161	2.000	33.339	86.161	2.139	35.658	86.161
3	0.707	11.791	97.952						
4	0.098	1.625	99.578						
5	0.025	0.422	100.000						
6	7.925E−17	1.321E −15	100.000						

4.2 Factor Analysis

Secondly, the study used SPSS to process and analyze the Scree Plot, as shown in Fig. 3. According to the shape trend of Scree Plot, the number of retention factors is determined, and the previous factor point when the curve changes from steep to gentle is considered as the maximum factor point to be extracted [19]. The main function of Scree Plot is to observe the changes of study factors, which is similar to Statistical Table of Variance. From the change trend in the figure, it can be seen that the change trend from factor 1 to factor 2 and factor 2 to factor 3 is relatively steep, and starting from factor 3, the subsequent curve changes gently, so it can be determined that factor 1 and factor 2 are significant factors in the research process.

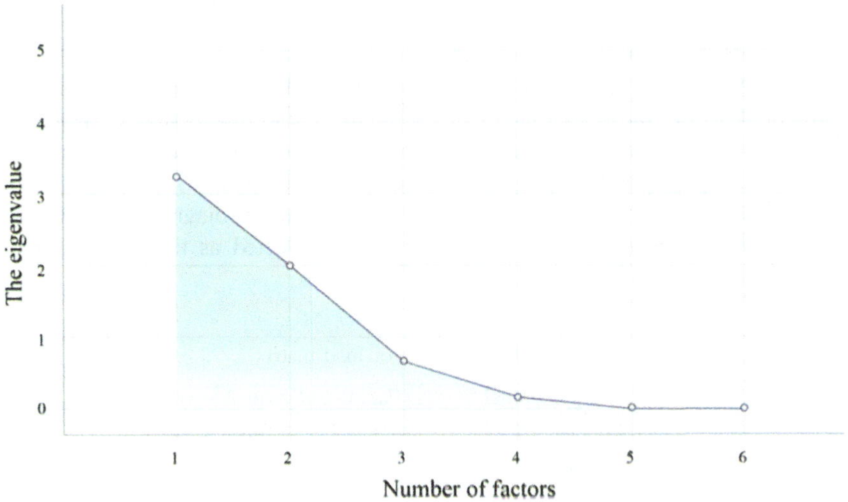

Fig. 3. Scree plot

The variance is used to illustrate the comparison of the statistical table of variance and the steep slope trend of the Scree Plot, and factors 1 and 2 are obtained and identified as significant factors in the research process. In the subsequent data analysis, these two factors will be used to establish the composition diagram and the initial factor load matrix in the rotation space.

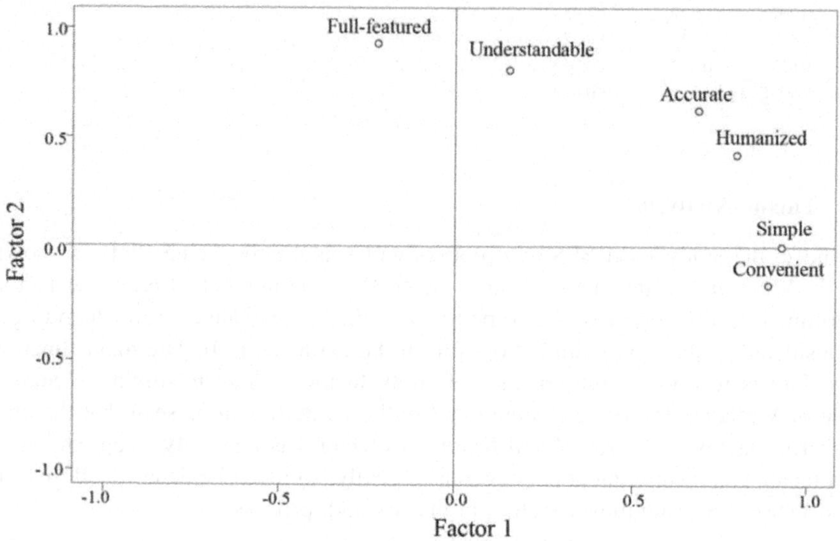

Fig. 4. Rotate the composition diagram in space

Based on the above two significant factors as the measuring scale, the composition diagram of perceptual words in the rotating space is obtained through data processing (Fig. 4). According to the figure, "Full-featured" and "Understandable" perceptual vocabularies are the most significant in the factor 1 coordinate scale, "Simple" and "Convenient" perceptual vocabularies are the most significant in the factor 2 coordinate scale, and other perceptual words are not prominent in the picture, and the significance is not great. Therefore, through the analysis of the component diagram, "Full-featured", "Understandable", "Simple" and "Convenient" are selected as the main component factors in the perceptual vocabularies.

Table 4. Initial factor load matrix

The perceptual words	Factor 1	Factor 2
Concise	0.325	−0.061
Full-featured	−0.155	0.473
Accurate	0.195	0.259
Understandable	0.016	0.352
Convenient	0.323	−0.191
Humane	0.257	0.152

In this study, factor 1 and factor 2 were taken as index items, and the perceptual vocabularies formed the initial factor load matrix (As shown in Table 4). The main perceptual factors were selected, and the load size occupied by the absolute value of the scoring coefficient of perceptual vocabularies under the factor was observed, and then the selection was made. According to the data statistics, the absolute value of the load ratio of "Simple" and "Convenient" perceptual words in factor 1 are relatively large. The absolute value of the load ratio of "Full-featured" and "easily understood" perceptual words are larger in factor 2. After comprehensive analysis, it is concluded that "Simple", "Convenient", "Full-featured" and "Understandable" are significant factors in the initial factor load matrix.

4.3 Result Analysis

Through mean statistics of perceptual images, factor analysis method, analysis of images and data, the following conclusions are drawn: Among the six perceptual words, "Simple", "Convenient", "Full-featured" and "Understandable" are significant perceptual words. The score of sample 3 in the above four factors is relatively high, so the combination of sample 3 can better meet the psychological needs of current users.

5 Discussion

The questionnaire's data statistics of the patients, combined with the observation and record of patients, allowed the study to finally discussed the data and phenomena obtained. In the initial factor loading matrix, "Simple" and "Convenient" have a higher proportion in factor 1, respectively 0.325 and 0.323, which indicates that patients attach more importance to these two perceptual words in the experience of interface use. In addition, according to the data in the mean table of perceptual intention of the interface samples, it was found that "Simple" and "Convenient" are positively correlated, and the higher the score of the samples in the "Simple" evaluation, the higher it will be in the "Convenient" evaluation. Based on the field investigation of the hospital, it was found that due to the problem of color contrast and block shape, the user could not easily identify the content of the icon and could not quickly find the required functional block. This phenomenon was also highlighted from the side, patient users in the "Simple" and "Convenient" level of demand.

In the initial factor loading matrix, "Full-featured" and "Understandable" account for a higher proportion in factor 2, respectively 0.473 and 0.352, which indicates that patients attach more importance to these two perceptual words in the experience of interface use. According to the data in the mean table of perceptual intention of the interface samples, it is found that there is no direct relation between them two. However, in the operation process of the interface, due to the age and education level of patients and other factors, the user's understanding of the service function was different, resulting in the phenomenon of difficult operation and psychological resistance to the service machine. Therefore, while satisfying the functional requirements, we should also improve the accessibility of the service interface.

In addition, according to the data in the mean table of perceptual intention of the interface samples, “Convenient” and “Full-featured” showed negative correlation. The higher the score of the interface samples in the “Convenient” evaluation, the lower it will be in the “Full-featured” evaluation. Combined with the field investigation of the hospital, it was found that the increase of interface function resulted in the problem of complicated information and large internal capacity. In the absence of relevant hints and guidance, patient users were prone to misoperation and repeated operation, resulting in low efficiency of the device. Therefore, optimizing the relationship between “Convenient” and “Full-featured” has become the key to improve the efficiency of self-service terminals.

Based on perceptual intention of hospital self-service machine interface and the research of problems in operation process, the article from the “Simple”, “Convenient”, “Understandable” and “Full-featured” perceptual perspective, proposed the improvements of the service interface design strategy: Improve interface readability, Optimize interface interaction logic, and Enhance interface guidance. The specific improvements are shown in Fig. 5.

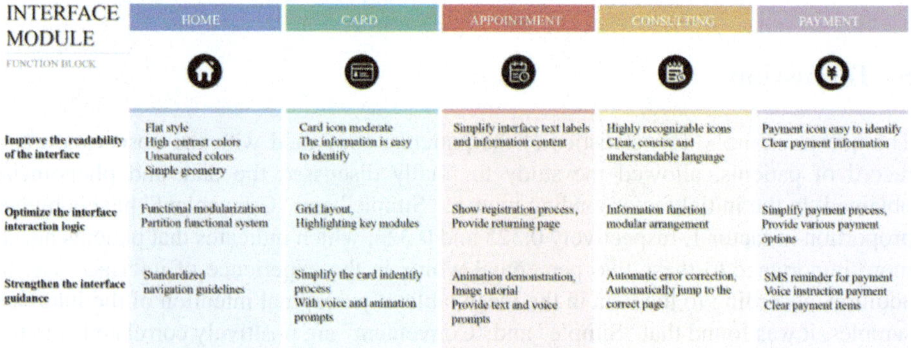

Fig. 5. Interface service improvement strategy

Based on the understanding of patients of different ages and knowledge backgrounds on the interactive content of self-service machine, this paper optimized the interface content from the following three aspects:

1. Improve the readability of the interface. The interface is flat style and the icons are simple, general geometric shapes that can be easily identified by users while illustrating functional information. In terms of color, high contrast color is used to highlight the high frequency function block. Unsaturated color is adopted to reduce the visual fatigue caused by color changes in the process of page jumping. In terms of text information, it simplifies the text label and information content of the interface, using clear and concise language to make it easy for users to understand.
2. Optimize interface logic. On the function panel, the service functions are modularized and integrated. While selecting the required functions, other functions are hidden and redundant information is omitted to improve the operation efficiency of

the interface. The function information is arranged in a grid according to the sorting logic, so that users can find the required function blocks conveniently while satisfying the multi-functions. In the process of interface operation, it provides the interface interaction mode that conforms to the user's operation preference, simplifies the functional operation steps, and gives timely feedback to the wrong operation; Display the level of the interface process, convenient for users to quickly select the required return page, improve the efficiency of operation.

3. Strengthen the interface guidance. For users who do not understand the operation process, the interface uses logical text interpretation, voice prompts and animation demonstration as the guide to assist the user to operate. When the user fails, the server actively corrects the wrong operation and provides correct guidance and feedback for the user.

6 Conclusion

Through the status of modern medical service, this paper conducts hospital self-service machine design research. Combining Kansei Engineering theory with Service design theory, this paper studies the interface of hospital self-service machine qualitatively and quantitatively. On the basis of the products already put into the market, this paper analyzes the preference of patients for the interface of self-service machine, summarizes its service characteristics, and extracts the four perceptual words of "Simple", "Convenient", "Understandable" and "Full-featured". Through field research, the paper summarizes the problems and difficulties in the operation process of self-service machine interface and puts forward the service demands for self-service function. Based on the interface service function block and key perceptual vocabularies, this paper summarizes the improvement strategies to improve the readability of the interface, optimize the interface interaction logic and strengthen the interface guidance.

References

1. Alnanih, R., Ormandjieva, O.: Mapping HCI principles to design quality of mobile user interfaces in healthcare applications. Proc. Comput. Sci. **94**, 75–82 (2016)
2. Pereira, A., Marins, F., Rodrigues, B., Portela, F., Santos, M.F., Machado, J., Rua, F., Silva, Á., Abelha, A.: Improving quality of medical service with mobile health software. Proc. Comput. Sci. **63**, 292–299 (2015)
3. Zhang, P., Ji, H., Jia, M., Sun, Z.: Develop self-service platform application system to improve the efficiency of outpatient medical service. Health Care China **34**(04), 259–260 +263 (2017)
4. Statistical information center of national health commission: A survey report on national health informatization, p. 81. People's Medical Publishing House, Beijing (2019)
5. Lyu, Y., Vincent, C.J., Chen, Y., et al.: Designing and optimizing a healthcare kiosk for the community. Appl. Ergon. **47**, 157–169 (2015)
6. Orlando, M.S., Rothman, R.E., et al.: Public health information delivery in the emergency department: analysis of a kiosk-based program. J. Emerg. Med. **50**(2), 223–227 (2016)

7. Mackrill, J., Marshall, P., Payne, S.R., et al.: Using a bespoke situated digital kiosk to encourage user participation in healthcare environment design. Appl. Ergon. **59**, 342–356 (2017)
8. Li, X., Zhou, D., Huang, C., Sun, S.: Interface usability design of self-service registration device based on eye tracking. Mach. Des. Manuf. (08), 145–148 (2018)
9. Nagamachi, M.: Kansei engineering: a new ergonomic consumer-oriented technology for product development. Int. J. Ind. Ergon. **15**(1), 3–11 (1995)
10. Zhou, M.: Kansei Design. Shanghai Science and Technology Press (2011)
11. Gummesson, E.: Organizing for strategic management—a conceptual model. Long Range Plan. **7**(2), 13–18 (1974)
12. Hu, F., Li, W.: Definition of "service design". Packag. Eng. **40**(10), 37–51 (2019)
13. Carreira, R., Patrício, L., Jorge, R.N., et al.: Development of an extended Kansei engineering method to incorporate experience requirements in product–service system design. J. Eng. Des **24**(10), 738–764 (2013)
14. Chen, M.-C., Chang, K.-C., Hsu, C.-L., Xiao, J.-H.: Applying a Kansei engineering-based logistics service design approach to developing international express services. Int. J. Phys. Distrib. Logist. Manage. **45**(6), 618–646 (2015)
15. Wang, Y.H., Lee, C.H., Trappey, A.J.C.: Service design blueprint approach incorporating TRIZ and service QFD for a meal ordering system: a case study. Comput. Ind. Eng. **107**, 388–400 (2017)
16. Osgood, C.E., et al.: The measurement of meaning. Audio-Vis. Commun. Rev. **2**(7), 503–504 (1957)
17. Liang, R., Zhang, L., Feng, W.: Psychological science education psychology multivariate statistics and SPSS software, vol. 3, p. 198. Beijing Institute of Technology Press (2005)
18. Thompson, B., Daniel, L.G.: Factor analytic evidence for the construct validity of scores: a historical overview and some guidelines. Educ. Psychol. Measur. **56**, 197–208 (1996)
19. Kong, M., Bian, R., Zhang, H.: The application of parallel analysis in exploratory factor analysis. Psychol. Sci. **30**(04), 924–925+918 (2007)

Kansei Evaluation Based on Quantifying the Identity of Smartphone Shape

Koki Chiba[1], Namgyu Kang[1(✉)], Jiuqiang Fu[2], Dongjin Seo[3], Khongthat Thongphun[4], Liu Jiehong[5], and Mohd Khairul Hussin[5]

[1] Future University Hakodate, 116-2 Kamedanakanocho, Hakodate, Hokkaido 0418655, Japan
{g2119025,kang}@fun.ac.jp

[2] Beijing Institute of Technology, 5 South Zhongguancun Street, Haidian District, Beijing 100081, China
fujq0306@126.com

[3] Sangmyung University, 20, Hongjimun 20gil, Jongno-gu, Seoul 03016, Korea
sdjin@smu.ac.kr

[4] Chiang Mai University, 239 Huaykaew Rd, Suthep, Mueang Chiang Mai District, Chiang Mai 50200, Thailand
kongtat.t@cmu.ac.th

[5] Nanyang Polytechnic, 180 Ang Mo Kio Ave 8, Singapore 569830, Singapore
{liu_jie_hong,mohd_khairul_hussin}@nyp.edu.sg

Abstract. Today, shapes of many smartphones are becoming similar to each other due to the fact that the shape has become minimal. The purpose of this research is to clarify the relationship between the identity of smartphones and their shapes. This research was conducted with the keywords of the nationality including GDP, share rate, and experience of using the smartphone. In an experiment, 185 participants of 5 different nationalities (Japan, Korea, China, Thailand, and Singapore) set the order of the 9 life-sized mock-up models (*iPhone X* and 8 *Android* smartphones) based on their similarity to the front shape of the *iPhone X*. As a result, the participants could distinguish the identity of the *iPhone X* to a certain degree from the information of only the shape. Moreover, the correct answer rate differs depending on the 3 conditions and experience of using the smartphone.

Keywords: *Kansei* evaluation · Identity · Smartphone · Shape

1 Introduction

Recent years, many users in diverse countries use a smartphone. (It means smartphone is demanding by people regardless of their wealth). Smartphone, without a physical keyboard, has been manufactured by many companies and sold in diverse countries since *Apple* introduced the *iPhone* in 2007. As a result, smartphones are becoming one of the essential tools in our lives [1]. There are many smartphone products, which can be held by one hand, are designed with liquid crystal screens covering most of the front of the main body. These shapes of many smartphones are becoming very similar to each other since the shape has become minimal with '*Notch Design*'. '*Notch Design*' means

G. Di Bucchianico et al. (Eds.): AHFE 2020, AISC 1202, pp. 689–695, 2020.
https://doi.org/10.1007/978-3-030-51194-4_89

the design that made the screen as large as possible except for a recessed part at the top of the screen for a speaker and a camera. This *Notch design* quickly became popular after *Essential Phone* had adopted this *Notch design* for the smartphone shape [2]. From that background, many companies are trying harder to attach their identities to their products for differentiating from others. However, there is a report of the case that the shape of the main body is similar even for products from different companies [3].

2 Previous Researches

2.1 Identity in Design

Ohkubo's research team conducted an experiment using 61 front-mask images of cars with 24 brands. They concluded that as a condition necessary for evaluating the identity of the design, a car brand must have a consistent shape or placement to some degree to maintain differentiated characteristics in shapes from those of others [4]. *Watami* reports that the identity of a product is connected, not only a matter of technology and appearance but also various experiences through the product [5]. Moreover, *Park* reported that consistency is the essential element for identity in smartphone and user experience [6]. In other words, as an element of smartphone identity, it was stated that consistent user experience from using a company's products is essential. From these previous researches, the importance of consistent experience, and required elements for the identity of the front mask shape of cars was revealed. However, there are almost no approaches to clarify the relationship between the shape of smartphones and identity yet.

2.2 Identity in Shape of Smartphone

Chiba and *Kang* conducted an experiment using 9 mock-up models, including the *iPhone X* and 8 *Android* smartphones, which had a similar shape and used the *Notch design* [7]. In the experiment, the participant looked at the printed nine mock-up models and ranked them in order from most likely *iPhone X* to most unlikely *iPhone X*. Those nine mock-up models had three different screen images that were prepared in the same conditions; 1) '*White condition*': All 9 models with the same white screen, 2) '*Android* condition': All 9 models with the same *Android* screen (*Huawei, P20 Pro*), and 3) '*iOS* condition': All 9 models with the same *iOS* screen (*Apple, iPhone X*). This experiment was conducted with various participants in Japan, Korea, and China. As a result, in participants' evaluation to recognize the shapes of the nine smartphones, the highest accuracy rate was in the '*White* condition', followed by '*Android* condition' and '*iOS* condition'. This research reports, even if the participants have experienced watching or touching the *iPhone X* so far, it was possible to recognize and distinguish, identify, to some extent the identity of the *iPhone X* from only the smartphone group of similar shapes. Still, it was difficult to identify as difficult in many cases.

3 Experiment

3.1 Purpose and Method

The purpose of this research is to clarify the relationship between the identity and shape of smartphones. This research was conducted with the keywords and the nationality, including GDP, share rate, and experience of using the smartphone. Therefore, the research needed to conduct in diverse countries to avoid the results become lean. Because the users' *Kansei* evaluation has been effected from their cultural background. Moreover, the share rate of smartphone brands is different depends on their country [8]. Based on this background, this research was conducted with the following 3 hypotheses; 1) Participants can accurately identify the shape of the front of the *iPhone X* regardless of the presence or type of information on the screen. 2) Participants can accurately identify the shape of the front of the *iPhone X* irrespective of their nationalities. 3) There is some correlation between the country whose participants can easily distinguish the shape of the front of the *iPhone X* and nominal GDP per capita/share rate.

In an experiment, the 185 participants looked and touched the 9 life-sized mock-up models, and ranked them in order from most likely *iPhone X* to most unlikely *iPhone X*. These 9 life-sized mock-up models consisted of the *iPhone X* and 8 *Android* smartphones, which had a similar shape with the *Notch design* (Fig. 1). The 9 similar shaped smartphones were available on the market in June 2018. The reason why we selected *iPhone X* as a standard to evaluate an identity in this experiment, the *iPhone X* had the highest market share in the world in 2018 [9].

Fig. 1. Various similarly shaped smartphones.

Based on the same size scale of each smartphone, the 9 mock-up models were made with 3 mm black acrylics. Each participant can check the size of each mock-up model and evaluate the identity of the shape of these 9 mock-up models while holding these mock-up models. All of 27 mock-up models were separated into 3 different screen images; 1) '*White* condition': All 9 models with the same white screen, 2) '*Android* condition': All 9 models with the same *Android* screen (*Huawei, P20 Pro*), and 3) '*iOS* condition': All 9 models with the same *iOS* screen (*Apple, iPhone X*). This experiment was conducted with various participants in the following 5 countries; 1) Japan, 2) China, 3) Korea, 4) Thailand, and 5) Singapore. In order to prevent the influence from the condition order, when participants conducted this experiment, each participant started from a condition picked randomly from the 3 conditions described above. This experiment was conducted with 185 students (93 males, 84 females, 8 no-answered) from Japan (*Hokkaido University of Education Hakodate School, and Future University Hakodate*), China (*Beijing Institute of Technology*), Korea (*Sangmyung University*), Thailand (*Chiang Mai University*) and Singapore (*Nanyang Polytechnic*) from 14th JUN 2018 to 25th JUN 2019.

3.2 Experiment Results

In '*White* condition', participants could identify *iPhone X* to a certain degree because *iPhone X* is scored in the second position (Fig. 2). In '*Android* condition', participants were less likely to identify the difference in shape between the mock-up models than in '*White* condition', because the scored position intervals of each smartphone were narrower than in '*White* condition' (Fig. 2). In '*iOS* condition', the range to identify of each smartphone is almost the same as '*Android* condition'. However, participants identified the shapes of the mock-up models into 3 big groups because of the scored results located around 0.5, 0.2, and −0.1 (Fig. 2). From those results, the identified score rate of smartphone shape depends on the type of information on the smartphone screen. Therefore, the first hypothesis was rejected.

In '*White* condition' of each country, *iPhone X* was concentrated around the scale value of 4.9 to 6.5, and *iPhone X* was identified similarly regardless of the country (Fig. 3). In '*Android* condition' of each country, identified accurately in Singapore and Korea but identified correctly in China (Fig. 4). In '*iOS* condition' of each country, identified accurately in Singapore and Japan but inaccurately in China (Fig. 5). From those results, the probability, which the participants identify *iPhone X*, has changed depending on the country, even though under the same conditions. Therefore, the second hypothesis was also rejected.

Finally, there was a strong correlation between the country whose participants can easily distinguish the shape of the front of the *iPhone* X and nominal GDP per capita/share rate [10] (Table 1). That reason is thought that because the price of the *iPhone X* is almost the same in various countries with different price levels. Moreover, in the case of except for the data of Singapore, it became clear that there was a negative correlation in '*White* condition', and there was a positive correlation in '*Android/iOS* condition'. This could be because the participants in Singapore have the same nationality but different cultural backgrounds. It means the correlation depends on the cultural background of participants. Therefore, the third hypothesis was supported.

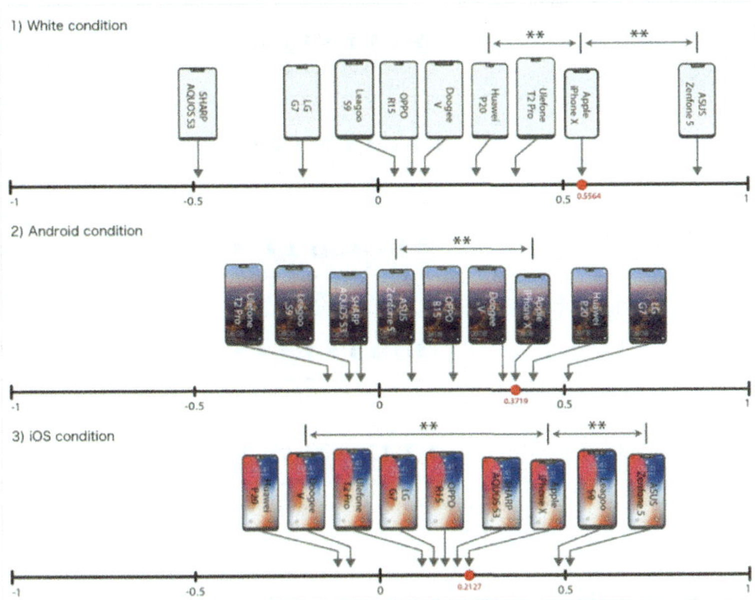

Fig. 2. Experiment results in three conditions.

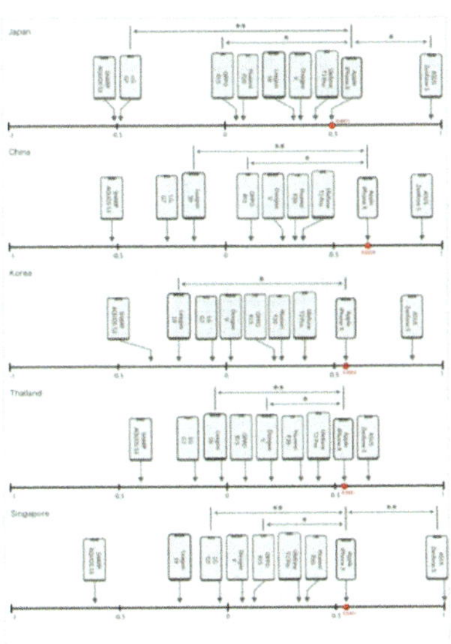

Fig. 3. Individual results in *White* condition

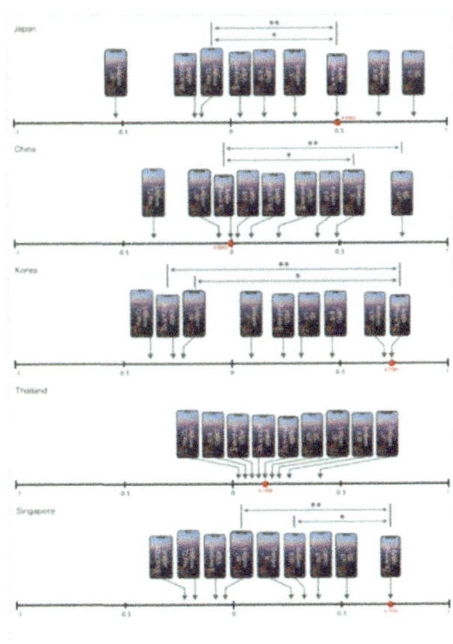

Fig. 4. Experiment results in *Android* condition

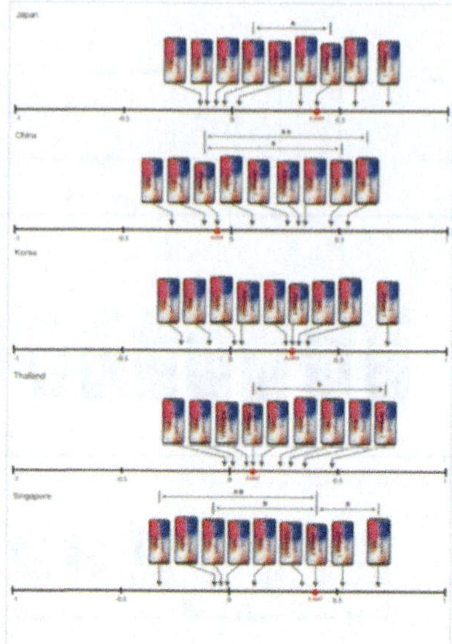

Fig. 5. Experiment results in *iOS* condition.

Table 1. Correlation coefficient of nominal GDP per capita/share rate.

	GDP/cpt.	Share rate	White cdt.	Android cdt.	iOS cdt.	
Japan	39,287	66.24	0.4911	0.4903	0.3889	
Korea	31,363	24.49	0.5601	0.7183	0.2803	
China	9,771	19.27	0.6524	0.0052	-0.054	
Thailand	7,274	16.72	0.5451	0.1458	0.0967	
Singapore	64,582	10.74	0.5471	0.7232	0.3987	
Correlation coefficient (excluded Singapore)	GDP		**-0.659**	**0.85**	**0.927**	**0.816**
	Share		**-0.706**	**0.421**	**0.778**	

4 Conclusion

As a result of this experiment, the participants could identify *iPhone X* to a certain degree from the information of only the shape. Moreover, the scoring rate of accurately identified differs depending on the 3 conditions and experience level of using the smartphone. In the case of except for the result of the participants in Singapore, there was a definite positive correlation (*0.816*) between GDP and share rate. The reason for except for Singapore, Singapore is a multi-cultural country, therefore the participants in Singapore have various cultural background comparing to the other 4 countries. In the case of the correlations with GDP and the result of the evaluation, there is a negative correlation (*−0.659*) in the *White* screen condition and a definite positive correlation in the *Android* (*0.85*), and *iOS* screen condition (*0.927*). As a consideration, these participants of a higher GDP country have many chances to see and touch the *iPhone X* and well know the identity of the shape from experiences of their daily life.

In conclusion, first, the scoring rate of accurately identified depends on the type of information on a smartphone screen or the difference of nationalities. Finally, the scoring rate of accurately identified depends on the nominal GDP per capita/share rate.

In this time, our experiment was conducted using mock-ups models that were reproduced the frontside shape of the smartphone. However, each smartphone has various backside shapes as a factor of identity. Moreover, a specific backside shape represents a particular brand also. Therefore, in the future, this research will be developed using the new keyword 'backside shape'. Additionally, our experiment will be conducted in more various countries.

Acknowledgements. In this research, we would like to thank to Prof. *Fu* at *Beijing Institute of Technology,* Prof. *Seo* at *Sangmyung University,* Prof. *Thongphun* at *Chiang Mai University,* Prof. *Jiehong* at *Nanyang Polytechnic, and* Prof. *Hussin* at *Nanyang Polytechnic.* And we also would like to thank to all participants in *Beijing Institute of Technology, Sangmyung University, Chiang Mai University, Nanyang Polytechnic, Hokkaido University of Education Hakodate School*, and *Future University Hakodate.*

References

1. Explosive popularization of smartphones seen by numbers (quantitative expansion over 5 years). http://www.soumu.go.jp/johotsusintokei/whitepaper/ja/h29/html/nc111110.html
2. Survey results that the number of notch phones sold in 2018 will be about 300 million units, 19% of the whole smartphone. https://japanese.engadget.com/2018/06/28/2018-3-19/
3. "Notch" storms that blow on Android devices. https://iphone-mania.jp/news-205315/
4. Shota, O., Hideyoshi, Y., Satoshi, N., Tadamasa, H., Hiroyuki, K.: Extraction of form features contributing to the recognition of the identity and discriminability of design. In: The 62nd Annual Conference of JSSD (2015)
5. Michihide, W.: On design issues concerning the formation of product identity. Hiroshima Univ. Manage. Rev. **3**, 137 (2003)
6. Park, Y., Jung, E.: A Basic study on extended brand identity model in product experience. In: International Design Trend Symposium & Forum, pp. 56–57 (2018)
7. Koki, C., Namgyu, K.: Exploring the relationship between shape of smartphone and identity. In: Advances in Affective and Pleasurable Design, pp. 358–368 (2019)
8. Major OS/model share status in 40 countries. Utilize share situation data for inbound Web promotion, March 2018. https://www.globalmarketingchannel.com/press/survey20180323
9. iPhone X now available in a total of 70 countries. https://www.cultofmac.com/515901/iphone-x-now-available-in-a-total-of-70-countries/
10. Visualized data of ASEAN-ASEAN Economic Statistics Basic Material. https://www.mofa.go.jp/mofaj/files/000127169.pdf

Proposal of System Based on *Kansei* Evaluation of Influence of the Margin Ratio of Plates

Shunsuke Ono(✉) and Namgyu Kang

Future University Hakodate, 116-2 Kamedanakano, Hakodate, Hokkaido, Japan
Shunn1348@gmail.com, kang@fun.ac.jp

Abstract. When we evaluate a dish, various factors affect the evaluation in a complicated way, it is not just the taste. In this study, we focus on the 'margin' of a plate with food. The 'margin' is one of the factors of the arrangement, and we aim to elucidate the effect of the margin of plate on people's emotions and sensibilities. The results were analyzed using factor analysis, one of the multivariate analyses. Furthermore, based on the results of the experiment, we proposed a *Kansei* system using the technology of projection mapping and a table with a translucent tabletop.

Keywords: Margin of plates · Factor analysis · *Kansei* system

1 Introduction

Food is not merely evaluated based on taste, and various factors have a compound effect. Therefore, various researches on food evaluation have been conducted from multiple perspectives. The assessment of food changes depending on the light color of illumination [1]. In Japanese food, the appetite changes depending on the proportion of blue in the dish [2]. According to the Japanese Cuisine Academy, Japanese food has a unique sensibility to beauty, such as "the beauty of margins" [3]. In Japanese food, the ratio of the margin ratio depending on the size of the plate and the purpose of the plate varies slightly, but it is said the average margin ratio of the plate is preferably about 2/3 of the area[4]. In addition, the area ratio between the fillet cutlet and the plate is preferably 4:10 [5]. From these previous studies, the optimal method and numerical value of the food presentation were performed. However, there are still few studies to clarify the effects of food presentation, the tableware like a plate, and the environment in which meals are served on the emotions and sensibilities of the recipient who are served. Therefore, the primary purpose of this study was to focus on the margin ratio of plates, which is one of the factors for serving dishes, and to clarify the effect of the margin ratio of plates on human sensitivity. Another purpose is to propose a system to support food presentation using IT based on the results.

G. Di Bucchianico et al. (Eds.): AHFE 2020, AISC 1202, pp. 696–702, 2020.
https://doi.org/10.1007/978-3-030-51194-4_90

2 Survey method

The purpose of this survey is to clarify the effect of the margin of plates on people's *Kansei*. In this study, we focused on the margin ratio of the plate only. Therefore, 15 evaluation samples with different sizes of the plate and the same food were used to the participants in the experiment. The result of the SD method was analyzed with factor analysis.

2.1 Selection of Evaluation Samples

In this experiment, 15 photos were used as evaluation samples, combining different 5 sizes of plates and 3 types of cakes (Fig. 1). All five plates are white and have the same design they differ only in size. The sizes of the five plates were 14 cm, 18 cm, 21 cm, 25 cm, and 30 cm from the smallest. The three types of cakes with different colors and shapes were used and were set as, Condition 1, Condition 2, and Condition 3, respectively. Two types of photos were presented together, one from directly above and the other from a low angle. The reason for selecting cakes and the white plates as the research subject is to avoid some research noises during the experiment.

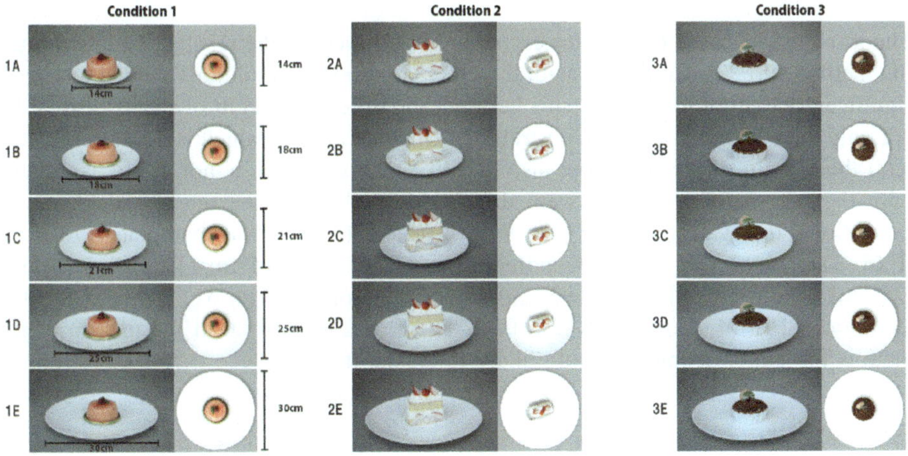

Fig. 1. 15 evaluation sample lists

2.2 Impression Evaluation Method

As a method to evaluate the impression of the margin of each plate, the SD method based on a 7-point rating was used with 12 adjective pairs as evaluation items (Table 1). The evaluation items were selected based on previou kinds of research [6] and specialized book [7].

Table 1. 12 adjective pairs as evaluation items

Expensive - Cheap	Mundane - Unique
Simple - Fancy	Lacking - Full
Tasteless - Flavorful	Constrained - Free
Sober - Brilliant	Repetitious - Fresh
Awkward - Graceful	Dislike - Like
Ugly - Beautiful	Frugal - Splendid

2.3 Impression Evaluation Method

This experiment was conducted from July 8 to 12, 2019, to a total of 36 men and women as a participant, mainly university students and university officials at Future University Hakodate. In addition, the order in which evaluation samples were presented was changed for each participant to prevent the order effect.

3 Results and discussion

3.1 SD Method Results

The three graphs were created for each condition based on the results obtained by the SD method (Fig. 2). Compared with the other conditions, Condition1 tended to have a significant variation in almost items between the different margin of plates. Primarily, not only the evaluation items "Constrained - Unconstrained", which is easily associated with the margin of the plate) but also "Dislike - Like" which is not easily associated with the margin of the plate, were differed depending on the margin of plates. In Condition1, the evaluated score of "Like" of plate C (21 cm) was than plate E (30 cm) ($F(4. 31) = 4.05$, $p < ,01$). In Condition 2 and 3, plate A (14 cm), B (18 cm), and C (21 cm) were preferable to plate E (30 cm) (Condition 2: ($F(4. 31) = 4.05$, $p < ,01$), Condition 3: ($F(4. 31) = 4.34$, $p < ,01$)). These results illustrate if the margin of the plate is too large, people are more likely to dislike it.

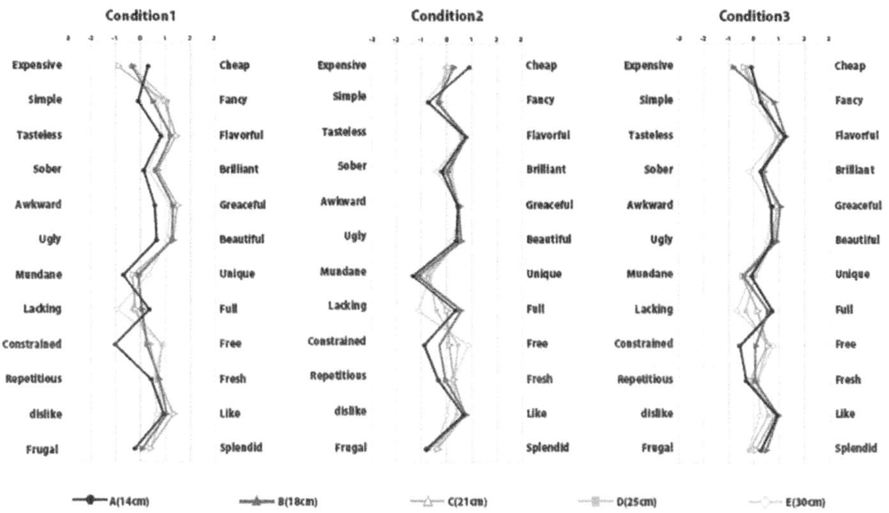

Fig. 2. The results of the evaluation with SD method

3.2 Factor Analysis Results

To explore the participants' impression of the margin of the plates, factor analysis was performed using the results of SD method. As a result, for the factors with eigenvalues of 1 or more, a total of 12 items with factor loadings of more 0.40 were extracted based on the Kaiser criterion and the screed plot, the three factors were extracted (Table 3). The first factor was interpreted as "Luxurious" because of the factor loading of scales such as "Fancy", "Splendid" and "Expensive" was strong. The second factor was interpreted as "Artistic" because of the factor load of the scales of "Grateful", "Fresh" and "Beautiful" was strong. The third factor was interpreted as "democratic" because the factor loading of scales such as "Like", "Full" and "Flavorful" was strong (Table 2).

Table 2. Factor interpretations

	Evaluation items	Extracted factor name
Factor1	Fancy, Splendid, Brilliant, Unique, Expensive	Luxurious
Factor2	Graceful, Fresh, Beautiful, Free	Artistic
Factor3	Like, Full, Flavorful	Democratic

Table 3. List of factor loadings

Evaluation Items		Factor1	Factor2	Factor3
Simple	Fancy	**0.729**	0.068	0.094
Frugal	Splendid	**0.700**	−0.004	0.197
Sober	Brilliant	**0.634**	0.051	0.289
Mundane	Unique	**0.606**	0.018	−0.283
Expensive	Cheap	**−0.572**	−0.079	0.118
Awkward	Graceful	0.123	**0.653**	0.293
Repetitious	Fresh	−0.091	**0.619**	0.110
Ugly	Beautiful	0.129	**0.588**	0.374
Constrained	Free	0.020	**0.456**	−0.083
Dislike	Like	−0.075	0.400	**0.699**
Lacking	Full	0.114	−0.104	**0.602**
Tasteless	Flavorful	0.052	0.384	**0.563**
Cumulative contribution ratio		0.182	0.323	0.459

3.3 Impression Evaluation Using Factor Scores

Each three average factor scores were obtained by the factor analysis (Fig. 3, Fig. 4, Fig. 5). "Luxurious" scores in Condition 2 were very different from the other Conditions. And "Artistic" scores in Condition 1 were very different from the other Conditions. These results illustrate that different cake type has more effect on "Luxurious" and "Artistic" than the margin of plates. Moreover, the factor score tends to draw a quadratic curve in "Luxurious". In other words, there is an adequate margin for each cake in the "Luxurious" factor, and the factor score tends to decrease as the distance from the adequate margin amount increases. In contrast, "Artistic" score graph tended to draw a positive linear function under all conditions; in other words, the larger margin, the more people tend to feel artistic. And Factor3 "Democratic" scores were highest for the smallest plate A (14 cm), and the factor score for the plate E (30 cm) is the lowest under all conditions, and the graph tends to draw a negative linear function.

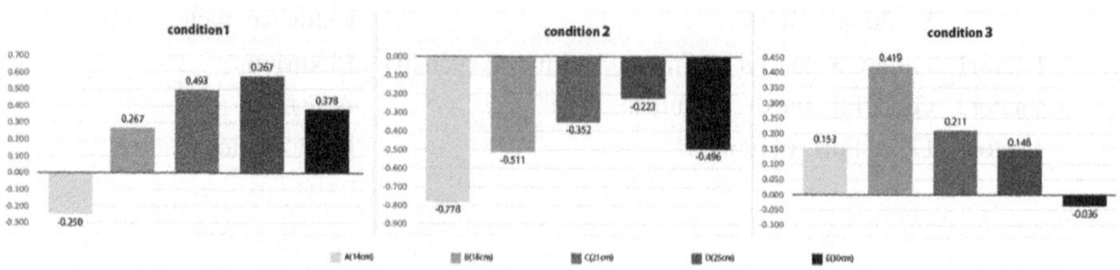

Fig. 3. The average scores of factor1 (Luxurious)

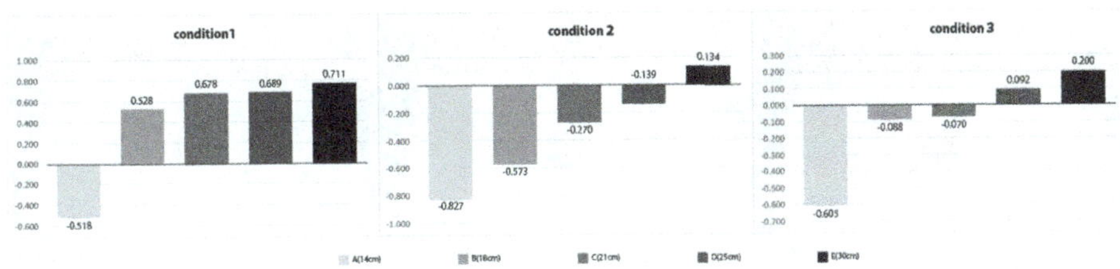

Fig. 4. The average scores of factor2 (Artistic)

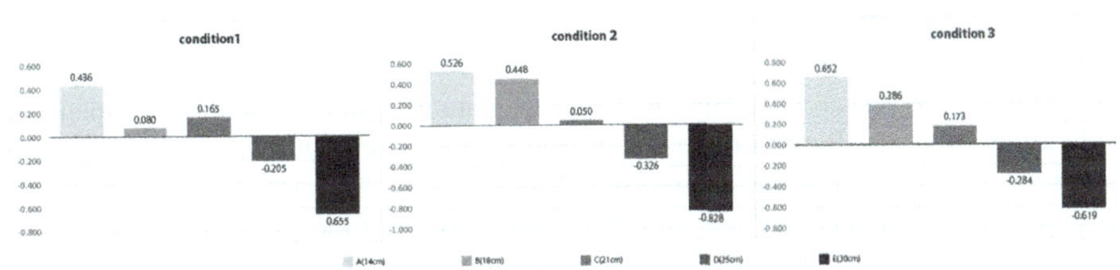

Fig. 5. The average scores of factor3 (Artistic)

4 Proposal of Support Tool

Kansei system using projection mapping was proposed based on the results of the experiment. The concept of this system is 'Freedom from preparing many plates, even though various foods are served'. This system provides an adequate plate size using projection mapping for served various foods. The system was consisted with a table with milky white acrylic plate and a single focus projector. This table-type projection mapping quickly changes the plate size depending on each food (Fig. 6). Figure 7 shows an example of projected various plates images with cake by the *Kansei* system.

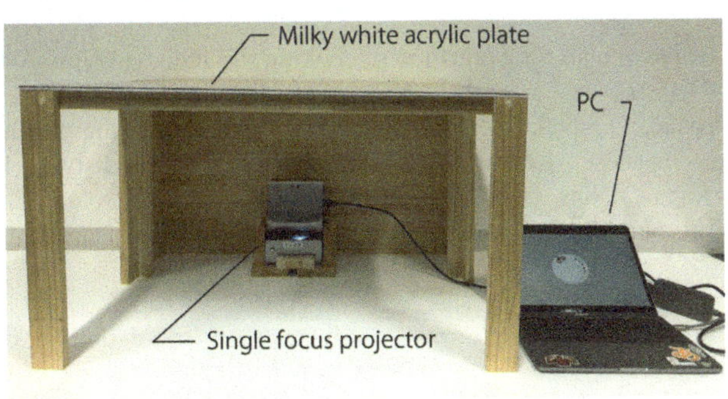

Fig. 6. System configuration

Fig. 7. Actual situation

5 Future Works

The results of this research clarified that the adequate margin of plate differs depending on the types of cakes. Based on the results, the *Kansei* system utilizing technology was proposed. In this research, the main research object was only cake, and the participant was only college students. However, the different types of foods and the characteristics of the participants may affect the impression of the food. Therefore, in the future, this research will be developed using various foods with various characteristics of the participants. Additionally, an experiment to clarify the efficiency of this *Kansei* system will be conducted.

References

1. Kobayashi, S.: Appetite for food illuminated with vivid color lights. J. Environ. Eng. **74**(637), 271–276 (2009)
2. Hino, K., Chieko, K.: Effect of the pattern of a blue plate on the appetite for Japanese food. J. Home Econ. Japan **67**(2), 66–68 (2016)
3. [nihonryouri(Japanese culinary art and culture)] no sekaimukeibunkaisannimuketa teiansyo (2012). http://www.let.uu.nl/~tanya.reinhart/personal/
4. Toshio, E.: Nihonryo-rimoritukesinan utukusiimoritukenotamenozissensido-syo, shibatasyoten (2006)
5. Yaeko, M.: A tentative plan for evaluation of cooking technique. (3) Jpn. Assoc. Home Econ. Educ. **13**, 55–63 (1972)
6. Yasuko, M., Hisae, O., Nakako, M., Enun, K.: To compare the Japanese and Korean conceptions of Japanese, Korean, Chinese, and western food-styles by SD: sematic differential. J. Kagawa Nutr. Univ. **36**, 85–93 (2005)
7. Toshihiko, H., Ryoko, H.: Ningenkougakugaido kanseiwokagakusuruhouhou, Saientexisutosya (2009)

Verification of Effectiveness with Sound Using Metaphor for Typing on Keyboard

Noriki Fukatsu(✉) and Namgyu Kang

Future University Hakodate, 116-2 Kamedanakano,
Hakodate, Hokkaido 0418655, Japan
{g2118034, kang}@fun.ac.jp

Abstract. The sounds effects have significant influences on our physical functions and emotions. Also, 'Metaphor' is one of the other ways to give a strong impression. Metaphor is one of the Metaphoric expression methods, which can express the word more simply by recalling a more specific image. Nowadays, Metaphor with sound is used in the fields such as electronic settlement and camera sounds. However, the utility of expressing Metaphor by sound has not been verified. Based on these backgrounds, in this research, we focused on "sound" to express Metaphor and the "time" as research keywords. That is, we use time as a gap and express a Metaphor using sound. For this research, we conducted two experiments by SD method. In the first experiment, the subjects did writing tasks with the two different conditions. In the second experiment, the subjects typed some sentences that we decided.

Keywords: Sounds · Metaphor · Emotional design

1 Introduction

Sound is widely known to have a significant influence on human physical functions and behaviors [1]. An example of the influence on human's physical functions is the behavior of the body along with the rhythm of the sounds. Another example of impacts on human behavior s is that various behavioral changes may occur, such as having fun or becoming sad [2]. These effects are widely applied to social life, such as music therapy and marketing. Besides, it has been reported that associating the sound with behavior can help lead the following desired behavior [3]. Therefore, it is possible to influence behavior by applying music in daily life. Metaphor is one of the other ways to give a strong impression to humans. Metaphor is one of the Metaphoric expression methods, which can express the word more by merely recalling a more specific image. The Metaphor is a kind of Metaphorical way, but it refers to a form that is not Metaphor, such as "she is a fairy" [4].

In the field of design, Metaphor has been applied widely in user interfaces. Furthermore, such a method of expression is said to have a potent Metaphor effect because odd ones are identification [5]. Also, there is the term 'Analogy', which is compared in conditions similar to Metaphor. Analogy is based on the fact that there is an essential similarity between two things, and that if one has a particular property, it infers that the other has the same property [6]. The difference between 'Metaphor' and 'Analogy' is

G. Di Bucchianico et al. (Eds.): AHFE 2020, AISC 1202, pp. 703–709, 2020.
https://doi.org/10.1007/978-3-030-51194-4_91

that Metaphor is based on similarities, such as the appearance of things. In contrast, Analogy is based on structural similarities, such as logic. In the field of design, Metaphor is a method of indirectly Metaphorizing an object that is not originality the same. Furthermore, Metaphors are classified into the following two types in the field of UI design; 1) 'Synecdoche' is the principles of icon design. A folder icon indicating a document is an example. 2) 'Metaphors, similes, analogies' that connect two different domains. A desktop Metaphor is an example. The desktop Metaphor is mapping between the real desktop world and trash Metaphors on desktop [7]. Nowadays, Metaphor with sound is used in the fields such as electronic settlement and camera sounds. However, the utility of expressing Metaphors by sound has not been verified yet. From that background, sound and Metaphor have a significant influence on our emotions. Therefore, we focus on one case of expressing a Metaphor using sound and verify whether it is useful or not.

2 Experiments

In this study, we conducted an evaluation experiment to verify the usefulness of expressing a Metaphor with sound. Therefore, this chapter describes an evaluation experiment to verify the usefulness of expressing a Metaphor with sound.

Besides, in the evaluation experiment, we used software that plays the typing sounds of a typewriter when typing on a keyboard as a product that expresses the Metaphor with sound. In the first experiment, the subjects did writing tasks with the two different conditions; The first condition is the one with the sound of a typewriter, and the other one is the one without the sound of a typewriter. In the second experiment, the subjects typed some sentences that we decided. The second experiment was also conducted with two conditions in the same way as the first experiment.

2.1 The Method of First Experiment

In the first evaluation experiment, participants typed two text tasks freely. After completing the two text tasks, the participants answered the evaluation sheet.

The two text task conditions were selected with the four experiment collaborators for making text easy to think. That is why participants typed the text contents freely. The two tasks are following two conditions; 1) Task A: "About the place you want to go", and 2) Task B: "About your hometown".

The participants were separated into two groups to avoid the effect of the order of the tasks. The first groups conducted the task A firstly, and the other one conducted the task B firstly. The adjective items in the first experiment were selected based on the reference to research using sounds with five experimental collaborators. Those were 'want to continue', 'fun', 'hard', 'interesting', 'heavy', 'pleasant' and 'comfortable'.

2.1.1 Implementation Schedule of First Experiment

This experiment was conducted with 24 students of Future University Hakodate on 14th and 15th June, 2019.

2.1.2 First Experimental Results and Discussion of Result

The result of the evaluation experiment was analyzed using factor analysis. Two factors were extracted according to the size of the initial eigenvalues and the attenuation situation. The first factor was interpreted as 'Sense of relief'. The elements of the first-factor were combined with the following items: pleasant, comfortable, and light. The second factor was interpreted as 'Sense of anticipation'. The elements of the second factor were combined with the following items; fun, interesting, pleasant, and want to continue.

From the graph of the first-factor score, there were significant differences between using the sound of a typewriter or not (Fig. 1). The averages factor scores when with typewriter sound and when without typewriter sound for each factor were compared using Analysis of variance.

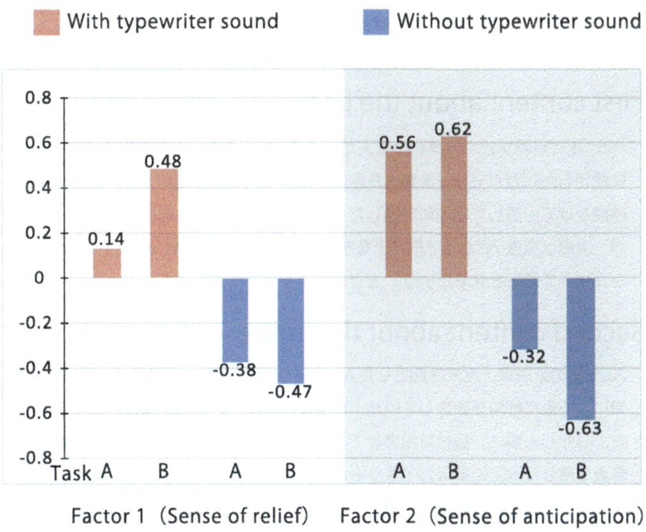

Fig. 1. The table of factor score

As a result, for the first-factor scores in task B, there was a significant difference between typewriter sound or not ($F(3, 8) = 3.57$, $p < .05$)). For the second-factor scores, there were significant differences also between typewriter sound or not ($F(3, 8) = 3.57$, $p < .05$)).

From the results of factor analysis, two factors, 'Sense of relief' and "Sense of anticipation" were obtained. Besides, there were significant differences between typewriter sound or not. Also, from the second factor, it is judged that the feeling of "fun" or "interesting" increases the subject's want to continue. Also, from the first factor, 'sense of relief' was extremely low without the typewriter sound in task B, whereas the 'sense of relief' was higher with the typewriter sound. Therefore, the utility of expressing Metaphor with sounds has become apparent.

2.2 The Method of Second Experiment

In this evaluation experiment, two text input tasks were used with the following two different conditions. After completing the writing each task, the participants answered using the evaluation sheet. Also, the text input tasks were prepared in two conditions with the four experiment collaborators consideration of gaps due to time series; 1) The first condition is the contents about the past, and 2) the second one is the contents about the future. Figure 2 illustrates the contents of two tasks. The participants were separated into two groups to avoid the effect of the order of the tasks. The first groups conducted the task A firstly, and the other one conducted the task B firstly. The used adjectives in the second experiment were also selected with five experimental collaborators. Those were 'want to continue', 'fun', 'hard', 'interesting', 'new', 'heavy', 'pleasant', 'stable', 'elegant' and 'comfortable'. Also, the time required for typing the task text was measured. Moreover, each participant's behavior to type the task text was recorded using a camera.

First content about the past

1961 年に当時ソビエト連邦の軍人だったガガーリンは，世界初の有人宇宙飛行に成功した．ガガーリンを乗せた宇宙船は，地球周回軌道に入り大気圏外を約 2 時間かけて一周して地球に帰還した．当初は宇宙船と共に着陸したとされていたが，実際は高度 7000m で飛行士を座席ごとカプセルから射出して，一人パラシュートで降下させるという，大きな危険を伴うものだった．

Second content about the future

2029 年の近未来，反乱を起こした人工知能スカイネットが指揮する機械軍と人類による最終戦争が勃発していた．機械軍により，人類は絶滅の危機を迎えていた．しかし人類は，抵抗軍指導者であるジョンコナーの指揮下，反撃に転じた．脅威を感じた人工知能スカイネットは，反乱軍を指揮するジョンコナーの誕生を事前に防ぐために，未来から現代へ殺人アンドロイドを送り込んだ．

Fig. 2. Contents of two tasks used in second experiment

2.2.1 Implementation Schedule of Second Experiment

This experiment was conducted for 28 students of Future University Hakodate on 29th October, 2019.

2.2.2 Second Experimental Results and Discussion of Result

The elements of the first principal component were combined with the following items: want to continue, fun, interesting, pleasant, and comfortable. Therefore, the first principal component was interpreted as 'Attractive typing'. That is why the users want to continue typing when they feel attracted to the typing experience. And, the elements of the second principal component were combined with the following items: heavy and awkward. The second principal component was interpreted as 'Unmotivated typing'. That is why the users feel heavier while they were typing, so they were typing without seriously. The elements of the third principal component were combined with the following factors: changeable and time is short. Therefore, the third principal component was interpreted as 'typing for efficiency'. That is why the time of typing is short, so work instability has occurred. The first principal component and the third principal component express '*Kansei* evaluation on typing text'. Therefore, these two principal components were focused in this research. Figure 4 showed when the users typed with the typewriter sound, and they felt the attractive feeling. Furthermore, the first principal component 'Attractive typing' scores between the two conditions with typewriter sound and without sound revealed a significant difference using analysis of variance. Also, there were significant differences in the third principal component 'Unmotivated typing' scores between the two conditions with typewriter sound and without sound (Fig. 3).

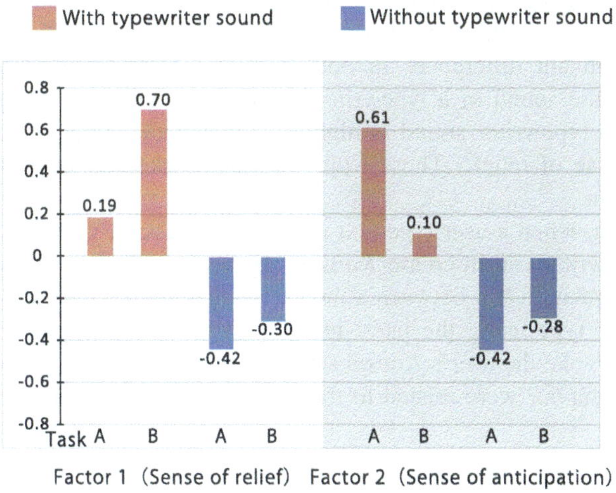

Fig. 3. Scatter plot of the first and third principal components

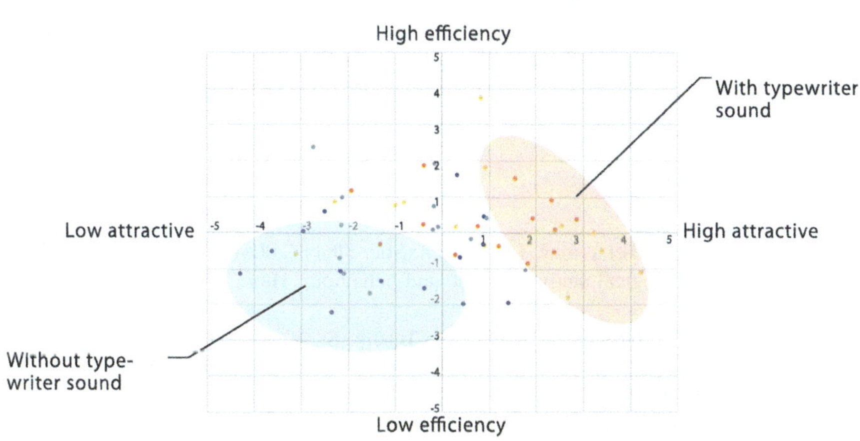

Fig. 4. Scatter plot of the first and third principal components

3 Conclusion

In the first experiment, the participants conducted typing text tasks freely with two different conditions with typewriter sound and without the sound. On the other hand, in the second experiment, the participants performed typing the already prepared text tasks with two different conditions as the first experiment. As a result of the first experiment, the following two factors were obtained using factor analysis; 1) 'Sense of relief' and 2) 'Sense of anticipation'. Also, from the first-factor' sense of relief' was extremely low on the condition without the typewriter sound in task B, whereas the second-factor 'sense of relief' was higher on the condition with the typewriter sound.

Therefore, the utility of expressing Metaphor with sounds has clarified based on the results of this research. Especially, as a result of the second experiment, we could obtain the following three principal components using Principal component analysis; 1) 'attractive typing', 2) 'unmotivated typing', and 3) 'typing for efficiency'. Moreover, there were significant differences in both principal component scores between the conditions with the sound of a typewriter and the one without the sound. From both experiments, the typewriter sound is highly evaluated the subject's 'Sense of anticipation' and 'Sense of relief'. These *Kansei* values evoke and increase the degree of users' satisfaction.

In conclusion, when a user type text using an expression with Metaphor using the sound of a typewriter can in-crease *Kansei* satisfaction. Therefore, users' *Kansei* satisfaction have a relationship between a time series gap by the time and sound. In other words, when we type using the latest technology with the typewriter sound as past technology can evoke the user's *Kansei* satisfaction. Therefore, the time series gap and the sound as Metaphor were related to the user's *Kansei* satisfaction deeply.

4 Future Works

From both experiments, the usefulness of expressing Metaphors with typewriter sound has been clarified. However, this is just one case study using a sound. Therefore, in the future, this research will be developed using various examples of expressing the Metaphor with sounds. Also, it is necessary to visualize which elements of Metaphor evoke a positive emotion on the user by expressing the Metaphor with sounds.

References

1. Kengo, O.: Music and behaviour, J. Biomech. Soc. **30**(1) (2006)
2. Teruo, Y.: A psychological study of music and behaviour. Hum. Sci. Res. Bull. **8**, 221–232 (2009)
3. Tomiei, I., Kenji, Y., Natsue, S., Akira, M., Tetsuo, S., Katsuyuki, S.: On the response of music and autonomic nerves-Consideration for using music for training -. Tokai Univ. Sports Med. J. **9**, 50–58 (1997)
4. World Art Dictionary, Monthly Art, Korea (1999)

5. Blackwell, A.F.: The reification of metaphor as a design tool. ACM Trans. Comput. Hum. Interact. (TOCHI) **13**(4), 490–530 (2006)
6. Selected Japanese language dictionary
7. Takashi, K.: The role of metaphor in cognitive interface. Inf. Res. Bull. **98-IM-9**, 103–112 (1993)

Proposal of *Kansei* Communication Tool for SNS Based on Combining Verbal and Non-verbal Communication

Sadahiro Suzuki(✉) and Namgyu Kang

Future University Hakodate,
116-2 Kamedanakano, Hakodate, Hokkaido 0418655, Japan
sadahiro0814@gmail.com, kang@fun.ac.jp

Abstract. In recent years, communication on chat via SNS has increased by the spread of information technology. Therefore, this research aims to verify the effectiveness and possibility of communication by converging Non-verbal and Verbal communication. To support more emotional communication, we propose an emotional communication tool combining drawing that is Nonverbal communication and voice that is Verbal communication. An experiment was conducted to clarify which degree of emotions sent by the messages on chat communication via SNS. And, the system of this research was proposed based on analyzing experimental results. In the future, this system will be essential for the users to communicate happily on SNS. Here is important to use them happily regardless of the ability of drawing and voice.

Keywords: Non-verbal communication · Verbal communication · *Kansei* system

1 Introduction

Nowadays, the usage rate of smartphones in Japan has increased to 80.4% due to the spread of smartphones. Besides, there are various social media services with the rapid development of information technology in Japan. Among Japanese social media services, the usage rate of *LINE* is highest at 75.8%, followed by *Facebook* at 31.9%, *Twitter* at 31.1%, and *Instagram* at 25.1%. In particular, the usage rate of *LINE* has increased significantly in the past five years, making it the most used social media service in Japan [1]. Thus, communication using SNS becomes active, and SNS promotes the connection between people has a positive effect on human relationships. The communication used when we communicate roughly divided into two types: Verbal and Non-verbal. Verbal communication is to communicate using words. Also, Non-verbal communication is to communicate in ways other than language, and the communication methods used vary widely depending on culture, gender, individual, and the situation at the place. It conveys intentions and emotions more powerfully than words [2]. Against this background, some researches have already shown the effectiveness of Non-verbal communication tools to

G. Di Bucchianico et al. (Eds.): AHFE 2020, AISC 1202, pp. 710–716, 2020.
https://doi.org/10.1007/978-3-030-51194-4_92

focus on drawing [3, 4]. Moreover, the combination of drawing and voices through a PC network at a childcare site has attracted people [5]. In particular, the act of drawing is an act of conveying thoughts at will, and it can be expected to be a powerful tool to develop and concentrate on expressiveness and sensitivity [6].

Also, people distinguish emotions such as anger, sadness, and joy from voice [7]. In other words, using drawing and voices in combination is expected to provide highly valuable communication by incorporating more diverse information visually, auditorily, and sensuously. Moreover, it makes it possible to stimulate people's emotions in various directions.

From these backgrounds, this research aims to clarify the degree of emotional transmission by messages on chats on SNS. Furthermore, we propose and produce a new communication tool that combines drawing, which is Non-verbal communication, and voice, which is Verbal communication. The final aim of this research is to verify the effectiveness and possibility of communication and to help everyone enjoy communication more freely and richly by converging drawing and voices. In this research, we first survey the current situation of SNS communication using a questionnaire method. Next, we verify the degree of emotional transmission by messages on chats on SNS. Then, based on the analysis results, we propose and produce a communication tool. The proposed system performs experiments to confirm its effectiveness.

2 Current Situation Survey

2.1 Method

This survey aims to explore new communication methods on SNS chats and to investigate the current situation of communication between users on SNS chats to avoid prejudice and guesswork. Therefore, a total of 49 men and women in their teens and twenties who are expected to have high SNS usage rates were surveyed using Google Forms. The contents of this questionnaire were questions about the frequency of use of each message on the chat. Each message was set in seven ways: text, voice, *EMOJI*, emoticon, stamp, photo, and video. *EMOJI* is a face message created by the text. The relative use of each message was rated on a four-point scale: low, slightly low, slightly high, and high. Each response was quantified as −2 for "low", −1 for "slightly low", −1 for "slightly high", and 2 for "high", and the average value and standard deviation for each message were calculated.

2.2 Results

The result of the question about the frequency of using each message on the chat is as follows. Comparing the average values of each message (Fig. 1), the text (M = 1.833, S.D. = 0.606) was the highest, followed by the stamps (M = 0.869, S.D. = 1.165). On the other hand, the voice (M = −1.958, S.D. = 0.198) was the lowest. Moreover, the difference between the average value of text and voice was huge.

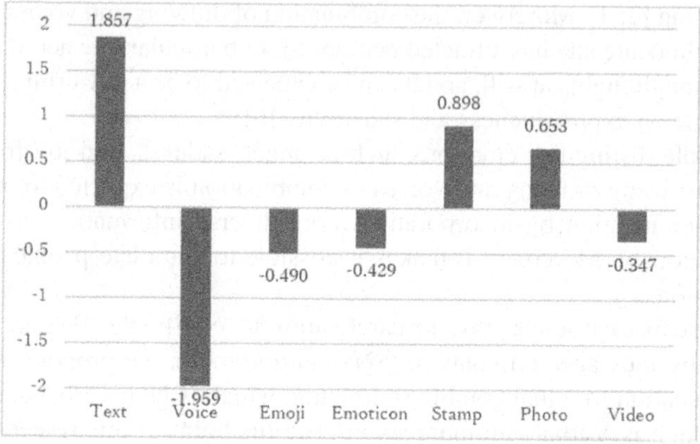

Fig. 1. The average values of frequency of each message on the chat

3 Evaluation Experiment

3.1 Method

The evaluation experiment aims to verify the degree of emotional transmission by each message in chat on SNS. This experiment was conducted with 15 male and female students in their twenties. They sent message using a smartphone and then evaluated it. They sent a message as a sender on *LINE* [1], the most used SNS in Japan, in two different scenes (Table 1). The two types of scenes were set by 10 experimental collaborators. After the experiment, a questionnaire was used to evaluate each message in terms of emotional communication. The questions were rated on a four-point scale. The degree was quantified by setting each answer from −2 to 2, and the average value and standard deviation of each message were calculated. Analysis of variance was performed.

Table 1. Set 2 types of scene

	Set scene	Emotion
Scene 1	A scene celebrating a friend's birthday	Joy
Scene 2	A friend does not come to meet and no contact	Anger

3.2 Results

The results of the degree of emotional transmission of each message on the chat are as follows (Fig. 2). Comparing the average values of each message way in Scene 1 (Joy), video (M = 1.200, S.D. = 1.373) was the highest, followed by voice (M = 0.733, S.D. = 1.335) and stamp (M = 0.733, S.D. = 1.486). On the other hand, text (M = −0.067, S.D. = 1.534) was the lowest. Comparing the average values of each message in Scene 2 (Anger), voice (M = 1.400, S.D. = 1.242) was the highest, followed by stamp

(M = 1.133, S.D. = 1.187). On the other hand, photo (M = −0.533, S.D. = 1.642) was the lowest score. Especially, the way using photo tends in the difference between the average values of Scene 1 and Scene 2 ($F(1.14) = 3.352$, $p < 0.10$). Furthermore, in the way using video, the difference between the average values of Scene 1 and Scene 2 was large ($F(1.14) = 7.485$, $p < 0.05$). That result means that the degree of using between a photo and a movie varied greatly depending on the Scene.

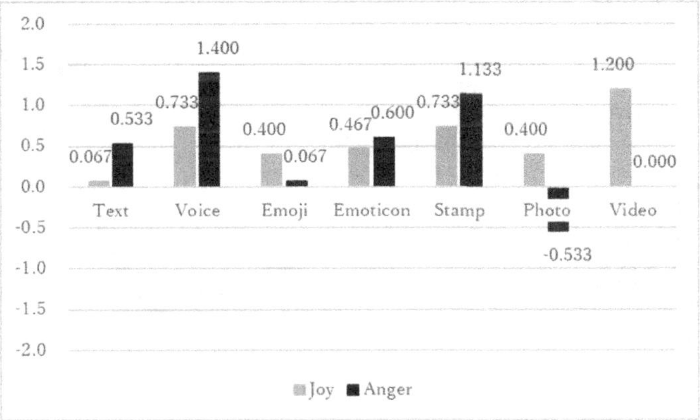

Fig. 2. The average values of emotional degree on 2 scenes

4 Proposal

There are existing messages with a voice that have already been added to SNS service. However, it is still impossible for users to create it by themselves. The act of drawing is divided into analog expression and digital expression. The analog expression is the act of drawing on paper using a pen, and the digital expression is the act of drawing on a screen using electronic devices.

In this proposal, we focused on the analog expressions, which many people are used to drawing, to make the act of drawing familiar. Furthermore, due to adding a voice will more affect the user's emotions and enable new and better communication. When users can manipulate drawing and voice more freely, they can be focusing on chat. That means users' usage rates to draw becomes more familiar into drawing have with the high. As a result, the enjoyment of drawing and achievement is promoted (Fig. 3).

The outline of the proposed new system for communication of SNS is as follows:

1. Acquires the user's own drawing with the smartphone camera.
2. Perform image processing to clarify the acquired drawing.
3. Acquire the user's own audio data with the proposed system.
4. Combine the image and audio to create a message for SNS communication.
5. Display and send the created stamp as a message.

In this research, as the first step, a prototype was created to verify the effectiveness of the created stamp by the proposed system. The development environment used Adobe Systems' prototyping creation tool, Adobe XD, to enable display on a smartphone (Fig. 4).

Fig. 3. Proposed system

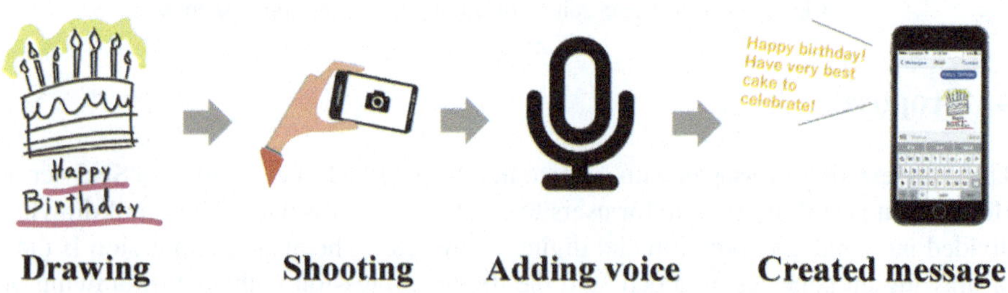

Fig. 4. Proposed system outline

5 Impression Evaluation

5.1 Method

The aim of the experiment using the prototype to verify the effectiveness of the created messages by the proposed system. Moreover, to clarify the users' impressions is the aim of the experiment also. That means this experiment is to understand the characteristics and value of the proposed system. In this evaluation experiment, the 41 participants used the prototype of the proposed system. Next, the SNS message, including the created stamp with drawing and voice by the proposed system, was evaluated using the SD (Semantic Differential) method. The 12 adjective pair items in the SD method were selected with ten experimental collaborators based on previous research [8].

After that, Factor analysis (maximum likelihood method, varimax rotation) using the results of the SD method was performed to clarify the characteristics of the created messages by the proposed system.

5.2 Results

A total of 10 adjective pair items were selected for main factors with an eigenvalue of 1 or more and a factor load of 0.4 or more (Table 2). And the following four factors were extracted based on the included content of the items in each factor. For the first factor, "Fun," "Bright," "Interesting," and "Happy" were extracted and interpreted as "Entertainment". As the second factor, "Progressive", "New" and "Free" were extracted and interpreted as "Futuristic". "Fancy" was extracted as the third factor and interpreted as "Decorative". As the fourth factor, "Clear" and "Continuous" were extracted and interpreted as "Readability" (Table 3). Finally, the three Factors with two or more adjectives were selected as characteristics of the created messages by the proposed system.

Table 2. Factor loading

Adjective pairs		Factor 1	Factor 2	Factor 3	Factor 4
Painful	Fun	0.89	0.07	-0.02	-0.08
Dark	Bright	0.69	0.12	0.17	-0.12
Boring	Interesting	0.64	0.29	-0.07	-0.27
Sad	Happy	0.52	0.20	-0.15	0.05
Regressive	Progressive	0.09	0.89	-0.27	0.35
Old	New	0.31	0.46	0.14	-0.07
Free	Cramped	-0.14	-0.40	-0.15	0.22
Sober	Fancy	0.12	0.13	0.96	-0.20
Clear	Vague	0.07	0.12	-0.19	-0.65
Temporary	Continuous	-0.09	0.04	-0.09	0.59
Soft	Hard	-0.29	-0.35	-0.08	0.24
Simple	Complex	-0.06	-0.01	0.27	0.13
Cumulative contribution		0.182	0.305	0.407	0.503

Table 3. Meaning of adjectives and factors

Factor	Adjective(s)	Factor name
Factor 1	Fun, Bright, Interesting, Happy	Entertainment
Factor 2	Progressive, New, Free	Futuristic
Factor 3	Fancy	Decorative
Factor 4	Clear, Continuous	Readability

6 Discussion

From the results of the factor analysis, three factors, "Entertainment", "Futurism", and "Readability" were selected as the factor of the created messages by the proposed system. This "Entertainment" can be interpreted as an improvement in the emotional mood for the SNS message. This result means the use of the user's own drawing and voice evokes a positive emotional feeling. "Futuristic" can be interpreted as an innovative futuristic communication tool for SNS messages. There is an existing message application using voice already, but the exist application is impossible for users to create messages by combining a user's own drawing and voice by themselves. Therefore, enabling the creation of messages with the user's own drawing and voice by the user himself/herself is very innovative. "Readability" illustrates the using the stamp with drawing and voice proposed system helps to understand the message and can be interpreted as this proposed system has a possibility that the user may use the messages not only once but also many times. Furthermore, the message with combined drawing and voice has a large amount of visual, auditory, and sensuous information, which leads to an improvement in understanding. Therefore, the features of the proposed system are "Entertainment", "Futurism", and "Readability".

In this research, we proposed a system for SNS communication that compose a message using drawing as Non-verbal communication, and voice as Verbal communication, to support better communication. Based on experiments and analysis, the potential of the proposed system was verified for this research. In other words, this system is expected as a *Kansei* communication tool for a chat on SNS.

References

1. Institute for Information and Communications Policy: 2017 Survey on information communication media usage time and information behavior (1981)
2. Marjorie, F.V.: An Introduction to Nonverbal Communication. Shinchosha, Shinjuku (1987). Translator, Tadashi, I
3. Lo, S.-K.: The nonverbal communication functions of emoticons in computer-mediated communication. CyberPsychology & Behavior **11**, 595–597 (2008)
4. Nobuyuki, H., Yusuke, U., Shuji, M.: Effects of emoticons and pictograms on communication of emotional states via mobile text-messaging. Jpn J. Res. Emot. **22**(1), 20–27 (2014)
5. Hirokazu, B., Yoko, O., Kota, O., Kazu, O.: Possibility of communication by drawing and voice via Pc network in childcare. Jpn J. Educ. Media Res. **17**(1), 37–47 (2010)
6. Brown, S.: The Doodle Revolution. Portfolio (2015)
7. Nobuaki, K., Hirotomo, A., Motoyuki, S., Shozo, M.: An investigation on discrimination among emotion expressions contained in speech. Inf. Process. Soc. Jpn. SLP **119**, 37–42 (2000)
8. Masaaki, I., Toshinobu, K.: The research domain and scale construction of adjective-pairs in a semantic differential method in Japan. Jpn. J. Educ. Psychol. **33**, 253–260 (1985)

Proposal and Kansei Evaluation of a Garbage Can to Change Human Behaviour Based on Fun Theory

Masayuki Takahashi(✉) and Namgyu Kang

Future University Hakodate, 116-2 Kamedanakanocho, Hakodate, Hokkaido 0418655, Japan
{g2119023, kang}@fun.ac.jp

Abstract. In general, people need to remove plastic bottle caps when they throw away plastic bottles. However, virtually half of the people do not remove it because it is troublesome. In previous research, a prototype was designed based on "Fun Theory," which is a theory to change human action with fun emotion, in order to encourage people to remove the cap. The prototype utilized a voting metaphor. It was observed that 73% of plastic bottles were thrown away without the cap. The prototype could encourage people to remove plastic bottle caps. However, it has not analyzed how people feel when they use the prototype yet. In this paper, a survey was conducted using word-pairs to clarify how people feel when they use the prototype. From the result of the survey, it was clarified that utilizing a voting metaphor as a garbage can for plastic bottle caps improved *Kansei* quality.

Keywords: *Kansei* engineering · Fun theory · Garbage can · Human behaviour

1 Introduction

1.1 Background

Garbage separation is one of the things that people need to do in daily life. However, it is one of the troublesome actions. People need to sort garbage when they throw it away, and the rule of garbage separation depends on the local government. It must be classified into various types, such as burnable garbage, non-burnable garbage, plastic, and plastic bottles.

According to a survey conducted by *Fumoto* and *Sakakibara* (2002), it was clarified that university students have slightly high knowledge about the environment. However, the percentage of students who can sort garbage correctly is low [1]. Furthermore, according to a survey by *Shinogi*, about 20% of the respondents answered that garbage separation was fun [2]. In other words, people feel bored when they sort garbage. Besides, an online questionnaire regarding garbage separation, which is answered by 3442 Japanese people, showed that 91.4% of the respondents sorted plastic bottles. On the other hand, only 41.1% of the respondents sorted plastic bottle caps [3]. Therefore, sorting plastic bottle caps is a particularly troublesome thing for people.

G. Di Bucchianico et al. (Eds.): AHFE 2020, AISC 1202, pp. 717–724, 2020.
https://doi.org/10.1007/978-3-030-51194-4_93

There are some researches to solve these problems by focusing on human *Kansei*. One of the research areas is *Kansei* engineering. *Ishihara* and *Nagamachi* defined *Kansei* engineering as "a collection of methodologies that support manufacturing from the perspective of human sensibilities" [4]. The Fun Theory is a concept advocated by the German company Volkswagen, stating that "fun with positive sensibility can change human behaviour." There are many products using the Fun Theory in the world. Piano stairs is an example. It is designed as a piano and encourages people to use stairs instead of escalators. Piano Stairs' attempts have been widely made worldwide, including Sweden, the United States, China, and Brazil. Piano stairs at the station in Sweden incorporate a system that plays the sound of a piano when people walk on the steps (Fig. 1). The piano stairs increased the number of people using the stairs instead of an escalator by more than 66% than usual [5]. Besides, people were going up and down the piano stairs and enjoying the sound of the piano. Going up the stairs is usually troublesome for people. However, it can be changed into a fun activity by changing the design.

Fig. 1. Piano stairs in Sweden

1.2 Previous Research

Definition of *Kansei* in System Design. Harada defines *Kansei* as "cognitive expression based on knowledge and experience in addition to innate properties" and "interaction between intuition and intellectual activity" [6].

According to the above definition of *Kansei*, we define *Kansei* quality as a quality that can intuitively change human behaviour by using information that has already been known from daily experience.

***Kansei* Garbage Separation Support System.** As a previous study, the authors focused on sorting plastic bottle caps among the behaviours that people feel troublesome. They designed a garbage can using a voting metaphor and evaluated its usefulness (Fig. 2). The experiment included installing the proposed prototype in a garbage can at Future University Hakodate, which is the place the evaluation was conducted for five days on weekdays. The sorting rate of plastic bottle caps was increased from 5% to 73% (Fig. 3) [7]. In other words, it was confirmed that the proposed prototype was highly useful as a garbage can that encourages the separation of plastic bottle caps. Although the effect of encouraging people to separate plastic bottle caps has been confirmed, the relationship between the user's *Kansei* evaluation and behaviour change has not been visualized yet.

Fig. 2. The proposed prototype

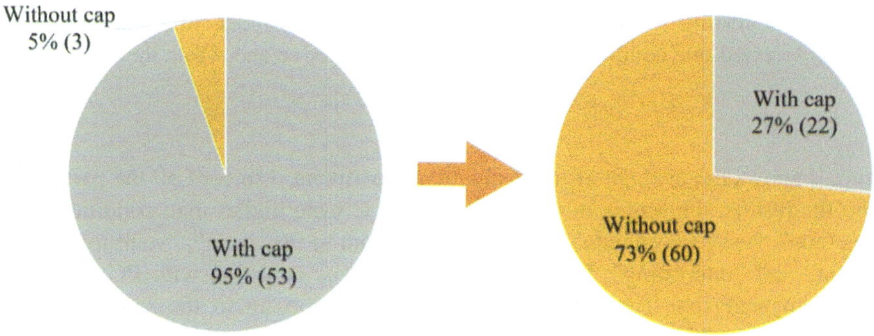

Fig. 3. Comparison between before placing prototype and after

2 Purpose

The purpose of this research is to evaluate *Kansei* quality of the prototypes produced in previous research and to visualize the way to improve the prototypes.

One of the ways to improve the prototypes is enhancing *Kansei* quality and adopting the concept of The Fun Theory. In other words, the way to design the prototype is to encourage people to separate plastic bottle caps intuitively and give a positive emotion to people.

3 Evaluation Experiment

3.1 Purpose of the Evaluation Experiment

The purpose of this evaluation experiment is to clarify how people feel when they separate plastic bottle caps under three conditions: one without the prototype (condition A), one with it without sound effect (condition B), and one with it with sound effect (condition C).

3.2 Method

This evaluation experiment was conducted for three days in 2019, targeting 24 university students. The participant was given a plastic bottle and instructed to "discard the plastic bottle." The participant discarded a plastic bottle under three conditions. The condition order was chosen randomly.

Besides, "Dog" and "Cat" were selected as voting items because they were considered familiar objects in daily life. Each animal sound was played using the mobile speakers connected to Bluetooth manually by the experimenter. After throwing plastic bottles in each condition, participants gave rankings on a five-grade evaluation for seven questionnaire items; 1) "not fun - fun", 2) "difficult to separate - easy to separate", 3) "troublesome - not troublesome", 4) "do not want to use again - want to use again", 5) "do not want to separate - want to separate", 6) "simple - complex" and 7) "boring - interesting". After discarding the plastic bottle under three conditions, participants selected one condition that is the most impressive and wrote the reason freely.

3.3 Result

Figure 4 shows the average of the evaluation experiment results of all the participants. From the results, the scores of conditions B and C were higher than condition A in 7 items "fun", "easy to separate", "easy to use", "want to use again", "want to separate", "simple" and "interesting". Moreover, the result was analyzed with PCA (principal component analysis). The cumulative contribution ratio up to the second principal component exceeds 70%, so the results of the first principal component and the second principal component were analyzed (Table 1). Two items were extracted in the first principal component (Table 2), and four items were extracted in the second principal component (Table 3) as items whose eigenvectors showed 0.40 or more.

In the first principal component, two values of interesting (*0.513*) and want to separate (*0.417*) are high, so the first principal component is interpreted as "*Kansei* quality." In the second principal component, four values of boring (*−0.516*), Not troublesome (*0.477*), easy to separate (0.428), and simple (*−0.411*) are high, so the second principal component is interpreted as "Usability quality." Furthermore, the

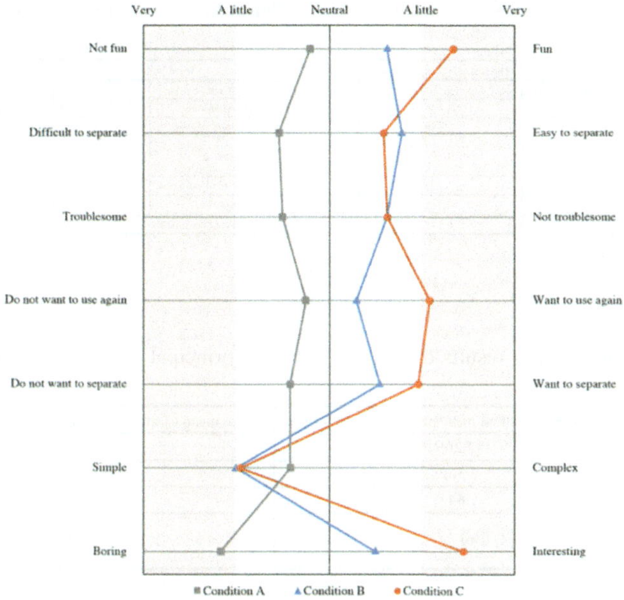

Fig. 4. The average of the evaluation experiment results

Table 1. The results of PCA

Word-pairs		The first principal component	The second principal component
Boring	Interesting	0.513	-0.516
Do not want to separate	Want to separate	0.417	0.016
Troublesome	Not troublesome	0.386	0.477
Difficult to separate	Easy to separate	0.367	0.428
Simple	Complex	-0.235	-0.411
Do not want to use again	Want to use again	0.356	-0.221
Not fun	Fun	0.312	-0.323
Eigenvalues		4.929	1.159
Proportion of Variance (%)		56.9%	13.4%
Cumulative Proportion (%)		56.9%	70.3%

results of PCA methods were analyzed with ANOVA (analysis of variance). The difference between the mean of the first principal component score and the mean of the second principal component score for each of the three conditions was analyzed and found that they had a significant difference (Table 4). In addition, the result turned out that only in the combination of the conditions A and B of the second principal component do not have a significant difference when examining the results among the three conditions by ANOVA for *Kansei* quality and Usability quality (Fig. 5).

Table 2. The first principal component

Word-pairs		The first principal component
Boring	Interesting	0.513
Do not want to separate	Want to separate	0.417
Troublesome	Not troublesome	0.386
Difficult to separate	Easy to separate	0.367
Simple	Complex	-0.235
Do not want to use again	Want to use again	0.356
Not fun	Fun	0.312

Table 3. The second principal component

Word-pairs		The second principal component
Boring	Interesting	-0.516
Troublesome	Not troublesome	0.477
Difficult to separate	Easy to separate	0.428
Simple	Complex	-0.411
Do not want to use again	Want to use again	-0.221
Not fun	Fun	-0.323
Do not want to separate	Want to separate	0.016

Table 4. The results of ANOVA of each principal component score

Conditions	The mean of the first principal component score	The mean of the second principal component score
Condition A	-2.206	0.257
Condition B	0.562	0.371
Condition C	1.644	-0.628
	F(2,46) = 59.70, p < 0.01	F(2,46) = 7.11, p < 0.01

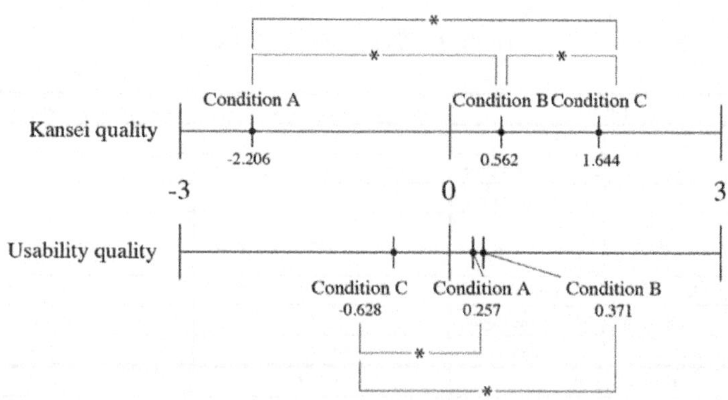

Fig. 5. The results of ANOVA in *Kansei* quality and usability quality

4 Discussion

From the graph of the average value of the questionnaire results shown in Fig. 4, the score of the condition with the prototype (condition B and condition C) is higher than the condition without the prototype (condition A) in 5 items "fun", "easy to separate", "easy to use", "simple" and "interesting ". In other words, the prototypes encouraged people to separate plastic bottle caps, and people could use the prototypes intuitively.

Therefore, it is useful to use a voting metaphor as a garbage can for separating plastic bottle caps. In terms of *Kansei* quality, the results of ANOVA confirmed that *Kansei* quality in the conditions B and C are higher than the condition A. Therefore, using a voting metaphor enhances *Kansei* quality. Moreover, the evaluation of *Kansei* quality in condition C was higher than condition B, so adding a sound effect to the voting metaphor could further enhance *Kansei* quality. In other words, using a voting metaphor with a sound effect can enhance *Kansei* quality the most. In terms of Usability quality, the evaluation of condition C was lower than conditions A and B. Therefore, adding a sound effect to the prototype decreased Usability quality. Although the evaluation of condition B in Usability quality is higher than condition A, there is no significant difference between them. In other words, even if the prototype using only a voting metaphor, Usability quality cannot be improved.

As shown in Fig. 5, there was a big significant difference among conditions A, B, and C in *Kansei* quality. However, there was a small significant difference among them in Usability quality. Condition C was able to enhance *Kansei* quality without extremely deteriorating Usability quality. Therefore, condition C gave suggestions as a way to enhance *Kansei* quality of the prototype.

In condition C, animal sounds were played as sound effects in response to the voting action. However, participants were confused because the sounds do not fit the context of the voting action. That is why Usability quality was impaired in condition C. Therefore, adding a voting metaphor and a sound reproduction function in a garbage can encourage the separation of plastic bottle caps, and that is useful for enhancing *Kansei* quality. However, sound effects should match the context of the voting action.

5 Future Works

From the results of this experiment, it is clarified that using a voting metaphor and a sound playing function for a garbage can is useful to improve *Kansei* quality. However, Usability quality was decreased. That means designers have to consider finding a sound relating to the design based on a logical process when we use a sound to the system design. In the future, this research will be developed to select a sound matched with the context of the voting action instead of the animal sounds to enhance not only *Kansei* quality but also Usability quality.

References

1. Fumoto S, Sakakibara N: On students' consciousness for segregating home trash. Kyoto University of Education Environmental Education Research Annual Report No. 10, pp. 19–28 (2002). (in Japanese)
2. Shinogi, M., Abe, K., Komatsu, Y.: The dilemma of social rationality between local governments and residents. Environ. Soc. Res. **17**, 19–34 (2011). (in Japanese)
3. DIMSDRIVE. http://www.dims.ne.jp/timelyresearch/2017/170314/

4. Ishihara, S., Nagamachi, M.: Evolution of Manufacturing and Kansei Engineering (Special Issue: Kansei and Manufacturing). Chugoku Reg. Res. Center **17**(2), 1–19 (2013). (in Japanese)
5. UNPROFESSIONAL DEVELOPMENT. https://unprofessionaldevelopment.org/portfolio/the-fun-theory/
6. Akira H: The Definition of *Kansei,* Report of Modeling the Evaluation Structure of *Kansei*, vol. 1, no. 1, pp. 41–47 (1998). (in Japanese)
7. Masayuki T, Namgyu K: Proposal of a bin to change human behavior with positive emotion. In: HCI International 2019 – Posters, pp. 394–401 (2019)

Human Factors for Textile and Society

The Importance of Textiles in the User Experience in Interior and Furniture Design

Rui Carreto[1,2(✉)] and Catarina Carreto[3]

[1] Escola Superior de Tecnologia e Gestão de Viseu, Instituto Politécnico de Viseu, Campus Politécnico, 3504-510 Viseu, Portugal
ruifccarreto@gmail.com

[2] Escola Superior de Tecnologia e Gestão, Instituto Politécnico da Guarda, Avenida Dr. Francisco Sá Carneiro, n.º 50, 6300-559 Guarda, Portugal

[3] Centro de Investigação em Arquitetura, Urbanismo e Design, Faculdade de Arquitetura, Universidade de Lisboa, Polo Universitário, Rua Sá Nogueira, Alto da Ajuda, 1349-055 Lisbon, Portugal
catarinaferreira.design@gmail.com

Abstract. Since Prehistory, the interior of human habitats is a reflection of their existence, therefore, man has adapted space to his needs and physical conditions. The adaptability of furniture and shelter spaces corresponded to sedentary or nomadic life forms.

During the Modern Era, from the 15th century Renaissance, and especially with the Baroque in the 17th and 18th centuries, furniture and decoration began to play a relevant role in the design of interiors, both religious and civil architecture. The relationship between architecture and design and its implications is a question that leads us to reflect on the nature of the disciplinary approach to design itself and its application or performance as interior design.

Thus, this article aims to emphasize the importance of textiles in the user's relationship with interior space and furniture. As evolutionary concepts, interiors and furniture have undergone profound changes and adaptations in response to the prevailing aesthetic power of symbolic character and to the comfort created by the needs of its users. In this sense, by reviewing the literature and presenting examples it is our intention to clarify concepts and modes of action.

Keywords: Interior design · Furniture design · Textiles · Human factors · User experience

1 Introduction

Since the beginning of civilizations, the search for well-being in shelter/housing spaces was something that was intended to be achieved, either through the comfort provided by the fire or by some utensils and equipment, or in the feeling of safety and protection of wild animals or of weather that the caves, for example, could assure them. Over time, man felt the need to settle in a place, to create roots and territorial and even social relations, which allowed him to cultivate a sense of belonging, of identification with his fellow men in sedentary communities, in a recognized physical space that could consider yours. The houses, in general, were multifunctional, as was the furniture,

G. Di Bucchianico et al. (Eds.): AHFE 2020, AISC 1202, pp. 727–732, 2020.
https://doi.org/10.1007/978-3-030-51194-4_94

which, in addition to being equally multifunctional, was also demountable. In this context, the chests used to store clothes and food also served as a seating unit, reducing the number of pieces due to their multifunctionality, thus facilitating their transport.

The first textiles were of natural origin made by hand. In furniture, linen and silk started to be the most used in Egyptian civilizations in beds and seats. Some houses had upper floors with galleries facing north, where people could sleep outdoors during heat periods. Decorative fabrics were hung on the walls or used as partitions in the compartments, and reed mats covered the floors.

From the end of the 18th century onwards, the Industrial Revolution brought new materials, new forms of construction, new production methods, which together with the appearance of the railways, the automobile and the replacement of human energy by driving energy were aspects advocating the development of new technical and constructive forms. The textile industry gained expression with great technical advances in production and finishing. Design also played an important role in this field, where the intervention of a designer added value to fabrics, raising profits. As a result, production increased, as did the quality of the fabrics, distinguishing it from products previously produced by artisanal systems. Nowadays, smart textiles are gaining more and more prominence, either in the monitoring of the human body through the fabrics used in clothing, in equipment or in furniture, through fabrics that heat or cool us in response to body temperature.

2 Interior and Furniture Design

It is relevant, for the understanding of our work, the definition and approach of interior design that shelters furniture design, considering that the history of interior design was also made through the history of furniture design. So, we start by saying that the term interior design came up to describe a group of projects, related to each other, involved in the design of any interior space, whatever the variety of human activities taking place in it, giving priority to the functionality of spaces instead of their decoration [1]. Gibbs [2] describes the profession as being multifaceted, so designers needs to be efficient, disciplined, have commercial qualities, be flexible and creative, but also have artistic sensitivity. It is a sector oriented to the public, which implies collaboration with customers and other professionals, specialists and suppliers, and it is vital that the designer is a good communicator. In this way, the function of the interior designer is to interpret the ideas and/or corporate identity of a company or institution to design environments in a creative way. The design process, from the initial contact with the customer until the end of the project, comprises several phases which include works and very different areas of specialization. They deal with the organization of spaces, layout and technical details [1]. Other areas of design, such as graphics, fashion or industrial, must consider function, aesthetics and the structure of the object as primary factors. Both, interior design and furniture design must also consider these aspects, adding the factor of the need to respond particularly to an interior space. According to Booth and Plunkett [3], without furniture there are no interiors, which is why the story of each one is not made without the other. Norman Foster [3] stated *"Furniture is like architecture in a microcosm"*. The authors emphasize that furniture should be studied with greater accuracy than the exterior of the building, since the people who live in an interior maintain direct, visual and physical contact with it. They immediately experience their practical efficiency and aesthetic language [3].

3 Textile User Experience in Interior and Furniture Design

Of the various materials that contribute to the design of an interior space, textiles stand out and occupy a very important place. They can provide a feeling of comfort, welcome and flexibility in a space. They appear in several colors, textures, quality and with an unlimited number of possibilities. Indoors, textiles can exist on the upholstery of chairs and sofas, on cushions, on beds, on tables, windows and without floors. According to Pile [1], textiles for internal use are divided into two groups: upholstery and drapery, where some can belong to the two types.

The history of textiles in interior and furniture design begins, according to Caan [4], in the discovery of the interior by man. It was based on the need to protect animals and atmospheric conditions, using animal skins and creating a controlled and more intimate environment, which allows greater comfort. Since prehistory, the interior of human habitats is a reflection of their presence, as it is man who adapts the space to his needs and conditions. As can be seen in the prehistoric histories of Lascaux and Altamira, zoomorphic representations were clearly intentional, reflecting a human need for artistic production, regardless of the purpose. In this way, we can indicate that even in this period, a relationship with space and its appropriation with a specific brand was indisputable. However, a discovery of the natural cave was not a unique long-term solution and the space was quickly adapted to new forms and demands of nomadic living.

The furniture adaptability and reference spaces correspond to the forms of life, sedentary or nomadic. Through the paintings existing in Greek vases or through the philosophical texts of classical thinkers, we can visualize the portrayed environment, an optical life form and ambiences of that time, in opposition to the campaign spaces that correspond to an itinerant and conquering life form. In this case, the campaign tents from the Roman Empire are examples of nomadism. Man moved from place to place, due to his wars and conquests', taking shelter and furniture with him, as was with the case of closing benches, which, due to their foldable nature, allowed him to move easily.

In Medieval times, civilian furniture almost disappeared. However, we can highlight that one of the emblematic furniture was the ark that served as a bed and table and where were transported the feudal lord goods who wandered through his properties. The spaces were simple and functional, taking into account the way of life and the use for which they were intended. Due to the roaming, many textiles are taken to provide greater thermal comfort and to delimit spaces, compartmentalizing the broadest.

During the Modern Period, from the Renaissance in the 15th century and, above all, with the Baroque in the 17th and 18th centuries, furniture and decoration came to play an important role in the design of interiors, both in religious architecture and in civil architecture. The relationship between architecture and design and its implications is an issue that leads us to reflect the design disciplinary approach nature and its application or performance as interior design. Although the definition of disciplinary territory is complex, difficult to define and the activity considered as a relatively recent profession, we can understand that interior design as a global and coherent interior design activity was practiced from an early age. Louis XIV's court in France, with a scenic or even expository character, undoubtedly provided an important milestone in the history of the discipline and profession of the interior designer. The Gobelins Manufacturing (or *Manufacture Royale des Meubles de la Couronne*) was responsible for everything that was used for the decoration of the royal palaces, more specifically for the supply of all the furniture that integrated the Palace of Versailles, for the profusion of textiles, inlaid walls and mirrored walls wrapped in marble. Upholstered furniture appears for the first time in response to the necessary comfort during waiting times in reception rooms [5].

The century, XVIII is marked by the Enlightenment that awakens a new mentality, more free and independent, with transformations in the economic and social level. The new society reacts against art that was too rigid and directed from the previous period. Therefore, the furniture multiplies to meet every need for life. They are simpler and lighter, continuing to obey the principles of symmetry. At this time, upholstered furniture appears with greater intensity, taking into account the different parts of the body that rest on the furniture, becoming curved and the seats having a more favorable height [6]. With the reign of Louis XV, the character of the rooms greatly influenced the dimensions and furniture types [5]. The manufacture of furniture was, at this time, a very specialized job. The seat furniture used thick cushions covered with tapestry on the seat, back and armrest, and often represented the work of painters [5].

In England, around 1620, the Tudor style was in force. The beds were usually made of oak wood and were the most important and most expensive pieces. Bed curtains were a very important element. The fabric considered most luxurious seems to have been taffeta and Turkish tapestry (imitation) was the most commonly used quality fabric. The tapestry and Turkish tapestry cheered up the dark interiors of the time. The paneled walls were ornamented with carved motifs imitating pleats of fabric. Over time, the straw gradually gave way to upholstered seats [5].

In 1754, Thomas Chippendale published the *Gentleman and Cabinet-Marker's Directory* Directory, a diverse collection of drawings of furniture pieces. This very practical book presented customers with possible ornamental solutions, resembling today's catalogs. Chippendale was a pioneer in creating large firms for the production of furniture, where he produced everything needed for a home. It employed numerous specialized artifices, from "cabinet-makers" (elite carpenters) to upholsters. For the first time, furniture is identified with a label, name and address, as was the norm of the French tradition, which until then was not found in England [7].

At the end of the 18th century, Neoclassicism prevailed in England, being Robert Adam one of its greatest exponents. We emphasize its importance for interior design, as according to Gibbs [2], Adam already thought the interiors as *"(...) intrinsic part of the project of the building, including designing the furniture in harmony with the complex decoration of the ceilings"* [2] (p. 16). In the 1770s, interior decoration became clearer and more delicate. Pastel colors started to be used. The motifs of the carpets reflected the painted and plastered decorations of the ceilings and the fixed furniture, consoles and mirrors, were designed as an integral part of the decoration [5].

The dramatic changes brought about by the Industrial Revolution affected various aspects of people's lives. The modern world of mechanized production has put artisan production aside, where design is an integral part of the product design process. *"During the 19th century, many interiors started to have a multifunctional character, being, therefore, much more hybrid in terms of ornamentation and content than their 18th century precedents"* [7] (p. 264).

In the twentieth century, from the 1980s, with the development of urban culture, the consumption of textile products in the interiors and in the furniture intensified and the offer corresponded with innovative solutions. In this context, fabrics are used especially for upholstery. New materials have great strength and firmness. Creating smooth and very resistant surfaces. While it is inevitable that certain modern materials will replace older ones, filled with foam instead of horsehair or coconut fiber, or staplers instead of nails and hammers for upholstery, traditional upholstery techniques remain equally important for the production of new parts quality and for the restoration of old parts.

Today textiles can incorporate digital technologies, light and heat sources, as well as react to the presence of users. More and more new materials appear quickly. In this field, concerns about the environment have also been valued as innovative solutions for the reuse and reuse of textiles for the design, not only of new fabrics, but also furniture design, as is the case with the various chairs of the designers Campana brothers or the designer Bernardita Marambio. This designer created the Demodé which consists of a new material that reuses textile waste, composed of both natural and plastic materials. The residues are agglomerated with a 100% biodegradable starch-based binder that provides structural resistance to the material. With this project, Bernardita Marambio intends to amortize, in large part, the high percentage of textile waste that ends up in landfills, from the concentrated Chilean industry. This material can be used in wall or floor coverings, and even molded in different formats for seats, benches and chairs.

In the interiors, smart fabrics are more prominent. Like many other products, they have their origin in the military and aerospace industry, due to the extreme conditions and the specific needs they are faced with. In this interior area, the Bloom project stands out, a sound-sensitive knitted fabric that opens and closes in response to ambient noise in order to provide the necessary sound attenuation. To avoid direct reflection of noise and, therefore, dampen sound, Bloom's surface opens to absorb sound waves. However, when the reverberating sound is desirable, the surface closes, allowing greater reflection of the sound in space. This smart textile is from the Yeadon Space Agency and is a collaboration with Jesse Asjes (Product Design Jsssjs) and Laura Wickesberg (WickesWerks LLC). Based on these evolutions in fibers, smart textiles, along with the recent development of technologies, are beginning to gain expression. Some specimens are able to measure our body temperature and send this information, through mobile applications, to the devices that regulate the temperature of our homes, so that they can adapt to our thermal needs, achieving an environment of total comfort.

4 Conclusions

Designers works to solve problems or issues, giving them form, functionality and creativity, always trying to improve and satisfy needs. Since design is something multidisciplinary and brings together project principles and attitudes, artistic creation can be born of everything. Interior designers deal with the space organization, layout and technical details and it should be noted that furniture design undoubtedly enters as a specific reason directed to living and interacting with the space.

Interior design and furniture design are areas of creative expression and, as such, their evolution and updating is permanent. Since the existence of man, he has sought for comfort in his habitats and the use of textiles has, to a certain extent, resolved some issues related to the thermal, acoustic, lighting or aesthetic of spaces and furniture in his relationship with the user.

The future of textiles, in its relationship with the user in interiors and furniture, is auspicious. More and more new projects are emerging with new technologies and new types of fabric focused on responding to an intelligent housing that monitors our body and our way of living. Thus, the importance of textiles in the user's experience in interior and furniture design will be, in fact, elementary in the life of the future.

References

1. Pile, J.F., Harry, N.: Interior Design. Abrams Publishers, New York (1995)
2. Gibbs, J.: Diseño de Interiores: Guía útil para estudiantes y profesionales. Editorial Gustavo Gili, Barcelona (2006)
3. Booth, S., Plunkett, D.: Muebles para diseñar Interiores. Blume, Barcelona (2014)
4. Caan, S.: Rethinking Design and Interiors: Human Beings in the built Environment. Laurence King, London (2011)
5. Brunt, A.: Guia dos Estilos de Mobiliário. Lisboa, Editorial Presença (1990)
6. Oates, P.: História do Mobiliário Ocidental. Lisboa, Editorial Presença (1991)
7. Kemp, M.: História da Arte no Ocidente. Editorial Verbo, Lisboa (2006). (coord.)

The Shape of Desire: Knitwear and the Body

Gianni Montagna[1(✉)], Cristina Carvalho[1], Carla Morais[1], and Maria Antonietta Sbordone[2]

[1] Ciaud, Lisbon School of Architecture, Universidade de Lisboa, Lisbon, Portugal
g.montagna@gmail.com, cristifig@gmail.com, carlamorais@gmail.com

[2] Università degli Studi della Campania Luigi Vanvitelli, Caserta, Italy
mariaantonietta.sbordone@unicampania.it

Abstract. If the body is the starting point for an unlimited journey of sharing signals of communication and recognition, it is also a point of arrival for desires and sensations that we live and share in our personal and public sphere.

This article aims to address the theme of body communication over time and its sharing with the textile knitted materials that throughout its development have helped to shape the objective and subjective factors that allow building a language of seduction and power.

This paper will present an overview of three different moments, throughout human history, where knitting and its characteristics were adopted as an integral part of the exercise for the development of aesthetic renewal and where technique and visual intersect in an exhibition and body projection.

Keywords: Knitwear · Body language · Seduction · Fashion communication

1 Introduction

The body is the privileged element of communication of any individual in relation to the context where he lives and develops. It communicates like a real object and adapts to the context and social situation where it belongs, communicating in an extremely complex way. Clothing enhances the communication of an image that we must understand as a social representation and that contributes to the construction of relationships [1]. As mentioned by Patel [2] different researches on body language reported that only 7% of the information human transmits to others is in the language we use; 38% of communication is in the quality of voice (accent, voice projection, emphasis, expression, pace, volume, pitch, etc.) and 55% through body language—posture, position, eye contact, facial expression, head and body movements, gestures, touch, etc.

If it is true that body communication is a constant element of our life and in the relationship with others, clothing represents the most natural and personal interface that interposes between the individual and the ecosystem where it is included. Fashion and clothing have the power to directly influence the society where we live, as it communicates visually and emotionally through symbols [3], which are constantly evolving and adapting. The fashion forms are almost inexhaustible and in constant adaptation. They provide new communicative constructions, whether material or immaterial,

G. Di Bucchianico et al. (Eds.): AHFE 2020, AISC 1202, pp. 733–740, 2020.
https://doi.org/10.1007/978-3-030-51194-4_95

objective or subjective, constantly reinventing themselves in the renewal of their expressive language. This language, on the one hand, allows a reading of fashion in its context and, on the other, it influences the social environment in which it operates, or the cultural world, as Ernst Harms [4] nicknamed, in a continuous exchange of influences and new adaptations. This communication and social construction game already identified and described for decades by Flügel [5], has existed over the centuries and in different times and social situations and continues to this day as a way of differentiating and affirming individuals and communities.

The individual's relationship with his geographical and social habitat is dynamic and interactive, giving rise to actions and reactions that build and shape the individual and the environment where he lives at the same time. It is a natural adaptation where the individual takes an objective and subjective posture of the environment, adapting the different internal and external dimensions in order to feel an integral and active part of society.

Textile material surrounds us throughout our lives. From birth until our death we are covered with textile materials that in different ways protect and represent us. They are materials that imbue us with sensations that cover a wide spectrum of our experience and that go from subjective to objective and vice versa, in a continuous interrelation of multidimensional sensations that integrate physical and psychological dimensions of the material itself and the individual.

The knitted material has specific characteristics that are unique in social and technical terms. Due to their characteristics, knit substrates present unique possibilities for representing their surface, where stories can be told and where specific drawings can be used to demonstrate belonging to an ethnic group or identity. The typical example of the "Fair Isle" drawings is paradigmatic of an identity representation of the homonymous island that gathers in the repetition of its coloured drawings common features of the population it represents. From a technical point of view the mesh substrates wefts present a high flexibility with a consequent improvement of comfort levels for the user. The weft mesh is developed by a single thread that, creating rows of loops that are attached to each other in the previous rows, produces an extremely flexible and elastic textile surface that has a high power of adaptation to different shapes such as those of the body. The knit substrate, when pulled at any point on its surface, dissipates the force applied at that point by the lower rows, allowing the loops that form the knit to adapt its shape to the new surface. This fact, taking into account that it does not concentrate the tension caused by the pressure exerted on the textile in a single point, allows to notably increase levels of comfort, even in situations of extreme stress of the textile substrate.

The characteristics of this type of substrate are probably the reasons that make this material one of the most used ever. In reality, some elements of clothing were probably created with this type of textile substrate and would not exist otherwise. The case of socks, which are probably the most anatomical pieces ever used on the male and female body, is the most paradigmatic of this reality.

The knitwear allowed a body exposure over time affecting the individual's behaviour from private to the social sphere in his identity and behaviour [6].

Knitted textile material is an element that, throughout its history, has allowed to potentiate a discourse of power and seduction so important that it has influenced peoples and shaped cultures. Its history is told through its use on the body and the combination of these two elements in continuous change and adaptation.

Three examples are presented below in order to understand its social, political, economic and technical context.

2 Socks and Tights

The origin of the knitting technique is not yet clear, but the Greeks already produced those that are now known as the first pair of knitted socks in history. Knitted socks have been found in some 3rd and 4th century Egyptian tombs believed to have been used to protect the feet of the pharaohs from entering sand in the middle of the fingers. Knitting technique started to be developed using a single needle and later evolved to two needles, in the middle of the sec. XIII. Probably imported from north Africa, this technique has been skillfully practiced in most Spanish houses, to produce wool and cotton socks for the people and for the making of silk socks for the aristocracy [7]. At that time, continues the author to explain, there was a commercial dispute between England and Spain as the production of Spanish silk stockings was highly acclaimed and this fact was emptying the English coffers. As related by Welters [8] "The knitting machine, invented by William Lee in 1589, resulted in stockings that provided a much better fit than those of woven fabric". From that moment on, socks became the element used in the legs, leaving behind the covers made with strips of cut fabric. Produced in planes open textile and subsequently made with a seam on the back of the leg [9], the socks had an important revolution with the invention of Marc Brunel that gave rise to the first circular knitting machine in 1816 [7].

In 1884 in France the new Rayon fiber was invented and came to revolutionize the market. Much cheaper than natural silk but with a similar appearance, the democratization of this type of feminine product has developed rapidly. In 1937 Dupon launched a new fiber on the market, Nylon, which even more like natural silk, floods the market and makes stockings an object of desire and provocation. It was between 1890 and 1900, refer Steele [7], the decade of the cancan and the Moulin Rouge, that the stockings managed to impose themselves with the sensual and provocative imagery that they will be able to maintain for several decades. If, on the one hand, this situation can expose women and their intimacy to the general public, it also places them as an element of social power through their desired body. Throughout history, tell us Caroline Cox [10], woman have skilled themselves in the art of seduction, one of the only routes to female power before the twentieth century.

The second decade of the twentieth century was probably the one where there was the greatest revolution in the exposure of female legs [11], with a great focus on footwear and especially on stockings that at that time were already being produced in nylon. Showing your legs from this moment on is no longer considered taboo and becomes an element of seduction and personal and social affirmation.

As reported by Chew [12] in his work, the exposition of short stockings transgresses the boundary between outerwear and underwear, [...] by displaying a part of dress that is deemed necessary to be kept hidden from public view.

Nowadays, tights are used both under clothing and in everyone's view, attracting attention to legs whose sexual message may be more or less evident. There are many iconic images of the use of body messages based on their sensuality, whether to mark historical moments and social resurrection, but also to sell an image of seduction available to all ages and all budgets, when buying products from specialized brands. such as La Perla, Wolford, Agent Provocateur, etc., among many others.

3 Lingerie

Vestiges demonstrate that underwear was already used in ancient Egypt and ancient Greece, even if its use was purely functional [6]. In many cases, underwear was used so that the dirt on the body did not contact the outer garments that were delicate, very rich in details and very expensive. It would then be easier to wash white linen between the dress and the body than to wash entire dresses.

Often used as a primary form of protection against the weather and the body's own dirt, it seems that from the middle ages its use began to be introduced into the habits of different western peoples. However, it seems a little obvious to say that hygiene habits at that time followed their own patterns and that they had nothing to do with those of today.

If with the forthcoming of the Renaissance individuals in general started to become an integral and central part of the universe and its creation and the body will begin to be represented in a more organic way and to show its forms. Leonardo da Vinci's studies of Vitruvius man are probably the maximum exposure of this turning point to the discovery of man as a central element of the world of the living.

With the Renaissance, the use of corsets will have the function of shaping the body and giving it a structured shape [11]. Iconic underwear, the corset has reached modern times where it is periodically used for a representation considered "perfect" of the female body and her ideal of beauty. It was probably from this moment on that the underwear began to have the connotation that still today are associated to it, with its multiple and complex meanings.

Used as a combat weapon, lingerie is a highly powerful seduction tool. Understood as a set of tangible and intangible elements, it uses symbols and their individual and social appearance to create in the individuals around it the expectation of desire or pleasure, as stated by Jean Baudrillard [13] in his work called Seduction.

It can be understood as a set of elements that constantly mix and remix in a logic of exploring sources of seduction and desire, constantly renewing themselves in the search for new languages. The basic elements of this language, the lingerie objects used on the body, respond more to requirements of form than function and represent and communicate a hidden intimacy reserved for the increase of what is a game of seduction and desire between individuals. Underwear can be perceived as the required interface between the body and clothing, but it can also constitute clothing's ultimate seduction [14].

As reported by Martin [14], clothing is a principal means by which we negotiate between private and public realms. In this continuous research and discovery of seduction, seducing and letting yourself be seduced, materials play a fundamental role, in the relationship with others but also in the relationship with the body itself.

A significant impulse to the refined decoration of underwear in the nineteenth century was its ability to share appearance with outer clothing [14], shortening the distances between outer and intimate clothing and paving the way for a public exhibition of the intimate plan.

Jersey was, refers Pajaczkowska [15], famously adopted by Coco Chanel as a modern, proto-feminist material for uniting the lightness and pliability required for freedom of movement with the softness and "feel" desirable for garments worn next to the skin.

Despite the recognized comfort offered by knitwear materials, whether in the increased possibility of movement or in the high possibility of adapting to the body and better showing its shapes, the knitwear material was used only very recently for the production of underwear and lingerie. In reality, until the 1950s, with few exceptions, underwear was mostly produced in orthogonal fabric with warp and weft and almost always in white due to its clear association with virginal purity.

With the development of knit textile materials and especially with the sexual revolution that had its maximum exponent between the 60s and 70s of the 20th century, the body frees itself from rigid and built structures to find more free forms of expression. The underwear gives up rigid materials that support the body to make space for the knitwear materials, more comfortable, elastic and with a more natural look.

Knit materials are nowadays the most used for the production of lingerie and those that best answer the needs of psychological and ergonomic comfort of this type of product. By adapting easily to the body, the knit materials helped to build the bridge of desire between the body shown and the fantasy.

4 The Body-Jewelry

The jewels have an intrinsic connection with the body"; this is what is reported in one of the tracks of the Journal of Jewelry Research, an international reference on research and critique on contemporary jewelry. The jewel was the first object created and conceived by the man in relation to the human body and above all to encourage commercial exchanges, creating the first economy based on luxury. Luxury linked to the symbolic dimension of the jewel, of which the technical perfection and the refinement of the materials are far from creating an object to satisfy the daily basic needs. "This intimate context means that jewels can function as a symbol of themselves, as something that becomes a witness to our experiences and as a signifier of aspects of identity and interpersonal relationships" [16]. The dynamic that is established between body and jewels leads us to different approaches: jewels "for the body" take on meaning and can be considered complete only if worn, therefore, in contact with the body. Website of the Journal of Jewelry Research, in its Chapter 7, called Jewelry and the body, refers that "demonstrative" jewels that rely on the "body as a point of reference or as a subject, rather than as a destination, suggest that in some cases the body is an accessory of the jewel". If we analyze the all-contemporary trend of Textile Jewelry, we move on to a third way which autonomously draws heavily from a manufacturing sector, in some ways unrelated to jewelry, properly so-called, which, however, fits into that thread of the handmade that it considers weaving as one of the first economies of everyday life [17]. Although the trend of handmade today is updated thanks to digital production; the technologies available or made available, ranging from laser cutting to chemical photoengraving, from 3D polymerization and sintering - in particular FDM (Fused Deposition Modeling), SLS (Selective Laser Sintering), DMLS (Direct Metal Laser Sintering) - up to inclusion of electronic components or other types of digital processes [18]. In a different way, these techniques allow the construction of complex models or very high precision engravings and mesh - such as the creation of internal undercuts and voids - which would be impossible to achieve using traditional bench techniques. It could be kicked that all digital technologies are difficult to cross with the timeless handmade craftsmanship of traditional jewelry [19] although contemporary literature speaks of "digital craftsmanship", hybridization could result either in the hybridization of languages, or vice versa, in production. The hybridization of languages is based on the mixture or juxtaposition of multiple techniques that find the right balance in the designer, and produces unique pieces; the other uses digital technology and through a design process based on a parametric development system manages the production of a single specimen in a series of multiples that differ from each other in a programmed manner. Self-produced and home-made: the model for the studio of jewelry in the economy of the 21st century is rooted in self-sufficiency, proximity distribution and hand intelligence [20]. The current model of the Textile Jewel that fully represents the 20th century adapts to other contemporary interests [21]: a collaborative project, nomadic structures and non-exclusive forms of knowledge (amateur), which all define new forms of "authenticity". How do these different narratives intersect and how do they differ? How are they redefining the role and perception of the professional of the contemporary atelier? An example is Frédérique Coomans whose experimental work combines two different processing techniques which explain in these terms: "These collections are born from the association of specific techniques to my two training areas: fashion design and contemporary jewelry. I tried to combine them trying to find a delicate balance through textile design. I experimented with knitting and other textile techniques that became outdated to honor them. I explore the boundaries between jewelry and textiles, fiber, clothing. While beads are strung, I dress forms and shapes with textiles. Like a second skin, I opted for a set of flexible mesh that reveals under their screens either organic or geometric shapes." The images shown below demonstrate how craft and contemporary techniques coexist in textile jewelry as elaborated by Coomans.

5 Final Remarks

Knit is a material that, due to its specific construction, has characteristics of elasticity and adaptability that other materials, whether textile or not, are rarely able to match. The concept applied to this material appeared in antiquity and very early on, it realized its characteristics and properties. The ancient Greeks, on the other hand, used socks made with this technique of a single thread that develops in loops around the previous rows and so on, giving rise to material with superior elastic characteristics that allow its adaptability to body shapes.

Even though in ancient times, in many cases, the use of this material was made only for its functional potential, as individuals gain recognition as being the center of life and its events and that began to value the body as a central element of real representation in his way of being, the knit allowed him to show the body beyond function, giving voice and shaping its aesthetic dimensions.

The knit material helps to expose the body and build the elements of desire created by showing and hiding the intrinsic beauty and pleasure elements of the body. If, on the one hand, this material allows showing forms, on the other hand it hides the body itself, leaving space for fantasy. It is a connecting element between the personal world and the social world that, constantly working on its different variables, presents renewed languages and social reinterpretations of our way of being and representing.

Acknowledgment. The authors of this paper wish to thank the Centre for Research in Architecture, Urbanism and Design (CIAUD) of the Lisbon School of Architecture of the University of Lisbon and FCT for founding this project.

References

1. Gomes, N.: Branding de Moda na Hipermodernidade Líquida, Revista Portuguesa de Mark. **17**(número 32), 9–19 (2014)
2. Patel, D.S.: Body language: an effective communication tool. IUP J. Eng. Stud. **9**(2), 90–95 (2014)
3. Barnard, M.: Fashion as Communication, 2nd edn. Routledge, London (2008)
4. Harms, E.: The Psychology of Clothes, New York (1938). https://www.worldcat.org/title/psychology-of-clothes/oclc/63981559
5. Flügel, C.: Psicologia dell'Abbigliamento. Franco Angeli Editore, Milan (1986)
6. Newman, K., Bressler, K., Proctor, G.: A Century of Lingerie: Icons of Style in the 20th Century. Chartwell Books, New Jersey (1997)
7. Steele, V. (ed.): Encyclopedia of Clothing and Fashion. Charles Scribner's Sons, New York (2005)
8. Welters, L.: Europe to 1700. In: Welters, L., Lillethun, A. (eds.) The Fashion Reader, 2nd edn, pp. 21–36. Berg, Oxford (2011)
9. Gleba, M., Mannering, U. (eds.): Textiles and Textile Production in Europe from Prehistory to AD 400. Oxbow Books, Oxford (2012)
10. Cox, C.: Seduction: A Celebration of Sensual Style. Octopus Publishing Group, London (2006). (n.d.)
11. Hennessy, K., DK Publishing, Inc., (eds.): Fashion: The Definitive Visual Guide. DK Publishing, London (2019). New edition
12. Chew, M.: The dual consequences of cultural localization: how exposed short stockings subvert and sustain global cultural hierarchy, Positions East Asia Cult. Critique, **11**, 479–509 (2003). https://doi.org/10.1215/10679847-11-2-479
13. Baudrillard, J.: Seduction. Macmillan, New York (1990)
14. Martin, R., Koda, H.: Infra-Apparel Metropolitan Museum of Art. H.N. Abrams, New York (1993)

15. Pajaczkowska, C.: book section, making known: nine types of textile thinking. In: Jefferies, J., Conroy, D.W., Clark, H. (eds.) The Handbook of Textile Culture, pp. 79–97. Bloomsbury, London (2015). ISBN 978-0-8578-775-0
16. Fiorani, E.: Leggere i materiali, Con l'antropologia, con la semiotica. Lupetti, Milano (2010)
17. Skinner, D.: Contemporary Jewelry in Perspective (In association with Art Jewelry Forum). Lark Jewelry & Beading, New York (2013)
18. Deckers, P.: Contemporary Jewellery in Context. Arnoldesche Art Publisher, Stuttgart (2017)
19. Musée de design et d'arts appliqués/contemporains (Lausanne): De main à main: Aprendre et transmettre dans le bijou contemporain Européen = From hand to hand: passing on skill and know-how in European contemporary jewellery, Mudac (2008)
20. Zilioli, E.M.: Open Space – Mind Maps. Positions in Contemporary Jewellery. Arnoldesche Art Publisher, Stockholm (2016)
21. Thompson, I., Stott, N., Kerridge, T.: Biojewellery. Oral and Maxillofacial Surgery, King's College, London (2006)

Ancient Activities Improved by Design

Isabel Bieger(✉), Cristina Carvalho, and Gianni Montagna

CIAUD, Faculty of Architecture, U. Lisbon, Lisbon, Portugal
isabelbieger@hotmail.com

Abstract. Each day Design is coming more important, especially when are interventions in order to improve objects or even systems. When design carry out their function, are able to develop new meanings to objects that already exists, enlighten them, improving lifestyles and giving the opportunity to turn products in a new goods.

Knowing the actual state of art of the bobbin lace and the glass factory in Portugal, we invite the design to reverse this situation. The bobbin lace at Peniche and the glass factory in Marinha Grande are two case studies that design brought solutions.

New methods were generated, and new goods were invented.

The bobbin lace that usually were made in cotton line, started to be produced by other materials like carbon fiber and glass fiber. The bobbin lace is no longer a totally traditional product and has accepted innovation. From these experiences emerged new ideas to new products to be developed.

Inviting to fulfill the actual design function, we transformed products and we transformed life.

Keywords: Bobbin lace · New materials · Design

1 Introduction

Design is crossing traditional boundaries giving people a better quality of life through the reorganization and requalification of products that were considered inopportune and often forgotten. This kind of activities are giving to the design the real importance, or even the real meaning that are named. Design has a function and a mission like to carry out with the people and with the world.

Associated with the direct observation of Peniche's bobbin lace and the finding of its decline as a craft activity that is mostly performed by women as a local tradition, in this way, we invite and insert design in search of better suggestions, in order to reverse this situation and make it more dynamic.

What design can do for the bobbin lace? How to streamline through design for the bobbin lace to be valued and even valued in terms of environmental, social and economic sustainability?

In Peniche (Portugal), the most used yarn for making bobbin lace is 100% cotton yarn. This yarn is chosen because it is a flexible yarn and therefore easy to twist and make the characteristic lace knots. The products made end up being the same over the years. In this context, the challenge of the present investigation is associated with new possibilities and new yarns capable of being used in the production of bobbin lace,

G. Di Bucchianico et al. (Eds.): AHFE 2020, AISC 1202, pp. 741–747, 2020.
https://doi.org/10.1007/978-3-030-51194-4_96

proposing an innovative contribution and aiming to obtain pieces different from those made to date. Therefore, with an active interventionist methodology, we intend to boost this activity.

The analysis of different materials and their properties is fundamental, the total alteration of materials involved in the production of income is crucial. In this way, we seek methods of contributions that design, and the designer can make in the innovation sector. Like this, associating the bobbin lace with other materials.

2 Contextualization

Bobbin lace is considered to be a typical handcraft from coastal regions. It is known as a traditional craft in the city of Peniche in Portugal. Peniche considers bobbin lace as part of its identity and culture.

"A piece of lace is an artistic composition expressed in twisted thread, just as a piece of woodcarving is the expression of the artist's idea in chiseled wood. Lace is not, like embroidery, an ornamental fabric; it is itself ornament. It is not the application of art to a craft; the whole pattern is the fabric, and the fabric is the pattern. It is this peculiarity that distinguishes lace from needlework and woven work" [1] (Fig. 1).

Fig. 1. Execution of bobbin lace. Source: author. Photograph captured on March 4, 2015

3 Action

For the author Peter Dormer, crafts have strong roots. It is considered a vigorous influence on the culture and society of the people, regardless of the relevance of some manual works, some handcrafts are considered cultural mainstream. "From earliest times, weaving was a craft reserved for women that gave them respect in the public realm; the hymn singles out crafts like weaving as practices that helped civilize the hunter-gatherer tribes" [2].

The search for materials was carried out in an intuitive way, trying to cover a wide typology of textile materials, but always paying particular attention to the mechanical-elastic property, flexibility, given the unalterable need to twist the thread in order to create the knots and thus originate the bobbin lace, as the lace execution technique would remain intact.

Tests were carried out, as well as the analysis of the results obtained in the tests of the use of new materials. Thus, the possibility of executing the points to be used was verified, as well as the difficulties encountered by the lace makers.

Analyzing the characteristics of the tested materials it was observed that many of them are resistant to high temperatures. The new samples made with the bobbin lace technique in different materials were now used in different experiments, small samples were used and fused to the glass.

Strategic changes are needed so that traditional handcrafts can remain present in history and especially in people's lives. For this to be a success, it is necessary to establish an important working relationship between the lace maker and the designer because the artisan must be seen as a protagonist and not an adjunct, thus enjoying his creative autonomy.

For the author Jean Poirier, a traditional object has the characteristic of never having completed its reification process. Some objects over the years retain their former status. According to Poirier, each of these elements is subject to change, given especially by the evolution of techniques [3].

It is through innovation that we will be able to upgrade things and products, both traditional and contemporary, innovating means having and introducing novelties, products never seen or commercialized. "innovation in a 'global' sense is defined to occur when new products, processes, raw materials, and management methods are first introduced to an existing or new market" [4].

As complexities, shorter product life cycles and increasingly variable market demands, the development of new products becomes essential [5]. To innovate is to think about the future, a future with a better quality of life. What interventions are necessary to be reconsidered for a better work practice. In this way, resulting in products with more meaning.

Please check that the lines in line drawings are not interrupted and have a constant width. Grids and details within the figures must be clearly legible and may not be written one on top of the other. Line drawings should have a resolution of at least 800 dpi (preferably 1200 dpi). The lettering in figures should have a height of 2 mm (10-point type). Figures should be numbered and should have a caption which should always be positioned *under* the figures, in contrast to the caption belonging to a table,

4 The Proposal Materials

The reuse of materials is one of the proposals of the author Blaine Brownell [6].

In this proposal the materials are readapted and used in other situations, thus redefining their use and taking advantage of all the functionalities of the materials, which can sometimes be less polluting and more economical, creating new and unexpected results. Another proposal by Blaine Blownell is the combination of different materials, changing them and at the same time provoking a symbiosis between them. It is part of this process to explore possibilities, guide the development of new materials to create new products, with the potential for innovation. This process aims to be able to reinterpret the opportunities and products presented by such developments and apply them in different and creative ways, thus facilitating the birth of a new product, with a different potential in the market.

Glass is considered a material with multiple work potentials. Reason why we chose it to be worked on and associated with our project. Its characteristics provide the creation of objects of singular peculiarity. "glass is a rigid liquid. Seems strange when you have a glass filed with water that both the container and the contained are one and the same. The molecules in the glass container are just thicker and appear to be solid" [7]. It is a material with unique characteristics combined with its functional and aesthetic possibilities that allow the exploration of its versatility [8].

The bobbin lace is made with 100% cotton thread. We propose to change it and explore new possibilities for using yarns. Here are some possibilities:

The carbon fiber "Has outstanding mechanical properties. Its tensile strength is 15 times greater than steel, but it is relatively cost-intensive. Suitable for high- performance structural elements. The black carbon fibers are manufactured from the petroleum derivate PAN (polyacrylonitrile) that is carbonized via pyrolysis ("charring" under oxygen-free conditions) at approx. 1,500 °C" [9].

According to the author Marie O'Mahony, carbon fiber is considered inorganic insulating fiber because it is resistant to high temperatures. "These lightweight fabrics are made of a blend of polyester and cotton and are designed to withstand repeated washings" [10].

Fiberglass is considered a resilient material; it also has the characteristic of insulating and fire-resistant properties. "Has extremely good chemical resistance and is thermally insulating and fireproof. Its mechanical properties are modest. Glass fiber is relatively cost-efficient" [9].

5 Results

About fiberglass yarn: Lubricated fiberglass, cross section: 1.5 mm.

The yarn presents the necessary flexibility for the execution of the bobbin lace stitches. This material leads us to suggest mixing and/or fusing with other materials where the use of high temperatures is an asset, such as glass.

The carbon fiber yarn is a yarn considered to have a large cross section for the execution of the lace. It is necessary to carry out a first operation, that is, try to separate the cables that initially constitute the carbon fiber yarn, in order to obtain a less dense yarn, with a smaller title, that is, a cross section/lower diameter. The difficulty presented in this operation is to maintain the uniformity of the cable throughout the process. The yarn has the necessary flexibility for making bobbin lace stitches. Its texture does not allow the presentation to be perfect, since the thread has frayed threads. The piece is very wavy on itself, does not remain in one plane as is characteristic of bobbin lace.

About fiberglass yarn: Is Lubricated fiberglass. Cross section: 1.5 mm.

The yarn has the necessary flexibility to execute bobbin lace stitches.

6 Creation of New Objects

Because it was detected that the materials had resistance to high temperatures, an improbable characteristic of the object of traditional study, for the development of new objects, we proposed to associate the result of this product with another material. We chose glass because it is a material with unique characteristics and likewise, it is produced at high temperatures. The city of Marinha Grande, neighboring Peniche, has an important glass production hub, known in international terms. Regarding the historical context, Marinha Grande glass suffers the same problems due to the decline in its demand compared to the bobbin lace.

Visually the pieces obtained are shown in some way as artistic pieces, unique, some of which may be considered functional, others not. However, they will always be unique and unrepeatable pieces, because although the same bobbin lace can be reproduced with the same material, the reaction when interconnected with the glass will always be different. The different factors that prevent the identical reproduction of a piece will always be taken into account. The glass reacts differently to each firing. The heat scale interferes with the burning of the pieces, which inevitably leads to obtaining exclusive pieces.

Glass, a material with high plasticity or flexibility when heated, takes on the shapes of the molds placed on them. The shapes are different and make an analogy with the bobbin lace. The delicacy of the lace is protected with glass (Fig. 2).

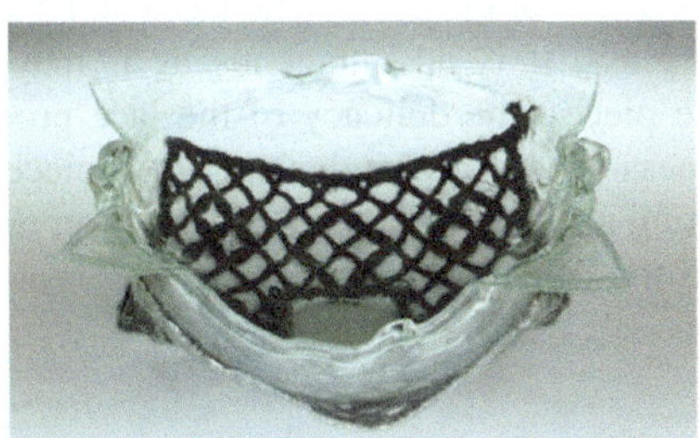

Fig. 2. Result of the merger of two float glass plates and bobbin lace in carbon fiber wire. Source: author. Captured August 2, 2017

Both pieces were made with two float glass plates and bobbin lace made with carbon fiber yarn. The carbon fiber yarn described here comes in a more rudimentary form. The yarn has an upper cross section, greater "thickness", that is, greater finesse, than that used so far, in the production of other bobbin lace tests. After burning, the piece presents the lace texture on the glass.

The larger piece acquired the concave shape corresponding to the mold. The contrast of carbon fiber with glass is very present. The yarn with the highest cross-section value of the carbon fiber allows to associate the bobbin lace. In this way, it will be possible to visualize some traditional points of bobbin lace from Peniche. However, the piece becomes more robust and striking. The higher the value of the cross section of the yarn to be used in making the lace, the greater the stiffness it will exhibit, thus losing flexibility. Factor that determines the shape of the object, because the upper glass cannot acquire the shape of the mold. This effect causes much higher air bubbles between glasses, which is why the glasses are unable to stick to each other. This piece can be considered both utilitarian and decorative (Fig. 3).

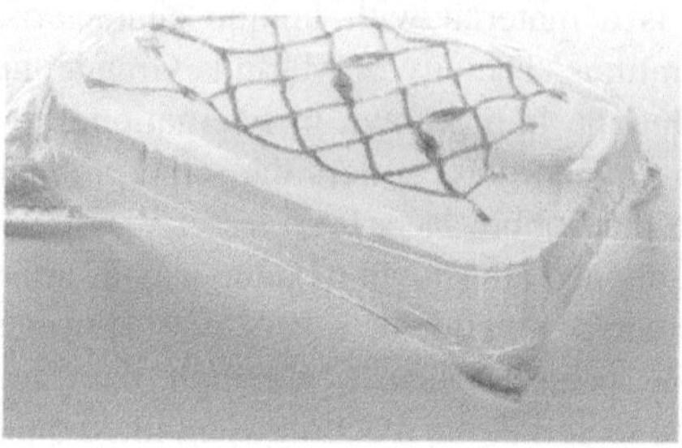

Fig. 3. Result of the merger of two float glass plates and bobbin lace in fiberglass yarn. Source: author. Captured August 2, 2017

For the elaboration of this piece, two float glass plates were used and among them is the bobbin lace made of fiberglass yarn. The glass of dimensions greater than the dimensions of the mold when melting slid again on the support table of the oven acquiring a support in the piece that refers to the idea of the piece having support feet, showing the solidity on this side of the piece. The lace can be seen on both sides, given the transparency of the glass, it is advantageous as it may facilitate the everyday utilitarian function of the piece. The delicacy of the lace produced in fiberglass yarn contrasts with the robustness of the glass at the end of the piece produced, there may be a duality of interest, creating a dispute of attentions, however the bobbin lace suggests a more intense conquest in the capacity of analysis or observation in the final result.

7 Conclusion

The intervention of design in this production activity, which is considered cultural and identity heritage, contributed to its consolidation and increased potential. The bobbin lace needs to establish strong relationships, to immediately abandon the preconceived idea of an obsolete piece and come to become the protagonist of history. The contributions that design can provide to bobbin lace will allow the tradition to continue. There is still much to explore, however, new paths for the bobbin lace of Peniche can be found.

In this way, the need for innovation adjusted to the realities of the products was combined. In both products, this is the characteristic glass of Marinha Grande and the Bobbin Lace of Peniche, present similar difficulty of survival. In this way, uniting two products with different characteristics and properties, it was intended to create new sensations in the observer. From these experiences, it becomes easier to understand that bobbin lace could acquire new meaning, ceasing to be a simple object and acquiring greater visibility.

The tests carried out with the glass, in addition to their results being always designated almost as a revelation, were another part of the creative development of the process.

The new pieces obtained using the technique of fusing glass with lace samples offer a new experience and strengthen the bond between the user/consumer and the object. They are pieces that naturally convey emotion, as they bring all the dedication and detail to the work of the artisan lace maker. This work can be considered eternal and closed between glass plates. The symbolism of the final piece refers to what remains kept and perpetuated, it can be both functional and ornamental. An exclusive and unique piece of added value.

References

1. Calado, M.: História da Renda de Bilros de Peniche. Gráfica Torriana, Peniche (2003). (Autor ed.)
2. Dormer, P.: The Culture of Craft - Status and Future. Manchester University Press, Manchester (1997)
3. Poirier, J.: História dos Costumes - O Homem e o Objeto, vol. 3. Editorial Estampa, Lisboa (1999)
4. Stoneman, P.: Soft Innovation - Economics, Product Aesthetics, and the Creative Industries. Oxford University Press, Oxford (2010)
5. Cooper, R., Aouad, G., Lee, A., Wu, S., Fleming, A., Kagioglou, M.: Process Management in Design and Construction. Blackweel, Oxford (2005)
6. Brownell, B.E.: Transmaterial. Princeton Architectural Press, New York (2006)
7. Kohler, L.: Glass - An Artist´s Medium. Krause, Iola (1988)
8. Beveridge, P., Doménech, I., Pascual, E.: O Vidro - Técnicas de Trabalho de Forno. Editora Estampa, Lisboa (2004)
9. Sauer, C.: Made of – New Materials Sourcebook for Architecture and Design. Gestalten, Berlin (2010)
10. O'Mahony, M.: Engineered textiles. In: Clarke, S.E., O'Mahony, M. (eds.) Techno Textiles, vol. 2, pp. 52–75. Thames & Hudson, London (2005)

Long-Line Composite Floating Modular System for Controlled Growth of Mytilus Galloprovincialis Mussels and the Development of Crassostrea Gigas Oysters

Alexandra Ene(✉) and Carmen Mihai

The National Research and Development Institute for Textiles and Leather, 16, Lucrețiu Patrașcanu Street, 030508 Bucharest, Romania
alexandra.ene@incdtp.ro

Abstract. The development of the composite materials for obtaining modular systems for the development of bio filter material is the central objective of the research and it responds to the provisions of the European directives “in force” - COM (2007) 0128, COM (2009) 0156 and COM (2012) 0670. The systemic approach of the problem aimed to the: principles of construction and functioning of the whole; analysis of the technical and functional characteristics of the components, as well as to the interdependencies between them; transfer of certain technical-functional properties (breaking resistance, shock, seawater, extreme temperature variations, etc.) to the whole; submission of the decision (from a managerial point of view) to specific problems of the whole system [1]. All these elements represented the basis for the elaboration of the mathematical model for the modular system for the development of the biofiltering material that enabled the prediction of the structural parameters of the composite materials used as subassemblies of the long-line system for the development of the biofiltering material.

Keywords: Bio filter · Coefficient of resistance losses · Buoyancy · Input data · Geometrical shape

1 Introduction

All aquatic organisms, from fresh or marine water, which live fixed on a natural substrate in the water (marine organisms) and which by their own vital functions contribute to maintaining the normal quality of the aquatic environment, form the so-called natural epibiontic filter.

Recent research has demonstrated the increased filtering capacity of marine water by epibiontic organisms, so that one kilogram of epibiontic filter material formed in 90% of mussels, in the juvenile state, have the capacity to clean within 24 h, a volume of approximately 73 m^3 of seawater, with a phytoplankton load of approx. 18 million cells/liter [2].

G. Di Bucchianico et al. (Eds.): AHFE 2020, AISC 1202, pp. 748–753, 2020.
https://doi.org/10.1007/978-3-030-51194-4_97

This technology for the production of marine epibiosis on artificial substrates is a purely ecological method, which aims to improve the quality of marine water in the coastal areas of tourism through local epibiontic bio filters [3].

Recovery of natural biological material is currently achieved through the development of complex systems for larval cultivation of epibiontic organisms. In this context, the solutions identified are aimed at developing suitable artificial media for fixing the epibiotic filter bodies [4].

The usage of the appropriate technical means will lead to the extension of the support area of the epibiontic organisms and, implicitly, will significantly increase the bio-cleaning capacity of the water from the formed epibiontic filter.

2 Materials and Methods

The hydro meteorological conditions characteristic for the open areas of the marine coast, respectively: H wave = 6 m; L wave = 25 m; V current = 1.5 m/s, the nature of the substrate (sandy areas) imposed the values of the design parameters of the modular system for the development of the bio filtering material and implicitly the typo-dimensional characteristics of the modules used as the floating superstructure (anchored in position) and of the artificial collectors (suspended in the water media) [5].

In order to design and develop the modular system, the following input data have been used: at the end of a biological cycle on a collector an amount of 50 kg bio filter material will be fixed; coefficient of resistance to roll forward cylinder: 0.63; bi-conical structure: 0.57; buoyancy for each floating module of the installation: min. 50kgf; the values of the load coefficients (safety, dynamics, asymmetry); the coefficients of loss of resistance (at joints, due to abrasion, fatigue and environmental factors) and of the ultimate load coefficient.

In addition, the following loading coefficients and associated values were considered:

- *Safe coefficient Cs* – as it was considered that the failure of an element would destroy the entire structure. In calculation of the floating subassemblies was considered a value of 1.5.

- *Dynamic loading coefficient Cd.* This coefficient has been introduced to include the specific effects of dynamic tasks. The value taken into account for this kind of application 1.2 ÷ 1.5.

- *Asymmetric load coefficient Ca*: it was appreciated that the forces are distributed evenly, regardless of the level of the perpendicular section on the axis of symmetry. The considered value: 1.5;

- *Final loading coefficient* $Ci = Cs \cdot Cd \cdot Ca$ – without having a concrete physical meaning, it was taken into account for the safety of the whole operation. From the mathematical point of view it is the algebraic product of the load coefficients.

In the practice of designing the systems used in the aeronautical or naval field, the loss coefficients of resistance are used to describe the behavior of textile materials in real working conditions.

If the load coefficients highlight the real load conditions, the resistance loss coefficients'role is to evidence the effect of the mechanical and environmental factors.

For this application the following coefficients were considered:

- *Coefficient for resistance in joints loss Cr* – used because composite structures are assembled with other materials (textile or other nature), usually the joint parts having a lower resistance than of the base materials. Thus, for the most efficient usage of the resistance of the composite materials throughout the entire system, it was adopted the concept of "equal strength joint" according to which the assembly must have the breaking resistance equal to that of the weakest resistant material in the assembly component. The value of the coefficient of loss of resistance in joints, obtained by the ratio between the average resistance to breaking for the joint and the breaking resistance of the weakest material in the assembly component was 2.3.

- *Coefficient for resistance due to abrasion loss Cf* – was taken into account the value 1.0, considering that the marine environment acts aggressively against the composite material.

- *Coefficient for resistance due to fatigue loss Co* – used in order to include the effects of multiple use. The value taken into account: 1.0.

- *Coefficient for resistance due to environmental factors loss Cm* – It was considered a value of 0.95, as the component composite structures of the system are subjected both to the aggressions of the solar radiation and sea water.

- *Total coefficient of resistance losses* $Cp = Cr \cdot Cf \cdot Co \cdot Cm$. From the mathematical point of view it is the algebraic multiplication of the coefficients for the resistance loss presented above.

The final load coefficient $C_U = Ci/Cp$ - that defines the level of resistance of the analysed assembly, encompassing both the effects of specific loading conditions and the effects of mechanical and environmental factors. The value of the final loading coefficient is obtained from the ratio between the final load coefficient and the total coefficient for the resistance loss. $C_U = 4.92$.

The composite structures were made of polyamide HT and polyester yarns. The resulting woven structures were coated with polyurethane resins used as a polyurethane emulsion in dimethylphthalate solvent, having an active substance content of 42%, as well as with vinyl polychloride which gives the material excellent abrasion resistance, very good resistance to microorganisms, good chemical resistance and possibilities of joining with adhesive or through thermal welding.

In order to increase the adhesion to the compound based on vinyl polychloride, the textile support was activated by immersion in isocyanate based products.

The lamination mixture was made of: PVC emulsion in approx. 45% in dioctyl phthalate solvent (approx. 22%), plasticizer (to increase the film's flexibility), chlorinated paraffin and tin trioxide (to increase the degree of fire resistance) and titanium dioxide (for resistance to solar radiation); the pigment used was gray.

The component elements are represented by the floating elements located offshore (max. 4000 m from shore) and in the area of shore (max. 1000 m); submersible cylinder for the floating elements destined to growth and development of biofilter material (mussels and oysters) fixation; central floating element for system support and growth - development of bio filter material (mussels and oysters); main and secondary anchors of mooring anchors type; floating elements for controlled growth of mussels; floating elements for controlled growth of mussels and oysters development; submerged enclosure for the development of oyster larvae; artificial collectors for the direct growth of mussels and the development of oysters; enclosure fixing systems for the development of oyster larvae.

The physical and mechanical characteristics of the composite materials and correspondence with functional models of the modular system are presented in Table 1 and Table 2, respectively.

Table 1. The physical and mechanical characteristics of the composite materials

Characteristic variant of the composite material	Mass, g/sqm	Breaking force, daN		Tearing resistance, daN	
		Warp	Weft	Warp	Weft
V1	437	412	410	38	42
V2	436	365	375	25	22
V3	303	204	217	25	29
V4	237	232	225	65	62
V5	312	287	172	12	13
V6	301	210	220	20	33
V7	1123	530	550	55	65

Table 2. The correspondence between variant of the composite material and functional models of the modular system

Variant of the composite material	Functional model of the modular system
V1: warp - 100% PA6.6, 660dtex/f100x1/350Z; weft - 100% PES, 1100dtex/f192x1/120Z; weft rep.	Submersible cylinder for the floating elements destined to growth and development of biofilter material (mussels and oysters) fixation
V2: warp -100% PA6.6, 940dtex/f140x1/160Z; weft - 100% PES 1100dtex/f192x1/120Z; weft rep.	Floating element located in the area of shore
V3 – warp, weft - 100%PES; 1100dtex/f192x1/120Z; warp rep.	Floating elements for controlled growth of mussels and oysters development
V4 - warp, weft - 100%PES; 167dtex/f32x2/250Z; basket.	Floating element located offshore
V5 - warp, weft - 100% PA6.6, 660dtex/f100x1/350Z; warp rep.	Submerged enclosure for the development of oyster larvae
V6 - warp, weft - 100%PES; 1100dtex/f192x1/120Z; warp rep.	Floating elements for controlled growth of mussels
V7 - warp, weft - 100%PES; 1100dtex/f192x1/120Z; twill 2/1.	Central floating element for system support and growth - development of bio filter material (mussels and oysters)

3 Results and Discussions

In order to obtain a minimum resistance in the sea current and to the action of the waves, the cylinder and the bi-cone have been chosen as geometrical shapes, that are efficient under the operating conditions of the system represented by: buoyancy, possibility to attach the catching and anchoring eyes, vertical stability etc.

In addition, in order to establish the values of the geometric parameters, it was considered that the surface strength of the modular system is increased proportionally with the fixed biological material growth that determine an increasement of the external forces on the anchors' action.

Table 3. Calculation of buoyancy of component elements

Geometrical shape	Input data, values and identification/coding elements
Bi-cone	Floating elements for the direct growth of mussels and the development of oysters. $\rho = 1.0245$ g/cm^3; r = 25 cm; h = 100 cm; g = 9.81 m/s^2; A = 0.3925 m^2; V = 0.13 m^3; W = 0.8 N; F_B = 1307 N; F = 1306 N
Cylinder	Floating elements for directed growth of mussels $\rho = 1.0245$ g/cm^3; r = 20 cm; h = 100 cm; g = 9.81 m/s^2; A = 1.38 m^2; V = 0.1256 m^3; W = 4 N; F_B = 1262 N; F = 1258 N
	Floating element located offshore $\rho = 1.0245$ g/cm^3; r = 50 cm; h = 230 cm; g = 9.81 m/s^2; A = 8.00 m^2; V = 1.805 m^3; W = 24 N; F_B = 18141 N; F = 18117 N
	Floating element located in the shore area $\rho = 1.0245$ g/cm^3; r = 40 cm; h = 180 cm; g = 9.81 m/s^2; A = 5.02 m^2; V = 0.90 m^3; W = 15.4 N; F_B = 9045.4 N; F = 9030 N
	Central floating element for system support and growth - development of bio filter material (mussels and oysters) $\rho = 1.0245$ g/cm^3; r = 22 cm; h = 130 cm; g = 9.81 m/s^2; A = 1.95 m^2; V = 0.20 m^3; W = 22 N; F_B = 2010 N; F = 1988 N

The buoyancy was calculated for the floating elements that ensure the integrity of the system throughout the growth and development period of the bio filter material (Table 3).

The solutions developed will contribute to the restoration of marine filters, by providing a breeding site for a variety of marine organisms, such as black mussels. At maturity, each medium reaches up to 50 L of seawater every day.

4 Conclusions

The process of slow and insignificant recovery of the natural bio filter in the coastal area requires the elaboration of ecological methods meant to increase the populations of epibiontic organisms which constitute in filters, able to accelerate the restoration of the quality of the marine environment in the tourist - recreational areas on the coast.

In order to improve the quality of the marine environment in coastal areas affected by the anthropogenic impact, it is recommended to build epibiontic bio filtration barriers throughout the water column, including on the affected sedimentary substrate.

Taking into account the input data, and the conditions of usage in the open sea, the modular system for the development of the bio filter material and seven floating modules together with the corresponding subassemblies were designed and realized.

The equipment was tested according to the biological cycle (April-November) and met the imposed requirements, ensuring an increased productivity of about 35%.

Acknowledgment. This scientific paper is funded by the Ministry of Research and Innovation within Program 1 - Development of the national RD system, Subprogram 1.2 - Institutional Performance - RDI excellence funding projects, Contract no. 6PFE/2018.

References

1. Irisarri, J., Fernández-Reiriz, M.J., Labarta, U., Cranford, P.J., Robinson, S.M.C.: Availability and utilization of waste fish feed by mussels Mytilus edulis in a commercial integrated multi-trophic aquaculture (IMTA) system: a multi-indicator assessment approach. Ecol. Ind. **48**, 673–686 (2015)
2. Brenner, M., Broeg, K., Frickenhaus, S., Buck, B.H., Koehler, A.: Multi-biomarker approach using the blue mussel (Mytilus edulis L.) to assess the quality of marine environments: season and habitat-related impacts. Mar. Environ. Res. **95**, 13–27 (2014)
3. Riisgård, H.U., Larsen, P.S., Pleissner, D.: Allometric equations for maximum filtration rate in blue mussel Mytilus edulis and importance of condition index. Helgol. Mar. Res. **68**, 193–198 (2014)
4. Delegrange, A., Vincent, D., Duretm, M., Amara, R.: The use of mussels for mitigating the noxious effect of phytoplankton spring blooms on farmed fish. Aquac. Eng. **66**, 52–61 (2015)
5. Azam, F., Malfatti, F.: Microbial structuring of marine ecosystems. Nat. Rev. Microbiol. **5**, 782–791 (2007)

Design Artisan and Art: Development of a Textile Collection in Contemporary Fashion

Ana Margarida Fernandes[1,2(✉)] and Isabele Lavado[1,2]

[1] Faculty of Architecture, CIAUD, University of Lisbon, Lisbon, Portugal
ana.design.fernandes@gmail.com
[2] Instituto Politécnico de Castelo Branco - IPCB, Castelo Branco, Portugal

Abstract. Promoting the articulation between design, artisan and textile art to develop concepts in contemporary fashion, here the focus is on developing joint projects between designers, artists and artisans, using different textile techniques for innovative productions. This approach drew on different types of materials and craft techniques, using these between designer, artists and artisans. Preserving the tradition in the technical execution of each textile piece is fundamental, where the articulation between the artist's design and the artisan developing a textile/fashion collection from knowledge capable of incorporating innovation. The articulation in the development of new products with design, satisfies different tastes where the culture and identity of certain regions can recover tradition and contribute to cultural dissemination. So, is the dynamics between designer, artist and artisan possible in the development of new products for the textile/fashion area?

Keywords: Design · Art · Artisan · Textile

1 Introduction

In a country where some handicraft traditions still proudly survive, it is urgent to revitalize and promote this knowledge. In traditional activities, reinventing is mandatory, says João Branco [1]. However, the cultural, social and economic weight of artisanal activities in Portugal still corresponds to a very fragile economic fabric. However, according to CEARTE - Professional Craft Training Center, the global situation in the field of handicrafts is no longer lacking, as was the case in the early 1980s. Today the outlook is much more encouraging, thanks to a set of changes in the last two decades that refer to affirmation as a social group and economic sector.

Today there is some recognition of handicrafts as a product now sought by an urban middle class looking for identification marks, either by the individualization of the products that result from the predominant manual characteristic of its manufacture or by its identification with a certain territory [1].

In a society that tends to invest in an increasingly globalized future, some seek to recover the past, preserve tradition, keep an identity alive through procedures that incorporate craftsmanship and design considering a partnership with the future [1].

G. Di Bucchianico et al. (Eds.): AHFE 2020, AISC 1202, pp. 754–760, 2020.
https://doi.org/10.1007/978-3-030-51194-4_98

A new impetus was introduced into handicrafts, expressed in the creation of new production units and in the appearance of people who, with increasing levels of education, with adequate professional preparation, often equipped with skills in the area of design, management and new technologies, became seduced by ceramics, glass, wood, textiles, and choose to value this knowledge [1]. Handicrafts seem to be in a phase of recovery from a decline while design seeks to consolidate itself in companies, institutions and consumers.

Over the past decade, design has become the most discussed management resource. It is seen as a key component in achieving quality standards and accepted as the main driver of innovation processes. Also, it is proposed as a way of giving products more added value and greater production profits [2].

As research in the field of design, it is important to clarify the concept of conceptual design, which distances itself from the "classic" design process and emphasizes its connection to artistic intentionality. Some authors [3] consider that the conceptual design is reflected in how to insert a new culture in objects within the scope of daily use and how a link can be created with the user that exceeds consumption and that is included in the recovery of the emotional and intellectual connection with the objectives. Here the notion of utilitarianism is overcome, against other aspects of the design project, such as the symbolic language of the object, so this estrangement takes on a greater relevance.

Conceptual design implies the materialization of the concept, to assert its intention and the objective of a project, although, sometimes, given the distant position in relation to the industry, this materialization can happen only in a prototype or in a small series [3].

2 Tradition and Modernity. Whomever Has a Past Has to Have a Future?

The future can be invented, given the impossibility of predicting. This indeterminacy of perspective also always contains the seed of hope, the constant search for happiness to which human beings are tied and makes their condition, at the same time incomplete and immense, the engine of their survival, as individuals and as a species [4].

In this line of thought, it is believed that the Portuguese clothing and textile sector can be the development paradigm when it serves as a vehicle to support traditional artefacts, in the preservation and reinterpretation of the national textile tradition, through contemporary products.

A study by the GEPE - Studies and Prospective Office of the Ministry of Economy in 2004, envisaged some strengths and opportunities in the clothing and textile sector in Portugal that reinforce the idea that this sector is privileging discovery and valuing traditional Portuguese culture, especially that which comes out of the hands and imagination of designers and artisans.

Intuitively, clothing makes it possible to make known, through textile work, other knowledge, other cultures, other Portuguese roots in addition to being a favored sector worldwide. Manuela Neves [2] in her work, mentions that it is an activity that easily attracts foreign investment, and because much of its technology is easy to transfer, it is generally one of the first industrial activities to grow in countries under development.

Consequently, the intelligent application of technology, concern and preservation of the environment, as well as new ideas for products, become the origin of creativity and innovation and consequently constitute a decisive factor in competitiveness and success. It is possible to produce goods that meet consumer expectations and that, at the same time, become competitive in the market [4].

3 Methodology

In the scope of this investigation, the interview as a data collection technique has an essential role because it is possible to gauge the perception of the interviewed designer/artisan. An interview consists of eliciting a set of individual responses, interpreting and generalizing them [5]. Starting from the initial research question, Is the dynamics between the designer and the artisan possible in the development of new products for the fashion/textile area?, this interview aims to confront the designer, mainly with the feasibility of articulating design/art and craftsman in applying new textile bases to contemporary garments.

In this sense, other questions that arise in order to answer the purpose of the investigation will be asked. The methodology for this process was: Description of the interviewee's profile; Selection of the population from the sample of individuals; Definition of the purpose of the interview; Establishment and description of the means of communication; Location and the time; Description of items or characteristics for the script: Questions related to the items, avoid influencing responses, the number of questions was established and ordered within each dimension.

Before the interview, a pre-test was carried out, and the script was adjusted to the interview. After being recorded in digital format, the information collected through the interview is treated with a view to its analysis; writing of results and conclusions (Table 1).

3.1 Procedures Followed During the Interview

The interview was conducted beginning with an explanation clarifying what the interviewer wants and the purpose of the interview. The confidentiality of the interviewee and the answers were guaranteed if they so wished. At this point, the need for collaboration was emphasized and a pleasant environment was created. Before starting, the interviewee was asked for authorization to make the recording. An audible distance was maintained between the interviewee and the interviewer while guaranteeing the audio recording. The interviewee's own words were recorded as were their gestures and expressions (noted). Despite the questions being presented orally and in writing, an informal conversational tone was used. There was no use of intonation or oral highlighting of words to avoided influencing the responses. The duration of the interview was 45 min. It ended as it had started in a friendly environment.

Table 1. Description of categories for the interview guide and elaboration of questions related to categories.

Analysis/Categorization Dimensions	Questions
Feasibility of revitalizing old techniques and craft knowledge	• *Do you consider it feasible to revitalize old techniques and craft knowledge?* • *What form do you think is most appropriate for this revitalization to be successful?*
Experiences of the designer/artist/artisan	• *From the experiences you have had in the evolution of your work and when you have gone abroad to receive your awards, what do you feel is that you carry a culture?*
Considerations about the work carried out	• *From the work that has been done, since Expo 98 in which several clothing or fashion designers were invited to participate in the exhibition (where Maria Helena also participated), how do her colleagues see her take clothes with these traditions, with which some experiences have already been made? How do you see these experiences?* • *How do you see the techniques for revitalizing this knowledge?*
Design issues	• *The type of customer that Maria Helena has in her store is a very specific customer, with some cultural basis?* • *How do you see the design framed in these pieces?* • *When Maria Helena thinks about a collection, does the methodology of a process enter?*

To reinforce and validate some of the research data, an individual and a unique interview was conducted with the Portuguese clothing and textile designer, Maria Helena Cardoso, as she was the only one who responded to the interviewee's profile.

The interviewee, taking into account the definition of the interview proposal and its objectives, was selected based on the characteristics previously mentioned.

Starting from the initial research question Is the dynamics between the designer and the artisan possible in the development of new products for the fashion/textile area?, this interview aims to confront the designer, mainly with the feasibility of articulating design/art and craftsman in applying new textile bases to contemporary garments.

In this sense, other questions that arise in order to answer the purpose of the investigation will be asked.

3.2 Selection of the Population from the Sample of Individuals to Be Interviewed

Among a very small sample of people identified for the interview, Maria Helena Cardoso was selected. Although one can count, at the national level, with other notable personalities, this was the one that best responded to the profile of the person to be interviewed. She mentioned, during the interview, another designer, Olga Rego, however, this is far from the focus area, so it was considered not to be an asset in terms of the final result.

Here, a small reference to the professional life that Maria Helena Cardoso shows the reason why she was selected to be the only interviewee. With a vast curriculum, mainly in the area of clothing and textiles, she has worked in the last 30 years with less conventional materials, applying them to contemporary garments. Maria Helena Cardoso became aware of the need to recover and value traditional knowledge, and it was in her eagerness to support these techniques that were on the brink of extinction, that she decided to carry out works, moving to locations. It has led to the recovery of the heritage, mainly forgotten textile, despised by the consumer society.

Her work begins with a process of raising awareness, learning and training in villages in the north of Portugal, to direct knowledge to the execution of current clothing, having the urban woman as a potential client.

Another of the fundamental points, in which she was successful, was the use of left over materials, in an intrinsic attitude of those who know that it is in taking advantage of the gain and that from waste new pieces can be made, thus combining the imagination with the acquired knowledge.

With a strong personality, Maria is also known for being an affable person, with an intense desire to help others when it comes to preserving what is typically Portuguese, she is a strong, persistent, agile and dynamic woman, wise, enthusiastic and well-disposed.

4 Registration of the Interviewee's Responses

4.1 Applicability of New Textile Bases

"Yes, right away, especially for a current market eager for the different (about the possibility of using textile bases in contemporary clothing). The things created en masse and the same things, from shopping centers, are having, I do not say an end (…). There are already a lot of people who prefer to go and look for the different because they start to miss and nostalgia for the different. (…) Now we can see that they are leaving them (shopping centers) and returning to the traditional (…)."

"Yes, from my experience, I know that these types of pieces have a way out, no doubt."

"In the case of their bases, they do not change, only with the Embroidery of Castelo Branco, you will see … this also affects the process of using silk, and there is such recycling and an adaptation that I think there are only positive vectors. The colors are not aggressive, they are natural and you move towards what is actually more natural; what a person wears and can pick up in a few years, never goes out of style, it's the great commercial value of these types of garments. We were able to make unique pieces because you get your hands dirty, which is a difficult feature to find elsewhere because you won't be able to do exactly the same. There is a registration and a mark of the person who made the piece."

"(…) It has to be thought that at this moment all people want are light pieces because of health problems. Linens, wools, silks have these characteristics that please people, as they are warm, comfortable and are not heavy. These fabrics only have advantages. (…) They are pieces that are the same, without being the same."

4.2 Feasibility of Revitalizing Old Techniques and Craft Knowledge

> "(…) It is viable, it is the path that this country has to follow, as Portugal is still the only country that has these memories and has structures and people that can do all these jobs and that do not lose those memories to be able to do them. (…) These works are an asset for Portugal and not a delay."

> "(…) Our country, with its "backwardness", becoming isolated and a very rural country, has kept knowledge and in fact techniques and materials (such as wool and linen) it is only us at this moment (…) we keep this ancestral knowledge of working with our hands, of making hand-crafted pieces."

> "(…) We try to "fit" these works into today's society and I think that it makes more and more sense (…), to make timeless pieces. We are in a world where we begin to discover that we have to monetize, recycle and women must also have clothing that is timeless because we will have to save."

"This is the way! First because we are taking things from ourselves, second because they are natural materials and third because it is to give life to a culture of a country that cannot be forgotten; a country is nothing if it lacks respect and dignity for all its culture, which comes from the roots, called popular culture (which is sometimes quite erudite) (…)."

"I think this project does have legs to stand on."

4.3 Experiences of the Designer/Artist/Artisan

"(…) We worked here a good few years ago (…) I see myself twenty-one years ago too, when I took things that nobody had ever thought to take and actually started to give them life, because they were dying."

> "(…) I feel that I have a culture (…)"

5 Conclusions Drawn from the Interview

The opinion of the designer/artist is clear and reflects the certainty of the feasibility of contemporary garments from different fabrics, created between craftsman and designer/craftsman. For Maria Helena Cardoso it is feasible to revitalize old techniques and traditional knowledge, being perhaps the only way to preserve them. As is the development of projects like the one presented to her, as Portugal is still one of the few countries that have these memories, structures and people who can do this type of work. For Portugal, this is an added value that will lead to progress, if well planned. Firstly, it draws on national details, then it works with natural materials and finally, it gives life to culture and what comes from Portuguese roots, the so-called popular culture that is sometimes quite erudite.

It is essential to look for a current market, avid for different pieces; according to the interviewee, everything that is created *en masse* and the same products from large shopping centers is having an apparent end and there is a desire to return to what is traditional.

It is possible to "fit in", as the designer says, this type of work presented in today's society because it is transformed into contemporary, timeless pieces, unique because everything is done manually. There are a registration and a mark of the person who performed the piece; according to the phrase of the designer "the pieces are the same, without being the same" [6], referring to the fact that it is possible to create identical pieces but that in the end are never the same because they are handmade works. We are also facing a world where we are beginning to discover that it is necessary to monetize, recycle and women must also have timeless clothing.

Another very important aspect that can throw a project to the ground if it is not considered is the price, which Maria Helena Cardoso stressed during the interview. It is necessary to think about pieces that serve most social classes, they will not be the most disadvantaged, but a considerable group. In addition to these aspects, she suggests thinking about pieces for a very specific customer, with some cultural basis as they end up buying a story, a differentiated image, pieces with presentation and distinct. It turns out to be a thoughtful and smarter purchase.

When approached about her professional experience in promoting the products she produces, the designer considers that they have immediate acceptance, regardless of whether they are in Portugal or other countries.

Any piece made from an existing cultural base must be worked with feeling and sensitivity; it is necessary to know the theme and the whole historical context behind it. There has to be collaborative work between those who design a piece and those who embroider and execute it (designers and craftsmen).

References

1. Branco, J.: CEARTE - Centro de Formação Profissional do Artesanato. Mãos 28 (2010)
2. Neves, M.: Panorâmica Geral. In: Desenho Têxtil - Tecidos, vol. I (2000)
3. Ferreira, P.C.: Design Conceptual na era pós-industrial: "A forma segue o conceito" In: dissertation to obtain a master's in product design, Faculty of Architecture Technical University of Lisbon (2010)
4. Vaz, P.: Tradição e modernidade. Quem tem passado tem obrigatoriamente futuro? In: P. Vaz, A tradição tem futuro? O sector têxtil e do vestuário português na mudança de paradigma de desenvolvimento p. 109. Braga: CENESTAP - Centro de Estudos Têxteis Aplicados (2004)
5. Coutinho, C.: Técnicas de Recolha de dados: Universidade do Minho (2011). http://claracoutinho.wikispaces.com/T%C3%A9cnicas+de+recolhas+de+dados
6. Cardoso, M.H.: Tecelagem Tradicional - Vestuário Contemporâneo. In: Atelier business card (2004)

Apparel Design and Textile Engineering

Hierarchical Model of "Feeling of Luxury:" Genuine and Artificial Leather Case Study

Shuhei Watanabe[1,2(✉)] and Takahiko Horiuchi[2]

[1] Innovation/Research and Development Division, Ricoh Company, Ltd., 2-7-1 Izumi, Ebina-Shi, Kanagawa 243-0460, Japan
shuuhei.sw.watanabe@jp.ricoh.com
[2] Department of Imaging Sciences, Graduate School of Science and Engineering, Chiba University, 1-33 Yayoicho, Inage-Ku, Chiba-Shi, Chiba 263-8522, Japan

Abstract. In this study, we used the physical properties of leather as a case study to quantify the "feeling of luxury" perceived by humans using a hierarchical model. We constructed our model with a physical property layer, an impression layer, and feelings layer. The physical property layer contains physical leather properties obtained from image measurement experiments. The experiments conducted multi-angle measurement because the appearance of leather changes depending on the observation angle. As described above, the remaining two layers include subjective scores obtained by subjective evaluation experiments. In particular, we determined the words used in the impression layer by an impression evaluation experiment in which we extracted the representative impressions perceived from leather. As a result, we were able to quantify the "feeling of luxury" perceived by the participants from leather by a simple estimation equation with two extracted representative impressions.

Keywords: Leather · Luxury · Perception · Modeling

1 Introduction

The appearance of material is an important aspect of product design. In recent years, methods for evaluating the appearance of various materials have been proposed [1, 2]. However, the appearance of material provides different impressions to viewers depending on the object [3, 4]. In particular, the appearance of genuine leather is considered to provide an impression of luxury to consumers. Although many product designers have a desire to use genuine leather for various products, such as automotive interiors and clothing, because of the high price of leather and animal rights issues involved, artificial leather is often used for such products. Artificial leather is made from polyurethane or vinyl chloride, and has the advantage of easy maintenance. However, the appearance of products made from artificial leather may provide different impressions to consumers compared with products made from genuine leather.

In the current study, we assumed that identifying the characteristics of leather perceived by consumers could be useful for the development of artificial leather products that induce the same impressions as genuine leather. Therefore, in the present study, we sought to investigate aspects of the perceptual impression of leather, and to

G. Di Bucchianico et al. (Eds.): AHFE 2020, AISC 1202, pp. 763–769, 2020.
https://doi.org/10.1007/978-3-030-51194-4_99

evaluate the subjective "feeling of luxury." To achieve this purpose, we constructed a hierarchy of subjective responses when considering visual perception mechanisms, and examined the relationships between each layer. The perception of the visual appearance of an object involves various factors, including surface color, gloss, and texture. In a previous study, Hiramatsu et al. demonstrated that brain activity in the initial visual cortex, such as V1 and V2, are strongly correlated with image features, whereas activity in the higher-order areas of the ventral pathway is strongly correlated with perceptual impressions [5]. These results indicate that the impression of a material is formed in the course of processing within the ventral path. A recent study in monkeys indicated that activity of the ventral visual pathway was involved in the conversion of image features into a perceptual impression [6].

Based on these previous studies, we confirmed the correlation between measurement values and subjective evaluation values, which were applied to construct a subjective hierarchy response, as shown in Fig. 1.

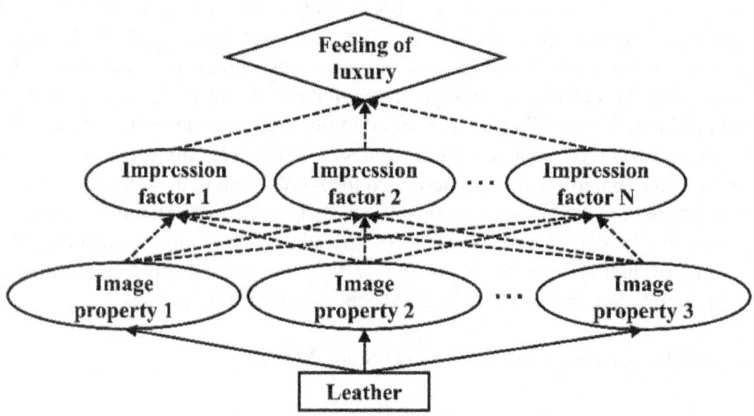

Fig. 1. Proposed subjective hierarchy response.

The structure of this paper is as follows: first, we describe the measurement methods. Then, we detail the methods used in the subjective experiment involving leather samples and present the results. Next, we report our examination of the correlation between subjective evaluation scores and measurement values. Finally, we provide some concluding remarks regarding the current findings.

2 Measurement Method

To develop an evaluation model for the "feeling of luxury", it was necessary to obtain image properties from the leather. Since the appearance of leather changes with the observation angle, we constructed a goniophotometer system. Two-measurement experiments (color measurement and surface measurement) were conducted to measure the image properties. Figure 2 shows a schematic illustration of each measurement system.

The color measurement system had a spectral camera and a light source. The spectral camera captured spectral images in 31 bands with a 10-bit depth. The size of the captured image was 600 × 600 pixels. The image resolution was approximately 1,000 dpi (25 μm/pixel). Because the camera recorded spectral images, these images could be converted into the CIE L*a*b* format. The spectral images were normalized by using spectral images of a standard white target. We performed the measurement under five-angle conditions as shown in Fig. 2(a).

The surface measurement system comprised a digital single-lens reflex (DSLR) camera (PENTAX K3-II, RICOH IMAGING COMPANY, LTD., Tokyo, Japan) and a lighting device. The raw images had a depth of 14 bits, and were converted into L* images. The raw images were then normalized by using raw images of a standard white target. The size of the captured image was trimmed to 3,000 pixels × 3,000 pixels. The image resolution was approximately 1000 dpi (25 μm/pixel). To obtain the surface property values, we calculated several surface characteristics. Integral values of the spatial frequency within the ranges 0–0.1, 0.1–1.0, and 1.0–4.0 cycles per millimeter were defined as surface characteristic values. The ranges were determined by measuring the grain size of the sample used in this study. These values corresponded to surface irregularity and large/small grains, respectively. We performed the measurement under four-angle conditions as shown in Fig. 2(b).

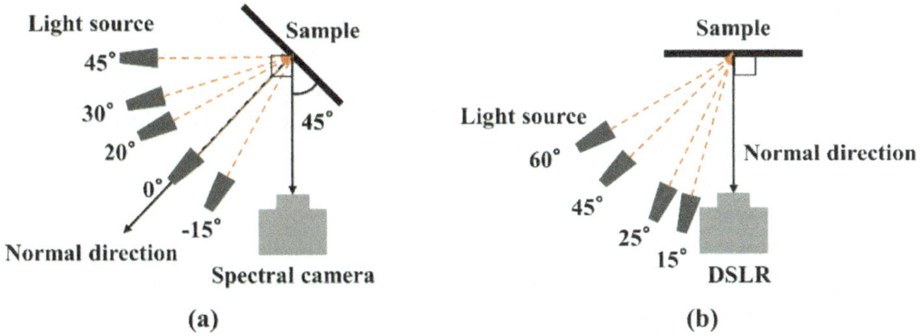

Fig. 2. Measurement method. (a) Color measurement. (b) Surface measurement.

3 Subjective Experiment

The subjective experiments were conducted to obtain impression factor and "feeling of luxury" for leather samples. Ten samples consisting of artificial or genuine leather were prepared for the experiments. Figure 3 shows the leather samples used in the experiments (sample 1–4 were genuine leather and sample 5–10 were artificial leather). The samples were black, and were approximately 210 × 300 mm in size.

3.1 Impression Factor

The impression factor values for the leather samples were obtained using factor analysis. The analysis estimated the impression factor for the appearance of leather samples.

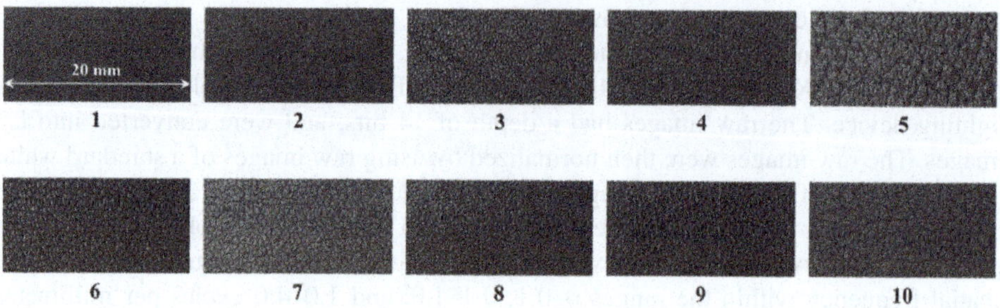

Fig. 3. Leather samples.

To determine appropriate evaluation adjectives for use in the analysis, we conducted three experiments: an impression-word extraction test, an impression-word appropriateness test, and a rating word distance measurement test.

Ten participants observed the leather samples. Since the appearance of the leather varies depending on the observation angle, participants observed the leather in a curved state. Since the appearance of leather varies depending on the observation angle, participants observed the leather in a curved state. This test was conducted using a standard light source (Spectra Light QC; X-rite, MI, USA). A standard illuminant D65 was used, and the luminance of the observation area was set to 1,270 lx. The observation distance was 300 mm.

Table 1 shows the results of the factor analysis. In the table, the words were selected by the above three consecutive experiments. The cumulative contribution ratio up to the second factor was approximately 0.89. The first factor contained "wrinkle," "granular," "sturdy," and "smooth." The second factor contained "texture" and "profound feeling." We determined the first factor as the "surface shape," because it corresponded to the surface conditions of the leather. The second factor was determined as "impression of stateliness" because it corresponded to a profound impression.

Table 1. Result of factor analysis.

Word name	Factor 1	Factor 2
Wrinkle	1.04	0.26
Granular	0.89	-0.10
Sturdy	-0.84	0.20
Smooth	-0.78	0.08
Texture	0.12	1.03
Profound feeling	-0.31	0.61

3.2 Feeling of Luxury

We used Scheffe's paired comparison method to obtain scores for "feeling of luxury" of the leather samples. A pair of test samples was chosen at random from the samples. Participants compared these samples and ranked them according to seven levels from "not felt at all" to "very strongly felt." The above comparison was conducted for all possible combinations of samples. Scheffe's paired comparison method can identify small differences between samples. The experiment was performed under the same observation conditions as the impression factor estimation experiments. The subjective scores were scaled using a correspondence analysis. Table 2 shows the results of the experiment. The magnitude of the score represents the strength of the feeling.

Table 2. Result of paired comparison method.

Sample No.	Subjective score
1	0.70
2	0.61
3	0.25
8	0.21
10	−0.02
5	−0.02
7	−0.12
6	−0.52
9	−0.54
4	−0.55

4 Results

In the current study, "feeling of luxury" was systematically evaluated by considering the underlying visual perception mechanisms. Therefore, we confirmed two types of correlations: those between the physical characteristics and the impression factors, and those between the impression factors and "feeling of luxury". We derived equations to estimate the impression factors using multiple regression analyses with a stepwise method. In these analyses, the objective variables are the factor scores of the impression factors, and the explanatory variables are the physical characteristics. The physical characteristic values have dissimilar ranges. Therefore, the standardized Z-scores of the physical characteristics were used as the explanatory variables.

First, "surface shape" and "impression of stateliness" were estimated from Eqs. (1) and (2), respectively. Here, x_{11} and x_{12} are the 60°/0° large grain and the 45°/45° L* value ($p < 0.05$). Also, x_{21}, x_{22}, and x_{23} are the 60°/0° surface irregularity, the 15°/0° surface irregularity, and the −15°/45° kurtosis value ($p < 0.05$), respectively. Figure 4 shows the correlations with the values estimated using Eqs. (1) and (2), and the subjective factor scores. We found a strong positive correlation and an R-squared value of 0.91 and 0.96 ($p < 0.001$).

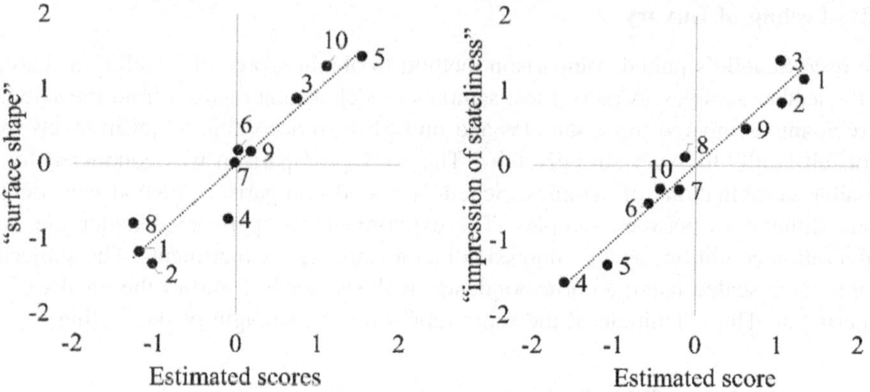

Fig. 4. Estimated values and subjective scores for impression factors.

Next, "feeling of luxury" was estimated from Eq. (3). Figure 5 shows the correlation with the values estimated using Eq. (3), and the subjective factor scores for the feeling. The results revealed the strong positive correlation, and an R-squared value of 0.65 ($p < 0.001$).

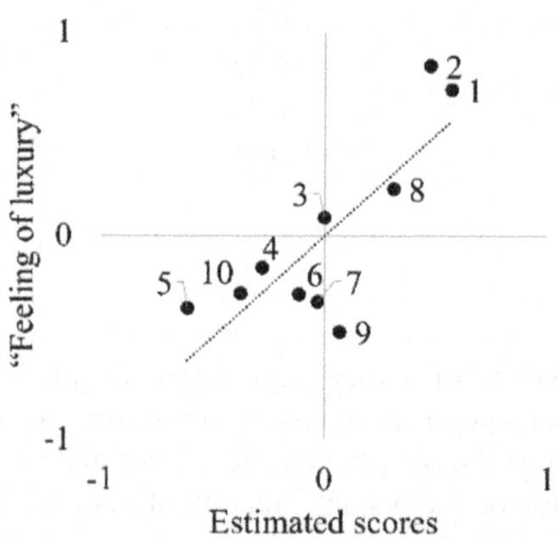

Fig. 5. Estimated values and subjective scores for "feeling of luxury".

$$\text{Surface shape} = 0.76 \cdot x_{11} - 0.31 \cdot x_{12}. \tag{1}$$

$$\text{Impression of stateliness} = 0.71 \cdot x_{21} + 0.24 \cdot x_{22} - 0.44 \cdot x_{23}. \tag{2}$$

$$\text{Feeling of luxury} = -0.55 \cdot \text{Surface shape} + 0.40 \cdot \text{Impression of stateliness}. \tag{3}$$

5 Conclusions

In the current study, we focused on "feeling of luxury" of the leather and investigated the hierarchical relationships between the image properties, impressions we perceived, and the feeling. As a result, we found that the participants perceived "surface shape" and "impression of stateliness" representatively from the leather. In the estimated equation, "surface shape" corresponded to the large grain under the shade (diffuse reflection far from the specular condition) angle. And, the grain was become invisible due to the high specular L* value. Whereas, "impression of stateliness" corresponded to the surface irregularity. Also, when the kurtosis value was high, the surface's L* was almost equally. That is, the surface had no surface irregularity. Therefore, the equation was negatively influenced by the kurtosis value.

Then, we found that "feeling of luxury" was estimated by the simple equation using only the two representative impressions. The equation was negatively influenced by "surface shape" and positively influenced "impression of stateliness". In the perception of leather, "feeling of luxury" was perceived as the leather surface had surface irregularity and no grain.

As the above results, we confirmed that "feeling of luxury" could estimate using our proposed hierarchy subjective response. In other words, the leather product's "feeling of luxury" can be controlled by the specified image properties through the representative impressions which participants perceive.

In the future, we plan to image editing of "feeling of luxury" based on the estimated model. Also, we will investigate the relationship between the specified image properties and design parameters of the artificial leather surface.

References

1. Fleming, R.W., Nishida, S., Gegenfurtner, K.R.: Perception of material properties (part I). Vis. Res. **109**(B), 123–236 (2015)
2. Fleming, R.W., Nishida, S., Gegenfurtner, K.R.: Perception of material properties (part II). Vis. Res. **115**(B), 157–302 (2015)
3. Watanabe, S.: One-shot multi-angle measurement device for evaluating the sparkle impression. J. Imaging Sci. Technol. **63**, 157–302 (2019)
4. Fujisaki, W., Tokita, M., Kariya, K.: Perception of the material properties of wood based on vision, audition, and touch. Vis. Res. **109**, 185–200 (2015)
5. Hiramatsu, C., Goda, N., Komatsu, H.: Transformation from image-based to perceptual representation of materials along the human ventral visual pathway. NeuroImage **57**(2), 482–494 (2011)
6. Goda, N., Tachibana, A., Okazawa, G., et al.: Representation of the material properties of objects in the visual cortex of nonhuman primates. J. Neurosci. **34**(7), 2660–2673 (2014)

Innovative Filtering Systems for High Temperature Chemical Processes

Carmen Mihai(✉) and Alexandra Ene

The National Research and Development Institute for Textiles and Leather, 16, Lucreţiu Patraşcanu Str., 030508 Bucharest, Romania
carmen.mihai@incdtp.ro

Abstract. The protection of the environment is a major problem of the planet, the legislation in this field being more and more severe and knowing an extension in an increasing number of countries. This required the development of both filtering facilities and filter materials, made today in a very varied range, corresponding to both normal conditions and special filtering conditions (high temperature, chemical aggression, high humidity, etc.).

The purpose of separating the components from the heterogeneous gas systems is multiple, being mainly represented by: the purification of the gas, for obtaining the gas phase in the most pure state, in order to avoid the pollution of the working section and for the total environment; powders recovery from the technological process, as an intermediate product - for later usage; separation of the powders involved in fluidization and pneumatic transport operations [1].

The researches in this field were focused on the design and accomplishment of a filtering material based on non-woven fabrics that nowadays plays an important role (in Europe over 90%), replacing the fabrics with very good results, due to the technical characteristics represented by: higher air permeability; lower tendency of clogging; high filtration efficiency; longer operating life [2].

Keywords: Filtration · Nonconventional technologies · High temperature processes

1 Introduction

Hot gas filtration is a physical-chemical process, where the high temperature, and other elements (chemical nature) can destroy the filtering material.

During the filtration process, the gas phase must overcome the resistance that the filter layer opposes to the fluid flow. However, the size of the pores of the filter layer and its resistance are important only at the initial moment of the process, because subsequently, the solid particles are gradually deposited on the surface of the filter layer. This layer, which increases during the process, is used as a filtration media, while the hydraulic resistance of the solid is minimized.

The reduction of the hydraulic resistance is achieved by continuous or periodic removal from the filtering layer of an important quantity of solid product.

G. Di Bucchianico et al. (Eds.): AHFE 2020, AISC 1202, pp. 770–776, 2020.
https://doi.org/10.1007/978-3-030-51194-4_100

The nature and thickness of the solid product layer deposited on the surface of the filter layer is, in most cases, one of the most important factors that determine the efficiency of the filter, respectively the filtration rate and the energy consumption for pushing the fluid through the filter.

Most of the filter layers have a low retention capacity at the beginning of the process. In addition, during the fine suspensions' filtration, the first portions of the filtrate almost always contain a certain number of suspended particles, which pass with the fluid through the pores of the filter.

This phenomenon is observed as long as on the surface of the filter layer a solid layer is formed, that having pores of much smaller size than the pores of the filter layer will completely retain all the solid particles.

The degree of complete separation depends largely on the pressure at which the filtration process takes place.

The filtrate is obtained clearly, if the filtration starts at low pressure, and the pressure increases, as the solid accumulates. In addition, the uniformity of pressure has a special importance. Thus, if the pressure is uniform and continuous, a complete separation can be achieved.

For such applications, non-woven filter materials made of interlaced individualized fiber layers, are currently used, usually with a reinforcement fabric used as an insertion. The fibrous layers play an important role in the filtration process, and the insertion confers resistance and dimensional stability. The insertion must have the same nature as the fibers of the non-woven support [3].

Due to the three-dimensional structure, the needle punched nonwoven filtering materials have a high filtration capacity, each fiber participating in the process of separating solid particles from air or gas.

Hot gas filtration requires the usage as raw materials of the fibers with special physical-mechanical characteristics, such as the high-tech fibers, that are resistant to high temperatures.

In general, the working conditions for the metallurgy and steel filtration installations are: working temperature: max. 280 °C; humidity: 0–10%; dust granulation: 0–1 mm (of which 50% below 5 μm); nature of the powders: iron oxides; coke; coal; graphite; ferroalloy (Fe Si, Fe Mg, Fe Cr); limestone; dolomite; dust content: at the entrance to the installation (min. 3–5 mg/m^3, max. 10–12 mg/m^3), at the exit from the installation: max. 20 mg/m^3.

2 Materials and Methods

The purpose of the research was the elaboration of a needle punched nonwoven textile material for filtration in continuous processes at temperatures of 150–200 °C.

The filtering material was made of three textile layers: two fibrous needle punched nonwoven beds reinforced with a woven insert distributed between the layers of nonwoven materials and reinforced by strong interlaced non-conventional process.

The physical and mechanical characteristics of the reinforcement woven are presented in Table 1.

Table 1. The physical and mechanical characteristics of the reinforcement woven

Characteristic	Value
Width, cm	129
Mass, g/sqm	160
Yarn density, warp/weft, no/10 cm	133/121
Breaking force, warp/weft, daN/5 cm	100/97
Elongation, warp/weft, %	36/38

As a raw material, an meta aramid fiber of 2.2 dtex/52 mm fineness was used, having high temperature resistance, good bending and abrasion behavior, resistance to acids, alkalis, fluorides and to media with low water vapor concentrations.

The main characteristics are: average length of the individual fibers: 52 mm; the relative breaking force of the fiber in the form of a loop, in conditioned state: 2.67 cN/dtex; frequency of corrugations: 6.1 corrugations/cm; the degradation temperature more than 370 °C (it does not melt, it is degraded by slow carbonization); oxygen limit index (L.O.I.) of 29–32% (does not light, L.O.I. is greater than 21% as the meta-aramid fiber is included in the difficult flammable group, avoiding the maintenance of combustion and therefore the propagation of fire); the temperature at which the fiber can be exposed for a long time, in continuous process is 200 °C, but it also supports short-term exposures at temperatures of 300–310 °C; at 200 °C and 10,000 h of operation it retains 80% of its initial resistance, and at 250 °C it retains 60% of its initial resistance.

The operation of cutting and mixing of the fibers was performed on a cutting plant - mixing (mixing wolf type Befama) with the purpose of unfolding, distributing as evenly as possible and making small volumes of fiber for carding operation facilitation.

The fibrous non-woven structure was made by carding by overlapping in approx. 20 layers, with the help of the plaiter (Fig. 1).

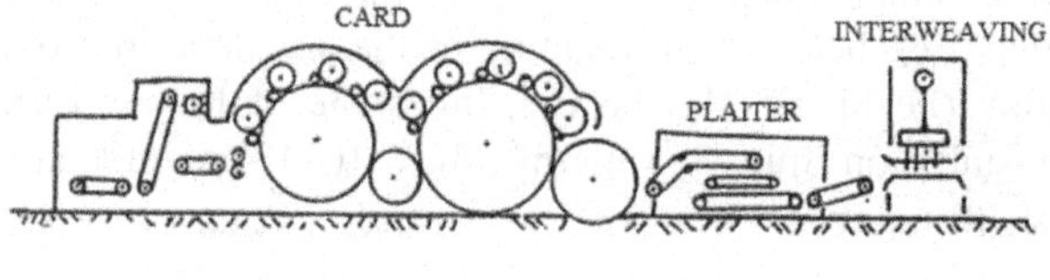

Fig. 1. Technological line for the production of interwoven filter materials

The deposition width was 1,6 m, and the speed of the bending band: 0.95 m/min. In addition, a simple, pre-interlace interlocking was performed, which provided the fibrous substrate with a pre-consolidation so that it could be rolled and transported to perform the interweaving operation with the reinforcement insert.

During the technological process for filtering material accomplishment, the needle punched nonwoven surfaces with the mass of 100 and 140 g/m^2 were used as outer layers and the interlaced fabric with the mass of 160 g/m^2 as the reinforcement support (insertion). The final product was obtained after six passages.

The main assembling and adjustment parameters used on the technological flow to obtain V_A and V_B variants of filter materials are:

- delivery speed (passage 1–6): 2,8-2,8-1,5-1,5-1,5-1,5 m/min.
- needle board frequency: 300 strokes/min.
- penetration depth (passage 1–6): 14-14-11,5-11,5-9-6,5 mm
- punching density (passage 1–6): 50-50-100-100-100-100 punches/cm^2.
- needles type: 15 × 18 × 40 × 3½ RB.

The variants of filtering materials V_A and V_B were subjected to a surface finishing treatment by thermo-plasticization in order to increase the regeneration capacity of the filter and to uniformize the air permeability due to the calibration of the material thickness (for variant V_B). The main physical-mechanical characteristics of the filter materials variants V_A and V_B are presented in the Table 2.

Table 2. The main physical-mechanical characteristics

Characteristics	V_A		V_B		
	Raw	Finished	Raw	Finished	Special surface treatment
Mass, g/sqm	385	386 ± 19	385	386 ± 19	739 ± 37
Thickness, mm	2.25	2.41 ± 0.12	2.25	2.41 ± 0.12	2,44 ± 0,12
Width, cm	128	127 ± 2	128	127 ± 2	125 ± 2
Air permeability, l/sqm/s	–	414 ± 66		414 ± 66	82 ± 21

The technologies and products have been experimented in the metallurgical field in the processes where heterogeneous gas-gas systems containing solid particles in suspension results from: mechanical operations (milling, screening, mixing, fluidization, etc.); contact operations between a gas and a powdery solid material; fine powders obtaining with increased efficiency of the filters by pneumatically driving the powders that have reached the desired fineness; chemical processes of combustion, frying, etc. or between two gases or between gases and a solid, with suspension formation.

3 Results and Discussions

In order to evaluate the filtering function of the obtained textile structure, the following parameters were considered: the dimensions of the particles that make up the filtration fluid: 5–45 μm, with the following particle size distribution: particles larger than 5 μm: 60–65%; particles smaller than 5 μm: 35–40%; apparent density: 2–2.5 g/cm^3; compacted density: 3.2–3.7 g/cm^3; temperature in the filter room: 80 °C; accidental temperature: 120 ° C; counter-current sieving time: 20 s; total filtration area: 260 m^2; depression created by the ventilation system: 0.2–0.35 PSI; installation capacity: 500 kg/h.

In order to determine the permeability to the air in the remnant clogged state, the specific loading σ of 100 m^3/m^2/h at a pressure difference of 580 Pa (standard condition for the laboratory simulation of the loading with material, that enable the determination of the permeability in the residual clogged state) was considered. In addition, was assumed that the sieving system starts to operate for $\sigma 1 = 300\ m^3/m^2/h$. The area of the measuring surface was 20 cm^2, and the sieving time was considered to be 20 s.

Considering that the air flow Q can be modified as the thickness of the material increases through clogging, the values of the permeability to the air in the remnant clogged state have been obtained for two variants of textile materials differentiated by thickness and permeability.

In order to simulate the industrial conditions, it was assumed that the sieving is performed with air jet in the opposite direction, and the flow of gas and powders subjected to the filtration process is constant. The specific loading $\sigma = 100\ m^3/m^2/h$ and the passage of a gas volume between 0,5 and 1 m^3 were considered until the pressure difference between the two sides of the material achieved a value of $\Delta P0 + 780$ Pa. Considering the particle size, sedimentation velocity and apparent density, it was possible to determine the mass of powder for the charge cycle. The filter was considered to be cleaned by blowing a reverse jet of air at a load of 300 $m^3/m^2/h$ for 2 s.

The filtration yield was determined as the ratio of the concentration of powders in gas before and after passing through the filtering material, the concentration being determined as a ratio between the mass of the powders from the gas on the front face and the volume of powdery from the gas that passed through the filter. The obtained data are presented in Table 3.

Table 3. Filtration yield for filtering materials V_A and V_B

Variant	Separation gradient for 100 $m^3/m^2/h$	Sedimentation velocity, mm/s		Filtration yield, %
		<5 μm	>5 μm	
V_A	0.439	0.38	4.6	99.939
V_B	0.962			99.997

The determination of air permeability and tear resistance was performed for filtering structures (Fig. 2) where the joints were made by simple and overlaid seams using a Necchi-type machine, with the step of 4 stakes per cm [4].

From the analysis of the obtained data were concluded the following:

- the value of air permeability determined on a joint in the form of simple seams is 15% lower than the value of the permeability of the ground material (experimentally determined to be of 265 $l/m^2/s$), for the variant V_A, and for V_B variant is higher by 19% (the value of the permeability of the ground material determined experimentally being of 18 $l/m^2/s$);
- in the case of the overlaid seam the value is lower by 22% for the variant V_A and higher for the variant V_B with 33%. It can be appreciated that both variants do not show significant variations for this parameter in industrial exploitation;

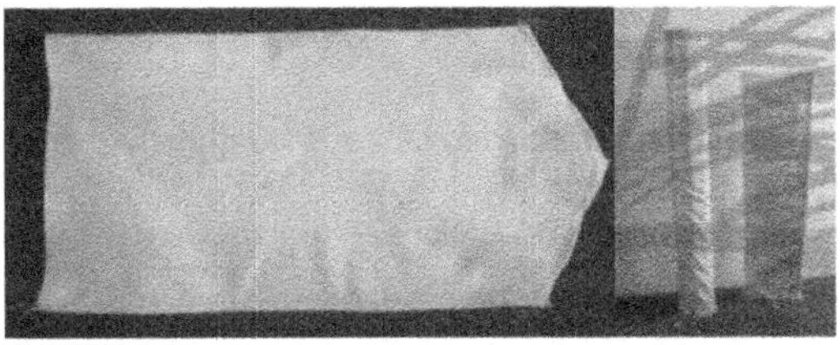

Fig. 2. Filtering materials prepared for tests

- the values of the breaking resistance are higher than the value of the ground material in the case of simple stitching with 39% for the variant V_A and 70% for the variant V_B (experimentally determined values for the background material: 45 daN for the variant V_A and 54 daN for the variant V_B);
- in the case of the overlapping stitches the values of the breaking resistance are higher than the value of the ground material with 59% for the variant V_A and 65% for the variant V_B, so there will be no material breaks in the situation in which the materials will be subjected to the specific operation conditions of the installation.

4 Conclusions

The planning of the test-evaluation activity played a very important role in ensuring the complete information on the system until its safe use and demonstrated the feasibility and estimation of the operational behavior mode.

The filtering material for high temperature processes is manufactured of meta - aramid fabric obtained through non-conventional technologies having the main parameters in accordance with the requirements of the process and installation.

The following technical and functional level of performances of the developed filtration system have been demonstrated:

- by clogging, the air permeability decreases by 50% compared to the value of the permeability in the clean state;

- the filtration efficiency is about 99,997%, higher than other usual systems;

- the values of the resistance to breaking are higher than the value of the ground material both in the case of the simple seam and of the overlaid one.

Acknowledgment. This scientific paper is funded by the Ministry of Research and Innovation within Program 1 - Development of the national RD system, Subprogram 1.2 - Institutional Performance - RDI excellence funding projects, Contract no. 6PFE/2018.

References

1. Lupion, M., Alonso-Fariñas, B., Rodriguez-Galan, M.: Modelling pressure drop evolution on high temperature filters. Chem. Eng. Process **66**, 12–19 (2013)
2. Arunangshu, M., Vipul, P., Kamal, D.: Effect of high temperature on the performance of filter fabric. J. Ind. Text. **45**(6), 1587–1602 (2016)
3. Rupayan, R., Soumili, C.: Development of a multi-component air filter by incorporating the density gradient structure in needle punched nonwoven. Fibers Polym. **19**, 2597–2603 (2018)
4. Priyal, D., Ishtiaque, S.M., Rupayan, R.: Influence of sequential punching in layered structure of needle punched nonwoven on the filtration behavior. Compos. Part B **182**, 1–11 (2019)

Unique Natural Fibers: Physicochemical Characteristics of Lignin-Cellulosic Fibers from *Aechmea Bromeliifolia* Leaves

Rayana Santiago de Queiroz[1(✉)], Ana Cristina Broega[2(✉)], and António Pedro Garcia Valadares Souto[2(✉)]

[1] Technical Textile and Protection Products Laboratory, Institute for Technological Research of São Paulo State, São Paulo, SP, Brazil
rayanasq@ipt.br

[2] Department of Textile Engineering, 2C2T – Centro de Ciência e Tecnologia Têxtil, University of Minho, Guimarães, Portugal
{cbroega, souto}@det.uminho.pt

Abstract. Ligno-cellulosic fibers are abundantly present in a diverse flora. Bromeliaceae is considered one of the most prominent vegetable family in terms of the availability of these fibrous plants, making for excellent textile leaf fibers. The objective of this work is to present a preliminary physicochemical characterization of the Brazilian-native *Aechmea Bromeliifolia* fiber. The fibers were manually extracted from the species' leaves to be studied in terms of their physical and chemical structure, and mechanical behavior. The results were compared to known fibers in the textile industry, identifying the fiber's potential for textile application. Nevertheless, further study is needed for property optimization and the improvement of extraction and treatment methods.

Keywords: Lignocellulosic fibers · Textiles · Physicochemical properties

1 Introduction

The textile activity is one of the oldest and largest in the world. Nowadays, the textile industry, associated with the prevailing logics of production and consumption of the fashion system and the rapid demographic and economic growth, has become a major factor in large-scale consumption and in the misuse of natural resources throughout its evolution. This background has resulted in an increasing discrepancy between fiber supply and demand, a phenomenon that Scheffer [1] refers to as a "fiber gap". In the present year of 2020, the global fibers consumption is expected to reach a total of 110 million tons, which gives fiber the second place in the scale of major items, following only to food sector [2].

Growing environmental concerns and the depletion of oil resources are increasingly forcing industries to seek alternatives among renewable resources. The United Nations (UN) and the Food and Agriculture Organization of the United Nations (FAO) consecrated 2009 as the International Year of Natural Fibers (IYNF). In a scenario in which the synthetic fibers market prevails over natural fibers, this milestone inaugurates a

G. Di Bucchianico et al. (Eds.): AHFE 2020, AISC 1202, pp. 777–784, 2020.
https://doi.org/10.1007/978-3-030-51194-4_101

prominent position of this theme at the international level, placing plant fibers in evidence among global trends [3–5].

Fiber is the primary unit of fabrics and other textile structures [6]. The natural fiber is entirely from nature, extracted from plants, animals, and minerals [7]. Plant fibers are natural fibers of lignocellulosic basis, obtained from stems, leaves, or seeds/fruits. They are advantageous from an environmental standpoint since they come from renewable sources and are fully biodegradable, and, when associated with good practices of production, its cultivation usually does not cause damages to ecosystems [8]. In addition to the environmental dimension, the benefits of natural fibers unfold in other aspects such as health and comfort, socioeconomic development of rural populations, and technological advancement, allowing the creation of new high-performance materials [9]. Compared to synthetic fibers, properties such as good air permeability, hygroscopicity and atoxity stand out [4].

Lignocellulose is the main component of the cell wall of vegetables and the most abundant organic substance on Earth [10]. It is estimated that the plant kingdom holds about 2,300 plants rich in lignocellulosic fibers, distributed in several botanic families [11]. Currently, hundreds of fibrous plants around the world are used as a fiber source for textiles, papermaking, and other manufactured goods [2]. In Brazil, which holds the most diverse flora of the planet [12, 13], there are several fibrous species with textile application potential. *Aechmea bromeliifolia* (Rudge) Baker, is an angiosperm from the Bromeliacea family that occurs from Central America to northwestern South America and Brazil [14, 15]. It is an epiphytic/rupicolous plant, rarely terrestrial, about 60 cm high and grows in Cerrado, Caatinga, Atlantic and Amazonian biomes, from the north to south regions of Brazil [16, 17]. The fibers of the leaves of *A. bromelii folia* can be used for varying purposes [11]. There is no specific study on the textile application of these fibers; however, there is outstanding potential for the plant's entire use, since not only do its leaves provide fibrous material, but it is also possible to extract coloring matter from its root.

This research carried out a preliminary investigation of the characteristics of the fibers obtained from *A. bromeliifolia* leaves, focusing on evaluating their applicability in textile materials. Basic properties related to its physical and chemical structure were determined, including analysis of morphology, linear density, and FT-IR and its mechanical behavior.

2 Materials and Methods

2.1 Fibers – Gathering and Processing

About fifty leaves of *A. bromeliifolia* were collected from three different individuals in a rural area in the countryside of Sao Paulo state (Brazil) located at 23°45′15.1″S 47°57′59.7″W.

The fibers were manually extracted from the fresh leaves by scraping their epidermal tissue with the aid of an ink scraper and tweezers to detach the fibers.

2.2 Characterization

Characteristics of the morphological and surface structure of *A. bromeliifolia* fibers were analyzed using an optical and a scanning electron microscopy (SEM). Their chemical structure was analyzed by Fourier transform-infra red spectroscopy (FT-IR), with 32 scans in the spectral range of 4000–400 cm^{-1} at a 4 cm^{-1} resolution. The linear density (tex) of fibers was determined gravimetrically based on the ISO 1973: 1995 standard test method, and its tensile properties (breaking force, breaking tenacity and elongation) were determined in a Constant-Rate-of-Extension (CRE) Type Tensile Testing Machine, based on standard test method ASTM D3822:2014. Twenty test specimens were tested with 30 mm of gauge length and 3 mm/min of extension rate.

3 Results and Discussion

3.1 General and Morphological Analysis Characteristics of the Fibers

The fibers obtained from the manual extraction process have a color that varies from cream-white to greenish-brown. They are shiny and with a soft touch, similar to that of linen. Figure 1 shows the general appearance of the obtained fibers. It was observed that their surface characteristics vary vastly according to the characteristics of the leaves. From the young leaves (Fig. 1b), very clear, shiny and malleable fibers are obtained, while the fibers of the more mature leaves (Fig. 1a) are darker and slightly more rustic.

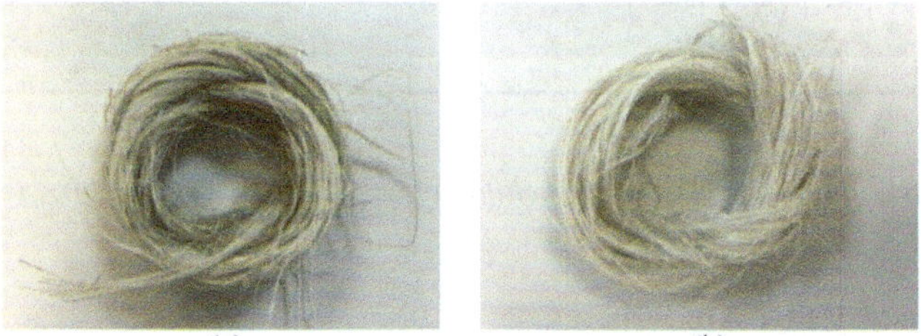

(a) (b)

Fig. 1. Fibers extracted from the *A. bromeliifolia* leaves: (a) mature leaves (b) young leaves

From the optical microscopy, it is possible to observe the characteristics of the fiber structure. Fiber's longitudinal shape is determined by the fiber's cross-section [18], which in turn also determines visual aspects, especially the gloss, and fiber hand. *A. bromeliifolia* fibers come in bundles, and their cross-section shape is polygonal; but not as pronounced as in sisal fibers, with a circular to oval lumen (Fig. 2a). Nodes are present in the longitudinal view of the fiber, a characteristic typically seen in bast fibers such as hemp and flax. Many similarities were observed with the abaca fiber, both in the longitudinal and in the cross section, when compared with other leaf fiber structures.

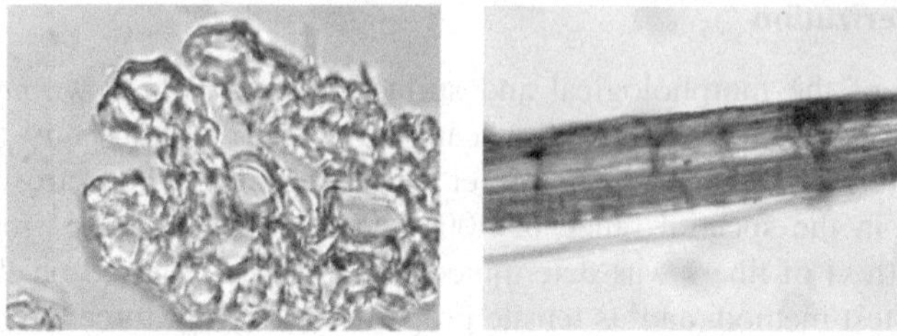

(a) (b)

Fig. 2. a) Cross-sectional view and b) Longitudinal view of *A. bromeliifolia* fiber by optical microscopy (500× magnification)

SEM images of *A. bromeliifolia* fibers (Fig. 3) also reveal fibers that are naturally arranged in bundles in an orderly longitudinal orientation. Figure 3b also shows residues of the epidermal tissue on the surface of the fiber bundle. This indicates the need to subject the samples to complementary chemical, enzymatic and/or physical treatment, such as cottonization, thus obtaining better individualization of the fibers and, consequently, cleaner and more uniform fibers, suitable for textile application [19, 20].

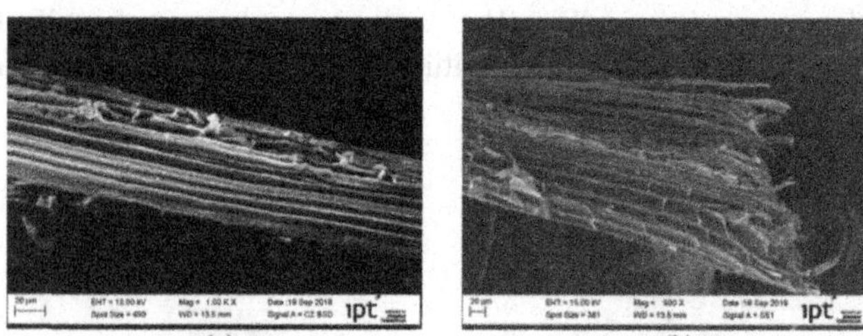

(a) (b)

Fig. 3. SEM longitudinal view of *A. bromeliifolia* fiber at 1000× (a) and 500× (b) magnification

3.2 Structural FTIR Analysis

FT-IR is the most widely available spectroscopic technique used to characterize fibers and investigate their chemical transformations at molecular level. The Fig. 4 shows the FT-IR spectra of *A. bromeliifolia* fibers, which exhibited the typical shape of lignocellulosic fibers [21].

The 3400 cm^{-1} band observed in the spectra corresponds to the hydrogen bonds for cellulose and hemicellulose [22, 23]. The peak wave number of 2850 cm^{-1} is attributed to the C–H stretching of saturated alkane for wax [22]. The peak at 1737 cm^{-1} corresponds to the acetyl groups and C–O bonds, characteristic of the hemicellulose, and also affected by the presence of pectin [22, 24]. The lignin is mainly characterized by the bands of 1234 cm^{-1} and 1035 cm^{-1} [22]. The observed bands of 1160 cm^{-1} and 890 cm^{-1} are related to the presence of cellulose and hemicellulose [22].

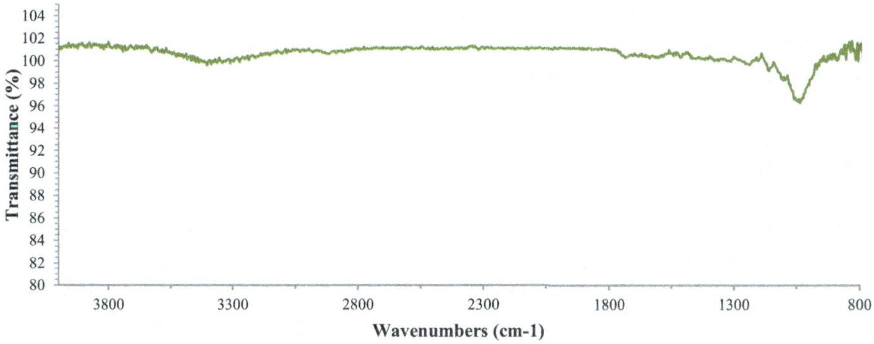

Fig. 4. FT-IR spectra of *A. bromeliifolia*

Comparing the spectra with other Bromeliaceae pant fibers studied by Sena Neto et al. [23] and de Queiroz et al. [24], *A. bromeliifolia* fibers have much less pronounced peaks, both attributed to lignin and waxes, and to cellulose and hemicelluloses. In fact, the fiber's tactile characteristic indicates a less rigid and lignified surface than other leaf fibers. However, it is also likely that part of chemical components is covered by the However, it is also likely that part of the chemical components was covered by the residual epidermal tissue of the fiber surface, keeping the fiber bundle structure closer.

3.3 Linear Density and Tensile Properties

Linear density is a property related to fiber fineness and its intrinsic mass. For natural vegetable fibers, linear density is the most useful and reliable mean of describing the fiber fineness [25]. The mechanical properties are also determinant in the evaluation of the fiber's textile application viability [26]. Table 1 shows the linear density and tensile properties of *A. bromeliifolia* fibers obtained from ripe and young leaves.

Table 1. Linear density and tensile properties of *A. bromeliifolia* fibers.

Sample	Linear density (dtex)	Breaking force (cN)	Tenacity (cN/tex)	Breaking elongation (%)
Ripe leaf fibers	58,8	52,13	8,87	1,59
Young leaf fibers	58,1	78,45	13,50	1,71

The younger leaf fibers have slightly lower linear density, which implies finer and more delicate fibers; however, the results can be considered statistically equivalent. On the other hand, breaking force and tenacity of young leaf fibers showed about 50% higher than those of more mature leaves. Regarding breaking elongation, there are no significant differences. Considering reference values for other textile fibers [25], the tenacity of *A. bromeliifolia* fibers is lower than that of vegetable fibers such as flax, jute, and hemp, being closer to fibers such as acetate and wool. As far as elongation is concerned, it is comparable to jute.

4 Conclusions

The environmental advantages of natural fibers as an alternative renewable source of biodegradable raw material have elevated attention to investigations on the textile uses of new sources of lignocellulosic fibers. In this paper, we presented a preliminary study on the textile applicability of the alternative fibers obtained from a native species from Brazil: *A. bromeliifolia*. Based on its structural morphological (microscopy) and chemical (FT-IT) characteristics, it can be concluded that the investigated fibers have the typical characteristics of vegetable textile fibers, compared to fibers such as flax and abaca. It was observed that fibers obtained from younger leaves present more favorable results in terms of appearance, handling and mechanical behavior. In general, the results of linear density (~60 dtex) revealed that the fineness of the fibers is also compatible with other vegetable textile fibers, but a slightly coarser than fibers such as linen and jute. Fiber tenacity (~9–14 cN/tex) and elongation (~1.6–1.7) were lower than most commercial vegetable fibers. These results indicate the possibility of using these fibers for textile purposes. Nonetheless, study advancement is advisable with a particular focus on alternative extraction methods and treatment of fibers to optimize their properties, such as cottonization.

Acknowledgments. The authors gratefully acknowledge the funding by Support Foundation of the Institute for Technological Research, FIPT, Brazil, under grants FLORATEXTIL I. "This work is also supported by FEDER funds through the Competitivity Factors Operational Programme - COMPETE and by national funds through FCT – Foundation for Science and Technology within the scope of the project POCI-01-0145-FEDER-007136".

References

1. Scheffer, M.R.: Trends in textile markets and their implications for textile products and processes. In: Shishoo, R. (ed.) The Global Textile and Clothing Industry: Technological Advances and Future Challenges, pp. 8–28. Woodhead Publishing, Cambridge (2012). https://doi.org/10.1533/9780857095626.8
2. Ramawat, K.G., Ahuja, M.R.: Fiber plants: an overview. In: Ramawat, K.G., Ahuja, M. (eds.) Fiber Plants, Sustainable Development and Biodiversity, pp. 3–15. Springer, Cham (2016). https://doi.org/10.1007/978-3-319-44570-0_1

3. Kozłowski, R.M., Muzyczek, M., Mackiewicz-Talarczyk, M., Barriga-Bedoya, J.: Quo vadis natural fibres in 21st century? Mol. Cryst. Liq. Cryst. **627**, 198–209 (2016). https://doi.org/10.1080/15421406.2015.1137675
4. Kozłowski, R.M., Muzyczek, M.: Introduction. In: Kozlowski, R.M., Muzyczek, M. (eds.) Natural Fibers: Properties, Mechanical Behavior, Functionalization and Applications, pp. vii–x. Nova Science Publishers, New York (2017)
5. Kozłowski, R.M., Baraniecki, P., Barriga-Bedoya, J.: Bast fibres (flax, hemp, jute, ramie, kenaf, abaca). In: Blackburn, R. (ed.) Biodegradable and Sustainable Fibres, pp. 36–88. Elsevier, Cambridge (2005). https://doi.org/10.1533/9781845690991.36
6. McIntyre, J.E., Daniels, P.N.: Textile Terms and Definitions. Taylor & Francis, Manchester (1995)
7. American Society for Testing and Materials: ASTM D123 - 17e1: Standard Terminology Relating to Textiles (2017). https://doi.org/10.1520/D0123-17.2
8. Kozłowski, R.M., Mackiewicz-Talarczyk, M.: Introduction to natural textile fibres. In: Kozlowski, R.M. (ed.) Handbook of Natural Fibres: Types, Properties and Factors Affecting Breeding and Cultivation, pp. 1–8. Woodhead Publishing, Cambridge (2012). https://doi.org/10.1016/B978-1-84569-697-9.50001-1
9. FAO: Why natural fibres? – International Year of Natural Fibres 2009. http://www.naturalfibres2009.org/en/iynf/index.html. Accessed 14 Jan 2019
10. Ramawat, K.G., Ahuja, M.R.: Fiber Plants: Biology, Biotechnology and Applications. (2016). https://doi.org/10.1007/978-3-319-44570-0
11. Medina, J.C.: Plantas Fibrosas da Flora Mundial (Exc.). Seção de Plantas Fibrosas, Instituto Agronômico, Campinas (1959)
12. Mittermeier, R.A., Robles-Gil, P., Mittermeier, C.G.: Megadiversity: Earth's Biologically Wealthiest Nations. CEMEX, Mexico (1997)
13. Myers, N., Mittermeier, R.A., Mittermeier, C.G., Fonseca, G.A.B., Kent, J.: Biodiversity hotspots for conservation priorities. Nature **403**, 853–858 (2000). https://doi.org/10.1038/35002501
14. Flora do Brasil 2020: Flora do Brasil 2020 em construção. http://floradobrasil.jbrj.gov.br/reflora/listaBrasil/PrincipalUC/PrincipalUC.do;jsessionid=819C0CCA767EDF022EA58AADFF38C777#CondicaoTaxonCP. Accessed 23 February 2018. https://doi.org/10.1590/2175-7860201566411
15. De Faria, A.P.G., Wendt, T., Brown, G.K.: A revision of Aechmea subgenus Macrochordion (Bromeliaceae) based on phenetic analyses of morphological variation. Bot. J. Linn. Soc. **162**, 1–27 (2010). https://doi.org/10.1111/j.1095-8339.2009.01019.x
16. Pio-Corrêa, M.: Dicionário de Plantas Úteis do Brasil e das Exóticas Cultivadas. Brasil - Ministério da Agricultura, Indústria e Comércio, Rio de Janeiro (1926)
17. Durigan, G., Baitello, J.B., Corrêa, G.A.D., de Siqueira, M.F.: Plantas do Cerrado Paulista: Imagens de uma paisagem ameaçada. Páginas & Letras, São Paulo (2004)
18. Markova, I.: Textile Fiber Microscopy: A Practical Approach. Wiley, Hoboken (2019)
19. Kozłowski, R.M., Mackiewicz-Talarczyk, M., Allam, A.M.: Bast fibres: flax. In: Kozłowski, R.M. (ed.) Handbook of Natural Fibres: Types, Properties and Factors Affecting Breeding and Cultivation, pp. 56–113. Woodhead Publishing, Cambridge (2012). https://doi.org/10.1533/9780857095503.1.56
20. Muzyczek, M.: The use of flax and hemp for textile applications. In: Kozłowski, R.M. (ed.) Handbook of Natural Fibres: Processing and Applications, pp. 312–328. Woodhead Publishing, Cambridge (2012). https://doi.org/10.1533/9780857095510.2.312
21. ASTM D276-12: Standard Test Methods for Identification of Fibers in Textiles (2012)

22. Gong, J., Zhang, Q., Lou, J., Zhang, T., Li, H., Li, Z., Li, Q., Zhang, J.: Investigation of the degradation of bio-recalcitrance in Apocynum venetum fiber biodegumming. J. Nat. Fibers **16**, 1–12 (2019). https://doi.org/10.1080/15440478.2017.1379043
23. Jabli, M., Tka, N., Ramzi, K., Saleh, T.A.: Physicochemical characteristics and dyeing properties of lignin-cellulosic fibers derived from Nerium oleander. J. Mol. Liq. **249**, 1138–1144 (2018). https://doi.org/10.1016/J.MOLLIQ.2017.11.126
24. Sena Neto, A.R., Araujo, M.A.M., Souza, F. V.D., Mattoso, L.H.C., Marconcini, J.M.: Characterization and comparative evaluation of thermal, structural, chemical, mechanical and morphological properties of six pineapple leaf fiber varieties for use in composites. Ind. Crops Prod. **43**, 529–537 (2013). https://doi.org/10.1016/j.indcrop.2012.08.001
25. Morton, W.E., Hearle, J.W.S.: Physical Properties of Textile Fibres. The Textile Institute, Manchester (1997)
26. Harwood, J., Harwood, R.: Testing of natural textile fibres. In: Handbook of Natural Fibres, pp. 345–390. Woodhead Publishing, Cambridge (2012)

Underwear in Personal Wardrobe – A Study About Consumption and Disposal

Carla Morais[1(✉)], Gianni Montagna[1], and Ana Sousa[2]

[1] Ciaud, Faculdade de Arquitectura, Rua Sá Nogueira, Polo Universitário, Alto da Ajuda, 1349-055 Lisbon, Portugal
carlota.morais@gmail.com, g.montagna@gmail.com

[2] Esart, Instituto Politécnico de Castelo Branco, Campus da Talagueira, Avenida do Empresário, 6000 Castelo Branco, Portugal
ana.filipa.j.sousa@gmail.com

Abstract. Clothing typologies and personal taste are determining factors that influence people to purchase and discard them in the same way. Design features, such as colour, shape, and texture, are characteristics that define the quality of the items that the market used to buy and use daily. However, regardless the quality concept, the comfort is a feature that has prevailed for decades and promotes the condition of the articles being perpetual in the wardrobe. Nowadays many people give clothes to family or friends that they do not use anymore or also discarded them in appropriate containers but not always these ever-growing practices still happen with all types of clothing. Underwear, or more precisely Lingerie, is being part of dress often overlooked by users when someone make questions, it is rarely both donated to and reused by others even still be bought and discarded alongside other types. The present study is the result of a literary review on the lingerie consumption in western women and about its practice of discarding it, aiming to envisage more sustainable design practices in the creation of this type of pieces, having account human, social and environmental factors.

Keywords: Underwear · Lingerie · Clothing · Bras · Briefs · Design

1 Introduction

According to a survey realized in 2019 to 25 higher academic students about what they kept inside their wardrobe, only 8% included accessories and anybody included suits, gym suits and underwear [1]. The open questionnaire was done with the purpose to give respondents freedom to illustrate what they really considered interesting in showing from their personal wardrobe. In addition to not talking about underwear, those female students had an interest for the same types of basic clothes and neutral colors, suggesting that the way Fashion System is currently based is completely out of step with the real consumer behavior.

After that and continuing reading about lingerie, is realized that in the developed world, lingerie brands have estimated the average number of pieces that women have in their wardrobe. Women buy tree, four bras and between five and nine pairs of briefs per year, so they have around eight to ten bras and ten to twenty pairs of briefs in their

G. Di Bucchianico et al. (Eds.): AHFE 2020, AISC 1202, pp. 785–791, 2020.
https://doi.org/10.1007/978-3-030-51194-4_102

closet [2]. Therefore, if the lingerie consumption is quite similar to the trend of global garment consumption, the number of items that may be discarded per year may also be high. Even more worrying is being expected if that discarded underwear no longer will be reused or recycled. With this, the present research begins with the aim that in foreseeable future Design will plan lingerie in a more sustainable way.

2 Literary Review

2.1 Lingerie and Underwear Consumption

The words underwear and lingerie are often used mutually as the same. Both also mean intimate clothing but there are a few differences between each other. Lingerie is a word taken directly from the French language that derives from old French, meaning 'linen', a fabric used under clothes. At the beginning of the 20th century women wore intimate clothes for three reasons: to alter their shape, for hygienic reasons and for modesty [3]. Then, women started feeling that having more practical undergarments and lingerie became easier to dress. Technically, all lingerie is underwear, but not all underwear is lingerie. Underwear is typically made from inexpensive cotton, it is worn for practicality and comfort by both genders, while lingerie (briefly bras and briefs) has been associated with more attractive and sensual underwear, made from main made fibers, in lace fabrics and typically worn by women.

According to Global Market Review of Lingerie Forecast 2023, EU is the largest importer in the world, representing 57% among USA, Asia and South América. The lingerie consumption almost follows the garment purchasing and is influenced by the personal disposable income. In Europe the leading brands are Hanes, Triumph, Calzedonia, ETam, H&M, Hunkemüller while in Asia are Cosmo-Lady, Embry, Maniform, Uniqlo and Wacoal. In USA Vitoria Secret and Vanity Fair are the leaders in spite of giant sportswear brands, Nike and Adidas, also having a significant share of the US lingerie/underwear market [2]. All these brands have estimated what quantities of lingerie the average women have in wardrobe and the amount between USA and Europe is alike. Women buy three to four bras and between five and nine pairs of briefs per year, having in her wardrobe around eight to ten bras and ten to twenty briefs.

In 2017, a London retail technology company (Edited) based on a sample of retailers in US, UK and Europe, realized that sales of push-up bras have fallen by 50% compared to a year ago while sales of triangle bras rose up to 120%. Since then many brands, such as Myla, Rossell, Marieyat Seamless, Les Girls Les Boys arose with sportier silhouettes, wire free, using the best quality fabrics. At the same time, the singer Rhianna was successful with the launch of her lingerie lines, reaching the most variety of body types in contrast to Victoria Secret that was widely criticised since its ex-chief marketing officer said to the Vogue magazine that had no interest in transgender models or a wider range of sizes and shapes [4].

There is an evidence that the lingerie sector is ready to change into something different, a trend has been followed by a new and old generation that have broken to the perfect body concept without fat or stretch marks and bring forth a new user, confident in his own body [5]. Therefore, sales of underwear has increased away from traditional wholesale model and digital brands are already building strong online communities often centred around the real body to engage costumers and lead them to interact directly and respond positively, with an extra like, if the body model shows 'a little bit of a roll on her tummy' [4].

2.2 End of Life and Clothing Management

Clothing for mass market is made in standard sizes, not facilitating variation and fit all types of the body's consumers. Sometimes users may keep their garment thinking they will wear it again when losing weight but probably it will never happen, times goes on and clothes stops being attractive lead them to discard those and buy others as replacement garment [1]. There are people who store such items as a significant extension of how they were used during their life and cannot discard them [6] but usually people put clothing out of use because «it is out of fashion», «no longer likes», «don't fit anymore» [7, 8].

Women can manage its appearance based on social networks, lifestyle and personal taste, even going in the opposite direction to what happens in fashion trends. Some garments play roles on their wardrobe but underwear they do not know which role they have. We only know that it is usually worn until not serving anymore. According to Laitalia 'no underwear, socks, stockings or nightwear were disposed of because of fashion [7] and the characteristics women find when they buy lingerie, special bras and briefs are «Function», «Tradition», «Sports Suitable», «Contemporary», «Extreme Glamour or Over-Sexy» [2].

Over the years, people have realised that their purchasing behaviour can have a direct impact on several elements of the environment [9] and the consumer recycling behaviour is a strong driver for donating clothing to charity [10].

There is a variety of options to discard clothing for reusing and recycling in Europe. In Uk, the most used is taken to charity shops (39%) while charity bags (18%) and bring banks (13%) remain important [11]. Re-sale is also another route for used clothing but if most of the people do not like to reuse second-hand clothing, they will never reuse lingerie. The reusing practise does not always happen for all stage of life because a study among Norwegians concluded that the amount of inheritance private clothing exchange among children clothing is greater than the formal market in Norway [12]. It is difficult to understand why adults do not reuse clothing as younger, but we think that in Portugal and in other western countries happens the same. According to Norris [13], the fear that people fells of contagion form wearing someone else's old clothing includes the fear of disease and - spreading germs. Therefore, the reuse of clothing depends on the certainty of hygiene practises of people who donate clothes as well as the care they have with clothing.

2.3 New Approach to Clothing Design and to Textile Recycling

Analysing each stage of garment's life from production to disposal, more sustainable practices need to be part of a join-up between designers, producers, retailers and consumers. A New Textile Economy aligned with the principles of a circular economy relies on: 'New textiles materials', 'transformation the way clothes are designed', 'sold and used according with their disposable nature', 'radically improve recycling by transforming design, collection and reprocessing clothes', and 'making effective use of resources moving on to renewable inputs' [14].

Relating to materials it needs ensure that the material input is safe and healthy to allow cycling and to avoid negative impacts during the producing, use, and after-use phases. Thus, designing and producing clothing to meet all sustainable challenges is more tangible than trying to counteract fashion change [7]. Especially because the reuse and recycling of materials integrated in outwear clothing is welcome, as long as everyone in the clothing supply chain feels the same.

Norris examined different research contexts focused upon strategies to re-introduced recycled wool back into fashion cycle, alongside the introducing of innovative new technologies and observed there are clear differences between sectors of the market to take different approaches. For example, *Skunkfunk* is a successful brand which always used waste as part of a mix of sustainable materials as the result of an overall collaboration with respected partners taking sustainability seriously while M&S brand prefers to do the hardworking of developing sustainable product very quietly. When M&S introduced a limited edition of women's coats made from recycled wool retail buyers were not convinced promoting the coat as recycled the best strategy because costumers could recoil. The same goes for a mentality of a conservative wool manufacture from Prato, Italy, which have built a reputation for fine fabrics and fancy yarns since the 1970s and when launched 'Cardato Regenerato', a sustainable certification for recycled wool, It was based on an assessment of the carbon footprint, water and energy requirements [13].

Although public perceptions of recycled textiles are of low quality, in fact often not even a lab test will reveal if fibres are recycled because it is possible to make high quality-fabrics of "mixed fibres", however that they are in the right amount way to guarantee the resistance attribute [15].

Even so, coordinated actions across the clothing value chain is required to align Design with recycling processes as the same implementing clothing collection at scale and stimulating market demand for recycled materials. For that all intervenient should also have a good knowledge about textiles materials processing to not make mistakes when choosing fabrics from and to recycled blends.

3 Methodology

3.1 Studying the Profile of Lingerie User

Once it is not clear which the preferred main disposal practices lead consumers awareness to discard their clothes and knowing that the post-purchase component of the clothing consumption process always involves whether clothing is re-used, recycled or simply discarded or destroyed [16] we would like to test how Portuguese women buy and discard their lingerie.

According to literature review there is no doubt that women buy more lingerie than they need as well as happen for clothing in general. Otherwise women would probably buy between three or four bras and six to eight briefs at any time until to rip or not fit, according to their washing and care cycle.

There are many underwear categories such as shapewear, swimwear, leisurewear that we will not include in the following survey. We prefer to be precisely in lingerie and asking people about briefs and bras to not confuse with shorts or panties, etc.

3.2 Survey

In order to reach users more quickly, we chose to carry out two satisfaction questionnaires, one about purchase and the other about disposal. We used *SURVIO* google software, for being easy to work, being free and adapted to different digital devices, both on the tablet and the mobile phone. The questions were made in order to help a Designer planning to design better according to the market needs. They mainly focused on the content related to shape, colour, materials, textures and embellishments. The language used is a common language, not a scientific one:

For purchasing profile follow the questions:

- How old are you?
- In the following list, what type of materials do you use in your lingerie? (Microfiber, Nylon Lace, Cotton, Polyester, Silk or Others)
- What kind of embellishments do you prefer using in your bras? (adjustable bras straps, Bras buttons, underwire, others)
- What kind of embellishments do you prefer using in your briefs? (seamless, not seamless, with cotton intimate coverage, others)
- What colours do you like more in lingerie? (Light colours, Pastel, Bright and strong, With patterns, Dark colours)
- What textures do you prefer? (Shine, Nourish, Lace, Fabric blend, Knit, Other)
- Which type of bras are more comfortable? (Classic cup, Triangle, balconette, strapless, push-up, sporty)
- Which brief type do you prefer? (shorts, classical brief, high waist, Brazilian, bikini, thong, G-string)
- What do you consider changing in order to satisfy the use of lingerie?

And for discarding practise follow the questions:

- How many pieces of lingerie do you usually buy per year? (None, 1–2 pieces, 3–4 pieces, 5–6 pieces, more than 6 pieces)
- What kind do you buy the most? (Briefs, Bras, Both types at the same time)
- Do you take care lingerie according to label rules? (No, Yes, Other)
- How many pieces of lingerie do you discard per year? (None, 1–2 pieces, 3–4 pieces, 5–6 pieces, more than 6 pieces)
- Which reasons lead you to discard lingerie? (Being out of Fashion, Tiredness, No longer fit, worn and damage, other)
- Where do you throw your lingerie away? (I pass on family or friends, give to Charity, Usual trash bin, other)
- Would you be available to wear second-hand lingerie? (Never, No, Perhaps, Yes. Other)
- Would you be available to wear modular lingerie with environmentally friendly materials? (Never, No, Perhaps, Yes. Other)

3.3 Results

Until the deadline of this article we may have some of the pre-conclusions in the universe of 70 people for each questionnaire besides we already realized mistakes after that.

Nobody filled in their age but the age range that inquired people is between 20 and 50. Most people prefer to wear lingerie in cotton, Lycra and in a mixture of fabrics (lace and plain fabric) in light or nude colors. In terms of shape, they prefer the classic-style bras over push-up and like thong-shaped briefs. Regarding the bra shape, there is a consensus that none of the usual fits well with the breast types, and they should be more practical and the cups adaptable to different body shapes.

According to discard practice, they buy more pieces than they discard accumulating items that they do not use in the wardrobe. Usually, they do not respect the rules of care and maintenance of their lingerie and when they throw them away, they do it in the common waste bin. They would never be willing wear second-hand lingerie, but they would be able to use modular underwear that will better adapt to the body of each person.

4 Conclusions

With reviewed literature and the survey results we confirmed the information about the consumption and disposal lingerie.

In consumption it was found that people are increasingly confident with their own body eliminating the thin body image. Massified clothing, namely bras, do not facilitate the adaptability of different types of bodies, in this case breasts, and almost all respondents would be willing to use modular lingerie.

On the other hand, we already confirmed that people will never reuse pieces of lingerie, concluding that the best way to move forward into a circular economy is recycling them.

Portuguese Textile waste management is not as advanced as that of the United Kingdom, with Doorstep Collectors, House Waste Recycling Centers or Clothing Banks, etc. [17] but it happens the same in private sector which take textile as resources to make lower quality products [15]. Nevertheless, the *Take back systems* around shops, such as H&M and Calzedonia, has increased as an easier routing discard possibility for people who throw away their lingerie.

Design also should take measures, not to prolong lifespan because lingerie is usually worn until be damaged but to be careful in the correct materials selection to be sure they will be recycled.

References

1. Morais, C., Carvalho, C., Montagna, G.: Dynamic Wardrobe from user's perspective- a previous survey study for female students. In: 10th International Conference on Applied Human Factors and Ergonomics, USA, 2019. Advances in Design for Inclusion, pp. 1–7. Springer Nature Switzerland AG (2020)

2. Newbery, M.: Global market review of lingerie forecasts to 2023. Aroq Limited, United Kingdom (2018)
3. Cox, C.: Lingerie: A Lexicon of Style. Scriptum Editions, London (2000)
4. Finnigan, K.: Underwear revolution: how lingerie grew up and put women's comfort first. The Guardian, The Observer, 09 December 2018)
5. Block, E.: If you want to wear shapewear, make it this shapewear. Who what wear, 14 June 2019. https://www.whowhatwear.co.uk/heist-bodysuits
6. Woodward, S., Miller, D.: Why Women Wear What They Wear. Berg, New York (2007)
7. Laitalia, K., Boks, C., Kleep, I.G.: Making clothing last: a design approach for reducing the environmental impacts. Int. J. Des. **9**(2), 93–107 (2005)
8. Morais, C., Carvalho, C., Broega, C.: A design tool to identify and measure the profile of sustainable conscious fashion costumer. In: Autex2011 - 11th World Textile Conference at ENSISA, Mulhouse, France, pp. 388–393 (2011)
9. Monto-Rios, F.J., Luque-Martinez, T., Fuentes-Moreno, F., Canadas-Soriano, P.: Improving attitudes toward brands with environmental associations: an experimental approach. J. Consum. Mark. **23**, 26–33 (2006)
10. Bianchi, C., Birtwistle, J.: Consumer clothing disposal behavior: a comparative study. Int. J. Consum. Stud. **36**, 335–341 (2014)
11. Waste and Resources Action Programme: Valuing our clothes: the cost of UK Fashion. WRAP Press, July 2017
12. Laitalia, K., Klepp, I.: Clothing Reuse: The potential in informal exchange (2017)
13. Norris, L.: Waste, dirt and desire: fashioning narratives of material regeneration. Sociol. Rev. Monogr. **67**(4), 886–907 (2019)
14. Ellen Macarthur Foundation: A New Textiles Economy: Redesigning Fashion's Future. Ellen Macarthur Foundation Press (2017)
15. Morais, C., Carvalho, C., Broega, C.: Clothing recycling as new value into Fashion cycle. In: Sustainable Development Symposium, Lisbon, Portugal (2015)
16. Ha-Brookshire, J.E., Hodges, N.N.: Socially responsible consumer behavior? Exploring used clothing donation behavior. Cloth. Text. Res. J. **27**(3), 179–196 (2009)
17. Morley, N.J., Bartlett, C., van Hemel, C.: Maximising Reuse and Recycling of UK Clothing and Textiles. Department for Environment, Food and Rural Affairs. Oakdene Hollins Ltd. (2009)

The Senior Target Audience: When the Body Does Not Keep Up with the Youthful Spirit, the Clothing Design Gives a Reply Through Comfort

Mariana Dias de Almeida[1(✉)], Ana Cristina da Luz Broega[2], and Marizilda dos Santos Menezes[3]

[1] Unisagrado, Bauru, Brazil
mari.ddalmeida@gmail.com
[2] Universidade do Minho, Guimarães, Brazil
[3] Universidade Estadual Paulista, Bauru, Brazil

Abstract. The new image of older women has demystified the profile of fragile ladies and with simple aesthetics, this woman has some new or the same yearnings, and the unpreparedness of the clothing sector towards these women may lead us to believe that fashion does not know the true image of the new elderly woman, who can be simple as well as modern and articulate, which will depend on the repertoire that she has built. Thus with methodological guidelines that provide a clothing project with experience of the user's reality, can provide greater inclusive direction of the elderly. Therefore, it is necessary to investigate the aspects of comfort in its in the all dimensions, well-being and valuation of clothing for elderly women from 65 to 85 years old. With the user-centred design methodologies and the aid of comfort assessment processes, we help to design apparel thought comfort for this increasingly demanding and purchasing power public target.

Keywords: Inclusive design · Comfort · User-centred design · Clothing design

1 Introduction

The work presented in this paper addresses the elderly woman (older adults) and her relationship with fashion, more specifically the way in which clothing has served this audience. It is easy verifiable that just as in Europe the Brazilian population is aging sharply, and that there is an urgent need to create solutions for the inclusion of this audience, and it should be emphasized that the country does not have a development that includes the elderly. Inclusion is discussed by various areas of society and is related to diversity and transversally. In clothing, it points to more specific perspectives, as in the case of the elderly woman - which is the direction of this study, where the interlocutor will be invited to explore and question the positioning of design in its contemporary requirements, where designers should be increasingly attentive to the human being and changes in society.

G. Di Bucchianico et al. (Eds.): AHFE 2020, AISC 1202, pp. 792–800, 2020.
https://doi.org/10.1007/978-3-030-51194-4_103

Senior women (elderly women) are an audience that has been gaining importance in the market and society. The construction of the image of these women also undergoes changes and the clothing constitutes a determining role in the characterization of the individual's identity, which can promote self-esteem, explore female identity and value this phase of life that should be that of "better age". Including the role that clothing plays in the way others see this person and relate to the woman at this stage. It is widely recognized that clothing and the entire fashion system can reinforce stereotypes and social stigmas, retract the elderly woman as a sick, incapable and fragile woman, and sometimes as a modern image reinforcing her healthy and active appearance, and sometimes even jovial and even artificial due to the need to adopt looks that were not designed for her, but that oblige her to make this choice due to the lack of commercial alternatives.

But these are the common places and these representations begin to be rejected by this target and by Brazilian society, because the "new elderly woman" has a rejuvenated aspect, with a modern character and breaks the senile figurative paradigms. As most of these women were not prepared for the longitude of age, fashion design as well as other sectors, also were not prepare to receive this audience as they needed. As much as the design is seeking to satisfy the needs of the elderly, clothing is still at a deficient stage, where there is still little concern and lack of preparation, which leads to the constitution of adapted rather than inclusive products based on user-centred principles. Immediate is the need for a change in the non-observance of the elderly woman, about her expectations regarding the role of clothing in her personality, as well as in the construction of her identity and physical and aesthetic well-being.

2 Comfort in Clothing

Comfort is one of the most important aspects of clothing, given its proximity to the body. A discomfort response is often immediate, however, the objective evaluation of comfort should be done accurately and quantitatively by the positive using Comfort Science methodologies, namely by Sensory Analysis for the evaluation of the touch of materials [1]. Unfortunately, this is a reality that is not observed in the clothing industry, but that is becoming increasingly necessary, especially when trying to develop more inclusive products.

On the other hand, aging is a process or set of processes that occur in living organisms naturally over time, and leaves people more predisposed to suffer from stress and anxiety due to decreased motor coordination and walking ability, loss of spatial perception, visual and auditory acuity, loss of bone and muscle resistance, etc.

However, it is worth emphasizing points highlighted by Piccoli [2], on the motor functions of the elderly person: when the individual is in good control of his functions, the construction of the basic notions, lead to a greater emotional enrichment social, physical, spiritual and intellectual, leads, therefore, to argue that when the elderly are in their basic function of motor skills, they can cover other factors in a positive way, therefore, when the elderly are able to perform their physical faculties a series of daily activities, their independence is preserved. Therefore, analysing the work of Pranke, Teixeira and Mota [3], in their research on studies involving the older adults (elderly's), it can be observed, that when the elderly develop a physical activity, their biomechanical functions have a more pronounced positive response than individuals who are institutionalized, more sedentary, that leads to a commitment for example of their march.

For Matsudo, Matsudo and Barros Neto [4], in their review of studies on the Impact of Aging on anthropometric, neuromotor and metabolic variables of physical fitness, the authors reflect on the main effects of aging, a statement stands out:

> As chronological age increases people become less active, their physical capabilities decrease and, with the psychological changes that accompany aging (feeling of old age, stress, depression), there is still a greater decrease in physical activity that consequently, facilitates the appearance of chronic diseases, which contributes to accelerate the aging process. More than chronic disease is the disuse of physiological functions that can create more problems [4].

The authors' statement demonstrates the close connection between physical and emotional conditions, one reflects on the other, this leads us to think that as one gets older, a chronological load is carried that affects the psychological, and this has a correlation with comfort, as previously explained. Comfort is related to psychological factors, therefore, the creation of a garment that causes positive sensations, and that provides satisfactory levels of comfort, can enable the elderly individual to increase their self-esteem, social, emotional and intellectual enrichment which contributes to their general well-being and happiness.

2.1 Inclusive Clothing Design and the Comfort of the Elderly

The researchers Simões and Bispo [5] argue that through inclusion it can create more balanced societies, which provide a better quality of life, not only for the disabled and the elderly people, but for society as a whole. In this perspective, inclusion in the clothing for the elderly is not just about adapting the appropriate measures for product development, but it is the clothing design itself, thinking about the physical and cultural diversity of this age group.

However, in the situation of unpreparedness in which the clothing industry is, it is not possible to glimpse the current scenario, leaving the elderly at the margin, limiting themselves only to making adjustments to what already exists. The *redesign* practiced is not enough anymore; a new attitude of innovation is needed. Thus, issues pertinent to the physiology of the elderly should be taken into consideration in the development of their clothing, as their comfort levels are correlated with clothing, as addressed previously. Based on the consulted literature, there are some points to be taken into account regarding the evolution of anthropometric variables, human motricity, etc., to be considered in the clothing design project. The aging process is accompanied by:

> '[…] increased body weight, especially from 40 to 60 years of age, with decrease after 70 years of age; gradual decreased in body height, explained, in large part, by bone mass loss; increased body fat, decreased resting metabolic rate, skeletal muscle mass and bone mass [4]'.

Yet according to the authors, there is still a tendency to "a decrease in height, over the years, because of vertebral compression" [4]. An important factor raised by the same study is that there seems to be a redistribution of body fat from the limbs to the trunk, translated into an increase in fat in the upper region of the body, especially in the abdomen, from which it appears that there will certainly be a lower demarcation of the waist (especially in the female body).

According to the study by Pranke, Teixeira and Mota [3] the changes most observed in the aging process, are reflected into the decline in the ability to move, the loss of acuity of fine motricity (which has to do with the "visuomanual" coordination when manipulating an object with the hand, for example writing, drawing, painting or even buttoning); the reduction in balance mechanisms, especially postural stability; decreased vibratory sensitivity and decreased muscle strength (due to loss of muscle tone). Other effects that have been consistently reported in the literature and are a consequence of the aging of the individual are: decrease in the number and size of neurons, with consequent decrease in nerve conduction speed, slower movement speed, decreased agility, coordination, flexibility, decreased in joint mobility and increased stiffness of cartilage, tendons and ligaments [4]. Thus there is an increasing demand for clothing products that meet the needs of users, especially the elderly and that respond to the needs of performing daily tasks that are vital to them, as is the simple case of dressing autonomously, as is the simple task that are vital to them, as is the case of simple dressing a coat without assistance. The Table 1 proposes some comfort requirements to be met in the design of a clothing product for the elderly user.

Table 1. Comfort requirements in clothing for the elderly

Comfort levels	Topics to attend
Thermophysiological	The user's thermo-physiological comfort can be improved through materials that allow an adequate thermal balance between the body and clothing The selected materials should allow good moisture management to the surface of the skin in thermal heat conditions. And they must be good thermal insulators in cold conditions These properties have to do with the porosity of textile structures (woven fabrics and knitted fabrics), i.e. with its structure more or less open With the typology of the yarns used in the construction of knitwear and fabrics, bulkier fabrics, allow to create more insulating textile substrates. Fabrics made from crepe yarns (larger twists), originate from more irregular surfaces facilitating air circulation [1] Fabrics made from natural fibers allow better moisture management, but their mixtures with special synthetic fibers (e.g. microfibers) or with special finishing tend to favour the thermal comfort properties of clothing. The option should always take into account its application to clothing for hot or cold environments The combination of porous textile surfaces made of natural fibers (or in certain mixtures with some synthetic) can promote thermal comfort

(*continued*)

Table 1. (*continued*)

Comfort levels	Topics to attend
Sensory "touch" (hand)	The sensory comfort of "touch" is related to the softness of the fabric that is the result of its composition and surface structure. Because the elderly tend to have a more dry skin, the pleasant "touch" of the fabric is critical to their feeling of comfort and avoid skin irritations. Softer textile materials with a "pleasant-touch" can be achieved with special finishes and/or functional blend of natural/synthetic fibers With a decrease in fine motricity skills there will be greater difficulty in buttoning or closing the garment, which implies the use of aviation's with more noticeable textures (in terms of size and texture), and even think different and innovative ways to tightening/adjusting clothing
Ergonomic	Ergonomic comfort is related to modelling (anthropometric measures of the elderly population) and the making of clothing, and should provide for the adequacy to the new configurations of the elderly's body, in order to allow the well-being and freedom of movements. This can be very conditioned (or enhanced) by the thermal and touch factors of the textile materials that make up clothing With the slow march, loss of muscle tone and the limitation of range of motion, clothing should be thought of in terms of ensuring greater range of motion, through clearances, new forms of openness, of modelling combined with the right choice of textile materials, parts should allow greater adaptability to the skeletal-motor reality of the elderly Modelling should also be attentive to the new postural dimensions of the elderly, at the levels of the corporate axis such as knees and elbows, where a tendency to flexion is denoted and, therefore, a tendency to excess of textile material in these areas, that can which can lead to discomfort in these areas of the body
Psycho-Aesthetic	The appearance of the object is a crucial factor so that the user does not feel excluded or stigmatized, therefore, the suitability of clothing to the user's lifestyle is fundamental here. Respecting the language of the individual with the environment in which he is inserted is fundamental By psycho-aesthetic comfort the elderly will follow confident insurance, will be able to express their personality, individuality, style The type of textile material applied, the conformation of clothing to the body, the dressing well in current lines valuing or hiding body parts will be valued; the range of colour within fashion trend palettes, the new materials (usually applied only to the youngest garments), will make them to feel safety in theirs clothing, which contributes to elevate there self-esteem and the quality of life

So when designing a garment, the first thing to consider is the user and his needs, in other words, the requirements to be taken into account, is the functionality, quality, comfort but also and not less important, the parameters of aesthetic order, in the sense of satisfying the user's needs and their inclusion in their social environment.

In the clothing product design, the object becomes of remarkable significance to be more inclusive, no longer being just a *redesign,* because when one knows the real needs of the user (namely the elderly), solutions can be developed that favour the relationship with the clothing object, witch provide more security, well-being and consequently more comfort, which involves aesthetic aspects, but also the formal and functional aspects through modelling and the fabric, befitting the needs of the elderly [6]. Completing these questions the "body-garment interactions (thermal and mechanical) perform very important roles in determining the wearer's comfort state, as well as the external environments, (physical, social and cultural)" [7].

According to Neves [8], the attendance of relevant factors of the movement of the elderly body, requires projects that facilitate dressing and undressing, which should take into account all aspects related to mobility, as they argue that some activities can be performed with more difficulty. Therefore, any elements relevant to clothing design, such as supplies (clothing construction accessories), textile surfaces, formal aspects of product development and modelling, may ensure greater success in clothing projects for the elderly public. Therefore, any elements relevant to clothing such as in the case of aircraft, textile surfaces, formal aspects of product development and modelling may ensure greater success in clothing projects for the elderly public, covering the requirements of these individuals (by their direct or indirect participation) raises more levels of comfort in clothing in general.

All these points and factors must be worked together with the end user, directly observing their difficulties and needs, which leads to rethinking the design of clothing in a more inclusive way, both in social terms and in providing the elderly with greater comfort and autonomy. Although these are factors that attend mainly to physical and physiological aspects, their direct correlation with psychological and emotional aspects will trigger direct responses of well-being that influence the daily life and the performance of daily activities of the elderly contributing to the increase of their self-esteem and social inclusion.

3 Design Guideline Focused on Comfort

When defining a project guideline some steps must be built in order to achieve an objective, in the case of this project it is to define the steps to develop an artefact of clothing that encompasses the public of elderly women, for this purpose, the first step was to develop interviews through a focus group, which seeks to understand the thoughts and emotions that externalizes what are their needs and desires [9], which for Pazmino [10] needs express reality and desires consumption.

The comfort within the elaborated guideline enters as a factor analysed in the interviews and subsequently applied in the quality analysis phase, that is, an evaluative indicator. Based on what has been raised in the theoretical bases, the four evaluative points can be understood: Thermo-physiological, Sensorial touch, ergonomic and psycho-aesthetic. Embodying this evaluative method, the project was based on the application of March and Smith's Design Research Methodology [11], which would classify this step as "Instantiations", that is to say, that the artefact being executed in its real scenario, because, it is in this scenario that its feasibility and effectiveness is

Table 2. Clothing comfort assessment

Age:__ Name:___________________			Date: ___/___/____		Temperature:_______	
Comfort criterions			piece 1 (top)	piece 2 (bottom)	piece 3 (coat)	Piece 4 (other)
Woven fabric / Knitted fabric	Touch	Natural				
		Artificial/ Synthetic				
		Softness				
		Rough				
		Warm touch				
		Cold touch				
		Hard				
		Flexible				
		Stocky/ Bodied				
		Fluids				
	Elast.	Fabric elasticity				
Garment	Weight	Light				
		Heavy				
	Ergonomics	Draping				
		Body-adjusted modelling				
		Facility of movement				
		Facility in dressing / undressing				
		Facility in tightening / buttoning				
		Seams Comfort				
	Thermal	Adequacy to climatic conditions				
		Freshness				
		Sweat absorption				
		Cold protection				
		Air permeability				
	Psycho-aesthetic	Fashion trends colour				
		Print / Pattern in fashion trends				
		Design piece (differentiated)				
		Durability / appearance of new				
		Quality of fit				
		Satisfaction in using				
		Security in using				
		Adjusted to personal style				
Totals		**Comfort Assessment Rate**				

Usability scoring legend:				
Very Unsatisfactory				Very Satisfactory
1	2	3	4	5

Legend temperature:				
1	2	3	4	5

shown, with the intent of the guideline to guide the application in industries, several factors should be added such as economic context, usability, market competitiveness and market insertion.

The application in the project must occur as follows:

1. Use of comfort viability checklist;
2. Investigation of market economic context;
3. Evaluation of strengths and weaknesses (product and economic context);
4. Application of a Decision Matrix.

After the quality evaluation process, the artefact goes into the factor that concludes what are the "propositions", because, through the collected data, it is possible to build a base so that solutions can be revisited for later problems [11].

The use of focus group methodology is used across the social sciences, providing access to participants in they own language, encouraging the production of more articulated attributes, identifying some key considerations, concepts and concerns.

For the creation of focus group, 11 women aged 65 to 85 years were selected and it had as intention of capturing the information that would lead to know the relationship of the elderly woman with aging and the role of Fashion in their lives. The use of focus group, also aims to highlight the difficulties experienced by the group in dressing and undressing their clothes, and offered an opportunity to observe their precessions about the comfort of their clothes. To this came the need to create a comfort assessment grid, which is presented in Table 2.

The parameters evaluated in this table originated from the research carried out in the literary review, uses a 5-point Likert Scale, represented by humour little faces, ranging from very unsatisfactory (0) to very satisfactory (5). A small scale of weather is also applied during evaluations. The target audience without much explanation easily understands this type of scales, by pictograms.

The comfort criteria in clothing as well as the scale were tested with this focus group. For each garment was made a detailed analysis was carried out to give to the design a tool for the evolution of clothing for this target audience.

4 Final Considerations

This paper has presented a clothing comfort assessment framework for a senior audience, based on specialized literature and an exploratory work with a focus group. This focus group of women is endowed with autonomy in their lives, it should be noted that these women with the age between 65 and 85 years, ended up presenting similar views regard the aging, in which they state that aging is not a problem, which is a natural evolution, and claim that they do not feel the pressure from society to remain young.

By observation (and photographic record) has verified the difficulties experienced by the group in dressing and undressing their clothes, in addition to allowing them to discuss their precessions regarding the comfort of their clothes. It has verified by observation (and photographic record) the difficulties experienced by the target audience (focus group) in dressing and undressing their clothes, and allows discuss their precession about the comfort of their clothes.

The comfort parameters as well as the scales were validated, however it still need to validate this process industrially. The next step is to insert this process into the industrial context of construction of the garment artefact, to create dynamism in the method, where designers will be the preponderant part of this phase.

Acknowledgments. "This work is supported by FEDER funds through the Competitivity Factors Operational Programme - COMPETE and by national funds through FCT – Foundation for Science and Technology within the scope of the project POCI-01-0145-FEDER-007136".

References

1. Broega, A.C.: Contribuição para a Definição de Padrões de Conforto de Tecidos Finos de Lã, Tese de doutoramento em Engenharia Têxtil, Ramo: Física Têxtil. Universidade do Minho, Guimarães (2008)
2. Piccoli, J.C.J., Santos, G.A., Ferrareze, M.E., Haas Junior, W.: Parâmetros motores e envelhecimento: um estudo de idosos de 60 a 83 anos de Ivoti, RS; Textos & Contextos (PA), vol. 8, núm. 2, pp. 306–318 (2009)
3. Pranke, G.I., Teixeira, C.S., Mota, C.B.: Contribuições biomecânicas ao público da terceira idade. Rev. Bras. Geriatr. Gerontol. **9**(2), 75–91 (2006)
4. Matsudo, S.M., Matsudo, V.K.R., Barros Neto, T.L., Impacto do envelhecimento nas variáveis antropométricas, neuromotoras e metabólicas da aptidão física; Rev. Bras. Ciên. e Mov. Brasília, vol. 8, no. 4, pp. 2 1–3 2, setembro 2000
5. Simões, J.F., Bispo, R.: Design inclusivo acessibilidade e usabilidade em produtos, serviços e ambientes. Centro Português de Design, Lisboa (2006)
6. Vianna, C., Quaresma, M.: Ergonomia: Conforto têxtil no vestuário do idoso. Recife, 15° Ergodesign/Usihc pp. 1–9 (2015)
7. Broega, A.C., Silva, M.E.C.: Ergonomia no vestuário: Conceito de conforto como valor agregado ao produto de moda. Buenos Aires, V Encuentro Latinoamericano de Diseño (2010)
8. Neves, E.P., Brigatto, A.C., Medola, F.O., Paschoarelli, L.C.: Biomecânica e Moda: Contribuições para a concepção de vestuário para os idosos. In: 6a Conferência Internacional sobre Factores Humanos Aplicados e Ergonomia (AHFE 2015) UNESP Bauru, Brasil (2015)
9. Pommer, C.P.C.R., Pommer, W.M.: A metodologia do grupo focal e a formação continuada do professor: um olhar interativo envolvendo a articulação cognição e emoção. Revista Itinerarius Reflectionis 5–21 (2014)
10. Pazmino, A.V.: Como se cria: 40 métodos para design de produtos. Blucher, São Paulo (2015)
11. Dresh, A., Lacerda, D.P., Antunes Junior, J.A.: Design Science Research: Métodos de Pesquisa para avanço da ciência e tecnologia. Bookman, Porto Alegre (2015)

Active Sportswear for Older Consumers

Laura Piccinini, Gianni Montagna(✉), and Cristina Carvalho

CIAUD, Lisbon School of Architecture, Universidade de Lisboa, Lisbon, Portugal
laurapiccinini@uol.com.br, g.montagna@gmail.com, cristifig@gmail.com

Abstract. The aging process determines significant alternations of different natures that lead to decreased physical or mental capacity to varying degrees, compromising their ability to exercise. Physical activity, as well as sports activity, regardless of the intensity with which it is practiced, is a psychophysiological need, which in addition to maintaining better health levels, also presents a high level of socialization.

Clothing adapted for the development of sport for seniors is beneficial for the wearer and has important comfort levels helping the wearer to develop in more active aging.

This research on clothing is based on a questionnaire answered by approximately 110 people over 60, active consumers in order to understand the particularities of their physical activities focusing on the study of comfort and preferences of types of garments and their characteristics. This article is part of a doctoral work that is being developed in order to better understand how clothing design and the textile materials used for its construction can be more valuable to users of this type of product.

Keywords: Knitwear · Sportwear · Older consumer · Sport for olders

1 Introduction

The advancement of research in science and technology in relation to the aging of the population has created a new paradigm for studies on the physical and psychological health of the elderly. The study of sports clothing design in this phase of life is aimed at improving the health and well-being for this age group of consumers.

In this sense, clothing should be adapted considering a multidimensional geriatric assessment orientation and multifunctional and interdisciplinary diagnostic processes, aimed to alleviating and improving medical and psychosocial problems.

Based on a preliminary research built on the observation that there was no offer of adapted clothing for the elderly in the market, a consultation was developed within this target audience that develops physical activity and about the objective and subjective dimensions of the specific equipment for sport movements. The auscultation carried out collected information from 110 people who regularly attend gyms and physical maintenance groups for the elderly over 60 years old and who develop recreational physical maintenance activities in a closed and controlled environment.

G. Di Bucchianico et al. (Eds.): AHFE 2020, AISC 1202, pp. 801–808, 2020.
https://doi.org/10.1007/978-3-030-51194-4_104

Clothing with an adapted design exists for different age groups, but it is scarce for this group of users, which is becoming increasingly important for their specific needs, due to the number of elderly people practicing maintenance sports, but also due to the renewed economic possibilities of this new economic class. Along with ease of use and ergonomic comfort, it is essential at this stage of life to provide some psychological comfort in order to ensure improvements in self-esteem and ensure better active and social aging.

The design of clothing includes the observation of all the multidimensional properties of the raw material, the construction processes, the pattern drafting and assembly of clothing, and others, which can provide good handling, facilitating the acts of dressing and undressing [1].

2 Aging Projections

Social and individual perspectives on what it means to be "Elderly" can be related to a variety of factors, including the age at which individuals stop actively participating in any job or civil society and how they have changed their social involvement or lost their functional capabilities. The definition of elderly is dynamic, specific to each country and has been adapted over time [2]. According to the same author, (2014, p. 60) "in 2010, the healthy life expectancy of the male world at birth was 59 years old and the female one was 63.2 years old, compared to 54.8 and 58.7 years, respectively, in 1990. According to Bernardes and Pinheiro [3] and Vincent and Velkoff [4], in recent years, there has been an increase in average life expectancy. In the USA, refer the same authors, it is estimated that in 2030 about 20% of the population is over 65, which will have worrying medical, economic and social implications.

According to Caldas [1] "In Brazil, people over 80 years of age are considered part of the very elderly segment, justified by the greater probability of being affected by the deterioration of their physical and mental condition, presenting, in general, large burden of chronic diseases and functional limitations. This classification is still in force, with no reformulation of the most recent categories found.

The aging process has progressive biological physical, sensitive, cognitive and coordination changes in the elderly who are more likely to suffer injuries through falls. Minimizing the risk of falling involves promoting the autonomy of the elderly. Fall mortality increases exponentially with age, verifying that individuals over 80 years old have a fall mortality rate six times higher than individuals between 65 and 79 years old.

3 Clinical Framework

The life span of women is about five years longer than that of men, due to less biological vulnerability and greater demand for medical assistance. Generally, female health problems are chronic, long-term or disabling, whereas male diseases tend to be short-term and fatal.

After 40 years of age, there is a gradual decrease in height, estimated at one centimeter in each decade; the factors most responsible for the stature decrease are: reduction of the plantar arch, alteration of the intervertebral discs, characterized by the decrease of its thickness and modification of its constitution and increase of the curvature of the spine. Generalized muscle atrophy occurs, and muscle mass is replaced by adipose tissue, loss of tone and muscle strength. Calcium is lost and the bones decrease in density. The joints become inflamed, deformed, constituting the most frequent chronic characteristic of aging [3].

Oliveira and Pinto [5] refers that chronic inflammatory activity associated with age is related to prolonged exposure to an antigenic load (a substance that, when introduced into the body, causes the production of antibodies). In fact, the increase in life expectancy provides more time for contact with antigens, which contributes to a change in the inflammatory response. Individuals genetically equipped with an effective inflammatory response mechanism and with an efficient control of inflammatory pathology, may be destined to be healthy centenarians".

Body fat increases with age, with a change in the location of the periphery to a more central, abdominal or visceral predominance. Adipose tissue acts as an active endocrine organ, capable of producing various cytokines and adipokines. Studies suggest that visceral fat, both in obese and non-obese elderly people, contributes specifically to inflammation. The decline in steroid sex hormones with age can also contribute to the increase in pro-inflammatory markers [5].

One of the main diseases is cardiovascular disease; corresponding to a group of disorders that occur in the heart and blood vessels [6].

Inactivity or inactivity in the elderly can cause or accentuate physical and non-physical problems ranging from impaired functionality, loss or decrease in muscle mass (Sarcopenia), loss of balance (ataxia), to emotional isolation and loss of self-esteem. Physical and mental health impairments can be attributed to extreme inactivity, which, in turn, can be caused or accentuated by health issues (real or not). There is a dialectical relationship between such factors [7].

Deformities in the spine occur because it is a delicate part and can also be congenital or result from different situations during the person's life, due to physical effort, infections, poor posture, among others. On the other hand, deformation of the spine may occur if the elderly remains seated incorrectly and in inappropriate places for too long, due to the deficiency of their supporting muscles [1].

In relation to posture, the elderly, while remaining in a sitting position, the abdominal muscles progress with flaccidity, resulting in an undesirable curvature in the back, from the dorsal to the lumbar area [8].

4 Old Age and Active Life

According to Bacha [9] about the factors that influence the active life of the Elderly is the "Immobilization Syndrome", defined as signs and symptoms resulting from the suppression of all joint movements, one of the main causes of immobilism such as: change in postural balance, falls, fear of falling, limited walking, loss of independence and autonomy, immobility in bed and dementia. All System are affected (Tegumentary, skeletal, Muscular, Cardiovascular, urinary, digestive and respiratory) That is, there is a global systemic involvement. Those who already have the syndrome should preferably move with the help and supervision of a professional in the specific area.

Pedersen [10] refers that is well known that cytokines, besides their immunoregulatory roles, are important players in metabolism and that it has become evident that skeletal muscles express several cytokines, which belong to distinct cytokine classes. The same author refers that IL-6 mediates anti-inflammatory effects by stimulating the production of anti-inflammatory cytokines and by suppressing TNFalpha (tumor necrosis factor alpha) production.

Regarding the psychological aspects of the elderly, successfully aging is trying to reduce the risks of diseases by preventing or reversing functional loss. Physical and social activities have therapeutic effects that prevent diseases. It is necessary to consider health in an expanded way, with some urgent change in the current context towards the production of a more favorable social and cultural environment for the elderly population, [...] in a safe and pleasant environment, consistent with the situation you are in right now [1].

Physical practice is important in this phase of life, it brings countless benefits to the elderly and relocates them independently to carry out their daily activities, makes them happier and more sociable, and combats one of the great bad of age, which is falls. Anyway, the Physical Education professional is a life changer, and with regard to this phase of life that we all aspire to reach, we can efficiently bring the greatest health benefits through physical activities. A clothing more suitable for the elderly, with easy handling to wear, comfortable, aesthetically pleasing and easy to wash is a way to help these individuals to be safer and happier in this phase of life that we all aim to reach, we have efficiently brought the greatest health benefits through physical activities.

5 Data Collection and Discussion

In this section, part of the data collected in the survey aimed at senior users who attend sports clubs and gyms will be presented in order to identify their preferences in terms of sport and clothing.

The first question tried to identify the biggest differences that the interviewees found in the development of sports practice as a senior compared to the development of their sports practice at a younger age. The answers, shown in image 1, show a clear change and deterioration in different physical aspects, compared to previous decades, changing their movement capacities for the execution of the exercises. Differences in impact and agility will be the factors that most differed over the years, followed by changes in performance, resistance, stretching and flexibility (Fig. 1).

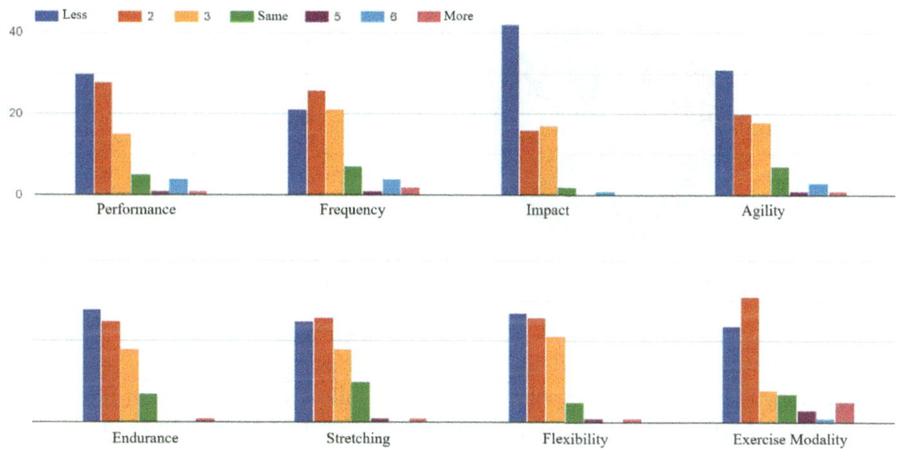

Fig. 1. Variations found in the development of physical activity in seniors compared to younger age. Source: Author.

Intensity of exercises practiced by the elderly was about physical activity. The answers to this question show, in Fig. 2, that almost 70% of the interviewees practice low and medium intensity activity and that only about 18% practice high intensity sports, such as weight lifting and cardio fitness, even when they are over 60 years old age.

Fig. 2. Intensity of exercises practiced by the elderly. Source: Author.

Bearing in mind that the majority of the interviewees practice light physical activity and that in most cases it implied a greater use of the lower limbs, it was asked what type of clothing they used for the lower body for the development of their physical activity. In the answers collected with this question, which can be seen in Fig. 3, more than 70% of the interviewees indicated the use of leggings and loose pants for use in gymnasiums and sports clubs.

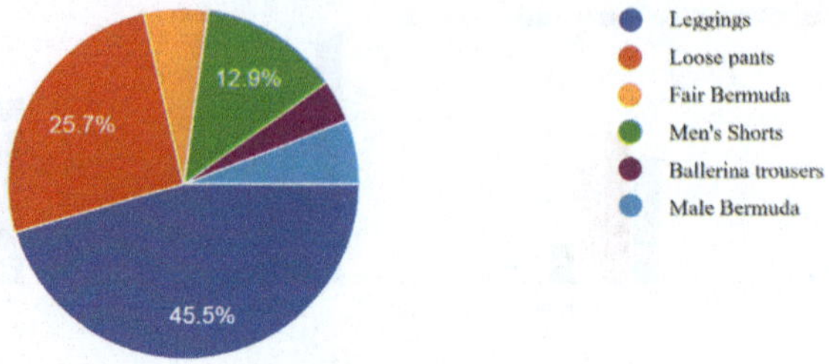

Fig. 3. Type of garment worn on lower limbs. Source: Author.

In the next question, respondents were asked to assign a level of comfort to the garments they normally use for sports. The results collected and shown in Fig. 4 show that, in general terms, the respondents consider themselves comfortable in using the existing equipment. Although all indicators have received a positive score, in the case of the chemical composition of the materials, about 30% of the auscultated people report some discomfort and lack of less adaptability to this type of use, while about 30% of the interviewees indicate needs improvements in terms of psychological comfort of this type of clothing, namely in the formal aspects.

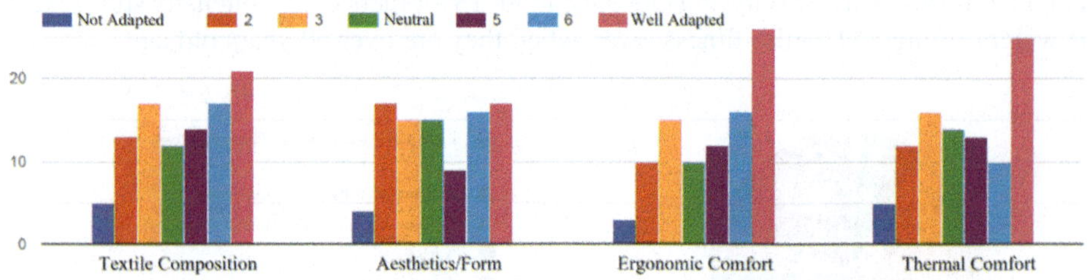

Fig. 4. Comfort levels of clothing normally used for general sports. Source: Author.

Taking into account the previous answers and the indoor sport practitioners in question, it was asked what type of underwear they would normally wear during the development of their maintenance exercises. Respondents had to indicate which types of interior parts they would normally use and what levels of comfort they attribute to them. The answers to this question, shown in Fig. 5, indicated a clear discomfort with most of the pieces of underwear used for sports, making this subject more deeply researched in the different levels of design of this type of products.

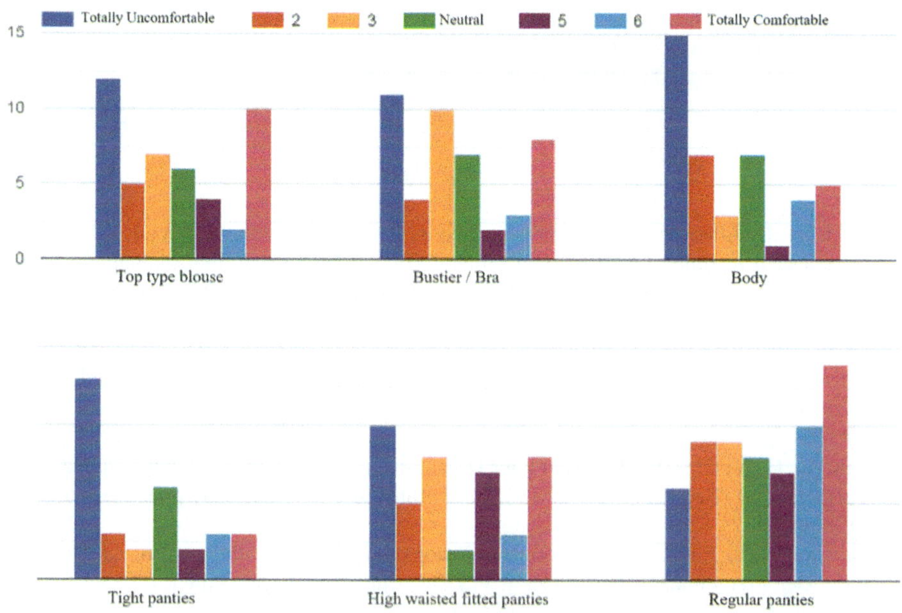

Fig. 5. Type of underwear and comfort levels. Source: Author.

6 Conclusions

A The study of sports clothing for the elderly raises specific questions with regard to different levels of comfort and in particular those that refer to ergonomic comfort and psychological comfort. It is a population group that has some motor impairment that in many cases can influence their psychological well-being and their experience in general terms. Active aging is an important way of maintaining the physical and psychological faculties of individuals in general and the elderly in particular. The general improvement that light sport can offer to this age group is very important and can help to develop and preserve the social relationships and integration that all individuals need.

Respondents' responses to the questions posed about sports clothing showed possible avenues for renewed clothing product developments. The interviewed users declared that they practice light sport and that they usually use generic clothing to practice this type of sport in a social environment. For the lower part of the body, they mostly use leggings or loose pants so that they can move and train in a comfortable way. When asked about the different comfort levels of their clothing, they are generally satisfied, although the sum of the different levels of dissatisfaction with ergonomic comfort and psychological comfort deserve to be taken into account for their level of expression. Particular attention should be paid to responses about the type of clothing worn under the sportswear. Most respondents consider most underwear to be uncomfortable for use in sports and their interaction with sportswear can be better analyzed and revised.

Acknowledgment. The authors of this paper wish to thank the Centre for Research in Architecture, Urbanism and Design (CIAUD) of the Lisbon School of Architecture of the University of Lisbon and FCT for founding this project.

References

1. Caldas, A.L.: Adequação do vestuário para idosas dependentes de cuidados, considerando a sua modificação anatómica. University of Minho (2017). http://repositorium.sdum.uminho.pt/
2. Ferreira, J.F.: Demografia e epidemiologia do envelhecimento; Geriatria Fundamental, Saber e praticar; Coordenação de Manuel Teixeira Veríssimo, Lidel, Edições Técnicas Ltda, Lisboa, PT (2014)
3. Bernardes, A., Pinheiro, S.: Anatomia do envelhecimento; Geriatria Fundamental, Saber e praticar; Coordenação de Manuel Teixeira Veríssimo, Lidel, Edições Técnicas Ltda, Lisboa, PT (2014)
4. Vincent, G., Velkoff, V.: The Next Four Decades: The Older Population in the United States: 2010 to 2050 (No. P25-1138; p. 16). (2018). https://www.census.gov/prod/2010pubs/p25-1138.pdf
5. Oliveira, C.R., Pinto, A.M.: Biologia e Fisiopatologia do Envelhecimento; capitulo 3 do livro Geriatria Fundamental. Saber e Praticar, Publisher: Lidel, Edições Técnicas, Lda., Editors: Manuel Teixeira Veríssimo (2014)
6. Martins, R.: Atividade física e envelhecimento; Geriatria Fundamental, Saber e praticar; Coordenação de Manuel Teixeira Veríssimo, Lidel, Edições Técnicas Ltda, Lisboa, PT (2014)
7. Barbosa, A.P., Concone, M.B.: Uma proposta de paradigma: capacidade e qualidade de vida de idosos; vários autores; O envelhecimento ativo e seus fundamentos; Suzana Carielo da Fonseca (organizadora), 1 edn. Portal Edições, São Paulo (2016)
8. Marques, S.: Discriminação da terceira idade. Fundação Manuel dos Santos, Lisboa (2011)
9. Bacha, et al.: Terceira idade: uma escala para medir atitudes em relação a lazer. In: Enanpad, 30, Salvador (2006)
10. Pedersen, B.: IL-6 signaling in exercise and disease. Biochem. Soc. Trans. **35**, 1295–1297 (2007). https://doi.org/10.1042/BST0351295

5D Virtual Model for Thermal Comfort and Ergonomics Assessment Applied in Clothing Industries

Consuelo Latorre-Sánchez(✉), Mateo Izquierdo(✉), Alejandro Conde(✉), Eduardo Parrilla(✉), Alfredo Ballester(✉), Jose Laparra-Hernández(✉), Juan Carlos Gonzalez(✉), and Sandra Alemany(✉)

Instituto de Biomecánica, Universidad Politécnica de Valencia, Edificio 9C. Camino de Vera s/n, 46022 Valencia, Spain
{Consuelo.Latorre, Mateo.Izquierdo, Alejandro.Conde, Eduardo.Parrilla, Alfredo.Ballester, Jose.Laparra, Juancarlos.Gonzalez, Sandra.Alemany}@ibv.org

Abstract. This paper describes the features and outcomes of a novel thermal 3D/4D scanner developed by IBV. The study is the continuation of the development of a relational model able to predict thermal comfort/discomfort from users' attributes and environmental conditions. The model is based on the "thermotype" concept, which synthesizes the climatic preferences, body shape and thermal response of an individual. The Digital Human Model generated by mapping thermal information into a 3D body representation has been evaluated with real subjects in a thermal chamber with different garments in order to validate the initial version of the model based in other type of products and to apply these techniques to Clothing industries.

Keywords: Human factors · Digital Human Model · Thermal avatar · Artificial intelligent · Thermal comfort · Biomechanics · Anthropometry · Temporal scanning 4D · Image processing · 3D body model · Health · HVAC · Clothing · Thermotype

1 Introduction

The Institute de Biomechanics of Valencia (IBV) is a Technology Center that studies the behavior of people in their relation to the products, environments and services they use. This study is the continuation of the development of a relational model framed in the COPET project. Preliminary results were presented in Latorre-Sánchez et al. [1].

Thermal comfort is a very important attribute especially in textile products with high interaction with users [2], conditioning the criteria design in garment or complements. The adaptation of clothing to the thermal characteristics of the environment and to the thermo physiological response of users has a great effect on the general comfort and satisfaction of the user. It is necessary for the clothing to allow regulation of the temperature and moisture of the human body within the limits of comfort. This aspect is critical in clothing for environments with extreme temperatures or moisture, in the case of high levels of activity or in special populations, such as children or seniors.

G. Di Bucchianico et al. (Eds.): AHFE 2020, AISC 1202, pp. 809–819, 2020.
https://doi.org/10.1007/978-3-030-51194-4_105

This research studies the mechanisms inherent to thermal comfort through the development of a relational model, capable of predicting thermal comfort based on objective and subjective measures on subjects [1]. It has been found that the classical relational model for thermal comfort [3], for user characterization parameters such as BMI or clothing insulation was not enough to predict the comfort. The parameter based on classification people according to their personal preferences in thermal comfort permits to generate models able to estimate thermal discomfort, based on simple questionnaires. This parameter has been named as "*thermotype*". It was characterized thanks to AI algorithms and a thermal avatar derived from the thermal body map and explains in a much better manner the obtained results [1].

Infrared thermography is technique, particularly valuable for the exploration of the topography or pattern of skin temperature across the body [4]. A protocol for IR data collection has been developed to minimize the issues associated with the uncertainties of this measure system [1, 4]. Additionally, 3D scanners can obtain accurate representations of the outer 3D body surface [6, 7]. The 2D/3D mapping of flat thermal images of human subjects to their 3D scans (herein, thermal avatars) can be used in product development [8, 9] and in health diagnostics [10–12].

Body thermal maps can be standardized to a common parameterization using a 3D human body template [13]. The point-to-point correspondence between thermal avatars improves reproducibility and facilitates the spatial, temporal or inter-individual comparisons of sets of data. Prior art uses different sensors to gather the thermal data and the 3D surface data. In this paper we propose a new method to create standardized thermal avatars from five thermal images of a person gathered using a single Long-Wave Infrared camera. In our method, 3D human surfaces are reconstructed from body outlines obtained from infrared images [14–16].

The main of this work was to provide a specific protocol for IR images collection and thermal scanner development to extract thermal and anthropometrical characteristics by generating a thermal avatar with objective and subjective information of the user [1]. *Move 4D* is a modular photogrammetry-based 4D body scanner developed by the Institute of Biomechanics of Valencia (IBV) [17]. The aim of this project is to integrate the thermal information in this system and increase its functionalities with very valuable data from the human body.

2 Methods

This section describes the methodology for the validation of the relational model in clothing sector. A brief summary of the experimentation with users and the equipment used to test and improve the data input of the model based on a thermal scanner prototype is presented. The protocol, data and image processing and analysis are also included.

2.1 Experimentation and Methodology

The first part of the project was based on a huge experimentation with more than 50 users balanced in gender, age and thermal profile, in five different scenarios in a range of temperature from −5 °C to +40 °C. This experimentation consisted in the evaluation of a heated car seat in a controlled environment simulating the interior of a vehicle and the air conditioning systems (HVAC) [1].

This experimentation feed and trained the relational model.

The validation of the model was carried out on clothing sector, based on testing different garment in controlled conditions. Experimentation was made ad-hoc with several clothing companies to test new fabrics and designs. All users wore identical costumes and size according to ISO 9920 [16] to predict thermal sensation and thermal comfort based on the surface temperature recorded in different parts of the body, and the estimation of its thermotype.

Four types of results were obtained: descriptive analysis of objective results from sensors, descriptive analysis of subjective results from questionnaires and thermal 2D maps from thermal images. Basic statistical analysis and subjective results separated per phases and per garments was analyzed.

Thermal maps before and after the experimentation provide information of the different level of insulation of the garments and the thermal behavior of the participants.

Table 1. Top: User with different garments. Bottom: 5 Users with same insulation clothing.

Garments with different insulation (same user)

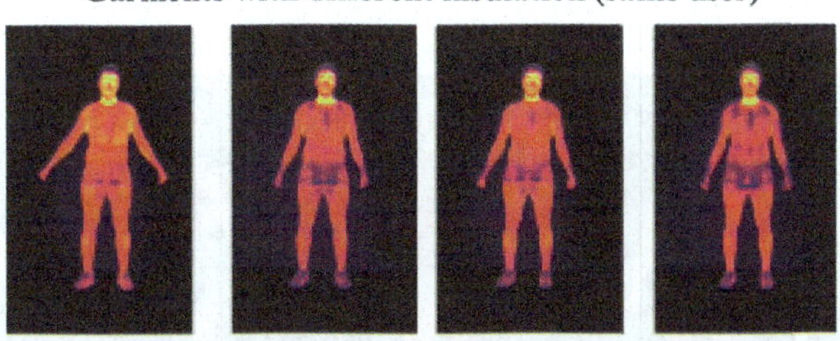

Users with different termotypes (same garment)

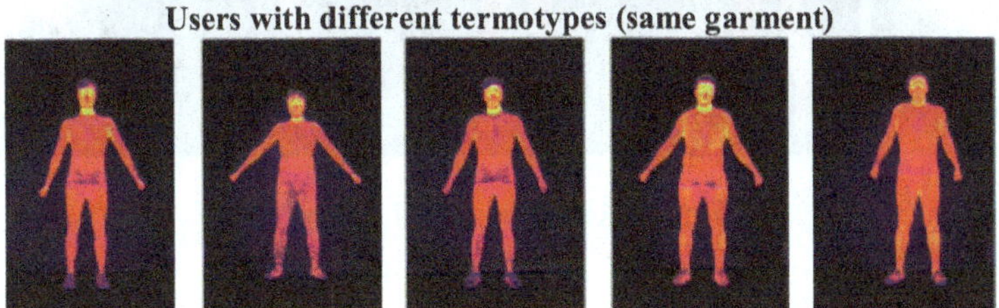

Table 1, shows some examples of thermograms taken for the same user with different garments (thermic long sleeve T-shirt) and five users with the same garment in a cold scenario at 0 °C.

All tests included a set of thermographs taken with the FLIR T650sc camera at 4.5 m and with the thermal scanner system (four FLIR A35 thermal cameras) at 2.5 m. This scanner is a system developed by IBV in order to integrate thermal layer information in the new IBV system MOVE 4D (https://new.ibv.org/tech/move4d/) [17].

2.2 Equipment

The experimentation was carried out in a climate chamber 3 m × 3 m × 3 m of capability (from −25 °C to 60 °C and moisture from 20% to 90%).

The objective measurements: body temperature and humidity were recorded with highly sensitive contact surface sensors along the whole test and with thermal images before and after the test (taking off the garment):

- Skin temperature sensors on several locations of the upper body to continuously measure the skin temperature.
- Microclimate sensors (temperature and humidity) on several of the upper body to continuously measure the temperature and humidity in the gap between the skin and the garment.
- Thermal camera images before and after the test to have a temperature mapping of the torso.

A tympanic thermometer recorded the internal temperature of the user during the experimentation, at the same instants that users answer the questionnaires.

Figure 1 shows a user with different parts of the torso with sensors of temperature and humidity. Several thermographs before and after the test show the thermoregulation effects on his body.

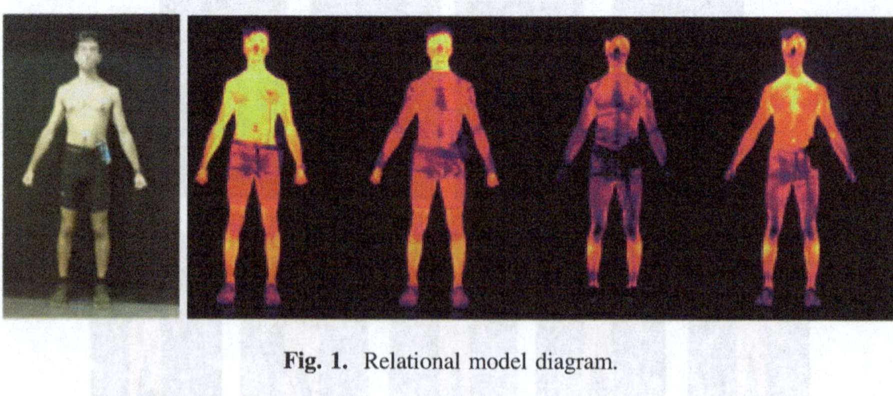

Fig. 1. Relational model diagram.

2.3 Protocol and Methodology

The protocol applied techniques for the objective measurement of temperature and moisture between clothing and other products and the body, as well as subjective tests to assess the thermal comfort provided by a garment in controlled conditions, to improve and evaluate different type of garments, as work clothes, thermal protection, smart clothing, textile fibers comparison, and other products with high interaction with the human body.

The subjective measurements were obtained with temporized surveys during the experimentation. Questionnaires follows ISO-10551[1] and ISO 9920:2007[2] standards. A set of questions related to thermal sensation (Fig. 2), humidity sensation and thermal comfort were collected before entering to the climate chamber, at 1, 3, 10, 15, 20 and 25 min. Out to the chamber the users completed another questionnaire about personal preferences and subjective feeling with each garment. Only the last 4 questionnaires (inside climate chamber in cold conditions) was considered to calculate the mean values and statistical results.

TEMPERATURE / HUMIDITY	BODY	CHEST	ABDOMEN	TRAPEZE	LUMBAR	ARMS/ ARMSPIT
Very cold	☐	☐	☐	☐	☐	☐
Cold	☐	☐	☐	☐	☐	☐
Cool	☐	☐	☐	☐	☐	☐
Neutral	☐	☐	☐	☐	☐	☐
Warm	☐	☐	☐	☐	☐	☐
Hot	☐	☐	☐	☐	☐	☐
Very hot	☐	☐	☐	☐	☐	☐

COMFORT	BODY	CHEST	ABDOMEN	TRAPEZE	LUMBAR	ARMPIT
Very comfortable	☐	☐	☐	☐	☐	☐
Comfortable	☐	☐	☐	☐	☐	☐
Neutral	☐	☐	☐	☐	☐	☐
Uncomfortable	☐	☐	☐	☐	☐	☐
Very uncomfortable	☐	☐	☐	☐	☐	☐

Fig. 2. Examples of questionnaires of thermal sensation and thermal comfort.

Figure 3 shows a typical example of the routine test to check the thermal comfort in dynamic conditions (treadmill inside the climate chamber) and the different phases of the experimentation.

[1] ISO 10551:2002. Ergonomics of the thermal environment – Assessment of the influence of the thermal environment using subjective judgement scales.

[2] ISO 9920:2007. Ergonomics of the thermal environment – Estimation of thermal insulation and water vapour resistance of a clothing ensemble.

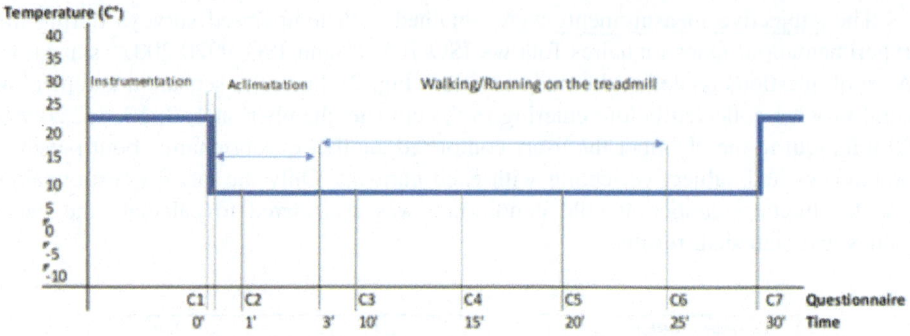

Fig. 3. Example of chronogram vs ambient temperature during the typical tests.

2.4 Image Processing and Data Analysis

ANOVA (Analysis of Variance) test, a parametric statistical technique used to compare datasets, was used to analyze the data from questionnaires. The value of questionnaires were also included in the relational model.

Thermographs were used to compare users and products. In cold conditions, the thermal behavior of body trying to compensate the losses of heat, this process is not homogenous in all humans.

The use of different thermotypes allows to test the garments for different thermal responses. Three thermotypes were selected to include in the experimentation the variability in the physiological thermal response of humans (chilly, neutral and warm people). Differences of temperature appeared in face, hands, neck and abdomen, typical for each thermal profile considered.

Figure 4 shows the same user and conditions with two different garments in different sessions and days.

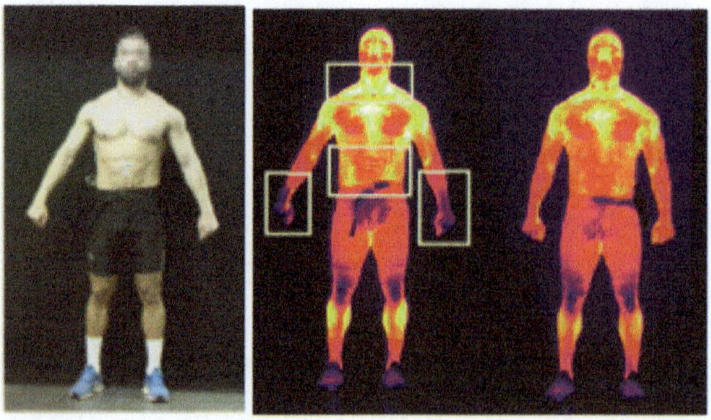

Fig. 4. Warm person after testing two different insulation garments.

3 Results and Discussion

Three types of results were obtained: the validation of the relational model [1] applied in clothing sector, thermal projection of thermal layer to 3D avatar from a sequence of thermal images (2D) [1] and thermal scanner development to improve the 2D to 3D reconstruction in a simultaneous capture.

3.1 Relational Objective-Subjective Model

The preliminary version of the relational model was exported to another more versatile programming platform such as ANACONDA (https://www.anaconda.com/), in Python programming language and codes of a new relational model integrated in a single convolutional neural network were implemented [1, 25–27].

After modifying the codes and expanding them with new functionalities, the assimilation of experimentation data of the new tests based on clothing carried out with the participating of several companies with innovative fabrics.

The new database served to train and validate the model based on simple neural networks as linear, with few layers and complex as convolutional, recursive and temporal networks.

Figure 5 shows the thermal images of one user used in the train model in different scenarios. Dark blue colors correspond to cooler areas.

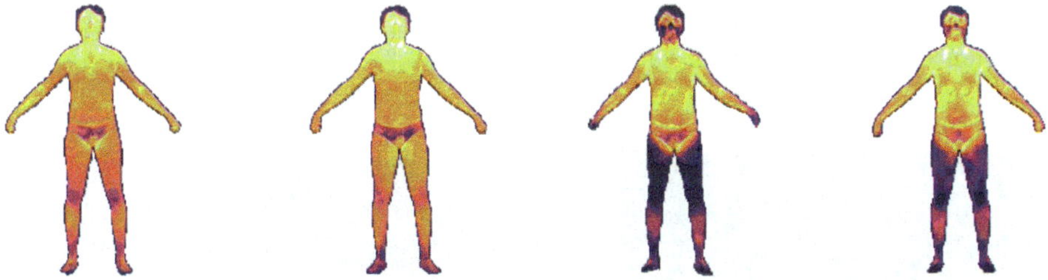

Fig. 5. User 34 at different scenarios (23 °C, 40 °C, −5 °C and −5 °C to 10 °C).

3.2 2D-3D Reconstruction

3D body reconstruction of full body was based on [16], which consists of estimating a combination of principal components by a Partial Least Square model from a set of parameters as measurements, age, body mass…etc. and then iteratively measure and modify the 3D shape within the principal components shape space until it converges [1, 16].

Figure 6 shows an example of user 15, female, 45 years old, in a sequence of different images from visible, infrared spectrum, 3D reconstruction and thermal projection.

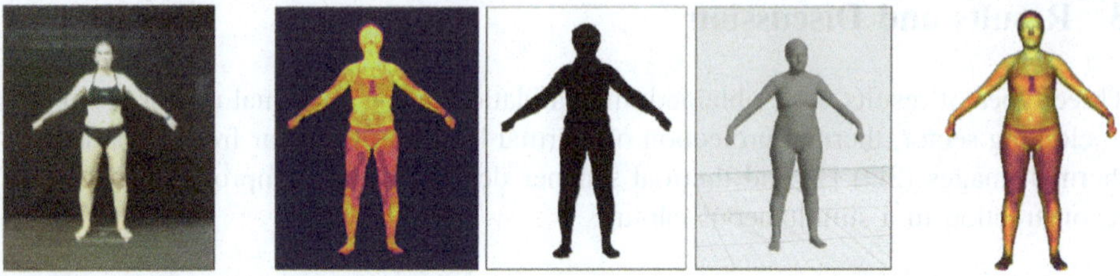

Fig. 6. Morpho-thermal avatar merging thermal and anthropometric information of a female body form infrared images.

3.3 Thermal Scanner

A thermal scanner prototype was developed to improve the thermal acquisition and reconstruction from 2D to 3D. Four thermal cameras FLIR A35 were selected to cover the total body with a good thermal and spatial resolution and integrable in a modular device.

These cameras were characterized, calibrated and synchronized to make a 3D reconstruction in a thermal avatar based on the anthropometry databases, as in the visible spectral. The synchronism trigger signal and its frequency, were conditioned by the scanner's signal in visible spectrum, in order to ensure its compatibility and future integration.

The cameras were distributed forming a 3 m^2 and the calibration was performed again through the intermediate space. Figure 7 shows the configuration of the cameras and their display after screen calibration.

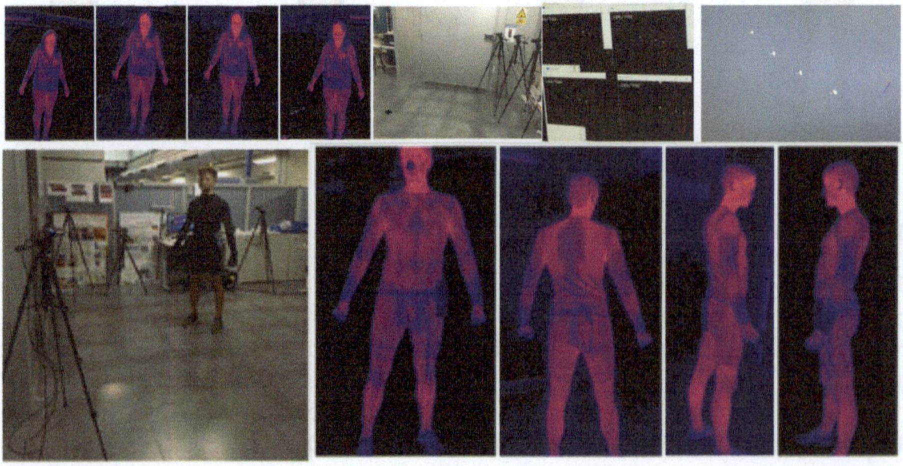

Fig. 7. Upper images: Cameras synchronization and spatial calibration. Down: Thermal scanner prototype.

4 Conclusion and Further Work

The main objective of this research work was predicting the thermal comfort from the user perception and validate the model and the thermotype concept. Additionally, the body scanning system MOTUS 4D has been implemented [17] by IBV. This system captures the 3D body surface in movement with an ultra-high-speed rate (90–180 frames per second). The processing software automatically provides continuous, watertight and intra-frame homologous meshes, including the rigging skeleton.

After the thermal scanner prototype, it is considered feasible to integrate a thermal layer in the MOVE 4D scanner through the projection obtained from a thermographic camera system in the current modules or in parallel (Fig. 8).

The next step, over the next few years, would be to integrate the infrared system into MOVE4D and its statistical correction with mathematical models.

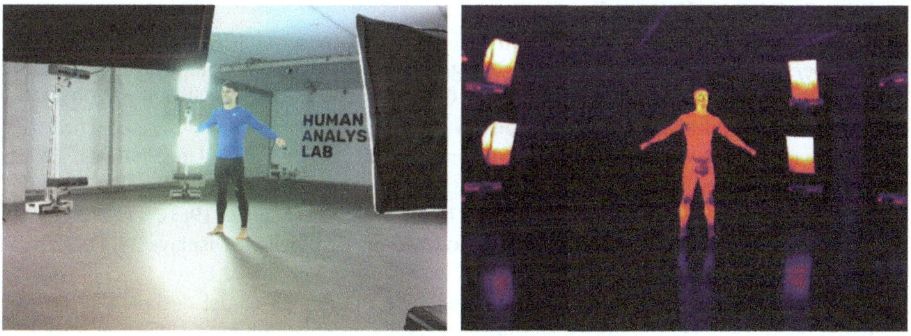

Fig. 8. Left: *Move 4D* at Human Analysis Lab (HAL) in the Institute of Biomechanics of Valencia. Right: Thermography at HAL with camera FLIR T650SC.

Ergonomics in clothing is also very important in human factors, fit assessment, and support to textile industry. Anthropometric department at IBV has developed a technology for the accurate and automated creation of avatars (closed homologous surfaces representing the human body shape equipped with a skeleton) from 3D raw scans. Since it is a fully automated process it can be applied massively to large 3D scan databases. The proposed thermal comfort models obtained in this study try to unify these two well-differentiated technical bases, thermal information with the 3D/4D models and image processing in movement applied to clothing industries.

Acknowledgements. This study is supported under the frame of the project IMDEEA/2019/14, financed by the IVACE within the framework of the program of grants directed to technological centers of the Valencian Community for the development of non-economic R & D projects carried out in cooperation with companies for the financial year 2018, co-financed by the European Development Fund Regional (ERDF) in a percentage of 50% through the ERDF Operational Program of the Comunitat Valenciana 2014–2020.

The IBV would like to thank companies such as MLS ELEBE 1992 S.L. (Lurbel) and Headway Solutions for their participation in providing samples to test and evaluate.

Conflict of Interest. The authors declare that there are no conflicts of interest.

References

1. Latorre-Sánchez, C., Soler, A., Parrilla, E., Ballester, A., Laparra-Hernández, J., Solaz, J.: Digital human updated: merging the thermal layers with the 3D anthropometric model. In: Di Nicolantonio, M., Rossi, E., Alexander, T. (eds.) Advances in Additive Manufacturing, Modeling Systems and 3D Prototyping, AHFE 2019. Advances in Intelligent Systems and Computing, vol. 975. Springer, Cham (2020)
2. M'Sellem, H., Alkama, D., Labidi, F.: An assessment of thermal comfort and users' "perceptions" in office buildings - case of arid areas with hot and dry climate. Energy Procedia **74**, 243–250 (2015)
3. García, A.C., Porcar, R., Bataller, A., Page, A., Martínez, A., González, J.C.: Functionality and personalization. Importance of design aspects and methodological approach. In: Presented at the World Congress on Mass Customization and Personalization (2001)
4. Fournet, D.: Skin temperature variations in the cold (2013)
5. Minkina, W., Dudzik, S.: Infrared Thermography. Errors and Uncertainties. Wiley, Chichester (2009)
6. Daanen, H.M., van de Water, G.J.: Whole body scanners. Displays **19**(3), 111–120 (1998)
7. Daanen, H.A., Ter Haar, F.B.: 3D whole body scanners revisited. Displays **34**(4), 270–275 (2013)
8. Chao, H., Luximon, A., Yeung, K.W.: Functional 3D human model design: a pilot study based on surface anthropometry and infrared thermography. Comput. Aided Des. Appl. **12** (4), 475–484 (2015)
9. Domina, T., Kinnicutt, P., MacGillivray, M.: Thermal pattern variations analyzed using 2D/3D mapping techniques among females. J. Text. Appar. Technol. Manag. **7**(1) (2011)
10. Chromy, A., Klima, O.: A 3D scan model and thermal image data fusion algorithms for 3D thermography in medicine. J. Healthc. Eng. **2017** (2017). 9 pages
11. Grubišić, I.: Medical 3D thermography system. Periodicum Biologorum **113**(4), 401–406 (2011)
12. Heuberger, R., Kinnicutt, P., Domina, T.: The relationship between thermal imaging and waist circumference in young adults. Health **4**(12), 1485 (2012)
13. Skala, K., Lipić, T., Sović, I., Grubišić, I.: Dynamic thermal models for human body dissipation. Periodicum Biologorum **117**(1), 167–176 (2015)
14. Ballester, A., Piérola, A., Parrilla, E., Uriel J., Ruescas, A., Perez, C., Durá, J.V., Alemany, S.: 3D human models from 1D, 2D & 3D inputs: reliability and compatibility of body measurements. In: Presented at the 9th International Conference on 3D Body Scanning Technologies, Lugano, Switzerland (2018)
15. Ballester, A., Parrilla, E., Piérola, A., Uriel, J., Perez, C., Piqueras, P., Alemany, S.: Data-driven three-dimensional reconstruction of human bodies using a mobile phone app. Int. J. Digit. Hum. **1**(4), 361–388 (2016)
16. Ballester, A., Parrilla, E., Vivas, J.A., Piérola, A., Uriel J., Puigcerver, S.A., Alemany, S.: Low-cost data-driven 3D reconstruction and its applications. In: Presented at the 6th International Conference on 3D Body Scanning Technologies, Lugano, Switzerland (2015)

17. ISO 9920:2007. Ergonomics of the thermal environment - Estimation of thermal insulation and water vapour resistance of a clothing ensemble
18. Parrilla, E., Ballester, A., Parra, P., Ruescas, A., Uriel, J., Garrido, D., Alemany, S.: MOVE 4D: accurate high-speed 3D body models in motion. In: Proceedings of 3D BODY TECH 2019, Lugano, Switzerland, 22–23 October 2019, pp. 30–32 (2019). https://doi.org/10.15221/19.030
19. Faul, F., Erdfelder, E., Lang, A.-G., Buchner, A.: G*Power 3: a flexible statistical power analysis program for the social, behavioral, and biomedical sciences. Behav. Res. Methods **39**, 175–191 (2007)
20. ISO 10551:2002. Ergonomics of the thermal environment – Assessment of the influence of the thermal environment using subjective judgement scales
21. Carroll, D., Finn, C., Judge, B., Gill, S., Sawyer, J.: A comparison of measurements from a temporal artery thermometer and a pulmonary artery catheter thermistor [abstract]. Am. J. Crit. Care **13**(3), 258 (2004)
22. Bouzas, J.: Circadian and gender differences in skin temperature in militaries by thermography. Infrared Phys. Technol. **71**, 322–328 (2015)
23. Kaiser, H.F.: The varimax criterion for analytic rotation in factor analysis. Psychometrika **23**, 187–200 (1958)
24. Eliasziw M.: Statistical methodology for the concurrent assessment of interrater and intrarater reliability: using goniometric measurements as an example. Phys. Ther. **74,** 777–788 (1994)
25. Schölkopf, B., Burges, J.C., Smola, A.: Advances in Kernel Methods: Support Vector Learning. MIT Press, Cambridge (1999)
26. ASHRAE Handbook Fundamentals. SI Edition, pp. 8.1–8.29 (2005)
27. Opitz, D., Maclin, R.: Popular ensemble methods: an empirical study. J. Artif. Intell. Res. **11** (11), 169–198 (1999)
28. Altman, N.S.: An introduction to kernel and nearest-neighbor nonparametric regression. Am. Stat. **46**(3), 175–185 (1992)
29. WO2014/053198. Simultaneous alignment and modeling of articulated 3D shapes
30. US20170053422. Mobile device human body scanning and 3d model creation and analysis
31. Allen, B., Curless, B., Popović, Z.: The space of human body shapes: reconstruction and parameterization from range scans. ACM Trans. Graph. (TOG) **22**(3), 587–594 (2003)
32. Alemany, S., Ballester, A., Parrilla, E., Pierola, A., Uriel, J., Nacher, B., Remon, A., Ruescas, A., Durá-Gil, J.V., Piqueras, P.: Solves, C.: 3D body modelling and applications. In: Presented at the International Ergonomics Association (2018)
33. Baydal-Bertomeu, J.M., Durá-Gil, J.V., Piérola-Orcero, Ana, Bernabé, E.P., Ballester, A., Alemany-Munt, A.: PCA-based bio-motion generator to synthesize new patterns of human running. PeerJ Comput. Sci. **2**, e102 (2016)

The Virtualization of the Fashion Product

Luís Ricardo Santos(✉), Gianni Montagna, and Maria João Pereira Neto

CIAUD - Lisbon School of Architecture, University of Lisbon, Rua Sá Nogueira, Alto da Ajuda, 1349-055 Lisbon, Portugal
luisricardosantos@gmail.com

Abstract. It was since the introduction of digital technology, at the end of XX century, that the social and cultural panorama begun to change exponentially. The ways used to communicate, to produce and to design shifted drastically, driven by a technology that doubles its processing power each eighteen months and that becomes increasingly monetarily accessible. With the dawn of the XXI century, this technology becomes ubiquitous and start transform behaviours, shaping the society in ways never seen before. With the rise of smartphones, often with as much computing power as a PC (Personal Computer), it became accessible to everyone to explore the internet—the global network of things—, in an easy, affordable and uninterrupted way. This integration of technology in the daily life, generated a duality in existence, splitting the individual into two kinds of simultaneous existence: the actual existence and the virtual existence.

Individuals could now explore a new kind of existence—the digital self—which gradually became natural and integrated with sociocultural behaviours. This virtualization of the individual generates shifts in the ways of life, that globally impacts the cultural and economic structure. New needs of expression emerge and with them, new products arise.

This digital culture that is solidified during the XXI century, not only virtualizes the individual but also all the images and products surrounding him. New ways of consuming emerge, and the acquisition of goods is increasingly made with the use of screens that portrait a virtualized product. The virtualization of products is a natural evolution of our society that is silently changing the way we create and consume. The study of these subjects is essential to understand the present and the future of design practices. The process of creation gradually evolves, with technology becoming intrinsic in it. The way we create and produce products and information becomes deeply connected with digital technology, and AI (Artificial Intelligence) becomes a major mediator of our daily lives. With the rise of an informational society, flooded by information coming from everyone from everywhere, things became constantly curated and sorted in our image, generating specific views of the world.

This paper is part of a PhD research on the influence of the digital tools and virtual environments in human creativity and concept development, in the specific case of fashion design products. The virtual environments induce a drastic change in the way research, concept development and project communication are made, changing forever the practice of design. The full understanding of the creative potential of digital languages is a real necessity to those who want to be prepared to the uncertain future of fashion design. With this paper we intend to comprehend how the fashion product has been virtualized,

G. Di Bucchianico et al. (Eds.): AHFE 2020, AISC 1202, pp. 820–830, 2020.
https://doi.org/10.1007/978-3-030-51194-4_106

and what is the impact of this virtualization, in society, in fashion design and in the individual.

Keywords: Virtualization · Society · Culture · Fashion · Design

1 Introduction

At the beginning of the 21st century, three-dimensional processes are introduced in the fashion industry, which are gradually being integrated into the lifecycle of products. Although 3D technology was already available at the end of the 20th century, difficulties in representing and constructing garments was a challenge only overcome by recent developments in software and hardware (mostly due to the gaming and film industries [1, 2]. The virtual representation of clothes (stretch, drape and gravity) is now portrayed in a more realistic way [3], allowing an easiest integration in the fashion business.

The emergence of these digital environments impacts not only the work processes but also the essence of fashion products [4], which acquire new characteristics and new forms of existence. A piece of clothing digitally created can originate a physical version of itself or can exist only in the digital environment. This way, garments can now exist in both worlds simultaneously - the actual and the virtual - sharing a feeling of symbiotic reality.

With the dematerialization of the fashion product, new forms of communication emerge, allowing a fast dissemination of products digitally [2] and in real time [5, 6]. Digital strategies and events start to appear and complement the more expensive physical approaches like photoshoots and catwalk shows [3].

An example of this approach can be seen in Louis Vuitton's Spring-Summer 2016 campaign, which shows characters from the "Final Fantasy" video game wearing 3D simulations of the brand's pieces (Fig. 1).

Fig. 1. Louis Vuitton Spring-Summer 2016 campaign. In https://www.slashgear.com/final-fantasy-xiii-character-is-louis-vuittons-newest-model-29420123/ [10 July 2018]

2 The Digital Product

2.1 The 3D Digital Environment

Designing virtually not only has huge potential for aesthetics and immersive experiences but can also help mitigate overproduction and over-consumption [3]. Digitally created products reduce the necessity of stocks and addresses made to order business models, allowing a bigger variety of options and explorations. These models are more sustainable in every aspect, reducing environmental impact and promoting social equity [7].

Business strategies rooted in digital products also leverage creativity and market variety by saving time and resources and thus reducing the economic risk of launching a fashion business. By not having the costs associated with physical store and products, designers can risk new solutions and establish new partnerships in the digital environment [8]. Through this, the digitalization of the fashion product has the potential to widen the fashion market, allowing the emergence of small creators with niche visions and markets. These immaterial products will also allow new ways of fashion design to emerge because clothes don't need to exist in the actual world [9].

It's this capacity to create and customize products without the dependency of inventory or stocks that allowed the growth of new business models in the digital economy [1]. According to Chavie Lieber from "Business of Fashion", we are at the dawn of a market in virtual fashion. Virtual products will allow users to keep up with trends in a more affordable and sustainable way.

The virtual product can interact with the human being in multiple ways:

I. *Augmented reality*
To let designers and consumers to virtually see (or "try on") products in a virtual mannequin or personal avatar [1].
One example of this kind of technological innovation is the company "Change of Paradigm" which innovates the traditional commerce mechanisms by using immersive and interactive experiences based solely in 3D simulated content (Fig. 2).

Fig. 2. The company Change of Paradigm promotes the 3D simulation as the future of fashion products promotion. In http://changeofparadigm.com [10 July 2018]

Another example is the "Total Appearance Capture (TAC) Ecosystem" from "X-Rite Pantone" that blurs the boundary between actual reality and virtual reality. The "Virtual Light Booth" (Fig. 3) is an immersive 3D system that shows virtual objects at their real size and in an extremely realistic way.

Fig. 3. Photo of the "Virtual Light Booth" from "X-Rite Pantone" company, taken by the author at Munich Fabric Start 2018.

II. *Virtual collections*
Pieces created to be worn in the virtual world.
One example of virtual collections is Carlings, a Scandinavian fashion brand that sold, in November 2019, 19 3D virtual pieces that are digitally applied to the shopper's photos (Fig. 4).

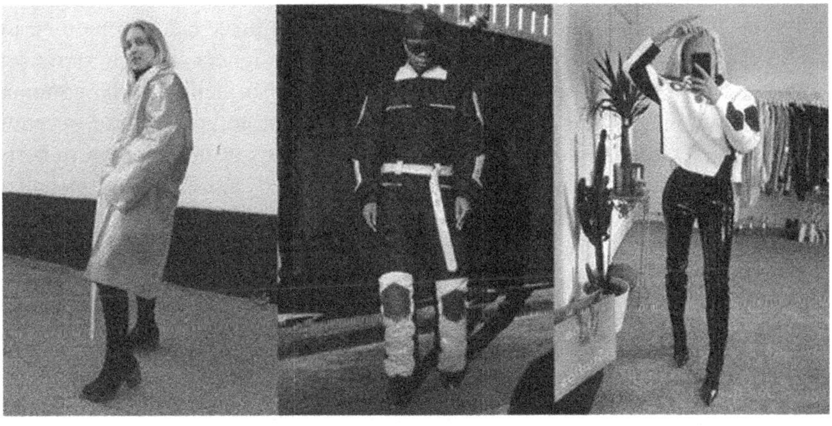

Fig. 4. Shoppers wearing looks from Carlings' virtual collection. In https://digitalcollection.carlings.com/ [14 February 2020]

III. *Virtual models*
A CGI (computer-generated imagery) character with a lifestyle and a digital presence in social media, acting as an influencer.
One example of a CGI influencer is Shudu, a virtual model with 100,000 followers (Fig. 5).

Fig. 5. Shudu, a CGI influencer, wearing a designer gown. https://www.instagram.com/shudu.gram/ [14 February 2020]

From a productive point of view, a 3D model allows a closer relationship with the construction techniques of the garments since the virtual piece can be modified in real time. This processed called "virtual prototyping" aids designers and pattern makers by presenting accurate digital simulations of products. It also facilitates the communication process and prevents misunderstandings during the prototyping phase, speeding up the process [1, 5] and enabling the shape, fit and graphics of the garment to be easily seen and corrected at any desired angle [2]. Specialized and advanced visualization tools integrated in three-dimensional software have the ability to monitor the mechanical properties of the garment, allowing a constant control of the degree of comfort and fit of the garment. There is also the possibility to test the garment in various bodies, postures or stages of an individual's life, enabling the creation of more informed and sustainable design responses [5]. By recurring to these tools, designers can produce more relevant fashion design solutions, targeted to specific markets and consumers [8].

One example of these processes is the CLO system which encompasses a 3D virtual prototyping software—CLO (Fig. 6), a collaborative working platform that facilitates communication between the various stakeholders during the product life cycle—CLOSET (Fig. 7) and a virtual fitting room that allows the end user to try on the different pieces of clothing created digitally (being able to compare different sizes and finishes in an avatar generated by him and with his body measurements)—Benefit by CLO (Fig. 8).

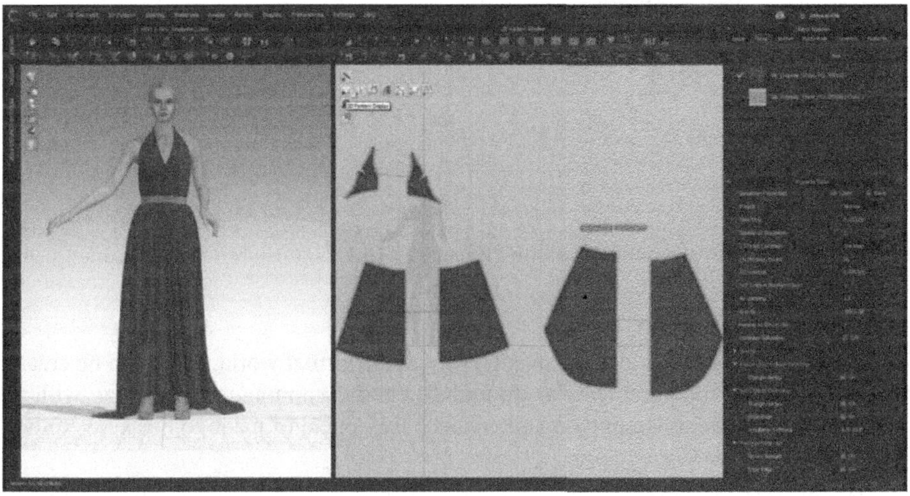

Fig. 6. Screenshot of the 3D fashion software "CLO". https://www.clo3d.com [26 June 2019]

Fig. 7. Screenshot of the online platform "CLOSET". https://www.clo-set.com [26 June 2019]

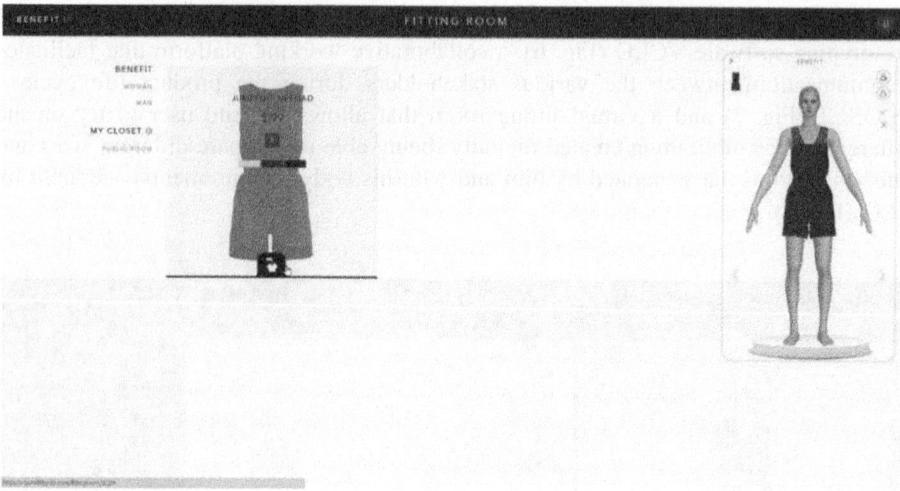

Fig. 8. Screenshot of the online platform "Benefit by CLO". https://benefitbyclo.com [26 June 2019]

The use of personal avatars is an entryway to the virtual world. They can be created by using personal measurements or through 3D body scanning, a technology with the potential to implement, strengthen and enhance the appeal of made to measure clothing [5, 10].

By using 3D body scanning technology, it becomes easier, quicker and more precise to extract body measurements. With the gathered data, fashion designers will be able to target specific niche markets and to create mass customisation business models [11].

2.2 Rapid Prototyping

One of the possibilities of three-dimensional product creation is its easy transposition into rapid prototyping processes. The spread of these technologies is making them increasingly accessible to designers and users, promoting a radical change in the consumer market. These new technological approaches allow the production of small series of customised products where changes in each unit produced have no associated economic penalty [12]. This customized response from production will increase the sustainability of the product by levering the emotional value these fashion items represent to the consumer [13], thus extending their life span.

In a near future these machines will be a commonplace and will be available in shopping centres and other locations where products can be made-to-order [13]. This whole process will inevitably become more demographic, agile, easily communicated, shared, manipulated, downloadable, and replicable in 2D, 3D and ultimately, with shape-changing potential, 4D [14].

2.3 Artificial Intelligence (AI)

Artificial Intelligence (AI) is increasingly imposing itself in the daily life of contemporary societies. Within this theme, Culturgest Lisbon (Portugal) organized in 2019 a series of conferences. These were some of the main points noted by the speakers:

- Artificial intelligence increases human intelligence and will never replace it. The tendency will be to increasingly happen a combination between human and AI, in order to increase our natural capabilities.
- AI can automate tasks and help make complex decisions. By doing this it carries inherent risks of misuse, loss of control or overuse and the consequent numbing of human skills.
- AI can generate fictitious worlds by creating non-real content. This affects the way we experience the actual world and places the individual in a utopic and disconnected existence from current reality.
- Current AI fakes human creativity quite well, but simply through recombinative processes.

3 The Digital Product in the Fashion Industry

During the workshop "What's Digital About Fashion Design?" hosted by the London College of Fashion, University of the Arts London in 2015, participants identified the three major needs in applying digital environments in the fashion product life cycle [8]:

I. *The challenge imposed by tradition*
The fashion industry is, of all the creative industries, the one that has made the least progress in its processes through digitalization. Traditional models are still imposed, and the use of new technologies continues to focus on product promotion and sales. The insertion of new technologies at an earlier stage in the process has the potential to infer changes not only in practice but also in existing business models. This integration will allow interdisciplinary and intersectoral support and learning.

II. *The design process*
Closer collaboration between fashion, technology and industry/production is needed. Therefore, in a first stage it is necessary to understand how this connection can respond to the problems current designers face in the product life cycle.

III. *New models*
There is a need for the gathering and dissemination of new projectual models used by contemporary fashion designers that use digital environments in their professional practices.

4 The Digital Product in the Fashion Design Teaching

In an educational context, three-dimensional representation allows a detailed view that students rarely capture through manual drawing. It is also a way to easily understand the interconnection between 2D and 3D of clothing (design vs. pattern making).

Since 2009 the Amsterdam Fashion Institute has been working with 3D software to carry out the virtual prototyping of 3rd and 4th grade students' collections. At first, the 3D simulation was only used for the final representation of the products, but, in time, the creative potential of this learning tool was understood when integrated in the process from the conceptual stage. Its use allowed a more fluid communication between the 3D representation of the garment and its pattern making, also facilitating the (visual) dialogue between student and teacher, which takes place in a more precise and defined way. With the use of this tool, students were able to take the experimental process further and were more incisive in choosing the pieces to prototype physically [2].

Also, at the University of Applied Sciences Hochschule Niederrhein (Germany) it was implemented 3D visualization at the undergraduate (BA) and master (MA) levels. There was a perception that the design process became faster and simpler because the software shows a model that is very close to the final product. The results end up being more experimental and bolder because the technical component is facilitated by the software, and there is no need for in-depth knowledge of the productive area (sewing and pattern making).

3D software is becoming an important tool for the fashion industry and, as such, it should be inserted in school curriculum. Like said by Taylor, A. & Unver, E., "*Current research has established that fashion learners need to be introduced to 3D digital technology quickly. They need to know -how to think and communicate in 3D, how to design using 3D CAD, how to link to 3D prototyping- 3D manufacture, and how to apply 3D/virtual marketing tools and 3D retail solutions to todays' technology focused apparel industries. [...] Encouraging undergraduate 3D awareness through experimentation with contemporary 3D digital design tools like portable 3D scan technology, is vital for digital innovation to emerge*" [10].

5 Conclusions

It is inevitable the proliferation of digital ambiences in the lifecycle of the fashion product. Humanity is now living in a hybrid existence between the virtual and the actual world and consequently all our cultural activities will be transformed, evolving into new kinds of expressions and existences.

The use of virtual products has many advantages and can help the fashion industry to find new business models that can respond to new social and cultural needs. It is therefore important to understand in what ways can these virtual environments support the design process in creating innovative products. These new approaches can boost creativity and sustainability in unprecedented ways, levering the emergence of real innovative solutions. This digitalization of the product also allows the intersection with other technologies such as the Artificial Intelligence, that can assist the creative process, and the Rapid Prototyping, that can help to automatically physically create fashion products.

The virtualization of the creative process can also facilitate the connection with the fashion market and industry and help both designers and students in addressing to real necessities using realistic resources and techniques.

The use of 3D laser scanning and 3D technological tools can provide real learning experiences for understanding and designing, anything imaginable on, in or around the human body [10]. It is therefore essential to understand how a new projectual posture using these new technologies can enhance learning and professional practice.

For Volino P., *"The next step will be to allow a more natural integration of the human gesture in the 3D creation process through the automatic recognition of shapes from sketches made by fashion designers [...] dressing a mannequin directly in the virtual world"* [5].

Acknowledgment. The authors wish to thank Research Centre for Architecture, Urbanism and Design (CIAUD) of ULisboa.

References

1. Spahiu, T., Shehi, E., Piperi, E.: Advanced CAD/CAM systems for garment design and simulation. In: 6th International Conference of Textile (2014)
2. Siersema, I.: The influence of 3D simulation technology on the fashion design process and the consequences for higher education. In: Proceedings of Digital Fashion Conference 2015, pp. 9–17. Digital Fashion Society, Seoul (2015)
3. WGSN: Strategy 2021 - Big Ideas Textiles (2019). www.wgsn.com
4. Treadaway, C.: Digital creativity: the impact of digital imaging technology on the creative practice of printed textile and surface pattern design. In: TEXTILE 2:3, pp. 256–273. Routledge (2004)
5. Volino, P., Cordier, F., Magnenat-Thalmann, N.: From early virtual garment simulation to interactive fashion design. Comput. Aided Des. **37**, 593–608 (2005)
6. Trivedi, V.: Int. J. Recent Res. Aspects (IJRRA) **2**(4), 25–29 (2014)
7. Papahristou, E., Bilalis, N.: Should the fashion industry confront the sustainability challenge with 3D prototyping technology? Int. J. Sustain. Eng. **10**(4–5), 207–214 (2017)
8. Black, S., Edwards, J., Miller, G.: What's Digital About Fashion Design? (2015). http://ualresearchonline.arts.ac.uk/9790/1/WHAT%27S%20DIGITAL%20ABOUT%20FASHION%20DESIGN.pdf
9. Lieber, C.: Would You Buy a Virtual Pair of Nikes? (2019). https://www.businessoffashion.com/articles/fashion-tech/virtual-fashion-digital-products
10. Taylor, A., Unver, E.: An experimental study to test a 3D laser Scanner for body measurement and 3D virtual garment design in Fashion education. In: Wearable Futures – Hybrid Culture in the Design and Development of Soft Technology, pp. 1–14. University of Wales (2005)

11. Apeagyei, P.: Application of 3D body scanning technology to human measurement for clothing fit. JDCTA **4**, 58–68 (2010)
12. Anderson, C.: Makers: The New Industrial Revolution. Random House Business Books, London ([2012] 2013)
13. Campbell, R.I., Bourell, D., Gibson, I.: Additive manufacturing - rapid prototyping comes of age. Rapid Prototyp. J. **18**(4), 255–258 (2012)
14. Treadaway, C.: Crafting textiles in the digital age - printed textiles. In: Nimkulrat, N., Kane, F., Walton, K. (eds.) Crafting Textiles in the Digital Age, pp. 17–33. Bloomsbury Academic, London (2016)

Effect of the Size of the Covered Area of the Chest Surface on Pressure Sensation

Tamaki Mitsuno(✉) and Tomomi Aruga

Faculty of Education, Shinshu University, 6-Ro, Nishinagano, Nagano, Japan
{mitsuno, 6e1602h}@shinshu-u.ac.jp

Abstract. Compression underwear improves the body's shape, but its effects on human health and stress remain controversial. Our goal is to develop underwear that applies a comfortable pressure sensation. The participants were eight women who evaluated the perceptible clothing pressure on the body surface in four planes (4th–8th ribs) while wearing a 2.5 cm wide elastic band. Their pressure perceptions were compared between two conditions: when the elastic band was tightened based on the feeling caused by adjusting the length in each plane singly and in all four planes simultaneously. When the clothing pressure was 3–4 hPa, they reported a perfect pressure sensation. When they were constricted on each plane singly, there was a positive correlation between the pressure sensation and the clothing pressure. When the size of covered area increased, the relation between the pressure sensation and the compressed area size was positively correlated ($P < 0.001$).

Keywords: Pressure sensation · Compression area · Clothing pressure · Elastic band

1 Introduction

The underwear worn in cold weather in Japan touches the skin directly. It therefore absorbs non-felt transpiration, sweat, and sebum secreted by the human body, thereby playing a role in maintaining skin cleanliness. It is also desirable that the underwear be designed such that it does not interfere with human body movements [1]. Underwear does not disturb movement if there is sufficient space, but during the winter, too much space results in the escape of radiated body heat, which makes it difficult to stay warm. Today, women's underwear with high fitting performance is commercially available. According to Ito et al., the pressure sense (e.g., in the chest, back, abdomen, and upper arm has a positive correlation with clothing pressure of the underwear [2]. Their report, however, showed that the pressure produced by the underwear was almost 1 hPa, which does not indicate good fitting performance. Therefore, our goal was to provide data for comfortable underwear that exerts optimal pressure sense. We thus focused our attention on women's chests, which have a complicated shape, to clarify the relations between pressure sense and the areas covered by the underwear.

T. Aruga—Formerly Faculty of Education, Shinshu University.

G. Di Bucchianico et al. (Eds.): AHFE 2020, AISC 1202, pp. 831–840, 2020.
https://doi.org/10.1007/978-3-030-51194-4_107

2 Experiment

2.1 Participants and Samples

The participants of the present study were eight young Japanese females of their twenties whose physical constitutions are shown in Table 1. The bodies of the participants ranged widely from thin to obese. All participants were non-smokers, with no history of systemic disease. They were explained their constitution was compared with Japanese female average of 20's whose height was 1.59 m, weight was 50.8 ± 5.9 kg, BMI was 20.1, so they were a bit of thick than the Japanese average. The study protocol for participants was approved by the ethics committees at Shinshu University and the Institute of Human and Social Sciences (permitted institutional review board number 173).

Table 1. Physical constitution of participants

Items	Age (Year)	Height (m)	Weight (kg)	Girth (cm) A	B	C	D	BMI (kg/m^2)
Average	21.9	1.58	52.5	83.4	84.6	85.1	75.2	21.0
SD	0.7	0.04	7.4	3.8	5.0	6.1	5.4	2.9

A typical example of the experimental band is shown in Fig. 1. The elastic band width was 2.5 cm; thickness 1.31 ± 0.03 mm, measured with a dial thickness gauge (Seimitsu Kogyo, Tokyo, Japan); and weight 0.08 ± 0.01 gf/cm, which was examined in a tensile test under conditions of a grab width of 100 mm and strain rate of 50 mm/min. The samples were strained to 60 mm (60%) and returned promptly by a tensile strength tester (Autograph S-100; Shimadzu, Tokyo, Japan). The force was 27.6 ± 5.6 gf/cm in 60% strain, which was easy to strengthen and shrink.

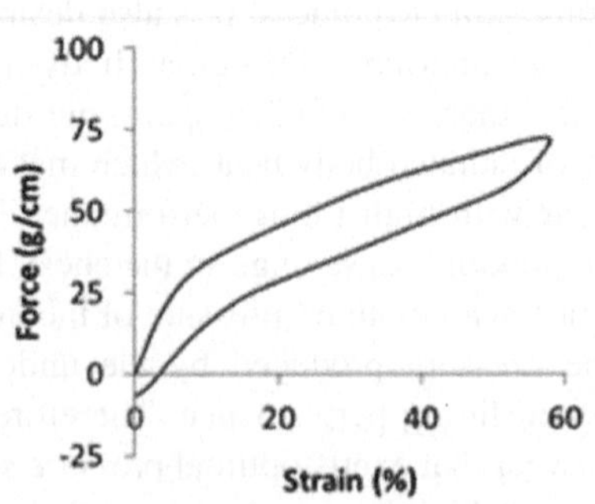

Fig. 1. Relationship between force and strain of experimental band

2.2 Experimental Schedule

The participants slept for 7 h during the night prior to the experiment and ate 2 h before the measurement. After entering a room with artificial climate control (temperature 24.5° ± 0.3 °C, relative humidity 50% ± 0.6%, air current 0.12 ± 0.17 m/s, illumination 829 ± 285 lx), the clothing pressure and pressure sense were measured for 1 h on the chest of each participant. The participants were able to evaluate a perfect pressure sense without sweating or shaking because they were in a comfortable environment at a neutral temperature. The participants rested for at least 1 h in the climate room before the experiment began.

They donned experimental clothes, including shorts, T-shirt, and half pants without a brassiere (5 × 5 × 1 cm cotton pads were placed on nipples using medical tape). We then evaluated the perceptible clothing pressure on the body surface in four planes (A–D) on the 4th, 5th, 6th, and 8th ribs, in that order, while they wore a 2.5 cm wide elastic band. Clothing pressure at 12 time points during 10 respirations were measured when the elastic band was tightened based on the feeling caused by adjusting the length in each plane singly. The nude girth, girth of the band, and band length were measured together (Experiment I). In Experiment II, while wearing the bands, participants' pressure perceptions were measured on all four planes simultaneously at the same lengths as in Experiment I. These experiments were carried out at their lower and higher body temperature phases during the same menstrual cycle (pressure sense and clothing pressure are changed by the menstrual cycle) [3, 4].

2.3 Measurement System for Clothing Pressure and Measurement Points on the Chest

Clothing pressure were measured using by a measuring system of hydrostatic pressure-balanced method at 12 points on a plane, which were shown in Fig. 2. The right figure shows the cross-section of each plane having 12 measuring points. F (B) is the cross-point between the anterior (posterior) middle line and the each girth line, whereas RA (LA) is the point of 3 cm to the right (left) of F on the each girth line, RB (LB) is the cross-point of the right (left) mammilla line and the each girth line, RC (LC) is the cross-point of the right (left) side line and the each girth line, RD (LD) is the cross-point between the right (left) shoulder line and the each girth line, and RE (LE) is the point of 3 cm to the right (left) of B on the each girth line.

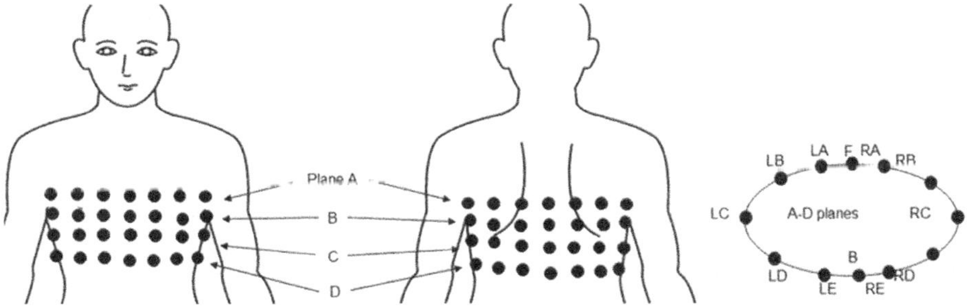

Fig. 2. Measuring points for clothing pressure

Left figure shows ventral and dorsal views and the right figure shows the cross section of each plane.

The horizontal planes from A to D planes showed on the 4th, 5th, 6th and 8th ribs in the order, and the heights of the band were decided according to JIS Z 8500:2002. Participants adjusted the length of the band to achieve "a perfect fit", corresponding to a comfortable pressure sense. The band length was fixed using a double clip exerting 5.4 gf by the experimenter. The participants prepared their posture standing in an upright position and evaluated the tightness pressure sense, and the clothing pressure generated between the skin and elastic band was then measured. First, the transducer was attached to a pole at the same height as the pressure sensor and the measuring plane, and zero pressure was confirmed. Second, the pressure sensor was inserted into the gap between the skin and band using a thin spatula that was inserted into the sensor pocket [3].

2.4 Evaluation of the Pressure Sense

Pressure sense was evaluated using a ratio scale [4]. The participants evaluated the pressure they felt when wearing the elastic band by completing a questionnaire (see Fig. 3). Before putting on the elastic band, participants estimated what they considered a "loose pressure sense" and "a perfect pressure sense." The participants were asked to

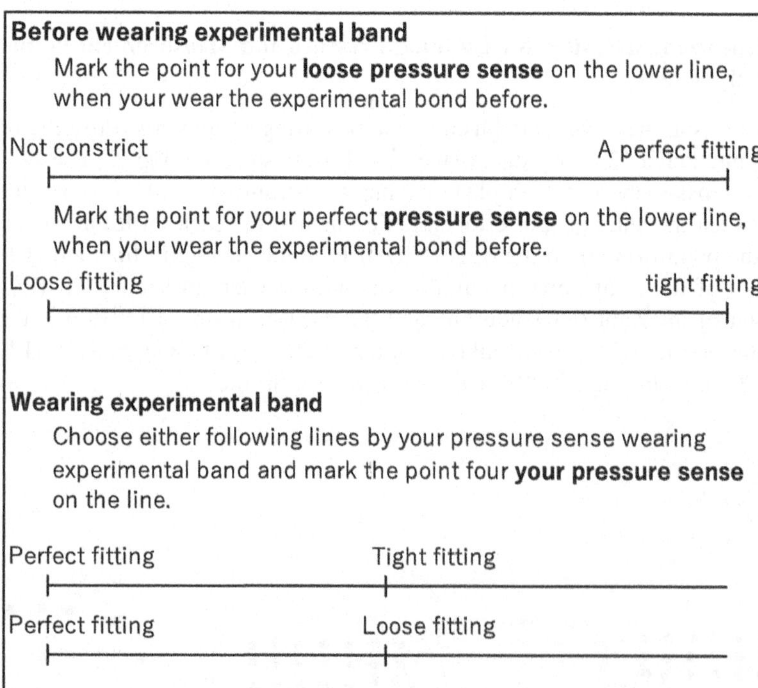

Fig. 3. Questionnaire for evaluation of pressure sense

indicate individual pressure sense under both conditions and to mark points on a line in the questionnaire corresponding to a loose pressure sense. Similarly, they were asked to mark points for a perfect pressure sense and tight pressure sense that could be used as baselines.

The participants then tightened the experimental band until they felt it fit perfectly and marked a point on a line in the questionnaire. They repeated the evaluations in Experiments I and II. The length of the line connecting the points marked by the participants to indicate the pressure senses were then measured. Numerical values for the pressure sense were obtained by supposing that the sense of no tightening scored zero and a perfect pressure sense scored 1.

2.5 Statistics

For both experiments, clothing pressure and pressure sense from it were calculated as significant differences using a paired t-test because of the large effect of individual differences. Correlation coefficients for the two experimental conditions while wearing one or more bands were calculated among nine items: perfect pressure sense; clothing pressure; band length; girth; nude girth; extension rate; compression rate on the body; constriction rate of the band; and the compressed area.

We defined some measuring data using equations. Band length (a) was measured as the length needed to constrict the band to achieve a perfect pressure sense with a single band on each plane. Girth (b) was measured by the length in each plane while wearing the band. Nude girth (c) was defined in Table 1 as the naked girth of each plane.

$$\text{Extension rate of the band} = b/a \times 100 - 100. \tag{1}$$

$$\text{Compression rate on the body} = b/c \times 100 - 100. \tag{2}$$

$$\text{Constricting rate of the band} = c/a \times 100 - 100. \tag{3}$$

$$\text{Compressed area} = 2.5 \times a. \tag{4}$$

Equation (4) was calculated on each plane with a single band in Experiment I. In Experiment II, calculations were for all four planes.

3 Results

3.1 Development of Clothing Pressure Under Two Pressure Senses While Wearing a Single Band

Development of clothing pressure with a single band under a perfect pressure sense is shown in Fig. 4 (left) and under tight pressure sense in Fig. 4 (right). Clothing pressure on the anteroposterior middle line was lower than on either side region because of the body shapes. Ventrally, the body regions on the former midline hollowed because of relief of the mammae, which rise and fall from the former midline to the mammillary line. Therefore, floating occurred, and pressure exertion was difficult. It was also because the participants were young women whose backs were indented at the vertebral column. The red fan areas seemed larger on the right than on the left. These results are summarized in Fig. 5 (left).

The clothing pressure in each plane was significantly higher with a tight pressure sense (Fig. 5, right) than with a perfect pressure sense. The correlation coefficients between the clothing pressure and its pressure sense were 0.387–0.600, all indicating a significantly positive correlation. The same relations between the clothing pressure and the pressure sense have been reported to develop at the waistband [5].

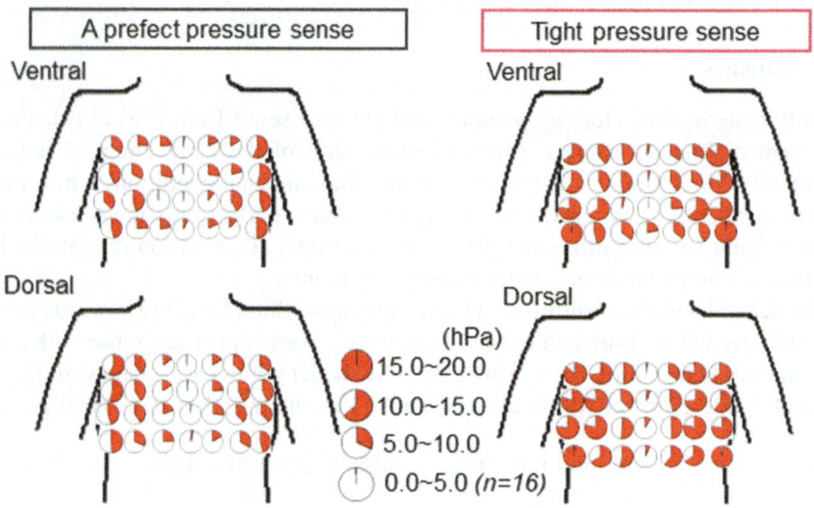

Fig. 4. Development of clothing pressure under the two types of pressure sense when wearing a single band

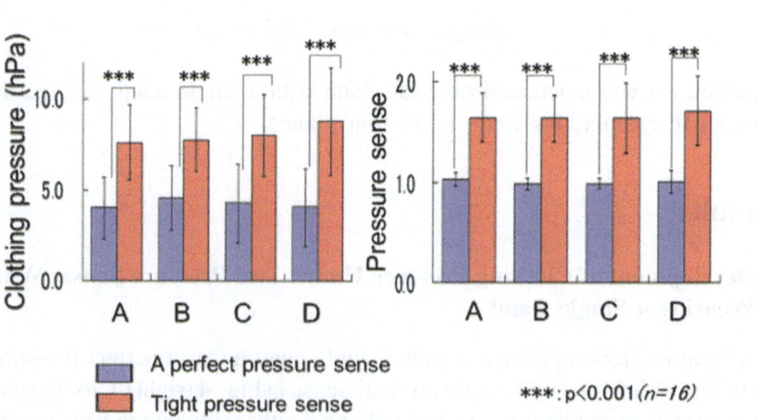

Fig. 5. Relations between clothing pressure and the pressure sense under two types of pressure sense

3.2 Development of Clothing Pressure on Two Compressed Areas and Their Pressure Sense

Development of clothing pressure on two compressed areas are shown in Fig. 6. The left figures show the presence of a single band with a perfect pressure sense (Experiment I). The right figures show the results of wearing all four bands simultaneously (Experiment II). Both conditions show almost the same development of clothing pressure (Fig. 7). The clothing pressure was 3–4 hPa, with the results in Experiment II slightly lower (but not significantly so) than those in Experiment I.

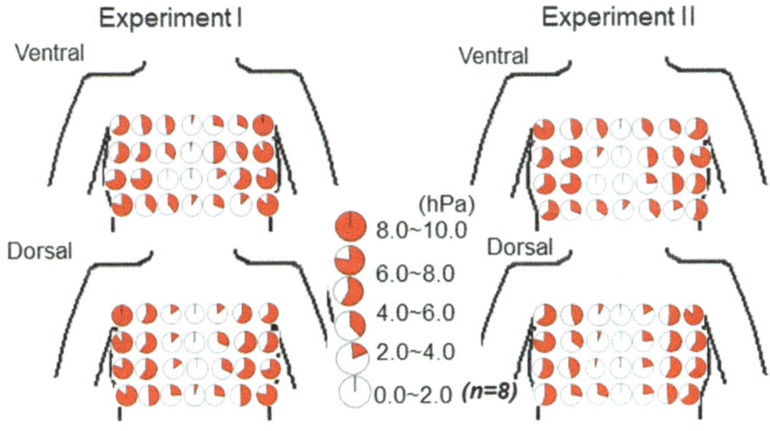

Fig. 6. Development of clothing pressure on two compressed areas

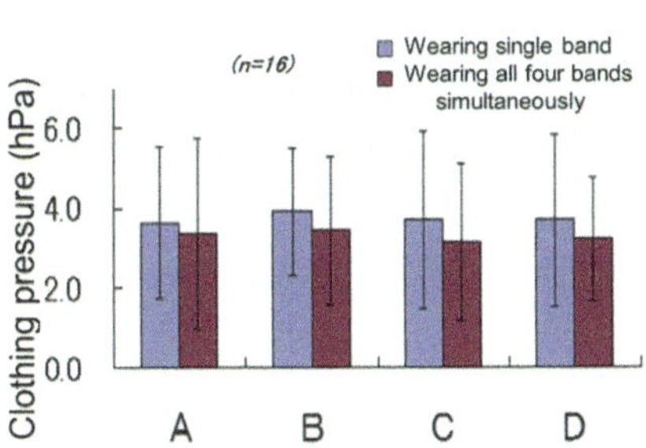

Fig. 7. Clothing pressure on two compressed areas

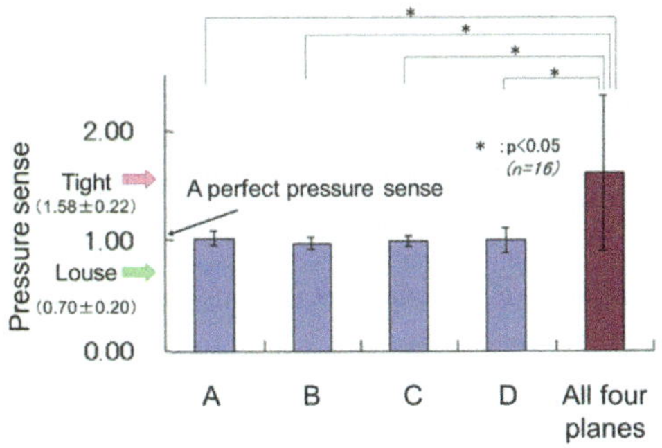

Fig. 8. Pressure sense on two compressed areas

Pressure sense for the two compressed areas is shown in Fig. 8. The participants recognized that the pressure sense in Experiment II was significantly greater than in each plane in Experiment I.

3.3 Compression Rate of the Band to the Body Surface

The compression rate of the band in the body on each plane [see Eq. (2)] is shown in Fig. 9. The compression rate on the body wearing all four bands simultaneously (Experiment II) was slightly larger than when wearing a single band (Experiment I), especially in plane A. The results of Experiment II were significantly greater than those of Experiment I. The surface tissue on planes B and C was pectoralis major muscle, whereas soft tissues were including on planes A and C. Plane C included adipose tissue of the mammae, and plane A included the armpit lymph nodes. Both planes had soft tissues, which are easily compressed, although plane A tissues were not likely to tighten. We therefore focused on planes B, C, and D and the relations between clothing pressure and pressure sense.

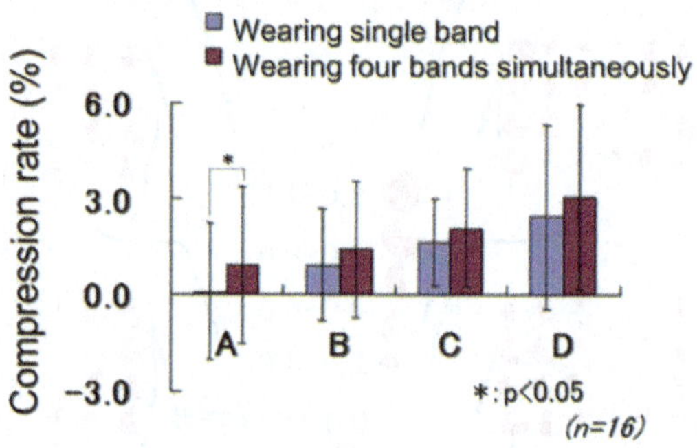

Fig. 9. Compression rates of the band to the body

3.4 Effect of the Compressed Area on Pressure Sense

We then calculated correlation coefficients between a perfect pressure sense and tight pressure sense in participants wearing a single band (condition I) and pressure sense wearing all four bands simultaneously (condition II) on planes B, C, and D for the

Table 2. Correlation coefficients between wearing a single band or four bands for a perfect pressure sensation

Items (Symbol)	Tight pressure sense with single band	Wearing four bands simultaneously
Clothing pressure	0.587 ***	0.093
Band length (a)	-0.218	0.302
Girth of wearing band (b)	-0.027	0.331
Nude girth (c)	0.132	0.350 *
Extension rate of band	0.513 **	0.054
Compression rate of the body	0.488 **	-0.092
Constricting rate of band	0.583 ***	-0.034
Compressed area (d)	-0.027	0.587 ***

*: $p < 0.05$, **: $p < 0.01$, ***: $p < 0.001$. a, b, c symbols are shown in 2.5 for detail.

items in Table 2. For condition I, significant positive correlations were obtained with not only clothing pressure but also the extension rate of the band/compression rate on the body and the constriction rate of the band. In contrast, significant positive correlations were lower for two items, nude girth and the compressed area. For condition I, participants recognized tightness when the band was stretched—the body was compressed and clothing pressure increased. For condition II, however, the participants recognized tightness when the nude length and compression area increased. Hence, the effects of items on the perfect pressure sense changed from clothing pressure to the compressed area.

On plane A, a multiple correlation coefficient was calculated between the perfect pressure sense and band length (a), girth while wearing a band (b), and compressed area (d). Equation (5) addresses these relations (multiple correlation coefficient 0.701), as follows.

$$\text{Perfect pressure sense} = -0.07\text{a} + 0.11\text{b} + 0.01\text{d} - 2.95 \tag{5}$$

The pressure sense on plane A was highly sensitized when the band length was shorter, and girth and the compressed area were larger, while wearing a band. Plane A thus does not cause a tightened pressure sense in all four planes.

4 Discussion

We aimed to design underwear using its clothing pressure and expected pressure sense as indices. In a previous study, the participants recognized their pressure sense as significantly tight (3–13 hPa) with higher clothing pressure. This phenomenon was the same as in our previous study [6]. The pressure sense on plane A has a significant multiple correlation explained by three items: band length (a), girth (b), and the compressed area (d). The participants recognized greater tightness on plane A than other planes because plane A had soft tissues (e.g., armpit lymph nodes) that are easily compressed and so not likely to tighten. Therefore plane A should not be tightened.

An area with a pressure sense of 3–4 hPa is affected by compression. If the distribution densities of the pressure sensor on four planes were approximately equal, the number of pressure sensors that accept a stimulus by tightening would increase in all four planes compared with that induced by a single band. Clothing pressure on planes B, C, and D under condition I was 3–13 hPa, whereas condition II had a low clothing pressure (3–4 hPa). In the same compressed area, the pressure sense increased because of increasing clothing pressure. Although condition II had almost the same clothing pressure, the pressure sense increased because of an increased compressed area caused by an increased number of stimulated pressure sensors. Whereas Ito reported that comfortable clothing pressure of underwear is about 1 hPa [2], we found that, from chest to under-bust, comfortable underwear has a higher pressure of 3–4 hPa.

Acknowledgments. Part of this work was supported by the Japanese Society for the Promotion of Science KAKENHI Grant Number 17H01954. We thank Nancy Schatken, BS, MT (ASCP), from Edanz Group (https://en-author-services.edanzgroup.com/), for editing a draft of this manuscript.

Trial Registration. This study was carried out after the review by the Ethical Review Board for Humans of Shinshu University (approval number 173).

References

1. Clothing Hygiene in Home Economics Society in Japan: Apparel and Health, Inoue shoin, pp. 65–73 (2012). (in Japanese)
2. Ito, N., Kotake, M., Santo, R., Yamada, T.: Effects of mean clothing pressure on compressive sensation of women's underwear. J. Home Econ. Jpn. **52**, 865–873 (2001). (English abstract and Japanese text). https://dl.ndl.go.jp/info:ndljp/pid/10582655/1
3. Niwa, H., Mitsuno, T.: Clothing pressure and pressure feeling developed by wearing yukata: in special reference to the phase of a seasonal and menstrual cycle in comparison with wearing waist-band. Jpn. Res. Assn. Text End-Uses **47**, 731–739 (2006). (English abstract and Japanese text). https://doi.org/10.11419/senshoshi1960.47.731
4. Mitsuno, T., Kai, A.: Distribution of the preferred clothing pressure over the whole body. Text. Res. J. **89**, 2187–2198 (2019). https://doi.org/10.1177/0040517518786272
5. Mitsuno, T., Hibino, A.: Which support pants are the most effective for burning body fat? J. Fiber Bioeng. Inf. **9**,187–199 (2016). https://doi.org/10.3993/jfbim00255
6. Mitsuno, T., Ueda, K.: The clothing pressure developed at waistband and pressure sense estimated in a ratio scale. J. Home Econo. Jpn. **48**, 989–998 (1997). (English abstract and Japanese text). https://doi.org/10.11428/jhej1987.48.989

Research on the Body Shape of Clothing for Chinese Adult

Linghua Ran(✉), Chaoyi Zhao, Hong Luo, Xin Zhang, and Taijie Liu

China National Institute of Standardization, Beijing, China
ranlh@cnis.ac.cn

Abstract. This study was based on the latest Chinese adult body size data collected by the China National Institute of Standardization from 2013 to 2018. Studies have shown that there are differences in body shape distribution between different regions and different age groups. Compared with the data of the first national body size survey in 1988, the body shape distribution of adults has also changed greatly. This study can provide data support for the revision of the current garment size national standard.

Keywords: Body shape · Clothing · Chinese adult

1 Introduction

In the current era of people-oriented consumption, the contradiction between clothing styles and their body shapes is increasingly prominent. According to the survey, half of consumers cannot find suitable ready-made clothes in the market, and 30% of returns of mail-order clothes are caused by inappropriate clothes. China is the world's most populous country and the world's largest producer and exporter of textile and apparel. Fully understanding the body shape of difference of people is extremely important for apparel manufacturers to produce clothes suitable for target consumer groups and consumers to choose clothes suitable for their own body shape.

From 2013 to 2018, China National Institute of Standardization measured the people's body size in 32 cities across the country, collected more than 26000 samples of human body size data in different regions, different age and different genders. This is the second nationwide data investigation since the national body dimension survey on the adult human body characteristics in 1988. Based on the newly collected national adult body size data, this paper makes statistics and analysis on the body size data used in clothing.

G. Di Bucchianico et al. (Eds.): AHFE 2020, AISC 1202, pp. 841–846, 2020.
https://doi.org/10.1007/978-3-030-51194-4_108

2 Sampling Method of Body Size Measurement

This survey sampling adopts stratified multi-order cluster sampling, which is stratified and graded according to the gender, age, region, urban and rural area and occupation of the sampling samples throughout the country. According to the size differences of Chinese population, the country is divided into six major regions, namely, northeast China, central and western China, middle reaches of the Yangtze river, lower reaches of the Yangtze river, guangdong, guangzhou, fujian and yunguichuan.

3 Body Type Classification

3.1 Body Type Classification Basis

In the national clothing size standard GB/T 1335.1-2008 "men of clothing size" [1] and GB/T 1335.2-2008 "women of clothing size" [2], the body size is divided according to the difference between chest circumference and waist circumference. The body shape classification as shown in Table 1.

Table 1. Body shape classification

Body shape	Chest and waist circumference difference for male (mm)	Chest and waist circumference difference for female (mm)
General body (type A)	120–169	140–189
Slightly obese (type B)	70–119	90–139
Fat body (type C)	20–69	40–89
Thin body (type Y)	170–220	190–250

3.2 Body Shape Distribution in Different Regions

The body shape distribution of male and female in different regions showed in Fig. 1 and Fig. 2.

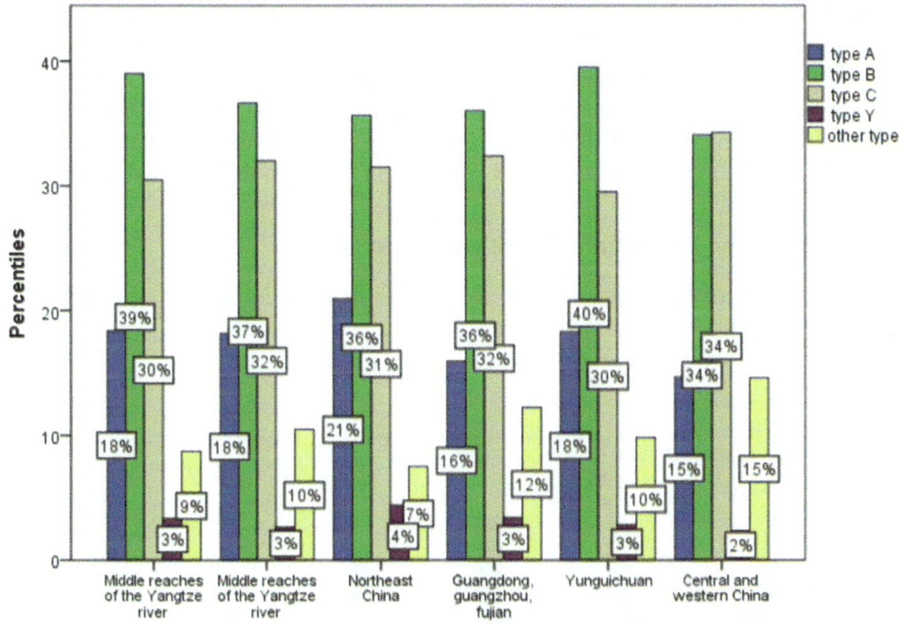

Fig. 1. The body shape distribution of male in different regions

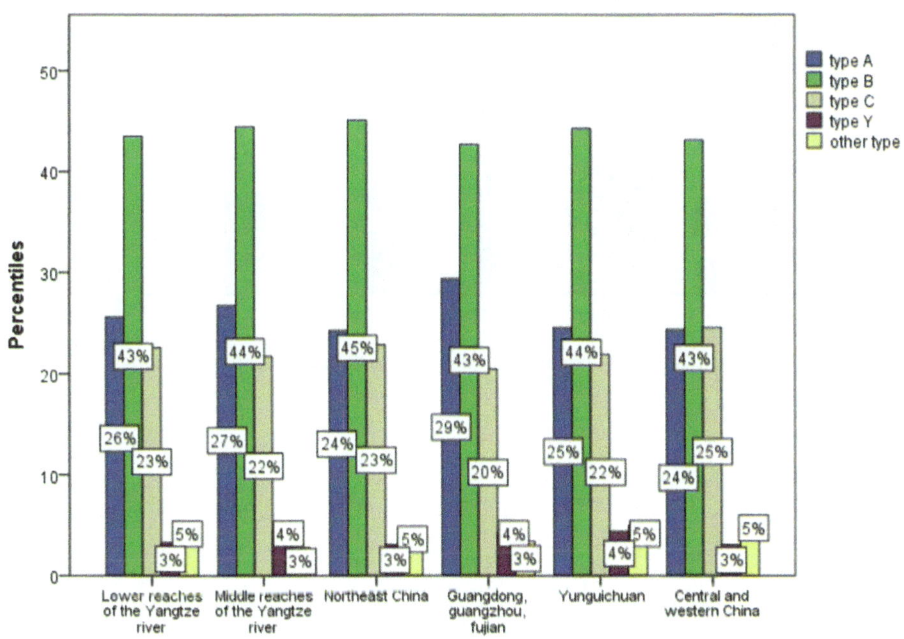

Fig. 2. The body shape distribution of female in different regions

The slightly obese type (type B) accounted for the highest proportion in each region, 34–40% for men and 24%–29% for women, followed by the obese type (type C), 30%–32% for men and 43–45% for women. The proportion of chest size, large waist size and thin body shape (type Y) was the lowest in each region, at 2%–4% for men and 3%–4% for women. Between 7% and 15% of the population in male regions and 3% and 5% of the population in female regions were not included in the four body types.

3.3 Body Shape Distribution in Different Age Groups

The body shape distribution of male and female in different age groups showed in Fig. 3 and Fig. 4.

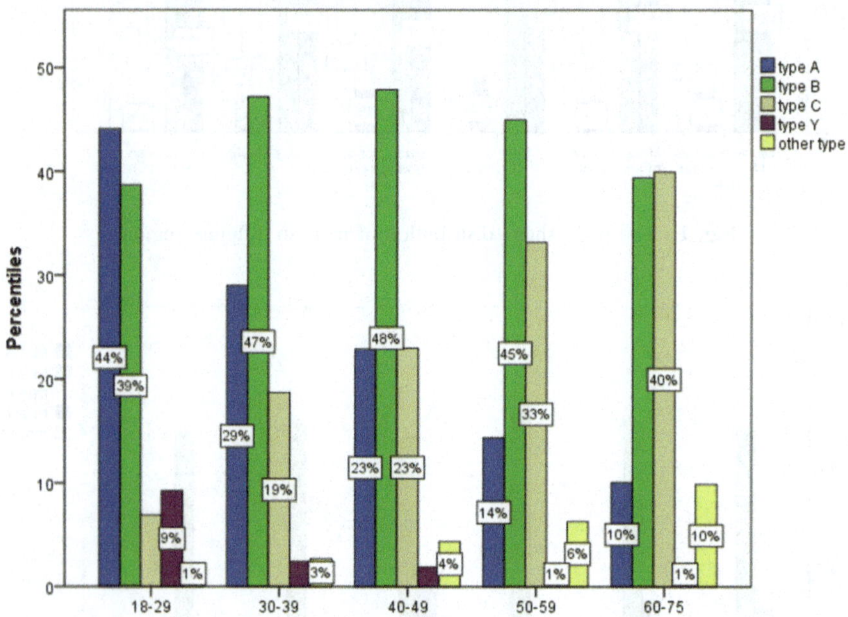

Fig. 3. The body shape distribution of male in different age groups

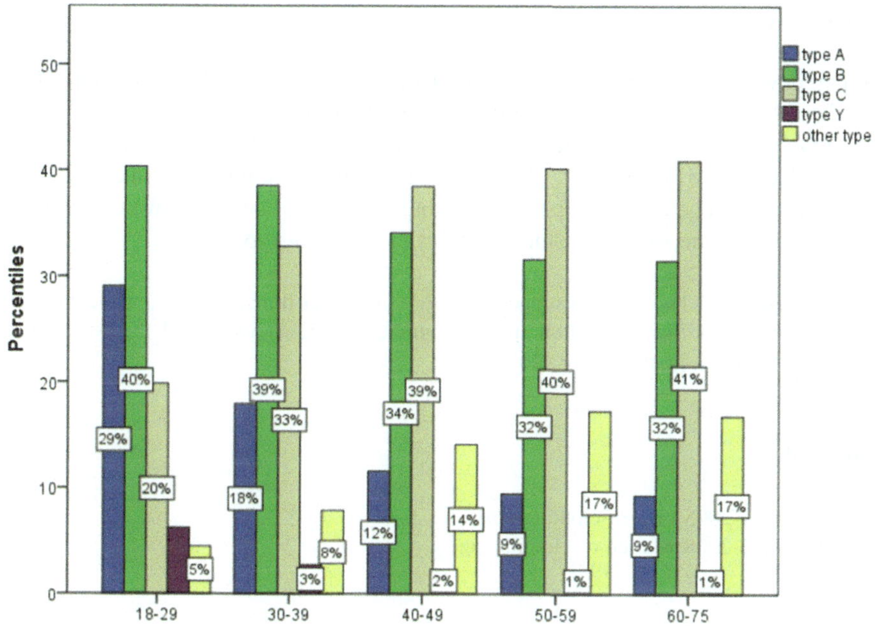

Fig. 4. The body shape distribution of female in different age groups

With the increase of age, the proportion of general body type (type A) decreased gradually. The proportion of men and women aged 18–29 was 20% and 34% higher than that of men and women aged 60–75, respectively. The percentage of obese (type C) people is increasing, with men and women aged 60–75 accounting for 21% and 33% more than men and women aged 18–29. The proportion of men with slight fat type (type B) gradually decreases with the increase of age, while the proportion of women first increases and then decreases. The proportion of men and women with Y body types gradually decreased, while the proportion of other body types gradually increased. It indicates that the older the age, the smaller the drop between the chest and waist, the less obvious the body shape, and the smaller the increment of chest circumference relative to the increment of waist circumference.

3.4 Changes Compared with 1988

Compared with 1988, the proportion of general body type (type A), chest waist thin type (type Y) of men of all ages decreased, the proportion of fat body type (type C) and other body types increased, the proportion of general body type (type A), micro fat body type (type B), chest waist thin type (type Y) and other body types of women of all ages decreased.

4 Conclusions

Compared with 1988, the body shape of Chinese adults has changed a lot. Due to China's vast territory and large population, there are significant differences in body shape among people of different ages. The coverage of age in China's current clothing size is relatively wide. In fact, there are significant differences in the body shape characteristics of each age group. The clothing size should be different in different age groups. It is very necessary to study and design clothing type by age.

Acknowledgments. This work is supported by Major project funded by the president of China National Institute of Standardization (project number 522018Y-5984) and the National Key Technology R&D Program (2014BAK01B02).

References

1. GB/T 1335.1-2008 men of clothing size
2. GB/T 1335.2-2008 women of clothing size

Smart Sportswear Design for Down Syndrome Patients

Ying-Chia Huang[1(✉)], Jen-Hau Chen[2], Guan-Yin Chen[1], and Kai Fang Tung[1]

[1] Department of Textiles and Clothing, College of Fashion and Textiles, Fu Jen Catholic University, Xinzhuang District, New Taipei City 24205, Taiwan (R.O.C.)
094537@mail.fju.edu.tw

[2] Department of Geriatric and Gerontology, National Taiwan University Hospital, Taipei City 10048, Taiwan (R.O.C.)

Abstract. The aim of this research was to develop 31 smart sportswear pieces for 15 males and 16 females with Down Syndrome. For the study, the researcher utilized a non-random sample comprising 31 people with Down Syndrome, 31 of their family members, and 31 case managers, to conduct semi-structured interviews to collect data and feedback. The garments were constructed using a hands-on method, involving manual body measurement, the invention patent No. I1621405 design procedure and apparel design to construct the garments. This research produced four results from the 93 interviews conducted. Top four colors of Down users expressed yellow, pink, black and blue color and style ideas regarding their preferred garment design. Two thirds of these unique users preferred pure pink for females and blue for males. Three quarters of the interviewees expressed similar textile, color and garment style preferences. They preferred wearing clothing with a 20% elastic functional textile, a round-neck and a short-sleeved blouse with cropped trousers.

Keywords: Smart clothing · Sportswear design · Down syndrome

1 Introduction

With the improved longevity of Down syndrome patients, the duration of long-term care has increased. Families of Down syndrome patients must strengthen the patients self-care capacity, to improve their quality of life and reduce family and caregiver stress. Over the last 80 years, the life expectancy of Down syndrome patients has increased by about 46 years. In 1929, the average life expectancy was 9 years; in 1949, 12 years; in 1982, 35 years; and in 2007, 55 years [1]. Chiang, Yang and Hsu [2] showed that in Taiwan, of the 3,684 Down syndrome patients using health insurance in 2011, 181 were over 50 years old (4.9% of the total sample size). These figures for Taiwan regarding the proportion of Down syndrome patients aged 50 years or older, is in sync with the rest of the world. This study also revealed that there are 1407 Down syndrome patients aged between 19 and 49 years old (comprising about 38% of the total sample).

G. Di Bucchianico et al. (Eds.): AHFE 2020, AISC 1202, pp. 847–855, 2020.
https://doi.org/10.1007/978-3-030-51194-4_109

For long-term care Down syndrome patients typically rely on family members. Although their number of medical visits is much lower than that of Downs patients in other age groups, Lin [3] pointed out that providing long-term care for Down syndrome patients exerts tremendous pressure, challenging families. The Department of Statistics at the Ministry of Health and Welfare in Taiwan indicated that in the third quarter of 2018, of those with mental disabilities, 101,799 have intellectual disabilities. In 2016, the government approved the Long-term Care 2.0 plan, defining categories for long-term care as relevant to long-term caregivers and services [3].

In recent years, studies evaluating the benefits of using smart wearable products to assist Down syndrome patients to maintain their health and quality of life have increased; one such product being smart garments. According to Ariyatum, Holland, Harrison and Kazi [4], smart garments were initially developed during the 1990s for use by the US and EU military. Since 2001, many companies have also begun to develop smart garments and products, such as the self-adapting shoes designed by Adidas, FM radio shirts from Gapkid's, MD & AMP jackets from Burton, the self-heating jacket by North Face, and smart shirts by Sensatex. In 2005, more smart apparel was developed; including iPod jeans from Levis, Bluetooth and solar jackets by Zegna, and solar apparel by Oakley. In ZoeTek's 2017 [5] smart heating wearable module, the smart thermal garment is a vest. The left pocket of the upper body also holds a mobile power supply of use to nursing home and elderly users.

In 2018, Lazar, Woglom, Chung, Schwartz, Hsieh, Moor, Crowley and Skotko [6] developed mobile phone application software for Down syndrome patients to address their nutritional decision-making problems at meal times. Their interface is user-centric, and the software was repeatedly revised in response to feedback from doctors, geneticists, psychologists and dietitians. designed a smart T-shirt with collision sensors on its shoulders for Down syndrome patients [7].

Kabel, Dimka and McBee-Black [8] discovered that people with disabilities require professional assistance in choosing their professional and social apparel, to ensure professional and social satisfaction. By using the monitoring method used in sportswear, wearable solutions can help healthcare professionals monitor the physiological and emotional indices of autism patients [9]. Kara, Güleç and Çağıltay [10] used wearable garments and animation to enable four intellectually disabled children in Turkey learn about the five organs of the human body. In a 2002 study by Carmeli, Kessel, Coleman and Avalon [11], adults with Down syndrome exercised 3 times a week on a treadmill for a period of 25 weeks. The result was that muscle endurance in their lower limbs improved significantly. Rimmer, Heller, Wang and Valerio [12] found that while muscle strength and muscular endurance training programs for Down syndrome patients did not significantly change their body weight, they do have a significant impact on reducing the health risks of Down syndrome patients, delaying deterioration in muscle strength and muscular endurance in adult Down syndrome patients, and maintaining independent bodily functions. In 2008, Shields, Taylor and Dodd [13] found that treadmill training three times a week for ten weeks significantly improved cardiopulmonary function, physical fitness, and the body fat ratio of children with Down syndrome.

The problems associated with garments for Down Syndrome patients in Taiwan are that existing patterns have poor accuracy and fit ability for these unique bodies.

2 Methods

Suh, Carroll, and Cassill [14] stated that 5 steps are involved in smart garment development, namely ideas generation, design, prototype development, evaluation and design refinement, and product planning (Fig. 1). This study focuses on the first three steps for the purpose of investigative research. The first step is ideas generation following semi-structured interviews with 16 females and 15 males with Down Syndrome, 31 of their family members and 31 case managers were engaged in pre-design observation and interviews to understand garment design issues. This study uses Panton Solid Coated Colors to determine the color preferences of individuals with Down Syndrome. The second step comprised of 8 items; i.e. hand drawn designs, smart electronics modules, computer drawn garment patterns and technical drawings, confirmation of design details, building 3D virtual human forms. According to human body data obtained from manual measurements, this study adopted Gerber computer pattern-making, and CLO 3D virtual garment try-on and preliminary pattern correction to create precise garment patterning and reduce the number of trials and errors in the construction of the first garment samples. The final step was prototype development using a Graphtec Cutting Plotter CE6000-120AP for printing the garment pattern, laser cutting the fabric, high-performance fitting, and an ISO 4915 stitch type for joining the pieces of the garment together [15].

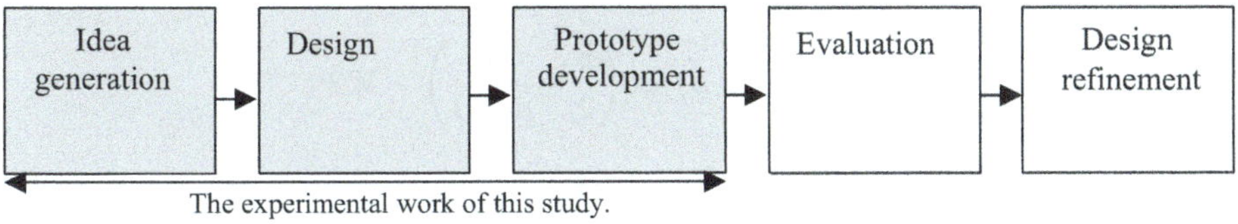

Fig. 1. This study takes the first three steps towards developing a smart garment.

In the second and third steps of this study, the author utilizes Patent No. I621405 3D Sculptural Garment Manufacturing Method and System Thereof [16] to generate the garment pattern. Before pattern-making, the researcher creates a design sketch on a scale grid. Then she enters variation data manually to establish a vertical model on CLO 3D (Fig. 2).

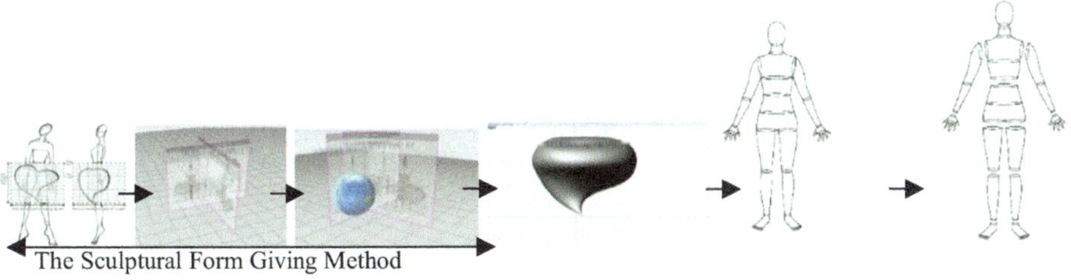

Fig. 2. Variation calculation for the pattern-making step in this research.

This patented method involves pattern-making in a square with the body figure being an oval shape. $\frac{b1}{a}$ and $\frac{b2}{a}$ are the parameters for pattern-making. Taking for example waist size, the data for the waist are $\frac{b1}{a} = \frac{17}{9.5}, \frac{b2}{a} = \frac{6}{9.5}$. Vertical and horizontal layers are then calculated to ensure garment variation.

For the vertical layers, *m* is the number of nodes and *n* is the sequence of nodes, as follows: $3 \leq \text{n} \leq m$. The pattern illustration reveals the number of nodes is greater than 3, and the smooth silhouette offers users a better garment fit when compared to fewer than 3. Th formula for the Sculptural Form Giving Method is listed below.

$$W1 = b2 + \frac{1}{m}(a - b) \times 1$$

$$W2 = b2 + \frac{1}{m}(a - b) \times 2$$

$$W3 = b2 + \frac{1}{m}(a - b) \times 3$$

$$Wn = b2 + \frac{1}{m}(a - b) \times n$$

$$BL1 = (b1 + B)\sec\left(\frac{\pi}{2} \times \frac{1}{m+1}\right) - W1$$

$$BL2 = (b1 + B)\sec\left(\frac{\pi}{2} \times \frac{1}{m+1}\right) - W2$$

$$BL3 = (b1 + B)\sec\left(\frac{\pi}{2} \times \frac{1}{m+1}\right) - W3$$

$$BLn = (b1 + B)\sec\left(\frac{\pi}{2} \times \frac{1}{m+1}\right) - Wn$$

In below are the four directions formulas:

$$BRn = (b1 + B)\sec\left(\frac{\pi}{2} \times \frac{n}{m+1}\right) - Xn$$

$$FRn = (b2 + F)\sec\left(\frac{\pi}{2} \times \frac{n}{m+1}\right) - Yn$$

$$BLn = (b1 + B)\sec\left(\frac{\pi}{2} \times \frac{n}{m+1}\right) - Wn$$

$$FLn = (b1 + F)\sec\left(\frac{\pi}{2} \times \frac{n}{m+1}\right) - Zn$$

For the horizontal layers, this study utilizes Excel to gain four data sets as illustrations (Fig. 3).

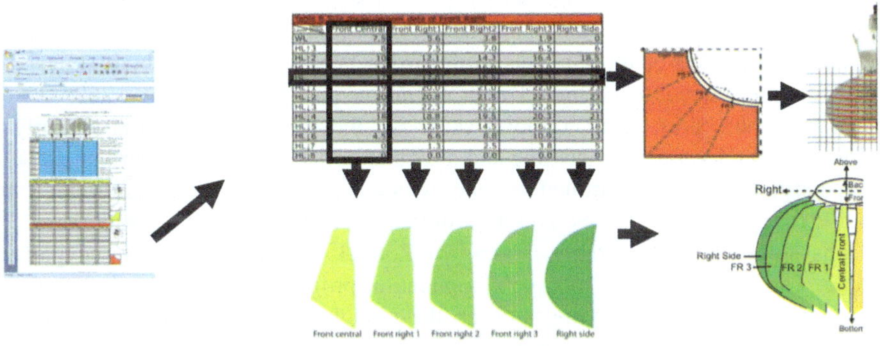

Fig. 3. Excel data sets are taken as vertical and horizonal layer illustrations.

2.1 Idea Generation

This step consists of observing and interviewing 16 females and 15 males with Down Syndrome, 31 of their family members and 31 case managers to understand garment design issues (Fig. 4). A question list of semi-structured interviews was conducted to collect data and feedback from the interviewees. The interview questions were designed to investigate the importance of the following features when designing smart sportswear focusing on Down Syndrome users' needs in summer. Details include users' background, textile thickness, hand feel, colors of functional textile, garment style and garment details.

Fig. 4. This study interviewed Down Syndrome users, their family members and case managers.

61% of the 31 Down Syndromes relied on their family members to commute between their homes and work. 38% of the 31 Down Syndrome patients take transport on their own daily to work. Even though their jobs are sedentary and include packing or making soap, they have a variety leisure activities, such as dancing, drumming and social networking.

When selecting the material, nearly half of all the interviewees selected a thin fabric with a 20% elastic functional textile, rather than the medium and the thick textiles. On the question of hand feel, 75% of all the interviewees preferred soft to the touch.

Almost all the interviewees stated that they intend to wear a mesh textile to combat the hot and humid weather of Taiwan summer. The top two-color preferences for female and male Down Syndrome patients, their family members and their case managers are yellow, pink, black and blue. For smart garment style, not only do all agree on short sleeves, but also 5% of the male data set suggest both short sleeved and sleeveless. 77%, 11% and 12% of all data sets prefer trousers, short pants, or both styles. The female data sets reveal that female Down Syndrome patients have a 33%, 26% and 6% need for a short skirt, dress, or long skirt. All the data sets show the majority of interviewees prefer a round neckline and a regular collar for Down Syndrome patients. Smart garment details mentioned include zippered and side pockets, easy care functional textiles, starched fabrics, and a garment style with a natural waistline.

This study uses a wearable module of ZoeTek to establish the initial garment sample. The features of the module's application program and operations include its 22 g weight, Bluetooth docking between the module and the smart phone interface, four temperature controls, GPS positioning, a fall alarm, SOS alarm, pedometer, and a calorie counter.

2.2 Smart Clothing Design Development

There are five steps involved in the smart clothing design, including body measurement (Fig. 5, Fig. 6 and Fig. 7), establishing a virtual avatar (Fig. 8) and pattern generation from garment technical drawing (Fig. 9). After the interviews, this study took the body data for Down Syndrome manually to garment pattern, via vertical avatar and computer aided design to do garment fitting (Fig. 10).

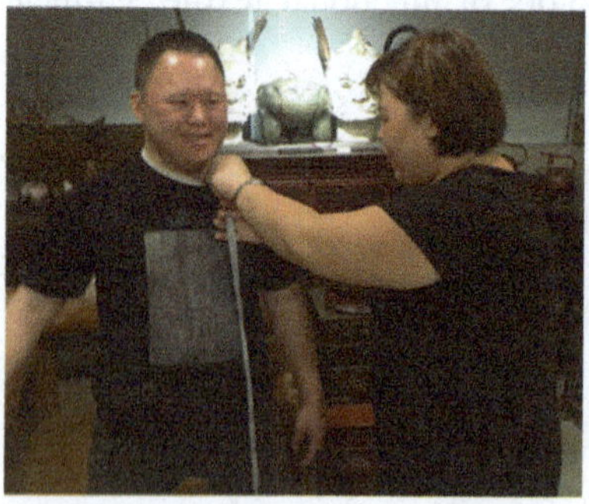

Fig. 5. Manual body measurement.

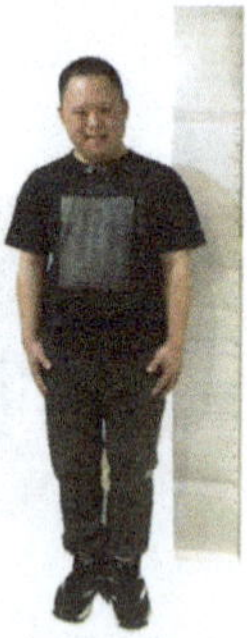

Fig. 6. Taking user's photo while measuring height.

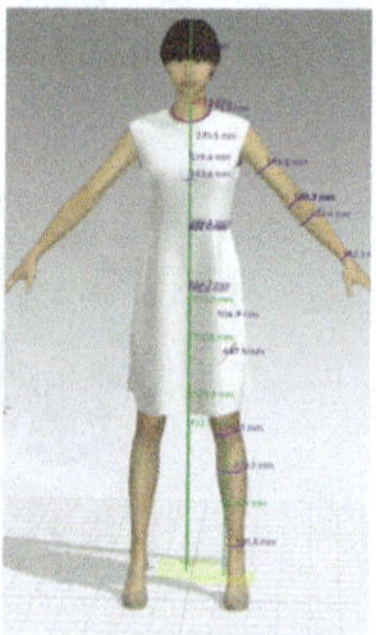

Fig. 7. 15 points of body measurement.

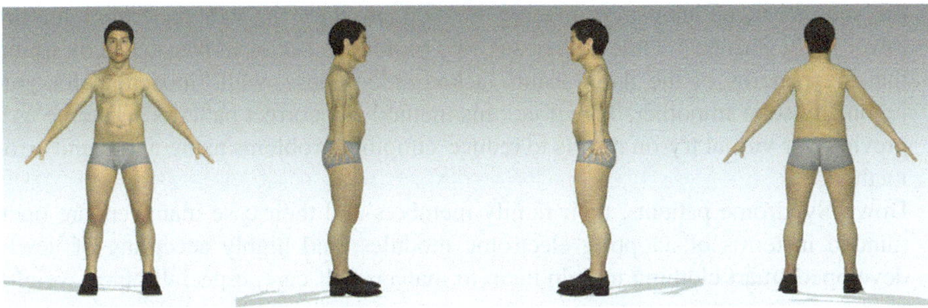

Fig. 8. Four views of male Down Syndrome by CLO 3D avatar.

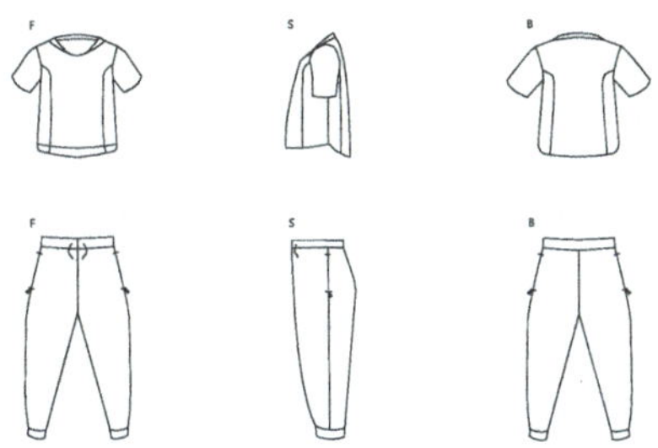

Fig. 9. Technical drawing of garment design.

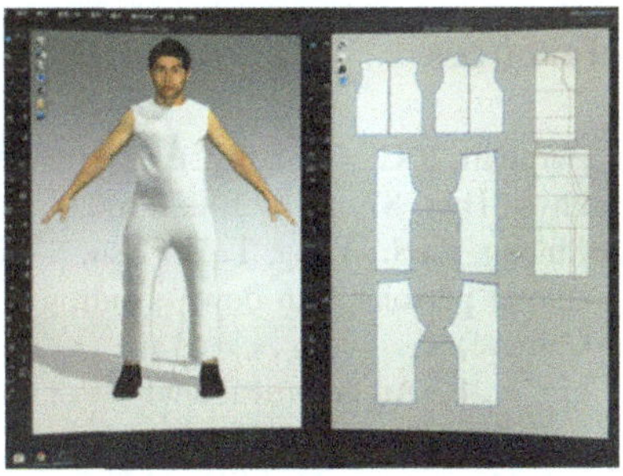

Fig. 10. CLO 3D and the garment pattern.

3 Conclusion

This research adopted 93 interviews to understand the garment design issues associated with Down Syndrome from 31 Down Syndrome patients, 31 family members, and 31 case managers. Then, taking CLO 3D, Gerber and appeal design through the steps of idea integration, design and garment prototyping. Three conclusions are listed below:

- Two thirds of Down Syndrome patients require summer season smart garment design features to include thin functional textiles, soft handling, 20% elastic fabric, mesh textile, yellow, pink, black, and blue textile colors with a garment style of a short sleeved blouse, cropped trousers, short skirt, dress, round neckline and a regular collar.
- Adopting a customized avatar helps individuals execute a high degree of accuracy in the garment pattern before sewing the first garment prototype. It is easy to modify the garment by varying the silhouette and other details. In this study, the use of CLO 3D clothing and virtual try-on software was associated with dress fitting problems before sewing the physical prototype. In terms of customized patterns,

the sculptural model pattern-making method provides smooth patterning for the lower body garments while addressing two problems. For example, the study found that the patterns of the flank seams lacked smoothness, while loosening for golf jupons must be smoother. Therefore, this method can correct patterns and serve as a preview for virtual try-on results to reduce sampling problems using a trial and error method.

- Down Syndrome patients, their family members and their case managers are open minded in terms of adopting electronic modules, and highly accepting of newly developed smart clothing to help them to manage self-care, especially their weight.

Acknowledgment. This research is founded by Taiwan Ministry of Science and Technology (MOST 108-2221-E-030-001).

References

1. Barnhart, R.C., Connolly, B.: Aging and down syndrome: implications for physical therapy. Phys. Ther. **87**, 1399–1406 (2007)
2. Chiang, P.-H., Yang, T.-H., Hsu, S.-W.: Trends in ambulatory care visits and expenditures among persons with down syndrome in Taiwan. J. Disabil. Res. **13**, 284–297 (2015). (in Chinese)
3. Lin, K.-P.: An empirical analysis of spatial distribution and facilities inventory of long-term care 2.0 – a case study of Kaohsiung City. Department of public policy and management, vol. Master, pp. 93. I-SHOU University, Kaohsiung City (2018). (in Chinese)
4. Ariyatum, B., Holland, R., Harrison, D., Kazi, T.: The future design direction of smart clothing development. J. Text. Inst. **96**, 199–210 (2005)
5. Zoetek Inc.: Smart Vest. Zoetek Inc., Taipei City (2017)
6. Lazar, J., Woglom, C., Chung, J., Schwartz, A., Hsieh, Y.G., Moore, R., Crowley, D., Skotko, B.: Co-design process of a smart phone app to help people with down syndrome manage their nutritional habits. J. Usability Stud. **13**, 73–97 (2018)
7. Semjonova, G., Vetra, J., Oks, A., Katashev, A.: Development of a new method to monitor shoulder girdle motion for ballerina with shoulder impingement syndrome based on DAid smart shirt application. In: World Congress on Medical Physics and Biomedical Engineering 2018, pp. 599–602. Springer, Cham (2019)
8. Kabela, A., Dimkab, E., McBee-Black, K.: Clothing-related barriers experienced by people with mobility disabilities and impairments. Appl. Ergon. **59**, 165–169 (2017)
9. Taj-Eldin, M., Ryan, C., O'Flynn, B., Galvin, P.: A review of wearable solutions for physiological and emotional monitoring for use by people with autism spectrum disorder and their caregivers. Sensors **18**, 4271 (2018)
10. Kara, E., Güleç, M., Çağıltay, K.: Teaching concepts with wearable technology: learning internal body organs. In: International Conference on Universal Access in Human-Computer Interaction 2018, pp. 542–550. Universal Access in Human-Computer Interaction. Methods, Technologies, and Users (2018)
11. Carmeli, E., Kessel, S., Coleman, R., Ayalon, M.: Effects of a treadmill walking program on muscle strength and balance in elderly people with down syndrome. J. Gerontol. **57**, M106–M110 (2002)

12. Rimmer, J.H., Braddock, D., Marks, B.: Health characteristics and behaviors of adults with mental retardation residing in three living arrangements. Res. Dev. Disabil. **16**, 489–499 (1995)
13. Shields, N., Taylor, N.F., Dodd, K.J.: Effects of a community-based progressive resistance training program on muscle performance and physical function in adults with Down syndrome: a randomized controlled trial. Arch. Phys. Med. Rehabil. **89**, 1215–1220 (2008)
14. Suh, M., Carroll, K.E., Cassill, N.L.: Critical review on smart clothing product development. Wilson Coll. Text. **6**, 1 (2010)
15. AMERICAN & EFIRD, INC.: ISO 4915 STITCH TYPES. ISO Stitch Terminology, pp. 4. AMERICAN & EFIRD, INC, North Carolina, United States (2006)
16. Huang, Y.-C.: 3D sculpturing garment manufacturing method and system thereof. In: Office, Ministry of Economic Affairs Intellectual Property Office. (ed.), vol. I1621405, Taiwan (2018). (in Chinese)

Effect of Fatigue of the Protective Clothing Wearer on Body Balance Performance Caused by Shorten or Prolonged Workload with Heat Exposure

Suyoung Son[1(✉)] and Ken Tokizawa[2]

[1] Department of Home economics Education, Teachers College, Kyungpook National University, Daehak-ro 80, Buk-gu, Daegu, South Korea
sonsy@knu.ac.kr

[2] National Institute of Occupational Safety and Health, Japan, Nagao, 6-21-1, Tama-ku, Kawaski 214-8585, Japan
tokizawa@h.jniosh.johas.go.jp

Abstract. This study aimed to investigate the effect of fatigue on body balance performance caused by shorten and prolonged workload with heat exposure. Shorten and prolonged experiments were conducted for the present study. Ten healthy men participated in the experiment for exploring of deterioration of balance ability by physiological and psychological fatigue due to heat exposure with wearing protective clothing for construction sites. The main finding of this study indicates that body balance performance was affected by shorten and prolonged workload with heat exposure. The core temperature and body weight loss measured during workload with heat exposure were significantly high. The body sway parameters showed a similar tendency, although the differences were not significant. Accumulation of fatigue from prolonged stress from workload and heat exposure seems to contribute to the decline in body balance performance.

Keywords: Body balance · Heat exposure · Workload test · Fatigue

1 Introduction

Japan has areas prone to high temperatures and relative humidity (Rh) during the summer season, which lead to increases in the incidence of heat stroke among outdoor workers. Between 2010 and 2012, approximately 86 deaths were caused by heat stroke were reported by the Ministry of Health, Labour and Welfare in Japan. And 40% of that report occurred at construction sites [1].

This study aimed to investigate the effect of fatigue on body balance performance caused by shorten and prolonged workload with heat exposure.

G. Di Bucchianico et al. (Eds.): AHFE 2020, AISC 1202, pp. 856–861, 2020.
https://doi.org/10.1007/978-3-030-51194-4_110

2 Methods

In the present study, Shorten and prolonged experiments were conducted for the present study (Fig. 1). In the first series, ten healthy men participated in the experiment for exploring of deterioration of balance ability by physiological and psychological fatigue due to heat exposure with wearing protective clothing for construction sites. The anthropometric characteristics of ten healthy participants for the present study were as below: (mean (SD): age, 31.9 (7.9) years; weight, 64.6 (6.7) kg; height, 171 (4.5) cm). The participants were required to wear a protective clothing for construction site, helmet, gloves, and boots, and performed workload test. The test chamber was maintained at four different combinations of temperature and humidity (CON1: 23 °C and 70% Rh, CON2: 27 °C and 40% Rh, CON3: 30 °C and 70% Rh, CON4: 34 °C and 40% Rh). The workload test consisted of a 1-h workload test (two times of 25 min walking at 5.5 km $\cdot$ h^{-1}, with a 10 min rest time), 1-h recovery time (sitting), and four times of balance tests (body sway test, functional reach test, and 2-steps test). The balance tests were performed before and after the workload test and during recovery time.

In the second series, eight healthy men were included in the experiment with a four time of workload with heat exposure (mean (SD): age 31.9 (4.1) years, weight, 62.5 (9.4) kg, height, 172.1 (7.8) cm). The test protocol consists of 45 min of workload test (walking at 4.5 km $\cdot$ h^{-1}) and 15 min of recovery (sitting). The participants performed these four times of tests on the same day. One hour of lunch break with a certain amount of meals were provided. The test chamber was maintained at different combinations of temperature and humidity (LOW: 25 °C and 50% Rh and HIGH: 34 °C and 50% Rh).

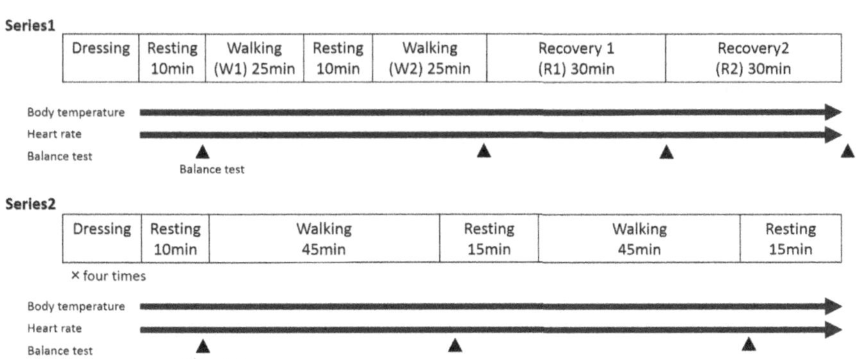

Fig. 1. Experimental protocols of two series.

Physiological responses and body balance performance were recorded during the experiment. Rectal temperature and 7-points skin temperature were measured and recorded every minute by thermistor sensors and data logger (Gram LT-8 series; Nikiso-Therm Co., Tokyo, Japan). Seven areas of the body temperature (forehead, abdomen, forearm, hand, thigh, calf, and foot) were measured, and the mean body temperature was calculated by the Hardy-Dubois formula [2]. Heart rate was monitored and recorded (RS400, Polar Electro Oy, Finland).

For the balance performance test, body sway test, functional reach test, and 2-steps test were carried out (Fig. 2). All measurement and analysis of body sway were performed using a force plate and system for measuring center of pressure (GS-3000Anima Co., Chou-Fu, Japan). Sway length and area when participants closed or opened their eyes were calculated. Participants stood up on the force plate with posture of upright, and asked to maintain their pose during 30 s with 'eyes open' or 'eyes closed'.

The functional reach test and 2-steps test were carried out as simple methods for measuring balance performance. The locations of the same side of hand and shoulder were set as measuring points for 'Functional reach test'. Participants stood up and raised their left arm straightly. They asked to extend their arm forward as par as they can do, however they should maintain their balance. The location of extended hand was recorded as the finish point, and was calculated those distances.

For the 2-steps test, participants stood up at the start line. When they gave a start sign, they take two steps forward as wide as possible. The distance of two steps was recorded.

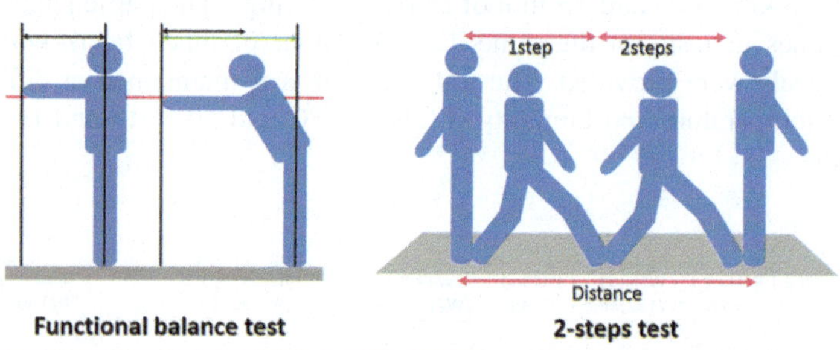

Fig. 2. Functional balance test (left) and 2steps test (right)

During the tests of two experimental series, participants were asked to subjectively evaluate rated perceived exertion (RPE), thermal sensation, thermal discomfort, fatigue, and thirty. For the subjective evaluation, visual analogue scale from 0 to 10 were used.

Results are presented as mean values and standard deviation. One-way ANOVA was carried out to analyze body weight loss. Two-way ANOVA was carried out to analyze the differences between body temperatures and balance performance (time × conditions). Tukey's post hoc test was used to the significant differences. In all the statistical analyses, $p < 0.05$ was considered as significant.

All of the experimental procedures were approved by the Human Research Ethics Committee of National Institute of Occupational Safety and Health, Japan.

3 Results

Figure 3 shows the results of Body weight loss of each condition from experimental series 1. The body weight loss of CON4 was 1.5% of participants' weight and that was significantly higher than other conditions.

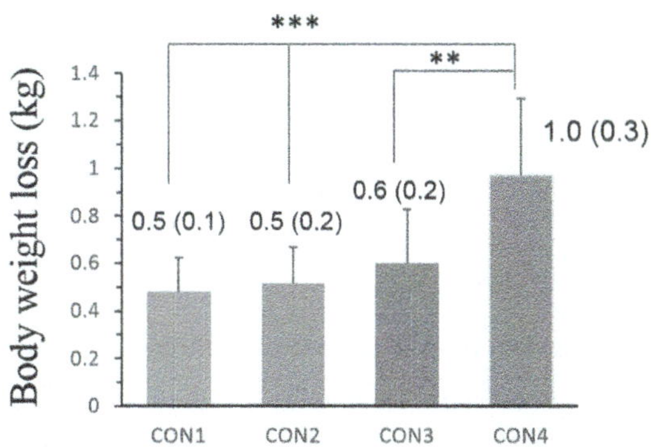

Fig. 3. Body weight loss of each condition from experimental series 1. ***$p < 0.001$, **$p < 0.01$

A similar tendency was shown in the results of body temperatures (Fig. 4). The rectal temperature of CON4 relatively increased from end of workload test, and it was significantly higher than other conditions. This tendency was maintained during one-hour recovery time. Also, mean skin temperature of CON4 extremely increased on CON3 and CON4. The mean values of changes of body temperature from rectal and 7 body parts are summarized in Table 1.

Almost of body sway data, such as sway length or area were largest among those under the second test (Table 2). Body sway length and area under the CON 2, 3, and 4 were significantly increased with eyes opened in experimental series.

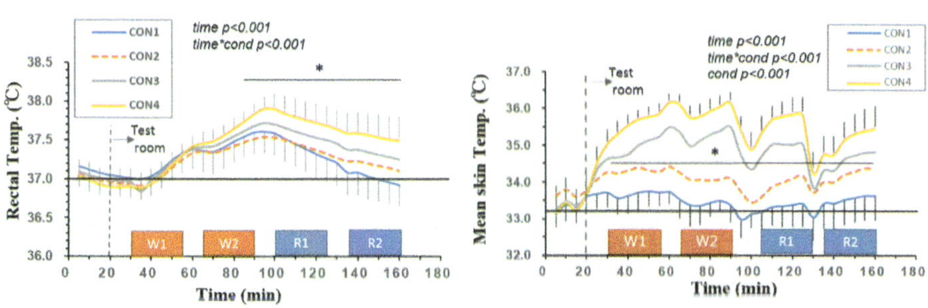

Fig. 4. Time course of rectal temperature (left) and mean skin temperature (right) of experimental series 1. *$p < 0.05$

Table 1. Changes of body temperature from rectal and 7 body parts ($T_{max} - T_{min}$).

Temperature from each body part	CON1	CON2	CON3	CON4
Tre	0.8 (0.1)	0.7 (0.1)	1.0 (0.4)	1.1 (0.2)
Tsk	1.9 (0.6)	1.9 (0.6)	2.3 (0.6)	2.8 (0.3)
T-head	2.8 (0.9)	2.4 (0.9)	2.7 (0.6)	3.4 (0.7)
T-abdomen	2.7 (0.7)	3.3 (1.7)	3.1 (0.8)	3.1 (0.7)
T-forearm	3.5 (0.6)	3.7 (1.1)	3.7 (1.9)	3.8 (0.6)
T-hand	3.2 (1.3)	2.6 (0.9)	2.7 (0.9)	3.3 (1.0)
T-calf	2.0 (0.5)	2.0 (0.3)	2.8 (0.7)	3.6 (0.4)
T-thigh	3.0 (1.1)	2.7 (0.6)	3.6 (1.1)	4.0 (0.7)
T-foot	3.9 (1.2)	4.1 (1.3)	4.1 (1.4)	5.4 (2.4)

Table 2. Body sway data from four times of test trials. * $P < 0.05$, † $p < 0.01$

		Body sway length (cm)		Body sway area (cm^2)	
		EO	EC	EO	EC
CON1	1st	34.0 ± 7.0	42.7 ± 8.9	2.4 ± 1.8	3.2 ± 1.9
	2nd	37.0 ± 9.7	48.9 ± 16.5	2.7 ± 1.3	4.8 ± 3.5
	3rd	35.2 ± 7	50.8 ± 14.7	3.4 ± 2.3	5.7 ± 3.9
	4th	34.9 ± 8.8	44.9 ± 14.2	3.2 ± 1.8	4.2 ± 3.8
CON2	1st	30.5 ± 6.7	41.5 ± 8.0	2.3 ± 1.6	3.0 ± 1.8
	2nd	39.7 ± 11.9*	53.3 ± 20.3	4.2 ± 4.0†	5.3 ± 4.1
	3rd	36.9 ± 8.2	48.8 ± 16.3	3.5 ± 2.2	5.1 ± 2.9
	4th	33.7 ± 7.1	46.5 ± 8.4	3.0 ± 1.3	4.7 ± 2.7
CON3	1st	32.2 ± 7.3	41.5 ± 11.0	1.9 ± 1.1	2.7 ± 1.1
	2nd	38.4 ± 11.9	47.3 ± 7.3	3.6 ± 2.3*	4.3 ± 2.4
	3rd	32.9 ± 7.6	46.8 ± 10.7	2.8 ± 1.6	4.7 ± 2.4
	4th	34.5 ± 5.8	47.2 ± 10.5	2.5 ± 1.0	4.6 ± 3.2
CON4	1st	31.6 ± 5.2	46.6 ± 10.0	2.1 ± 0.8	5.2 ± 5.5
	2nd	39.9 ± 10.3*	58.1 ± 24.0	4.7 ± 4.3*	6.1 ± 4.5
	3rd	36.3 ± 8.6	56.2 ± 10.3	3.0 ± 1.7	6.4 ± 4.5
	4th	35.7 ± 5.1	48.5 ± 9.4	3.0 ± 1.3	4.3 ± 1.0

The mean values of functional balance test and 2-steps test decreased by 1-h workload, and values of CON4 more decreased at the middle of the recovery period (Fig. 5). However, we could not observe significant differences in time courses and among the conditions.

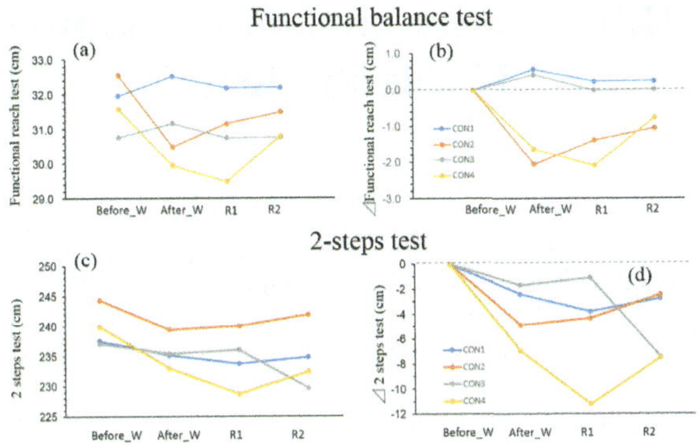

Fig. 5. Body balance performance test results from experimental series 1: (a) Functional balance test, (b) Delta values of functional balance test, (c) 2-steps test. (d) Delta values of 2-steps test.

4 Discussion and Conclusion

In this study, we investigated the variation in physical performance, especially body balance performance caused by workload and heat exposure. The main finding of this study indicates that balance ability and physiological responses were affected by physical workload and heat exposure. For the physiological responses, change of rectal temperature and body weight loss of high temperature conditions were significantly higher than low temperature conditions. Even though body sway parameters did not show significantly big differences among conditions, body sway parameters increase with workload or heat exposure. Moreover, body balance performance deteriorated with workload and more until the middle of the recovery period with heat exposure. It could not be recovered enough during 1-h recovery period. In conclusion, this study demonstrated that balance ability is affected by physical workload and heat exposure. Based on our results, to have enough recovery time (less than 30 min) after around 1-h workload with heat exposure is required for maintaining workers balance performance and their safety. For this study, we designed two experimental series. However, the test for experimental series 2 is now in progress and the results will be reported soon.

Acknowledgement. This research was supported by Kyungpook National University Research Fund, 2017.

References

1. A research of occurrence situation of heat strain in Japan, Ministry of Health, Labour and Welfare in Japan (2012). (Japanese). http://www.mhlw.go.jp/bunya/roudoukijun/anzeneisei51/
2. Hardy, J.D., DuBois, E.F.: The technic of measuring radiation and convection. J. Nutr. **15**, 461–475 (1938)

Textile Historical Sustainability and Innovative Textile Products

Eleonora Trivellin[1(✉)], Giuseppe Lotti[2], Marco Marseglia[2], Elisa Matteucci[2], Margherita Vacca[2], Irene Fiesoli[2], Claudia Morea[1], Alessio Tanzini[1], Francesco Tanzini[2], and Lu Ji[1]

[1] Dipartimento di Architettura, Università di Ferrara, Ferrara, Italy
eleonora.trivellin@unife.it
[2] Dipartimento di Architettura, Università di Firenze, Firenze, Italy

Abstract. Textile surfaces have expressed a great deal of changes related to the meanings they are able to transmit as a communication layer and also as a technological product.

We can read this path in an exemplary way with the analysis of the largest textile district in Europe, the one that belongs to the city of Prato in Tuscany. The typical product of this area - regenerated carded - has seen its nature transform from a low quality and value product to a technological and environmental excellence.

Keywords: Textile · Innovation · Prato · Carded wool · Industry 4.0

1 Introduction

The strategic importance of textile products manifests itself in the various production sectors and not only in the clothing sector.

With the value of "communication layer" they were able to take on not only technological but also communicative and symbolic meanings in the strictest sense [1].

The strength of these products lies in being able to express strongly consolidated meanings together with others of great innovation.

Somehow, during their evolution, what we can see defined by the first industrial revolution forward has been confirmed, that is to say, it is a field of great experimentation for the application of new forms of energy, new design methods, new applications of mechanics and the standardization of products and, at the same time, remain closely linked to local resources and, more generally, to territorial skills [2].

The text was thought of authors all together. In particular Eleonora Trivellin wrote paragraphe no. 4 Carded: manufacturing cycle and no 5 The Manteco company; Giuseppe Lotti wrote paragraphe no. 1 Introduction; Elisa Matteucci wrote paragraphe no. 3 The production structure; Marco Marseglia wrote the first paragraph of paragraphe no. 2 Territory and society; Claudia Morea wrote the second paragraph of paragraphe no. 2 Territory and society; Francesco Cantini wrote the third paragraph of paragraphe no. 2 Territory and society; Alessio Tanzini wrote the fourth and fifth paragraphs of paragraphe no. 2 Territory and society; Margherita Vacca edited the bibliography and the notes; Lu Ji found and edited the images.

G. Di Bucchianico et al. (Eds.): AHFE 2020, AISC 1202, pp. 862–869, 2020.
https://doi.org/10.1007/978-3-030-51194-4_111

The fact that man, from the dawn of civilization until today, has used most of his energies to nourish himself and to cover his body, giving these activities also strongly symbolic values, reminds us of the reason, precisely on the actions of intertwining a lot of attention has always been focused both from an operational and theoretical point of view. Furthermore, the link between nourishment, and therefore breeding and agriculture and weaving has always been very strong and strategic for the development of archaic societies. With the great attention paid today to sustainability, the relationship with the food supply chain, and agriculture in general, has returned to being very topical.

Contextualizing our study on the Prato district, the elements to be taken into consideration are the conformation of the territory and urban development; the economy; the organization and transformation of the company, the process and the resulting product. We will necessarily focus on the last two factors but, it would not be possible to understand the originality and peculiarity of carded and regenerated wool without at least mentioning the other factors as they are all closely connected.

In order to exemplify, we will go on to state the example of an emblematic company in the Prato area.

2 Territory and Society

The urbanized countryside [3], which characterizes the Prato area around the urban center, is made up of a fabric that holds industrial and civil buildings together; In Prato there are still the houses with the adjoining *stanzone* in which small artisan companies, often of a family nature, operate on behalf of third parties for the woolen clients.

Great production fragmentation but, at the same time, coordination for waste management: since the 1980s, a cutting-edge centralized system for the purification of civil and industrial waste water has been created thanks to public-private investments. In Prato's society, work is at the center of life and entrepreneurship is considered an attitude based more on knowing how to do than on knowing so much that Prato, at least until 2008 the year of the great crisis, precisely because of the high percentage of employees, had a very low rate of graduates compared to the rest of Tuscany. The rarefied urban fabric of Prato which develops in many small smaller towns, as well as favored the development of a predominantly horizontal economy, has been conducive to the establishment of Chinese workers who have been setting up their community since the 1990s [4].

It is from the beginning of the millennium that the demographic increase has made itself felt in a very incisive way up to become the largest Chinese Italian community and the third European after London and Paris[1]. The Prato manufacture and that relating to the Chinese community traveled in many ways on different tracks with parallel skills; with due exceptions, the Chinese deal with fast fashion for medium-low markets, and Prato with increasingly high quality fabrics, the former increasing their

[1] The factors that influenced the growth of the community were China's entry into the WTO and the end of the Multifibre agreement which no longer allowed to regulate manufacturing imports from countries with non-advanced economies.

turnover a lot, the latter, instead, downsizing it. So the spaces left free by Prato in many cases were occupied by the Chinese.

In fact, there are very few elements of contact between these two realities: Italian staff with the skills of fashion designer and accountant is present in Chinese companies and as far as the processes are concerned, printing works and dyers have both Italian and Chinese customers [5].

The shortening of the design/processing times that led from fast fashion to fast fashion has exacerbated certain mechanisms and even more marks the separation between the two realities.

3 The Production Structure

From the beginning of the twentieth century until the Second World War, two production circuits were substantially established: that of vertically integrated companies with standardized productions (plaids, military blankets, etc.) oriented towards exports, whose greatest strength was given by the large numbers of production; that of the teams of companies with phase processing with a production more oriented towards the clothing markets.

Large full cycle companies have had a prevalent role only for fifty years, that is, from the second decade of the 1900s to the 1960s, after which they will undergo the pulverization process.

In some respects the war destruction triggered this phenomenon which then determined the identity and characteristics of the Prato model: "according to some estimates, 33% of the looms, 80% of the spinning mills, seventy percent of the assortments in operation at the beginning of the war were destroyed or damaged [6][2].

Giampiero Nigro writes, "Starting from the 1960s, with the evolution of the role of the small business and the tumbler, the flexibility characteristics were enriched with new meanings, above all in terms of rapid adaptation of the production to the requests for textile products less and less standardized, more personalized and qualitatively better" [7].

Taking as a firm point the competitive market price deriving from the use of recycled material, production was then differentiated thanks to the production structure. So even knitwear, combed products and patterned fabrics, intended for women's clothing, became part of the Prato products. As a result, an ever closer relationship developed with the fashion industry [8].

---

[2] It is said that some companies managed to save their frames from the destruction of the German army by dismantling and burying them. These episodes, which are still being told, are reported without documentation, also by Edoardo Nesi in the novel Storia della mia gente (2010).

In June 1944 at the station of Carmignano, Alighiero and Bogardo Buricchi, Ariodante Naldi and bruno Spinelli blew up eight German railway wagons loaded with TNT destined for the destruction of Prato's industries.

4 Carded: Manufacturing Cycle

From the nineteenth century, with precision between 1850–60, spinning machines began to be developed with materials obtained from the scraps and scraps of used fabrics and clothing, starting the processing of regenerated wool [9].

Carding is a specific fiber processing technique. The yarns are produced with the use of virgin fibers or through the reuse of the fibers obtained from the recycling of fabrics and from new and used knitwear cuttings.

The characteristic of this process is the use of short and even non-homogeneous fibers, mixed together in the most varied compositions such as to give the yarn a particular physiognomy that distinguishes it from the other type represented by the hairstyle.

For over a century, regenerated wool has represented an important opportunity for development and affirmation of the Prato textile district on markets all over the world.

Recycling techniques [...] collect and synthesize a culture of attention and sensitivity towards the environment, the tradition of man and his professional skills [10].

The operations that are part of the carded fabric cycle are sorting, carbonizing, tearing, carding, spinning, warping, weaving, fulling, dyeing, finishing, recovery of processing waste and new start of the cycle. They are the first four and the last, the actions that characterize it [11].

The figure that performs the actions of differentiation and selection of textile materials is the sorting machine or, more generally, the *cenciaiolo* or *stracciaiolo*. Those who still carry out this job have a highly developed visual and tactile sensitivity that allows you to quickly distinguish the rags by composition and color and which for their speed and precision did not seem replaceable with an equally efficient machine. On the whole territory there are no more than forty. There is a generational change for this particular profession: they are young people from Morocco, India who today acquire this professionalism that allows you to distinguish the composition of the fabric at a touch and the nuance of the colors with your eyes (Fig. 1).

Fig. 1. Pieces of fabric during the selection operation

We are therefore in the presence of a productive culture that is not only based but rooted in recovery, an element which, perhaps, has not yet been given the cultural recognition it deserves.

In this direction, the initiative promoted by the Prato Chamber of Commerce, in collaboration with the Consortium for the Enhancement of Carded Textile Products, Pratese Industrial Union, CNA National Confederation of Craft and *Confartigianato*, (trade association) which established the *Cardato* brand and the *Cardato Riciclato* brand, and even if today there are not many member companies, there are some that, while applying the principles, however company policy preferred not to join the brand[3].

If in the past the Prato companies had a certain homogeneity, today the companies that work on this reality are very different. In general, the awareness of wanting to enhance the historical industrial culture is growing. It is the smallest and youngest companies that manage most effectively that manage to superimpose the product on effective and very dynamic communication.

Among these, the reality Rifò can be mentioned, which also had attention in the national press despite being a small company. A network of four companies has been created with the Re-think your jeans project: used jeans are collected through the Naturasì food store network; Recooper social cooperative deals with the selection of materials; Pinori filati produces yarn from recycled material and, finally, Rifò transforms the yarn into clothing products (Fig. 2).

Fig. 2. Control operation on frayed fabrics

[3] The certification is guaranteed by SGS, the international certification body, which also guarantees a connection with fashion brands (http://www.cardato.it/it/cardato/il-progetto/).

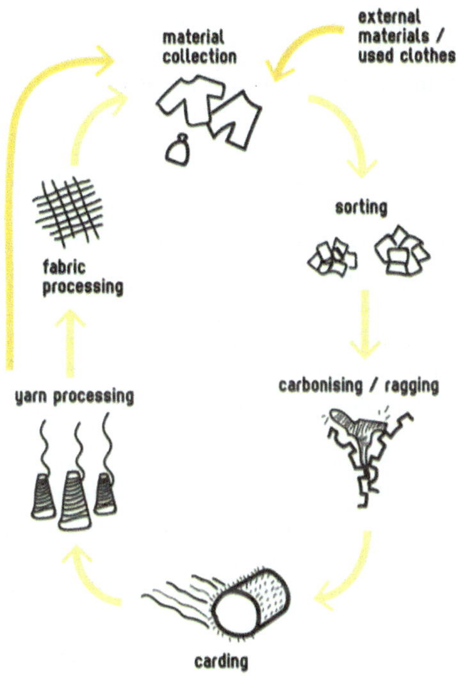

Fig. 3. Scheme of the carded regenerated cycle

5 The Manteco Company

Young growing companies are important to understand the creative potential that exists in the area, however our attention is right that it focuses on large companies that face the challenges of the present with a vision that knows how to keep historical heritage and future challenges together sales, for which the company purchases the "mechanical wool" spun in the various stocks and then be woven and finished according to the expected qualities; the second line consists of a wide range of fabrics produced with carded and combed virgin wool fibers; the third line is made up of fabrics of variable composition obtained through the use of worsted wool, cotton, linen and their blends. In this case the yarns are directly purchased from preferably European first-rate spinning mills.

Applications of enabling technologies, attention to the sustainability of corporate value communication are some of the most visible actions that come into contact with the company.

Let's start with the company archive set up in the last year, which organized and made the 40,000 samples accessible. Through the inventory, cataloging, digitization, conservation and display of the company's historical heritage, stylists, researchers, customers, students and journalists can consult the material on site. Each folder that collects the historical material is marked with a quarcode that communicates the complete content to the visitor without having to take it from the shelf.

Next to the physical archive, a digital archive was set up in which there are photos of the collections of items developed over the years starting from the collections of the 1970s. Sending via e-mail photos of fabrics as an alternative to samples sent by express courier leads to high energy savings.

The sustainability of the company is not expressed only through the use of regenerated material. By way of example, we would like to mention Project 43 which is taking more and more field. This is the collection of all the fabric production scraps that occur during the cutting of the fabric by the packer. These fabric scraps, suitably selected for color and composition, are used again so as to create new fabric. It should also be added that the company is "Zero Waste" in fact.

The production waste (false selvedges, test samples, piece heads, spinning and twisting scraps, etc. …) are grouped and re-entered in the production cycle.

The commitment to sustainability and the circular economy has been recognized and rewarded: it was among the five Italian companies to receive the Radicalgreen award in 2018, within the third edition of the Green Economy Festival [12]; presented its proposals to Premier Vision in September 2019 within the section/platform dedicated to companies that work in favor of responsible innovation; has been included among the 1000 companies that inspire Europe.

They then obtained the Cribis Prime Company certification[4].

There are also many projects that focus on the physico-chemical characteristics of the product. One last work carried out was the verification of the presence of bacteria in the recycled raw materials which over time deteriorated the fibers, decreasing their color fastness to light. As a result of this study, the Bacteria Free system was developed which sanitizes the fibers before spinning, making them more sanitized and leading to evident product improvements [13].

We are therefore in the presence of a territory and companies that are transforming traditional knowledge that has developed thanks to resourcefulness and know-how in a model capable of giving answers to the needs of our time.

It is no coincidence that Prato will be a "pilot city" of a national project, together with Milan and Bari, becoming a real laboratory for promoting and experimenting innovative initiatives on the circular economy and on all issues with high environmental impact on waste. The themes of circularity have always been historically linked to the city of Prato, since reuse has been the basis of the development of the textile district with the creation of recycled carded wool.

References

1. Dal Curto B., Fiorani E., Passaro C.: La pelle del design. Lupetti, Milano (2010)
2. Vv.Aa.: Sustainable Thinking, Exhibition Catalog, Florence 2019. Electa, Milan (2019)
3. Becattini G.: Lo sviluppo economico della Toscana: con particolare riguardo all'industrializzazione leggera. IRPET, Firenze (1975)
4. Bonomi, A., Masiero, R.: Dalla Smart City alla Smart Land. Marsilio, Venezia (2014)

[4] Award issued to companies that simultaneously have Tating 1 (maximum level of reliability) and Paydex® >70.

5. Etnografia del pronto moda: I laboratory cinesi nel distretto di Prato. https://journals.openedition.org/qds/379
6. Nigro G.: Il "caso" Prato in Storia d'Italia. Le regioni dall'Unità ad oggi. La Toscana, pp. 823–868. Einaudi, Torino (1986)
7. Nigro G.: Il "caso" Prato in Storia d'Italia. Le regioni dall'Unità ad oggi. La Toscana, pp. 823–868. Einaudi, Torino (1986)
8. Cardato made in Prato. http://www.cardato.it/it/cardato/il-progetto/
9. Prato arte e storia. http://www2.po-net.prato.it/artestoria/storia/memorie-tessile/dizionario/pagina448.html. Accessed 30 Jan 2020
10. Magi F., Ceccarelli F.,: Il cardato pratese. La spola, Prato (2002)
11. Ricchetti, M. (ed.): Neomateriali nell'economia circolare. Moda. Edizioni Ambiente, Milan (2017)
12. Greenweek festival della green economy. https://www.greenweekfestival.it/premier-gentiloni-premio-radical-green-aprono-la-green-week/
13. Mantellassi Compagnia tessile. https://www.manteco.com/prodotti-eco-friendly/\

Coordinating for Emotional Satisfaction

From Control to Coordination

Shuichi Fukuda(✉)

Keio University, 4-1-1, Hiiyoshi, Kohoku-ku, Yokohama 223-8526, Japan
shufukuda@gmail.com

Abstract. As the environments and situations change frequently, extensively and in an unpredictable manner, engineering is quickly shifting from control-based to coordination-focused. In control, we can apply rational approaches in a straightforward manner and the goal is clear from the first. But to cope with the changes today, we need to decide which parts we should use to cope with the situation or the context. We need to best coordinate our resources to overcome the changes to achieve our goal. Although this is a very much challenging job, it provides us with the psychological satisfaction we have never experienced in control-based engineering.

Keywords: Control to coordination · Unpredictable changes · Increasing complexity and complicatedness · Movement · Individual to team · Adapt to overcome · Self-determination · Mental satisfaction

1 Control-Based Engineering

Our engineering has been control-based up to now. As changes were small and smooth, we could predict the future. Engineers could foretell the operating conditions and as the products were produced on an individual basis, engineers could focus on the functions of their products. The world was small and closed with boundary, so that the world yesterday was explicit, and engineers could apply rational approaches in a straightforward manner. But our world expanded rapidly, and it became an open world without boundary (Fig. 1).

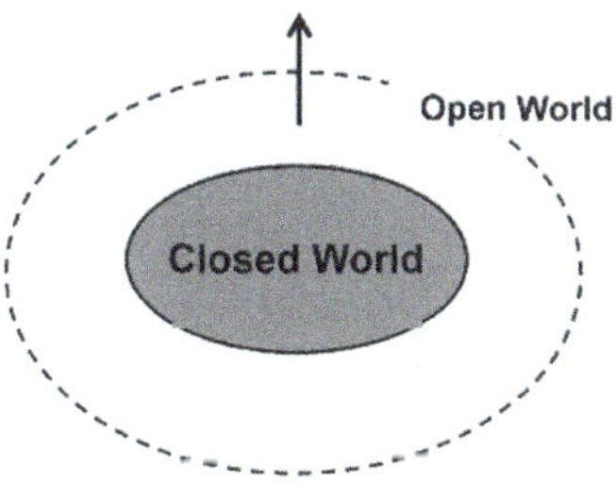

Fig. 1. Closed World to Open World

G. Di Bucchianico et al. (Eds.): AHFE 2020, AISC 1202, pp. 873–878, 2020.
https://doi.org/10.1007/978-3-030-51194-4_112

Therefore, we could not apply rational approaches anymore as we did yesterday. So, engineers developed System Identification Approach. This approach is basically the same idea as we identify the name of a river. If we look at the flow, it changes every minute and we cannot identify its name. But if we look around, we can find mountains, forests, etc. that do not change. Using these feature points, we can identify its name (Fig. 2).

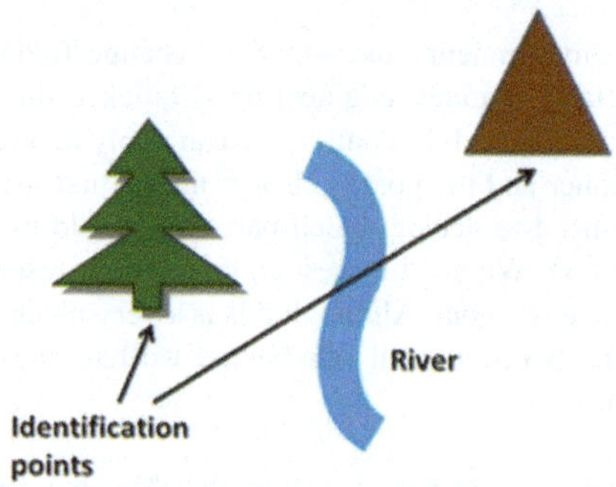

Fig. 2. How we identify the name of a river

By finding such feature points, we can identify the system and we can identify the parameters. Thus, although our world expanded very quickly, we expanded the rational world and established the controllable world (Fig. 3).

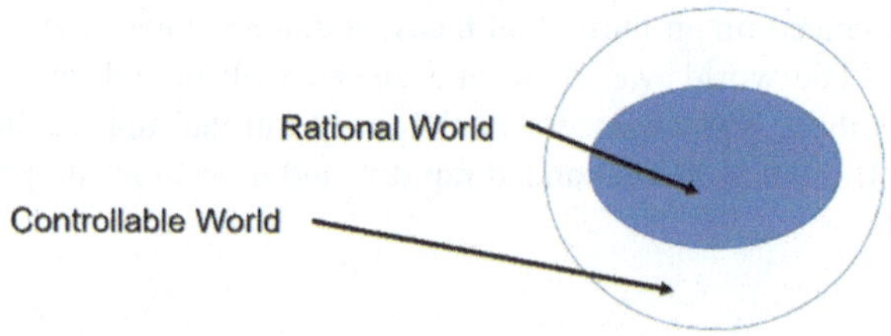

Fig. 3. Rational World to Controllable World

2 Changes Are Changing

Although we succeeded in expanding the rational world and established the controllable world, changes are changing. Changes yesterday were smooth, so that we could differentiate them mathematically, i.e., we could predict the future. But changes today are sharp, so we cannot differentiate them. We cannot predict the future anymore (Fig. 4).

Changes are taking place more and more frequently and extensively, but what makes the problem very difficult is they become unpredictable.

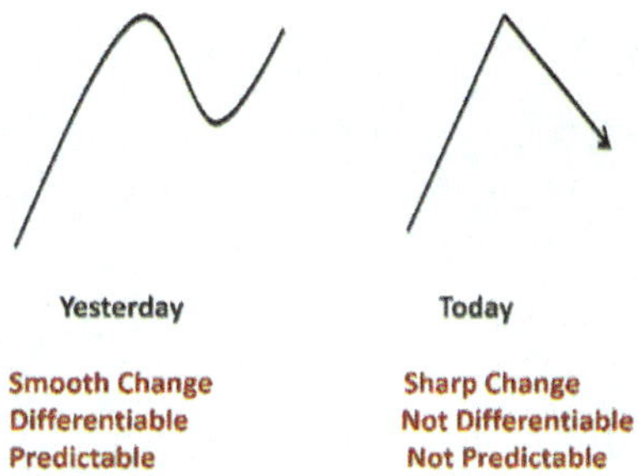

Fig. 4. Changes Yesterday and Today

3 Increasing Complexity and Complicatedness

In addition to the shift of changes from predictable to unpredictable, another problem emerged. That is Complexity and Complicatedness. There is much discussion on complexity, but they pay only small attention to complicatedness. The discussion about complexity is based on digital thinking or discrete world. How discrete elements can be put together into a system. But we must remember that when we discuss reliability, the time to break in and the time to failure or how products deteriorate are very important. These are issues related to analog world. In fact, our world is composed of digital and analog elements. We must pay more attention to analog phenomena and we should make efforts to integrate digital and analog world.

4 Movement: Core of Engineering

Nobody would object to the idea that movement constitutes the core of engineering. Machines play a central role in engineering and they are robots in the broad sense. Karel Chapek [1] coined the word "robota" in his science fiction play "R.U.R." [2] and it meant "forced labor" and labor itself comes from the Latin and it means "toil", "trouble", i.e., physical work. So, it may be said that the primary objective of engineering is to reduce such physical works. Thus, movement plays an important role in engineering.

But up to now, when we talk about the movement of robots or machines, we always take up movements that can be analyzed rationally. There is such a word as humanoid robots, but their movement is rational. They reproduce the rational movements, because engineering up to now has focused on control.

However, the world has changed tremendously, and changes become unpredictable. We are now in the flow, which changes every minute. As the fact that there is still no swimming robot developed indicates, we need to find another direction to move forward.

Robots were modeled on humans. So, let us study how humans move. Nikolai Bernstein, famous human motion control researcher, made clear that our movement trajectories vary widely from time to time at the early stages, but as we approach the target, our trajectories are fixed, so we can apply rational approaches [3, 4] (Fig. 5).

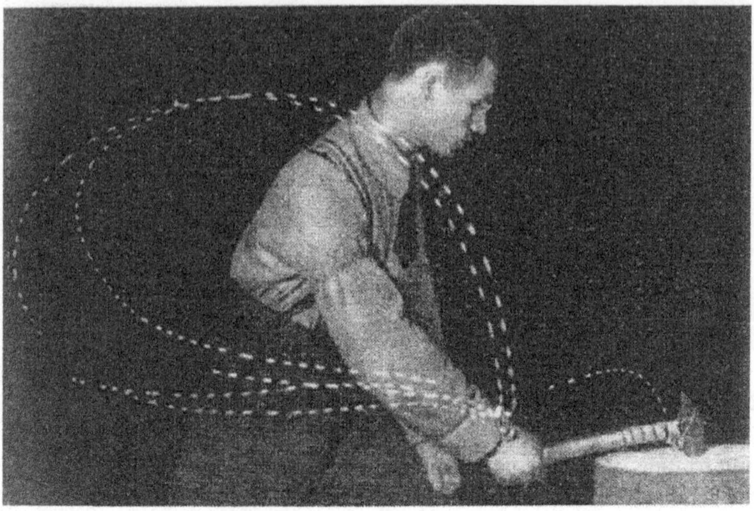

Fig. 5. Cyclogram of hammering

Once we come close to the target, we reduce the number of degrees of freedom to control our trajectories. But at the early stage, the number of degrees of freedom is tremendously large so we have no other way to make a decision on how to move by trial and error.

In other words, we coordinate our many body parts and balance our body. This may be said that we need teamworking of our body parts to balance and the formation of this team varies from time to time, depending on the context. And as our body builds are different, movements vary from person to person. Thus, we are now entering the age of coordination. How we can team up our body parts becomes most important.

5 Individual to Team

The importance of teamworking is emphasized. But we should remember that this teamworking is no more that of yesterday. Teamworking itself is changing rapidly.

To understand it easily, let us take up an example of soccer. Soccer is well known for its formation. But yesterday, the game did not change appreciably, so the formation did not change during the game. Every player was expected to do his best at his position. But today, the game changes so frequently and widely. So, the formation changes continuously. And the number of players in the formation changes. We need flexible and viable capabilities for organizing formation.

As the environments and situations come to change so much today, "Adaptability" becomes the keyword today. We do not care anymore about reproducibility. But we should note that Adaptability is not enough. We need to overcome the situation to win. In other words, we need to outwit the enemy to win. We need to betray enemy's anticipation.

In soccer, to catch up with the frequent and wide changes in the game, midfielders are now taking the position of a playing manger. Other players move proactively to respond. So, the formation may be 3-member team, or 4-member team, depending on the situation.

6 Adapt to Overcome

Thus, we need to "overcome" the situation to achieve what we expect. Adaptability is just letting yourself go with the flow. If we wish to achieve our goal, we need to swim against the flow. Thus, the ability to overcome is what is needed most today. In fact, if we come to think about the evolution, we expanded the world of our species by this way. We need to realize that we are playing the game with the real world. If we adapt to it, we will be carried away with the flow.

7 Summary

Table 1 summarize what are discussed in this paper. It should be emphasized that such shifts are deeply related to IoT, Internet of Things. Up to now, humans and machines are different. Humans are living and machines are nonliving. Their worlds are different. Humans are outside of the machine world and gave instructions. And machines responded. They are master and slave.

But today, we are not separated anymore. We work together on the same pitch and machines and humans communicate. Both are things and we work together as things member on the same team. Thus, every member must be context and situational aware and must be proactive.

Table 1. Engineering yesterday and tomorrow

	Yesterday	Tomorrow
Basis of engineering	Individual control	Team coordination
World	Closed with boundary Simple	Open without boundary Complex and complicate
Changes	Predictable	Unpredictable
How to deal with the situation	Adapt	Overcome
Satisfaction	Physical	Mental

Finally, it should be emphasized that moving from control to coordination will bring us psychological or mental satisfaction. Edward Deci and Richard Ryan made clear in their Self-Determination Theory [5, 6] that if we are self-motivated and act, then we will get the maximum satisfaction and happiness, which external reward cannot provide.

Therefore, in addition to the benefits mentioned above, shifting from control to coordination implies that we will be moving from processing things externally to processing things internally, which opens the door to the world of much greater satisfaction.

References

1. https://en.wikipedia.org/wiki/Karel_Capek
2. https://en.wikipedia.org/wiki/R.U.R
3. https://en.wikipedia.org/wiki/Nikolai_Bernstein
4. Bernstein, N.: The Co-ordination and Regulation of Movements. Pergamon Press, Oxford (1967)
5. https://en.wikipedia.org/wiki/Self-determination_theory
6. Deci, E.L., Ryan, R.: Intrinsic Motivation and Self-determination in Human Behavior. Plenum, New York (1985)

Color in Glossolalia: Green in Western Culture (Part I)

Leonor Ferrão[1,2(✉)]

[1] Lisbon School of Architecture, CIAUD, University of Lisbon, Rua Sá Nogueira, Polo Universitário, Alto da Ajuda, 1349-063 Lisbon, Portugal
[2] CITAD, Universidade Lusíada de Lisboa, Rua da Junqueira, 188-198, 1349-001 Lisbon, Portugal
lferrao@fa.ulisboa.pt

Abstract. All meanings of the word *glossolalia* (from the Greek words *glossa*, meaning tongue or language, and *laleō*, meaning to speak, talk, chat, or to make a sound) raise a knowing smile because they can be applied to what happens with words that circulate in design discourse. Since color is intrinsic to the creation of any design product and is essential in the recognition, perception and communication process, it determines the semantics of a given object in whatever form concerns the 'design object'. In establishing the meanings of color, the aspects addressed encompass a broad gamut that includes neurology, vision and illumination and the interactions between light and color in understanding the nature of design at stake and the cultural context. At a more detailed scale, a multidisciplinary approach to color allows other qualities to be addressed, including those produced by variations in surface and materiality, elements of composition, psycho-physiological effects of both color and light, synaesthetic reactions to color, communication variables, color terminology, aesthetics of color, as well as color's historical and cultural background. In this article, I will discuss some aspects related to GREEN in Western European culture. From paintings to fashion and product design, GREEN suggests, among many other things, creativity, but also a certain kind of consciousness. The word GREEN also circulates in common discourse, sometimes as a noun, sometimes as an adjective, and sometimes in idiomatic expressions. Given the vastness of contexts in which GREEN is used, I will focus on some aspects of its meanings in Western European culture, as it is impossible to remove from the material and immaterial culture the symbolic force that results from being GREEN.

Keywords: Human factors · Color · Western European culture · Semantics · Green

1 Introduction

This article is part of the Glossolalia project [1] and seeks to review the presence of the color green in Western culture. However, it is not exhaustive, and therefore a second article will also be dedicated to green.

G. Di Bucchianico et al. (Eds.): AHFE 2020, AISC 1202, pp. 879–884, 2020.
https://doi.org/10.1007/978-3-030-51194-4_113

Saying that green falls between yellow and blue in the visible spectrum means very little, because it merely expresses a characteristic unknown until the work of Isaac Newton (1643–1727) on light (published in 1704). The social and symbolic dimensions of this color remain to be clarified, and they are much more interesting and varied. As with other colors, green has experienced many changes over time in the European cultural space, depending on the contexts in which it has been used and circulated.

Reviewing the etymology, in ancient languages, such as Greek, there is no word equivalent to green. The words that does exist can refer to blue, gray or even a material, such as glass. This omission has had several interpretations. The possibility that it played a limited role in social relations, religious life, symbolically and in the imagination of the Greeks seems to be the most credible [2]. Unlike the Greeks, the Romans appreciated green (although not in the same way they loved other colors), and had a word for it - *virĭdis*, a word that is the root of *verde*, which means green in Portuguese, Italian, Spanish and Romanian, and *vert*, in French. This Latin word is at the root of others, which ended up producing a kind of symbolic contamination. By this I mean *virere* (which means being green but also being vigorous), *vis* (which means strength), *vir* (which means man and is the source of virile), *virtus* (which means courage, or virtue, depending on the context).

Northern barbarians loved this color, even in their clothing, so there was a time when wearing green was synonymous with not being Roman. Unlike the Greeks, they had a specific word for this color. The origin of the word is proto-German *grōni* (*grün* in modern German), which branched into *green* (in English) *grøn* (in Danish), *groen* (in Dutch), *grön* (in Swedish) and *grønn* (in Norwegian).

The Romans learned to like green, not because of the influence of Germans, but possibly because of the Egyptians, people they admired and conquered: they transported many artifacts of their material culture to Rome and assimilated many aspects of their immaterial culture.

However, with some exceptions (of which Nero's taste for this color is an example), green and blue colored the two rival teams in horse racing at the *Circus Maximus* in Rome. The aristocracy used to support the blue team; the people supported the green team. This sociological nuance did not continue into the High Middle Ages, perhaps because Christian morality disapproved of gambling and the notion of "honest" and "dishonest" colors emerged as a result of this and other disapprovals. The first great transformation was brought about by the future Pope Innocent III (1198–1216): in a text dating from the end of the 12th century, the prelate links the color green with hope and eternal life. It was a short hop from here to green beginning to be used for liturgical clothing. While for Islam the association between green and nature is "natural", for Christians, green did not allow this relationship, at least until the 18th century.

From antiquity to contemporaneity, the color green has made a very interesting symbolic journey [3]: i) it is the color of fate, good luck, bad luck, wealth, money and hope; ii) nature, ecology, hygiene, health and freshness; iii) youth and debauchery; iv) freedom; v) poison [4]. The instability of the pigments that make it possible to obtain the color green and the high toxicity of some, such as arsenic, if nothing else, justify the ambivalence of this color in Western culture. Below, I will focus on the most important aspects that still remain today.

2 The Color of Negativity

The association between the color green and negative passions is abundant in literature. Let us look at three texts. In Shakespeare's *Othello* (Act 3, Scene 3), Iago warns the Moor not to surrender to the green-eyed monster, which has no other result than to increase the sting of jealousy: "O, beware, my lord, of jealousy; // It is the *green-eyed monster* which doth mock // The meat it feeds on; that cuckold lives in bliss // Who, certain of his fate, loves not his wronger" [5]. In *The Merchant of Venice*, it appears in a speech by Portia (Act III, Scene 2): "How all the other passions fleet to air, // As doubtful thoughts, and rash-embraced despair, // And shuddering fear, and *green-eyed jealousy*! O love, // Be moderate; allay thy ecstasy, // In measure rein thy joy; scant this excess. // I feel too much thy blessing: make it less, // For fear I surfeit" [6]. In the satire "A Connecticut Yankee in King Arthur's Court" (1889), the color green is associated with envy: "It turned Brer Merlin green with envy and spite, which was a great satisfaction to me. But there was one thing I couldn't understand — nobody had asked for an autograph" [7].

William Langham's book is one of the oldest printouts in the English language that identifies the green sickness; it provides the plant-based recipe and how to take it [8]. It is, after all, a type of anemia (Hypochromic anemia), resulting from a lack of iron in the blood, which translates into reduced production of red blood cells and a greenish skin color, among other symptoms [9]. In this topic we could also include the health hazards resulting from the use of copper derivatives to obtain a green color in fabrics. It was an important and widely discussed issue in the 19th century, a century in which this color was very much appreciated, even by social elites.

In the fable of "The fox and the grapes", green is the color of the disguised disdain of estrangement, a feeling as negative and destructive as that of jealousy or envy. There are several versions of Aesop's fable. La Fontaine's version (1668) maintains the sharpness: "Certain Renard Gaston, d'autres disent Normand, // Mourant Presque de faim, vit au haut d'une treille // Des raisins mûrs apparemment, // Et couverts d'une peau vermeile. // Le galante em eût fait volontiers un repas. // Mais comme il n'y pouvoit attendre; // *Ils font trop verds*, dit-il, & bons pour des goujats. // Fit-il pas mieux que de se plaindre?" [10]. The version illustrated in 1900 by Walter Crane (1845–1915), a major figure in the Arts and Crafts movement, is very concise: "This fox has a longing for grapes, // He jumps, but the bunch still // escapes. // So he goes away sour; // And, 'tis said, to his hour // Declares that he's no taste for grapes" [11].

The color green is also associated with poison and a drink that was reputed to be hallucinogenic, absinthe, also known as "the green fairy". This drink, discovered at the end of the 18th century, was very popular with intellectuals and bohemians in the late 19th century [12]. Finally, but not in last place, the Devil may have a black, red, or green body – as in *The Devil Presenting St. Augustine with the Book of Vices* (1471–1475), by Michael Pacher (c. 1435–1498).

So, for many reasons, in Western culture, there are ancient motives for detesting the color green. And yet…

3 Green Is the Color of Youth, Nature and Health

At least since the 13th century, green has been associated with youth – in Shakespeare, however, youth is associated with a lack of experience. I am referring specifically to a speech by Cleopatra (Act I, Scene 5) [13]. However, we had to wait for romanticism to associate the color green with nature. The "discovery" of nature is manifested in the "Le hameau de la reine", a farm that Marie Antoinette (1755–1793) ordered to be built adjacent to the Petit Trianon (in two phases, one in 1777 and the other in 1786): it expresses the idealization of a supposed simplicity and authenticity of country life, bringing together several styles (although the result has a great deal of unity) [14]. The elites' love for nature results from their repulsion for the effects of industrialization on the urban landscape, concomitant with other social transformations, no less disturbing and threatening to the social order of the *ancien régime*.

The link between the color green and healthy living has not ceased to be made. Even today, green is the color for ecological parties and sustainability. In the common sense, parks and gardens are the "green lungs" of cities. In the more technical sense of architecture and urbanism, they are called "green spaces" or "green infrastructure". The adjective green, when attached to architecture or design, designates a practice that seeks to minimize environmental footprint, saving water and energy and abstaining from using non-renewable raw materials or manufacturing processes that consume fossil fuels.

A concern with losses in green areas on a global scale is the basis of the concept of a sustainable and self-sufficient city in terms of energy: a *Smart Forest City* is also self-sufficient in terms of food production and is equipped with an electric and semi-automated transport system [15]. The summary description of the project is silent on the matter of building materials, but it is not difficult to imagine that they will also be environmentally friendly.

In design, the 1973 oil crisis put the dangers of industrial economies' dependence on fossil fuels on the agenda. Since then, criticism of consumerism and environmental concerns have been expressed in various ways, according to theoretical and, above all, ideological filters – from design nihilism – a dangerous abstinence and, therefore, something evil to fight [16] – to design criticism [17] and finally radical design [18]. The object for sitting in Fig. 1 is an ironic example, reusing remnants of previous chairs in a tree trunk to take the green statement to extremes of paroxysm (if we look at the cost and size of the environmental footprint because of packaging and transport) [19].

Fig. 1. *Tree-Trunk Bench* by Dutch designer Jürgen Bey (b.), edited by Vitra

As a very critical voice of the concept of hot design would say, the criticism of consumption and waste that is at the root of its project does not exorcise the evils of capitalism, but rather replicates them [20].

Extending products' useful life (or postponing their disposal – which can be done in many ways that are not explained here) allows us to have some hope for the future: at worst, if we continue to dismantle natural resources as we have until now, today's products, at least some of them, even if they lose their function of use, may be the only receptacles of raw materials for new uses we have in the not too distant future [21].

4 Can the Color Green Be Merely Instrumental or Is It Always Symbolic?

In Western painting, green was used in small details to highlight other colors. Examples of this are abundant. They include the *St. Columba Altarpiece* (c. 1455) by Rogier van der Weyden (1400–1464) and a beautiful watercolor and gouache on vellum by Albrech Dürer (1471–1528) – the painter used green primed papers, as in *Green Passion: Christ before Caiaphas* (1504). As a matter of fact, figures dressed entirely in green are rare in paintings of the Modern Era, but the color is abundant in a) backgrounds, as in *Portrait of a Man* (c.1530) by Jan Gossaert (c.1478–1532), b) in curtains, as in *Sir Thomas More* (1527) and *The Ambassadors* (1533) by Hans Holbein (1497–1543), or in *Girl Reading a Letter at an Open Window* (1657) by Johannes Vermeer (1632–1675), c) and in garments, as in *The Magdalene Reading* (c.1445) by Rogier van der Weyden (1400–1464). Paolo Veronese (1528–1588) was one of the Italian painters who loved the color green… The examples chosen do not provide a complete answer to the question posed, perhaps because these works can be stripped of the symbolic dimension they had in their time to be enjoyed today, simply as beautiful exercises in pure visual delight.

References

1. Ferrão, L.: All about words [on design]: on glossolalia project. In: Procedia Manufacturing, 6th International Conference on Applied Human Factors and Ergonomics (AHFE 2015) and the Affiliated Conferences, vol. 3, pp. 6253–6257. Springer, Heidelberg (2015). https://doi.org/10.1016/j.promfg.2015.07.774
2. Pastoureau, M.: Vert, histoire d'une couleur. Éditions du Seuil, Paris (2013)
3. Pastoureau, M., Simonnet, D.: Couleurs, le grand livre, p. 65. Éditions Panama, Paris (2008)
4. Pastoureau, M.: Dictionnaire des couleurs de notre temps: symbolique et société, pp. 163–164. Christine Bonneton, Paris (2007)
5. Shakespeare, W.: Othello. http://shakespeare.mit.edu/othello/othello.3.3.html
6. Shakespeare, W.: The Merchant of Venice. http://shakespeare.mit.edu/merchant/full.html
7. Twain, M.: A Connecticut Yankee in King Arthur's Court. https://www.gutenberg.org/files/86/86-h/86-h.htm
8. Langham, W.: The Garden of Health (By the deputies of Christopher Barker), London (1597)

9. Ali, M.A.M.: The Hypochromic Anemias. https://www.ncbi.nlm.nih.gov/pmc/articles/PMC2378373/pdf/canfamphys00321-0044.pdf
10. La Fontaine, J.: Fables choisies mises en vers par… tome premier, p. 110. Charles-Antoine Jombert, Paris (1755)
11. Baby's Own Aesop. http://www.waltercrane.com/illustrations/BabysOwnAesop/
12. Padosch, S.A., Lachenmeier, D.W., Kröner, L.U.: Absinthism: a fictitious 19th century syndrome with present impact. Subst. Abuse Treat. Prev. Policy **1**, 14 (2006). https://doi.org/10.1186/1747-597X-1-14
13. Shakespeare, W.: Antony and Cleopatra. http://shakespeare.mit.edu/cleopatra/full.html
14. Château de Versailles, Le hameau de la reine. http://www.chateauversailles.fr/decouvrir/domaine/domaine-trianon/hameau-reine
15. Boeri, S.: Smart Forest City. https://www.stefanoboeriarchitetti.net/en/project/smart-forest-city-cancun/
16. Maldonado, T.: La speranza progettuale: ambiente e società. Einaudi, Torino (1970)
17. Papanek, V.: The green imperative: ecology and ethics in design and architecture. Thames and Hudson, London (1995)
18. Branzi, A.: The Hot House: Italian New Wave Design. Thames and Hudson, London (1984)
19. Bay, J.: Tree-Trunk Bench. https://www.droog.com/project/tree-trunk-bench-jurgen-bey
20. Selle, G.: There is no Kitch, there is only design! Des. Issues **1**(1), 41–52 (1984)
21. Fry, T.: Remakings: Ecology, Design, Philosophy. Envirobook, Sydney (1994). Design futuring: sustainability, ethics and new practice. Berg, Oxford (2008)

Predicting Attitudes Towards Advertisement Strategy Based on Personality

Ezekiel B. Skovron(✉) and Kim-Phuong Vu

California State University, Long Beach, CA 90840, USA
Kim.vu@csulb.edu

Abstract. A theory developed by Myers, Sen, and Alexandrov in 2010 proposes that individuals who possess higher levels of certain personality dimensions will show a more favorable attitude towards different types of advertisement (ad) strategy. This study investigated, via an online survey, relationships between the Big Five personality dimensions and attitudes toward two advertisement strategy clusters: transformational versus informational, and comparative versus non-comparative. Results from a multivariate regression analysis revealed a variety of significant relationships, some of which supported the propositions of the Myers and associates theoretical framework. Individuals scoring high on extraversion favored all types of ad strategy, those scoring high on agreeableness favored both types of informational ads, those scoring high on conscientious favored transformational non-comparative ads, and those scoring high on neuroticism favored informational non-comparative and transformational comparative ads. However, openness was not related to any of the ad strategies. Research should continue to investigate the relationship between the more affective components of the personality dimensions and their relationship with ad strategy. In addition, the creativity and quality of an advertisement could be what attracts individuals with higher levels of openness to experience.

1 Introduction

According to an advertisement forecast performed by Magna in September 2018, digital ad spending is expected to increase 11.8% by the end of 2019 [6]. With a larger budget comes larger opportunity, thus forcing businesses to expend more on digital marketing campaigns to remain competitive. Many marketers are now investing a sizable portion of their advertising on consumer insights (i.e., collecting data from consumers to inform marketing decisions). Consumer insights range from basic demographics of a consumer population, to actual behaviors and attitudes. These data are used to create more compatible experiences for users, much like how Amazon's algorithm learns user interests based on their behaviors. For marketers to remain competitive, they must understand their consumers the best they can, but consumer personality is often neglected from consumer insight data. It has already been revealed that there are distinct, significant relationships between certain human personality traits and brand characteristics; however, research is also beginning to show significant relationships between the way a product is marketed towards a consumer, and the personality dimension levels of consumers. This suggests that marketers are underutilizing personality as a key consumer insight, as focusing on

G. Di Bucchianico et al. (Eds.): AHFE 2020, AISC 1202, pp. 885–891, 2020.
https://doi.org/10.1007/978-3-030-51194-4_114

personality could not only help make the ad more successful, but also create a more personalized user experience.

In a pivotal study, Hirsh and associates [4] discovered a significant relationship between human personality and persuasive messaging. Five advertisements were created featuring a picture of a cell phone and a few lines of text. The text was manipulated so that each advertisement appealed to the *motivational concerns* of one of the Big Five personality traits. For example, extraverts are especially sensitive to rewards and social attention [5]. Thus, the "extraverted" advertisement text said the following: "with XPhone, you'll always be where the excitement is" [4]. Participants rated the effectiveness of each advertisement via a series of Likert scales. Some examples of the statements presented were "I would purchase this product after seeing this advertisement," and "I find this advertisement to be persuasive." Results showed significant positive correlations between the individual's personality dimension and the ratings given to the compatible advertisement [4]. These results confirm that there is a definite relationship between how a product is marketed, and the personality of the consumer.

More recent research has focused on the relationship between the Big Five personality traits and social media marketing on platforms like Facebook or Instagram. In a study conducted by Ryan and Xenos in 2011, a combination of Facebook users and non-Facebook users were given an online questionnaire to determine how they score on each of the Big Five personality dimensions, as well as their levels of narcissism, shyness, and loneliness. Results of this study showed that Facebook users tend to be more extraverted and narcissistic, but also less conscientious and socially lonely [9]. The previous research conducted by Hirsh and associates [4] established a definite connection between human personality and how messages are communicated, and Ryan and Xenos's research was able to create a *persona* of typical Facebook users based on personality. Such findings suggest that more "extraverted" advertising may be more compatible with Facebook users.

To better understand this, Clark and Calli [1] had participants take a short personality survey, called the Ten Item Personality Inventory (TIPI), and showed them two classes of Facebook advertisements to see how participants would respond [3]. One class of ads were promotional posts (referred to as "sponsored stories") that appeared based on a friend's activity (i.e., liking a particular product or service) and the other class of ads were actual paid ads shown on Facebook (referred to as "page post ads"). Participants were asked to predict the likelihood of any actions they might take as a result (share, comment, etc.), based on a 5-point Likert scale. Results found that participants with higher levels of Extraversion were more inclined to share sponsored stories and like page post ads [1]. As previously mentioned, Ryan and Xenos [9] discovered the typical Facebook user to be extraverted, so these findings are consistent with the view that extraverts have high engagement with the presented advertisements.

From past research, it is clear the following marketing factors are related to personality: the specific way a message is communicated, if the ad is being displayed via social media, and the affect of the advertisement (positive versus negative) [1, 4, 7]. This research is integral for understanding the relationship between personality and advertising, but raises an interesting problem. The brand personality of products greatly influences how a message can be communicated and its affect. It also is now known that certain human personality traits are significantly related to certain brand personality traits. So, how can a brand create advertising that can be manipulated to appeal to all individuals excelling at one or more of the personality dimensions? For example, an alarm system brand will likely appeal to someone who is more neurotic, as alarm systems are intended for those individuals who are uneasy and worried about individuals breaking into their property. Thus, it is difficult for alarm system brands to appeal to extroverts who crave excitement. Despite the clear relationship that exists between human personality and marketing, there is still a limitation of brand personality, and this makes it difficult for companies to create compatible digital advertising experiences for web users.

To address this problem, Susan D. Myers and colleagues have created a theoretical framework hypothesizing the relationship between the Big Five personality traits and higher-level advertising strategy [8]. Any brand can manipulate an advertising strategy without going against their personality and compromising relationships with its consumers. The framework introduces two strategies routinely used by marketers: Transformational versus Informational and Comparative versus Non-Comparative. She hypothesized that individuals with a higher degree of extraversion will have a more favorable attitude toward transformational non-comparative ads, individuals with higher levels of agreeableness will prefer transformational and non-comparative advertisements, individuals high in conscientiousness will show a more favorable attitude towards both informational comparative and informational non-comparative ads, individuals with higher levels of neuroticism will show a more favorable attitude towards both transformational non-comparative and comparative advertisements, and individuals high in openness to experience should show a more favorable attitude towards non-comparative ads.

These hypotheses serve as the foundation for the present research. In addition to better understanding the relationships between human personality and specific advertising strategy, this study will help to determine if there is an ad strategy that is holistically favorable in terms of attitude.

2 Methods

2.1 Participants

A survey was administered to 216 participants. However, 10 participants were excluded for failing to answer a quality control question. Thus, the sample for the present study consisted of 206 participants (143 males, 63 females). All participants were adults, ranging from 18 to 65 years old. Over half of the participants were between the ages of 25 and 34, with the next highest age group being 35 to 44 years of age. Participants were recruited via Amazon's Mechanical Turk (AMT) [2]. Participants who completed the survey were paid $1 USD.

2.2 Materials

Advertisement stimuli were created using Adobe Photoshop CC 2015 and were 660 pixels by 560 pixels in size. Eight fictional products were created: the CafMate Coffee

Maker, Happy Detergent, the Ultraphone X, Roaders Running Shoes, Delight Cola, ADF Speakers, Live Fresh Toothpaste, and the Vaclfify Vacuum. Transformational stimuli showed one or multiple persons enjoying the product and a positive affect tagline. Informational stimuli showed the product on its own on a neutral background, a list of facts about the product, and a tagline such as "Just the facts". Transformational comparative stimuli showed someone having a positive experience with the advertised product and another person having a negative experience with a competitor product. Informational comparative stimuli put the advertised product and competitor product side by side and listed the facts of each on the same neutral background as the informational advertisement.

2.3 Procedure

Participants signed up for the study via AMT and were provided with an informed consent. Participants then completed a 30 question personality inventory called the Big Five Inventory-2 Short Form (BFI-2-S) and brief demographics questionnaire. Advertisement stimuli were presented to participants and attitudes toward them were measured via responses to rating scale questions used by Hirsh and associates [4]. Individual responses to rating scale items were averaged to obtain an "attitude score" for each type of advertisement strategy. As there were 6 rating scale items for each advertisement, and each participant saw an advertisement type twice, a total of 12 rating scale responses made up an attitude score. As there were four advertisement types, each participant had a total of four attitude scores (i.e., informational non-comparative, transformational non-comparative advertisements, informational comparative advertisements, and transformational comparative). Thus, there were four dependent variables in this study.

3 Results

A multivariate regression analysis was run to study the underlying patterns of relationships between each of the Big Five dimensions of personality and attitudes toward the four key advertisement strategy types (see Table 1). This section will break down the results for each hypothesis that was tested.

The multivariate regression analysis yielded a significant predictive relationship, between extraversion and attitudes towards transformational non-comparative ads, with individuals possessing higher extraversion scores showing a more favorable attitude towards transformational non-comparative advertisements. Thus, Hypothesis 1 was supported. Extraversion also showed several significant predictive relationships with the remaining three advertisement strategies. Similar to transformational non-comparative ads, extraversion was highly predictive of scores for transformational comparative advertisements, with individuals possessing higher extraversion scores showing a more favorable attitude towards transformational comparative advertisements. A smaller effect was observed between extraversion and attitude towards both types of informational advertisements. For extraversion and informational non-comparative ads, individuals possessing higher extraversion scores showed more favorable attitudes toward informational non-comparative advertisements. Similarly, for extraversion and informational comparative ads, individuals possessing higher extraversion scores showed more favorable attitudes toward informational comparative advertisements.

The second hypothesis was that individuals with higher levels of agreeableness would yield high attitude scores for transformational non-comparative ads. The multivariate regression analysis did not yield a significant predictive relationship between agreeableness and attitudes towards transformational non-comparative advertisements. Thus, Hypothesis 2 was not supported.

Table 1. Multivariate regression analysis of big five personality dimensions and advertisement strategy.

Advertisement strategy type	Parameter (personality type)	*B*	*t*	*Model* R^2
Informational non-comp				.114
	Extraversion	.535**	3.91	
	Agreeableness	.356*	2.53	
	Conscientiousness	−.319	−1.79	
	Neuroticism	.348*	2.42	
	Openness	−.136	−.99	
Transformational non-comp				.136
	Extraversion	.716**	4.89	
	Agreeableness	.063	.42	
	Conscientiousness	−.394*	−2.06	
	Neuroticism	.269	1.74	
	Openness	−.208	−1.42	
Informational comparative				.093
	Extraversion	.429**	3.28	
	Agreeableness	.396**	2.95	
	Conscientiousness	−.137	−.80	
	Neuroticism	.235	1.71	
	Openness	−.092	−.71	
Transformational comparative				.141
	Extraversion	.719**	4.68	
	Agreeableness	.098	.62	
	Conscientiousness	−.378	−1.88	
	Neuroticism	.416*	2.57	
	Openness	−.270	−1.76	

*Significant at $p < .05$, **Significant at $p < .01$

However, there was a significant predictive relationship between agreeableness and the two types of Informational ads. For informational non-comparative advertisements, individuals possessing higher agreeableness scores showed more favorable attitudes toward informational non-comparative advertisements. There was also a significant predictive relationship between agreeableness and informational comparative advertisements, with individuals possessing higher agreeableness scores showing a more favorable attitude towards informational comparative advertisements.

The third hypothesis was that individuals with higher levels of conscientiousness will yield high attitude scores for both informational non-comparative advertisements and informational comparative advertisements. The multivariate regression analysis did not yield a significant relationship between conscientiousness and attitude towards informational non-comparative advertisements and informational comparative advertisements. Thus, Hypothesis 3 was not supported. However, there was a significant predictive relationship between conscientiousness and attitude towards transformational non-comparative advertisements, with individuals possessing higher conscientiousness scores showing a less favorable attitude towards transformational non-comparative advertisements. There were no other significant relationships observed between conscientiousness and the other advertisement strategies.

The fourth hypothesis was that individuals with higher levels of neuroticism will yield high attitude scores for both transformational non-comparative advertisements and transformational-comparative advertisements. The multivariate regression analysis yielded a significant predictive relationship between neuroticism and attitude towards transformational comparative advertisements, with individuals possessing higher levels of neuroticism showing a more favorable attitude towards transformational comparative advertisements. However, there were no significant relationships observed between neuroticism and attitude towards transformational non-comparative advertisements. Thus, Hypothesis 4 was only partially supported. Although not hypothesized, Neuroticism was predictive of attitude scores for informational non-comparative advertisements, with individuals possessing higher neuroticism scores showing a more favorable attitude towards informational non-comparative advertisements.

The fifth hypothesis was that individuals with higher levels of openness to experience will yield high attitude scores for transformational non-comparative advertisements. The multivariate regression analysis yielded no significant predictive relationships between openness to experience and attitude towards any of the four advertisement strategy types. Thus, Hypothesis 5 was not supported.

4 Discussion

Being that this study was a survey, it was subject to some limitations. Self-selection bias is a major limitation, as there are undoubtedly some individuals on the internet who are more likely than others to complete an online survey [10]. In addition, relatively little is known about the characteristics of people in online communities, aside from basic demographics. There is no guarantee that the responses provided by participants are truly accurate. In addition, market researchers frequently provide incentives to individuals for completing surveys, such as a chance to win a prize or gift certificate, which can appear suspicious. Participants were offered compensation for completing the survey, but based on prior experience, a participant could be reluctant to participate. Sampling issues make it difficult to generalize study findings, since the data collected from the sample may not actually be representative of the general population.

This study was able to contribute to research investigating at the relationship between personality and attitude toward advertisement strategy, rather than the content of advertisements alone. The results show that there was a relationship between personality and attitude toward the four advertisement strategies used in this study. However, the affective components of personality may play a larger role in attitude than originally anticipated. Thus, there is much more to learn regarding this relationship, and further research should be done to add to this body of research. There appears to be a clear link between individuals scoring higher on neuroticism and transformational comparative advertising, which could be beneficial to marketers that want to further personalize their advertising campaigns. In addition, when building marketing personas, it could be helpful to consider the affective components of personality. Although affect was not measured in the present study, it could be the case that affective components of personality may be able to explain why individuals scoring high on extraversion engage regardless of advertisement strategy, and individuals scoring high on agreeableness favor more informational advertisements. Further examining the affective components of personality can help marketers better understand how to use personality to its full potential when targeting consumers.

References

1. Clark, L., Çalli, L.: Personality types and Facebook advertising: an exploratory study. J. Direct Data Digit. Mark. Pract. **15**(4), 327–336 (2014)
2. Crump, M.J., McDonnell, J.V., Gureckis, T.M.: Evaluating Amazon's mechanical turk as a tool for experimental behavioral research. PLoS ONE **8**(3), e57410 (2013)
3. Gosling, S.D., Rentfrow, P.J., Swann Jr., W.B.: A very brief measure of the Big-Five personality domains. J. Res. Pers. **37**(6), 504–528 (2003)
4. Hirsh, J.B., Kang, S.K, Bodenhausen, G.V.: Personalized persuasion: tailoring persuasive appeals to recipients' personality traits. Psychol. Sci. (2012)
5. Lucas, R.E., Fujita, F.: Factors influencing the relation between extraversion and pleasant affect. J. Pers. Soc. Psychol. (2000)
6. Magna Advertising Forecasts. http://magnaglobal.com/magna-advertising-forecasts-fall-update-executive-summary/
7. Mooradian, T.A.: Personality and ad-evoked feelings: the case for extraversion and neuroticism. J. Acad. Mark. Sci. **24**(2), 99–109 (1996)
8. Myers, S.D., Sen, S., Alexandrov, A.: The moderating effect of personality traits on attitudes toward advertisements: a contingency framework. Manag. Mark. **5**(3), 3–20 (2010)
9. Ryan, T., Xenos, S.: Who uses Facebook? An investigation into the relationship between the Big Five, shyness, narcissism, loneliness, and Facebook usage. Comput. Hum. Behav. **27**(5), 1658–1664 (2011)
10. Wright, K.B.: Researching Internet-based populations: advantages and disadvantages of online survey research, online questionnaire authoring software packages, and web survey services. J. Comput.-Med. Commun. **10**(3), 1034 (2005)

Research and Practice on the Design Method of Cultural Output for Small Commodities Based on "Without Thought Design"

Junnan Ye(✉), Siyao Zhu, Menglan Wang, Jingyang Wang, and Xu Liu

East China University of Science and Technology, Shanghai 200237, People's Republic of China
yejunnan971108@qq.com, 17708479775@163.com, 847284821@qq.com, 1435498591@qq.com, 2529540842@qq.com

Abstract. Small commodity economy is an important branch of China's market economy. However, in the design process of small commodities, there is a lack of consideration of cultural attributes and other issues. Without thought design is a design method that pays attention to the user's psychological feeling, which can help small commodities to enhance their cultural value. This paper summarizes the methods of without thought design by combing the methods of without thought design and combining with the methods of "looking for relevance" in the theory of without thought design. Through two small commodity design cases, it shows how to integrate cultural experience into small commodity design through the without thought design method. Finally, it verifies the feasibility and effectiveness of this method through questionnaire. This study will enhance the cultural brand competitiveness and commercial value of small commodities through design theory and method.

Keywords: Small commodities · Without thought design · Cultural output · Design method

1 Introduction

Small commodity is to point to those variety various, the value is relatively low small department store, certain daily life articles and partial cultural articles and so on. The small commodity industry is an important driving force in the development of China's manufacturing industry. It rapidly expands the market with small profits and large sales, but it also forms the impression of "low price and low quality" in the world. And now facing consumption upgrade, small commodities are also facing urgent transformation and upgrading [1]. At the same time, as a country with profound cultural deposits, China's cultural characteristics are a unique advantage, which has not been exploited by the current small commodity industry. On the other hand, without thought, as a design method focusing on life, is more suitable for the transformation of small commodity industry. This paper studies the theory of without thought design to help small commodities use the power of culture for brand building, so as to further export Chinese characteristic culture [2].

G. Di Bucchianico et al. (Eds.): AHFE 2020, AISC 1202, pp. 892–898, 2020.
https://doi.org/10.1007/978-3-030-51194-4_115

2 Current Situation of Small Commodity Market

Small commodities are not only closely related to People's Daily life, but also promote the national economy. Commodity industry broad prospects, is determined by a number of national policy, is run by its flexible form, low barriers to entry. But due to its low barriers to entry, most practitioners do not have system design experience, so have the homogeneity serious, commodity quality, design, lack of cultural attribute, failed to set up the brand value, such problems as lack of innovative spirit and service consciousness.

2.1 Serious Homogenization

The operation of Yiwu small commodity market is faced with the serious problem of homogenization [3]. Compulsory small commodity city because of its high visibility. And in the field of foreign trade achievements, so easy to become others to imitate the object. With the technology radiation and talent flow, less developed areas gradually gain the ability to enter the small commodity industry. At the same time, less developed areas have lower labor costs, so it is very easy to make the small commodity industry into the trap of blindly imitation, this is the source of the homogenization crisis.

2.2 Lack of Cultural Attributes

The design of small commodities generally lacks cultural connotation. Small commodities are usually very close to People's Daily life, and relatively easy to wear and tear, often need to replace, so the demand is great. Mass manufacturing means that products must use as few complex molds as possible, and products are designed with more conventional parts than personalized ones [4]. Under this premise, the backward thinking that small commodities only seek utility and not quality has been formed. Traditional small commodities have given up thinking about the spiritual needs of users at a higher level, thus abandoning the cultural attributes of products [5].

2.3 Lack of Systematic Design Methods

The lack of systematic design method makes small commodities less innovative and difficult to form an effective service system. Sellers of small commodities generally lack the consciousness of providing good service to customers. This is because in the early development of the small commodity industry, because of the market relationship between demand and supply, so that sellers formed this habit, do not know the user-centered, it is difficult to form a systematic design method to improve the quality of small commodities [6].

3 A Study of Without Thought Design Methods

3.1 The Definition of Without Thought Design

Austrian psychiatrist sigmund Freud founded the "iceberg theory". He divided people's psychological behavior into conscious and unconscious, which can be divided into preconscious and unconscious. In the "iceberg theory" model, the unconscious occupies the dominant position [7]. Unconsciousness is the experience, knowledge, memory and habit that people have accumulated for a long time, which are all some instinctive reactions. It exists at the bottom of people's psychological activities. For the most part, the unconscious dominates people's behavior, such as people unconsciously stretch after sitting for a long time, and people are accustomed to leaning their umbrellas against the wall.

Without thought design was first put forward by Japanese designer Naoto Fukasawa, is a kind of "visible" unconscious action can be converted to the design concept, namely by designers to observe people's unconscious behavior, and then use it to guide design, create a fair and reasonable design, and in use process conforms to the people's existing experience or habits, don't need to think too much, so as to improve the user experience [8].

Without thought design makes all kinds of information of products easy to understand through rational thinking, and decorates life with refined and simple design elements. Making users feel comfortable and close is the meaning of unconscious design.

3.2 Methods and Processes of Unconscious Design

The first thing to make clear is that without thought design is not simply looking for similarities or connections between things, which is wrong, because there is no connection between things. If we look for similarities between things, the result will only be boring and naive.

Things send out all kinds of information, which is read by people's five senses, thus forming the unconscious form of experience, memory and so on. In general, there are two categories of unconscious design, including four methods:

First, The Objective Sketch

Objective sketching is the direct and intuitive use of the information of things, and ultimately serve the things themselves. [9] Direct perception is an innate human response to stress, not an accumulated experience or memory, such as a fear of heights or blood. Mr. Fukazawa calls the design 'no-brainer design,' and it's comfortable to use.

Like the grooved umbrella designed by Naoto Fukasawa, he observed that people habituated to hang shopping bags on the handle of the umbrella upside down and dig a groove out of the handle, because this behavior was in line with the user's instinct and the user naturally understood the function of the groove.

Second, Look for Correlation

Different from objective sketch, all the "search for correlation" is the process of information and without thought exchange between two things, and in this process, a logical connection is formed, thus making the design result more natural and reasonable.

1. Without thought **grafting**
   The unconsciousness of thing B serves the information of thing A, thus producing additional functions or perceptions in the use of thing A. Murata Tomomi designed the 'gum eraser,' which is inspired by chewing gum. The thin surface of the gum gives the eraser the ability to erase both large and small areas.
2. **Subject perception substitution**
   The designer will extract an unconscious of thing B and magnify it until it can replace the overall perception of thing A. At this point, the physical properties of thing A still exist, but the overall experience has been replaced by the unconsciousness of thing B. The most famous work is Naoto Fukasawa's CD player, which is very similar to a ventilator in both appearance and interaction, giving the user a completely new experience.
3. **Strengthen individual concept**
   The unconsciousness of thing A merges with the unconsciousness of thing B to form A new experience-object C. For example, the ripples left by a ship when it is moving are like an unzipper. Therefore, Japanese designer Suzuki Yasunori designed a zipper boat, in which the zipper and the boat are combined on an unconscious level, which is closer to installation art in terms of expression (Fig. 1).

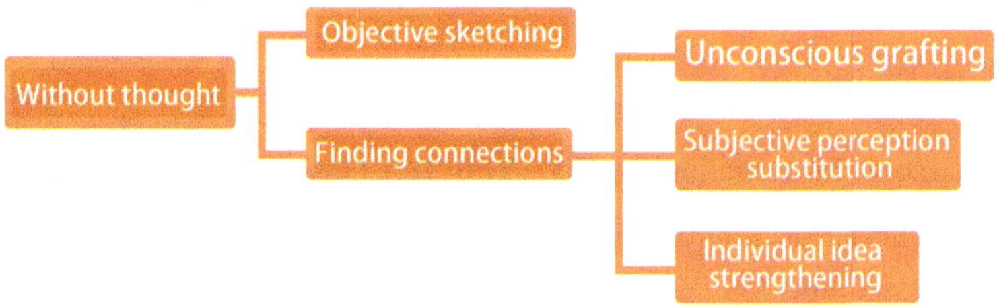

Fig. 1. Methods and processes of unconscious design.

3.3 Construct a Methodology for Without Thought Design of Small Commodities

Small commodities are closely related to people's lives. Only by using without thought design, which focuses on life, can we better solve various problems in the small commodity industry.

Objective sketching is a kind of "without thought design", which relies more on human instinct, while culture needs precipitation. On the other hand, the method of searching for correlation is to find the logical connection between two different things, which is easier to integrate culture into the design. After experiencing the five senses, the culture will form the unconsciousness. Later, the designer applies the unconsciousness of this culture to the design of small commodities through the three methods of "finding the correlation" to increase the cultural connotation of the products, which is also conducive to the cultural output.

When designing small commodities, designers should start from two aspects of service function or service perception, and then use the method of unconscious design to abstract specific small commodities and rise to the level of unconsciousness, and then use the method of finding correlation to further materialize the design intention. If it starts from the service function, it can use the method of symbol and homophony. If we start from perception, we can use experience, memory and other methods to achieve specific results (Fig. 2).

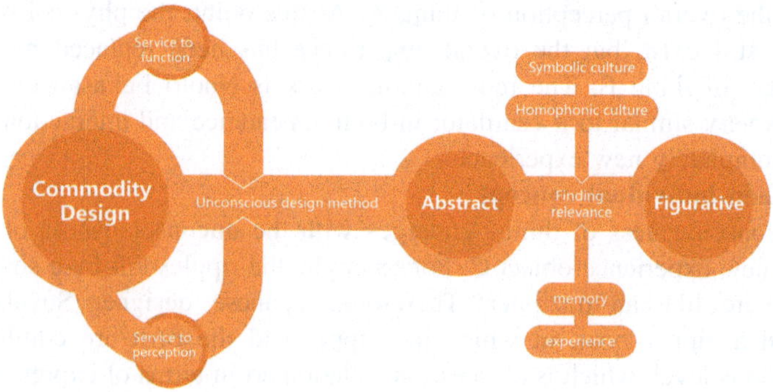

Fig. 2. Construct a methodology for unconscious design of small commodities.

4 Design Practice

4.1 Start from the Function – "Give You Top Gas" Pen Bag Design

"Give You Top Gas" is a on the test scenarios pen bag products, its use crowd mainly between 10 to 22 years old, this age group is also known as Z, Z crowd has obvious tendency of brand, for them, brand and product is a kind of channels and means of their social, they, too, in order to reveal their attitude.

The design of this product starts from the function of examination and combines the method of unconscious grafting to integrate the Chinese champion culture into the design. Symbolism is a way of materializing the design intention. When people see the image of the top scorer, they will associate it with the auspicious implication of the exam and give a positive psychological hint. As the main visual communication image of this design, the reverse image of "losing the champion" is more likely to be popular among the target population, and the slogan of "give you the champion" directly conveys a positive implication. When users use the pen bag in the exam room, the unconscious champion culture will naturally help them to keep a positive mood for the exam (Fig. 3).

Fig. 3. "Give You Top Gas" pen bag design.

4.2 Starting from Perception – The Packaging Design of "Golden Age Looking Back" Gift Box

As a New Year gift box packaging design, "Golden Age Looking Back" series with the tang style aesthetic tone, in order to improve the quality of the gift box at the same time, highlight the identity of the tang culture. This product starts from perception and USES people's memories of the flourishing tang dynasty to create the packaging design of this gift box [10].

Festival gift boxes should first make consumers feel the festive atmosphere of joy and peace. Therefore, the information about prosperity in the tang dynasty is abstracted from the patterns of cultural relics in the tang dynasty and finally implemented in the packaging design. The appearance of the rich and elegant objects in the tang dynasty carries people's memories of the prosperous times. After receiving such a gift box, users will feel the strong auspicious atmosphere and intuitively feel the results of the inheritance of traditional culture. The integration of tang dynasty culture into packaging design also reflects the great potential of small commodities in promoting cultural connotation (Fig. 4).

Fig. 4. The packaging design of "Golden Age Looking Back" gift box.

5 Questionnaire Validation

This paper will conduct a survey on the perception of cultural elements of two products through questionnaires to verify the feasibility and effectiveness of the unconscious design method of small commodities. For the two products, the main questions are "whether to experience cultural elements", "the psychological feelings brought by the product" and "the degree of perception of Chinese culture in the product".

A total of 37 samples were collected in the survey. The survey results show that 83.78% of users perceive the examination culture and 59.46% perceive the no. 1 scholar culture in the pen bag products. Products that are interesting reach 67.57%, while those that are funny reach 56.76%. The average score for perception of Chinese culture was 5.22. Among the gift box products, 94.59% of the users perceive traditional culture and festival culture, 91.89% think that the products are happy, 56.76% are happy, and the average score of perception of Chinese culture is 7.76.

Through the questionnaire results, it can be concluded that more than half of the users correctly perceive the cultural elements and atmosphere that the products want to convey, thus verifying that the unconscious design method of small commodities is feasible and effective.

6 Conclusion and Prospect

According to design, discusses the without thought design, to build the design method of small commodity unconscious cultural exports, namely according to the service function and perception two paths, use for correlation of unconscious design method to find the correlation between products and culture, thus further promote the competitiveness of the commodity culture of the brand and business value.

China has a long and profound history and profound cultural background, these cultural characteristics not only need to export the carrier of small commodities, but also the source of inspiration behind the design of small commodities, small commodity market in the future will have unlimited potential. At the same time, I also hope that more designers can continue to explore and enrich the design methods of unconscious cultural export of small commodities, continuously improve the cultural value and commercial value of small commodities, and continuously export the high-quality Chinese cultural characteristics.

References

1. Chen, Z.: Enlightenment of "kingdom of small commodities" – report from Yiwu small commodities market, Zhejiang Province. J. Outlook Wkly (04), 22–24 (1990)
2. Chen, X., Ren, X.: On human care and unconscious design. J. Art Sci. Technol. **32**(11), 172–173 (2019)
3. Guo, Z., Li, Q.: Economic development status and SWOT analysis of Yiwu small commodity market. J. Tax Payment (09), 112 (2017)
4. Qiao, W.: Status quo and bottleneck of international development of China's small commodity market – a case study of Yiwu. J. Chin. Mark. (42), 187–188 (2016)
5. Liu, J.: Present situation and development prospect of China's small commodity market. J. Sci. Technol. Inf. (2008)
6. Ji, R., Teng, H.: Status quo and countermeasures of China's commodity standard system. J. Stand. Sci. (2) (2015)
7. Shi, L. (ed.): Freud: Freud on Self-awareness. Tianjin Academy of Social Sciences Press (2014)
8. Gu, L., Wang, S.: Application of the concept of "unconscious design" in product design. J. Packag. Eng. (10), 162–166 (2018)
9. Goto, T., Sasaki, M., Fukasawa, N.: Ecology of Design, p. 101. Guangxi Normal University Press, Guilin (2016)
10. Chen, P.: Shaping cultural symbols and telling good Chinese stories. J. (2019)

The Creation of Aesthetic Conception on Landscape Space Through Information Technology

Shuwang Zhao(✉)

School of Design, Shanghai Jiao Tong University,
International Center for Innovation Design, 800 Dongchuan Road,
Minhang District, Shanghai 200240, China
236992069@qq.com

Abstract. The pan visualization of contemporary landscape space is developing together with the era of information media. With the rapid spread of digital information and the wide range of communication, the pan visual expression of visual and image works has become an important form of the main creation of landscape space form. In addition to its own physical function, landscape spatial form is perceived and consumed by the public in the form of visual image. "We have been living in the era of visual elements, and it is appropriate for today's landscape to express and present visual images as the formal language of landscape." According to Jean Nouvel [1], a famous contemporary French architect. The contemporary landscape spatial form is dominated by vision, and its pan visual aesthetic experience is different from the traditional space. This is a kind of spatial aesthetic experience which is suitable for multi-dimensional, diversified and full of complexity. The creation of aesthetic conception of human in landscape space has broken through the "gap" between subject and object, and based on the concept of interactive aesthetics, it has moved towards interaction and integration.

Keywords: Complex situation · Spatial experience · Artistic conception

1 Introduction

1.1 Research Background

Virtual reality technology has not only updated people's cognitive system, but also changed people's aesthetic habits. Based on the landscape space form, the aesthetic consciousness has been improved unprecedentedly. People began to pay attention to the form beauty of physical landscape, but also to the combination of virtual reality. The concept of "truth" in "truth", "goodness" and "beauty", as the ideal realm pursued in human thought and practice, has gradually expanded into "the combination of truth and virtual" [2].

The application of new technology in the construction of landscape form not only changes the aesthetic object of landscape, but also permeates the generation of aesthetic significance into the virtual non-material field. Through the network display platform,

G. Di Bucchianico et al. (Eds.): AHFE 2020, AISC 1202, pp. 899–906, 2020.
https://doi.org/10.1007/978-3-030-51194-4_116

landscape designers gradually pay attention to the public's generation and development of landscape form. The resulting rich experience of landscape and the change of psychological level make the traditional inherent landscape The cognitive system of aesthetic perception has been strongly impacted. The so-called aesthetic of eternal meaning is no longer eternal. On the contrary, it shows the value level concern for dynamic beauty. From the perspective of the history of aesthetic research, the classical aesthetics of the Renaissance mainly pursues the preciseness of order and the sanctity of form, and mainly examines the beauty and ugliness of things through the golden section ratio as the measurement standard. The aesthetics entering the industrial period is the applicability and usability of the leading function, advocating the aesthetic thought of pragmatism. The aesthetic thought of natural ecologism is determined to get along with nature harmoniously, pursue the unity of man and nature and the overall development of natural ecosystem.

Nowadays, the aesthetic way of people in landscape space has broken through the "gap" between the subject and object, and based on the concept of interactive aesthetics, it has moved towards interaction and integration.

1.2 Research Methods

This paper uses literature analysis and comprehensive induction to analyze and compare the interactive and aesthetic relationship of landscape space in the information digital era, and produces several new aesthetic tendencies. At the same time, it explores the landscape space with the visual aesthetic scene as the leading role, summarizes the aesthetic perception experience after the form of Pan visualization, and produces three kinds of spatial aesthetic perception experiences that are suitable for multi-dimensional, diversified and full of complexity.

2 Landscape Space Aesthetics and Artistic Conception

2.1 Comparison of Several Aesthetic Types

In the information age, the focus of aesthetic development is the unity of information and content, and the rationality of content distribution, such as complexity and simplicity of information, more and less, emptiness and reality, etc., resulting in a new information interaction aesthetics (as shown in Fig. 1).

Therefore, under the influence of the present aesthetic thought of information interaction aesthetics, a series of new aesthetic tendencies have been constructed in the landscape space, which is based on the concept of interaction aesthetics, towards interaction and integration.

2.2 Synaesthesia of Complex Situation

The interactive experience of landscape space in the era of synaesthesia and mood information of complex situation is rich and colorful, especially the role of media digital technology has led the public to produce a series of new psychological changes

Aesthetic connotation	Aesthetic types			
	Classical aesthetics	Naturalistic aesthetics	Machine aesthetics	Information aesthetics
Aesthetic ideal	Form harmony	Natural harmony	Functional harmony	Information harmony
Aesthetic standard	Mathematical norm	Environmental standard	Functional standard	Quality and quantity standards
Constitutive characteristics	Form art beauty	Environment art beauty	Function - artistic beauty	Interaction - artistic beauty
Formal characteristics	Static equilibrium	Natural organic	Dynamic equilibrium	Virtual reality balance

Fig. 1. Aesthetic type and connotation

caused by perception in the landscape scene, which not only exist in the perception dimension, but also rise to the emotional level of resonance. According to Deleuze, the philosopher, body instinct is a kind of "flow" of desire. He first proposed the concept of "body without organ", thus breaking the theory that the body is subject to consciousness, redefining the connection mode between active behavior and body organs, and forming a multiple and open "Synaesthesia" between subconscious perception and body organs. In today's landscape form, the public is easier to obtain the perception experience directly from the landscape space. On this basis, combined with their own cognitive background and the information elements that designers want to express, it rises to an immersive mental state, which can also be called the synaesthesia mood of complex emotions. The traditional landscape space also has such a synaesthesia space, but the degree is different from that in the information age. Under the background of technology development, designers gradually realize that creating a kind of synaesthesia mood has a very important significance in landscape space design, so in the specific space aesthetic experience, synaesthesia mood should be interpreted in two different ways:

1. The contemporary landscape space presents the aesthetic perception of the complex situation from non-image to synaesthesia. With the help of the means of complexity science and technology, it emphasizes the multiple perception of the non-organ body in the urban space, and carries on the instantaneous comprehensive experience of the space with the flow of "encounter". At this time, the perceptual experience is a kind of encounter characteristic, which is a kind of encounter after people pass

through different fields such as building space, volume, form, etc., and forms synesthesia resonance of different levels, different times, different events. From the perspective of Deleuze's body perception without organs, we can see that he has a negative attitude towards the body dominated by consciousness. Therefore, in the process of perception, Deleuze emphasizes a kind of encounter rather than recognition. He believes that recognition is a process in which rational cognition and prior logical structure coincide with experience in the process of perception, that is, the assimilation of perception in consciousness. The real artistic creation is a kind of encounter, which is the shock and stimulation of the senses of the body without organs [3]. The perception of landscape space emphasizes instantaneity, which is the result of the body breaking through the body structure. In the "concept movement 2005" interactive device space designed by Olaf ariason, according to the free movement of the viewer in the space, the wooden floor in the space can convert the energy of movement into the signal of projection and present it through the wall (as shown in Fig. 2). Different motion sensors transmit energy signals and produce different random patterns, just like waves on the water surface, which are sometimes rapid and sometimes slow. The game of light in space dissolves the walls of space, and generates synaesthesia through the intuitive induction and interactive stimulation of the body. The viewer seems to be on the vast water surface, and the ripples on the water surface break through the boundaries of space and extend to all sides.

Fig. 2. Olaf Ellison concept movement interaction space

2. In the complex situation aesthetic perception of landscape space, the form of the body without organs can be presented in the space form of complex situation with the dynamic state of mutual penetration of body desire flow. Through the body perception flow of synaesthesia, the image of landscape space can be presented in the way of vision through the primitivity in the sense, so as to give the viewer a multi-sensory image visualization Perception and experience [4]. In this process, it reflects the original state of unconsciousness, the sense of body without organs. The first step in this process is to establish a domain, and the well establishes a relationship between the domain and all the forces that pass through it. In the landscape, the fields and forces here refer to vision, hearing, touch, smell, etc. Therefore, when

perceiving the landscape space in the way of synaesthesia, the space form of landscape brings us visual, physical and sensory stimulation and impact. It emphasizes the desire flow of the body, the instantaneous perceptual power of instinct, so as to abandon physiology and biology and take the way of body perception without organs as the landscape form to the emergence of "body image" [5].

2.3 Nomadic Mood of Space Experience

In the process of space experience of landscape, if the experiencer is in a happy mood, he can completely immerse himself in the space scene, thus creating a state of mind called the unity of man and nature. Specifically speaking, this mood contains two meanings: one is the experience mood between human and space, and the other is the feedback mood generated in the experience. In the traditional landscape space shaping, designers are very keen to place their good will in the landscape, which is called the integration of emotion, scenery and human, which is a representation of the primary nomadic mood. In the space dimension of modern landscape construction, nomadism is more manifested in the pursuit of whether space and psychology fit, and the expectation of freedom and tranquility. In his book "pleats: Leibniz and Baroque style", the philosopher Deleuze expounds his deep interpretation of the formation and evolution of space. He believes that the emergence of nomadic thought is based on the organizational structure in space, and its most important characteristics are the integration of multiple elements, the homogeneity of different elements and the openness of all dimensions. It is not necessary to deliberately add or remove new elements in space construction It is to redefine the sense of order by adjusting the arrangement of space forms, to form a space order that follows certain internal logic, and to change correspondingly according to the changes of surrounding environment.

On the other hand, the use of landscape media technology is more to stimulate the public to produce spiritual pleasure, which is a form of expression to meet spiritual demands. The first is to construct another dimension of landscape space scene through virtual reality, parameterization and other technical means, and the second is to obtain a high sense of immersion based on a new cognitive communication means, which may exist in the real space, or may be superior to the real space, showing a sense of freedom. At the same time, as an experiencer, the public will have a resonance with the space, as if they are part of the landscape space form, so as to obtain the surprise and pleasure of the experience. It is true that there are some limitations in the application of technology, especially a series of short-term problems that can not be solved, and the creation of scene space can not achieve full-scale virtualization. However, the production of nomadic mood is not entirely driven by technology, but also determined by the personality, cultural heritage, experience cognition and other factors of the experiencer. With the help of technology, the experiencer will also give birth to different levels of nomadic mood in the space experience [6].

The nomadic emotion in the landscape space is more detached from the shackles of reality. With the upgrading of the space atmosphere, it produces the maximum spiritual release and the double effects of Transcendentalism and deepening of experience. In addition, the creation of nomadic mood is more of a kind of spiritual self-release. The concept of time and space seems to be blurred. Walking in a space full of connotation can produce the most in-depth resonance with the designer's thinking. A very classic design case was completed by the design team in Shenzhen, China. They widely used the technology of virtual reality to create a wonderful panoramic view of the starry sky and realize the full-scale interactive aesthetic experience between the public, nature and artificial environment. The project is named "starry sky", covering an area of about 400 m^2 and the total height of the dome is 9 m. The initial hope of the design team is to simulate the distribution of stars in the northern hemisphere through this project, so as to show people the vastness of the universe (as shown in Fig. 3). We have realized these two characteristics in technology means and design experience means.

Fig. 3. Artificial star experience space

2.4 The Symbiotic Artistic Conception of the Whole Space

Landscape creation belongs to man-made natural landscape, which is different from pure natural landscape but complements each other. When the experiencer can't perceive the difference between the two in the landscape space and fully integrate into the atmosphere created by the space, a kind of symbiotic spiritual mood is generated. The so-called symbiosis, originated from the theory of biology, mainly said that there is a kind of interdependence and correlation between cells and other cells. As we all know, Darwin's evolutionary biological chain, no one can exist without the whole, because without the relevance of the biological chain, the so-called symbiosis is more of a paper battle. It can be seen that symbiosis refers to a holistic mechanism system, in which all elements are interrelated, influenced and restricted. The harmonious coexistence of elements and elements together forms the establishment of symbiosis system. Since ancient China, people have been pursuing the "unity of man and nature" in the construction of gardens, combining the living environment with the natural landscape organically, achieving the goal of moving scenery step by step and being scenery everywhere. It can be seen that landscape design should be based on the form of nature, but also higher than nature, and its core goal is to reflect people-oriented. With the development and progress of technology, landscape creation faces a series of new challenges and opportunities. This is because people's aesthetic tendency has changed in the complex and changeable social forms. Before creating, designers should first make a full investigation and Research on the space scene, and pay attention to the internal logical relationship between different monomers in addition to the formation of monomers. So symbiotic artistic conception is also the expression way of landscape space. In specific space construction, we need to focus on the following dimensions:

Connection and interaction of space nodes: in landscape creation, considering the relevance of form and space, the landscape nodes in the space unit are often focused on in the design. As a "grey zone" connecting different spaces in series, the landscape node not only plays a role in connecting each space, but also can't break the overall presentation effect of the landscape space because its design is too "characteristic". Therefore, when we introduce complex technical forms into a landscape space which is inclined to natural form, we must use the non-linear thinking analysis method to optimize the packaging of landscape space. At the same time, because the relationship between the monomer and the monomer in the landscape space is also complex, we also need to excavate the internal systematic relationship to form an integrated parametric model to organize the spatial form layout of the generating unit.

The integration of spatial presentation: the spatial form of landscape has the characteristics of opening and changing, whether it is a single spatial unit, landscape node and spatial sequence follow an internal self-regulation mechanism. In addition, landscape space does not exist independently. It has a close relationship with natural environment and human environment. In order not to make the construction of landscape space and surrounding environment seem incompatible, it is necessary to carefully analyze and reference the relationship between them in the early stage of design, because natural environment is not only the basis of landscape creation, but also the basic element for landscape to show its design charm. As a technical factor of shaping landscape form, it not only realizes the multiple media means of landscape, but also realizes the perfect integration of technology and nature, which is the premise and development goal of media technology integrating into the process of landscape creation. The existence of technology should increase the degree of contact between the public and nature, and make the viewing and experiencing landscape form more natural. In addition, the accumulation of technology is not necessarily able to achieve the ideal creation and presentation effect. All designs have a so-called measurement standard, which is the symbiotic artistic conception of landscape. Only with this concept as the core guidance, we can always pay attention to the complementary relationship between nature and man in the process of space creation [7].

"Body" of the experiencer in the space: no matter what type of landscape creation, the most fundamental core element lies in the study of human beings. The process of creation is also to deal with the organic unity between human beings and nature. It should not only be dominated by the subjective needs of human beings, but also excessively break the original landscape natural environment, that is, to find the balance between the two. The confliction between the two focuses on whether to focus on the needs of the experiencer or the environment of the natural original ecology. Based on this, there are two different creative thoughts: people-oriented and nature restoration. People oriented needs to highlight people's personality in the design. No matter the space form, unit, experience, interest and other links of the landscape, people's interest should be aroused to create a "fragmented" aesthetic experience. The importance of the experiencer is displayed in the space and becomes the "center" of the landscape. The design concept of restoring nature starts from the naturalization of landscape spatial form, which weakens the individual behavior of the experiencer and integrates it into the overall landscape ecosystem. If we want to unify the two, we can first abstract and refine the information elements of the landscape, and transform the materialized information elements into the virtualized information elements. In the process of landscape spatial form design, a series of complete story clues can be formed by several different factors, and the natural space and technical factors can be linked together by analogy to achieve the unity of opposites and strengthen the multi integration of information.

3 Conclusion

A series of new aesthetic perception elements are generated in the participation experience of the landscape space created by the digital multimedia. These new aesthetic modes, along with the transformation of the space experience mode, derive three kinds of artistic conception, namely, the synaesthesia of the complex situation, the nomadic artistic conception of the space experience and the symbiotic artistic conception of the whole space.

References

1. Nouvel, J.: Works and Thoughts, p. 8. Editorial Department of Master Series. China Electric Power Press (2006)
2. Cai, L., Zeng, J.: Philosophical connotation and theoretical development of information architecture aesthetics. Urban Arch. (2) (2005)
3. Han, G.: On the creative value of body feeling: an interpretation of Deleuze's logic of feeling. J. Nanjing Normal Univ. (Soc. Sci. Ed.) (03), 40–45 (2010)
4. Deleuze, G., Jiang, Y. (Trans.): Capitalism and Schizophrenia: Thousand Highlands (1985)
5. Jiang, Y.: Research on Deleuze's Body Aesthetics, p. 132. East China Normal University Press, Shanghai (2007)
6. Zhang, Z.: Aesthetics and intuitionism. Tangdu J. (3), 63–68 (2013)
7. Jia, W.: Interaction of architectural design in the information age, p. 98. Doctoral dissertation, Tianjin University (2008)

Motor Control of Hand Force for Visual Indicator Without Hand Displacement

Takashi Oyama(✉) and Teruaki Ito

Okayama Prefectural University, Kuboki 111, Soja, Okayama, Japan
{ooyama, tito}@ss.oka-pu.ac.jp

Abstract. This study assessed the hand force exerted to a fixed controlled object under three conditions with the dominant and non-dominant hand. The position, velocity, and acceleration of a visual indicator were determined by the exerted hand force and the participants were asked to move the indicator to a target position. Our findings indicated that the non-dominant hand had an advantage when an appropriate magnitude and time interval was required while the dominant hand had an advantage in the pseudo-natural condition. These findings could be applied to the designing of devices and interfaces.

Keywords: Motor control · Force control · Man-machine interface · Laterality

1 Introduction

This study investigated hand force control in humans to contribute to improvement of the usability of haptic interfaces. There has been recent development of many devices and gadgets that are controlled using somatosensory interfaces. For example, multiple smartphone systems use a touch screen as the tactile interface, and some virtual reality systems use a motion sensor as the proprioceptive interface. Therefore, it is important to determine the human somatosensory characteristic in the interaction between humans and devices to develop convenient systems.

There have been many previous studies on human movements with respect to motor control [1–5]. Contrastingly, there have been few studies on human force control without concrete movements [6], e.g., using a pointing stick on a laptop. In this study, the aspect of hand force control without hand displacement was investigated. This study was approved by the ethic committee of Okayama Prefectural University.

2 Experiment

2.1 Measurement Experiment

The subject of this study was 4 right-handed students. The subject was instructed to grasp an almost fixed stylus of a haptic device (Touch X, 3D Systems) and exert force to the stylus without hand displacement. Subsequently, the subject was instructed to move an indicator circle displayed on an LCD monitor toward the target position. The distance between the start position of the indicator circle and the target position was set

G. Di Bucchianico et al. (Eds.): AHFE 2020, AISC 1202, pp. 907–912, 2020.
https://doi.org/10.1007/978-3-030-51194-4_117

to 20 cm. The position of the indicator circle changed depending on the hand force exerted to the stylus by the participant. The position of the indicator circle was controlled by hand force in three different aspects; namely, its position (FP condition), velocity (FV condition), and acceleration (FA condition). The manual asymmetry of force control was accessed by asking the subject to use both dominant and non-dominant hands, respectively. This was based on previous findings on differences in the posture preservation between the dominant and non-dominant hand [4]. 6 experimental conditions (3 force conditions $\times$ 2 hand conditions) were set up, and 45 trials for each condition were conducted. The arrangement of start and target positions flipped depending on the using hand. The position of the indicator circle and the hand force exerted to the stylus at 60 Hz were measured.

2.2 Simulation Experiment

Many previous studies have reported computational models of human motor control based on optimal theories [1–3, 5]. Flash & Hogan [1] showed that humans produce movements under the condition that allowed optimized movement smoothness. The smoothness of a movement trajectory $x(t)$ is quantified as shown in Eq. (1):

$$J = \int_0^{t_f} \left(\frac{d^3x}{dt^3}\right)^2 dt \qquad (1)$$

Where, t_f denotes the movement duration. A computed trajectory where value J in Eq. (1) is minimized reflects actual human movement with unconstrained extremities. Using the variation method, Eq. (1) can be solved under the boundary conditions $x(0) = \dot{x}(0) = \ddot{x}(0) = 0, t_f = 1, x(t_f) = 1, \dot{x}(t_f) = \ddot{x}(t_f) = 0$; subsequently, the optimized solution $x(t)$ is expressed as the fifth-order polynomial function shown in Eq. (2). The trajectory $x(t)$ is known as the minimum jerk (third derivative of position) trajectory.

$$x(t) = 6t^5 - 15t^4 + 10t^3 \qquad (2)$$

According to a method of variational calculus, if $(\ddot{x})^2$ and $(\dot{x})^2$ are specified as functionals in Eq. (1), the optimal functions are expressed as their third- and first-order polynomials, respectively.

When the movement path in a task is constrained, e.g., when a human grips a handle and rotates a crank, the optimized function is expressed as Eq. (3) [7,8].

$$L = \int_0^{t_f} \dot{F}^2 dt \qquad (3)$$

Where F denotes the hand force exerted to the controlled object. This optimal function suggests that human motor control is responsible for hand force smoothness when using a device. The optimal function of Eq. (3) is associated with the optimal function using the derivative of joint torques as the smoothness criterion [2]. These previous studies speculated that a human motor control system accounts for the dynamics of arms and controlled objects when required.

As aforementioned, hand force exerted to the fixed stylus moved the indicator in three different aspects; namely, position, velocity, and acceleration. If a motor control system takes into account the dynamics of arms and controlled objects, the strategies of hand force exertion vary based on the relationship between hand force and indicator movement. However, it is possible that a motor control system rarely accounts for dynamics information but rather predominantly considers visual and kinematics information given the priority of visual information in making movements [9]. This simulation experiment aimed to investigate whether a motor control system accounts for dynamics when controlling a visual indicator using hand force. This study assumed that if dynamics affected motor control, the indicator trajectory would differ due to the force conditions; contrastingly, if it did not, the trajectories would be invariant regardless of the conditions.

Based on the optimal function Eq. (3), the optimized trajectories of the indicator could be calculated. The trajectories were expressed as first-, third-, and fifth-order polynomials for the FP, FV, and FA conditions. This study analyzed the mean differences between the actual indicator trajectory and the calculated optimal trajectory.

3 Result

3.1 Measurement Experiment

This study analyzed the error of movement endpoint, which was defined as the distance between the movement endpoint and the target position (EE), as well as the movement duration, which was defined as the time interval between the movement start and the movement end (MD). Figure 1 shows the means and standard deviations.

A two way analysis of variance (Force condition × hand) revealed a significant main effect of Force condition in EE and MD (both $< .001$), no significant main effect of Hand in EE ($p = .283$) and MD ($p = .294$), and a significant interaction between Force condition and Hand in EE ($p < .05$) and MD ($p < .001$). Multiple comparisons with Tukey's HSD (honestly significant difference) test revealed no significant difference between the dominant and non-dominant hands in EE (FP: $p = .063$, FV: $p = .967$, and FA: $p = .509$). The MD of the non-dominant hands was shorter when the hand force corresponded to the indicator velocity ($p < .001$), and that of the dominant hands was shorter when it corresponded to the indicator acceleration ($p < .001$).

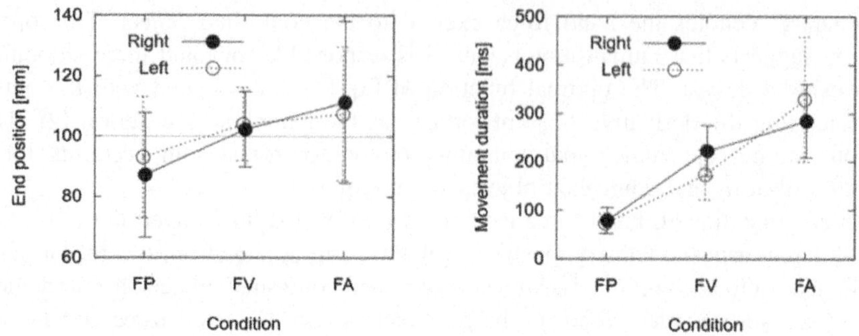

Fig. 1. The endpoint and movement duration of the indicator trajectories under each condition (FP, FV, and FA).

3.2 Simulation Experiment

We applied the first-, third-, and fifth-order polynomials based on the optimal theory to the measured trajectories in each condition. Generally, high-order polynomial approximation is better than low-order; therefore, it is inappropriate to inadvertently evaluate high-order polynomial approximation. The residual difference between n-th and m-th polynomials ($n > m$) was defined as shown in Eq. (4):

$$R_{n,m} = \frac{1}{N}\sum_{i=0}^{N-1}\left(p_i - \sum_{j=0}^{n} a_j t^j\right)^2 - \frac{1}{N}\sum_{i=0}^{N-1}\left(p_i - \sum_{k=0}^{m} b_k t^k\right)^2 \tag{4}$$

Where N denotes the data size of indicator movement, $t = i\Delta t$, and $p_0, p_1, \cdots, p_{N-1}$ denote the time-series of indicator movement. Figure 2 presents the means and standard errors of $R_{5,3}$ and $R_{5,1}$ for each condition.

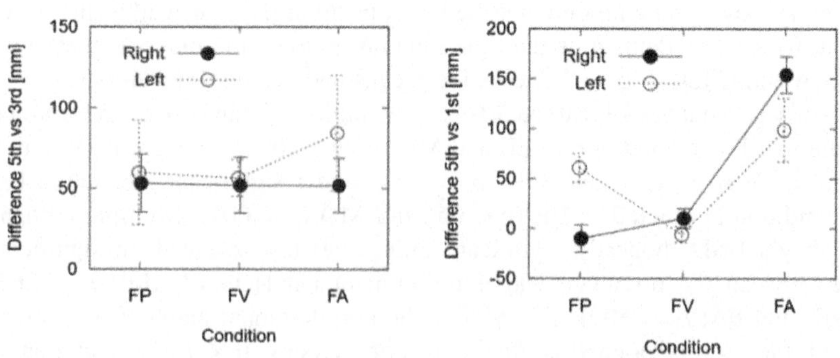

Fig. 2. Residual differences $R_{5,3}$ and $R_{5,1}$ for each condition FP, FV, and FA.

This study performed a two-way analysis of variance (Force condition × Hand) for $R_{5,3}$ and $R_{5,1}$. In $R_{5,3}$, there were no significant main effect of Force condition ($p = .838$) and Hand ($p = .455$), as well as no significant interaction between Force condition and Hand ($p = .778$). In $R_{5,1}$, there was a significant main effect of Force condition ($p < .001$), no significant main effect of Hand ($p < .001$), and a significant interaction between Force condition and Hand ($p = .989$). Multiple comparisons revealed a significant difference between the condition of FP and FV condition for the left hand ($p < .05$), FV and FA condition for both hands (both $p < .001$), and the right and left hand for the FP condition ($p < .05$).

4 Discussion

This study investigated the manual laterality and motor control in a task involving hand force exertion to a fixed device. It was found that the motor control systems of both dominant and non-dominant hands produce movements with similar accuracy. However, there was a significant difference in the movement durations. Given the speed-accuracy tradeoff [10], a shorter movement duration could indicate a higher accuracy. The FA condition was a pseudo-natural condition since accelerations are usually proportional to forces in the real world. Therefore, the finding that dominant hands could perform better than non-dominant hands in the FA condition seems reasonable. In the FV condition, non-dominant hands performed better than dominant hands. Previous studies have reported an advantage of non-dominant hands in some types of task [4, 11, 12], which have been assumed to involve posture maintenance and impedance control [13] for the target position. A profile of the measured hand force formed a roughly trapezoid-like shape. This reflects three phases of force control as follows: initial increase, maintaining of a constant magnitude over an appropriate time interval, and final decrease. This maintenance of a constant magnitude is speculated to be similar to posture maintenance, which explains the better performance of the non-dominant hand in the FV condition. Contrastingly, there was no significant difference in EE between the dominant and non-dominant hands in the FP condition. This could be attributed to lack of manual asymmetry in the exertion of a constant force magnitude without applying a specific time interval.

The simulation experiment results revealed differences in the polynomial approximation between the FV and FA conditions. This reflects differences in the indicator movement trajectories depending on the force conditions; moreover, indicator movement trajectories in the FA condition should have a higher order polynomial approximate than that in other conditions. This suggests that a motor control system does not consider visual information concerning the position of a controlled object as the optimal criterion. This study adopted the optimal function expressed as Eq. (3); however, other previously reported optimal theories are also plausible. As aforementioned, the profile of exerted hand force formed a trapezoid-like shape. Therefore, it is likely that a theoretically optimal motor control system produces hand force based on bang-bang control and minimization of force transition [14]. Given the assumption that a motor control system prefers fewer hand force transitions, the task in the FP condition was easier than that in the other conditions.

Our findings suggest that there is hand preference for force control without hand displacements and provide insights that could improve the design of devices and interfaces.

Acknowledgments. This work was supported by JSPS KAKENHI Grant Number 15K21301.

References

1. Flash, T., Hogan, N.: The coordination of arm movements: an experimentally confirmed mathematical model. J. Neurosci. **5**, 1688–1703 (1985)
2. Uno, Y., Kawato, M., Suzuki, R.: Formation and control of optimal trajectory in human multipoint arm movement. Biol. Cybern. **61**, 89–101 (1989)
3. Richardson, M.J.E., Flash, T.: Comparing smooth arm movements with the two-thirds power law and the related segmented-control hypothesis. J. Neurosci. **22**, 8201–8211 (2002)
4. Wang, J., Sainburg, R.L.: The dominant and nondominant arms are specialized for stabilizing different features of task performance. Exp. Brain Res. **178**, 565–570 (2007)
5. Ben-Itzhak, S., Karniel, A.: Minimum acceleration criterion with constraints implies bang-bang control as an underlying principle for optimal trajectories of arm reaching movements. Neural Comput. **20**, 779–812 (2008)
6. Tsuji, T., Miki, Y., Ito, K.: Hand force during constrained arm movements and virtual trajectory hypothesis. SICE **28**, 366–373 (1992)
7. Ota, K., Svinin, M.M., Luo, Z., Hosoe, S., Laboissière, R.: Optimal trajectory formation of constrained human arm reaching movements. Biol. Cybern. **91**, 23–36 (2004)
8. Ota, K., Svinin, M.M., Luo, Z., Hosoe, S.: Optimal trajectory formation of human reaching movement in crank rotation task. IEICE D-II **87**, 1707–1717 (2004)
9. Flanagan, J.R., Rao, A.K.: Trajectory adaptation to a nonlinear visuomotor transformation: evidence of motion planning in visually perceived space. J. Neurophysiol. **74**, 2174–2178 (1995)
10. Fitts, P.M.: The information capacity of the human motor system in controlling the amplitude of movement. J. Exp. Psychol. **47**, 381–391 (1954)
11. Bagesteiro, L.B., Sainburg, R.L.: Nondominant arm advantages in load compensation during rapid elbow joint movements. J. Neurophysiol. **90**, 1503–1513 (2003)
12. Oyama, T., Sako, A.: Difference in bi-articular muscle during postural fixation between dominant and non-dominant arms. In: IIAI-AAI 2015, p. 187 (2015)
13. Yadav, V., Sainburg, R.L.: Handedness can be explained by a serial hybrid control scheme. Neurosci. **278**, 385–396 (2014)
14. Karniel, A.: The minimum transition hypothesis for intermittent hierarchical motor control. Front. Comput. Neurosci. **7**, 12 (2013)

Style and Color Design of Technology Packaging

Lung-Wen Kuo[1], Chih-Chun Lai[2(✉)], and Ming-Chi Hsieh[3]

[1] Department of Multimedia Design, Sanming University, Fujian 365004, China
longwen0425@gmail.com
[2] Department of Industrial Design, Tatung University, Taipei 104, Taiwan
lai@ttu.edu.tw
[3] Graduate Institute of Design Science, Tatung University, Taipei 104, Taiwan
art8studio@gmail.com

Abstract. There are various types of packaging designs with different design styles. However, in terms of overall packaging materials, packaging mostly used cartons, plastics, iron boxes and textiles, of which cartons is the most commonly used material. For consumers, the product's packaging design will first be noticed in a store in the aim to attract consumers, and a good packaging design can highlight the importance of products. This study is based on two experiments color box design of the computer mouse and best visual effect of the packaging design of computer mouse. From the results of the color box design experiment, we concluded that yellow and red packages are the most commonly used, followed by black, while green is the least used. From the results of the best visual effects packaging design experiment, the visual effect on transparent window design of 60%:60% (114 mm × 90 mm) is considered to be ideally suited for the computer mouse, whereas the visual effect on transparent window packaging design of 75%:75% (127 mm × 75 mm) is considered to be very unsuitable.

Keywords: Product packaging design · Color design · Technology products

1 Introduction

Packaging design can be classified into industrial packaging and commercial packaging according to the product. Industrial packaging has the mainly purpose of protecting products from being damaged during transportation, whereas commercial packaging is mainly for attractions in stores, which considers the color, style and product of the packaging design.

The basic conditions of packaging design is to protect the product content from damage by external forces, make the product stand out more and accurately contain the product [1]. Scholar Xu Xingrong believes that packaging design must have an attractive style, easily opened and closed, and convenient for store displays [2]. Product packaging is based on materials and styles that each have a representative design for the package or container. Packaging design shall be considered as part of the product. In addition to the function of protection, it must also stimulate consumers' desires [3].

G. Di Bucchianico et al. (Eds.): AHFE 2020, AISC 1202, pp. 913–919, 2020.
https://doi.org/10.1007/978-3-030-51194-4_118

The most important functions of graphics in packaging design are the representation of the product, creating brand image and decoration [4]. Presentation methods of packaging graphic design include concrete graphics, semi-concrete graphics and abstract graphics. Concrete graphics are represented by drawings or images, semi-figurative graphics can be represented by illustrations, and abstract graphics can be divided into geometric, non-geometric and incidental graphics [5]. Abstract graphics can give consumers a stronger visual impression and easily express a unique appearance.

Product packaging design can become the symbolic meaning of a brand [6]. The visual coherence of packaging design can raise product and brand awareness [7]. In the study of cognitive linguistics, design elements in visual space include the function of specific metaphors and image patterns [7] and the impression brought by the appearance of packaging may be transferred to other sensory impressions [8]. Products must focus on successful packaging design methods that can attract the consumers' visual attention. There are many concepts and techniques of various packaging design materials and structures, which comprehensively apply unique methods to stimulate new thinking directions.

2 Discussion

In general, the structural composition method of packaging design includes shapes of the color box such as square, rectangular, circular, heart-shaped or polygonal. The shape and materials of the packaging design is the most important in visual psychology to be ideally accepted under the reasonable range of packaging cost, and must first be discussed and studied when designing. For the selection of packaging materials, the price of packaging materials varies greatly and the choice of paper seems very essential. The printing method also affects the cost and the most reasonable design method is to match the positioning and pricing of the product. A lot of money can be spent to make advanced and complicated colored boxes for the packaging of high-priced products, whereas the cost must be carefully considered for low-priced products.

3 Research Methods

For the purpose of research and analysis, the survey and statistics of the packaging color for the commercially available packaging color boxes are based on the color and shape of the packaging design and selected a total of 100 color box styles to be statistically analyzed for the basis of research on future packaging designs.

3.1 Survey on Styles and Colors of Color Boxes

This experiment on the size of computer mouse has a total of 30 participants aged 18–22, which were randomly selected college students, including 15 males and 15 females. The experimental results show that Black C (80) is the most popular color for color boxes, followed by 877C silver (67), whereas 877 silver packaging (18) with color box style 5 is the most popular shape and color of the color box design, which will be deemed as having the best visual effects on computer mouse packaging design for experiment 5 to design the style and color of color boxes.

According to the experimental results on the emotional characteristics of colors, the eight most commonly used packaging styles, coupled with 10 commonly used colors for color boxes, are combined into 80 packaging design test samples for the experiment and the packaging styles and colors with the most choice are selected as the best style and color of computer mouse packaging color box (Fig. 1).

color	1	2	3	4	5	6	7	8	Total	Sort

	10	6	6	8	7	7	9	8	62	3
	11	10	12	9	5	12	10	9	80	1
	2	3	5	3	3	5	4	5	29	7
	6	2	4	7	2	4	3	4	32	6
	3	5	6	4	5	6	6	7	42	4
	1	0	1	2	0	0	1	1	6	10
	4	8	3	5	5	2	6	3	36	5
	2	2	4	3	2	3	2	4	22	9
	3	3	3	4	3	2	3	2	24	8
	7	11	6	5	18	9	6	7	67	2
Total	50	50	50	50	50	50	50	50	400	

Fig. 1. Statistics on style and color of color boxes

The experimental results show that Black C (80) is the most popular color for color boxes, followed by 877C silver (67), whereas 877 silver packaging (18) with color box style 5 is the most popular shape and color of the colored box design.

3.2 The Experiment on Best Visual Effects for Computer Mouse Packaging

For the experiment on best visual effects for computer mouse packaging, we use a 64.5 × 95 × 35.5 mm mouse (No. 2 mouse) as basis for designing 5 types of size ratio of transparent windows for computer mouse color box packaging, in order to obtain the computer mouse packaging style with the best visual effect through experimental methods (Figs. 2 and 3).

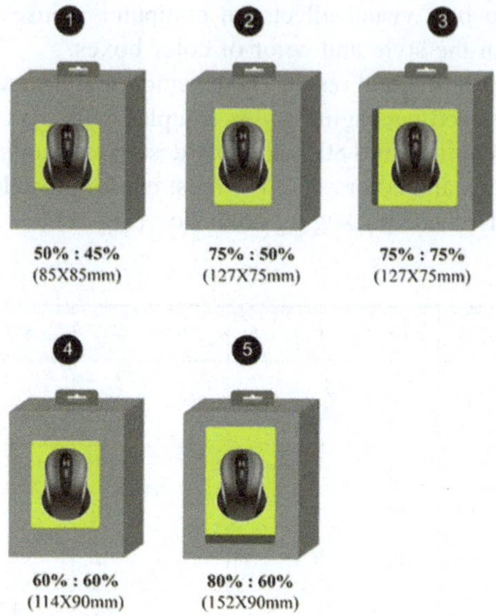

Fig. 2. Five sizes of transparent window designs are used for the experiment on best visual effects for computer mouse packaging

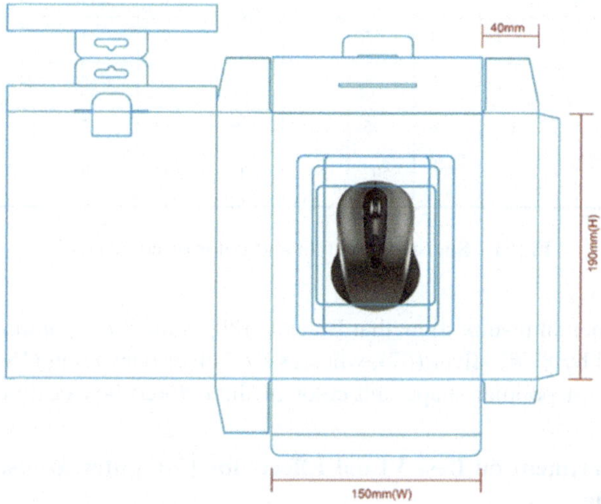

Fig. 3. Extended drawing of color box (knife mold) size is used for the experiment on best visual effects for computer mouse packaging

The size of the colored box is 190 mm (length) × 150 mm (height) × 40 mm (height). The size ratio of transparent window design are as follows: color box 1 of 50%:45% (85 mm length × 85 mm width), color box 2 of 75%:50% (127 mm length × 75 mm width), color box 3 of 75%:75% (127 mm length × 75 mm width), color box 4 of 60%:60% (114 mm length × 90 mm width) and color box 5 of 80%:60% (152 mm length × 90 mm width) (Fig. 4).

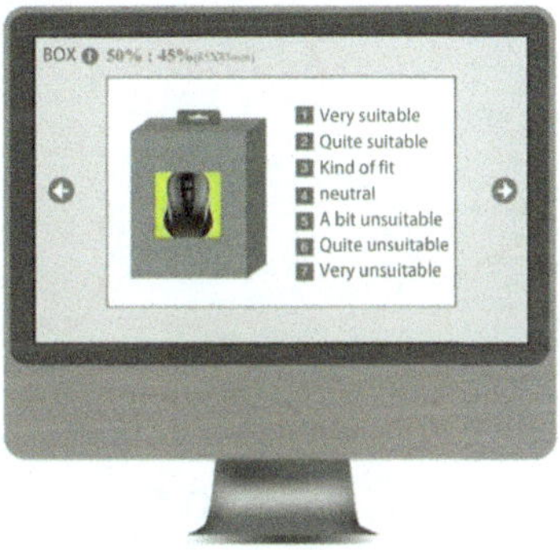

Fig. 4. User interface used in the best visual effects experiment for computer mouse packaging

Step 1. After viewing the package on the screen, the test subject selects the adjective for emotional characteristic of the color at the right.

Step 2. Select answers of 1–7 in the corresponding position on the paper.

Step 3. Press the next page button on the right side of the screen for the next question and continue on until the survey is completed.

(Testing equipment: 27″iMac with Retina display.)

3.3 Experimental Results of the Best Visual Effects for Computer Mouse Packaging

The best visual effects experiment for computer mouse packaging is based on 5 size ratios of the transparent windows of color boxes. The subjects of the experiment are 50 randomly selected college students, including 23 males and 27 females (Table 1, Fig. 5).

Table 1. Average and standard deviation of the best visual effects for computer mouse packaging

Mouse no.	Number	M	SD
1	50	2.98	1.515
2	50	2.23	1.23
3	50	6.88	1.02
4	50	1.28	.905
5	50	4.59	1.33

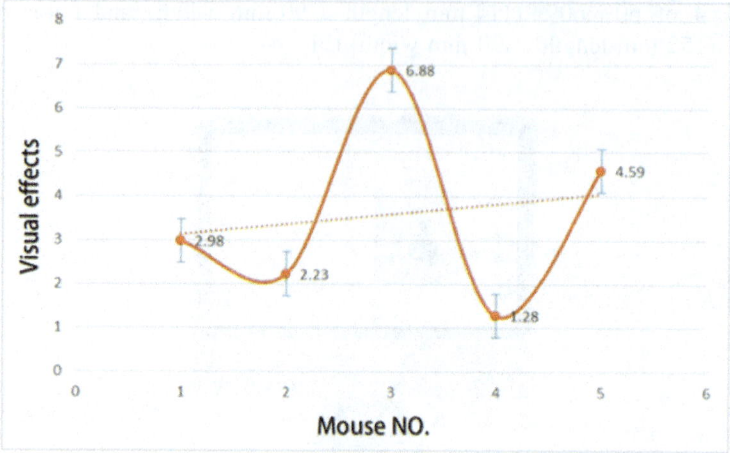

Fig. 5. Experimental chart of the best visual effects for computer mouse packaging

Experimental results show that the visual effect of transparent window packaging design No. 4 of 60%:60% (114 mm × 90 mm) is considered to be ideally suited (1.28) for the computer mouse, whereas the visual effect on transparent window packaging design No. 3 of 75%:75% (127 mm × 75 mm) is considered to be very unsuitable (6.88).

4 Discussion and Conclusions

For the experiment on best visual effects for computer mouse packaging, we use a 64.5 × 95 × 35.5 mm mouse (No. 2 mouse) as basis for designing 5 types of size ratio of transparent windows for the computer mouse colored box packaging, in order to obtain the computer mouse packaging style with the best visual effect through experimental methods. The ranking of the best visual effects for computer mouse packaging as well as the average and standard deviation (SD) and ranking results of the five samples are shown in Table 2.

Table 2. Statistical ranking of the best visual effects for computer mouse packaging

No.	M	SD	Significance	Sort
1	2.98	1.515	.000*	3
2	2.23	1.23	.000*	2
3	6.88	1.02	.012*	5
4	1.28	.905	.001*	1
5	4.59	1.33	.058	4

Table 2 shows the visual effect of transparent window packaging design No. 4 of 60%:60% (114 mm × 90 mm) is considered to be ideally suited (1.28) for the computer mouse, whereas the visual effect on transparent window packaging design No. 3 of 75%:75% (127 mm × 75 mm) is considered to be very unsuitable (6.88). The best visual effects design of computer mouse No. 4 and No. 2 are 0.001 and 0.000 respectively, with P value of less than 0.001, indicating that the suitability of "best visual effects for computer mouse packaging" is ranked by subjects to have a consistent significant level. Through variation analysis, the results of the above rankings indicate that the best visual effect suitability of computer mouse packaging was $F = 25.849$, $p < 0.001$, which reached significant difference, as shown in Table 3.

Table 3. Variation analysis of the best visual effects for computer mouse packaging

	Sum of square	Degrees of freedom	Mean square sum	F test	Distinctiveness
Visual effects	318.352	5	64.283	25.849	.000*

*$p < 0.05$

References

1. Lin, D., Zhang, L.: Package Design. New Image Publishing Corporation, New Taipei City (1994)
2. Xu, X.: Modern Commercial Packaging Science Theory, Concept, Practice. Video Culture Limited Company, New Taipei City (2003)
3. Zhu, C.: Package Design. Jin Guan Publishing, Taipei City (1996)
4. Huang, R.: Research on graphic design of commercial packaging. Taichung City: a large social and cultural enterprise (1990)
5. Li, T.: CI Visual Design Packing Pointine Surface PART1. New Image Publishing Business Has Limited Company, Taipei City (1992)
6. Long,, D.: Commercial Packaging Design. Lemon, Yellow Cultural Enterprise, Taipei City (1992)
7. Van Rompay, T.J.L., Pruyn, A.T.H., Tieke, P.: Symbolic integration in design and its impact on product and brand evaluation. Int. J. Des. **3**(2), 19–26 (2009)
8. Van Rompay, T.J.L., Pruyn, A.T.H.: When visual product features use the same language: the impact of shape font consistency on brand perception and price expectations. Prod. Innov. Manag. J. **28**(4), 599–610 (2011)
9. Johnson, M.: The Meaning of the Body: The Aesthetics of Human Understanding. University of Chicago Press, Chicago (2007)
10. Schifferstein, H.N.J., Spence, C.: Multi-sensory Product Experience. Product experience at HNJ Schifferstein and PPM Hekkert (eds.), pp. 133–161. Elsevier, Amsterdam (2008)

A Methodology for Designing Meaningful Interactions in Public Space

Liang Tan(✉)

Guangzhou Academy of Fine Arts, Guangzhou, Guangdong, China
tanliang@gzarts.edu.cn

Abstract. As digital computing and embedded technologies are increasingly pervasive and accessible in public spaces, interactions with technology-mediated physical environments become the focus of a growing amount of research in human-computer interaction (HCI). This study develops a new methodology towards designing meaningful interactions in public spaces. Based on the theories of embodied cognition, we conducted case analyses, experience tests on an interactive prototype, and design ideations. The findings show that the methodology is useful for the design of meaningful interaction through linking physical interactions with embodied metaphorical expressions. Related design principles and research limitations are summarized.

Keywords: Embodied interaction · Public space · Interaction design

1 Introduction

While digital computing and sensor technology are becoming increasingly accessible in public space, people's everyday experiences have been extending from interacting with digital products to interacting with mediated physical environments [2, 11]. Many studies suggest that sensory-motor mechanism of the physical body has a key role in experiencing digital media [1, 3–5] and product experience [12]. However, less work has been done on exploring approaches to the design of embodied interactions in the context of public spaces. Building on the prior studies [15, 16], this study aims to develop an embodied design approach for ambient media design.

We first reviewed the concepts and studies related to embodied interaction. Then, an embodied approach to designing ambient media was developed based on interpretive and empirical studies. The approach proposes that everyday environments can be transformed into interactive ambient media, which provides physical affordances to facilitate bodily engagement. Next, we organized a workshop of ambient media design to evaluate whether the design strategies derived from the prior researches could assist designers to generate new design works towards meaning making in public spaces. The research results show that these design methods are useful in developing ambient media works.

G. Di Bucchianico et al. (Eds.): AHFE 2020, AISC 1202, pp. 920–926, 2020.
https://doi.org/10.1007/978-3-030-51194-4_119

2 Theoretical Background and Related Work

2.1 Embodied Interaction and Embodied Metaphor

Building on the insights from phenomenology, Dourish [7] regards embodiment as a key foundation of understanding HCI. He uses the term embodied interaction to describe that people make and share meaning through both physical and social interactions, and embodied experience is closely connected with bodily behaviors and social interactions. The embodiment of experience can be understood in terms of three types: body's innate capacities, bodily skills, and cultural habits, which play a central role in understanding how meaning is generated in the world [8]. Embodied interaction is usually contrasted with cognitivist approaches that consider human thinking model as internal information processing. Besides, embodiment is a central idea in cognitive linguistics that believes language reflects the way how people think [9]. Many researchers in HCI have been inspired by the insights from cognitive semantics to design embodied interactions with interactive systems [2, 3, 5]. "Embodied metaphors" means the metaphors that are closely rooted in bodily experiences.

2.2 Experiencing Ambient Media

The concept "ambient" refers to encompassing atmosphere and has philosophical roots in embodiment, as Merleau-Ponty described that human body is a physical basis for engaging in the world [13, 17, 18]. Ambient medium is defined as anything in public space that can be used for delivering brand or product information in advertising field [15]. In ubiquitous computing, the most significant feature of ambient media is that interactive technologies penetrate people's everyday spaces and provide physical affordances for interactions [1, 11]. We define ambient media as a novel media form that can engage embodied interaction and facilitate meaning making through utilizing existing physical environments public spaces. Three main properties of ambient media were identified: The spatial features of ambient media (Spatiality) are drawn upon to attract audience's attention, and audience gain unexpected experience (Unexpectedness) while they are engaged in the active interaction with ambient media (Engagement) [15].

2.3 Embodied Approaches in HCI Research

Although embodied thinking provides HCI community with valuable insights for understanding user experience, only a few studies have focused on developing embodied methodologies for guiding design practice. Loke and Robertson [10] presented a methodology called "Moving and Making Strange" for designing movement-based interactions. Building on embodied interaction, Millard and Soylu [14] introduced an embodied approach to designing interaction in pervasive computing environments by focusing on tangible interaction and multi-modal interaction. These studies apply the theory of embodiment into more specific interaction design contexts. In context of experience with ambient media, the main objectives of this paper are to propose an embodied approach to the design of ambient media toward facilitating meaningful experience, and to conduct design experiments for evaluating the approach.

3 Developing a Methodology for Designing Meaningful Interactions

Building on prior work [5, 15, 16], we developed a methodology for ambient media design towards meaningful interactions. In regard to the attributes of ambient media, a range of strategies, models, and tools were developed the design of evaluation of embodied interactions and experiences.

3.1 Design Strategies

Case analyses [15] on interactive ambient media works indicate that the interaction in public spaces can be understood as a complex process. The affordances in the physical environments can be perceived by people, and they invite people to act according to the specific environmental features. Embodied metaphors reflect the cognitive patterns in people's long-term bodily experience. The in-depth study on audience experience with Piano Staircase (Fig. 1) summarized the major experiential elements in interaction with public spaces, based on empirical and interpretive studies [6, 16].

Fig. 1. Piano staircase.

Bodily engagement plays a key role in facilitating meaningful experience and giving rise to metaphorical meanings. Passersby noticed that the staircase was designed as a piano keyboard that forms the visual mapping of similarity in shapes. With the physical affordances of the staircase, people can walk up or walk down. These bodily actions generated different musical effects. Then, people further explored the staircase through walking up and down, which belongs to the schema of Orientation. This schema projects the abstract concept of musical notes onto another cognitive concept of spatial relationship of the stairs. Based on the empirical studies, four design strategies were summarized:

1) Embedding the interactive installation into public environments: Existing public environments are employed as basis and provide affordances for designing embodied interaction and engaging people's bodily actions.
2) Designing I/O relations for eliciting unexpected experience: When audience performs bodily actions (Input), it triggers expected feedbacks (Output). The appearance of ambient media resembles the other object or space out of the current context.

3) Mapping various metaphorical domains: The source domain (bodily interactions) is projected onto a target domain (abstract concepts).
4) Linking people's bodily movements to perceptual responses: In spite of unexpectedness, the relationship between bodily movements and sensory responses should correspond with the metaphorical meaning (e.g., walking triggering music is like playing piano).

3.2 Relationship Between Spatial Characteristics and Image Schemas

Regarding case analyses [15] and image schema theory [9] we devise a model reflecting the relationship between spatial characteristics and image schemas. As shown in Fig. 2, the characteristics of physical environments provide corresponding potentials of bodily movements, which are in line with a series of cognitive patterns (image schemas) generating from people's physical interactions with various environments. For example, a long path indicates the possibility of moving along a certain track. This is organized by the schema of Path, which is a recurrent cognitive mode grounded in the mind.

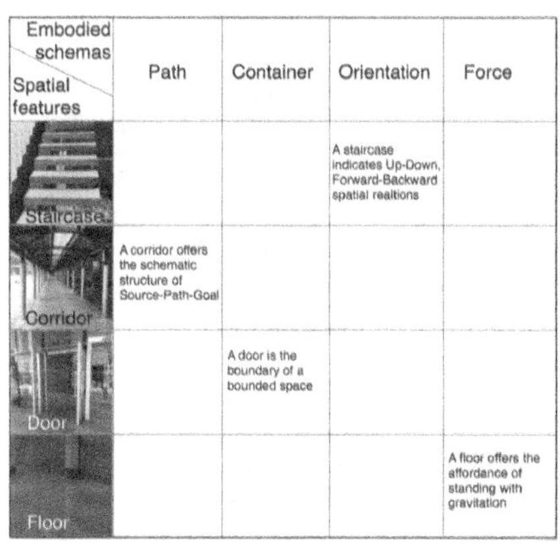

Fig. 2. Relationship between spatial characteristics and image schemas.

Path schema includes an underlying mode with a start position and an end position, which constructs a process of movements. Human's everyday interactions with different corridors, paths and roads construct the cognitive pattern (Path schema). This would inspire new interaction design ideas related to the meanings of life journey and temporal concepts. The real-time sensory-motor actions are tightly related to embodied schemas. Designers can employ this model to bridge everyday movements with spatial characteristics, which is suggested to help ideations of meaningful interaction design in public spaces.

3.3 The Model of Embodied Meaning Making

Figure 3 shows a model demonstrating how the meaning is generated through bodily interactions in public spaces [16]. This model offers a tool for analyzing audience experience and evaluating embodied interaction with ambient media works. People's physical (sensory and motor) experiences in growing up process construct different cognitive modes (Embodied schemas). Line 2 indicates that the conceptual metaphor is activated by the physical interactions with the public spaces, which projects real-time bodily experience onto abstract domain. Line 3 means that the recurring cognitive pattern (embodied schema) can structure embodied metaphors (specific metaphorical expressions). Embodied schema-based imaginations are also triggered by embodied metaphors, which can be physical objects, certain brands, and past moments. Embodied schema also regulates embodied imagination, which is activated by bodily actions. Embodied metaphor and embodied imagination have a similar form of mapping. However, embodied metaphor is normally a projection from a concrete domain (e.g., bodily movement) to an abstract domain (concept), while embodied imagination is a mapping between an abstract domain and a concrete environment.

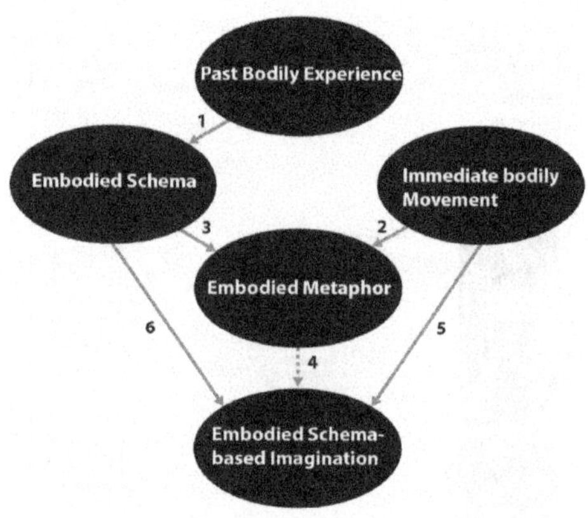

Fig. 3. Embodied meaning making model.

4 Designing Meaningful Interaction in Public Spaces

In order to further explore an embodied design methodology, a design experiment was conducted to evaluate how the design strategies and methods can assist designers to ideate new interaction designs in public spaces. We invited four students in interaction design program to join the experiment. The embodied design methods and models (the design strategies and the matrix in Sect. 3) were introduced to the students in the first phase. Then, the designers identified a design topic related to environmental protection, and the researcher suggested them to ideate design proposals in regard to the four design strategies. In the design experiment, some designers felt that the concept of embodied schema was hard to comprehend. The researcher mainly focused on observing to what extent the participants can effectively use the embodied methodology to ideate ideas. The design was less consumption of towel paper. Based on the embodied design methodology, the participants generated a design proposal as shown in Fig. 4. The embodied metaphor is formed through a mapping from the disappearing of forest onto the bodily behavior of consuming the toilet paper. The interactive ambient work was expected to be installed in the public restrooms. There are two embodied mappings: tearing a piece of paper from the paper box metaphorically leads to cutting down of the trees on the screen mirror. Secondly, water dropping down caused by shaking hands metaphorically results in growing up of the trees. These metaphors raise the awareness of sustainable lifestyles and reducing wastes.

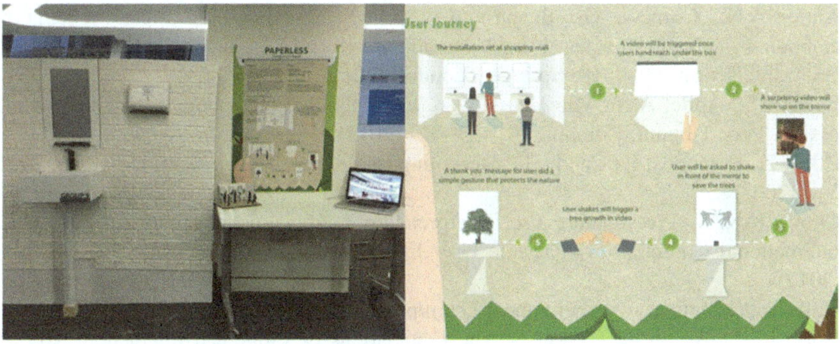

Fig. 4. Paperless.

The experiment of interaction design shows that the embodied methodology can effectively serve as guidance for the developing interaction design ideas in public spaces. In regard to an embodied schema, aligning bodily movements with an embodied metaphor is a valuable ideation method in public integration design.

5 Conclusion

Based on a well-defined framework, this study proposes a methodology for designing meaningful interaction in public spaces. This methodology comprises a range of models and strategies, which provide a powerful guidance in understanding and designing embodied interactions with physical spaces. An experiment on embodied interaction design was carried out to test whether this methodology is applicable in guiding designers to design meaningful interactions with public spaces. The results of this experiment indicate that this methodology can effectively generate design ideas on embodied meaning making. Some research limitations will remind the researchers to further investigate and explore this design methodology. For instance, some designers may have difficulties in constructing embodied metaphors when designing for a certain topic, which needs to have a deep understanding of the relationship between the features of physical environments and embodied schemas. Future research will focus on how to make this approach more applicable to interaction designers. In addition, more empirical studies are needed to evaluate the ideations. We will also explore other modalities besides visual and auditory feedbacks.

Acknowledgments. This study is supported by the grant from Guangdong Philosophical and Social Science Planning Fund Program (GD17XYS16).

References

1. Alves Lino, J., Salem, B., Rauterberg, M.: Responsive environments: user experiences for ambient intelligence. J. Amb. Intel. Smart.En. **2**, 347–367 (2010)
2. Antle, A.N., Corness, G., Bevans, A.: Springboard: designing image schema based embodied interaction for an abstract domain. In: Karat, J., Vanderdonckt, J. (eds.) Whole Body Interaction, pp. 7–18. Springer, London (2011)
3. Bakker, S., Antle, A.N., Van Den Hoven, E.: Embodied metaphors in tangible interaction design. Pers. Ubiquit. Comput. **16**, 433–449 (2012)
4. Bilda, Z., Edmonds, E., Candy, L.: Designing for creative engagement. Des. Stud. **29**, 525–540 (2008)
5. Chow, K.K.N.: Investigating user interpretation of dynamic metaphorical interfaces. In: International Conference of Design, User Experience, and Usability, pp. 45–59. Springer (2017)
6. Volkswagen. http://www.thefuntheory.com/piano-staircase
7. Dourish, P.: Where the Action Is: The Foundations of Embodied Interaction. MIT Press, Cambridge (2001)
8. Dreyfus, H.L.: The current relevance of Merleau-Ponty's phenomenology of embodiment. Electron. J. Anal. Philos. **4**, 1–16 (1996)
9. Lakoff, G.: Explaining embodied cognition results. Top. Cogn. Sci. **4**, 773–785 (2012)
10. Loke, L., Robertson, T.: Moving and making strange: an embodied approach to movement-based interaction design. ACM Trans. Comput.-Hum. Interact. **20**, 7 (2013)
11. Lugmayr, A.: Connecting the real world with the digital overlay with smart ambient media—applying Peirce's categories in the context of ambient media. Multimed. Tools Appl. **58**, 385–398 (2012)
12. McCarthy, J., Wright, P.: Technology as Experience. MIT Press, Cambridge (2004)
13. Merleau-Ponty, M.: Phenomenology of Perception. Routledge & Kegan Paul, London (1962)
14. Millard, M.O., Soylu, F.: An embodied approach for engaged interaction in ubiquitous computing. In: International Conference on Human-Computer Interaction, pp. 464–472. Springer (2009)
15. Tan, L., Chow, K.K.N.: An embodied interaction framework for facilitating audience experience with ambient media. In: Fourth International Conference on Design Creativity (4th ICDC), pp. 1–8. The Design Society (2016)
16. Tan, L., Chow, K.K.N.: Piano staircase: exploring movement-based meaning making in interacting with ambient media. In: IFIP Conference on Human-Computer Interaction (Interact 2017), pp. 282–291. Springer (2017)
17. Varela, F., Thompson, E., Rosch, E.: The Embodied Mind: Cognitive Science and Human Experience. MIT Press, Cambridge, Mass (1991)
18. Wilson, M.: Six views of embodied cognition. Psychon. Bull. Rev. **9**, 625–636 (2002)

Case Study: How Design Students Tailor Their Crowdfunding Presentation to Different Audiences

Jose Rivera-Chang(✉)

California State University, Long Beach, USA
Jose.Rivera@csulb.edu

Abstract. Popular crowdfunding websites such as Kickstarter and IndieGogo have transformed the practice of Industrial Design and have had a significant impact on Industrial Design education as well. Students and professionals use these online platforms to communicate their ideas and to crowdfund their own projects. The goals are to launch new product campaigns and to test new ideas.

Since most investors, project donors or backers want to understand the mission of a new project or campaign, Industrial Design students have to make sure that their crowdfunding video has a good level of engagement and is conveyed in a manner that is easy to understand by all audiences. How they choose to engage their target audience matters. Investors, donors and backers may have both similar and different expectations.

This case study will present the different strategies used by Industrial Design students to engage their target audiences using video as their primary communication format.

Keywords: Crowdfunding · Investor · Donor · Backer · Online platform

1 Online Crowdfunding Platforms

Online crowdfunding platforms such as IndieGogo and Kickstarter were launched around 2009 during a time of economic recession. Both were part of the first wave of internet crowdfunding websites. They both allow people to solicit funds (investment) for an idea or project. These crowdfunding platforms circumvent traditional avenues of investment. After only one year of online presence, traditional publishers such as *The New York Times* and *Time Magazine* recognized them and featured them as some of the best websites available. Their notoriety helped fund ideas and projects in investment categories that are traditionally considered difficult to support financially such as Art, Dance, Design, Fashion, Film, Photography, Music, Games, Theater and Writing. These categories were considered high risk and thus avoided by traditional investors such as banks or Wall Street institutions.

Within these investment categories avoided by traditional investors but promoted by websites such as IndieGogo and Kickstarter, Design represents a much smaller proportion of the total number of projects, below 4% [1], and within this 4%, it is not clear which ones are considered Industrial Design or Product Design projects. These

G. Di Bucchianico et al. (Eds.): AHFE 2020, AISC 1202, pp. 927–931, 2020.
https://doi.org/10.1007/978-3-030-51194-4_120

websites do not make a distinction between Industrial Design or Interior Design or Graphic Design or any other design discipline. Nevertheless, six out of the ten largest successfully completed Kickstarter projects by total funds pledged are in the Product Design category (Fig. 1).

<table>
<tr><th>Rank</th><th>Total USD</th><th>Project name</th><th>Creator</th><th>Category</th><th>% funded</th><th>Backers</th><th>Closing date</th></tr>
<tr><td>1</td><td>20,338,986</td><td>Pebble Time – Awesome Smartwatch, No Compromises[105]</td><td>Pebble Technology</td><td>Product design</td><td>4,067</td><td>78,471</td><td>2015-03-27</td></tr>
<tr><td>2</td><td>13,285,226</td><td>Coolest Cooler: 21st Century Cooler that's Actually Cooler[106]</td><td>Ryan Grepper</td><td>Product design</td><td>26,570</td><td>62,642</td><td>2014-08-30</td></tr>
<tr><td>3</td><td>12,779,843</td><td>Pebble 2, Time 2 + All-New Pebble Core[107]</td><td>Pebble Technology</td><td>Product design</td><td>1,277</td><td>66,673</td><td>2016-06-30</td></tr>
<tr><td>4</td><td>12,393,139</td><td>Kingdom Death: Monster 1.5[108]</td><td>Kingdom Death/Adam Poots</td><td>Tabletop games</td><td>12,393</td><td>19,264</td><td>2017-01-07</td></tr>
<tr><td>5</td><td>12,143,435</td><td>Travel Tripod by Peak Design[109]</td><td>Peak Design</td><td>Product design</td><td>2,429</td><td>27,168</td><td>2019-12-13</td></tr>
<tr><td>6</td><td>11,385,449</td><td>Critical Role: The Legend of Vox Machina Animated Special[110]</td><td>Critical Role</td><td>Film</td><td>1,518</td><td>88,887</td><td>2019-04-19</td></tr>
<tr><td>7</td><td>10,266,845</td><td>Pebble: E-Paper Watch for iPhone and Android[111]</td><td>Pebble Technology</td><td>Product design</td><td>10,266</td><td>68,929</td><td>2012-05-18</td></tr>
<tr><td>8</td><td>9,192,055</td><td>The World's Best Travel Jacket with 15 Features || BAUBAX[112]</td><td>BAUBAX LLC</td><td>Product design</td><td>45,960</td><td>44,949</td><td>2015-09-03</td></tr>
<tr><td>9</td><td>8,782,571</td><td>Exploding Kittens[113]</td><td>Elan Lee</td><td>Playing cards</td><td>87,825</td><td>219,382</td><td>2015-02-20</td></tr>
<tr><td>10</td><td>8,596,474</td><td>OUYA: A New Kind of Video Game Console[114]</td><td>Ouya Inc.</td><td>Video games</td><td>904</td><td>63,416</td><td>2012-08-09</td></tr>
</table>

Fig. 1. List of ten largest Kickstarter projects by total funds pledged.

IndieGogo and Kickstarter do not distinguish whether the crowdfunding campaigns featured on their websites are run by professionals, students or amateurs. They are all considered *creators*. California State University Long Beach encourage senior-level students in the Industrial Design program to create promotional videos that could be shown to promote their product design ideas. These videos follow a format that is similar to the format used in the crowdfunding websites but they are not necessarily intended to be uploaded or posted to those websites. The video is only one of several *deliverables* that the students had to submit as part of a 10-week project called *Students as Entrepreneurs*. The goal for each student was to create a business idea based on a simple product. This product must be simple enough to allow a working prototype to be developed. The prototype was the basis for a promotional video and a website.

2 Industrial Design and Entrepreneurship Awareness

While Industrial Design programs around the country often emphasize the development of technical skills for the design industry, the Industrial Design program at California State University Long Beach (CSULB) places additional emphasis on entrepreneurial awareness. This emphasis was designed to provide senior-level students with a broader view of Industrial Design beyond the discipline itself and more as an enterprise or entrepreneurial activity that engages a broad audience of stakeholders in a series of activities that explore not only design but also marketing, promotion, business and investment. This emphasis on entrepreneurship teaches CSULB students to consider the business side of Industrial Design.

In traditional business or management schools, students learn that typical categories of investment are gold, real state and bonds [2]. Traditional business investors often regard these as proper investment categories. On the other hand, Design is not considered an investment category. Since 2009 online crowdfunding has changed or expanded the perception of investment categories. Investors, backers or donors can now invest in non-traditional categories such as Design. According to the Kickstarter website, as of February 2020, there were 474,946 launched projects and the total amount pledged by investors over a period of ten years was more than $4.76 billion [3].

3 Crowdfunding Video Presentation

Most investors, project donors or backers want to understand the mission of a new project or campaign. The final objective for the CSULB Industrial Design senior students was to develop a crowdfunding campaign in video format where the video is supposed to promote a new business-product-idea and to encourage or solicit investment funds or *seed money*. The format was similar to the ones seen on the Kickstarter and IndieGogo websites. The video would feature the prototype developed during earlier stages of the project.

Investors are typically more interested in the project's business plan and the potential return on investment. Donors and backers are more interested in finding unique features or emotional appeal. The challenge for the Industrial Design students was to develop a video that was compelling, relatable and engaging to a variety of audiences. Investors, donors and backers may have both similar and different expectations.

When Industrial Design students develop a video presentation they usually rely on a combination of tools such as digital drawings, 3D model renderings, color animations and working prototypes. They were also given a basic storyboard to help them understand the sequence of frames and time needed for the video (Fig. 2).

Fig. 2. Basic storyboards examples given to students.

Since investors, donors and backers have both similar and different expectations, it became necessary for students to tailor their crowdfunding presentations to these different audiences.

4 Investors, Donors and Backers

Investor is a business term used to refer to a person or organization who commits capital with the expectation of getting a return or achieving a profit. This is a more conservative type of person that will take the time to review a business plan and make sure that the idea promoted by the entrepreneur or creator has real potential.

Traditional investors need significant documentation and proof that the work has been done otherwise they would not commit to the project. Their investment funds are typically much larger than those of donors or backers. They prefer a more personalized or dedicated type of presentation. To tailor to this type of audience, design students should pay more attention to the type of business or industry the investor belongs to. Numbers are important to them so the video presentation should focus on those numbers and how it benefits their bottom line. Maybe the suggestion of a partnership in the new business is more appealing to them. In other cases, more aggressive investors will contact the owner of the videos online and try to buy part of their business.

In the case of donors and backers, they don't necessarily commit capital with the expectation of achieving a profit. They more likely want to support the project because they like the entrepreneur or creator of the idea, or, they just want to be the first one to benefit from the product or idea. Depending on the website or venue, donor and backers are considered the same person but in other venues a donor could be considered a person that contributes to the project without expecting anything in return. Most of the people who contribute to crowdfunding websites are donors and backers. Design students looking to post their video on a crowdfunding website should consider donors and backers as their target audience. To tailor to this type of audience the video presentation must be entertaining and should connect at a more emotional level, making emphasis on product features and/or creator's personality (Fig. 3).

Fig. 3. Student example of video presentation for donors and backers.

5 Conclusions

The project *Students as Entrepreneurs* was an ambitious and intense 10-week project that covered several phases of design development and topics such as market research, business plan, concept development, prototype development, website delivery and video production. The video production phase (the subject of this case study) took about two weeks to complete and students only had a chance to revise their video twice. The emphasis on the video presentation was on the content rather than technical achievement. Minor technical difficulties such as audio or voice over edits were common but overlooked in favor of more compelling content where the mission of the project was clearly stated to the target audience.

By tailoring their crowdfunding presentation to specific audiences, Industrial Design students were able to not only understand the design process but also understand the design business model and generated awareness on how design projects get funded.

References

1. Ultimate List of Crowdfunding and Fundraising Websites. Crowd 101. https://www.crowd101.com/list-crowdfunding-and-fundraising-websites. Accessed 2 Feb 2020
2. 6 types of investments: RuleOne. https://www.ruleoneinvesting.com/blog/how-to-invest. Accessed 1 Feb 2020
3. Kickstarter Stats: Kickstarter. https://www.kickstarter.com/help/stats. Accessed 1 Feb 2020

Research on the Application of Traditional Cultural Elements in the Design of Bank Card Surface

Yun-chen Xi(✉) and Xin-xiong Liu

Huazhong University of Science and Technology, Wuhan 430074, Hubei, China
2655518427@qq.com

Abstract. Objective: design the bank card surface with the characteristics of traditional culture and explore the application methods and design principles of traditional cultural elements in the design of bank card surface. Method: collect bank cards of traditional culture that have been issued in the domestic market to explore the application status of traditional cultural elements in the design of bank card surface. The methods and principles of integrating traditional cultural elements into bank card surface design are analyzed and summarized. Taking the bank card surface design of the Drawing on five ethics series as an example, the traditional cultural elements are integrated into the bank card surface design to explore the application of traditional cultural elements in the bank card surface design. Conclusion: The application of traditional cultural elements to the bank card surface design, combined with research and practice. It provides ideas and methods for the integration of traditional cultural elements into the bank card surface design, which can promote the spread of traditional Chinese culture while improving the card surface delicacy.

Keywords: Bank card · Card surface design · Traditional cultural elements · Visual design

1 Introduction

Bank cards are a payment tool that people often use. With the development of the times, people's requirements for bank cards are not limited to professional technology and services. The card surface design of bank cards has also become a necessary factor favored by people. This makes the design of bank cards tend to be aesthetic [1, 2]. Among them, many designers choose to use traditional culture as the design element of the card surface, hoping to improve the fineness of the card surface of the bank card and promote the development of banking business, while helping to spread the traditional culture. How to better integrate traditional culture with modern financial technology, and let traditional culture enter contemporary households with contemporary finance, this requires thinking and research [3].

G. Di Bucchianico et al. (Eds.): AHFE 2020, AISC 1202, pp. 932–937, 2020.
https://doi.org/10.1007/978-3-030-51194-4_121

2 Development Status

At present, the traditional culture type bank cards that have been issued in the domestic market can be roughly divided into four categories: cultural relics appreciation, innovation, entertainment and interaction.

The cultural relics viewing bank card mainly focuses on the display of related cultural relics, and usually displays the cultural relics on the card surface of the bank card in the form of original photos. By displaying the original image of the cultural relic, the purpose of publicizing the cultural relic and popularizing the public knowledge is achieved [4]. This type of bank card is also the most common traditional culture bank card on the market. This kind of bank card face design is not easy to be ambiguous in terms of cultural popularization, not easy to make mistakes, and has good educational significance and value for viewing collections. But its design and innovation are insufficient. In addition to the user groups who like the cultural relics to buy and collect, its user group is relatively small.

Innovative bank cards focus on the manifestation of cultural connotations, extracting cultural elements from ancient cultural relics, and presenting them to users in a modern art form through a combination with modern design [5]. This type of card surface design can reflect the connotation of traditional culture and the aesthetics of modern design. At the end of 2018, a set of Forbidden City bank cards officially issued by the Forbidden City and Industrial and Commercial Bank of China set off a wave of discussion on the Internet and received wide public praise (Fig. 1 shows an example). This type of bank card face is highly innovative and novel in design, and its expression is more popular. The design extracts the most core elements of ancient relics, which can help users understand the cultural connotation behind the cultural relics. It has the value of cultural dissemination. However, this type of bank card has higher requirements for the designer, and the designer's understanding and grasp of the cultural connotation is very important. If it is not grasped, it will lead to controversy after it is issued. Therefore, the card must be confirmed by strict examination Whether there is any ambiguity in the design content.

Fig. 1. ICBC & Forbidden City-Debit card (*Spring*).

Entertainment bank cards refer to the extracted cultural elements in a fun, entertaining, and popular way. In this way, the user's attention is attracted, and the user has a desire to buy. For example, the sign series can enrich the card surface by adding popular words and other methods on the card surface without affecting the transmission of traditional cultural elements (Fig. 2 shows an example). This type of bank card is loved by young people. But in the design, we need to pay attention to whether the spread of its culture is affected, and we must grasp the degree of entertainment.

Fig. 2. CMBC-Symbol Card Series.

The interactive bank card refers to the interaction with the user in the card surface design. In general, you can deepen its interactivity through DIY, senses, and so on. The winter card of the Four Seasons Card issued by the Industrial and Commercial Bank of China is presented in the form of a red envelope, with a blank golden circle on the card surface, in which the cardholder can freely fill in numbers or auspicious words. It gives us a feeling of personal customization (Fig. 3 shows an example).

Fig. 3. ICBC & Forbidden City-Debit card (*Winter*).

Most of the traditional cultural bank cards currently on the market are cultural relics. However, with the issuance of co-branded cards between the Bank and the Forbidden City, innovative, entertainment and interactive traditional cultural bank card faces have become the current development trend of such bank card face designs.

2.1 Design Methods and Principles

Direct Reference. The classic cultural relic works are displayed on the bank card surface in the form of photos and copy. Through direct display, it shows the charm of traditional culture and gives bank cards higher product value. The credit cards pay tribute to the Dunhuang series of credit cards. Elements from the Dunhuang frescoes, showing the charm of the characters in the Dunhuang frescoes and the infinite charm of the murals (Fig. 4 shows an example).

Fig. 4. CMBC National Treasure Series Theme Credit Card-Tribute to Dunhuang Series.

Extraction Element. Extract the representative elements of the cultural relics and give them the theme of auspicious meaning. The spirit and connotation of traditional Chinese culture is revealed through the extraction of the elements in the classic cultural relics and the reconstruction of the theme.

Craft. Designers and craftsmen have begun to unite and continue to break through the traditional form of card surface. The card surface has been used reasonably [6]. The Twelve Beauties Card displays the original Meticulous painting works, added 3D raster three-dimensional craftsmanship, incorporating modern popular elements. At the same time as the Twelve Beauties showing traditional culture, it also showed interesting and cute. It enriches the visual expression of the card surface and also adds to the fun of the card (Fig. 5 shows an example).

Craft performance has become a design factor in the visual appearance of bank card faces. Reasonable use of craft can help highlight the layering of themes and pictures, and enhance the texture and exquisiteness of bank card faces.

Fig. 5. CMBC—Twelve Beauties Card.

2.2 Design Principles

First of all, it is necessary to ensure the accuracy of the card surface content and the legibility. Bank cards are popular products. Therefore, when designing the card surface, you must pay attention to the place where there is no ambiguity on the card surface. At the same time, you need to ensure that the public can understand the design.

Secondly, when designing the card face of traditional cultural bank cards, it is important to pay attention to the morality and auspiciousness of the theme. The traditional concept of the Chinese nation pays great attention to everything. From the cases analyzed above, it can be seen the theme of traditional cultural bank cards is mostly auspicious, which is also the traditional concept of the Chinese nation for thousands of years.

In addition, in the visual expression of the card surface, it is necessary to pay attention to the problems of color selection and clear visual hierarchy. Color is an indispensable part of traditional Chinese culture. Different colors have different meanings. Adding colors from traditional Chinese culture to the color matching of traditional cultural bank card faces can add a lot of Chinese flavor. In addition, when designing the card surface, you need to pay attention to the fact that the extracted elements cannot be too single or complicated. A single will make the card surface simple and lack a sense of hierarchy. Complexity will make the card surface elements messy and the main body not prominent. Therefore, the extracted elements need to be divided into primary and secondary, the main elements need to be prominent, and the secondary elements need to be weakened. This can make the visual center of the card surface more obvious and the overall theme more prominent. It will also look fine on the card surface, with a sense of layering.

3 Design Application

Taking the Drawing on five ethics as a design prototype, a set of innovative bank card faces about the Drawing on five ethics was designed. Suitable for use and collection, the target population is the general public. The element extraction method was used to extract the elements of the Drawing on five ethics, giving each card an auspicious meaning, and combining the craftsmanship to enrich the card surface level and improve the card surface refinement (Fig. 6 shows an example).

Fig. 6. Design plan.

References

1. Zhang, S.: Explore the visual language in the design of bank card face. Art Panorama (1), 153 (2012)
2. Li, J.: Creative card design and bank card business development. China Credit Card **22**, 40–43 (2006)
3. Zhu, Y.: Listen to the Dean Shan Ji-xiang said the palace museum. Mod. Commerc. Bank. (22), 122–123 (2018)
4. Wang, J.: The research about traditional culture APPS. Jiangnan University, Jiangsu (2016)
5. Bi, F.: Application of traditional culture elements in modern packaging design. Packag. Eng. **31**(18), 69–72 (2010)
6. Liang, L.: Contemporary bank card design and technological innovation. China Credit Card **8**, 66–67 (2012)

The Construction of Story-Telling Illustration in Tea Packaging Design

Ren Long, Lu Geng(✉), and Chenyue Sun

School of Mechanical Science and Engineering,
Huazhong University of Science and Technology,
Wuhan, People's Republic of China
1562582561@qq.com

Abstract. With the improvement of the economic level and the prosperity of the commodity market, consumers' demand for commodities is not only at the material level. The packaging design of the story illustration has strong readability. On the basis of satisfying the visual needs of consumers, it can also show the cultural background behind the tea products through illustration. Through the research on the feasibility and implementation method of story graphics in packaging design, it is confirmed that story graphics is a direct and efficient packaging graphics language, which has certain significance for the improvement of product quality and the shaping of brand image and can provide new ideas and methods for the development of packaging design.

Keywords: Story-telling illustration · Tea packaging design · Illustration design

1 The Situation and Demand of Tea Packaging Design

In the past, the demand for tea packaging mainly served the storage, transportation and sales functions of tea, but with the development of commodity economy, consumers generally increased their aesthetic demand for goods, and the traditional tea packaging is not enough to meet the buyer's market. Businesses need packages with publicity properties to attract consumers. Packaging with unique cultural added value can not only meet the needs of consumers in visual aesthetics, but also improve the spread of the brand, and develop the cultural media connecting the brand and consumers.

At present, there are many kinds of tea packaging in the market, but the quality of the design is different. There are many problems in the design of tea beverage packagings, such as the lack of distinct culture, the complexity and disunity of packaging style, the lack of design sense, the monotonous or mutual imitation of packaging elements, and the lack of brand identification. Good tea packaging design requires designers to open their minds, not stick to the widespread packaging style on the market, understand the tea culture and the advantages of tea brand, and design a bright package for the transferor.

G. Di Bucchianico et al. (Eds.): AHFE 2020, AISC 1202, pp. 938–944, 2020.
https://doi.org/10.1007/978-3-030-51194-4_122

2 Overview of Story-Telling Illustration Packaging

2.1 The Concept of Story-Telling Illustration

The story can be expressed through a variety of carriers. The ways we are familiar with every day are written words, dynamic images, comics, storytelling, etc. The story can be spread not only in sound mode but also in silent modes such as image symbols, gestures and actions. Generally, the human visual system is more complex than the auditory system and can receive more information. So the same story, visual information is more easily captured and processed quickly. It is more intuitive to use the visual image as the medium of spreading story information. Story illustration can be used as a means to spread ideological and cultural connotations.

Story illustration is different from ordinary illustration and decorative painting. It uses narrative techniques to construct narrative images that can be perceived and understood by the audience through character actions, expression, scene interaction, element exchange, and other aspects. The audience can perceive the narrative intention of the work from the static illustration. The illustrator planarizes the story scene in his mind, connects the audience with the information conveyed by the illustration, and completes a silent plot narrative. Compared with ordinary illustration and decorative painting, story illustration is more interesting and readable. When it is used in packaging design, it can also spread brand stories or specific culture to the public through story descriptions.

2.2 Advantages of Story-Telling Illustration Tea Packaging

Express clearly and attract consumers: in order to convey story information to consumers, story illustration packaging will be more vivid and interesting. This kind of packaging is easier to stand out in a variety of shelves.

Highlight the cultural connotation of tea brand: There are many tea brands, and those who can stand firm in the high-intensity brand competition undoubtedly have distinct brand characteristics and culture. Story-telling illustration can effectively tell the brand culture and build the information bridge between brand and consumers through packaging. Consumers enjoy tea culture and brand culture while enjoying tea, thus opening a new tea-drinking experience.

Diversified styles: The style of illustration is changeable, which is suitable for cultural tea under various sales modes. Analysis of brand attributes, tea characteristics, and consumer groups can be combined with different styles of story-telling illustration. For example, for mature consumer groups, we can choose realistic illustration performance; for young consumer groups, we can choose abstract illustration performance of fashion personality; for children consumer groups, we can choose comic illustration performance.

3 The Construction of Story-Telling Illustration Based on Tea Culture

3.1 Frame

We should first understand the brand culture and tea culture, analyze the purpose and connotation of the brand, and understand the attributes of tea.

In the process of cultural analysis, we need to screen out representative cultural factors, which can come from the brand, product, tea, region, and other aspects. Different kinds of cultural factors constitute the core of the story. The story formed in this way has a unique cultural brand in theme emotion, scene construction, and interactive relationship. Then combined with the product target market, select the appropriate style to draw, and finally combine the illustration and text information to produce a unique story illustration package. The construction framework is shown (Fig. 1).

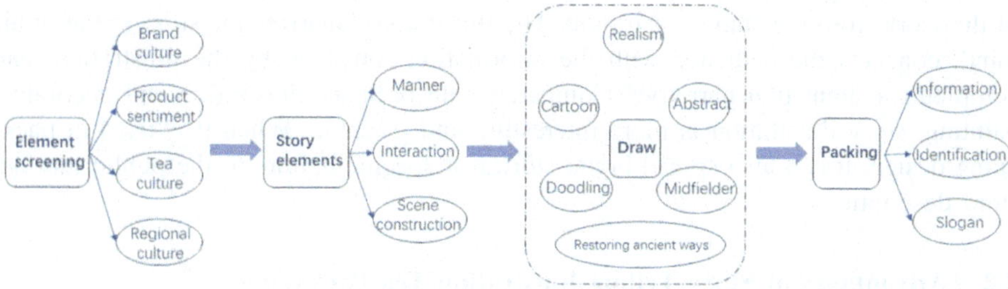

Fig. 1. Research framework

The story-telling illustration in the packaging ultimately needs to serve the brand, products, and consumers. Refining brand culture, product emotion, tea culture, and regional culture, forming the elements of the story, attracting consumers with unique tea regional culture, imperceptibly letting consumers experience the brand culture and product emotion, increasing brand memory and user stickiness.

The four basic elements of a story are time, place, characters and events. However, the expression of illustration is static. It needs to create a "sense of story" through the action and expression of the protagonist and the interaction with the things in the scene, to inspire the imagination of consumers and let consumers perceive the content of the story.

Product positioning, target population, and other factors determine the style of illustration. For example, product positioning is an urban youth. This kind of consumer group pursues fashion, convenience, and individuality. The style of illustration can choose Abstract geometry and graffiti, while the middle-aged and old people prefer realistic and national style illustration.

Illustration serves products, and the ultimate goal is to promote products and brands. The integration of story illustration and packaging needs to consider many aspects. Whether brand information and product information are eye-catching, whether the combination of illustration and packaging box is appropriate, etc. need to be balanced through design.

3.2 Integrate Cultural Elements into the Story

The core of illustration is the story, Consumers actually understand cultural connotation through "story". The elements that make up the story are the protagonist (character image, animal image, a virtual image), space scene, interactive relationship, etc. consumers often rely on these elements when they understand the cultural connotation of the story. Therefore, designers need to combine the culture of product demand and intention expression to naturally integrate elements and stories.

Three years ago, NONGFU SPRING company produced a series of fruit tea drinks – "茶 π" for young consumers. When it was first introduced, the packaging was in the form of a flat illustration (Fig. 2). Although the bright color illustration is vivid and interesting, and the illustration elements can also reflect the taste characteristics of tea, it is not easy to stand out in the same type of tea products on the market. So in 2019, they launched a new package (Fig. 3). It adopts the way of narrative hand-painted illustration to publicize the taste of products more emotionally. For example, rose and litchi black tea. The taste of which is aimed at young women captures the characteristics of women's preference for sweetness and roses. There are girls, horses, flowers and birds, mermaids and other elements in the initial packaging, but there is no interaction between these elements, and each element seems to be in an independent space, so it does not create a sense of story. The new package uses the china familiar story, depicting a beautiful woman in the Tang Dynasty costume who came here in a big bird and an elephant on the other side was hunching litchi. Although there are not many kinds of elements in the picture, the interaction between elements creates a story illustration. Although there is no detailed expression of the story content, imagination initiated by the picture is actually the thinking bridge between the designer's intention and the consumer's construction. The integration of cultural elements into story design increases the sense of identity of consumers and promotes the tradition in a modern way.

Fig. 2. Old packaging

Fig. 3. New packaging

3.3 Artistic Conception and Emotion

The aesthetic of consumers is constantly improving; a simple expression can not meet the aesthetic needs of modern consumers. Designers create different artistic conception to express different emotions and use a variety of illustration language to convey ideas to consumers. The creation of artistic conception is related to color, expression, and visual intention. The artistic conception of the package affects the emotional attributes of the transferred emotion. For example, two series of the same brand have selected the unique animals in the rain forest: elephant, peacock, monkey, and hornbill. However, the forms of expression are flat cartoon illustration and realistic illustration, showing different artistic conception. The plane cartoon illustration abstracts and refines the animal form, which is full of modern sense and fashionable and young. The realistic illustration expresses the elephant image, but the elephant body combines the natural environment in the rainforest to create a quiet and harmonious sense of natural ecology. Designers create a different atmosphere to let consumers experience different emotions from the two kinds of packaging (Fig. 4).

Fig. 4. Different painting styles packaging

4 Tea Brand Packaging Design Case

Through the research and exploration of story illustration packaging, a set of tea packaging with regional characteristics is designed by using the form of story illustration. Extract story elements are based on product demand, brand positioning, corporate culture and other directions (Fig. 5).

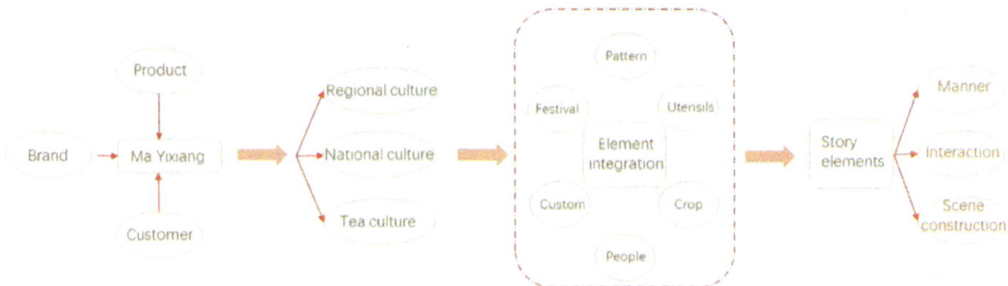

Fig. 5. Illustration element extraction

The brand of “Ma Yixiang” comes from Ma Yidui village, Lincang City, Yunnan Province. It was called “Yi Ting” in ancient times. It has a long history of tea horse ancient road culture and fresh ethnic characteristics. Therefore, in the design process of the story, the corporate culture of “drinking a cup of tea together in the world”, the regional characteristics of Yunnan style, and the ever-changing national style are integrated. Based on the creative background of “Ma Yidui” village, five short stories are created. Each of these five stories depicts a piece of local people’s life. Then, the color factor, pattern factor and scene factor are extracted, and the story is imaged by hand-painted illustration (Fig. 6). Finally applied to tea packaging (Fig. 7).

Fig. 6. Story-telling illustration

Fig. 7. Packaging effect drawing

References

1. Zhang, Z., Lu, J.: Product narrative design based on regional cultural themes. Packag. Eng. (2016)
2. Yang, X., Liu, M.: Application of story-telling graphics in packaging design. Packag. Eng. (2019)
3. Gobé, M.: Let's brandjam to humanize our brands. In: Design Thinking Integrating Innovation, Customer Experience, and Brand Value, pp. 127–139 (2016)
4. Ji, T., Chen, J., Guo, Y.: Product-service system design of food culture based on dong minority's oil-tea. Pack. Eng. (2019)
5. Li, L.: Artistic features of commercial illustration and its application in packaging design. Packag. Eng. (2008)
6. Lu, Q.: The construction of image narration in art jewelry. Sinogram Cult. (2019)

Affective Value and Kawaii Engineering

Analysis of Factors Affecting Luxury of Images of Cooked Dishes

Yuri Hamada[1,2(✉)], Yumi Takei[1,2], and Hiroko Shoji[1,2]

[1] Chuo University, 1-13-27 Kasuga, Bunkyo-Ku, Tokyo 112-8551, Japan
{yhamada, hiroko}@kc.chuo-u.ac.jp
[2] Graduate School of Chuo University,
1-13-27 Kasuga, Bunkyo-Ku, Tokyo 112-8551, Japan

Abstract. This study aims to identify the essential features of photos that make food look "luxury". The authors consider as Kansei evaluation "Looks luxury" as a hierarchical structure constructed from the impression layer and the physical layer, and modeled it using the impression evaluation and physical features of the image. First, the impression evaluation and a principal component analysis was performed. As a result, four principal components were extracted as the impression layer: glamorous, friendliness, moist feeling and moderate portion. The physical layer used six image features: hue, saturation, brightness, number of pixels, presence of background, and presence of background blur. Next, a "looks luxury" hierarchical structure model was constructed using multiple regression analysis. This made it possible to structure and explain the factors that affect the luxury of the images of cooked dishes.

Keywords: Luxury · Impression evaluation · Multiple regression analysis

1 Introduction

In recent years, with the widespread use of the Internet, various services are now available, and a huge amount of information is being shared. In particular, opportunities to acquire and share information via images are increasing. Thus, it becomes more important to make effective use of photographs when advertising services and products to consumers. This is especially true on gourmet sites, where consumers can check the atmosphere and food in a restaurant by means of images. Moreover, it is necessary to know what kind of image gives a good impression to the viewer.

This study focuses on images of cooked dishes. In order to convey the image and atmosphere from the image, it is necessary to clarify what kind of image gives what impression. There have been studies on how to take pictures and the structure of cooking images, but few studies have clarified how impressions are received by engineering approaches. Therefore, the authors are working on the analysis of the factors that affect the evaluation by modeling the sensitivity evaluation in images of cooked dishes. In this paper, we report on the results of the analysis of "luxury".

G. Di Bucchianico et al. (Eds.): AHFE 2020, AISC 1202, pp. 947–952, 2020.
https://doi.org/10.1007/978-3-030-51194-4_123

2 Approach of This Study

In this study, images of spaghetti were used. Referring to the hierarchical structure of Kansei proposed by Nagata et al. [1], the hierarchical structure of "Looks luxury" in an image of spaghetti as shown in Fig. 1, was considered. The value layer was taken as "Looks luxurious - looks no luxury", the impression layer, as "Impression of spaghetti", and the physical layer, as "Image feature quantities". By modeling this hierarchical structure in the study, the physical layer feature quantities that affect the evaluation in the value layer are clarified.

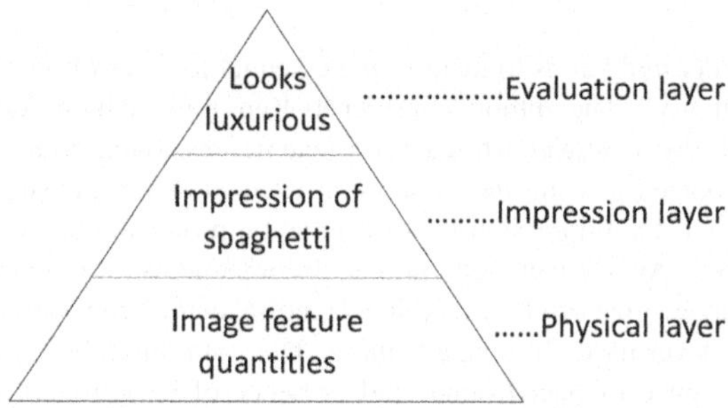

Fig. 1. Hierarchical structure of looks luxurious of the study

3 Data Used in the Study

The data used in modeling the structure of "Looks luxurious" are described for each layer below.

For the "Image feature quantities" of the physical layer, "hue", "saturation", "brightness", "number of pixels", "presence of background", "presence of background blur", "ratio of pasta to the image", "ratio of plate to the image", "ratio of yellow/green/red/black/white color to the image", "presence of other elements", "presence of white dish" and "number of representative colors" were acquired. For "hue", "saturation", and "brightness", the average values of one whole image were obtained using OpenCV [5], an image processing library of Python which is a programming language. In OpenCV, the hue is output between 0° and 180°. In this study, 165° to 180° and 0° to 15° were defined as "red", 15° to 45° as "yellow", 45° to 75° as "green", 75° to 105° as "light blue", 105° to 135° as "blue", and 135° to 165° as "purple". In the images used in this study, the values averaged on any one whole image did not fall into "blue" nor "purple" as defined in the above. Therefore, in this study, the four colors, "red", "yellow", "green" and "light blue" were used as dummy variables. Images with "presence with background" and/or "presence with background blur" were taken as 1, and those without, as 0. Image feature quantities were normalized and then used for analysis. The number of representative colors was calculated using the representative color extraction method of Takahashi et al. [3].

For the value layer, i.e., "Looks luxurious - Looks no luxury", and for the impression layer, i.e., "Impression of spaghetti", impression evaluation values obtained by the SD method were used. Table 1 shows adjective pairs used in the impression evaluation experiment. The impression evaluation experiment is described in Sect. 4.

Table 1. Adjectives used in the impression evaluation experiment

Layer	Adjective pairs
Evaluation layer	Looks luxury – Looks no luxury
Impression layer	Boring – Fashionable
	Clude – Delicate
	Dark – Bright
	Simple – Gorgeous
	Vulgar – Refind
	Not like – Like
	Traditional – Contemporary
	Mediocre – Unique
	Looks unappetizing – Looks delicious
	Looks heavy – Looks light
	Looks dry – looks moist
	Looks hard – Looks soft
	Looks too much – Looks too little

4 Impression Evaluation Experiment

To obtain evaluation values for the value layer and the impression layer, an impression evaluation experiment was performed using the SD method. The subjects were 20 university students and graduate students. In the experiment, while looking at one image of spaghetti, all 14 items in Table 1 were evaluated in 7 stages. The images used in the experiment were 60 images of an Italian restaurant pasta dish with a budget set at 5,000 yen in Tokyo and were collected by GURUNAVI API [3].

In order to adjust the differences of the evaluation scale among subjects, the experimental results for each subject were normalized. In this study, in order to better visualize trends, the evaluation, "Looks luxurious - Looks no luxury" was analyzed with the bottom 25% as low and the top 25% as high. The number of low evaluation data was 300, and the number of high evaluation data was 331.

The dimensions of the impression evaluation results for the impression layer were reduced using principal component analysis. Table 2 shows the factor loading and contribution rate up to the 4th principal component. Since the cumulative contribution rate up to the 4th principal component was 73% from a result of the analysis, subsequent analysis was performed using these principal components in this study. In addition, the 1st principal component was named as "brilliance", the 2nd, "friendly", the 3rd, "moist", and the 4th, "moderate portion".

Table 2. Factor loading and contribution rate

	1st principal component	2nd principal component	3rd principal component	4th principal component
Boring - Fashionable	0.369652	0.037414	0.025453	-0.030983
Clude - Delicate	0.348061	0.047425	0.022534	0.154601
Dark - Bright	0.263431	0.207361	0.078119	-0.170271
Simple - Gorgeous	0.35436	-0.298645	0.007292	-0.325256
Vulgar - Refind	0.312907	0.076404	0.015441	0.017596
Not like - Like	0.283951	0.442256	0.032608	-0.183571
Traditional - Contemporary	0.313978	-0.408881	-0.04499	0.162977
Mediocre - Unique	0.322429	-0.489133	0.020347	-0.187935
Looks unappetizing - Looks delicious	0.282742	0.417282	0.01406	-0.216963
Looks heavy - Looks light	0.117111	0.204228	0.541232	0.341381
Looks dry - looks moist	0.087803	0.113166	-0.648362	0.024929
Looks hard - Looks soft	0.093346	0.161469	-0.523601	0.208987
Looks too much - Looks too little	0.236816	-0.039862	-0.035875	0.732499
Contribution rate	0.444823	0.128581	0.094082	0.070887
Cumulative contribution rate	0.444823	0.573404	0.667486	0.738373

5 Modeling

Multiple regression analysis was performed for the value layer and the impression layer, and the impression layer and the physical layer, respectively, using the factor scores up to the fourth factor obtained from the principal component analysis and the physical quantities obtained from the images.

First, the relationship between the value layer and the impression layer was analyzed. The correlation coefficients of the brilliance, friendly, moist, moderate portion were 0.459, −0.172, −0.001 and −0.031, respectively. The correlation coefficient was 84%, which was significant for the two variables of brilliance and friendly. From the regression coefficient, it was found that increasing the brilliance and decreasing the other three variables increased the luxury.

Next, the relationship between the brilliance and the physical layer was analyzed. The correlation coefficient was 56%, which was significant for the ratio of pasta, ratio of plate, ratio of yellow, presence of background, presence of background blur, presence of other elements, presence of white dish and number of representative colors. The correlation coefficients were −0.043, 0.023, −0.022, −1.014, −0.948, 0.823, and 0.320.

Next, the relationship between the friendly and the physical layer was analyzed. The correlation coefficient was 34%, which was significant for brightness, ratio of yellow, ratio of black, presence of white dish and number of pixels. The correlation coefficients were 0.010, 0.018, 0.062, −0.465 and 0.000000632.

Next, the relationship between the moist and the physical layer was analyzed. The correlation coefficient was 28%, which was significant for presence of other elements and number of representative colors. The correlation coefficients were −0.493 and 0.122.

Next, the relationship between the moderate portion and the physical layer was analyzed. The correlation coefficient was 52%, which was significant for the ratio of pasta, ratio of plate, ratio of yellow, ratio of green, and number of representative colors. The correlation coefficients were −0.030, 0.012, 0.020, 0.052 and −0.127.

A summary of the above results is shown in Fig. 2. In this figure, only the significant elements are connected by lines. It was also found that setting the physical layer values as shown in Table 3 enhances the luxurious.

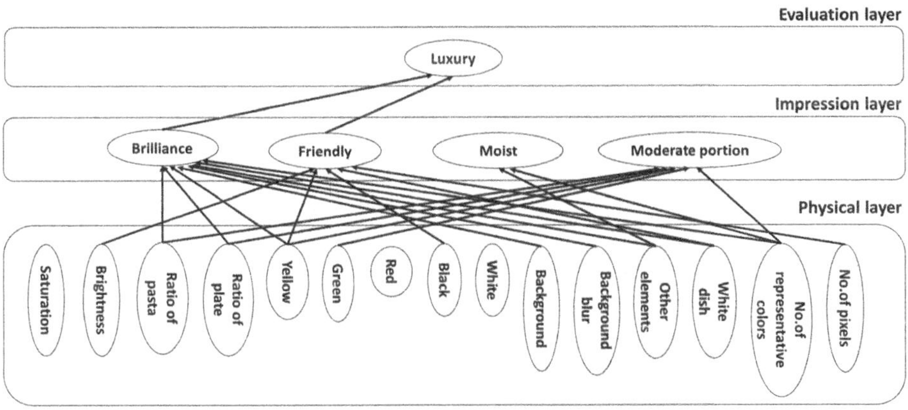

Fig. 2. Model reflecting multiple regression analysis result

Table 3. Physical feature quantities to enhance the luxurious

Brightness	Ration of pasta	Ratio of plate	Yellow	Green	Black	Background	Background blur	Other elements	White dish	No.of representative colors	No.of pixcels
Low	Small	Large	Small	Small	Small	None	Presence	None	Presence	Large	Low

6 Summary

In this study, we modeled the hierarchical structure of "Looks luxurious" in images of cooked dishes, and analyzed the physical feature quantities that affect the evaluation. As a result of modeling by Multiple regression analysis, it was suggested that "brilliance" and "friendly" has an influence especially on "Looks luxurious". In addition, the physical feature quantities which had an effect on "Looks luxurious" were revealed.

In the future, we will evaluate the validity of the model by conducting evaluation experiments using high-quality images and low-quality images based on the models obtained from the analysis.

References

1. Katahira, K., Muto, K., Hashimoto, S., Tobitani, K., Nagata, N.: Evaluation hierarchy in Kansei measurement using the SD method - focusing on the ambiguity of evaluation control factors in the EPA structure. J. Jpn. Soc. Kansei Eng. **17**(4), 453–463 (2018)
2. OpenCV. https://opencv-python-tutroals.readthedocs.io/en/latest/index.com
3. GURUNAVI Web Service. http://api.gnavi.co.jp/api/
4. Takahashi, N., Sakamoto, T., Kato, T.: Method of representative color extraction from images based on perception modeling and data analysis, and its experimental evaluation. J. Color Sci. Assoc. Jpn. **42**(4), 170–179 (2018)

Differences of Image Words for Clothes Depending on Fashion Brands

Yuri Hamada(✉), Takahiro Tanaka, and Hiroko Shoji

Chuo University, 1-13-27 Kasuga, Bunkyo-ku, Tokyo 112-8551, Japan
{yhamada, hiroko}@kc.chuo-u.ac.jp

Abstract. This study aims to investigate how the nuances of image words in the fashion field differ depending on the fashion brand. First, the authors extracted fashion brands that represent feminine, mode, and street types, and collected fashion images on the brand site and explanations for each image. Next, we extracted the image words contained in the collected explanations and investigated their frequency. As a result, it was found that the three image words "simple", "casual", and "cute" appear frequently among the three brands. Next, in order to investigate whether the nuances and usage of image words differ depending on the brand, we investigated words that co-occur in frequent image words. In addition, we compared co-occurring words by brand. As a result, the authors clarified that the concept represented by the image word differs depending on each fashion brand.

Keywords: Co-occurrence network · Text mining · Fashion

1 Introduction

In recent years, as we have become materially rich, our lives have become richer. While the food, clothing and shelter responsible for life are becoming richer, various information is intricate, making it difficult to differentiate the images of things we see and pick up. Above all, in fashion related to clothing, there are many similar items because products spread according to trends and times. This has made it very difficult to differentiate the image of apparel products. In today's information society, SNS and the Internet are the main media for obtaining information. Therefore, many consumers use apparel sales sites to collect fashion information. In this study, using the product introduction sentence on the apparel sales site as the subject, we clarify what image words are used in the introductions by fashion genre and what differences there are in the image words.

2 Related Studies

Tomobe et al. [1] conducted an impression evaluation experiment on the suitability of the image words for the coordination, using 40 snap shots of fashion coordination and 11 image words. As a result, there was variation in the image words selected for the fashion style. In particular, image words such as feminine, girly, modern, and elegant,

G. Di Bucchianico et al. (Eds.): AHFE 2020, AISC 1202, pp. 953–956, 2020.
https://doi.org/10.1007/978-3-030-51194-4_124

which are frequently cited as fashion image words, were selected for a single fashion style. From this, it can be seen that various image words are applied to the fashion style and are widely quoted. However, it is not clear how image words differ for each fashion style. In this study, we clarify the differences between fashion genres.

3 Experimental Methods

In this study, we used product introductions by three fashion genres. Three fashion genres were selected: feminine, mode and street. The data used in this study was selected with reference to the ranking site [2]. Images from the selected genres and product introductions are quoted from the fashion site Wear [3]. From the data quoted, images and sentences with zero characters in the product introduction sentence were removed, resulting in 1386 feminine, 2047 mode, and 1764 street items. Experiments were conducted on these data. The appearance frequency of image words was visualized using Text Mining Studio [4] for the introduction sentences of each brand. At that time, seven parts of speech, including adjectives, adjective verbs, and adjective verb stems, were selected as possible parts of speech as image words.

4 Experimental Result and Discussion

Table 1 shows the appearance frequencies of the image words of each brand.

Table 1. Adjectives used in the impression evaluation experiment ("〇〇" in the table is the brand name of the quoted brand)

Ranking	Feminine brand	Mode brand	Street brand
1	Feminine	Simple	Good
2	Good	Womanly	Simple
3	Simple	Good	Usual
4	Womanly	Beautiful	Graphic
5	Typocal of 〇〇	Casual	Standard
6	Moderate	Fashionable	Cute
7	Casual	Cute	Large
8	Cute	Various	Basic
9	Long	Long	Big(in Japanese)
10	Classy	Suspecious	Exactly
11	Adult feminine	Moderate	Casual
12	Glamorous	New	Original
13	Bold	Happy	Big(inEnglish)
14	Delicate	Basic	Nonchalant
15	Compact	High	Typocal of 〇〇

Table 1 shows that among the image words that appear frequently, there are image words (simple, cute, casual) that are common to all three brands. A co-occurrence network is created for the three words to clarify the differences between image words for each brand.

First, focus on image words that appear frequently in each genre. From Table 1, it can be seen that feminine, cute, womanly, classy, etc. appear frequent-ly in the feminine brand. In general, feminine has a strong connection with cute and feminine images, so these words have a high frequency of appearance, and are used to image feminine in the introduction. It can be seen that the appearance frequency of womanly, beautiful, casual, etc. is high in the mode brand. The mode is a cool and unique fashion system. Therefore, it is considered that the frequency of casual appearance is high. In addition, women often tend to be conscious of beautiful silhouettes and womanly dress up, so it is thought that the frequency of appearance of beautiful and womanly is high.

In the street brand, large, big, casual, etc. appeared frequently. In recent years, oversized fashion is prevalent among both of men and women. Therefore, it is considered that the frequency of appearance of words that image a sense of large size such as large and big may have been high. In general, it is considered that the frequency of casual appearance is high because street elements generally have a strong casual element.

Next, consider an image word having a high appearance frequency commonly in each genre. The words common to each genre are simple, cute, and casual. Simple, cute, and casual are used as fashion image words on a daily basis, so it seems natural that they appear frequently. Also, in this study, the word "cute" is likely to appear more frequently because the subject is a product introduction sentence for lady's fashion. Then, we created a co-occurrence network for image words using Text Mining Studio to visualize what common words co-occur with each genre. Table 2 shows the results of co-occurrence terms of image words classified by genre. From Table 2, it can be said that simple, cute, and casual have different genres because co-occurring words are different for each image word.

Table 2. Co-occurring words by genre and image word

	Feminine brand	Mode brand	Street brand
Simple	Smart	Slit style Healthy	Over Fancy Psychedelic Warm
Cute	Peplum style		Over Typocal of ○○
Casual	Mannish Cool Not too rough Loose	Girly Womanly	

First, think about simple. The word smart co-occurred in the feminine brand. This suggests that it may be related to the calm impression and polite image. A mode brand co-occurred with a slit style, and a street brand with overs. This reflects the trend of ladies' fashion. Next, think about cute. The word peplum style co-occurred in feminine brand. Peplum is a trend in women's fashion, expressing a silhouette that feels feminine. Therefore, it seems that "cute" in feminine brand is strongly related to trends and womanly. The word "over" co-occurred in the street brand. In recent years, clothes with a large size have become popular, and are often called big size or over-size. For this reason, it is thought that there is a strong relationship with words related to recent trends such as big and large. Finally, think about casual. The feminine brand was co-occurring with the masculine meaning of mannish, cool, not too rough, and loose. This suggests that it is strongly related to masculine style, seasonality, and trends. Girly and womanly co-occurred in the mode brand. Girly has a cute image like a girl, and womanly has a mature image. This suggests that there is a relationship between cute images and mature images.

From the above, it was found that words similar to the image of each fashion genre are often used. In addition, it became clear that image words commonly used in the three fashion genres have different co-occurring words for each genre, and there are differences.

5 Summary

In this study, it was clarified that there is a difference in the image words used in product introduction sentences for each fashion genre. Image words reflecting the image of each genre were often used. In addition, it was found that co-occurring words were different for image words that had a high frequency of appearance in common, and that there were differences between them.

The problem is that there were few data such as the number of fashion genres and the amount of product introduction sentences used in this study. It is thought that more reliable results can be obtained by increasing the amount of product introduction sentences using various genres.

References

1. Tomobe, N., Yanagida, Y.: A consideration of the fitness of fashion image terminology to fashion styles. Trans. Jpn. Soc. Kansei Eng. **13**(1), 137–144 (2014)
2. Lady's Me. http://design-dtp.net/
3. WEAR. https://wear.jp/
4. Text Mining Studio. https://www.msi.co.jp/tmstudio/

Preliminary Study on Color Therapy Effect Evaluation by the Emotion Estimation Method with Biological Signals

Yohei Kawakami, Makoto Komazawa, Chen Feng(✉), and Midori Sugaya

Shibaura Institute of Technology, Tokyo, Japan
{al15021,nb18109,doly}@Shibaura-it.ac.jp,
makoto.komazawa@winfrontier.com

Abstract. While the number of people with mild depression and emotional disorders is increasing, it is an issue that there is little support for them. These mild illness people, it is not very easy to understand the emotional status of themselves and tends to lose control of emotion. There are several therapies approach to control psychological status. However, there are little objectively observed approaches to evaluate them. In recent years, an objective method has been proposed to recognize the psychological changes of people by analyzing the biological signals with sensors. We assume this approach will be effective in detecting the emotional status objectively. Based on the idea, we preliminary propose a tool that can objectively detect the emotional status using the biological signals such as EEG and heart rate sensors with the emotion estimation method and present a therapy color which aims to control their emotion toward the target emotional state. Then evaluate the effect of emotion by the method. In this paper, we describe the assumption and proposed method and developed tool, and preliminary experiment by using them. Then, we discussed the effectiveness of our approach.

Keywords: Heart rate · pNN50 · Electroencephalogram · Color therapy · Stabilization of emotion · Support for assessment · Support for counseling

1 Introduction

According to a survey of patients with mental illness [1], there was an increase of approximately 1.88 million patients with mental illness. Among the people, the proportion of "Emotional disorders, including depression" is the highest, accounting for 28% of the total. But there is an inadequate of the clinical psychologists and the prescription [2]. The number of mild illness other than these statistical results are widely spreading. However, the techniques and methods for quantitatively measuring emotions have not been discussed sufficiently.

Regarding emotion estimation, a variety of methods has been proposed based on human behavior such as facial, languages, and behavior [4, 5]. However, these approaches sometimes do not estimate emotion accurately since the behavior information includes arbitrary actions based on individual cognition. On contrast, biological

G. Di Bucchianico et al. (Eds.): AHFE 2020, AISC 1202, pp. 957–963, 2020.
https://doi.org/10.1007/978-3-030-51194-4_125

signals are controlled by the autonomic nerve, which is the involuntary nervous system that has been presented [6, 7]. These approaches provide the method to evaluate the biological signals collected by the EEG (Electroencephalography), and heart rate sensors directly from the human body.

Usually, several therapy approaches are applied to control the emotional status. However, there are a few approaches that can objectively observe and evaluate. On the contrary, the emotion estimation method with biological signals can objectively observe and evaluate people's psychological states. We consider that the emotion estimation method with biological signals would be valid for the people who are difficult to understand their emotional state. Based on the emotion estimation method with biological signals, it is possible to apply the quantitative evaluation of their emotional status.

Since psychological researches show that different color can affect emotion status [8], we proposed a system that can objectively detect the user's emotional status with biological signals such as EEG and pulse sensors. The system will present a therapy color that fits with the target emotion. The system aims to help users to control their emotions toward the target emotion.

This paper consists of the following sections: Sect. 2 describes the relevant research, Sect. 3 describes the objective and proposal, Sect. 4 describes the experiments, Sect. 5 contains the discussions, and Sect. 6 states the conclusion and future challenges.

2 Related Work

There are several therapies approach to control emotional status. Within these therapies, psychotherapy, such as behavioral therapy [9], cognitive therapy [10] is widely used for the treatment. Behavioral therapy focuses on behavioral problems such as present habits, mindset, etc. Cognitive therapy, both logically and in practice of treatment, focuses on the thoughts and visual images consciously perceived by the examinees [11]. These therapies would be applied to the patient who has problematic behavior or cognitive problems in the brain.

In contrast, color therapy [12] is applied to a more casual case of controlling the emotional status. It uses the biological effects of colors and provides psychiatric treatment by presenting appropriate colors to the examinees. The colors presented are based on color psychology, and colors that are necessary or not sufficient for people are selected using a model in which the biological effects of colors are mapped on a hue circle.

Since various impacts of colors on emotions have been pointed out, recent years, impacts of colors on emotions have applied in color therapy and industrial products that use the effect of colors. To regard the impact of colors on the body, Zhang et al. evaluated the heart rate and found that the colors green and red affect the heart rate [13]. Also, Mishima et al. evaluated brainwaves and found that the colors yellow and blue affect brainwaves [14]. In color therapy, the correction of mental balance by the presentation of color is called the stabilization of emotions. Several color therapies are performed to this end; however, a method of formulation has not been explicitly established for the therapeutic purpose.

We consider the therapy would be applied to mild depression and emotion disorders; however, the effect of the therapy has not shown. Although many color therapies have been used in psychology treatment, there is still no specific formulation method that has been established for improvement. Therefore, it is necessary to verify and discuss the effect of color therapy.

In order to measure the emotional state objectively, these days, there are approaches to detect emotion by biological sensors. So, in order to increase the validity of the stimulus effects of the previous research, we consider using the combination of the EEG and the heart rate signal to evaluate the effect. Ikeda et al. classified and estimated emotions [7] by associating brainwaves and heart rate with arousal and valence axes, respectively. The values obtained from two sensors, EEG sensor and heart rate sensor, were assigned based on Russell's circumplex model [15]. However, there is little research to apply this method to evaluate the emotional status of them.

3 Proposal

3.1 Purpose of the Research

As described to the consider emotion estimation method with biological signals would be effective for people for whom it is difficult to understand their emotional state, and it is also possible to apply the quantitative evaluation of their emotional status to control their emotion toward the target emotion.

Since our objective is to propose a system to support to detect emotional status objectively and control the emotion easily, it was considered that appropriate to use the effect on emotion by the presentation of stimuli such as color and music, rather than language. Also, from the fact that among the perceptions by the five senses, information through visual organs constitutes 80% [14], we considered color therapy using colors that are associated with the sense of vision would be the most effective.

Based on the idea, we preliminary proposed a tool to apply both objective emotions estimate method and color therapy. This tool can detect the user's emotional status objectively with biological sensors based on the emotion estimate method and present a target color to help the user to control their emotional status.

It is necessary to estimate emotions quantitatively. Therefore, we considered a method to classify emotions using biological signals acquired from EEG and heart rate sensors based on the emotion estimation method. As the primary step, we will examine a method that attempts to control emotions by presenting colors based on the emotions classified with the assessment support system.

3.2 Proposed System

To recognize the emotion more objectively with sensors, we use the Ikeda et al. approach. They applied EEG and heart rate (pNN50) to Russell's circumplex model and classify the emotion using the result of the sensors. In this approach, the heart rate variability values assigned to the X-axis, and the result of the EEG assigned to the Y-axis. We consider the methods defining the positive region of the X-axis as "Valence/Pleasant", the negative region as "Unpleasant", the positive region of the Y-axis as "Arousal", and the negative region as "Sleepiness" have been found to yield results in various studies.

Based on these researches, we a mechanism to classify user's emotion and used color to target emotion.

In this system, a tool that maps the emotions obtained by the biological emotion estimation method based on the values of the EEG and heart rate sensor onto the emotion model created by referring to the Russell's Model emotion model and outputs it to the screen The emotions are output on the screen in the same manner as above, and the emotions are expressed in 10 levels of intensity using eight characters: "Happy", "Excited", "Alarmed", "Angry", "Gloomy", "Tired", "Sleepy", and "Relaxed". In the color.

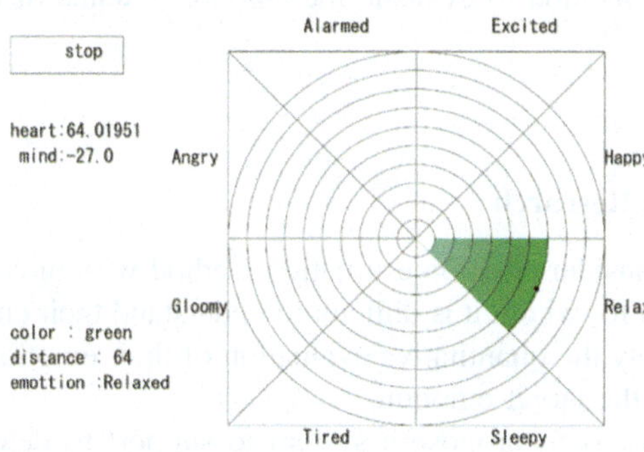

Fig. 1. Emotion mapping on the emotion visualizer

In the method used by Ikeda et al., the X-axis values are calculated using a sensor that obtains the heart rate from the heart rate. The sensor measures the heart rate from the heart rate by photo plethysmography and uses pNN50 as an evaluation index. pNN50 is an index for time domain analysis of RR interval fluctuation, and it is an index to determine the state from the rate of fluctuation over a fixed time; however, it is difficult to consider the balance of sympathetic and parasympathetic nerves which can be obtained by frequency analysis.

We used the Attention and Meditation values obtained from Mind Wave Mobile of NeuroSky's for Y-axis values. These values remove noise from the α and β-waves, respectively [16], and are considered to indicate concentration and rest, however, because the difference is made to correspond to the Y-axis, it is not possible to accurately reflect the feature value.

In this system, the Emotion Color Model (Fig. 2. Left) was created by first arranging colors that have an effect on emotions based on the Russell Model. Zhang et al. Evaluated the heart rate on the effect of color on the body, and determined that green and red had an effect on the heart rate [13]. Mishima et al. Evaluated the EEG, and yellow and blue evaluated the effect on the EEG [14]. Based on this, in this system, the red color, which is dominant in the sympathetic nervous system to the left of the origin of the X axis, and the green color, which is dominant in the parasympathetic nervous system to the right of the origin, are used. The higher the value, the more arousal is yellow, and the lower the value, the blue. In color therapy, the purpose is to correct the mental balance by presenting colors, and to control emotions, the colors are point-symmetric with the current emotions (right). It was designed to be controllable.

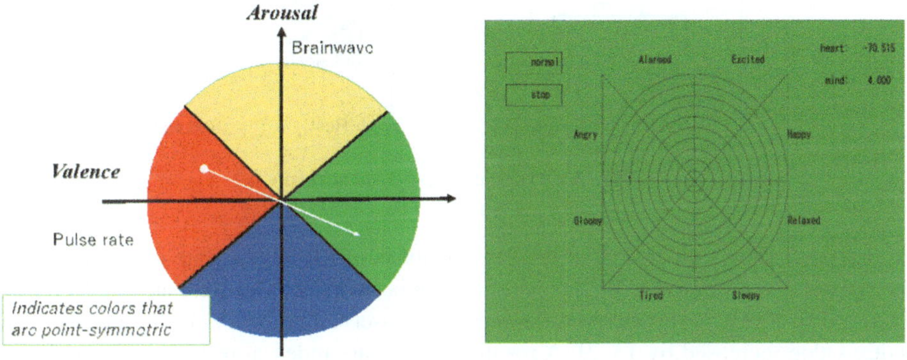

Fig. 2. Emotion color model (left) and select therapy color presenting (right)

4 Preliminary Experiment

To evaluate the changes of the emotional status by presenting the colors through our system, we conducted a preliminary experiment. To analyze the changes of the body, we use EEG and heart rate during rest and the color presentation.

First, we observed the effect of presenting the red color stimulus to the experimental co-workers whose emotions during resting for one minute were "sleepy" and "relaxed". The experiment collaborators performed on 15 women in their 60 s.

Experiment Procedure

1. Attach the electroencephalograph and heart rate meter to the experiment cooperator to get used to the discomfort.
2. The living body is measured for 1 min at rest.
3. The color to be presented is selected according to the measured biological information, and the color stimulus is presented for one minute.

   The biological measurement at that time is performed.

4. Repeat steps 2 and 3 three times.

5 Results

Figure 3 shows the results of analyzing changes in biological information when presenting a red color stimulus to an experimental co-worker who spent a lot of time in the coordinate area within the emotional area of "full" in the resting minute. The result was obtained. These results indicate that the red stimulus tends to stabilize emotions in the "full" state, indicating the validity of red placement.

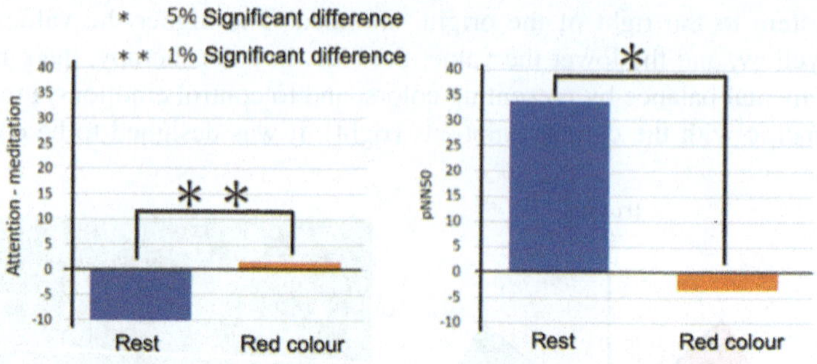

Fig. 3. Average of experiment result

Figure 3 shows the results of the t-test analysis of EEG/ heart rate at rest and when the red color was presented by the experimental co-workers who were at rest. As for the EEG index at rest, a significant difference was observed at the 1% level, and the EEG index value increased by 13.29. A result of heart rate index at rest was analyzed with respect to experiment co-workers is a pleasant state of 0 to 100 with respect to heart rate, a significant difference was observed at the 5% level, the value of the heart rate index is 36.99 decrease.

From these results, when red stimuli were presented to subjects whose emotions at rest were "full", the emotions moved toward the target emotion. It shows the possibility of the effectiveness of our approach.

6 Conclusion

In this study, we proposed a method to support objectively observed the change of the emotion by biological sensors, and applying the color therapy method, evaluate the effect of presenting a color through our proposed system as a preliminary study. The EEG and heart rate (pNN50) indices and proposed a method to display colors according to the emotion classified by the prototype support system.

The result show that the red color showed a significant difference for both arousal and valance in the experiment. For emotion classification models, since individual differences of the experimental collaborators cannot be considered in the experiment, it is necessary to prepare the emotion classification model for each individual which depends on information at rest. It is considered that the emotion model should also be associated with colors based on the biological signals at rest.

References

1. Mental disease data, Ministry of Health, Labor and Welfare. Disability health welfare department, Ministry of Health, Labor and Welfare, 17 January 2019
2. Transition of persons with "Clinical Psychologist" qualification, Foundation of the Japanese Certification Board for Clinical Psychologists, 17 January 2019

3. Natsukari, I.: How do patients and their families assess the "Psychiatric physicians consultation attitude" - findings from approximately 6,000 respondents and recommendations based on it. Psychiatria et neurologia Japonica (10), 868–886 (2018)
4. Faulhaber, G.R.: Design of service systems with priority reservation. In: Conference Recreation: IEEE International Conference Communications, pp. 3–8 (1995)
5. Doyle, W.D.: Magnetization reversal in films with biaxial anisotropy. In: 1987 Proceedings of INTERMAG Conference, pp. 2.2–12.2-6 (1987)
6. Soraia, M., et al.: Emotions recognition using EEG signals a survey. IEEE Trans. Affect. Comput. **3**, 1–12 (2017)
7. Ikeda, Y., Horie, R., Sugaya, M.: Estimate emotion with biological information for robot interaction. In: 21st International Conference on Knowledge-Based and Intelligent Information & Engineering Systems (KES 2017), Marseille, France, 6–8, September (2017). Procedia Computer Science, vol. 112, pp. 1589–1600
8. Ou, L.C., Luo, M.R., Woodcock, A., Wright, A.: A study of colour emotion and colour preference. Part I: colour emotions for single colours. Color Res. Appl. **29**(3), 232–240 (2004)
9. Kingdon, D.G., Turkington, D.: Cognitive-Behavioral Therapy of Schizophrenia. Guilford Press, New York (1994)
10. Beck, J.S., Beck, A.T.: Cognitive Therapy: Basics and Beyond (No. Sirsi i9780898628470). Guilford Press, New York (1995)
11. Japan Association for Cognitive Therapy. What is cognitive therapy? http://jact.umin.jp/introduction/. Accessed 17 Jan 2019
12. Daiwa Pharmaceutical Co., Ltd. Color Therapy. http://www.daiwa-pharm.com/info/holistic/18/. Accessed 17 Jan 2019
13. Zhen, Z.: The Effects of the environmental color change on the ability of the human movement. Hum. Dev. Stud. (5), 23–29 (2014)
14. Mishima, K.: Fundamental studies on the characteristics of electroencephalogram inspecting. Sev. Colors **54**(5), 108–113 (1990)
15. Russell, J.A.: A circumplex model of affect. J. Pers. Soc. Psychol. **39**(6), 1161–1178 (1980)
16. Lee, C.F., Katsuura, T., Shibata, S., Ueno, Y., Ohta, T., Hagimoto, S., Sumita, K., Okada, A., Harada, H., Kikuchi, Y.: Responses of electroencephalogram to different odors. Ann. Physiol. Anthropol. **13**(5), 281–291 (1994)

Comparing Subjective and Objective Evaluations of the Ratio of Visible Green

Momoko Kawakami[1](✉), Masayoshi Tanishita[1], and Michiko Ohokura[2]

[1] Advances in Civil and Environmental Engineering, Chuo University, Tokyo 112-8551, Japan
{a15.akya,matanishita.45e}@g.chuo-u.ac.jp
[2] SIT Research Laboratory, Shibaura Institute of Technology, Tokyo 135-8548, Japan
ohkura@sic.shibaura-it.ac.jp

Abstract. Ratio of visible green is known as one of the important indicators of comfort of the city. The ratio of the green area in the image, generally calculated by counting green pixels in the image of a digital camera. We named the ratio as "objective ratio of the visible green" in contrast to the ratio named "subjective ratio of the visible green" which is answered by the person who evaluated the ratio of green amount by looking at the image. Our interests are the difference between the two and the individual differences of subjective ratio of visible green. We conducted an experiment to present multiple images with different ratio of visible green for 30 people. The subjective ratio of visible green tends to be evaluated greater than the objective one. In addition, using multiple regression analysis, we showed that characteristics of the image of a digital camera and individual characteristics affect the differences.

Keywords: Objective ratio of visible green · Subjective ratio of visible green · Evaluation · Multiple regression analysis

1 Introduction

From the viewpoint of improving health and quality of life, the development of human friendly towns and urban spaces has become increasingly important. From the study on the relaxing effect of forest therapy and green in the city, it became clear that green gives a relaxing effect to people [1, 2]. So, the ratio of visible green became recognized as an index of the comfort of the city. We focused on the relationship between ratio of visible green and the relaxing effect of urban space. Then we conducted an experiment to measure heart rates of walking participants in four spaces with different ratio of visible green. It is important how to define and measure the ratio of visible green to estimate the effect of green properly.

Ratio of visible green are used as an environmental evaluation index in several municipalities in Japan [3]. In general, it is calculated by the ratio of the green area which is occupied by the image taken by a digital camera.

Various studies have been conducted on how to measure the ratio of visible green, which captures green as a large group. Fujii et al. have created a CG model of the

G. Di Bucchianico et al. (Eds.): AHFE 2020, AISC 1202, pp. 964–970, 2020.
https://doi.org/10.1007/978-3-030-51194-4_126

university campus and calculated the ratio of visible green [4]. Hirasawa et al. proposed method to estimate the ratio of visible green by defining that of trees as the ratio of the three-dimensional angle of the tree-lined to the unit sphere [5]. The calculation of the ratio of visible green as a physical quantity has been studied by various methods but does not consider the relationship with the amount of green actually felt by humans.

We named the ratio of visible green calculated from the digital image "objective ratio of visible green", and answered by a person looked at the image and answered about how much green in the images "subjective ratio of visible green". The purpose of this paper is to compare these two criteria values and examine how to set the visible green rate when it is intended to be used for spatial evaluation for humans. In addition, we clarify the factors that influence the evaluation value of the subjective visible green rate.

The first hypothesis is that the objective ratio of visible green and the subjective ratio of visible green do not always match because people see visible green as a unit. The second hypothesis, person's perception of the amount of green may have individual differences influenced by their perceptions and preferences. Therefore, it is thought that the subjective ratio of visible green can be an important index not only for evaluating the amount of green but also for understanding the sensitivity of the participants.

2 Preliminary Experiment

The experiment was conducted to confirm that the objective ratio of visible green and the subjective ratio of visible green have different values.

2.1 Method

Ten participants (six men and four women) joined the experiment. They watched sixteen images with different objective ratio of visible green. After 20-s watching, the participants answered the amount of green of each image in 10 s. The image was presented by projecting on a white wall about five meters in width and four meters in height. The images used in the experiment, and those objective visible green rates are shown in Fig. 1.

2.2 Data

Subjective ratio of visible green obtained by experiment (n = 10).
Objective visible green rate obtained by image analysis.
(The hue range for judging green was 60° to 170°).

2.3 Results and Discussion

Figure 2 shows the objective ratio of visible green and the subjective ratio of visible green answered by the participants. Those two kinds of ratio did not necessarily match, and it was confirmed that the subjective values differ up to 50% between individuals in the same image. In 14 out of 16 images, the subjective ratio of visible green exceeded the objective ratio of visible green. And the differences of other two images are less

Fig. 1. 16 photos presented in the experiment. The number at the upper left of each image is the ID, and the number at the below is the objective green rates (%).

than 3%. In other words, the subjective ratio of visible green is evaluated roughly as the same or higher than the objective ratio of visible green measured this time. The reason for this result is considered to be that people tend to perceive the green of plants as a group. The gap between the leaves where the building or the sky are observed is recognized not to be green in the objective evaluations, whereas it is likely to be evaluated as a group of green in subjective evaluations.

The differences between subjective and objective ratio are less than 10% for images when the objective ratio of visible green is less than 20% and more than 50%. This is probably due to the characteristics of the image, such as the presence of green clumps. On the other hand, a presented image in which green is mottled over the entire screen including the upper part of the screen has a high evaluation of green viewing rate and is easily estimated.

It is probable that there was a large error in the measurement of the objective ratio of visible green for the images of No. 4, 5, 8, 11, which had a particularly large difference. When calculating the objective green viewing rate, it is necessary to pay attention brightness or saturation of the image set to adjust them, and this is a future topic.

Next, we examine the tendency of individual differences in subjective ratio of visible green in the same image. The difference between the maximum and minimum values ranged from 18% to 50%. It is necessary to increase the number of participants and survey items and analyze the major factors in the evaluation of ratio of visible green.

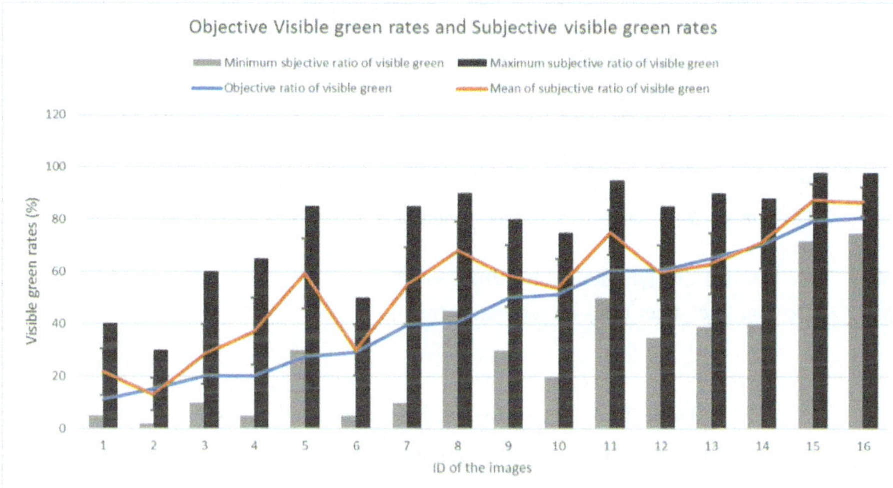

Fig. 2. Objective ratio of visible green and subjective ratio of visible green of each participant

3 Comparing Objective and Subjective Ratio of Visible Green and the Factor Analysis of Individual Differences

Next, we examine the tendency of individual differences in subjective ratio of visible green in the same image. The difference between the maximum and minimum values ranged from 18% to 50%. It is necessary to increase the number of participants and survey items and analyze the major factors in the evaluation of ratio of visible green.

3.1 Method

The experiment was performed on 27 participants (15 men and 12 women). In total, 37 people (21 men and 16 women) participated in the experiment. The method of presenting the image and answering the ratio of visible green is the same as in the preliminary experiment. After the presentation of the images, they answered some question about thoughts on their background, green environment and the presented images. The numbers of options are two (yes/no) or five (strongly agree/agree a little/neither agree or disagree/disagree a little/strongly disagree). The items of the question are shown in Table 1.

3.2 Method of the Analysis

Multiple regression analysis is performed to clarify the factors that cause the subjective ratio of visible green to fluctuate. The explanatory variables consist of the characteristics of the photograph and the individual characteristics of the participants. The variables are shown in Table 2.

Table 1. The question items

No.	Contents of the question	Number of choices
Q1-1	Do you like each 16 view?	2
Q1-2	Have you seen in 16 views?	2
Q2	Was your place of childhood rich in nature?	5
Q3	Do you go to green places (parks, mountains, etc.) for leisure now?	5
Q4	Do you prefer more natural or urban areas?	5
Q5	Do you like street trees on the main street?	5

Table 2. Variables for multiple regression analysis

Type	Name of the variable	Contents of the variable
Image	green1	Ratio of visible green for the entire image
	green2	Ratio of visible green of outer half of image
	Sky	Whether the sky is not hidden
	Human	Whether the person is in the picture
	Building	Whether the building is visible in the image
Person	Q1	1: Yes, 0: No
	Q2	1: strongly agree, 2: agree a little, 3: neither agree or disagree, 4: disagree a little, 5: strongly disagree
	Q3	1: Often go, 2: Sometimes go, 3: Sometimes go with someone, 4: Do not go often, 5: Hardly go
	Q4	1: Nature, 2: Comparative nature, 3: Not either, 4: Comparative City, 5: City
	Q5	1: strongly agree, 2: agree a little, 3: neither agree or disagree, 4: disagree a little, 5: strongly disagree

3.3 Data

Subjective ratio of visible green obtained by experiment (n = 37).
Objective visible green rate obtained by image analysis.

3.4 Results and Discussions

Figure 3 shows boxplot of the relationship the difference and image with/without building. If buildings are in the image, people tend to evaluate similar value with objective ratio of visible green. Figure 4 shows boxplot of the relationship the

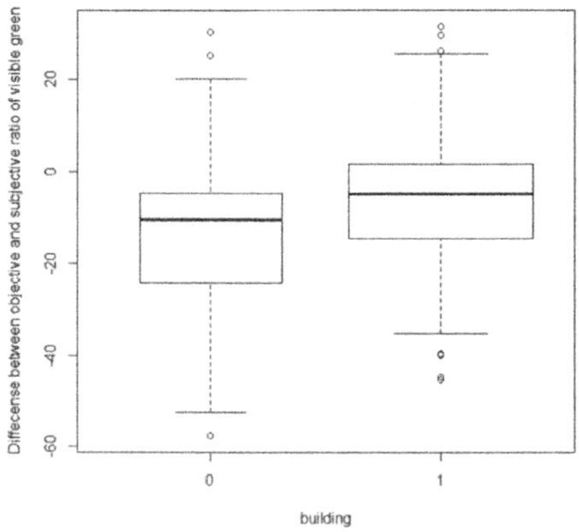

Fig. 3. Boxplot of the relationship the difference and image with/without building

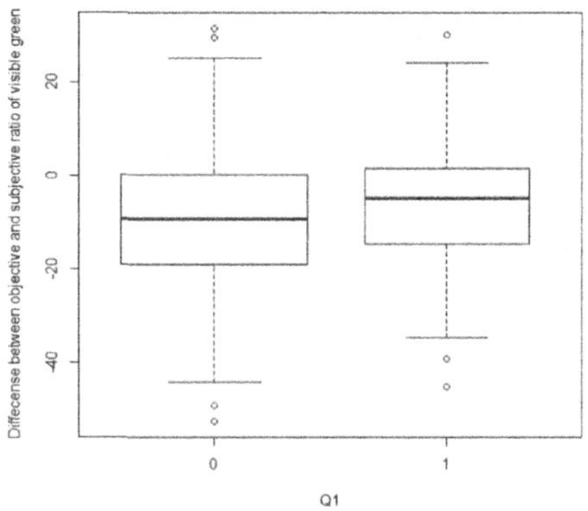

Fig. 4. Boxplot of the relationship the difference and the answer of Q1.

Table 3. Estimation results

Dependent variable: subjective green rate (%)				
		Estimate	t value	
	(Intercept)	25.59	7.39	***
Image	green1	1.64	12.91	***
	I(green1^2)	−0.01	−7.98	***
	Sky	7.32	1.98	***
	Human	−10.77	−5.33	*
	Building	−23.66	−12.30	***
	Sky: green2	−0.51	−3.86	***
	Human: building	19.63	7.21	***
Individual characteristics	Q1 (yes)	4.87	4.46	***
	ifelse (Q2 > 3, 0, 1)	2.48	2.28	*
	ifelse (Q3 == 4, 1, 0)	3.84	2.54	*
	ifelse (Q4 == 3, 1, 0)	4.54	3.11	**
	ifelse (Q5 < 3, 1, 0)	−5.86	−4.17	***
Coefficient of determination		0.82		
The number of samples		432		

difference and the answer of Q1. If the person likes the image, he/she tends to evaluate similar value with objective ratio of visible green.

Table 3 shows the estimation results of multiple regression analysis. First, the relationship between the characteristics of the image and the subjective ratio of visible green will be discussed. Focusing on human, it can be seen that the presence of people in the picture indicates that there is little green. And the presence of buildings tends to underestimate the amount of green. If both human and building are present, the ratio of visible green is answered about 15% smaller. The image in which the sky looks good, the ratio of visible green tends to be highly evaluated, but when green 2 (the objective ratio of visible green surrounding the image) is high, the evaluation value is small.

Focusing on individual characteristics, it can be seen that the objective ratio of visible green is smaller for favorite photos. It was also found that participants who answered negatively whether they spent their childhood in nature-rich areas tended to appreciate the ratio of visible green in the images.

4 Conclusion

Focusing on the subjective evaluation of participants, the subjective ratio of visible green was measured, and the effects of images and personal attributes on the evaluation were analyzed. The newly obtained knowledge this time is as follows;

- The objective ratio of visible green and the subjective ratio of visible green do not always match. In many cases, the subjective evaluation is greater than the objective evaluation.
- The value of the subjective ratio of visible green is determined from both the characteristics of the image and the characteristics of the individual. The features of the image are the sky, people, and the presence of buildings. It was suggested that individual characteristics such as the richness of nature in places where they spent their childhood, and whether they like the natural environment and street trees, affect the subjective ratio of visible green.

Along with the ratio of visible green calculated by image analysis, it may be an important index for evaluating a space in which a person feels comfortable. This time, the amount of green in the 2D image was extracted by determining the hue range. The future task is to compare multiple ratio of visible green calculated by various methods, as in previous studies, with the subjective ratio of visible green.

References

1. Matsuba, N., Lee, J., Park, B.J., Lee, M., Song, C., Miyazaki, Y.: Physical effects of walking in Shinjuku Gyoen, a large-scale urban green area. Japan. J. Physiol. Anthropol. **16**(3), 133–139 (2018). (in Japanese)
2. Lanki, T., Siponen, T., Ojala, A., Korpela, K., Pennanen, A., Tiittanen, P., Tsunetsugue, Y., Kagawae, T., Tyrväinen, L.: Acute effects of visits to urban green environments on cardiovascular physiology in women: a field experiment. Environ. Res. **159**, 176–185 (2017)
3. Osaka Prefecture: Ryokushiritsu tyousa gaidorainn (Survey guidelines for ratio of visible green), 2–24 (2013). (in Japanese)
4. Fujii, T., Yamada, S., Hirose, T., Oikawa, K.: Measurement of the ratio of visible green spaces in the omnidirectional field of vision using CG models and their potential applications. Architect. Inst. Jpn. J. Technol. Des. **19**(43), 1067–1072 (2013). (in Japanese)
5. Hirasawa, Y., Ukai, T., Kurita, O.: Ratio of visible green associated with pedestrian location and visual line-solid angle calculation introducing sphere model. J. City Plan. Inst. Jpn. **52**(3) (2017). (in Japanese)

Color Analysis of Popular Lipstick Brands

Peeraya Sripian[1(✉)], Keiko Miyatake[2], Kejkaew Thanasuan[3], and Michiko Ohkura[1]

[1] Shibaura Institute of Technology, 3-7-5, Toyosu, koto-ku, Tokyo 135-8548, Japan
{peeraya, ohkura}@sic.shibaura-it.ac.jp

[2] Kyoritsu Women's University, 2-2-1 Hitotsubashi, Chiyoda, Tokyo 101-8437, Japan
kmiyatake@kyoritsu-wu.ac.jp

[3] King Mongkut's University of Technology Thonburi, 49 Soi Thian Thale 25, Bangkhuntien, Bangkok 10150, Thailand
kejkaew.tha@mail.kmutt.ac.th

Abstract. Lipstick is one of the significant makeup tools for females of all ages. The current trend of lipsticks among Japanese undergraduate female is found to revolve around global contemporary luxury brands and also around more affordable brands with Kawaii-images. The goal of this study is to classify the characteristics of lipstick brands using color features in order to clarify differences in lipstick colors and brands. We collected a total of 294 lip colors from the trending brands and extracted RGB and HSV colors as features and developed the classification models on the color spaces using decision tree. The classification results successfully showed the brand characteristics with their color features of lipsticks.

Keywords: Cosmetic · Lipstick · Color analysis

1 Introduction

"Kawaii" has been officially recognized and used as Japan's foreign policy tool since the Ministry of Foreign Affairs appointed "Kawaii Ambassadors" in 2009. Kawaii concept mainly includes street fashion such as the unique Harajuku "Kawaii" type fashion that can be distinguishably observed in Harajuku [1–4]. Fashion comprises not only clothing combination but also facial cosmetic, hairstyle, accessories, poses, and so on. To investigate "Kawaii" fashion, we are also interested in looking at the cosmetics application. In this research, we focus on the lipstick colors of popular brands in Japan.

According to the second author's preliminary survey among Fashion-design major university female students, she found that the current trend of lipstick is revolving around global contemporary luxury brand like CHANEL and Yves Saint Laurent (YSL), MAC Cosmetics, to brands with Kawaii images like Etude House and Canmake. The global luxury brands such as CHANEL, YSL, and MAC have always been ranked as one of the top lipstick brands in Japan [5]. Especially for YSL lipstick, which has been ranked as the best cosmetics award from cosme.net, a major Japanese cosmetic ranking website [6].

G. Di Bucchianico et al. (Eds.): AHFE 2020, AISC 1202, pp. 971–977, 2020.
https://doi.org/10.1007/978-3-030-51194-4_127

On the other hand, Korean cosmetic brands are popular for their affordable price, Kawaiiness and high quality [7, 8]. Etude House is one of the top Korean cosmetic brands that acquired its popularity in Japan, mainly due to its Kawaiiness [9]. Finally, Canmake is a Japanese cosmetic brand that positioned itself as Kawaii, reasonable price and high quality.

We hypothesized that lipstick colors of contemporary luxury brands are somehow different from teenagers-target brands. We verified this hypothesis by the following methods;

- Data collection of lipstick color images
- Data analysis by
  - Color feature extraction using image processing
  - Analyze the extracted features using a decision tree

2 Data Collection of Lipstick Color Images

We collected a total of 294 lipstick color images that are downloadable as sample lip color images from the five cosmetic brand's websites; CHANEL, YSL, MAC, Etude House and Canmake. Several product lines for each brand were carefully selected as shown in Table 1. The examples of the collected data are shown in Fig. 1.

Table 1. Collected Lipstick brands (brand) separated by their product lines (Product Line) with the number of products (N).

Brand	Product line	N	Total per brand
CHANEL	Le Rouge Crayon	13	63
CHANEL	Rouge Allure	23	
CHANEL	Rouge Coco	27	
YSL	Rouge Volupté Shine	47	67
YSL	Volupté Plump-in-Color	8	
YSL	Volupté Tint-In-Balm	12	
MAC	Matte Lipstick	42	94
MAC	Powder Kiss Lipstick	24	
MAC	Creamsheen Lipstick	28	
Etude House	Dear My Glossy	10	50
Etude House	Better Lips Talk	30	
Etude House	Dear My Matt	10	
Canmake	Melty Luminous Rouge	3	20
Canmake	Stay-On Balm Rouge	14	
Canmake	Crayon Matte Lip	3	

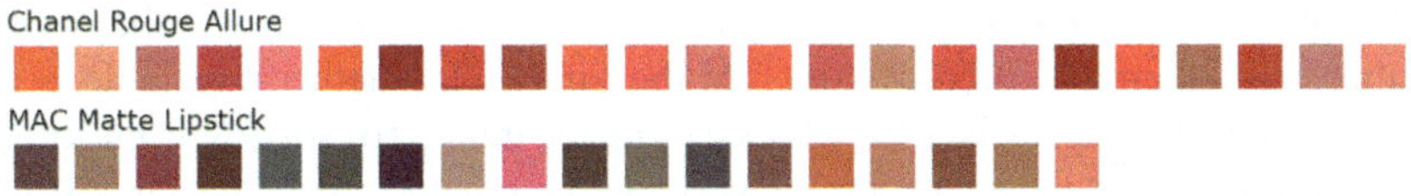

Fig. 1. Example of the collected lip color data.

3 Data Analysis

3.1 Color Feature Extraction Using Image Processing

For data analysis, we extracted RGB and HSV color values from every lip color image. The RGB color value is obtained by calculating the average value of red, green and blue channel from all pixels in the image. In our work, the RGB value is stored in the range of 0–255, with 0 indicates the lowest value and 255 is the maximum value.

The HSV color value, represented by 3 color parameters; hue (H), saturation (S) and intensity (V), can be displayed as a hexacone with the central vertical axis shows the intensity (V), the hue H is given by the angle about the vertical axis, while the saturation S varies between 0 to 1. In this work, we converted the image to HSV (Hue, Saturation and Brightness) color space before proceeding with the average value for hue, saturation and intensity to obtain HSV color value. The HSV color value is stored in the range of 0–1, with 0 indicates the lowest value and 1 is the maximum value for that component.

3.2 Analyze the Extracted Features Using a Decision Tree

In this study, we applied decision tree models on two color spaces, which are RGB and HSV in order to classify the lipstick brands (i.e., categorical variables) including CHANEL, Etude House, Canmake, YSL, and MAC. Then, we reported model performances and decision trees of each color space.

4 Result and Discussion

We implemented a decision tree model for each color space including RGB and HSV. The classification results were probabilities of each lip brand of each observation. Then, the results were used to evaluate the model performances.

4.1 Result Obtained from Decision Tree by RGB Features

The decision tree of RGB model is presented in Fig. 2A. Each branch shows a decision rule for splitting data to the next level of the tree. For example, the decision rule at the top-level is determining whether r-variable, which indicates red color, of each lipstick is greater than or equal to 216. If the answer is yes, that observation will be classified into the left branch and will be judged by the next decision rule at the next level. Otherwise, it will go to the right branch and will be judged by another decision rule. Each leaf node shows the classification results, which are the lipstick brands on the first line and a proportion of the data categorized at that branch on the second line.

For example, if a CHANEL lipstick has r-variable = 236, g-variable = 46 and b-variable = 84, then it will be classified at the top level, whether r-variable is greater than 216 or not. The result is 236, which is greater than 216, so it will go to another classification rule which is whether r-variable is greater than 246 and it is not. Therefore, it should go the right branch of the second level. The next step is to check whether b-variable is less than 128 and 84 is less than 128, so it is classified to the next rule, which is whether r-variable is greater than 224. The r-variable is greater than 224. Thus, the next classification is to check whether g-variable is less than 55 and it is true (46 < 55). The final result of the process will show the lipstick brand; in this case it is CHANEL. The model is able to categorize the RGB shade of lipstick color. Furthermore, the variable importance scores showed that r and g are the most significant classification variables.

From the decision tree obtained by RGB features, we found that CHANEL and MAC were distinct and able to be classified from the other brands. Meanwhile, YSL, on the other hand, contains many variations so it can be difficult to create a cluster of this brand. Canmake was found to be classified similar to CHANEL, while Etude house was found that it could be classified using many possible criteria like YSL.

4.2 Result Obtained from Decision Tree by HSV Features

Additionally, the decision tree of HSV model is presented in Fig. 2B. The lipstick brand was categorized based on H, S, V variables, and the most significant variables were V and S. Using saturation and intensity value, the decision tree obtained by HSV features classification shows that it is possible to classify MAC. As seen from the rightmost node of the decision tree that 18% of MAC is classified using S < 0.62 and V < 0.8 combination criteria. For CHANEL, it is still possible to classify using the combination of S, V and H as seen from the decision tree. On the other hand, YSL can be classified with many possible criteria, similar to Etude House.

4.3 Comparison Between Model Performances Obtained from RGB and HSV

The overall accuracy of RGB and HSV models, which indicated the models successfully classified the brand of each lip using only the color-space variables, were 57.82% and 58.5%, respectively. It also showed that both models were able to categorize MAC (RGB model: 67.02% and HSV model: 75.53%) and CHANEL (RGB model: 61.9% and HSV model: 63.49%) correctly more than the others. This may suggest that lipstick colors of MAC and CHANEL were distinct and able to be classified from the other brands using both RGB and HSV models.

4.4 Discussion

In this work, we successfully implemented decision tree based on lipstick's color feature to visualize five lipstick brand's characteristic. It is well known for middle-aged females that CHANEL is very distinct in the product colors, as their lipstick shades somehow contain more "red" color than other brands, especially comparing to those of

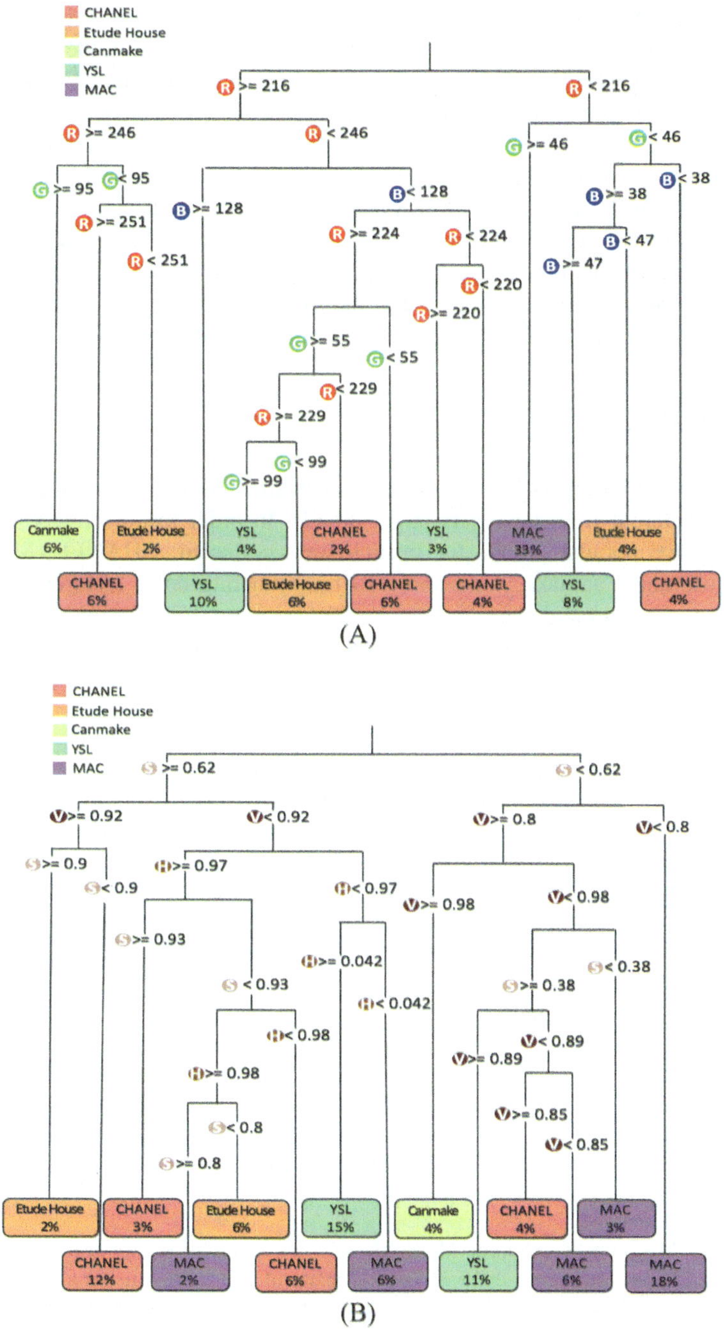

Fig. 2. Decision trees from RGB model (A) and HSV model (B). Each leaf node shows a classification result, which is a lipstick brand on the 1st line, and the proportion of the entire data that was classified at that branch on the 2nd line.

MAC. While YSL's position could be quite uncertain. On the other hand, more affordable brands like Canmake and Etude House seem to position themselves similar to the popular high-end brands. From the classification result, Chanel contains higher shades of red as seen from the result of the decision tree obtained from RGB that R >= 251 and G < 95 is classified as CHANEL, while for MAC is a mixture of other shades and its red color will definitely not be greater than 216 (R < 216 and G >= 46 is classified as MAC). Meanwhile, YSL contains many different possible color combinations, therefore there are many possible classification criteria for YSL. Canmake, as seen from the classification result, is classified similar to Chanel while Etude house can be classified using many possible criteria like YSL. However, the limited number of products for Canmake could affect the classification results.

5 Conclusion

We hypothesized based on the experience that lipstick colors of contemporary luxury brands are different from teenagers-target brands. In order to objectively verify such hypothesis, five popular brands for lipstick were collected; CHANEL, MAC, YSL, Etude House and Canmake, were analyzed by extracting color features using image processing techniques and classified using a decision tree. In this work, we showed the classification results of decision tree obtained by RGB color features and HSV color features. These results follow the hypothesis. Future work will include adding more data of each brand and explore other classification methods.

Acknowledgments. We thank Ms. Hitomi Kiuchi, a student in the Department of Textile and Clothing, Faculty of Home Economics, Kyoritsu Women's University for lip color data collection and also thank Ms. Nithita Riansumrit, a student in Media Technology, King Mongkut's University of Technology Thonburi for creating decision tree figures.

References

1. Miyatake, K.: Japanese Kawaii fashion-historic changes in trends and styles. In: Spring Conference of Japan Society of Kansei Engineering, Nagoya (2018)
2. Miyatake, K., Sripian, P., Ohkura, M.: Study on style of Harajuku Kawaii fashion (1st Report) - design features. In: The 20th Annual Meeting of Japan Society of Kansei Engineering, Tokyo (2018)
3. Sripian, P., Miyatake, K., Ohkura, M.: Study on style of Harajuku Kawaii fashion (2nd report) - color features. In: The 20th Annual Meeting of Japan Society of Kansei Engineering, Tokyo (2018)
4. Sripian, P., Miyatake, K., Ohkura, M.: Study on color feature of Harajuku Kawaii fashion. In: The 14th Spring Conference 2019 of Japan Society of Kansei Engineering (2019)
5. Brand Lipstick Cosme Goods Present Ranking. https://bestpresent.jp/cosme/lipstick-cosme/. Accessed 19 July 2019. (in Japanese)
6. @Cosme. https://www.cosme.net/. Accessed 19 July 2019

7. Park, H.: Preliminary Study on Transition of Korean Cosmetic Industry and Sources of Sustained Competitive Advantage of Cosmetic Manufacturers: Focusing on Amore Pacific Case. Fukuoka Jo Gakuin University Bulletin. Faculty of International Career Development, 1, 45–68 (2015). (in Japanese)
8. Nohara, U.: Recommended 11 Korean Lip. https://lipscosme.com/articles/1295. Accessed 19 July 2019. (in Japanese)
9. Etude House that everyone bought is for instant. Kawaii. https://matome.naver.jp/odai/2150819610043777701. Accessed 19 July 2019. (in Japanese)

Sensory Engineering and Emotion

Comparison of Object-Based Three-Dimensional Sound and Ambient Sound for Straight Walking

Keiichi Muramatsu[1,2(✉)], Shigekimi Tsukada[1], Kazunori Kaede[1,2], and Keiichi Watanuki[1,2]

[1] Graduate School of Science and Engineering, Saitama University, 255 Shimo-okubo, Sakura-ku, Saitama-shi, Saitama 338-8570, Japan
{muramatsu, kaede, watanuki}@mech.saitama-u.ac.jp

[2] Advanced Institute of Innovative Technology, Saitama University, 255 Shimo-okubo, Sakura-ku, Saitama-shi, Saitama 338-8570, Japan

Abstract. At present, visually impaired people have difficulty using mobility support facilities due to their dependence on the surrounding environment, such as sound and snowfall. In addition, people who later became visually impaired may find it difficult to use these facilities. The current study aimed to develop an easy-to-use walking guidance system that uses three-dimensional (3D) sounds, regardless of the surrounding environment, and examined the characteristics of the guidance sounds produced by the 3D sounds. We implemented this system using a head-mounted display capable of generating three-dimensional sound. Participants were asked to walk in a visually restricted environment to either a real guidance sound or a 3D guidance sound with three types of ambient sounds, and their walking trajectory was evaluated. As a result, we found that the 3D guidance sound showed the opposite tendency to deflect from the ambient sound. However, the guidance accuracy of the 3D guidance sound was similar to that of the real guidance sound.

Keywords: Three-dimensional sound · Ambient sound · Walking guidance

1 Introduction

In Japan, there are currently about 310,000 visually impaired people who are unable to use their eyes to collect information from the environment, accounting for about 9% of the Japanese population with disabilities [1]. These visually impaired people need to be able to walk for their daily activities. The visual information required for walking, however, is highly dependent on vision, and when visual impairment occurs, gait impairment also occurs at the same time. Braille blocks, tactile guidance lines (called escort zones introduced by the Ehime Prefectural Police in 1997) and acoustic traffic signals have been widely installed in Japan to help visually impaired people walk more safely and independently. Many studies have been reported in Japan on how to establish escort zones for the visually impaired and how to cross the road [2, 3]. However, in Tokyo, there is concern that about 0.6% of crosswalks have escort zones and about 2% of crosswalks have acoustic traffic signals [4]. These devices can be

G. Di Bucchianico et al. (Eds.): AHFE 2020, AISC 1202, pp. 981–986, 2020.
https://doi.org/10.1007/978-3-030-51194-4_128

difficult to use because they have the inherent problem of being dependent on the surrounding environment. For instance, when it is snowing, it is difficult to recognize the uneven surfaces of braille blocks and escort zones that are affected by environmental changes. In addition, if an acoustic signal is installed at an intersection, it may not be operable due to the need to consider the impact on residents close to the installation site. Therefore, the current study proposes an object-based three-dimensional acoustic guidance method for solitary walking, which is designed for the visually impaired. The purpose of the current study is to examine the inductive effects of three-dimensional sounds and the effects of ambient sounds under visual constraints.

2 Implementation of Object-Based 3D Sound

In the current study, we used HoloLens, a head-mounted display (HMD) wearable computer developed by Microsoft. Figure 1 shows the position of the HoloLens speaker playing the three-dimensional sound. The reasons for using HoloLens for the reproduction device is that it is wearable and can be fixed on the user's head. The speakers built into the left and right sides of the HoloLens (Fig. 1) play the three-dimensional sound. As the speakers do not cover the ears, they do not interfere with other surrounding sounds during actual use. In addition, since the HoloLens is wire-free, it allows for natural walking, and its position tracking system can record data on the user's position and rotation. The positioning data accuracy is 0.1 mm. HoloLens adopts the left-hand coordinate system (the positive x, y, and z axes are right, up, and front, respectively), and the activated position is the origin. The development environment used was Unity, which is a multi-platform game engine that supports HoloLens. There are individual differences in the head-related transfer function (HRTF), however, it is not practical to prepare HRTF individually. Therefore, we used the MS HRTF Spatializer with the sound source in Unity, in order to reproduce the three-dimensional sound.

Fig. 1. HoloLens and speaker position.

3 Experiment on Walking Guidance by 3D Sound

The purpose of the walking experiments was to verify the effect of the three-dimensional sound guidance as well as the effect of the surrounding sound under visual restriction by comparing the real sound and the three-dimensional sound as sounds to guide walking.

Participants were 12 normally sighted and hearing men. They were university and graduate students between 20 and 25 years of age (average age 23 years). The experiment was performed in a large and quiet room where a sound-absorbing material was installed. The task of the experiment was to walk toward the guidance sound. At that time, the water sound that was in a previous study [5] was used as the guidance sound. Two conditions were set for the reproduction: "real sound," in which speakers were actually placed, and "3D sound," which was reproduced using HoloLens and Unity as mentioned above. In addition to the guidance sound, three conditions were set as ambient sound conditions: "control" which did not reproduce any ambient sound, and "right speaker" and "left speaker," which reproduced ambient sounds from the right or left speakers. The background noise was assumed to be noise heard when people walked on sidewalks and pedestrian crossings. The noise contains frequency in a wide band. In this experiment, six conditions consisting of a combination of the above-mentioned conditions were randomly implemented, and one condition was performed twice for one participant. No feedback was given for each trial; therefore, we assumed that there was no effect from repetition.

After explaining the outline of the experiment, participants were given time to listen to the water sound and to get used to the guidance sound. Next, after wearing the eye mask and entering the soundproof room, the walking practice was performed within a visually restricted environment. During the experiment, participants were asked to wear HoloLens and were allowed to stand facing the guidance sound. At that time, they were only informed that the guidance sound was ahead.

Figure 2 shows an overview of the soundproof room where the experiment was conducted. The starting point of walking was 6 m from the center wall of the stage. A speaker (RUH-5, Ashida Sound Co., Ltd.), which produced a guidance sound, was placed in the front direction, and speakers (CM1MW, Bowers & Wilkins) that produced the ambient sound were installed around the walking path. The sound pressures from the speakers were adjusted to 60 dB (A characteristic) at a height of 1.5 m beginning at the walking starting point.

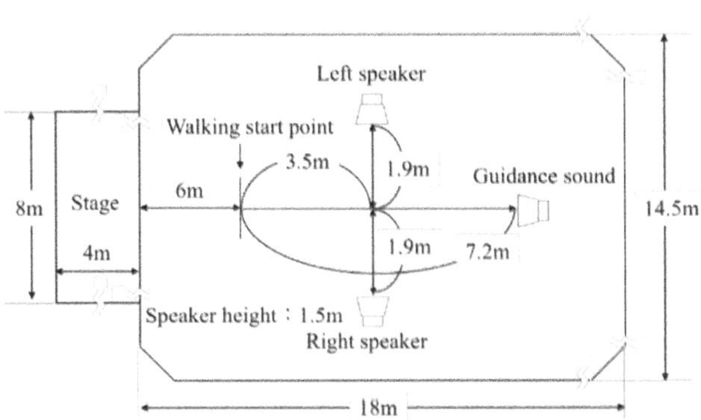

Fig. 2. Setup of speakers in the soundproof room.

As an objective evaluation, the walking locus was measured using the HoloLens. The walking locus is a plot of HoloLens coordinate data, and the time resolution is

1/60 s. The degree of veering and the amount of veering in the walking locus as well as the veering angle and the average veering distance were defined within the quantitative analysis. As shown in Fig. 3 (left), the angle of veering was defined as the angle between the ideal trajectory and the regression line of the actual trajectory from 0 to 6 m by using the least-squares method. The positive and negative angles indicated rightward and leftward veering, respectively. Figure 3 also shows the definition of the veering distance for each meter, based on the ideal trajectory from 0 to 6 m. The overall veering distance is the average value of the veering distance at each meter.

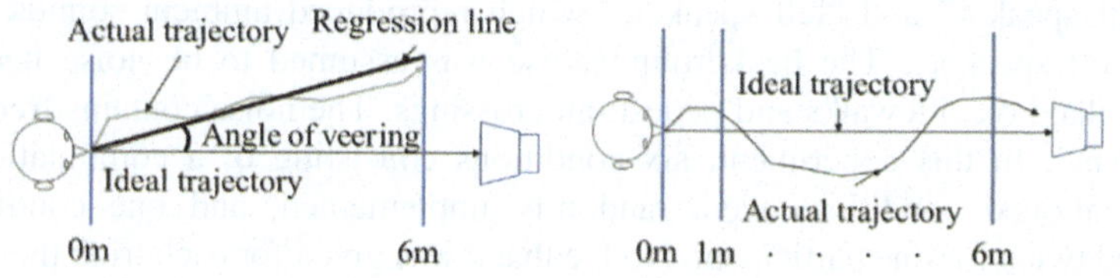

Fig. 3. Definition of angle of veering and averaged veering distance.

4 Results and Discussion

Figure 4 shows the walking trajectory recorded by the HoloLens under the six conditions. Each walking trajectory covered about 1 to 1.5 m. In addition, they reached the point of 6 m without deviating left and right on the way.

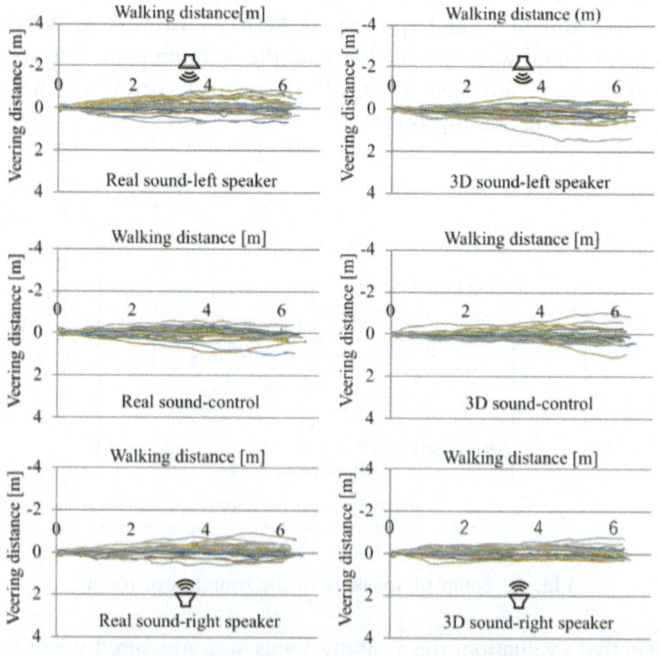

Fig. 4. Walking locus of each condition.

Figure 5 shows the average and standard deviation of the veering angle under each condition of real and three-dimensional sound. Results from multiple comparisons using the Tukey-Kramer procedure for each of the guidance methods using ambient sound factors showed no significant differences regarding the real sound conditions. On the other hand, a significant difference was found in the 3D sound condition between the left and right speakers.

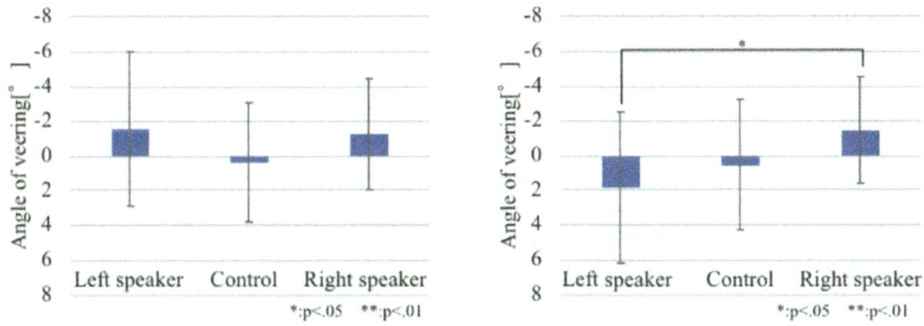

Fig. 5. Angle of veering for each condition of real (left) and 3D (right) sound.

Figure 6 shows the average veering distance and standard deviation under the six conditions. Results from the *t*-test between the ambient sound conditions of real sound and stereophonic sound showed no significant differences. Considering the average value of each condition, it was less than 0.3 m and the standard deviation was less than 0.5. The guidance of straight walking by 3D sound had the same accuracy even under ambient sound conditions when compared to the guidance by real sound. This suggests that people can be guided by 3D sound keeping them out of the roadway, assuming a sidewalk width of 3.5 m.

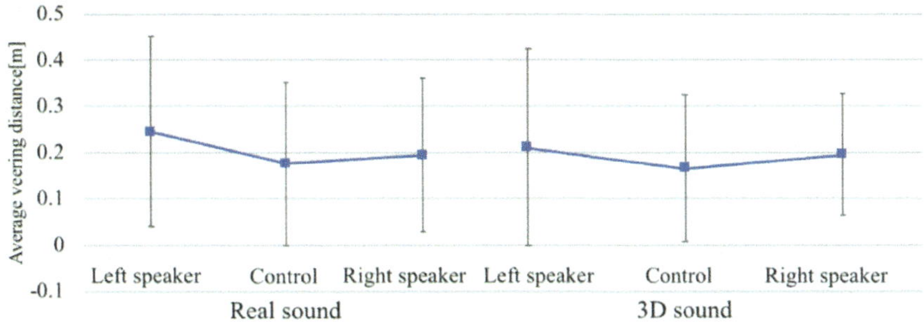

Fig. 6. Average veering distance under each condition.

5 Conclusion

In this study, an experiment was conducted aimed at verifying the guidance effect and the influence of surrounding sounds using the guidance sound of stereophonic sounds as a clue to walking under visual restriction. Results revealed that when the water sound was used as the guidance sound, the real sound was not affected by the surrounding sound, and the 3D sound occurred on the side opposite to the surrounding sound. In addition, 3D sound could be used as a signal to guide people with visual impairment into walking straight with similar accuracy as real sound.

References

1. Department of Health and Welfare for Persons with Disabilities: Survey on Persons with Physical Disability 2006. Ministry of Health, Labor and Welfare, Japan (2008)
2. Ohkura, M., Urakawa, R., Teramoto, H., Sato, J., Kondo, G., Ikegami, A., Nakagawa, Y., Yada, Y., Tauchi, M., Murakami, T., Kitano, M.: Guidelines for installing tactile guiding lines to assist blind pedestrians in crossing intersections (Part 1). J. Sci. Labour **78**(6), 310–316 (2002)
3. Ohkura, M., Ogawa, H., Furukawa, S., Tauchi, M., Kitano, M.: Guidelines for installing tactile guiding lines to assist blind pedestrians in crossing intersections (Part 2). J. Sci. Labour **80**(4), 160–170 (2004)
4. Statistics Bureau: Social Indicators by Prefecture 2015. Ministry of Internal Affairs and Communications, Japan (2015)
5. Muramatsu, K., Tsukada, S., Kaede, K., Watanuki, K.: Kansei evaluation of object-based three-dimensional sound for visual impairments. In: IEEE International Conference on Systems, Man and Cybernetics (SMC), pp. 3185–3189 (2019)

Chaotic Analysis of Finger Plethysmogram During Aroma Exposure for Evaluating Emotions and Examination of Optimal Analysis Method

Tomofumi Sakata[1(✉)], Keiichi Watanuki[2], Kazunori Kaede[2], and Keiichi Muramatsu[2]

[1] Graduate School of Science and Engineering, Saitama University, 255 Shimo-okubo, Saitam-ku, Saitama-shi, Saitama 338-8570, Japan
t.sakata.448@ms.saitama-u.ac.jp
[2] Advance Institute of Innovative Technology, Saitama University, 255 Shimo-okubo, Saitam-ku, Saitama-shi, Saitama 338-8570, Japan

Abstract. Fragrance influences the psychophysiological state of a person as the volume of fingertip plethysmogram changes during one's aspiration of aroma. Recent studies have used aroma in various settings, such as for medical care, relaxation, and productivity-enhancing environments. However, no study has elucidated the kind of behavior shown on the fingertip plethysmograph near the stimulation and discrimination threshold for the sense of smell when the odor intensity of aroma changes. Although our research suggests that emotions can be evaluated using the Lyapunov exponent of chaos analysis, no study has clarified the appropriate method for chaos analysis. Therefore, in this study, we performed a chaotic analysis of fingertip plethysmograms for evaluating emotional changes near the stimulation and discrimination thresholds during aroma inhalation. Results were evaluated using a correlation dimension and Lyapunov exponent. The optimal analysis method was examined thereafter.

Keywords: Fragrance · Chaos analysis · Lyapunov exponent · Correlation dimension

1 Introduction

Olfaction is the modality of the cerebral limbic system of the brain that is responsible for causing emotions. Among the five senses, olfaction is the only one that is directly connected to the limbic system of the cerebrum [1]. Therefore, it is said that "smell" has a major influence on emotion and concentration. In recent years, research has focused on elucidating how olfaction perceives "smell" [2]. Attempts have been made to use aroma in various settings; for example, it has been used in medical care, relaxation, and improving work environments to enhance productivity. However, if the odors used are too strong, individuals may feel uncomfortable, even if the odor is pleasant. When classifying odors in various situations, it is essential to evaluate the emotion caused during aspiration. The fingertip plethysmogram is closely related to the

G. Di Bucchianico et al. (Eds.): AHFE 2020, AISC 1202, pp. 987–992, 2020.
https://doi.org/10.1007/978-3-030-51194-4_129

psychophysiological state of humans, and studies have revealed fluctuations owing to emotional changes and stress loads [3]. The finger plethysmogram has a chaotic nature, and it has been reported that it moves subtly in response to unpleasant emotions [4]. In this study, we quantitatively measured the change in pleasant/unpleasant emotions using finger plethysmograms. However, chaos analysis has various useful findings, there are a variety of indices to measure, and there is little knowledge to determine a superior analysis method. In particular, chaos analysis of studies using biological indicators still has many uncertainties. Therefore, evaluations and comparisons were performed by analyzing the correlation dimension and Lyapunov exponent, which are typically used in chaos analysis.

2 Purpose of the Experiment

For this study, we built a fragrance generator to confirm the stimulus threshold, discrimination threshold, and acclimatization time of smell in accordance with a previous study by Sotoike [5]. The fragrant gas was randomly changed by the fragrance generator and sent directly to the nose of 10 research participants (aged 20–25 years) [6], who had not applied any fragrance (perfume, deodorant). The participants were asked to make subjective evaluations of aroma intensity, comfortableness (comfort/discomfort), and aroma frequency in real-time. A multi-sensor physiological measurement system (NeXus-10 MKII, Mindmedia) measured the fingertip plethysmograph at rest and during aspiration, as a physiological index.

3 Experimental Environment

An odorless room was created in a laboratory. The odorless room was a 1.7-m rectangular-parallelepiped-shaped space, created using a pole draped with a sealed vinyl sheet. A deodorizing machine, humidifier, and a dehumidifier were permanently installed in this odorless room. The temperature inside the odorless room was set at 25 °C with 50% humidity. The odorless room was kept odorless. During the experiment, the experimenter gave directions to the participant from outside the laboratory to avoid air interference.

4 Experiment Device

For this experiment, we created a fragrance generator, as shown in Fig. 1. First, air was supplied through an air pump, then dehydrated through the silica gel tank and deodorized by passing through an activated carbon tank. After deodorization, an air filter was used to remove the dirt and dust in the air. Thereafter, the airflow rate was controlled at 4 L/min using a flow meter. The temperature of the air sent to the odor cell in the water tank was set at 45 °C, and a cannula was used to send the air directly to the participants' nose. Odor cells with several different concentrations were prepared, and the cannula was manually changed between them. Odorless air was supplied by passing air through purified water instead of the odor cells.

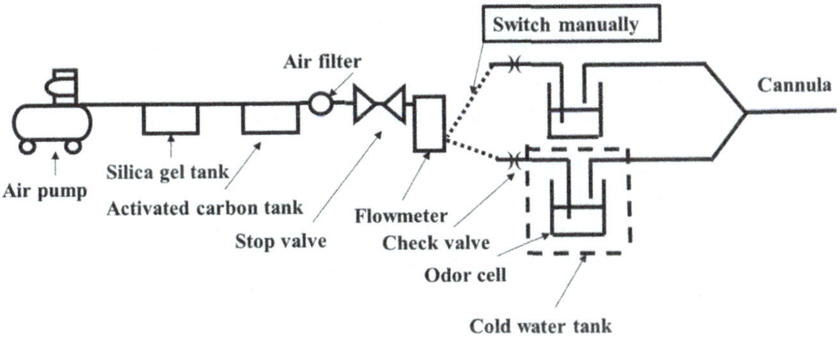

Fig. 1. The fragrance generator

5 Experimental Method

In this experiment, the odor of the odorless room was regarded as the basic smell. The participants waited until their sense of smell acclimatized to the odor of the odorless room before beginning the experiment. The flow of the experiment is shown in Fig. 2. The concentration of the released odor cell was randomly selected. Participants were asked to provide subjective evaluations of not only the "smell strength" using a slider but also of "pleasure/discomfort" using a joystick, in real-time when the odor was released. A multi-sensor physiological measurement system (NeXus-10 MKII, Mind-media) measured the fingertip plethysmograph at rest and during odor aspiration as a physiological index.

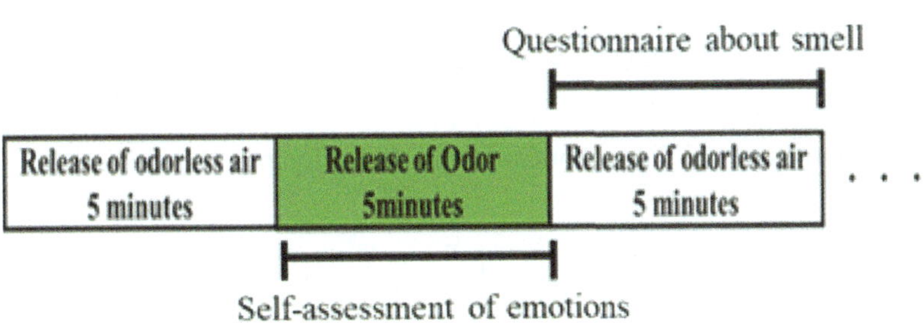

Fig. 2. Time protocol

6 Analysis Method

Chaos is a waveform that has repeated self-similarity shapes; it is often considered a biological signal. Even if one part is enlarged (i.e., a complicated, similar shape appears), the nonlinear part exists everywhere, and it is difficult to analyze it using differential equations [7]. Therefore, the chaos dimension analysis method is used to quantify and evaluate this chaos. This method may reveal changes that do not appear in frequency analysis. In addition, because it has the merit of being able to eliminate the influence of noise, it is considered suitable for the analysis of complex signals.

6.1 Measurement of Fluctuation of Attractor by Lyapunov Index

The Lyapunov exponent was used as an index for quantitatively evaluating the fluctuation of the attractor, a characteristic of chaos [8]. The Lyapunov exponent is an index that quantitatively evaluates the enlargement rate of the attractor's trajectory based on the difference between the initial values. In this study, the Sano-Sawada method [8] was used to calculate the Lyapunov exponent. In this method, one point of the reconstructed attractor and its neighboring point are evaluated. We examined the difference between the two pairs after τ time. We defined $d(t)$, the distance between $P_t(m)$ and the nearest neighbor point of $Q_t(m)$, on a reconstructed attractor.

$$\mathrm{d}(t) = \mathrm{a}Q_t(m) - P_t(m)\mathrm{a} \quad (1)$$

The distance $\mathrm{d}'(t + I?)$ developed after τ time was defined as

$$\mathrm{d}'(t + I?) = \mathrm{a}Q'_{t+I?}(m) - P_{t+I?}(m)\mathrm{a} \quad (2)$$

The distance ratio between $\mathrm{d}(t)$ and $\mathrm{d}'(t + I?)$ was defined as

$$\Lambda_m(t, \tau) = \frac{\mathrm{d}'(t + \tau)}{\mathrm{d}(t)} \quad (3)$$

In many respects, distance ratios were taken for the nearest pairs.

$$\langle log\, \Lambda_m(t, \tau)\rangle = \frac{1}{N}\sum_{i=1}^{N} log\, \Lambda_m(t, \tau) \quad (4)$$

Therefore, the amount $\lambda(t, \tau)$ corresponding to the Lyapunov exponent at $I?$ was as follows:

$$\lambda_m(t, \tau) = \frac{1}{\tau}\langle \log \Lambda_m(t, \tau)\rangle \quad (5)$$

We performed this process for each dimension below m found in the previous section. The largest of these was regarded as the largest Lyapunov exponent (LLE) and used as the physiological index.

6.2 Measurement of Fluctuation of Attractor by Correlation Dimensionl

By calculating the correlation integral, a correlation dimension, which is a kind of fractal dimension, can be calculated. The correlation integral is a cumulative frequency function of the distance between two points and is defined by the following equation.

$$C(r) = \frac{1}{N^2}\sum_{i,j=1}^{N}\theta\left(\varepsilon - \left|P_{ij}\right|\right) \tag{6}$$

ε is a point P_i on the attractor in m-dimensional space. Increase the radius of the m-dimensional hypersphere at the center. The correlation integration counts the number of remaining (N − 1) points excluding the point P_i that falls within the m-dimensional hypersphere of radius r, centered on the point P_i. Repeat the work to find P_i.

7 Experimental Results

Comparison of LLE and correlation dimension using DTW is shown in Fig. 3 and Table 1. LLE was superior at all odor concentrations.

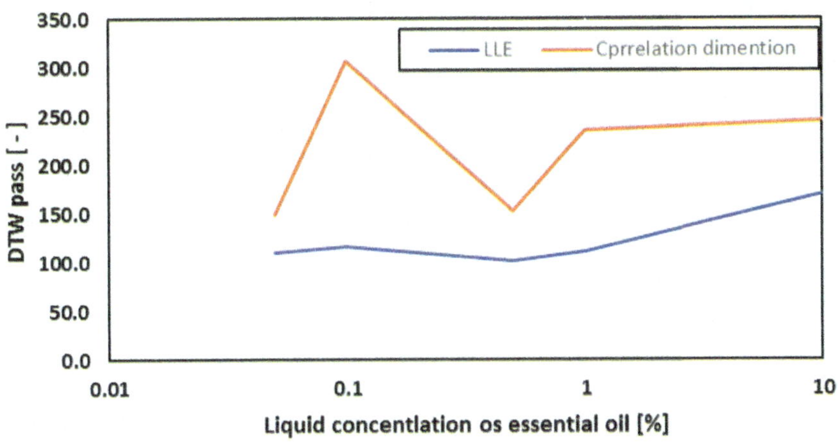

Fig. 3. Comparison of LLE and correlation dimension

Table 1. Comparison of LLE and correlation dimension

	Liquid concentration of essential oil [%]				
	0.05	0.1	0.5	1	10
LLE	110.4	116.8	102.5	111.5	170.7
Correlation dimension	150.3	307.0	153.1	236.8	246.3

8 Discussion and Conclusion

This study confirmed that the fingertip plethysmogram did not fluctuate in the absence of stimulation. Additionally, as the odor concentration increased, the degree of fit tended to decrease. It was considered that the fluctuation increased when the participant felt uncomfortable due to the odor stimulation becoming stronger. This was thought to be the case as fluctuations occurred more frequently during stronger odor concentrations than during the neutral state. It is further considered that the value needed to return to the neutral state is large. The correlation dimension may be unsuitable for small fluctuations. This study suggests that the fluctuation of the finger plethysmograms makes it possible to determine the minute changes in emotion that occur during aroma inhalation. This indicates the possibility of visualizing the human mental state with a simpler device.

References

1. Higashihara, K.: Olfactory odor acceptance mechanism. Otorhinolaryngol. Soc. Bull. Jpn. **118**, 1072–1075 (2018)
2. Konagai, C., Koga, Y.: The effect of coffee aroma on human cognitive function: the psychophysiological evaluation of differences in effect of aroma induced by various kinds of coffee beans. Jpn. J. Taste. Smell. Res. **9**, 561–564 (2002)
3. Sakata, T., Watanuki, K., Muramatsu, K., Kaede, K.: Chaotic analysis of finger plethysmograms for evaluation of emotion induced by exposure to fragrance. In: IEEE SMC, 1135–1139 (2019)
4. Suzuki, K., Okada, Y.: Evaluation of driver's, mental workload in terms of the fluctuation of finger pulse. Proc. Jpn. Soc. Mech. Eng. **74**, 85–94 (2008)
5. Oyama, M., Imanishi, A.: Relations between the different parameters of chaos analysis and the largest Lyapunov exponent in finger plethysmograms. Jpn. J. Ergon. **45**, 141–147 (2019)
6. Sakata, T., Watanuki, K., Muramatsu, K., Kaede, K.: Evaluation of psychological emotions during fragrance exposure by chaotic analysis of fingertip plethysmography. AHFE **2019**, 167–176 (2019)
7. Sotoike, M.: Olfactory stimulator. Inst. Elec. Comput. Sci. **53**, 1266–1275 (1989)
8. Ohyama, T.: Methods of measurements in sensation and perception. Ergonomics **4**, 37–47 (1968)

Exploring User Experience of Music Social Mode - Take NetEase Cloud Music as an Example

Han Wang and RongRong Fu(✉)

College of Art Design and Media,
East China University of Science and Technology, 130 Meilong Road,
Xuhui District, Shanghai 200237, People's Republic of China
731450310@qq.com, muxin789@126.com

Abstract. NetEase Cloud Music, one of the most popular music streaming services in China, innovated music social networking mode by launching the "One Song, One Encounter" campaign. However, it failed to deliver ideal user experience due to deficiency in key factors. This research identified and verified eight key factors that influence user experience, e.g. interaction mode, secondary social networking etc., and created the User Experience Model of Influence Factors. By the principle of in-depth integration of GUI with media content, multi-functional linkage to creative immersive scenarios, an optimization scheme was proposed to address the key factors and problems—"creating scenarios by emotional linkage of online elements, extending music social scenarios to offline channels, making it customary by the regular timing". The scheme effectively improved the targeted experience, and it's expected to provide reference for immersive UX design in music social mode.

Keywords: Music social mode · User experience · NetEase Cloud Music

1 Introduction

The mode of "Music + Social" is one of the most important future trends for digital music streaming services. In the Experience Economy, emotional satisfaction and psychological experience in consumer behaviors are the research focuses for service providers. As one of the most popular music streaming services in China, NetEase Cloud Music's users had reached 800 million by 2019, with a year-on-year growth of 50% and a 135% growth on paid members. Around the Chinese New Year of 2020, NetEase Music launched the "One Song, One Encounter" campaign. By leveraging the special timing, matching algorithm and immersive situational socializing, it innovated the music social mode. In usage testing and analysis, however, deficiency was found in the scenario of yearly personal music report, failures were identified to address user needs in matching and social scenarios. Moreover, a closed loop was not formed in the music social networking process. Deficiencies in key factors as aforementioned led to poor user experience.

G. Di Bucchianico et al. (Eds.): AHFE 2020, AISC 1202, pp. 993–999, 2020.
https://doi.org/10.1007/978-3-030-51194-4_130

There have been no efficient solutions to improve, nor reliable model to explain the user experience in music social mode in current studies. This research proposed the following questions targeting at user experience in NetEase music social mode.

What are the key factors in user experience of NetEase music social mode? How do these factors influence user experience? What solutions could there be to optimize the user experience? Can these solutions optimize the user experience in practice?

2 Research on User Experience of NetEase Music Social Mode

NetEase music social mode consists of four sessions including yearly personal music report, user matching, situational social and sharing. The mechanism works in this flow: Learn your music listening pattern of the past year, enter the user matching interface, get matched with another user (who shares similar music taste), converse in the scenario accompanied by the same and simultaneously played piece of music, and afterwards, share to SNS such as WeChat or start a new round. The key factors liable to influence user experience in this process will be discussed in Sect. 2.1.

2.1 Key Factors Extracted from Music Social User Experience

In the yearly music report session, users can have a sense of belonging stimulated by the tailored content. The background illustrations create immersive scenarios. And with multi-sensory elements e.g. music, the experience become multi-dimensional. Moreover, storytelling on a special timing tend to be more touching and provoking.

In the user matching session, there is a tagline that manifests the common point of the both sides and the tone of this social activity. The dynamic graphics and background music are soothing and relaxing. User profiles are actually users' social assets, indicating what can she/he can offer to the user on the other side.

In the situational social session, the shared song is a common ground, also serves the function of agenda-setting. Social time setting can avoid boredom, also improve the efficiency of communication. The virtual facilitator influences user experience mainly in three aspects: image, interaction and content. The background graphics indicates the content and topics, and present multi-sensory experience in combination with background music. A survey in China has shown 95.5% participants frequently use stickers when chatting online [1]. Easy-to-use, entertaining and expressive, stickers have the potential to optimize user experience in music social as well.

The sharing mechanism is featured by internal and external sharing. Diversified channels of communication leads to diversified experience in terms of cognition, interpretation, learning, memorizing and feedback [2]. Secondary socializing and extending social after sharing action is the key to transforming weak solidarity to strong solidarity, also key to success, since it has direct impact on the experience.

11 potential influence factors were identified by analyzing the music social mode of NetEase Cloud Music and touch points: H1-Background Music, H2-Background Graphics, H3-Taglines, H4-User Profiles, H5-Interaction Mode, H6-Timesetting, H7-Sharing (Internal & External), H8-Music Recommendation, H9-Stickers, H10-Virtual Facilitator, H11-Secondary Socializing (Fig. 1).

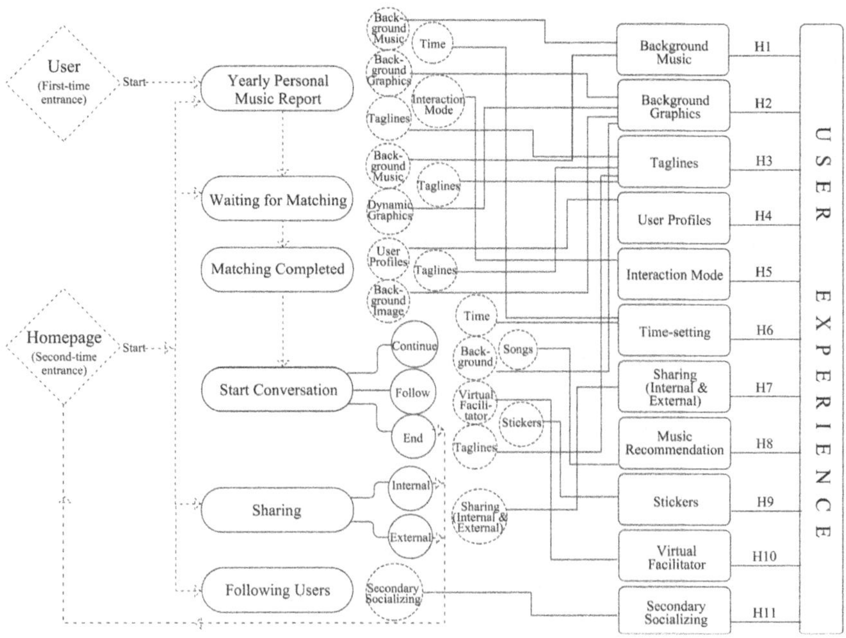

Fig. 1. Influence factors in NetEase music social mode

2.2 User Experience Evaluation Model and Questionnaire

A model was developed to evaluate the targeted user experience based on three components of attitude theory (Rosenberg and Hovland). This research is user-focused, so whether the user needs are satisfied determines the user experience quality. Therefore, the cognitive component in the theory was adapted to need component. The three levels in NetEase music social mode are needs, behaviors and emotions. And two specific indicators were proposed on each of the three levels (Fig. 2).

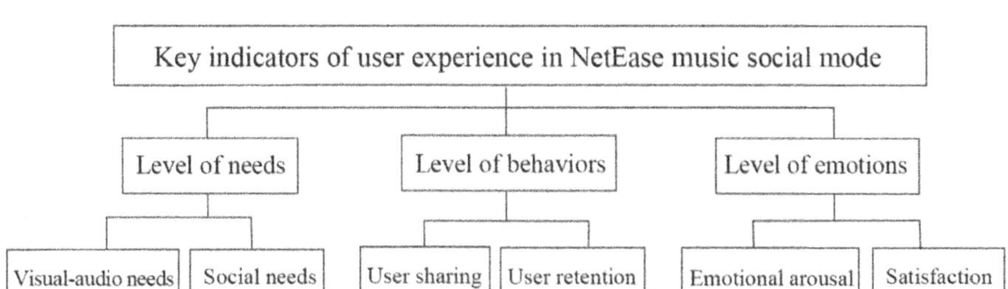

Fig. 2. User experience evaluation model for music social mode

Based on the influence factors of user experience and user experience evaluation model, a questionnaire was designed. Questions 1–5 are aimed to collect user information. In Question 6–17, with influence factors as independent variables and evaluation model as dependent variables, the questions and answers form a mapping between questions and variables, and thus could be used to determine the value of variables through the answers.

2.3 User Experience Model of Influence Factors

181 valid questionnaires were retrieved. By analyzing the answers to questions 1–5, it was found among the targeted users, there were more females than males; Most of them were under the age of 35, well-educated students and white-collars. This group has strong identities and willingness to share on social medias.

A regression analysis was carried out with influence factors H1–H11 (questions 6–16) as independent variables, and visual-audio needs in user experience evaluation model (question 17) as dependent variables. As shown in Fig. 3, for Q9, 12 and 13, Sig value < 0.05, which indicates H4-User Profile, H7-Sharing, H8-Music Recommendation are significant for satisfying visual-audio needs under 95% level of significance. For Q16, Sig < 0.01, which indicates H11 is significant for satisfying visual-audio needs under 99% level of significance (Table 1).

Table 1. Regression analysis on influence factors for user retention

Model		Non-standard coefficient		Standard coefficient	t	Sig.	Nonlinear statistic	
		B	Standard deviation	Beta version			Tolerance	VIF
1	(constant)	.584	.527		1.108	.269		
	Q6	−.026	.080	−.025	−.331	.741	.451	2.219
	Q7	−.035	.085	−.030	−.410	.683	.486	2.057
	Q8	.037	.080	.033	.460	.646	.494	2.024
	Q9	−.161	.074	−.156	−2.170	.031	.494	2.026
	Q10	.071	.083	.068	.846	.399	.399	2.506
	Q11	.105	.079	.103	1.329	.186	.427	2.340
	Q12	.162	.070	.157	2.312	.022	.554	1.804
	Q13	.192	.092	.173	2.100	.037	.378	2.644
	Q14	.099	.069	.103	1.444	.150	.506	1.978
	Q15	.116	.072	.121	1.623	.106	.462	2.165
	Q16	.348	.069	.378	5.024	.000	.452	2.213

a. dependent variable. Q17

With the same method, regression analysis was carried out with social need on the level of needs, user sharing and user retention on the level of behavior, emotional arousal and satisfaction on the level of emotions.

Analysis has shown, 3 out of the 11 influence factors don't have significant influence on any of the three levels of evaluation model—H2-Background Graphics, H3-Taglines, H9-Stickers failed the verification. The other 8 factors have varied degrees of influence on user experience on three levels. The user experience model of influence factor is as shown in Fig. 3.

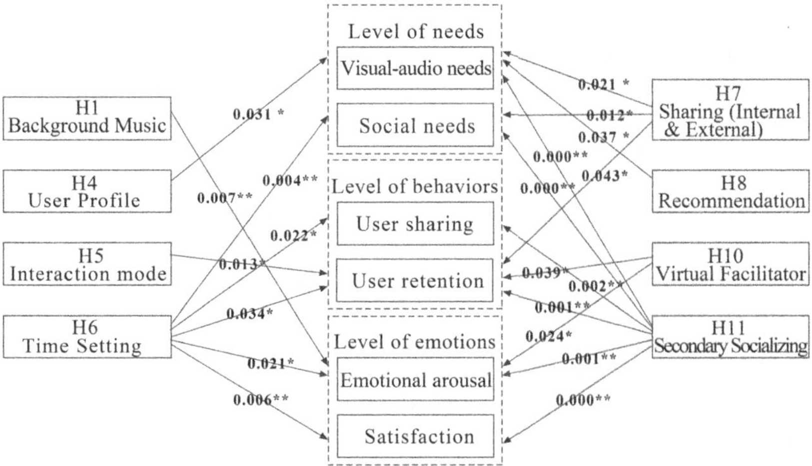

Fig. 3. User experience model of influence factors

3 Optimizing Music Social User Experience

Based on user experience model of influence factors, by the design principle of "in-depth integration of GUI with media content, multi-functional linkage to creative immersive scenarios", an optimization scheme was proposed to improve the user experience in music social mode:

On the choosing of background music, it should leverage the trend of times and new year festive elements, at the same time, be matched with user interaction. The background music should be available in varied genres and customized, so that the boredom of waiting period can be relieved. In order to meet the users' visual-audio needs, classification and recommendation algorithm should be improved.

On the user profiles, service providers should, on the one hand, protect user privacy and interpersonal relationship to maintain the advantage of social networking, and on the other hand, avoid the social risk of information loss by empowering the users by acquiring more social assets for them. Service providers should give back the initiative of matching to the users, for example, use filters, to compensate the deviation of algorithm

The interaction of yearly music report should be easy to operate and efficient, creating immersive experience, turning "reading a report" to "experiencing a story".

NetEase Music users tend to be enthusiastic about new year activities. Service providers can make a buzz on social platforms to attract attention and create a festive atmosphere before launching yearly music report at zero o'clock sharply as well as the music social campaign. By taking advantage of the special timing and the regular online buzz, the yearly music report and music social can be turned into a new year custom. A chatting time "free refill" function should be available and visible on the chatting interface. One part can be notified of the other part's willingness to prolong the conversation. Such design can encourage users to seize the opportunity, but can also extend the time spent on music social mode.

On the external sharing, service providers should ensure the channels are commonplace and platforms are diversified, the experience is accessible from the external platforms. On the internal sharing, service providers should produce tailored content by taking advantage of music life's topics and the musicians' influence; encourage and protect UGC, stimulate originality, cater to users' self-awareness.

On the virtual facilitator, its expression should be rich in emotions. It should assist the users to propose desirable questions. The accesses to functions should be easy-to-use and diversified. In order to compensate for the deviation of recommendation algorithm, Song-on-demand playlist could be added which consists of the shared "liked" songs. The mascot of the brand of NetEase Cloud Music—reindeer characters that is warm and affectionate could be applied to the virtual facilitator, to promote the brand IP and culture, concentrate branding effort and reduce the cost.

Service providers should also expand the online mode to online + offline mode. So that the music social service becomes a holistic experience.

By implementing the optimization scheme as aforementioned, the music social experience is expected to improve in aspects of precision, convenience and interest.

In the testing, a clickable mock-up is produced following the proposed design strategies. For comparison, the same group of interviewees were given a trial on the mock-up. In reference to the same user experience model, the comparison result has shown all average values of the indicators of social music mode were improved after the experience is optimized. The data is as shown in Table 2.

Table 2. Comparison of user experience before/after optimization

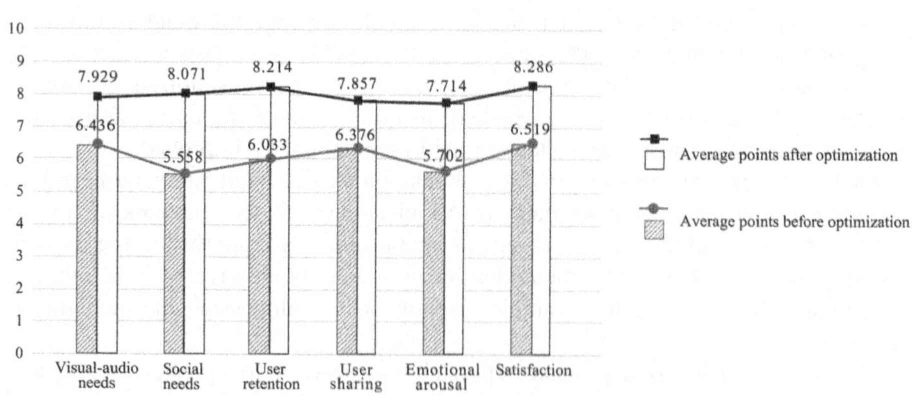

4 Conclusion

This research constructed user experience model of influence factors for NetEase Cloud Music social mode. It is expected to serve as a reference for the immersive experience design in music social mode.

References

1. Yuancheng, W., Qi, Z.: A primary research of the communication power of stickers. Radio TV J. **12**, 119–120 (2017)
2. Chen, Yu.: The impact of big data on mass communication studies——and the interaction between brand image communication and identity. J. Lang. Lit. **5**, 31–36 (2018)
3. Schedl, M., Zamani, H., Chen, C.-W., Deldjoo, Y., Elahi, M.: Current challenges and visions in music recommender systems research. Int. J. Multimedia Inf. Retrieval **7**, 95–116 (2018)
4. Tanaka, A.: Interaction, experience and the future of music. In: O'Hara, K., Brown, B. (eds.) Consuming Music Together. Computer Supported Cooperative Work, vol. 35, pp. 267–288. Springer, Dordrecht (2006)
5. Xuehui, Z.: Music social platform construction based on emotions—NetEase Music as an example. Radio TV J. **1**, 128–129 (2019)
6. Xambó, A.: Embodied music interaction: creative design synergies between music performance and HCI. In: Broadhurst, S., Price, S. (eds.) Digital Bodies. Palgrave Studies in Performance and Technology. Palgrave Macmillan, London (2017)
7. Håkansson, M., Rost, M., Holmquist, L.E.: Gifts from friends and strangers: a study of mobile music sharing. In: Bannon, L.J., Wagner, I., Gutwin, C., Harper, R.H.R., Schmidt, K. (eds.) ECSCW 2007. Springer, London (2007)
8. Ya, L.: Research on NetEase Cloud Music user interaction behavior. He Bei University (2018)
9. Xin, L., Zouqiannan, L., Yonghai, Y.: Innovative exploration of "music socialization" based on mobile application. J. Design **1**, 82–83 (2016)
10. Jie, J.: Analysis on the NetEase music social ecosystem. J. News Res. **19**, 205–206, 208 (2018)
11. Xu, W.: Analysis on the interaction ritual chain. J. News Res. **19**, 96, 219 (2017)
12. Ping, Y.: Self-presentation and identity of online social networking from the perspective of interaction ritual chain—from NetEase Cloud Music yearly personal music report. J. New Media Res. **5**, 29–31 (2018)

Evaluation of the Attention Effect of the Fraser-Wilcox Illusion in a Visual Discrimination Task

Kota Tokonabe[1(✉)], Keiichi Watanuki[1,2], Kazunori Kaede[1,2], and Keiichi Muramatsu[1,2]

[1] Graduate School of Science and Engineering, Saitama University, 255 Shimo-okubo, Sakura-ku, Saitama-shi, Saitama 338-8570, Japan
k.tokonabe.220@ms.saitama-u.ac.jp, {watanuki,kaede,muramatsu}@mech.saitama-u.ac.jp
[2] Advanced Institute of Innovative Technology, Saitama University, 255 Shimo-okubo, Sakura-ku, Saitama-shi, Saitama 338-8570, Japan

Abstract. This study aimed to propose a display that incorporates the Fraser-Wilcox illusion with a higher attention effect. Fraser-Wilcox illusion is a type of visual illusion in which a part of a still image appears to be moving. We compared the attention effects of shapes that induce the Fraser-Wilcox illusion and those that do not. As an evaluation method of attention recall effect, the visual discrimination task, which was an experimental technique to measure the characteristics of visual attention, was used. The correlation between the strength of visual illusion and the target reaction time was examined. As a result, a strong negative correlation was observed. The larger the strength of visual illusion, the faster the detection of the target. These results suggest that the Fraser-Wilcox illusion has an alerting effect on visual illusion.

Keywords: Attention · Cognition · Fraser–Wilcox illusion · Visual discrimination task

1 Introduction

Visual caution markings are used to notify danger in various daily situations such as in construction sites and factories. Such alerts are called safety markings. On the other hand, tumble accident is one of the most frequent occupational accidents. According to the work-related injury and disease reports, the number of victims of the tumble accidents in 2017 was 28,310 (24% of all occupational accidents), while in 2008, it was 24,792 (19% of all occupational accidents) [1]. This result shows that the number of tumble accidents has been increasing. To reduce accidents and disasters by using safety markings, it is necessary to display a stronger attention recall ability.

There is an illusion that utilizes the property of vision. The motion illusion is the illusion in which the human eye appears to be moving whereas it is static. By incorporating this moving effect into the caution markings, it can improve the attention recall ability. The purpose of this study is to propose a display with high attention recall ability that incorporates motion illusion. This study compares the alerting effect of the

G. Di Bucchianico et al. (Eds.): AHFE 2020, AISC 1202, pp. 1000–1006, 2020.
https://doi.org/10.1007/978-3-030-51194-4_131

display that induces and does not induce motor illusions by the visual discrimination task.

In this research, the Fraser–Wilcox illusion was adopted, which is a kind of motion illusion. The Fraser–Wilcox illusion repeats a dark to light gradient of "Black → dark gray → white → light gray → black" that appears to move in the direction of the gradient [2].

2 Method

The visual discrimination task was used as an experimental method for measuring the characteristics of visual attention. In this method, the participants were given the task of making decisions on the target as quickly as possible by pressing a key. The reaction time was measured from the time the target was presented until the participant pressed the key. During that time, a leading cue (Cue) was presented to indicate the location of the target just before it was presented. There were two types of prior cues: a match condition that correctly indicated the location where the target was presented (Cued), and a mismatch condition that incorrectly indicated the location (Uncued). Generally, the reaction time of the target was shortened when the preceding cue correctly indicated the position of the target [3].

The attraction of attention to the place of the stimulation by the external stimulation regardless of the intention as when the light suddenly appears in the peripheral visual field is called the exogenous attention recall. When the target appears at the position from 100 ms to 150 ms after the presentation of the cue stimulus, the detection and processing of the target become faster. This may be because attention has already shifted to that location by the presentation of a cue stimulus.

The process of such attention arousal is said to be fast, automatic, and reflective. In these experimental paradigms, the time course from the cue stimulus to the target presentation is deeply related to the nature of the attention calling process. This process is called Stimulus Onset Asynchronous (SOA) [3].

3 Experiment 1

In this experiment, to evaluate the attention calling effect of the Fraser Wilcox illusion using the visual discrimination task, the illusion image used as a precedent clue is presented to the participant at the same time as the SOA before the experiment using the visual discrimination task, and the amount of the illusion at that time is acquired.

3.1 Design and Procedure

The stimulus was presented on a 55-in liquid crystal display (BenQ ST 550 K) using stimulus presentation software (PsychoPy3). In the experimental environment as shown in Fig. 1, the figure of the Fraser Wilcox illusion used as a precedent clue in the evaluation experiment of the attention calling effect of the illusion display, which is the next experiment carried out on the display 1000 mm away from the participant, was respectively presented 3 times for 100, 300, 500 ms. At that time, the questionnaire was carried out on how much the figure seemed to move by the 5 step evaluation as shown in Fig. 2. The participants were five men in their twenties. The obtained sensitivity evaluation was made to be the intensity of the illusion effect of the preceding clue of the condition with the illusion in this study, that is, the amount of the illusion.

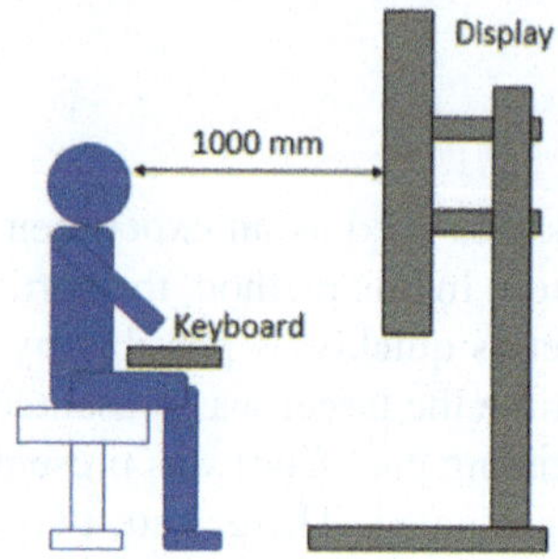

Fig. 1. Experimental environment. The participants observed the display at a distance of 1000 mm and answered by the keyboard.

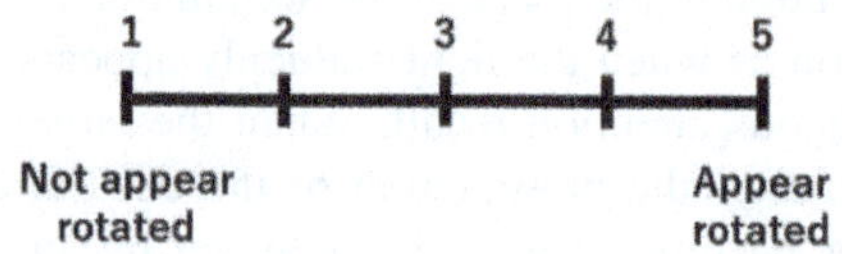

Fig. 2. The intensity of the illusion is evaluated by a 5-stage subjective evaluation scale.

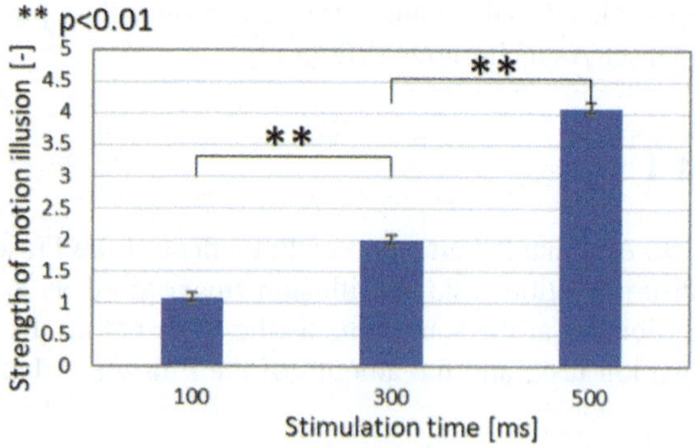

Fig. 3. Strength of motion illusion. The higher the value on the vertical axis, the greater the strength of the illusion at the stimulation time.

3.2 Results

The results of the questionnaire are shown in Fig. 3. When a one-way analysis of variance (one-way ANOVA) was performed at a significance level of 1%, significant differences were found between 100 ms and 300 ms and between 300 ms and 500 ms, respectively.

4 Experiment 2

4.1 Design and Procedure

The stimulus was presented on a 55-in liquid crystal display (BenQ ST 550 K) using stimulus presentation software (PsychoPy3). A lead cue (Cue) and a target were presented on a display 1000 mm away from the participants, and the reaction time and the number of errors were measured. The participants were five men in their twenties. As a preceding clue, the visual illusion which induces the rotational motion and the figure which does not induce the visual illusion are presented as Fig. 4. The configuration of one trial is shown in Fig. 5. After the fixation point was presented in the center of the screen for 1000 ms, the preceding clue was presented in 8 directions with the fixation point as the origin. After a lapse of 150 ms, targets of Landolt rings with cuts on either side were randomly presented at any of 8 points on the circumference.

The presented position of the cue and target was 12° from the fixation point, the size of the preceding cue was 9 × 9, and the size of the target was 1 × 1.

Fig. 4. Illustration for visual stimulation. Control (left) and Motion illusion (right) are compared.

The participants were instructed to determine as quickly and accurately as possible whether the nicks were on the left or right side of the target displayed on the display. Responses were requested by pressing keys on the keyboard. In all, 64 trials for 1clue were made to be 1 block, and type and SOA (100, 300, 500 ms Write) of the precedence clue were changed in every block, and 6 blocks were done for 1 participant. The number of conditions (Uncued) where the presentation position of the clue in 1 block does not coincide with the presentation position of the target (Cued) was constituted equally.

4.2 Results

Figure 6 shows the average reaction time of the target detection of the match condition and the mismatch condition in each preceding cue. The detection reaction time of the target was shortened in comparison with the condition in which the coincidence condition was discordant for both visual illusion figure and figure which does not induce visual illusion, and there was the clue coincidence effect. There was no significant difference in the shortening or prolongation of the reaction time on the existence of the optical illusion effect of the preceding clue.

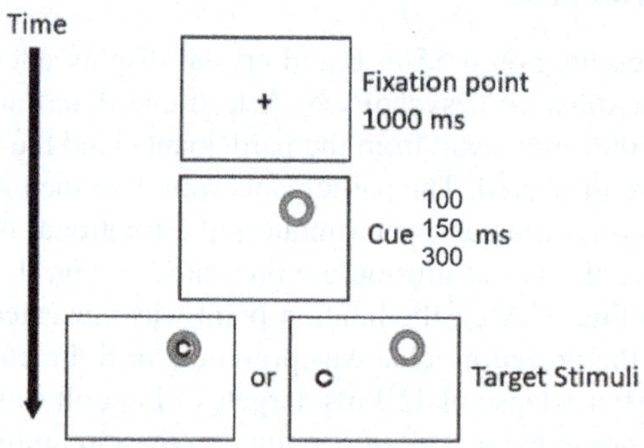

Fig. 5. Construction of one trial. 3 kinds of cues were presented.

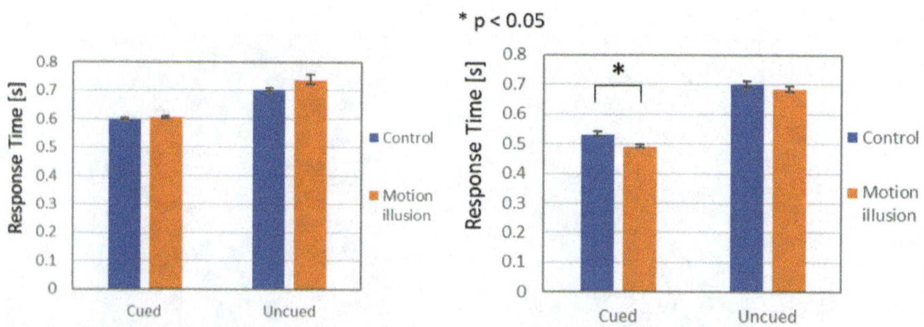

Fig. 6. Mean response time of Cued and Uncued (left) and Mean response time of a group of experimental participants

Figure 6 shows the reaction time at SOA: 500 ms in the group of participants in which the target reaction time was shortened in the preceding cue with visual illusion effect under the condition of agreement (Cued). In the condition with visual illusion, the reaction time significantly decreased at the significance level of 5%. On the other hand, there was no significant difference in the reaction rate in the discordant condition.

The correlation coefficient of the visual illusion quantity for each presentation time obtained by the visual illusion quantity measurement of the Fraser Wilcox visual illusion and the average reaction time for each SOA obtained by the evaluation experiment of the caution markings was −0.72, and there was the strong negative correlation.

Additionally, the analysis was carried out when the participant pressed the wrong key in the detection problem (Fig. 7). In the matching condition (Cued), the reaction time did not change even when an error occurred. In the mismatch condition (Uncued), the reaction time was shortened when the error occurred. No change in reaction time was observed under the concordance conditions.

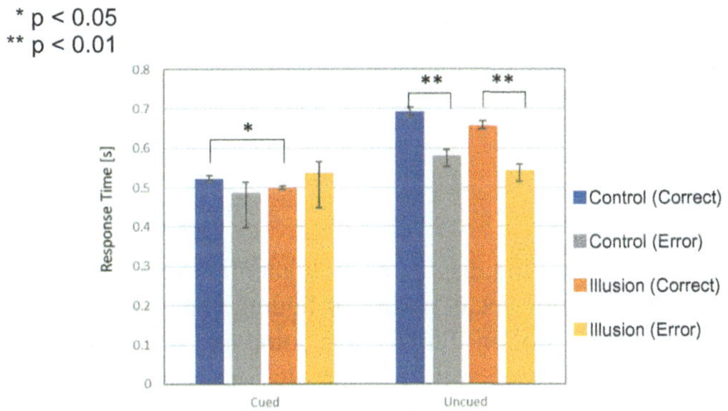

Fig. 7. In the comparison between reaction error and correct, the time of reaction error was shortened in the Uncued condition.

5 Discussion

From the result of the questionnaire of the optical illusion quantity measurement of the Fraser Wilcox illusion, the optical illusion quantity of the Fraser Wilcox illusion showed a significant increase at the significance level of 5% between 100 ms and 300 ms and between 300 ms and 500 ms, respectively. These results suggest that the longer the presentation time, the greater the amount of Fraser-Wilcox illusion in the short presentation time, which affects the automatic and reflective arousal.

In Experiment 1, on the participant group in which the shortening of the reaction time of the target was observed in the precedent clue of the illusion effect under the coincidence condition (Cued), the clue coincidence effect was observed in SOA: 300 and 500 ms. In this case, the visual illusion effect of the preceding clues prompted attention by the long time display of large strength of visual illusion, and the target detection seemed to be accelerated.

The correlation coefficient of the visual illusion quantity for each presentation time obtained by the visual illusion quantity measurement of the Fraser Wilcox visual illusion and the average reaction time of the target for each SOA obtained by Experiment 1 was −0.72, and there was the strong negative correlation. It is considered that the larger the strength of visual illusion, the faster the detection of the target, and the higher the attention calling effect.

There was no significant difference in error rate between the two groups. In the previous study using the visual discrimination task, the analysis of errors was rarely performed because the number of errors was small [4]. In this study, the number of errors varied greatly among individuals, and no significant difference was found. This suggests that the presence or absence of the optical illusion effect of the preceding cue did not affect the performance of the visual discrimination task.

In the analysis when the wrong key was pressed, there was no change of the reaction time in the coincidence condition, and it seemed to carry out the promotion of the attention when the error occurred.

6 Conclusion

This study aimed to propose the display of high attention recall ability combined with motion illusion. The attention recall ability of figures that induce motion illusion and figures that do not induce motion illusion was compared by the visual discrimination task. As a result, the target reaction time of the illusion figure was shortened under the concordance condition in a part of the participant group. Also, by examining the correlation between the strength of visual illusion and the target reaction time, it was shown that an increase in the strength of visual illusion may enhance the attention recall ability.

References

1. Ministry of Health, Labour and Welfare: Occupational accident occurrence status. https://www.mhlw.go.jp/bunya/roudoukijun/anzeneisei11/rousai-hassei/
2. Fraser, A., Wilcox, K.J.: Perception of illusory movement. Nature **281**, 565–566 (1979)
3. Ogishi, M.: Spatial attention effect in visual discrimination task. Kanazawa Univ. Sch. Educ. J. Humanit. Soc. Sci. **52**, 1–18 (2003)
4. Hashimoto, Y., Utsugi, S.: Visual attention with human gaze and arrow symbols. Japan. J. Ergon. **41**(6), 337–344 (2005)

Visualization of Features in Multivariate Gait Data: Use of a Deep Learning for the Visualization of Body Parts and Their Timing During Gait Training

Yusuke Osawa[1(✉)], Keiichi Watanuki[1,2], Kazunori Kaede[1,2], and Keiichi Muramatsu[1,2]

[1] Graduate School of Science and Engineering, Saitama University, Shimo-Okubo 255, Sakura-ku, Saitama-shi 338-8570, Japan
y.osawa.949@ms.saitama-u.ac.jp, {watanki,kaede, muramatsu}@mech.saitama-u.ac.jp
[2] Advanced Institute of Innovative Technology, Saitama University, Shimo-Okubo 255, Sakura-ku, Saitama-shi 338-8570, Japan

Abstract. In this study, we aimed to examine the usefulness of gait classification and feature visualization based on multivariate data for the development of a gait feedback training system capable of considering the physical differences among the trainees. The multivariate data considered in this study were the joint angles and the ground reaction forces. In addition, all multivariate gait data were labeled as gait "rarely associated with stumbling" or "frequently associated with stumbling". A convolutional neural network was used to learn the gait features. Furthermore, the feature parts of the multivariate gait data used for classification were visualized on a heat map created using Grad-CAM. As the results indicate, a heatmap is able to show the feature parts of a gait frequently associated with stumbling, through which the trainee can adjust their gait.

Keywords: Healthcare · Gait training · Motion analysis

1 Introduction

As numbers of elderly and stroke patients in developed countries have increased, so has the importance of gait training in ensuring a good quality of life. Gait training is applied as a form of physical therapy, which consists of the use of walkers, walking sticks, and handrails under the guidance of a physical therapist. Several training methods have been proposed to avoid the need for constant professional assistance [1]. Such methods provide the trainee with measured and target gait values in real time; in particular, the target values are based on ideal gaits. Notably, people who undergo gait training (e.g., the elderly) have weak muscles, whereas people who can walk with an ideal gait (e.g., young people) have strong muscles. For this reason, a target gait based on the movements of young people is unsuitable for the elderly, and target gaits need to be

G. Di Bucchianico et al. (Eds.): AHFE 2020, AISC 1202, pp. 1007–1013, 2020.
https://doi.org/10.1007/978-3-030-51194-4_132

tailored to the physical characteristics of the trainee (e.g., height, range of motion, and muscle strength).

The purpose of our study is to develop a gait feedback training system capable of considering the physical differences among trainees [2]. In this study, gaits rarely or frequently associated with stumbling were defined as "ideal" or "non-ideal," respectively; in fact, most falls among the elderly are caused by stumbles. Our system uses a deep learning model that is constructed by learning the gait features from both ideal and non-ideal multivariate gait data. The trainee data are measured and input into the model during training to classify the gait in real time; furthermore, the feature parts of the data used for classification are visualized and presented to the trainees using a method for a visualization of the influence of the input or intermediate layer on the convolutional neural network (CNN) output, allowing the users to adjust their gait.

In this paper, we aim to examine the usefulness of gait classification and feature visualization based on multivariate gait data for gait training. The multivariate gait data considered in this study are the joint angles and ground reaction forces (GRF). In addition, all multivariate gait data are labeled as either a "gait rarely associated with stumbling" or a "gait frequently associated with stumbling". A CNN model learns the gait features using multivariate gait data as the input and the labels as the output. Furthermore, the feature parts of the multivariate gait data used for classification are visualized in a heat map created by Grad-CAM (i.e., a method for a visualization of the influence of the input or intermediate layer on the CNN output) [3].

2 Construction of Multivariate Gait Data

2.1 Measurement of Multivariate Gait Data

The participants were eight Japanese men with an average age of 23.9 $\pm$ 1.0 years. The experiment applied two conditions: (1) normal walking or (2) walking with a load applied to the muscle and limited joint movement. The cameras of the optical 3D motion analyzer were placed around a treadmill with a built-in force plate. In addition, markers were placed on the body of the participant, as shown in Fig. 1 (left). Under condition (2), the participants wore weights at the front of their upper limbs and their wrists and ankles, as well as braces to restrict the movement of the joints at the elbows and knees, as shown in Fig. 1 (right). The participants put themselves in a reference posture for 5 s and then walked on a treadmill for 120 s under each condition. We measured the 3D coordinates of each marker and the GRF (lateral, longitudinal, and vertical). The reference posture was defined as the back held straight, the width of the foot stance the same as the width of the shoulders, and the hand kept away from the body by the width of a closed fist.

2.2 Preprocessing for Learning

The 3D coordinate data were calculated by applying a cubic spline interpolation method. From the marker coordinates, the angles of the joints (the trunk, right and left shoulders, elbows, hips, knees, and ankles) in the sagittal and frontal planes were

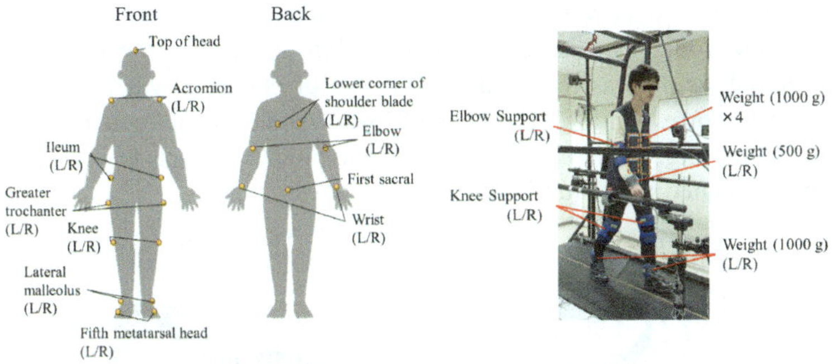

Fig. 1. Left: marker placement; Right: walking with load applied to muscle and limited joint movement.

calculated. In addition, the thumb-to-ground distance (TGD) was calculated. The changes in these data were calculated by subtracting them while walking from the reference posture because such data depend on the placement of the markers. The GRF were normalized, and thus the weight of each participant was 100%. To create input data for the CNN, each change during a 120 s period was split into data beginning from the moment of one right foot contact with the ground until the next right foot contact. The foot contact was defined as the moment when the vertical GRF of the foot reached 5% or more. These data were arranged in the row direction, and the matrix was defined as multivariate gait data. In addition, the left foot contact was detected in the same way, and the data were divided. To be able to treat the data split by the right foot contact and the data by the left foot contact as similar, the left and right rows of the data split by the left foot contact were replaced, and the signs (plus/minus) of the angle of the frontal plane were switched. As a result, 1,548 multivariate gait data were created. These data were normalized using the maximum and minimum values of each variable for all data. In addition, all data were normalized in the time direction with the walking cycle at 100%. Finally, a bandpass filter was applied to the data with a passband of 30 to 220 Hz.

3 Labeling Gait Focusing on Ease of Stumbling Using K-Shape Clustering

For classification using a CNN, correct labels corresponding to the input data are required. In this study, the TGD was clustered based on the k-shape [4], and all multivariate gait data were labeled as either a "gait rarely associated with stumbling" or a "gait frequently associated with stumbling" based on a cluster of the corresponding TGD because stumbling was defined as a movement associated with contact between the toe and ground or obstacle during the swing phase. The k-shape is a clustering method applied using unsupervised learning, particularly for time-series data. The number of clusters was determined using an elbow method.

The number of clusters was set to three based on the elbow method. As a result, the TGD of the target foot was clustered as shown in Fig. 2. Because the TGD of clusters 2 during the initial swing was the smallest and that of cluster 1 was the largest, multivariate gait data corresponding to clusters 2 were labeled as a “gait frequently associated with stumbling,” whereas those corresponding to cluster 1 were labeled as a “gait rarely associated with stumbling.”

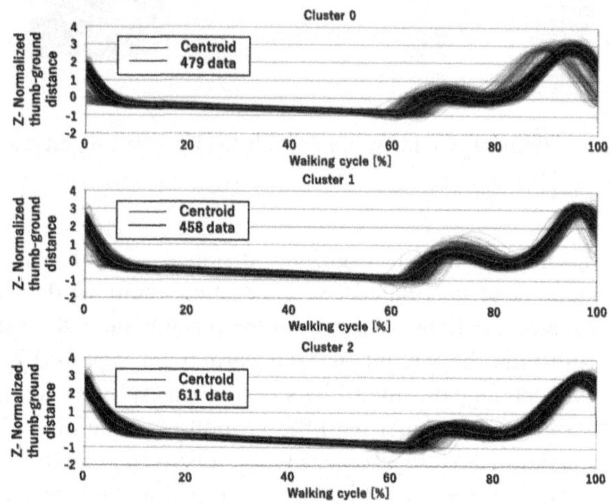

Fig. 2. Clustering TGD for labeling multivariate gait data using k-shape.

4 Learning of Multivariate Gait Data Using CNN and Visualization of Features Through Introduction of Grad-CAM

The CNN learned the gaits using the input multivariate gait data, and outputs the correct answer labels. The structure of the CNN model generated is shown in Table 1. The number of epochs was 1000, the optimizer was an SGD, and the objective function was categorical cross entropy. Data for verification of the feature visualization were randomly selected five for each label from 1069 data. In addition, 80% of the 1,059 data excluding the ten data were used as the training data, and 20% were used as validation data. The model with the lowest validation loss of 1000 epochs was saved using the model check point function. The gait of the data for verification of the feature visualization were determined using the CNN model. In addition, Grad-CAM outputs the feature position of a gait rarely associated with stumbling in the multivariate gait data of each label as a heat map.

The accuracy of the validation data was 91.98%. Figure 3 shows the Grad-CAM output of the “gait frequently associated with stumbling” feature of the data for verification of the feature visualization of each label that had the highest softmax function output, as well as the difference between them. The locations where data with the label “gait frequently associated with stumbling” have a stronger feature corresponding to

Table 1. Structure of CNN mode.

Layer name	Layer description
Input	28 × 400 × 1 Two-dimensional data
Conv2d_1	filters: 32, kernel size: (1,50), stride: (1,1), padding: valid, activation: relu
Max pooling2d_1	pool size: (1,3), padding: valid
Conv2d_2	filters: 64, kernel size: (1,20), stride: (1,1), padding: valid, activation: relu
Max pooling2d_2	pool size: (1,3), padding: validpool size: (1,3), padding: valid
Dropout_1	Rate: 0.25
Flatten_1	–
Dense_1	units: 128, activation: relu
Dropout_2	Rate: 0.25
Dense_3	units: 64, activation: relu
Dense_4	units: 2, activation: softmax

this label are shown in red, whereas the locations where the data of the label "gait rarely associated with stumbling" have a stronger feature corresponding with "gait frequently associated with stumbling" are shown in blue. The "gait frequently associated with stumbling" feature are shown on the data of label "gait frequently associated with stumbling". In particular, the data of the label "gait frequently associated with stumbling" has strong features in the elbow angle, the frontal plane angle of the right hip joint, and the GRF of right foot.

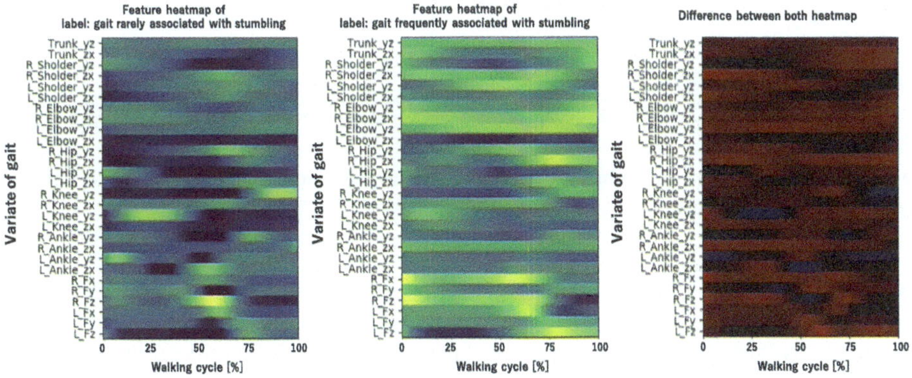

Fig. 3. Feature parts as "gait frequently associated with stumbling of data" for verification of feature visualization of each label when applying Grad-CAM.

5 Consideration

Comparing the heat maps of the two labels, there was a marked difference in the lateral GRF of right foot and the frontal plane angle change of the right hip joint. Therefore, these data used for verification of each label are shown in Fig. 4. The "gait frequently associated with stumbling" has a larger lateral GRF than the "gait rarely associated with stumbling", and at 0%–40% and 60%–100%, the "gait frequently associated with stumbling" has a larger frontal plane angle change of the right hip joint than the "gait rarely associated with stumbling". In particular, during the latter half of the walking cycle, which is the swing phase, the features in frontal plane angle change of the right hip joint was large, and the large frontal hip angle during the swing phase means that the knee cannot be sufficiently flexed when the leg is swung forward, and that the walker has a circumduction gait. In addition, the lateral movement of the center of gravity caused by the circumduction gait increases the ground force. Due to the circumduction gait, the data for verification had a low TGD, and the gait was the "gait frequently associated with stumbling". Therefore, it can be considered that this CNN model was able to learn the features of a gait frequently associated with stumbling. In addition, it was considered that the trainee can adjust their gaits using Grad-CAM to visualize the feature parts.

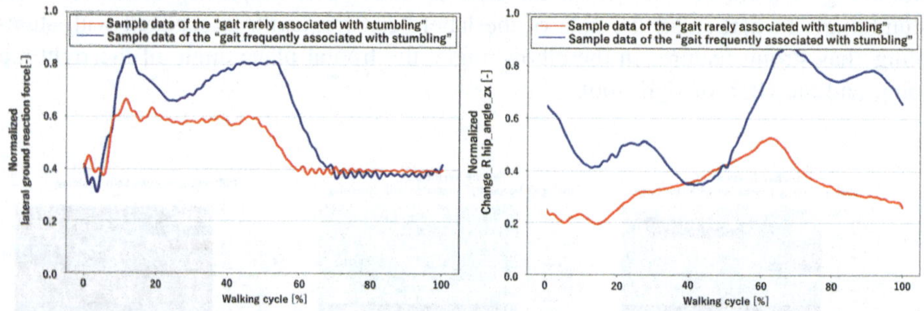

Fig. 4. The lateral GRF of right foot (left) and the frontal plane angle change of the right hip joint (right) of the data used for verification of each label.

6 Conclusion

In this study, we aimed at examining the usefulness of gait classification and feature visualization based on multivariate data for gait training. All multivariate gait data were labeled as a "gait rarely associated with stumbling" or as a gate "frequently associated with stumbling". A CNN model was used to learn the gait features. Furthermore, the feature parts of the multivariate gait data used for the classification were visualized in a heat map created using Grad-CAM. As a result, the CNN model was able to learn the features as a "gait frequently associated with stumbling". In addition, it was considered that the heatmap of the Grad-CAM output is able to show the feature parts of a "gait frequently associated with stumbling," and that the trainee can adjust their gait using the heatmap.

References

1. Stanton, R., Ada, L., Dean, C.M., Preston, E.: Biofeedback improves activities of the lower limb after stroke: a systematic review. J. Physiother. **57**(3), 145–155 (2011)
2. Yusuke, O., Keiichi, W., Kazunori, K., Keiichi, M.: Study on gait discrimination method by deep learning for biofeedback training optimized for individuals. In: Karwowski, W., Ahram, T. (eds.) Intelligent Human Systems Integration 2019. Advances in Intelligent Systems and Computing, vol. 903, pp. 155–161. Springer, Cham (2019)
3. Ramprasaath R., Selvaraju, M.C., Abhishek, D., Ramakrishna, V., Devi, P., Dhruv, B.: Grad-CAM: visual explanations from deep networks via gradient-based localization. arXiv:1610.02391v3 (2017)
4. Paparrizos, J., Gravano, L.: K-shape: efficient and accurate clustering of time series. ACM SIGMOD Record **45**(1), 69–76 (2016)

A Comparative Study on the Communication Effect of Stage Performance Promotion Platforms Based on User Experience

Jingyi Ji, Rongrong Fu(✉), and Jianxin Cheng

School of Art Design and Media, East China University of Science and Technology, Shanghai 200030, China
chumeocho@qq.com, muxin789@126.com,
13901633292@163.com

Abstract. With young people are away from traditional stage theaters, resulting in the loss of media for large stage performances. However, due to the innovation of digital media technology and the prevalence of consumerism, the new generation of users rely more on digital platforms to obtain entertainment information. Therefore, how to promote stage performances with the help of digital promotion platforms has become the key to media and even to enterprise research. At present, most stage performances in China rely on third-party platforms for promotion. However, they have problems including miscellaneous performance business, messy platform construction, etc., and thus hindering the further development of the platform. Therefore, this paper takes user experience as the entrance, uses experimental methods of cognitive neurology, and to establish a set of optimized communication strategies in line with user experience for the platforms.

Keywords: Communication effect · Stage performance · Promotion platforms · User experience

1 Introduction

With the continuous development of China's cultural industry, the new generation of users are more and more aware of payment and increasingly rely on digital platforms to obtain entertainment information. The quality of the communication effect has a direct impact on the attraction and acceptance of the works to the audience. Due to the complexity of the performance business, the lack of concern for user experience reduces the audience's recognition and acceptance of the platform, thus hindering the further development of the platform.

The integration of cognitive neuroscience and communication science has grasped the cognitive response in the communication process with a dynamic vision, and it is also conducive to objectively grasp the psychological mechanism of communicators and receivers, and research tools for the research of communication effect [1].

In view of the current problems, this article selects two mainstream domestic stage performance promotion platforms for comparative analysis. Comprehensively use user experience research methods, apply cognitive neurological experimental methods to this

G. Di Bucchianico et al. (Eds.): AHFE 2020, AISC 1202, pp. 1014–1021, 2020.
https://doi.org/10.1007/978-3-030-51194-4_133

field, and combine user satisfaction tables to test the user's memory level, willingness to use [2] and acceptance behavior of the platforms. Finally, to mine the significant elements that affect the communication effect of stage performance promotion platform and explore the role of user experience in its communication effect (Fig. 1).

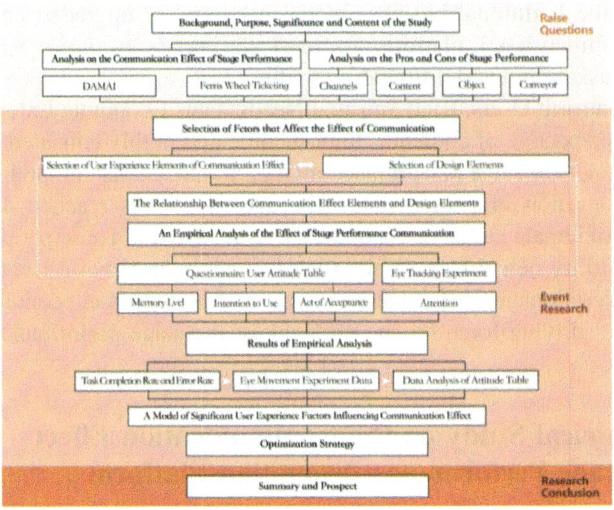

Fig. 1. Technical circuit diagram (author's original drawing)

2 Current Situation of Stage Performance Promotion Platform Communication

According to the annual comprehensive analysis of China's on-site entertainment ticketing market 2018, it can be seen that two platforms with the highest industry share and industry penetration are Damai and Motianlun ticketing [3].

Stage performance promotion platform has become an increasingly important part of the performance market. However, the domestic stage performance promotion platform has been suffering from the following problems: the overall scale is small, the degree of marketization is insufficient, the gray areas are all over the place, and the whole industrial chain is not integrated.

1. High overlap in content, accelerate platform integration. The overall consumption market is characterized by high consumption, youth and consumption preference, and the overlap of different platform projects is very high. The content of the performance promotion is too single and inconsistent with the reality, the lack of innovation in the way of interaction. Therefore, users' attention and stickiness tend to focus on the platform with the most comprehensive brand coverage, high brand awareness and the most complete supporting services. In addition, some small and medium-sized platforms will face integration under the competition of multiple factors.

2. Formulas many communication channels, poor transparency, single mode and the platform should meet the needs of the industry. With the increasing demand of the performance market, the performance activities are often random, real-time, scarce, high price and low frequency. There are many communication channels for performance works, but the profit space is low. It is difficult for a single ticketing model to achieve long-term development in a multi competitive environment. Therefore, the traditional business model needs to be upgraded and transformed. The multi-dimensional platform to meet the needs of users will assume the important task of industrial digital upgrading.
3. Younger Platform Users, Rich Spiritual Needs, Fans Economic Extension Platform. From the perspective of audience, high-income and highly educated women are the main consumers of stage performance market. In the annual comprehensive analysis [3], the proportion of people under the age of 30 has reached 42.4%, and the proportion of female audience accounts for 61% in 2017. These people love life and are willing to try new things. While focusing on the quality of life, the demand for spiritual entertainment is also increasing. In addition, the fan economy has become an important driving force for the outbreak of the stage performance industry.

3 An Empirical Study on the Communication Effect of the Stage Performance Promotion Platform

3.1 Selection of User Experience Elements for Communication Effects and Design Elements

Through the analysis of the content and characteristics of the stage performance promotion platform and the interview and analysis of the target users, combined with the five element model of Arphippainen and Tahti [4], the first level elements and the second level user experience elements and design elements are determined (Fig. 2).

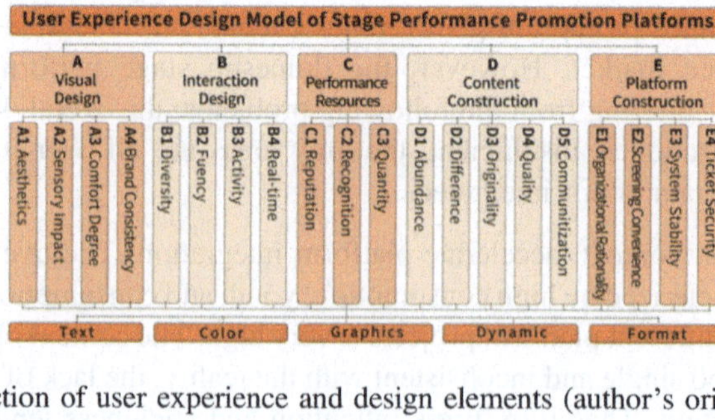

Fig. 2. Selection of user experience and design elements (author's original drawing)

3.2 Eye Movement Tracking Experiment and Satisfaction Table

According to the attributes of the core user group of the stage performance promotion platform, it is proposed to select 30 audiences (12 males and 18 females) as the test objects. Two groups of attention tests were conducted through eye movement tracking experiment. Damai and Motianlun app were given to users as two groups of experimental materials, and five tasks corresponding to user experience elements were arranged. At the end of the study, the Richter 7-point satisfaction scale was completed [5] (Fig. 3).

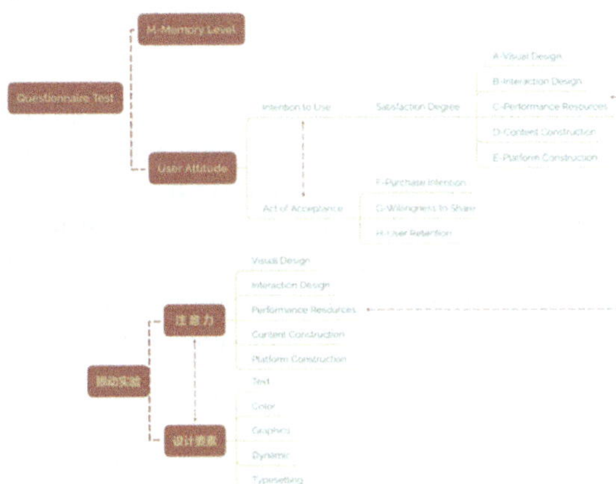

Fig. 3. Scale test and eye movement experiment (author's original drawing)

4 Analysis and Results

4.1 Task Completion Rate and Error Rate

From Table 1, it can be seen that task completion rate of both platforms is over 85%, which proves that the function design corresponding to the task is effective. Task 4 has the highest error rate. By observing the mistakes of the subjects in task 4 operation, they mainly focused on the selection of position and the change of ticket price. Therefore, this paper considers that the content setting of Damai and Motianlun ticketing is too cumbersome and the labeling is not clear, which leads to poor user experience.

Table 1. Completion rate (left) and Error rate per capita (right) (author's original drawing)

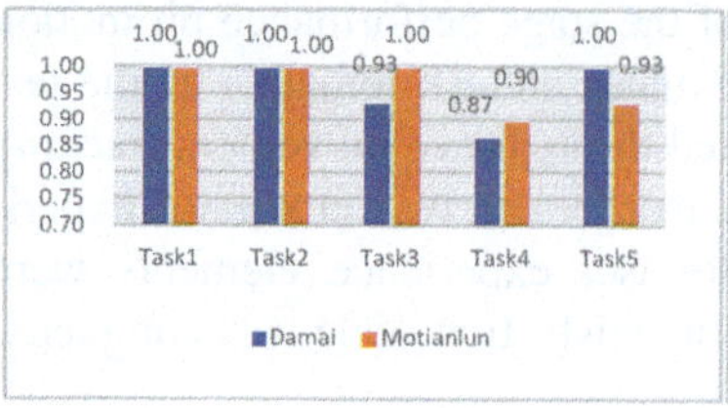

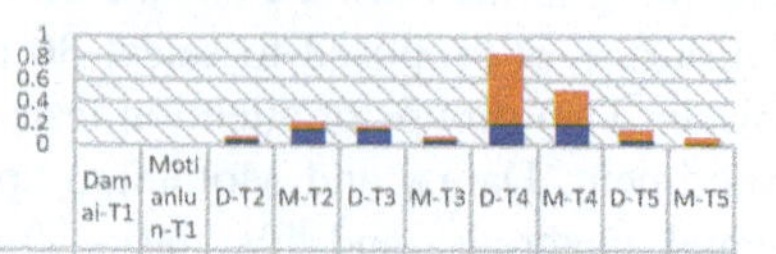

	Damai-T1	Motianlun-T1	D-T2	M-T2	D-T3	M-T3	D-T4	M-T4	D-T5	M-T5
Error Rate	0	0	0.03	0.07	0.03	0.03	0.63	0.3	0.1	0.07
Error Deviation Rate	0	0	0.07	0.17	0.17	0.07	0.23	0.23	0.07	0.03

4.2 Task Completion Rate and Error Rate

In this study, SPSS20.0 software was used to analyze the scale. The questionnaire of satisfaction scale not only has reasonable item setting of each dimension, but also shows that the scale has good overall effect and can be used for further analysis.

Through the descriptive statistical results of the scale, we can see that users' overall satisfaction with Motianlun is higher than Damai. From the overall average point of view, Damai needs to improve the interaction real-time and content originality. The interaction design of the two platforms has little difference for users (Table 2).

Table 2. Average score of questionnaire questions (author's original drawing)

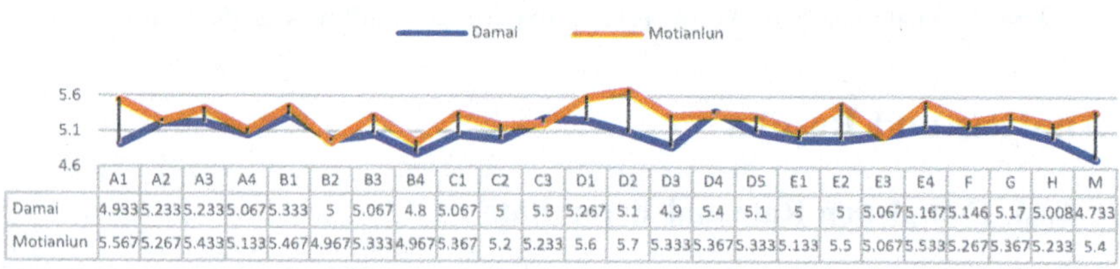

	A1	A2	A3	A4	B1	B2	B3	B4	C1	C2	C3	D1	D2	D3	D4	D5	E1	E2	E3	E4	F	G	H	M
Damai	4.933	5.233	5.233	5.067	5.333	5	5.067	4.8	5.067	5	5.3	5.267	5.1	4.9	5.4	5.1	5	5	5.067	5.167	5.146	5.17	5.008	4.733
Motianlun	5.567	5.267	5.433	5.133	5.467	4.967	5.333	4.967	5.367	5.2	5.233	5.6	5.7	5.333	5.367	5.333	5.133	5.5	5.067	5.533	5.267	5.367	5.233	5.4

Next, by using Pearson correlation analysis method, this paper analyzes the correlation among purchase intention, sharing intention and user retention. It shows that there is correlation among the influencing factors of purchase intention, sharing intention and user retention, which is suitable for regression analysis.

In the regression analysis of Damai, A2, A3, A4, B2 and B4 had no effect on F; D4 and D5 had no effect on G. In regression analysis of Motianlun ticketing, A2, A3 and A4 have no effect on F; C1, C2, C3, D1, D3 and D5 have no effect on G. After deleting the influence factors that are not established, to get the final stage performance promotion platform user experience design model, which is summarized as follows:

- Purchase Intention of Damai = 1.786 + 0.188 * Visual Aesthetics + 0.218 * Interaction Diversity + 0.250 * Interactivity

- Purchase Intention of Motianlun = 0.844 + 0.264 * Visual Aesthetics + 0.134 * Interaction Diversity + 0.162 * Interaction Fluency + 0.121 * Cross-activity + 0.167 * Interactive Real time
- Sharing Intention of Damai = 0.377 + 0.139 * Resource Visibility + 0.130 * Resource Recognition + 0.244 * Quantity of Resources + 0.163 * Content Abundance + 0.148 * Content Difference + 0.109 * Content Originality
- Sharing Intention of Motinaliun = 0.967 + 0.575 * Content Difference + 0.255 * Content Quality
- User Retention of Damai = 0.419 + 0.207 * Platform Organization Rationality + 0.239 * Convenience of Platform Screening + 0.204 * Platform System Stability + 0.256 * Security of Platform Ticket Purchase
- User Retention of Motianlun = 0.568 + 0.183 * Platform Organization Rationality + 0.253 * Convenience of Platform Screening + 0.324 * Platform System Stability + 0.169 * Security of Platform Ticket Purchase

4.3 Eye Movement Data

The correlation analysis between the eye movement experiment and the satisfaction scale showed that there was a positive correlation between fixation time and fixation points, fixation points and memory degree, fixation time and purchase intention.

According to the scanning path chart, the result analysis of the corresponding scale: in Damai, the user's error point in task 4 ticket purchase is seat selection, and the platform sets different ticket prices corresponding to different colors. The selection of ticket file is contrary to the user's usual behavior of selecting gray position, so that the user has a high error rate and poor user experience in this step. In Motianlun, there is no corresponding position area of ticket file in task 4 (a), which increases the page turning rate for users in the options of seat selection and vote file, and because they can't choose the position, they can only choose the area, so the first choice of the page of ticket file will make the platform lose many users who pay attention to the experience of watching (Fig. 4).

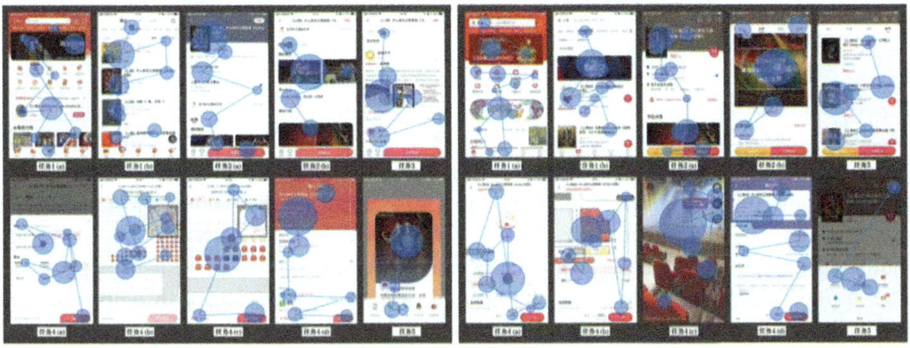

Fig. 4. Scanning path map of Damai (left) and Motianlun (right) (author's original drawing)

5 The References Section Optimization Strategy of Stage Performance Promotion Platform

1. With the development of media, the huge audience is no longer a collection of individual characteristics. On the contrary, they diverge and divide into numerous minority groups. It emphasizes subdividing the audience group and launching a highly targeted communication mode according to their individual needs. Relying on information and digital technology, different users can "meet" their own content when consuming their products. This firmly grasps the masses, so as to improve the user experience effect.
2. Pay attention to the content setting of the platform to meet the content needs of different users. Based on the use of different forms of text, audio, video content, in the first time to accurately and timely convey product information to users, this is an effective way to improve user satisfaction by establishing emotional ties between users and the platform. We should pay attention to the construction of content, so as to produce a deep impression. It is also necessary to attract users' consumption behavior, establish emotional connection with users and finally achieve the goal of brand loyalty.
3. Establish a community platform to enhance the user's use dependence. Users in the community often join the community because of a certain interest, which helps to improve the loyalty of users, and also enhances the dependence of consumers on the community. The increase of users' dependence on the community will affect more people and attract more non-users. The increase of the number of community users and the increase of loyalty will further promote the consumption behavior of the platform, expand the scope of influence, and e-commerce business, forming a virtuous circle.
4. Combine online platform with offline service to stimulate users' consumption behavior. The purpose of the establishment of online platform is to improve the efficiency of offline service, so as to improve the user's interest in the platform, increase the coverage of information dissemination, and enhance the popularity of the platform. For the stage performance work promotion platform, the ultimate goal is to promote the works to users and stimulate users' consumption behavior.

6 Conclusion

China's stage performance promotion market has long been in a state of low-level operation and inefficient operation. Breaking this barrier has become the top priority of the development of performance market. This paper focus on how to study the communication effect of stage performance promotion platforms based on user experience. The application of this research to the digital promotion platforms will help to clarify the factors to be concerned about and the issues to be avoided in the design and dissemination of this type of promotion platform, and to establish a set of optimized communication strategies in line with user experience for the stage performance promotion platforms, so as to make up for the defects of the current lack of targeted stage performance promotion platforms, providing target users with movie-viewing service experience scene, and finally achieving the changes and breakthroughs of difficult spreading of traditional stage performances among young groups.

References

1. Yang, H.: The fusion of cognitive neuroscience and human-computer interaction: new trends in human-computer interaction research. Packag. Eng. **40**(22) (2019)
2. Yu, G.: Cognitive neurocommunications. Paradigm innovation and research framework. J. Zhejiang Inst. Media Commun. **25**(01), 9–13+140 (2018)
3. Li, H.: Comprehensive analysis of China's on-site entertainment ticketing market in 2018 (2019). https://www.analysys.cn/article/analysis/detail/20019227
4. Arhippainen, L., Tahti, M.: Empirical evaluation of user experience in two adaptive mobile application prototypes. In: Ollila, M., Rantzer, M. (eds.) Proceedings of the 2nd International Conference on Mobile and Ubiquitous Multimedia, pp. 27–34 (2003)
5. Zheng, W.L., et al.: Emotion meter: a multimodal framework for recognizing human emotions. IEEE Trans. Cybern. **49**(3), 1110–1122 (2019)

The Influence of Video Barrage on the Emotion of Young Users

Shanang He[✉] and Fumie Muroi

School of Design, Shanghai Jiao Tong University, Shanghai 200240, China
shanang@sjtu.edu.cn

Abstract. This article will explore the impact of the barrage in entertainment videos on the emotions of young Chinese video users to explore how to improve the viewing experience of video users through the setting of the barrage. According to China's most famous barrage website—Bilibili, and the popularity of barrage has become a unique way to record changes in the times. This article will select entertainment videos with a large number of barrage in station B, and study the impact of barrage in entertainment videos on the emotions of young Chinese video users. Research was conducted through questionnaires and personas. The results show that browsing with a barrage is more conducive to improving users 'positive emotions than without a barrage; when barrage can be sent, participating in barrage interactions can not only reduce users' negative emotions, but also improve users to watch videos Positive emotions.

Keywords: Entertainment video · Barrage video · Emotion · User experience

1 Introduction

This article will explore the impact of barrage in entertainment videos on the emotions of young Chinese video users. With the continuous development of science and technology and the Internet, live broadcast and video platforms in China are rapidly emerging. In online video, "barrage" has become a new way for users to interact with video. "Barrage" is an emerging feature on online video sites in China and Japan, where viewers can comment while watching a video, and these real-time comments are flying like bullets across the screen. Bilibili (Station B) is the pioneer of Chinese video barrage. It has the most advanced video barrage technology on the Chinese content platform and the most widely used barrage user group [1, 2].

According to the B Station Report 2018 jointly released by Bilibili and the Chinese Academy of Social Sciences, as an important interaction mode and characteristic culture of B station, the popularity of barrage has not only overturned the way of social and viewing, it has also become a record of the changing times A unique way. According to official data from station B, 38.5% of users at station B in 2019 are young people under 24 years old, and barrage is particularly popular among young people. The popularity of the barrage reflects the young people's unique perspective and perspective on the world and themselves, and reflects the young people's quest for the resonance of reality.

G. Di Bucchianico et al. (Eds.): AHFE 2020, AISC 1202, pp. 1022–1028, 2020.
https://doi.org/10.1007/978-3-030-51194-4_134

In recent years, the use of barrage on video has become a functional trend of Chinese video platforms. Due to the relatively short rise of barrage, there is no scientific research in China on the impact of barrage on the mood of video users. Therefore, this article will fill the gap in this research content and explore the impact of barrage on young video users (under 24 years). In the selection of video types, this study selected entertainment videos. Entertainment video is one of the most pop-up video types in station B. This article explores whether the barrage video can affect the mood of the user watching the video, thereby further expanding the barrage interaction method, providing suggestions for improving the user experience design, helping users better adjust their emotions, relax their mood, and relieve stress [3].

2 Research

2.1 Questionnaire Survey

The questionnaire method is a widely used method in social surveys. Researchers use this kind of controlled measurement to measure the problems they are studying to collect reliable data [4]. The research in this step is mainly to quickly understand the viewing habits of entertainment video users, especially the barrage habits. According to the results of the questionnaire, users are divided into different types, which is convenient for the subsequent use of the persona [5].

During the questionnaire survey period, a total of 300 valid questionnaires were collected. The participants in the questionnaire survey were limited based on the descriptions of user portraits of young users (under 24 years old) in the official data of station B. These descriptions included age, gender, region, and occupation. The results of the questionnaire show that the proportions of non-opening barrage, opening barrage, but not participating in interaction, and engaging in barrage interaction are 25%, 49%, and 26%, respectively. There are no extremely low types. Therefore, three types Should be studied. In addition, the results of the questionnaire show that users watch entertainment videos for a maximum of 0–10 min. And through the questionnaire, we got the most popular entertainment video bloggers.

2.2 Persona

Sixteen people were selected from the three types of people in the questionnaire results (no barrage, barrage but no barrage interactions, and barrage with barrage interactions), to conduct user interviews. The content of the interviews included the use of station B, the Screen usage and daily entertainment life. Through interview results, the characteristics and habits of these three types of users are clearer [6].

Focused and Excited users: without opening the barrage, the self-centered experience is the center of the movie, and they think that the barrage will affect their subjective feelings. Such people will more carefully choose the content of the entertainment video, not just watching the entertainment mentality. Video, also hope to benefit from it. Indifferent and Introverted users: open the barrage but hardly participate in interaction. Such people are usually more cautious or not immersed. They browse the Internet, but do not easily post content on the Internet. Enthusiastic and Interactive users: participate in barrage interactions. Such people are enthusiastic and generous in interaction.

2.3 Experimental Method

In the experimental method stage, the PANAS scoring table was mainly used to investigate the effect of the presence or absence of barrage and the participation of barrage interaction on the user's emotions, and an attempt was made to explore the emotional differences among the three populations.

Experimental Subject. The experimental subjects in the experimental method were 48 subjects in user interviews, including 26 males and 22 females, aged between 18 and 24, mainly college students. All subjects had no visual obstacles and problems, and had not seen the video of the experiment before the experiment. Before the experiment, make sure the subject's emotional state is normal and stable.

Experimental Video. In the B station, select 10 sets of videos of the same series of popular entertainment video bloggers. In order to ensure the enthusiasm and patience of users to participate in the experiment, the 10 groups of videos are all shorter entertainment videos of about 2 min. The playback volume of these videos was more than 400,000 in the experimental stage, and the number of the barrage was about 1500. Based on the video's degree of interest, content of video, barrage's degree of interest, content of barrage, and influence of barrage as the evaluation factors, 10 users from station B outside the experiment were selected to rate the video. Finally, three groups of videos with similar scores were selected as experimental videos.

Experiment. Phase one: Fatigue intervention experiment. Before the start of the formal experiment, perform a fatigue intervention experiment on the subject through a word-finding game, and ensure that the subject understands the basic form of the experiment [7]. Phase two: Before the start of the formal experiment, the subjects fill out the PANAS emotional questionnaire. Phase three: Watch the video and fill out the PANAS questionnaire. In the formal experiment phase, the subjects went through three times of watching the video-filling the questionnaire process. Participants first watched the video without barrage, and then completed the PANAS emotional questionnaire after watching it. Rest for 30 min. After the break, the participants watched the second barrage video (but did not send the barrage). After watching the video, they filled out the PANAS emotional questionnaire. Rest for 30 min. After the break, the participants watched the third barrage video and sent a barrage interaction. After watching it, they filled out the PANAS emotional questionnaire and the experiment ended.

The PANAS questionnaire set 1–5 points for the degree of each option. After the test, the total score obtained from each option was used for statistics. Finally, all experimental data were analyzed by SPSS software.

3 Research Results

3.1 Experimental Evaluation

In the experiment, PANAS negative and positive emotion questionnaire was used to measure the emotional state of the subjects before and after the test. The reliability of the questionnaire results was analyzed by the α-reliability coefficient method, and the validity of the experiment was supported by the validity test of the experimental results [8].

Reliability Analysis. The reliability of this experiment is monitored and analyzed by the alpha reliability coefficient method. The evaluation criteria are as follows: $\alpha > 0.8$ high reliability; $0.7 > \alpha > 0.8$ good reliability; $0.6 > \alpha > 0.7$ acceptable reliability; $\alpha < 0.6$ The reliability is not good. If the CITC value is lower than 0.3, consider deleting this item. Through the first round of analysis, the less relevant options are deleted, and the reliability correction analysis table of the sample is obtained [9].

According to the reliability correction table, it can be seen that the reliability coefficient value of the research data is higher than 0.8, and the reliability coefficient value will not be significantly improved after deleting the items, which shows that the reliability of the data is high and can be used for further analysis.

Validity Analysis. The validity is mainly judged by KMO, commonality, variance interpretation rate value, factor load coefficient value and other indicators using factor analysis to evaluate the validity level of the data to determine the rationality of the analysis item design.

Generally, it is reasonable to correspond to 4–7 research items in one dimension. According to the SPSS software automatic identification, factor analysis is performed in 4 dimensions. The results of research and analysis show that the values of the variance interpretation rates of the four factors are 29.795%, 20.713%, 19.131%, 6.565%, and the cumulative variance interpretation rate after rotation is 76.205% > 50%. It means that the data validity is high, and the information amount of the research items can be effectively extracted. In addition, the KMO value is 0.723, which is greater than 0.6, which means the data has validity.

3.2 Experimental Results 1

Barrage in entertainment videos is more conducive to positive emotions.

Chi-square test (cross analysis) was used to study the relationship between the barrage situation when watching the video and the positive and negative emotions of 20 items.

Samples of the barrage situation when watching the video showed significant significance for a total of 20 items of positive and negative emotions, that is, the sample of the barrage situation when the subjects watched the video showed a difference for a total of 20 items of positive and negative emotion. The analysis results are shown in Table 1.

Table 1. Cross chi-square analysis of the subject's emotion questionnaire

Analysis project	Option	Barrage when you watch the videos			X^2	p
		No barrage	With barrage	Barrage interaction		
Positive emotion	1-little	53 (33.13)	4 (2.50)	5 (3.13)	44.4	0.000**
	2-a little	63 (39.38)	9 (5.63)	9 (5.63)		
	3-smoe	42 (26.25)	36 (22.50)	41 (25.63)		
	4-much	2 (1.25)	75 (46.88)	52 (32.50)		
	5-a great deal of	0 (0.00)	36 (22.50)	53 (33.13)		
Total		160	160	160		
Negative emotion	1-little	27 (16.88)	95 (59.38)	86 (53.75)	34.825	0.000**
	2-a little	33 (20.63)	59 (47.20)	57 (35.63)		
	3-smoe	68 (42.50)	6 (3.75)	12 (7.50)		
	4-much	30 (18.75)	0 (0.00)	5 (3.13)		
	5-a great deal of	2 (1.25)	0 (0.00)	0 (0.00)		
Total		160	160	160		

$*p < 0.05$ $**p < 0.01$

According to the statistical analysis of the results, the following results were obtained: The barrage situation when watching the video showed a 0.01 level significance (Chi = 44.40, P = 0.00 < 0.01) for positive emotions. According to the percentage comparison difference, it is known that when "1 almost 33.13% of "not" will be significantly higher than the average of 12.92%. The proportion of "2 less" without barrage selection is 39.38%, which is significantly higher than the average of 16.88%. The percentage of "4 more" with barrage selection is 46.88%, which will be significantly higher than the average of 26.87%. The percentage of "5 very many" in the barrage interactive selection is 33.13%, which is significantly higher than the average of 18.54%.

The barrage situation when watching the video showed a level of significance of 0.01 for negative emotions (Chi = 34.83, P = 0.00 < 0.01). It can be seen from the percentage contrast difference that the ratio of "1 almost nothing" to 53.75% for barrage interaction selection will be obvious. 43.33% above the average. The percentage of "2 less" with barrage selection is 47.20%, which is significantly higher than the average of 31.04%. The proportion of "3 general" without barrage selection is 42.50%, which is significantly higher than the average of 17.92%. The proportion of "4 more" without barrage selection is 18.75%, which is significantly higher than the average of 7.29%.

In general, there is no barrage. In the case of barrage, the score of positive emotions is higher and the score of negative emotions is lower. In the premise of barrage, the score of positive emotions is participated in the interaction of barrage. Higher, lower negative emotion scores.

3.3 Experimental Results 2

There are differences in the emotional situation of the three groups of people.

In the experimental results, for different types of people, there are differences between positive and negative emotions in different situations. among them:

Focused and Excited users: In the process of changing from non-barrage to non-interaction with barrage and then to barrage interaction, their positive and negative emotions have increased. Among them, the scores of "proud" and "firm" in positive emotions gradually increased; the scores of "easy to be distracted" and "hostile" in negative emotions gradually increased, and the "easy to be distracted" increased significantly.

Introverted and Indifferent users: This group has relatively small emotional changes; the emotional changes are mainly reflected in positive emotions. In the case where there is no barrage and there is a barrage, the "spiritual" and "enthusiastic" in positive emotions have risen significantly.

Enthusiastic and Interactive users: In the process of changing from a non-barrage to a barrage but not interacting, and then to a barrage interaction, its positive emotions gradually increase and negative emotions gradually decrease. In the absence of a barrage, the scores of "nervous", "annoyed", and "guilt" in negative emotions are high; in the case of barrage interaction, the "spiritual" and "susceptible" in positive emotions are high. "Inspiring", "enthusiastic", and "proactive" all increased significantly.

4 Discussion

The experimental results show that in entertainment videos, the appearance and interaction of the barrage really helps to increase the positive emotions and reduce the negative emotions of users. This result has important links with the nature of entertainment video. Entertainment videos are popular, relaxed, and funny, and users' viewing motivation is mainly to relax themselves, get pleasure, and improve their mood. Therefore, most of the barrage is to express that the video is amused, get interesting points and hope Other users spread this interesting content, and these barrage contents often make the movie-viewing users more resonated and fun. Even in some entertainment videos, the barrage is more laughable than the video content. At the same time, it is easier for users to reduce negative emotions in the easy interaction with others.

According to the above experimental results, the user experience design of the video barrage can be inspired. For example, a video site can make a barrage screening mechanism for different groups of people. For the Focused and Excited people, screening for repeated mood barrage (for example, repeated 'hahaha') may increase their chances of participating in barrage interactions, and they Participation in speech can improve the quality of the barrage to a certain extent. In addition, you can set the parameters such as the transparency, size, color, and position of the barrage. For the Introverted and Indifferent people, they can adjust the layout of the barrage and video content to watch the video and barrage at the same time. After all, both The meaning in their eyes is almost the same. For the Enthusiastic and Interactive people, you can also set whether to temporarily pause the video when publishing the barrage, so that they can better edit the content of the barrage.

References

1. Yu, H., Dongyu, X.: On the culture of "cyber barrage". J. Commun. Res. **2**(021)(09), 183 (2018). (in Chinese)
2. Zhuoling, H.: Research on development of barrage video website. Value Eng. **37**(494)(18), 261–262 (2018). (in Chinese)
3. Zifei, L., Xiaoyu, H.: Experimental study on the influence of internet emoticons and emotional words on emotional priming. J. Gannan Normal Univ. **4**,125–130 (2019). (in Chinese)
4. Ellard, J.H., Rogers, T.B.: Teaching questionnaire construction effectively: the ten commandments of question writing (1993)
5. Linong, D.: Design and Research on the Children Furniture of Shanghai Middle-and-upper Income Families Based on Persona (2017). (in Chinese)
6. Chapman, C.N., Milham, R.P.: The personas' new clothes: methodological and practical arguments against a popular method. Hum. Factors Ergon. Soc. Ann. Meet. Proc. **50**, 634–636 (2014)
7. Rupp, M.A., Sweetman, R., Sosa, A.E., Smither, J.A., Mcconnell, D.S.: Searching for affective and cognitive restoration: examining the restorative effects of casual video game play. Hum. Factors 001872081771536 (2017)
8. Wittchen, H.U.: Reliability and validity studies of the WHO-Composite International Diagnostic Interview (CIDI): a critical review. J. Psychiatr. Res. **28**(1), 57–84 (1994)
9. Guo, Q., Wang, G., Gao, J., et al.: Reliability analysis of weapon systems based on SPSS. In: 2012 Proceedings of International Conference on Modelling, Identification & Control (ICMIC). IEEE (2012)

Author Index

G. Di Bucchianico et al. (Eds.): AHFE 2020, AISC 1202, pp. 1029–1032, 2020.
https://doi.org/10.1007/978-3-030-51194-4

The manufacturer's authorised representative in the EU is Springer Nature Customer Service Centre GmbH, Europaplatz 3, 69115 Heidelberg, Germany. If you have any concerns regarding our products, please contact ProductSafety@springernature.com

Printed and bound by CPI Group (UK) Ltd, Croydon, CR0 4YY
30/09/2026
02225441-0005